# CAD/CAM THEORY AND PRACTICE

# McGraw-Hill Series in Mechanical Engineering

## Consulting Editors

**Jack P. Holman,** *Southern Methodist University*
**John R. Lloyd,** *Michigan State University*

Also Available from McGraw-Hill

## Schaum's Outline Series in Mechanical Engineering

Most outlines include basic theory, definitions, and hundreds of solved problems and supplementary problems with answers.

Titles on the Current List Include:

*Acoustics*
*Basic Equations of Engineering*
*Continuum Mechanics*
*Engineering Economics*
*Engineering Mechanics, 4th edition*
*Fluid Dynamics, 2d edition*
*Fluid Mechanics & Hydraulics, 2d edition*
*Heat Transfer*
*Introduction to Engineering Calculations*
*Lagrangian Dynamics*
*Machine Design*
*Mathematical Handbook of Formulas and Tables*
*Mechanical Vibrations*
*Operations Research*
*Statics and Mechanics of Materials*
*Strength of Materials, 2d edition*
*Theoretical Mechanics*
*Thermodynamics, 2d edition*

## Schaum's Solved Problems Books

Each title in this series is a complete and expert source of solved problems containing thousands of problems with worked out solutions.

Related Titles on the Current List Include:

*3000 Solved Problems in Calculus*
*2500 Solved Problems in Differential Equations*
*2500 Solved Problems in Fluid Mechanics and Hydraulics*
*1000 Solved Problems in Heat Transfer*
*3000 Solved Problems in Linear Algebra*
*2000 Solved Problems in Mechanical Engineering Thermodynamics*
*2000 Solved Problems in Numerical Analysis*
*700 Solved Problems in Vector Mechanics for Engineers: Dynamics*
*800 Solved Problems in Vector Mechanics for Engineers: Statics*

Available at your College Bookstore. A complete list of Schaum titles may be obtained by writing to: Schaum Division, McGraw-Hill, Inc., Princeton Road, S-1, Hightstown, NJ 08520

# CAD/CAM THEORY AND PRACTICE

**Ibrahim Zeid**

*Department of Mechanical Engineering*
*Northeastern University*

**McGraw-Hill, Inc.**

New York   St. Louis   San Francisco   Auckland   Bogotá   Caracas
Hamburg   Lisbon   London   Madrid   Mexico   Milan   Montreal   New Delhi
Paris   San Juan   São Paulo   Singapore   Sydney   Tokyo   Toronto

This book was set in Times Roman.
The editors were John J. Corrigan and John M. Morriss;
the production supervisor was Friederich W. Schulte.
The cover was designed by Rafael Hernandez.
Project supervision was done by Santype International Ltd.
R. R. Donnelley & Sons Company was printer and binder.

**CAD/CAM THEORY AND PRACTICE**

2 3 4 5 6 7 8 9 0 DOC DOC 9 0 9 8 7 6 5 4 3 2 1

ISBN 0-07-072857-7

Library of Congress Cataloging-in-Publication Data

Zeid, Ibrahim.
    CAD/CAM theory and practice/Ibrahim Zeid.
      p.        cm.
    ISBN 0-07-072857-7
     1. CAD/CAM systems.    I. Title.
TS155.6.Z45    1991           90-42281
670'.285—dc20

# ABOUT THE AUTHOR

**Ibrahim Zeid** is an Associate Professor of Mechanical Engineering at Northeastern University. He received his Ph.D. in Mechanical Engineering from the University of Akron. Professor Zeid has an international background. He received his B.S. (with highest honor) and M.S., both in Mechanical Engineering, from Cairo University in Egypt. He has received various honors and awards both in Egypt and the United States. He is the recipient of both the Northeastern University Excellence in Teaching Award (1983) and the SAE Ralph R. Teetor National Educational Award (1984).

Professor Zeid is credited with initiating and establishing the instructional and research CAD/CAM program and its related laboratory at Northeastern University. He is the author or coauthor of many publications in journals and conference proceedings. He has been active in research in design and manufacturing. He has received various grants to support his research activities. He is a consultant for various companies. He is active in professional societies and most recently was elected the Chairman of the Executive Committee of the ASME Boston Section.

# CONTENTS

## Part II   Geometric Modeling

# Chapter 7  Types and Mathematical Representations of Solids

# Part III   Two- and Three-Dimensional Graphics Concepts

# Part IV   Interactive Tools

# Part V   Design Applications

# PREFACE

Engineering design and manufacturing form the core of the engineering profession. The engineering curricula and the engineering educational process attempt to provide today's students, tomorrow's engineers, with a sufficient number of sciences and tools to perform, among other things, design and manufacturing. Engineering sciences are well established and most often include physics, engineering mechanics, mechanical behavior and processes of materials, and thermal fluids. Mathematics, computers and computational techniques, communication methods, and drafting skills are among the essential tools a designer needs. In contrast to engineering sciences some of these tools, in particular computers and drafting skills, have been changing quite often to reflect changes and advances in manufacturing and technology. Over the past thirty years, engineering has changed from using mathematical tables, to slide rules, to pocket calculators, to personal computers. In the past fifteen years the interactive computer graphics and CAD/CAM technology have been impacting the drafting, design, and manufacturing tools significantly. It is because of these important impacts that this book has been written.

In an attempt to write a meaningful book with enough subject depth and breadth in the area of CAD/CAM, a focus for the book must be defined. Among the many available choices, this book focuses on presenting a balanced mix on the theory and practice of the CAD/CAM concepts. Throughout the book, the influence of the theoretical and practical aspects of CAD on CAM is also presented. The late chapters of the book, such as Chapters 16 and 20, discuss the integration of CAD and CAM databases. It is believed that the true integration between CAD and CAM forms the bottleneck for achieving automation. It is hoped that the "A" in CAD/CAM will mean automated instead of aided.

The purpose of this book is to present CAD/CAM principles and tools in generic and basic forms with enough depth and breadth. These principles are supplemented with engineering and design applications as well as problems. The presentation of these principles and tools maintains a balance between both theory and practice. The book is concerned with developing the proper attitudes and approaches to utilizing the existing CAD/CAM technology in engineering. It attempts to expand the reader's imagination beyond just creating interactive graphics. Therefore, Parts IV, V, and VI illustrate how geometric modeling and graphics concepts covered in previous parts can be applied to engineering and design applications. Whenever new tools and applications become available in the future, these three parts can be updated without affecting the book organization. This is important for those who adopt a book in a rapidly developing field such as CAD/CAM. Throughout the book, examples, applications, and computer algorithms are covered independently of any specific hardware or programming languages. However, it is assumed that the reader is familiar with computers and has a basic background in engineering and computer programming.

The book is targeted at students, engineers, and professionals who are interested in the CAD/CAM technology and its applications to design. Most often, this group utilizes, in one form or another, a CAD/CAM system. It may be a fully commercial system or a low-end PC-based system. In either case, the user is faced with understanding the same basic concepts and principles underlying the system. Failure of such understanding often results in user frustration and a significant decline in productivity and utilization of the system relative to manual procedures. Manuals and documentation which are typically provided with CAD/CAM systems tend to concentrate on the user interface and the syntax associated with it. They usually assume that the user has the proper theoretical background which this book attempts to provide. Such a background helps the user a great deal in understanding the various jargon and terminology encountered in the system documentation as well as enabling the user to deal with system errors more intelligently.

The material in the book can be used in various ways. As a textbook, it could be used at either the advanced undergraduate or first graduate level. A two quarter-long or a one semester-long undergraduate course is adequate to cover most of the material and allows time for a project which is a valuable experience for students. The book provides a complete menu of topics. The depth and choice of topic coverage and projects may vary based on a particular curriculum. A graduate course should be designed to cover all the book material and allow for a comprehensive project. A course with an interactive computer graphics focus may cover Chapters 2, 3, 5, 6, 9, 10, 13, and 15. A course with a geometric modeling focus may cover Chapters 2, 3, 5, 6, 7, 9, and 15. A course with a CAD/CAE focus may cover Chapters 1 to 9, 11 to 15, and 17 to 19. A course with a CAM focus may cover Chapters 1 to 9, 11 to 15, 16, and 20 (supplemental material to these chapters may be provided by the instructor). Courses with other foci can easily be designed in a similar fashion. Many instructors may prefer to supplement Parts IV, V, and VI of the book with their own experience and/or their applications. If an engineering curriculum does not offer separate CAD/CAM courses, this book is then ideal as a reference for outside reading by the students. The book can also serve as a reference for the CAD/CAM industry. Training courses typically offered by CAD/CAM vendors to engineers and professionals concentrate on system syntax and documentation.

To write a book in the very rapidly changing CAD/CAM field is perhaps the most challenging endeavor an individual can undertake. The book design and organization has taken this observation into consideration. The book has been divided into six integrated parts which can be updated in the future to reflect new trends, tools, and applications when they evolve without changing the book organization. For example, if a new application subject becomes available in the future, it can be added as a new chapter in Part V or VI. Future updates will always be made taking into consideration the book size. The author has tried to collect as much material from the literature as possible into this book with a unified notation. This represents a major task. The author would be grateful to receive any suggestions, opinions, ideas, and advice regarding the book. The author would also appreciate receiving any errors which went undetected in this edition, and will acknowledge them by name and institution in subsequent editions.

A final word regarding the book organization and style. The book is organized and written in such a way to be suitable for self-study. There are enough details about each subject. Instructors using the book do not have to cover all these details in class. Instead, they can assign some of these details as out-of-class reading exercises. In this way the class time can be utilized effectively by both students and instructors to discuss design projects and applications or issues related to using a particular CAD/CAM system. With this style, engineers and professionals should also find the book material handy to use and easy to understand.

The author is indebted to all the people who helped directly or indirectly to make this book idea a reality. Without their assistance this project would never have been completed. The author would like to thank the following reviewers for their valuable comments, suggestions, encouragement, and sound advice throughout the project: Abdulsamad Ata, University of Detroit; Samir B. Billatos, University of Connecticut; Richard G. Budynas, Rochester Institute of Technology; Jan Evans, University of Tennessee at Chattanooga; Herbert Freeman, Rutgers University; Gary A. Gabriele, Rennselear Polytechnic Institute; Gary L. Kinzel, Ohio State University; Michael B. McGrath, Colorado School of Mines; Charles Mischke, Iowa State University; John J. Moskwa, University of Wisconsin-Madison; Albert P. Pisano, University of California, Berkeley; Donald R. Riley, University of Minnesota; Eric Teicholz, Graphics Systems, Inc.; and Robert O. Warrington, Jr., Louisiana Tech University. The author has made every possible effort to take advantage of their suggestions. The author is also indebted to the many CAD/CAM vendors and their personnel who provided photographs and slides for the book.

A book cannot be published without the help of many people. I would like to thank all my students and colleagues who contributed directly and indirectly through their constructive criticism in the evolution and preparation of the book manuscript. Special thanks are due to Ms. Sohela Shafai, Ms. Leslie Schreiter, and others who typed the manuscript. Thanks are also due to McGraw-Hill staff for their patience and professional help. The diligences and encouragement of

Scott Stratford and Anne C. Duffy in the early stages of the project were very valuable. The valuable experience and vision of John Corrigan, the book editor, has permitted the successful completion of the manuscript. His phone calls and visits maintained the steady progress of the manuscript. In addition, his efforts during the production phase of the book were invaluable to its completion. I would also like to thank Karen Jackson, John Morriss, Fred Schulte, and others for their efforts.

Last, but not least, very special thanks are due to my family and friends for their constant love and support. The patience, understanding, and encouragement of my wife and my children are greatly appreciated.

*Ibrahim Zeid*

# CAD/CAM THEORY AND PRACTICE

# PART
# I

# OVERVIEW OF
# CAD/CAM
# SYSTEMS

# CHAPTER
# 1

# INTRODUCTION

## 1.1 CAD/CAM CONTENTS AND TOOLS

In engineering practice, CAD/CAM has been utilized in different ways by different people. Some utilize it to produce drawings and document designs. Others may employ it as a visual tool by generating shaded images and animated displays. A third group may perform engineering analysis of some sort on geometric models such as finite element analysis. A fourth group may use it to perform process planning and generate NC part programs. In order to establish the scope and definition of CAD/CAM in an engineering environment and identify existing and future related tools, a study of a typical product cycle is necessary. Figure 1-1 shows a flowchart of such a cycle.

The product begins with a need which is identified based on customers' and markets' demands. The product goes through two main processes from the idea conceptualization to the finished product: the design process and the manufacturing process. Synthesis and analysis are the main subprocesses that constitute the design process. Synthesis is as crucial to design as analysis. The philosophy, functionality, and uniqueness of the product are all determined during synthesis. The major financial commitments to turn the conceived product idea into reality are also made. Most of the information generated during the synthesis subprocess is qualitative and consequently is hard to capture in a computer system. Expert and knowledge-based systems have made a great deal of progress in this regard and the interested reader should refer to the corresponding literature. The end goal of the synthesis subprocess is a conceptual design of the prospective product. Typically, this design takes the form of a sketch or a layout drawing that shows the relationships among the various product parts, as well as any surrounding constraints. It is also employed during brainstorming discussions among various design teams and for presentation purposes.

The analysis subprocess begins with an attempt to put the conceptual design in the context of the abstracted engineering sciences to evaluate the per-

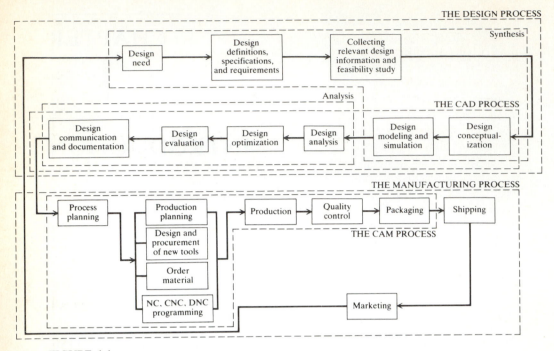

**FIGURE 1-1**
Typical product cycle.

formance of the expected product. This constitutes design modeling and simulation. The quality of the results and decisions involved in the activities to follow such as design analysis, optimization, and evaluation is directly related to and limited by the quality of the chosen design model. It is the responsibility of the designer to ensure the adequacy of a chosen model to a particular design. An important characteristic of the analysis subprocess is the "what if" scenario, which is usually valuable in design situations where analytical solutions do not exist. A computer environment where various design alternatives can be investigated is ideal to make better design decisions in shorter periods of time. Algorithms for both design analysis and optimization can be implemented and utilized. While design optimization may be embedded in design analysis, it is identified as a separate phase in Fig. 1-1 to emphasize its significance to the design process as a whole. Once the major elements of the design have been analyzed and their nominal dimensions determined, the design evaluation phase starts. Prototypes can be built in a laboratory or a computer to test the design. More often computer prototypes are utilized because they are less expensive and faster to generate. They also help the designer determine other dimensions of the product that are not analyzed, as well as finalize those that result from analysis by employing commonsense design rules. The designer can also generate bills of materials, specify tolerances, and perform cost analyses. The last phase of the analysis subprocess is the design communication and documentation which involves preparations of drawings, reports, and presentations. Drawings are utilized to produce blueprints to be passed to the manufacturing process.

The main phases of the manufacturing process are shown in Fig. 1-1. It begins with the process planning and ends with the actual product. Process planning is considered the backbone of the manufacturing process since it attempts to determine the most efficient sequence to produce the product. A process planner must be aware of the various aspects of manufacturing to plan properly. The planner works typically with blueprints and may have to communicate with the design department of the company to clarify or request changes in the final design to fit manufacturing requirements. The outcome of process planning is a production plan, tools procurement, material order, and machine programming. Other special manufacturing needs such as design of jigs and fixtures are planned. Process planning to the manufacturing process is analogous to synthesis to the design process; it involves considerable human experience and qualitative decisions. This makes it difficult to computerize. However, CAPP (computer aided process planning) has progressed significantly. In addition to a centralized CAD/CAM database for CAPP, geometric models that are accessed must be unambiguous. Solid models possess such a characteristic and are used in CAPP development.

Once the process planning phase is complete, the actual production of the product begins. The produced parts are inspected and usually must pass certain standard quality control (assurance) requirements. Parts that survive inspection are assembled, packaged, labeled, and shipped to customers. Market feedbacks are usually valuable in enhancing the products. These feedbacks are usually incorporated into the design process. With the market feedback, a closed-loop product cycle results, as shown in Fig. 1-1.

The phases of the design and manufacturing processes shown in Fig. 1-1 serve as the basis to define the design and manufacturing contents and consequently the tools that a CAD/CAM system must provide for engineers. To identify these tools properly, a CAD process and a CAM process have been defined in relation to the other processes. The CAD process is a subset of the design process. Similarly, the CAM process is a subset of the manufacturing process. The implementation of the CAD process on current systems takes the generic flow presented in Fig. 1-2. Once a conceptual design materializes in the designer's mind, the definition of a geometric model starts via the user interface provided by the relevant software. The choice of a geometric model to CAD is analogous to the choice of a mathematical model to engineering analysis. It depends directly on the type of analysis to be performed. For example, finite element analysis might require a different model than kinematic analysis. A valid geometric model is created by the CAD/CAM system through its definition translator which converts the designer input into the proper database format. In order to apply engineering analysis to the geometric model, interface algorithms are provided by the system to extract the required data from the model database to perform the analysis. In the case of finite element analysis, these algorithms form the finite element modeling package of the system. Design testing and evaluation may require changing the geometric model before finalizing it. When the final design is achieved the drafting and detailing of the models starts, followed by documentation and production of final drawings.

Table 1.1 relates the CAD tools to the various phases of the design process.

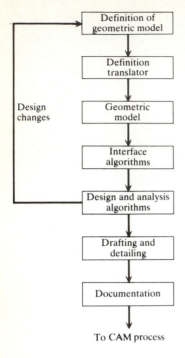

**FIGURE 1-2**
Implementation of a typical CAD process on a CAD/CAM system.

The core of the CAD tools are geometric modeling and graphics applications. Aids such as color, grids, geometric modifiers, and group facilitate structuring geometric models. Manipulations include transformation of the model in space so it can be viewed properly. Visualization is achieved via shaded images and animation procedures which help design conceptualization, communication, and interference detections in some cases. The tools for design modeling and simulation are well diversified and are closely related to the available analysis packages.

**TABLE 1.1**
**CAD tools required to support the design process**

| Design phase | Required CAD tool(s) |
|---|---|
| Design conceptualization | Geometric modeling techniques; graphics aids, manipulations, and visualization |
| Design modeling and simulation | Same as above; animation; assemblies; special modeling packages |
| Design analysis | Analysis packages; customized programs and packages |
| Design optimization | Customized applications; structural optimization |
| Design evaluation | Dimensioning; tolerances; bill of materials; NC |
| Design communication and documentation | Drafting and detailing; shaded images |

**FIGURE 1-3**
Implementation of a typical CAM process on a CAD/CAM system.

Optimization CAD tools are also available. Some FEM (finite element modeling) packages provide some form of shape and structural optimization. Even though CAD tools for design evaluations are hard to identify, they may include the proper sizing of the model after the analysis is performed to ensure engineering practices such as gradual change in dimensioning and avoidance of stress concentrations. Adding tolerances, performing tolerance analysis, generating a bill of materials, and investigating the effect of manufacturing on the design by utilizing NC packages are also valuable tools that are available to designers.

The implementation of the CAM process on CAD/CAM systems is shown in Fig. 1-3. The geometric model developed during the CAD process forms the basis of the CAM activities. Various CAM activities may require various CAD information. Interface algorithms are usually utilized to extract such information from CAD databases. In the case of process planning, features that are utilized in manufacturing (e.g., holes, slots, etc.) must be recognized to enable efficient planning of manufacturing. NC programs, along with ordering tools and fixtures, result from process planning. Once parts are produced, CAD software can be used to inspect them. This is achieved by superposing an image of the real part with a master image stored in its model database. After passing inspection, CAM software can be utilized to instruct robot systems to assemble the parts to produce the final product.

Table 1.2 relates the CAM tools to the previous phases of the manufacturing process. CAPP techniques include variant, generative, and hybrid approaches. Various part programming languages are supported by most CAM software. These include APT, COMPACT II, SPLIT, etc. Inspection software

**TABLE 1.2**
**CAM tools required to support the manufacturing process**

| Manufacturing phase | Required CAM tool(s) |
|---|---|
| Process planning | CAPP techniques; cost analysis, material and tooling specification |
| Part programming | NC programming |
| Inspection | Inspection software |
| Assembly | Robotics simulation and programming |

utilizes CMMs (coordinate measuring machines) which compares the coordinates of the actual parts with those of the master database. The robotics software supports robot simulation, offline programming, and image processing and vision applications.

## 1.2   HISTORY OF CAD/CAM DEVELOPMENT

The roots of today's CAD/CAM go back to the beginning of civilization when graphics communication was acknowledged by engineers of ancient Egypt, Greece, and Rome. Some of the existing drawings on Egyptian tombs can be considered as technical drawings. Available work and notes of Leonardo da Vinci show the use of today's graphics conventions such as isometric views and cross-hatching. Orthographic projection which we practice today was invented by the French mathematician Gaspard Monge (1746–1818) who was employed as a designer by his government. This method of projection was made available for public engineers at the beginning of the nineteenth century after the military kept it as a secret for thirty years. The inventions of computers and xerography later in that century have given graphics, and consequently CAD/CAM, their current dimensions and power.

CAD/CAM has gone through four major phases of developments in the past four decades. The first phase spanned the decade of the 1950s and can be characterized as the era of conceiving interactive computer graphics. Developments during the first half of the decade were slowed down by the expense and inadequacy of computers of that period for interactive use. MIT was able to produce simple pictures by interfacing a television-like CRT (cathode ray tube) with a Whirlwind computer in 1950. In 1952, MIT's Servo Mechanisms Laboratory demonstrated the concept of numerical control (NC) on a three-axis milling machine. Passive graphics, displayed on CRTs, were used in the mid 1950s to solve military command and control problems. The second half of the 1950s had witnessed the conception of the lightpen. Such a conception was related to the project SAGE (semi-automatic ground environment) Air Defense System developed out of MIT's Lincoln Lab. During the late 1950s, APT (automatically programmed tools) was developed, and in 1959, GM began to explore the potential of interactive graphics.

The decade of the 1960s represents the most critical research period for interactive computer graphics. The fact that the computer came out of research laboratories helped spark the development in this decade. The milestone of research achievements was the development of the Sketchpad system by Ivan Sutherland, which was published in 1962 as his thesis. The Sketchpad system was a dramatic event because it demonstrated that it was possible to create drawings and alterations of objects interactively on a CRT. By the mid 1960s, large computer graphics research was initiated by various groups. The term "computer aided design," or CAD, started to appear and was used. The term implied computer graphics with the word "design" extending it beyond basic drafting concepts. General Motors announced their DAC-1 system (design augmented by computers) in 1964. In 1965, Lockheed Aircraft initiated CADAM and Bell Telephone Laboratories announced their GRAPHIC 1 remote display system. In the late 1960s direct view storage tubes became available commercially and storage tube-based "turnkey" systems began to evolve.

During the decade of the 1970s, the research efforts of the 1960s in computer graphics had begun to be fruitful and the important potential of interactive computer graphics in improving productivity was realized by industry, government, and academia. Various lectures and courses were organized by interested groups. In 1974, the first national SIGGRAPH Conference was held in Boulder, Colorado, and a few years later the National Computer Graphics Association (NCGA) was formed and held its first meeting in 1980 in Washington D.C. Other important developments include the initiation of IGES (Initial Graphics Exchange Specification) in 1979. The decade of the 1970s can also be characterized as the golden era for computer drafting and the beginning of ad hoc instrumental design applications. Turnkey systems supplied draftsmen and/or designers with three-dimensional centralized databases primarily for modeling and drafting purposes. Some of these systems were slow 16-bit machines and the majority of them supported wireframe modeling in large and, on a limited basis, some surface modeling. Due to limitations and restrictions of modeling, only basic design applications were available. Such applications were mostly manual and far from being able to handle real industrial design problems. Mass property calculations, finite element modeling, NC tape generation and verification, and integrated circuits were, and still are, the most well-developed applications available.

The management in various industries began to realize the impact of the then new CAD/CAM technology on improving productivity in the late 1970s. Engineers have been stretching the technology beyond drafting since then. They have demanded various design and manufacturing applications from CAD/CAM vendors who have been responding successfully within the existing limits of hardware, software, and the basic theories underlying the field. Consequently, the decade of the 1980s can be identified as the CAD/CAM heady years of research. New theories and algorithms have evolved. An essential goal for this decade is to integrate and/or automate the various elements of design and manufacturing to achieve the factory of the future. The major research focus is to expand CAD/CAM systems beyond three-dimensional geometric design and provide more engineering applications. Accurate representations of sculptured surfaces based on Coons, Bezier, and Gordon as well as B-spline surfaces have advanced the mass property calculations, the NC milling, and the finite element applications that had existed a decade ago. Other applications such as mechanisms and robotics analysis and simulation, injection molding design and analysis, front-end tools to automate conceptual design, and many others are examples of the breadth of the CAD/CAM field and its development.

Another significant achievement is the acceptance and growing credibility of the solid modeling theory. The fundamental potential of solid modeling lies in the fact that it provides unique and unambiguous geometric representations of solids which, in turn, help automate and/or support design and manufacturing applications. Major solid modeling systems now exist such as GMSolid (General Motors), Romulus (ShapeData), PADL-2 (University of Rochester), SynthaVision-based (Applicon), and Solidesign (Computervision). The hardware has kept pace with the software and applications developments. In addition to developments of special computer hardware, improved displays, almost real-time simulation hardware, and microcomputer-based and workstation-based CAD/CAM systems have been emerging rapidly into the market. Most recent systems are as capable as most of the mainframe-based systems that appeared a decade ago.

While the CAD/CAM field has come a long way in four decades thus far, its future certainly holds many challenges. Extrapolating this existing history reveals that the decade of the 1990s and beyond will represent the age where the fruits of the current research efforts in integrating and automating design and manufacturing applications will mature. It is anticipated that new design and manufacturing algorithms and capabilities will become available. These applications will be supported by better and faster computing hardware, and efficient networking and communication software.

## 1.3   CAD/CAM MARKET TRENDS[1]

The CAD/CAM market has always been in a state of flux since it began. New hardware configurations and software concepts are continuously developed. Chapters 2 and 3 cover the details and definitions of existing concepts. It is most likely that the market will continue to change rapidly over the next decade. The emergence of microcomputers and engineering workstations have contributed to the decline in price which make CAD/CAM systems more affordable by small businesses. In constant 1985 dollars, the U.S. market for CAD/CAM systems is expected to grow from about $3 billion in 1985 to $8 billion by 1992. In current dollar units, the 1992 sales figures may approach $12 billion, assuming an average inflation rate of 5.2 percent per year for the period 1985–1992. The average yearly growth rate in real (constant dollar) terms is expected to be 15 percent per year for the period; that is, the real growth is to slow down from an initial 21 percent per year in 1985 to about 10 percent per year in 1992. In current dollars, the average yearly growth is 21 percent per year. Growth is

---

[1] This section includes excerpts from the Frost and Sullivan Report 1564.

expected to decline over the seven-year span, 1985–1992, from 27 percent per year in 1985 to 17 percent per year in 1992.

In the next few years, it is expected that purchases of large numbers of personal computers and workstations will be made. Traditional turnkey systems will continue to be sold but not at the rate seen in the past. These will be aimed at the project group which works together, at the drafting and drawing archival environment, and at others where sharing a system does not seriously impede the productivity of coworkers. Turnkey systems will continue to offer high levels of software and peripheral capability and can be equipped with the same types of software tools and graphics terminals as their stand-alone workstation counter-parts. Chapter 2 discusses the various types of CAD/CAM systems.

Forecasts by end-user industry are shown in Table 1.3 and Fig. 1-4. The largest users of CAD/CAM are the electronics, aerospace, and automotive industries. The relative rankings of the top three industries are expected to be preserved until 1992. The growth rates of industries' purchases of CAD/CAM systems vary. The fastest growth is seen in the construction, electronics, and chemicals market segments. The rapid growth in construction is due to the com-bination of a relatively small installed base and the development of CAD/CAM technology to a point where constructions can productively use it on a large scale. The electronics segment is forecast to grow rapidly due to the rapid growth

**TABLE 1.3**
**CAD/CAM market, United States—sales by end-user industry**

| Industry | Sales by industry, $ million | | | | | Growth per year 1985–1992, % |
|---|---|---|---|---|---|---|
| | 1984 | 1986 | 1988 | 1990 | 1992 | |
| Aerospace | 361 | 494 | 634 | 764 | 876 | 11.0 |
| Automotive | 312 | 448 | 596 | 737 | 860 | 12.6 |
| Chemicals | 65 | 99 | 144 | 196 | 254 | 17.9 |
| Construction | 111 | 182 | 290 | 444 | 649 | 24.5 |
| Electrical machinery | 159 | 237 | 321 | 398 | 462 | 13.1 |
| Electronics | 432 | 679 | 1012 | 1417 | 1869 | 19.4 |
| Food | 47 | 62 | 80 | 97 | 113 | 11.1 |
| Glass, stone | 23 | 32 | 41 | 51 | 61 | 12.3 |
| General machinery | 151 | 207 | 270 | 333 | 391 | 12.0 |
| Fabricated metals | 151 | 207 | 266 | 322 | 371 | 11.2 |
| Petroleum, gas | 207 | 292 | 391 | 493 | 591 | 13.3 |
| Paper | 20 | 29 | 40 | 53 | 65 | 15.3 |
| Rubber and plastics | 31 | 49 | 70 | 92 | 111 | 16.2 |
| Primary metals | 13 | 17 | 21 | 23 | 25 | 7.7 |
| Textile, apparel | 32 | 46 | 59 | 68 | 75 | 9.7 |
| Other manufacturing | 85 | 129 | 180 | 234 | 284 | 15.3 |
| Architecture and engineering | 110 | 157 | 205 | 248 | 282 | 11.5 |
| Research | 52 | 76 | 104 | 132 | 158 | 14.0 |
| Other services | 121 | 184 | 265 | 357 | 455 | 17.3 |
| Total "other" | 369 | 546 | 754 | 971 | 1180 | 14.8 |
| Total | 2484 | 3626 | 4990 | 6460 | 7951 | 14.9 |

*Source:* Frost and Sullivan Report 1564.

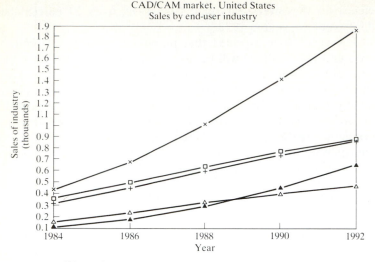

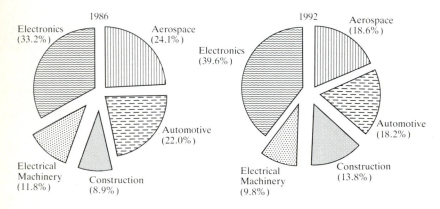

**FIGURE 1-4**
CAD/CAM market by end-user industry.

of the use of CAD/CAM within it, as well as continued overall rapid growth of the industry.

The expected growth of the CAD/CAM industry is shown segmented by application type in Table 1.4 and Fig. 1-5. The largest application area is mechanical. This includes mechanical numerical control and robot programming also. This segment is about $1.2 billion in size in 1985 and is forecast to grow to 2.9 billion in 1985 dollars by 1992. The corresponding average annual growth rate in that period is 13.3 percent.

The future of CAD/CAM is bright. The price of hardware is continually decreasing while its speed and performance is increasing. Thus, the price/

**TABLE 1.4**
**CAD/CAM market, United States—sales by application type**

| | 1984 | 1986 | 1988 | 1990 | 1992 | Growth per year 1985–1992, % |
|---|---|---|---|---|---|---|
| Sales by application type, 1985 $ millions | | | | | | |
| Mechanical | 1030 | 1426 | 1891 | 2390 | 2898 | 13.3 |
| Electronics | 502 | 827 | 1234 | 1688 | 2149 | 18.7 |
| Electrical | 100 | 133 | 167 | 197 | 223 | 9.9 |
| AEC | 404 | 621 | 891 | 1197 | 1519 | 17.2 |
| Mapping | 160 | 205 | 246 | 278 | 303 | 7.6 |
| Technical illustration | 141 | 207 | 285 | 369 | 452 | 14.9 |
| Other | 146 | 207 | 275 | 343 | 407 | 12.9 |
| Total | 2484 | 3626 | 4990 | 6460 | 7951 | 14.9 |
| Percent shares | | | | | | |
| Mechanical | 41.5 | 39.3 | 37.9 | 37.0 | 36.5 | −1.4 |
| Electronics | 20.2 | 22.8 | 24.7 | 26.1 | 27.0 | 3.3 |
| Electrical | 4.0 | 3.7 | 3.3 | 3.0 | 2.8 | −4.4 |
| AEC | 16.3 | 17.1 | 17.9 | 18.5 | 19.1 | 1.9 |
| Mapping | 6.4 | 5.7 | 4.9 | 4.3 | 3.8 | −6.4 |
| Technical illustration | 5.7 | 5.7 | 5.7 | 5.7 | 5.7 | 0.0 |
| Other | 5.9 | 5.7 | 5.5 | 5.3 | 5.1 | −1.8 |
| Total | 100.0 | 100.0 | 100.0 | 100.0 | 100.0 | |
| Growth rates | | | | | | |
| Mechanical | | 18.0 | 14.3 | 11.8 | 9.4 | −8.3 |
| Electronics | | 27.7 | 20.6 | 15.9 | 11.6 | −12.2 |
| Electrical | | 15.3 | 10.9 | 8.1 | 6.0 | −12.2 |
| AEC | | 23.9 | 18.7 | 15.1 | 11.6 | −9.9 |
| Mapping | | 13.0 | 8.5 | 5.7 | 4.1 | −15.7 |
| Technical illustration | | 21.1 | 16.3 | 13.0 | 9.9 | −10.1 |
| Other | | 19.0 | 14.2 | 10.9 | 8.2 | −11.7 |
| Total | | 20.9 | 16.3 | 13.1 | 10.1 | −9.8 |

*Source:* Frost and Sullivan Report 1564.

the future" hinges on the successful integration and automation of various CAD/CAM functions.

## 1.4 DEFINITION OF CAD/CAM TOOLS

In Sec. 1.1 we have defined CAD and CAM as subsets of the design and manufacturing processes respectively. Tables 1.1 and 1.2 list some CAD and CAM tools. In this section, we define CAD, CAM, and CAD/CAM tools. These definitions are based on practical and industrial use of the CAD/CAM technology. The definitions are broad enough to encompass many of the details that readers may wish to add.

Employing their constituents, CAD tools can be defined as the intersection of three sets: geometric modeling, computer graphics, and the design tools.

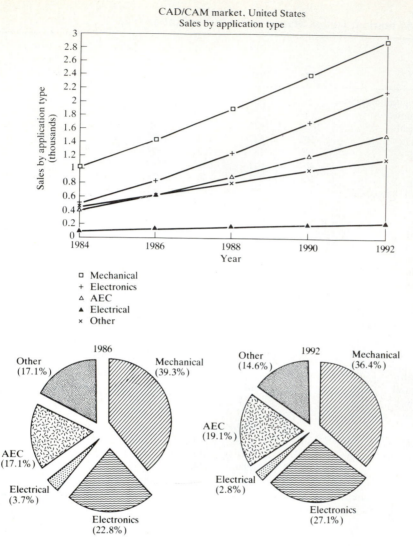

**FIGURE 1-5**
CAD/CAM market by application type.

performance ratio of hardware is always decreasing. CAD/CAM is rewarding itself. Designing better chips in less time and cost (via the existing CAD/CAM technology) enables better graphics hardware to be produced and more extensive graphics algorithms to be embedded (in these chips). The resulting firmware will dramatically improve the response time. Consequently, real-time simulations and interactions will be a reality. A key factor to the future success of CAD/CAM is the development of versatile tools for design and manufacturing applications. While progress has been made, CAD/CAM still needs further development to achieve full design and manufacturing automation. The concept of the "factory of

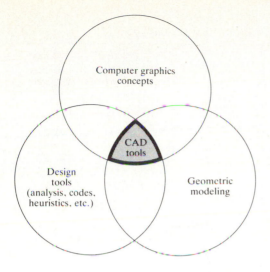

**FIGURE 1-6**
Definition of CAD tools based on their constituents.

Figure 1-6 shows such definition. As can be perceived from this figure, the abstracted concepts of geometric modeling and computer graphics must be applied innovatively to serve the design process. Based on implementation in a design environment, CAD tools can be defined as the design tools (analysis codes, heuristic procedures, design practices, etc.) being augmented by computer hardware and software throughout its various phases to achieve the design goal efficiently and competitively as shown in Fig. 1-7. The level of augmentation determines the design capabilities of the various CAD/CAM systems and the effectiveness of the CAD tools they provide. Designers will always require tools that provide them with fast and reliable solutions to design situations that involve iterations and testings of more than one alternative. CAD tools can vary from geometric tools, such as manipulations of graphics entities and interference checking, on one extreme, to customized applications programs, such as developing analysis and optimization routines, on the other extreme. In between these two extremes, typical tools currently available include tolerance analysis, mass property calculations, and finite element modeling and analysis—to name a few.

CAD tools, as defined above, resemble a guidance to the user of CAD technology. The definition should not, and is not intended to, represent a restriction

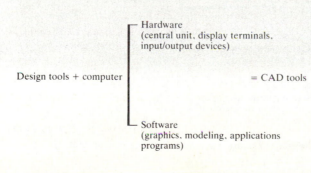

Design tools + computer = CAD tools

Hardware
(central unit, display terminals, input/output devices)

Software
(graphics, modeling, applications programs)

**FIGURE 1-7**
Definition of CAD tools based on their implementation in a design environment.

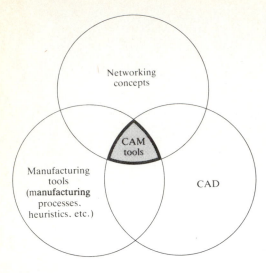

**FIGURE 1-8**
Definition of CAM tools based on their constituents.

on utilizing it in engineering design and applications. The principal purposes of this definition are the following:

1. To extend the utilization of current CAD/CAM systems beyond just drafting and visualization.
2. To customize current CAD/CAM systems to meet special design and analysis needs.
3. To influence the development of the next generation of CAD/CAM systems to better serve the design and manufacturing processes.

Similar to the definition of CAD tools, CAM tools can be defined as the intersection of three sets: CAD tools, networking concepts, and the manufacturing tools as shown in Fig. 1-8. This definition enforces the link between CAD and CAM as well as database centralization. The main elements to implement CAM into a manufacturing environment are shown in Fig. 1-9. There are two main factors that determine the success of this implementation. First, the link

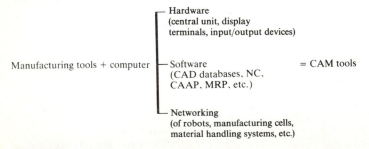

**FIGURE 1-9**
Definition of CAM tools based on their implementation in a manufacturing environment.

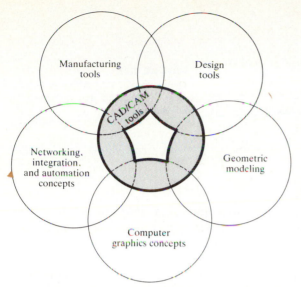

**FIGURE 1-10**
Definition of CAD/CAM tools based
on their constituents.

between CAD and CAM must be a two-way route. CAD databases must reflect manufacturing requirements such as tolerances and features. Designers must think in terms of CAM requirements when finalizing their designs. On the other hand, CAD databases and their limitations must be conveyed to manufacturing engineers who plan to utilize them in process planning and other manufacturing functions. It should be pointed out that not all manufacturing processes are, or need to be, computer driven.

The second factor that decides the success of CAM is the hardware and software networking of the various CAM elements to automate the manufacturing process. The factory of the future and its levels of automation are directly related to the soundness of the networking concepts. Timely synchronization among robots, vision systems, manufacturing cells, material handling systems, and other shop-floor tasks is one of the most challenging networking problems that face the implementation of CAM.

With the definitions of CAD and CAM tools in hand, the reader might speculate on the definition of CAD/CAM. Extending the philosophy of these former definitions, Fig. 1-10 presents the definition of CAD/CAM tools as the intersection of five sets: the design tools, the manufacturing tools, geometric modeling, computer graphics concepts, and networking concepts. If one argues that the ultimate goal of an engineering task is a consumer product, CAD/CAM can be defined as a subset of the product cycle augmented by computer, as shown in Fig. 1-11.

## 1.5 INDUSTRIAL LOOK AT CAD/CAM

Historically, thus far, CAD/CAM is a technology (both hardware and software) and applications driven field. Aerospace, automotive, and shipbuilding industries have influenced, to a great extent, the development of lofted and sculptured sur-

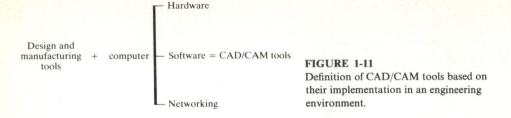

**FIGURE 1-11**
Definition of CAD/CAM tools based on their implementation in an engineering environment.

faces. Therefore, understanding the utilization and implementation of the CAD/CAM technology in an industrial environment helps to close the gap between creating the technology, managing it, using it, and more importantly learning it. Figure 1-12 shows, in a general sense, how a typical CAD/CAM system is utilized in a typical industrial environment. The figure shows the major components or packages that exist. The detailed capabilities and functions of each package as well as the various types of existing user interface are what makes these systems look entirely different. As a matter of fact, practical experience has proven that learning one system is sufficient to learn another one at a much faster pace. This faster pace is attributed to dealing with the same functions. All the user has to do is to adjust to the user interface and the management hierarchy of the new system. One might conclude that learning the generic basic concepts behind these systems does not only speed up the training curve of perspective users but it also helps them utilize the technology productively.

The principal packages available consist of geometric modeling and graphics, design, manufacturing, and programming software. The three available types of modeling are wireframes, surfaces, and solid modeling. The underlying theories of these modeling types are presented in Chaps. 5, 6, and 7 respectively. A wide variety of geometric entities or items are accessible by the designer under each modeling technique. Graphics encompass such functions as geometric transformations, drafting and documentation, shading, coloring, and layering. The design applications package includes mass property calculations, finite element modeling and analysis, tolerance stack analysis, mechanisms modeling, and interference checking. If a design or manufacturing application is encountered where the system's standard software can not be utilized, a customized software may be developed using the programming language provided. These languages are typically either system dependent or independent. Once the design is complete, drafting and documentation are performed on the model database. The model is now ready for CAM applications such as process planning, tool path generation and verification, inspection, and assembly.

This text is written with the above industrial look at CAD/CAM in mind. A coherent realization of the current CAD/CAM tools and their relationships to one another form an essential core to the learning process. Thus learning the basis of existing tools enhances both the utilization of current systems and the development of new design and manufacturing applications.

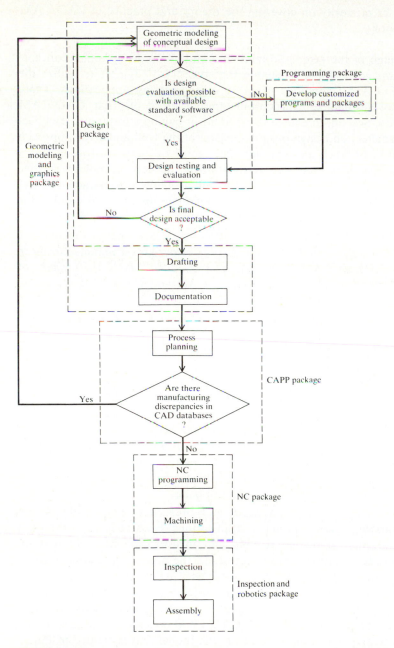

**FIGURE 1-12**
Typical utilizations of typical CAD/CAM systems in an industrial environment.

## 1.6   BOOK APPROACH

This book focuses primarily on covering the theory and practice of CAD/CAM concepts. The book:

1. Presents enough of the geometric modeling and computer graphics concepts to achieve two goals. First, to enable the student or engineer to utilize the existing CAD/CAM technology properly in engineering applications. There are always overestimated expectations among users of what CAD/CAM systems can do. Second, to enable the student or engineer to predict the effect of input parameters on design outputs instead of the trial-and-error approach that is widely used. Such an approach often results in user frustration and a significant decline in productivity when utilizing the new technology relative to manual procedures. The user should be able to know the characteristics of a B-spline surface, or a Bezier surface, for example, in order to control the surface behavior in a particular design application.

2. Presents enough CAM concepts to achieve two goals: first, to enable the student or engineer to recognize the influence of design on manufacturing in general and CAD on CAM in particular; second, to enable the student or engineer to appreciate the subtleties facing CAD and CAM integration and automation.

3. Relates the book topics to the current practice of the CAD/CAM technology. Each topic in the book is covered first from a theoretical point of view and then followed by its implementation, if possible, on CAD/CAM systems. This approach is extended to the "Problems" section at the end of each chapter. This section is divided into three parts. Problems of the first part will test the concepts introduced in the chapter. Problems of the second part will require a CAD/CAM system. Third-part problems will be on programming. This approach of mixing theory and applications will better prepare students and will help engineers evaluate CAD/CAM systems. It also gives instructors additional flexibility in choosing the types of problems that reflect their teaching philosophy and emphasis. Some might emphasize first and second parts, first and third parts, second and third parts, or all parts.

4. Presents the material in enough depth and breadth to help students fully utilize the various concepts in actual applications. Each chapter of the book has a bibliography for further reading and investigation in order to fulfill specialized needs of certain groups that utilize the book. These bibliographies are essential for an emerging area such as CAD/CAM. The book also attempts, as much as possible, to point out the integration among the various concepts. Thus, the student will be able to see the return on investment of spending time to perform design on CAD/CAM systems. For example, the student will be required to utilize a geometric model that is created in Part II in Parts III, IV, V, and VI of the book.

The development of the book has taken into consideration the wide variety of hardware and software configurations that various schools utilize in conjunction with their CAD/CAM curricula. The book material is suitable for any configuration as long as it provides students with three basic packages: modeling and graphics, applications, and programming. The difference comes in the degree of emphasizing one topic versus another and the types of problems that students can handle. For example, if a configuration does not have solid modeling in the modeling and graphics package, the instructor will adjust accordingly when solid modeling material is covered.

## 1.7 BOOK ORGANIZATION

The book material is organized to reflect the definitions of CAD/CAM tools, the book approach, and the current practice of CAD/CAM technology. The book presents ideas and strategies that are effective for a large class of engineering problems. It also organizes these ideas in a creative fashion. Some of the highlights of the organization are the sections on design and engineering applications, the problems, and the bibliographies. These sections present generic solutions and approaches that have been developed and tested on typical CAD/CAM systems for various design problems. It is recommended that the student utilizes and adapts these solutions to a particular system.

The book is organized in a logical order into six major parts. Part I presents an overview of CAD/CAM systems—both hardware and software. It sets the stage for the following parts and establishes the relative relationships between the various topics of the book. It also provides the necessary background and terminology underlying existing CAD/CAM systems. Moreover, the knowledge gained in this part is related later on to the theoretical concepts covered in other chapters. For example, the raster scan displays and hardware resolution covered in Chap. 2 will be related to the topic of shading and coloring in Chap. 10. Similarly, Chap. 9 will show how the homogeneous geometric transformation relates both the database and the working coordinate systems covered in Chap. 3 to each other.

The motivation to include Part I in the book stems from the fact that there are many similarities among current various CAD/CAM systems which, if they are understood correctly, will make the utilization and evaluation of these systems more effective. This first chapter is an introduction which provides the rationale for learning about the CAD/CAM field and its relation to product design and manufacturing. Chapter 2 presents a generic description of available hardware. Various existing CAD/CAM systems are classified based on their hardware configurations. Advantages and disadvantages of each configuration are presented. Input and output devices are also covered in the chapter.

Chapter 3 complements the material of Chap. 2 and extends its philosophy to software. The major goal of this chapter is to show that CAD/CAM systems are very similar to each other and conceptually utilize the same concepts. To achieve this goal, basic definitions most often utilized by CAD/CAM software are introduced. In addition, the major software modules available to designers and engineers are provided with their available capabilities. Details of the basic concepts governing these modules are covered in later chapters of the book. Chapter 4 is dedicated to PC-based CAD/CAM which is a viable solution to many organizations and institutions.

Part II deals with the subject of geometric modeling. This part presents the available three types of modeling techniques: wireframes (curves), surfaces, and solids. Chapters 5, 6, and 7 cover these techniques respectively in a consistent fashion. Each chapter deals first with the various geometric entities provided by each modeling technique. This is followed by the manipulations' functions such as displaying, trimming, intersection, and projection. Examples are provided throughout the chapters. In addition, there is the design and applications section in each chapter which combines the material in each chapter into relevant engineering problems. Chapter 8 ends this part by covering the exchange of CAD/CAM databases among various CAD/CAM systems.

Part III is a logical flow of Part II. It treats the basic graphics concepts which when applied to geometric models result in the versatile and basic visual tools CAD/CAM systems provide. Chapter 9 introduces the concepts of homogeneous transformation. Chapter 10 covers algorithms for hiding surfaces and solids, shading, and coloring.

While the first three parts of the book deal with the fundamentals of CAD/CAM, the last three parts harness the acquired knowledge and cover the related CAD and CAM tools, and applications. Part IV deals with the interactive tools that are typically available to designers on all today's CAD/CAM systems. These tools facilitate the creation and management of CAD/CAM databases.

Chapter 11 presents the available graphics aids. Some of these aids such as geometric modifiers and grids help minimize the amount of calculations to create a geometric model. Others such as names and groups help manage and manipulate the model. This chapter also introduces the concept of layering which is important for managing graphics of large and extensive models. Chapter 12 deals with manipulating graphics of geometric models for creation, visualization, and calculation purposes. Chapter 13 covers animation procedures. Examples of how to employ animation in engineering studies are included. Chapter 14 treats the concepts of mechanical assemblies and its merits to designing products consisting of more than one component. Modeling and representation schemes of assemblies are covered. Part IV ends with Chapter 15 which presents interactive programming as a useful tool. In many engineering problems, there is a need to customize the available tools. In the case of the CAD/CAM subject, it is desired to customize CAD/CAM systems to meet particular needs and increase productivity. Examples and problems are covered in the chapter.

Part V describes the most widely available CAD applications on CAD/CAM systems. The basic concepts of each application, its implementation into software, and its utilization in design problems are covered. Chapter 16 deals with the basics and applications of mechanical tolerancing. Chapter 17 covers the mass property calculations. Chapter 18 deals with finite element modeling and analysis and discusses the issues related to interfacing modeling and analysis. Chapter 19 discusses how design projects with CAE focus can be developed to reflect the capabilities of CAD/CAM systems.

Part VI discusses the issue of integrating CAD and CAM databases. Chapter 20 describes the available prominent CAM applications such as process planning and part programming. The chapter also discusses the influence of CAD on CAM.

# PROBLEMS

## Part 1: Theory

**1.1.** CAD/CAM is an interdisciplinary and dynamic field. Many magazines and journals exist or appear frequently. Go to your library and make a list of available journals, magazines, and newsletters in CAD, CAM, CAE (computer aided engineering), and CIM (computer integrated manufacturing). Refer to the bibliography section in this chapter as a start.

**1.2.** The development of the Sketchpad and APT are considered to be the turning point in establishing the CAD and CAM fields respectively. Describe, in detail, the objectives and developments of each project when they were initiated in the 1950s and 1960s.

**1.3.** Most often when you talk to designers who utilize CAD/CAM systems in their design work, they mention that these systems should provide them with more design capabilities. Why? Do you have any suggestions to help them? Make a wish list of your favorite design capabilities which you would like to see provided by the CAD/CAM technology.

**1.4.** Most often when you talk to managers of CAD/CAM systems, in industry, about the evaluation of these systems, they feel that productivity gains promised to them by using the systems are hard to achieve. Why? Identify the various factors that influence these gains.

**1.5.** Make a list of some of the engineering projects you were involved with and apply the definitions of CAD, CAM, and CAD/CAM covered in Sec. 1.4 to them.

**1.6.** Apply the design and CAD tools defined in Sec. 1.1 to some of the design projects you identified in Prob. 1.5. Draw the corresponding flowchart for each one. What are the specific CAD tools that each project requires?

**1.7.** Some design applications would be better with CAD than others. Why? What characterizes an application as a good candidate for CAD?

**1.8.** Discuss the benefits of CAD/CAM to engineering design as compared to conventional methods.

## Part 2: Laboratory

**1.9.** Familiarize yourself with the CAD/CAM system you will use to do the laboratory assignments at the end of each chapter. Make sure you can use the user interface and documentation provided by the system.

## Part 3: Programming

**1.10.** Familiarize yourself with the programming languages available to you on the CAD/CAM system. Make sure you can deal with files (creation, editing, deletion, etc.), and compile and execute computer programs.

# BIBLIOGRAPHY

Chang, T. C., and R. A. Wysk: *An Introduction to Automated Process Planning Systems*, Prentice-Hall, 1985.
Chasen, S. H.: "Historical Highlights of Interactive Computer Graphics," *Mech. Engng*, vol. 103, no. 11, 1981.

Childs, J. J.: *Principles of Numerical Control*, Industrial Press, New York, 1969.

Dieter, G.: *Engineering Design*, McGraw-Hill, 1983.

Everett, R. R.: "The Whirlwind I Computer," *Joint AIEE-IRE Conf., 1952*, Rev. Electron. Digital Comput., February 1952.

Foley, J. D., and A. van Dam: *Fundamentals of Interactive Computer Graphics*, Addison-Wesley, 1982.

Frost and Sullivan, Inc.: "CAD/CAM Market in the United States," Report 1564, New York, 1986.

Haug, E. H., and J. S. Arora: *Applied Optimal Design; Mechanical and Structural Systems*, John Wiley, 1979.

Jacks, E.: "A Laboratory for the Study of Man-Machine Communication," AFIPS, FJCC 26, pt. 1, 1964.

Newman, W. M., and R. F. Sproull: *Principles of Interactive Computer Graphics*, 2d ed., McGraw-Hill, New York, 1979.

Ninke, W.: "GRAPHIC 1—A Remote Graphical Display Console System," AFIPS, FJCC 22, pt. 1, 1965.

Pusztai, J., and M. Sava: *Computer Numerical Control*, Reston, A Prentice-Hall Company, Reston, Va., 1983.

Requicha, A. A. G., and H. B. Voelcker: "Solid Modeling: A Historical Summary and Contemporary Assessment," *IEEE Computer Graphics and Applic.*, vol. 2, no. 2, March 1982.

Requicha, A. A. G., and H. B. Voelcker: "Solid Modeling: Current Status and Research Directions," *IEEE Computer Graphics and Applic.*, vol. 3, no. 7, October 1983.

Shigley, J. E., and L. D. Mitchell, *Mechanical Engineering Design*, 4th ed., McGraw-Hill, 1983.

Steidel, Jr., R. F., and J. M. Henderson: *The Graphic Language of Engineering*, John Wiley, New York, 1983.

Sutherland, I. E.: "SKETCHPAD: A Man-Machine Graphical Communication System," *Spring Joint Computer Conf.*, Spartan, Baltimore, Md., 1963.

Teicholz, E.: *CAD/CAM Handbook*, McGraw-Hill, 1985.

Vanderplaats, G. N.: *Numerical Techniques for Engineering Design: With Applications*, McGraw-Hill, 1984.

Zeid, I.: "Understanding Turnkey CAD/CAM System Capabilities: Overview of the Computervision CDS 4000," *J. Engng Computing and Applic.*, vol. 1, no. 2, Winter 1987.

Zeid, I., and T. Bardasz: "The Role of Turnkey CAD/CAM Systems in the Development of the 'Graphysis' Concept," *Proc. 1985 ASME International Computers in Engineering Conf. and Exhibit*, Boston, Mass., August 4–8, 1985.

# CHAPTER
# 2

## CAD/CAM
## HARDWARE

## 2.1 INTRODUCTION

CAD/CAM systems are usually known for their fast interactive response and graphics display. It is a common question to ask: what are the differences between a CAD/CAM system and a conventional computer system? The answer is related to the two basic elements of the system, that is, hardware and software. The differences in hardware lie, in general, in the specialized input and output devices typically required by a CAD/CAM system to handle interactive graphics inputs and displays. Input devices may include alphanumeric keyboards, programmed-function keyboards, digitizing tablets, lightpens, electronic styluses, trackballs, mouse systems, and touch input devices. Output devices may include conventional plotters, hardcopy printers, and more importantly graphics display terminals. These terminals are quite different from the conventional video display, or video terminals, usually referred to as computer terminals which generally display only text. Graphics display terminals possess various local processors and controllers that can perform various graphics functions, such as transformations and graphics generation, locally at the hardware level to reduce the interactive time response between the user and the system.

The unique characteristics of software that contribute to the apparent distinct look of CAD/CAM systems are the integration and user interfaces. Such software typically provides the user with a full range of functions and modules that all relate to a centralized database. While the core of CAD/CAM software is written in conventional programming languages such as FORTRAN or C, various user interfaces vary distinctively from one system to another and quite sharply from conventional interfaces such as data files or user dialogues encountered in typical engineering software.

CAD/CAM hardware has progressed steadily from slow specialized systems to fast standard ones. In the 1970s and early 1980s, the majority of commercially available CAD/CAM systems were based on 16-bit word minicomputers. It was typical then to find turnkey vendors who designed and manufactured their own hardware to run their software. As a result, users of such systems were faced with the problem of dealing with nonstandard operating systems which most often resulted in system isolation. In addition, interfacing and networking these systems with other computers were not feasible. Other vendors have envisioned the importance of standard hardware and have consequently adapted it to fit the architecture and configuration of their systems. Bundled systems were common at that time. A bundled system is a packaged hardware and software that is sold and maintained by a single vendor. It is also known as a turnkey system. With the constantly increasing demand in performance and diversity of utilizations, CAD/CAM systems have migrated to 32-bit word minicomputers to provide the accuracy and support calculation intensity required by applications such as finite element analysis and solid modeling. These systems are commonly unbundled and offer users the flexibility to choose the optimum hardware and software configurations to meet their needs.

The majority of today's CAD/CAM systems utilizes open hardware architecture and standard operating systems. Open hardware architecture implies that CAD/CAM vendors no longer design and manufacture their own hardware platforms. Instead the CAD/CAM industry relies upon the giant general-purpose computer companies and smaller firms that specialize in engineering workstations. Thus users can network the CAD/CAM systems to other computer systems as well as hardwire them to various manufacturing cells and facilities. They can also run third party software to augment the analysis capabilities typically provided by CAD/CAM vendors. With the advancements in IC (integrated circuit), PC (printed circuit), and VLSI (very large scale integration) technology, current CAD/CAM systems are based on the workstation concept. Such a concept provides both single-user and timesharing environments. These advancements have resulted in reducing the developing and manufacturing costs and time of chips. It has therefore become feasible to develop firmware by embedding calculations and graphics-intensive algorithms into chips to speed up their executions instead of developing conventional software.

The microcomputer (PC)-based CAD systems have been developing remarkably in the past few years. The conventional problems of memory size, processing speed, and memory-accessing speed seem to be going away. Similarly, peripheral storage has been enhanced by developing high-capacity fixed (hard) disks with high access speeds. User-interaction techniques have also been developing rapidly.

CAD/CAM systems based on either the microcomputer or workstation concept represent a distinct philosophy or trend in hardware technology which is based on a distributed (stand-alone) but networked (linked) environment. Workstations can be linked together as well as to mainframes dedicated to numerical computations. Other processors may exist in the network to control other types of hardware such as file and print servers. These distributed systems are able to perform major graphics functions locally at the workstations, and operations that require more power are sent to the mainframe. The communication between devices in this distributed design and manufacturing environment becomes an important part of the system configuration and design.

The dynamics and rapid changes in the hardware technology have created an absorption problem at the user's part. There are always various types and configurations of CAD/CAM systems to choose from. To choose and implement a system in an industrial or educational environment requires the development of a set of guidelines that must address both hardware and software requirements. Section 2.3 discusses how to choose a system. A key factor in a system evaluation is the capabilities and integration of its software which influence the productivity rate directly.

Managing a CAD/CAM system (as other computer systems) covers both hardware and software. CAD/CAM managers are typically responsible for day-to-day operation and maintenance of the system, developing training programs for users, and keeping informed of the latest hardware and software trends. Day-to-day operation involves developing a sign-up scheme because there are never enough terminals for unlimited access, reporting hardware problems and software bugs to the vendor, dealing with users' problems and questions, and keeping a record of system utilization. Developing a training program for users is an important function of the manager. Another function of a manager is to keep informed of the latest products and trends. This typically requires attending vendor and/or users' group meetings, as well as trade shows/exhibits and conferences. Thus the manager is always in a position to recommend future upgrades of the existing system.

The main objective of this chapter is to provide generic descriptions of CAD/CAM hardware and its related terminology. We mainly cover the types of architectures of CAD/CAM systems, input devices, and output devices.

## 2.2   TYPES OF SYSTEMS[1]

The types of CAD/CAM systems available to end users are quite diverse. Within each type, there exist various configurations and options. A generic classification of these systems and a general description of each type and its features are beneficial to understand hardware trends. The classification criteria are subjective and can vary significantly from one occasion to another. One might classify systems on the basis of their hardware configuration and performance. A second person might choose software capabilities as a criterion or systems may be categorized on the basis of their main geometric modeling technique, whether it is wireframe, surface, or solid modeling.

In this section, CAD/CAM systems are classified based on their hardware. More specifically, the type of host computer that drives the system is the major factor in the classification. While various configurations and peripherals may exist within each type, this is less important here. Some of the types described below are of less use than others due to the rapid changes in the hardware technology or due to the newness of the hardware concept itself. Nevertheless, it is interesting and informative to be aware of the development of this technology.

---

[1] This section includes excerpts from the Frost and Sullivan report 1564.

## 2.2.1 Mainframe-Based Systems

At one point all CAD/CAM systems had been mainframe-based since that was the only type of computer available. A typical mainframe-based CAD/CAM system consists of one or more design/drafting station. Each of these include, at a minimum, a graphics display, an alphanumeric control display (possibly integrated with the graphics screen), and a keyboard. The designer/drafting station is linked to the mainframe computer. To the latter are also linked plotters, printers, storage devices, digitizing boards, alphanumeric terminals, and possibly other output devices. Design/drafting stations may be equipped with input devices such as lightpens, joysticks, or other cursor-control components. They may also have the programmed-function keyboard (PFK) instead of a conventional keyboard or a digitizing tablet to enter commands.

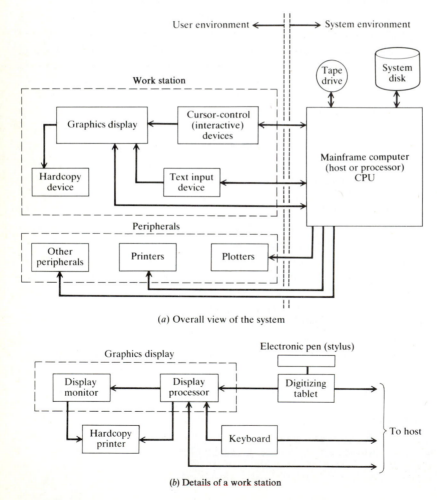

(a) Overall view of the system

(b) Details of a work station

**FIGURE 2-1**
Schematic diagram of a typical mainframe-based CAD/CAM system.

Figure 2-1 shows a schematic of the overall system components as well as details of a work station. The computer environment is typically divided into two: the user environment and the system environment. The first signifies the components and the area to which the user can have access. These include primarily work stations and peripherals. There is always a maximum number of work stations that a host can support to avoid degradation of response time between users and the system. Users spend most if not all of their time on work stations to perform their work. They normally know how to use and operate the various components of the stations and peripherals. A typical work station consists mainly of two major segments, input and output devices. The former includes cursor control devices for graphics input and text input devices. The cursor can be controlled via a lightpen, joystick, mouse, electronic pen (stylus) with a digitizing tablet, thumbwheel, or trackball. Text can be input through a keyboard which may have programmed-function keys. Output devices consist of a graphics display with a hardcopy printer (usually a dot matrix printer) to provide convenient raster plots of full screen contents. Figure 2-1*b* shows the details of a work station with specific input and output devices. In this figure, the cursor-control device is an electronic pen and a digitizing tablet and the keyboard is the text input device. The graphics display is shown as a display monitor and processor.

Figures 2-2 and 2-3 show examples of existing commercial CAD/CAM systems which are based on the mainframe concept. The mainframe-based

**FIGURE 2-2**
Mainframe-based CAD/CAM system. (*Courtesy of McDonnell Douglas Information Systems Group.*)

**FIGURE 2-3**
Mainframe-based CAD/CAM system. (*Courtesy of Computervision Corp.*)

CAD/CAM solution is suitable to integrate with the existing mainframe in a particular organization. Thus CAD/CAM users have access to accounting, planning, and management information systems (MIS) databases. A requirement when using a mainframe for CAD/CAM is that the system provides high priority to CAD/CAM users so that the response time is adequate. Should a large job tie the computer down, the ability of CAD/CAM users to concentrate on their work would be impaired. This is particularly the case when there are random fluctuations in system load, affecting the CAD/CAM operators. Of course, placing too many CAD/CAM stations on the mainframe can have the same effect.

### 2.2.2 Minicomputer-Based Systems

The development of LSI (large scale integrated) circuits, and now VLSI, has changed the basic principles of computer architecture and has directly led to the proliferation of minicomputers during the 1970s. Minicomputers began to take over from mainframe computers in the 1970s as it was possible to configure them at costs less than the basic cost of communication networks with a mainframe. Early versions of minis were 16-bit word, slow, and limited-storage computers. The DEC (Digital Equipment Corporation) PDP series offers a typical example.

In the late 1970s, the arrival of superminicomputers, such as the VAX 11/780, with 32-bit word and virtual memory operating systems, boosted CAD/CAM applications and helped decentralize them from mainframes. Minicomputers have enabled the rapid growth of the CAD/CAM industry. Their low-cost end-user control, generally trouble-free programming, and small size have all been significant factors in getting vendors and customers interested. The superminicomputers of today are available with speeds, accuracy, and storage that are more adequate for sophisticated CAD/CAM, including computations. Additionally, users of many minis have the option of employing devices that speed up certain computations, the latter ranging from board-level products to host computers. Array processors and other special-purpose hardware are increasingly utilized to speed up array manipulations.

Minicomputer-based CAD/CAM systems look similar to mainframe-based ones, only the computer is smaller. Figure 2-1 can serve as a schematic diagram of a typical minicomputer-based CAD/CAM system by replacing the host or the processor with a mini. The number of design/drafting stations is typically smaller with minis. It is common to configure a mini-based CAD/CAM system with a dedicated mini which may be networked to other MIS computers. This helps keep the interactive time as small as possible.

Most of the supermini-based CAD/CAM systems are sold as turnkey systems. A turnkey system is defined as a computer hardware and software configuration that is provided by one supplier. The configuration depends on the range of applications that users are interested in and on the financial investment to be made. Turnkey systems' suppliers, or turnkey vendors for short, offer a wide range of postsales support, hardware and software maintenance, technical training, documentation, and consultation.

Turnkey systems can be classified into two types based on the supermini utilized in their configurations. In the first type, the vendor designs and manufactures its own proprietary superminicomputer as well as develops the OS (operating system) necessary to run this specialized hardware. This type existed in the 1970s and early 1980s when adequate computers were not available for CAD/CAM. A typical example is Computervision's Designer series and CDS 4000 system. Figure 2-4 shows the various components of the CV Designer V system. The second type of turnkey system was the most popular in the 1980s. Vendors typically acquire a variety of industry standard hardware platforms (such as those offered by SUN, HP, DEC, and IBM) and, therefore, standard OSs as the basis of their systems. Vendors usually offer their software on more than one platform in an attempt to attract potential buyers. Figures 2-5 to 2-7 and 2-9 show examples of some existing configurations. Figure 2-8 shows a schematics of a typical enhanced standard minicomputer where graphics boards (GPU) are added to improve system response.

### 2.2.3   Microcomputer-Based Systems

Microcomputers, originally popularized by the Apple Computer, have impacted the CAD/CAM field greatly. The advent of the IBM Personal Computer (PC) provided the first significant impetus for CAD on micros. Currently, there exists a

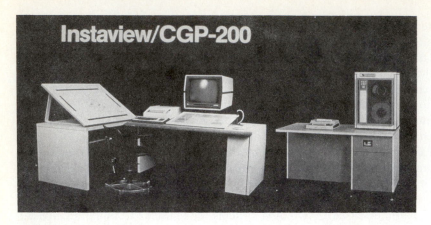

**FIGURE 2-4**
Typical configuration of a turnkey CAD/CAM system based on a specialized superminicomputer. (*Courtesy of Computervision.*)

wealth of CAD software for PCs ranging from two-dimensional drafting to three-dimensional modeling and applications. There are two main factors for the popularity and fast emergence of micro-based systems. First, the speed, size, and accuracy problems are going away. Microcomputers of a 32-bit word length are available with enough memory size, disk storage, and speed for CAD/CAM applications. Second, various applications programs have matured and cover most, if not all, users' needs.

Micro-based CAD/CAM systems, such as IBM PS/2 and Macintosh IICx, generally utilize one computer per graphics terminal. This allows rapid and predictable response to the operator's commands. If there are several systems in an

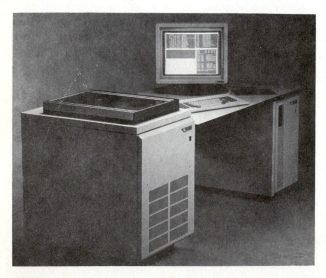

**FIGURE 2-5**
Typical configuration of a turnkey CAD/CAM system based on SUN 3. (*Courtesy of Computervision.*)

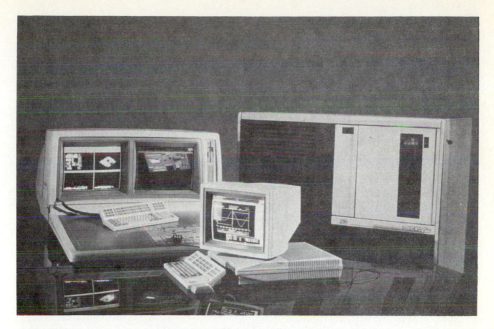

**FIGURE 2-6**
Typical configuration of a turnkey CAD/CAM system based on Digital MicroVAX II. (*Courtesy of Intergraph.*)

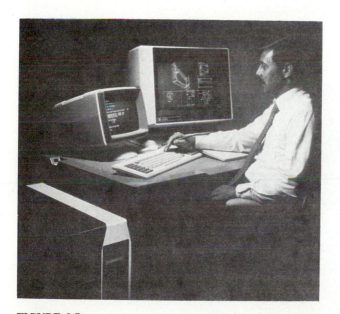

**FIGURE 2-7**
Typical configuration of a turnkey CAD/CAM system based on Apollo DN570A. (*Courtesy of GE (U.S.A.) Calma.*)

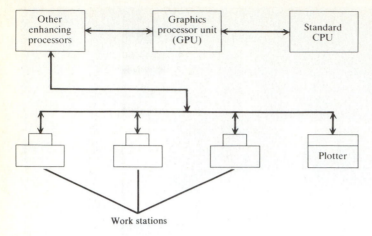

**FIGURE 2-8**
Schematic of a typical enhanced minicomputer.

environment, one is generally designated as the server. This unit may control the plotter and other unique devices. The other units are networked to the server to be able to access the peripherals and possibly software and data. Figure 2-10 shows an example of a micro-based system that uses a PC-AT with an EGA (enhanced graphics adaptor) display. Figure 2-11 shows an example of other systems that upgrade the PC-AT hardware and replace the EGA displays with higher resolution and better performance ones.

**FIGURE 2-9**
Typical configuration of a turnkey CAD/CAM system based on an enhanced VAX 11/780. (*Courtesy of Intergraph.*)

**FIGURE 2-10**
Typical microcomputer-based CAD/CAM system. (*Courtesy of Auto-trol Technology.*)

**FIGURE 2-11**
Typical enhanced microcomputer-based CAD/CAM system. (*Courtesy of Adage, Inc.*)

## 2.2.4 Workstation-Based Systems

The workstation concept underlying these systems has emerged from the simple downward evolution of well-established systems technology into a single-user or an office environment. For CAD/CAM applications, the concept offers significant advantages over the timesharing, central computing facility accessed through graphic display terminals offered by mainframe- or supermini-based systems. Among these advantages offered by workstations are their availability, portability, the ability to dedicate them to a single task without affecting other users, and their consistency of time response. By supplying engineering professionals with their own dedicated computing resources, workstations have proven their effectiveness in shortening time to complete typical engineering tasks. Studies have shown that a user's train of thought is broken and productivity suffers when the system's response to a user's command is not within one second. More specifically, an IBM study of system response time showed that each one-tenth-second decrease in this time led to a four-minute reduction in time required to complete the experimental design task.

Graphics terminals attached to mainframes, minis, or PCs do not qualify as workstations. These terminals may be referred to as "work stations" (two words). A workstation (sometimes also referred to as a personal, technical, or engineering workstation) can be defined as a "work station" with its own computing power to support major software packages, multitasking capabilities demanded by increased usage and complex tasks, and networking potential with other computing environments. A set of standards have been adopted since 1981 to guide the development of workstations and to differentiate them from low-cost 16-bit PCs and graphics terminals. The most important of these standards are the 32-bit architecture, Unix operating system, and Ethernet local area network. However, the rapid development of graphics terminals, PCs, and workstations is making the dichotomy between them difficult to sustain. One might look at the workstation concept as a trend that is placing downward pressure on the PCs and upward pressure on the high-end graphics terminals, obviously suggesting the diminishment of both.

The workstation concept seems to form the basis of the next generation of CAD/CAM systems. The style of distributed computing offered by CAD/CAM workstations is preferred over that of central computing offered by mainframe- or superminicomputer-based CAD/CAM systems. The argument for such preference is typically based on the inadequate performance of multiuser environments provided by the latter systems, the networking potential of workstations, and low entry costs. Figures 2-12 to 2-16 show examples of various workstations that are currently available.

## 2.3 CAD/CAM SYSTEMS EVALUATION CRITERIA

The various types of CAD/CAM systems have been discussed in the previous section. The implementation of these types by various vendors, software developers, and hardware manufacturers result in a wide variety of systems, thus making the selection process of one rather difficult. CAD/CAM selection committees find themselves developing long lists of guidelines to screen available

**FIGURE 2-12**
Workstation, based on Appolo DN 3000 workstation, for Auto-trol software. (*Courtesy of Auto-trol Technology.*)

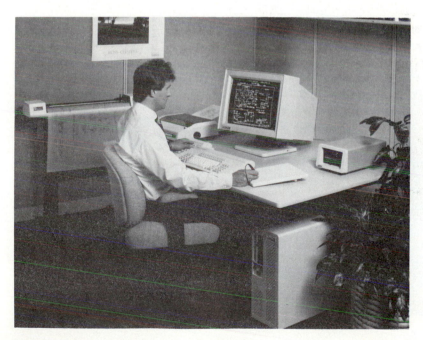

**FIGURE 2-13**
Workstation, based on Adage hardware, for CADAM software. (*Courtesy of Adage, Inc.*)

**FIGURE 2-14**
Tektronix graphics workstations. (*Courtesy of Tektronix, Inc.*)

**FIGURE 2-15**
Workstation, based on DEC VAXstation II/GPX, for Auto-trol software. (*Courtesy of Auto-trol Technology.*)

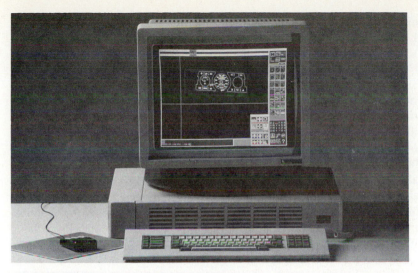

**FIGURE 2-16**
Workstation, based on SUN hardware, for CADDS 4X software. (*Courtesy of Computervision.*)

choices. These lists typically begin with cost criteria and end with sample models or benchmarks chosen to test system performance and capabilities. In between comes other factors such as compatibility requirements with in-house existing computers, prospective departments that plan to use the system, and credibility of CAD/CAM systems' suppliers.

In contrast to many selection guidelines that may vary sharply from one organization to another, the technical evaluation criteria are largely the same. They are usually based on and are limited by the existing CAD/CAM theory and technology. These criteria can be listed as follows.

### 2.3.1   System Considerations

**2.3.1.1   HARDWARE.**   One of the main issues to consider is how open and standard the involved computer is. Some of the standard computers such as VAX or SUN Workstations utilized in CAD/CAM systems are enhanced by the vendors to improve the performance of their software. In doing so, they might disable some of the functions of the original hardware. This might impair using the CAD/CAM computer to support third party software. The open architecture of the hardware is crucial to system expandability and future networking needs. The types of peripherals supported by the system are also important to the documentation phase of design work performed on the system.

Two popular hardware configurations are disked and diskless workstations. In the former, each workstation has enough local disk space and memory to be truly stand-alone. In the latter, each workstation is connected to a central computer, called the server, which has enough large disk and memory to store users' files and applications programs as well as executing these programs. In the diskless configuration, it is recommended that each workstation has enough swapping disk to avoid degrading the response time, which is crucial when CAD/CAM is involved.

**2.3.1.2 SOFTWARE.** Three major contributing factors are the type of operating system the software runs under, the type of user interface (syntax), and the quality of documentation. The first factor is important to perform CAE work which typically involves running third party software. In addition, it is psychologically easier to teach new CAD/CAM users a standard operating system, if they do not already know it, than a nonstandard one. Standard systems also become important when in-house software developments are needed.

In evaluating the type of user interface, care should be given to whether the interface can accommodate both experienced and new users. For example, while the menu-driven systems are preferred by beginners, they represent an uncomfortable rigid structure for day-to-day advanced operators. Some systems might offer both menu-driven and non-menu-driven interfaces. Furthermore, the ability to create customized menus should be investigated.

The quality of documentation of software is perhaps the easiest criterion to evaluate by simply reading the relevant manuals. The most important issue to investigate is how easy it is to use the documentation and if there are online help functions available or not.

**2.3.1.3 MAINTENANCE.** Repair of hardware components and software updates comprise the majority of typical maintenance contracts. The annual cost of these contracts is substantial (about 5 to 10 percent of the initial system cost) and should be considered in deciding on the cost of a system in addition to the initial capital investment. Various vendors offer various types of contracts ranging from an immediate service to an on-call basis service.

**2.3.1.4 VENDOR SUPPORT AND SERVICE.** Vendor support typically includes training, field services, and technical support. Most vendors provide training courses, sometimes on-site if necessary. The timely response of the vendor response centers to customer's technical questions is important during the start-up time when no in-house technical expertise is available.

## 2.3.2 Geometric Modeling Capabilities

**2.3.2.1 REPRESENTATION TECHNIQUES.** The geometric modeling module of a CAD/CAM system is its heart. The applications module of the system is directly related to and limited by the various representations it supports. Wireframes, surfaces, and solids are the three types of modeling available. Most commercial CAD/CAM systems provide them. However, it is important to consider the various entities supported by each representation. The integration between the various representations and the applications they support is an essential issue.

**2.3.2.2 COORDINATE SYSTEMS AND INPUTS.** In order to provide the designer with the proper flexibility to generate geometric models, various types of coordinate systems and coordinate inputs ought to be provided. A working coordinate system is an example of the former. The ability of the designer to define the proper planes of constructions is valuable to create faces that are not parallel to the standard orthogonal planes (front, top, right). Coordinate inputs can take the form of cartesian $(x, y, z)$, cylindrical $(r, \theta, z)$, and spherical $(\theta, \phi, z)$.

**2.3.2.3  MODELING ENTITIES.** The fact that a system supports a representation scheme is not enough. It is important to know the specific entities provided by the scheme. The ease to generate, verify, and edit these entities should be considered during evaluation.

**2.3.2.4  GEOMETRIC EDITING AND MANIPULATION.** It is essential to ensure that these geometric functions exist for the three types of representations. Editing functions include intersection, trimming, and projection, and manipulations include translation, rotation, copy, mirror, offset, scaling, and changing attributes.

**2.3.2.5  GRAPHICS STANDARDS SUPPORT.** If geometric models' databases are to be transferred from one system to another, both systems must support exchange standards. These standards are valuable if various systems exist in one organization or if design models are shipped to outside vendors for various reasons such as tools, jigs, and fixture designs, or generating NC part programs. It should be mentioned here that graphics standards may introduce inconsistencies and errors into CAD/CAM databases.

### 2.3.3  Design Documentation

**2.3.3.1  GENERATION OF ENGINEERING DRAWINGS.** After a geometric model is created, standard drafting practices are usually applied to it to generate the engineering drawings or the blueprints. Various views (usually top, front, and right side) are generated in the proper drawing layout. Then dimensions are added, hidden lines are eliminated and/or dashed, tolerances are specified, general notes and labels are added, etc. These activities are time-consuming. To generate an engineering drawing, it typically takes as much as two to three times as long as it takes to generate the geometric model.

### 2.3.4  Applications

**2.3.4.1  ASSEMBLIES OR MODEL MERGING.** Generating assemblies and assembly drawings from individual parts is an essential process. The two issues that are worth investigation are the assembly procedure itself and the clean-up of the resulting assembly.

**2.3.4.2  DESIGN APPLICATIONS.** There are design packages available to perform applications such as mass property calculations, tolerance analysis, finite element modeling and analysis, injection modeling analysis, and mechanism analysis and simulation. What should be evaluated are the capabilities of these packages, their integration and interfaces with geometric databases, and the representation techniques they utilize. Some packages might require clumsy user input while others might lack the proper way to display results.

**2.3.4.3 MANUFACTURING APPLICATIONS.** The common packages available are tool path generation and verification, NC part programming, postprocessing, computer aided process planning, group technology, CIM applications, and robot simulation. It is essential to ensure that the CAD and CAM applications that are provided by the system are truly integrated.

**2.3.4.4 PROGRAMMING LANGUAGES SUPPORTED.** It is vital to look into the various levels of programming languages a system supports. Attention should be paid to the syntax of graphics commands when they are used inside and outside the programming languages. If this syntax changes significantly between the two cases, user confusion and panic should be expected.

All of the above evaluation criteria are covered in full detail throughout the book. To gain a better understanding of any one of these criteria, the reader is advised to refer to the corresponding part, chapter, or section in the book.

## 2.4 INPUT DEVICES

The user of a CAD/CAM system spends much time sitting at a workstation communicating and interacting with the computer to develop a particular engineering design. As shown in Fig. 2-1, the user utilizes both input and output devices that comprise a workstation to achieve the design task. These devices are universal and mostly independent of the types of CAD/CAM systems discussed in Sec. 2.2. Software drivers are required to enable the host application program, i.e., the CAD/CAM software, to interpret the information received from input devices as well as send information to output devices. In the past, most of these drivers used to be hardware-dependent. However, greater acceptance of graphics standards, such as GKS, enables programmers to write device-independent codes.

A number of input devices are available. These devices are used to input the two possible types of information: text and graphics. Text-input devices are the alphanumeric (character-oriented) keyboards. There are three classes of graphics-input devices: locating devices, digitizers, and image-input devices. Locating devices, or locators, provide a position or location on the screen. These include lightpens, mice, digitizing tablets and styluses, joysticks, trackballs, thumbwheels, touchscreens, and touchpads. The keyboard arrow (cursor direction) keys are inadequate for most graphics applications and therefore are not considered here. Locating devices typically operate by controlling the position of a cursor on the screen. Thus, they are also referred to as cursor-control devices. The popularity of window and icon-oriented user interfaces has stimulated the further development and enhancement of locating devices. Normally, locators do not only have positioning functionality; they also provide other functions and graphical input modes such as picking and choosing, tracing, and sketching. Picking or pointing means selecting a displayed item or entity on the screen. A drawing can be input to the computer by carefully tracing over its graphics entities. Line segments can be traced by simply digitizing their endpoints. Sketching, sometimes referred to as painting, involves the freehand generation of a drawing. All locating devices, with the exception of the joystick which may provide three-dimensional input, are two-dimensional input devices. Three-axis input devices have been demonstrated using acoustic and mechanical techniques. However, they remain expensive and are not employed in CAD/CAM systems.

Another class of graphics input devices, besides locating devices, is digitizer boards or tablets, or simply digitizers. They are considered as electronic drafting boards. A digitizer consists of a large synthesized electronic board with a movable stylus called the cursor. It is a two-dimensional input device with high resolution and accuracy. Typical sizes are 36 × 48 and 48 × 72 inches. Available resolution and accuracy are up to 0.001 and 0.003 inch respectively. Digitizers can be divided into three kinds relative to the mode of operation of the cursor. They are free-cursor, constrained-cursor, and motor-cursor digitizers. In the first kind, the cursor is attached to the end of a flexible chord, in the second it slides along a gantry that traverses the entire digitizing board area, and in the third kind the cursor motion is accomplished by motors driven by an operator-controlled joystick. Each kind has advantages and disadvantages. The first kind provides greater ease of moving the cursor. The cursor is restricted in the second kind but the digitizer can be used in an upright position. Motorized digitizers are expensive but combine the best features of the first two.

Image-input devices such as video frame grabbers and scanners comprise the third class of graphics-input devices. Electronic imaging is an area of relevance to image processing. This area may become significant to the CAD/CAM field if robot vision systems are to be driven by CAD/CAM databases. Video digitizers is another area where digitizers are connected to a video source, whether it is a video camera or recorder. They can utilize standard NTSC (national television system committee) composite input or color input with separate RGB (red, green, blue) inputs. The resolution of video digitizers is determined by that of the original source. Image scanners are another form of image-input devices. They are used to convert flat paper drawings or plots into digital bit-map form. They range in size and have been applied to help in the conversion of existing engineering blueprints into CAD databases. Some scanners incorporate optical character recognition capability for common typewriter fonts. Additional details regarding image-input devices are beyond the scope of this text.

There are four relevant parameters to measure the performance of graphics-input devices. These are resolution, accuracy, repeatability, and linearity. Some may be more significant to some devices than others. The resolution of a device is defined as the smallest distance the device requires to recognize two adjacent points as spatially separate or addressable. For example, if a digitizer has a resolution of 0.001 inch, the user cannot digitize or represent two points that are less than 0.001 inch apart. Resolution becomes important if a driver for the device is to be written. Typically, the resolution of digitizers is higher than that for digitizing tablets. Accuracy is defined as the error in the measurement of actual data by the input device. This parameter applies more to digitizers. The accuracy of a digitizer is measured by how closely a point is reported as to its actual location on the digitizer grid. Repeatability measures the device ability to return to a given position. For example, if the same point is digitized many times, how close are the coordinates of the resulting point? Linearity measures the response of a device to the user hand movements. How does the device, for example, increase or decrease the input coordinates in accordance with the user hand movements? These parameters are frequently used very loosely and should be carefully defined within the context of each input device of interest.

While the above part presents an overview of input devices and their related issues, the remainder of the section discusses in more detail the most frequently used devices. In Secs. 2.4.2 to 2.4.5 that follow the two words "locating" and "positioning" are used interchangeably as well as the two words "picking" and "pointing."

### 2.4.1  Keyboards

Conventional keyboards are text-only devices and form an essential and basic input device. They are typically employed to create/edit programs or to perform word processing functions. These keyboards have been modified to perform graphics tasks by adding special function keys or attaching graphics-input devices such as mice to them. The programmable function keyboard (PFK) is another type that typically has pushbuttons that are programmed to eliminate extensive typing of commands or entering coordinate information. The pushbuttons are controlled by the software and may be assigned different functions at different phases of the software. PFK may be built as a separate unit, or buttons may just be integrated with a conventional keyboard. Figures 2-17 to 2-19 show examples of modified keyboards.

In CAD/CAM systems, information entered through keyboards should be displayed back to the user on a screen for verification, as in the case with conventional computers. However, the situation is different in CAD/CAM due to the simultaneous display of text and interactive graphics. Various vendors have devised different alternatives. A workstation may have a dual screen. One is used as an exclusive alphanumeric terminal to display input text, commands, and user prompts. The other screen is used to display graphics. Thus, no loss of graphics display area is suffered. Figure 2-20 shows a GE (U.S.A.) Calma as an example of a dual screen system. Other systems such as the Computervision workstation

**FIGURE 2-17**
Alphanumeric keyboard with a trackball and function keys. (*Courtesy Tektronix, Inc.*)

**FIGURE 2-18**
Alphanumeric keyboard with a mouse and function keys. (*Courtesy Tektronix, Inc.*)

(Fig. 2-5) utilizes one screen for both text and graphics by reserving a small portion in the bottom of the display for text. The user has the option to display more text by letting it overlap with the graphics. Irrespective of the text/graphics display organization, users prefer to have a means to obtain hard copies of the text as it is input for various reasons. Hardcopy printers may be provided for this purpose. If they are not available, software functions may be available to generate session files or hard files of sessions' inputs, which may be printed later. Users

**FIGURE 2-19**
Alphanumeric keyboard with function keys. (*Courtesy Intergraph Corp.*)

**FIGURE 2-20**
A dual screen CAD/CAM system. (*Courtesy of GE (U.S.A.) Calma, Inc.*)

should be warned that these files are usually hard to read, especially when functions such as backspaces, control keys, or digitizes are encountered in the files.

How does the keyboard communicate with the CAD/CAM software or the main application program? How is the software interrupted to receive the keyboard input? Each keyboard or input device, in general, is connected to the computer by means of registers whose contents can be read by the computer. A keyboard has typically two registers, one to set a status bit when a key has been struck, the other to identify the key by its character code. The value of the status bit is monitored in a continuous repetitive manner by the software via a programming technique known as "polling." When the user hits a key, the status bit is set and the application program is consequently interrupted to clear the status bit, followed by reading the corresponding code of the key character. The loop is repeated every time the user strikes a key. Keyboard characters are identified by their ASCII (American Standard Code for Information Interchange) codes. ASCII codes for alphanumeric characters are 7-bit codes. Therefore, a character register has seven bits. The character code for a capital letter, say A, is different from that for a small letter, say a. ASCII codes for A and a are 1000001 and 1100001 respectively, the difference being the replacement of the first zero from the left by one which is the case for all the other characters. This is why some programs may require text input as either lower case or upper case. EBCDIC (Extended Binary Coded Decimal Interchange Codes) for alphanumeric characters are 8-bit codes and do not have this distinction (11000001 is the code for A); that is, capital letters are always used.

## 2.4.2  Lightpens

The lightpen is intrinsically a pointing or picking device that enables the user to select a displayed graphics item on a screen by directly touching its surface in the vicinity of the item. The application program processes the information generated from the touching to identify the selectable item to operate on. The lightpen, however, does not typically have hardware for tracking, positioning, or locating in comparison to a digitizing tablet and stylus. Instead, these functions are performed by utilizing the hardware capabilities of the graphics display at hand. The lightpen itself does not emit light but rather detects it from graphics items displayed on the screen. Using the emitted light as an input, it sends an interrupt signal to the computer to determine which item was seen by the pen.

A generic lightpen structure is shown in Fig. 2-21. The pen operation as a pointing device can be described as follows. The user moves the pen to the vicinity of the item to be selected and then depresses the pushbutton to allow the emitted light from the item to reach the pen internal circuitry and hence to signal to the computer to select the item. The application program can then identify the segment of the data structure corresponding to the item detected and operate on it promptly. The lightpen is not sensitive to the ambient room light or the prolonged phosphor typically used with graphics displays. It only detects the sharp intensity of the fluorescent light emitted when the electron beam of the display is drawing the graphics items during any refresh cycle. It is important to synchronize the lightpen response with the speed of the graphics display processor to avoid identifying the wrong item.

The lightpen normally operates as a logical pick in conjunction with a vector refresh display, as explained above, or as a locator on a scan display. In the latter, the $(x, y)$ address of the detected pixel provides a location that is made

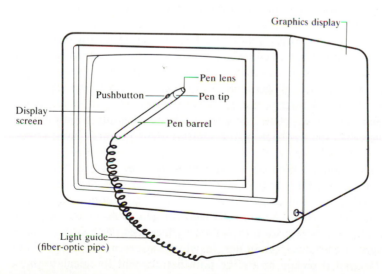

**FIGURE 2-21**
Sketch of a typical lightpen.

available to the application program through the image display system. The lightpen cannot be used with the storage tube display due to the lack of a refresh cycle. However, its functionality can be extended to perform positioning, in the case of a vector refresh display, and pointing, in the case of a raster scan display, by writing the corresponding proper software. In this case, the cursor of the display device is utilized with the lightpen to track its movement and consequently its position on the surface of the display. The cursor may be a cross, cross-hairs, or an arrow.

The lightpen is no longer popular as an input device as it used to be in the late 1960s and early 1970s. During that period, the vector refresh displays that were popular provided high-quality display systems such that the lightpen could be efficiently used to pick out segments in a picture. With the availability of modern technologies, the lightpen is no longer useful in this respect. Another factor that adds to the unpopularity of the lightpen is the need to hold it in an elevated position at the display surface for long periods of time which is tiring for some users.

### 2.4.3  Digitizing Tablets

A digitizing tablet is considered to be a locating as well as a pointing device. It is a small, low-resolution digitizing board often used in conjunction with a graphics display. The tablet is a flat surface over which a stylus or a puck (a hand-held cursor to differentiate it from a display screen cursor) can be moved by the user. The close resemblance of the tablet and stylus to paper and pencil contributes to its popularity as an input device in computer graphics. The stylus is shaped like a pen, and a puck is a little hand-held box. The puck contains a rectile and at least one pushbutton. The rectile's engraved cross-hairs help locate a point for digitizing. Pressing the pushbutton sends the coordinates at the cross-hairs to the computer. Additional buttons may be available on the puck and may be programmed by the software for other functions than digitizing locations such as selecting alphanumeric font sizes or electronic symbols. Sizes of digitizing tablets range from $11 \times 11$ to $36 \times 36$ inches and perhaps larger. A tablet's typical resolution is 0.005 inch or 200 dots per inch. If the digitizing tablet and stylus are primarily used to position a screen cursor, its parameters (resolution, accuracy, linearity, and repeatability) are of less concern since no accurate positioning is needed and the user has the feedback of the screen cursor to guide hand movements.

The tablet operation is based on sensitizing its surface area to be able to track the pointing element (stylus or puck) motion on the surface. Several sensing methods and technologies are used in tablets. The most common sensing technology is electromagnetic, where the pointing element generates an out-of-phase magnetic field sensed by a wire grid in the tablet surface (the pad). Other tablets are based on magnetorestrictive techniques, in which a current pulse sent from the wire grid in the pad is picked up by the pointing element to calculate the $x$ and $y$ coordinates of the desired position. A third type of tablet is based on sonic or acoustic sensing. The pointing element in these tablets has a sound generator with strip microphones along the edges of the tablet to pick up the periodic

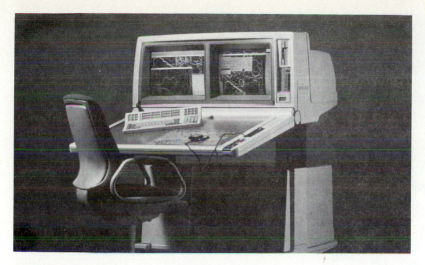

**FIGURE 2-22**
A built-in digitizing tablet with a puck. (*Courtesy Intergraph Corp.*)

sound to determine the $x$ and $y$ values. Figures 2-4 and 2-22 to 2-24 show various tablet configurations with either a stylus or a puck.

The operational concept of the digitizing tablet is simple. The user moves the pointing element to the desired position and then interrupts the computer to accept the coordinate value of this position. In the case of a stylus, the user presses it against the tablet surface or depresses a pushbutton near its tip. This, in turn, activates a switching mechanism inside the stylus that picks (or sends) signals from (to) the tablet position and sends them to the stylus decoding logic

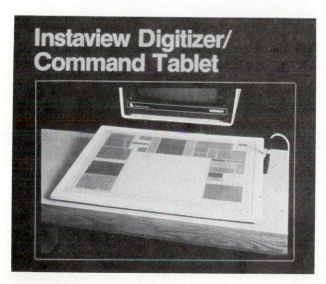

**FIGURE 2-23**
A digitizing tablet with a stylus. (*Courtesy Computervision Corp.*)

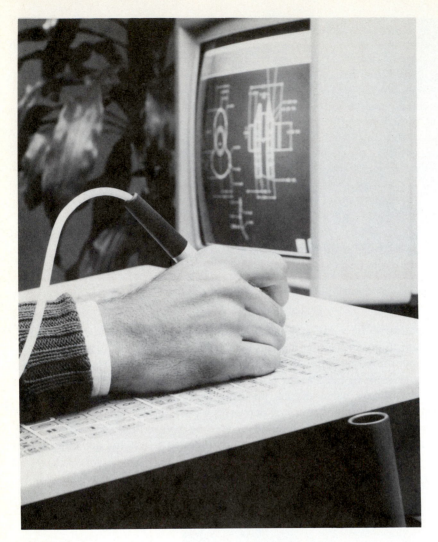

**FIGURE 2-24**
A digitizing tablet with a tablet menu in use. (*Courtesy Computervision Corp.*)

which stores the corresponding coordinates in the tablet's buffer registers. These registers are read by the application program as soon as it is interrupted by the depression of the stylus or its pushbutton. All digitizing tablets provide an origin, typically in the lower left corner, for the $x$ and $y$ coordinates, the $x$ axis being horizontal and the $y$ axis being vertical. Some tablets let the user move the origin to any point on the tablet's area.

Most tablets have designated areas to input digitizes (usually called graphics areas). The remaining area of the tablet surface is used to input menu commands, therefore called the menu area. The screen of the graphics display which the tablet is connected to is mapped to the graphics area via the tablet calibration procedure. Such a procedure is done automatically by the software or the user is requested to do it only once, when the tablet is used for the first time. In the latter case, the user can always recalibrate the tablet as desired. The mapping process (Fig. 2-25) lets the motion of the screen cursor follow that of the pointing element. If the pointing element leaves the boundaries of the graphics area on the tablet, the screen cursor always disappears from the screen.

There are three basic modes of utilizing a digitizing tablet. The first is to employ it to digitize graphics information. The second is to use it to select menu items and commands. Menus can be on-screen menus displayed on the graphics display or printed menus laid on the tablet surface. The third mode enables the user to select different types of data-output operations. Depending on the tablet, select point, stream, switched-stream, polled, and incremental operations may be available. In the point mode, the user locates each point with the pointing element and presses a button to send the coordinates to the computer. In the

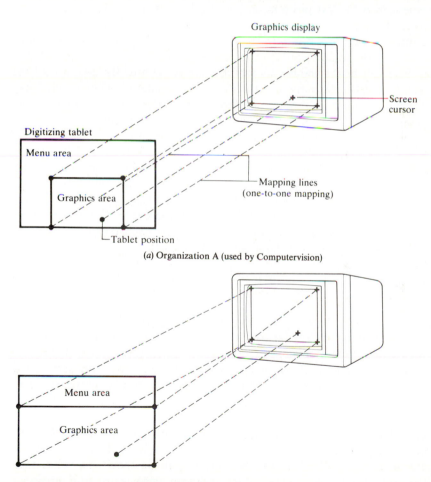

(a) Organization A (used by Computervision)

(b) Organization B (used by GE (U.S.A.) Calma)

**FIGURE 2-25**
Mapping between a tablet graphics area and a display screen.

stream mode, a tablet continually transmits the cursor's coordinates to the computer. In the switched-stream mode, the tablet transmits coordinates only for as long as the user continues to press the button. In the polled mode, the application program requests specific information about the cursor's location and the status of the buttons of the pointing element. The incremental mode allows new positional information to be sent to the computer only after the cursor moves a predefined distance on the tablet surface.

### 2.4.4 Mouse Systems

The mouse was invented in the late 1960s as a location device but has only recently become fairly popular due to its convenient use with icons and pop-up and pull-down menus. Unlike the digitizing tablet, the mouse measures its relative movement from its last position, rather than where it is in relation to some fixed surface. There are two basic types of mice available: mechanical and optical. The mechanical mouse is a box with two metal wheels or rollers on the bottom whose axes are orthogonal in order to record the mouse motion in the $X$ and $Y$ directions. The roll of the mouse on any flat surface causes the rotation of the wheel which is encoded into digital values via potentiometers. These values may be stored, when a mouse pushbutton is depressed, in the mouse registers accessible by the application program either immediately or during the computer interrupt every refresh cycle. Using these values, the program can determine the direction and magnitude of the mouse movement. Unlike the mechanical one, the optical mouse is used in conjunction with a special surface (the mouse pad). Movements over this surface are measured by a light beam modulation and optical encoding techniques. The light source is located at the bottom and the mouse must be in contact with the surface for the screen cursor to follow its movements. Pushbuttons may be mounted on top of the mouse and programmed to various functions. Figure 2-26 shows an optical mouse.

### 2.4.5 Joysticks, Trackballs, and Thumbwheels

These are less popular locating devices than the tablet or the mouse. Their concept of operation is very similar to that of the mechanical mouse discussed in the above section. Potentiometers record and encode the movements of the device. Figures 2-27 and 2-28 show the sketches of a joystick and a trackball respectively while Fig. 2-29 shows thumbwheels utilized with a Tektronix workstation.

The joystick works by pushing its stick backward or forward or to the left or the right. The extreme positions of these directions correspond to the four corners of the screen. A joystick may be equipped with a rotating knob on the top, as shown in Fig. 2-27, which can be used to enter a third axis value, thus making the joystick a three-dimensional input device. Springs are often provided to return the joystick to its center position. Joysticks are seldom used for graphics input because a slight movement of the stick by the user causes the screen cursor to change significantly due to an amplification factor. However,

**FIGURE 2-26**
A typical optical mouse. (*Courtesy Computervision Corp.*)

they are desirable to control velocity or force in some simulation applications where fast responses are required.

A trackball is similar in principle to a joystick but it allows more precise fingertip control. The ball rotates freely within its mount. Both the joystick and the trackball have been used historically in radar and flight control systems. Both are used to navigate the screen display cursor. The user of a trackball can learn quickly how to adjust to any nonlinearities in its performance.

Two thumbwheels are usually required to control the screen cursor, one for its horizontal position and the other for its vertical position. Each position is

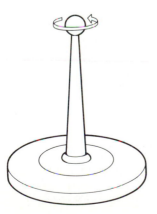

**FIGURE 2-27**
Sketch of a joystick.

**FIGURE 2-28**
Sketch of a trackball.

**FIGURE 2-29**
Thumbwheels as input device with a workstation. (*Courtesy Tektronix, Inc.*)

indicated on the screen by a cross-hair. Thumbwheels are usually mounted on the keyboard as shown in Fig. 2-29.

### 2.4.6 Other Input Devices

In addition to the above input devices, there are other nonmainstream devices. These include three-dimensional input devices such as some digitizers, touch, and

**TABLE 2.1**
**Advantages and disadvantages of cursor-control devices**

| Lightpen | Tablet/stylus | Mouse, joystick, trackball |
|---|---|---|
| Advantages | | |
| Ideal for tracking moving objects | Multipurpose input modes | Can be attached to keyboard |
| | High resolution | High resolution |
| Error rate is low when picking displayed items | Resembles paper and pencil | Comfortable to use |
| | Comfortable to use | |
| Good for simple drawing | | |
| Disadvantages | | |
| Not a pen | Requires extra work surface | Poor for freehand input |
| Low resolution | | Mouse requires extra work surface |
| Tiring to use | High cost | |
| Obstructs part of screen | | Poor control of the cursor |
| Does not work well with small targets | | |

**TABLE 2.2**
**Comparison of input devices**

| Function | Lightpen | Tablet/stylus | Tablet/puck | Mouse | Joystick | Trackball | Touchscreen |
|---|---|---|---|---|---|---|---|
| | | | | Device | | | |
| Resolution | Low | High | High | Medium | Medium | Medium | Low |
| Response to hand movement (positioning speed | High | High | High | High | High | High | High |
| Locating (positioning) | Poor | Very good | Very good | Very good | Good | Good | — |
| Picking (pointing) | Good | Very good | Very good | Very good | Good | Good | — |
| Digitizing | — | Good | Very good | — | — | — | — |
| Traking | — | Very good | Good | — | — | — | — |
| Sketching | — | Very good | Good | Good | Poor | Poor | Fair |
| Fatigue | High | Low | Low | Medium | Low | Low | Low |

voice devices. Touch and voice input devices are attempts to satisfy the great demand for easy-to-use man-machine interfaces. The touch devices are more often used to select items or buttons from an on-screen menu (sometimes called soft buttons) by touching the screen. Touch entry systems incorporate various types of touch-sensing methods such as resistive membranes, optical sensing, acoustic sensing, and capacitance sensing. Details of concepts governing the use of these devices are beyond the scope of this book.

Table 2.1 lists the advantages and disadvantages of the most popular devices when used for cursor movement control. Table 2.2 gives a comparison of the general functionality of these devices. The table reflects the sensitivity and reliability of the tablet and the mouse which contribute to their popularity as input devices. The low cost of the mouse in comparison to the tablet and its use with icon-oriented user interfaces have been important in its acceptance. Joysticks, trackballs, thumbwheels, and touchscreens are considered rare devices.

## 2.5 OUTPUT DEVICES

Output devices form the other half of a CAD/CAM workstation, the first being the input devices. While CAD/CAM applications require the conventional output devices such as alphanumeric (video) displays (terminals) and hardcopy printers, they require output devices to display graphics to the user. Graphics output devices can be divided into soft and hard devices. The former refer to the graphics displays or terminals which only display information on a screen. Hard output devices refer to hardcopy printers and plotters that can provide permanent copies of the displayed information.

### 2.5.1 Graphics Displays

The graphics display of a workstation is considered its most important component because the quality of the displayed image influences the perception of generated designs on the CAD/CAM system. In addition to viewing images, the graphics display enables the user to communicate with the displayed image by adding, deleting, blanking, and moving graphics entities on the display screen. As a matter of fact, this communication process is what gives interactive graphics its name to differentiate it from passive graphics, as in the case of a home television set, that the user cannot change.

Various display technologies are now available to the user to choose from. They are all based on the concept of converting the computer's electrical signals, controlled by the corresponding digital information, into visible images at high speeds. Among the available technologies, the CRT (cathode ray tube) is the most dominating and has produced a wide range of extremely effective graphics displays. Other technologies utilize laser, flat panel displays, or plasma panel displays. In the first, a laser beam, instead of an electron beam, is used to trace an image in a film. In the second, a liquid crystal display (LCD) and light-emitting diodes (LEDs) are used to generate images. The plasma display uses small neon bulbs arranged in a panel which provides a medium resolution display. Thus far, none of these display technologies has been able to displace the CRT as the dominant graphics display device.

Figure 2-30 shows a schematic diagram of a typical CRT. The operation of the CRT is based on the concept of energizing an electron beam that strikes the phosphor coating at very high speed. The energy transfer from the electron to the phosphor due to the impact causes it to illuminate and glow. The electrons are

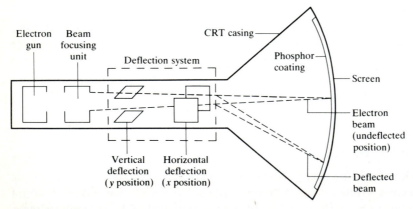

**FIGURE 2-30**
Schematic diagram of a CRT.

generated via the electron gun that contains the cathode and are focused into a beam via the focusing unit shown in Fig. 2-30. By controlling the beam direction and intensity in a way related to the graphics information generated in the computer, meaningful and desired graphics can be displayed on the screen. The deflection system of the CRT controls the $x$ and $y$, or the horizontal and vertical, positions of the beam which in turn are related to the graphics information through the display controller, which typically sits between the computer and the CRT. The controller receives the information from the computer and converts it into signals acceptable to the CRT. Other names for the display controller are the display processor, the display logical processor, or the display processing unit. The major tasks that the display processor performs are the voltage-level convergence between the computer and the CRT, the compensation for the difference in speed between the computer and the CRT (by acting as a buffer), and the generation of graphics and texts. More often, display processors are furnished with additional hardware to implement standard graphics software functions into hardware to improve the speed of response. Such functions include transformations (scaling, rotation, and translation) and shading. Figure 2-1 shows the display processor in relation to the components of the workstation.

The graphics display can be divided into two types based on the scan technology used to control the electron beam when generating graphics on the screen. These are random and raster scan. In random scan (also referred to as stroke writing, vector writing, or calligraphic scan), graphics can be generated by drawing vectors or line segments on the screen in a random order which is controlled by the user input and the software. The word "random" indicates that the screen is not scanned in a particular order. On the other hand, in the raster scan system, the screen is scanned from left to right, top to bottom, all the time to generate graphics. This is similar to the home television scan system, thus suggesting the name "digital scan." Figure 2-31 shows the two techniques of scans. The three existing CRT displays that are based on these techniques are the refresh (calligraphic) display, direct view storage tube, and the raster display. The first two are vector displays based on the random scan technique and the last is based on the raster scan technique. The details of each display are discussed below.

**2.5.1.1 REFRESH DISPLAY.** Early displays in the 1960s were refresh vector displays. Figure 2-32 shows the main components of a typical refresh display. In comparison to Fig. 2-30, the deflection system of the CRT is controlled and driven by the vector and character generators and digital-to-analog converters. The refresh buffer stores the display file or program, which contains points, lines, characters, and other attributes of the picture to be drawn. These commands are interpreted and processed by the display processor. The electron beam accordingly excites the phosphor, which glows for a short period. To maintain a steady flicker-free image, the screen must be refreshed or redrawn at least 30 to 60 times per second, that is, at a rate of 30 to 60 Hz.

The display file shown in Fig. 2-32 is generated by the CAD/CAM software and is considered a data structure which must be updated and constructed according to the needs of the application program, that is, the software. Changes

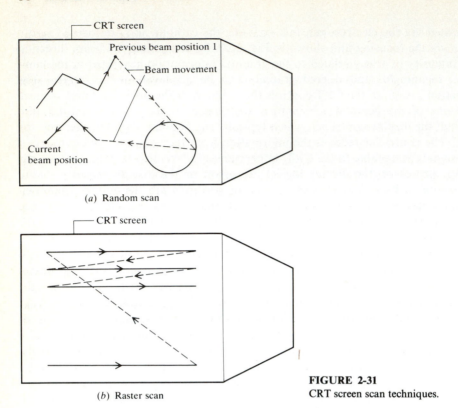

(a) Random scan

(b) Raster scan

**FIGURE 2-31**
CRT screen scan techniques.

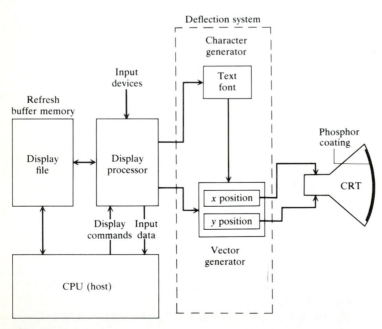

**FIGURE 2-32**
Refresh display.

made to the display file by the software must be synchronized with the display refresh cycle to prevent the display of an incomplete picture. If the software updates the file fast enough, then it is possible to use the dynamic techniques such as animation to simulate movements as well as developing responsive user interfaces.

The principal advantage to the refresh display is its high resolution (4096 × 4096) and thus its generation of high-quality pictures. However, the need to refresh the picture places a limit on the number of vectors that can be displayed without flicker. The present limit is many thousands of lines which is adequate for many users. This, though, requires expensive circuitry to achieve. In addition, being a binary display, the refresh display is able to generate only two levels of color intensity. In some displays, the intensity of the electron beam can vary to provide better color capabilities.

**2.5.1.2  DIRECT VIEW STORAGE TUBE (DVST).** Refresh displays were very expensive in the 1960s, due to the required refresh buffer memory and fast display processor, and could only display a few hundred vectors on the screen without flicker. Consequently, at the end of the 1960s the DVST was introduced by Tektronix as an alternative and inexpensive solution. It is believed that the emergence of the DVST in that time had a significant impact on making CAD/CAM systems affordable for both users and programmers. The DVST eliminates the refresh processors completely and, consequently, the refresh buffer used with the refresh display, as shown in Fig. 2-33. It also uses a special type of phosphor that

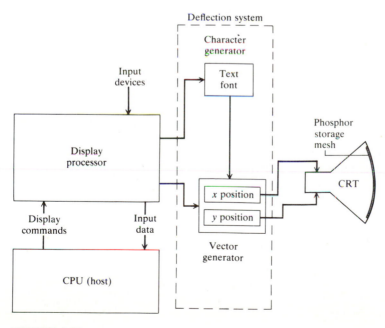

**FIGURE 2-33**
Direct view storage tube.

has a long-lasting glowing effect. The phosphor is embedded in a storage tube. In addition, the speed of the electron beam in the DVST is slower than in the refresh display due to elimination of the refresh cycle.

In the DVST, the picture is stored as a charge in the phosphor mesh located behind the screen's surface. Therefore, complex pictures could be drawn without flicker at high resolution. Once displayed, the picture remains on the screen until it is explicitly erased. This is why the name "storage tube" was suggested. New picture items can be added and displayed rapidly. However, if a displayed item is erased, the entire screen must be cleared and the new picture displayed (typically by using a "repaint" command) to reflect the removal of the item.

In addition to the lack of selective erasure, the DVST cannot provide colors, animation, and use of a lightpen as an input device. Due to its main advantages of inexpensive price and high resolution, early turnkey CAD/CAM systems used storage tubes for their displays. DVST can now be found with a local intelligence and display file to provide selective erasure without the need to refresh the picture.

**2.5.1.3 RASTER DISPLAY.** The inability of the DVST to meet the increasing demands by various CAD/CAM applications for colors, shaded images, and animation motivated hardware designers to continue searching for a solution. During the late 1970s raster displays based on the standard television technology began to emerge as a viable alternative. The drop in memory price due to advances in solid states made large enough refresh buffers available to support high-resolution displays. A typical resolution of a raster display is 1280 × 1204 with a possibility to reach 4096 × 4096 as the DVST. Raster displays are very popular and nearly all recent display research and development focus on them.

In raster displays, the display screen area is divided horizontally and vertically into a matrix of small elements called picture elements or pixels, as shown in Fig. 2-34. A pixel is the smallest addressable area on a screen. An $N \times M$ resolution defines a screen with $N$ rows and $M$ columns. Each row defines a scan line. A rasterization process is needed in order to display either a shaded area or graphics entities. In this process, the area or entities are converted into their

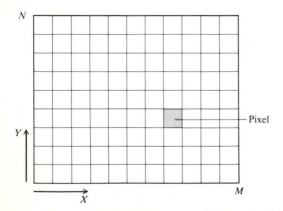

FIGURE 2-34
Typical pixel matrix of a raster display.

corresponding pixels whose intensity and color are controlled by the image display system.

Figure 2-35 shows a schematic of a typical color raster display. Images (shaded areas or graphics entities) are displayed by converting geometric information into pixel values which are then converted into electron beam deflection through the display processor and the deflection system shown in the figure. If the display is monochrome, the pixel value is used to control the intensity level or the gray level on the screen. For color displays, the value is used to control the color by mapping it into a color map.

The creation of raster-format data from geometric information is known as scan conversion or rasterization. A rasterizer that forms the image-creation system shown in Fig. 2-35 is mainly a set of scan-conversion algorithms. Due to the universal need for these algorithms, the scan conversion or rasterization process is now hardware implemented and is done locally in the workstation. As an example, there are standard algorithms such as the DDA (digital differential analyzer) and Bresenham's method which are used to draw a line by generating pixels to approximate the line. Similar algorithms exist to draw arcs, text, and surfaces. This is why it is possible to create images with different colors and hollow areas on raster displays.

The values of the pixels of a display screen that result from the scan-conversion process are stored in an area or memory called frame buffer or bit

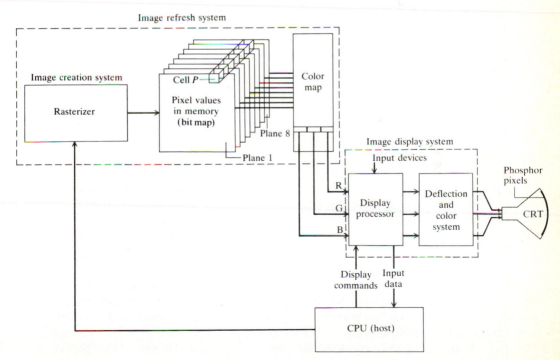

**FIGURE 2-35**
Color raster display with eight planes.

map refresh buffer (bit map, for short), as shown if Fig. 2-35. Each pixel value determines its brightness (gray level) or most often its color on the screen. There is a one-to-one correspondence between every cell in the bit map memory and every pixel on the screen. The display processor maps every cell into its corresponding screen pixel brightness or color. In order to maintain a flicker-free image on the screen, the screen must be refreshed at the rate of 30 or 60 Hz, as in the case of refresh displays. The refresh process is performed by passing the pixel values in the bit map to the display processor every refresh cycle regardless of whether these values represent the image or the background. Therefore the refresh process is independent of the complexity of the image and the number of its graphics items. Thus there is no chance of a flicker problem with the increased complexity of the image as in the case of refresh displays.

To understand the performance of raster displays and to evaluate them, one must ask the following question: how many bits are required in the bit map to adequately represent the intensity of any one pixel on the display screen? The trivial answer of one bit/pixel produces only a two-level image (bright or dark) which is very unsatisfactory to basic applications. The practice suggests that 8 bits/pixel are needed to produce satisfactory continuous shades of gray for monochrome displays. For color displays, 24 bits/pixel would be needed: 8 bits for each primary color red, blue, and green. This would provide $2^{24}$ different colors, which are far more than needed in real applications. Typically, 4 to 8 bits/pixel are adequate for both monochrome and color displays utilized in most engineering applications. Specialized image processing applications may require more than that. The bit map memory is arranged conceptually as a series of planes, one for each bit in the pixel value. Thus an eight-plane memory provides 8 bits/pixel, as shown in Fig. 2-35. This provides $2^8$ different gray levels or different colors that can be displayed simultaneously in one image. The number of bits per pixel directly affects the quality of its display and consequently its price.

The value of a pixel in the bit map memory is translated to a gray level or a color through a lookup table (also called a color table or color map for a color display). The pixel value is used as an index for this lookup table to find the corresponding table entry value which is then used by the display system (display processor and beam deflection system) to control the gray level or color. Figure 2-36 shows how the pixel value is related to the lookup table in an eight-plane display. If cell $P$ in the bit map corresponds to pixel $P$ at the location $P(x, y)$ on the screen, then the gray level of this pixel is 50 (00110010) or its corresponding color is 50.

Figure 2-36 shows raster displays with what is called direct-definition systems in which the lookup table always has as many entries as there are pixel values in the bit map. For the eight-plane display shown in the figure, the lookup table has 256 ($2^8$) entries, which correspond to all possible values a pixel may have. For color displays this may imply that the number of bits per pixel must be increased to increase the number of entries in the color map, and therefore increase the available number of colors to the user. This, however, is not true and leads to increasing the size of the bit map memory and the cost of the display. Thus, how can the number of color indexes in the color map increase while keeping the pixel definition (number of bits per pixel) in the bit map to a

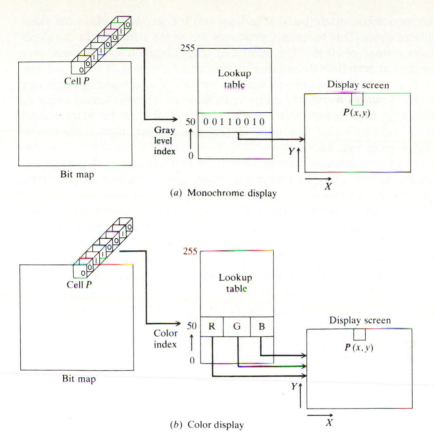

(a) Monochrome display

(b) Color display

**FIGURE 2-36**
Relationship between pixel value and a lookup table.

minimum? For example, how can a display have 4 bits/pixel with 24 bits of color output ($2^{24}$ different colors)? This is achieved by designing a color map with $2^{24}$ (16.7 million) available color indexes. The 4 bits/pixel provides 16 ($2^4$) simultaneous colors, in an image, which can be chosen from the color map. A pixel value (0 to 15) can be used to set the value of the color index which corresponds to the proper color to be displayed. This scheme, in this example, provides 16 simultaneous colors from a palette of 16.7 million. To the user, the color map is made available where colors are chosen and the application program relates the chosen color to the proper pixel value. For example, if the user chooses the color purple for an image element, the corresponding program sets the corresponding pixels to reflect the color purple.

While raster displays are now a standard offering from nearly all CAD/CAM vendors, the quality of the displayed images is affected by flicker and aliasing problems. The flicker of an image is reduced by simply reducing the time of the refresh cycle. The image refresh system (Fig. 2-35) may use an interlaced scan of two fields. In the interlaced scan (as in the home television), the refresh cycle of $\frac{1}{30}$ second is divided into two subcycles each lasting $\frac{1}{60}$ second. The first subcycle displays the odd-numbered scan lines and the second displays the even-numbered scan lines. This technique produces an image with almost a refresh rate of 60 Hz instead of 30 Hz. The interlaced scan scheme does not work very well if the adjacent scan lines do not display similar information. Another scheme is to use a noninterlaced scan of one field by operating at a higher refresh rate such as 60 Hz. In this scheme, the entire scan lines are refreshed once every $\frac{1}{60}$ second. This high rate means more and faster accesses to the bit map (refresh buffer) per second and higher bandwidth deflection amplifiers used in the deflection system shown in Fig. 2-35.

The aliasing problem is directly related to the resolution of the display which determines how good or bad is the raster approximation of geometric information. The jaggedness of lines at angles other than multiples of 45 degrees, assuming square pixels, is the feature of a raster display known as aliasing. Various methods of antialiasing exist which use various intensity levels to soften the edges of the lines or shades. Of course, the aliasing problem diminishes as the resolution increases and is only related to the screen image and not to the geometric representation in the computer or to drawings plotted on paper.

**Example 2.1.** An eight-plane raster display has a resolution of 1280 horizontal × 1024 vertical and a refresh rate of 60 Hz noninterlaced. Find:

(a) The RAM size of the bit map (refresh buffer).
(b) The time required to display a scan line and a pixel.
(c) The active display area of the screen if the resolution is 78 pixels (dots) per inch.
(d) The optimal design if the bit map size is to be reduced by half.

*Solution*
(a) The RAM size of the bit map = 8 × 1280 × 1024 = 1.3 Mbytes.
(b) The times to display a scan line $t_s$ and a pixel $t_p$ are given by

$$t_s = \frac{1/60}{1024} = 16 \text{ microseconds}$$

$$t_p = \frac{16}{1280} = 12 \text{ nanoseconds}$$

(c) The active display area = 1280/78 horizontal × 1024/78 vertical = 16.4 × 13.1 inch.
(d) Assuming there is only one bit map available, the two solutions are to reduce the number of planes by half and keep the resolution as it is or vice versa. Thus the two choices are a four-plane 1280 × 1024 display or an eight-plane 640 × 512 display. The first choice is preferred, especially if 16 simultaneous colors are adequate for most applications that utilize the display.

**Example 2.2.** What is a reasonable resolution of an eight-plane display refreshed from a bit map of 256 kbytes of RAM?

*Solution.* Bit map size per plane = $\dfrac{256 \times 1000 \times 8}{8}$ = 256,000 bits

This could support a display with a resolution of 505 × 505, 640 × 400 or other combinations. If four planes are used instead, then the resolution can be 715 × 715, 640 × 800, or other combinations of 512,000.

**Example 2.3.** How can you draw a 500 pixel wide square on a 1280 horizontal × 1024 vertical screen whose aspect ratio is 4 : 3?

**Solution.** There are two possibilities to draw the square. The first is to assume a horizontal value of 500. The corresponding vertical value is given by 500 × 1024/1280 × 4/3 = 533. In pixel value, the square corners become (assuming the bottom left corner is located at pixel (h, v):

$$(h, v + 533) \qquad\qquad (h + 500, v + 533)$$

$$533$$

$$(h, v) \qquad\qquad (h + 500, v)$$

$$500$$

The second possibility is to assume a vertical value of 500. The horizontal value becomes 500 × 1280/1024 × 3/4 = 469. The square corners become:

$$(h, v + 500) \qquad\qquad (h + 469, v + 500)$$

$$500$$

$$(h, v) \qquad\qquad (h + 469, v)$$

$$496$$

This section has covered the details of the three major types of graphics displays. It is left as an exercise, at the end of the chapter, for the reader to compare the capabilities, advantages, and disadvantages of these types. The general trend in display hardware design is to increase the local capabilities and intelligence of the display. This is possible by the advances in the VLSI field and consequently the reduction of prices of RAM. In addition, window memory techniques are available to provide multiwindow screens without windows overlapping.

## 2.5.2 Hardcopy Printers and Plotters

Output devices of both printers and plotters are available to CAD/CAM systems for purposes such as creating checkplots for offline editing and producing final drawings and documentation on paper. Relative to the drawing rate on screen, they are slow. Printers usually provide hard copies of text as well as graphics. Hardcopy devices, in general, employ one of two methods of plotting: vector or raster plotting. The two methods are very close in concept to refresh and raster displays respectively. Vector plotting can employ either absolute or incremental plotting depending on whether the coordinates of the current point are measured relative to the absolute origin of the plot or to the last point plotted. In general, absolute plotting is preferred because it eliminates any errors or noise from the surrounding environment. Raster (or dot matrix) plotting is based on using the presence or absence of dots to draw lines and other geometric information present in the document. The resolution of the plotter may cause jaggedness in the hard copy similar to raster displays. For many years, plotters were the only means of obtaining high-quality outputs from a computer. Currently, various printers are available that provide either black and white or multicolor pictures. The hardcopy devices that are available include impact and nonimpact devices. Impact devices include dot matrix printers and typically produce an image on paper by hammering a ribbon onto the surface of the paper—hence the name "impact printers." Nonimpact devices utilize other methods and include pen, photographic, electrophotographic, electrostatic, thermal-transfer, and inkjet plotters and/or printers. While most of these devices may be central and shared by all workstations, each workstation usually has its own hardcopy unit. The basic operations of some of these devices are described below.

**2.5.2.1 PEN PLOTTERS.** There are two common types of conventional pen plotters: flat-bed and drum. In the flat-bed plotter, the paper is stationary and the pen-holding mechanism can move in two axes. In the drum plotter, the paper is attached to a drum that rotates back and forth, thereby providing movement in one axis. The pen mechanism moves in the transverse direction to provide movement along the other axis. Pen plotters are considered vector plotters and employ multiple pens (four to eight) to provide varying line widths or colors. Pens may be of wet ink, ball-point, or felt-tip type. The plotting speed is typically limited by the speed of the writing pen used. Pen plotters require supervision to oversee pen performance and are relatively inexpensive to acquire and operate.

**2.5.2.2 ELECTROSTATIC PLOTTERS.** They are considered dot matrix or raster plotters. The image in vector form, as lines, arcs, characters, and symbols, has to be converted into raster form and sorted. Then these rows of dots can be printed across the width of the paper or plastic film as it slowly moves through the plotter. Typically the plotter resolution is 200 dots/inch or more and each dot is arranged to overlap adjacent ones. This provides a relatively high quality image. Electrostatic plotters have the virtue of being quiet, usually trouble free, undemanding of the operator's time, and about an order of magnitude faster than pen plotters. However, they are also an order of magnitude more expensive than pen plotters. It is important to know where the vector-to-raster conversion and sorting is done due to the time they take. If they are performed at the central or the host computer, then the response time at workstations is expected to degrade. Separate processors can be provided specifically for this conversion task. Electrostatic plotters are normally monochrome, that is, black images on a white, translucent, or transparent medium. Color plotters are available but are quite expensive.

**2.5.2.3 HARDCOPY UNITS.** Each workstation in a CAD/CAM system should have its own hardcopy unit to produce quick low-quality copies of screen images (screen dump) and to provide a copy of user input and system output information which is always useful in tracking errors and mistakes. These units are particularly useful when a central plotter is located some distance from workstations. Hardcopy units can take the form of small electrostatic plotters with a relatively coarse raster grid, impact dot matrix printers using normal paper, or other devices that use light-sensitive paper. A drawback with such paper is that it darkens with time.

**2.5.2.4  INKJET PLOTTERS.** They utilize the dot matrix method of plotting. As with electrostatic plotters, they produce a raster image. Each dot is, however, created by impelling a tiny jet of ink on the surface of the paper. The jets are switched on and off at high speeds to create multicolor plots. Typical applications include color plots of solid models, shaded images, and contour plots.

**2.5.2.5  BLACK-AND-WHITE PRINTERS.** The major two types of black-and-white printers are dot matrix and laser printers. Dot matrix printers are inexpensive but slow. Their resolution is typically 75 dpi (dots per inch). It is possible to obtain a resolution of 300 dpi from dot matrix printers. Laser printers are the most popular printers. They are more expensive but faster and better than dot matrix printers. Their typical resolution is 300 dpi (and may reach 600 dpi). Their speed, measured in pages per minute, is typically 8 (the range is 6 to 12). The laser printing mechanism is simple. A laser source is turned on and off. The laser beam bounces off a spinning mirror and exposes a photosensitive drum covered with an even charge. When the drum is exposed to the beam, the charge on the drum surface changes which attracts the toner. The toner is then transferred to the paper to produce the image.

**2.5.2.6  COLOR PRINTERS.** The demand for color printers has increased since the color displays have become affordable and popular. The major six types of color printers available are impact, photographic, electrophotographic, electrostatic, thermal-transfer, and inkjet printers. Color impact dot matrix printers operate similarly to black-and-white dot matrix printers. Dots of ink are transferred from an inked ribbon onto paper through the printhead wires or hammers. Color can be placed on ribbon or multiple ribbons, one per color, can be used. Typically, four colors (black, red, green, and blue) are available. The electrophotographic color printer works in two steps. First, the negative images of the primary colors (red, green, and blue) are scanned one at a time and selectively discharged onto a charged photoconductor drum. Second, each latent image corresponding to each primary color is transferred to paper by developing it with cyan-colored toner. The thermal-transfer printer (wax printers belong to this type) has a thermal print-read unit which consists of tiny resistors. By controlling the temperature of these transistors, selectively melted dots of colored wax can be transferred onto paper. This transfer process is repeated for each color primary in addition to the fourth cycle for black to generate a full-color image. The electrostatic printer works in a similar way to the electrophotographic printer by producing a charged latent image. The latent image is produced in the printer by generating discharges through the control of voltage potentials on wire printing "nibs." The latent image is then developed with the proper colored liquid toner. This process is repeated four times for each full-colored image. The inkjet color printer works in a similar way to the inkjet plotter. It ejects ink onto the paper medium as it revolves on a drum beneath the carriage of the ink jets. The jets are

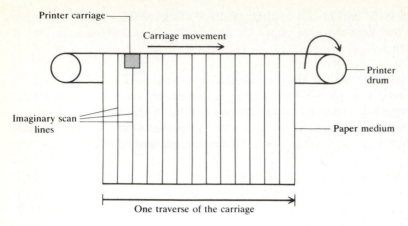

**FIGURE 2-37**
Basic movements in inkjet color printer.

controlled by digital data received from the image to be printed. The carriage contains four jets to produce full-color images. The carriage movement is synchronized with the drum revolution such that it advances one scan line for each revolution; hence the image is produced in one carriage traverse, as shown in Fig. 2-37. The photographic printer produces the highest quality image by receiving video signals from the color terminal and transferring it to conventional photographic media. The printer separates the primary color components of the image, outputs them to a monochrome monitor, and then passes them through corresponding color filters onto the photographic film. Table 2.3 shows a comparison between the various types of color printers.

**TABLE 2.3**
**Comparison of color printers**

| | | | Printer | | | |
|---|---|---|---|---|---|---|
| Factor | Impact | Electro-static | Inkjet | Photo-graphic | Electro-photographic | Thermal-transfer |
| Image quality | Poor | Good | Good | Very good | Fair | Fair |
| Supplies cost | Reasonable | Reasonable | Reasonable | Expensive | Inexpensive | Reasonable |
| Image production speed | Reasonable | Reasonable | Reasonable | Slow | Very fast | Slow |
| Noise level | Noisy | Reasonable | Reasonable | Quiet | Acceptable (less than reasonable) | Reasonable |
| Maintenance and reliability | Good | Good | Good | Good | Fair | Good |

## 2.6 HARDWARE INTEGRATION AND NETWORKING

As discussed in the previous section, there exists various CAD/CAM systems and input and output devices. The integration and networking between the various components and peripherals of a system ensures the success of CAD/CAM installations. The need for hardware integration and networking in CAD/CAM are manifold. CAD/CAM is interdisciplinary by nature and therefore its functions are distributed among various departments, such as design and manufacturing, in many organizations. The hardware components in these departments must communicate together and have access to common databases. Another need for networking is to share common resources and peripherals such as plotters and printers. Stand-alone workstations are most often networked together and to central computing facilities. The need to expand a CAD/CAM system by adding new workstations in an incremental fashion necessitates networking. It is also common to have a need to network devices that complement each other. An example is connecting a high-end CAD/CAM system to a low-end system to allow database transfer.

Local area networks (LANs) are the main communication technology available. A LAN is a data communication system that allows various types of digital devices to talk to each other over a common transmission medium. Low-cost and low-performance LANs use a twisted-pair cable. Shielded coaxial cables and fibre-optic connections are used for higher speed communications. The three most popular LAN configurations available are star, ring, and bus LANs. The star LAN (Fig. 2-38a) consists of a central computer (sometimes referred to as a file server) to which several workstations (sometimes referred to as diskless nodes) and central peripherals are attached. This configuration is typical for workstations supported by mainframes or superminicomputers and which do not have disks—thus the name "diskless"—for storage or computations. The major advantage of the star configuration is that it provides a central database that is accessible by all users. However, its main disadvantage is that the whole network fails with the failure of the central computer. The ring, or closed-loop, LAN shown in Fig. 2-38b is well suited when devices are mostly similar such as stand-alone workstations discussed in Sec. 2.2. Examples of ring LANs are Prime Ringnet and the Apollo Domain network. One advantage of the ring network is that databases or files on one workstation can be shared by others in the network. Another advantage is that if one workstation goes down, the rest remain operational. The bus LAN (Fig. 2-38c) is an open-loop system which may take the form of a main bus or branched or tree systems. These are particularly suited when devices to be connected are mostly dissimilar. The Xerox Ethernet system is the best-known bus LAN. Typical configurations of CAD/CAM laboratories may utilize various types of LANs to provide users with as much access of existing computing facilities as possible. Figure 2-39 shows an example of using the star and bus configuration together. The central mainframe serves as the host for the control of the star LAN and is typically used for number-crunching and control of central devices such as plotters. Workstations are connected to the mainframe in a star fashion with either low-speed asynchronous lines such as RS-232 or a high-speed synchronous line. The RS-232 links are adequate for

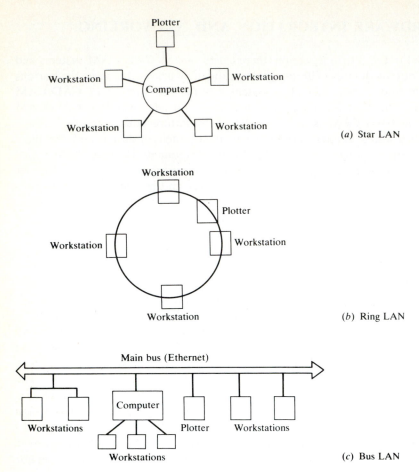

**FIGURE 2-38**
Star, ring, and bus LAN configurations.

short distances. The workstations themselves can also communicate with one another over an Ethernet bus LAN which is considered a high-speed communication network.

A LAN performance is directly related to the efficiency and ease of use of the related operating system, communication speeds, and communication protocols supported. An efficient operating system incorporates virtual memory management to provide any node on the network with the sum of all disk capacity on it, and supports a wide range of protocols and file-access procedures. Unix is an example of existing operating systems.

Communication speeds are related to the system of transmission utilized in the network, whether it is synchronous or asynchronous. Standard speeds include 300, 1200, 4800, 9600, and 15,600 baud (one baud is equivalent to one byte per second). High-speed LANs utilize synchronous transmission (data transmission is based on a predefined timing pattern or signal) at speeds of 4800 baud and up.

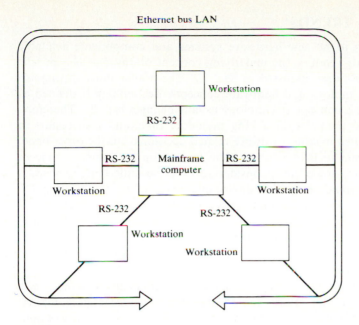

Ethernet bus LAN

**FIGURE 2-39**
Combined star and bus LAN configuration.

Low-speed LANs utilize asynchronous transmission (data transmission is *not* based on a predefined timing pattern or signal) at relatively low speeds, the most popular being 300 baud. Many slow-speed devices including keyboards, digitizers, and plotters operate at 300 baud. CRT display may operate on higher-speed asynchronous networks to improve the response speed. Most low-speed asynchronous lines are interlaced via RS-232 cables which are the standard for connecting low-speed devices.

The communication protocol that a CAD/CAM system's LAN supports is important to prevent system isolation from other computing facilities. The protocol is simply the format or language used by the network to transmit the desired information through the network cables or lines. TCP/IP is the most popular protocol supported by various operating systems such as Unix, VMS, and MS-DOS. It is also adopted by the United States government as the standard protocol. The network file system (NFS) protocol is considered one step higher than TCP/IP. It is developed to address the one serious drawback of Unix: only one user at a time can copy the network's original files. The NFS code works with any kind of medium, workstation, and protocol. It also allows transparent file access by multiple users simultaneously.

LANs are "local" in extent, but this in practice can range over a distance of six miles depending on circumstances. Long-distance communications would normally use modems and either leased telephone lines or a public packet-switched network, depending on the amount of usage intended. Dial-up facilities using the telephone network are seldom fast enough for most CAD/CAM work.

## 2.7 HARDWARE TRENDS[2]

Although there seems to be new hardware systems and components available every day from various vendors, the underlying concept of any one system falls under one of the four types discussed in Sec. 2.2. The reader must distinguish between hardware technology and hardware concepts. Technology is defined as the implementation of a concept. Technology usually changes rapidly. Therefore, it becomes difficult to predict over a long period of time, and also results in confusing end users when making hardware-related decisions. On the other hand, hardware concepts seem to take a longer time to develop and, once a concept evolves, it lasts for quite some time. Considering PCs, the underlying concept is the single-user or personal computing environment. However, the existing PC technology includes infinite models and types and keeps changing every day. It should also be realized that concepts and technology influence each other and both originate from the end user and criticism of existing technology. A methodology for technology prediction based on the available hardware concepts is presented at the end of this section.

It is expected that engineering design and manufacturing usages will increase dramatically in the future due to the steady decrease in computer response time and steady decrease in hardware and software costs. For example, MCAE (mechanical computer aided engineering) systems (see Figs. 2-40 and 2-41) are dedicated to design functions. Hardware networking and true integration between CAD, CAM, and CAD and CAM software will play a major role in increasing these usages.

Another future trend is the increasing modularity of CAD/CAM and increasing standardization of many CAD/CAM components. These standards will provide users with more options and the ability to configure their systems to

**FIGURE 2-40**
Desktop MCAE system based on an enhanced IBM PC/AT. (*Courtesy of Aries Technology, Inc.*)

---

[2] This section includes excerpts from the Frost and Sullivan report 1564.

**FIGURE 2-41**
MCAE system based on an enhanced
**IBM PC/AT** or MicroVax II.
(*Courtesy of Cognition, Inc.*)

their needs. As technology advances and user needs change, new standards will replace the old. It is therefore not necessary to have permanent hardware standards. What is happening and will continue to evolve is that every component of CAD/CAM systems, from operating systems to LANs, will be standardized among some key manufacturers.

Upgradability of existing CAD/CAM systems seems and will continue to be difficult. Technology progresses too fast to allow major vendors to invest into providing fully upgradable systems. They generally provide only an ability to convert data from one upgrade to the other. Other than data, hardware, software, and training investment become obsolete.

Other trends can be summarized as follows. There has been a shift toward more open architecture. Prices will continue to decline. There will be a continuation of the trend toward more desktop systems. Mainframes will be used to support centralized management of software and data, rapid computations, archival storage, and network management.

Tables 2.4 and 2.5 summarize forecasts of CAD/CAM systems by configuration type in dollars and units terms respectively for the period 1984–1992. Measured in 1985 U.S. dollar values, as shown in Table 2.4, CAD/CAM systems sales are expected to increase from $2484 million in 1984 to $7951 million in 1992 at a growth rate of about 15 percent per year. The table also shows the clear trend that workstation-based and microcomputer-based CAD/CAM systems will become prevailing concepts in the years to come. Figures 2-42 to 2-44 present Table 2.4 in terms of line graphs and pie charts. As seen from Table 2.5, the growth rate of workstations is expected to be the largest, at an average yearly rate of 27.4 percent in the period 1985–1992. Personal computer-based, mainframe-based, and turnkey systems sales are to grow by 20.1, 16.5, and 15.1 percent respectively. When CAD/CAM sales are measured in the number of graphics seats (the maximum number of simultaneous interactive graphics users), as shown in Table 2.5, workstations are followed in growth by turnkey

TABLE 2.4
**CAD/CAM market, United States—distribution of sales by system configu-ration type**

| | 1984 | 1986 | 1988 | 1990 | 1992 | Growth per year 1985–1992, % |
|---|---|---|---|---|---|---|
| Sales by configuration type, 1985 $ millions | | | | | | |
| Turnkey system | 1548 | 1947 | 2465 | 3080 | 3761 | 11.8 |
| Workstation | 320 | 548 | 838 | 1169 | 1511 | 20.1 |
| Personal computer | 124 | 366 | 594 | 777 | 914 | 208 |
| On mainframe host | 492 | 765 | 1093 | 1434 | 1765 | 16.3 |
| Total | 2484 | 3626 | 4990 | 6460 | 7951 | 14.9 |
| Shares of configurations, % of total | | | | | | |
| Turnkey system | 62.3 | 53.7 | 49.4 | 47.7 | 47.3 | −2.7 |
| Workstation | 12.9 | 15.1 | 16.8 | 18.1 | 19.0 | 4.5 |
| Personal computer | 5.0 | 10.1 | 11.9 | 12.0 | 11.5 | 5.1 |
| On mainframe host | 19.8 | 21.1 | 21.9 | 22.2 | 22.2 | 1.1 |
| Total | 100.0 | 100.0 | 100.0 | 100.0 | 100.0 | |
| Growth rates, % per year | | | | | | |
| Turnkey system | | 13.1 | 11.7 | 11.4 | 10.1 | −1.6 |
| Workstation | | 30.4 | 22.1 | 16.9 | 12.4 | −12.3 |
| Personal computer | | 50.8 | 24.6 | 13.3 | 6.4 | −32.1 |
| On mainframe host | | 24.5 | 18.4 | 13.6 | 10.1 | −12.2 |
| Total | | 20.9 | 16.3 | 13.1 | 10.1 | −9.8 |

*Source:* Frost and Sullivan Report 1564.


CAD/CAM system seats, personal computer-based units, and mainframe-based seats respectively. Figures 2-45 to 2-48 present Table 2.5 as line graphs and pie charts.

In order to predict long-term changes in computer hardware a method-ology for technology prediction is described here. The method is based on the observation that parameters governing computer advances in technology from year to year are relatively constant. These parameters, for example, include the cost per chip and cost per unit weight. The method presents a computer tier model based on the following axioms:

1. Regardless of their size, the cost of computer hardware is about $200/lb when adjusted for inflation.
2. Weight is the dominant factor in deciding computer use independently of time. Other factors such as memory, disk space, and speed of execution seem to be less dominant. For example, a person is expected to use 50 lb of computer in the office at all times.
3. Based on axiom 2, there exists seven tiers, as shown in Table 2.6. There is a factor of ten in weight separating the tiers. The same factor exists in price based on axiom 1.

**TABLE 2.5**
**CAD/CAM market, United States—distribution of sales by system configuration type in units**

| | 1984 | 1986 | 1988 | 1990 | 1992 | Growth per year 1985–1992, % |
|---|---|---|---|---|---|---|
| **Sales by configuration type, number of complete systems†** | | | | | | |
| Turnkey system | 3,658 | 4,569 | 6,050 | 8,101 | 10,810 | 15.1 |
| Workstation | 3,752 | 7,224 | 12,461 | 19,581 | 28,504 | 27.4 |
| Personal computer | 7,926 | 23,073 | 36,930 | 47,710 | 55,418 | 20.1 |
| On mainframe host | 452 | 707 | 1,014 | 1,337 | 1,653 | 16.5 |
| Total | 15,788 | 35,573 | 56,455 | 76,729 | 96,385 | 20.9 |
| **Sales by configuration type expressed in number of graphics seats‡** | | | | | | |
| Turnkey system | 17,560 | 25,588 | 37,512 | 54,278 | 76,754 | 20.3 |
| Workstation | 3,752 | 7,224 | 12,461 | 19,581 | 28,504 | 27.4 |
| Personal computer | 7,926 | 23,073 | 36,930 | 47,710 | 55,418 | 20.1 |
| On mainframe host | 6,927 | 11,240 | 16,743 | 22,917 | 29,419 | 18.7 |
| Total | 36,165 | 67,125 | 103,646 | 144,486 | 190,095 | 20.9 |
| **Sales by configuration type, number of complete systems** | | | | | | |
| Turnkey system | 48.6 | 38.1 | 36.2 | 37.6 | 40.4 | −0.4 |
| Workstation | 10.4 | 10.8 | 12.0 | 13.6 | 15.0 | 5.5 |
| Personal computer | 21.9 | 34.4 | 35.6 | 33.0 | 29.1 | −0.6 |
| On mainframe host | 19.1 | 16.7 | 16.2 | 15.8 | 15.5 | −1.8 |
| Total | 100.0 | 100.0 | 100.0 | 100.0 | 100.0 | |
| **Growth rates of number of graphics seats** | | | | | | |
| Turnkey system | | 21.8 | 20.2 | 19.9 | 18.4 | −0.9 |
| Workstation | | 38.4 | 29.6 | 24.1 | 19.3 | −9.6 |
| Personal computer | | 49.8 | 23.8 | 12.6 | 5.7 | −33.1 |
| On mainframe host | | 27.1 | 20.9 | 16.0 | 12.4 | −10.8 |
| Total | | 33.0 | 22.7 | 17.3 | 13.6 | −14.1 |
| **Terminals/average system** | | | | | | |
| Turnkey system | 4.8 | 5.6 | 6.2 | 6.7 | 7.1 | 4.5 |
| Workstation | 1.0 | 1.0 | 1.0 | 1.0 | 1.0 | 0.0 |
| Personal computer | 1.0 | 1.0 | 1.0 | 1.0 | 1.0 | 0.0 |
| On mainframe host | 15.3 | 15.9 | 16.5 | 17.1 | 17.8 | 1.9 |

† Note that the above figures are for CAD/CAM systems sold as complete systems to end users.

‡ The number of graphics seats is a measure of how much interactive user the system can support simultaneously.

*Source:* Frost and Sullivan Report 1564.

4. A factor-of-ten improvement is achieved every seven years. This is suggested by the fact that there is about 35 percent per year aggregate technology improvement across the tiers. This implies that technology migrates from one tier to the next less-capable tier in seven years.

5. The transition from one tier to another introduces qualitative changes in computing usage which are usually the most difficult to predict. For example, while the tier model could have predicted the emergence of workstations 20 years ago, it would have been unable to predict the need and role of the LAN

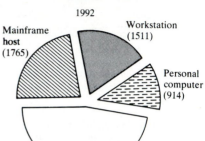

1986

Workstation
(548)

Mainframe
host
(765)

Personal
computer
(366)

Turnkey systems
(1947)

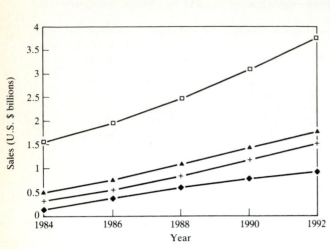

FIGURE 2-42
CAD/CAM market sales.

□ Turnkey system
▲ On mainframe host
+ Workstation
◆ Personal computer

1992

Mainframe
host
(1765)

Workstation
(1511)

Personal
computer
(914)

Turnkey systems
(3761)

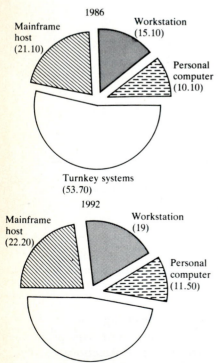

1986

Mainframe
host
(21.10)

Workstation
(15.10)

Personal
computer
(10.10)

Turnkey systems
(53.70)

1992

Mainframe
host
(22.20)

Workstation
(19)

Personal
computer
(11.50)

Turnkey systems
(47.30)

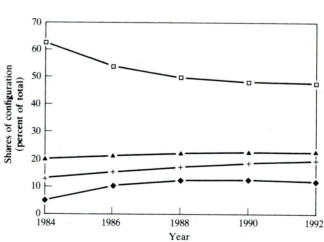

□ Turnkey system
▲ On mainframe host
+ Workstation
◆ Personal computer

FIGURE 2-43
Shares of CAD/CAM systems.

**76**

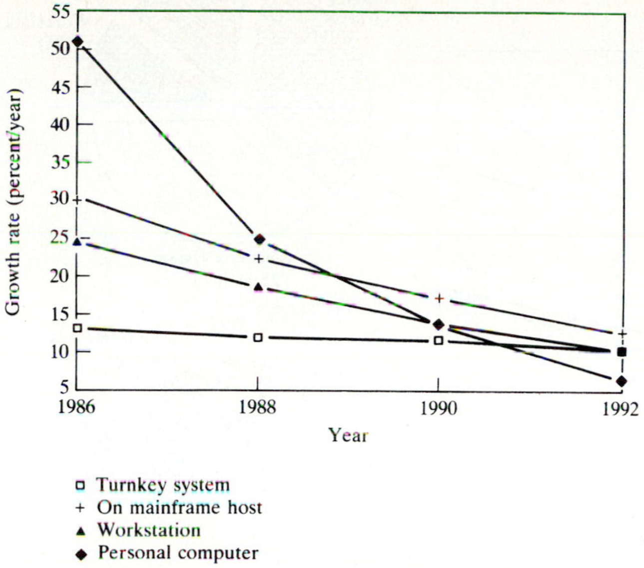
55
50
45
40
35
30
25
20
15
10
5
Growth rate (percent/year)
1986
1988
1990
1992
Year
Turnkey system
On mainframe host
Workstation
Personal computer

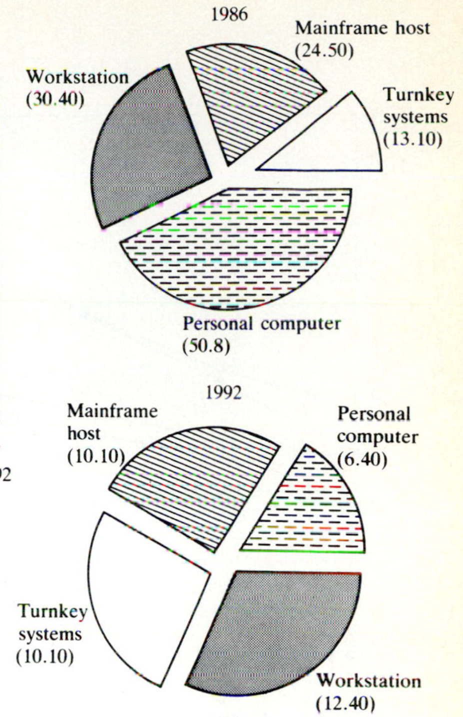
1986
Mainframe host (24.50)
Workstation (30.40)
Turnkey systems (13.10)
Personal computer (50.8)
1992
Mainframe host (10.10)
Personal computer (6.40)
Turnkey systems (10.10)
Workstation (12.40)

**FIGURE 2-44**
Growth rate of CAD/CAM systems.

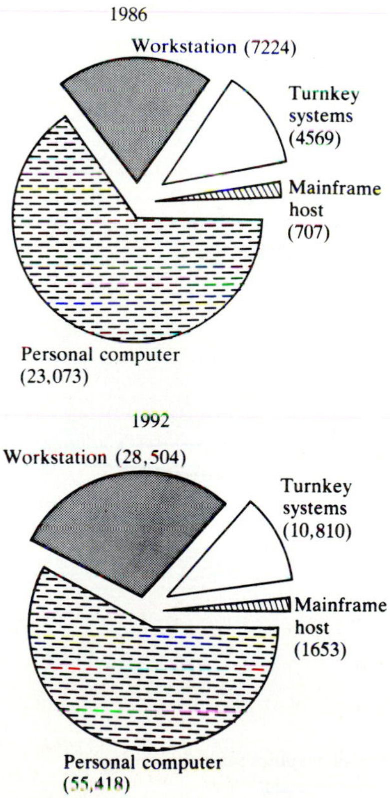
1986
Workstation (7224)
Turnkey systems (4569)
Mainframe host (707)
Personal computer (23,073)
1992
Workstation (28,504)
Turnkey systems (10,810)
Mainframe host (1653)
Personal computer (55,418)

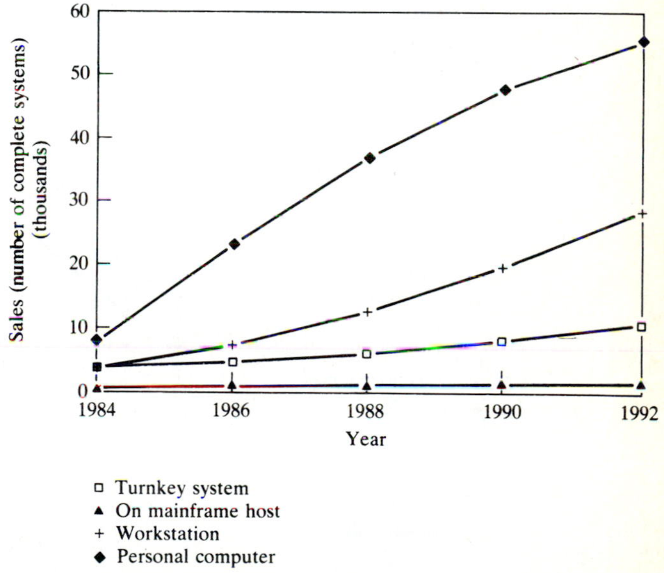
60
50
40
30
20
10
0
Sales (number of complete systems) (thousands)
1984
1986
1988
1990
1992
Year
Turnkey system
On mainframe host
Workstation
Personal computer

**FIGURE 2-45**
CAD/CAM market sales (number of complete systems).

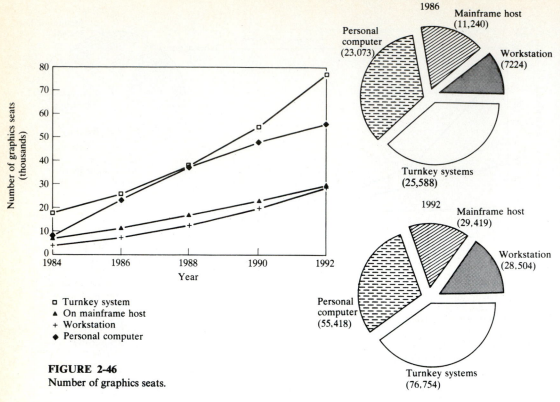

**FIGURE 2-46**
Number of graphics seats.

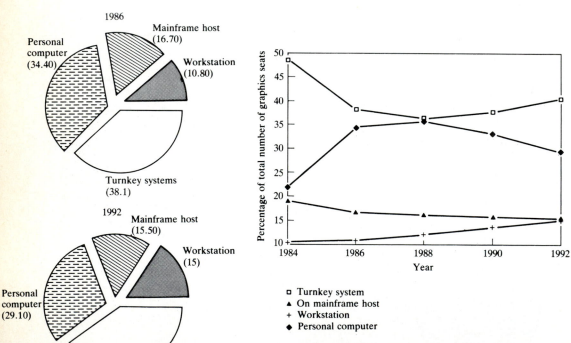

**FIGURE 2-47**
Growth of number of graphics seats (percentage of total growth number of seats).

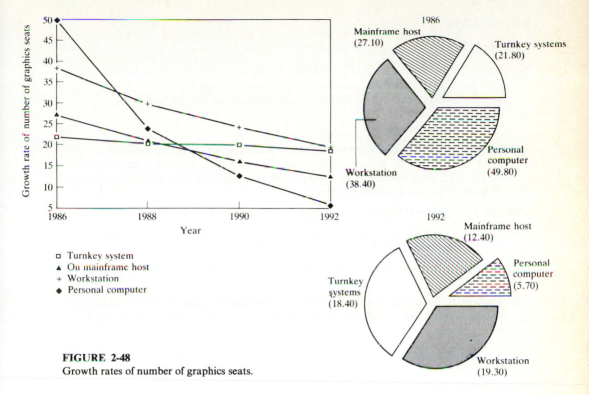

**FIGURE 2-48**
Growth rates of number of graphics seats.

concept. The model also cannot predict what is coming beyond the seventh tier.

The tier model identifies and correlates existing hardware concepts and can predict when and how (mode of uses) they are used. Applying such a model to CAD/CAM hardware, one would expect supercomputer power to be available at the engineer's desk in about 28 years (to migrate from tier 7 to tier 4).

**TABLE 2.6**
**Computer tiers**

| Tier | Location | Name | Price, $ | Weight, lb |
|------|----------|------|----------|------------|
| 1 | Wallet | Calculator | 10 | 0.05 |
| 2 | Pocket | Special function | 100 | 0.5 |
| 3 | Briefcase | Kneetop | 1K | 5 |
| 4 | Office | Workstation | 10K | 50 |
| 5 | Department | Shared mini | 100K | 500 |
| 6 | Center | Mainframe | 1M | 5000 |
| 7 | Region | Supercomputer | 10M | 50,000 |

*Source:* Reprinted with permission from D. L. Nelson and C. G. Bell, "The Evolution of Workstations," *IEEE Circuits and Devices Mag.*, pp. 12–16, July 1986. © 1986 IEEE.

# PROBLEMS

## Part 1: Theory

**2.1.** Compare the existing types of CAD/CAM systems considering the following issues:
   (a) Cost: capital investment, running cost, and maintenance cost
   (b) Capabilities: geometric modeling, design applications, and programming
   (c) Convenience of use
   (d) Future improvements

**2.2.** The first step in implementing the CAD/CAM technology in an engineering environment is to justify its benefits to management. Assuming you are a member of a CAD/CAM select committee, develop guidelines to justify it. Also develop guidelines to evaluate various available systems.

**2.3.** What are the advantages and disadvantages of the various interactive input devices?

**2.4.** Can a lightpen be used with a storage tube display? Why? How can a locating or pointing function be performed on this type of display?

**2.5.** Compare the capabilities, advantages, and disadvantages of the three types of displays.

**2.6.** Explain the following statement: "A typical raster display would have a resolution of $1280 \times 1024$, capable of presenting 256 simultaneous colors selected from a palette of 16.7 million."

**2.7.** The following specifications are read in one of the catalogues of a CAD/CAM system:

   Display monitor:
   | | |
   |---|---|
   | Resolution | 78 pixels (dots) per inch |
   | Format | 1024 horizontal × 864 vertical |
   | Active display area | 11.10 × 13.95 in |
   | Refresh | 60 Hz noninterlaced |
   | Colors | 8 IN/24 OUT |

   Keyboard:
   105 sculptured keys, typewriter-style main array, editing keypad, numerical keypad, 20 special-function keys
   | | |
   |---|---|
   | Profile | 1.2 in from palm rest to home row |
   | Cord length | 12.0 ft coiled |
   | Weight | 4 lb |

   Mouse:
   | | |
   |---|---|
   | Output | 100 pulses per inch (X and Y relative displacement) |
   | Function buttons | 3 |

   Explain, in terms of the technical background covered in this chapter, what these specifications mean. If you were a member of a CAD/CAM select committee, would you approve the above configuration of the workstation? Explain your answer.

**2.8.** Which of the following graphics displays would you choose for better performance?
   (a) Two displays with the same specifications except the refresh rate. One has 30 Hz interlaced and the other has 60 Hz noninterlaced.
   (b) Same as (a) but one has 60 Hz interlaced and the other has 60 Hz noninterlaced.
   (c) Two displays with the same specifications except one has colors of 8 IN/12 OUT and the other 12 IN/8 OUT.

   Justify your choice. If you have to compromise between performance and cost, which display do you think is less expensive in each case?

**2.9.** For the display monitor in Prob. 2.7, calculate the size of its bit map (RAM), number of pixels accessed per second, and the access time per pixel.

**2.10.** A lookup table is defined as an optional RAM memory (distinct from pixel memory) that allows a user to display any given pixel at an arbitrary shade or color. The lookup table stores user-defined intensity indexes and values for each pixel. The intensity index is used in the bit map and the intensity value is used for the beam deflection and energy. Below is a lookup table for three plane monochrome and color displays:

| Intensity index | | Intensity value | | | Red | Green | Blue |
|---|---|---|---|---|---|---|---|
| 0 | (000) | 0 | Black | 0 | | | |
| 1 | (001) | 27 | | 1 | | | |
| 2 | (010) | 73 | | 2 | | | |
| 3 | (011) | 109 | | 3 | | | |
| 4 | (100) | 145 | | 4 | | | |
| 5 | (101) | 181 | Medium bright | Yellow 5 | 255 | 255 | 0 |
| 6 | (110) | 217 | | 6 | | | |
| 7 | (111) | 255 | White | 7 | | | |

Monochrome display                                    Color display

How can you display a horizontal medium bright line on the monochrome and a horizontal yellow line on the color display between pixels (100, 100) and (500, 100)?

**2.11.** How long does it take a dot matrix printer with a speed of 250 characters/second to print text only displayed on a 1024 horizontal × 1280 vertical monitor? Assume that a 5 horizontal × 7 vertical grid (pattern of dots) is required to generate one character as shown in Fig. P2-11.

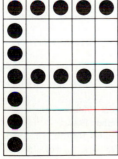

Character F          **FIGURE P2-11**

**2.12.** How long does it take to plot a drawing of size $8\frac{1}{2} \times 11$ on an electrostatic plotter with 9 inch wide paper, a resolution of 200 dots/inch in each direction, and paper speed of 600 dots/second. What is the rate of transfer of dots (bits) per second?

## Part 2: Laboratory

**2.13.** What type of CAD/CAM system is available to you? What are the specifications of the system drive(s)?

**2.14.** With the help of the in-house CAD/CAM system manager, generate a sheet of the specifications of the existing input and output devices. Do you see any possible future improvements?

**2.15.** What is the type of LAN that supports the in-house CAD/CAM system available to you? Obtain data about it and relate it to system performance. Suggest some improvements.

**2.16.** If you would have the responsibility to upgrade the in-house CAD/CAM system, what directions would you take and recommendations would you make? Why?

## Part 3: Programming

**2.17.** What does the term 16-bit or 32-bit machine or computer mean?

**2.18.** What is meant by single or double precision? How does that affect engineering and design calculations?

**2.19.** What is meant by fixed-point and floating-point numbers?

**2.20.** What is a bit, byte, kbyte, and Mbyte?

**2.21.** What is the difference between RAM and ROM?

## BIBLIOGRAPHY

Allan, R.: "Networking," *Electronic Des.*, pp. 152–158, January 1986.

Aronson, R. B.: "Printer and Printer Proliferation," *CAE*, pp. 46–56, April 1986.

"Built-in Processor for Interface of Graphics Workstations," *IBM Tech. Disclosure Bull.*, vol. 28, no. 12, pp. 5524–5527, May 1986.

Cowan, D.: "Survey of the CAD Field," *Computer Aided Des. J.*, vol. 18, no. 1, pp. 5–9, 1986.

Cragg, R.: "The Graphics Terminal Market Is Moving Toward More Intelligent, Less Expensive Terminals," *Hardcopy*, pp. 116–117, August 1986.

"Display Definition Control for Graphics Displays," *IBM Tech. Disclosure Bull.*, vol. 28, no. 12, pp. 5603–5604, May 1986.

"Drawing a Bead on High-Performance Graphics," *Hardcopy*, pp. 108, March 1986.

Foley, J. D., and A. Van Dam: *Fundamentals of Interactive Computer Graphics*, Addison-Wesley, Reading, Mass., 1982.

Frost and Sullivan, Inc.: "Computer Aided Design Systems in Europe," Report E712, New York, 1984.

Frost and Sullivan, Inc.: CAD/CAM Market in the United States," Report 1564, New York, 1986.

Groover, M. P., and E. W. Zimmers: *CAD/CAM: Computer-Aided Design and Manufacturing*, Prentice-Hall, Englewood Cliffs, N.J., 1984.

Hearn, E. D.: "Digitizers for Data Entry," *BYTE*, pp. 261–266, November 1986.

"High-Powered Desktops Handle CAD/CAM," *CAE*, pp. 11–12, November 1984.

Hordeski, M. F.: *CAD/CAM Techniques*, Reston, A Prentice-Hall Company, Reston, Va., 1986.

Hubbold, R. J.: "Computer Graphics and Displays," *Computer Aided Des. J.*, vol. 16, no. 3, pp. 127–132, 1984.

Hughs, G.: "Price/Performance Trends in High Resolution, Large Format Plotting," *Computer Graphics '86, NCGA*, vol. III, pp. 431–435, 1986.

Jafe, M.: "Evaluating CAD/CAM Systems: Priorities and Integration," *Infosystems*, pp. 56–57, June 1985.

Kacala, J.: "Hardcopy Output for CAD/CAM," *CAE*, pp. 36–42, November 1985.

Killmon, P.: "Tailored Design Match Workstations to Applications," *Computer Des.*, pp. 49–66, June 1986.

Knox, C. S.: *CAD/CAM Systems; Planning and Implementation*, Marcel Dekker, New York, 1983.

Krouse, J. K.: *What Every Engineer Should Know about Computer-Aided Design and Computer-Aided Manufacturing*, Marcel Dekker, New York, 1982.

Kulinski, T.: "Mice and Digitizers Gain Respectability and Market Share," *Hardcopy*, pp. 114, August 1986.

Machover, C.: "Hardware Directions," *Computer Graphics '86, NCGA*, vol. III, pp. 436–441, 1986.

Maekawa, M.: "Multiwindow Screens without Window Overlapping," in *Microprocessing and Microprogramming*, vol. 18, pp. 539–546, North-Holland, 1986.

Mills, R. B.: "Devices for Faster Data Entry," *CAE Mag.*, pp. 70–73, December 1985.

Mills, R. B.: "CAD/CAM Drives Display Technology," *CAE*, p. 34, September 1986.

Myers, W. (Contributing Ed.): "Computer Graphics: The Next 20 Years," *IEEE Computer Graphics and Applic.*, vol. 5, no. 8, pp. 69–76, 1985.

Nelson, D. L., and C. G. Bell: "The Evolution of Workstations," *IEEE Circuits and Devices Mag.*, pp. 12–16, July 1986.

Newman, W. M., and R. F. Sproull: *Principles of Interactive Computer Graphics*, 2d ed., McGraw-Hill, New York, 1979.

Rippiner, H.: "Workstations: Present and Future Trends," *Computer Aided Des. J.*, vol. 18, no. 1, pp. 17–21, 1986.

Scrivener, S. A. R. (Ed.): *Computer-Aided Design and Manufacture; State-of-the-Art Report*, Pergamon Infotech, Maidenhead, Berkshire, 1985.

Stover, R. N.: *An Analysis of CAD/CAM Applications, with an Introduction to CIM*, Prentice-Hall, Englewood Cliffs, N.J., 1984.

Teicholz, E. (Editor-in-Chief): *CAD/CAM Handbook*, McGraw-Hill, New York, 1985.

Titus, J.: "Digitizing Tablets Offer Choices of Formats, Operating Modes, and Pointers," *EDN*, pp. 69–74, April 1986.

Vaughan, J.: "Solids Modeling: Hardware Grabs the Spotlight," *Digital Des.*, pp. 50–53, August 1986.

Wang, P. C. C. (Ed.): *Advances in CAD/CAM Workstations; Case Studies*, Kulwer Academic, Hingham, Mass., 1986.

Williams, T.: "Graphics Processing Migrates from Host to Workstation," *Computer Des.*, pp. 49–57, July 1985.

Yoshikawa, K., H. Yamaguchi, and T. Asano: "Captive Touch Panel Using Uniform Resistive Film," *Fujitsu Sci. Tech. J.*, vol. 22, no. 2, pp. 124–131, 1986.

Zyda, M. J.: "Workstation Graphics Capabilities for the 1990s and Beyond," *Computer Graphics '86, NCGA*, vol. III, pp. 442–453, 1986.

# CHAPTER
# 3
# CAD/CAM SOFTWARE

## 3.1 INTRODUCTION

CAD/CAM has been acknowledged as the key to improving manufacturing productivity and the best approach for meeting the recent critical design requirements. The National Science Foundation's Center for Productivity has stated that "CAD/CAM has more potential to radically increase productivity than any development since electricity." All these promises and dreams are directly related to the quality and capabilities of CAD/CAM software. Such software provides engineers with the tools needed to perform their technical jobs efficiently and free them from the tedious and time-consuming tasks that require little or no technical expertise. Experience has shown that CAD/CAM software speeds the design process, therefore increasing productivity, innovation, and creativity of designers. Furthermore, in some design cases thus far, such as VLSI, CAD/CAM software has provided the only means to meet the new technological design and production requirements of increased accuracy and uniformity. The need for the software in the future will be even greater due to the expected intricate design and manufacturing requirements and the increasing industrial competition.

Similar to hardware, CAD/CAM software has progressed steadily since the development of interactive computer graphics in the 1960s. As in the case of computing in general, software has emerged from working out techniques, followed by developing sound algorithms, to placing more emphasis on basic principles and less on techniques that were quite inadequate when compared to the demands of actual graphics and engineering applications. As a result, the need to develop sound algorithms to support both graphics and applications was apparent.

The future potential of CAD/CAM has stimulated various investments which resulted in the creation of a wide variety of CAD/CAM systems that provided a large spectrum of capabilities and applications. Although these systems may seem greatly different due to the format of user interface and different implementations offered by each one, they all operate on the same theory and offer similar functionalities. Thus, knowledge about one system and its software can be related to other systems.

An investigation of existing CAD/CAM software in general reveals that it has common characteristics regardless of the hardware it runs on. It is an interactive program typically written in a standard programming language: FORTRAN, Pascal, or C. It is hardware-dependent and seems different to the user from conventional software due to the user interface. The database structure and database management system of the software determines its quality, speed, and ease of information retrieval. Users of the software are usually faced with learning its related semantics and syntax of its user interface. Semantics specifies how the software functions and what information is needed for each operation on an object. For example, a block requires three lengths and an orientation to create. Semantics are usually bounded by the principles and theories underlying a given field. Syntax defines the format of inputs and outputs. It is considered the grammar of the software. It specifies the rules that users must follow to achieve the desired semantics.

Performance is another common characteristic of software. The larger the number of interactive users, the longer the interactive response time. The software occasionally "locks" and ceases to respond to or accept user commands. This is typically referred to as a "system crash." When this happens, the user loses the work performed after the last filing or save command is issued and rebooting the system is required. This is why users are always advised to file or save their work frequently.

The most important characteristic of CAD/CAM software is its fully three-dimensional, associative, centralized, and integrated database. Such a database is always rich in information needed for both the design and manufacturing processes. The centralized concept implies that any change in or addition to a geometric model in one of its views is automatically reflected in the existing views or any views that may be defined later. The integrated concept implies that a geometric model of an object can be utilized in all various phases of a product cycle, as discussed in Chap. 1. The associativity concept implies that input information can be retrieved in various forms. For example, if the two endpoints of a line are input, the line length and its dimension can be output.

CAD/CAM software is typically a large, complex program that has been developed over the years. Typical users do not, and should not, have access to the source code. Therefore, they need professional training on the semantics and syntax of the software to become productive experienced users. During the training period, two main observations should be kept in mind. First, the learning curve is usually slow at the beginning and then speeds up towards the end, eventually leveling off. Second, the highest level of confusion occurs at the beginning, thus accounting for the slow learning period. Further training is usually easier and faster. Figure 3-1 shows a typical trend of the learning curve. A key to the success of a CAD/CAM installation in an industrial environment is the support of the company management during and after the training period. The posttraining support is very crucial because the trainee's early productivity on the

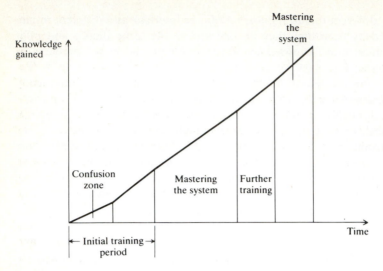

**FIGURE 3-1**
Typical learning curve during CAD/CAM training.

CAD/CAM system to complete a design task may be less than that needed to complete the same task utilizing conventional ways.

Users of CAD/CAM software can be classified into three groups: software operators, applications programmers, and system programmers. The majority of users including engineers and designers fall into the operators category. The main concern of this group is to master using the software so that the anticipated productivity increases are achieved. A typical operator tends to specialize in one or two modules of the software. For example, a designer may master geometric modeling and finite element modeling and analysis modules while a draftsman concentrates primarily on the drafting and detailing module. Operators are usually assisted by their system manager and vendors' customer response centers and hot lines to resolve any problems and answer any questions they may encounter in their day-to-day use of the software. Applications programmers can develop new programs and link them with the software, but they are not allowed to modify the existing source code. Such a need arises when a CAD/CAM system is to be customized by adding special applications and means to its software. These programmers are also experienced operators of the software with extensive programming background. They are usually familiar with the programming module of the software. In contrast to applications programmers, system programmers have the privilege to change the source code. In essence, they are the developers of the software itself. They are usually knowledgeable of the internal organization of the software, its database structure, and its database management system. They also know how to modify the user interface and usually possess backgrounds in computer graphics, engineering analysis, and computer science. They usually work for either turnkey vendors or R&D groups and centers.

Management of CAD/CAM software mirrors that of software in general and more. System managers usually report any software bugs to the vendor, develop backup procedures, and ensure that new revisions of software are installed with minimum interruptions to the system schedule. The manager is always seeking feedback from users about the software behavior and response for evaluation purposes. The manager is also responsible for developing a file-naming scheme to be used in conjunction with the software. Users are usually requested to use this scheme to name the files of the parts and models they create on the system. Such a scheme helps identify these parts by projects, groups, and/or years. Another software-related management issue is the selection of the most efficient procedures to achieve specific goals. For example, sometimes users can access already setup parts files to start their geometric modeling. The setup usually entails the various types of views and their organizations that can be employed. Another example is related to assemblies where a choice between the top-down and bottom-up approaches is made and adhered to.

The main objective of this chapter is twofold. First, definitions of the most common terms related to the software are introduced. Second, generic and common functions of CAD/CAM software are covered. The descriptions of these functions should enable readers to utilize documentation of their in-house CAD/CAM software effectively. It should be noted that the material covered here applies to all systems.

## 3.2   GRAPHICS STANDARDS

CAD/CAM software may be perceived as an application program supported by a graphics system as shown in Fig. 3-2a. The graphics system performs all related graphics techniques. In the actual source code of the application program, the graphics system is embedded in the form of subroutine calls. Therefore, software becomes inevitably device-dependent. If input/output devices change or become obsolete, its related software becomes obsolete as well unless significant resources are dedicated to modify such software. This approach was very costly to both

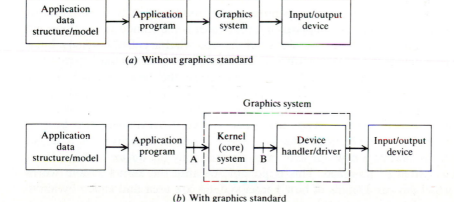

(a) Without graphics standard

(b) With graphics standard

**FIGURE 3-2**
Organization of a typical CAD/CAM software.

CAD/CAM vendors as well as users. This approach characterized the period from 1963 to 1974 due to the undeveloped nature of computer graphics during this period.

The needs for graphics standards were obvious and were acknowledged by the CAD/CAM community—both vendors and users. The following are some of these needs:

1. Application program portability. This avoids hardware dependence of the program. For example, if the program is written originally for a DVST display, it can be transported to support a raster display with minimal effort.
2. Picture data portability. Description and storage of pictures should be independent of different graphics devices.
3. Text portability. This ensures that text associated with graphics can be presented in an independent form of hardware.
4. Object database portability. While the above needs concern CAD/CAM vendors, transporting design and manufacturing (product specification) data from one system to another is of interest to CAD/CAM users. In some cases, a company might need to ship a CAD database of a specific design to an outside vendor to manufacture and produce the product.

With the above needs in mind, the search for standards began in 1974. Since then, various organizations and interest groups have been developing standard practices to encourage the more general and cost-effective use of CAD/CAM software development. In 1974, the GSPC (Graphics Standards Planning Committee) was formed to address the standards issue. The focus of standards is that the application program should be device-independent and should interface to any input device through a device handler and to any graphics display through a device driver. This leads to the conceptual organization of CAD/CAM software as shown in Fig. 3-2b. Here, the graphics system is divided into two parts: the kernel (core) system, which is hardware-independent, and the device handler/driver, which is naturally hardware-dependent. The kernel system, therefore, acts as a buffer between the application program and the specific hardware to ensure the independence and portability of the program. At interface A in the figure, the application program calls the standard functions and subroutines provided by the kernel system through what is called language bindings. These functions and subroutines, in turn, call the device handler/driver functions and subroutines at interface B to complete the task required by the application program. Figure 3-2b shows the benefits of standards. CAD/CAM software can now serve several hardware generations. It is also portable from one graphics system to another. Application and system programmers also become portable and can move from one system to another. Moreover, if a device becomes obsolete or a new one is to be supported, only the device handler/driver is to be written or modified. This is possible because the kernel system works with virtual devices. Details of how kernel systems are used and called by application programs are beyond the scope of this book.

The search for standards that began in 1974 continued both at the USA and international levels. In 1977 and 1979 the ACM (Association for Computing

Machinery) SigGraph group published two landmark reports (not formal standards) on the "core system." Core was never standards but influenced many related efforts. In 1981 the GSPC disbanded, and the ANSI (American National Standards Institute) has formed the Technical Committee on Computer Graphics Languages, X3H3, to produce a standardized core of device-independent computer graphics functions. At the international level, similar efforts to that of the GSPC were directed by the ISO (International Standards Organization). The technical work was led by the German Standards Institute (DIN) and resulted in the GKS (graphics kernel system). GKS has been adopted by the USA with the ANSI version having four output levels instead of three.

As a result of these worldwide efforts, various standards functioning at various levels of the graphics system shown in Fig. 3-2 exist. These are:

1. GKS is an ANSI and ISO standard. It is device-independent, host-system-independent, and application-independent. It supports both two-dimensional and three-dimensional data and viewing. It interfaces the application program with the graphics support package (interface A in Fig. 3-2b).

2. PHIGS (programmer's hierarchical interactive graphics system) is intended to support high function workstations and their related CAD/CAM applications. The significant extensions it offers beyond GKS-3D are in supporting segmentation used to display graphics and the dynamic ability to modify segment contents and relationships. PHIGS operates at the same level as GKS (interface A).

3. VDM (virtual device metafile) defines the functions needed to describe a picture. Such description can be stored or transmitted from one graphics device to another. It functions at the level just above device drivers. VDM is now called CGM (computer graphics metafile).

4. VDI (virtual device interface) lies between GKS or PHIGS and the device handler/driver code (interface B in Fig. 3-2b). Thus VDI is the lowest device-independent interface in a graphics system. It shares many characteristics with CGM. VDI is designed to interface plotters to GKS or PHIGS. It is not suitable to interface intelligent workstations. It is also not well matched to a distributed or network environment. VDI is now called CGI (computer graphics interface).

5. IGES (initial graphics exchange specification) was approved in September 1981 as the ANSI Standard Y14.26M. It enables an exchange of model databases among CAD/CAM systems. IGES functions at the level of the object database or application data structure.

6. NAPLPS (North American presentation-level protocol syntax) was accepted by Canada and ANSI in 1983. It describes text and graphics in the form of sequences of bytes in ASCII code.

Various CAD/CAM users and application or system programmers may be interested in one or more of the above standards. Awareness of these standards can be used as a guideline in evaluating various CAD/CAM systems. For example, mechanical design requires three-dimensional modeling. Therefore a system that supports GKS-3D or PHIGS is required. However, for two-dimensional applications such as VLSI design, GKS-2D is adequate. In addition, the future needs of the system must also be considered to avoid locking the system into software that will be unnecessarily difficult to upgrade over the coming years. Finally, knowledge of these standards and their functions might stimulate engineers to think of developing design and manufacturing standards, through engineering organizations, and enforce them on CAD/CAM vendors.

## 3.3   BASIC DEFINITIONS

This section introduces some of the basic terms and their definitions. The knowledge gained in this section and the remainder of the chapter should enable the readers to understand how their in-house CAD/CAM systems are structured, how they work, and how to go about using them effectively in an engineering environment.

### 3.3.1   Data Structure

Formally a data structure is defined as a set of data items or elements that are related to each other by a set of relations. Applying these relations to the elements of the set results in a meaningful object. From a CAD/CAM point of view, a data structure is a scheme, logic, or a sequence of steps developed to achieve a certain graphics, non-graphics, and/or a programming goal.

   As an example consider the object shown in Fig. 3-3. Three different types of data structures have been identified to construct the object. They are based on edges, vertices, or blocks. Within the context of the above formal definition of a data structure, the set of edges, vertices, or blocks is the set of data items for each type, and edges, vertices, or blocks are the data items themselves. Furthermore, the connectivity vertices for the first type, the edge information for the second, and the set operators for the third form the set of relations required by each type. As an example, $\boxed{1, A \& B}$ in Fig. 3-3b indicates that vertex 1 is shared by edges A and B while in Fig. 3-3c, $\boxed{A, 1 \& 4}$ indicates that edge A has the two vertices 1 and 4.

### 3.3.2   Database

The term "database" is commonly used and may mean different things to different users. Casually, it is synonymous with the terms "files" and "collection of files." Formally, a database is defined as an organized collection of graphics and nongraphics data stored on secondary storage in the computer. It could, therefore, be viewed as the art of storing or the implementation of data structure into the computer. Hence, it is a repository for stored data. From a software development point of view, a decision on the data structure has to be made first, followed by a choice of a database to implement such a structure. There may exist more than one alternative of database to implement a given data structure.

   The objective of a database is to collect and maintain data in a central storage so that it will be available for operations and decision-making. The

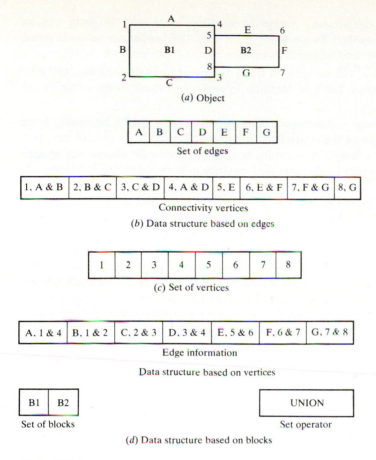

(a) Object

Set of edges

Connectivity vertices

(b) Data structure based on edges

(c) Set of vertices

Edge information

Data structure based on vertices

Set of blocks                    Set operator

(d) Data structure based on blocks

**FIGURE 3-3**
Various data structures of an object.

advantages that accrue from having centralized control of the data, or a central-ized database, is manyfold:

1. **Eliminate redundancy.** This is important for integrated CAD/CAM functions and CIM applications. The database should be rich enough to support all various phases of product design and manufacturing. If both design and manufacturing departments, for example, have access to the same database, inconsistent and conflicting decisions are inherently eliminated, and data is shared by all applications. Thus, engineering "assets" and experiences of a company can be captured in a database and modified for new product designs.

2. **Enforce standards.** With central control of the database, both national and international standards are followed. Dimensioning and tolerancing are exam-ples. In addition, a company can develop its own internal standards required by various departments. Standards are desirable for data interchange or migration between systems.

3. Apply security restrictions. Access to sensitive data and projects can be checked and controlled by assigning each user the proper access code (read, write, delete, copy, and/or none) to various parts of the database.
4. Maintain integrity. The integrity of the database ensures its accuracy. Integrity precedes consistency. Lack of database integrity can result in inputting inconsistent data.
5. Balance conflicting requirements. Compromises can easily be made when designing a model of the centralized database to provide its overall best performance. If, for example, a software is designed solely for design and modeling, one would expect inadequate performance in manufacturing functions.

CAD/CAM databases must be able to store pictorial data in addition to textural and alphanumeric data typically stored in conventional databases. A brief description of the popular database models is provided below:

1. Relational database. Data is stored in tables, called relations, that are related to each other. The relations are stored in files which can be accessed sequentially or in a random access mode. Sequential access files are widely used. As an example, the relations needed to describe the object in Fig. 3-3 are shown in Fig. 3-4. The object is represented by the three relations POINT, LINE/CURVE, and SURFACE. A particular data structure shown in Fig. 3-3 determines which relations are to be entered by the user and which are to be calculated automatically. One of the disadvantages of the relational database is that it requires substantial sorting, which might result in slowing the system response to user commands.
2. Hierarchical database. In this model, data is represented by a tree structure. The top of the tree is usually known as the "root" and the superiority, or hierarchy, of the tree levels relative to each other descends from the root

| Point | $x$ | $y$ |
|-------|-----|-----|
| 1 | $x_1$ | $y_1$ |
| 2 | $x_2$ | $y_2$ |
| 3 | $x_3$ | $y_3$ |
| 4 | $x_4$ | $y_4$ |
| 5 | $x_5$ | $y_5$ |
| 6 | $x_6$ | $y_6$ |
| 7 | $x_7$ | $y_7$ |
| 8 | $x_8$ | $y_8$ |

Relation POINT

| Line | Start point | End point |
|------|-------------|-----------|
| A | 1 | 4 |
| B | 1 | 2 |
| C | 2 | 3 |
| D | 3 | 4 |
| E | 5 | 6 |
| F | 6 | 7 |
| G | 7 | 8 |

Relation LINE/CURVE

| Surface | Line/curve | Type |
|---------|-----------|------|
| | A | Line |
| | B | Line |
| 1 | C | Line |
| | D | Line |
| | E | Line |
| | F | Line |
| 2 | G | Line |
| | D | Line |

Relation SURFACE

**FIGURE 3-4**
Sample relational database of object shown in Fig. 3-3.

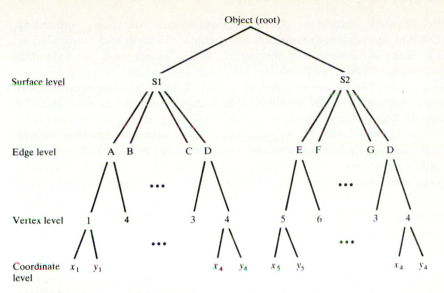

**FIGURE 3-5**
Sample hierarchical database of object shown in Fig. 3-3.

down. Figure 3-5 shows a hierarchical database of the object shown in Fig. 3-3. Four levels are required to represent the object completely. One of the drawbacks of the hierarchical approach is the asymmetry of the tree structure, which forces database programmers to devote time and effort to solving problems, introduced by the hierarchical approach, which are not intrinsic to the object modeling itself.

3. Network database. The network approach permits modeling of many-to-many correspondence more directly than the hierarchical approaches. Figure 3-6 shows a network database of the object shown in Fig. 3-3. The prime disadvantage of the network approach is its undue complexity both in the database structure itself and in the associated programming of it.

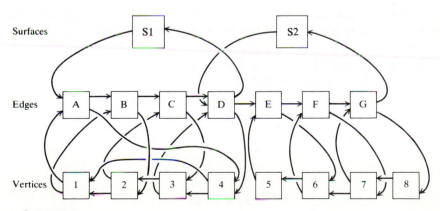

**FIGURE 3-6**
Sample network database of object shown in Fig. 3-3.

4. Object-oriented database. Unlike conventional database processing, CAD/CAM applications require object-oriented accessing and manipulation; that is, units of retrieval and storage are design objects and not individual records in files. These design objects also form the basis for ensuring database integrity upon the insertion, deletion, or modification of component objects. The object-oriented model should be able to capture all the relevant semantics of objects. This, in turn, results in a "rich," well-integrated, and complete database readily accessible for applications. Object-oriented database models include the entity relationship model, complex object representation, molecular object representation, and abstract data model. The abstract data model is close to solid modeling databases. It employs abstract objects as primitives in the design of the database. Figure 3-7 shows an example of this database. Primitives are constructed from input data and form the lowest field or record of storage in the database.

Object-oriented databases seem to be ideal for CAD/CAM applications. Hybrid database models may also be useful. The following are some of the functional requirements and specifications that CAD/CAM databases must support:

1. Multiple engineering applications from conceptual design to manufacturing operations.
2. Dynamic modification and extension of the database and its associativity.
3. The iterative nature of design. This nature is not common in business data processing. CAD/CAM database management systems must support the tentative, iterative, and evolutionary nature of the design process.
4. Design versions and levels of detail. CAD databases must provide a capability for storage and management of multiple design solutions that may exist for a particular design. There is seldom a unique solution to a design problem and there may exist several optimal solutions.
5. Concurrent and multiple users must be supported from the database. Large design projects usually involve multiple designers working simultaneously on multiple aspects of a project.
6. Temporary database support. Due to the iterative nature of design, earlier generated data may not be committed to the database until the design process is completed.
7. Free design sequence. The database system should not impose constraints on the designer to follow because different designs require different sequences.
8. Easy access. Application programs requiring data from a CAD/CAM database should not require extensive knowledge of the database structure to extract

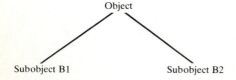

Object

Subobject B1                    Subobject B2

**FIGURE 3-7**
Sample object-oriented database of object shown in Fig. 3-3.

the data needed. This is important in customizing CAD/CAM systems for specific design and manufacturing procedures.

### 3.3.3 Database Management System (DBMS)

A DBMS is defined as the software that allows access to use and/or modify data stored in a database. The DBMS forms a layer of software between the physical database itself (i.e., stored data) and the users of this database as shown in Fig. 3-8a. DBMS protects the database from user's abuse. It also shields users from having to deal with hardware-level details by interpreting their input commands and requests from the database. For example, a command such as retrieve a line could involve few lower-level steps to execute. In general, a DBMS is responsible for all database-related activities such as creating files, checking for illegal users of the database, and synchronizing user access to the database.

DBMSs designed for commercial business systems are too slow for CAD/CAM. The handling of graphics data is an area where the conventional DBMSs tend to break down under the shear volume of data and the demand for quick display. By contrast, data handled in the commercial realm is mostly alphanumeric and the objects described are usually not very complex. A DBMS is directly related to the database model it is supposed to manage. For example, relational DBMSs require relatively large amounts of CPU time for searching and sorting data stored in the relations or tables. Therefore, the concept of database machines exists where a DBMS is implemented into hardware that can lie between the CPU of a computer and its database disks, as shown in Fig. 3-8b.

The requirements of a DBMS for CAD/CAM are fundamentally different from those required by commercial data processing applications. Therefore, current DBMS and techniques that are originally designed to support business data processing are not directly applicable to CAD/CAM databases. The latter

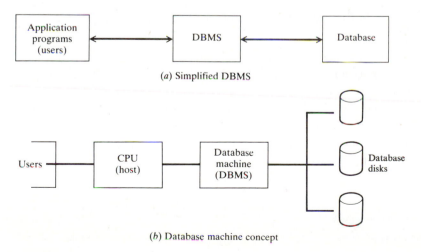

(a) Simplified DBMS

(b) Database machine concept

**FIGURE 3-8**
A typical DBMS.

are characterized by many different data types and also large numbers of instances of each type; that is, these databases exhibit very large, but static, database structures. In addition, CAD/CAM databases must support complex relationships between data items in contrast to business databases that are designed for record keeping and modeling of relatively simple relationships between data types. Another fundamental difference is that business databases are relatively stable over time. However, CAD/CAM databases must reflect the iterative nature of design and manufacturing. Object-oriented databases and their related DBMSs seem to be ideal for CAD/CAM and can result in a significant improvement of CAD/CAM systems.

### 3.3.4 Database Coordinate System

Three types of coordinate systems are needed in order to input, store, and display model geometry and graphics. These are the working coordinate system (WCS), the model coordinate system (MCS), and the screen coordinate system (SCS) respectively. Other names for MCS are database or world coordinate system. Another name for SCS is device coordinate system. Throughout this book, MCS, WCS, and SCS are used.

The model coordinate system is defined as the reference space of the model with respect to which all the model geometrical data is stored. It is a cartesian system which forms the default coordinate system used by a particular software. The $X$, $Y$, and $Z$ axes of the MCS can be displayed on the graphics display by using the proper software command. The origin of the MCS can be arbitrarily chosen by the user while its orientation is established by the software.

In order for the user to communicate properly and effectively with a model database, the relationships between the MCS orthogonal planes and the model views must be understood by the user. Typically, the software chooses one of two possible orientations of the MCS in space. As shown in Fig. 3-9a, the $XY$ plane is the horizontal plane and defines the model top view. As an example, this convention is adopted in Computervision's software. The front and right side views

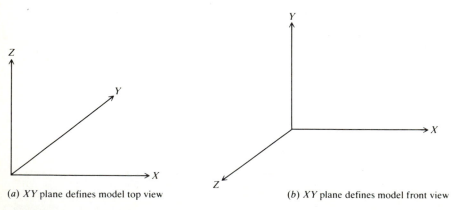

(a) $XY$ plane defines model top view          (b) $XY$ plane defines model front view

**FIGURE 3-9**
Possible orientations of MCS in space.

are consequently defined by the $XZ$ and $YZ$ planes respectively. Figure 3-9b shows the other possible orientation of the MCS where the $XY$ plane is vertical and defines the model front view. GE Calma software utilizes this convention. As a result, the $XZ$ and the $YZ$ planes define the top and the right side views respectively. Moreover, software of existing CAD/CAM systems uses the MCS as the default WCS (see next section). In both orientations, the $XY$ plane is the default construction plane. If the user utilizes such a plane, the first face to be constructed of a model becomes the top or front view depending on which MCS is used. This applies to wireframe and surface modeling only. For solid modeling, this plane determines the orientation of the input primitives.

The MCS is the only coordinate system that the software recognizes when storing or retrieving geometrical information in or from a model database. Many existing software packages allow the user to input coordinate information in cartesian $(x, y, z)$, cylindrical $(r, \theta, z)$, and/or spherical $(r, \theta, \phi)$ systems. However, this input information is transformed to $(x, y, z)$ coordinates relative to the MCS before being stored in the database. Obtaining views is a form of retrieving geometrical information relative to the MCS. If the MCS orientation does not match the desired orientation of the object being modeled (see Example 3.1), users become puzzled and confused. Another form of retrieving information is entity verification. Coordinates of points defining the entity are given relative to MCS by default. However, existing software allows users to obtain the coordinates relative to another system (WCS) by using the proper commands or modifiers.

> **Example 3.1.** Figure 3-10 shows a geometric model that is to be utilized for design and manufacturing applications. The center of hole B is the center of the inclined plane and is $\frac{1}{2}$ inch deep. Two designers construct the model by using point $P_2$ as the origin of the MCS and creating the front face $P_1 P_2 P_3 P_4 P_5$ first. Each designer uses one of the MCSs shown in Fig. 3-9. What are the top, front, right side, and isometric views each designer obtains assuming the default construction planes are active? How do you advise them if both are to obtain identical views and yet still use different software?

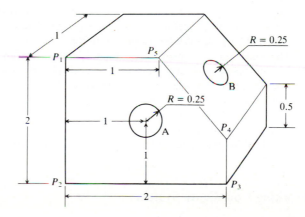

All dimensions
in inches

**FIGURE 3-10**
Geometric model of an object.

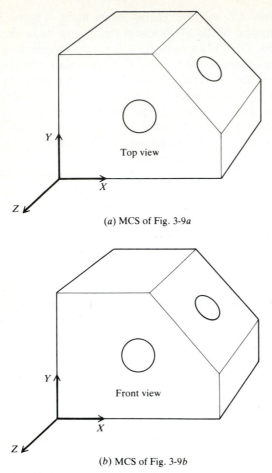

(a) MCS of Fig. 3-9a

(b) MCS of Fig. 3-9b

**FIGURE 3-11**
Orientation of MCS relative to the model.

*Solution.* Under the above construction conditions, the orientations of the two MCSs relative to the model are shown in Fig. 3-11. The corresponding views are then generated by each software, as shown in Fig. 3-12.

The above disagreement of the model views as given by the two software packages has resulted because the top view ($XY$ plane) as defined by the MCS of Fig. 3-9a does not coincide with the top view of the object being modeled, but instead coincides with its front view. Therefore, the designer using the corresponding software to Fig. 3-9a is advised to activate the construction plane that corresponds to the $XZ$ plane; that is, the front view defined by the software, before beginning construction. This guarantees obtaining identical results to those obtained by the other designer.

### 3.3.5  Working Coordinate System

It is often convenient in the development of geometric models and the input of geometrical data to refer to an auxiliary coordinate system instead of the MCS.

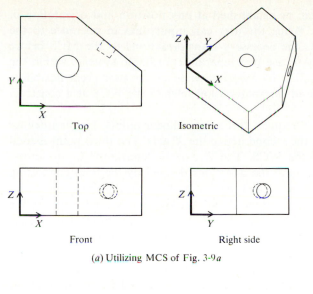

Top          Isometric

Front          Right side

(a) Utilizing MCS of Fig. 3-9a

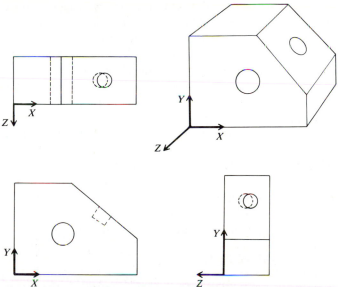

(b) Utilizing MCS of Fig. 3-9b

**FIGURE 3-12**
Views of object shown in Fig. 3-10.

This is usually useful when a desired plane (face) of construction is not easily defined as one of the MCS orthogonal planes, as in the case of inclined faces of a model. The user can define a cartesian coordinate system whose $XY$ plane is coincident with the desired plane of construction. That system is the working coordinate system (WCS). It is a convenient user-defined system that facilitates geometric construction. It can be established at any position and orientation in space that the user desires. While the user can input data in reference to the WCS, the software performs the necessary transformations to the MCS before storing the data. The ability to use two separate coordinate systems within the same model database in relation to one another gives the user great flexibility. Some software such as those of Computervision refer to the WCS as a construction plane.

The definition of a WCS requires three noncollinear points. The first defines the origin and the first with the second define the $X$ axis. The third point is used to define the $XY$ plane of the WCS. The $Z$ axis is determined as the cross-product of the two unit vectors in the directions defined by the lines connecting the first and the second (the $X$ axis), and the first and the third points. The $Y$ axis is determined as the cross product of the $Z$ and $X$ unit vectors (see Prob. 3.3). We will use the subscript $W$ to distinguish the WCS axes from those of the MCS. The $X_W Y_W$ plane becomes the active construction (working) plane if the user defines a WCS. In this case, the WCS and its corresponding $X_W Y_W$ plane override the MCS and the default construction plane respectively. As a matter of fact, the MCS with its default construction plane can be viewed by the user as the default WCS with its $X_W Y_W$ plane. All CAD/CAM software packages provide users with three standard WCSs that correspond to the three standard views of front, top, and right sides. Other WCSs can be defined by the users.

There is only one active WCS at any one time. If the user defines few WCSs in one session during a model construction, the software recognizes only the last one and stores it with the model database if the user stores the model by filing the session work. When retrieved later, the user is advised to display the axes of the current WCS before beginning construction to check its origin and orientation. If confused, the user can simply set the WCS back to the MCS by using the same command that defines a WCS but with the proper modifiers.

Once a WCS is defined, user coordinate inputs are interpreted by the software in reference to this system. At the mean time, the software calculates the corresponding homogeneous transformation matrix between the WCS and the MCS to convert these inputs into coordinates relative to the MCS before storing them in the database. The transformation equation can be written as

$$P = [T]P_W \tag{3.1}$$

where $P$ is the position vector of a point relative to the MCS and $P_W$ is the vector of a point relative to the active WCS. Each vector is given by

$$P = [x \quad y \quad z \quad 1]^T \tag{3.2}$$

The matrix $[T]$ is the homogeneous transformation matrix. It is a $4 \times 4$ matrix and is given by

$$[T] = \begin{bmatrix} t_{11} & t_{12} & t_{13} & t_{14} \\ t_{21} & t_{22} & t_{23} & t_{24} \\ t_{31} & t_{32} & t_{33} & t_{34} \\ \hline 0 & 0 & 0 & 1 \end{bmatrix} = \begin{bmatrix} {}^M_W[R] & {}^M P_{W,\text{org}} \\ \hline 0 \quad 0 \quad 0 & 1 \end{bmatrix} \tag{3.3}$$

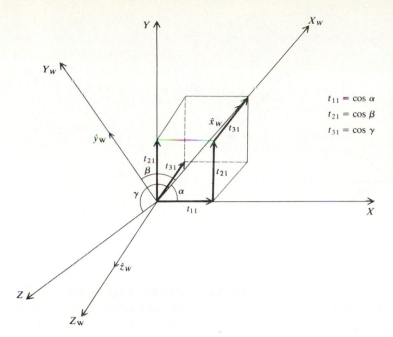

**FIGURE 3-13**
Direction cosines of WCS relative to MCS.

where $^M_W[R]$ is the rotation matrix that defines the orientation of the WCS relative to the MCS and $^M P_{W, \text{org}}$ is the position vector that describes the origin of the WCS relative to the MCS. The columns of $^M_W[R]$ give the direction cosines of the unit vectors in the $X_W$, $Y_W$, and $Z_W$ directions relative to the MCS, as shown in Fig. 3-13.

The WCS serves another function during geometric construction. Its $X_W Y_W$ plane is used by the software as the default plane of circles. A circle plane is usually not defined using its center and radius. In addition, the $Z_W$ axis of a WCS can be useful in defining a projection direction which may be helpful in geometric construction.

**Example 3.2.** Write a procedure to construct the holes shown in the model used in Example 3.1. Use the MCS shown in Fig. 3-9a.

**Solution.** Let us assume that the user has defined the $(WCS)_1$ as shown in Fig. 3-14 to construct the model without the holes. The procedure to construct the holes becomes:

1. With the $(WCS)_1$ active, construct circle A with center (1, 1, 0) and radius 0.25.
2. Construct hole A by projecting circle A at a distance of $-1.0$ (in the opposite direction to $Z_{W1}$).
3. Define $(WCS)_2$ as shown by using points $E_1$, $E_2$, and $E_3$.
4. Construct circle B with center $(x_c, y_c)$ and radius 0.25. The center can easily be found implicitly as the midpoint of line $E_2 E_3$.
5. Repeat step 2 but with a distance $-0.5$ (in the opposite direction to $Z_{W2}$).

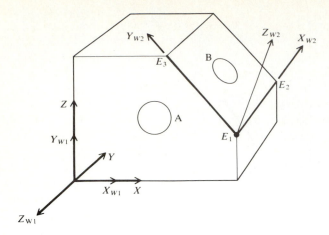

**FIGURE 3-14**
WCSs required to construct holes A and B.

The coordinates in step 1 are given relative to $(WCS)_1$ which is active at the time of construction. With reference to Fig. 3-14, these coordinates are (1, 0, 1) relative to the MCS and these are the values that are stored in the model database. To verify this, using Eq. (3.3) we can write:

$$C = \begin{bmatrix} 1 & 0 & 0 & \vdots & 0 \\ 0 & 0 & -1 & \vdots & 0 \\ 0 & 1 & 0 & \vdots & 0 \\ \hdashline 0 & 0 & 0 & \vdots & 1 \end{bmatrix} \begin{bmatrix} 1 \\ 1 \\ 0 \\ 1 \end{bmatrix} = \begin{bmatrix} 1 \\ 0 \\ 1 \\ 1 \end{bmatrix} \tag{3.4}$$

A similar approach can be followed to find the center of the hole B and is left as an exercise for the reader at the end of the chapter (see Prob. 3.4).

### 3.3.6  Screen Coordinate System

In contrast to the MCS and WCS, the SCS is defined as a two-dimensional device-dependent coordinate system whose origin is usually located at the lower left corner of the graphics display, as shown in Fig. 3-15. The physical dimensions

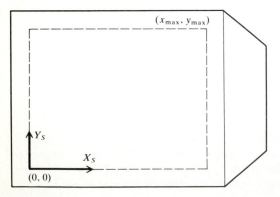

**FIGURE 3-15**
Typical SCS.

of a device screen (aspect ratio) and the type of device (vector or raster) determine the range and the measurement unit of the SCS. The SCS is mostly used in view-related digitizes such as definitions of view origin and window or digitizing a view to select it for graphics operations.

The range and the measurement unit of an SCS can be determined in three different methods. For raster graphics displays, the pixel grid serves as the SCS. A 1024 × 1024 display has an SCS with a range of (0, 0) to (1024, 1024). The center of the screen has coordinates of (512, 512). This SCS is used by the CAD/CAM software to display relevant graphics by converting directly from MCS coordinates to SCS (physical device) coordinates. This approach of defining SCSs is appropriate if the software supports or drives only one type of graphics display. For software packages that must drive multiple display units, it is convenient to define a normalized coordinate system that can be utilized to represent an image. Such representation can be translated by device-dependent codes to the appropriate physical device coordinates. In such a case, the range of the SCS can be chosen from (0, 0) to (1, 1). The third method of defining an SCS is by using the drawing size that the user chooses. If a size A drawing is chosen, the range of the SCS becomes (0, 0) to (11, 8.5) while size B produces the range (0, 0) to (17, 11). The rationale behind this method stems from the conventional drawing board so that the drafting paper is represented by the device screen.

A transformation operation from MCS coordinates to SCS coordinates is performed by the software before displaying the model views and graphics. Typically, for a geometric model, there is a database to store its geometric data (relative to MCS) and a display file to store its display data (relative to SCS).

**Example 3.3.** A view is typically defined by a view origin and a view window. Use the possible three methods (pixel grid, normalized values, drawing size) to define the four views shown in Fig. 3-16. The origins of the top, front, and right views ($O_T$, $O_F$, $O_R$) must line up as shown and the origin of the isometric view $O_I$ is assumed in the middle of its window. Assume a 1000 × 1000 pixel grid, a maximum normalized value of 1, and size A drawing for the three methods respectively.

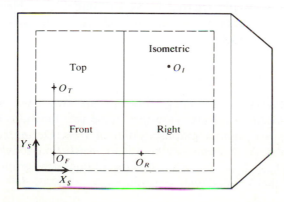

**FIGURE 3-16**
A typical screen layout.

*Solution.* A view window (viewport) is usually defined by one of its diagonals. Assume that the lower left $(x_{V,\,min}, y_{V,\,min})$ and the top right $(x_{V,\,max}, y_{V,\,max})$ corners of a view window are used to define such a window. The coordinates required to define the above views become

| | | Method | |
| --- | --- | --- | --- |
| | | Pixel grid | |
| View | $(x_0, y_0)$ | $(x_{V,\,min}, y_{V,\,min})$ | $(x_{V,\,max}, y_{V,\,max})$ |
| Front | 10, 10 | 0, 0 | 500, 500 |
| Top | 10, 510 | 0, 500 | 500, 1000 |
| Right | 510, 10 | 500, 0 | 1000, 500 |
| Isometric | 750, 750 | 500, 500 | 1000, 1000 |
| | | Normalized values | |
| View | $(x_0, y_0)$ | $(x_{V,\,min}, y_{V,\,min})$ | $(x_{V,\,max}, y_{V,\,max})$ |
| Front | 0.01, 0.01 | 0, 0 | 0.5, 0.5 |
| Top | 0.01, 0.51 | 0, 0.5 | 0.5, 1.0 |
| Right | 0.51, 0.01 | 0.5, 0 | 1.0, 0.5 |
| Isometric | 0.75, 0.75 | 0.5, 0.5 | 1.0, 1.0 |
| | | Drawing size | |
| View | $(x_0, y_0)$ | $(x_{V,\,min}, y_{V,\,min})$ | $(x_{V,\,max}, y_{V,\,max})$ |
| Front | 0.5, 0.5 | 0, 0 | 5.5, 4.25 |
| Top | 0.5, 4.75 | 0, 4.25 | 5.5, 8.5 |
| Right | 6.0, 0.5 | 5.5, 0 | 11, 4.25 |
| Isometric | 8.25, 6.375 | 5.5, 4.25 | 11, 8.5 |

An investigation of the coordinates of the origins of the top, front, and right views shows that these origins line up so that the conventional rules of orthographic views are followed. Similar results can be obtained by using the GRID function provided by CAD/CAM software. Grids are covered in Chap. 11 in more detail. In a production environment, the screen layout shown in Fig. 3-16 is usually defined and stored as a master file that is accessible by designers for modeling purposes. This file is usually referred to as an "initial" or "blank" model. This approach does not only save time but also enforces company standards during the definition of the model.

## 3.4  MODES OF GRAPHICS OPERATIONS

Constructing geometric models and producing drawings are two basic and popular functions of a CAD/CAM software. However, these two functions are contradictory in nature from a computer graphics point of view. During construction, it is required to keep the associativity between construction activities and the model database. These activities may include adding new graphics entities to the database, deleting or modifying existing entities, and/or defining new views of the model. Contrarily, producing drawings necessitates the disso-ciativity between drafting activities and the model database. Changing entity fonts and adding dimensions and/or text are typical activities that should not be reflected in all model views if performed in a particular view.

Software provides users with two basic graphics modes to enable them to perform the two functions. These are the model and drafting (detailing or drawing) modes. Only one mode can be active at any one time. There is always one command with the proper modifiers to turn on or off either mode. Both model and drafting modes utilize existing information in the model database. However, if the model mode is active, the result of every CAD/CAM operation or command is recorded to the model database. Consequently, work in one view of the model is reflected in the other current or future defined views. If, on the other hand, the drafting mode is active, the results of the commands are only local to the view they are applied to and do not affect the model database. This is a direct outcome of the fact that drafting work is mainly done for appearance (line fonts) and documentation (dimensions and text) purposes. Conceptually, the relation between the two modes and the model database is shown in Fig. 3-17. The switch shown in the loop of the drafting mode indicates that work done in this mode is still stored in the model database, although it does not affect it centrally.

While the main purpose of the model mode is to construct a model geometry, the main purpose of the drafting mode is to generate engineering draw-ings. The major three activities of the drafting mode are model clean-up, docu-mentation, and plotting. Model clean-up is the most tedious and time-consuming activity a user can engage in. Starting with the three standard orthographic views from the model database, the user should hide or change fonts of entities accord-ing to the standard drafting rules. The user is usually faced with two main prob-lems during model clean-up. The first is overlapping entities. This problem appears if the user has to blank (hide) these entities or change their fonts to dashed font. Reliable automatic hidden line removal algorithms based on wire-frame modeling are helpful in model clean-up. The typical time ratio between model creation and its clean-up is between $1:2$ and $1:3$, depending on the com-plexity of the model and its clean-up work. The second problem is that clean-up work is usually not recoverable. If the user destroys a view during the clean-up process, the only alternative is to start from scratch by repeating the work on a "fresh" view and discarding any previous clean-up work. It is also possible that the CAD/CAM system might crash in the middle of a drafting session which

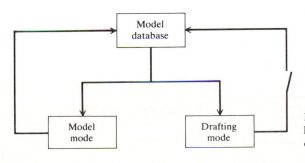

**FIGURE 3-17**
Relationship between model and drafting modes and model database.

results in loss of the drafting work also. The general advice is that users must file
their work more frequently during an active session.

Once the clean-up work is completed, model documentation (detailing)
includes adding dimensions and tolerances, adding text notes, generating drawings, and producing bills of materials. Users are assumed to know the standard
rules of dimensioning and tolerancing. The majority of CAD/CAM systems
provide users with rich menus and commands for dimensioning and tolerance
purposes. Obtaining high-quality plots of the generated drawings is usually
achieved by issuing the proper plot command. The standard drawing sizes are
supported by all systems. These are (in inches)

$$
\begin{array}{ll}
\text{SIZE A} & 8\frac{1}{2} \times 11 \\
\text{SIZE B} & 17 \times 11 \\
\text{SIZE C} & 17 \times 22 \\
\text{SIZE D} & 34 \times 22 \\
\text{SIZE E} & 34 \times 44 \\
\end{array}
$$

In addition to these standard sizes, user-defined drawing sizes are always provided. There is always one of these sizes that is used as default by the software. The
mapping between the orientation of a drawing on the display screen and that of
the drafting paper mounted on the plotter must be understood by the user to
avoid surprises. There is a modifier that can be used with the plotter command to
rotate the drawing orientation on the screen by 90 degrees before being plotted
on the paper.

## 3.5   USER INTERFACE

A user interface is defined as a collection of commands that users can use to
interact with a particular CAD/CAM system. User-interface or man-machine dialogue represents the only means of communication between users and
CAD/CAM software. The language of the interface should be simple enough for
the user to understand. It should also be efficient and complete and should have
a natural grammar, that is, the minimum number of easy-to-grasp rules. This
helps minimize user training and allows the user to concentrate on the problem
to be solved. The user interface should allow the user to undo mistakes if needed.
The design of the user-computer dialogue is a separate subject by itself.

Regardless of the type of user interface, the generic structure of a
CAD/CAM command consists of two parts as shown in Fig. 3-18. The user communication part includes the dialogue that the user follows to achieve specific
goals. The database part includes the geometrical data input to or retrieved from

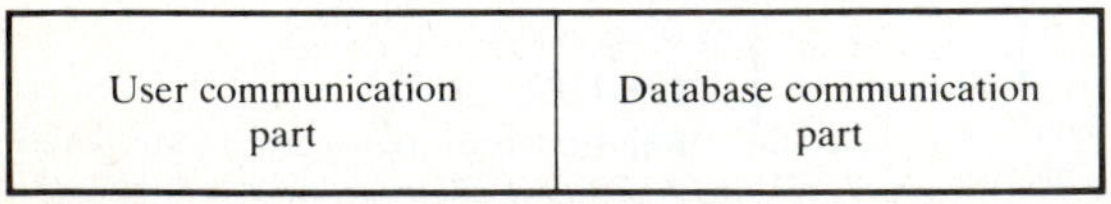

| User communication part | Database communication part |
|---|---|

**FIGURE 3-18**
Generic structure of a CAD/CAM command.

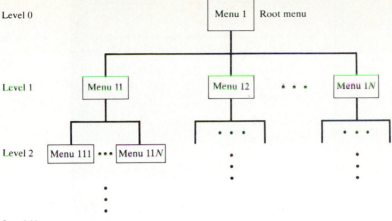

Level 0

Menu 1    Root menu

Level 1

Menu 11          Menu 12      · · ·     Menu 1*N*

Level 2

Menu 111  · · ·  Menu 11*N*

Level *N*

**FIGURE 3-19**
A typical menu structure.

the database. For example, consider creating a line between the two points (1, 2, 0) and (3, 5, 0) using the command "LINE." LINE is considered the first part of the command and the coordinates are the database part. Figure 3-19 shows a typical menu structure while Figs. 3-20 to 3-25 show examples of menu-based user interfaces.

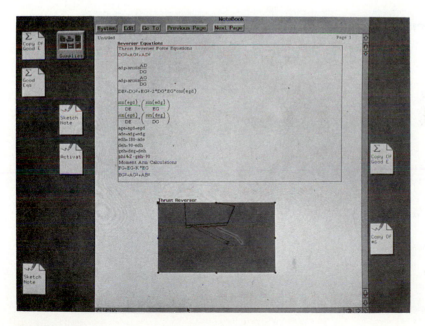

**FIGURE 3-20**
Reviewing design equations via menu interface. (*Courtesy of Cognition, Inc.*)

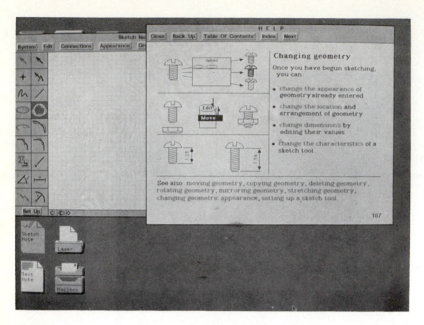

**FIGURE 3-21**
HELP information superposed on menu structure. (*Courtesy of Cognition, Inc.*)

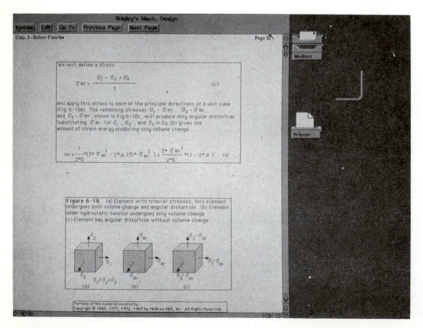

**FIGURE 3-22**
Storing design information as part of menu structure. (*Courtesy of Cognition, Inc.*)

**FIGURE 3-23**
Icon user interface under Unix
operating system. (*Courtesy of
Computervision Corporation.*)

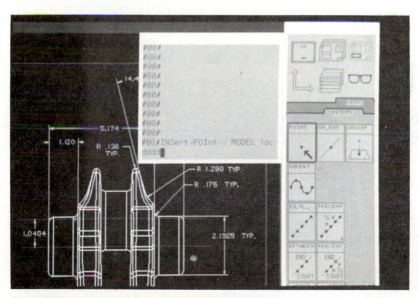

**FIGURE 3-24**
Provoking point definition icon. (*Courtesy of Computervision Corporation.*)

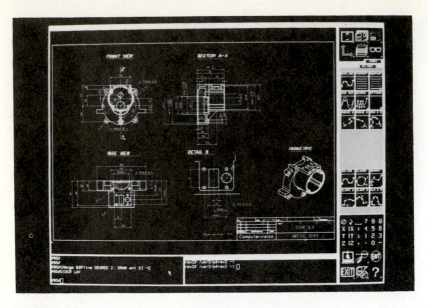

**FIGURE 3-25**
Provoking splines icon. (*Courtesy of Computervision Corporation.*)

## 3.6   SOFTWARE MODULES

There are considerable numbers of software packages for the various types of
CAD/CAM systems discussed in Chap. 2. Each package has its own strengths
and uniqueness and is usually targeted toward a specific market and group of
users. For example, there are mechanical, electrical, and architectural CAD and
CAM software for the respective users. Investigating an existing software on
various systems reveals that it has a generic structure and common modules.
Awareness of such structure and modules enables users to better understand
system function for both evaluation and training purposes. The major available
modules are discussed below.

### 3.6.1   Operating System (OS) Module

This module provides users with utility and system commands that deal with
their accounts and files. Typical functions such as file manipulations (delete, copy,
rename, etc.), managing directories and subdirectories using text editors, programming, and accounts setups are supported by the OS module. Files that are
generated in a user's CAD/CAM account by the OS can be classified into two
groups. The first group includes all conventional files (text files). The second
group includes graphics-related files. A model geometry and its shaded images
are stored in these files.

   Due to the distinction between the OS and graphics functions on a
CAD/CAM system, two working levels are available to the user. These are the
OS and graphics levels. The user can easily invoke one level from the current one.
The software usually provides its users with a command or procedure to go back

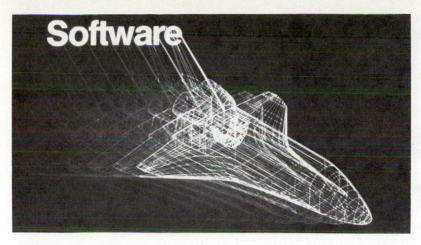

**FIGURE 3-26**
A typical surface model. (*Courtesy of Computervision Corporation.*)

and forth between the two levels to achieve maximum flexibility and increase user's productivity.

### 3.6.2 Graphics Module

This module provides users with various functions to perform geometric modeling and construction, editing and manipulation of existing geometry, drafting and documentation. The typical graphics operations that users can engage in are model creation, clean-up, documentation, and plotting. Figures 3-26 to 3-29 show

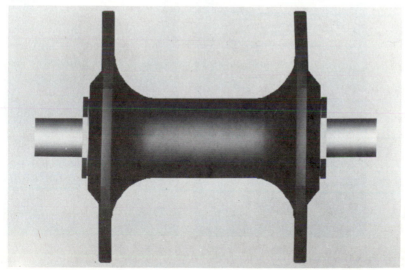

**FIGURE 3-27**
A solid model (shaded) of a bicycle hub. (*Courtesy of Adage, Inc.*)

**FIGURE 3-28**
A shaded image of a terminal frame. (*Courtesy of GE (U.S.A.) Calma.*)

some typical models. Shaded images can be generated as part of model documentation. Figure 3-30 shows a shaded image of a universal joint for both visualization and documentation purposes.

### 3.6.3  Applications Module

The creation of a geometric model of an object represents a means and not a goal to engineers. Their ultimate goal is to be able to utilize the model for design and manufacturing purposes. This module varies from one software to the other.

(a)                              (b)                              (c)

**FIGURE 3-29**
Solid model of a glass. (*Courtesy of McDonnell Douglas Manufacturing and Engineering Systems Company.*)

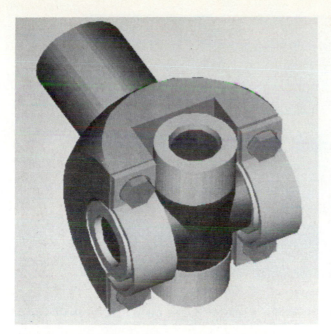

**FIGURE 3-30**
A shaded image of a universal joint. (*Courtesy of Adage, Inc.*)

However, there are common applications shared by most packages. Mechanical applications include mass property calculations, assembly analysis, tolerance analysis, sheet metal design, finite element modeling and analysis, mechanisms analysis, animation techniques, and simulation and analysis of plastic injection molding. Manufacturing applications include process planning, NC, CIM, robot simulation, and group technology. Figures 3-31 to 3-42 show samples of these applications.

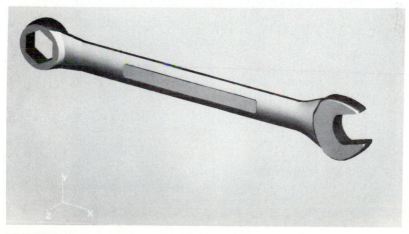

**FIGURE 3-31**
A wrench model used for finite element analysis. (*Courtesy of Tektronix, Inc.*)

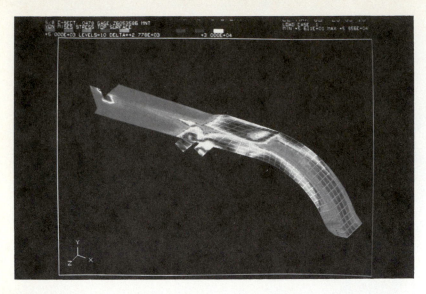

**FIGURE 3-32**
Stress contours of a finite element analysis of a motorcycle chain guard. (*Courtesy of GE (U.S.A.) Calma.*)

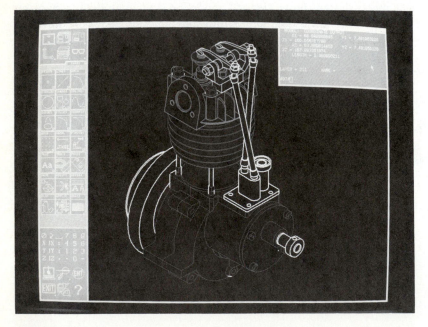

**FIGURE 3-33**
An engine assembly. (*Courtesy of Computervision Corporation.*)

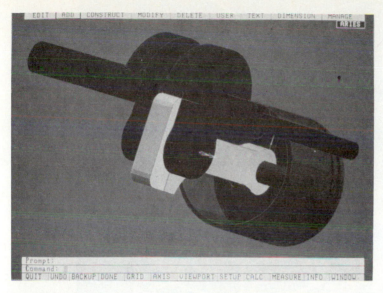

**FIGURE 3-34**
Piston assembly (crankshaft, connecting rod, and a piston). (*Courtesy of Aries Technology, Inc.*)

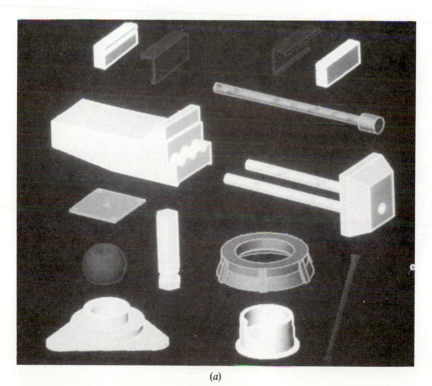

(*a*)

**FIGURE 3-35**
Vise assembly. (*Courtesy of McDonnell Douglas Manufacturing and Engineering Systems Company.*)

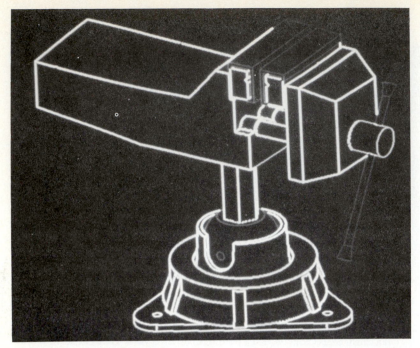

(b)

(c)

**FIGURE 3-35**
Vise assembly. (*Courtesy of McDonnell Douglas Manufacturing and Engineering Systems Company.*)

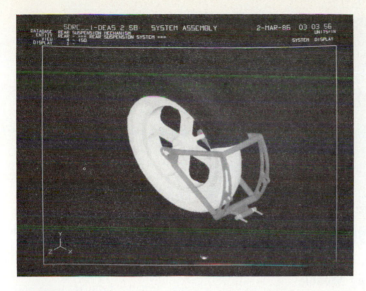

**FIGURE 3-36**
Kinematic study of motorcycle rear wheel assembly. (*Courtesy of GE (U.S.A.) Calma.*)

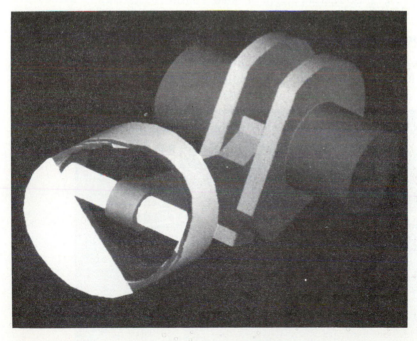

**FIGURE 3-37**
A piston assembly with piston head clipped for animation applications. (*Courtesy of McDonnell Douglas Manufacturing and Engineering Systems Company.*)

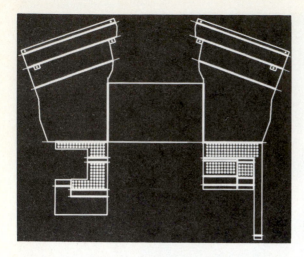

**FIGURE 3-38**
Flat sheet metal pattern of a terminal frame. (*Courtesy of GE (U.S.A.) Calma.*)

### 3.6.4 Programming Module

Typically, this module provides users with system-dependent and standard programming languages. The former is provided for graphics purposes while the latter is used for analysis and calculations. As an example of system-dependent languages, Computervision, GE Calma, and McDonnell Douglas provide VARPRO2 and CVMAC, DAL, and GRIP respectively. Programming as applied to design problems is covered in Chap. 15.

### 3.6.5 Communications Module

This module is crucial if integration is to be achieved between the CAD/CAM system, other computer systems, and manufacturing facilities. It is common to

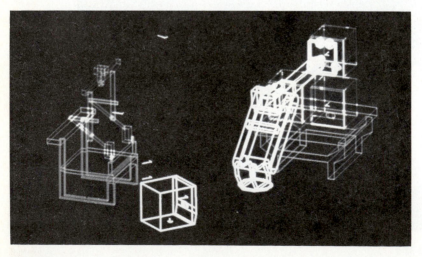

**FIGURE 3-39**
Robot simulation via offline programming. (*Courtesy of GE (U.S.A.) Calma.*)

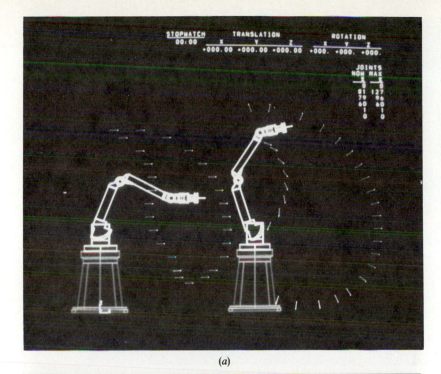

(a)

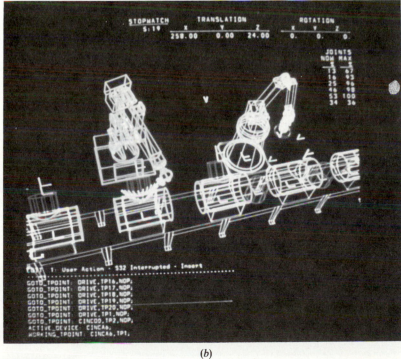

(b)

**FIGURE 3-40**
Robot simulation and programming. (*Courtesy of McDonnell Douglas Manufacturing and Engineering Systems Company.*)

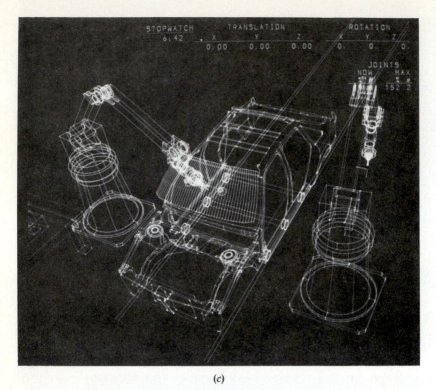

(c)

**FIGURE 3-40**
Robot simulation and programming. (*Courtesy of McDonnell Douglas Manufacturing and Engineering Systems Company.*)

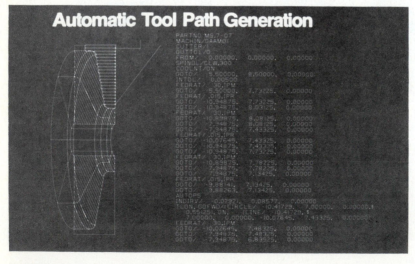

**FIGURE 3-41**
NC tool path generation. (*Courtesy of Computervision Corporation.*)

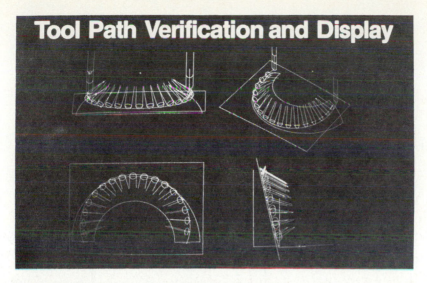

**FIGURE 3-42**
NC tool path verification. (*Courtesy of Computervision Corporation.*)

network the system to transfer the CAD database of a model for analysis purposes or to transfer its CAM database to the shop floor for production. This module also serves the purpose of translating databases between CAD/CAM systems using graphics standards such as IGES.

## 3.7 MODELING AND VIEWING

Modeling is the art of abstracting or representing a phenomenon and geometric modeling is no exception. Geometric modeling and simulation via computers have reached a level to replace the real-life prototypes or tests. A geometric model is defined as the complete representation of an object that includes both its graphical and nongraphical information. Objects can be classified into three types from a geometric construction point of view. These are two-and-a-half dimensional, three dimensional, or a combination of both. Figures 3-43 and 3-44 show some examples. As Fig. 3-43 shows, two-and-a-half-dimensional objects are classified to have uniform cross sections and thicknesses in directions perpendicular to the planes of the cross sections. Constructing such an object via wireframe modeling requires only constructing the proper entities (faces), projecting them along the proper directions by the thickness value, and then creating the proper edges along these directions. This is a much more efficient way of construction than calculating and inputing the coordinates of all the corner points of the model. The construction of a true three-dimensional object requires the coordinate input of key points and then connecting them with the proper types of entities. Figure 3-44 shows a phone model. The model is two-and-a-half dimensional although it may seem to be three dimensional at first glance.

Irrespective of the specific syntax of any software package, a particular model setup procedure is usually needed by the package to organize the model

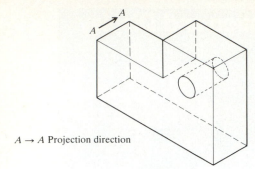

$A \rightarrow A$ Projection direction

Two-and-a-half dimensional

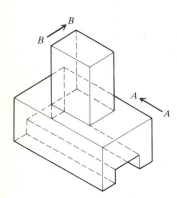

Two-and-a-half dimensional

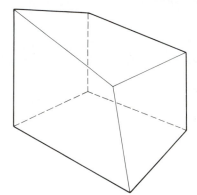

Three dimensional

$A \rightarrow A$
$B \rightarrow B$ Projection directions

**FIGURE 3-43**
Two-and-a-half-dimensional and three-dimensional geometric models.

database before the user is allowed to construct geometry. The procedure can be listed in a generic form in the following order:

**1.** INITIATE NEW MODEL.
**2.** CHOOSE A SCREEN LAYOUT.
**3.** DEFINE THE WINDOWS OF THE LAYOUT AS THE MODEL DESIRED VIEWS.
**4.** SELECT THE PROPER CONSTRUCTION PLANE OR WCS.

In the first step, the model name becomes the file name that stores the geometric model information. The setup procedure results in a model database with the hierarchy shown in Fig. 3-45. Thus, the user cannot define views before choosing a screen layout. The user can choose the layouts and define their views at any time during construction and not only during the model setup. Dealing with a centralized database has two consequences on software response to user's commands regarding screen layouts and views. First, whenever the user defines a

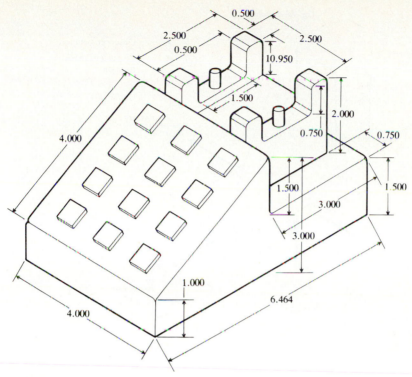

**FIGURE 3-44**
A telephone geometric model.

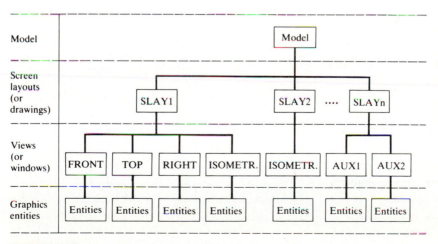

**FIGURE 3-45**
Typical hierarchy of a geometric model database.

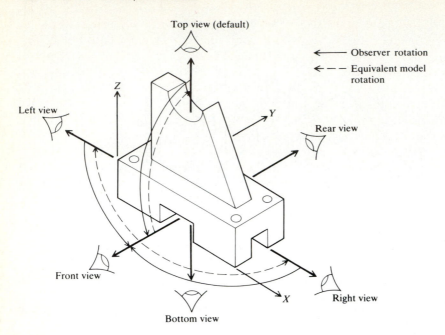

**FIGURE 3-46**
Viewing standard angles.

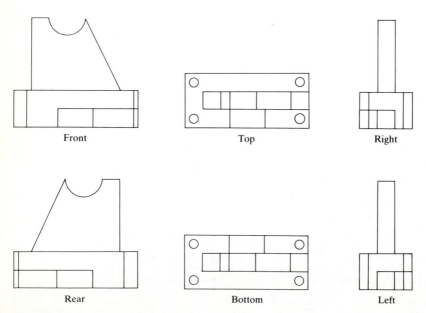

**FIGURE 3-47**
Standard two-dimensional views of a typical model.

new view in a layout, existing model geometry in the model database is automatically transformed and displayed in the view window (viewpart). Second, if a view is deleted from a layout or an entire layout is deleted from the database, graphics entities making the view or the layout are not deleted. Only the view or the layout display disappear from the screen. As a matter of fact, the user can delete all views and all screen layouts from the database and still have the model entities invisible to the user. Entities are deleted only if the user does so explicitly.

While model construction involves the creation of the model database, viewing affects the way the model is displayed on the screen. Modeling and viewing are interrelated but yet display techniques and approximations do not affect the object representation inside the database. Views are defined by the various angles from which a model can be observed. Essentially, the observer changes position in the MCS, while the model maintains its original orientation. However, it appears to the viewer as if the model moves, when in fact its MCS simply moves with it. Figure 3-46 shows the equivalence between the two interpretations. The first interpretation is easier and more convenient to define mathematically and is used by CAD/CAM software to define views. An infinite number of views can be defined for a model. Most software provides commands for the standard views. Figures 3-47 and 3-48 show the standard two-dimensional and three-dimensional views of a model. The " + " designates the origin of the MCS.

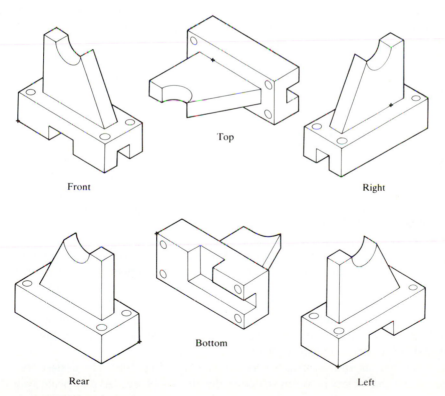

**FIGURE 3-48**
Standard three-dimensional views of a typical model.

## 3.8 SOFTWARE DOCUMENTATION

Documentation represents the only formal source of information users can refer to in order to learn about software capabilities. The source code of CAD/CAM software is usually very long and not available to users. Even if it is, it is too intricate and complex to understand. Understanding the organization and general theme of the documentation helps users utilize the software to its full capacity.

Two types of documents may be identified. The first type describes the semantics and theoretical background of the software (sometimes called user reference manuals). The second type describes the syntax and user interfaces (sometimes called user guides or user manuals). Online documentation and help functions provided by CAD/CAM software usually replace the second type of documentation.

## 3.9 SOFTWARE DEVELOPMENT

CAD/CAM software as delivered by vendors is seldom complete to meet all users' specific needs. Customizing existing software and/or developing new programs are inevitable. In general, there are two levels of developing software by users. The first and most popular level does not require knowledge of the database structure of the software. During execution, these programs usually require user interaction to input information and/or digitize graphics entities. The second and less popular level requires modifying and/or accessing the database. This level always requires extensive knowledge of the database structure of the software. While programs are more difficult to develop utilizing the second level of programming, they are usually more efficient to run.

## 3.10 EFFICIENT USE OF CAD/CAM SOFTWARE

CAD/CAM software is usually a complex software that requires training and understanding of its philosophy and underlying principles. Once this is achieved the user should develop the habit of devising a strategy to construct geometric models or achieve other goals before logging into the system. The following recommendations may be helpful to users:

1. Develop an efficient planning strategy. A good strategy for complex models can result in great time savings to create the model database. The user must decide first on the type of object at hand, if it is two-and-a-half-dimensional or three-dimensional and the type of geometric modeling desired, if it is wireframes, surfaces, or solids. The key faces should be identified for two-and-a-half-dimensional objects. If more than one alternative exists, choose the one that makes it easier to construct the model.

   Consider, as an example, the model shown in Fig. 3-49. The easiest way to construct this object is by constructing the entities of face (A) and projecting them back along the direction shown and then creating the holes. Other alternatives exist but they are not optimal. For both two-and-a-half-dimensional

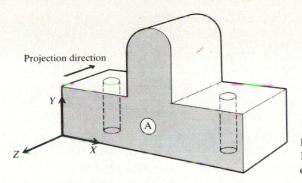

Projection direction

Y

Z

X

A

**FIGURE 3-49**
Plan to construct a two-and-a-half-dimensional object.

and three-dimensional objects, the user must develop the key coordinates required to construct the model relative to the chosen MCS and possible WCSs. The general rule here is that the user should avoid excessive coordinate calculations. The software usually provides users with many tools in this regard, such as WSCs, geometric modifiers, and graphic manipulations. Chapters 11 and 12 cover these topics in detail.

Planning strategy should also include the choice of the MSC origin and orientation, the screen layout, views, and colors. Typically, a screen layout with either an isometric view or four views (front, top, right, and isometric) are adequate to begin construction. Additional layouts and views can be defined later. Figure 3-49 shows one possibility for the MSC.

An important geometric characteristic of an object that usually facilitates construction is symmetry. If an object is symmetric with respect to one plane or more, the user can only construct, say, half of the model and then use the "mirror" command to create the full model. The "translate" and "rotate" commands can also be used.

2. Prepare the initial command sequence away from the workstation. Once the planning strategy is finished, the user should write the command sequence needed to implement such a strategy. The sequence does not have to be complete, but more or less a skeleton to remind the user of what to do at the workstation. If the user is not sure of a command or likes to try new commands or syntax, this uncertainty could be resolved during an active session.

3. Use syntax to its fullest capacity. Every software provides its users with syntax of various capabilities.

4. Document every step of constructing the model for future reference. Headings can be added to the initial command sequence to label the various sections of the sequence. If programming is used, comment statements are always available. Saving the planning strategy is also helpful.

5. Use available menus to increase productivity during session work. Tablet and/or on-screen menus are available to the user in addition to the keyboard. Input via typing on the keyboard is typically slow.

6. Program the command sequence to construct the object. This approach is very useful in design applications if the user should investigate different geometry or other design parameters. Programming the command sequence

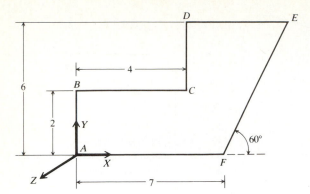

**FIGURE 3-50**
A typical two-dimensional model.

should be as simple as stacking the sequence in a file (macro) that can be executed later. The program can be thought of as a backup. In case the model database becomes corrupted or lost due to the system sudden crash, the program can simply be executed to create another one.

7. Use online documentation as much as possible. Once the user is in an interactive session, the use of the online documentation becomes more efficient than referring to a manual.

8. Avoid unnecessary calculations. This requires knowledge and experience of the available graphics aids. It is always possible to avoid calculations necessitated by manual procedures if the associativity and centralized concepts of the database are utilized. Geometric modifiers and editing functions are usually helpful in this regard.

As an example, consider constructing the model shown in Fig. 3-50. The user should not attempt to calculate the coordinates of point $E$ using trigonometric relationships. Instead, after the obvious construction of lines $AB$, $BC$, $CD$, and $AF$, the user creates a horizontal line passing by the endpoint $D$ of an arbitrary length of, say, 10. Another line can be created passing by endpoint $F$ at an angle of 60° (use cylindrical coordinates input mode) of arbitrary length of, say, 10. The user completes the construction by trimming the two lines to their point of intersection. Verifying point $E$ can provide its explicit coordinates.

## 3.11   SOFTWARE TRENDS

The demand and expectation from CAD/CAM software has been increasing steadily. Efforts to develop applications software is decisive in meeting users' demands. Table 3.1 and Fig. 3-51 reflect the commitment of CAD/CAM vendors to software development. Software sales, including the value of bundled software, is expected to grow from $281 million in 1984 to about $1.8 billion by 1992 at an average growth rate of 25.2 percent during this period. This is the highest rate among the other components of CAD/CAM.

Software, in general, is bounded by the basic principles of its field as well as the development efforts and therefore tends to change more slowly than

**TABLE 3.1**

## Sales of CAD/CAM systems by components (in millions of 1985 dollars)

| | 1984 | 1986 | 1988 | 1990 | 1992 | Growth per year 1985–1992 |
|---|---|---|---|---|---|---|
| Computers | 617 | 895 | 1205 | 1505 | 1720 | 12.7 |
| Peripherals | | | | | | |
|    Graphics station | 859 | 1211 | 1610 | 1997 | 2339 | 12.6 |
|    External storage | 316 | 459 | 628 | 802 | 967 | 14.3 |
|    Other | 411 | 585 | 784 | 981 | 1158 | 13.1 |
| Software | 281 | 477 | 763 | 1174 | 1767 | 25.2 |
| Total | 2484 | 3627 | 4990 | 6459 | 7951 | 14.9 |

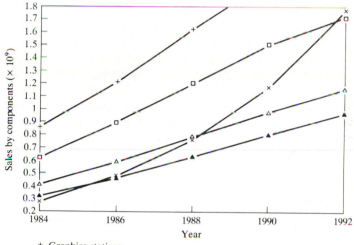

+ Graphics stations
□ Computers
▲ External storage
× Software
△ Other

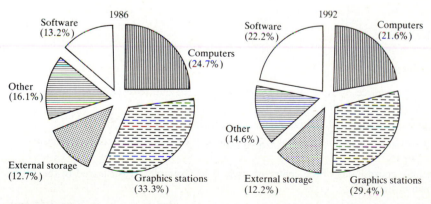

**FIGURE 3-51**
Sales of CAD/CAM systems by component.

hardware. CAD/CAM software modules discussed in this chapter have been progressing steadily. However, some modules are more mature than others. Most existing commercial software is strong in graphics applications.

Integration and automation of design procedures are important characteristics of CAD/CAM software. Typically, analysis programs can be written or finite element analysis can be used to test the design. These programs tend to draw their geometric input from databases. If the design procedure is performed with considerable user interaction, it becomes an integrated procedure. On the other hand, an automated procedure would require minimum or no user intervention. Consider the classical example of finite element analysis. An integrated procedure would require the users to create the geometric model database, create the finite element model, perform the analysis, and finally display the results. An automated procedure would require the same user to only create the geometric model with the physical finite element data. The procedure, then, processes the data automatically and displays the results. The automation procedure is much better suited for designers who, most of the time, are not interested in the analysis techniques themselves but rather how to use them. Another important benefit of automation is that it guarantees making the proper assumptions for the proper analysis. Therefore the designer can have more confidence in the results of the software.

The achievement of true integration and automation of design and manufacturing is a primary goal of various software developments. This depends mainly on database structure and management as well as on the completeness and uniqueness of the geometric modeling techniques utilized to develop these databases. Among the various available techniques, solid modeling seems to be the key technique to automate and integrate the CAD and CAM functions.

Adaptive analysis and optimization are other characteristics that CAD/CAM software should possess to better serve design and engineering applications. These characteristics reflect the iterative nature of design. Instead of leaving the burden of adaptation and optimization to the user, it is very beneficial if software can provide it. Expert and learning systems as well as AI (artificial intelligence) techniques are useful in this regard.

## PROBLEMS

### Part 1: Theory

**3.1.** Give an example of how the centralized integrated database concept can help the "what if" situations that arise during the design process.

**3.2.** Discuss the contents of a database for a line, a circle, and an arc.

**3.3.** Can you define a nonorthogonal WCS? How is the three-point definition interpreted by software?

**3.4.** Find the coordinates of the center of circle B in Example 3.1 measured in the MCS shown in Fig. 3-9a.

**3.5.** What are the most efficient ways to construct the models shown in Figs. 3-44 and 3-46? Assume a wireframe modeling technique is used.

## Part 2: Laboratory

Use your in-house CAD/CAM system to answer the questions in this part.

**3.6.** How can you define a WCS? What is the command syntax and modifiers?

**3.7.** How can you display the orientation of the active WCS and the MCS on the screen?

**3.8.** Solve Prob. 3.4 using the entity verification command. You can also solve the problem by using the angle measurement command to find the matrix $[T]$. Compare the answer with the answer to Prob. 3.4. What are the conclusions you can draw?

**3.9.** What are the parameters that are required to define views?

**3.10.** Which of the three methods discussed in Sec. 3.3.6 is used by your particular system?

**3.11.** Design and implement a user-interface menu that can perform beam stress analysis (call it the beam design menu).

**3.12.** What are the software modules that are available on your system? List the capabilities of each module.

**3.13.** What is the setup procedure required by your system before you can construct the geometry? Write the corresponding command sequence.

**3.14.** Find the commands that define the six standard three-dimensional views and the six standard two-dimensional views.

**3.15.** Figure P3-15 shows a model view with different locations of its MCS. Inside each viewport provided below each view, select the proper view origin such that the constructed geometry lies completely within the viewport to avoid clipping.

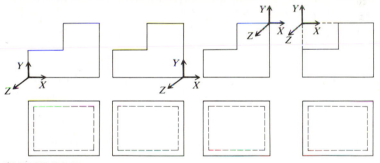

**FIGURE P3-15**

**3.16.** Find the optimum procedure for using the documentation of your system. How can you invoke the online documentation during a session?

**3.17.** Develop the planning strategy required to construct the two models used in Prob. 3.5.

**3.18.** Develop the most efficient command sequence for Prob. 3.17.

## Part 3: Programming

**3.19.** Develop a macro of the setup procedure developed in Prob. 3.13.

**3.20.** Develop a macro to create an "initial" or "blank" model.

**3.21.** Develop a macro to generate the six standard three-dimensional views.

**3.22.** Develop a macro to generate the six standard two-dimensional views.

## BIBLIOGRAPHY

Batory, D. S., and W. Kim: "Modeling Concepts for VLIS CAD Objects," ACM *Trans. on Database System*, vol. 10, no. 3, pp. 322–346, September 1985.

Beasant, C. B.: *Computer Aided Design and Manufacturing*, Ellis Horwood, West Sussex, 1980.

Bradley, J.: *Introduction to Data Base Management in Business*, CBS College Publishing, 1983.

Brannigan, M.: "GKS: The New Graphics Standard," *Computer Language*, pp. 26–30, May 1986.

Buchmann, A.: "Current Trends in CAD Databases," *Computer Aided Des. J.*, vol. 16, no. 3, pp. 123–126, 1984.

Carden, J.: "The Structural Stability of Corporate and Devolved Databases," *The Computer J.*, vol. 29, no. 4, pp. 361–367, 1986.

Date, C. J.: *An Introduction to Database Systems*, 3d ed., vol. 1, Addison-Wesley, Reading, Mass., 1982.

Date, C. J.: *An Introduction to Database Systems*, vol. 11, Addison-Wesley, Reading, Mass., 1983.

Encarnacao, J., and E. G. Schlechtendahl: *Computer Aided Design: Fundamentals and System Architectures*, Springer-Verlag, Germany, 1983.

Falk, H.: "Databases Add Functions to Meet Computer Aided Design Needs," *Computer Des.*, pp. 33–35, October 1984.

Freeman, H. (Ed.): *Interactive Computer Graphics*, IEEE Computer Society, Los Alamitos, Calif., 1980.

Giloi, W. K.: *Interactive Computer Graphics; Data Structures, Algorithms, Languages*, Prentice-Hall Inc., Englewood Cliffs, N.J., 1978.

Hatfield, L., and B. Herzog: "Graphics Software—From Techniques to Principles," *IEEE Computer Graphics and Applic.*, vol. 2, no. 1, pp. 59–80, January 1982.

Kalay, Y. E.: "A Database Management Approach to CAD/CAM Systems Integration," in *IEEE 22nd Design Automation Conf.*, Las Vegas, Nevada, pp. 111–116, June 23–26, 1985.

Kobori, K., Y. Nagata, Y. Sato, K. Jones, and I. Nishioka: "A 3-D CAD/CAM System with Interactive Simulation Facilites," *IEEE Computer*, pp. 14–20, December 1984.

Law, K. H., M. K. Jouaneh, and D. L. Spooner: "Abstraction Database Concept for Engineering Modeling," *Engng with Computers*, vol. 2, pp. 79–94, 1987.

Machover, C., and W. Myers: "Interactive Computer Graphics," *IEEE Computer*, pp. 145–161, October 1984.

Mills, R. B.: "Database Machines Tackle the CAD/CAM Information Glut," *CAE Mag.*, pp. 61–64, October 1985.

Panasuk, C.: "Focus on Graphics Workstations," *Electronic Des.*, pp. 157–161, August 1985.

Scrivener, S. A. R. (Ed.): *Computer-Aided Design and Manufacture: State-of-the-Art Report*, Pergamon Infotch, Maidenhead, Berkshire, 1985.

Shenoy, R. S., and L. M. Patnaik: "Data Definition and Manipulation Languages for a CAD Database," *Computer Aided Des. J.*, vol. 15, no. 3, pp. 131–134, 1983.

Shepherd, B. A.: "Graphics Standards: Status and Impact," in *Advances in CAD/CAM Workstations* (Ed. C. C. Wang), Kluwer Academic, Hingham, Mass., pp. 209–214, 1986.

Shepherd, B. A.: "Graphics Related Standards," in *Advances in CAD/CAM Workstations* (Ed. C. C. Wang), Kluwer Academic, Hingham, Mass., pp. 215–223, 1986.

Staley, S. M., and D. C. Anderson: "Functional Specification for CAD Databases," *Computer Aided Des. J.*, vol. 18, no. 3, pp. 132–138, April 1986.

Stempson, G.: "Database Management; New-Generation Systems Speed Graphics' Role," *Computer Graphics World*, pp. 51–56, April 1986.

Stover, R. N.: *An Analysis of CAD/CAM Applications, with an Introduction to CIM*, Prentice-Hall, Englewood Cliffs, N.J., 1984.

Ullman, J. D.: *Principles of Database Systems*, 2d ed., *Computer Science*, Maryland, 1982.

Warner, J.: "CAD Update with Portable Software," *CAE Mag.*, p. 97, March 1985.

Warner, J.: "Standard Graphics Software for High-Performance Applications," *IEEE Computer Graphics and Applic.*, pp. 74–79, March 1985.

Wiederhold, G.: "Databases," *IEEE Computer*, pp. 211–223, October 1984.

Yao, S. B. (Ed.): *Principles of Database Design*, vol. 1: *Logical Organization*, Prentice-Hall, Englewood Cliffs, N.J., 1985.

Zeid, I.: "CAD on Computervision System," Unpublished Document, CAD/CAM Laboratory, Northeastern University, Boston, Mass., 1984.

Zeid, I.: "CAD on G.E. Calma System," Unpublished Document, CAD/CAM Laboratory, Northeastern University, Boston, Mass., 1985.

# MICROCOMPUTER-BASED CAD/CAM[1]

## 4.1 INTRODUCTION

What is micro-based CAD? Micro-based CAD, also referred to as microCAD and PC CAD, is defined simply as computer aided design and drafting on a microprocessor-based computer known as a microcomputer. Microcomputers are also referred to as personal computers (PCs). Figure 4-1 shows a typical microCAD system.

The cost of a microCAD system is at the low end of the spectrum compared to conventional large-scale mini- and mainframe-based CAD systems. Most industry experts agree with the 80/20 rule; that is, microCAD gives you about 80 percent of the functionality of large-scale mainframe CAD at approximately 20 percent of the cost. As time passes, this rule will change as the cost of large-scale CAD declines due to the decline in prices of hardware.

MicroCAD's popularity comes mainly from its performance/cost ratio. The prominence of the microcomputer and its ease of use has also led to widespread acceptance. Thousands of small firms, even one- and two-person companies, now consider CAD as an alternative to laborious and time-consuming manual methods.

Architects, civil engineers, land developers, facility planners, and others use microCAD. In addition, the manufacturing industry realizes its potential benefits and has converted or is in the process of converting their manual environments. In addition, experienced organizations are expanding their installed base of large-scale CAD systems with microCAD.

---

[1] Written by Terry Wohlers of Wohlers Associates, Fort Collins, Colorado.

**FIGURE 4-1**
A typical microCAD system. (*Courtesy of Versacad Corporation.*)

In comparison, the features and capabilities of today's microCAD technology are similar to those found in large-scale CAD. In the past, mini- and mainframe computers offered powerful, multitasking, and networking capabilities far beyond the scope of the microcomputer. Today, though, microcomputers offer these capabilities. These capabilities often have been scaled down to have fewer options when moved to microCAD software. With fewer options, it becomes inherently easier to learn and use microCAD software.

## 4.2   GENERAL FEATURES

MicroCAD has general features which contribute to its popularity. Some of these features may be shared by mini- or mainframe CAD. Some of these features are:

*Modeling and drafting.* The majority of microCAD systems provide two-dimensional and three-dimensional modeling capabilities. Some systems are dedicated to two-dimensional drafting.

*Ease of use.* Users find microCAD relatively easy to learn and use, especially in comparison to large-scale CAD. Because of microCAD's popularity, a range of courses, texts, and support groups are available to help the learner.

*System flexibility.* MicroCAD systems are flexible in two ways. First, users have the flexibility of applying many of the general-purpose software packages for a variety of applications. For example, one user may choose to use it for mechanical part design while another may use it for facility layout. Another may choose to construct flowcharts and organizational charts. Second, microCAD provides flexibility when configuring the hardware. Hundreds of computers, display systems, expansion boards, input and output components are compatible and configurable with popular microCAD software.

*Standards/modularity.* Standards develop as the technology develops. A good example is the IBM PC standard. Other examples are the 5.25 and 3.5 inch standard diskette sizes and the standard RS-232 serial port. Standard input and output devices connect to the serial port, making the systems modular in nature, much like components that make up a hi-fidelity stereo system. As yet another example, modular printed circuit boards, also referred to as cards, mount in standard slots inside standard microcomputers.

*Low maintenance.* Generally, microCAD software and hardware is very durable. Little maintenance is needed to keep the system fully functional. When parts do malfunction, the repair or replacement costs are very low in comparison to mini- and mainframe computers.

## 4.3   SYSTEM IMPLEMENTATION

It is important to realize that not all microCAD installations are successful. Several variables affect the level of the success or failure of the installation. Personnel issues are usually more difficult to overcome than those issues directly associated with the hardware and software.

The individuals or department that are implementing the microCAD system(s) must gain the support from management. The management should have a general awareness of the techniques involved in operating a microCAD system as well as the system's capabilities and limitations.

Often management has high expectations. They may expect much more than the system and individual are able to deliver. The management may not realize that the system does not produce results automatically. A learning curve is involved (see Fig. 3-1), and at first it may take longer to produce the work on the CAD system than on the drafting board. Months may pass before the user(s) and system(s) are up to speed.

In-house standards evolve as the users become familiar with the need for standards. Hardware and software upgrades are identified and implemented as the users discover their requirements. Each step of change takes time and affects productivity in the short term. Training is critical. It can come from an educational institution, independent consultant, or from an in-house expert. Regardless, the learning process should be well planned, and the instruction must be current, accurate, and made interesting for the learners. On-going training should be parallel with changes in the technology.

A knowledgeable and ambitious person within an organization should assume responsibility, or be made responsible, for overseeing the microCAD system. This supervisor should work closely with all other microCAD users in the organization. Hardware, software, and standards, as well as communication among all users, are his/her responsibility. Without such a person, each user will move in a different direction and the system implementation is likely to fail.

## 4.4 HARDWARE COMPONENTS AND CONFIGURATION

The quality, functionality, and affordability of microCAD hardware have improved dramatically over the last several years. Affordable microcomputers offering the power of mainframe computers of the past are available. Micro-computer display systems produce bright, large, and high-resolution screens. Fast and compact storage devices are able to store enormous amounts of graphics information. Input devices are becoming easier to use, more functional, and very affordable. A range of fast and high-quality, color output options are also avail-able to suit the needs of microCAD users.

The general trend in computers indicates a continued reduction of size and cost, and a continued increase in power, capacity, and array of applications. In short, users are able to purchase more with less money.

This section covers hardware details related to microCAD. Other details shared by mini- and mainframe-based systems have already been covered in Chap. 2.

### 4.4.1 Microcomputer Components

As mentioned above, microcomputer components are very modular. A typical microcomputer is comprised of several components, each having its own set of characteristics, features, and limitations:

*System board.* The system board, also referred to as the mother board, is the main unit on which other components connect. The CPU, coprocessor, RAM, and ports (discussed here) connect directly to the system board. Figure 4-2 shows a typical system board.

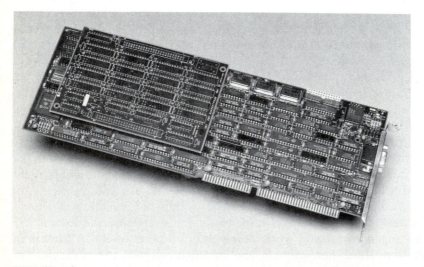

**FIGURE 4-2**
Microcomputer system board. (*Courtesy of Compaq Computer Corporation. All rights reserved.*)

*Central processing unit (CPU)*. The CPU, also referred to as a processor or microprocessor, is like the brain of the computer. Generally, the faster the CPU, the more powerful the computer. The standard IBM PC which became popular in the early 1980s contains an Intel 8088 8-bit microprocessor. A faster chip—the 8086— is also found in low-end microcomputers. It offers a faster processing rate than the 8088 chip.

The Intel 80286 16-bit chip is used in the IBM PC/AT and compatibles. The 80286 is typically the minimum speed chip for most microCAD applications. The 8088 and 8086 chips run at approximately 5 to 8 megahertz (MHz) clock speed, while the 80286 runs at a range of 6 to 12 MHz.

A popular platform for serious CAD users with stiff requirements is a computer based on the Intel 80386 of 80486 (Fig. 4-3). The 80386 is a 32-bit microprocessor that handles a much larger quantity of data because of its 32-bit data path between it and the memory (RAM). In addition, it provides fast clock speeds of 16 to 20 MHz or more, and a performance rate of about 4 to 5 million instructions per second (MIPS). IBM, Compaq microcomputers, and compatibles use the 80386 and other Intel series of microprocessors. Motorola's 68020 and 68030 (Fig. 4-4) are 32-bit microprocessors comparable to the Intel 80386 chip. The popular Apollo, Apple MAC II, MicroVAX, and Sun workstations use the 68020 chip. Table 4.1 shows a comparison between popular chips.

Apple Corp. products as represented by the Macintosh computer line have always offered a visual interface superior to that of PC's. Finer pixel count and graphical rather than textual displays give the user an easier visual environment. The OS enabled the user to do more and do it easier and faster. For many reasons, too numerous or dense to list here, the CAD industry did not take this product line seriously enough to write software for this hardware platform until recently. More work was done on producing PC-CAD (versus MAC-CAD) and workstation software.

**TABLE 4.1**
**Microcomputer Chip Comparison**

| CPU | Bus bits | Speed MHz | MEM | CO-PRO | Best video |
|-----|----------|-----------|-----|--------|------------|
| | | IBM | | | |
| 8088 | 8 | 4.7 | 640 KB | 8087 | EGA |
| 8086 | 16 | 4.7 | 640 KB | 8087 | EGA |
| 80286 | 16 | 6–20 | 16 MB | 80287 | VGA |
| 80386 | 32 | 16–33 | 32 MB | 80387 | 24 BIT |
| 80486 | 32 | 16–50 | 1 GB+ | NA | 24 BIT |
| 80586* | | | | | |
| | | APPLE | | | |
| 68000 | 16 | 16 | 4 MB | 68880 | MONO |
| 68020 | 32 | 16 | 4 MB | 68881 | 8 BIT |
| 68030 | 32 | 16–25 | 32 MB | 68882 | 8 BIT |
| 68030 | 32 | 40 | 128 MB | 68882 | 24 BIT |
| 68040* | | | | | |

* Future.

The 386SX™ microprocessor from Intel features a 100-lead plastic surface mount package. Compared here with the 132-lead pin grid array package of the 386™ microprocessor, the 386SX microprocessor allows for low-cost system manufacturing.

(a)

(b)

**FIGURE 4-3**
Intel microprocessors. (*Courtesy of Intel Corporation.*)

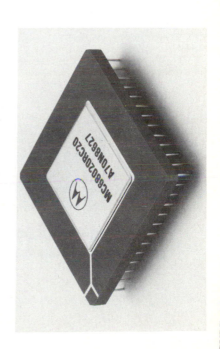

(a)

(b)

**FIGURE 4-4**
Motorola microprocessors. (*Courtesy of Motorola Inc.*)

139

*Numeric coprocessors.* Coprocessors, also referred to as math chips, provide dramatic speed increases with math-intensive CAD packages. Speed improvements are usually 3 to 5 times or more over microCAD systems without a coprocessor. The Intel 8087 coprocessor works with the Intel 8088 and 8086 CPUs, while the Intel 80287 coprocessor works with the 80286 and the Intel 80387 coprocessor with the 80386. The Motorola 68881 coprocessor works with the 68020 CPU.

*Random access memory (RAM).* The RAM is the computer's temporary but fast memory storage. MicroCAD systems contain RAM of 640 K (thousand) bytes, minimum, to 8 mega (million) bytes or more. Generally, microCAD systems containing several megabytes of RAM are faster and more capable than those containing less. This is due mainly because the system accesses the available RAM instead of the slower fixed (hard) disk or floppy disk. RAM chips are also rated at different speeds (in nanoseconds) and sizes (8 KB, 256 KB, and up to 1 meg chips).

*Ports.* Input devices, such as mice and digitizers, and output devices, such as pen plotters, connect to the standard 25-pin or 9-pin RS-232 serial port. Telephone modems and some networks also use the serial port for communication purposes. Most dot matrix and letter-quality printers connect to the standard 25-pin parallel port, also found at the rear of the computer. Figure 4-5 shows sample communication ports.

*Bus expansion slots.* Most microcomputers contain three or more bus expansion slots. Commercially available boards fit into these slots for a variety of purposes. As examples, certain add-on boards increase the computer's RAM,

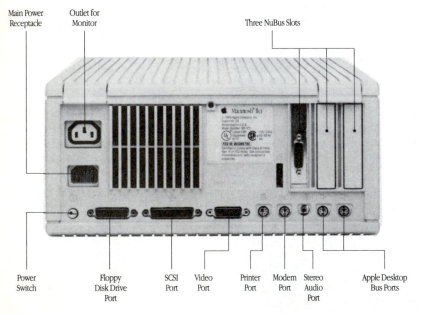

**FIGURE 4-5**
Communication ports. (*Courtesy of Apple Computer, Inc.*)

while others connect to the monitor for enhanced color or monochrome graphics. Accelerator boards are also available to boost the power of slow computers.

### 4.4.2   Display Systems

*Monitors.* A typical display system is comprised of a monitor and display board. Monitors vary greatly in size, resolution, and cost. MicroCAD systems use 12- to 20-inch monitors. Generally large, high-resolution color monitors are best for CAD, though small, medium-resolution monochrome monitors work satisfactorily for those companies on a restricted budget. Low-resolution monitors are approximately 300 pixels horizontally × 200 pixels vertically, while medium-resolution monitors are around 640 × 480. High-resolution monitors, which cost more than low-resolution monitors, are 1024 × 768 pixels and up. Some microCAD systems use two monitors (dual monitors): a large, high-resolution color monitor for the graphics and a small monochrome monitor for the text (commands and menus) information. Other microCAD systems combine the graphics and text into a single color monitor, thereby eliminating the extra cost of a second monitor and the added use of desk space.

*Display boards.* Display boards, also referred to as graphics boards or display adapters or display controllers, are mounted in one of the computer's bus slots. Like monitors, they too range in color and resolution capabilities as well as cost. Controllers are technically different from adapters because they contain added graphics capabilities and are usually faster. Consequently, they cost more. Some computers incorporate the graphics board's functions in the computer's system board, which is the case with the IBM Personal System/2 family. Regardless, its purpose is to provide graphics to the monitor. Its color and resolution capabilities must be as good or better than that of the monitor in order for the monitor to display its color and resolution capabilities. Sophisticated controllers offer extremely fast hardware pan and zoom capabilities. The display pan and zoom functions are actually performed within the controller instead of with the CPU, coprocessor, and RAM.

### 4.4.3   Storage Devices

*Floppy disks.* These units are used for permanent storage of files. Most microcomputers contain at least one floppy disk drive. The most popular sizes are 5.25- and 3.5-inch disk drives. The 5.25-inch diskette contains between 360 KB and 1.2 MB of storage space depending on the computer on which the diskette is being formatted. The 3.5-inch diskette contains between 700 KB and 1.44 MB of storage space.

*Fixed disks.* Fixed-disk drives (Fig. 4-6), also referred to as hard disks, have much faster file access capabilities and they store far greater amounts of information. For example, 20, 40, and 70 MB hard disks are popular among microCAD users. Capacities of up to 300 MB and beyond are also available. Hard-disk access speeds range from 15 to 70 milliseconds.

*Backup systems.* Large-capacity tape backup systems are available for producing backup copies of numerous, large files. Low-cost floppy diskettes are

(a) Internal hard disk. (*Courtesy of Seagate Technology.*)

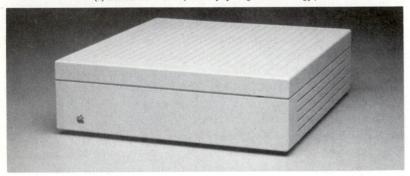

(b) External hard disk. (*Courtesy of Apply Computer Inc.*)

**FIGURE 4-6**
Fixed-disk drive units.

often used for backup purposes in lieu of the cartridge system, although floppy diskettes are more cumbersome and time-consuming to use.

## 4.5  MICRO-BASED CAD SOFTWARE

MicroCAD software packages offer a range of commands and features for a variety of applications. Some are two-dimensional drafting packages while others provide powerful three-dimensional modeling and engineering analysis capabilities in addition to two-dimensional drafting. Each package supports a different set of computers, display systems, and input and output devices. Also, these packages have different disk storage space and RAM requirements for their installation. For these reasons, the software is usually chosen before the computer and peripherals.

MicroCAD packages vary in speed, ease of use, menus, and customization features. Sluggish systems are constantly being improved to compete with fast software. The computer largely impacts the speed of the system, but fast software further increases user productivity.

Certain packages are easier to learn and easier to use than others. At least three to six months of learning are typically expected before achieving productivity gains over manual drafting methods. This varies depending on the application, software, and hardware as well as the individual. Some users catch on quickly while others constantly struggle.

The microCAD user interfaces the system using the keyboard, digitizer or mouse, and monitor. Most software packages contain screen menus which contain commands and other functions. Some contain digitizer/tablet menus in addition to the screen menus. When used in combination, they offer fast input of commands and functions.

Some screen menus offer pull-down menus, also referred to as pop-up menus, and dialogue boxes. For example, a hatching command is chosen from the menu; a box temporarily appears on the display screen containing numerous hatch patterns from which to choose. After the pattern is picked with the pointing device, the pull-down menu disappears. Pull-down menus of this type provide a fast and simple method of input for new and experienced users.

## 4.6 CUSTOMIZING THE SOFTWARE

Certain microCAD packages allow for customization. This may involve something as simple as writing a two-item command macro or as sophisticated as writing 300 lines of code. The level of customization varies based on need, application, and ability of the user. Regardless, the result can be a more powerful, useful, and productive program.

Several keystrokes and CAD commands are grouped to form a macro. This macro is invoked by a single keystroke or by a pick from the screen or tablet menu. Time and effort is saved over performing each operation separately.

Here is an example of a simple two-item macro: Zoom Window. A function key or item in a menu is assigned to invoke macro. When the key is pressed or a menu item is picked, the computer performs a zoom operation and prompts the user to choose a window. Without the macro, this is a two-step operation. When several operations are combined, the benefit of the macro increases. Some micro-CAD packages include a user programming language to produce supplement macros.

Ambitious users, software developers, and programmers often create numerous macros and include them in a custom screen and tablet menus. The menu often fulfils a specific need. Menus of this nature are available and sold commercially while others are designed for in-house use only.

## 4.7 FILE TRANSLATION

Exchange of files from one system to another is important in the corporate environment. Often, CAD systems are mixed within an organization, or organizations using one system must exchange files with another organization using a different system. Their compatibility becomes a critical issue. Two file formats have emerged as industry standards for moving CAD files from one system to another. The first is IGES which is used frequently across the wide-ranging levels of CAD and computer graphics technology. The Drawing Interchange Format (DXF) is a second popular file format. DXF (developed by Autodesk, Inc.) is considered a defacto standard.

IGES and DXF are very important to users who need to transfer a CAD file from one system to another. However, both are not without their drawbacks. Standards on layers/levels, linetypes, dimensioning, text, symbols, and others are not defined across all CAD systems. Therefore these items are potential problem areas when using IGES or DXF. For example, a CAD file may contain layers/ levels by name, that is, OBJECTS, DIM, and NOTES. If this file is transferred to a system that uses numbers only for layers/levels, then these names OBJECTS, DIM, and NOTES turn into arbitrary numbers such as 1, 2, and 3. Potential problems also exist with linetypes, text, symbols, and many other non-standard CAD methodologies.

Despite the current limitations of both DXF and IGES, they play an important role in the industry-wide trend toward sending and receiving electronic CAD data. When drawing files are compatible with other CAD systems, they are transferred freely and often sent thousands of miles through telephone lines in just minutes.

## 4.8 OPERATING SYSTEMS

In simplistic terms, the operating system is the software bridge between the computer and the microCAD software. Certain microCAD packages run under different operating systems, but of the several available, the first and most popular among microCAD users is MS-DOS. MS-DOS, short for Microsoft Disk Operating System, was developed by Microsoft Corporation during the early 1980s. This effort was supported by IBM, who at the time introduced the industry defacto standard IBM Personal Computer. The IBM version of MS-DOS, called PC-DOS, set a standard for future operating systems on IBM PC compatible hardware and software.

MicroCAD software adopted this standard too. Consequently, most micro-CAD packages run under MS-DOS. However, limitations to MS-DOS prohibit microCAD taking full advantage of the hardware capabilities. The most important example is MS-DOS's inability to address more than 640 KB of RAM. Since microCAD packages are typically large and sophisticated, especially in comparison to an average word processing program, the MS-DOS memory limitation is even more critical to microCAD users.

Microsoft and IBM were quite aware of the problem and need for a more sophisticated operating system. As a result, Operating System/2 (OS/2) was introduced in 1987 and made available in 1988. OS/2 addresses 16 MB of RAM which is 25 times that of MS-DOS. Now, microCAD programs can run resident in RAM, minimizing the number of times the computer needs to read and write to disk. Increased speed and productivity is the result.

Since OS/2 takes advantage of this large amount of memory, it also provides multitasking capabilities. This means that a user may run several tasks or programs simultaneously. For example, a microCAD program may be repainting the image in one window on the screen while the user is working in another window writing a specification with a word processor. Furthermore, a drawing may be plotting in the background.

## 4.9  MECHANICAL APPLICATIONS

### 4.9.1  Two-Dimensional Drafting

Two-dimensional drafting is by far the most common use of microCAD by mechanical drafters, designers, and engineers. The reason in large part comes from the traditional use of the drafting board. Drawings generated on the board are two-dimensional in nature. This mindset carries over when converting to microCAD.

For many mechanical design applications, two-dimensional drafting sufficiently and accurately describes the part geometry. Geometric dimensioning and tolerancing, feature control symbols, and detailed specifications usually communicate the part or product to the manufacturing engineer and NC part programmer.

### 4.9.2  Symbol Libraries

Symbols avoid repetitious drawing of shapes and details used mainly in two-dimensional work. A symbol is one or more drawing elements combined to form a specific shape, and is given a name and stored on disk. Symbols are retrieved by name, scaled, rotated, and inserted into drawings as needed. Groups of symbols are stored in libraries. An example is a set of feature control symbols.

All elements contained in a symbol are treated as one element. For instance, only one selection is necessary to pick and move a symbol. A symbol can be decomposed into its individual entities with the explode function. After a symbol is exploded, the entities within the symbol can be edited one by one.

Symbol libraries, also referred to as component libraries, is a collection of predefined symbols. When used properly, they dramatically speed the design and drafting process. Once the component is stored in the library, it is a simple matter to insert the component into the drawing. Several tool design libraries are commercially available.

### 4.9.3  Report Generation

Most microCAD packages offer a facility for report and bill of materials generation. Attributes are assigned to various drawing elements and symbols such as those mentioned above. These attributes are extracted to form a report. Attributes are text information such as cost, model number, color, and material. The attributes can be made visible on the graphics screen, but typically they are kept invisible.

The microCAD software contains a facility to extract and collect the attributes to form a file. This attribute file is then read and manipulated by another program, written in BASIC, LISP, or some other language, to create a report. The sophistication of the report usually depends on the sophistication of the program. Sometimes these programs are supplied as part of the microCAD package while others are available commercially.

Attribute files are often linked with word processors, spreadsheets, and database programs for special applications. The file is brought into a word processor for text editing. Reports are formatted; column headings and notes/specifications are added or changed as desired, applying the strengths of the word processor.

Spreadsheets, such as Lotus 1-2-3, display the attribute information in cells made up of rows and columns. Formulas are assigned to values, such as cost of materials, contained in the spreadsheet. Cost estimates are calculated, printed, and attached to the drawing. Database programs, such as Reflex and dBase III, are strong in handling large amounts of data. They sort and filter the data according to specific needs and provide specialized reports.

### 4.9.4  Parametric Design

With parametric design, a drawing is stored on disk and the dimensions of the object remain variable. This means that when the object, such as a screw, is inserted on the screen, the user is prompted to enter design parameters such as the type of head, length, diameter, pitch, etc. An unlimited number of variations of the same basic screw design can be produced. Storage of the object is extremely efficient, too, because only the parametric form is stored on disk as opposed to hundreds of variations.

### 4.9.5  Three-Dimensional Functions

Certain CAD software packages are two dimensional, some are two-and-a-half dimensional, while others store their information in a full three-dimensional database. Two-dimensional systems store coordinate data with $x$ and $y$ values only. Three-dimensional systems, on the other hand, store graphic data using $x$, $y$, and $z$ coordinates. Three-dimensional systems offer more capability, but are typically complex and more difficult to learn. Wireframe, surface, and solid modelings are supported by three-dimensional systems. Wireframe is the representation of an object with lines that form a wire mesh. These systems usually allow for hidden line removal. Surface modeling is the placement of skin on the wireframe object, making the object appear more real. The skin may appear in a variety of colors and shades depending on the capabilities of the system. Solid modeling is sometimes confused with surface modeling because both appear identical in certain instances. True solid modeling treats the object as a real solid object. For instance, if you were to slice through the solid model, it would not be hollow like a surface model. Complicated solid models require substantial CPU power and memory. Shaded images provide realistic views of three-dimensional objects.

### 4.9.6  Finite Element Analysis (FEA)

FEA is the solution of stress, both structural and thermal, within a mechanical member. FEA software calculates the amount of stresses that is exerted on a mechanical member due to externally applied loads. MicroCAD FEA software is usually a subset of mini- and mainframe-based software. It usually has a limit on the size (number of nodes) of the finite element model. This limitation is a result of the memory size and processing speeds of the PCs. The maximum number of nodes that can be handled by microCAD FEA software ranges from 500 to 2000. This software is typically useful at early stages of design. For example, a designer may first run a crude (using a coarse mesh) FEA on a PC to test various design alternatives. Once a final design is chosen, a more refined FEA can be performed on a mini- or mainframe-based system.

### 4.9.7  Manufacturing

MicroCAM is a subset of mainframe CAM. For example, microCAM software may support two-axis to three-axis machining compared to three-axis to five-axis machining offered by mainframe CAM. However, for some users, it provides satisfactory performance at an affordable cost. To beginners, microCAM is more easily understood in its separate components than as a complete system. These components include the microCAD software, NC software, microcomputer and its peripherals, and a computerized NC machine such as a vertical mill or lathe.

The NC software is the link between the microCAD software and the NC machine. It saves time by eliminating the creation of new geometry for machine tool paths. The NC software generates the tool path from the microCAD drawing, creating the path in NC code. A wide variety of NC packages are available for NC programming. They range in capabilities; some support two-axis machining while others support three-axis machining. Some NC software has the capability of checking the tool path at the machine tool one step at a time. The NC machine tool is typically the most expensive component of the micro-based CAM system. Table-top NC machines are well suited for prototyping and training. These machines can cut a variety of materials such as aluminum, brass, wood, plastics, and wax to an accuracy of 0.01 mm.

## 4.10  MICROCAD TRENDS

The microcomputer industry will continue to experience improvements in the range of operating systems. MicroCAD users will migrate from MS-DOS of the early 1980s to systems such as OS/2, Xenix, and Unix. These sophisticated systems take full advantage of the hardware memory capabilities and will offer attractive multitasking and networking capabilities.

Also in the future, microCAD software will become compatible with a broader range of software. For example, most desktop publishing packages will be able to accept microCAD-produced drawings. As another example, painting packages will capture microCAD drawings and techniques such as cut and paste, and special hatching, background colors and text fonts will be applied to embellish the drawing.

MicroCAD packages will incorporate some of these packages too. They may include a built-in word processor or text editor to ease the development and correction of lengthy notes and specifications in drawings. The same is possible and also occurs with spreadsheet and database management capabilities.

As microCAD software becomes more integrated, its power and functionality increases. Parametric design, described earlier in this chapter, will be a common function of the software too. Built-in finite element analysis capabilities will provide the mechanical and structural engineer with a convenient and low-cost means for analyzing the capacity of structural members within a design.

Product design and manufacturing engineers will combine their efforts by using a design and NC package that provides for both. The single database will result in the design and production of a higher quality product in a shorter span of time.

Companies separated by thousands of miles will send and receive CAD files using communication software, modems, and standard telephone lines. These drawings and their databases will reach clients and subcontractors in minutes instead of days. A substantial saving in time and money is the payoff.

The hardware performance/cost ratio will continue to improve. Low-cost microcomputers will perform better than mainframes of the past. Display technology will improve too with bigger and higher resolution color displays. Storage systems will store huge amounts of data such as 500 MB or more, and this information will be stored and retrieved at a faster rate. Libraries will use CD-ROM devices to store thousands of photographic images and drawings. Each will be indexed and displayed on the screen in a matter of seconds.

## 4.11  PRODUCT DISTRIBUTION TRENDS

In the past, CAD/CAM systems were bundled (hardware and software purchased from one vendor). Today, they are unbundled and as many as three or more suppliers can contribute to configuring a system. In addition, most microCAD suppliers are not bound only to sales of microCAD-related products.

Mail order companies are pioneering an unusual way of distributing unbundled microcomputer and microCAD products. Over the phone and through the mail, users are able to buy hardware and software, saving substantial amounts on the purchase price.

Integrators, often called system houses, are also on the rise. They are companies that buy the microCAD software and hardware components, assemble the system, and sell it. Often they add value to the package by including software such as a tablet menu or symbol library they have developed in-house.

## PROBLEMS

**4.1.** What led to microCAD's popularity?

**4.2.** Name at least three advantages of using a microCAD system.

**4.3.** Name and describe the main components that make up a microCAD system (e.g., the computer is one component).

**4.4.** Name and briefly describe the function of at least two drawing commands and two editing commands found in typical microCAD software.

**4.5.** Explain differences between two-dimensional and three-dimensional functions.

**4.6.** Remove the cover from a microcomputer, such as an IBM PS/2 or a MAC, and identify each of the following components: system board, CPU, math coprocessor, RAM chips, power supply, and expansion slots.

## PUBLICATIONS

*CADalyst*, CADalyst Publications, Vancouver, B.C., Canada; covers techniques, tips, applications, developments, and news for AutoCAD users and developers.

*CADENCE*, Ariel Communications, Inc., Austin, Texas; covers techniques, tips, applications, developments, and news for AutoCAD users and developers.

*Computer Graphics Today*, New York; covers news, applications, developments, and events on the computer graphics industry.

*Computer Graphics World*, Littleton, Massachusetts; covers applications, developments, trends, news and events on the computer graphics industry.

*MicroCAD News*, Ariel Communications, Inc., Austin, Texas; covers microCAD hardware and software developments, applications trends, and events.

*PC Week*, Boston, Massachusetts; covers news, developments, and trends on the IBM PC-standard computing industry.

*Technology and Business Communications*, Sudbury, Massachusetts; covers trends, opinions, and developments on the computer graphics industry.

*Versatility*, Ariel Communications, Inc., Austin, Texas; covers techniques, applications, and developments for VersaCAD users and developers.

# PART
# II

GEOMETRIC
MODELING

# CHAPTER
# 5

# TYPES AND MATHEMATICAL REPRESENTATIONS OF CURVES

## 5.1 INTRODUCTION

CAD tools have been defined in Chap. 1 as the melting pot of three disciplines: design, geometric modeling, and computer graphics. While the latter discipline is covered in Part III, this part discusses geometric modeling and its relevance to CAD/CAM. Early CAD/CAM systems focused on improving the productivity of draftsmen. More recently, they have focused on modeling engineering objects. As a result, geometric models that once were more than adequate for drafting purposes are not acceptable for engineering applications. A basic requirement, therefore, is that a geometric model should be an unambiguous representation of its corresponding object. That is to say, the model should be unique and complete to all engineering functions from documentation (drafting and shading) to engineering analysis to manufacturing.

A geometric model of an object and its related database have been defined in Chap. 3. The three types of geometric models, wireframes, surfaces, and solids, are covered in Chaps. 5, 6, and 7 respectively. Each chapter presents the available types of entities of the modeling technique and their related mathematical representations to enable good understanding of how and when to use these entities in engineering applications. Users usually have to decide on the type of modeling technique based on the ease of using the technique during the construction phase and on the expected utilization of the resulting database later in the design and manufacturing processes. Regardless of the chosen technique, the user constructs a geometric model of an object on a CAD/CAM system by inputting the object data as required by the modeling technique via the user interface provided by the software. The software then converts such data into a mathematical representation which it stores in the model database for later use. The user may retrieve and/or modify the model during the design and/or manufacturing processes.

To convey the importance of geometric modeling to the CAD/CAM process, one may refer to other engineering disciplines and make the following analogy. Geometric modeling to CAD/CAM is as important as governing equilibrium equations to classical engineering fields as mechanics and thermal fluids. From an engineering point of view, modeling of objects is by itself unimportant. Rather, it is a means (tool) to enable useful engineering analysis and judgment. As a matter of fact, the amount of time and effort a designer spends in creating a geometric model cannot be justified unless the resulting database is utilized by the applications module discussed in Chap. 3.

The need to study the mathematical basis of geometric modeling is manyfold. From a strictly modeling point of view, it provides a good understanding of terminology encountered in the CAD/CAM field as well as CAD/CAM system documentation. It also enables users to decide intelligently on the types of entities necessary to use in a particular model to meet certain geometric requirements such as slopes and/or curvatures. In addition, users become able to interpret any unexpected results they may encounter from using a particular CAD/CAM system. Moreover, those who are involved in the decision-making process and evaluations of CAD/CAM systems become equipped with better evaluation criteria.

From an engineering and design point of view, studying geometric modeling provides engineers and designers with new sets of tools and capabilities that they can use in their daily engineering assignments. This is an important issue because, historically, engineers cannot think in terms of tools they have not learned to use or been exposed to. The tools are powerful if utilized innovatively in engineering applications. It is usually left to the individual imagination to apply these tools usefully to applications in a new context. For example, the mere fact that CAD/CAM databases are centralized and associative provides great capabilities that are utilized in Sec. 5.8 of this chapter. These capabilities are usually more efficient than writing analyses and plotting programs on conventional computers.

Having established the need for geometric modeling, what is the most useful geometric model to engineering applications? Unfortunately, there is no direct answer to this question. Nevertheless, the following answer may be offered. In this book, the answer has two levels. At one level, engineers may agree that some sort of geometry is required to carry engineering analysis. The degree of geometric detail depends on the analysis procedure that utilizes the geometry. Engineers may also agree that there is no model that is sufficient to study all behavioral aspects of an engineering component or a system. A machine part, for example, can be modeled as a lumped mass rigid body on one occasion or as a distributed mass continuum on another occasion.

At the second level, the adequacy of geometry or a geometric model to an analysis procedure is decided by its related useful attributes to that procedure. Attributes of geometry is never an issue for manual procedures because the engineer's mind coordinates all the related facts and information. In computer-based modeling and analysis, attributes attached to geometric models determine their relevance to design, analysis, and manufacturing. Current geometric models offered by CAD/CAM systems seem to have adequate and enough geometric and visualization (colors and shades) attributes. Based on these attributes, they are utilized successfully in applications such as mass property calculations, mechanism analysis, finite element modeling, and NC. If these attributes are insufficient or must be reorganized for other applications, then new software must be written based on the existing database structure of the geometric model which may be modified to accept new attributes; or a completely new structure may have to be developed. In conclusion, the study of existing geometric models provides designers and engineers with the capabilities and limitations of these models and paves the road for them to utilize the attributes of these models or create new ones to benefit new engineering applications.

This chapter covers the available types and most useful mathematical representations of curves. Sections 5.1 through 5.7 cover the basic related topics. Section 5.8 applies these topics to design and engineering applications to demonstrate their usefulness.

## 5.2   WIREFRAME MODELS

A wireframe model of an object is the simplest, but mose verbose, geometric model that can be used to represent it mathematically in the computer. It is sometimes referred to as a stick figure or an edge representation of the object. The word "wireframe" is related to the fact that one may imagine a wire that is bent to follow the object edges to generate the model. Typically, a wireframe model consists entirely of points, lines, arcs and circles, conics, and curves. Wireframe modeling is the most commonly used technique and all commercial CAD/CAM systems are wireframe-based.

Early wireframe modeling techniques developed in the 1960s were strictly two dimensional and were designed to automate drafting and simple NC. Two-dimensional wireframe models contained enough useful information to perform the NC work. Users had to construct geometry in the desired various views independently due to the lack of centralization and associativity of the resulting database. Later in the early 1970s, the centralized associative database concept enabled modeling of three-dimensional objects as wireframe models that can be subject to three-dimensional transformations. Creating geometry in one view is automatically projected and displayed in other views. This represents a substantial saving and flexibility over manual design and drafting.

In constructing a wireframe model on a CAD/CAM system, the user should follow the modeling guidelines discussed in Chap. 3. The detailed step-by-step procedures to create models may vary according to system capabilities and the user's individual habits. In addition to the commands required to create the common wireframe entities, users are provided with other tools that facilitate the model construction. Part IV of the book covers these tools in detail. These tools, in general, help users to manage geometry as well as to avoid unnecessary calculations. For example, typical CAD/CAM systems provide users with possibly three modes to input coordinates: cartesian, cylindrical, or spherical. Each mode has explicit or implicit inputs. Explicit input could be absolute or incremental coordinates. Implicit input involves user digitizes. Another example is the geometric modifiers which automatically identify specific locations such as end- or midpoints of entities that are convenient to access once these entities are created by the system.

Despite its many disadvantages, the major advantages of wireframe modeling is its simplicity to construct. Therefore, it does not require as much computer time and memory as does surface or solid modeling. However, the user or terminal time needed to prepare and/or input data is substantial and increases rapidly with the complexity of the object being modeled. Wireframe modeling is considered a natural extension of traditional methods of drafting. Consequently, it does not require extensive training of users; nor does it demand the use of unusual terminology as surfaces and solids. Wireframe models form the basis for surface models. Most existing surface algorithms require wireframe entities to generate surfaces (refer to Chap. 6). Lastly, the CPU time required to retrieve, edit, or update a wireframe model is usually small compared to surface or solid models.

The disadvantages of wireframe models are manyfold. Primarily, these models are usually ambiguous representations of real objects and rely heavily on human interpretation. A wireframe model of a box offers a typical example where the model may represent more than one object depending on which face(s) is assumed to exist. Models of complex designs having many edges become very confusing and perhaps even impossible to interpret. To overcome this confusion, lines can be hidden, dashed, or blanked. If done manually, these operations are very tedious, error-prone, and can result in "nonsense" objects. Automatic hidden line removal algorithms based on wireframe modeling are usually helpful. Another disadvantage is the lack of visual coherence and information to determine the object profile. The obvious example is the representation of a hole or a curved portion of the object. In most systems, the hole is displayed as two parallel circles separated by the hole length. Some systems may connect a line between the two circles on one side of the hole. In many cases, users add edges of the hole for appearance purposes at the drafting mode or may use a cylindrical surface to represent the hole which introduces problems later on during the model clean-up phase. In adding the edges, inexperienced users tend to attempt to create tangent lines beweeen the hole circles which obviously does not work. Figure 5-1 shows possible cases to display holes and/or curved ends of objects. Representing the intersection of plane faces with cylinders, cylinders with cylinders, or tangent surfaces in general is usually a problem in wireframe modeling and requires user manipulations.

Wireframe models are also considered lengthy or verbose when it comes to the amount of defining data and command sequence required to construct them. For example, compare the creation of a simple box as a wireframe and as a solid. In the latter, the location of one corner, the length, width, and height are the required input while in the former the coordinates of at least four corners of one face, the depth, and the edge connectivity are required, considering the box as a two-and-a-half-dimensional object. In other words, both topological and geometrical data are needed to construct wireframe models while solids require only

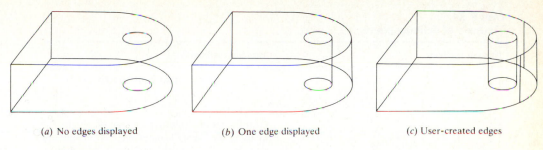

(*a*) No edges displayed          (*b*) One edge displayed          (*c*) User-created edges

**FIGURE 5-1**
Displaying holes and curved ends in wireframe models.

geometrical data (refer to Chap. 7 for the difference between topology and geometry). In addition, WCSs and construction planes always need to be defined to facilitate model construction.

From an application, and consequently engineering, point of view, wireframe models are of limited use. Unless the object is two-and-a-half dimensional, volume and mass properties, NC tool path generation, cross-sectioning, and interference detections cannot be calculated. The model can, however, be used in manual finite element modeling and tolerance analysis.

Despite the above-mentioned limitations, wireframe models are expected to last and may extend to certain classes of solid modeling. For example, some solid modelers (such as Medusa) are based on wireframe input. The simplicity of the geometrical concepts based on wireframe modeling makes them attractive to use to introduce users to the CAD/CAM field. In addition, at early design stages, designers might just need a sketchpad to try various ideas. Wireframes are ideal to provide them with such a capability. From an industrial point of view, wireframe models may be sufficient to many design and manufacturing needs. From a practical point of view, many companies have large amounts of wireframe databases that are worth millions of dollars and man-hours and therefore make it impossible to get rid of wireframe technology.

## 5.3   WIREFRAME ENTITIES

All existing CAD/CAM systems provide users with basic wireframe entities which can be divided into analytic and synthetic entities. Analytic entities are points, lines, arcs and circles, fillets and chamfers, and conics (ellipses, parabolas, and hyperbolas). Synthetic entities include various types of spline [cubic spline, B-spline, $\beta$(beta)-spline, $v$(nu)-spline] and Bezier curves. The mathematical properties of each entity and how it is used in engineering applications or converted into a user interface are covered in the remainder of the chapter. It is quite common for a user to be faced with many modifiers that can be used to create a particular entity on a particular CAD/CAM system. Knowledge of the basics of such an entity can only increase user productivity. Choosing the proper modifier for a given input can save unnecessary calculations. Also, knowing a curve behavior in relation to its input data can save time trying to achieve the impossible.

**TABLE 5.1**
**Methods of defining points**

| Explicit methods | Implicit methods |
|---|---|
| 1. Absolute cartesian co-ordinates | A digitize $d$ (with or without an active grid) |

$+ P(x,y,z)$

$+ d$

Coordinates of resulting point can be obtained by using the "verify" command. Coordinates are measured relative to the MCS as discussed in Chap. 3

---

2. Absolute cylindrical coordinates

$+ P(R, \theta, z)$

(Spherical coordinates are seldom used in practice)

Endpoint of an existing entity†

$E_2$

$E_1$

Line

$E_1, E_2$

Circle

$E_1 \quad E_2$

Arc

$E_1 \quad E_2$

Curve

---

3. Incremental cartesian coordinates

$P(x + \Delta x, y + \Delta y, z - \Delta z)$

$\Delta z$

$\Delta x \quad \Delta y$

Reference point $P_0 (x,y,z)$

Centerpoint (origin) of an existing entity†

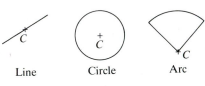

Line

Circle

Arc

$C$

---

4. Incremental cylindrical coordinates

$P(R + \Delta R, \theta + \Delta \theta)$

$\Delta R \quad \Delta \theta$

Reference point $P_0 (R, \theta)$

(Some CAD/CAM systems require moving the current WCS to the reference point $P_0$ to avoid unexpected results)

Intersection point of two existing entities†

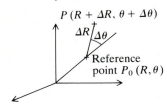

$I_1 \quad I_2$

---

† More details of these methods are covered in Sec. 11.2, Chap. 11.

**TABLE 5.2**
**Methods of defining lines**

| Method | Illustration |
|---|---|
| 1. Points defined by any method of Table 5.1 | |
| 2. Horizontal (parallel to the $X$ axis of the current WCS) or vertical (parallel to the $Y$ axis of the current WCS) | |
| 3. Parallel or perpendicular to an existing line | Reference line |
| 4. Tangent to existing entities | One of the four possibilities is obtained depending on user digitizes of the two circles One of the two possibilities is obtained |

Tables 5.1 to 5.5 show the most common methods utilized by CAD/CAM systems to create wireframe entities. Readers are advised to compare and/or modify these methods to what their respective systems offer and what user interfaces of these systems require. As an example, defining a point using the endpoint method requires the " END" or "PND (point end)" modifier on the Computervision or Calma system respectively. Such similarities among systems exist in all

software modules, simply because they all share the same theory. Readers are also advised to find commands corresponding to these tables on their particular CAD/CAM systems. The following are some examples to illustrate using wire-frame modeling techniques. These examples are independent of any system or user interface and readers can simply convert them into a command sequence of their choice.

**TABLE 5.3**
**Methods of defining arcs and circles**

| Method | Illustration |
|---|---|
| 1. Radius or diameter and center. In the case of an arc, beginning and ending angles $\theta_1$ and $\theta_2$ are required | |
| 2. Three points defined by any method of Table 5.1 | |
| 3. Center and a point on the circle | |
| 4. Tangent to line, pass through a given point, and with a given radius | |

**TABLE 5.4**
**Methods of defining ellipses and parabolas**

| Methods | Illustration |
|---|---|
| 1. *Ellipses* | |
| (a) Center and axes lengths | |
| (b) Four points | |
| (c) Two conjugate diameters | |
| 2. *Parabolas* | |
| (a) Vertex and focus | |
| (b) Three points | |

**TABLE 5.5**
**Methods of defining synthetic curves**

| Method | Illustration |
|---|---|
| 1. *Cubic spline*<br><br>A given set of data points and start and end slopes | 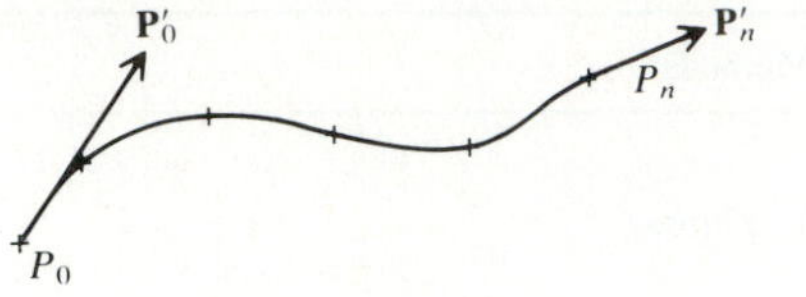 $P_0'$ ... $P_n'$, $P_n$, $P_0$ |
| 2. *Bezier curves*<br><br>A given set of data points | 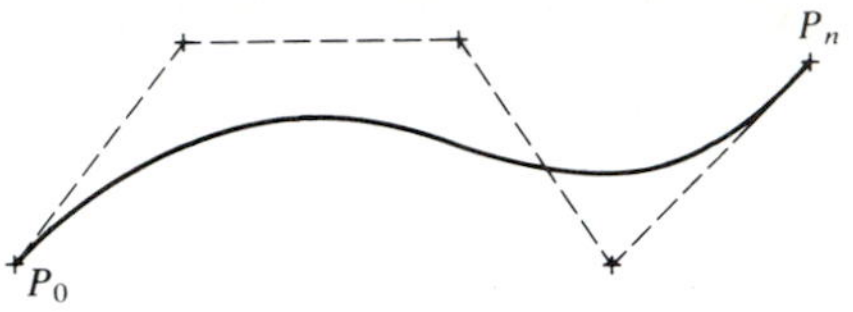 $P_n$, $P_0$ |
| 3. *B-spline curves*<br><br>(a) Approximate a given set of data points<br><br><br>(b) Interpolate a given set of data points | |

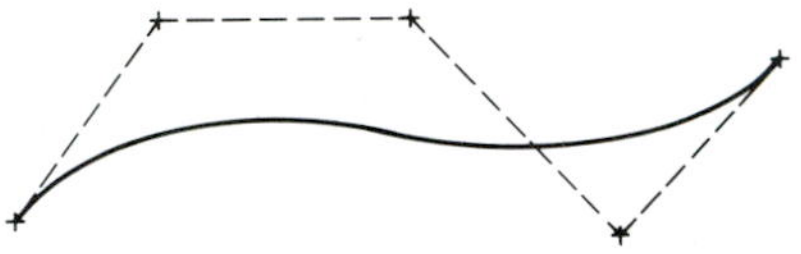

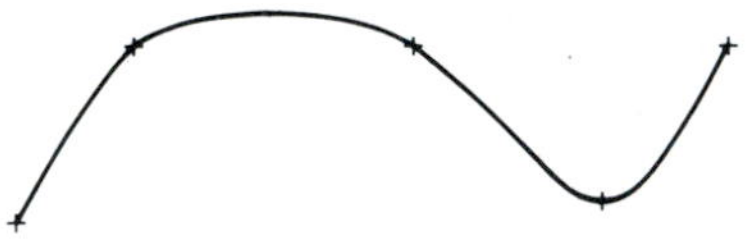

**Example 5.1.** For the guide bracket shown in Fig. 5-2:

(a) Create the model database utilizing a CAD/CAM system.
(b) Obtain the orthographic views of the model.
(c) Obtain a final drawing of the model.

*Solution*

(a) An efficient planning strategy is required before logging into the CAD/CAM system. Examining Fig. 5-2 reveals that the geometric model of the guide bracket is a two-and-a-half-dimensional model. It is symmetric with respect to the model central vertical plane if the cut on the top right corner is ignored. Let us choose the origin of the MCS at point $A$ shown in the figure. The orientation of the MCS is chosen properly, as discussed in Chap. 3, so that the front view of the bracket is defined as shown. Assume the orientation shown to facilitate discussion. The following steps may be followed to construct the model:

  1. Follow the part setup procedure of the CAD/CAM system. Define an iso-metric view to begin constructing the model.

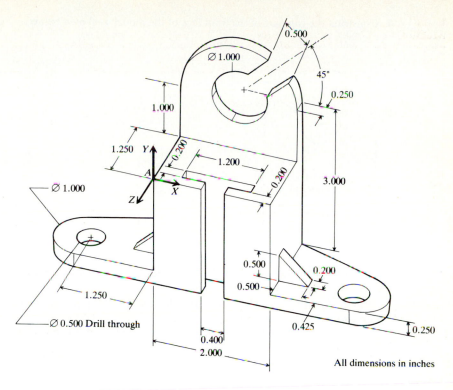

**FIGURE 5-2**
Guide bracket.

2. Begin by constructing the right face using a line command with the following sequence of point input:

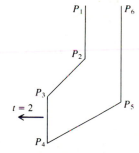

| Point | Coordinate input relative to MCS | | |
|-------|------|------|------|
|       | $x$  | $y$  | $z$  |
| $P_1$ | $2^\dagger$ | 1.00 | $-1.25$ |
| $P_2$ | $-^\ddagger$ | $\Delta y = -1.00^\S$ | $-$ |
| $P_3$ | $-$ | $-$ | 0 |
| $P_4$ | $-$ | $y = -2$ | $-$ |
| $P_5$ | $-$ | $-$ | $z = -1.5$ |
| $P_6$ | $-$ | $y = 3.00$ | $-$ |

   † Absolute coordinate input.
   ‡ System defaults to the last input value.
   § Relative (incremental) coordinate input.
   Incremental coordinate input is very efficient
   in this case.

3. Project that face a distance of 2 inches in the direction shown above ($X$
   axis).

4. Construct the top part of the front face of the model as shown below:

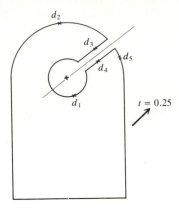

Trimming, offsetting, circle, and arc commands are useful. Cylindrical input is needed to create the line at 45°.

5. Project the entities labeled by digitizes $d_1$ to $d_5$ a distance of 0.25 inch in the direction shown above ($Z$ axis).

6. Create the slot in the top face using a line command with the point sequence shown below:

| Point | Coordinate input relative to MCS | | |
|-------|------|------|------|
| | $x$ | $y$ | $z$ |
| $P_1$ | 1.2 | 0 | 0 |
| $P_2$ | — | — | $\Delta z = -0.2$ |
| ⋮ | ⋮ | ⋮ | ⋮ |
| $P_8$ | — | — | 0 |

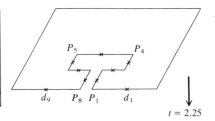

Trim the front horizontal line between points $P_1$ and $P_8$.

7. Project the entities labeled by digitizes $d_1$ to $d_9$ a distance of 2.25 inches in the direction shown above ($Y$ axis).

8. Construct the top face of the right flange of the bracket as shown below:

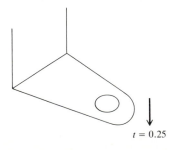

First construct the two circles and then the two tangents. Trim the outer circle to finish the construction.

9. Project that top face a distance of 0.25 inch in the direction shown ($Y$ axis).

10. Construct the front face of the right support (wedge) using a line command with the following point sequence:

| Point | Coordinate input relative to MCS | | |
|---|---|---|---|
| | $x$ | $y$ | $z$ |
| $P_1$ | 2 | $-2$ | $-0.425$ |
| $P_2$ | $\Delta x = 0.5$ | — | — |
| $P_3$ | 2 | $-1.5$ | — |
| $P_4$ | — | $-2$ | — |

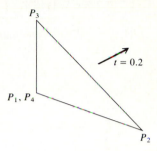

11. Project that face a distance of 0.2 inch in the direction shown (Z axis).
12. Using the mirror command, copy the entities that resulted in steps 8 to 11 to create the left flange of the model. The mirror plane is the vertical central plane of the model. To facilitate using this command, a different layer (see Chap. 11) should be used for steps 8 to 11 to isolate the entities to be mirrored from the others.
13. Connect any missing lines and obtain a hard copy or a plot of the model.
14. File the model to save it.

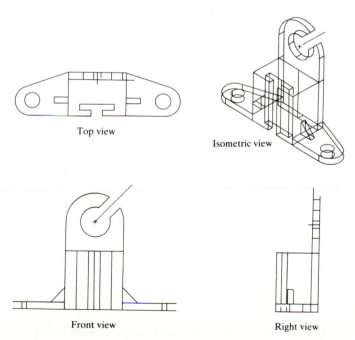

Top view

Isometric view

Front view

Right view

**FIGURE 5-3**
Orthographic views of guide bracket.

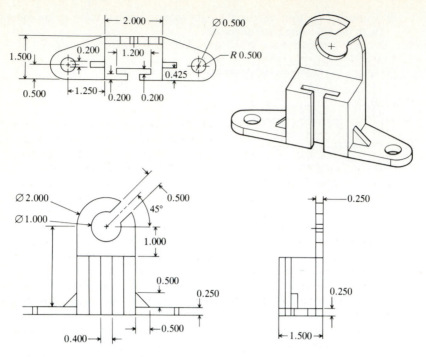

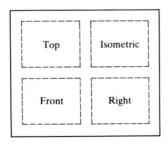

**FIGURE 5-4**
Final drawing of guide bracket.

(b) Using a new drawing or screen layout, define views as shown below:

| | |
|---|---|
| Top | Isometric |
| Front | Right |

Follow all discussions in Chap. 3 to define these views. The result is shown Fig. 5-3.

(c) Using the model clean-up procedure on the system, the model views are cleaned according to the drafting conventions and the drawing of the model is shown in Fig. 5-4.

This example illustrates most of the experiences encountered in creating two-and-a-half-dimensional wireframe models on major CAD/CAM systems with their related drafting work.

**Example 5.2.** For the stop block shown in Fig. 5-5:

(*a*) Create the model database utilizing a CAD/CAM system.
(*b*) Obtain the isometric view of the model.
(*c*) Obtain the final drawing of the model.

*Solution*

(*a*) Unlike Example 5.1, the geometric model of the stop block is given by its three
   orthographic views. All existing mechanical designs are available in this form
   where their related information are stored in blueprints. While scanning techniques exist to convert blueprints into CAD databases automatically, or at least
   semi-automatically, this example illustrates how to create a geometric model
   database from its three orthogonal views. The planning strategy in this case
   consists of choosing the proper location and orientation of the database MCS
   and of trying to construct the model by working in four views simultaneously.
   The general guiding rule should be to minimize the amount of calculations by
   taking advantage of the centralization and associativity property of the database. For example, the part of the block that is oriented at 30° should be constructed in the front view first and then finished off in the top and right views.
   Assuming an MCS as shown, the following steps may be followed to create the
   model database:
   1. Follow the part setup procedure of the CAD/CAM system. Define four views
      (front, top, right, and isometric) to construct the model.

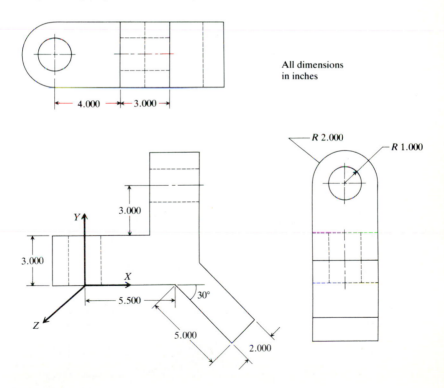

All dimensions
in inches

**FIGURE 5-5**
Stop block.

2. Begin by constructing the front view as shown below:

| Point | Coordinate input relative to MCS | | |
|---|---|---|---|
| | $x$ | $y$ | $z$ |
| $P_1$ | 0 | 0 | 0 |
| $P_2$ | 5.5 | — | — |
| $P_3$ | $R = 5$ | $\theta = -30$ | — |
| $P_4$ | 0 | 3 | 0 |
| $P_5$ | $\Delta x = 4$ | — | — |
| $P_6$ | — | $\Delta y = 5$ | — |
| $P_7$ | $\Delta x = 3$ | — | — |
| $P_8$ | — | $\Delta y = -7$ | — |

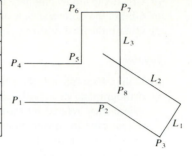

Connect $P_1$, $P_2$, and $P_3$ using a line command. Connect $P_4$, $P_5$, $P_6$, $P_7$, and $P_8$ using a line command. Create line $L_1$ with a perpendicular line command. Create line $L_2$ with a parallel line command. Trim lines $L_2$ and $L_3$ to their intersection point.

3. Complete the creation of the top view. Notice that all entities created in the front view are properly projected and displayed in the other three views. An existing entity can be chosen (located) by digitizing it in any view where it is most feasible and accessible. This usually makes construction much simpler.

4. Complete the creation of the right side view. The isometric view is now automatically completed and the four views are shown in Fig. 5-6.

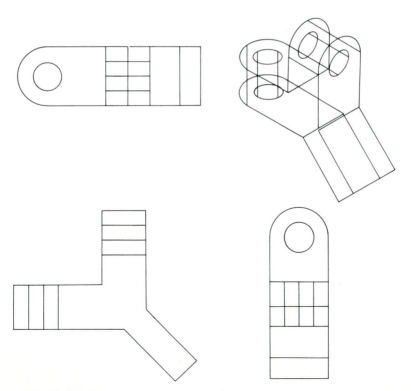

**FIGURE 5-6**
Views of stop block.

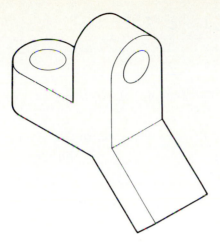

**FIGURE 5-7**
Isometric view of stop block.

(b) Access a new drawing or screen layout and define the whole screen as an isometric view. Clean up the view to obtain Fig. 5-7.

(c) Using the clean-up procedure as discussed in Example 5.1, the final drawing of the model is shown in Fig. 5-8.

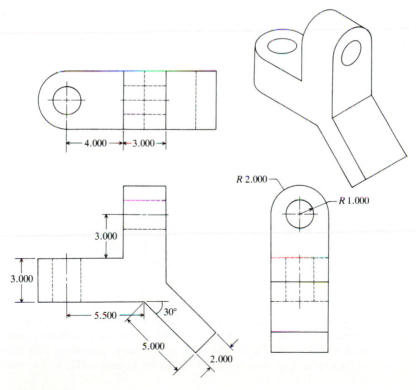

**FIGURE 5-8**
Final drawing of stop block.

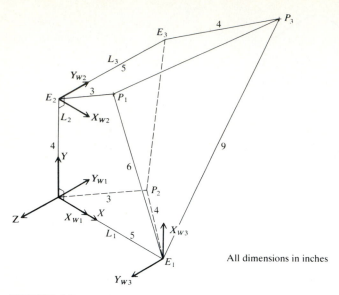

**FIGURE 5-9**
Adjustment block.

**Example 5.3.** Create the database of the three-dimensional model shown in Fig. 5-9.

*Solution.* This is a three-dimensional geometric model. Its construction does not usually follow any general guidelines. Typically, coordinates of its corners are required for construction. However, the general rule of minimizing calculations and utilizing capabilities of various CAD/CAM functions must always be followed. The shortest way to construct this model is discussed below:

1. After the part setup procedure, define an isometric view. The MCS is chosen as shown in Fig. 5-9.
2. Construct line $L_1$ between points (0, 0, 0) and (5, 0, 0) and $L_2$ between points (0, 0, 0) and (0, 4, 0).
3. Construct two circles to find the corner $P_1$. The first circle has a center at $E_1$ (endpoint of $L_1$) and a radius of 6, and the second has a center at $E_2$ (end of line $L_2$) and a radius of 3. The intersection of the two circles defines $P_1$.
4. Connect $E_1$ and $P_1$ and $E_2$ and $P_1$ using the proper line command.
5. Define a (WCS)$_1$ such that its $XY$ plane coincides with the bottom of the model which is in turn assumed to lie in the $XZ$ plane of the MCS.
6. Find point $P_2$ and construct the bottom edges in the same way as in steps 3 and 4.
7. Construct $L_3$ by connecting $E_2$ and point (0, 5, 4). The coordinates of the point are measured with respect to the active (WCS)$_1$.
8. Construct two circles to find the corner $P_3$. Define a (WCS)$_2$ such that its $XY$ plane coincides with the top view of the model. Construct a circle with center $E_3$ and radius 4. Define another (WCS)$_3$ with its $XY$ plane coincident with the right view of the model. Construct a circle with center $E_1$ and radius 9. The intersection of the two circles defines $P_3$.
9. Connect $E_1$ and $P_3$, $E_3$ and $P_3$, $P_1$ and $P_3$, and $P_2$ and $E_3$ to complete the model construction as shown in Fig. 5-10.

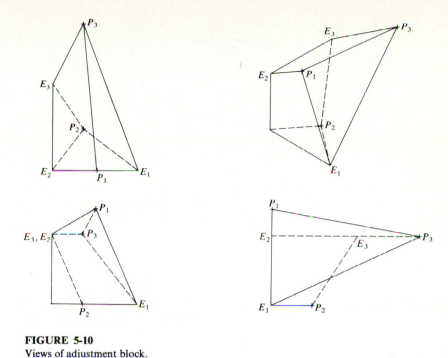

**FIGURE 5-10**
Views of adjustment block.

## 5.4  CURVE REPRESENTATION

Examples in the previous section have been chosen to require only simple geometric entities (lines and circles) to provide a good understanding of issues encountered in practicing CAD/CAM technology. However, many applications (automotive and aerospace industries) require other general curves to meet various shape constraints (continuity and/or curvature). The remainder of this chapter covers the basics of these curves.

A geometric description of curves defining an object can be tackled in several ways. A curve can be described by arrays of coordinate data or by an analytic equation. The coordinate array method is impractical for obvious reasons. The storage required can be excessively large and the computation to transform the data from one form to another is cumbersome. In addition, the exact shape of the curve is not known, therefore impairing exact computations such as intersections of curves and physical properties of objects (e.g., volume calculations). From a design point of view, it becomes difficult to redesign shapes of existing objects via the coordinate array method. Analytic equations of curves provide designers with information such as the effect of data points on curve behavior, control, continuity, and curvature.

The treatment of curves in computer graphics and CAD/CAM is different from that in analytic geometry or approximation theory. Curves describing engineering objects are generally smooth and well-behaved. In addition, not every available form of a curve equation is efficient to use in CAD/CAM software due to either computation or programming problems. For example, a curve

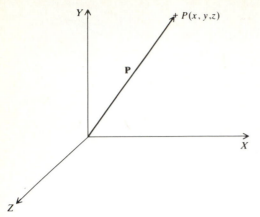

**FIGURE 5-11**
Position vector of point $P$.

equation that results in a division by zero while calculating the curve slope causes overflow and errors in calculations. Similarly, if the intersection of two curves is to be found by solving their two equations numerically, the forms of the two equations may be inadequate to program due to the known problems with numerical solutions. In addition, considering that most design data of objects are available in a discrete form, mainly key points, the curve equation should be able to accept points and/or tangent values as input from the designer.

Curves can be described mathematically by nonparametric or parametric equations. Nonparametric equations can be explicit or implicit. For a nonpara-metric curve, the coordinates $y$ and $z$ of a point on the curve are expressed as two separate functions of the third coordinate $x$ as the independent variable [see Eq. (5.1)]. This curve representation is known as the nonparametric explicit form. If the coordinates $x$, $y$, and $z$ are related together by two functions [see Eq. (5.2)], a nonparametric implicit form results. For a parametric curve, on the other hand, a parameter is introduced and the coordinates $x$, $y$, and $z$ are expressed as functions of this parameter [see Eq. (5.3)].

Explicit nonparametric representation of a general three-dimensional curve takes the form:

$$\mathbf{P} = [x \quad y \quad z]^T = [x \quad f(x) \quad g(x)]^T \tag{5.1}$$

where $\mathbf{P}$ is the position vector of point $P$ as shown in Fig. 5-11. Equation (5.1) is a one-to-one relationship. Thus, this form cannot be used to represent closed (e.g., circles) or multivalued curves (e.g., parabolas). The implicit nonparametric representation can solve this problem and is given by the intersection of two surfaces as

$$F(x, y, z) = 0$$
$$G(x, y, z) = 0 \tag{5.2}$$

However, the equation must be solved to find its roots ($y$ and $z$ values) if a certain value of $x$ is given. This may be inconvenient and lengthy. Other limitations of nonparametric representations of curves are:

1. If the slope of a curve at a point is vertical or near vertical, its value becomes infinity or very large, a difficult condition to deal with both computationally and programming-wise. Other ill-defined mathematical conditions may result.

2. Shapes of most engineering objects are intrinsically independent of any coordinate system. What determines the shape of an object is the relationship between its data points themselves and not between these points and some arbitrary coordinate system.

3. If the curve is to be displayed as a series of points or straight line segments, the computations involved could be extensive.

Parametric representation of curves overcomes all of the above difficulties. It allows closed and multiple-valued functions to be easily defined and replaces the use of slopes with that of tangent vectors, as will be introduced shortly. In the case of commonly used curves such as conics and cubics, these equations are polynomials rather than equations involving roots. Hence, the parametric form is not only more general but it is also well suited to computations and display. In addition, this form has properties that are attractive to CAD/CAM and the interactive environment, as will be seen later in this chapter.

In parametric form, each point on a curve is expressed as a function of a parameter $u$. The parameter acts as a local coordinate for points on the curve. The parametric equation for a three-dimensional curve in space takes the following vector form:

$$\mathbf{P}(u) = [x \quad y \quad z]^T = [x(u) \quad y(u) \quad z(u)]^T, \qquad u_{min} \leq u \leq u_{max} \qquad (5.3)$$

Equation (5.3) implies that the coordinates of a point on the curve are the components of its position vector. It is a one-to-one mapping from the parametric space (euclidean space $E^1$ in $u$ values) to the cartesian space ($E^3$ in $x$, $y$, $z$ values), as shown in Fig. 5-12. The parametric curve is bounded by two parametric values $u_{min}$ and $u_{max}$. It is, however, convenient to normalize the parametric variable $u$ to have the limits 0 and 1. The positive sense on the curve is the sense in which $u$ increases (Fig. 5-12).

The parametric form as given by Eq. (5.3) facilitates many of the useful related computations in geometric modeling. To check whether a given point lies on the curve or not reduces to finding the corresponding $u$ values and checking whether that value lies in the stated $u$ range. Points on the curve can be computed by substituting the proper parametric values into Eq. (5.3). Geometrical transformations (as discussed in Chap. 9) can be performed directly on parametric equations. Parametric geometry can be easily expressed in terms of vectors and matrices which enables the use of simple computation techniques to solve complex analytical geometry problems. In addition, common forms for curves which are extendable to surfaces can be found. For example, a cubic polynomial, and a cubic polynomial for surfaces, can describe a three-dimensional curve sufficiently in space. Furthermore, curves defined by Eq. (5.3) are inherently bounded and, therefore, no additional geometric data is needed to define boundaries. Lastly, the parametric form is better suited for display by the special graphics hardware. Numerical values of coordinates of points on a curve can control the deflection of the electron beam of a graphics display, as discussed in Chap. 2.

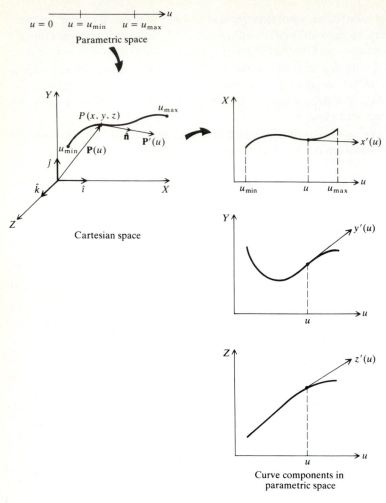

**FIGURE 5-12**
Parametric representation of a three-dimensional curve.

To evaluate the slope of a parametric curve at an arbitrary point on it, the concept of the tangent vector must be introduced. As shown in Fig. 5-12, the tangent vector is defined as vector $\mathbf{P}'(u)$ in the cartesian space such that

$$\mathbf{P}'(u) = \frac{d\mathbf{P}(u)}{du} \tag{5.4}$$

Substituting Eq. (5.3) into Eq. (5.4) yields the components of the tangent vector in the parametric space as

$$\mathbf{P}'(u) = [x' \quad y' \quad z']^T = [x'(u) \quad y'(u) \quad z'(u)]^T, \qquad u_{min} \le u \le u_{max} \tag{5.5}$$

where $x'(u)$, $y'(u)$, and $z'(u)$ are the first parametric derivatives (with respect to $u$) of the position vector components $x(u)$, $y(u)$, and $z(u)$ respectively. The slopes of the curve are given by the ratios of the components of the tangent vector:

$$\frac{dy}{dx} = \frac{dy/du}{dx/du} = \frac{y'}{x'}$$

(5.6)

$$\frac{dz}{dy} = \frac{z'}{y'} \quad \text{and} \quad \frac{dx}{dz} = \frac{x'}{y'}$$

The tangent vector has the same direction as the tangent to the curve—hence the name "tangent vector." The magnitude of the vector is given by

$$|\mathbf{P}'(u)| = \sqrt{x'^2 + y'^2 + z'^2}$$

(5.7)

and the direction cosines of the vector are given by

$$\hat{\mathbf{n}} = \frac{\mathbf{P}'(u)}{|\mathbf{P}'(u)|} = n_x \hat{\mathbf{i}} + n_y \hat{\mathbf{j}} + n_z \hat{\mathbf{k}}$$

(5.8)

where $\hat{\mathbf{n}}$ is the unit vector (Fig. 5-12) with cartesian space components $n_x$, $n_y$, and $n_z$. The magnitude of the tangent vectors at the two ends of a curve affects its shape and can be used to control it, as will be seen later.

There are two categories of curves that can be represented parametrically: analytic and synthetic. Analytic curves are defined as those that can be described by analytic equations such as lines, circles, and conics. Synthetic curves are the ones that are described by a set of data points (control points) such as splines and Bezier curves. Parametric polynomials usually fit the control points. While analytic curves provide very compact forms to represent shapes and simplify the computation of related properties such as areas and volumes, they are not attractive to deal with interactively. Alternatively, synthetic curves provide designers with greater flexibility and control of a curve shape by changing the positions of the control points. Global as well as local control of the shape can be obtained as discussed in Sec. 5.6.

**Example 5.4.** The nonparametric implicit equation of a circle with a center at the origin and radius $R$ is given by $x^2 + y^2 = R^2$. Find the circle parametric equation. Using the resulting equation, find the slopes at the angles 0, 45, and 90°.

*Solution.* In general, the parametric equation of a curve is not unique and can take various forms. For a circle, let

$$x = R \cos 2\pi u, \quad 0 \le u \le 1$$

Substituting into the circle equation gives $y = R \sin 2\pi u$ and the parametric equation of the circle becomes

$$\mathbf{P}(u) = [R \cos 2\pi u \quad R \sin 2\pi u]^T, \quad 0 \le u \le 1$$

and the tangent vector is

$$\mathbf{P}'(u) = [-2\pi R \sin 2\pi u \quad 2\pi R \cos 2\pi u]^T, \quad 0 \le u \le 1$$

The parameter $u$ takes the values 0, 0.125, and 0.25 at the angles 0, 45, and 90° respectively, and the corresponding tangent vectors are given by

$$\mathbf{P}'(0) = [0 \quad 2\pi R]^T$$

$$\mathbf{P}'(0.125) = [-2\pi R/\sqrt{2} \quad 2\pi R/\sqrt{2}]^T$$

$$\mathbf{P}'(0.25) = [-2\pi R \quad 0]^T$$

For a two-dimensional curve, the slope is given by $y'/x'$. The slope at the given angles are calculated as $\infty$, $-1$, and 0.

**Example 5.5.** The parametric equation of the helix shown in Fig. 5-13 with a radius $a$ and a pitch $b$ is given by

$$\mathbf{P}(u) = [a \cos 2\pi u \quad a \sin 2\pi u \quad 2b\pi u]^T, \qquad 0 \le u \le 1$$

Find the nonparametric equation of the helix.

*Solution.* The helix equation gives

$$x = a \cos 2\pi u \qquad y = a \sin 2\pi u \qquad z = 2b\pi u$$

Solving the first equation for $u$ and substituting in the other two gives the following explicit nonparametric equation [compare with Eq. (5.1)]:

$$\mathbf{P} = \left[ x \quad a \sin\left( \cos^{-1}\left(\frac{x}{a}\right) \right) \quad b \cos^{-1}\left(\frac{x}{a}\right) \right]^T$$

Solving the Z component for $u$ and substituting in the $x$ and $y$ equations gives the implicit nonparametric equation as [compare with Eq. (5.2)]

$$x - a \cos\left(\frac{z}{b}\right) = 0$$

$$y - a \sin\left(\frac{z}{b}\right) = 0$$

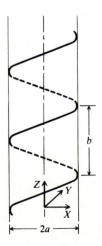

FIGURE 5-13
A helix curve.

## 5.5 PARAMETRIC REPRESENTATION OF ANALYTIC CURVES

This section covers the basics of the parametric equations of analytic curves that are most widely utilized in wireframe modeling. These developments are related, whenever possible, to the common practice encountered on CAD/CAM systems. This should enable users to fully realize the input parameters they deal with and usually find in system documentation. It should also help developers who may be interested in writing their own CAD/CAM software.

Parametric equations and their developments are presented in vector form. The benefits of the vector form include developing a unified approach and consistent notation to treat both two-dimensional and three-dimensional curves in addition to yielding concise equations that are more convenient to program. The following section provides a quick review of the most relevant vector algebra and analysis needed here. Additional equations can be found in standard linear algebra textbooks.

### 5.5.1 Review of Vector Algebra

Let $\mathbf{A}$, $\mathbf{B}$, and $\mathbf{C}$ be independent vectors, $\hat{\mathbf{i}}$, $\hat{\mathbf{j}}$, and $\hat{\mathbf{k}}$ be unit vectors in the $X$, $Y$, and $Z$ directions respectively, and $K$ be a constant.

**1.** Magnitude of a vector is

$$|\mathbf{A}| = \sqrt{A_x^2 + A_y^2 + A_z^2}$$

where $A_x$, $A_y$, and $A_z$ are the cartesian components of the vector $\mathbf{A}$.

**2.** The unit vector in the direction of $\mathbf{A}$ is

$$\hat{\mathbf{n}}_A = \frac{\mathbf{A}}{|\mathbf{A}|} = n_{Ax}\hat{\mathbf{i}} + n_{Ay}\hat{\mathbf{j}} + n_{Az}\hat{\mathbf{k}}$$

The components of $\hat{\mathbf{n}}_A$ are also the direction cosines of the vector $\mathbf{A}$.

**3.** If two vectors $\mathbf{A}$ and $\mathbf{B}$ are equal, then

$$A_x = B_x \qquad A_y = B_y \qquad \text{and} \qquad A_z = B_z$$

**4.** The scalar (dot or inner) product of two vectors $\mathbf{A}$ and $\mathbf{B}$ is a scalar value given by

$$\mathbf{A} \cdot \mathbf{B} = \mathbf{B} \cdot \mathbf{A} = A_x B_x + A_y B_y + A_z B_z = |\mathbf{A}||\mathbf{B}| \cos \theta$$

where $\theta$ is the angle between $\mathbf{A}$ and $\mathbf{B}$. Therefore the angle $\theta$ between two vectors is given by

$$\cos \theta = \frac{\mathbf{A} \cdot \mathbf{B}}{|\mathbf{A}||\mathbf{B}|}$$

The scalar product can give the component of a vector $\mathbf{A}$ in the direction of another vector $\mathbf{B}$ as

$$\mathbf{A} \cdot \hat{\mathbf{n}}_B = |\mathbf{A}| \cos \theta$$

Other properties of the scalar product are:

$$\mathbf{A} \cdot \mathbf{A} = |\mathbf{A}|^2$$

$$\mathbf{A} \cdot \mathbf{B} = \mathbf{B} \cdot \mathbf{A}$$

$$\mathbf{A} \cdot (\mathbf{B} + \mathbf{C}) = \mathbf{A} \cdot \mathbf{B} + \mathbf{A} \cdot \mathbf{C}$$

$$(K\mathbf{A}) \cdot \mathbf{B} = \mathbf{A} \cdot (K\mathbf{B}) = K(\mathbf{A} \cdot \mathbf{B})$$

$$\hat{\mathbf{i}} \cdot \hat{\mathbf{i}} = \hat{\mathbf{j}} \cdot \hat{\mathbf{j}} = \hat{\mathbf{k}} \cdot \hat{\mathbf{k}} = 1 \qquad \hat{\mathbf{i}} \cdot \hat{\mathbf{j}} = \hat{\mathbf{j}} \cdot \hat{\mathbf{k}} = \hat{\mathbf{k}} \cdot \hat{\mathbf{i}} = 0$$

5. The vector (cross) product of two vectors $\mathbf{A}$ and $\mathbf{B}$ is a vector perpendicular to the plane formed by $\mathbf{A}$ and $\mathbf{B}$ and is given by

$$\mathbf{A} \times \mathbf{B} = \begin{vmatrix} \hat{\mathbf{i}} & \hat{\mathbf{j}} & \hat{\mathbf{k}} \\ A_x & A_y & A_z \\ B_x & B_y & B_z \end{vmatrix}$$

$$= (A_y B_z - A_z B_y)\hat{\mathbf{i}} + (A_z B_x - A_x B_z)\hat{\mathbf{j}} + (A_x B_y - A_y B_x)\hat{\mathbf{k}}$$

$$= (|\mathbf{A}| |\mathbf{B}| \sin \theta)\hat{\mathbf{I}}$$

where $\hat{\mathbf{I}}$ is a unit vector in a direction perpendicular to the plane of $\mathbf{A}$ and $\mathbf{B}$ and a sense determined by the advancement of the tip of a right-hand screw when it is rotated from $\mathbf{A}$ to $\mathbf{B}$ (the right-hand rule). Utilizing both the scalar and vector products, the angle $\theta$ between two vectors can be written as

$$\tan \theta = \frac{|\mathbf{A} \times \mathbf{B}|}{\mathbf{A} \cdot \mathbf{B}}$$

The vector product can give the component of a vector $\mathbf{A}$ in a direction perpendicular to another vector $\mathbf{B}$ as

$$|\mathbf{A} \times \hat{\mathbf{n}}_B| = |\mathbf{A}| \sin \theta$$

Other properties of the vector product are:

$$\mathbf{A} \times \mathbf{B} = -\mathbf{B} \times \mathbf{A}$$

$$|\mathbf{A} \times \mathbf{B}|^2 = |\mathbf{A}|^2 |\mathbf{B}|^2 \sin^2 \theta$$

$$\mathbf{A} \times (\mathbf{B} + \mathbf{C}) = \mathbf{A} \times \mathbf{B} + \mathbf{A} \times \mathbf{C}$$

$$(K\mathbf{A}) \times \mathbf{B} = \mathbf{A} \times (K\mathbf{B}) = K(\mathbf{A} \times \mathbf{B})$$

$$\hat{\mathbf{i}} \times \hat{\mathbf{i}} = \hat{\mathbf{j}} \times \hat{\mathbf{j}} = \hat{\mathbf{k}} \times \hat{\mathbf{k}} = 0$$

$$\hat{\mathbf{i}} \times \hat{\mathbf{j}} = \hat{\mathbf{k}}$$

$$\hat{\mathbf{j}} \times \hat{\mathbf{k}} = \hat{\mathbf{i}}$$

$$\hat{\mathbf{k}} \times \hat{\mathbf{i}} = \hat{\mathbf{j}}$$

$$\mathbf{A} \times (\mathbf{B} \times \mathbf{C}) = \mathbf{B}(\mathbf{C} \cdot \mathbf{A}) - \mathbf{C}(\mathbf{A} \cdot \mathbf{B}) \qquad \text{(vector triple product)}$$

$$\mathbf{A} \cdot (\mathbf{B} \times \mathbf{C}) = \mathbf{B} \cdot (\mathbf{C} \times \mathbf{A}) = \mathbf{C} \cdot (\mathbf{A} \times \mathbf{B})$$

$$= \begin{vmatrix} A_x & A_y & A_z \\ B_x & B_y & B_z \\ C_x & C_y & C_z \end{vmatrix} \qquad \text{(scalar triple product)}$$

**6.** Two vectors **A** and **B** are parallel if and only if

$$\hat{\mathbf{n}}_A \cdot \hat{\mathbf{n}}_B = 1 \qquad \text{or} \qquad |\hat{\mathbf{n}}_A \times \hat{\mathbf{n}}_B| = 0$$

**7.** Two vectors **A** and **B** are perpendicular if and only if

$$\hat{\mathbf{n}}_A \cdot \hat{\mathbf{n}}_B = 0 \qquad \text{or} \qquad |\hat{\mathbf{n}}_A \times \hat{\mathbf{n}}_B| = 1$$

The conditions for parallelism and perpendicularity are written as shown above to reflect the expected skew symmetry of the two properties. For calculation purposes, it might be useful to replace $|\hat{\mathbf{n}}_A \times \hat{\mathbf{n}}_B| = 0$ by $\hat{\mathbf{n}}_A \times \hat{\mathbf{n}}_B = \mathbf{0}$ or $\mathbf{A} \times \mathbf{B} = \mathbf{0}$ for parallel lines and replace $\hat{\mathbf{n}}_A \cdot \hat{\mathbf{n}}_B = 0$ by $\mathbf{A} \cdot \mathbf{B} = 0$ for perpendicular lines.

### 5.5.2 Lines

Basic vector parametric equations of straight lines are derived here with two questions in mind. First, how is a line equation converted by the CAD/CAM software into the line database which is at a minimum at the two endpoints of the line? Second, how are the mathematical requirements of an equation correlated with various modifiers available with line commands offered by common user interfaces? Consider the following two cases:

**1.** A line connecting two points $P_1$ and $P_2$, as shown in Fig. 5-14. Define a parameter $u$ such that it has the values 0 and 1 at $P_1$ and $P_2$ respectively. Utilizing the triangle $OPP_1$, the following equation can be written:

$$\mathbf{P} = \mathbf{P}_1 + (\mathbf{P} - \mathbf{P}_1) \tag{5.9}$$

However, the vector $(\mathbf{P} - \mathbf{P}_1)$ is proportional to the vector $\mathbf{P}_2 - \mathbf{P}_1$ such that

$$\mathbf{P} - \mathbf{P}_1 = u(\mathbf{P}_2 - \mathbf{P}_1) \tag{5.10}$$

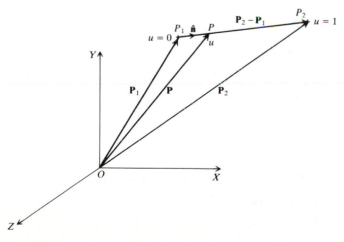

**FIGURE 5-14**
Line connecting two points $P_1$ and $P_2$.

Thus the equation of the line becomes

$$\mathbf{P} = \mathbf{P}_1 + u(\mathbf{P}_2 - \mathbf{P}_1), \qquad 0 \le u \le 1 \tag{5.11}$$

In scalar form, this equation can be written as

$$\left.\begin{array}{l} x = x_1 + u(x_2 - x_1) \\ y = y_1 + u(y_2 - y_1) \\ z = z_1 + u(z_2 - z_1) \end{array}\right\} \qquad 0 \le u \le 1 \tag{5.12}$$

Equation (5.11) defines a line bounded by the endpoints $P_1$ and $P_2$ whose associated parametric values are 0 and 1 respectively. Any other point on the line or its extension has a certain value of $u$ which is proportional to the point location, as Fig. 5-15 shows. The coordinates of any point in the figure are obtained by substituting the corresponding $u$ value in Eq. (5.11).

The tangent vector of the line is given by

$$\mathbf{P}' = \mathbf{P}_2 - \mathbf{P}_1 \tag{5.13}$$

or, in scalar form,

$$\begin{array}{l} x' = x_2 - x_1 \\ y' = y_2 - y_1 \\ z' = z_2 - z_1 \end{array} \tag{5.14}$$

The independence of the tangent vector from $u$ reflects the constant slope of the straight line. For a two-dimensional line, the known infinite (vertical line) and zero (horizontal line) slope conditions can be generated from Eq. (5.14).

The unit vector $\hat{\mathbf{n}}$ in the direction of the line (Fig. 5-14) is given by

$$\hat{\mathbf{n}} = \frac{\mathbf{P}_2 - \mathbf{P}_1}{L} \tag{5.15}$$

where $L$ is the length of the line:

$$L = |\mathbf{P}_2 - \mathbf{P}_1| = \sqrt{(x_2 - x_1)^2 + (y_2 - y_1)^2 + (z_2 - z_1)^2} \tag{5.16}$$

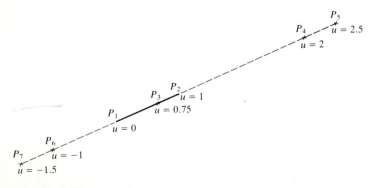

**FIGURE 5-15**
Locating points on an existing line.

Regardless of the user input to create a line, a line database stores its two endpoints and additional information such as its font, width, color, and layer. Equations (5.11) and (5.13) show that the endpoints are enough to provide all geometric properties and characteristics of the line. They are also sufficient to construct and display the line. For reference purposes, CAD/CAM software usually identifies the first point input by the user during line construction as $P_1$, where $u = 0$. These two equations can be programmed into a subroutine that can reside in a graphics library of the software and which can be invoked, via the user interface, to construct lines. Point commands (or definitions) on most systems provide users with a modifier to specify a $u$ value relative to an entity to generate points on it. In the case of a line, the value is substituted into Eq. (5.11) to find the point coordinates.

2. A line passing through a point $P_1$ in a direction defined by the unit vector $\hat{n}$ (Fig. 5-16). Case 1 is considered the basic method to create a line because it provides the line database directly with the two endpoints. This case and others usually result in generating the endpoints from the user input or given data, as discussed below.

To develop the line equation for this case, consider a general point $P$ on the line at a distance $L$ from $P_1$. The vector equation of the line becomes (see triangle $OP_1P$)

$$\mathbf{P} = \mathbf{P}_1 + L\hat{n}, \qquad -\infty \le L \le \infty \qquad (5.17)$$

and $L$ is given by

$$L = |\mathbf{P} - \mathbf{P}_1| \qquad (5.18)$$

$L$ is the parameter in Eq. (5.17). Thus, the tangent vector is $\hat{n}$.

Once the user inputs $\hat{\mathbf{P}}_1$, $\hat{n}$, and $L$, the point $P$ is calculated using Eq. (5.17), and the line has the two endpoints $P_1$ and $P$ with $u$ values of 0 and 1 as discussed in case 1.

The following examples show how parametric equations of various line forms can be developed. The examples relate to the most common line commands offered by CAD/CAM software.

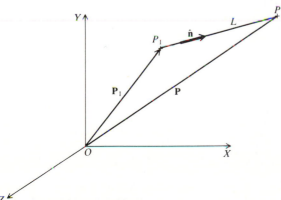

**FIGURE 5-16**
Line passing through $P_1$ in direction **n**.

**Example 5.6.** Find the equations and endpoints of two lines, one horizontal and the other vertical. Each line begins at and passes through a given point and is clipped by another given point.

*Solution.* Horizontal and vertical lines are usually defined in reference to the current WCS axes. Horizontal lines are parallel to the $X$ axis and vertical lines are parallel to the $Y$ axis. Figure 5-17 shows a typical user working environment where the WCS has a different orientation from the MCS. In this case, the WCS is equivalent to the coordinate system used to develop the line equations. Once the endpoints are calculated from these equations with respect to the WCS, they are transformed to the MCS before the line display or storage.

Assume that three points $P_1$, $P_2$, and $P_3$ are given. The vertical line passes through $P_1$ and ends at $P_2$ while the horizontal line passes through $P_1$ and ends at $P_3$. In general, the two lines cannot pass through points $P_2$ or $P_3$. Therefore, the ends are determined by projecting the points onto the lines as shown. Using Eq. (5.17), the line equations are

Vertical: $$\mathbf{P} = \mathbf{P}_1 + L\hat{\mathbf{n}}_1, \qquad 0 \le L \le L_1$$

or

$$x_W = x_{1W}$$

$$y_W = y_{1W} + L$$

$$z_W = z_{1W}$$

where

$$L_1 = y_{2W} - y_{1W}$$

and the endpoints are $(x_{1W}, y_{1W}, z_{1W})$ and $(x_{1W}, y_{1W} + L_1, z_{1W})$

Horizontal: $$\mathbf{P} = \mathbf{P}_1 + L\hat{\mathbf{n}}_2, \qquad 0 \le L \le L_2$$

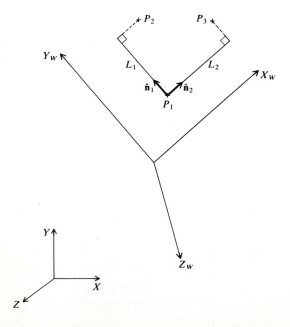

**FIGURE 5-17**
Horizontal and vertical lines.

or
$$x_W = x_{1W} + L$$
$$y_W = y_{1W}$$
$$z_W = z_{1W}$$

where
$$L_2 = x_{3W} - x_{1W}$$

and the endpoints are $(x_{1W}, y_{1W}, z_{1W})$ and $(x_{1W} + L_2, y_{1W}, z_{1W})$.

**Example 5.7.** Find the equation and endpoints of a line that passes through a point $P_1$, parallel to an existing line, and is trimmed by point $P_2$ as shown in Fig. 5-18.

**Solution.** To minimize confusion, the differentiation between the orientations of the WCS and the MCS is ignored and the position vectors of the various points are omitted from the figure. Assume that the existing line has the two endpoints $P_3$ and $P_4$, a length $L_1$, and a direction defined by the unit vector $\hat{n}_1$. The new line has the same direction $\hat{n}_1$, a length $L_2$, and endpoints $P_1$ and $P_5$. $P_5$ is the projection of $P_2$ onto the line.

The equation of the new line is found by substituting the proper vectors into Eq. (5.17). The unit vector $\hat{n}_1$ is given by

$$\hat{n}_1 = \frac{\mathbf{P}_4 - \mathbf{P}_3}{L_1}$$

where $L_1$ is a known value [Eq. (5.16)]. $L_2$ can be found from the following equation:

$$L_2 = \hat{n}_1 \cdot (\mathbf{P}_2 - \mathbf{P}_1)$$

$$= \frac{\mathbf{P}_4 - \mathbf{P}_3}{L_1} \cdot (\mathbf{P}_2 - \mathbf{P}_1)$$

$$= \frac{(x_4 - x_3)(x_2 - x_1) + (y_4 - y_3)(y_2 - y_1) + (z_4 - z_3)(z_2 - z_1)}{\sqrt{(x_4 - x_3)^2 + (y_4 - y_3)^2 + (z_4 - z_3)^2}}$$

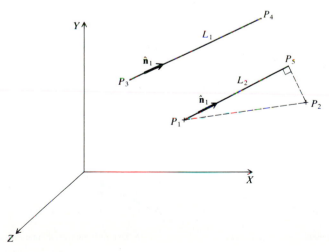

**FIGURE 5-18**
Line parallel to an existing line.

The equation of the new line becomes

$$\mathbf{P} = \mathbf{P}_1 + L\hat{\mathbf{n}}_1, \qquad 0 \le L \le L_2$$

and its two endpoints are $P_1(x_1, y_1, z_1)$ and $P_5(x_1 + L_2 n_{1x}, y_1 + L_2 n_{1y}, z_1 + L_2 n_{1z})$. As numerical examples, consider the simple two-dimensional case of constructing lines parallel to existing horizontal and vertical ones. Verify the above equations for the horizontal lines case $P_1(3, 2)$, $P_2(7, 4)$, $P_3(2, 3)$, and $P_4(5, 3)$. Repeat for the vertical lines case $P_1(8, 9)$, $P_2(9, 2)$, $P_3(5, 3)$, and $P_4(5, 9)$. Readers can also utilize the corresponding command to this example to construct the line on their CAD/CAM systems, verify the line to obtain its length and endpoints, and then compare with the results obtained here.

**Example 5.8.** Relate the following CAD/CAM commands to their mathematical foundations:

(a) The command that measures the angle between two intersecting lines.
(b) The command that finds the distance between a point and a line.

*Solution*

(a) If the endpoints of the two lines are $P_1$, $P_2$ and $P_3$, $P_4$ (Fig. 5-19a), the angle measurement command uses the equation

$$\cos \theta = \frac{(\mathbf{P}_2 - \mathbf{P}_1) \cdot (\mathbf{P}_4 \cdot \mathbf{P}_3)}{|\mathbf{P}_2 - \mathbf{P}_1||\mathbf{P}_4 - \mathbf{P}_3|}$$

$$= \frac{(x_2 - x_1)(x_4 - x_3) + (y_2 - y_1)(y_4 - y_3) + (z_2 - z_1)(z_4 - z_3)}{\sqrt{[(x_2 - x_1)^2 + (y_2 - y_1)^2 + (z_2 - z_1)^2][(x_4 - x_3)^2 + (y_4 - y_3)^2 + (z_4 - z_3)^2]}}$$

(b) Figure 5-19b shows the distance $D$ from point $P_3$ to the line whose endpoints are $P_1$ and $P_2$ and direction is $\hat{\mathbf{n}}$. Its length is $L$. $D$ is given by

$$D = |(\mathbf{P}_3 - \mathbf{P}_1) \times \hat{\mathbf{n}}| = \left|(\mathbf{P}_3 - \mathbf{P}_1) \times \frac{(\mathbf{P}_2 - \mathbf{P}_1)}{|\mathbf{P}_2 - \mathbf{P}_1|}\right| = \begin{Vmatrix} \hat{\mathbf{i}} & \hat{\mathbf{j}} & \hat{\mathbf{k}} \\ x_3 - x_1 & y_3 - y_1 & z_3 - z_1 \\ \dfrac{x_2 - x_1}{L} & \dfrac{y_2 - y_1}{L} & \dfrac{z_2 - z_1}{L} \end{Vmatrix}$$

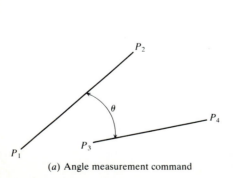

(a) Angle measurement command

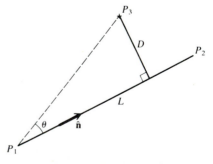

(b) Distance measurement command

**FIGURE 5-19**
Data related to line commands.

(where, on the furtherest right-hand side, the inner vertical bars indicate the determinant and the outer vertical bars indicate the magnitude of the resulting vector) or

$$D = \frac{1}{L} \text{SQRT}\{[(y_3 - y_1)(z_2 - z_1) - (z_3 - z_1)(y_2 - y_1)]^2$$

$$+ [(z_3 - z_1)(x_2 - x_1) - (x_3 - x_1)(z_2 - z_1)]^2$$

$$+ [(x_3 - x_1)(y_2 - y_1) - (y_3 - y_1)(x_2 - x_1)]^2\}$$

where SQRT is the square root. For a two-dimensional case, this equation reduces to

$$D = \frac{1}{L} [(x_3 - x_1)(y_2 - y_1) - (y_3 - y_1)(x_2 - x_1)]$$

**Example 5.9.** Find the unit tangent vector in the direction of a line:

(a) Parallel to an existing line.
(b) Perpendicular to an existing line.

*Solution.* The conditions of parallelism or perpendicularity of two lines given in the vector algebra reviewed in Sec. 5.5.1 are useful if the vector equations of the two lines exist. If one equation is not available, which is mostly the case in practical problems, the conditions should be reduced to find the unit tangent vector of the missing line in terms of the existing one. Figure 5-20 shows the existing line as $L_1$ with a known unit tangent vector $\hat{n}_1$. The unit tangent vector $\hat{n}_2$ is to be found in terms of $\hat{n}_1$.

(a) For $L_1$ and $L_2$ to be parallel (Fig. 5-20a),

$$\hat{n}_2 = \hat{n}_1$$

or
$$[n_{2x} \quad n_{2y} \quad n_{2z}]^T = [n_{1x} \quad n_{1y} \quad n_{1z}]^T$$

This equation defines an infinite number of lines in an infinite number of planes in space. Additional geometric conditions are required to define a specific line. This equation is equivalent to the condition of the equality of the two slopes of

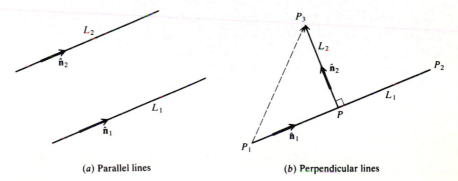

(a) Parallel lines                    (b) Perpendicular lines

**FIGURE 5-20**
Unit tangent vectors of various lines.

the two lines in the two-dimensional case. Neglecting the $Z$ component, the above equation gives the condition for this case as

$$[n_{2x} \quad n_{2y}]^T = [n_{1x} \quad n_{1y}]^T$$

or

$$\frac{n_{2y}}{n_{2x}} = \frac{n_{1y}}{n_{1x}}$$

or

$$m_2 = m_1$$

where $m_1$ and $m_2$ are the slopes of the lines.

(b) For $L_1$ and $L_2$ to be perpendicular (Fig. 5-20b),

$$\hat{\mathbf{n}}_2 \cdot \hat{\mathbf{n}}_1 = 0$$

or

$$n_{1x} n_{2x} + n_{1y} n_{2y} + n_{1z} n_{2z} = 0 \qquad (5.19)$$

Additional equations are needed to solve for $\hat{\mathbf{n}}_2$. The following two equations can be written:

$$|\hat{\mathbf{n}}_1 \times \hat{\mathbf{n}}_2| = 1$$

or

$$(n_{1y} n_{2z} - n_{1z} n_{2y})^2 + (n_{1z} n_{2x} - n_{1x} n_{2z})^2 + (n_{1x} n_{2y} - n_{1y} n_{2x})^2 = 1 \quad (5.20)$$

and

$$\hat{\mathbf{n}}_2 \cdot \hat{\mathbf{n}}_2 = |\hat{\mathbf{n}}_2|^2$$

or

$$n_{2x}^2 + n_{2y}^2 + n_{2z}^2 = 1 \qquad (5.21)$$

However, only two equations out of the above three are independent. For example, squaring Eq. (5.19) and adding it to Eq. (5.20) results in Eq. (5.21) if the identity $n_{1x}^2 + n_{1y}^2 + n_{1z}^2 = 1$ is used. Therefore, only two equations are available to solve for $n_{2x}$, $n_{2y}$, and $n_{2z}$, which implies that only two of these components can be obtained as functions of the third. This situation results from the fact that the perpendicular line $L_2$ to $L_1$ at point $P$ shown in Fig. 5-20 is not unique. A plane perpendicular to $L_1$ at $P$ defines the locus of $L_2$.

As a more defined case, assume that point $P_3$ is known and $L_2$ is the perpendicular line from $P_3$ to $L_1$. Using the above equations to solve for $\hat{\mathbf{n}}_2$ is usually cumbersome. Instead, utilizing the triangle $P_1 P P_3$, the following equation can be written:

$$\mathbf{P}_1\mathbf{P}_3 = \mathbf{P}_1\mathbf{P} + \mathbf{P}\mathbf{P}_3$$

or

$$\mathbf{P}_3 - \mathbf{P}_1 = [(\mathbf{P}_3 - \mathbf{P}_1) \cdot \hat{\mathbf{n}}_1]\hat{\mathbf{n}}_1 + D\hat{\mathbf{n}}_2 \qquad (5.22)$$

where $D$ is the perpendicular distance between $P_3$ and $L_1$ and is given in the previous example. Equation (5.22) gives $\hat{\mathbf{n}}_2$ as

$$\hat{\mathbf{n}}_2 = \frac{1}{D} \{\mathbf{P}_3 - \mathbf{P}_1 - [(\mathbf{P}_3 - \mathbf{P}_1) \cdot \hat{\mathbf{n}}_1]\hat{\mathbf{n}}_1\}$$

In scalar form,

$$n_{2x} = \frac{1}{D} [(x_3 - x_1)(1 - n_{1x}^2) - (y_3 - y_1)n_{1x} n_{1y} - (z_3 - z_1)n_{1x} n_{1z}]$$

$$n_{2y} = \frac{1}{D} [(y_3 - y_1)(1 - n_{1y}^2) - (x_3 - x_1)n_{1x} n_{1y} - (z_3 - z_1)n_{1y} n_{1z}]$$

$$n_{2z} = \frac{1}{D} [(z_3 - z_1)(1 - n_{1z}^2) - (y_3 - y_1)n_{1y} n_{1z} - (x_3 - x_1)n_{1x} n_{1z}]$$

For the two-dimensional case, Eq. (5.22) becomes

$$x_3 - x_1 = [(x_3 - x_1)n_{1x} + (y_3 - y_1)n_{1y}]n_{1x} + Dn_{2x}$$

$$y_3 - y_1 = [(x_3 - x_1)n_{1x} + (y_3 - y_1)n_{1y}]n_{1y} + Dn_{2y}$$

Multiplying the first and second equations by $n_{1x}$ and $n_{1y}$ respectively and adding them, utilizing the identity $n_{1x}^2 + n_{1y}^2 = 1$, we get

$$D(n_{2x}n_{1x} + n_{2y}n_{1y}) = 0$$

or

$$\frac{n_{2y}}{n_{2x}} = -\frac{n_{1x}}{n_{1y}}$$

or

$$\frac{n_{2y}}{n_{2x}}\frac{n_{1y}}{n_{1x}} = -1$$

or

$$m_1 m_2 = -1$$

which is a known result from two-dimensional analytic geometry. In the two-dimensional case $\hat{n}_2$ can be chosen such that

$$n_{2x} = n_{1y} \qquad n_{2y} = -n_{1x}$$

or

$$n_{2x} = -n_{1y} \qquad n_{2y} = n_{1x}$$

*Note:* it is left to the reader to show that Eq. (5.22) can be reduced to the condition $\hat{n}_1 \cdot \hat{n}_2 = 0$.

Other useful line cases are assigned as problems at the end of the chapter.

### 5.5.3  Circles

Circles and circular arcs are among the most common entities used in wireframe modeling. Circles and circular arcs together with straight lines are sufficient to construct a large percentage of existing mechanical parts and components in practice. Besides other information, a circle database stores its radius and center as its essential geometric data. If the plane of the circle cannot be defined from the user input data as in the case of specifying a center and a radius, it is typically assumed by the software to be the $XY$ plane of the current WCS at the time of construction. Regardless of the user input information to create a circle, such information is always converted into a radius and center by the software. This section presents some cases to show how such conversion is possible. Other cases are left to the reader as problems at the end of the chapter.

The basic parametric equation of a circle can be written as (refer to Fig. 5-21)

$$\left.\begin{aligned} x &= x_c + R \cos u \\ y &= y_c + R \sin u \\ z &= z_c \end{aligned}\right\} \qquad 0 \le u \le 2\pi \qquad (5.23)$$

assuming that the plane of the circle is the $XY$ plane for simplicity. In this equation, the parameter $u$ is the angle measured from the $X$ axis to any point $P$ on the circle. This parameter is used by commercial software to locate points at certain

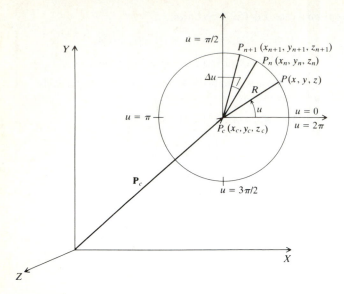

**FIGURE 5-21**
Circle defined by a radius and center.

angles on the circle for construction purposes. Some certain values of $u$ are shown in Fig. 5-21.

For display purposes, Eq. (5.23) can be used to generate points on the circle circumference by incrementing $u$ from 0 to 360. These points are in turn connected with line segments to display the circle. However, this is an inefficient way due to computing the trigonometric functions in the equation for each point. A less computational method is to write Eq. (5.23) in an incremental form. Assuming there is an increment $\Delta u$ between two consecutive points $P(x_n, y_n, z_n)$ and $P(x_{n+1}, y_{n+1}, z_{n+1})$ on the circle circumference, the following recursive relationship can be written:

$$x_n = x_c + R \cos u$$
$$y_n = y_c + R \sin u$$
$$x_{n+1} = x_c + R \cos (u + \Delta u) \qquad (5.24)$$
$$y_{n+1} = y_c + R \sin (u + \Delta u)$$
$$z_{n+1} = z_n$$

Expanding the $x_{n+1}$ and $y_{n+1}$ equation gives

$$x_{n+1} = x_c + (x_n - x_c) \cos \Delta u - (y_n - y_c) \sin \Delta u$$
$$y_{n+1} = y_c + (y_n - y_c) \cos \Delta u + (x_n - x_c) \sin \Delta u \qquad (5.25)$$
$$z_{n+1} = z_n$$

Thus, the circle can start from an arbitrary point and successive points with equal spacing can be calculated recursively. Cos $\Delta u$ and sin $\Delta u$ have to be calculated only once, which eliminates computation of trigonometric functions for each point. This algorithm is useful for hardware implementation to speed up the circle generation and display.

Circular arcs are considered a special case of circles. Therefore, all discussions covered here regarding circles can easily be extended to arcs. A circular arc equation can be written as

$$
\left.
\begin{aligned}
x &= x_c + R \cos u \\
y &= y_c + R \sin u \\
z &= z_c
\end{aligned}
\right\} \qquad u_s \le u \le u_e \qquad (5.26)
$$

where $u_s$ and $u_e$ are the starting and ending angles of the arc respectively. An arc database includes its center and radius, as a circle, as well as its starting and ending angles. Most user inputs offered by software to create arcs are similar to these offered to create circles. In fact, some software packages do not even offer arcs, in which case users have to create circles and then trim them using the proper trimming boundaries. As indicated from Eq. (5.26) and Fig. 5-21, the arc always connects its beginning and ending points in a counterclockwise direction. This rule is usually the default of most CAD/CAM packages when the user input is not sufficient to determine the arc position in space. For example, Fig. 5-22 shows the two possibilities to create an arc given two input points $P_1$ and $P_2$ to define its diameter. In this case, the arc is obviously half a circle and $P_1$ is always its starting point as it is input first by the user.

Following are some examples that show how various geometric data and constraints, which can be thought of as user inputs required by software packages, can be converted to a radius and center before its storage by the software in the corresponding circle database. They also show that three constraints (points and/or tangent vectors) are required to create a circle except for the obvious case of a center and radius. In all these examples, the reader can verify the resulting equations by comparing their results with those of an accessible CAD/CAM software package.

**Example 5.19.** Find the radius and the center of a circle whose diameter is given by two points.

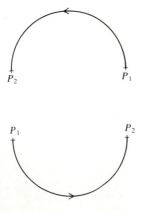

**FIGURE 5-22**
Arc creation following counterclockwise direction.

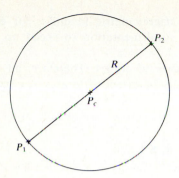

**FIGURE 5-23**
Circle defined by diameter $P_1P_2$.

*Solution.* Assume the circle diameter is given by the two points $P_1$ and $P_2$ as shown in Fig. 5-23. The circle radius and center are

$$R = \tfrac{1}{2}\sqrt{(x_2 - x_1)^2 + (y_2 - y_1)^2 + (z_2 - z_1)^2}$$

$$P_c = \tfrac{1}{2}(P_1 + P_2)$$

or

$$[x_c \quad y_c \quad z_c]^T = \left[\frac{x_1 + x_2}{2} \quad \frac{y_1 + y_2}{2} \quad \frac{z_1 + z_2}{2}\right]^T$$

**Example 5.11.** Find the radius and the center of a circle passing through three points.

*Solution.* Figure 5-24 shows the three given points $P_1$, $P_2$, and $P_3$ that the circle must pass through. The circle radius and center are shown as $P_c$ and $R$ respectively. From analytic geometry, $P_c$ is the intersection of the perpendicular lines to the chords $P_1P_2$, $P_2P_3$, and $P_1P_3$ from their midpoints $P_4$, $P_5$, and $P_6$ respectively. The unit vectors defining the directions of these chords in space are known and given by

$$\hat{n}_1 = \frac{P_2 - P_1}{|P_2 - P_1|} \qquad \hat{n}_2 = \frac{P_3 - P_2}{|P_3 - P_2|} \qquad \hat{n}_3 = \frac{P_1 - P_3}{|P_1 - P_3|}$$

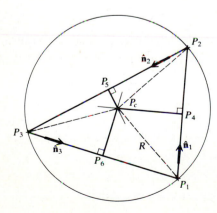

**FIGURE 5-24**
Circle passing through three points.

To find the center of the circle, $P_c$, the following three equations can be written:

$$(\mathbf{P}_c - \mathbf{P}_1) \cdot \hat{\mathbf{n}}_1 = \frac{|\mathbf{P}_2 - \mathbf{P}_1|}{2}$$

$$(\mathbf{P}_c - \mathbf{P}_2) \cdot \hat{\mathbf{n}}_2 = \frac{|\mathbf{P}_3 - \mathbf{P}_2|}{2} \qquad (5.27)$$

$$(\mathbf{P}_c - \mathbf{P}_3) \cdot \hat{\mathbf{n}}_3 = \frac{|\mathbf{P}_1 - \mathbf{P}_3|}{2}$$

Each of these vector equations implies that the component of a vector radius in the direction of any of the three chords is equal to half the chord length. These equations can be used to solve for $P_c$ $(x_c, y_c, z_c)$. Expanding and rearranging the equations in a matrix form yields

$$\begin{bmatrix} n_{1x} & n_{1y} & n_{1z} \\ n_{2x} & n_{2y} & n_{2z} \\ n_{3x} & n_{3y} & n_{3z} \end{bmatrix} \begin{bmatrix} x_c \\ y_c \\ z_c \end{bmatrix} = \begin{bmatrix} b_1 \\ b_2 \\ b_3 \end{bmatrix} \qquad (5.28)$$

where

$$b_1 = \frac{|\mathbf{P}_2 - \mathbf{P}_1|}{2} + (x_1 n_{1x} + y_1 n_{1y} + z_1 n_{1z})$$

$$b_2 = \frac{|\mathbf{P}_3 - \mathbf{P}_2|}{2} + (x_2 n_{2x} + y_2 n_{2y} + z_2 n_{2z})$$

$$b_3 = \frac{|\mathbf{P}_1 - \mathbf{P}_3|}{2} + (x_3 n_{3x} + y_3 n_{3y} + z_3 n_{3z})$$

The matrix equation (5.28) has the form $[A]\mathbf{P}_c = \mathbf{b}$. Therefore,

$$\mathbf{P}_c = [A]^{-1}\mathbf{b} = \frac{\text{Adj} ([A])}{|A|}\mathbf{b}$$

where Adj $([A])$ and $|A|$ are the adjoint matrix and determinant of $[A]$ respectively. Adj $([A])$ is the matrix $[C]^T$ where $[C]$ is the matrix formed by the cofactors $C_{ij}$ of the elements $a_{ij}$ of $[A]$. The cofactor $C_{ij}$ is given by

$$C_{ij} = (-1)^{i+j} M_{ij}$$

where $M_{ij}$ is a unique scalar associated with the element $a_{ij}$ and is defined as the determinant of the $(n-1) \times (n-1)$ matrix obtained from the $n \times n$ matrix $[A]$ by crossing out the $i$th row and $j$th column. Thus

$$\mathbf{P}_c = \frac{[C]^T}{|A|}\mathbf{b}$$

and $\quad |A| = n_{1x}(n_{2y}n_{3z} - n_{2z}n_{3y}) - n_{1y}(n_{2x}n_{3z} - n_{2z}n_{3x}) + n_{1z}(n_{2x}n_{3y} - n_{2y}n_{3x})$

$$[C] = \begin{bmatrix} C_{11} & C_{12} & C_{13} \\ C_{21} & C_{22} & C_{23} \\ C_{31} & C_{32} & C_{33} \end{bmatrix}$$

The elements of $[C]$ are given by

$$C_{11} = n_{2y}n_{3z} - n_{2z}n_{3y} \qquad C_{12} = n_{2z}n_{3x} - n_{2x}n_{3z} \qquad C_{13} = n_{2x}n_{3y} - n_{2y}n_{3x}$$

$$C_{21} = n_{1z}n_{3y} - n_{1y}n_{3z} \qquad C_{22} = n_{1x}n_{3z} - n_{1z}n_{3x} \qquad C_{23} = n_{1y}n_{3x} - n_{1x}n_{3y}$$

$$C_{31} = n_{1y}n_{2z} - n_{1z}n_{2y} \qquad C_{32} = n_{1z}n_{2x} - n_{1x}n_{2z} \qquad C_{33} = n_{1x}n_{2y} - n_{1y}n_{2x}$$

The coordinates of the center can now be written as

$$x_c = \frac{1}{|A|}(C_{11}b_1 + C_{21}b_2 + C_{31}b_3)$$

$$y_c = \frac{1}{|A|}(C_{12}b_1 + C_{22}b_2 + C_{32}b_3)$$

$$z_c = \frac{1}{|A|}(C_{13}b_1 + C_{23}b_2 + C_{33}b_3)$$

The radius $R$ is the distance between $P_c$ and any of the three data points, that is,

$$R = |P_c - P_1| = |P_c - P_2| = |P_c - P_3|$$

or, for example,

$$R = \sqrt{(x_c - x_1)^2 + (y_c - y_1)^2 + (z_c - z_1)^2}$$

For the two-dimensional case, only two of the three relationships shown in Eq. (5.27) are sufficient to find the center $P_c(x_c, y_c)$. The third equation can be used to check the results. Using the first two relationships, the following matrix equations can be written:

$$\begin{bmatrix} n_{1x} & n_{1y} \\ n_{2x} & n_{2y} \end{bmatrix}\begin{bmatrix} x_c \\ y_c \end{bmatrix} = \begin{bmatrix} b_1 \\ b_2 \end{bmatrix}$$

where

$$b_1 = \frac{|P_2 - P_1|}{2} + (x_1 n_{1x} + y_1 n_{1y})$$

$$b_2 = \frac{|P_3 - P_2|}{2} + (x_2 n_{2x} + y_2 n_{2y})$$

Similar to the three-dimensional case, the center is given by

$$x_c = \frac{n_{2y}b_1 - n_{1y}b_2}{n_{1x}n_{2y} - n_{1y}n_{2x}}$$

$$y_c = \frac{n_{1x}b_2 - n_{2x}b_1}{n_{1x}n_{2y} - n_{1y}n_{2x}}$$

**Example 5.12.** Find the center of a circle that is tangent to two known lines with a given radius.

*Solution.* This case is shown in Fig. 5-25. The two existing lines are defined by the point pairs $(P_1, P_2)$ and $(P_3, P_4)$. The unit vectors $\hat{n}_1$ and $\hat{n}_2$ define the directions of

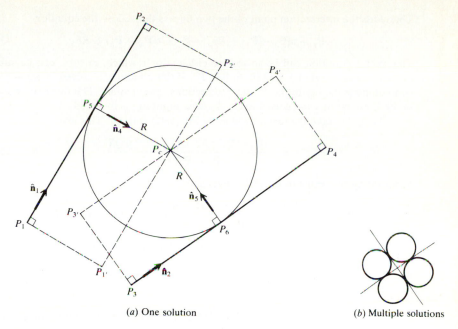

(a) One solution                                    (b) Multiple solutions

**FIGURE 5-25**
Circle tangent to two lines with given radius $R$.

the lines in space. The center of the circle $P_c$ is the intersection of the two normals to
the two lines at the tangency points. The unit vectors $\hat{n}_4$ and $\hat{n}_5$ define the directions
of these two normals in space.

Conceptually, $P_c$ can be found by finding the intersection of the two normals
by solving their two vector equations. However, this route is cumbersome. Instead,
$P_c$ is found as the intersection of the two lines $P_{1'}P_{2'}$ and $P_{3'}P_{4'}$ that are parallel to
$P_1P_2$ and $P_3P_4$ at distance $R$ respectively. The following development is based on
the observation that the two lines $P_1P_2$ and $P_3P_4$ define the plane of the circle and
all the other lines and points shown in Fig. 5-25 lie in this plane.

The unit vectors can be defined as

$$\hat{n}_1 = \frac{P_2 - P_1}{|P_2 - P_1|} \qquad \hat{n}_2 = \frac{P_4 - P_3}{|P_4 - P_3|} \qquad \hat{n}_3 = \frac{\hat{n}_1 \times \hat{n}_2}{|\hat{n}_1 \times \hat{n}_2|}$$

$$\hat{n}_4 = \hat{n}_3 \times \hat{n}_1 \qquad \hat{n}_5 = \hat{n}_2 \times \hat{n}_3$$

where $\hat{n}_3$ is the unit vector perpendicular to the plane of the circle. The endpoints of
the parallel lines are given by

$$P_{1'} = P_1 + R\hat{n}_4 \qquad P_{2'} = P_2 + R\hat{n}_4$$

and $\qquad\qquad\qquad\qquad P_{3'} = P_3 + R\hat{n}_5 \qquad P_{4'} = P_4 + R\hat{n}_5$

Thus, the parametric vector equations of these parallel lines becomes

$$P = P_1 + u(P_2 - P_1) + R\hat{n}_4 \tag{5.29}$$

$$P = P_3 + v(P_4 - P_3) + R\hat{n}_5 \tag{5.30}$$

Therefore, the intersection point of the two lines is defined by the equation

$$\mathbf{P}_1 + u(\mathbf{P}_2 - \mathbf{P}_1) + R\hat{\mathbf{n}}_4 = \mathbf{P}_3 + v(\mathbf{P}_4 - \mathbf{P}_3) + R\hat{\mathbf{n}}_5 \qquad (5.31)$$

This vector equation can be solved for either $u$ or $v$ which, in turn, can be substituted into Eq. (5.29) or Eq. (5.30) to find $P_c$. If this equation is solved in scalar form, two components, say in the $X$ and $Y$ directions, give $u$ and $v$. The third component, in the $Z$ direction, can be used to check the computations involved.

To find $u$, take the scalar product of Eq. (5.31) with $\hat{\mathbf{n}}_5$. This gives

$$u = \frac{(\mathbf{P}_3 - \mathbf{P}_1) \cdot \hat{\mathbf{n}}_5 + (1 - \hat{\mathbf{n}}_4 \cdot \hat{\mathbf{n}}_5)R}{(\mathbf{P}_2 - \mathbf{P}_1) \cdot \hat{\mathbf{n}}_5} \qquad (5.32)$$

Substituting this value in Eq. (5.29) gives

$$\mathbf{P}_c = \mathbf{P}_1 + \left[\frac{(\mathbf{P}_3 - \mathbf{P}_1) \cdot \hat{\mathbf{n}}_5 + (1 - \hat{\mathbf{n}}_4 \cdot \hat{\mathbf{n}}_5)R}{(\mathbf{P}_2 - \mathbf{P}_1) \cdot \hat{\mathbf{n}}_5}\right](\mathbf{P}_2 - \mathbf{P}_1) + R\hat{\mathbf{n}}_4 \qquad (5.33)$$

which can easily be written in scalar form to yield the coordinates $x_c$, $y_c$, and $z_c$ of the centerpoint $P_c$. The two-dimensional case exhibits no special characteristics from the three-dimensional case.

The above development was not concerned with the existence of the four multiple solutions shown in Fig. 5-25b if the two known lines, and not their extensions, intersect. In such a case, the software package often follows a certain convention or requests the user to digitize the quadrant where the circle is to reside. One common convention is that the circle becomes tangent to the closest line segments chosen by the user while digitizing the two known lines to identify them. For all solutions, the above development is valid.

A point of interest here is the possible graphical solution to find the point $P_c$ which is useful to verify the above results if the case of constructing a circle tangent to two lines is not provided by the software, but the other simpler case of a radius and center is. In terms of a typical CAD/CAM software package, the graphical solution consists of defining a WCS utilizing the endpoints of the two known lines. The $XY$ plane of WCS becomes the plane of the lines. Using the parallel line command, the user can construct the two lines $P_{1'}P_{2'}$ and $P_{3'}P_{4'}$. The circle can now be constructed with the radius $R$ and $P_c$ as the intersection of these two parallel lines. Using the verify command the user can obtain the coordinates of $P_c$ and compare with the results of Eq. (5.33).

Two special cases related to Eqs. (5.32) and (5.33) are discussed here. First, consider the case when the two known lines are parallel, as shown in Fig. 5-26a. In this case $\hat{\mathbf{n}}_1 = \hat{\mathbf{n}}_2$, $\hat{\mathbf{n}}_4 = -\hat{\mathbf{n}}_5$, $\hat{\mathbf{n}}_4 \cdot \hat{\mathbf{n}}_5 = -1$, and $(\mathbf{P}_2 - \mathbf{P}_1) \cdot \hat{\mathbf{n}}_5 = 0$. Therefore, $u \to \infty$ and $P_c$ is not defined. However, the locus of $P_c$ is defined as the parallel line to the known lines at the middle distance between them, as shown in the figure. In this case the software can override the user input of $R$ and replaces it by half the perpendicular distance between the two lines. Such a distance can be computed as shown in Sec. 5.5.2. The two endpoints of the locus of $P_c$ are given by

$$\mathbf{P}_{L1} = \frac{\mathbf{P}_1 + \mathbf{P}_3}{2} \qquad \mathbf{P}_{L2} = \frac{\mathbf{P}_2 + \mathbf{P}_4}{2}$$

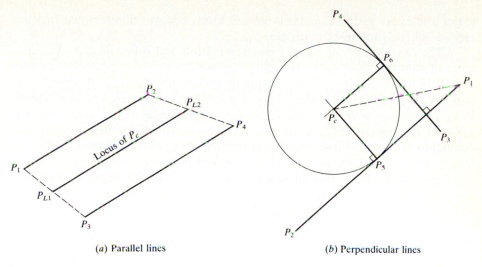

(a) Parallel lines                                    (b) Perpendicular lines

**FIGURE 5-26**
Circles tangent to parallel and perpendicular lines.

An infinite number of circles exists with centers on the locus. The software can either display a warning message to the user or choose point $P_{L1}$ or $P_{L2}$ as a default center.

The second case is shown in Fig. 5-26b, where the two known lines are perpendicular to each other. In this case $\hat{\mathbf{n}}_4 = \hat{\mathbf{n}}_2$, $\hat{\mathbf{n}}_5 = \hat{\mathbf{n}}_1$, and $\hat{\mathbf{n}}_4 \cdot \hat{\mathbf{n}}_5 = 0$. Equations (5.33) and (5.32) become

$$u = \frac{(\mathbf{P}_3 - \mathbf{P}_1) \cdot \hat{\mathbf{n}}_1 + R}{|\mathbf{P}_2 - \mathbf{P}_1|}$$

$$\mathbf{P}_c = \mathbf{P}_1 + [(\mathbf{P}_3 - \mathbf{P}_1) \cdot \hat{\mathbf{n}}_1 + R]\hat{\mathbf{n}}_1 + R\hat{\mathbf{n}}_2$$

This equation for $\mathbf{P}_c$ is also obvious if we consider the triangle $P_1 P_c P_5$ and write the vector equation $\mathbf{P}_1\mathbf{P}_c = \mathbf{P}_1\mathbf{P}_5 + \mathbf{P}_5\mathbf{P}_c$.

The above development can be extended to fillets connecting lines as follows. Once $\mathbf{P}_c$ is known from Eq. (5.33), the two points $P_5$ and $P_6$ (Fig. 5-25a) are given by

$$\mathbf{P}_5 = \mathbf{P}_c - R\hat{\mathbf{n}}_4$$

$$\mathbf{P}_6 = \mathbf{P}_c - R\hat{\mathbf{n}}_5$$

These two points define the beginning and the end of the fillet and can be used to construct the proper part of the circle that forms the fillet.

## 5.5.4 Ellipses

Mathematically the ellipse is a curve generated by a point moving in space such that at any position the sum of its distances from two fixed points (foci) is constant and equal to the major diameter. Each focus is located on the major axis of the ellipse at a distance from its center equal to $\sqrt{A^2 - B^2}$ ($A$ and $B$ are the major and minor radii). Circular holes and forms become ellipses when they are viewed obliquely relative to their planes.

The development of the parametric equation and other related characteristics of ellipses, elliptic arcs, and fillets are similar to those of circles, circular arcs, and fillets. However, four conditions (points and/or tangent vectors) are required to define the geometric shape of an ellipse as compared to three conditions to define a circle. The default plane of an ellipse, as in a circle, is the $XY$ plane of the current WCS at the time of construction if the user input is not enough to define the ellipse plane, as in the case of inputing the center, half of the length of the major axis, and half of the length of the minor axis. The database of an ellipse usually stores user input as a centerpoint, half the length of the major axis, half the length of the minor axis, and other information (orientation, starting and ending angles, layer, font, name, color, etc.).

Figure 5-27 shows an ellipse with point $P_c$ as the center and the lengths of half of the major and minor axes are $A$ and $B$ respectively. The parametric equation of an ellipse can be written as

$$\left.\begin{array}{l} x = x_c + A \cos u \\[6pt] y = y_c + B \sin u \\[6pt] z = z_c \end{array}\right\} \quad 0 \le u \le 2\pi \qquad (5.34)$$

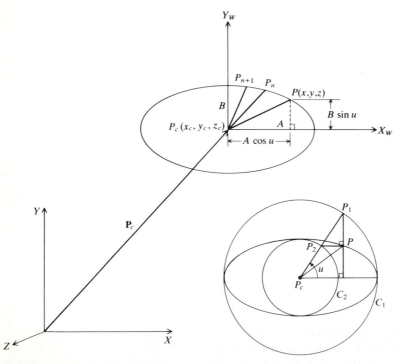

**FIGURE 5-27**
Ellipse defined by a center, major and minor axes.

assuming the plane of the ellipse is the $XY$ plane. The parameter $u$ is the angle as in the case of a circle. However, for a point $P$ shown in the figure, it is not the angle between the line $PP_c$ and the major axis of the ellipse. Instead, it is defined as shown. To find point $P$ on the ellipse that corresponds to an angle $u$, the two concentric circles $C_1$ and $C_2$ are constructed with centers at $P_c$ and radii of $A$ and $B$ respectively. A radial line is constructed at the angle $u$ to intersect both circles at points $P_1$ and $P_2$ respectively. If a line parallel to the minor axis is drawn from $P_1$ and a line parallel to the major axis is drawn from $P_2$, the intersection of these two lines defines the point $P$.

Similar development as in the case of a circle results in the following recursive relationships which are useful for generating points on the ellipse for display purposes without excessive evaluations of trigonometric functions:

$$x_{n+1} = x_c + (x_n - x_c) \cos \Delta u - \frac{A}{B} (y_n - y_c) \sin \Delta u$$

$$y_{n+1} = y_c + (y_n - y_c) \cos \Delta u + \frac{A}{B} (x_n - x_c) \sin \Delta u \qquad (5.35)$$

$$z_{n+1} = z_n$$

If the ellipse major axis is inclined with an angle $\alpha$ relative to the $X$ axis as shown in Fig. 5-28, the ellipse equation becomes

$$\left.\begin{aligned} x &= x_c + A \cos u \cos \alpha - B \sin u \sin \alpha \\ y &= y_c + A \cos u \sin \alpha + B \sin u \cos \alpha \\ z &= z_c \end{aligned}\right\} \qquad 0 \le u \le 2\pi \qquad (5.36)$$

Equations (5.36) cannot be reduced to a recursive relationship similar to what is given by Eqs. (5.35). Instead these equations can be written as

$$x_{n+1} = x_c + A \cos (u_n + \Delta u) \cos \alpha - B \sin (u_n + \Delta u) \sin \alpha$$

$$y_{n+1} = y_c + A \cos (u_n + \Delta u) \sin \alpha + B \sin (u_n + \Delta u) \cos \alpha \qquad (5.37)$$

$$z_{n+1} = z_n$$

where $u_n = (n - 1)u$. The first point corresponds to $n = 0$ which lies at the end of the major axis. In addition, $\cos (u_n + \Delta u)$ and $\sin (u_n + \Delta u)$ are evaluated from the double-angle formulas for cosine and sine. If the calculations from the previous point for $\sin u_n$ and $\cos u_n$ are stored temporarily, $\cos (u_n + \Delta u)$ and $\sin (u_n + \Delta u)$ can be evaluated without calculating trigonometric functions for each point. Cos $\Delta u$ and sin $\Delta u$ need to be calculated only once. Therefore, computational savings similar to the circle case can be achieved.

**Example 5.13.** Find the center, the lengths of half the axes, and the orientation in space of an ellipse defined by:

(a) Its circumscribing rectangle.
(b) One of its internal rectangles.

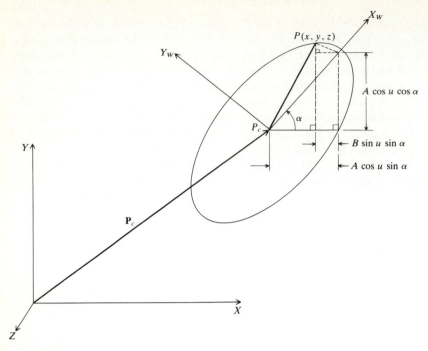

**FIGURE 5-28**
An inclined ellipse.

**Solution.** Both cases are equivalent to defining an ellipse by four points. The user interface can let the user input the three corner points of the rectangle or one corner point, the lengths of the rectangle sides, and an orientation.

(*a*) From Fig. 5-29*a*:

$$P_c = \tfrac{1}{2}(P_1 + P_3) = \tfrac{1}{2}(P_2 + P_4)$$

$$P_H = \tfrac{1}{2}(P_2 + P_3)$$

$$P_v = \tfrac{1}{2}(P_3 + P_4)$$

$$A = |P_c - P_4| = \sqrt{(x_c - x_4)^2 + (y_c - y_4)^2 + (z_c - z_4)^2}$$

$$B = |P_c - P_v| = \sqrt{(x_c - x_v)^2 + (y_c - y_v)^2 + (z_c - z_v)^2}$$

The orientation of the ellipse in space can be defined by the unit vectors $\hat{n}_1$, $\hat{n}_2$, and $\hat{n}_3$ instead of the angle $\alpha$. These vectors are given by

$$\hat{n}_1 = \frac{P_2 - P_1}{|P_2 - P_1|} \qquad \hat{n}_2 = \frac{P_4 - P_1}{|P_4 - P_1|} \qquad \hat{n}_3 = \hat{n}_1 \times \hat{n}_2$$

To display the ellipse, points on its circumference can be computed in the local (WCS) system $X_W Y_W Z_W$ utilizing the equation

$$\left.\begin{array}{l} x_W = A \cos u \\[4pt] y_W = B \cos u \\[4pt] z_W = 0 \end{array}\right\} \qquad 0 \le u \le 2\pi$$

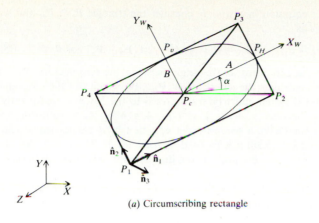

(a) Circumscribing rectangle

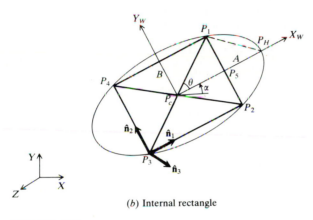

(b) Internal rectangle

**FIGURE 5-29**
An ellipse defined by four points.

These points can then be transformed to the MCS utilizing Eq. (3.3) as follows:

$$[x \quad y \quad z \quad 1]^T = \begin{bmatrix} n_{1x} & n_{2x} & n_{3x} & x_c \\ n_{1y} & n_{2y} & n_{3y} & y_c \\ n_{1z} & n_{2z} & n_{3z} & z_c \\ 0 & 0 & 0 & 1 \end{bmatrix} \begin{bmatrix} x_W \\ y_W \\ 0 \\ 1 \end{bmatrix}$$

The transformed points can be input to the graphics display driver to display the ellipse.

(b) $P_c$, $\hat{\mathbf{n}}_1$, $\hat{\mathbf{n}}_2$, and $\hat{\mathbf{n}}_3$ can be calculated as in case (a). To find $A$ and $B$, one can write

$$\mathbf{P}_5 = \tfrac{1}{2}(\mathbf{P}_2 + \mathbf{P}_3)$$

In reference to the WCS coordinate system, the $x$ and $y$ coordinates of point $P_3$ are $|\mathbf{P}_5 - \mathbf{P}_c|$ and $|\mathbf{P}_3 - \mathbf{P}_5|$ respectively. Substituting into the ellipse equation,

$$\frac{|\mathbf{P}_5 - \mathbf{P}_c|^2}{A^2} + \frac{|\mathbf{P}_3 - \mathbf{P}_5|^2}{B^2} = 1 \tag{5.38}$$

To write another equation in $A$ and $B$, consider the triangle $P_c P_3 P_H$ and write the law of cosines as

$$|\mathbf{P}_3 - \mathbf{P}_H|^2 = |\mathbf{P}_3 - \mathbf{P}_c|^2 + A^2 - 2A|\mathbf{P}_3 - \mathbf{P}_c| \cos \theta \qquad (5.39)$$

The angle $\theta$ can be computed as the angle between the two vectors $(\mathbf{P}_5 - \mathbf{P}_c)$ and $(\mathbf{P}_3 - \mathbf{P}_c)$. To calculate the length $|\mathbf{P}_3 - \mathbf{P}_H|$, $\mathbf{P}_H$ which has the coordinates $(A, 0, 0)$ in the WCS system must be transformed to the MCS. Consequently its MCS coordinates become $(x_c + n_{1x} A,\ y_c + n_{1y} A,\ z_c + n_{1z} A)$. Substituting these coordinates into Eq. (5.39), a second-order equation in $A$ results which can be solved for $A$. Then Eq. (5.38) can be used to solve for $B$. The remainder of the solution (finding the ellipse orientation and displaying it) is identical to case $(a)$.

**Example 5.14.** Find the center, the lengths of half the axes, and the orientation of an ellipse given two of its conjugate diameters.

*Solution.* This is a case of defining an ellipse by four points that form two of its conjugate diameters. Figure 5-30 shows the two conjugate diameters and the relevant unit vectors. The center of the ellipse is given by

$$\mathbf{P}_c = \tfrac{1}{2}(\mathbf{P}_1 + \mathbf{P}_3) = \tfrac{1}{2}(\mathbf{P}_2 + \mathbf{P}_4)$$

The unit vectors $\hat{\mathbf{n}}_1$, $\hat{\mathbf{n}}_2$, and $\hat{\mathbf{n}}_3$ are given by

$$\hat{\mathbf{n}}_1 = \frac{\mathbf{P}_3 - \mathbf{P}_1}{|\mathbf{P}_3 - \mathbf{P}_1|} \qquad \hat{\mathbf{n}}_2 = \frac{\mathbf{P}_4 - \mathbf{P}_2}{|\mathbf{P}_4 - \mathbf{P}_2|} \qquad \hat{\mathbf{n}}_3 = \frac{\hat{\mathbf{n}}_1 \times \hat{\mathbf{n}}_2}{|\hat{\mathbf{n}}_1 \times \hat{\mathbf{n}}_2|}$$

Utilizing the ellipse equation for points $P_2$ and $P_3$, the following two equations can be written:

$$\frac{(L_1 \cos \theta_1)^2}{A^2} + \frac{(L_1 \sin \theta_1)^2}{R^2} = 1 \qquad (5.40)$$

$$\frac{(L_2 \cos \theta_2)^2}{A^2} + \frac{(L_2 \sin \theta_2)^2}{B^2} = 1 \qquad (5.41)$$

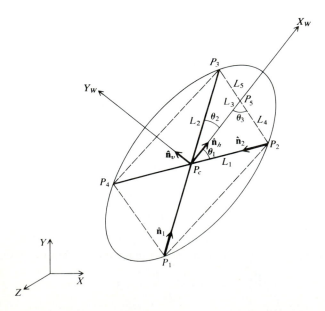

**FIGURE 5-30**
An ellipse defined by two conjugate diameters.

where

$$L_1 = |\mathbf{P}_2 - \mathbf{P}_c| \qquad L_2 = |\mathbf{P}_3 - \mathbf{P}_c|$$

The angles $\theta_1$ and $\theta_2$ can be related together as follows:

$$\hat{\mathbf{n}}_1 \cdot \hat{\mathbf{n}}_2 = \cos(\theta_1 + \theta_2)$$

or

$$\theta_1 + \theta_2 = K \qquad (5.42)$$

where $K$ is a known value. In addition, applying the law of sines to the triangles $P_c P_3 P_5$ and $P_c P_2 P_5$, we can write

$$\frac{L_1}{\sin \theta_3} = \frac{L_4}{\sin \theta_1} \qquad (5.43)$$

$$\frac{L_2}{\sin(180 - \theta_3)} = \frac{L_5}{\sin \theta_2} \qquad (5.44)$$

where

$$L_4 = |\mathbf{P}_2 - \mathbf{P}_5| \qquad L_5 = L_6 - L_4 \qquad L_6 = |\mathbf{P}_3 - \mathbf{P}_2|$$

Dividing Eq. (5.44) by (5.43) and using (5.42) gives

$$\frac{\sin \theta_1}{\sin(K - \theta_1)} = \frac{L_2 L_4}{L_1(L_6 - L_4)} \qquad (5.45)$$

Applying the law of cosines to the same triangles gives

$$L_4^2 = L_1^2 + L_3^2 - 2L_1 L_3 \cos \theta_1 \qquad (5.46)$$

$$(L_6 - L_4)^2 = L_2^2 + L_3^2 - 2L_2 L_3 \cos \theta_2 \qquad (5.47)$$

Equations (5.40), (5.41), (5.42), (5.45), (5.46), and (5.47) form six equations to be solved for $A$, $B$, $\theta_1$, $\theta_2$, $L_3$, and $L_4$. Subtracting Eqs. (5.47) and (5.46) and using (5.42) gives

$$L_3 = \frac{L_2^2 - L_1^2 - L_6^2 + 2L_4 L_6}{2[L_2 \cos(K - \theta_1) - L_1 \cos \theta_1]} \qquad (5.48)$$

If Eq. (5.48) is substituted into (5.46) and the resulting $L_4$ is substituted into (5.45), a nonlinear equation in $\theta_1$ results which can be solved for $\theta_1$. Consequently $\theta_2$ can be found from Eq. (5.42). Equations (5.40) and (5.41) can therefore be solved for $A$ and $B$.

The orientation of the ellipse can be found by determining the unit vectors $\hat{\mathbf{n}}_h$ and $\hat{\mathbf{n}}_v$ that define the directions of the major and the minor axes respectively. To find $\hat{\mathbf{n}}_h$, the following three equations can be written:

$$\hat{\mathbf{n}}_1 \cdot \hat{\mathbf{n}}_h = \cos \theta_2$$

$$\hat{\mathbf{n}}_2 \cdot \hat{\mathbf{n}}_h = \cos \theta_1$$

$$\hat{\mathbf{n}}_3 \cdot \hat{\mathbf{n}}_h = 0$$

These equations can be solved for the components $n_{hx}$, $n_{hy}$, and $n_{hz}$ using the matrix approach utilized in Example 5.11. The unit vector $\hat{\mathbf{n}}_v$ can be found as

$$\hat{\mathbf{n}}_v = \hat{\mathbf{n}}_3 \times \hat{\mathbf{n}}_h$$

With $\hat{n}_h$, $\hat{n}_v$, and $\hat{n}_3$ known, the orientation of the ellipse is completely defined in space and transformation of points on the ellipse from the WCS system to the MCS can be performed for display and plotting purposes.

*Note:* case (b) of Example 5.13 is a special case of this example.

**Example 5.15.** Find the tangent to an ellipse from a given point $P_1$ outside the ellipse.

*Solution.* Two tangents can be drawn to the ellipse from $P_1$ as shown in Fig. 5-31. Assume the tangency point is $P_T$. First, transform $P_1$ from the MCS to the ellipse local WCS system using the equation

$$\mathbf{P}_1 = [T]\mathbf{P}_{1W} \tag{5.49}$$

The transformation matrix $[T]$ is known because the orientation of the ellipse is known. $P_{1W}$ holds the local coordinates of $P_1$ which should be $[x_{1W} \quad y_{1W} \quad 0]$. Therefore:

$$\mathbf{P}_{1W} = [T]^{-1}\mathbf{P}_1$$

The tangent vector to the ellipse is given by

$$\mathbf{P}' = [-A \sin u \quad B \cos u \quad 0]^T$$

At point $P_T$, this vector becomes

$$\mathbf{P}' = [-A \sin u_T \quad B \cos u_T \quad 0]^T$$

and the slope of the tangent is given by

$$S = -\frac{B \cos u_T}{A \sin u_T}$$

This slope $S$ can also be found using the two points $P_1$ and $P_T$ as

$$S = \frac{y_{1W} - B \sin u_T}{x_{1W} - A \cos u_T} \tag{5.50}$$

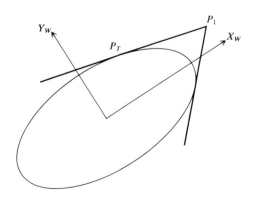

**FIGURE 5-31**
Tangent to an ellipse from an outside point.

Therefore,

$$\frac{y_{1W} - B \sin u_T}{x_{1W} - A \cos u_T} = -\frac{B \cos u_T}{A \sin u_T}$$

which gives

$$x_1 B \cos u_T + y_1 A \sin u_T = AB$$

or

$$K_1 \cos u_T + K_2 \sin u_T = AB$$

where $K_1 = x_1 B$ and $K_2 = y_1 A$. Define an angle $\gamma$ such that $\gamma = \tan^{-1}(K_1/K_2)$. Thus, the above equation can be rewritten as

$$\sin(\gamma + u_T) = \frac{AB}{\sqrt{K_1^2 + K_2^2}}$$

or

$$u_T = \sin^{-1}\left(\frac{AB}{\sqrt{K_1^2 + K_2^2}}\right) - \gamma$$

The arcsine function gives two angles that result in two tangents.

Once $u_T$ is known, the local coordinates of $P_T$ become

$$\mathbf{P}_{TW} = [x_{TW} \quad y_{TW} \quad 0]^T = [A \cos u_T \quad B \sin u_T \quad 0]^T$$

$\mathbf{P}_{TW}$ can be substituted in Eq. (5.49) to obtain $\mathbf{P}_T$. Therefore, the tangent is defined and stored in the database by the two endpoints $P_1$ and $P_T$. In practice, the CAD/CAM software can ask the user to digitize close to the desired tangent so that the other one is eliminated.


## 5.5.5  Parabolas

The parabola is defined mathematically as a curve generated by a point that moves such that its distance from a fixed point (the focus $\mathbf{P}_F$) is always equal to its distance to a fixed line (the directrix) as shown in Fig. 5-32. The vertex $P_v$ is the intersection point of the parabola with its axis of symmetry. It is located midway between the directrix and the focus. The focus lies on the axis of symmetry. Useful applications of the parabolic curve in engineering design include its use in parabolic sound and light reflectors, radar antennas, and in bridge arches.

Three conditions are required to define a parabola, a parabolic curve, or a parabolic arc. The default plane of a parabola is the $XY$ plane of the current WCS at the time of construction. The database of a parabola usually stores the coordinates of its vertex, distances $y_{HW}$ and $y_{LW}$, that define its endpoints as shown in Fig. 5-32, the distance $A$ between the focus and the vertex (the focal distance), and the orientation angle $\alpha$. Unlike the ellipse, the parabola is not a closed curve. Thus, the two endpoints determine the amount of the parabola to be displayed.

Assuming the local coordinate system of the parabola as shown in Fig. 5-32, its parametric equation can be written as

$$\left. \begin{array}{l} x = x_v + Au^2 \\ y = y_v + 2Au \\ z = z_v \end{array} \right\} \quad 0 \le u \le \infty \qquad (5.51)$$

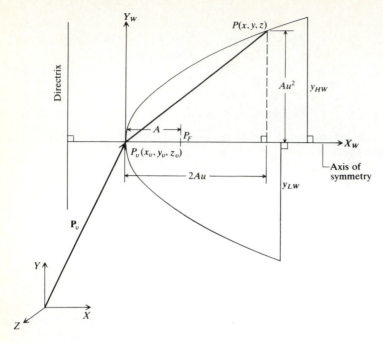

**FIGURE 5-32**
Basic geometry of a parabola.

If the range of the $y$ coordinate is limited to $y_{HW}$ and $y_{LW}$ for positive and negative values respectively, the corresponding $u$ values become

$$u_H = \frac{y_{HW}}{2A}$$

$$u_L = \frac{y_{LW}}{2A}$$

(5.52)

The recursive relationships to generate points on the parabola are obtained by
substituting $u_n + \Delta u$ for points $n + 1$. This gives

$$x_{n+1} = x_n + (y_n - y_v)\,\Delta u + A(\Delta u)^2$$

$$y_{n+1} = y_n + 2A\,\Delta u$$

(5.53)

$$z_{n+1} = z_n$$

If the parabolic axis of symmetry is inclined with an angle $\alpha$ as shown in
Fig. 5-33, its equation becomes

$$x = x_v + Au^2 \cos \alpha - 2Au \sin \alpha$$

$$y = y_v + Au^2 \sin \alpha + 2Au \cos \alpha$$

(5.54)

$$z = z_v$$

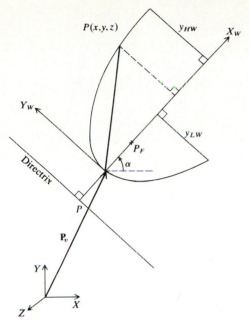

**FIGURE 5-33**
An inclined parabola.

and the recursive relationships reduce to

$$x_{n+1} = x_n \cos \alpha + (1 - \cos \alpha)x_v + (\Delta u \cos \alpha - \sin \alpha)(y_n - y_v)$$
$$+ A \, \Delta u(\Delta u \cos \alpha - 2 \sin \alpha)$$
$$y_{n+1} = (\cos \alpha + \Delta u \sin \alpha)y_n + (1 - \cos \alpha - \Delta u \sin \alpha)y_v \qquad (5.55)$$
$$+ (x_n - x_v) \sin \alpha + A \, \Delta u(\Delta u \sin \alpha + 2 \cos \alpha)$$
$$z_{n+1} = z_n$$

An alternative to Eqs. (5.55) would be to use Eqs. (5.53) to generate the points in the WCS local coordinate system of the parabola and then utilize Eq. (3.3) to transform these points to the MCS before displaying or plotting them. If the $X_W Y_W$ and the $XY$ planes coincide, the transformation matrix becomes

$$[T] = \begin{bmatrix} \cos \alpha & \sin \alpha & 0 & x_v \\ -\sin \alpha & \cos \alpha & 0 & y_v \\ 0 & 0 & 1 & z_v \\ \hline 0 & 0 & 0 & 1 \end{bmatrix} \qquad (5.56)$$

If the two planes are different, then the unit vectors $\hat{n}_{xW}$, $\hat{n}_{yW}$, and $\hat{n}_{zW}$ that define the orientation of the parabolic local coordinate system relative to the MCS must be calculated before using Eq. (3.3).

**Example 5.16.** Find the focal distance and the orientation in space of a parabola that passes through three points, one of which is the vertex.

**Solution.** Figure 5-34 shows two cases in which point $P_1$ is the vertex. In Fig. 5-34a, the other two points $P_2$ and $P_3$ define a line perpendicular to the $X_W$ axis of the

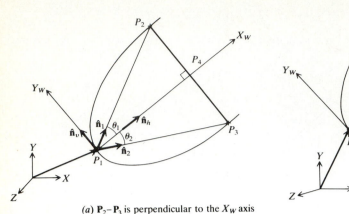

(a) $\mathbf{P}_2 - \mathbf{P}_3$ is perpendicular to the $X_W$ axis

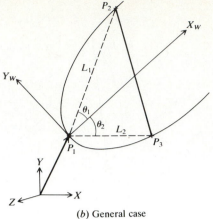

(b) General case

**FIGURE 5-34**
A parabola passing through three points.

parabola while Fig. 5-34b shows the general case. The solution for the first case can be found as follows. The angles $\theta_1$ and $\theta_2$ in this case are equal. Thus,

$$\mathbf{P}_4 = \frac{\mathbf{P}_2 + \mathbf{P}_3}{2}$$

The angle $\theta_1$ can be found from

$$\tan \theta_1 = \frac{|\mathbf{P}_2 - \mathbf{P}_4|}{|\mathbf{P}_4 - \mathbf{P}_1|}$$

Using the parabolic equation $y_W^2 = 4Ax_W$, the focal distance $A$ can be written as

$$A = \frac{|\mathbf{P}_2 - \mathbf{P}_4|^2}{4|\mathbf{P}_4 - \mathbf{P}_1|}$$

The orientation of the parabola is determined by the unit vectors $\hat{\mathbf{n}}_h$ and $\hat{\mathbf{n}}_v$ along the $X_W$ and $Y_W$ axes and the vector $\hat{\mathbf{n}}_3$ which is perpendicular to its plane. These vectors are given as

$$\hat{\mathbf{n}}_h = \frac{\hat{\mathbf{n}}_1 + \hat{\mathbf{n}}_2}{|\hat{\mathbf{n}}_1 + \hat{\mathbf{n}}_2|}$$

$$\hat{\mathbf{n}}_3 = \frac{\hat{\mathbf{n}}_2 \times \hat{\mathbf{n}}_1}{|\hat{\mathbf{n}}_2 \times \hat{\mathbf{n}}_1|} = \frac{\hat{\mathbf{n}}_2 \times \hat{\mathbf{n}}_1}{\sin(\theta_1 + \theta_2)}$$

$$\hat{\mathbf{n}}_v = \hat{\mathbf{n}}_3 \times \hat{\mathbf{n}}_h$$

where the unit vectors $\hat{\mathbf{n}}_1$ and $\hat{\mathbf{n}}_2$ can easily be computed from the given points.

Unlike the above case, the angles $\theta_1$ and $\theta_2$ are not equal for the second case (Fig. 5-34b). Applying the parabolic equations to points $P_2$ and $P_3$, we can write respectively

$$L_1 \sin^2 \theta_1 = 4A \cos \theta_1 \tag{5.57}$$

$$L_2 \sin^2 \theta_2 = 4A \cos \theta_2 \tag{5.58}$$

where $L_1 = |\mathbf{P}_2 - \mathbf{P}_1|$ and $L_2 = |\mathbf{P}_3 - \mathbf{P}_1|$. In addition,

$$\hat{\mathbf{n}}_1 \cdot \hat{\mathbf{n}}_2 = \cos\,(\theta_1 + \theta_2)$$

or

$$\theta_1 + \theta_2 = K \qquad\qquad (5.59)$$

where

$$K = \cos^{-1}\,(\mathbf{n}_1 \cdot \hat{\mathbf{n}}_2)$$

The solution of Eqs. (5.57), (5.58), and (5.59) gives $A$, $\theta_1$, and $\theta_2$. If we divide Eq. (5.57) by (5.58) and use (5.59), we obtain

$$\frac{L_1}{L_2}\left[\frac{\sin\,\theta_1}{\sin\,(K - \theta_1)}\right]^2 = \frac{\cos\,\theta_1}{\cos\,(K - \theta_1)}$$

which is a nonlinear equation in $\theta_1$. One way to solve it would be to plot both the left- and right-hand sides as functions of $\theta_1$ ($0 \leq \theta_1 \leq 90$) and find the intersection point using the intersection modifier provided by the CAD/CAM software. Once $\theta_1$ is found, Eq. (5.57) can be solved for $A$ and (5.59) solved for $\theta_2$. To find the orientation of the parabola, write

$$\hat{\mathbf{n}}_1 \cdot \hat{\mathbf{n}}_h = \cos\,\theta_1$$

$$\hat{\mathbf{n}}_2 \cdot \hat{\mathbf{n}}_h = \cos\,\theta_2$$

$$\hat{\mathbf{n}}_3 \cdot \hat{\mathbf{n}}_h = 0$$

which can be solved for the components of $\hat{\mathbf{n}}_h$. Then

$$\hat{\mathbf{n}}_v = \hat{\mathbf{n}}_3 \times \hat{\mathbf{n}}_h$$

### 5.5.6  Hyperbolas

A hyperbola is described mathematically as a curve generated by a point moving such that at any position the difference of its distances from the fixed points (foci) $F$ and $F'$ is a constant and equal to the transverse axis of the hyperbola. Figure 5-35 shows the geometry of a hyperbola.

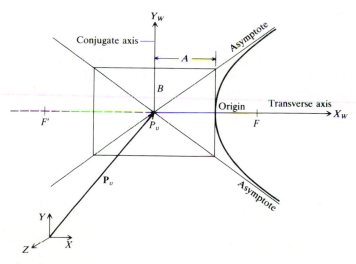

**FIGURE 5-35**
Hyperbola geometry.

The parametric equation of a hyperbola is given by

$$x = x_v + A \cosh u$$

$$y = y_v + B \sinh u \qquad (5.60)$$

$$z = z_v$$

This equation is based on the nonparametric implicit equation of the hyperbola which can be written as

$$\frac{(x - x_v)^2}{A^2} - \frac{(y - y_v)^2}{B^2} = 1 \qquad (5.61)$$

by utilizing the identity $\cosh^2 u - \sinh^2 u = 1$.

Similar to the ellipse developments, equations of an inclined hyperbola and recursive relationships can be derived. Also, the examples covered in the ellipse section can be extended to hyperbolas.

### 5.5.7 Conics

Conic curves or conic sections form the most general form of quadratic curves. Lines, circles, ellipses, parabolas, and hyperbolas covered in the previous sections are all special forms of conic curves. They all can be generated when a right circular cone of revolution is cut by planes at different angles relative to the cone axis—thus the derivation of the name conics. Straight lines result from intersecting a cone with a plane parallel to its axis and passing through its vertex. Circles result when the cone is sectioned by a plane perpendicular to its axis while ellipses, parabolas, and hyperbolas are generated when the plane is inclined to the axis by various angles.

The general implicit nonparametric quadratic equation that describes a planar conic curve has five coefficients if the coefficient of the term $x^2$ is made equal to one; that is, normalize the equation by dividing it by this coefficient if it is not one. Thus, five conditions are required to completely define a conic curve. As presented in the previous sections, these reduce to two conditions to define lines, three for circles and parabolas, and four for ellipses and hyperbolas.

The conic parametric equation can be developed if five conditions are specified. Two cases are discussed here: specifying five points or three points and two tangent vectors. The development is based on the observation that a quadratic equation can be written as the product of two linear equations. Figure 5-36 shows a conic curve passing through points $P_1$ to $P_5$. Define the two pairs of lines $(L_1, L_2)$ and $(L_3, L_4)$ shown in the figure. Their equations are

$$L_1 = 0 \qquad L_2 = 0 \qquad L_3 = 0 \qquad L_4 = 0 \qquad (5.62)$$

The four intersection points of these pairs are given by points $P_1$ to $P_4$. Let us define the two conics:

$$L_1 L_2 = 0 \qquad L_3 L_4 = 0 \qquad (5.63)$$

Each one of these conics passes through points $P_1$ to $P_4$ but not necessarily through point $P_5$. However, any linear combination of these two conics rep-

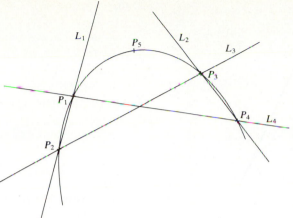

**FIGURE 5-36**
A conic curve defined by five points.

resents another conic (since it is quadratic) which passes through their intersection points $P_1$ to $P_4$. For example, the equation

$$aL_1L_2 + bL_3L_4 = 0 \qquad (5.64)$$

represents such a conic. For the conic to pass through point $P_5$ its coordinates are substituted into Eq. (5.64) to find the ratio $b/a$. A more convenient form of Eq. (5.64) is

$$(1 - u)L_1L_2 + uL_3L_4 = 0, \qquad 0 \le u \le 1 \qquad (5.65)$$

Equation (5.65) gives the parametric equation of a conic curve with the parameter $u$. Changing the value of $u$ results in a family (or pencil) of conics, two of which are $L_1L_2 = 0$ ($u = 0$) and $L_3L_4 = 0$ ($u = 1$). To use Eq. (5.65), four data points are used to find the equations for the lines $L_1$ to $L_4$ and the fifth point is used to find the $u$ value.

The case of a conic defined by three points and two tangent vectors is considered as an adaptation of the above case. If we make lines $L_1$ and $L_2$ tangent to the conic curve, points $P_1$ and $P_2$ become one point (the tangent point) and $P_3$ and $P_4$ become another point. Consequently, the two lines $L_3$ and $L_4$ are merged into one line. Figure 5-37 shows the conic geometry in this case. The definition of this conic curve is equivalent to a four-point definition: the two points of tangency $P_1$ and $P_2$, the intersection point $P_4$ of the two tangents $L_1$ and $L_2$, and a fourth point $P_3$, known as the shoulder point. $P_3$ must always be chosen inside the triangle $P_1P_2P_4$ to ensure the continuity of the conic curve segment that lies inside the triangle between $P_1$ and $P_2$. Equation (5.65) is then reduced for this case to

$$(1 - u)L_1L_2 + uL_3^2 = 0, \qquad 0 \le u \le 1 \qquad (5.66)$$

Similar to the first case, points $P_1$, $P_2$, and $P_4$ determine the equations for the lines $L_1$ to $L_3$ and the point $P_3$ is used to find the $u$ value.

The choice of the position of $P_3$ determines the type of resulting conic curve. If $P_3$ is the midpoint of the line joining the midpoints of the tangents $L_1$

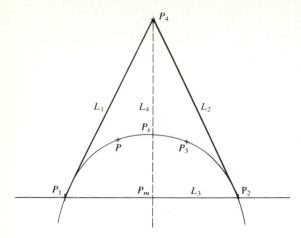

**FIGURE 5-37**
A conic curve defined by three points
and two tangent vectors.

and $L_2$, then the conic is a parabola. $P_3$ becomes the vertex of the parabola and
the line connecting $P_3$ and $P_4$ becomes its axis of symmetry. It is also obvious
that the distance between $P_3$ and $P_4$ is equal to the distance between $P_3$ and the
intersection point of the axis of symmetry and $L_3$ (a known characteristic of
parabolas). If $P_3$ lies inside the parabola and line $L_3$, the resulting conic is an
ellipse. If it is outside the parabola, a hyperbolic curve results. To check which
type of conic curve results from a given input data, let $P_m$ be the midpoint of the
line $L_3$ and let the line $L_4$ intersect the conic curve at the point $P_s$. The para-
metric equation of line $L_4$ is $\mathbf{P} = \mathbf{P}_m + u(\mathbf{P}_4 - \mathbf{P}_m)$. Let the parameter $u$ take the
value $s$ at $P_s$. Then $(\mathbf{P}_s - \mathbf{P}_m) = s(\mathbf{P}_4 - \mathbf{P}_m)$. Therefore, the conic curve is parabolic if $s = \frac{1}{2}$, elliptic if $s < \frac{1}{2}$, and hyperbolic if $s > \frac{1}{2}$. A further test is necessary to
determine whether an eliptic curve is circular. If $|\mathbf{P}_4 - \mathbf{P}_1| = |\mathbf{P}_2 - \mathbf{P}_4|$ and

$$\frac{s^2}{1 - s^2} = \frac{|\mathbf{P}_2 - \mathbf{P}_1|^2}{4|\mathbf{P}_4 - \mathbf{P}_1||\mathbf{P}_4 - \mathbf{P}_2|} \tag{5.67}$$

then the arc is circular.

**Example 5.17.** Find the equation of a conic curve defined by five points $P_1$ to $P_5$.

**Solution.** Equation (5.65) must be reduced further to be able to utilize it to generate
points on the conic curve for display purposes. The approach taken in this example
is to create a local (WCS) coordinate system $(X_W Y_W Z_W)$ in the plane of the conic
curve (Fig. 5-38), write Eq. (5.65) in this system, generate points, and lastly transform these points to the global MCS $(XYZ)$. The input point $P_2$ is taken as the
origin of the local system and the vector $(\mathbf{P}_4 - \mathbf{P}_2)$ as its $X_W$ axis. To define the
system, the unit vectors along the axes are calculated as

$$\hat{\mathbf{n}}_1 = \frac{\mathbf{P}_4 - \mathbf{P}_2}{|\mathbf{P}_4 - \mathbf{P}_2|} \qquad \hat{\mathbf{n}}_4 = \frac{\mathbf{P}_1 - \mathbf{P}_2}{|\mathbf{P}_1 - \mathbf{P}_2|}$$

$$\hat{\mathbf{n}}_3 = \frac{\hat{\mathbf{n}}_1 \times \hat{\mathbf{n}}_4}{|\hat{\mathbf{n}}_1 \times \mathbf{n}_4|} \tag{5.68}$$

$$\hat{\mathbf{n}}_2 = \hat{\mathbf{n}}_3 \times \hat{\mathbf{n}}_1$$

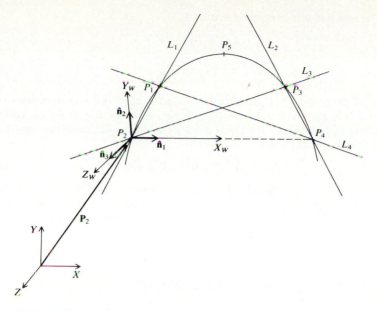

**FIGURE 5-38**
Local coordinate system of a conic curve.

At this point, it might be useful to ensure that the five points the user has inputted lie in one plane (the conic curve plane). This can simply be achieved by checking whether the scalar (dot) products of the vector $\hat{n}_3$ with the vectors $(P_3 - P_2)$ and $(P_3 - P_5)$ are zeros or not.

The points $P_1$ to $P_5$ can be transformed from the MCS to the local system using the equation

$$[x \quad y \quad z \quad 1]^T = [T] \begin{bmatrix} x_W \\ y_W \\ z_W \\ 1 \end{bmatrix} \tag{5.69}$$

or

$$[x_W \quad y_W \quad z_W \quad 1]^T = [T]^{-1} \begin{bmatrix} x \\ y \\ z \\ 1 \end{bmatrix} \tag{5.70}$$

where

$$[T] = \begin{bmatrix} n_{1x} & n_{2x} & n_{3x} & x_2 \\ n_{1y} & n_{2y} & n_{3y} & y_2 \\ n_{1z} & n_{2z} & n_{3z} & z_2 \\ \hline 0 & 0 & 0 & 1 \end{bmatrix} \tag{5.71}$$

Notice that the $Z_W$ components of all the points will be zeros.

The equation of a line can be written as

$$Ax_W + By_W + C = 0$$

or
$$L = x_W + \frac{B}{A} y_W + \frac{C}{A} = 0 \qquad (5.72)$$

For each of the lines $L_1$ to $L_4$, the local coordinates of its two known points can be substituted into the above equation to find the two ratios $B/A$ and $C/A$. Utilizing the resulting equations with Eq. (5.65) gives an implicit equation of the conic curve that has the general form

$$f(x_W, y_W, u) = 0, \qquad 0 \le u \le 1 \qquad (5.73)$$

Substituting the local coordinates of $P_5$ into this equation gives the value of $u$. Now, this equation can be used to generate points on the conic curve. These points can be transformed to the MCS using Eq. (5.69) for display, plotting, and storage purposes. Notice that this approach can also apply to Eq. (5.66).

## 5.6   PARAMETRIC REPRESENTATION OF SYNTHETIC CURVES

Analytic curves, described in the previous section, are usually not sufficient to meet geometric design requirements of mechanical parts. Products such as car bodies, ship hulls, airplane fuselage and wings, propeller blades, shoe insoles, and bottles are a few examples that require free-form, or synthetic, curves and surfaces. The need for synthetic curves in design arises on two occasions: when a curve is represented by a collection of measured data points and when an existing curve must change to meet new design requirements. In the latter occasion, the designer would need a curve representation that is directly related to the data points and is flexible enough to bend, twist, or change the curve shape by changing one or more data points. Data points are usually called control points and the curve itself is called an interpolant if it passes through all the data points.

Mathematically, synthetic curves represent a curve-fitting problem to construct a smooth curve that passes through given data points. Therefore, polynomials are the typical form of these curves. Various continuity requirements can be specified at the data points to impose various degrees of smoothness of the resulting curve. The order of continuity becomes important when a complex curve is modeled by several curve segments pieced together end to end. Zero-order continuity ($C^0$) yields a position continuous curve. First ($C^1$)- and second ($C^2$)-order continuities imply slope and curvature continuous curves respectively. A $C^1$ curve is the minimum acceptable curve for engineering design. Figure 5-39 shows a geometrical interpretation of these orders of continuity. A cubic polynomial is the minimum-order polynomial that can guarantee the generation of $C^0$, $C^1$, or $C^2$ curves. In addition, the cubic polynomial is the lowest-degree polynomial that permits inflection within a curve segment and that allows representation of nonplanar (twisted) three-dimensional curves in space. Higher-order polynomials are not commonly used in CAD/CAM because they tend to oscillate about control points, are computationally inconvenient, and are uneconomical of storing curve and surface representations in the computer.

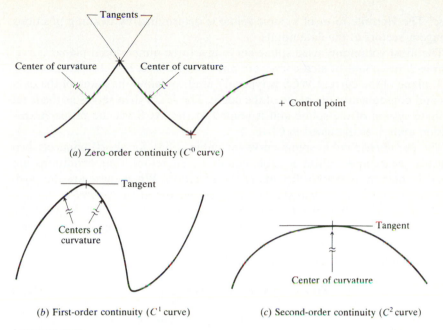

(a) Zero-order continuity ($C^0$ curve)

(b) First-order continuity ($C^1$ curve)

(c) Second-order continuity ($C^2$ curve)

**FIGURE 5-39**
Various orders of continuity of curves.

The type of input data and its influence on the control of the resulting synthetic curve determine the use and effectiveness of the curve in design. For example, curve segments that require positions of control points and/or tangent vectors at these points are easier to deal with and gather data for than those that might require curvature information. Also, the designer may prefer to control the shape of the curve locally instead of globally by changing the control point(s). If changing a control point results in changing the curve locally in the vicinity of that point, local control of the curve is achieved; otherwise global control results.

Major CAD/CAM systems provide three types of synthetic curves: Hermite cubic spline, Bezier, and B-spline curves. The cubic spline curve passes through the data points and therefore is an interpolant. Bezier and B-spline curves in general approximate the data points, that is, they do not pass through them. Under certain conditions, the B-spline curve can be an interpolant, as will be seen in Sec. 5.6.3. Both the cubic spline and Bezier curves have a first-order continuity and the B-spline curve has a second-order continuity. The formulation of each curve is discussed below.

## 5.6.1  Hermite Cubic Splines

Parametric spline curves are defined as piecewise polynomial curves with a certain order of continuity. A polynomial of degree $N$ has continuity of derivatives of order $(N - 1)$. Parametric cubic splines are used to interpolate to given data, not to design free-form curves as Bezier and B-spline curves do. Splines draw their name from the traditional draftsman's tool called "French curves or splines." The Hermite form of a cubic spline is determined by defining positions and tangent vectors at the data points.

The most commonly used spline curve is a three-dimensional planar curve. The three-dimensional twisted curves are not covered here. For the planar curve, the $XY$ plane of the current WCS is typically used to define the plane of the data points and consequently the plane of the curve. The WCS then serves as the local coordinate system of the spline and is related to the MCS via the proper transformation matrix, as discussed in Chap. 3.

The parametric cubic spline curve (or cubic spline for short) connects two data (end) points and utilizes a cubic equation. Therefore, four conditions are required to determine the coefficients of the equation. When these are the positions of the two endpoints and the two tangent vectors at the points, a Hermite cubic spline results. Thus the Hermite spline is considered as one form of the general parametric cubic spline. The reader is encouraged to extend the forthcoming development of the Hermite cubic spline to a cubic spline defined by four given data points. The parametric equation of a cubic spline segment is given by

$$P(u) = \sum_{i=0}^{3} C_i u^i, \qquad 0 \le u \le 1 \tag{5.74}$$

where $u$ is the parameter and $C_i$ are the polynomial (also called algebraic) coefficients. In scalar form this equation is written as

$$
\begin{aligned}
x(u) &= C_{3x} u^3 + C_{2x} u^2 + C_{1x} u + C_{0x} \\
y(u) &= C_{3y} u^3 + C_{2y} u^2 + C_{1y} u + C_{0y} \\
z(u) &= C_{3z} u^3 + C_{2z} u^2 + C_{1z} u + C_{0z}
\end{aligned}
\tag{5.75}
$$

In an expanded vector form, Eq. (5.74) can be written as

$$P(u) = C_3 u^3 + C_2 u^2 + C_1 u + C_0 \tag{5.76}$$

This equation can also be written in a matrix form as

$$P(u) = U^T C \tag{5.77}$$

where $U = [u^3 \quad u^2 \quad u \quad 1]^T$ and $C = [C_3 \quad C_2 \quad C_1 \quad C_0]^T$. $C$ is called the coefficients vector.

The tangent vector to the curve at any point is given by differentiating Eq. (5.74) with respect to $u$ to give

$$P'(u) = \sum_{i=0}^{3} C_i i u^{i-1}, \qquad 0 \le u \le 1 \tag{5.78}$$

In order to find the coefficients $C_i$, consider the cubic spline curve with the two endpoints $P_0$ and $P_1$ shown in Fig. 5-40. Applying the boundary conditions ($P_0$, $P_0'$ at $u = 0$ and $P_1$, $P_1'$ at $u = 1$), Eqs. (5.74) and (5.78) give

$$
\begin{aligned}
P_0 &= C_0 \\
P_0' &= C_1 \\
P_1 &= C_3 + C_2 + C_1 + C_0 \\
P_1' &= 3C_3 + 2C_2 + C_1
\end{aligned}
\tag{5.79}
$$

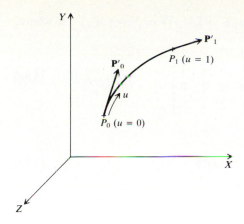

**FIGURE 5-40**
Hermite cubic spline curve.

Solving these four equations simultaneously for the coefficients gives

$$\mathbf{C}_0 = \mathbf{P}_0$$
$$\mathbf{C}_1 = \mathbf{P}'_0$$
$$\mathbf{C}_2 = 3(\mathbf{P}_1 - \mathbf{P}_0) - 2(\mathbf{P}'_0 - \mathbf{P}'_1)$$
$$\mathbf{C}_3 = 2(\mathbf{P}_0 - \mathbf{P}_1) + \mathbf{P}'_0 + \mathbf{P}'_1$$

(5.80)

Substituting Eqs. (5.80) into Eq. (5.76) and rearranging gives

$$\mathbf{P}(u) = (2u^3 - 3u^2 + 1)\mathbf{P}_0 + (-2u^3 + 3u^2)\mathbf{P}_1$$
$$+ (u^3 - 2u^2 + u)\mathbf{P}'_0 + (u^3 - u^2)\mathbf{P}'_1, \qquad 0 \le u \le 1 \qquad (5.81)$$

$\mathbf{P}_0, \mathbf{P}_1, \mathbf{P}'_0$, and $\mathbf{P}'_1$ are called geometric coefficients. The tangent vector becomes

$$\mathbf{P}'(u) = (6u^2 - 6u)\mathbf{P}_0 + (-6u^2 + 6u)\mathbf{P}_1$$
$$+ (3u^2 - 4u + 1)\mathbf{P}'_0 + (3u^2 - 2u)\mathbf{P}'_1, \qquad 0 \le u \le 1 \qquad (5.82)$$

The functions of $u$ in Eqs. (5.81) and (5.82) are called blending functions. The first two functions blend $\mathbf{P}_0$ and $\mathbf{P}_1$ and the second two blend $\mathbf{P}'_0$ and $\mathbf{P}'_1$ to produce the left-hand side in each equation.

Equation (5.81) can be written in a matrix form as

$$\mathbf{P}(u) = \mathbf{U}^T[M_H]\mathbf{V}, \qquad 0 \le u \le 1 \qquad (5.83)$$

where $[M_H]$ is the Hermite matrix and $\mathbf{V}$ is the geometry (or boundary conditions) vector. Both are given by

$$[M_H] = \begin{bmatrix} 2 & -2 & 1 & 1 \\ -3 & 3 & -2 & -1 \\ 0 & 0 & 1 & 0 \\ 1 & 0 & 0 & 0 \end{bmatrix} \qquad (5.84)$$

$$\mathbf{V} = [\mathbf{P}_0 \quad \mathbf{P}_1 \quad \mathbf{P}'_0 \quad \mathbf{P}'_1]^T \qquad (5.85)$$

Comparing Eqs. (5.77) and (5.83) show that $\mathbf{C} = [M_H]\mathbf{V}$ or $\mathbf{V} = [M_H]^{-1}\mathbf{C}$ where

$$[M_H]^{-1} = \begin{bmatrix} 0 & 0 & 0 & 1 \\ 1 & 1 & 1 & 1 \\ 0 & 0 & 1 & 0 \\ 3 & 2 & 1 & 0 \end{bmatrix} \tag{5.86}$$

Similarly, Eq. (5.82) can be written as

$$\mathbf{P}'(u) = \mathbf{U}^T[M_H]^u\mathbf{V} \tag{5.87}$$

where $[M_H]^u$ is given by

$$[M_H]^u = \begin{bmatrix} 0 & 0 & 0 & 0 \\ 6 & -6 & 3 & 3 \\ -6 & 6 & -4 & -2 \\ 0 & 0 & 1 & 0 \end{bmatrix} \tag{5.88}$$

Equation (5.81) describes the cubic spline curve in terms of its two end-points and their tangent vectors. The equation shows that the curve passes through the endpoints ($u = 0$ and 1). It also shows that the curve's shape can be controlled by changing its endpoints or its tangent vectors. If the two endpoints $P_0$ and $P_1$ are fixed in space, the designer can control the shape of the spline by changing either the magnitudes or the directions of the tangent vectors $\mathbf{P}'_0$ and $\mathbf{P}'_1$. The change of both the magnitudes and the directions is, of course, per-missible. However, for planar splines tangent vectors can be replaced by slopes. In this case, a default value, say one, for the lengths of the tangent vectors might be assumed by the software to enable Eq. (5.81) to be used. For example, if the slope at $P_0$ is given as 30°, then $\mathbf{P}'_0$ becomes [cos 30   sin 30   0]. It is obvious that the slope angle and the components of $\mathbf{P}'_0$ are given relative to the axes of the WCS that is active at the time of creating the spline curve.

Equation (5.81) can also be used to display or plot the spline. Points can be generated on the spline for different values of $u$ between 0 and 1. These points are then transformed to the MCS for display or plotting purposes.

Equation (5.81) is for one cubic spline segment. It can be generalized for any two adjacent spline segments of a spline curve that are to fit a given number of data points. This introduces the problem of blending or joining cubic spline seg-ments which can be stated as follows. Given a set of $n$ points $P_0$, $P_1$, ..., $P_{n-1}$ and the two end tangent vectors $\mathbf{P}'_0$ and $\mathbf{P}'_{n-1}$ (Fig. 5-41) connect the points with a cubic spline curve. The spline curve is created as a blend of spline segments connecting the set of points starting from $P_0$ and ending at $P_{n-1}$. Tangent vectors at the intermediate points $P_1$ through $P_{n-2}$ are needed as shown in Eq. (5.81) to compute these segments. To eliminate the need for these vectors, the continuity of curvature at these points can be imposed. To illustrate the procedure, consider eliminating $\mathbf{P}'_1$ between the first two segments that connect points $P_0$, $P_1$, and $P_2$. For curvature continuity between the first two segments, we can write

$$\mathbf{P}''(u_1 = 1) = \mathbf{P}''(u_2 = 0) \tag{5.89}$$

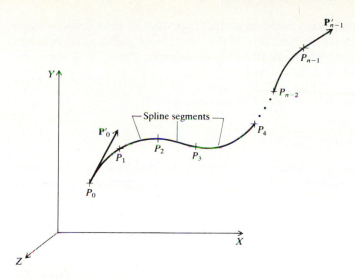

**FIGURE 5-41**
Hermite cubic spline curve.

where the subscripts of $u$ refer to the segment number. Differentiating Eq. (5.82) and using the result with Eq. (5.89), we obtain

$$\mathbf{P}'_1 = -\tfrac{1}{4}(3\mathbf{P}_0 + \mathbf{P}'_0 - 3\mathbf{P}_2 + \mathbf{P}'_2) \tag{5.90}$$

For more than two segments, a matrix equation can result from repeating this procedure, which can be solved for the intermediate tangent vectors in terms of the data points and the two end tangent vectors $\mathbf{P}'_0$ and $\mathbf{P}'_{n-1}$. Thus, the geometric information of a cubic spline database consists of the set of the data points and the two end tangent vectors.

The use of the cubic splines in design applications is not very popular compared to Bezier or B-spline curves. The control of the curve is not very obvious from the input data due to its global control characteristics. For example, changing the position of a data point or an end slope changes the entire shape of the spline which does not provide the intuitive feel required for design. In addition,

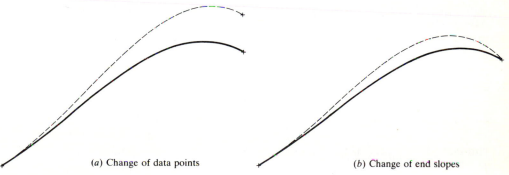

(a) Change of data points          (b) Change of end slopes

**FIGURE 5-42**
Control of cubic spline curve.

the order of the curve is always constant (cubic) regardless of the number of data points. In order to increase the flexibility of the curve, more points must be input, thus creating more splines which are all still of cubic order. Figure 5-42 shows the control aspects of the cubic spline curve.

**Example 5.18.** What shape of a cubic spline curve results if:

(a) $\mathbf{P}_0 = \mathbf{P}_1$, $\mathbf{P}'_1 = \mathbf{P}'_0$?
(b) $\mathbf{P}_0 = \mathbf{P}_1$, $\mathbf{P}'_1 = -\mathbf{P}'_0$?

**Solution.** This example illustrates how to create a closed curve using cubic splines. Consider only one segment in this example. Therefore the two endpoints $\mathbf{P}_0$ and $\mathbf{P}_1$ are always identical.

(a) If we substitute the given end conditions into Eqs. (5.81) and (5.82), we get

$$\mathbf{P}(u) = (2u^3 - 3u^2 + u)\mathbf{P}'_0 + \mathbf{P}_0$$

$$\mathbf{P}'(u) = (6u^2 - 6u + 1)\mathbf{P}'_0$$

This spline passes by $\mathbf{P}_0$ at $u = 0$, $\frac{1}{2}$, 1. Figure 5-43a shows the curve for $\mathbf{P}_0 = [0 \ \ 0 \ \ 0]^T$ and a slope of 45°, that is, $\mathbf{P}'_0 = [1/\sqrt{2} \ \ 1/\sqrt{2} \ \ 0]^T$. The spline is a straight line in the cartesian space because the slope $y'/x' = y'_0/x'_0$ is constant. Points 1, 2, 3, ..., 11 shown in the figure correspond to values of $u$ equal to 0, 0.1, 0.2, ..., 1 respectively. The spline has two extreme points 3 and 9 at $u = 0.2$ and 0.8 respectively. The extreme points can be obtained from solving the equation $\mathbf{P}'(u) = \mathbf{0}$.

(b) Similar to (a) we obtain

$$\mathbf{P}(u) = (-u^2 + u)\mathbf{P}'_0 + \mathbf{P}_0$$

$$\mathbf{P}'(u) = (-2u + 1)\mathbf{P}'_0$$

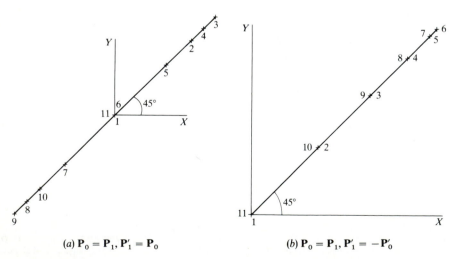

(a) $\mathbf{P}_0 = \mathbf{P}_1$, $\mathbf{P}'_1 = \mathbf{P}_0$　　　　　　　(b) $\mathbf{P}_0 = \mathbf{P}_1$, $\mathbf{P}'_1 = -\mathbf{P}'_0$

**FIGURE 5-43**
Closed cubic spline curves.

This spline has an extreme point at $u = \frac{1}{2}$. It begins the tangent to $\mathbf{P}'_0$ and then overlaps to become the tangent to $\mathbf{P}'_1$. Figure 5-43$b$ shows the curve for $\mathbf{P}_0 = [0 \quad 0 \quad 0]^T$ and slopes of 45 and 225°. The same analysis made above in ($a$) can be applied here. The reader may attempt to solve this example on a CAD/CAM system and compare the results using the "verify" command.

### 5.6.2   Bezier Curves

Cubic splines discussed in the previous section are based on interpolation techniques. Curves resulting from these techniques pass through the given points. Another alternative to create curves is to use approximation techniques which produce curves that do not pass through the given data points. Instead, these points are used to control the shape of the resulting curves. Most often, approximation techniques are preferred over interpolation techniques in curve design due to the added flexibility and the additional intuitive feel provided by the former. Bezier and B-spline curves are examples based on approximation techniques.

Bezier curves and surfaces are credited to P. Bezier of the French car firm Regie Renault who developed (about 1962) and used them in his software system called UNISURF which has been used by designers to define the outer panels of several Renault cars. These curves, known as Bezier curves, were also independently developed by P. DeCasteljau of the French car company Citroen (about 1959) which used it as part of its CAD system. The Bezier UNISURF system was soon published in the literature; this is the reason that the curves now bear Bezier's name.

As its mathematics show shortly, the major differences between the Bezier curve and the cubic spline curve are:

1. The shape of Bezier curve is controlled by its defining points only. First derivatives are not used in the curve development as in the case of the cubic spline. This allows the designer a much better feel for the relationship between input (points) and output (curve).
2. The order or the degree of Bezier curve is variable and is related to the number of points defining it; $n + 1$ points define an $n$th degree curve which permits higher-order continuity. This is not the case for cubic splines where the degree is always cubic for a spline segment.
3. The Bezier curve is smoother than the cubic spline because it has higher-order derivatives.

The Bezier curve is defined in terms of the locations of $n + 1$ points. These points are called data or control points. They form the vertices of what is called the control or Bezier characteristic polygon which uniquely defines the curve shape as shown in Fig. 5-44. Only the first and the last control points or vertices of the polygon actually lie on the curve. The other vertices define the order, derivatives, and shape of the curve. The curve is also always tangent to the first and last polygon segments. In addition, the curve shape tends to follow the polygon shape. These three observations should enable the user to sketch or

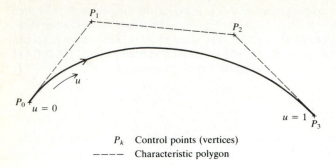

$P_k$   Control points (vertices)
- - - -   Characteristic polygon

**FIGURE 5-44**
Cubic Bezier curve (nomenclature).

predict the curve shape once its control points are given as illustrated in Fig. 5-45. The figure shows that the order of defining the control points changes the polygon definition which changes the resulting curve shape consequently. The arrow depicted on each curve shows its parametrization direction.

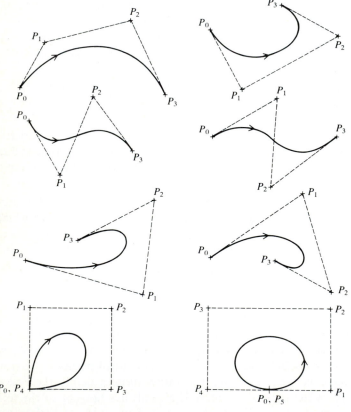

**FIGURE 5-45**
Cubic Bezier curves for various control points.

Mathematically, for $n + 1$ control points, the Bezier curve is defined by the following polynomial of degree $n$:

$$P(u) = \sum_{i=0}^{n} P_i B_{i,n}(u), \qquad 0 \leq u \leq 1 \tag{5.91}$$

where $P(u)$ is any point on the curve and $P_i$ is a control point. $B_{i,n}$ are the Bernstein polynomials. Thus, the Bezier curve has a Bernstein basis. The Bernstein polynomial serves as the blending or basis function for the Bezier curve and is given by

$$B_{i,n}(u) = C(n, i)u^i(1 - u)^{n-i} \tag{5.92}$$

where $C(n, i)$ is the binomial coefficient

$$C(n, i) = \frac{n!}{i!(n-i)!} \tag{5.93}$$

Utilizing Eqs. (5.92) and (5.93) and observing that $C(n, 0) = C(n, n) = 1$, Eq. (5.91) can be expanded to give

$$P(u) = P_0(1 - u)^n + P_1 C(n, 1)u(1 - u)^{n-1} + P_2 C(n, 2)u^2(1 - u)^{n-2}$$
$$+ \cdots + P_{n-1}C(n, n-1)u^{n-1}(1 - u) + P_n u^n, \qquad 0 \leq u \leq 1 \tag{5.94}$$

The characteristics of the Bezier curve are based on the properties of the Bernstein polynomials and can be summarized as follows:

1. The curve interpolates the first and last control points; that is, it passes through $P_0$ and $P_n$ if we substitute $u = 0$ and 1 in Eq. (5.94).
2. The curve is tangent to the first and last segments of the characteristic polygon. Using Eqs. (5.91) and (5.92), the $r$th derivatives at the starting and ending points are given by respectively:

$$P^r(0) = \frac{n!}{(n-r)!} \sum_{i=0}^{r} (-1)^{r-i} C(r, i)P_i \tag{5.95}$$

$$P^r(1) = \frac{n!}{(n-r)!} \sum_{i=0}^{r} (-1)^i C(r, i)P_{n-i} \tag{5.96}$$

Therefore, the first derivatives at the endpoints are

$$P'(0) = n(P_1 - P_0) \tag{5.97}$$

$$P'(1) = n(P_n - P_{n-1}) \tag{5.98}$$

where $(P_1 - P_0)$ and $(P_n - P_{n-1})$ define the first and last segments of the curve polygon. Similarly, it can be shown that the second derivative at $P_0$ is determined by $P_0$, $P_1$, and $P_2$; or, in general, the $r$th derivative at an endpoint is determined by its $r$ neighboring vertices.
3. The curve is symmetric with respect to $u$ and $(1 - u)$. This means that the sequence of control points defining the curve can be reversed without change of the curve shape; that is, reversing the direction of parametrization does not

change the curve shape. This can be achieved by substituting $1 - u = v$ in Eq. (5.94) and noticing that $C(n, i) = C(n, n - i)$. This is a result of the fact that $B_{i,n}(u)$ and $B_{n-i,n}(u)$ are symmetric if they are plotted as functions of $u$.

4. The interpolation polynomial $B_{i,n}(u)$ has a maximum value of $C(n, i) (i/n)^i (1 - i/n)^{n-i}$ occurring at $u = i/n$ which can be obtained from the equation $d(B_{i,n})/du = 0$. This implies that each control point is most influential on the curve shape at $u = i/n$. For example, for a cubic Bezier curve, $\mathbf{P}_0$, $\mathbf{P}_1$, $\mathbf{P}_2$, and $\mathbf{P}_3$ are most influential when $u = 0$, $\frac{1}{3}$, $\frac{2}{3}$, and 1 respectively. Therefore, each control point is weighed by its blending function for each $u$ value.

5. The curve shape can be modified by either changing one or more vertices of its polygon or by keeping the polygon fixed and specifying multiple coincident points at a vertex, as shown in Fig. 5-46. In Fig. 5-46a, the vertex $P_2$ is pulled to the new position $P_2^*$ and in Fig. 5-46b, $P_2$ is assigned a multiplicity $K$. The higher the multiplicity, the more the curve is pulled toward $P_2$.

6. A closed Bezier curve can simply be generated by closing its characteristic polygon or choosing $P_0$ and $P_n$ to be coincident. Figure 5-45 shows examples of closed curves.

7. For any valid value of $u$, the sum of the $B_{i,n}$ functions associated with the control points is always equal to unity for any degree of Bezier curve. This fact can be used to check numerical computations and software developments.

Thus far, only one Bezier curve segment is considered. In practical applications, the need may arise to deal with composite curves where various curve

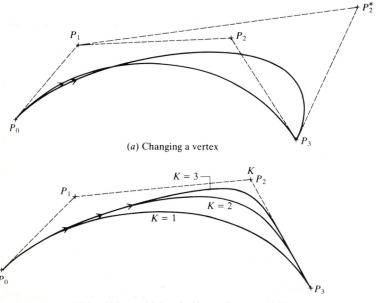

(a) Changing a vertex

(b) Specifying multiple coincident points at a vertex

**FIGURE 5-46**
Modifications of cubic Bezier curve.

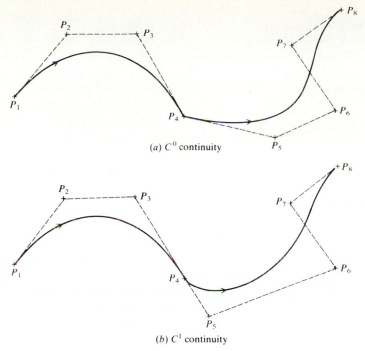

(a) $C^0$ continuity

(b) $C^1$ continuity

**FIGURE 5-47**
Blending Bezier curve segments.

segments are blended or joined together. In these applications maintaining continuity of various orders between the segments might be desired. Figure 5-47
shows two curve segments defined by the two sets of points $P_1, P_2, P_3, P_4$ and
$P_4, P_5, P_6, P_7, P_8$. To achieve a zero-order ($C^0$) continuity, it is sufficient to
make one of the end control points of the segments common, e.g., $P_4$ in Fig. 5-
47a. To achieve a first-order ($C^1$) continuity, the end slope of one segment must
equal the starting slope of the next segment; that is, the corresponding tangent
vectors are related to each other by a constant. This condition requires that the
last segment of the first polygon and the first segment of the second polygon form
a straight line. With regards to Fig. 5-47b, the three points $P_3, P_4$, and $P_5$ must
be collinear. Utilizing Eqs. (5.97) and (5.98), we can write

$$\mathbf{P_4} - \mathbf{P_3} = \tfrac{4}{3}(\mathbf{P_5} - \mathbf{P_4}) \tag{5.99}$$

A most desirable feature for any curve defined by a polygon such as the
Bezier curve is the convex hull property. This property relates the curve to its
characteristic polygon. This is what guarantees that incremental changes in
control point positions produce intuitive geometric changes. A curve is said to
have the convex hull property if it lies entirely within the convex hull defined by
the polygon vertices. In a plane, the convex hull is a closed polygon and in three
dimensions it is a polyhedron. The shaded area shown in Fig. 5-48 defines the
convex hull of a Bezier curve. The hull is formed by connecting the vertices of the
characteristic polygon.

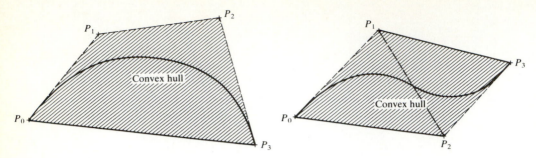

**FIGURE 5-48**
The convex hull of a Bezier curve.

Curves that possess the convex hull property enjoy some important consequences. If the polygon defining a curve segment degenerates to a straight line,
the resulting segment must therefore be linear. Thus a Bezier curve may have
locally linear segments embedded in it, which is a useful design feature. Also, the
size of the convex hull is an upper bound on the size of the curve itself; that is,
the curve always lies inside its convex hull. This is a useful property for graphics
functions such as displaying or clipping the curve. For example, instead of testing
the curve itself for clipping, its convex hull is tested first and only if it intersects
the display window boundaries should the curve itself be examined. A third con-
sequence of the convex hull property is that the curve never oscillates wildly
away from its defining control points because the curve is guaranteed to lie
within its convex hull.

From a software point of view, the database of a Bezier curve includes the
coordinates of the control points defining its polygon stored in the same order as
input by the user. Other information which may obviously be stored include
layer, color, name, font, and line width of the curve.

While a Bezier curve seems superior to a cubic spline curve, it still has some
disadvantages. First, the curve does not pass through the control points which
may be inconvenient to some designers. Second, the curve lacks local control. It
only has the global control nature. If one control point is changed, the whole
curve changes. Therefore, the designer cannot selectively change parts of the
curve.

**Example 5.19.** The coordinates of four control points relative to a current WCS are
given by

$$\mathbf{P}_0 = [2 \ \ 2 \ \ 0]^T, \quad \mathbf{P}_1 = [2 \ \ 3 \ \ 0]^T, \quad \mathbf{P}_2 = [3 \ \ 3 \ \ 0]^T, \quad \text{and} \quad \mathbf{P}_3 = [3 \ \ 2 \ \ 0]^T$$

Find the equation of the resulting Bezier curve. Also find points on the curve for
$u = 0, \frac{1}{4}, \frac{1}{2}, \frac{3}{4}$, and 1.

*Solution.* Equation (5.91) gives

$$\mathbf{P}(u) = \mathbf{P}_0 B_{0,3} + \mathbf{P}_1 B_{1,3} + \mathbf{P}_2 B_{2,3} + \mathbf{P}_3 B_{3,3}, \quad 0 \le u \le 1$$

Using Eqs. (5.92) and (5.93), the above equation becomes

$$\mathbf{P}(u) = \mathbf{P}_0(1-u)^3 + 3\mathbf{P}_1 u(1-u)^2 + 3\mathbf{P}_2 u^2(1-u) + \mathbf{P}_3 u^3, \quad 0 \le u \le 1$$

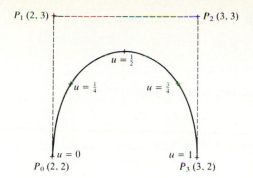

**FIGURE 5-49**
Bezier curve and generated points.

Substituting the $u$ values into this equation gives

$$\mathbf{P}(0) = \mathbf{P}_0 = [2 \quad 2 \quad 0]^T$$

$$\mathbf{P}\left(\frac{1}{4}\right) = \frac{27}{64}\mathbf{P}_0 + \frac{27}{64}\mathbf{P}_1 + \frac{9}{64}\mathbf{P}_2 + \frac{1}{64}\mathbf{P}_3 = [2.156 \quad 2.563 \quad 0]^T$$

$$\mathbf{P}\left(\frac{1}{2}\right) = \frac{1}{8}\mathbf{P}_0 + \frac{3}{8}\mathbf{P}_1 + \frac{3}{8}\mathbf{P}_2 + \frac{1}{8}\mathbf{P}_3 = [2.5 \quad 2.75 \quad 0]^T$$

$$\mathbf{P}\left(\frac{3}{4}\right) = \frac{1}{64}\mathbf{P}_0 + \frac{9}{64}\mathbf{P}_1 + \frac{27}{64}\mathbf{P}_2 + \frac{27}{64}\mathbf{P}_3 = [2.844 \quad 2.563 \quad 0]^T$$

$$\mathbf{P}(1) = \mathbf{P}_3 = [3 \quad 2 \quad 0]^T$$

Observe that $\sum_{i=0}^{3} B_{i,3}$ is always equal to unity for any $u$ value. Figure 5-49 shows the curve and the points.

**Example 5.20.** A cubic spline curve is defined by the equation

$$\mathbf{P}(u) = \mathbf{C}_3 u^3 + \mathbf{C}_2 u^2 + \mathbf{C}_1 u + \mathbf{C}_0, \quad 0 \leq u \leq 1 \tag{5.100}$$

where $\mathbf{C}_3$, $\mathbf{C}_2$, $\mathbf{C}_1$, and $\mathbf{C}_0$ are the polynomial coefficients [see Eq. (5.76)]. Assuming these coefficients are known, find the four control points that define an identical Bezier curve.

*Solution.* The Bezier equation is

$$\mathbf{P}(u) = \mathbf{P}_0 B_{0,3} + \mathbf{P}_1 B_{1,3} + \mathbf{P}_2 B_{2,3} + \mathbf{P}_3 B_{3,3} \tag{5.101}$$

where

$$B_{0,3} = 1 - 3u + 3u^2 - u^3 \qquad B_{1,3} = 3u - 6u^2 + 3u^3$$

$$B_{2,3} = 3u^2 - 3u^3 \qquad\qquad B_{3,3} = u^3$$

Substituting all these functions into Eq. (5.101) and rearranging, we obtain

$$\mathbf{P}(u) = (-\mathbf{P}_0 + 3\mathbf{P}_1 - 3\mathbf{P}_2 + \mathbf{P}_3)u^3 + (3\mathbf{P}_0 - 6\mathbf{P}_1 + 3\mathbf{P}_2)u^2$$

$$+ (-3\mathbf{P}_0 + 3\mathbf{P}_1)u + \mathbf{P}_0 \tag{5.102}$$

Comparing the coefficients of Eqs. (5.100) and (5.102) gives

$$\mathbf{P}_0 = \mathbf{C}_0$$
$$\mathbf{P}_1 = \tfrac{1}{3}\mathbf{C}_1 + \mathbf{C}_0$$
$$\mathbf{P}_2 = \tfrac{1}{3}(\mathbf{C}_2 + 2\mathbf{C}_1 + 3\mathbf{C}_0)$$
$$\mathbf{P}_3 = \mathbf{C}_3 + \mathbf{C}_2 + \mathbf{C}_1 + \mathbf{C}_0$$

*Note:* to check these results, observe that the Bezier curve passes through points $P_0$ and $P_3$ where $u = 0$ and 1 respectively. These two points are obtained from Eq. (5.100) for the same value of $u$.

### 5.6.3   B-Spline Curves

B-spline curves provide another effective method, besides that of Bezier, of generating curves defined by polygons. In fact, B-spline curves are the proper and powerful generalization of Bezier curves. In addition to sharing most of the characteristics of Bezier curves they enjoy some other unique advantages. They provide local control of the curve shape as opposed to global control by using a special set of blending functions that provide local influence. They also provide the ability to add control points without increasing the degree of the curve.

B-spline curves have the ability to interpolate or approximate a set of given data points. Interpolation is useful in displaying design or engineering results such as stress or displacement distribution in a part while approximation is good to design free-form curves. Interpolation is also useful if the designer has measured data points in hand that must lie on the resulting curve. This section covers only B-spline curves as used for approximation.

In contrast to Bezier curves, the theory of B-spline curves separates the degree of the resulting curve from the number of the given control points. While four control points can always produce a cubic Bezier curve, they can generate a linear, quadratic, or cubic B-spline curve. This flexibility in the degree of the resulting curve is achieved by choosing the basis (blending) functions of B-spline curves with an additional degree of freedom that does not exist in Bernstein polynomials. These basis functions are the B-splines—thus the name B-spline curves.

Similar to Bezier curves, the B-spline curve defined by $n + 1$ control points $P_i$ is given by

$$\mathbf{P}(u) = \sum_{i=0}^{n} \mathbf{P}_i N_{i,k}(u), \qquad 0 \le u \le u_{\max} \tag{5.103}$$

$N_{i,k}(u)$ are the B-spline functions. Thus B-spline curves have a B-spline basis. The control points (sometimes called deBoor points) form the vertices of the control or deBoor polygon. There are two major differences between Eqs. (5.103) and (5.91). First, the parameter $k$ controls the degree $(k - 1)$ of the resulting B-spline curve and is usually independent of the number of control points except as restricted as shown below. Second, the maximum limit of the parameter $u$ is no longer unity as it was so chosen arbitrarily for Bezier curves. The B-spline functions have the following properties:

Partition of unity: $\sum\limits_{i=0}^{n} N_{i,k}(u) = 1$

Positivity: $N_{i,k}(u) \geq 0$

Local support: $N_{i,k}(u) = 0 \qquad \text{if } u \notin [u_i, u_{i+k+1}]$

Continuity: $N_{i,k}(u)$ is $(k-2)$ times continuously differentiable

The first property ensures that the relationship between the curve and its defining control points is invariant under affine transformations. The second property guarantees that the curve segment lies completely within the convex hull of $P_i$. The third property indicates that each segment of a B-spline curve is influenced by only $k$ control points or each control point affects only $k$ curve segments. It is useful to notice that the Bernstein polynomial, $B_{i,n}(u)$, has the same first two properties mentioned above.

The B-spline function also has the property of recursion which is defined as

$$N_{i,k}(u) = (u - u_i)\frac{N_{i,k-1}(u)}{u_{i+k-1} - u_i} + (u_{i+k} - u)\frac{N_{i+1,k-1}(u)}{u_{i+k} - u_{i+1}} \tag{5.104}$$

where

$$N_{i,1} = \begin{cases} 1, & u_i \leq u \leq u_{i+1} \\ 0, & \text{otherwise} \end{cases} \tag{5.105}$$

Choose $0/0 = 0$ if the denominators in Eq. (5.104) become zero. Equation (5.105) shows that $N_{i,1}$ is a unit step function.

Because $N_{i,1}$ is constant for $k = 1$, a general value of $k$ produces a polynomial in $u$ of degree $(k-1)$ [see Eq. (5.104)] and therefore a curve of order $k$ and degree $(k-1)$. The $u_i$ are called parametric knots or knot values. These values form a sequence of nondecreasing integers called the knot vector. The values of the $u_i$ depend on whether the B-spline curve is an open (nonperiodic) or closed (periodic) curve. For an open curve, they are given by

$$u_j = \begin{cases} 0, & j < k \\ j - k + 1, & k \leq j \leq n \\ n - k + 2, & j > n \end{cases} \tag{5.106}$$

where

$$0 \leq j \leq n + k \tag{5.107}$$

and the range of $u$ is

$$0 \leq u \leq n - k + 2 \tag{5.108}$$

Relation (5.107) shows that $(n + k + 1)$ knots are needed to create a $(k - 1)$ degree curve defined by $(n + 1)$ control points. These knots are evenly spaced over the range of $u$ with unit separation $(\Delta u = 1)$ between noncoincident knots. Multiple (coincident) knots for certain values of $u$ may exist.

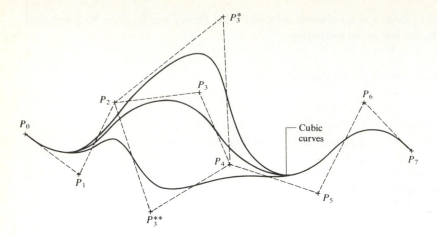

**FIGURE 5-50**
Local control of B-spline curves.

While the degree of the resulting B-spline curve is controlled by $k$, the range of the parameter $u$ as given by Eq. (5.108) implies that there is a limit on $k$ that is determined by the number of the given control points. This limit is found by requiring the upper bound in Eq. (5.108) to be greater than the lower bound for the $u$ range to be valid, that is,

$$n - k + 2 > 0 \tag{5.109}$$

This relation shows that a minimum of two, three, and four control points are required to define a linear, quadratic, and cubic B-spline curve respectively.

The characteristics of B-spline curves that are useful in design can be summarized as follows:

1. The local control of the curve can be achieved by changing the position of a control point(s), using multiple control points by placing several points at the same location, or by choosing a different degree $(k - 1)$. As mentioned earlier, changing one control point affects only $k$ segments. Figure 5-50 shows the local control for a cubic B-spline curve by moving $P_3$ to $P_3^*$ and $P_3^{**}$. The four curve segments surrounding $P_3$ change only.

2. A nonperiodic B-spline curve passes through the first and last control points $P_0$ and $P_{n+1}$ and is tangent to the first $(P_1 - P_0)$ and last $(P_{n+1} - P_n)$ segments of the control polygon, similar to the Bezier curve, as shown in Fig. 5-50.

3. Increasing the degree of the curve tightens it. In general, the less the degree, the closer the curve gets to the control points, as shown in Fig. 5-51. When $k = 1$, a zero-degree curve results. The curve then becomes the control points themselves. When $k = 2$, the curve becomes the polygon segments themselves.

4. A second-degree curve is always tangent to the midpoints of all the internal polygon segments (see Fig. 5-51). This is not the case for other degrees.

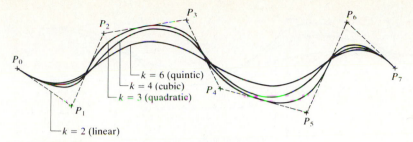

**FIGURE 5-51**
Effect of the degree of B-spline curve on its shape.

5. If $k$ equals the number of control points $(n + 1)$, then the resulting B-spline curve becomes a Bezier curve (see Fig. 5-52). In this case the range of $u$ becomes zero to one [see Eq. (5.108)] as expected.

6. Multiple control points induce regions of high curvature of a B-spline curve. This is useful when creating sharp corners in the curve (see Fig. 5-53). This effect is equivalent to saying that the curve is pulled more towards a control point by increasing its multiplicity.

7. Increasing the degree of the curve makes it more difficult to control and to calculate accurately. Therefore, a cubic B-spline is sufficient for a large number of applications.

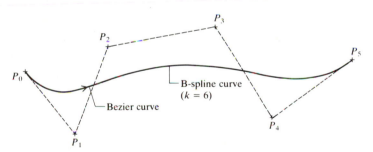

*(a)* No multiple control points

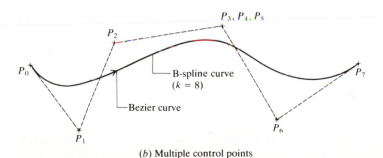

*(b)* Multiple control points

**FIGURE 5-52**
Identical B-spline and Bezier curves.

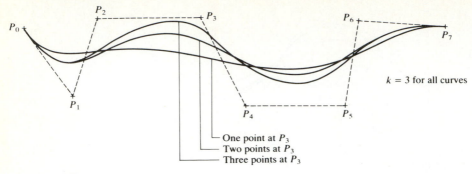

One point at $P_3$
Two points at $P_3$
Three points at $P_3$

**FIGURE 5-53**
Multiple control point B-spline curves.

Thus far, open or nonperiodic B-spline curves have been discussed. The same theory can be extended to cover closed or periodic B-spline curves. The only difference between open and closed curves is in the choice of the knots and the basic functions. Equations (5.106) to (5.108) determine the knots and the spacing between them for open curves. Closed curves utilize periodic B-spline functions as their basis with knots at the integers. These basis functions are cyclic translates of a single canonical function with a period (interval) of $k$ for support. For example, for a closed B-spline curve of order 2 ($k = 2$) or a degree 1 ($k - 1$), the basis function is linear, has a nonzero value in the interval (0, 2) only, and has a maximum value of one at $u = 1$, as shown in Fig. 5-54. The knot vector in this case is [0  1  2]. Quadratic and cubic closed curves have quadratic and cubic basis functions with intervals of (0, 3) and (0, 4) and knot vectors of [0  1  2  3] and [0  1  2  3  4] respectively.

The closed B-spline curve of degree ($k - 1$) or order $k$ defined by ($n + 1$) control points is given by Eq. (5.103) as the open curve. However, for closed curves Eqs. (5.104) to (5.108) become

$$N_{i,k}(u) = N_{0,k}((u - i + n + 1) \bmod (n + 1)) \tag{5.110}$$

$$u_j = j, \qquad 0 \le j \le n + 1 \tag{5.111}$$

$$0 < j \le n + 1 \tag{5.112}$$

and the range of $u$ is

$$0 \le u \le n + 1 \tag{5.113}$$

The mod ($n + 1$) in Eq. (5.110) is the modulo function. It is defined as

$$A \bmod n = \begin{cases} A, & A < n \\ 0, & A = n \\ \text{remainder of } A/n, & A > n \end{cases} \tag{5.114}$$

For example, 3.5 mod 6 = 3.5, 6 mod 6 = 0, and 7 mod 6 = 1. The mod function enables the periodic (cyclic) translation [mod ($n + 1$)] of the canonical basis function $N_{0,k}$. $N_{0,k}$ is the same as for open curves and can be calculated using Eqs. (5.104) and (5.105).

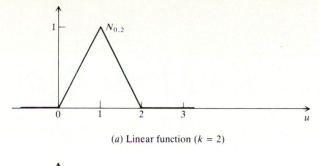

(a) Linear function ($k = 2$)

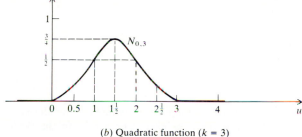

(b) Quadratic function ($k = 3$)

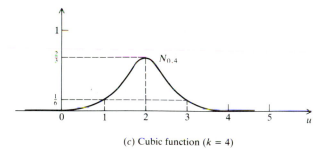

(c) Cubic function ($k = 4$)

**FIGURE 5-54**
Periodic B-spline basis functions.

Like open curves, closed B-spline curves enjoy the properties of partition of unity, positivity, local support, and continuity. They also share the same characteristics of the open curves except that they do not pass through the first and last control points and therefore are not tangent to the first and last segments of the control polygon. In representing closed curves, closed polygons are used where the first and last control points are connected by a polygon segment. It should be noticed that a closed B-spline curve can not be generated by simply using an open curve with the first and last control points being the same (coincident). The resulting curve is only $C^0$ continuous, as shown in Fig. 5-55. Only if the first and last segments of the polygon are colinear does a $C^1$ continuous curve result as in a Bezier curve.

Based on the above theory, the database of a B-spline curve includes the type of curve (open or closed), its order $k$ or degree ($k - 1$), and the coordinates of the control points defining its polygon stored in the same order as input by the user. Other information such as layer, color, name, font, and line width of the curve may be stored.

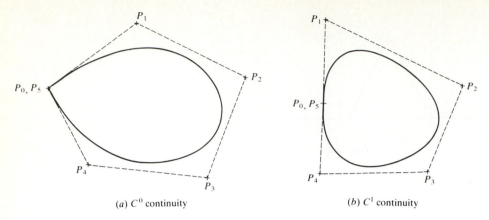

(a) $C^0$ continuity          (b) $C^1$ continuity

**FIGURE 5-55**
An open B-spline curve with $P_0$ and $P_n$ coincident.

**Example 5.21.** Find the equation of a cubic B-spline curve defined by the same control points as in Example 5.19. How does the curve compare with the Bezier curve?

**Solution.** This cubic spline has $k = 4$ and $n = 3$. Eight knots are needed to calculate the B-spline functions. Equation (5.106) gives the knot vector

$$[u_0 \quad u_1 \quad u_2 \quad u_3 \quad u_4 \quad u_5 \quad u_6 \quad u_7] \quad \text{as} \quad [0 \quad 0 \quad 0 \quad 0 \quad 1 \quad 1 \quad 1 \quad 1]$$

The range of $u$ [Eq. (5.108)] is $0 \le u \le 1$. Equation (5.103) gives

$$P(u) = P_0 N_{0,4} + P_1 N_{1,4} + P_2 N_{2,4} + P_3 N_{3,4}, \qquad 0 \le u \le 1 \qquad (5.115)$$

To calculate the above B-spline functions, use Eqs. (5.104) and (5.105) together with the knot vector as follows:

$$N_{0,1} = N_{1,1} = N_{2,1} = \begin{cases} 1, & u = 0 \\ 0, & \text{elsewhere} \end{cases}$$

$$N_{3,1} = \begin{cases} 1, & 0 \le u \le 1 \\ 0, & \text{elsewhere} \end{cases}$$

$$N_{4,1} = N_{5,1} = N_{6,1} = \begin{cases} 1, & u = 1 \\ 0, & \text{elsewhere} \end{cases}$$

$$N_{0,2} = (u - u_0) \frac{N_{0,1}}{u_1 - u_0} + (u_2 - u) \frac{N_{1,1}}{u_2 - u_1} = \frac{u N_{0,1}}{0} + \frac{(-u) N_{1,1}}{0} = 0$$

$$N_{1,2} = (u - u_1) \frac{N_{1,1}}{u_2 - u_1} + (u_3 - u) \frac{N_{2,1}}{u_3 - u_2} = \frac{u N_{1,1}}{0} + \frac{(-u) N_{2,1}}{0} = 0$$

$$N_{2,2} = (u - u_2) \frac{N_{2,1}}{u_3 - u_2} + (u_4 - u) \frac{N_{3,1}}{u_4 - u_3} = \frac{u N_{2,1}}{0} + \frac{(1 - u) N_{3,1}}{1} = (1 - u) N_{3,1}$$

$$N_{3,2} = (u - u_3) \frac{N_{3,1}}{u_4 - u_3} + (u_5 - u) \frac{N_{4,1}}{u_5 - u_4} = u N_{3,1} + \frac{(1 - u) N_{4,1}}{0} = u N_{3,1}$$

$$N_{4,2} = (u - u_4) \frac{N_{4,1}}{u_5 - u_4} + (u_6 - u) \frac{N_{5,1}}{u_6 - u_5} = (u - 1) \frac{N_{4,1}}{0} + \frac{(1 - u)N_{5,1}}{0} = 0$$

$$N_{5,2} = (u - u_5) \frac{N_{5,1}}{u_6 - u_5} + (u_7 - u) \frac{N_{6,1}}{u_7 - u_6} = \frac{(u - 1)N_{5,1}}{0} + \frac{(1 - u)N_{6,1}}{0} = 0$$

$$N_{0,3} = (u - u_0) \frac{N_{0,2}}{u_2 - u_0} + (u_3 - u) \frac{N_{1,2}}{u_3 - u_1} = u \frac{0}{0} + (-u) \frac{0}{0} = 0$$

$$N_{1,3} = (u - u_1) \frac{N_{1,2}}{u_3 - u_1} + (u_4 - u) \frac{N_{2,2}}{u_4 - u_2} = u \frac{N_{1,2}}{0} + \frac{(1 - u)N_{2,2}}{1} = (1 - u)^2 N_{3,1}$$

$$N_{2,3} = (u - u_2) \frac{N_{2,2}}{u_4 - u_2} + (u_5 - u) \frac{N_{3,2}}{u_5 - u_3} = uN_{2,2} + (1 - u)N_{3,2} = 2u(1 - u)N_{3,1}$$

$$N_{3,3} = (u - u_3) \frac{N_{3,2}}{u_5 - u_3} + (u_6 - u) \frac{N_{4,2}}{u_6 - u_4} = u^2 N_{3,1} + (1 - u) \frac{N_{4,2}}{0} = u^2 N_{3,1}$$

$$N_{4,3} = (u - u_4) \frac{N_{4,2}}{u_6 - u_4} + (u_7 - u) \frac{N_{5,2}}{u_7 - u_5} = (u - 1) \frac{N_{4,2}}{0} + (1 - u) \frac{N_{5,2}}{0} = 0$$

$$N_{0,4} = (u - u_0) \frac{N_{0,3}}{u_3 - u_0} + (u_4 - u) \frac{N_{1,3}}{u_4 - u_1} = (1 - u)^3 N_{3,1}$$

$$N_{1,4} = (u - u_1) \frac{N_{1,3}}{u_4 - u_1} + (u_5 - u) \frac{N_{2,3}}{u_5 - u_2} = 3u(1 - u)^2 N_{3,1}$$

$$N_{2,4} = (u - u_2) \frac{N_{2,3}}{u_5 - u_2} + (u_6 - u) \frac{N_{3,3}}{u_6 - u_3} = 3u^2(1 - u)N_{3,1}$$

$$N_{3,4} = (u - u_3) \frac{N_{3,3}}{u_6 - u_3} + (u_7 - u) \frac{N_{4,3}}{u_7 - u_4} = u^3 N_{3,1}$$

Substituting $N_{i,4}$ into Eq. (5.115) gives

$$\mathbf{P}(u) = [\mathbf{P}_0(1 - u)^3 + 3\mathbf{P}_1 u(1 - u)^2 + 3\mathbf{P}_2 u^2(1 - u) + \mathbf{P}_3 u^3]N_{3,1}, \qquad 0 \le u \le 1$$

Substituting $N_{3,1}$ into this equation gives the curve equation as

$$\mathbf{P}(u) = \mathbf{P}_0(1 - u)^3 + 3\mathbf{P}_1 u(1 - u)^2 + 3\mathbf{P}_2 u^2(1 - u) + \mathbf{P}_3 u^3, \qquad 0 \le u \le 1$$

This equation is the same as the one for the Bezier curve in Example 5.19. Thus the cubic B-spline curve defined by four control points is identical to the cubic Bezier curve defined by the same points. This fact can be generalized for a $(k - 1)$-degree curve as mentioned earlier.

There are two observations that are worth mentioning here. First, the sum of the two subscripts $(i, k)$ of any B-spline function $N_{i,k}$ cannot exceed $(n + k)$. This gives a control on how far to go to calculate $N_{i,k}$. In this example six functions of $N_{i,1}$, five of $N_{i,2}$, and four of $N_{i,3}$ were needed such that $(6 + 1)$ for the first, $(5 + 2)$ for the second, and $(4 + 3)$ for the last are always equal to 7 $(n + k)$. Second, whenever the limits of $u$ for any $N_{i,1}$ are equal, the $u$ range becomes one point.

**Example 5.22.** Find the equation of a closed (periodic) B-spline curve defined by four control points.

***Solution.*** This closed cubic spline has $k = 4$, $n = 3$. Using Eqs. (5.111) to (5.113), the knot vector $[u_0 \quad u_1 \quad u_2 \quad u_3 \quad u_4]$ is the integers $[0 \quad 1 \quad 2 \quad 3 \quad 4]$ and the range of $u$ is $0 \le u \le 4$. Equation (5.103) gives the curve equation as

$$P(u) = P_0 N_{0,4} + P_1 N_{1,4} + P_2 N_{2,4} + P_3 N_{3,4}, \qquad 0 \le u \le 4 \qquad (5.116)$$

To calculate the above B-spline functions, use Eq. (5.110) to obtain

$$N_{0,4}(u) = N_{0,4}((u + 4) \bmod 4)$$

$$N_{1,4}(u) = N_{0,4}((u + 3) \bmod 4)$$

$$N_{2,4}(u) = N_{0,4}((u + 2) \bmod 4)$$

$$N_{3,4}(u) = N_{0,4}((u + 1) \bmod 4)$$

In the above equations, $N_{0,4}$ on the right-hand side is the function for the open curve and on the left-hand side is the periodic function for the closed curve. Substituting these equations into Eq. (5.116) we get

$$P(u) = P_0 N_{0,4}((u + 4) \bmod 4) + P_1 N_{0,4}((u + 3) \bmod 4)$$

$$+ P_2 N_{0,4}((u + 2) \bmod 4) + P_3 N_{0,4}((u + 1) \bmod 4), \qquad 0 \le u \le 4 \quad (5.117)$$

In Eq. (5.117), the function $N_{0,4}$ has various arguments, which can be found if specific values of $u$ are used. To find $N_{0,4}$, similar calculations to the previous example 5.21 are performed using the above knot vector as follows:

$$N_{0,1} = \begin{cases} 1, & 0 \le u \le 1 \\ 0, & \text{elsewhere} \end{cases}$$

$$N_{1,1} = \begin{cases} 1, & 1 \le u \le 2 \\ 0, & \text{elsewhere} \end{cases}$$

$$N_{2,1} = \begin{cases} 1, & 2 \le u \le 3 \\ 0, & \text{elsewhere} \end{cases}$$

$$N_{3,1} = \begin{cases} 1, & 3 \le u \le 4 \\ 0, & \text{elsewhere} \end{cases}$$

$$N_{0,2} = (u - u_0) \frac{N_{0,1}}{u_1 - u_0} + (u_2 - u) \frac{N_{1,1}}{u_2 - u_1} = u N_{0,1} + (2 - u) N_{1,1}$$

$$N_{1,2} = (u - u_1) \frac{N_{1,1}}{u_2 - u_1} + (u_3 - u) \frac{N_{2,1}}{u_3 - u_2} = (u - 1) N_{1,1} + (3 - u) N_{2,1}$$

$$N_{2,2} = (u - u_2) \frac{N_{2,1}}{u_3 - u_2} + (u_4 - u) \frac{N_{3,1}}{u_4 - u_3} = (u - 2) N_{2,1} + (4 - u) N_{3,1}$$

$$N_{0,3} = (u - u_0) \frac{N_{0,2}}{u_2 - u_0} + (u_3 - u) \frac{N_{1,2}}{u_3 - u_1} = \frac{1}{2} u N_{0,2} + \frac{1}{2}(3 - u) N_{1,2}$$

$$= \tfrac{1}{2} u^2 N_{0,1} + \tfrac{1}{2}[u(2 - u) + (3 - u)(u - 1)] N_{1,1} + \tfrac{1}{2}(3 - u)^2 N_{2,1}$$

$$N_{1,3} = (u - u_1) \frac{N_{1,2}}{u_3 - u_1} + (u_4 - u) \frac{N_{2,2}}{u_4 - u_2} = \frac{1}{2}(u - 1) N_{1,2} + \frac{1}{2}(4 - u) N_{2,2}$$

$$= \tfrac{1}{2}(u - 1)^2 N_{1,1} + \tfrac{1}{2}[(u - 1)(3 - u) + (u - 2)(4 - u)] N_{2,1} + \tfrac{1}{2}(4 - u)^2 N_{3,1}$$

$$N_{0,4} = (u - u_0)\frac{N_{0,3}}{u_3 - u_0} + (u_4 - u)\frac{N_{1,3}}{u_4 - u_1} = \frac{1}{3}uN_{0,3} + \frac{1}{3}(4 - u)N_{1,3}$$

$$= \tfrac{1}{6}\{u^3 N_{0,1} + [u^2(2 - u) + u(3 - u)(u - 1) + (4 - u)(u - 1)^2]N_{1,1}$$

$$+ [u(3 - u)^2 + (4 - u)(u - 1)(3 - u) + (4 - u)^2(u - 2)]N_{2,1} + (4 - u)^3 N_{3,1}\}$$

or

$$N_{0,4} = \tfrac{1}{6}[u^3 N_{0,1} + (-3u^3 + 12u^2 - 12u + 4)N_{1,1} + (3u^3 - 24u^2 + 60u - 44)N_{2,1}$$

$$+ (-u^3 + 12u^2 - 48u + 64)N_{3,1}]$$

Due to the non-zero values of the functions $N_{i,1}$ for various intervals of $u$, the above equation can be written as

$$N_{0,4}(u) = \begin{cases} \tfrac{1}{6}u^3, & 0 \le u \le 1 \\ \tfrac{1}{6}(-3u^3 + 12u^2 - 12u + 4), & 1 \le u \le 2 \\ \tfrac{1}{6}(3u^3 - 24u^2 + 60u - 44), & 2 \le u \le 3 \\ \tfrac{1}{6}(-u^3 + 12u^2 - 48u + 64), & 3 \le u \le 4 \end{cases} \qquad (5.118)$$

To check the correctness of the above expression of $N_{0,4}(u)$, one would expect to obtain Fig. 5-54c if this function is plotted. Indeed, this figure is the plot of $N_{0,4}$. If $u = 0, 1, 2, 3,$ and $4$ are substituted into this function, the corresponding values of $N_{0,4}$ that are shown in the figure are obtained.

Equations (5.117) and (5.118) together can be used to evaluate points on the closed B-spline curve for display or plotting purposes. As an illustration, consider the following points:

$$P(0) = P_0 N_{0,4}(4 \bmod 4) + P_1 N_{0,4}(3 \bmod 4)$$

$$+ P_2 N_{0,4}(2 \bmod 4) + P_3 N_{0,4}(1 \bmod 4)$$

$$= P_0 N_{0,4}(0) + P_1 N_{0,4}(3) + P_2 N_{0,4}(2) + P_3 N_{0,4}(1)$$

$$= \tfrac{1}{6}P_1 + \tfrac{2}{3}P_2 + \tfrac{1}{6}P_3$$

Similarly,

$$P(0.5) = P_0 N_{0,4}(0.5) + P_1 N_{0,4}(3.5) + P_2 N_{0,4}(2.5) + P_3 N_{0,4}(1.5)$$

$$= \frac{1}{48}P_0 + \frac{1}{48}P_1 + \frac{23}{48}P_2 + \frac{23}{48}P_3$$

$$P(1) = P_0 N_{0,4}(1) + P_1 N_{0,4}(0) + P_2 N_{0,4}(3) + P_3 N_{0,4}(2)$$

$$= \tfrac{1}{6}P_0 + \tfrac{1}{6}P_2 + \tfrac{2}{3}P_3$$

$$P(2) = P_0 N_{0,4}(2) + P_1 N_{0,4}(1) + P_2 N_{0,4}(0) + P_3 N_{0,4}(3) = \tfrac{2}{3}P_0 + \tfrac{1}{6}P_1 + \tfrac{1}{6}P_3$$

$$P(3) = P_0 N_{0,4}(3) + P_1 N_{0,4}(2) + P_2 N_{0,4}(1) + P_3 N_{0,4}(0) = \tfrac{1}{6}P_0 + \tfrac{2}{3}P_1 + \tfrac{1}{6}P_2$$

$$P(4) = P_0 N_{0,4}(0) + P_1 N_{0,4}(3) + P_2 N_{0,4}(2) + P_3 N_{0,4}(1) = \tfrac{1}{6}P_1 + \tfrac{2}{3}P_2 + \tfrac{1}{6}P_3$$

In the above calculations, notice the cyclic rotation of the $N_{0,4}$ coefficients of the control points for the various values of $u$ excluding $u = 0.5$. Notice also the effect of the canonical (symmetric) form of $N_{0,4}$ on the coefficients of the control points. If the $u$ values are 0.5, 1.5, 2.5, and 3.5, or other values separated by unity, a similar cyclic rotation of the coefficients is expected. Finally, notice that $P(0)$ and $P(4)$ are equal, which ensures obtaining a closed B-spline curve.

The theory of the B-spline has been extended further to allow more control of the curve shape and continuity. For example, $\beta$-spline (beta-spline) and $v$-spline (nu-spline) curves provide manipulation of the curve shape and maintain its geometric continuity rather than its parametric continuity as provided by B-spline curves. The $\beta$-spline (sometimes called the spline in tension) curve is a generalization of the uniform cubic B-spline curve. The $\beta$-spline curve provides the designer with two additional parameters: the bias and the tension to control the shape of the curve. Therefore, the control points and the degree of the $\beta$-spline curve can remain fixed and yet the curve shape can be manipulated.

Although the $\beta$-spline curve is capable of applying tension at each control point, its formulation as piecewise hyperbolic sines and cosines makes its computation expensive. The $v$-spline curve is therefore developed as a piecewise polynomial alternative to the spline in tension.

### 5.6.4 Rational Curves

A rational curve is defined by the algebraic ratio of two polynomials while a nonrational curve [Eq. (5.103) gives an example] is defined by one polynomial. Rational curves draw their theories from projective geometry. They are important because of their invariance under projective transformation; that is, the perspective image of a rational curve is a rational curve. Rational Bezier curves, rational B-spline and $\beta$-spline curves, rational conic sections, rational cubics, and rational surfaces have been formulated. The most widely used rational curves are NURBS (nonuniform rational B-splines). A brief description of rational B-spline curves is given below.

The formulation of rational curves requires the introduction of homogeneous space and the homogeneous coordinates. This subject is covered in detail in Chap. 9 (Sec. 9.2.5). The homogeneous space is four-dimensional space. A point in $E^3$ with coordinates $(x, y, z)$ is represented in the homogeneous space by the coordinates $(x^*, y^*, z^*, h)$, where $h$ is a scalar factor. The relationship between the two types of coordinates is given by Eq. (9.59).

A rational B-spline curve defined by $n + 1$ control points $P_i$ is given by

$$\mathbf{P}(u) = \sum_{i=0}^{n} \mathbf{P}_i R_{i,\,k}(u), \qquad 0 \le u \le u_{\max} \tag{5.119}$$

$R_{i,\,k}(u)$ are the rational B-spline basis functions and are given by

$$R_{i,\,k}(u) = \frac{h_i N_{i,\,k}(u)}{\sum_{i=0}^{n} h_i N_{i,\,k}(u)} \tag{5.120}$$

The above equation shows that $R_{i,\,k}(u)$ are a generalization of the nonrational basis functions $N_{i,\,k}(u)$. If we substitute $h_i = 1$ in the equation, $R_{i,\,k}(u) = N_{i,\,k}(u)$. The rational basis functions $R_{i,\,k}(u)$ have nearly all the analytic and geometric characteristics of their nonrational B-spline counterparts. All the discussions covered in Sec. 5.6.3 apply here.

The main difference between rational and nonrational B-spline curves is the ability to use $h_i$ at each control point to control the behavior of the rational B-splines (or rational curves in general). Thus, similarly to the knot vector, one can define a homogeneous coordinate vector $H = [h_0 \quad h_1 \quad h_2 \quad h_3 \quad \cdots \quad h_n]^T$ at the control points $P_0, P_1, \ldots, P_n$ of the rational B-spline curve. The choice of the $H$ vector controls the behavior of the curve.

A rational B-spline is considered a unified representation that can define a variety of curves and surfaces. The premise is that it can represent all wireframe, surface, and solid entities. This allows unification and conversion from one modeling technique to another. Such an approach has some drawbacks, including the loss of information on simple shapes. For example, if a circular cylinder (hole) is represented by a B-spline, some data on the specific curve type may be lost unless it is carried along. Data including the fact that the part feature was a cylinder would be useful to manufacturing to identify it as a hole to be drilled or bored rather than a surface to be milled.

## 5.7 CURVE MANIPULATIONS

Analytic and synthetic curves essential to wireframe modeling have been presented in Secs. 5.5 and 5.6. The effective use of these curves in a design and manufacturing environment depends mainly on their manipulation to achieve goals in hand. A user might want to blend Bezier and B-spline curves with a certain continuity requirement, or the intersection of two curves in space might provide the coordinates of an important point to engineering calculations or modeling of a part. The next sections cover some useful features of curve manipulations. Refer to Chap. 12 for more details on the implementation of curve manipulations in CAD/CAM softwear.

### 5.7.1 Displaying

Displaying curves provides the designer with a means of visualizing geometric models within the limits of the wireframe modeling technique. Various colors can be assigned to various curves for identification and other purposes. To display a curve many closely spaced points on it are generated for various permissible values of the parameter $u$. The curve parametric equation is obviously used to generate the coordinates of these points. The display processing unit (see Chap. 2) receives these coordinates and changes them to two-dimensional coordinates relative to the device coordinate system of the display terminal. The resulting points are displayed and are connected by short-line segments. Line-generation hardware exists (refer to Chap. 2) for creating and displaying these line segments.

To display a straight line, the two endpoints are fed to the line-generation hardware to generate intermediate points and connect them by short-line segments. Equations of curves, such as circles and conic sections, that involve trigonometric functions are usually rewritten to minimize computation time to generate points on these curves for display purposes [see Eqs. (5.25), (5.35), (5.37), and (5.55)]. As mentioned in Chap. 2, displaying a curve and representing it in its database are two different things.

## 5.7.2  Evaluating Points on Curves

Points on curves are generated for display purposes as discussed previously or for other purposes. For example, finite element modeling techniques (see Chap. 18) requires generating nodal points (nodes) on the model to be used later in finite element analysis. In most CAD/CAM systems, the function of evaluating points on curves is made available to the user via a "generate point on" command or its equivalent.

It is also mentioned in the previous section that a curve parametric equation is used to evaluate points on it. Evaluation methods must be efficient and fast for interactive purposes as well as capable of producing enough points to display a smooth curve. The obvious method of calculating the coordinates of points on a curve by substituting successive values of its parameter into its equation is inefficient. Incremental methods proved more efficient.

The forward difference technique to evaluate a curve polynomial equation at equal intervals of its parameter is the most common incremental method. Consider applying the method to a cubic polynomial as an example. If the curve is given by

$$\mathbf{P}(u) = \mathbf{a}u^3 + \mathbf{b}u^2 + \mathbf{c}u + \mathbf{d}, \qquad 0 \le u \le 1 \tag{5.121}$$

and if $(n + 1)$ equally spaced points are to be evaluated for the $u$ range, then $\Delta u = 1/n$. The following equations can be used to initialize the method:

$$
\begin{aligned}
\mathbf{P}(0) &= \mathbf{d} \\
\Delta_1 \mathbf{P}_0 &= \mathbf{a}(\Delta u)^3 + \mathbf{b}(\Delta u)^2 + \mathbf{c}(\Delta u) \\
\Delta_2 \mathbf{P}_0 &= 6\mathbf{a}(\Delta u)^3 + 2\mathbf{b}(\Delta u)^2 \\
\Delta_3 \mathbf{P}_0 &= 6\mathbf{a}(\Delta u)^3
\end{aligned}
\tag{5.122}
$$

Any successive point can be evaluated from

$$
\begin{aligned}
\mathbf{P}_{i+1} &= \mathbf{P}_i + \Delta_1 \mathbf{P}_i, \qquad 0 \le i \le n \\
\Delta_1 \mathbf{P}_{i+1} &= \Delta_1 \mathbf{P}_i + \Delta_2 \mathbf{P}_i \\
\Delta_2 \mathbf{P}_{i+1} &= \Delta_2 \mathbf{P}_i + \Delta_3 \mathbf{P}_i
\end{aligned}
\tag{5.123}
$$

These equations show that only three additions are needed to generate any point and $3n$ additions generate the $n$ points. In general, for a polynomial of degree $n$, $n$ additions are required to calculate one point.

Evaluating a point on a curve given by its parametric value, $u$, is sometimes called the direct point solution. It entails evaluating three polynomials in $u$, one for each coordinate of the point. The inverse problem is another form of evaluating a point on a curve. Given a point on or close to a curve in terms of its cartesian coordinates $x$, $y$, and $z$, find the corresponding $u$ value. This problem arises, for example, if tangent vectors are to be evaluated at certain locations on the curve. The solution of this problem is called the inverse point solution and requires the solution of a nonlinear polynomial in $u$ via numerical methods (see Prob. 5.18 at the end of the chapter).

### 5.7.3 Blending

The construction of composite curves from the various types of parametric curve segments covered in Secs. 5.5 and 5.6 forms the core of the blending problem which can be stated as follows. Given two curve segments $P_1(u_1)$, $0 < u_1 < a$, $P_2(u_2)$, $0 < u_2 < b$, find the conditions for the two segments to be continuous at the joint. Notice that the upper limits on $u_1$ and $u_2$ are taken to be $a$ and $b$, and not 1, for generality (B-spline curves have an upper limit that is not unity in general). Three classes of continuity at the joint can be considered (see Fig. 5-39). The first is $C^0$ continuity; that is, the ending point of the first curve and the starting point of the second curve are the same. This gives

$$P_1(a) = P_2(0) \tag{5.124}$$

If the two segments are to be $C^1$ continuous as well, they must have slope continuity, that is,

$$P'_1(a) = \alpha_1 T$$
$$P'_2(0) = \alpha_2 T \tag{5.125}$$

where $\alpha_1$ and $\alpha_2$ are constants and $T$ is the common unit tangent vector at the joint. It has already been shown how to use curve characteristics for blending purposes as in the case with Bezier curves. Another example is the blending of a Bezier curve and an open B-spline curve. For slope continuity at the joint, the last segment of the control polygon of the former and the first segment of the control polygon of the latter must be colinear.

The third useful class of continuity is if the curvature ($C^2$ continuity) is to be continuous at the joint in addition to position and slope. To achieve curvature continuity is less straightforward and requires the binormal vector to a curve at a point. Figure 5-56 shows the tangent unit vector $T$, the normal unit vector $N$, the center of curvature $O$, and the radius of curvature $\rho$ at point $P$ on a curve

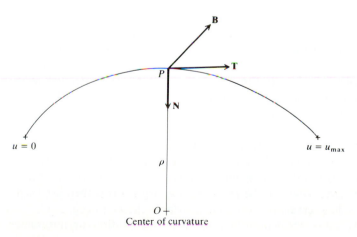

**FIGURE 5-56**
Binormal vector to a curve.

segment. The curvature at $P$ is defined as $1/\rho$. The binormal vector $\mathbf{B}$ is defined as

$$\mathbf{B} = \mathbf{T} \times \mathbf{N} \tag{5.126}$$

The curvature is related to the curve derivatives through the vector $\mathbf{B}$ by the following equation:

$$\frac{1}{\rho} \mathbf{B} = \frac{\mathbf{P}' \times \mathbf{P}''}{|\mathbf{P}'|^3} \tag{5.127}$$

where $\mathbf{P}''$ is the second derivative with respect to the parameter $u$. The following condition can then be written for curvature continuity at the joint:

$$\frac{\mathbf{P}_1'(a) \times \mathbf{P}_1''(a)}{|\mathbf{P}_1'(a)|^3} = \frac{\mathbf{P}_2'(0) \times \mathbf{P}_2''(0)}{|\mathbf{P}_2'(0)|^3} \tag{5.128}$$

Substituting Eqs. (5.125) into the above equation gives

$$\mathbf{T} \times \mathbf{P}_1''(a) = \left(\frac{\alpha_1}{\alpha_2}\right)^2 \mathbf{T} \times \mathbf{P}_2''(0) \tag{5.129}$$

This equation can be satisfied if

$$\mathbf{P}_2''(0) = \left(\frac{\alpha_2}{\alpha_1}\right)^2 \mathbf{P}_1''(a) \tag{5.130}$$

or, in general, if

$$\mathbf{P}_2''(0) = \left(\frac{\alpha_2}{\alpha_1}\right)^2 \mathbf{P}_1''(0) + \gamma \mathbf{P}_1'(a) \tag{5.131}$$

where $\gamma$ is an arbitrary scalar which is chosen as zero for practical purposes.


### 5.7.4 Segmentation

Segmentation or curve splitting is defined as replacing one existing curve by one or more curve segments of the same curve type such that the shape of the composite curve is identical to that of the original curve. Segmentation is a very useful feature for CAD/CAM systems; it is implemented as a "divide entity" command. Model clean-up for drafting and documentation purposes is an example where an entity (curve) might be divided into two at the line of sight of another (normally the two entities do not intersect in space). One of the resulting segments is then removed or blanked out. Another example is when a closed curve has to be split for modeling purposes. Consider the case of creating a ruled surface using a closed rectangle and a circle as rails of the surface (see Chap. 6). In this case the circle is split into four segments at the intersections of the rectangle diagonals with the circle. Each of the resulting four segments is then used with the proper rectangle side to create the four ruled surfaces shown in Fig. 5-57.

Mathematically, curve segmentation is a reparametrization or parameter transformation of the curve. Splitting lines, circles, and conics is a simple problem. To split a line connecting two points $P_0$ and $P_1$ at a point $P_2$, all that

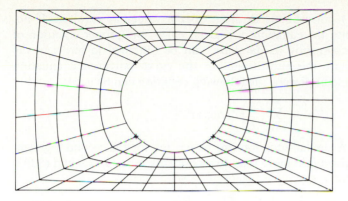

**FIGURE 5-57**
Segmentation of a circle for modeling purposes.

is needed is to define two new lines connecting the point pairs $(P_0, P_2)$ and $(P_2, P_1)$. For circles and conics, the angle corresponding to the splitting point together with the starting and ending angles of the original curve defines the proper range of the parameter $u$ (in this case the angle) for the resulting two segments. In other words, the parameter transformation in the case of lines, circles, and conics is trivial.

Polynomial curves such as cubic splines, Bezier curves, and B-spline curves require a different parameter transformation. If the degree of the polynomial defining a curve is to be unchanged, which is the case in segmentation, the transformation must be linear. Let us assume that a polynomial curve is defined over the range $u = u_0, u_m$. To split the curve at a point defined by $u = u_1$ means that the first and the second segments are to be defined over the range $u = u_0, u_1$ and $u = u_1, u_m$ respectively. A new parameter $v$ is introduced for each segment such that its range is $v = 0, 1$ (see Fig. 5-58). The parameter transformation takes the form:

$$u = u_0 + (u_1 - u_0)v \qquad \text{for the first segment}$$

$$(5.132)$$

and $\qquad u = u_1 + (u_m - u_1)v \qquad \text{for the second segment}$

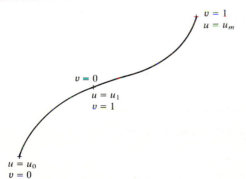

**FIGURE 5-58**
Reparametrization of a segmented curve.

It is clear that $v = 0, 1$ corresponds to the proper $u$ values as required. If Eq. (5.132) is substituted into the equation of a given curve, the proper equation of each segment, and consequently its database, can be obtained in terms of the parameter $v$. To facilitate this substitution in the curve polynomial equation where $u$ is raised to different powers, the following equation can be used:

$$u^n = \sum_{r=0}^{n} \binom{n}{r} u_0^r (\Delta u_0 \, v)^{n-r} \tag{5.133}$$

where $\Delta u_0 = u_1 - u_0$. A similar equation can be written for the second segment. If the curve is to be divided into more than two segments, Eqs. (5.132) and (5.133) can be applied successively.

**Example 5.23.** A cubic spline connecting two points is to be divided by a user into two segments. Find the endpoints and slopes for each segment.

**Solution.** Instead of substituting Eq. (5.133) into Eq. (5.81) of the spline directly and reducing the result, it is more efficient to use the matrix form given by Eq. (5.83). The vector **U** in this equation can be written in terms of the parameter $v$ using Eq. (5.133) as:

$$[u^3 \quad u^2 \quad u \quad 1] = [v^3 \quad v^2 \quad v \quad 1] \begin{bmatrix} \Delta u_0^3 & 0 & 0 & 0 \\ 3u_0 \, \Delta u_0^2 & \Delta u_0^2 & 0 & 0 \\ 3u_0^2 \, \Delta u_0 & 2u_0 \, \Delta u_0 & \Delta u_0 & 0 \\ u_0^3 & u_0^2 & u_0 & 1 \end{bmatrix}$$

or in a matrix form

$$\mathbf{U}^T = \mathbf{v}^T[T]$$

Substituting this equation into Eq. (5.83), we obtain

$$\mathbf{P}(u) = \mathbf{v}^T[T][M_H]\mathbf{V} = \mathbf{P}^*(v)$$

This equation can be rewritten in a similar form to Eq. (5.83) as

$$\mathbf{P}^*(v) = \mathbf{v}^T[M_H][M_H]^{-1}[T][M_H]\mathbf{V} = \mathbf{v}^T[M_H]\mathbf{V}^*$$

Therefore, the modified geometry, or boundary conditions, vector of the first spline segment is given by

$$\mathbf{V}^* = [M_H]^{-1}[T][M_H]\mathbf{V}$$

Expanding this equation and utilizing Eqs. (5.84) to (5.86) and $\Delta u_0 = u_1 - u_0$, we get

$$\mathbf{P}_0^* = (2u_0^3 - 3u_0^2 - 1)\mathbf{P}_0 + (-2u_0^3 + 3u_0^2)\mathbf{P}_1$$
$$+ (u_0^3 - 2u_0^2 + u_0)\mathbf{P}_0' + (u_0^3 - u_0^2)\mathbf{P}_1'$$

$$\mathbf{P}_1^* = (2u_1 - 3u_1^2 - 1)\mathbf{P}_0 + (-2u_1^3 + 3u_1^2)\mathbf{P}_1$$
$$+ (u_1^3 - 2u_1^2 + u_1)\mathbf{P}_0' + (u_1^3 - u_1^2)\mathbf{P}_1'$$

$$\mathbf{P}_0^{*\prime} = (u_1 - u_0)[(6u_0^2 - 6u_0)\mathbf{P}_0 + (-6u_0^2 + 6u_0)\mathbf{P}_1$$
$$+ (3u_0^2 - 4u_0 + 1)\mathbf{P}_0' + (3u_0^2 - 2u_0)\mathbf{P}_1']$$

$$\mathbf{P}_1^{*\prime} = (u_1 - u_0)[(6u_1^2 - 6u_1)\mathbf{P}_0 + (-6u_1^2 + 6u_1)\mathbf{P}_1$$
$$+ (3u_1^2 - 4u_1 + 1)\mathbf{P}_0' + (3u_1^2 - 2u_1)\mathbf{P}_1']$$

where $\mathbf{P}_i^{*\prime} = d\mathbf{P}_i^*/dv$ and $\mathbf{P}_i^{\prime} = d\mathbf{P}_i/du$. Comparing the above four equations with Eqs. (5.81) and (5.82) reveals that

$$\mathbf{P}_0^* = \mathbf{P}_0 \qquad\qquad \mathbf{P}_1^* = \mathbf{P}(u_1)$$

$$\mathbf{P}_0^{*\prime} = (u_1 - u_0)\mathbf{P}_0^{\prime} \qquad \mathbf{P}_1^{*\prime} = (u_1 - u_0)\mathbf{P}^{\prime}(u_1)$$

These equations conclude that the first spline segment has the endpoints $\mathbf{P}_0$ and $\mathbf{P}(u_1)$ which lie on the original curve and its end tangent vectors are related to the end vectors of the original curve by the factor $(u_1 - u_0)$. The boundary conditions for the second segment can, therefore, be written as

$$\mathbf{P}_0^* = \mathbf{P}(u_1) \qquad\qquad \mathbf{P}_1^* = \mathbf{P}_1$$

$$\mathbf{P}_0^{*\prime} = (u_m - u_1)\mathbf{P}^{\prime}(u_1) \qquad \mathbf{P}_1^{*\prime} = (u_m - u_1)\mathbf{P}_1^{\prime}$$

While the endpoints of each segment are logically understood, the above equations show that the magnitudes of the tangent vectors are scaled by the range $(u_1 - u_0)$ [or $(u_m - u_1)$] of the parametric variable to preserve the directions of the tangent vectors and thus the shape of the curve. This scale factor can be derived in another way as follows:

$$\mathbf{P}^{*\prime} = \frac{d\mathbf{P}^*}{dv} = \frac{d\mathbf{P}}{dv} = \frac{d\mathbf{P}^*}{du}\frac{du}{dv} = \frac{d\mathbf{P}}{du}\frac{du}{dv}$$

$$= \frac{du}{dv}\mathbf{P}^{\prime}$$

Substituting Eq. (5.132) gives the same result as above.

Similar development can be applied to split a Bezier curve or a B-spline curve. Refer to the problems at the end of the chapter.

### 5.7.5 Trimming

Trimming curves or entities is a very useful function provided by all CAD/CAM systems. Trimming can truncate or extend a curve. Trimming is mathematically identical to segmentation covered in the previous section. The only difference between the two is that the result of trimming a curve is only one segment of the

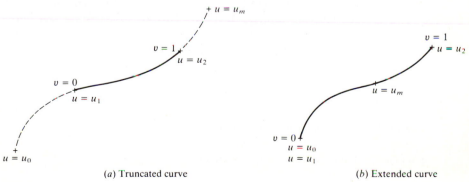

(a) Truncated curve                    (b) Extended curve

**FIGURE 5-59**
Reparametrization of a trimmed curve.

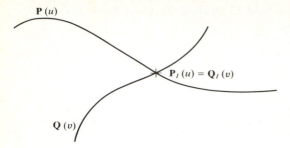

P (u)

$P_I(u) = Q_I(v)$

Q (v)

**FIGURE 5-60**
Intersection of two parametric curves in space.

curve bounded by the trimming boundaries. All the mathematical treatment of the segmentation applies here. The trimming function requires evaluating the equation of the desired segment of the curves, deleting the original curve, and then storing and displaying the desired segment. Figure 5-59 shows the reparametrization of a trimmed curve. Extending general curves (Fig. 5-59b) such as cubic splines, Bezier curves, and B-spline curves may not be recommended because the curve behavior outside its original interval $u_0 < u < u_m$ may not be predictable.

### 5.7.6 Intersection

The intersection point $P_I$ (see Fig. 5-60) of two parametric curves $\mathbf{P}(u)$ and $\mathbf{Q}(v)$ in three-dimensional space requires the solution of the following equation in the parameters $u$ and $v$:

$$\mathbf{P}(u) - \mathbf{Q}(v) = 0 \tag{5.134}$$

This equation represents three scalar equations that take the polynomial nonlinear form generally in the two parameter unknowns. One way to find $u$ and $v$ is to solve the $X$ and $Y$ components of the equation simultaneously, that is,

$$P_x(u) - Q_x(v) = 0$$
$$P_y(u) - Q_y(v) = 0 \tag{5.135}$$

and then use the $Z$ component, $P_z(u) - Q_z(v) = 0$, to verify the solution.

The roots of Eq. (5.134) or (5.135) can be found by numerical analysis methods such as the Newton-Raphson method. This method requires an initial guess which can be determined interactively. The user is usually asked by the CAD/CAM software to digitize the two curves whose intersection is required. These digitizes could be possibly changed to an initial guess to start the solution. If more than one intersection point exists, the user must digitize close to the desired intersection point.

**Example 5.24.** Find the intersection point of two lines.

*Solution.* Assume the equations of the lines are

$$\mathbf{P}(u) = \mathbf{P}_0 + u(\mathbf{P}_1 - \mathbf{P}_0) \tag{5.136}$$

$$\mathbf{Q}(v) = \mathbf{Q}_0 + v(\mathbf{Q}_1 - \mathbf{Q}_0) \tag{5.137}$$

The intersection point is given by the solution of the equation

$$\mathbf{P}_0 + u(\mathbf{P}_1 - \mathbf{P}_0) = \mathbf{Q}_0 + v(\mathbf{Q}_1 - \mathbf{Q}_0) \tag{5.138}$$

Taking the dot product of the above equation with the vector $(\mathbf{Q}_0 \times \mathbf{Q}_1)$ we obtain

$$(\mathbf{Q}_0 \times \mathbf{Q}_1) \cdot [\mathbf{P}_0 + u(\mathbf{P}_1 - \mathbf{P}_0)] = 0$$

since $(\mathbf{Q}_0 \times \mathbf{Q}_1)$ is perpendicular to the plane of $\mathbf{Q}_0$ and $\mathbf{Q}_1$ and hence to the two vectors. Solving the above equation for $u$ gives

$$u \doteq -\frac{(\mathbf{Q}_0 \times \mathbf{Q}_1) \cdot \mathbf{P}_0}{(\mathbf{Q}_0 \times \mathbf{Q}_1) \cdot (\mathbf{P}_1 - \mathbf{P}_0)} \tag{5.139}$$

Thus, $u$ is given in terms of the endpoints of the two lines. Substituting Eq. (5.139) into (5.136) results in the coordinates of the intersection point which are usually calculated in terms of the MCS of the given model or part. Equation (5.138) could have been solved for $v$ instead of $u$ to give

$$v = -\frac{(\mathbf{P}_0 \times \mathbf{P}_1) \cdot \mathbf{Q}_0}{(\mathbf{P}_0 \times \mathbf{P}_1) \cdot (\mathbf{Q}_1 - \mathbf{Q}_0)}$$

The above approach can be extended to find the intersection of a line given by $\mathbf{P}_0 + u(\mathbf{P}_1 - \mathbf{P}_0)$ and a curve given by $\mathbf{Q}(v)$ such as a Bezier or B-spline curve. In this case $u$ should be eliminated from the two equations to obtain

$$(\mathbf{P}_0 \times \mathbf{P}_1) \cdot \mathbf{Q}(v) = 0$$

### 5.7.7 Transformation

Manipulation or transformation of geometric entities during model construction or creating the model database offers a distinct advantage of CAD/CAM technology over traditional drafting methods. With transformation techniques, the designer can project, translate, rotate, mirror, and scale various entities. Section 5.3 shows how two-and-a-half-dimensional objects can be constructed easily using a "project" command. It also shows how the "mirror" command helps construct symmetric objects. Transformation is also useful in studying the motion of mechanisms, robots, and other objects in space. Animation techniques (Chap. 13) may be based on transforming an object into various positions and then replaying these positions continuously.

Simple transformations such as those mentioned above are usually referred to as rigid-body transformations. They can be directly applied to parametric representations of geometric models. To translate a model of a car, all points, lines, curves, and surfaces forming the model must be translated.

Homogeneous transformation, as discussed in Chap. 9, offers a concise matrix form to perform all rigid-body transformations as matrix multiplications, which is a desired feature from the software development point of view. Equation (3.3) gives the general homogeneous transformation matrix $[T]$. The proper choice of the elements of this matrix produces the various rigid-body transformations.

One of the main characteristics of rigid-body transformations is that geometric properties of curves, surfaces, and solids are invariant under these transformations. For example, originally parallel or perpendicular straight lines remain so after transformations. Intersection points of curves are transformed into the new intersection points, that is, one-to-one transformation. Chapter 9 discusses in more detail the subject of geometric transformation.

## 5.8   DESIGN AND ENGINEERING APPLICATIONS

This section presents some examples that show how theories covered in this chapter are applied to design and engineering problems. Consequently, it shows how existing CAD/CAM systems can be stretched beyond just using them for drafting, geometric modeling, and beyond what application modules of these systems offer. The reader is advised to work these examples on an actual CAD/CAM system.

**Example 5.25.**  For the state of plane stress shown in Fig. 5-61, determine:

(a) The principal stresses.
(b) The state of stress exerted on plane a-a.

*Solution.* The solution of this typical problem is to use Mohr's circle which is a graphical method. The centralization of CAD/CAM databases can be used to sim-plify the use of the method. The designer follows the part setup procedure covered in Chap. 3 and utilizes the default view and construction plane. Then two lines are created, one horizontal and one vertical, passing through the origin of the MCS. These form the $\sigma$ and $\tau$ axes (see Fig. 5-62). The stresses at the $X$ and $Y$ planes can be used to form the endpoints $P_1$ and $P_2$ of a line. These points have the coordi-nates $(-400, -600, 0)$ and $(800, 600, 0)$. Now a circle can be constructed whose center is the intersection of the line with the $\sigma$ axis and whose diameter is that line. The intersection points $P_3$ and $P_4$ of the circle with the $\sigma$ axis are the principal stresses. Their coordinates are $(\sigma_{min}, 0, 0)$ and $(\sigma_{max}, 0, 0)$, and $\sigma_{max}$ and $\sigma_{min}$ can simply be obtained using the "verify" command available on the CAD/CAM system. The orientation of the principal planes can be found by evaluating the angle $\alpha$ between the $\sigma$ axis and the line using the "measure angle" command.

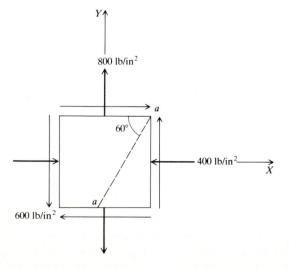

**FIGURE 5-61**
Stress state.

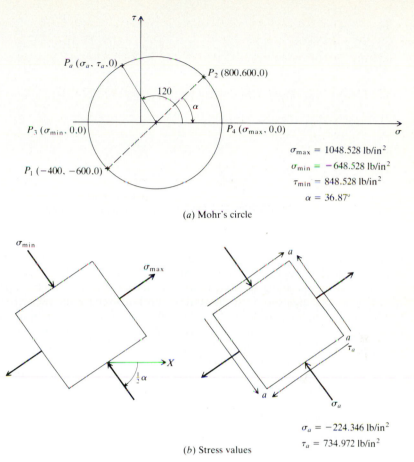

$\sigma_{max} = 1048.528 \ \text{lb/in}^2$

$\sigma_{min} = -648.528 \ \text{lb/in}^2$

$\tau_{min} = 848.528 \ \text{lb/in}^2$

$\alpha = 36.87°$

(a) Mohr's circle

$\sigma_a = -224.346 \ \text{lb/in}^2$

$\tau_a = 734.972 \ \text{lb/in}^2$

(b) Stress values

**FIGURE 5-62**
Stress calculations via Mohr's circle.

To find the state of stress on plane *a-a*, a line that passes through the circle center and has an angle of 120° with the $\sigma$ axis can be constructed. Its intersection point with the circle gives the state of stress on this plane. The complete solution is shown in Fig. 5-62.

The above procedure can be automated by writing a macro (refer to Chap. 15). The ease with which the designer can learn macro programming versus learning a full programming language makes this procedure attractive.

**Example 5.26.** The bar *AB* shown in Fig. 5-63 moves with its ends in contact with the horizontal and vertical walls. Assuming a plane motion, find the locus of point *C* on the bar for $D = 6$ inches and $L = 20$ inches. What is the locus if *C* is the centerpoint of the rod? Prove your answer.

*Solution.* The planning strategy to solve this problem begins by trying to generate enough points on the locus and then interpolating these points by a B-spline curve. Follow a similar part setup procedure as in the previous example. Create two lines,

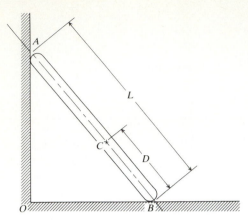

**FIGURE 5-63**
Bar *AB* in plane motion.

horizontal and vertical, passing through the origin of the MCS to represent the two walls. The bar *AB* is represented by a line of length 20 inches (see Fig. 5-64).

The extreme positions of point *C* are at distances *D* and $L - D$ from point *O* when the bar *AB* coincides with the vertical and horizontal walls respectively. To find any other point on the locus, let us start from the vertical position of *AB* and increment its motion until it becomes horizontal. Thus, we start when point *A* is at the location (0, *L*, 0). Assuming that ten points are enough to generate the locus, point *A* has the coordinates (0, $L - m \, \Delta L$, 0) where $\Delta L = L/10$ and $0 < m < 10$. To find point *C* for any position of the bar, create a circle with center at *A* and radius equal to *L*. The intersection point between the circle and the horizontal line gives point *B* and, therefore, the orientation of the bar. Point *C* can be located on *AB* using a point command with a parameter *u* value equal to $D/L$ or $(1 - D/L)$ depending on whether point *B* or *A* is input first in the line command used to create the line. The algorithm to generate points on the locus can therefore be written as

```
LOOP C1 = circle with Pc = A, R = L
     L1 = Line connecting P0 = A and P1 = Intersection of C1 and horizontal line
     C = point at u = 1 − D/L
     A = point at (0, L − m ΔL, 0)
     when m is equal to 10 exit
     Go to LOOP
```

Once all the points are generated, they are connected with a B-spline curve via a B-spline command. The locus is shown in Fig. 5-64. To facilitate the management of all graphics entities during construction, it is recommended that all possible aids such as layers, colors, and fonts be used (see Chap. 11). This method lends itself to macro programming which is useful in parametric design studies.

If point *C* is the center of *AB*, the locus becomes a circle with center at *O* and radius equal to the distance *OC*. Indeed, the locus of *C* is an ellipse in general with center at *O* and major and minor axes of lengths equal to $(L - D)$ and *D* if $D < L/2$ and vice versa if $D > L/2$. To prove this, consider Fig. 5-65, which gives the following coordinates of point *C*:

$$x = (L - D) \cos \theta \qquad y = D \sin \theta \qquad z = 0$$

which is an equation of an ellipse. If $D = L/2$, an equation of a circle results.

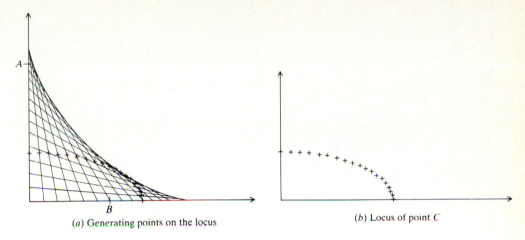

(a) Generating points on the locus

(b) Locus of point $C$

**FIGURE 5-64**
Locus of point $C$ on bar $AB$.

**Example 5.27.** Minimize the function

$$\phi(x, y) = (x - 3)^2 + (y - 3)^2$$

subject to the constraints

$$\phi_1 \equiv x \geq 0$$

$$\phi_2 \equiv y \geq 0$$

$$\phi_3 \equiv x + y - 4 \leq 0$$

*Solution.* After following the part setup, construct the horizontal and vertical lines passing through the origin of the MCS. The above constraint set is the shaded triangle shown in Fig. 5-66. The function $\phi$ is a circle with a center at $C(3, 3, 0)$ and has a minimum value at that point ignoring the constraints. With the constraints, the minimum corresponds to the circle that is tangent to the line $\phi_3$. The tangency point is the intersection point $P$ between $\phi_3$ and the line connecting the origin $O$ and the center $C$. Verifying this point gives its coordinates as $(2, 2, 0)$. Therefore, $\phi_{\min} = \phi(2, 2) = 2$.

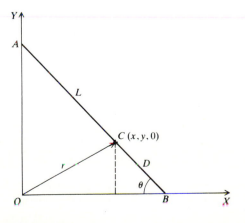

**FIGURE 5-65**
Analytic development of locus of point $C$.

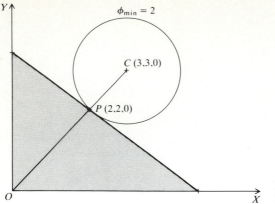

**FIGURE 5-66**
Minimum of the function $\phi(x, y)$.

**Example 5.28.** A four-bar mechanism is shown Fig. 5-67. The input angular velocity of the link $AB$ is $\omega_1 = 200$ r/min clockwise. Point $E$, the center of the link $CB$, is connected to a valve that is not shown in the figure. The mechanism is to be redesigned such that:

      *Design criterion.* The maximum linear velocity of point $E$, $v_{E,\,max}$, must be greater than its current value by at least 5 inches/s.

      *Design constraints.* (1) Only $L$ and $LL$ lengths can change and (2) $AB$ must rotate the full 360°.

*Solution.* The solution to this design problem requires the velocity analysis of mechanisms. Either the relative velocity or instantaneous center concept can be utilized. The latter is used here because it lends itself to CAD/CAM techniques. The velocity analysis is shown in Fig. 5-68. The instantaneous center of the mechanism is the intersection point $I$ of links $AB$ and $CD$ for any configuration of the mechanism. The velocity analysis gives

$$v_B = \omega_1 L = \omega_2 L_1$$

$$\omega_2 = \omega_1 L/L_1 \qquad\qquad (5.140)$$

$$v_E = \omega_2 L_2 = \omega_1 LL_2/L_1$$

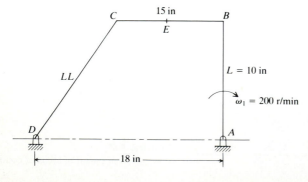

**FIGURE 5-67**
Four-bar mechanism.

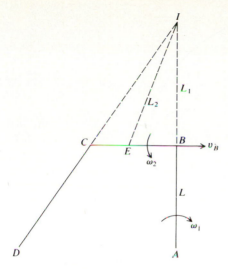

**FIGURE 5-68**

Velocity analysis of mechanism shown in Fig. 5-67.

The last equation gives the velocity of point $E$ in terms of the input velocity and the lengths $L$, $L_1$, and $L_2$. To obtain $v_{E,\,\max}$, rotate the link $AB$ around point $A$ an angle $\theta$, construct the mechanism in the new position, find $L_1$ and $L_2$, and substitute in Eqs. (5.140) to find $v_E$. Repeat for the admissible range of the angle $\theta$. Connect the resulting points [each point has coordinates $(\theta, v_E, 0)$] with a B-spline curve. Find $v_{E,\,\max}$ from the curve. To construct the mechanism at any angle $\theta$, rotate $AB$ about $A$, the required angle. This defines point $B$. Point $C$ is the intersection of two circles. The first has a center at $B$ and a radius of 15 in and the second has a center at $D$ and a radius equal to $LL$. Thus, the admissible range of angle of rotation $(\theta)$ of link $AB$ for a certain set of dimensions is found when the above two circles do not intersect.

With the above strategy in mind, the designer can choose various values for $L$ and $LL$ in an attempt to achieve the design goal and the given design constraints. The major commands needed are "rotate, measure distance, B-spline" commands in addition to layer and utility (delete. . .) commands.

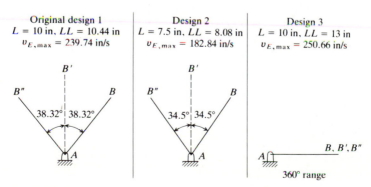

| Original design 1 | Design 2 | Design 3 |
|---|---|---|
| $L = 10$ in, $LL = 10.44$ in | $L = 7.5$ in, $LL = 8.08$ in | $L = 10$ in, $LL = 13$ in |
| $v_{E,\max} = 239.74$ in/s | $v_{E,\max} = 182.84$ in/s | $v_{E,\max} = 250.66$ in/s |

**FIGURE 5-69**

Results for three design iterations.

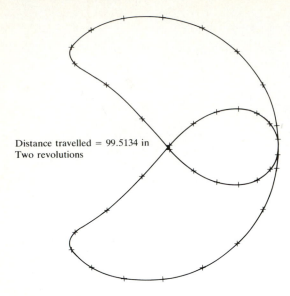

Distance travelled = 99.5134 in
Two revolutions

**FIGURE 5-70**
Locus of point $E$ for design 3.

Figure 5-69 shows the results for two designs in addition to the original. Design 3 is the final design. Figure 5-70 shows the locus of point $E$ for this design and Fig. 5-71 shows the velocity curves of $E$ for the three designs. Notice that for the final design $v_{E, max} = 250.656$ inch/s, which meets the design goal.

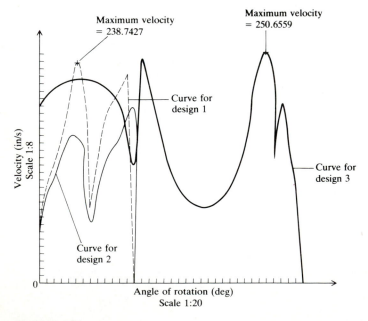

Maximum velocity = 238.7427

Maximum velocity = 250.6559

Curve for design 1

Curve for design 3

Velocity (in/s)
Scale 1:8

Curve for design 2

0

Angle of rotation (deg)
Scale 1:20

**FIGURE 5-71**
Velocity curves of designs 1, 2, and 3.

# PROBLEMS
## Part 1: Theory

**5.1.** Find the equation and endpoints of each of the following lines:

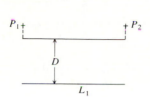

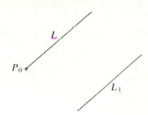

(a) Parallel to $L_1$ at distance $D$ and bounded by points $P_1$ and $P_2$

(b) Parallel to $L_1$, passes through $P_0$, and has length $L$

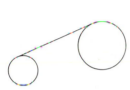

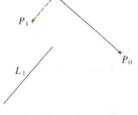

(c) Tangent to two given circles

(d) Perpendicular to $L_1$, passes through $P_0$, and bounded by $P_1$

(e) Tangent to a given circle from a given point $P_0$

(f) Tangent to a given ellipse from a given point $P_0$

**5.2.** Find the length of the common perpendicular to two skew lines.

**5.3.** Find the radius and the center of the following circles:
   (a) Tangent to a given circle and a given line with a given radius.
   (b) Tangent to two lines and passing through a given point.
   (c) Passing through two points and tangent to a line.
   (d) Tangent to a line, passing through a point, and with a given radius.

**5.4.** Find the center and the major and minor radii of an ellipse defined by two points and two slopes.

**5.5.** Find the intersection point of two tangent lines at two known points on an ellipse.

**5.6.** Find the tangent to an ellipse:
   (a) At any given point on its circumference.
   (b) From a point outside the ellipse.

**5.7.** As Prob. 5.6 but for a parabola.

**5.8.** Figure P5-8 shows a cubic spline curve consisting of two segments 1 and 2 with end conditions $\mathbf{P}_0$, $\mathbf{P}_0'$, $\mathbf{P}_1$, $\mathbf{P}_{11}'$, and $\mathbf{P}_1$, $\mathbf{P}_{12}'$, $\mathbf{P}_2$, $\mathbf{P}_2'$. The second subscripts in the tangent vectors at $P_1$ refer to the segment number. If $\mathbf{P}_{11} = \mathbf{R}$ and $\mathbf{P}_{12} = K\mathbf{R}$, where $K$ is a constant, prove that the curve can only be $C^1$ continuous at $P_1$ and $C^2$ elsewhere.

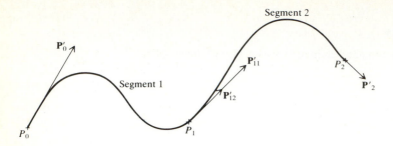

**FIGURE P5-8**

**5.9.** Find the normal vector to a cubic spline curve at any of its points.

**5.10.** For a cubic Bezier curve, carry a similar matrix formulation to a cubic spline. Compare $[M_B]$ and $V$ for the two curves.

**5.11.** Find the condition that a cubic Bezier curve degenerates to a straight line connecting $\mathbf{P}_0$ and $\mathbf{P}_3$.

**5.12.** Derive a method by which you can force a Bezier curve to pass through a given point in addition to the starting and ending points of its polygon. Achieve that by changing the position of only one control point, say $\mathbf{P}_2$.

**5.13.** Investigate the statement "each segment of a B-spline curve is influenced by only $k$ control points or each control point affects only $k$ curve segments." Use $n = 3$, $k = 2$, 3, 4.

**5.14.** Find the equation of an open quadratic B-spline curve defined by five control points.

**5.15.** For an open cubic B-spline curve defined by $n$ control points, carry a similar matrix formulation to a cubic spline. Compare $[M_S]$ and $V$ with those of the cubic spline and Bezier curves.

    *Hint:* Start with $n = 5$, that is, six control points, and then generalize the resulting $[M_S]$.

**5.16.** As Prob. 5.14 but for a closed B-spline curve.

**5.17.** Derive Eqs. (5.122) and (5.123).

**5.18.** Given a point $Q$ and a parametric curve in the cartesian space, find the closest point $P$ on the curve to $Q$.

    *Hint:* Find $P$ such that $(\mathbf{Q} - \mathbf{P})$ is perpendicular to the tangent vector.

**5.19.** A cubic Bezier curve is to be divided by a designer into two segments. Find the modified polygon points for each segment.

## Part 2: Laboratory

**5.20.** Obtain the three orthographic views (front, top, and right side) and a perspective view looking along the $Z$ axis at a distance 20 inches from the parts shown in Fig. P5-20 (all dimensions in inches). Obtain final drawings of these views. Follow the model clean-up and documentation procedure on your particular CAD/CAM system. Obtain the standard six isometric views of each model. Clean up each view.

**5.21.** Using your CAD/CAM system, investigate the line, circle, ellipse, parabola, conics, cubic spline, Bezier curve, and B-spline curve commands and their related modifiers. Relate these modifiers to their theoretical background covered in this chapter.

**5.22.** Choose a mechanical element such as a gear and generate its geometric model.

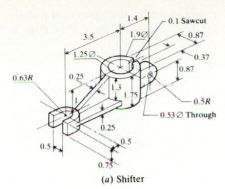

(a) Shifter

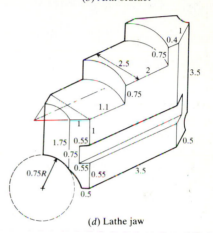

(b) Arm bracket

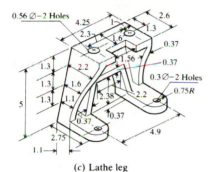

(c) Lathe leg

(d) Lathe jaw

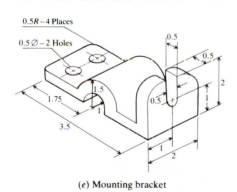

(e) Mounting bracket

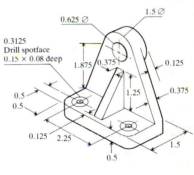

(f) Pipe bracket

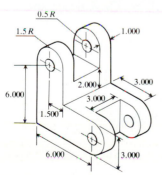

(g) Support bracket　　**FIGURE P5-20**

**5.23.** In Fig. P5-23, the similar links $AB$ and $CD$ rotate about the fixed pins at $A$ and $D$. Find the locus of point $P$ for $0 < \theta < 180$ for $L = 7$ and 3.5 inches. How does $L$ affect the locus?

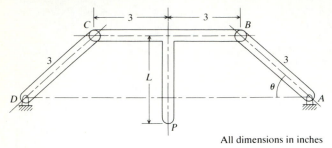

All dimensions in inches

**FIGURE P5-23**

**5.24.** The slider crank mechanism shown in Fig. P5-24 is part of a machine. The sensor comes in contact with the centerpoint of the connecting rod $BC$ only when this point is in its extreme position. The sensor has to change position to meet the following design requirements:

  Design objective (goal): sensor should move 0.5 in outward.
  Design constraints: $a + b$ must be kept to a minimum to minimize the mechanism weight ($a$ and $b$ have same cross section and same material).

Find the new lengths $a$ and $b$.

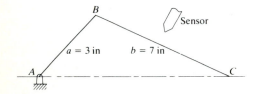

**FIGURE P5-24**

## Part 3: Programming

**5.25.** Based on the theory presented in this chapter, write a program that takes the proper user input and display lines, circles, ellipses, and parabolas.

**5.26.** An ellipse at a general orientation can be generated and displayed in two ways:
  (a) Use Eq. (5.37) to generate points ready for display directly.
  (b) Use the equations $x_L = A \cos u$, $y_L = B \sin u$, $z_L = 0$ (see Fig. 5-28) and use the transformation

$$
[x \quad y \quad z \quad 1]^T = \left[ \begin{array}{ccc:c} \cos \alpha & \sin \alpha & 0 & x_c \\ -\sin \alpha & \cos \alpha & 0 & y_c \\ 0 & 0 & 1 & z_c \\ \hdashline 0 & 0 & 0 & 1 \end{array} \right] \left[ \begin{array}{c} x_L \\ y_L \\ 0 \\ 1 \end{array} \right]
$$

Program both ways and compare the CPU times of both cases.

**5.27.** As Prob. 5.26 but for a parabola [use Eq. (5.55)].

**5.28.** Write a program for a Hermite cubic spline connecting two points.

**5.29.** As Prob. 5.28 but for a cubic Bezier curve.

**5.30.** As Prob. 5.28 but for both open and closed B-spline curves.

# BIBLIOGRAPHY

Abhyankar, S. S., and C. Bajaj: "Automatic Parametrization of Rational Curves and Surfaces 1: Conics and Conicoids," *Computer Aided Des.*, vol. 19, no. 1, pp. 11–14, 1987.

Barnhill, R. E., and R. F. Riesenfeld (Eds.): *Computer Aided Geometric Design*, Academic Press, New York, 1974.

Beer, F. P., and E. R. Johnston: *Mechanics of Materials*, McGraw-Hill, New York, 1981.

Boehm, W.: "A Survey of Curve and Surface Methods in CAGD," *Computer Aided Geometric Des. (CAGD) J.*, vol. 1, pp. 1–60, 1984.

Boehm, W.: "Triangular Spline Algorithms," *CAGD J.*, vol. 2, pp. 61–67, 1985.

Boehm, W.: "Curvature Continuous Curves and Surfaces," *CAGD J.*, vol. 2, pp. 313–323, 1985.

Boehm, W.: "Multivariate Spline Methods in CAGD," *Computer Aided Des. J.*, vol. 18, no. 2, pp. 102–104, 1986.

Boehm, W.: "Curvature Continuous Curves and Surfaces," *Computer Aided Des. J.*, vol. 18, no. 2, pp. 105–106, 1986.

Casen, S. H.: *Geometric Principles and Procedures for Computer Graphic Applications*, Prentice-Hall, Englewood Cliffs, N.J., 1978.

Chang, T. C., and R. A. Wysk: *An Introduction to Automated Process Planning Systems*, Prentice-Hall, Englewood Cliffs, N.J., 1985.

Cohen, E., T. Lyche, and L. L. Schumaker: "Algorithms for Degree-Raising of Splines," *ACM Trans. on Graphics*, vol. 4, no. 3, pp. 171–181, 1985.

Cohen, E., and L. L. Schumaker: "Rates of Convergence of Control Polygons," *CAGD J.*, vol. 2, pp. 229–235, 1985.

Dakken, T.: "Finding Intersections of B-Spline Represented Geometrics Using Recursive Subdivision Techniques," *CAGD J.*, vol. 2, pp. 189–195, 1985.

Delvos, F. J.: "Bernoulli Functions and Periodic B-Splines," *Computing*, vol. 38, pp. 23–31, 1987.

Encarnacao, J., and E. G. Schlechtendahl: *Computer Aided Design; Fundamentals and System Architectures*, Springer-Verlag, New York, 1983.

Farin, G.: "Some Remarks on V2-Splines," *CAGD J.*, vol. 2, pp. 325–328, 1985.

Faux, I. D., and M. J. Pratt: *Computational Geometry for Design and Manufacture*, Ellis Horwood (John Wiley), Chichester, West Sussex, 1979.

Foley, J. D., and A. Van Dam: *Fundamentals of Interactive Computer Graphics*, Addison-Wesley, 1982.

Forrest, A. R.: "Curves and Surfaces for Computer-Aided Designs," Ph.D.Thesis, University of Cambridge, 1968.

Gardan, Y., and M. Lucas: *Interactive Graphics in CAD*, Kogan Page, London, 1984.

Giloi, W. K.: *Interactive Computer Graphics; Data Structures, Algorithms, Languages*, Prentice-Hall, Englewood Cliffs, N.J., 1978.

Goldman, R. N.: "Vector Elimination: A Technique for the Implicitization, Inversion, and Intersection of Planar Parametric Rational Polynomial Curves," *CAGD J.*, vol. 1, pp. 327–356, 1984.

Goldman, R. N.: "The Method of Resolvents: A Technique for the Implicitization, Inversion, and Intersection of Non-Planar, Parametric, Rational Cubic Curves," *CAGD J.*, vol. 2, pp. 237–255, 1985.

Goldman, R. N., and D. J. Filip: "Conversion from Bezier Rectangles to Bezier Triangles," *Computer Aided Des. J.*, vol. 19, no. 1, pp. 25–27, 1987.

Goodman, T. N. T., and K. Unsworth: "Manipulating Shape and Producing Geometric Continuity in Spline Curves," *IEEE CG&A*, pp. 50–56, February 1986.

Groover, M. P., and E. W. Zimmers: *CAD/CAM; Computer-Aided Design and Manufacturing*, Prentice-Hall, Englewood Cliffs, N.J., 1984.

Hagen, H.: "Geometric Spline Curves," *CAGD J.*, vol. 2, pp. 223–227, 1985.

Haug, E. J., and J. S. Arora: *Applied Optimum Design*, John Wiley, New York, 1979.

Lasser, D.: "Bernstein-Bezier Representation of Volumes," *CAGD J.*, vol. 2, pp. 145–149, 1985.

Lee, E. T. Y.: "Some Remarks Concerning B-Splines," *CAGD J.*, vol. 2, pp. 307–311, 1985.

Luzadder, W. J.: *Innovative Design with an Introduction to Design Graphics*, Prentice-Hall, Englewood Cliffs, N.Y., 1975.

Maccallum, K. J., and J. M. Zhang: "Curve Smoothing Techniques Using B-Splines," *The Computer J.*, vol. 29, no. 6, pp. 564–571, 1986.

Mortenson, M. E.: *Geometric Modeling*, John Wiley, New York, 1985.

Newman, W. M., and R. F. Sproull: *Principles of Interactive Computer Graphics*, 2d ed., McGraw-Hill, New York, 1979.

Nielson, G. M.: "Rectangular Splines," *IEEE CG&A*, pp. 35–40, February 1986.

Patterson, R. R.: "Projective Transformation of the Parameter of a Bernstein-Bezier Curve," *ACM Trans. on Graphics*, vol. 4, no. 4, pp. 276–290, 1985.

Piegl, L.: "Recursive Algorithms for the Representation of Parametric Curves and Surfaces," *Computer Aided Des. J.*, vol. 17, no. 5, pp. 225–229, 1985.

Piegl, L.: "Infinite Control Points, A Method for Representing Surfaces of Revolution Using Boundary Data," *IEEE CG&A*, pp. 45–55, March 1987.

Piegl, L.: "Interactive Data Interpolation by Rational Bezier Curves," *IEEE CG&A*, pp. 45–58, April 1987.

Rogers, D. F., and J. A. Adams: *Mathematical Elements for Computer Graphics*, McGraw-Hill, New York, 1976.

Sablonniere, P.: "Bernstein-Bezier Methods for the Construction of Bivariate Spline Approximations," *CAGD J.*, vol. 2, pp. 29–36, 1985.

Sederberg, T. W.: "Degenerate Parametric Curves," *CAGD J.*, vol. 1, pp. 301–307, 1984.

Vandoni, C. E. (Ed.): *Eurographics '85*, Elsevier Science, New York, 1985.

Wilson, P. R.: "Conic Representations for Shape Description," *IEEE CG&A*, pp. 23–30, April 1987.

# CHAPTER

# 6

# TYPES AND MATHEMATICAL REPRESENTATIONS OF SURFACES

## 6.1 INTRODUCTION

Shape design and representation of complex objects such as car, ship, and airplane bodies as well as castings cannot be achieved utilizing wireframe modeling covered in the previous chapter. In such cases, surface modeling must be utilized to describe objects precisely and accurately. Due to the richness in information of surface models, their use in engineering and design environments can be extended beyond just geometric design and representation. They are usually used in various applications such as calculating mass properties, checking for interference between mating parts, generating cross-sectioned views, generating finite element meshes, and generating NC tool paths for continuous path machining.

Creating surfaces in general has some quantitative data, such as a set of points and tangents, and some qualitative data, such as intuition of the desired shape and smoothness. Quantitative and qualitative data can be thought of as hard and soft data respectively. Surface formulation must provide the designer with the flexibility to use both types of data in a simple form that is suitable for interactive use. Similar to curves, available surface techniques can interpolate or approximate the given hard data. As will be seen in this chapter, the Bezier surface is a form of approximation and the B-spline surface is a form of interpolation.

In addition to using surfaces to model geometric objects, they can also be used to fit experimental data, tables of numbers, and discretized solutions of differential equations. In all these cases, the multidimensional surface problem arises. The problem can be stated as follows. Given positional data (points in three-dimensional space) and a variable value at each point, how can the surface representing the variable distribution be constructed? Consider, for example, constructing the pressure surface, or distribution, on an oblique airplane wing or constructing the stress distribution in a mechanical part. The construction of these four-dimensional surfaces usually involves finding the proper contours of the given variable (pressure or stress) and then displaying them in a color-coded form.

The choice of the surface form depends upon the application; that is, there is no single solution for all problems. For example, the surface form used to model a car body may not be adequate to model the human heart. The choice of surface form may depend on manufacturing methods needed to produce the surface. It is usually preferred to choose a surface that can be produced using three-axis machining instead of five-axis machining to reduce the manufacturing cost. However, all surface forms must be easy to differentiate to determine surface tangents, normals, and curvatures. Polynomial functions are an obvious choice. Polynomials of higher orders are not appropriate from a surface design point of view due to their large number of coefficients which may make it difficult to control the resulting surface or may introduce unwanted oscillations in the surface. For most practical surface applications, cubic polynomials are sufficient.

It is desirable, if possible, to choose a surface mathematical description or form that is applicable to both surface design and surface representation. Surface representation involves using given data to display and view the surface. It can take place in $n$-space, as the pressure surface problem cited above shows. Surface design involves using key given data and making interactive changes to obtain the desired surface. Surface design usually takes place in three-dimensional space. In this chapter, we concern ourselves with surface design only.

The most obvious and inefficient way to describe a surface is to list sufficient points on it. This approach is cumbersome and cannot be used to derive any surface properties. Instead, it is common to employ some form of interpolation scheme and use fewer points. It is even more convenient if analytic forms of surfaces exist all the time. When analytic forms are not available, surfaces are defined in patches that are connected together similarly to curves, which can be defined in a piecewise manner.

Surface creation on existing CAD/CAM systems usually requires wireframe entities as a start. A system might request two boundary entities (rails of the surface) to create a ruled surface or might require one entity (generator curve) to create a surface of revolution. All analytic and synthetic wireframe entities covered in the previous chapter can be used to generate surfaces. In order to visualize surfaces on a graphics display, a mesh, say $m \times n$ in size, is usually displayed. The mesh size is controllable by the user. CAD/CAM systems must provide their users with a wide range of surfaces and surface manipulations as discussed in this chapter.

This chapter covers both theoretical and practical aspects of surfaces. Throughout the chapter, related issues to constructing surfaces on CAD/CAM systems are covered. Sections 6.2 and 6.3 are dedicated to some of these issues which may be helpful to users. Sections 6.4 to 6.7 cover the mathematical representations of most popular surfaces. Section 6.8 applies the chapter material to design and engineering applications.

## 6.2 SURFACE MODELS

A surface model of an object is a more complete and less ambiguous representation than its wireframe model. It is also richer in its associated geometric contents, which makes it more suitable for engineering and design applications. A surface model can be used, for example, to drive the cutter of a machine tool while a wireframe model cannot. Surface models take the modeling of an object one step beyond wireframe models by providing information on surfaces connecting the object edges. Typically, a surface model consists of wireframe entities that form the basis to create surface entities. Surface description is usually tackled as an extension to the wireframe representation. Analytic and synthetic surface entities are available and provided by most CAD/CAM systems.

Surface modeling has been developing rapidly due to the shortcomings and inconveniences of wireframe modeling. The former is considered an extension of the latter. In general, a wireframe model can be extracted from a surface model by deleting or blanking all surface entities. Databases of surface models are centralized and associative. Thus, manipulating surface entities in one view is automatically reflected in other views. These entities can also be subjected to three-dimensional geometric transformations.

Despite their similar look, there is a fundamental difference between surface and solid models. Surface models define only the geometry of their corresponding objects. They store no information regarding the topology of these objects. As an example, if there are two surface entities that share a wireframe entity (edge), neither the surfaces nor the entity store such information. There is also a fundamental difference between creating surface and solid models. To create a surface model, the user begins by constructing wireframe entities and then connecting them appropriately with the proper surface entities. For solids based on boundary representation, either faces, edges, and vertices are created or solid primitives are input which are converted internally by the software to faces, edges, and vertices. Refer to the next chapter on solid modeling.

In constructing a surface model on a CAD/CAM system, the user should follow the modeling guidelines discussed in Chap. 3. All the design tools provided by CAD/CAM systems and covered in Part IV of the book are applicable to surface models. Only geometric modifiers are meaningless and cannot be applied to surface entities. From a practical point of view, it is more convenient to construct a surface model in an isometric view to enable clear display and visualization of its entities. Visualization of a surface is aided by the addition of artificial fairing lines (called mesh) which criss-cross the surface and so break it up into a network of interconnected patches. If wireframe entities are to be digitized to generate surfaces the user must do so in an ordered manner to ensure that the right surfaces are created. Figure 6-1 shows how the wrong ruled surface is created if the surface rails are digitized near the wrong ends. The +'s in the figure indicates the digitized locations. Another practical tip in constructing surface models is the change in mesh size of a surface entity. The most obvious way is to delete the surface entity and then reconstruct it with the desired mesh size. A better solution is to simply change the size of the mesh and then regenerate the surface display. Figure 6-2 shows surfaces of revolutions with a mesh size of $4 \times 6$ and Fig. 6-3 shows the regeneration of the surfaces with a $20 \times 20$ mesh

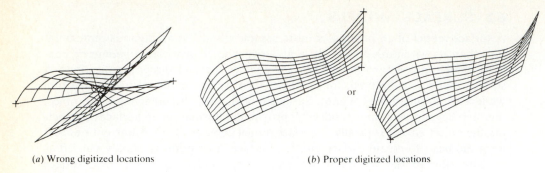

(a) Wrong digitized locations          or          (b) Proper digitized locations

**FIGURE 6-1**
Construction of improper and proper ruled surfaces.

size. It should be mentioned here that the finer the mesh size of surface entities in a model, the longer the CPU time to construct the entities and to update the graphics display, and the longer it takes to plot the surface model. Finally, some CAD/CAM systems do not permit their users to delete wireframe entities used to create surface entities unless the latter are deleted first.

Surface models have considerable advantages over wireframe models. They are less ambiguous. They provide hidden line and surface algorithms to add realism to the displayed geometry. Shading algorithms are only available for surface and solid models. From an application point of view, surface models can be utilized in volume and mass property calculations, finite element modeling, NC path generation, cross sectioning, and interference detections.

Despite the above-cited advantages, surface models have few disadvantages. Surface modeling does not lend itself to drafting background. It therefore requires more training and mathematical background on the user's part. Surface

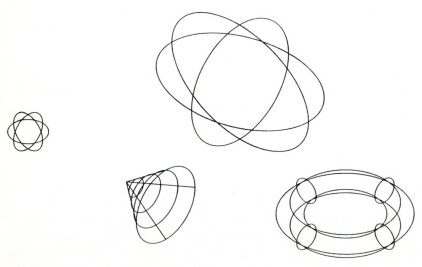

**FIGURE 6-2**
Surfaces of revolution with a 4 × 4 mesh size.

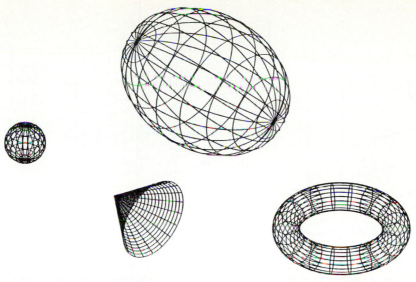

**FIGURE 6-3**
Surfaces of revolution with a 20 × 20 mesh size.

models are generally more complex and thus require more terminal and CPU time and computer storage to create than wireframe models. They are still ambiguous in some applications, as is the case when determining which of an object's surfaces define its volume. Surface models are sometimes awkward to create and may require unnecessary manipulations of wireframe entities. Consider the example of a surface with holes or cuts in it. Surface patches may have to be created with the aid of intermediate wireframe entities to create the full surface. Refer to Fig. 5-57 where the circle has to be divided to create the ruled surfaces.

Surfaces and wireframes form the core of all existing CAD/CAM systems. Surface descriptions and capabilities are generalized to provide modeling flexibilities. For example, Gordon surfaces are considered alternatives, with generalization and improvements, to Bezier and Coons surfaces. Triangular Bezier and Coons patches are available for the case of arbitrarily located input data. Additional triangular patches also exist. As a result, new surface possibilities are created for those situations in which four-sided topology cannot be assumed.

## 6.3   SURFACE ENTITIES

Similar to wireframe entities, existing CAD/CAM systems provide designers with both analytic and synthetic surface entities. Analytic entities include plane surface, ruled surface, surface of revolution, and tabulated cylinder. Synthetic entities include the bicubic Hermite spline surface, B-spline surface, rectangular and triangular Bezier patches, rectangular and triangular Coons patches, and Gordon surface. The mathematical properties of some of these entities are covered in this chapter for two purposes. First, it enables users to correctly choose the proper surface entity for the proper application. For example, a ruled

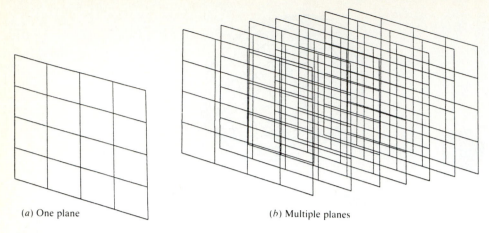

(a) One plane                        (b) Multiple planes

**FIGURE 6-4**
Plane surface.

surface is a linear surface and does not permit any twist while a B-spline surface is a general surface. Second, users will be in a position to better understand CAD/CAM documentation and the related modifiers to each surface entity command available on a system. The following are descriptions of major surface entities provided by CAD/CAM systems:

1. *Plane surface*. This is the simplest surface. It requires three noncoincident points to define an infinite plane. The plane surface can be used to generate cross-sectional views by intersecting a surface model with it, generate cross sections for mass property calculations, or other similar applications where a plane is needed. Figure 6-4 shows a plane surface.

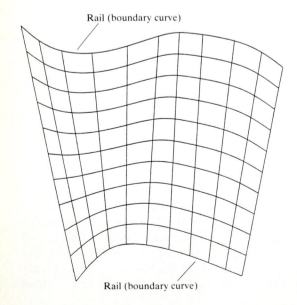

Rail (boundary curve)

Rail (boundary curve)

**FIGURE 6-5**
Ruled surface.

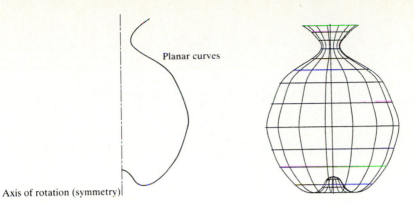

Planar curves

Axis of rotation (symmetry)

**FIGURE 6-6**
Surface of revolution.

2. *Ruled (lofted) surface.* This is a linear surface. It interpolates linearly between two boundary curves that define the surface (rails). Rails can be any wireframe entity. This entity is ideal to represent surfaces that do not have any twists or kinks. Figure 6-5 gives some examples.

3. *Surface of revolution.* This is an axisymmetric surface that can model axisymmetric objects. It is generated by rotating a planar wireframe entity in space about the axis of symmetry a certain angle (Fig. 6-6).

4. *Tabulated cylinder.* This is a surface generated by translating a planar curve a certain distance along a specified direction (axis of the cylinder) as shown in Fig. 6-7. The plane of the curve is perpendicular to the axis of the cylinder. It

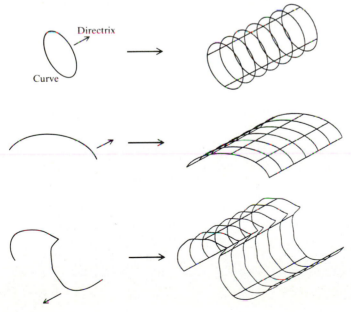

Directrix

Curve

**FIGURE 6-7**
Tabulated cylinder.

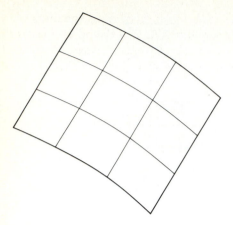

**FIGURE 6-8**
Bezier surface.

is used to generate surfaces that have identical curved cross sections. The word "tabulated" is borrowed from the APT language terminology.

5. *Bezier surface.* This is a surface that approximates given input data. It is different from the previous surfaces in that it is a synthetic surface. Similarly to the Bezier curve, it does not pass through all given data points. It is a general surface that permits, twists, and kinks (Fig. 6-8). The Bezier surface allows only global control of the surface.

6. *B-spline surface.* This is a surface that can approximate or interpolate given input data (Fig. 6-9). It is a synthetic surface. It is a general surface like the Bezier surface but with the advantage of permitting local control of the surface.

7. *Coons patch.* The above surfaces are used with either open boundaries or given data points. The Coons patch is used to create a surface using curves that form closed boundaries (Fig. 6-10).

8. *Fillet surface.* This is a B-spline surface that blends two surfaces together (Fig. 6-11). The two original surfaces may or may not be trimmed.

9. *Offset surface.* Existing surfaces can be offset to create new ones identical in shape but may have different dimensions. It is a useful surface to use to speed

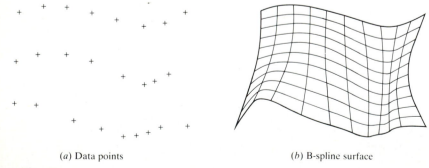

(a) Data points  (b) B-spline surface

**FIGURE 6-9**
B-spline surface.

up surface construction. For example, to create a hollow cylinder, the outer or inner cylinder can be created using a cylinder command and the other one can be created by an offset command. Offset surface command becomes very efficient to use if the original surface is a composite one. Figure 6-12 shows an offset surface.

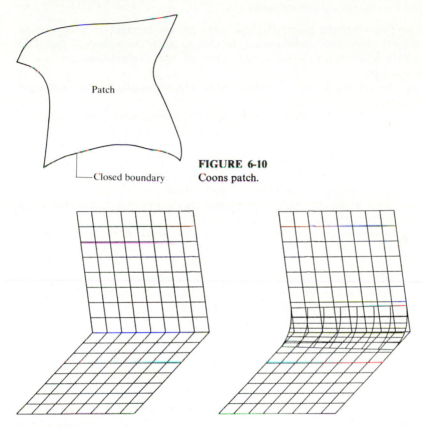

Patch

Closed boundary

**FIGURE 6-10**
Coons patch.

**FIGURE 6-11**
Fillet surface.

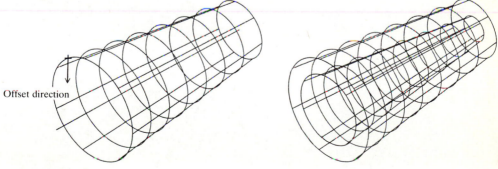

Offset direction

**FIGURE 6-12**
Offset surface.

**Example 6.1.** Create the surface model of the guide bracket shown in Fig. 5-2.

*Solution.* Before creating the surface model, the wireframe model of the bracket created in Example 5.1 is required and its database is assumed to be available for this example. A quick look at Fig. 5-2 reveals that ruled surfaces and tabulated cylinders are sufficient to construct the surface model. The following steps may be followed to construct the surface model:

1. Retrieve the wireframe model database of the bracket. It might be more practical to copy the wireframe model and use the copy to create the surfaces so that if the database is corrupted during surface creation, the original database can be copied again.
2. Select new layer(s) for surface entities to facilitate managing surface and wireframe entities.
3. Create surfaces on the right face of the model by using a ruled surface command. Referring to Fig. 6-13a, the line $P_5 P_6$ is divided first into two entities at the intersection point ($P_7$) of lines $P_2 P_3$ and $P_5 P_6$. The ruled surface command is used twice to create two surfaces using lines $P_3 P_4$ and $P_5 P_7$ (identified by digitizes $d_1$ and $d_2$ in Fig. 6-13a) as rails for the first surface and lines $P_1 P_2$ and $P_6 P_7$ ($d_3$ and $d_4$ in the figure) as rails for the other surfaces.
4. Use a mirror or duplicate command to copy these surfaces to create surfaces on the left face of the model.
5. Create ruled surfaces on the front face of the top part of the model as shown in Fig. 6-13a. Five surfaces are constructed. These surfaces use entities $d_5$ and $d_6$; $d_7$ and $d_8$; $d_9$ and $d_{10}$; $d_{11}$ and $d_{12}$; and $d_{13}$ and $d_{14}$ as rails. The small circle has to be divided into five arc segments to create these surfaces.
6. Create the surface of the hole shown in Fig. 6-13a. Use a tabulated cylinder command with the circle as the cylinder generator.
7. Follow a similar approach to construct all the surfaces of the model. During construction, the user encounters either dividing existing entities or creating new ones to define the appropriate rails of the various ruled surfaces. Figure 6-13b shows the completed surface model while Fig. 6-13c shows a shaded image of the surface model for better visualization.

The above example illustrates most of the experiences encountered in creating surface models on major CAD/CAM systems. The long construction time and user inconveniences due to dividing or creating entities make surface models

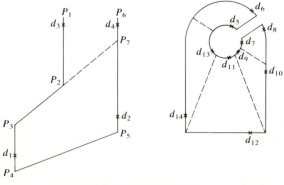

(a) Intermediate construction

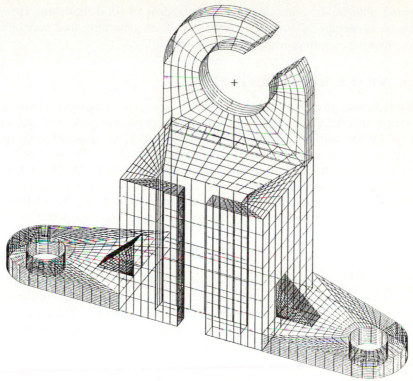

(b) Surface model

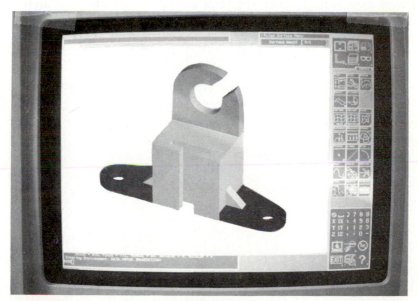

(c) Shaded image

**FIGURE 6-13**
Surface model of the guide bracket of Example 5.1.

somewhat difficult to create and give more appeal to solid modeling. However, the major advantage of surface modeling is in its generality and capability to handle any kind of surface including sculptured surfaces.

## 6.4  SURFACE REPRESENTATION

Surface representation is considered, in many aspects, an extension of curve representation covered in the previous chapter. The nonparametric and parametric forms of curves can be extended to surfaces as will be covered later in this chapter. Similarly, the treatment of surfaces in computer graphics and CAD/CAM requires developing the proper equations and algorithms for both computation and programming purposes. Moreover, surface description is usually related to machining requirements to manufacture the surface. The surface description must successfully drive a tool to generate its path. Chapter 20 covers more on machining.

Surfaces can be described mathematically in three-dimensional space by nonparametric or parametric equations. There are several methods to fit nonparametric surfaces to a given set of data points. These fall into two categories. In the first, one equation is fitted to pass through all the points while in the second the data points are used to develop a series of surface patches that are connected together with at least position and first-derivative continuity. In both categories, the equation of the surface or surface patch is given by

$$\mathbf{P} = [x \quad y \quad z]^T = [x \ y \ \ f(x, y)]^T \tag{6.1}$$

where $\mathbf{P}$ is the position vector of a point on the surface as shown in Fig. 6-14. The natural form of the function $f(x, y)$ for a surface to pass through all the given data points is a polynomial, that is,

$$z = f(x, y) = \sum_{m=0}^{p} \sum_{n=0}^{q} a_{mn} x^m y^n \tag{6.2}$$

where the surface is described by an $XY$ grid of size $(p + 1) \times (q + 1)$ points.

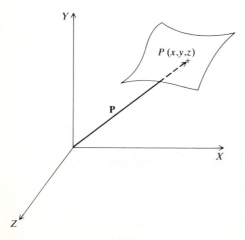

**FIGURE 6-14**
Point $P$ on a nonparametric surface patch.

The nonparametric surface representation suffers from all the disadvantages, when compared with parametric surface representation, that nonparametric curves suffer from when compared with parametric curves. However, nonparametric surfaces do have some advantages when it comes to solving surface intersection problems, but these advantages do not warrant their use in CAD/CAM.

The parametric representation of a surface means a continuous, vector-valued function $P(u, v)$ of two variables, or parameters, $u$ and $v$, where the variables are allowed to range over some connected region of the $uv$ plane and, as they do so, $P(u, v)$ assumes every position on the surface. The function $P(u, v)$ at certain $u$ and $v$ values is the point on the surface at these values. The most general way to describe the parametric equation of a three-dimensional curved surface in space is

$$P(u, v) = [x \quad y \quad z]^T = [x(u, v) \quad y(u, v) \quad z(u, v)]^T,$$

$$u_{min} \leq u \leq u_{max}, v_{min} \leq v \leq v_{max} \quad (6.3)$$

As with curves, Eq. (6.3) gives the coordinates of a point on the surface as the components of its position vector. It uniquely maps the parametric space ($E^2$ in $u$ and $v$ values) to the cartesian space ($E^3$ in $x$, $y$, and $z$), as shown in Fig. 6-15. The parametric variables $u$ and $v$ are constrained to intervals bounded by minimum and maximum values. In most surfaces, these intervals are [0, 1] for both $u$ and $v$.

Equation (6.3) suggests that a general three-dimensional surface can be modeled by dividing it into an assembly of topological patches. A patch is considered the basic mathematical element to model a composite surface. Some surfaces may consist of one patch only while others may be a few patches connected together. Figure 6-16 shows a two-patch surface where the $u$ and $v$ values are [0, 1]. The topology of a patch may be rectangular or triangular as shown in Fig. 6-17. Triangular patches add more flexibility in surface modeling because they do not require ordered rectangular arrays of data points to create the surface as the rectangular patches do. Both triangular and rectangular Coons and Bezier patches are available.

Analogous to curves, there are analytic and synthetic surfaces. Analytic surfaces are based on wireframe entities and include the plane surface, ruled surface, surface of revolution, and tabulated cylinder. Synthetic surfaces are formed from a given set of data points or curves and include the bicubic, Bezier, B-spline, and Coons patches. There are few methods to generate synthetic surfaces such as the tensor product method, rational method, and blending method. The rational method develops rational surfaces which is an extension of rational curves. The blending method approximates a surface by piecewise surfaces.

The tensor product method is the most popular method and is widely used in surface modeling. Its widespread use is largely due to its simple separable nature involving only products of univariate basis functions, usually polynomials. It introduces no new conceptual complications due to the higher dimensionality of a surface over a curve. The properties of tensor product surfaces can easily be deduced from properties of the underlying curve schemes. The tensor product formulation is a mapping of a rectangular domain described by the $u$ and $v$

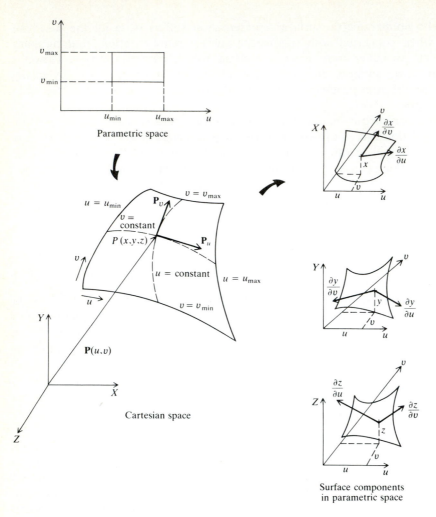

**FIGURE 6-15**
Parametric representation of a three-dimensional surface.

values; e.g., $0 \leq u \leq 1$ and $0 \leq v \leq 1$. Tensor product surfaces fit naturally onto rectangular patches. In addition, they have an explicit unique orientation (triangulation of a surface is not unique) and special parametric or coordinate directions associated with each independent parametric variable.

There is a set of boundary conditions associated with a rectangular patch. There are sixteen vectors and four boundary curves as shown in Fig. 6-18. The vectors are four position vectors for the four corner points $P(0, 0)$, $P(1, 0)$, $P(1, 1)$, and $P(0, 1)$; eight tangent vectors (two at each corner); and four twist vectors at the corner points (see below for the definition of a twist vector). The four boundary curves are described by holding one parametric variable fixed at one of its limiting values and allowing the other to change freely. The boundary curves are then defined by the curve equations $u = 0$, $u = 1$, $v = 0$, and $v = 1$.

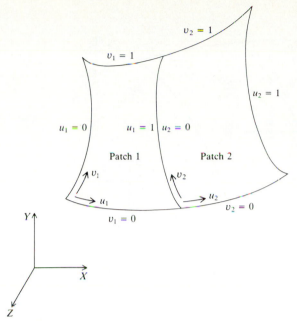

**FIGURE 6-16**
Two-patch parametric surface.

To generate curves on a surface patch, one can fix the value of one of the parametric variables, say $u$, to obtain a curve in terms of the other variable, $v$. By continuing this process first for one variable and then for the other using a certain set of arbitrary values in the permissible domain, a network of two parametric families of curves are generated. Only one curve of each family passes through any point $P(u, v)$ on the surface. The positive sense of any of these curves is the sense in which its nonfixed parameter increases. As mentioned earlier, the user can specify a mesh size, say $m \times n$, to display a surface on the graphics display. That mesh size determines the number of curves in each family. The curves of each family are usually equally spaced in the permissible interval of the corresponding parametric variable.

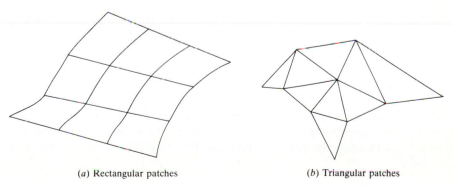

(a) Rectangular patches          (b) Triangular patches

**FIGURE 6-17**
Surfaces composed of rectangular and triangular patches.

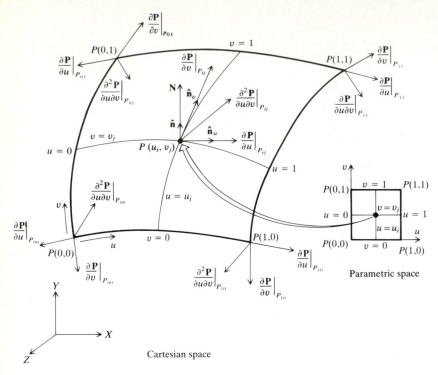

**FIGURE 6-18**
A parametric surface patch with its boundary conditions.

Geometric surface analysis forms an important factor in using surfaces for other purposes than just displaying them. For example, knowing the tangent vectors to a surface enables driving a cutting tool along the surface to machine it and knowing the normal vectors to the surface provides the proper directions for the tool to approach and retract from the surface. Differential geometry plays a central role in the analysis of surfaces. The measurements of lengths and areas, the specification of directions and angles, and the definition of curvature on a surface are all formulated in differential terms. The parametric surface $P(u, v)$ is directly amenable to differential analysis. There are intrinsic differential characteristics of a surface such as the unit normal and the principal curvatures and directions which are independent of parametrization. These characteristics require introducing few parametric derivatives.

The tangent vector concept that is introduced in Chap. 5 can be extended to surfaces. The tangent vector at any point $P(u, v)$ on the surface is obtained by holding one parameter constant and differentiating with respect to the other. Therefore, there are two tangent vectors, a tangent to each of the intersecting curves passing through the point as shown in Fig. 6-18. These vectors are given by

$$\mathbf{P}_u(u, v) = \frac{\partial \mathbf{P}}{\partial u} = \frac{\partial x}{\partial u}\hat{\mathbf{i}} + \frac{\partial y}{\partial u}\hat{\mathbf{j}} + \frac{\partial z}{\partial u}\hat{\mathbf{k}}, \qquad u_{\min} \leq u \leq {}_{\max}, v_{\min} \leq v \leq v_{\max} \quad (6.4)$$

along the $v =$ constant curve, and

$$\mathbf{P}_v(u, v) = \frac{\partial \mathbf{P}}{\partial v} = \frac{\partial x}{\partial v} \hat{\mathbf{i}} + \frac{\partial y}{\partial v} \hat{\mathbf{j}} + \frac{\partial z}{\partial v} \hat{\mathbf{k}}, \qquad u_{\min} \le u \le u_{\max}, \; v_{\min} \le v \le v_{\max} \quad (6.5)$$

along the $u =$ constant curve. These two equations can be combined to give

$$\begin{bmatrix} \mathbf{P}_u \\ \mathbf{P}_v \end{bmatrix} = \begin{bmatrix} \dfrac{\partial x}{\partial u} & \dfrac{\partial y}{\partial u} & \dfrac{\partial z}{\partial u} \\ \dfrac{\partial x}{\partial v} & \dfrac{\partial y}{\partial v} & \dfrac{\partial z}{\partial v} \end{bmatrix} \qquad (6.6)$$

These components of each unit vector are shown in Fig. 6-15. If the dot product of the two tangent vectors given by Eqs. (6.4) and (6.5) is equal to zero at a point on a surface, then the two vectors are perpendicular to one another at that point. Figure 6-18 shows the tangent vectors at the corner points of a rectangular patch and at point $P_{ij}$. The notation $\partial \mathbf{P}/\partial u|_{P_{ij}}$, for example, means that the derivative is calculated at the point $P_{ij}$ defined by $u = u_i$ and $v = v_j$. Tangent vectors are useful in determining boundary conditions for patching surfaces together as well as defining the motion of cutters along the surfaces during machining processes. The magnitudes and unit vectors of the tangent vectors are given by

$$|\mathbf{P}_u| = \sqrt{\left(\frac{\partial x}{\partial u}\right)^2 + \left(\frac{\partial y}{\partial u}\right)^2 + \left(\frac{\partial z}{\partial u}\right)^2}$$

$$|\mathbf{P}_v| = \sqrt{\left(\frac{\partial x}{\partial v}\right)^2 + \left(\frac{\partial y}{\partial v}\right)^2 + \left(\frac{\partial z}{\partial v}\right)^2} \qquad (6.7)$$

and

$$\hat{\mathbf{n}}_u = \frac{\mathbf{P}_u}{|\mathbf{P}_u|} \qquad \hat{\mathbf{n}}_v = \frac{\mathbf{P}_v}{|\mathbf{P}_v|} \qquad (6.8)$$

The slopes to a given curve on a surface can be evaluated (refer to Chap. 5), although they are less significant and seldom used in surface analysis.

The twist vector at a point on a surface is said to measure the twist in the surface at the point. It is the rate of change of the tangent vector $\mathbf{P}_u$ with respect to $v$ or $\mathbf{P}_v$ with respect to $u$, or it is the cross (mixed) derivative vector at the point. Figure 6-19 shows the geometric interpretation of the twist vector. If we increment $u$ and $v$ by $\Delta u$ and $\Delta v$ respectively and draw the tangent vectors as shown, the incremental changes in $\mathbf{P}_u$ and $\mathbf{P}_v$ at point $P$, whose position vector is $\mathbf{P}(u, v)$, are obtained by translating $\mathbf{P}_u(u, v + \Delta v)$ and $\mathbf{P}_v(u + \Delta u, v)$ to $P$ and forming the two triangles shown. The incremental rate of change of the two tangent vectors become $\Delta \mathbf{P}u/\Delta v$ and $\Delta \mathbf{P}v/\Delta u$, and the infinitesimal rate of change is given by the following limits:

$$\underset{\Delta v \to 0}{\text{Limit}} \; \frac{\Delta \mathbf{P}_u}{\Delta v} = \frac{\partial \mathbf{P}_u}{\partial v} = \frac{\partial^2 \mathbf{P}}{\partial u \, \partial v} = \mathbf{P}_{uv}$$

$$\underset{\Delta u \to 0}{\text{Limit}} \; \frac{\Delta \mathbf{P}_v}{\Delta u} = \frac{\partial \mathbf{P}_v}{\partial u} = \frac{\partial^2 \mathbf{P}}{\partial u \, \partial v} = \mathbf{P}_{uv} \qquad (6.9)$$

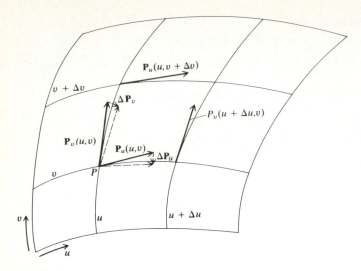

**FIGURE 6-19**
Geometric interpretation of twist vectors.

The twist vector can be written in terms of its cartesian components as

$$\mathbf{P}_{uv} = \left[ \frac{\partial^2 x}{\partial u\, \partial v}\ \frac{\partial^2 y}{\partial u\, \partial v}\ \frac{\partial^2 z}{\partial u\, \partial v} \right]^T = \frac{\partial^2 x}{\partial u\, \partial v}\ \hat{\mathbf{i}} + \frac{\partial^2 y}{\partial u\, \partial v}\ \hat{\mathbf{j}} + \frac{\partial^2 z}{\partial u\, \partial v}\ \hat{\mathbf{k}},$$

$$u_{min} \le u \le u_{max}, \; v_{min} \le v \le v_{max} \qquad (6.10)$$

The twist vector depends on both the surface geometric characteristics and its parametrization. Due to the latter dependency, interpreting the twist vector in geometrical terms may be misleading since $\mathbf{P}_{uv} \neq 0$ does not necessarily imply a twist in a surface. For example, a flat plane is not a twisted surface. However, depending on its parametric equation, $\mathbf{P}_{uv}$ may or may not be zero.

The normal to a surface is another important analytical property. It is used to calculate cutter offsets for three-dimensional NC programming to machine surfaces, volume calculations, and shading of a surface model. The surface normal at a point is a vector which is perpendicular to both tangent vectors at the point (Fig. 6-18); that is,

$$\mathbf{N}(u, v) = \frac{\partial \mathbf{P}}{\partial u} \times \frac{\partial \mathbf{P}}{\partial v} = \mathbf{P}_u \times \mathbf{P}_v \qquad (6.11)$$

and the unit normal vector is given by

$$\hat{\mathbf{n}} = \frac{\mathbf{N}}{|\mathbf{N}|} = \frac{\mathbf{P}_u \times \mathbf{P}_v}{|\mathbf{P}_u \times \mathbf{P}_v|} \qquad (6.12)$$

The order of the cross-product in Eq. (6.11) can be reversed and still defines the normal vector. The sense of $\mathbf{N}$, or $\hat{\mathbf{n}}$, is chosen to suit the application. In machining, the sense of $\hat{\mathbf{n}}$ is usually chosen so that $\hat{\mathbf{n}}$ points away from the surface being machined. In volume calculations, the sense of $\hat{\mathbf{n}}$ is chosen positive when pointing toward existing material and negative when pointing to holes in the part.

The surface normal is zero when $\mathbf{P}_u \times \mathbf{P}_v = \mathbf{0}$. This occurs at points lying on a cusp, ridge, or a self-intersecting surface. It can also occur when the two derivatives $\mathbf{P}_u$ and $\mathbf{P}_v$ are parallel, or when one of them has a zero magnitude. The latter cases correspond to a pathological parametrization which can be remedied.

The calculation of the distance between two points on a curved surface forms an important part of surface analysis. In general, two distinct points on a surface can be connected by many different paths, of different lengths, on the surface. The paths that have minimum lengths are analogous to a straight line connecting two points in euclidean space and are known as geodesics. Surface geodesics can, for example, provide optimized motion planning across a curved surface for numerical control machining, robot programming, and winding of coils around a rotor. The infinitesimal distance between two points $(u, v)$ and $(u + du, v + dv)$ on a surface is given by

$$ds^2 = \mathbf{P}_u \cdot \mathbf{P}_u \, du^2 + 2\mathbf{P}_u \cdot \mathbf{P}_v \, du \, dv + \mathbf{P}_v \cdot \mathbf{P}_v \, dv^2 \qquad (6.13)$$

Equation (6.13) is often called the first fundamental quadratic form of a surface and is written as

$$ds^2 = E \, du^2 + 2F \, du \, dv + G \, dv^2 \qquad (6.14)$$

where

$$E(u, v) = \mathbf{P}_u \cdot \mathbf{P}_u \qquad F(u, v) = \mathbf{P}_u \cdot \mathbf{P}_v \qquad G(u, v) = \mathbf{P}_v \cdot \mathbf{P}_v \qquad (6.15)$$

$E$, $F$, and $G$ are the first fundamental, or metric, coefficients of the surface. These coefficients provide the basis for the measurement of lengths and areas, and the specification of directions and angles on a surface.

The distance between two points $P(u_a, v_a)$ and $P(u_b, v_b)$, shown in Fig. 6-20, is obtained by integrating Eq. (6.14) along a specified path $\{u = u(t), v = v(t)\}$ on the surface to give

$$S = \int_{t_a}^{t_b} \sqrt{Eu'^2 + 2Fu'v' + Gv'^2} \; dt \qquad (6.16)$$

where $u' = du/dt$ and $v' = dv/dt$. The minimum $S$ is the geodesic between the two points.

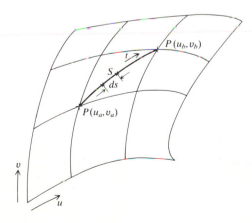

**FIGURE 6-20**
Surface geodesics.

The first fundamental form gives the distance element $ds$ which lies in the tangent plane of the surface at $P(u, v)$ and, therefore, yields no information on how the surface curves away from the tangent plane at that point. To investigate the surface curvature, another distance perpendicular to the tangent plane at $P(u, v)$ is introduced and given by

$$\tfrac{1}{2} \, dh^2 = \mathbf{\hat{n}} \cdot \mathbf{P}_{uu} \, du^2 + 2\mathbf{\hat{n}} \cdot \mathbf{P}_{uv} \, du \, dv + \mathbf{\hat{n}} \cdot \mathbf{P}_{vv} \, dv^2 \qquad (6.17)$$

Similarly to Eq. (6.13), Eq. (6.17) is often called the second fundamental quadratic form of a surface and is written as

$$\tfrac{1}{2} \, dh^2 = L \, du^2 + 2M \, du \, dv + N \, dv^2 \qquad (6.18)$$

where

$$L(u, v) = \mathbf{\hat{n}} \cdot \mathbf{P}_{uu} \qquad M(u, v) = \mathbf{\hat{n}} \cdot \mathbf{P}_{uv} \qquad N(u, v) = \mathbf{\hat{n}} \cdot \mathbf{P}_{vv} \qquad (6.19)$$

$L$, $M$, and $N$ are the second fundamental coefficients of the surface and form the basis for defining and analyzing the curvature of a surface.

A surface curvature at a point $P(u, v)$ is defined as the curvature of the normal section curve that lies on the surface and passes by the point. A normal section curve is the intersection curve of a plane passing through the normal $\mathbf{\hat{n}}$ at the point and the surface. Obviously, there can exist a family of planes, and therefore a family of normal section curves, that can contain $\mathbf{\hat{n}}$. The surface curvature at a point on a normal section curve given by the form $\{u = u(t), \, v = v(t)\}$ can be written as

$$\mathcal{K} = \frac{Lu'^2 + 2Mu'v' + Nv'^2}{Eu'^2 + 2Fu'v' + Gv'^2} \qquad (6.20)$$

and the radius of curvature at the point is $\rho = 1/\mathcal{K}$. The sense of curvature must be chosen. One convention is to choose the sign in Eq. (6.20) to give positive curvature for convex surfaces and negative curvature for concave surfaces.

Few types of surface curvatures exist. The gaussian curvature $K$ and mean curvature $H$ are defined by

$$K = \frac{LN - M^2}{EG - F^2}$$

$$H = \frac{EN + GL - 2FM}{2(EG - F^2)} \qquad (6.21)$$

Equation (6.20) gives the surface curvature in any direction at point $P(u, v)$. However, it can be used to obtain the principal curvatures which are the upper (maximum) and lower (minimum) bounds on the curvature at the point:

$$\mathcal{K}_{\text{max}} = H + \sqrt{H^2 - K}$$

$$\mathcal{K}_{\text{min}} = H - \sqrt{H^2 - K} \qquad (6.22)$$

The gaussian and mean curvatures can be written in terms of Eqs. (6.22) as

$$K = \mathcal{K}_{\text{max}} \mathcal{K}_{\text{min}}$$

$$H = \tfrac{1}{2}(\mathcal{K}_{\text{max}} + \mathcal{K}_{\text{min}}) \qquad (6.23)$$

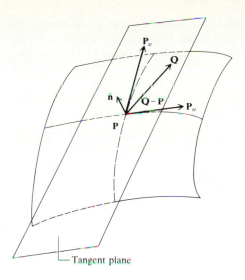

Tangent plane

**FIGURE 6-21**
Tangent plane to a surface.

The gaussian curvature can be positive (as in a hill), negative (as in a saddle point), or zero (as in ruled or cylindrical surfaces) depending on the signs at $\mathscr{K}_{max}$ and $\mathscr{K}_{min}$. Surfaces that have zero gaussian curvature everywhere are called developable, that is, they can be laid flat on a plane without stretching, tearing, or distorting them. Some CAD/CAM systems display gaussian curvature contour maps to convey information about local surface shape.

From a practical point of view, the principal curvatures are of primary interest. For example, to machine a surface with a spherical cutter, it is important to ensure that the cutter radius is smaller than the smallest concave radius of curvature of the surface if gouging is to be avoided.

Analogous to the family of planes that are perpendicular to a surface and contain the normal $\hat{\mathbf{n}}$, a tangent plane to a surface at a point can be defined. It is the common plane on which all the tangent vectors to the surface at a point lie. To develop the equation of a tangent plane, consider a point $q$ in a tangent plane to a given surface at a given point $P$, as shown in Fig. 6-21. The vectors $\mathbf{P}_u$, $\mathbf{P}_v$, and $\mathbf{Q} - \mathbf{P}$ lie on the plane. The normal $\hat{\mathbf{n}}$ is normal to any vector in the plane. Thus, we can write

$$\hat{\mathbf{n}} \cdot (\mathbf{Q} - \mathbf{P}) = (\mathbf{P}_u \times \mathbf{P}_v) \cdot (\mathbf{Q} - \mathbf{P}) = 0 \qquad (6.24)$$

which gives the equation of the tangent plane.

**Example 6.2.** The parametric equation of a sphere of radius $R$ and a center at point $P_0(x_0, y_0, z_0)$ is given by

$$x = x_0 + R \cos u \cos v$$

$$y = y_0 + R \cos u \sin v \qquad -\frac{\pi}{2} \leq u \leq \frac{\pi}{2}, 0 \leq v \leq 2\pi$$

$$z = z_0 + R \sin u$$

Find the sphere implicit equation.

*Solution.* Equation (6.1) suggests a useful and systematic way of developing the implicit equation of a surface from its parametric form. Implicit equations of surfaces are useful when solving surface intersection problems. Based on Eq. (6.1), the $z$ coordinate of a point on the sphere is to be rewritten as a function of $x$ and $y$. Squaring and adding the first two equations give

$$\frac{(x - x_0)^2}{R^2} + \frac{(y - y_0)^2}{R^2} = \cos^2 u = 1 - \sin^2 u$$

which gives

$$\sin u = \sqrt{1 - \frac{(x - x_0)^2}{R^2} - \frac{(y - y_0)^2}{R^2}}$$

By substituting $\sin u$ into the $z$ coordinate equation, $f(x, y)$ for a sphere becomes

$$z = z_0 + \sqrt{R^2 - (x - x_0)^2 - (y - y_0)^2}$$

The sphere implicit equation can now be written as

$$\mathbf{P} = [x \quad y \quad z_0 + \sqrt{R^2 - (x - x_0)^2 - (y - y_0)^2}]^T$$

It is obvious that the square of the $z$ coordinate gives the classical sphere equation:

$$(x - x_0)^2 + (y - y_0)^2 + (z - z_0)^2 = R^2$$

## 6.5  PARAMETRIC REPRESENTATION OF ANALYTIC SURFACES

This section covers the basics of the parametric equations of analytic surfaces most often encountered in surface modeling and design. The background provided in this section and the next one should enable users of CAD/CAM technology to realize the limitations of each surface available to them for modeling and design as well as cope with the documentation of surface commands.

The distinction between the WCS and MCS has been ignored in the development of the parametric equations of the various surfaces to avoid confusion. The transformation between the two systems is obvious and has already been discussed in Chap. 5. In addition, the terms "surface" and "patch" are used interchangeably in this chapter. However, in a more general sense a surface is considered the superset since a surface can contain one or more patch.

### 6.5.1  Plane Surface

The parametric equation of a plane can take different forms depending on the given data. Consider first the case of a plane defined by three points $P_0$, $P_1$, and $P_2$ as shown in Fig. 6-22. Assume that the point $P_0$ defines $u = 0$ and $v = 0$ and the vectors $(\mathbf{P}_1 - \mathbf{P}_0)$ and $(\mathbf{P}_2 - \mathbf{P}_0)$ define the $u$ and $v$ directions respectively. Assume also that the domains for $u$ and $v$ are $[0, 1]$. The position vector of any point $P$ on the plane can be now written as

$$\mathbf{P}(u, v) = \mathbf{P}_0 + u(\mathbf{P}_1 - \mathbf{P}_0) + v(\mathbf{P}_2 - \mathbf{P}_0), \qquad 0 \le u \le 1, 0 \le v \le 1 \quad (6.25)$$

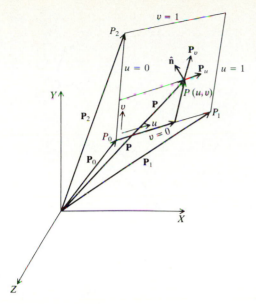

**FIGURE 6-22**
A plane patch defined by three points.

The above equation can be seen as the bilinear form of Eq. (5.11). Utilizing Eqs. (6.4) and (6.5), the tangent vectors at point $P$ are

$$\mathbf{P}_u(u, v) = \mathbf{P}_1 - \mathbf{P}_0 \qquad \mathbf{P}_v(u, v) = \mathbf{P}_2 - \mathbf{P}_0, \qquad 0 \leq u \leq 1, 0 \leq v \leq 1 \quad (6.26)$$

and the surface normal is

$$\hat{\mathbf{n}}(u, v) = \frac{(\mathbf{P}_1 - \mathbf{P}_0) \times (\mathbf{P}_2 - \mathbf{P}_0)}{|(\mathbf{P}_1 - \mathbf{P}_0) \times (\mathbf{P}_2 - \mathbf{P}_0)|}, \qquad 0 \leq u \leq 1, 0 \leq v \leq 1 \quad (6.27)$$

which is constant for any point on the plane. As for the curvature of the plane, it is equal to zero [see Eq. (6.20)] because all the second fundamental coefficients of the plane are zeros [see Eq. (6.19)].

Another case of constructing a plane surface is when the surface passes through a point $P_0$ and contains two directions defined by the unit vectors $\hat{\mathbf{r}}$ and $\hat{\mathbf{s}}$ as shown in Fig. 6-23. Similar to the above case, the plane equation can be written as

$$\mathbf{P}(u, v) = \mathbf{P}_0 + uL_u\hat{\mathbf{r}} + vL_v\hat{\mathbf{s}}, \qquad 0 \leq u \leq 1, 0 \leq v \leq 1 \quad (6.28)$$

This equation is also considered as the bilinear form of Eq. (5.17). The equation assumes a plane of dimensions $L_u$ and $L_v$ that may be set to unity.

The above two cases can be combined to provide the equation of a plane surface that passes through two points $P_0$ and $P_1$ and is parallel to the unit vector $\hat{\mathbf{r}}$. In this case, we can write

$$\mathbf{P}(u, v) = \mathbf{P}_0 + u(\mathbf{P}_1 - \mathbf{P}_0) + vL_v\hat{\mathbf{r}}, \qquad 0 \leq u \leq 1, 0 \leq v \leq 1 \quad (6.29)$$

The last case to be considered is for a plane that passes through a point $P_0$ and is perpendicular to a given direction $\hat{\mathbf{n}}$. Figure 6-24 shows this case. The vector $\hat{\mathbf{n}}$ is normal to any vector in the plane. Thus,

$$(\mathbf{P} - \mathbf{P}_0) \cdot \hat{\mathbf{n}} = 0 \qquad (6.30)$$

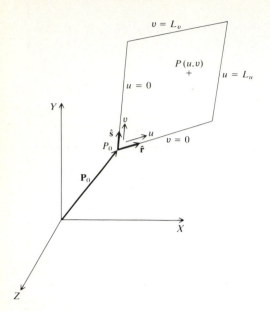

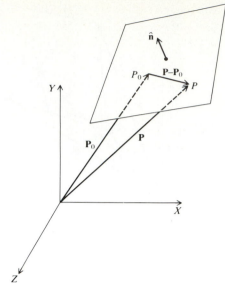

**FIGURE 6-23**
A plane patch defined by a point and two directions.

**FIGURE 6-24**
A plane patch passing through point $P_0$ and normal to $\hat{\mathbf{n}}$.

which is a nonparametric equation of the plane surface. A parametric equation can be developed by using Eq. (6.30) to generate two points on the surface which can be used with $P_0$ in Eq. (6.25). Planes that are perpendicular to the axes of a current WCS are special cases of Eq. (6.30). For example, in the case of a plane perpendicular to the X axis, $\hat{\mathbf{n}}$ is (1, 0, 0) and the plane equation is $x = x_0$.

A database structure of a plane surface can be seen to include its unit normal $\hat{\mathbf{n}}$, a point on the plane $P_0$, and $u$ and $v$ axes defined in terms of the MCS coordinates. For example, if a plane passes through the points $P_0(0, 0, 0)$, $P_1(2, 0, 0)$, and $P_2(0, 0, 2)$ as shown in Fig. 6-25, the verification of the plane

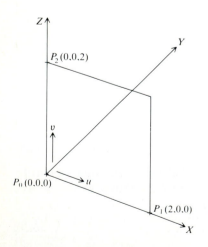

**FIGURE 6-25**
Plane surface construction.

surface entities shows the entity is a plane passing through $P_0(0, 0, 0)$ and has a unit normal of $(0, -1, 0)$. In addition, the $u$ and $v$ axes are defined by the coordinates $(1, 0, 0)$ and $(0, 0, 1)$ respectively.

**Example 6.3.** Find the minimum distance between a point in space and a plane surface.

**Solution.** The minimum distance between a point and a plane is also the perpendicular distance from the point onto the plane. Let us assume the plane equation is given by

$$\mathbf{P} = \mathbf{P}_0 + u\hat{\mathbf{r}} + v\hat{\mathbf{s}}, \qquad 0 \le u \le 1, 0 \le v \le 1 \tag{6.31}$$

This assumption can be made in the light of the database structure described above. From Fig. 6-26, it is obvious that the perpendicular vector from point $Q$ to the plane is parallel to its normal $\hat{\mathbf{n}}$. Thus, we can write

$$\mathbf{P} = \mathbf{Q} - D\hat{\mathbf{n}} \tag{6.32}$$

By using Eq. (6.31) in (6.32), we get

$$\mathbf{P}_0 + u\hat{\mathbf{r}} + v\hat{\mathbf{s}} = \mathbf{Q} - D\hat{\mathbf{n}} \tag{6.33}$$

Equation (6.33) can be rewritten in a matrix form as

$$\begin{bmatrix} r_x & s_x & n_x \\ r_y & s_y & n_y \\ r_z & s_z & n_z \end{bmatrix} \begin{bmatrix} u \\ v \\ D \end{bmatrix} = \begin{bmatrix} x_Q - x_0 \\ y_Q - y_0 \\ z_Q - z_0 \end{bmatrix} \tag{6.34}$$

where $r_x$, $r_y$, and $r_z$ are the components of the unit vector $\hat{\mathbf{r}}$. Similarly, the components of the other vectors are given in Eq. (6.34). Solving Eq. (6.34) (see Chap. 5) gives the normal distance $D$ and $u$ and $v$, which can give the point $P$.

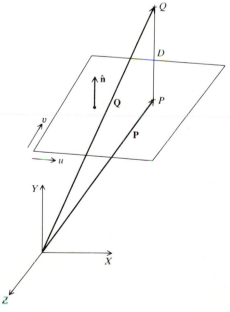

**FIGURE 6-26**
Minimum distance between a point and plane surface.

### 6.5.2 Ruled Surface

A ruled surface is generated by joining corresponding points on two space curves (rails) $G(u)$ and $Q(u)$ by straight lines (also called rulings or generators), as shown in Fig. 6-27. The main characteristic of a ruled surface is that there is at least one straight line passing through the point $P(u, v)$ and lying entirely in the surface. In addition, every developable surface is a ruled surface. Cones and cylinders are examples of ruled surfaces and the plane surface covered in Sec. 6.5.1 is considered the simplest of all ruled surfaces.

To develop the parametric equation of a ruled surface, consider the ruling $u = u_i$ joining points $G_i$ and $Q_i$ on the rails $G(u)$ and $Q(u)$ respectively. Using Eq. (5.11), the equation of the ruling becomes

$$P(u_i, v) = G_i + v(Q_i - G_i) \tag{6.35}$$

where $v$ is the parameter along the ruling. Generalizing Eq. (6.35) for any ruling, the parametric equation of a ruled surface defined by two rails is

$$P(u, v) = G(u) + v[Q(u) - G(u)] = (1 - v)G(u) + vQ(u),$$
$$0 \le u \le 1, 0 \le v \le 1 \tag{6.36}$$

Holding the $u$ value constant in the above equation produces the rulings given by Eq. (6.35) in the $v$ direction of the surface, while holding the $v$ value constant

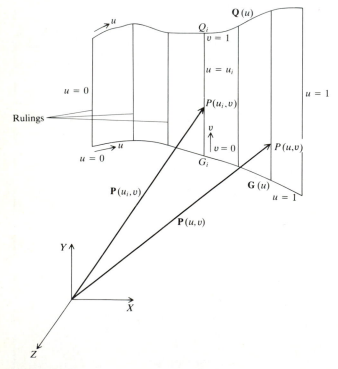

**FIGURE 6-27**
Parametric representation of a ruled surface.

yields curves in the $u$ direction which are a linear blend of the rails. In fact, $\mathbf{G}(u)$ and $\mathbf{Q}(u)$ are $\mathbf{P}(u, 0)$ and $\mathbf{P}(u, 1)$ respectively. Therefore, the closer the value of $v$ to zero, the greater the influence of $\mathbf{G}(u)$ and the less the influence of $\mathbf{Q}(u)$ on the $v =$ constant curve. Similarly, the influence of $\mathbf{Q}(u)$ on the ruled surface geometry increases when the $v$ value approaches unity (see Fig. 6-1).

Based on Fig. 6-27 and Eq. (6.35), it is now obvious why digitizing the wrong ends of the rails produces the undesirable ruled surface as shown in Fig. 6-1a. In addition, Eq. (6.36), together with Eqs. (6.15), (6.19), and (6.20), shows that a ruled surface can only allow curvature in the $u$ direction of the surface provided that the rails have curvatures. The surface curvature in the $v$ direction (along the rulings) is zero and thus a ruled surface cannot be used to model surface patches that have curvatures in two directions.

## 6.5.3 Surface of Revolution

The rotation of a planar curve an angle $v$ about an axis of rotation creates a circle (if $v = 360$) for each point on the curve whose center lies on the axis of rotation and whose radius $r_z(u)$ is variable, as shown in Fig. 6-28. The planar curve and the circles are called the profile and parallels respectively while the various positions of the profile around the axis are called meridians.

The planar curve and the axis of rotation form the plane of zero angle, that is, $v = 0$. To derive the parametric equation of a surface of revolution, a local coordinate system with a $Z$ axis coincident with the axis of rotation is assumed as shown in Fig. 6-28. This local system shown by the subscript $L$ can be created as follows. Choose the perpendicular direction from the point $u = 0$ on the profile as the $X_L$ axis and the intersection point between $X_L$ and $Z_L$ as the origin of the local system. The $Y_L$ axis is automatically determined by the right-hand rule. Now, consider a point $\mathbf{G}(u) = \mathbf{P}(u, 0)$ on the profile that rotates an angle $v$ about $Z_L$ when the profile rotates the same angle. Considering the shaded triangle which is perpendicular to the $Z_L$ axis, the parametric equation of the surface of revolution can be written as

$$\mathbf{P}(u, v) = r_z(u) \cos v \hat{\mathbf{n}}_1 + r_z(u) \sin v \hat{\mathbf{n}}_2 + z_L(u) \hat{\mathbf{n}}_3,$$

$$0 \le u \le 1, 0 \le v \le 2\pi \qquad (6.37)$$

If we choose $z_L(u) = u$ for each point on the profile, Eq. (6.37) gives the local coordinates $(x_L, y_L, z_L)$ of a point $P(u, v)$ as $[r_z(u) \cos v, r_z(u) \sin v, u]$. The local coordinates are transformed to MCS coordinates before displaying the surface using Eq. (3.3) where the rotation matrix is formed from $\hat{\mathbf{n}}_1$, $\hat{\mathbf{n}}_2$, and $\hat{\mathbf{n}}_3$, and the position of the origin of the local system is given by $\mathbf{P}_L$ (see Fig. 6-28).

The database of a surface of revolution must include its profile, axis of rotation, and the angle of rotation as starting and ending angles. Whenever the user requests the display of the surface with a mesh size $m \times n$, the $u$ range is divided equally into $(m - 1)$ divisions and $m$ values of $u$ are obtained. Similarly, the $v$ range is divided equally into $n$ values and Eq. (6.37) is used to generate points on the surface.

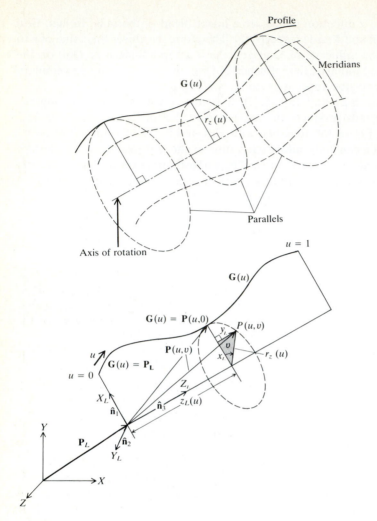

**FIGURE 6-28**
Parametric representation of a surface of revolution.

### 6.5.4 Tabulated Cylinder

A tabulated cylinder has been defined as a surface that results from translating a space planar curve along a given direction. It can also be defined as a surface that is generated by moving a straight line (called generatrix) along a given planar curve (called directrix). The straight line always stays parallel to a fixed given vector that defines the $v$ direction of the cylinder as shown in Fig. 6-29. The planar curve $\mathbf{G}(u)$ can be any wireframe entities. The position vector of any point $P(u, v)$ on the surface can be written as

$$\mathbf{P}(u, v) = \mathbf{G}(u) + v\hat{\mathbf{n}}_v, \qquad 0 \le u \le u_{\max}, 0 \le v \le v_{\max} \qquad (6.38)$$

From a user point of view, $\mathbf{G}(u)$ is the desired curve the user digitizes to form the cylinder, $v$ is the cylinder length, and $\hat{\mathbf{n}}_v$ is the cylinder axis. The repre-

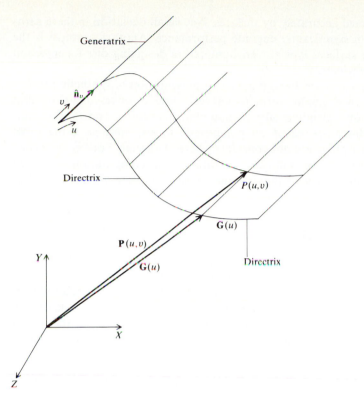

**FIGURE 6-29**
Parametric representation of a tabulated cylinder.

sentation of $G(u)$ is already available in the database at the time of creating it. The cylinder length $v$ is input in the form of lower and higher bounds where the difference between them gives the length. A zero value of the lower bound indicates the plane of the directrix. The user inputs the cylinder axis as two points that are used to determine $\hat{n}_v$ which is the unit vector along the axis.

As seen from Eq. (6.38), the database of a tabulated cylinder includes its directrix, the unit vector $\hat{n}_v$, and the lower and upper bounds of the cylinder. The display of a tabulated cylinder with a mesh $m \times n$ follows the same approach as discussed with surfaces of revolution.

## 6.6 PARAMETRIC REPRESENTATION OF SYNTHETIC SURFACES

The arguments regarding the needs for synthetic curves discussed in Sec. 5.6 of Chap. 5 apply here. Synthetic surfaces provide designers with better surface design tools than analytic surfaces. Consider the design of blade surfaces in jet aircraft engines. The design of these surfaces is usually based on aerodynamic and fluid-flow simulations, often incorporating thermal and mechanical stress deformation. These simulations yield ordered sets of discrete streamline points which must then be connected accurately by surfaces. Any small deviation in these aero-dynamic surfaces can significantly degrade performance. Another example is the creation of blending surfaces typically encountered in designing dies for injection molding of plastic products.

For continuity purposes, the parametric representation of synthetic surfaces is presented below in a similar form to curves. Surfaces covered are bicubic, Bezier, B-spline, Coons, blending offset, triangular, sculptured, and rational sur-faces. All these surfaces are based on polynomial forms. Surfaces using other forms such as Fourier series are not considered here. Although Fourier series can approximate any curve given sufficient conditions, the computations involved with them are greater than with polynomials. Therefore, they are not suited to general use in CAD/CAM.

### 6.6.1   Hermite Bicubic Surface

The parametric bicubic surface patch connects four corner data points and uti-lizes a bicubic equation. Therefore, 16 vector conditions (or 48 scalar conditions) are required to find the coefficients of the equation. When these coefficients are the four corner data points, the eight tangent vectors at the corner points (two at each point in the $u$ and $v$ directions), and the four twist vectors at the corner points, a Hermite bicubic surface patch results. The bicubic equation can be written as

$$\mathbf{P}(u, v) = \sum_{i=0}^{3} \sum_{j=0}^{3} \mathbf{C}_{ij} u^i v^j, \qquad 0 \le u \le 1, 0 \le v \le 1 \tag{6.39}$$

This equation can be expanded in similar ways, as given by Eqs. (5.75) and (5.76). Analogous to Eq. (5.77), the matrix form of Eq. (6.39) is

$$\mathbf{P}(u, v) = \mathbf{U}^T[C]\mathbf{V}, \qquad 0 \le u \le 1, 0 \le v \le 1 \tag{6.40}$$

where $U = [u^3 \quad u^2 \quad u \quad 1]^T$, $V = [v^3 \quad v^2 \quad v \quad 1]^T$, and the coefficient matrix $[C]$ is given by

$$[C] = \begin{bmatrix} \mathbf{C}_{33} & \mathbf{C}_{32} & \mathbf{C}_{31} & \mathbf{C}_{30} \\ \mathbf{C}_{23} & \mathbf{C}_{22} & \mathbf{C}_{21} & \mathbf{C}_{20} \\ \mathbf{C}_{13} & \mathbf{C}_{12} & \mathbf{C}_{11} & \mathbf{C}_{10} \\ \mathbf{C}_{03} & \mathbf{C}_{02} & \mathbf{C}_{01} & \mathbf{C}_{00} \end{bmatrix} \tag{6.41}$$

In order to determine the coefficients $\mathbf{C}_i$, consider the patch shown in Fig. 6-18. Applying the boundary conditions into Eq. (6.40), solving for the coeffi-cients, and rearranging give the following final equation of a bicubic patch:

$$\mathbf{P}(u, v) = \mathbf{U}^T[M_H][B][M_H]^T\mathbf{V}, \qquad 0 \le u \le 1, 0 \le v \le 1 \tag{6.42}$$

where $[M_H]$ is given by Eq. (5.84) and $[B]$, the geometry or boundary condition matrix, is

$$[B] = \begin{bmatrix} \mathbf{P}_{00} & \mathbf{P}_{01} & \mathbf{P}_{v00} & \mathbf{P}_{v01} \\ \mathbf{P}_{10} & \mathbf{P}_{11} & \mathbf{P}_{v10} & \mathbf{P}_{v11} \\ \mathbf{P}_{u00} & \mathbf{P}_{u01} & \mathbf{P}_{uv00} & \mathbf{P}_{uv01} \\ \mathbf{P}_{u10} & \mathbf{P}_{u11} & \mathbf{P}_{uv10} & \mathbf{P}_{uv11} \end{bmatrix} \tag{6.43}$$

The matrix $[B]$ is partitioned as shown above to indicate the grouping of the similar boundary conditions. It can also be written as

$$[B] = \begin{bmatrix} [P] & [P_v] \\ [P_u] & [P_{uv}] \end{bmatrix} \tag{6.44}$$

where $[P]$, $[P_u]$, $[P_v]$, and $[P_{uv}]$ are the submatrices of the corner points, corner $u$-tangent vectors, corner $v$-tangent vectors, and the corner twist vectors respectively.

The tangent and twist vectors at any point on the surface are given by

$$\mathbf{P}_u(u, v) = \mathbf{U}^T[M_H]^u[B][M_H]^T\mathbf{V} \tag{6.45}$$

$$\mathbf{P}_v(u, v) = \mathbf{U}^T[M_H][B][M_H]^{vT}\mathbf{V} \tag{6.46}$$

$$\mathbf{P}_{uv}(u, v) = \mathbf{U}^T[M_H]^u[B][M_H]^{vT}\mathbf{V} \tag{6.47}$$

where $[M_H]^u$ or $[M_H]^v$ is given by Eq. (5.88).

Similar to the cubic spline, the bicubic form permits $C^1$ continuity from one patch to the next. The necessary two conditions are to have the same curves ($C^0$ continuity) and the same direction of the tangent vectors ($C^1$ continuity) across the common edge between the two patches. The magnitude of the tangent vectors does not have to be the same.

Before writing the continuity conditions in terms of the $[B]$ matrix, let us expand Eqs. (6.42) and (6.45) to (6.47) to see what influences the position and tangent vectors. Equations (6.42) and (6.45) to (6.47) give

$$\mathbf{P}(u, v) = [F_1(u) \quad F_2(u) \quad F_3(u) \quad F_4(u)][B]\begin{bmatrix} F_1(v) \\ F_2(v) \\ F_3(v) \\ F_4(v) \end{bmatrix} \tag{6.48}$$

$$\mathbf{P}_u(u, v) = [G_1(u) \quad G_2(u) \quad G_3(u) \quad G_4(u)][B]\begin{bmatrix} F_1(v) \\ F_2(v) \\ F_3(v) \\ F_4(v) \end{bmatrix} \tag{6.49}$$

$$\mathbf{P}_v(u, v) = [F_1(u) \quad F_2(u) \quad F_3(u) \quad F_4(u)][B]\begin{bmatrix} G_1(v) \\ G_2(v) \\ G_3(v) \\ G_4(v) \end{bmatrix} \tag{6.50}$$

$$\mathbf{P}_{uv}(u, v) = [G_1(u) \quad G_2(u) \quad G_3(u) \quad G_4(u)][B]\begin{bmatrix} G_1(v) \\ G_2(v) \\ G_3(v) \\ G_4(v) \end{bmatrix} \tag{6.51}$$

where

$$F_1(x) = 2x^3 - 3x^2 + 1$$
$$F_2(x) = -2x^3 + 3x^2$$
$$F_3(x) = x^3 - 2x^2 + x$$
$$F_4(x) = x^3 - x^2$$

$$(6.52)$$

and

$$G_1(x) = 6x^2 - 6x$$
$$G_2(x) = -6x^2 + 6x$$
$$G_3(x) = 3x^2 - 4x + 1$$
$$G_4(x) = 3x^2 - 2x$$

$$(6.53)$$

For $u = 0$ and $u = 1$ edges these equations become

$$\begin{bmatrix} \mathbf{P}(0, v) \\ \mathbf{P}(1, v) \\ \mathbf{P}_u(0, v) \\ \mathbf{P}_u(1, v) \end{bmatrix} = [B] \begin{bmatrix} F_1(v) \\ F_2(v) \\ F_3(v) \\ F_4(v) \end{bmatrix}$$

$$(6.54)$$

Similarly, for $v = 0$ and $v = 1$ edges we can write

$$\begin{bmatrix} \mathbf{P}(u, 0) \\ \mathbf{P}(u, 1) \\ \mathbf{P}_v(u, 0) \\ \mathbf{P}_v(u, 1) \end{bmatrix} = [B] \begin{bmatrix} F_1(u) \\ F_2(u) \\ F_3(u) \\ F_4(u) \end{bmatrix}$$

$$(6.55)$$

Equation (6.54) shows that the corner $v$-tangent and twist vectors affect the position and tangent vectors respectively all along the $u = 0$ and $u = 1$ edges except at $v = 0$ and 1 where $F_3$ and $F_4$ are both zero.

To write the blending conditions for $C^1$ continuity between two patches, consider the surface shown in Fig. 6-16. These conditions to connect patch 1 and patch 2 along the $u$ edges are

$$[\mathbf{P}(0, v)]_{\text{patch 2}} = [\mathbf{P}(1, v)]_{\text{patch 1}} \qquad C^0 \text{ continuity}$$
$$[\mathbf{P}_u(0, v)]_{\text{patch 2}} = K[\mathbf{P}_u(1, v)]_{\text{patch 1}} \qquad C^1 \text{ continuity}$$

$$(6.56)$$

where $K$ is a constant. Equation (6.56) can be interpreted as blending an infinite number of cubic spline segments on each patch (each corresponding to a particular $v$ value) with $C^1$ continuity across the $u$ edge. Utilizing Eq. (6.54), Eq. (6.56) can be expressed in terms of the rows of the $[B]$ matrix as shown in Fig. 6-30. The figure shows the constrained elements of each matrix only. Empty elements of each matrix are unconstrained and can have arbitrary values. It is left to the reader to find similar matrices to join the two patches at the $v = 1$ and $v = 0$ edges or any other combination.

The Ferguson surface (also called the F-surface patch) is considered a bicubic surface patch with zero twist vectors at the patch corners, as shown in

$[B]_{\text{patch 1}}$          $[B]_{\text{patch 2}}$

| | | | | $P_{10}$ | $P_{11}$ | $P_{v10}$ | $P_{v11}$ |
|---|---|---|---|---|---|---|---|
| $P_{10}$ | $P_{11}$ | $P_{v10}$ | $P_{v11}$ | | | | |
| | | | | $K\,P_{u10}$ | $K\,P_{u11}$ | $K\,P_{uv10}$ | $K\,P_{uv11}$ |
| $P_{u10}$ | $P_{u11}$ | $P_{uv10}$ | $P_{uv11}$ | | | | |

**FIGURE 6-30**
Constrained elements of boundary matrix $[B]$ to blend two bicubic patches along a $u$ edge.

Fig. 6-31. Thus, the boundary matrix for the F-surface patch becomes

$$[B] = \left[\begin{array}{cc|cc} P_{00} & P_{01} & P_{v00} & P_{v01} \\ P_{10} & P_{11} & P_{v10} & P_{v11} \\ \hline P_{u00} & P_{u01} & 0 & 0 \\ P_{u10} & P_{u11} & 0 & 0 \end{array}\right] \qquad (6.57)$$

This special surface is useful in design and machining applications. The tangent vectors at the corner points can be approximated in terms of the corner positions using the direction and the length of chord lines joining the corner points. Hence, the designer does not have to input tangent vector information and the computations required to calculate the surface parameters are simplified. This is useful if tool paths are to be generated to mill the surface.

The characteristics of the bicubic surface path are very similar to those of the cubic spline. The patch can be used to fit a bicubic surface to an array of data

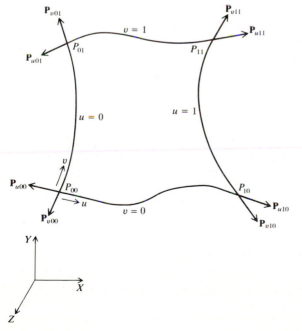

**FIGURE 6-31**
The F-surface patch.

points homomorphic to an $m \times n$ rectangular grid. The control of the resulting surface is global and is not intuitively based on the input data. In addition, the requirement of tangent and twist vectors as input data does not fit very well the design environment because the intuitive feeling for such data is usually not clear.

**Example 6.4.** Show that a bicubic surface patch degenerates to a cubic spline if the four corner points of the patch are collapsed to two.

**Solution.** Consider the surface patch shown in Fig. 6-18. Let us assume that the $v = 1$ edge coincides with the $v = 0$ edge. In this case, the corners $P_{00}$ and $P_{01}$ coincide and so do the corners $P_{10}$ and $P_{11}$. All the derivatives with respect to $v$ are set to zero and the $u$-tangent vectors at the coincident corners are equal to one another. Finally let us choose the value of $v$ to equal 1. Substituting all these values into Eq. (6.48), we obtain

$$P(u, 1) = [F_1(u) \quad F_2(u) \quad F_3(u) \quad F_4(u)] \begin{bmatrix} P_{00} & P_{00} & 0 & 0 \\ P_{10} & P_{10} & 0 & 0 \\ P_{u00} & P_{u00} & 0 & 0 \\ P_{u10} & P_{u10} & 0 & 0 \end{bmatrix} \begin{bmatrix} 1 \\ 0 \\ 0 \\ 0 \end{bmatrix}$$

Expanding the equation gives

$$P(u, 1) = (2u^3 - 3u^2 + 1)P_{00} + (-2u^3 + 3u^2)P_{10} + (u^3 - 2u^2 + u)P_{u00}$$
$$+ (u^3 - u^2)P_{u10}$$

Dropping the reference to the $v$ variable in this equation gives

$$P(u) = (2u^3 - 3u^2 + 1)P_0 + (-2u^3 + 3u^2)P_1 + (u^3 - 2u^2 + u)P_{u0} + (u^3 - u^2)P_{u1}$$

which is identical to Eq. (5.81) developed in Chap. 5. Also, notice that substituting $v = 1$ in Eq. (6.39) results in Eq. (5.74).

**Example 6.5.** Sometimes it is useful to reformulate a given surface equation in terms of the bicubic form given by Eq. (6.42). What is the equivalent bicubic patch to a plane given by Eq. (6.28)?

**Solution.** The general procedure to achieve reformulation of a surface form to a bicubic form is to use Eq. (6.39) and its derivatives to find $P(u, v)$, $P_u(u, v)$, $P_v(u, v)$, and $P_{uv}(u, v)$ in terms of the $C_{ij}$ coefficients. If the boundary values of $u$ and $v$ (0 and 1) are substituted into the resulting expressions, the boundary conditions (elements of $[B]$) of the patch can be written in terms of these coefficients. The resulting general functions take the form:

$$B_{mn} = f(C_{ij})$$

where $B_{mn}$ is an element of $[B]$.

If the equation of a given surface is compared to Eq. (6.39), the corresponding set of $C_{ij}$ coefficients are obtained which can in turn be substituted into the above equation to provide $[B]$ of the equivalent bicubic patch.

Let us apply the above general procedure to the plane given by Eq. (6.28). Comparing this equation to Eq. (6.39) gives

$$C_{00} = P_0$$

$$C_{10} = L_u \hat{r}$$

$$C_{01} = L_v \hat{s}$$

and all the other coefficients are zeros. The $[B]$ matrix then becomes

$$[B] = \begin{bmatrix} P_0 & P_0 + L_v \hat{s} & L_v \hat{s} & L_v \hat{s} \\ P_0 + L_u \hat{r} & P_0 + L_u \hat{r} + L_v \hat{s} & L_v \hat{s} & L_v \hat{s} \\ \hline L_u \hat{r} & L_u \hat{r} & 0 & 0 \\ L_u \hat{r} & L_u \hat{r} & 0 & 0 \end{bmatrix}$$

**Example 6.6.** A bicubic surface patch passes through the point $P_{00}$ and has tangent vectors at the point as $P_{u00}$ and $P_{v00}$. Show that the patch is planar if the other corner vectors are defined as linear functions of $P_{u00}$ and $P_{v00}$ and the corner twist vectors are zeros.

*Solution.* Figure 6-32 shows the above defined bicubic patch. The corner position and tangent vectors can be defined by the following linear functions:

$$\begin{bmatrix} P_{10} \\ P_{01} \\ P_{11} \\ P_{u10} \\ P_{u01} \\ P_{u11} \\ P_{v10} \\ P_{v01} \\ P_{v11} \end{bmatrix} = \begin{bmatrix} 1 & a_1 & b_1 \\ 1 & a_2 & b_2 \\ 1 & a_3 & b_3 \\ 0 & a_4 & b_4 \\ 0 & a_5 & b_5 \\ 0 & a_6 & b_6 \\ 0 & a_7 & b_7 \\ 0 & a_8 & b_8 \\ 0 & a_9 & b_9 \end{bmatrix} \begin{bmatrix} P_{00} \\ P_{u00} \\ P_{v00} \end{bmatrix} \qquad (6.58)$$

Substituting the above equation into Eq. (6.43) we get

$$[B] = \begin{bmatrix} P_{00} & P_{00} + a_2 P_{u00} + b_2 P_{v00} & P_{v00} & a_8 P_{u00} + b_8 P_{v00} \\ P_0 + a_1 P_{u00} + b_1 P_{v00} & P_{00} + a_3 P_{u00} + b_3 P_{v00} & a_7 P_{u00} + b_7 P_{v00} & a_9 P_{v00} + b_9 P_{v00} \\ \hline P_{u00} & a_5 P_{u00} + b_5 P_{v00} & 0 & 0 \\ a_4 P_{u00} + b_4 P_{v00} & a_6 P_{u00} + b_6 P_{v00} & 0 & 0 \end{bmatrix}$$

$$(6.59)$$

If the bicubic patch whose geometry matrix $[B]$ is given by Eq. (6.59) is planar, the following equation can be written for any point on the patch:

$$N_{00} \cdot [P(u, v) - P_{00}] = 0 \qquad (6.60)$$

which states that the normal vector at point $P_{00}$ must be perpendicular to any vector $[P(u, v) - P_{00}]$ in the plane, as shown in Fig. 6-32. To prove that Eq. (6.60) is

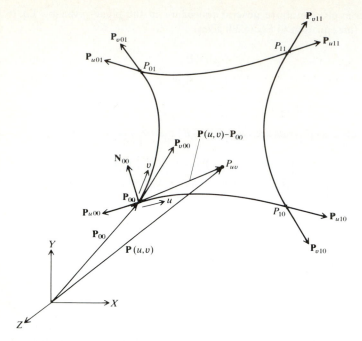

**FIGURE 6-32**
A bicubic plane patch.

valid for any point on the patch under investigation, substitute Eq. (6.59) into Eq. (6.48) and reduce the result to obtain

$$
\begin{aligned}
\mathbf{P}(u, v) = \ &\mathbf{P}_{00}[F_1(u)F_1(v) + F_2(u)F_1(v) + F_1(u)F_2(v) + F_2(u)F_2(v)] \\
&+ \mathbf{P}_{u00}[a_1 F_2(u)F_1(v) + F_3(u)F_1(v) + a_4 F_4(u)F_1(v) + a_2 F_1(u)F_2(v) \\
&\qquad + a_3 F_2(u)F_2(v) + a_5 F_3(u)F_2(v) + a_6 F_4(u)F_2(v) \\
&\qquad + a_7 F_2(u)F_3(v) + a_8 F_1(u)F_4(v) + a_9 F_2(u)F_4(v)] \\
&+ \mathbf{P}_{v00}[b_1 F_2(u)F_1(v) + b_4 F_4(u)F_1(v) + b_2 F_1(u)F_2(v) + b_3 F_2(u)F_2(v) \\
&\qquad + b_5 F_3(u)F_2(v) + b_6 F_4(u)F_2(v) + F_1(u)F_3(v) + b_7 F_2(u)F_3(v) \\
&\qquad + b_8 F_1(u)F_4(v) + b_9 F_2(u)F_4(v)]
\end{aligned}
$$

The coefficient of $\mathbf{P}_{00}$ in the above equation is equal to unity if Eq. (6.52) is used. The coefficients of $\mathbf{P}_{u00}$ and $\mathbf{P}_{v00}$ are functions of $u$ and $v$ only. Assuming these functions are $f(u, v)$ and $g(u, v)$ respectively, the above equation can be written as

$$
\mathbf{P}(u, v) - \mathbf{P}_{00} = f(u, v)\mathbf{P}_{u00} + g(u, v)\mathbf{P}_{v00} \tag{6.61}
$$

The normal vector $\mathbf{N}_{00}$ can be written as

$$
\mathbf{N}_{00} = \mathbf{P}_{u00} \times \mathbf{P}_{v00} \tag{6.62}
$$

Substitute Eqs. (6.61) and (6.62) into the left-hand side of Eq. (6.60). This gives

$$
(\mathbf{P}_{u00} \times \mathbf{P}_{v00}) \cdot [f(u, v)\mathbf{P}_{u00} + g(u, v)\mathbf{P}_{v00}]
$$

$$
= f(u, v)(\mathbf{P}_{u00} \times \mathbf{P}_{v00}) \cdot \mathbf{P}_{u00} + g(u, v)(\mathbf{P}_{u00} \times \mathbf{P}_{v00}) \cdot \mathbf{P}_{v00} \tag{6.63}
$$

The right-hand side of this equation is equal to zero regardless of the values of $u$ and $v$. Therefore, Eq. (6.60) is valid for any point and the bicubic patch is planar.

The above technique can be generalized to test for planarity/nonplanarity of a bicubic patch based on its geometry matrix $[B]$. Let us define the matrix $[S]$ as

$$[S] = [M_H][B][M_H]^T \tag{6.64}$$

If a bicubic patch is to be planar, the elements $s_{ij}$ of $[S]$ must satisfy the following equation:

$$N_{x00} \sum_{i=1}^{4} \sum_{j=1}^{4} \frac{s_{ijx}}{(5-i)(5-j)} + N_{y00} \sum_{i=1}^{4} \sum_{j=1}^{4} \frac{s_{ijy}}{(5-i)(5-j)}$$

$$+ N_{z00} \sum_{i=1}^{4} \sum_{j=1}^{4} \frac{s_{ijz}}{(5-i)(5-j)} - K = 0 \tag{6.65}$$

where $N_{x00}$, $N_{y00}$, and $N_{z00}$ are the components of the normal vector $N_{00}$ and $K$ is defined by

$$K = \mathbf{N}_{00} \cdot \mathbf{P}_{00} \tag{6.66}$$

The practical implication of this example is the ability to construct planes with curved boundaries as opposed to straight boundaries, discussed in Sec. 6.5.1.

### 6.6.2  Bezier Surface

A tensor product Bezier surface is an extension of the Bezier curve in two parametric directions $u$ and $v$. An orderly set of data or control points is used to build a topologically rectangular surface as shown in Fig. 6-33. The surface equation

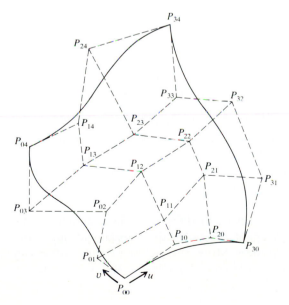

**FIGURE 6-33**
A 4 × 5 Bezier surface.

can be written by extending Eq. (5.91); that is,

$$P(u, v) = \sum_{i=0}^{n} \sum_{j=0}^{m} P_{ij} B_{i,n}(u)B_{j,m}(v), \qquad 0 \le u \le 1, 0 \le v \le 1 \qquad (6.67)$$

where $P(u, v)$ is any point on the surface and $P_{ij}$ are the control points. These points form the vertices of the control or characteristic polyhedron (shown dashed in Fig. 6-33) of the resulting Bezier surface. The points are arranged in an $(n + 1) \times (m + 1)$ rectangular array, as seen from the above equation. In comparison with Eq. (5.94) of the Bezier curve, expanding Eq. (6.67) gives

$$
P(u, v) = P_{00}(1 - u)^n(1 - v)^m + P_{11}C(n, 1)C(m, 1)uv(1 - u)^{n-1}(1 - v)^{m-1}
$$

$$
\left.
\begin{array}{l}
+ P_{22}\,C(n, 2)C(m, 2)u^2v^2(1 - u)^{n-2}(1 - v)^{m-2} + \cdots \\[6pt]
+ P_{(n-1)(m-1)}C(n, n - 1)C(m, m - 1)u^{n-1}v^{m-1}(1 - u)(1 - v) \\[6pt]
+ P_{nm}\,u^n v^m
\end{array}
\right\}
\begin{array}{c}\text{Symmetric}\\ \text{terms in}\\ u \text{ and } v\end{array}
$$

$$
\begin{array}{c}\text{Nonsymmetric}\\ \text{terms in}\\ u \text{ and } v\end{array}
\left[
\begin{array}{l}
+ P_{10}\,C(n, 1)u(1 - u)^{n-1}(1 - v)^m + P_{01}C(m, 1)v(1 - u)^n(1 - v)^{m-1} \\[6pt]
+ P_{20}\,C(n, 2)u^2(1 - u)^{n-2}(1 - v)^m + P_{02}\,C(m, 2)v^2(1 - u)^n(1 - v)^{m-2} \\[6pt]
+ \cdots + P_{n0}\,u^n(1 - v)^m + P_{0m}\,v^m(1 - u)^n \\[6pt]
+ \cdots + P_{(n-1)(m-2)}\,C(n, n - 1)C(m, m - 2)u^{n-1}v^{m-2}(1 - u)(1 - v)^2 \\[6pt]
+ P_{(n-2)(m-1)}C(n, n - 2)C(m, m - 1)u^{n-2}v^{m-1}(1 - u)^2(1 - v) \\[6pt]
+ P_{(n-1)m}\,C(n, n - 1)u^{n-1}v^m(1 - u) + P_{n(m-1)}u^n v^{m-1}(1 - v)
\end{array}
\right.
\qquad (6.68)
$$

The characteristics of the Bezier surface are the same as those of the Bezier curve. The surface interpolates the four corner control points (see Fig. 6-33) if we substitute the $(u, v)$ values of $(0, 0)$, $(1, 0)$ $(0, 1)$, and $(1, 1)$ into (6.68). The surface is also tangent to the corner segments of the control polyhedron. The tangent vectors at the corners are:

$$
\begin{array}{llll}
P_{u00} = n(P_{10} - P_{00}) & \quad P_{un0} = n(P_{n0} - P_{(n-1)0}) & \quad \text{along } v = 0 \text{ edge} \\[6pt]
P_{u0m} = n(P_{1m} - P_{0m}) & \quad P_{unm} = n(P_{nm} - P_{(n-1)m}) & \quad \text{along } v = 1 \text{ edge} \\[6pt]
P_{v00} = m(P_{01} - P_{00}) & \quad P_{v0m} = m(P_{0m} - P_{0(m-1)}) & \quad \text{along } u = 0 \text{ edge} \\[6pt]
P_{vn0} = m(P_{n1} - P_{n0}) & \quad P_{vnm} = m(P_{nm} - P_{n(m-1)}) & \quad \text{along } u = 1 \text{ edge}
\end{array}
\qquad (6.69)
$$

In addition, the Bezier surface possesses the convex hull property. The convex hull in this case is the polyhedron formed by connecting the furthest control points on the control polyhedron. The convex hull includes the control polyhedron of the surface as it includes the control polygon in the case of the Bezier curve.

The shape of the Bezier surface can be modified by either changing some vertices of its polyhedron or by keeping the polyhedron fixed and specifying multiple coincident points of some vertices. Moreover, a closed Bezier surface can be generated by closing its polyhedron or choosing coincident corner points as illustrated in Fig. 6-34.

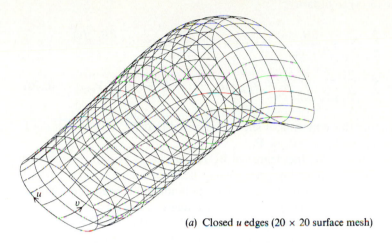

(a) Closed $u$ edges ($20 \times 20$ surface mesh)

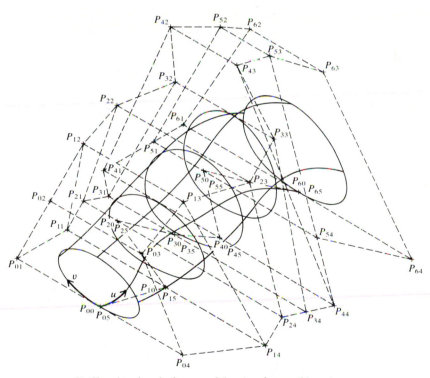

(b) Closed $v$ edges (polygon and $4 \times 4$ surface mesh)

**FIGURE 6-34**
Closed Bezier surface.

The normal to a Bezier surface at any point can be calculated by substituting Eq. (6.67) into (6.11) to obtain

$$
N(u, v) = \sum_{i=0}^{n} \sum_{j=0}^{m} P_{ij} \frac{\partial B_{i,n}(u)}{\partial u} B_{j,m}(v) \times \sum_{k=0}^{n} \sum_{l=0}^{m} P_{kl} B_{k,n}(u) \frac{\partial B_{l,m}(v)}{\partial v}
$$

$$
= \sum_{i=0}^{n} \sum_{j=0}^{m} \sum_{k=0}^{n} \sum_{l=0}^{m} \frac{\partial B_{i,n}}{\partial u} B_{j,m}(v) B_{k,n}(u) \frac{\partial B_{l,m}(v)}{\partial v} P_{ij} \times P_{kl} \tag{6.70}
$$

When expanding this equation, it should be noted that $P_{ij} \times P_{kl} = 0$ if $i = k$ and $j = l$, and that $P_{ij} \times P_{kl} = -P_{kl} \times P_{ij}$.

As with the Bezier curve, the degree of Bezier surface is tied to the number of control points. Surfaces requiring great design flexibility need a large control point array and would, therefore, have a high polynomial degree. To achieve required design flexibility while keeping the surface degree manageable, large surfaces are generally designed by piecing together smaller surface patches of lower degrees. This keeps the overall degree of the surface low but requires a special attention to ensure that appropriate continuity is maintained across patch boundaries. A composite Bezier surface can have $C^0$ (positional) and/or $C^1$ (tangent) continuity. A positional continuity between, say, two patches requires that the common boundary curve between the two patches must have a common boundary polygon between the two characteristic polyhedrons (see Fig. 6-35a). For tangent continuity across the boundary, the segments, attached to the common boundary polygon, of one patch polyhedron must be colinear with the corresponding segments of the other patch polyhedron, as shown in Fig. 6-35b. This implies that the tangent planes of the patches at the common boundary curve are coincident.

In a design environment, the Bezier surface is superior to a bicubic surface in that it does not require tangent or twist vectors to define the surface. However, its main disadvantage is the lack of local control. Changing one or more control point affects the shape of the whole surface. Therefore, the user cannot selectively change the shape of part of the surface.

Common boundary polygon

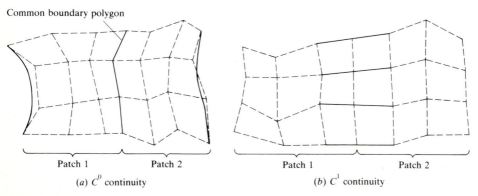

Patch 1          Patch 2                     Patch 1                    Patch 2

(a) $C^0$ continuity                          (b) $C^1$ continuity

**FIGURE 6-35**
Composite Bezier surface.

**Example 6.7.** Find the equivalent bicubic formulation of a cubic Bezier surface patch.

*Solution.* This equivalence is usually useful for software development purposes where, say, one subroutine can handle the generation of more than one surface. It is also useful in obtaining a better understanding of the surface characteristic.

Figure 6-36 shows a cubic Bezier patch. Substituting $n = 3$ and $m = 3$ into Eq. (6.67), the patch equation is

$$\mathbf{P}(u, v) = \sum_{i=0}^{3} \sum_{j=0}^{3} \mathbf{P}_{ij} B_{i,3}(u) B_{j,3}(v), \qquad 0 \le u \le 1, 0 \le v \le 1 \tag{6.71}$$

This equation can be expanded to give

$$\mathbf{P}(u, v) = \sum_{i=0}^{3} B_{i,3}(u)[\mathbf{P}_{i0} B_{0,3}(v) + \mathbf{P}_{i1} B_{1,3}(v) + \mathbf{P}_{i2} B_{2,3}(v) + \mathbf{P}_{i3} B_{3,3}(v)]$$

$$= B_{0,3}(u)[\mathbf{P}_{00} B_{0,3}(v) + \mathbf{P}_{01} B_{1,3}(v) + \mathbf{P}_{02} B_{2,3}(v) + \mathbf{P}_{03} B_{3,3}(v)]$$

$$+ B_{1,3}(u)[\mathbf{P}_{10} B_{0,3}(v) + \mathbf{P}_{11} B_{1,3}(v) + \mathbf{P}_{12} B_{2,3}(v) + \mathbf{P}_{13} B_{3,3}(v)]$$

$$+ B_{2,3}(u)[\mathbf{P}_{20} B_{0,3}(v) + \mathbf{P}_{21} B_{1,3}(v) + \mathbf{P}_{22} B_{2,3}(v) + \mathbf{P}_{23} B_{3,3}(v)]$$

$$+ B_{3,3}(u)[\mathbf{P}_{30} B_{0,3}(v) + \mathbf{P}_{31} B_{1,3}(v) + \mathbf{P}_{32} B_{2,3}(v) + \mathbf{P}_{33} B_{3,3}(v)]$$

This equation can be written in a matrix form as

$$\mathbf{P}(u, v) = [B_{0,3}(u) \quad B_{1,3}(u) \quad B_{2,3}(u) \quad B_{3,3}(u)] \begin{bmatrix} \mathbf{P}_{00} & \mathbf{P}_{01} & \mathbf{P}_{02} & \mathbf{P}_{03} \\ \mathbf{P}_{10} & \mathbf{P}_{11} & \mathbf{P}_{12} & \mathbf{P}_{13} \\ \mathbf{P}_{20} & \mathbf{P}_{21} & \mathbf{P}_{22} & \mathbf{P}_{23} \\ \mathbf{P}_{30} & \mathbf{P}_{31} & \mathbf{P}_{32} & \mathbf{P}_{33} \end{bmatrix} \begin{bmatrix} B_{0,3}(v) \\ B_{1,3}(v) \\ B_{2,3}(v) \\ B_{3,3}(v) \end{bmatrix}$$

$$\tag{6.72}$$

or $\quad \mathbf{P}(u, v) = [(1 - u)^3 \quad 3u(1 - u)^2 \quad 3u^2(1 - u) \quad u^3][P] \begin{bmatrix} (1 - v)^3 \\ 3v(1 - v)^2 \\ 3v^2(1 - v) \\ v^3 \end{bmatrix}$

or $\quad \mathbf{P}(u, v) = \mathbf{U}^T[M_B][P][M_B]^T \mathbf{V}$ $\tag{6.73}$

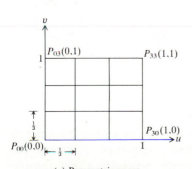

(a) Parametric space

(b) Cartesian space

**FIGURE 6-36**
A cubic Bezier patch.

where the subscript $B$ denotes Bezier and

$$[P] = \begin{bmatrix} \mathbf{P}_{00} & \mathbf{P}_{01} & \mathbf{P}_{02} & \mathbf{P}_{03} \\ \mathbf{P}_{10} & \mathbf{P}_{11} & \mathbf{P}_{12} & \mathbf{P}_{13} \\ \mathbf{P}_{20} & \mathbf{P}_{21} & \mathbf{P}_{22} & \mathbf{P}_{23} \\ \mathbf{P}_{30} & \mathbf{P}_{31} & \mathbf{P}_{32} & \mathbf{P}_{33} \end{bmatrix}$$

and

$$[M_B] = \begin{bmatrix} -1 & 3 & -3 & 1 \\ 3 & -6 & 3 & 0 \\ -3 & 3 & 0 & 0 \\ 1 & 0 & 0 & 0 \end{bmatrix} \qquad (6.74)$$

and the $\mathbf{U}$ and $\mathbf{V}$ vectors are $[u^3 \quad u^2 \quad u \quad 1]^T$ and $[v^3 \quad v^2 \quad v \quad 1]^T$ respectively. Notice that $[M_B]$ given by Eq. (6.74) is the same matrix for the cubic Bezier curve (see Prob. 5.10 in Chap. 5).

Equating Eq. (6.73) with (6.42) gives

$$\mathbf{U}^T[M_H][B][M_H]^T\mathbf{V} = \mathbf{U}^T[M_B][P][M_B]^T\mathbf{V}$$

or

$$[M_H][B][M_H]^T = [M_B][P][M_B]^T$$

Solving for $[B]$ gives

$$[B] = [M_H]^{-1}[M_B][P][M_B]^T[M_H]^{T-1}$$

Using Eq. (5.86) for $[M_H]^{-1}$, this equation can be reduced to give

$$[B] = \begin{bmatrix} \mathbf{P}_{00} & \mathbf{P}_{03} & 3(\mathbf{P}_{01} - \mathbf{P}_{00}) & 3(\mathbf{P}_{03} - \mathbf{P}_{02}) \\ \mathbf{P}_{30} & \mathbf{P}_{33} & 3(\mathbf{P}_{31} - \mathbf{P}_{30}) & 3(\mathbf{P}_{33} - \mathbf{P}_{32}) \\ 3(\mathbf{P}_{10} - \mathbf{P}_{00}) & 3(\mathbf{P}_{13} - \mathbf{P}_{03}) & 9(\mathbf{P}_{00} - \mathbf{P}_{10} - \mathbf{P}_{01} + \mathbf{P}_{11}) & 9(\mathbf{P}_{02} - \mathbf{P}_{12} - \mathbf{P}_{03} + \mathbf{P}_{13}) \\ 3(\mathbf{P}_{30} - \mathbf{P}_{20}) & 3(\mathbf{P}_{33} - \mathbf{P}_{23}) & 9(\mathbf{P}_{20} - \mathbf{P}_{21} - \mathbf{P}_{30} + \mathbf{P}_{31}) & 9(\mathbf{P}_{22} - \mathbf{P}_{23} - \mathbf{P}_{32} + \mathbf{P}_{33}) \end{bmatrix}$$

$$(6.75)$$

Comparing this equation with Eq. (6.43) for the bicubic patch reveals that the tangent and twist vectors of the Bezier surface are expressed in terms of the vertices of its characteristic polyhedron.

*Note:* equation (6.72) offers a concise matrix form of Eq. (6.67) for a cubic Bezier surface. This form can be extended to an $(n + 1) \times (m + 1)$ surface as follows:

$$\mathbf{P}(u, v) = [B_{0,n}(u) \quad B_{1,n}(u) \quad \cdots \quad B_{n,n}(u)] \begin{bmatrix} \mathbf{P}_{00} & \mathbf{P}_{01} & \cdots & \mathbf{P}_{0m} \\ \mathbf{P}_{10} & \mathbf{P}_{11} & \cdots & \mathbf{P}_{1m} \\ \vdots & \vdots & & \vdots \\ \mathbf{P}_{n0} & \mathbf{P}_{n1} & \cdots & \mathbf{P}_{nm} \end{bmatrix} \begin{bmatrix} B_{0,m}(v) \\ B_{1,m}(v) \\ \vdots \\ B_{n,m}(v) \end{bmatrix}$$

$$(6.76)$$

### 6.6.3  B-Spline Surface

The same tensor product method used with Bezier curves can extend B-splines to describe B-spline surfaces. A rectangular set of data (control) points creates the surface. This set forms the vertices of the characteristic polyhedron that approximates and controls the shape of the resulting surface. A B-spline surface can approximate or interpolate the vertices of the polyhedron as shown in Fig. 6-37.

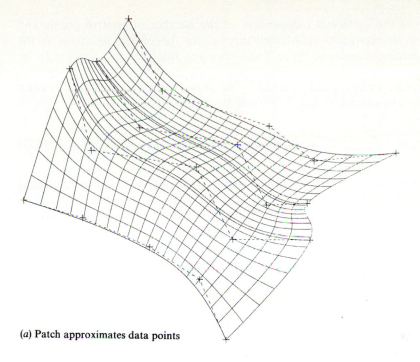

(*a*) Patch approximates data points

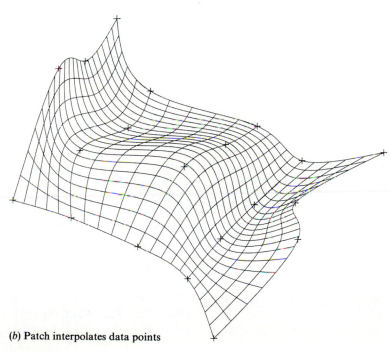

(*b*) Patch interpolates data points

**FIGURE 6-37**
4 × 5 B-spline surface patches.

The degree of the surface is independent of the number of control points and continuity is automatically maintained throughout the surface by virtue of the form of blending functions. As a result, surface intersections can easily be managed.

A B-spline surface patch defined by an $(n + 1) \times (m + 1)$ array of control points is given by extending Eq. (5.103) into two dimensions:

$$\mathbf{P}(u, v) = \sum_{i=0}^{n} \sum_{j=0}^{m} \mathbf{P}_{ij} N_{i,k}(u) N_{j,l}(v), \qquad 0 \le u \le u_{max}, \ 0 \le v \le v_{max} \quad (6.77)$$

All the related discussions to Eqs. (5.104) and (5.105) apply to the above equation. Equation (6.77) implies that knot vectors in both $u$ and $v$ directions are constant but not necessarily equal. Other formulations could allow various knot vectors in a given direction to increase the flexibility of local control.

B-spline surfaces have the same characteristics as B-spline curves. Their major advantage over Bezier surfaces is the local control. Composite B-spline surfaces can be generated with $C^0$ and/or $C^1$ continuity in the same way as composite Bezier surfaces.

**Example 6.8.** Find the equivalent bicubic formulation of an open and closed cubic B-spline surface.

*Solution.* Most of the results obtained in Examples 5.21 and 5.22 can be extended to a cubic B-spline surface. First, let us find the matrix form of Eq. (6.77). This equation is identical in form to the Bezier surface equation (6.67). Thus, by replacing the Bernstein polynomials in Eq. (6.76) by the B-spline functions yields the matrix form of Eq. (6.77) as

$$\mathbf{P}(u, v) = [N_{0,k}(u) \quad N_{1,k}(u) \quad \cdots \quad N_{n,k}(u)] \begin{bmatrix} \mathbf{P}_{00} & \mathbf{P}_{01} & \cdots & \mathbf{P}_{0m} \\ \mathbf{P}_{10} & \mathbf{P}_{11} & \cdots & \mathbf{P}_{1m} \\ \vdots & \vdots & & \vdots \\ \mathbf{P}_{n0} & \mathbf{P}_{n1} & \cdots & \mathbf{P}_{nm} \end{bmatrix} \begin{bmatrix} N_{0,l}(v) \\ N_{1,l}(v) \\ \vdots \\ N_{m,l}(v) \end{bmatrix} \quad (6.78)$$

or

$$\mathbf{P}(u, v) = [N_{0,k}(u) \quad N_{1,k}(u) \quad \cdots \quad N_{n,k}(u)][P] \begin{bmatrix} N_{0,l}(v) \\ N_{1,l}(v) \\ \vdots \\ N_{m,l}(v) \end{bmatrix} \quad (6.79)$$

where $[P]$ is an $(n + 1) \times (m + 1)$ matrix of the vertices of the characteristic polyhedron of the B-spline surface patch. For a $4 \times 4$ cubic B-spline patch, Eq. (6.78) becomes

$$\mathbf{P}(u, v) = [N_{0,4}(u) \quad N_{1,4}(u) \quad N_{2,4}(u) \quad N_{3,4}(u)] \begin{bmatrix} \mathbf{P}_{00} & \mathbf{P}_{01} & \mathbf{P}_{02} & \mathbf{P}_{03} \\ \mathbf{P}_{10} & \mathbf{P}_{11} & \mathbf{P}_{12} & \mathbf{P}_{13} \\ \mathbf{P}_{20} & \mathbf{P}_{21} & \mathbf{P}_{22} & \mathbf{P}_{23} \\ \mathbf{P}_{30} & \mathbf{P}_{31} & \mathbf{P}_{32} & \mathbf{P}_{33} \end{bmatrix} \begin{bmatrix} N_{0,4}(v) \\ N_{1,4}(v) \\ N_{2,4}(v) \\ N_{3,4}(v) \end{bmatrix}$$

For the open patch, the B-spline functions are the same as the Bernstein polynomials of the previous example (refer to Example 5.21) and the equivalent bicubic formulation results in Eq. (6.75). For, say, a $5 \times 6$ open cubic B-spline patch, we get

$$P(u, v) = [N_{0,4}(u) \quad N_{1,4}(u) \quad N_{2,4}(u) \quad N_{3,4}(u) \quad N_{4,4}(u)][P] \begin{bmatrix} N_{0,4}(v) \\ N_{1,4}(v) \\ N_{2,4}(v) \\ N_{3,4}(v) \\ N_{4,4}(v) \\ N_{5,4}(v) \end{bmatrix}, \quad \begin{matrix} 0 \le u \le 2, \\ 0 \le v \le 3 \end{matrix}$$

where

$$[P] = \begin{bmatrix} P_{00} & P_{01} & P_{02} & P_{03} & P_{04} & P_{05} \\ P_{10} & P_{11} & P_{12} & P_{13} & P_{14} & P_{15} \\ P_{20} & P_{21} & P_{22} & P_{23} & P_{24} & P_{25} \\ P_{30} & P_{31} & P_{32} & P_{33} & P_{34} & P_{35} \\ P_{40} & P_{41} & P_{42} & P_{43} & P_{44} & P_{45} \end{bmatrix}$$

All the B-spline functions shown in this equation are calculated by following the procedure described in Example 5.21. After this is done, the above equation can be reduced to

$$\begin{bmatrix} P_{11}(u, v), & P_{12}(u, v), & P_{13}(u, v), \\ 0 \le u \le 1, & 0 \le u \le 1, & 0 \le u \le 1, \\ 0 \le v \le 1 & 1 \le v \le 2 & 2 \le v \le 3 \\ \hline P_{21}(u, v), & P_{22}(u, v), & P_{23}(u, v), \\ 1 \le u \le 2, & 1 \le u \le 2, & 1 \le u \le 2, \\ 0 \le v \le 1 & 1 \le v \le 2 & 2 \le v \le 3 \end{bmatrix}$$

$$= \begin{bmatrix} U^T[M_S][P_{11}][M_S]^T V & U^T[M_S][P_{12}][M_S]^T V & U^T[M_S][P_{13}][M_S]^T V \\ U^T[M_S][P_{21}][M_S]^T V & U^T[M_S][P_{22}][M_S]^T V & U^T[M_S][P_{23}][M_S]^T V \end{bmatrix}$$

$$\text{(6.80)}$$

where $[P_{11}]$, $[P_{12}]$, ..., $[P_{23}]$ are partitions of $[P]$. Each partition is $4 \times 4$ and is different from its neighbor (moving in the row direction of $[P]$) by one row or by one column (moving in the column direction of $[P]$). The general form of any partition is given by

$$[P_{ij}] = \begin{bmatrix} P_{(i-1)(j-1)} & P_{(i-1)j} & P_{(i-1)(j+1)} & P_{(i-1)(j+2)} \\ P_{i(j-1)} & P_{ij} & P_{i(j+1)} & P_{i(j+2)} \\ P_{(i+1)(j-1)} & P_{(i+1)j} & P_{(i+1)(j+1)} & P_{(i+1)(j+2)} \\ P_{(i+2)(j-1)} & P_{(i+2)j} & P_{(i+2)(j+1)} & P_{(i+2)(j+2)} \end{bmatrix}$$

The matrix $[M_S]$ is the same as for cubic B-spline curves (see Prob. 5.15 in Chap. 5). It is given by

$$[M_S] = \frac{1}{6} \begin{bmatrix} -1 & 3 & -3 & 1 \\ 3 & -6 & 3 & 0 \\ -3 & 0 & 3 & 0 \\ 1 & 4 & 1 & 0 \end{bmatrix} \qquad \text{(6.81)}$$

Equation (6.80) can be written in a more concise form as

$$P_{ij}(u, v) = U^T[M_S][P_{ij}][M_S]^T V$$

$$1 \le i \le 2, 1 \le j \le 3; (i-1) \le u \le i, (j-1) \le v \le j \qquad \text{(6.82)}$$

Equating the above equation with the bicubic equation, the equivalent $[B_{ij}]$ matrix becomes

$$[B_{ij}] = [M_H]^{-1}[M_S][P_{ij}][M_S]^T[M_H]^{T-1} \tag{6.83}$$

Notice that the above procedure can be extended to an $n \times m$ cubic B-spline surface.

A similar procedure can be followed for a closed cubic B-spline patch. The difference comes in the form of the B-spline functions. For a $4 \times 4$ cubic B-spline patch closed in the $u$ direction, the $v$ direction, or both directions, the following three equations can be written respectively:

$$\mathbf{P}(u, v) = [N_{0,\,4}((u + 4) \bmod 4),\ N_{0,\,4}((u + 3) \bmod 4),$$
$$N_{0,\,4}((u + 2) \bmod 4),\ N_{0,\,4}((u + 1) \bmod 4)]$$

$$\times [P] \begin{bmatrix} N_{0,\,4}(v) \\ N_{1,\,4}(v) \\ N_{2,\,4}(v) \\ N_{3,\,4}(v) \end{bmatrix}$$

$$\mathbf{P}(u, v) = [N_{0,\,4}(u)\quad N_{1,\,4}(u)\quad N_{2,\,4}(u)\quad N_{3,\,4}(u)][P] \begin{bmatrix} N_{0,\,4}((v + 4) \bmod 4) \\ N_{0,\,4}((v + 3) \bmod 4) \\ N_{0,\,4}((v + 2) \bmod 4) \\ N_{0,\,4}((v + 1) \bmod 4) \end{bmatrix}$$

and $\mathbf{P}(u, v) = [N_{0,\,4}((u + 4) \bmod 4),\ N_{0,\,4}((u + 3) \bmod 4),$
$$N_{0,\,4}((u + 2) \bmod 4),\ N_{0,\,4}((u + 1) \bmod 4)]$$

$$\times [P] \begin{bmatrix} N_{0,\,4}((v + 4) \bmod 4) \\ N_{0,\,4}((v + 3) \bmod 4) \\ N_{0,\,4}((v + 2) \bmod 4) \\ N_{0,\,4}((v + 1) \bmod 4) \end{bmatrix}$$

The closed B-spline functions have been evaluated in Example 5.22. Investigating Eqs. (5.118) reveals that there are four different expressions for the matrix $[B]$ [see Eq. (6.83) above] depending on the value of $u$, $v$, or $u$ and $v$. For, say, a $5 \times 6$ closed cubic B-spline patch, the above procedure is repeated but with the functions $[N_{0,\,4}((u + 5) \bmod 5),\ N_{0,\,4}((u + 4) \bmod 5),\ N_{0,\,4}((u + 3) \bmod 5),\ N_{0,\,4}((u + 2) \bmod 5),\ N_{0,\,4}((u + 1) \bmod 5)]$ and $[N_{0,\,4}((u + 6) \bmod 6),\ N_{0,\,4}((u + 5) \bmod 6),\ N_{0,\,4}((u + 4) \bmod 6),\ N_{0,\,4}((u + 3) \bmod 6),\ N_{0,\,4}((u + 2) \bmod 6),\ N_{0,\,4}((u + 1) \bmod 6)]$. The control point matrix $[P]$ is a $5 \times 6$ matrix as in the case of the open patch.

### 6.6.4  Coons Surface

All the surface methods introduced thus far share one common philosophy; that is, they all require a finite number of data points to generate the respective surfaces. In contrast, a Coons surface patch is a form of "transfinite interpolation" which indicates that the Coons scheme interpolates to an infinite number of data points, that is, to all points of a curve segment, to generate the surface. The Coons patch is particularly useful in blending four prescribed intersecting curves which form a closed boundary as shown in Fig. 6-38. The figure shows the given four boundary curves as $\mathbf{P}(u, 0)$, $\mathbf{P}(1, v)$, $\mathbf{P}(u, 1)$, and $\mathbf{P}(0, v)$. It is assumed that $u$

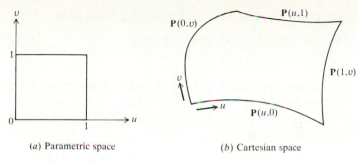

(a) Parametric space          (b) Cartesian space

**FIGURE 6-38**
Boundaries of Coons surface patch.

and $v$ range from 0 to 1 along these boundaries and that each pair of opposite boundary curves are identically parametrized. Development of the Coons surface patch centers around answering the following question: what is a suitable well-behaved function $P(u, v)$ which blends the four given boundary curves and which satisfies the boundary conditions, that is, reduces to the correct boundary curve when $u = 0$, $u = 1$, $v = 0$, and $v = 1$?

Let us first consider the case of a bilinearly blended Coons patch which interpolates to the four boundary curves shown in Fig. 6-38. For this case, it is useful to recall that a ruled surface interpolates linearly between two given boundary curves in one direction as shown by Eq. (6.36). Therefore, the superposition of two ruled surfaces connecting the two pairs of boundary curves might satisfy the boundary curve conditions and produces the Coons patch. Let us investigate this claim. Utilizing Eq. (6.36) in the $v$ and $u$ directions gives respectively

$$P_1(u, v) = (1 - u)P(0, v) + uP(1, v) \tag{6.84}$$

and
$$P_2(u, v) = (1 - v)P(u, 0) + vP(u, 1) \tag{6.85}$$

Adding these two equations gives the surface

$$P(u, v) = P_1(u, v) + P_2(u, v) \tag{6.86}$$

The resulting surface patch described by Eq. (6.86) does not satisfy the boundary conditions. For example, substituting $v = 0$ and 1 into this equation gives respectively

$$P(u, 0) = P(u, 0) + [(1 - u)P(0, 0) + uP(1, 0)] \tag{6.87}$$

$$P(u, 1) = P(u, 1) + [(1 - u)P(0, 1) + uP(1, 1)] \tag{6.88}$$

These two equations show that the terms in square brackets are extra and should be eliminated to recover the original boundary curves. These terms define the boundaries of an unwanted surface $P_3(u, v)$ which is embedded in Eq. (6.86). This surface can be defined by linear interpolation in the $v$ direction, that is,

$$P_3(u, v) = (1 - v)[(1 - u)P(0, 0) + uP(1, 0)] + v[(1 - u)P(0, 1) + uP(1, 1)] \tag{6.89}$$

Subtracting $\mathbf{P}_3(u, v)$ (called the "correction surface") from Eq. (6.86) gives

$$\mathbf{P}(u, v) = \mathbf{P}_1(u, v) + \mathbf{P}_2(u, v) - \mathbf{P}_3(u, v) \tag{6.90}$$

or

$$\mathbf{P}(u, v) = \mathbf{P}_1(u, v) \oplus \mathbf{P}_2(u, v) \tag{6.91}$$

where $\oplus$ defines the "boolean sum" which is $\mathbf{P}_1 + \mathbf{P}_2 - \mathbf{P}_3$. The surface $\mathbf{P}(u, v)$ given by the above equation defines the bilinear Coons patch connecting the four boundary curves shown in Fig. 6-38. Figure 6-39 shows the graphical representation of Eq. (6.91) and its matrix form is

$$\mathbf{P}(u, v) = -\begin{bmatrix} -1 & (1-u) & u \end{bmatrix} \begin{bmatrix} 0 & \vdots & \mathbf{P}(u, 0) & \mathbf{P}(u, 1) \\ \hline \mathbf{P}(0, v) & \vdots & \mathbf{P}(0, 0) & \mathbf{P}(0, 1) \\ \mathbf{P}(1, v) & \vdots & \mathbf{P}(1, 0) & \mathbf{P}(1, 1) \end{bmatrix} \begin{bmatrix} -1 \\ 1-v \\ v \end{bmatrix} \tag{6.92}$$

The left column and the upper row of the matrix represent $\mathbf{P}_1(u, v)$ and $\mathbf{P}_2(u, v)$ respectively while the lower right block represents the correction surface $\mathbf{P}_3(u, v)$. The functions $-1$, $1 - u$, $u$, $1 - v$, and $v$ are called blending functions because they blend together four separate boundary curves to give one surface.

The main drawback of the bilinearly blended Coons patch is that it only provides $C^0$ continuity (positional continuity) between adjacent patches, even if their boundary curves form a $C^1$ continuity network. For example, if two patches are to be connected along the boundary curve $\mathbf{P}(1, v)$ of the first patch and $\mathbf{P}(0, v)$ of the second, it can be shown that continuity of the cross-boundary derivatives

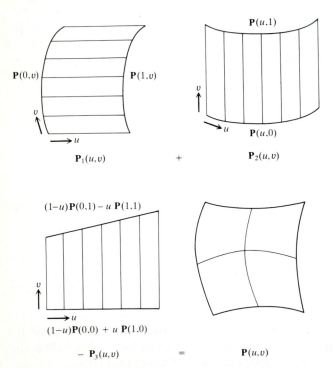

**FIGURE 6-39**
Bilinearly blended Coons patch (boolean sum).

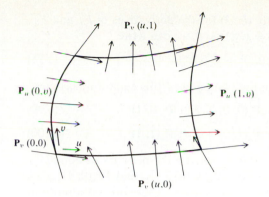

**FIGURE 6-40**
Cross-boundary derivatives.

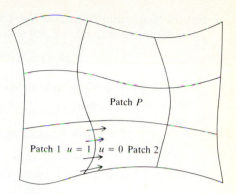

**FIGURE 6-41**
A composed Coons surface formed by a network of $C^1$ boundary curves.

(Fig. 6-40) $\mathbf{P}_u(1, v)$ and $\mathbf{P}_u(0, v)$ of the two patches cannot be made equal (see Prob. 6.10 at the end of the chapter).

Gradient continuity across boundaries of patches of a composite Coons surface is essential for practical applications. If, for example, a network of $C^1$ curves is given as shown in Fig. 6-41, it becomes very desirable to form a composite Coons surface which is smooth or $C^1$ continuous, that is, it provides continuity of cross-boundary derivatives between patches. Investigation of Eq. (6.92) shows that the choice of blending functions controls the behavior of the resulting Coons patch. If the cubic Hermite polynomials $F_1(x)$ and $F_2(x)$ given in Eq. (6.52) are used instead of the linear polynomials $(1 - u)$ and $u$ respectively, a bicubically blended Coons patch results. This patch guarantees $C^1$ continuity between patches. Substituting $F_1(x)$ and $F_2(x)$ into Eqs. (6.84) and (6.85) gives

$$\mathbf{P}_1(u, v) = (2u^3 - 3u^2 + 1)\mathbf{P}(0, v) + (-2u^3 + 3u^2)\mathbf{P}(1, v) \tag{6.93}$$

$$\mathbf{P}_2(u, v) = (2v^3 - 3v^2 + 1)\mathbf{P}(u, 0) + (-2v^3 + 3v^2)\mathbf{P}(u, 1) \tag{6.94}$$

Similar to the bilinear patch, the boolean sum $P_1 \oplus P_2$ can be formed and the matrix equation of the bicubic Coons patch becomes

$$\mathbf{P}(u, v) = -[-1 \quad F_1(u) \quad F_2(u)] \begin{bmatrix} 0 & \vdots & \mathbf{P}(u, 0) & \mathbf{P}(u, 1) \\ \hdashline \mathbf{P}(0, v) & \vdots & \mathbf{P}(0, 0) & \mathbf{P}(0, 1) \\ \mathbf{P}(1, v) & \vdots & \mathbf{P}(1, 0) & \mathbf{P}(1, 1) \end{bmatrix} \begin{bmatrix} -1 \\ F_1(v) \\ F_2(v) \end{bmatrix} \tag{6.95}$$

As can easily be seen from Eqs. (6.92) and (6.95), the correction surface is usually formed by applying the blending functions to the corner data alone.

To check continuity across patch boundaries, let us consider patch 1 and patch 2 in Fig. 6-41. For $C^0$ and $C^1$ continuity we should have

$$[\mathbf{P}(0, v)]_{\text{patch 2}} = [\mathbf{P}(1, v)]_{\text{patch 1}} \tag{6.96}$$

$$[\mathbf{P}_u(0, v)]_{\text{patch 2}} = [\mathbf{P}_u(1, v)]_{\text{patch 1}} \tag{6.97}$$

$C^0$ continuity is automatically satisfied between the two patches because they share the same boundary curve. For $C^1$ continuity, differentiating Eq. (6.95) with respect to $u$, using $G_1(x)$ and $G_2(x)$ in Eq. (6.53) for the derivatives of $F_1$ and $F_2$, and noticing that $G_1(0) = G_2(0) = G_1(1) = G_2(1) = 0$, we can write

$$\mathbf{P}_u(u, v) = F_1(v)\mathbf{P}_u(u, 0) + F_2(v)\mathbf{P}_u(u, 1) \tag{6.98}$$

At the common boundary curve between the two patches, this equation becomes

$$[\mathbf{P}_u(0, v)]_{\text{patch 2}} = F_1(v)\mathbf{P}_u(0, 0) + F_2(v)\mathbf{P}_u(0, 1) \tag{6.99}$$

and
$$[\mathbf{P}_u(1, v)]_{\text{patch 1}} = F_1(v)\mathbf{P}_u(1, 0) + F_2(v)\mathbf{P}_u(1, 1) \tag{6.100}$$

Based on Eqs. (6.99) and (6.100), Eq. (6.97) is satisfied if the network of boundary curves is $C^1$ continuous because this makes $[\mathbf{P}_u(0, 0)]_{\text{patch 2}}$ and $[\mathbf{P}_u(0, 1)]_{\text{patch 2}}$ equal to $[\mathbf{P}_u(1, 0)]_{\text{patch 1}}$ and $[\mathbf{P}_u(1, 1)]_{\text{patch 1}}$ respectively. Therefore, continuity of cross-boundary derivatives is automatically satisfied for bicubic Coons patches if the boundary curves are $C^1$ continuous.

The bicubic Coons patch as defined by Eq. (6.95) is easy to use in a design environment because only the four boundary curves are needed. However, a more flexible composite $C^1$ bicubic Coons surface can be developed if, together with the boundary curves, the cross-boundary derivatives $\mathbf{P}_u(0, v)$, $\mathbf{P}_u(1, v)$, $\mathbf{P}_v(u, 0)$, and $\mathbf{P}_v(u, 1)$ are given (see Fig. 6-40). Note that at the corners these derivatives must be compatible with the curve information. Similar to Eq. (6.93) and (6.94), we can define the following cubic Hermite interpolants:

$$\mathbf{P}_1(u, v) = F_1(u)\mathbf{P}(0, v) + F_2(u)\mathbf{P}(1, v) + F_3(u)\mathbf{P}_u(0, v) + F_4(u)\mathbf{P}_u(1, v) \tag{6.101}$$

and $\quad \mathbf{P}_2(u, v) = F_1(v)\mathbf{P}(u, 0) + F_2(v)\mathbf{P}(u, 1) + F_3(v)\mathbf{P}_v(u, 0) + F_4(v)\mathbf{P}_v(u, 1) \tag{6.102}$

where the functions $F_1$ to $F_4$ are given by Eq. (6.52). Forming the boolean sum $(\mathbf{P}_1 \oplus \mathbf{P}_2)$ of the above two equations results in the bicubic Coons patch that incorporates cross-boundary derivatives. The introduction of the cross-boundary derivatives causes the twist vectors at the corners of the patch to appear (see Prob. 6.11 at the end of the chapter). The matrix equation of this Coons patch can be written as

$$\mathbf{P}(u, v) = -[-1 \quad F_1(u) \quad F_2(u) \quad F_3(u) \quad F_4(u)]$$

$$\times \begin{bmatrix} 0 & \mathbf{P}(u, 0) & \mathbf{P}(u, 1) & \mathbf{P}_v(u, 0) & \mathbf{P}_v(u, 1) \\ \mathbf{P}(0, v) & \mathbf{P}(0, 0) & \mathbf{P}(0, 1) & \mathbf{P}_v(0, 0) & \mathbf{P}_v(0, 1) \\ \mathbf{P}(1, v) & \mathbf{P}(1, 0) & \mathbf{P}(1, 1) & \mathbf{P}_v(1, 0) & \mathbf{P}_v(1, 1) \\ \mathbf{P}_u(0, v) & \mathbf{P}_u(0, 0) & \mathbf{P}_u(0, 1) & \mathbf{P}_{uv}(0, 0) & \mathbf{P}_{uv}(0, 1) \\ \mathbf{P}_u(1, v) & \mathbf{P}_u(1, 0) & \mathbf{P}_u(1, 1) & \mathbf{P}_{uv}(1, 0) & \mathbf{P}_{uv}(1, 1) \end{bmatrix} \begin{bmatrix} -1 \\ F_1(v) \\ F_2(v) \\ F_3(v) \\ F_4(v) \end{bmatrix} \tag{6.103}$$

The upper $3 \times 3$ matrix determines the patch defined previously by Eq. (6.95). The left column and the upper row represent $\mathbf{P}_1(u, v)$ and $\mathbf{P}_2(u, v)$ respectively, while the lower right $4 \times 4$ matrix represents the bicubic tensor product surface discussed in Sec. 6.6.1 [see Eq. (6.43)]. Equation (6.103) shows that every bicubically blended Coons surface reproduces a bicubic surface. On the other hand, bicubically blended Coons patches cannot be described in general by bicubic tensor product surfaces. Thus, Coon formulation can describe a much richer variety of surfaces than do tensor product surfaces.

**Example 6.9.** Show that if the boundary curves of a bilinear Coons patch are coplanar, the resulting patch is also planar.

*Solution.* We need to show that the surface normal at any point on the surface is normal to the plane of the boundary curves. Based on Eq. (6.90), the surface tangent vectors are

$$\mathbf{P}_u(u, v) = [\mathbf{P}_u(u, 0) + \mathbf{P}(1, v) - \mathbf{P}(0, v) + \mathbf{P}(0, 0) - \mathbf{P}(1, 0)]$$

$$+ v[\mathbf{P}_u(u, 1) - \mathbf{P}_u(u, 0) + \mathbf{P}(1, 0) - \mathbf{P}(0, 0) - \mathbf{P}(1, 1) + \mathbf{P}(0, 1)]$$

$$= \mathbf{A} + v\mathbf{B}$$

and $\quad \mathbf{P}_v(u, v) = [\mathbf{P}_v(0, v) + \mathbf{P}(u, 1) - \mathbf{P}(u, 0) + \mathbf{P}(0, 0) - \mathbf{P}(0, 1)]$

$$+ u[\mathbf{P}_v(1, v) - \mathbf{P}_v(0, v) - \mathbf{P}(0, 0) + \mathbf{P}(0, 1) + \mathbf{P}(1, 0) - \mathbf{P}(1, 1)]$$

$$= \mathbf{C} + u\mathbf{D}$$

It can easily be shown that $\mathbf{P}_u(u, v)$ and $\mathbf{P}_v(u, v)$ lie in the plane of the boundary curves. Considering $\mathbf{P}_u(u, v)$, the vectors $\mathbf{P}_u(u, 0)$, $\mathbf{P}(1, v) - \mathbf{P}(0, v)$ and $\mathbf{P}(0, 0) - \mathbf{P}(1, 0)$ lie in the given plane. By investigating the coefficient of $v$, the vectors $\mathbf{P}_u(u, 1)$, $\mathbf{P}_u(u, 0)$, $\mathbf{P}(1, 0) - \mathbf{P}(0, 0)$, and $\mathbf{P}(0, 1) - \mathbf{P}(1, 1)$ also lie in the given plane. Therefore, vectors $\mathbf{A}$ and $\mathbf{B}$ lie in the plane. Consequently, the tangent vector $\mathbf{P}_u(u, v)$ to the surface at any point $(u, v)$ lies in the plane of the boundary curves. The same argument can be extended to $\mathbf{P}_v(u, v)$.

The surface normal is given by

$$\mathbf{N}(u, v) = \mathbf{P}_u \times \mathbf{P}_v = (\mathbf{A} + v\mathbf{B}) \times (\mathbf{C} + u\mathbf{D})$$

which is perpendicular to the plane of $\mathbf{P}_u$ and $\mathbf{P}_v$ and, therefore, the plane of the boundary curves. Thus for any point on the surface, the direction of the surface normal is constant (the magnitude depends on the point) or the unit normal is fixed in space. Knowing that the plane surface is the only surface that has a fixed unit normal, we conclude that a bilinear Coons patch degenerates to a plane if its boundary curves are coplanar. Thus, this patch can be used to create planes with curved boundaries similar to the bicubic surface covered in Example 6.6. Note, however, that a bicubic Coons patch or any other patch that has nonlinear blending functions does not reduce to a plane when all its boundaries are coplanar.

## 6.6.5  Blending Surface

This is a surface that connects two nonadjacent surfaces or patches. The blending surface is usually created to manifest $C^0$ and $C^1$ continuity with the two given patches. The fillet surface shown in Fig. 6-11 is considered a special case of a blending surface. Figure 6-42 shows a general blending surface. A bicubic surface can be used to blend patch 1 and patch 2 with both $C^0$ and $C^1$ continuity. The corner points $P_1$, $P_2$, $P_3$, and $P_4$ of the blending surface and their related tangent and twist vectors are readily available from the two patches. Therefore, the $[B]$ matrix of the blending surface can be evaluated. A bicubic blending surface is suitable to blend cubic patches, that is, bicubic Bezier or B-spline patches.

For patches of other orders, a B-spline blending surface may be generated in the following scenario. A set of points and their related $v$-tangent vectors

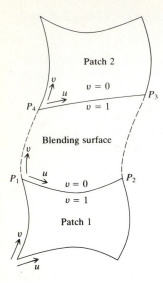

Patch 2

$v = 0$

$P_3$

$v = 1$

Blending surface

$P_1$

$v = 0$

$P_2$

$v = 1$

Patch 1

**FIGURE 6-42**
A blending surface.

beginning with $P_1$ and ending with $P_2$ can be generated along the $v = 1$ edge of patch 1. Similarly, a corresponding set can be generated along the $v = 0$ edge of patch 2. Cubic spline curves can now be created between the two sets. These curves can be used to generate an ordered rectangular set of points that can be connected with the B-spline surface which becomes the blending surface. Some CAD/CAM systems allow users to connect a given set of curves with a B-spline surface directly. In the case of the fillet surface shown in Fig. 6-11, a fillet radius is used to generate the surface. Here, the rectangular set of points to create the B-spline surface can be generated by creating fillets between corresponding $v = $ constant curves on both patches. In turn, points can be generated on these fillets.

### 6.6.6 Offset Surface

If an original patch and an offset direction are given as shown in Fig. 6-12, the equation of the resulting offset patch can be written as:

$$\mathbf{P}(u, v)_{\text{offset}} = \mathbf{P}(u, v) + \hat{\mathbf{n}}(u, v) \, d(u, v) \qquad (6.104)$$

where $\mathbf{P}(u, v)$, $\hat{\mathbf{n}}(u, v)$, and $d(u, v)$ are the original surface, the unit normal vector at point $(u, v)$ on the original surface, and the offset distance at point $(u, v)$ on the original surface respectively. The unit normal $\hat{\mathbf{n}}(u, v)$ is the offset direction shown in Fig. 6-12. The distance $d(u, v)$ enables generating uniform or tapered thickness surfaces depending on whether $d(u, v)$ is constant or varies linearly in $u, v$ or both.

### 6.6.7 Triangular Patches

Triangular patches are useful if the given surface data points form a triangle or if a given surface cannot be modeled by rectangular patches only and may require at least one triangular patch. In tensor product surfaces, the parameters are $u$ and

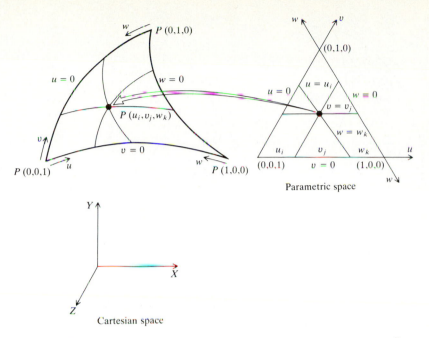

**FIGURE 6-43**
Representation of a triangular patch.

$v$ and the parametric domain is defined by the unit square of $0 \le u \le 1$ and $0 \le v \le 1$. In triangulation techniques, three parameters $u$, $v$, and $w$ are used and the parametric domain is defined by a symmetric unit triangle of $0 \le u \le 1$, $0 \le v \le 1$, and $0 \le w \le 1$, as shown in Fig. 6-43. The coordinates $u$, $v$, and $w$ are called "barycentric coordinates." While the coordinate $w$ is not independent of $u$ and $v$ (note that $u + v + w = 1$ for any point in the domain), it is introduced to emphasize the symmetry properties of the barycentric coordinates.

The formulation of triangular polynomial patches follows a somewhat similar pattern to that of tensor product patches. For example, a triangular Bezier patch is defined by

$$\mathbf{P}(u, v, w) = \sum_{i, j, k} \mathbf{P}_{ijk} B_{i, j, k, n}(u, v, w), \qquad 0 \le u \le 1, 0 \le v \le 1, 0 \le w \le 1 \quad (6.105)$$

where $i, j, k \ge 0$, $i + j + k = n$ and $n$ is the degree of the patch. The $B_{i, j, k, n}$ are Bernstein polynomials of degree $n$:

$$B_{i, j, k, n} = \frac{n!}{i!\,j!\,k!}\, u^i v^j w^k \qquad (6.106)$$

The coefficients $\mathbf{P}_{i, j, k}$ are the control or data points that form the vertices of the control polygon. The number of data points required to define a Bezier patch of degree $n$ is given by $(n + 1)(n + 2)/2$. Figure 6-44 shows cubic and quartic triangular Bezier patches with their related Bernstein polynomials. The order of inputting data points should follow the pyramid organization of a Bernstein polynomial shown in the figure. For example, 15 points are required to create a

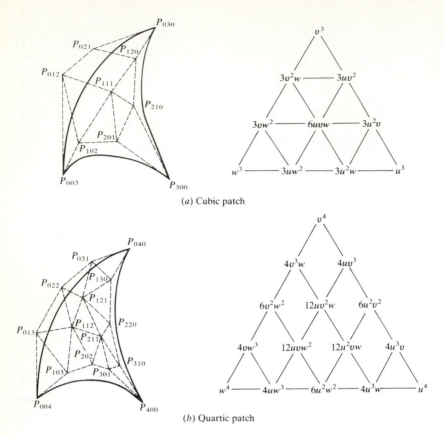

(a) Cubic patch

(b) Quartic patch

**FIGURE 6-44**
Triangular Bezier patches.

quartic Bezier patch and must be input in five rows. The first row has five points $(n + 1)$ and each successive row has one point less than its predecessor until we reach the final row that has only one point. This pattern of input can be achieved symmetrically from any direction as shown. Note that the degree of the triangular Bezier patch is the same in all directions, in contrast to the rectangular patch which can have $n$- and $m$-degree polynomials in $u$ and $v$ directions respectively. However, all the characteristics of the rectangular patch hold true for the triangular patch.

A rectangular Coons patch can be modified in a similar fashion as described above to develop a triangular Coons patch. It is left to the reader, as an exercise, to extend the above formulation to a triangular Coons patch.

### 6.6.8 Sculptured Surface

Any of the surface patches introduced thus far is seldom enough by itself to model complex surfaces typically encountered in design and manufacturing applications. These complex surfaces are known as sculptured or free-form surfaces.

They arise extensively in automotive die and mold-making, aerospace, glass, cameras, shoes, and appliance industries—to name a few.

A sculptured surface is defined as a collection or sum of interconnected and bounded parametric patches together with blending and interpolation formulas. The surface must be susceptible to APT, or other machining languages, processing for NC machine tools. The analytic and synthetic patches described thus far can be used to create a sculptured surface. From a modeling viewpoint, a sculptured surface can be divided into the proper patches which can be created to produce a $C^0$ or $C^1$ continuous surface. The two-patch surfaces (e.g., Fig. 6-16) discussed throughout this chapter thus far are considered sculptured surfaces.

### 6.6.9 Rational Parametric Surface

The rational parametric surface is considered a very general surface. The shape of the surface is controlled by weights (scalar factors) assigned to each control point. A rational surface is defined by the algebraic ratio of two polynomials. A rational tensor product surface can be represented as

$$P(u, v) = \frac{\sum\limits_{i=0}^{n} \sum\limits_{j=0}^{m} P_{ij} h_{ij} F_i(u) F_j(v)}{\sum\limits_{i=0}^{n} \sum\limits_{j=0}^{m} h_{ij} F_i(u) F_j(v)} \tag{6.107}$$

where $h_{ij}$ are the weights assigned to the control points. Rational bicubic, Bezier, and B-spline surfaces are available. The discussion regarding rational curves covered in Sec. 5.6.4 can be extended to rational surfaces.

## 6.7 SURFACE MANIPULATIONS

As with curves, surface manipulation provides the designer with the capabilities to use surfaces effectively in design applications. For example, if two surfaces are to be welded by a robot, their intersection curve provides the path the robot should follow. Therefore, the robot trajectory planning becomes very accurate. This section covers some useful features of surface manipulations.

### 6.7.1 Displaying

The simplest method to display surfaces is to generate a mesh of curves on the surface by holding one parameter constant at a time. Most surface commands on CAD/CAM systems have a mesh size modifier. Using this size and the surface equation, the curves of the mesh are evaluated. The display of these curves follows the same techniques discussed in Sec. 5.7.1. For example, to display a Bezier surface of mesh size $3 \times 4$, three curves that have constant $v$ values ($v = 0, \frac{1}{2}, 1$) and four curves that have constant $u$ values ($u = 0, \frac{1}{3}, \frac{2}{3}, 1$) are evaluated using Eq. (6.67) and displayed. This method is not very efficient as fine details of the surface may be lost unless the mesh is very dense, in which case it becomes slow and expensive to generate, display, and/or update the surface. This method of displaying a surface by a mesh of curves is sometimes referred to as a wire-frame display of a surface.

To improve the visualization of a surface, surface normals can be displayed as straight lines in addition to the wireframe display. If the lengths of these straight lines are made long enough, any small variation in surface shape becomes evident. Another good method to use to display surfaces is shading. Shading techniques are discussed in more detail in Chap. 10. Various surfaces can be shaded with various colors. Surface curvatures and other related properties can also be displayed via shading. If surfaces are nested, inner shaded surfaces can be viewed by using X-ray and transparency techniques.

### 6.7.2   Evaluating Points and Curves on Surfaces

Generating points on surfaces is useful for applications such as finite element modeling and analysis. The "generate point on" command or its equivalent, provided by CAD/CAM systems and mentioned in Sec. 5.7.2, is also available for surfaces. The obvious method of calculating points on a surface is by substituting their respective $u$ and $v$ values into the surface equation. The forward difference technique for evaluating points on curves described by Eq. (5.121) can be extended to surfaces by using the equation for constant values of $v$. If the specific $v$ value is substituted into the surface equation, Eq. (5.121) results. Therefore, Eqs. (5.122) and (5.123) can be applied to generate the points for the given $v$ value. This procedure can be repeated for all $v$ values to generate the desired points on the surface.

Similarly to curves, evaluating a point on a surface for given $u$ and $v$ values is the direct point solution, which consists of evaluating three polynomials in $u$ and $v$, one for each coordinate of the point. The inverse problem, which requires finding the corresponding $u$ and $v$ values if the $x$, $y$, and $z$ coordinates of the point are given, arises if tangent vectors at the point are to be evaluated or if two coordinates of the point are given and we must find the third coordinate. The solution of this problem (the inverse point solution) requires solving two nonlinear polynomials in $u$ and $v$ simultaneously via numeric methods (see Prob. 6.12 at the end of the chapter).

### 6.7.3   Segmentation

The segmentation problem of a given surface is identical to that of a curve, which is discussed in Sec. 5.7.4. Surface segmentation is a reparametrization or parameter transformation of the surface while keeping the degree of its polynomial in $u$ and $v$ unchanged.

Let us assume a surface patch is defined over the range $u = u_0$, $u_m$ and $v = v_0$, $v_m$ and it is desired to split it at a point $P_1$ defined by $(u_1, v_1)$ as shown in Fig. 6-45. If the point is defined by its cartesian coordinates, the inverse point solution must be found first to obtain $u_1$ and $v_1$. Let us introduce the new parameters $u^1$ and $v^1$ such that their ranges are $(0, 1)$ for each resulting subpatch (see Fig. 6-45).

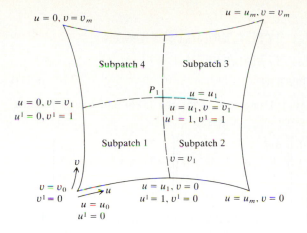

**FIGURE 6-45**
Reparametrization of a segmented surface patch.

The parameter transformation takes the form:

$$u^1 = u_0 + (u_1 - u_0)u \atop v^1 = v_0 + (v_1 - v_0)v \Bigg\} \quad \text{for subpatch 1} \qquad (6.108)$$

Similar equations can be written for the other subpatches. Notice that $u^1 = 0$, 1 and $v^1 = 0$, 1 correspond to the proper values of $u$ and $v$ for the subpatch. If the surface is described by a polynomial, Eq. (5.133) and a similar one for $v$ can be used to facilitate substitutions in the surface equation.

If the surface is to be divided into two segments or subpatches along the $u = u_1$ or $v = v_1$ curve instead of four segments, Eq. (6.108) becomes

$$u^1 = u_0 + (u_1 - u_0)u \atop v^1 = v_0 + (v_m - v_0)v \Bigg\} \quad \text{for the first segment} \qquad (6.109)$$

if segmentation is done along the $u = u_1$ curve, and

$$u^1 = u_0 + (u_m - u_0)u \atop v^1 = v_0 + (v_1 - v_0)v \Bigg\} \quad \text{for the first segment} \qquad (6.110)$$

if segmentation is done along the $v = v_1$ curve.

**Example 6.10.** A bicubic surface patch is to be divided by a designer into four subpatches. Find the boundary conditions of each of the resulting subpatches.

**Solution.** Using Eq. (5.133) in its matrix form found in Example 5.23, Eq. (6.42) can be written as

$$\mathbf{P}(u, v) = \mathbf{U}^{1T}[T]^u[M_H][B][M_H]^T[T]^{v^T}\mathbf{V}^1 = \mathbf{P*}(u^1, v^1)$$

This equation can be rewritten in the form of Eq. (6.42) as

$$\mathbf{P*}(u^1, v^1) = \mathbf{U}^{1T}[M_H][M_H]^{-1}[T]^u[M_H][B][M_H]^T[T]^{v^T}[M_H]^{T^{-1}}[M_H]^T\mathbf{V}^1$$

$$= \mathbf{U}^{1T}[M_H][B]^*[M_H]^T\mathbf{V}^1$$

where

$$[B]^* = [M_H]^{-1}[T]^u[M_H][B][M_H]^T[T]^{v^T}[M_H]^{-1^T}$$

$$= [M_H]^{-1}[T]^u[M_H][B][[M_H]^{-1}[T]^v[M_H]]^T$$

or $\qquad [B]^* = [A]^u[B][A]^{v^T}$ \hfill (6.111)

where $[A] = [M_H]^{-1}[T][M_H]$. Using Eqs. (5.84) and (5.86) and the $[T]$ matrix from Example 5.23, $[A]^u$ is given by

$$[A]^u = \begin{bmatrix} 0 & 0 & 0 & 1 \\ 1 & 1 & 1 & 1 \\ 0 & 0 & 1 & 0 \\ 3 & 2 & 1 & 0 \end{bmatrix} \begin{bmatrix} \Delta u_0^3 & 0 & 0 & 0 \\ 0 & \Delta u_0^2 & 0 & 0 \\ 0 & 0 & \Delta u_0 & 0 \\ 0 & 0 & 0 & 1 \end{bmatrix} \begin{bmatrix} 2 & -2 & 1 & 1 \\ -3 & 3 & -2 & -1 \\ 0 & 0 & 1 & 0 \\ 1 & 0 & 0 & 0 \end{bmatrix}$$

The above diagonal $[T]$ matrix resulted from assuming $u_0 = 0$. Similarly, we assume $v_0 = 0$. These assumptions do not restrict the solution method but they rather simplify the matrix manipulations (the reader can re-solve Example 5.23 with $u_0 = 0$ for clarification of this point if necessary). Performing the above matrix multiplications and utilizing Eqs. (6.52) and (6.53) in the result, we obtain

$$[A]^u = \begin{bmatrix} 1 & 0 & 0 & 0 \\ F_1(\Delta u_0) & F_2(\Delta u_0) & F_3(\Delta u_0) & F_4(\Delta u_0) \\ 0 & 0 & \Delta u_0 & 0 \\ \Delta u_0\, G_1(\Delta u_0) & \Delta u_0\, G_2(\Delta u_0) & \Delta u_0\, G_3(\Delta u_0) & \Delta u_0\, G_4(\Delta u_0) \end{bmatrix}$$

The matrix $[A]^{v^T}$ is obtained from the above matrix by replacing $\Delta u_0$ by $\Delta v_0$ and transposing the result to get

$$[A]^{v^T} = \begin{bmatrix} 1 & F_1(\Delta v_0) & 0 & \Delta v_0\, G_1(\Delta v_0) \\ 0 & F_2(\Delta v_0) & 0 & \Delta v_0\, G_2(\Delta v_0) \\ 0 & F_3(\Delta v_0) & \Delta v_0 & \Delta v_0\, G_3(\Delta v_0) \\ 0 & F_4(\Delta v_0) & 0 & \Delta v_0\, G_4(\Delta v_0) \end{bmatrix}$$

Substituting $[A]^u$ and $[A]^{v^T}$ into Eq. (6.111) gives the geometry or boundary condition matrix for subpatch 1 shown in Fig. 6-45. To reduce the results into a useful form, the cubic polynomial interpolation given by Eqs. (5.81) and (5.82) must be recognized during matrix multiplications. In Eq. (6.111), assume $[A]^u[B] = [D]$. Thus, the elements of $[D]$ are

$$d_{11} = P_{00}$$

$$d_{12} = P_{01}$$

$$d_{13} = P_{v00}$$

$$d_{14} = P_{v01}$$

$$d_{21} = P_{00} F_1(u_1) + P_{10} F_2(u_1) + P_{u00} F_3(u_1) + P_{u10} F_4(u_1)$$

We have substituted $u_1$ for $\Delta u_0$ based on the assumption that $u_0 = 0$. The right-hand side of the above equation is a cubic spline of the form given by Eq. (5.81) evaluated at $u_1$. It interpolates the boundary conditions at the corners $P_{00}$ and $P_{10}$ of the original patch. Therefore:

$$d_{21} = P(u_{1,0})$$

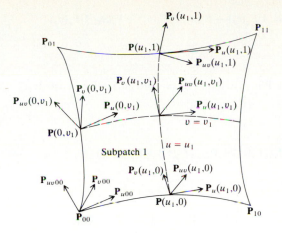

**FIGURE 6-46**
Related vectors to subpatch 1.

This point is shown in Fig. 6-46. Repeating this interpretation process and referring to Fig. 6-46, we obtain

$$\mathbf{d}_{22} = \mathbf{P}_{01}F_1(u_1) + \mathbf{P}_{11}F_2(u_1) + \mathbf{P}_{u01}F_3(u_1) + \mathbf{P}_{u11}F_4(u_1) = \mathbf{P}(u_1, 1)$$

$$\mathbf{d}_{23} = \mathbf{P}_{v00}F_1(u_1) + \mathbf{P}_{v10}F_2(u_1) + \mathbf{P}_{uv00}F_3(u_1) + \mathbf{P}_{uv10}F_4(u_1) = \mathbf{P}_v(u_1, 0)$$

In the above equation, it is obvious that the cubic polynomial interpolates the $v$-tangent vector between the corners $\mathbf{P}_{00}$ and $\mathbf{P}_{10}$.

$$\mathbf{d}_{24} = \mathbf{P}_{v01}F_1(u_1) + \mathbf{P}_{v11}F_2(u_1) + \mathbf{P}_{uv01}F_3(u_1) + \mathbf{P}_{uv11}F_4(u_1) = \mathbf{P}_v(u_1, 1)$$

$$\mathbf{d}_{31} = \Delta u_0\, \mathbf{P}_{u00}$$

$$\mathbf{d}_{32} = \Delta u_0\, \mathbf{P}_{u01}$$

$$\mathbf{d}_{33} = \Delta u_0\, \mathbf{P}_{uv00}$$

$$\mathbf{d}_{34} = \Delta u_0\, \mathbf{P}_{uv01}$$

$$\mathbf{d}_{41} = \Delta u_0[\mathbf{P}_{00}\, G_1(u_1) + \mathbf{P}_{10}\, G_2(u_1) + \mathbf{P}_{u00}\, G_3(u_1) + \mathbf{P}_{u10}\, G_4(u_1)]$$

Using Eq. (5.82), we can write

$$\mathbf{d}_{41} = \Delta u_0\, \mathbf{P}_u(u_1, 0)$$

$$\mathbf{d}_{42} = \Delta u_0[\mathbf{P}_{01}G_1(u_1) + \mathbf{P}_{11}G_2(u_1) + \mathbf{P}_{u01}G_3(u_1) + \mathbf{P}_{u11}G_4(u_1)] = \Delta u_0\, \mathbf{P}_u(u_1, 1)$$

$$\mathbf{d}_{43} = \Delta u_0[\mathbf{P}_{v00}\, G_1(u_1) + \mathbf{P}_{v10}\, G_2(u_1) + \mathbf{P}_{uv00}\, G_3(u_1) + \mathbf{P}_{uv10}\, G_4(u_1)] = \Delta u_0\, \mathbf{P}_{uv}(u_1, 0)$$

$$\mathbf{d}_{44} = \Delta u_0[\mathbf{P}_{v01}G_1(u_1) + \mathbf{P}_{v11}G_2(u_1) + \mathbf{P}_{uv01}G_3(u_1) + \mathbf{P}_{uv11}G_4(u_1)] = \Delta u_0\, \mathbf{P}_{uv}(u_1, 1)$$

Returning to Eq. (6.111), the elements of $[B]^*$ result from multiplying $[D]$ and $[A]^{v^T}$:

$$\mathbf{b}_{11}^* = \mathbf{P}_{00}$$

$$\mathbf{b}_{12}^* = \mathbf{P}_{00}F_1(v_1) + \mathbf{P}_{01}F_2(v_1) + \mathbf{P}_{v00}F_3(v_1) + \mathbf{P}_{v01}F_4(v_1) = \mathbf{P}(0, v_1)$$

$$\mathbf{b}_{13}^* = \Delta v_0\, \mathbf{P}_{v00}$$

$$\mathbf{b}_{14}^* = \Delta v_0[\mathbf{P}_{00}\, G_1(v_1) + \mathbf{P}_{01}G_2(v_1) + \mathbf{P}_{v00}\, G_3(v_1) + \mathbf{P}_{v01}G_4(v_1)] = \Delta v_0\, \mathbf{P}_v(0, v_1)$$

$$\mathbf{b}_{21}^* = \mathbf{P}(u_1, 0)$$

$$\mathbf{b}_{22}^* = P(u_1, 0)F_1(v_1) + P(u_1, 1)F_2(v_1) + P_v(u_1, 0)F_3(v_1) + P_v(u_1, 1)F_4(v_1)$$

$$= P(u_1, v_1) = \mathbf{P}_1$$

$$\mathbf{b}_{23}^* = \Delta v_0 P_v(u_1, 0)$$

$$\mathbf{b}_{24}^* = \Delta v_0[P(u_1, 0)G_1(v_1) + P(u_1, 1)G_2(v_1) + P_v(u_1, 0)G_3(v_1) + P_v(u_1, 1)G_4(v_1)]$$

$$= \Delta v_0 P_v(u_1, v_1)$$

$$\mathbf{b}_{31}^* = \Delta u_0 P_{u00}$$

$$\mathbf{b}_{32}^* = \Delta u_0[P_{u00} F_1(v_1) + P_{u01} F_2(v_1) + P_{uv00} F_3(v_1) + P_{uv01} F_4(v_1)]$$

$$= \Delta u_0 P_u(0, v_1)$$

$$\mathbf{b}_{33}^* = \Delta u_0 \Delta v_0 P_{uv00}$$

$$\mathbf{b}_{34}^* = \Delta u_0 \Delta v_0[P_{u00} G_1(v_1) + P_{u01} G_2(v_1) + P_{uv00} G_3(v_1) + P_{uv01} G_4(v_1)]$$

$$= \Delta u_0 \Delta v_0 P_{uv}(0, v_1)$$

$$\mathbf{b}_{41}^* = \Delta u_0 P_u(u_1, 0)$$

$$\mathbf{b}_{42}^* = \Delta u_0[P_u(u_1, 0)F_1(v_1) + P_u(u_1, 1)F_2(v_1) + P_{uv}(u_1, 0)F_3(v_1) + P_{uv}(u_1, 1)F_4(v_1)]$$

$$= \Delta u_0 P_u(u_1, v_1)$$

$$\mathbf{b}_{43}^* = \Delta u_0 \Delta v_0 P_{uv}(u_1, 0)$$

$$\mathbf{b}_{44}^* = \Delta u_0 \Delta v_0[P_u(u_1, 0)G_1(v_1) + P_u(u_1, 1)G_2(v_1) + P_{uv}(u_1, 0)G_3(v_1) + P_{uv}(u_1, 1)G_4(v_1)]$$

$$= \Delta u_0 \Delta v_0 P_{uv}(u_1, v_1)$$

The boundary condition matrix $[B]^*$ for subpatch 1 takes the form:

$$[B]^* = \left[\begin{array}{cc:cc}
\mathbf{P}_{00} & P(0, v_1) & \Delta v_0 \mathbf{P}_{v00} & \Delta v_0 P_v(0, v_1) \\
P(u_1, 0) & P(u_1, v_1) & \Delta v_0 P_v(u_1, 0) & \Delta v_0 P_v(u_1, v_1) \\
\hdashline
\Delta u_0 \mathbf{P}_{u00} & \Delta u_0 P_u(0, v_1) & \Delta u_0 \Delta v_0 \mathbf{P}_{uv00} & \Delta u_0 \Delta v_0 P_{uv}(0, v_1) \\
\Delta u_0 P_u(u_1, 0) & \Delta u_0 P_u(u_1, v_1) & \Delta u_0 \Delta v_0 P_{uv}(u_1, 0) & \Delta u_0 \Delta v_0 P_{uv}(u_1, v_1)
\end{array}\right]$$

This result is consistent with Example 5.23. The corner points of subpatch 1 are those evaluated from the original patch. However, the subpatch tangent vectors have a factor of $\Delta u_0$ or $\Delta v_0$ multiplying those obtained from the original patch. There is also a factor of $\Delta u_0 \Delta v_0$ associated with the twist vectors. These factors ensure the adherence to the shape of the original patch. It is obvious that surface normals do not reverse directions. The $[B]$ matrices for the other three subpatches can now easily be written by substituting the proper $\Delta u$, $\Delta v$, and the vectors into the above matrix.

If the original patch is to be divided into two segments along either the $u = u_1$ or $v = v_1$ curve, the two boundary condition matrices can be written in a similar way.

## 6.7.4 Trimming

Trimming of surface entities is useful for engineering applications. It helps eliminate unnecessary calculations on the user's part. Surface trimming can be treated as a segmentation or intersection problem. If the surface is to be trimmed between two point trimming boundaries, we then have a segmentation problem

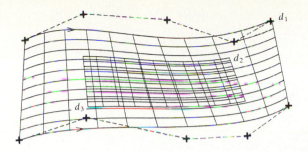

**FIGURE 6-47**
Surface trimming.

at hand. If, on the other hand, a surface is to be trimmed to another surface, the intersection curve of the two surfaces must be found first and then the desired surface is trimmed to it. Figure 6-47 shows an example of trimming surfaces. A Bezier surface (identified by the digitize $d_1$) is trimmed between the two points identified by the two digitizes $d_2$ and $d_3$.

### 6.7.5 Intersection

The intersection problem involving surfaces is complex and nonlinear in nature. It depends on whether it is a surface/curve or surface/surface intersection as well as on the representation, parametric or implicit, of the involved surfaces and/or curves. In surface-to-curve intersection, the intersection problem is defined by the equation

$$P(u, v) - P(w) = 0 \qquad (6.112)$$

where $P(u, v) = 0$ and $P(w) = 0$ are the parametric equations of the surface and the curve respectively. Equation (6.112) is a system of three scalar equations, generally nonlinear, in three unknown $u$, $v$, and $w$. There are efficient iterative methods that can solve Eq. (6.112) such as the Newton-Raphson method. The detailed solution of Eq. (6.112) is left to the reader as an exercise.

The problem of surface-to-surface intersection is defined by the equation

$$P(u, v) - P(t, w) = 0 \qquad (6.113)$$

where $P(u, v) = 0$ and $P(t, w) = 0$ are the equations of the two surfaces. The vector equation (6.113) corresponds to three scalar equations in four variables $u$, $v$, $t$, and $w$. To solve this overspecified problem, some methods hold one of the unknowns constant and therefore reduce the original surface/surface intersection problem into a surface/curve intersection. Other methods introduce a new con-strained function of the four variables. In both approaches, the solutions are iterative and may yield any combination of curves (closed or open) and isolated points. Solutions for sculptured surfaces usually employ curve-tracing, subdivi-sion (divide-and-conquer method), or a combination of both techniques. Some of these solutions assume a given class of surface types, such as Bezier or B-spline representation. Figure 6-48 shows an example of surface/surface intersection. Some CAD/CAM systems implement the special case of a plane intersecting a surface as a "cut plane" command.

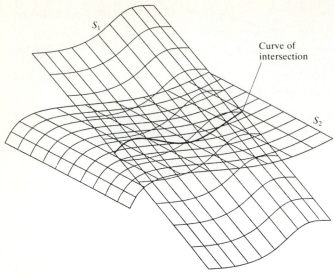

**FIGURE 6-48**
Surface/surface intersection.

## 6.7.6  Projection

Projecting an entity onto a plane or a surface is useful in applications such as determining shadows or finding the position of the entity relative to the plane or the surface. Entities that can be projected include points, lines, curves, or surfaces. Projecting a point onto a plane or a surface forms the basic problem shown in Fig. 6-49a. Point $P_0$ is projected along the direction $r$ onto the given plane. It is desired to calculate the coordinates of the projected point $Q$. The plane equation, based on Eq. (6.28), can be written as follows:

$$\mathbf{P}(u, v) = \mathbf{a} + u\mathbf{b} + v\mathbf{c} \qquad (6.114)$$

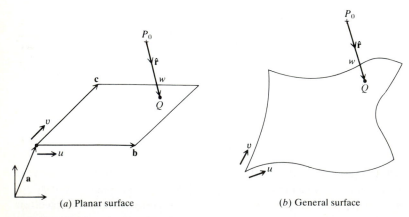

(a) Planar surface                    (b) General surface

**FIGURE 6-49**
Projecting a point onto a surface.

where the vectors **a**, **b**, and **c** are shown in Fig. 6-49$a$. The equation of the projection line is given by

$$P(w) = P_0 + w\hat{r} \tag{6.115}$$

The projection point $Q$ is the intersection point between the line and the plane; that is, the following equation must be solved for $u$, $v$, and $w$:

$$P(u, v) - P(w) = 0 \tag{6.116}$$

or

$$a + ub + vc = P_0 + w\hat{r} \tag{6.117}$$

To solve for $w$, dot-multiply both sides of the above equation by $(b \times c)$ to get

$$(b \times c) \cdot a = (b \times c) \cdot (P_0 + w\hat{r}) \tag{6.118}$$

since $(b \times c)$ is perpendicular to both **b** and **c**. Equation (6.118) gives

$$w = \frac{(b \times c) \cdot (a - P_0)}{(b \times c) \cdot \hat{r}} \tag{6.119}$$

Similarly, we can write

$$u = \frac{(c \times \hat{r}) \cdot (P_0 - a)}{(c \times \hat{r}) \cdot b} \tag{6.120}$$

and

$$v = \frac{(b \times \hat{r}) \cdot (P_0 - a)}{(b \times \hat{r}) \cdot c} \tag{6.121}$$

The projection point $Q$ results by substituting Eq. (6.119) into (6.115) or Eqs. (6.120) and (6.121) into (6.114).

If point $P_0$ is to be projected onto a general surface as shown in Fig. 6-49$b$, Eq. (6.114) is replaced by the surface equation and Eq. (6.116) becomes a nonlinear equation similar to Eq. (6.112).

The above approach can be extended to projecting curves and surfaces onto a given surface as shown in Figs. 6-50 and 6-51. The projection of a straight line passing by the two endpoints $P_0$ and $P_1$ (see Fig. 6-50$a$) along the direction $\hat{r}$ involves projecting the two points using Eqs. (6.119) to (6.121) and then connecting $Q_0$ and $Q_1$ by a straight line which, of course, must lie in the given plane.

The projection of a general curve $P(s)$ onto a plane (Fig. 6-50$b$) or a general surface (Fig. 6-50$c$) requires repetitive solution of Eq. (6.112). One simple strategy

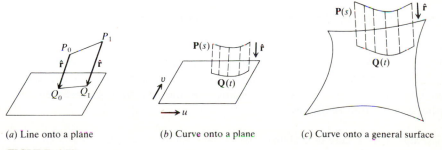

(a) Line onto a plane          (b) Curve onto a plane          (c) Curve onto a general surface

**FIGURE 6-50**
Projecting a curve onto a surface.

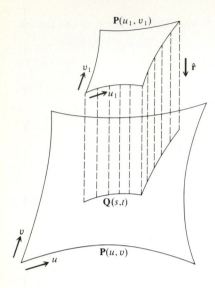

**FIGURE 6-51**
Projecting a surface onto a surface.

is to generate a set of points on $P(s)$, project them onto the given plane or surface, and then connect them by a B-spline curve to obtain the projection curve $Q(t)$.

The projection of a surface $P(u_1, v_1)$ onto a surface $P(u, v)$ (Fig. 6-51) can be seen as an extension of the above strategy. A set of points is first generated on the surface $P(u_1, v_1)$. Utilizing Eq. (6.112), this set is projected onto the surface $P(u, v)$. The projection points are then connected by a B-spline surface to produce the projection surface $Q(s, t)$. Note here that the greater the number of projection points, the closer the surface $Q(s, t)$ to the surface $P(u, v)$.

### 6.7.7 Transformation

The homogeneous transformation of surfaces is an extension of those of curves. Transformations offer very useful tools to designers while creating surface models. Functions such as translation, rotation, mirror, and scaling are offered by most CAD/CAM systems and are based on transformation concepts. To transform a surface, points on the surface may be evaluated first, transformed, and then the surface is redisplayed in the new transformed position and/or orientation. Other methods may exist. More on transformation is covered in Chap. 9.

### 6.8 DESIGN AND ENGINEERING APPLICATIONS

The following examples show how surface functions offered by CAD/CAM systems and their related theory covered in this chapter are used for engineering and design applications. The reader can work these examples on a CAD/CAM system. The reader is also advised to expand the line of thinking presented here into other applications of interest.

**Example 6.11.** Figure 6-52 shows a set of data points whose coordinates are measured directly from a given closed surface. Reconstruct the surface from the data points.

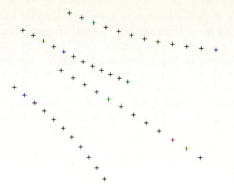

**FIGURE 6-52**
Data points.

*Solution.* A B-spline surface is the proper surface to connect the above given set of data points. "B-spline surface" commands provided by most CAD/CAM systems allow users to create B-spline surfaces by connecting a set of data points as described in Sec. 6.6.3 or by connecting a set of B-spline curves. In the latter case, the CAD/CAM system is actually using the points that define the spline curves for the surface definition. Therefore, the splines must have been created with the same number of points and in the same order, that is, connect points from left to right or right to left; otherwise the resulting surface will be twisted.

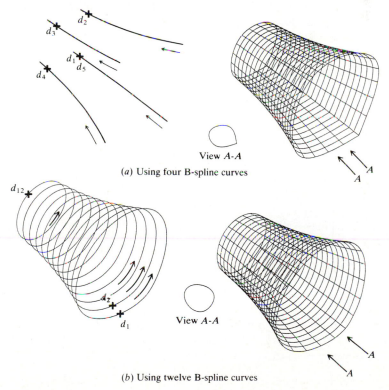

View A-A
(a) Using four B-spline curves

View A-A
(b) Using twelve B-spline curves

**FIGURE 6-53**
Constructing a B-spline surface from a given set of points.

The data points shown in Fig. 6-52 can be seen to form four B-spline curves which can then be used to reconstruct the surface (see Fig. 6-53a). They can also be interpreted as points on 12 cross sections of the surface, four points per cross section. In this case, the surface is shown in Fig. 6-53b. In both figures, the points are connected in the order shown by arrows to create the splines and the splines are digitized in the order shown to create the surface. In Fig. 6-53b we used the same tangents at the beginning and end points of each spline to force $C^1$ continuity. If we had not used it, we would have obtained an identical surface to that shown in Fig. 6-53a.

**Example 6.12.** Figure 6-54 shows a surface whose cross sections are circular. Construct the surface.

*Solution.* This is an axisymmetric surface. The most efficient way to construct it is to choose an MCS as shown in Fig. 6-54 and create points $a$, $b$, $c$, and $d$. These points are then copied using a mirror command to generate points $b'$, $c'$, and $d'$. The seven points are connected with a B-spline curve. Utilizing a surface of revolution command, this curve can be rotated 360° about the $Y$ axis to generate the required surface (see Fig. 6-55).

Another way of constructing the surface is to replace the circular cross sections by B-spline curves which can be connected with a B-spline surface. Figure 6-56 shows the B-spline surface where 36 points are used to represent each circle.

Both ways of constructing the surface are compared by superposing them as shown in Fig. 6-57. As can be seen, while each surface has its unique database, both ways yield identical surfaces.

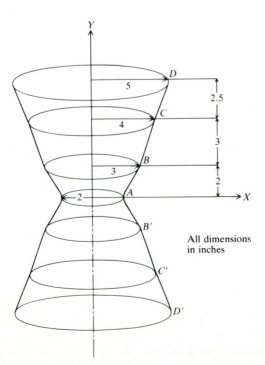

All dimensions
in inches

**FIGURE 6-54**
Circular cross sections of a surface.

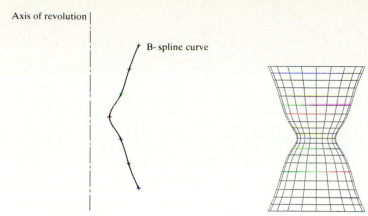

Axis of revolution

B-spline curve

**FIGURE 6-55**
Surface of revolution of Example 6.12.

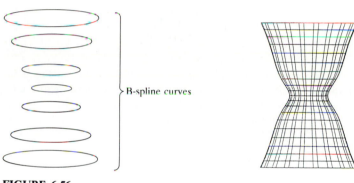

B-spline curves

**FIGURE 6-56**
B-spline surface of Example 6.12.

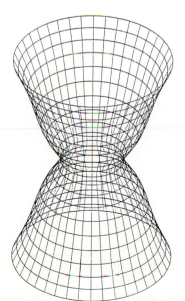

**FIGURE 6-57**
Comparison of both ways of constructing the surface of
Example 6.12.

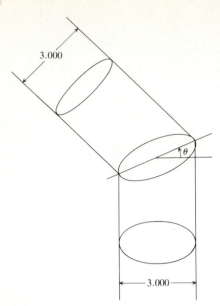

**FIGURE 6-58**
Pipeline design.

**Example 6.13.** Two pipes whose axes form an angle $\theta$ are shown in Fig. 6-58. Each pipe has a radius of 1.5 inch. The two pipes are to be butt welded along the common ellipse of intersection. The area of the ellipse must be 10 in$^2$ ± 1 percent for pressure drop considerations inside the pipes at the intersection cross section. Find the angle $\theta$ for which the resulting ellipse has the required area. Find also the lengths of the ellipse axes. For welding purposes, find also the perimeter of the ellipse. Ignore the thickness of the pipes for simplicity.

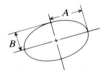

$\theta = 15°$
$A = 1.553$ in, $B = 1.5$ in
Area $= 7.318$ in$^2$
Perimeter $= 9.592$ in

$\theta = 25°$
$A = 1.655$ in, $B = 1.5$ in
Area $= 7.799$ in$^2$
Perimeter $= 9.918$ in

$\theta = 35°$
$A = 1.831$ in, $B = 1.5$ in
Area $= 8.628$ in$^2$
Perimeter $= 10.490$ in

$\theta = 45°$
$A = 2.121$ in, $B = 1.5$ in
Area $= 9.995$ in$^2$
Perimeter $= 11.460$ in

**FIGURE 6-59**
Characteristics of intersection ellipse of two intersecting pipes.

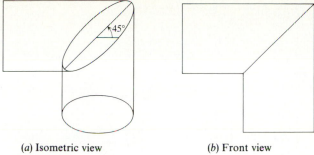

(*a*) Isometric view         (*b*) Front view

**FIGURE 6-60**
Final design of pipeline of Fig. 6-58.

*Solution.* The strategy to solve this problem is to construct the vertical pipe shown in Fig. 6-58 as a tabulated cylinder with a radius of 1.5 inch and an arbitrary length, say 6 inches, cut it with a plane at various values of $\theta$ ranging from 0 to 90°, verify the resulting ellipse entity to obtain the lengths of the ellipse axes, and calculate its area using a "calculate area" command provided by the particular CAD/CAM system. The ellipse that produces the given area is the desired one and the corresponding angle $\theta$ is the solution. The ellipse perimeter can be evaluated using the "measure length" command.

Figure 6-59 shows some of the intermediate results with the solution and Fig. 6-60 shows the final design of the pipeline. To check the system accuracy, compare the obtained results with those obtained using the following equations:

$$\text{Ellipse area } a = \pi AB$$

$$\text{Ellipse perimeter } p = 2\pi \sqrt{\frac{A^2 + B^2}{2}} = 4aE \qquad \text{(approximate)}$$

where $E$ is the complete elliptic integral at $\mathscr{K} = \sqrt{A^2 - B^2}/A$. $E$ can be obtained from the table of elliptic integrals available in CRC standard mathematical tables. For this problem $E$ is found to be 1.3506.

Some CAD/CAM systems may recognize the intersection curve between the plane and the cylinder as a closed B-spline curve. In this case, the area and the perimeter can be calculated using the "calculate area" and "measure length" commands as above. To check the accuracy, the B-spline intersection curve can be replaced by an ellipse using $B = 1.5$ and $A = B/\cos \theta$. The reader is encouraged to re-solve this example on a CAD/CAM system. To find out whether your system treats the intersection curve as an ellipse or a B-spline curve, verify it using the "verify entity" command. *Note:* instead of using a plane to intersect the vertical cylinder, the two cylinders can be created and intersected.

**Example 6.14.** Figure 6-61 shows a duct of an air-conditioning system. The 4-inch diameter pipe is connected to a 4-inch diameter elbow. The elbow is joined to a truncated cone having a 4-inch diameter and 6-inch diameter ends. The 6-inch diameter end is increased to a $10 \times 10$ inch square end. If the thickness of the duct is ignored:

(*a*) Develop a surface model of the duct.
(*b*) Find the duct cross section at the valve location shown.

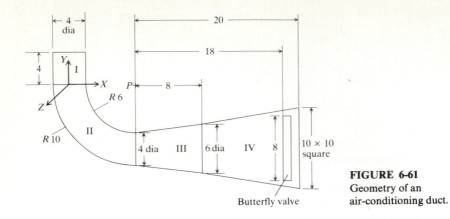

**FIGURE 6-61**
Geometry of an
air-conditioning duct.

*Solution*

(a) Figure 6-61 also shows the four different surfaces required to create the duct. Surface I is a tabulated cylinder with a 4-inch diameter and 4-inch length. Surface II is a surface of revolution. It is created by rotating a horizontal circle located at the end of surface I about an axis passing through point $P$ an angle of 90°. Surfaces III and IV can be created as ruled surfaces. Figure 6-62 shows the top, front, right, and isometric views of the duct.

(b) If the duct geometric model is cut by a vertical plane at the valve location, the resulting duct cross section is shown in Fig. 6-63. The area of the cross section is 85.229 in². The figure also shows cross sections 2 inches apart for surface IV. Notice that the right (first) cross section is square and the left (last) cross section

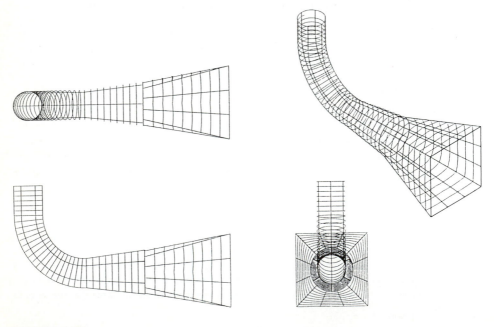

**FIGURE 6-62**
Duct geometric model.

Duct cross section
at valve location

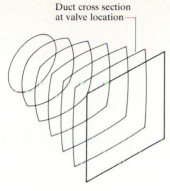 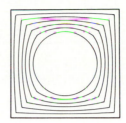

**FIGURE 6-63**
Cross sections of duct surface IV.

is circular as expected and gradual change in shape occurs in between. The areas, centroids, and perimeters of these cross sections can easily be calculated using the "calculate area" and "measure length" commands.

## PROBLEMS

### Part 1: Theory

**6.1.** Equation (6.36) gives the equation of a ruled surface joining two space curves. Find the equation of a ruled surface that passes through one space curve along a given set of direction vectors defining the surface rulings.

**6.2.** How can you implement Eq. (6.37) into a surface package?

**6.3.** Derive the parametric equations of an ellipsoid, a torus, and a circular cone.

**6.4.** Find the tangent and normal vectors to a surface of revolution in terms of its profile equation.

**6.5.** Repeat Prob. 6.4 but for a tabulated cylinder in terms of its directrix.

**6.6.** Prove that the curvature of a circular cylinder is zero. What is the radius of curvature at any point on its surface?

**6.7.** Derive Eq. (6.65) of Example 6.6 which states the planar condition of a bicubic surface patch.

**6.8.** Develop the relationships between the position, tangent, and twist vectors at the corner points of a bicubic surface patch and $C_{ij}$ coefficients used in Eq. (6.39).

**6.9.** Derive the conditions for $C^0$ and $C^1$ continuity of a cubic Bezier composite surface of two patches.

**6.10.** Show that a bilinear Coons patch cannot provide $C^1$ continuity between adjacent patches even if their boundary curves form a $C^1$ continuity network.

**6.11.** Derive Eq. (6.103) for a bicubic Coons patch whose boundary curves and cross-boundary derivatives are given.

**6.12.** Given a point $Q$ and a parametric surface in cartesian space (Fig. P6-12), find the closest point $P$ on the surface to $Q$.

  *Hint:* Find $P$ such that $(Q - P)$ is perpendicular to the tangent vectors $\mathbf{P}_u$ and $\mathbf{P}_v$ at $P$.

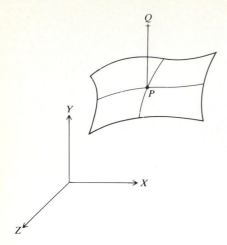

**FIGURE P6-12**

**6.13.** Find the minimum distance between:
   (*a*) A point and a surface
   (*b*) A curve and a surface
   (*c*) Two surfaces
**6.14.** Refer to the published literature and find algorithms for the intersection between:
   (*a*) A curve and a surface
   (*b*) Two surfaces

## Part 2: Laboratory

**6.15.** Create the surface models of the wireframe models of Prob. 5.20 in Chap. 5. Make a copy of each model database for surface construction.
   *Hint:* Use layers and colors provided by your CAD/CAM system in order to manage the amount of graphics displayed on the screen and to facilitate identifying each surface-generating curve.
**6.16.** Create the surface models of the additional geometric models shown in Fig. P6-16. Follow the part setup procedure on your CAD/CAM system. All dimensions are in inches.
**6.17.** Create the quadric (quadratic) surfaces shown in Fig. P6-17 on your CAD/CAM system.

## Part 3: Programming

Write a computer program to generate a:

**6.18.** Plane surface that passes through three points.
**6.19.** Ruled surface that connects two given rails.
**6.20.** Surface of revolution.
**6.21.** Tabulated cylinder.
**6.22.** Bicubic surface.
**6.23.** Cubic rectangular Bezier patch.
**6.24.** Open cubic B-spline surface patch.

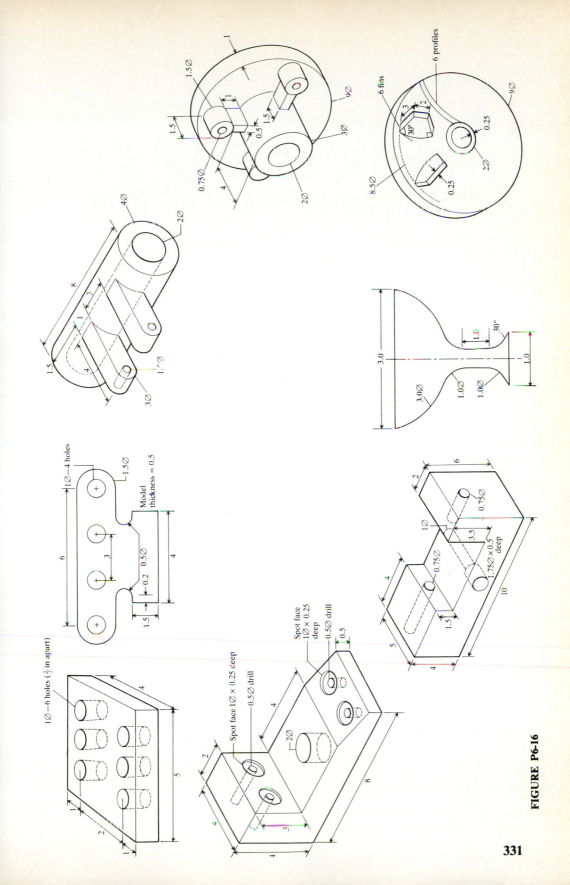

**FIGURE P6-16**

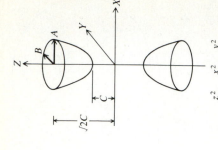

$$\frac{z^2}{C^2} - \frac{x^2}{A^2} - \frac{y^2}{B^2} = 1$$

Hyperboloid of two sheets

$$x^2 + y^2 + z^2 = R^2$$

Sphere

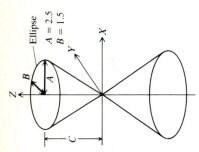

$$\frac{x^2}{A^2} + \frac{y^2}{B^2} = Cz$$

Elliptic paraboloid

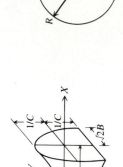

$$\frac{x^2}{A^2} - \frac{y^2}{B^2} = Cz$$

Hyperbolic paraboloid

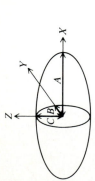

Ellipse

$A = 2.5$
$B = 1.5$

$A = 2.5, B = 1.5, C = 3$

$$\frac{x^2}{A^2} + \frac{y^2}{B^2} - \frac{z^2}{C^2} = 0$$

Elliptic cone

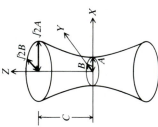

$$\frac{x^2}{A^2} + \frac{y^2}{B^2} - \frac{z^2}{C^2} = 1$$

Hyperboloid of one sheet

$A = 5, B = 1, C = 2$

$$\frac{x^2}{A^2} + \frac{y^2}{B^2} + \frac{z^2}{C^2} = 1$$

Ellipsoid

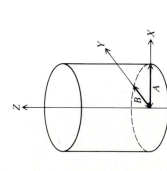

$$\frac{x^2}{A^2} + \frac{y^2}{B^2} = 1 \text{ for any } z \text{ value}$$

Elliptic cylinder

**FIGURE P6-17**

## BIBLIOGRAPHY

Barnhill, R. E.: "A Survey of the Representation and Design of Surfaces," *IEEE Computer Graphics and Applic. Mag.*, pp. 9–16, October 1983.

Barnhill, R. E.: "Surfaces in Computer Aided Geometric Design: A Survey with New Results," *Computer Aided Geometric Des. J.*, vol. 2, pp. 1–17, 1985.

Barnhill, R. E., and W. Boehm (Eds.): *Surfaces in CAGD'84*, Elsevier Science, 1985.

Beck, J. M., R. T. Farouki, and J. K. Hinds: "Surface Analysis Methods," *IEEE CG&A*, pp. 18–36, December 1986.

Bohm, W., G. Farin, and J. Kahmann: "A Survey of Curve and Surface Methods in CAGD," *CAGD J.*, vol. 1, pp. 1–60, 1984.

Brunet, P.: "Increasing the Smoothness of Bicubic Spline Surfaces," *CAGD J.*, vol. 2, pp. 157–164, 1985.

CAM-I, Inc.: "Common Normals for Arbitrary Parametric Surfaces," Document no. R-64-APT-03, Arlington, Tex., 1964.

CAM-I, Inc.: "Final Report on Parametric Surfaces," Document no. R-66-APT-02, Arlington, Tex., 1966.

CAM-I, Inc.: "Internal Data Structure Design for Sculptured Surfaces SS × 2 System," Document no. R-71-SS-02, Arlington, Tex., 1971.

CAM-I, Inc.: "Mathematical Specifications for Sculptured Surfaces Patches," Document no. R-71-SS-03, Arlington, Tex., 1971.

CAM-I, Inc.: "Sculptured Surfaces Phase IV," Document no. R-75-SS-01, Arlington, Tex., 1975.

CAM-I, Inc.: "Demonstration Test of Sculptured Surfaces," Document no. R-75-SS-03, Arlington, Tex., 1975.

CAM-I, Inc.: "Proceedings on Sculptured Surfaces," Document no. P-80-SS-01, Arlington, Tex., 1980.

CAM-I, Inc.: "Proceedings on Sculptured Surfaces," Document no. P-83-SS-01, Arlington, Tex., 1983.

CAM-I, Inc.: "Extended Geometric Facilities for Surface and Solid Modelers," Document no. R-83-GM-02.1, vols. 1, 2, 3, Arlington, Tex., 1983.

Chang, G., and Y. Feng: "An Improved Condition for the Convexity of Bernstein-Bezier Surfaces Over Triangles," *CAGD J.*, vol. 1, pp. 279–283, 1984.

Chang, G., and B. Su: "Families of Adjoint Patches for a Bezier Triangle Surface," *CAGD J.*, vol. 2, pp. 37–42, 1985.

Chasen, S. H.: *Geometric Principles and Procedures for Computer Graphics Applications*, Prentice-Hall, Englewood Cliffs, N.J., 1978.

Cohen, E., T. Lyche, and R. Riesenfeld: "Discrete Box Splines and Refinement Algorithms," *CAGD J.*, vol. 1, pp. 131–148, 1984.

Cottingham, M. S.: "A Compressed Data Structure for Surface Representation," *Computer Graphics Forum*, vol. 4, pp. 217–228, 1983.

Dahmen, W., and C. A. Micchelli: "Subdivision Algorithms for the Generation of Box Spline Surfaces," *CAGD J.*, vol. 1, pp. 115–129, 1984.

Dahmen, W., and C. A. Micchelli: "Line Average Algorithm: A Method for the Computer Generation of Smooth Surfaces," *CAGD J.*, vol. 2, pp. 77–85, 1985.

Dodd, S. L., D. F. McAllister, and J. A. Roulier: "Shape-Preserving Spline Interpolation for Specifying Bivariate Functions on Grids," *IEEE CG&A*, pp. 70–79, September 1983.

Encarnacao, J., and E. G. Schlechtendahl: *Computer Aided Design; Fundamentals and System Architectures*, Springer-Verlag, New York, 1983.

Faux, I. D., and J. J. Pratt: *Computational Geometry for Design and Manufacture*, Ellis Horwood (John Wiley), Chichester, West Sussex, 1979.

Foley, J. D., and A. van Dam: *Fundamentals of Interactive Computer Graphics*, Addison-Wesley, 1982.

Forrest, A. R.: "Curves and Surfaces for Computer-Aided Design," Ph.D. Thesis, University of Cambridge, 1968.

Giloi, W. K.: *Interactive Computer Graphics; Data Structures, Algorithms, Languages*, Prentice-Hall, Englewood Cliffs, N.J., 1978.

Miller, J. R.: "Sculptured Surfaces in Solid Models: Issues and Alternative Approaches," *IEEE CG&A*, pp. 37–48, December 1986.

Mortenson, M. E.: *Geometric Modeling*, John Wiley, New York, 1985.

Newman, W. A., and R. F. Sproull: *Principles of Interactive Computer Graphics*, 2d ed., McGraw-Hill, New York, 1979.

Piegl, L.: "Representation of Quadratic Primitives by Rational Polynomials," *CAGD J.*, vol. 2, pp. 151–155, 1985.

Piegl, L.: "The Sphere as a Rational Bezier Surface," *CAGD J.*, vol. 3, pp. 45–52, 1986.

Pratt, M. J.: "Smooth Parametric Surface Approximation to Discrete Data," *CAGD J.*, vol. 2, pp. 165–171, 1985.

Rogers, D. F., and J. A. Adams: *Mathematical Elements for Computer Graphics*, McGraw-Hill, New York, 1976.

Rogers, D. F., S. G. Satterfield, and F. Rodriguez: "Ship Hulls, B-Spline Surfaces, and CAD/CAM," *IEEE CG&A*, pp. 37–43, December 1983.

Satterfield, S. G., and D. F. Rogers: "A Procedure for Generating Contour Lines from a B-Spline Surface," *IEEE CG&A*, pp. 71–75, April 1985.

Sederberg, T. W.: "Piecewise Algebraic Surface Patches," *CAGD J.*, vol. 2, pp. 53–59, 1985.

Worsey, A. J.: "A Modified C2 Coon's Patch," *CAGD J.*, vol. 1, pp. 357–360, 1984.

# CHAPTER
# 7

# TYPES AND MATHEMATICAL REPRESENTATIONS OF SOLIDS

## 7.1 INTRODUCTION

Wireframe and surface geometric modeling techniques have been presented in Chaps. 5 and 6 respectively. This chapter presents the third modeling technique available to designers on a CAD/CAM system, that is, solid modeling. The use of solid modeling in design and manufacturing is increasing rapidly because of the reduced computing costs, fast computing hardware, improved user interfaces, increased capabilities of solid modeling itself, and software improvements. Some twenty modelers are commercially available in the United States, and a body of theory and technology continues to develop. It is forecast[1] that the total solid modeling market will grow steadily over the next few years reaching $2.9 billion by 1991 in comparison to $0.73 billion in 1985. This constitutes a growth in the solid software revenues from $45 million in 1985 to $892 million in 1991. Hardware running solid modeling software is projected to grow from $284 million in 1985 to $2 billion in 1991.

Solid modeling has been acknowledged as the technological solution to automating and integrating design and manufacturing functions. Indeed, the complete definition of part shape (geometry and topology) through solids models has been called a key to CIM. Programmable or flexible automation could very well be achieved via developing application algorithms that operate directly on

---

[1] Solid modeling market forecast is based on The Merrit Company's *Solid Modeling Today*, vol. 1, no. 1, May 1986.

solid modeling databases. Most of the solid modeling systems in the design/ manufacturing environment were installed to test the feasibility of integrating these two functions. However, recent use of these systems has started shifting toward the design and manufacturing of actual parts and assemblies. It is expected, though, that the original goal of automation and integration set for solid modeling will eventually be achieved.

Solid modeling techniques are based on informationally complete, valid, and unambiguous representations of objects. Simply stated, a complete geometric data representation of an object is one that enables points in space to be classified relative to the object, if it is inside, outside, or on the object. This classification is sometimes called spatial addressability. If completeness, validity, and unambiguity are not achieved formally by the geometric modeling technique, the technique has no other option but to depend on users to verify the creations of models interactively. Therefore, automation and integration of tasks such as interference analysis, mass property calculations, finite element modeling, computer aided process planning (CAPP), machine vision, and NC machining are not possible to achieve.

Solid modelers store more information (geometry and topology) than wireframe or surface modelers (geometry only). Both wireframe and surface models are incapable of handling spatial addressability as well as verifying that the model is well formed, the latter meaning that these models cannot verify whether two objects occupy the same space.

Other disadvantages of wireframe modeling have been discussed in Chap. 5. While surface models provide a precise definition of surfaces and can handle complex geometries, they are slow to render, are computationally intensive, and do not further CAD/CAM automation and integration goals. A shaded surface model is by no means considered a solid model. On the other hand, solid modeling produces accurate designs, provides complete three-dimensional definition, improves the quality of design, improves visualization, and has potential for functional automation and integration. However, solid modeling has some limitations. For example, it cannot automatically create other models from the solid definition and neither can it automatically use data created in other models to create a solid. In addition, solid modeling has not been proven for large-scale production applications. Other limitations such as slow rendering and computations as well as poor user interface are fading away with the rapid enhancement of both hardware and software.

Solid modeling (sometimes called volumetric modeling) techniques began to develop in the late 1960s and early 1970s. Early solid modeling projects appeared at this period of time in Europe, Japan, and the United States. Build-1 system, and later Build-2, was developed by Braid's CAD group in Cambridge, England, in 1973. TIPS-1 from Hakkaido University was publicized in 1973. In the mid 1970s, GLIDE-1 was developed by Eastman's group at Carnegie-Mellon University. Baumgart, from Stanford University, introduced Euler operators and a winged-edge polyhedron structure for boundary representation. The Production Automation Project (PAP) was founded at the University of Rochester in 1972. The PAP group headed by Voelcker and Requicha launched the research that led to the PADL-1 and PADL-2 systems. By the late 1970s, solid modeling had gained enough credibility to penetrate the commercial market. In 1980, Evans and Sutherland began to market Romulus; in 1981 Applicon and Computervision announced their SynthaVision-based and Solidesign systems respectively.

Until recently, solid modeling has been confined to mainframes and large minicomputers (early runs were in batch mode) primarily because of the extensive computations necessary to produce and render solids. By improving solid modeling algorithms, and with the design of frame buffers especially for solid modeling, microcomputer-based solid modeling systems are now available and expected to grow. These buffers are capable of storing enough information for two complete screen images (for animation) and can support 512 × 512 resolution with 16 bits per pixel. These bits can be partitioned as four bits for cursor and text and 12 bits of colors, thus allowing $4096(2^{12})$ colors to be displayed concurrently from a palette of 16.8 million, assuming 256 intensities for each primary color (refer to Chap. 2). In addition, the buffers do not have processors for computations which can, instead, be performed by the host microcomputer to keep hardware costs down.

User input required to create solid models on existing CAD/CAM systems depends on both the internal representation scheme used by each system as well as the user interface. It is crucial to distinguish between the user interface and the internal data representation of a given CAD/CAM system. The two are quite separate aspects of the systems and can be linked together by software that is transparent to the user. For example, a system that has a B-rep (boundary representation) internal data representation may use a CSG (constructive solid geometry)-oriented user interface; that is, input a solid model by its primitives. Most systems use the building-block approach (CSG oriented) and sweep operations as the basis for user interface. Some early user interfaces were based on the boundary representation scheme and used commands such as "make edge/face," "kill edge/face," etc. Such interfaces are not efficient. Object-oriented user interfaces (input a solid by its features) are more acceptable by users. For example, a user can create a hole in a block using a command such as "create hole" instead of the "subtract cylinder" command. In order to best visualize solids on a graphics display, a shaded image is usually displayed. Some systems, especially those based on the boundary representation, are also available to display a mesh on the solid as in the case of surface models.

This chapter covers the theoretical and practical aspects of solids. Throughout the chapter, related issues to constructing solids on CAD/CAM systems are covered. Sections 7.2 to 7.5 are directly related to practicing solid modeling theory. Sections 7.6 to 7.14 discuss the mathematical representations of solids and other important topics. Section 7.15 applies the chapter material to design and engineering applications.

## 7.2 SOLID MODELS

A solid model of an object is a more complete representation than its surface model. It is unique from the latter in the topological information it stores which potentially permits functional automation and integration. For example, the mass property calculations or finite element mesh generation of an object can be performed fully automatically, at least in theory, without any user intervention. Typically, a solids model consists of both the topological and geometrical data of its corresponding object.

Defining an object with a solid model is the easiest of the available three modeling techniques (curves, surfaces, and solids). Solid models can be quickly created without having to define individual locations as with wireframes. In many cases, solid models are easier to build than wireframe or surface models. For example, representing the intersection of two cylinders using wireframe modeling is not possible unless points on the intersection curve are evaluated in order to be input to a CAD/CAM system and connected with a B-spline curve. Another example is shown in Fig. 7-1. Using solid modeling, the object shown can be created as a block with six cylinders subtracted from it. It is a cumbersome and lengthy process to create the surface model of the same object although only ruled surfaces are needed (refer to Prob. 6.16).

The completeness and unambiguity of solid models are attributed to the information that the related databases of these models store. Unlike wireframe and surface models, which contain only geometric data, solid models contain both geometric data and topological information of the corresponding objects. The difference between geometry and topology is illustrated in Fig. 7-2. Geometry (sometimes called metric information) is the actual dimensions that define the entities of the object. The geometry that defines the object shown in Fig. 7-2 is the lengths of lines $L_1$, $L_2$, and $L_3$, the angles between the lines, and the radius $R$ and the center $P_1$ of the half-circle. Topology (sometimes called combinatorial structure), on the other hand, is the connectivity and associativity of the object entities. It has to do with the notion of neighborhood; that is, it determines the relational information between object entities. The topology of the object shown in Fig. 7-2b can be stated as follows: $L_1$ shares a vertex (point) with $L_2$ and $C_1$, $L_2$ shares a vertex with $L_1$ and $L_3$, $L_3$ shares a vertex with $L_2$ and $C_1$, $L_1$ and $L_3$ do not overlap, and $P_1$ lies outside the object. Based on these definitions, neither geometry nor topology alone can completely model objects. Wireframe and surface models deal only with geometrical information of objects, and are therefore considered incomplete and ambiguous. From a user point of view, geometry is visible and topology is considered to be nongraphical relational information that is stored in solid model databases and are not visible to users.

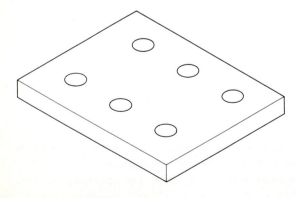

**FIGURE 7-1**
A typical solid model.

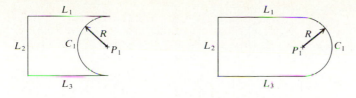

(a) Same geometry but different topology

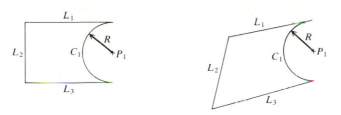

(b) Same topology but different geometry

**FIGURE 7-2**
Difference between geometry and topology of an object.

For automation and integration purposes, solid models must be accurate. Although accurate models are not a necessity during conceptual design, they are needed for analysis and application algorithms that work off the solid model. Accuracy and speed of creation of a solid model depend directly on the representation scheme, and consequently the data stored in the database of the model. The various available schemes are discussed later in this chapter. Each of those schemes has its own advantages and disadvantages, depending on the application. For example, B-rep modelers can better represent general shapes but usually require more processing time. In contrast, CSG models are easier to build and better suited for display purposes. However, it may be difficult to define a complex shape.

In constructing a solid model on a CAD/CAM system, the user should follow the modeling guidelines discussed in Chap. 3. All the design tools provided by these systems and covered in Part IV of the book, excluding the geometric modifiers, are applicable to solid models. Practically, it might be more convenient to construct solid models in isometric views to enable clear display and visualization of the solid as it is being constructed. It is also recommended that solid entities (primitives) as well as intermediate solids be placed on different layers to allow convenient reference to them during the construction process. A mesh similar to that used with surface models can be added to B-rep-based solid models after they are created. However, solid models are better visualized via shading. Finally, it should be noted that most user interfaces available to input solids have compatibility for CSG input. Such compatibility does not reflect the internal core representation scheme implemented in a particular solid modeling package, and users must consult with the package developers if they wish to know that information.

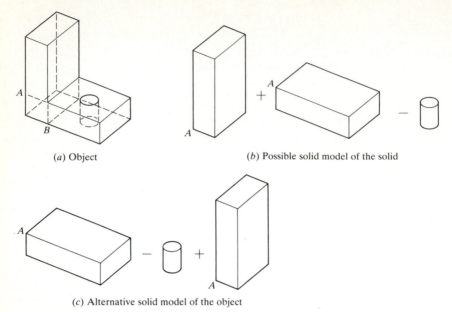

(a) Object                                    (b) Possible solid model of the solid

(c) Alternative solid model of the object

**FIGURE 7-3**
Nonuniqueness of solid model of an object.

While solid models are complete and unambiguous, they are not unique. An object may be constructed in various ways. Consider the solid shown in Fig. 7-3. One can construct the solid model of the object shown by extending the horizontal block to point $A$, add two blocks, and subtract a cylinder as shown in Fig. 7-3$b$. Another alternative is shown in Fig. 7-3$c$, where the subtraction is performed first followed by the addition. Other possibilities exist including extending the vertical block to point $B$ instead and repeating the same two alternatives. Regardless of the order and method of construction as well as the representation scheme utilized, the resulting solid model of the object is always complete and unambiguous. However, there will always be a more efficient way than others to construct the solid models as in the case with wireframe and surface models.

Users are now more aware of the potential benefits of solid models. Consequently CAD/CAM vendors are investing more resources into developing solid modeling. However, most existing CAD/CAM systems offer solid modeling as packages that are not linked to wireframe or surface capabilities offered by these systems. It is expected, though, that the next generation of these systems will be based on solid modeling if it matures and proves useful in the production environment.

## 7.3 SOLID ENTITIES

Most commercially available solid modeling packages have a CSG-compatible user input and therefore provide users with a certain set of building blocks, often called primitives. Primitives are simple basic shapes and are considered the solid modeling entities which can be combined by a mathematical set of boolean operations to create the solid. Primitives themselves are considered valid "off-the-shelf" solids. In addition, some packages, especially those that support sweeping operations, permit users to utilize wireframe entities to create faces that are swept later to create solids. The user usually positions primitives as required before applying boolean operations to construct the final solid.

There is a wide variety of primitives available commercially to users. However, the four most commonly used are the block, cylinder, cone, and sphere. These are based on the four natural quadrics: planes, cylinders, cones, and spheres. For example, the block is formed by intersecting six planes. These quadrics are considered natural because they represent the most commonly occurring surfaces in mechanical design which can be produced by rolling, turning, milling, cutting, drilling, and other machining operations used in industry. Planar surfaces result from rolling, chamfering, and milling; cylindrical surfaces from turning or filleting; spherical surfaces from cutting with a ball-end cutting tool; conical surfaces from turning as well as from drill tips and countersinks. Natural quadrics are distinguished by the fact that they are combinations of linear motion and rotation. Other surfaces, except the torus, require at least dual axis control.

From a user-input point of view and regardless of a specific system syntax, a primitive requires a set of location data, a set of geometric data, and a set of orientation data to define it completely. Location data entails a primitive local coordinate system and an input point defining its origin. Geometrical data differs from one primitive to another and are user-input. Orientation data is typically used to orient primitives properly relative to the MCS or WCS of the solid model under construction. Primitives are usually translated and/or rotated to position and orient them properly before applying boolean operations. Following are descriptions of the most commonly used primitives (refer to Fig. 7-4):

1. *Block.* This is a box whose geometrical data is its width, height, and depth. Its local coordinate system $X_L Y_L Z_L$ is shown in Fig. 7-4. Point $P$ defines the origin of the $X_L Y_L Z_L$ system. The signs of $W$, $H$, and $D$ determine the position of the block relative to its coordinate system. For example, a block with a negative value of $W$ is displayed as if the block shown in Fig. 7-4 is mirrored about the $Y_L Z_L$ plane.

2. *Cylinder.* This primitive is a right circular cylinder whose geometry is defined by its radius (or diameter) $R$ and length $H$. The length $H$ is usually taken along the direction of the $Z_L$ axis. $H$ can be positive or negative.

3. *Cone.* This is a right circular cone or a frustum of a right circular cone whose base radius $R$, top radius (for truncated cone), and height $H$ are user-defined.

4. *Sphere.* This is defined by its radius or diameter and is centered about the origin of its local coordinate system.

5. *Wedge.* This is a right-angled wedge whose height $H$, width $W$, and base depth $D$ form its geometric data.

6. *Torus.* This primitive is generated by the revolution of a circle about an axis lying in its plane ($Z_L$ axis in Fig. 7-4). The torus geometry can be defined by the radius (or diameter) of its body $R_1$ and the radius (or diameter) of the centerline of the torus body $R_2$, or the geometry can be defined by the inner radius (or diameter) $R_I$ and outer radius (or diameter) $R_O$.

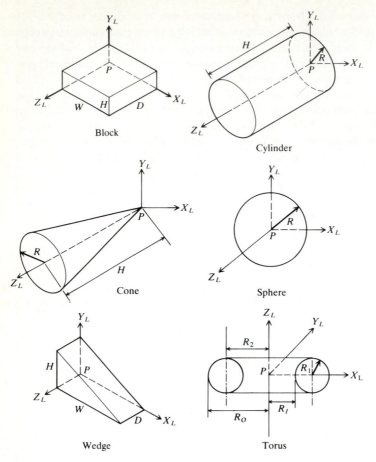

**FIGURE 7-4**
Most common primitives.

For all the above primitives, there are default values for the data defining their geometries. Most packages use default values of 1. In addition, the local coordinate systems for the various primitives shown in Fig. 7-4 may change from one package to another. Some packages assume that the origin, $P$, of the local coordinate system is coincident with that of the MCS or WCS and require the user to translate the primitive to the desired location, thus eliminating the input of point $P$ by the user.

Two or more primitives can be combined to form the desired solid. To ensure the validity of the resulting solid, the allowed combinatorial relationships between primitives are achieved via boolean (or set) operations. The available boolean operators are union ($\cup$ or $+$), intersection ($\cap$ or I), and difference ($-$). The union operator is used to combine or add together two objects or primitives. Intersecting two primitives gives a shape equal to their common volume. The difference operator is used to subtract one object from the other and results in a shape equal to the difference in their volumes. Figure 7-5 shows boolean operations of a block $A$ and a cylinder $B$.

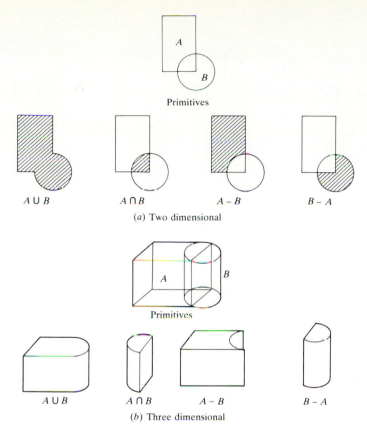

**FIGURE 7-5**
Boolean operations of a block $A$ and cylinder $B$.

**Example 7.1.** Create the solid model of the guide bracket shown in Fig. 5-2.

**Solution.** The creation of the solid model of the guide bracket is much simpler than its wireframe and surface models created in Examples 5.1 and 6.1 respectively. In fact, combinations of blocks and cylinders are all that is needed to create the solid model. While translational sweep can be used to create the solid model, it is not discussed in this example and is left to the reader as an exercise. The following steps may be followed to construct the solid model:

1. Follow the setup procedure discussed in Chap. 3.
2. To create the upper part of the object, create a block of size $2 \times 1 \times 0.25$, and two cylinders of sizes $R = 1.0$, $H = 0.25$ and $R = 0.5$, $H = 0.25$. Create another block of size $0.5 \times 0.5 \times 0.25$ and rotate it $45°$ about the $Z$ axis (assuming the MCS shown in Fig. 5-2 is used here). These primitives are combined to produce the upper part as shown by branch 1 of the tree shown in Fig. 7-6a. The locations and orientations of these primitives can be easily done and are not discussed here.
3. In a similar fashion, branches 2, 3, and 4 of the tree show how to create the lower part of the object, the left flange, and the right flange respectively.
4. The union of all the four branches produces the final solid model.

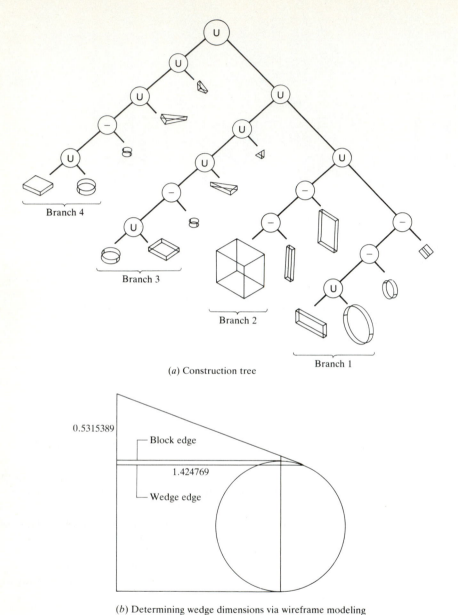

(a) Construction tree

(b) Determining wedge dimensions via wireframe modeling

**FIGURE 7-6**
Solid model of the guide bracket of Example 5.1.

In the above steps, the dimensions of the horizontal wedges used to create the flange are obtained using a wireframe construction, as shown in Fig. 7-6b. Figure 7-6c and d shows the primitives used to create the model and the final solid model respectively.

It is useful in practice to place primitives, intermediate solids, and the final solid in the above example on separate layers. This makes management of creating the solid model easier. For example, if two blocks are added and the result is placed

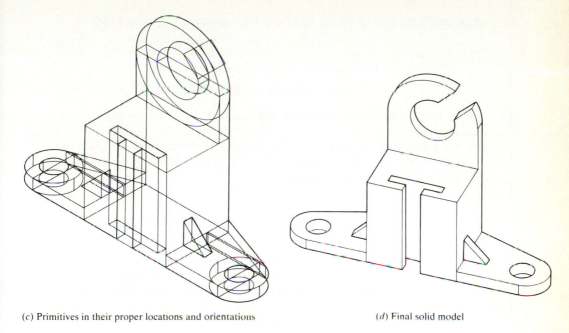

($c$) Primitives in their proper locations and orientations          ($d$) Final solid model

**FIGURE 7-6 (*continued*)**

on the same layer, distinguishing between the three solids becomes difficult and requires some of them to be blanked. A need to reuse a primitive after repositioning it might arise, in which case having it on a different layer is very helpful.

The above example illustrates most of the experiences encountered in creating solid models that do not have sculptured surfaces. If the reader were to create the guide bracket on a CAD/CAM system and keep track of the amount of time and effort needed to create the wireframe, surface, and solid models of the object, the latter would clearly require the minimum of both time and effort. This results from the richness of information embedded in defining the primitives and boolean operations. Consequently, solid models appeal strongly to engineering, design, and manufacturing applications.

## 7.4  SOLID REPRESENTATION

Solid representation of an object can support reliably and automatically, at least in theory, related design and manufacturing applications due to its informational completeness. Such representation is based fundamentally on the notion that a physical object divides an $n$-dimensional space, $E^n$, into two regions: interior and exterior separated by the object boundaries. A region is defined as a portion of space $E^n$ and the boundary of a region is a closed surface, as in the case of a sphere, or a collection of open surfaces connected at proper edges, as in the case of a box.

In terms of the above notion, a solid model of an object is defined mathematically as a point set $S$ in three-dimensional euclidean space ($E^3$). If we denote the interior and boundary of the set by iS and bS respectively, we can write

$$S = iS \cup bS \qquad (7.1)$$

and if we let the exterior be defined by cS (complement of S), then

$$W = iS \cup bS \cup cS \qquad (7.2)$$

where $W$ is the universal set, which in the case of $E^3$ is all possible three-dimensional points.

The solid definition given by Eq. (7.1) introduces the concept of geometric closure which implies that the interior of the solid is geometrically closed by its boundaries. Thus, Eq. (7.1) can be rewritten as

$$S = kS \qquad (7.3)$$

where kS is the closure of the solid or point set S, and is given by the right-hand side of Eq. (7.1); that is, $kS = iS \cup bS$.

Figure 7-7 shows the geometric explanation of Eqs. (7.1) to (7.3). It should be noted here that both wireframe and surface models lack geometric closure which is the main reason for their incompleteness and ambiguity. Based on Eq. (7.1), an object is represented by bS (its boundary) only in both modeling techniques. In the wireframe technique, bS represents $E^3$ curves that occupy one-dimensional parametric regions while it represents $E^3$ surfaces that occupy two-dimensional regions in surface modeling.

The foundations of formalizing the solid modeling theory have been well established by the Requicha and Voelcker research group (PADL-1 and PADL-2 authors and developers) and others. The successful representation of solid models in computers and their utilization in engineering applications depend on their properties as well as the properties of the schemes representing them. In the context of the solid modeling theory, the solid model (sometimes called the

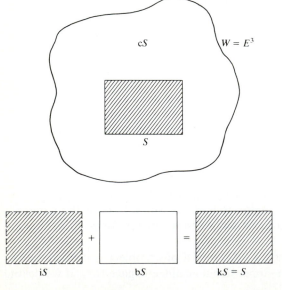

**FIGURE 7-7**
Solid and geometric closure definitions.

abstract solid) is considered the mathematical model of the real object (sometimes called the physical solid). The properties of this mathematical model determine its behavior when geometric algorithms manipulate its related data structure. The importance of these properties can perhaps be realized if we relate to other classical engineering fields. As a dynamic model of a car dictates the relevance of the associated equilibrium equation and its results, a mathematical model of an object decides the class of algorithms that can be applied to it and the level of their automation.

The properties that a solid model or an abstract solid should capture mathematically can be stated as follows:

1. Rigidity. This implies that the shape of a solid model is invariant and does not depend on the model location or orientation in space.
2. Homogeneous three-dimensionality. Solid boundaries must be in contact with the interior. No isolated or dangling boundaries (see Fig. 7-8) should be permitted.
3. Finiteness and finite describability. The former property means that the size of the solid is not infinite while the latter ensures that a limited amount of information can describe the solid. The latter property is needed in order to be able to store solid models into computers whose storage space is always limited. It should be noted that the former property does not include the latter and vice versa. For example, a cylinder which may have a finite radius and length may be described by an infinite number of planar faces.
4. Closure under rigid motion and regularized boolean operations. This property ensures that manipulation of solids by moving them in space or changing them via boolean operations must produce other valid solids.
5. Boundary determinism. The boundary of a solid must contain the solid and hence must determine distinctively the interior of the solid.

The mathematical implication of the above properties suggests that valid solid models are bounded, closed, regular, and semi-analytic subsets of $E^3$. These subsets are called r-sets (regularized sets). Intuitively, r-sets are "curved

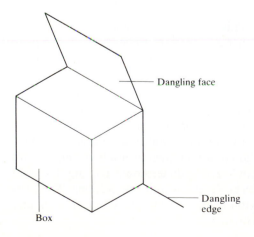

Dangling face

Dangling edge

Box

**FIGURE 7-8**
Example of isolated boundaries.

polyhedra" with "well-behaved" boundaries. The point set $S$ that defines a solid model and is given by Eq. (7.1) is always an r-set. Intuitively, a "closed regular set" means that the set is closed and has no dangling portions, as shown in Fig. 7-8, and a "semi-analytic set" means that the set does not oscillate infinitely fast anywhere within the set. The concept of "semi-analytic set" is important in choosing equations to describe surfaces or primitives of solid models. For example, the point set that satisfies $\sin(x) < 0$ is a semi-analytic set while the set that satisfies $\sin(1/x) < 0$ is not because the function $\sin(1/x)$ oscillates fast when $x$ approaches zero.

Having discussed the desired properties of solid models, let us discuss the properties of representation schemes that usually operate on point sets or r-sets to produce valid solid models. A representation scheme is defined as a relation that maps a valid point set into a valid model. For example, a CSG scheme maps valid primitives into valid solids via boolean operations. Informally, a representation scheme is unambiguous or complete, and unique if one model produced by the scheme represents one and only one object, that is, one-to-one mapping. A scheme is unambiguous or complete, but not unique, if more than one model can represent the object (refer to Fig. 7-3). On the other hand, a scheme is ambiguous or incomplete if one model can represent more than one object, as in the case of wireframe models. Figure 7-9 shows these various schemes.

The formal properties of representation schemes which determine their usefulness in geometric modeling can be stated as follows:

1. Domain. The domain of a representation scheme is the class of objects that the scheme can represent or it is the geometric coverage of the scheme.
2. Validity. The validity of a representation scheme is determined by its range, that is, the set of valid representations or models it can produce. If a scheme produces an invalid model, the CAD/CAM system in use may crash or the model database may be lost or corrupted if an algorithm is invoked on the model database. Validity checks can be achieved in three ways: test the resulting databases via a given algorithm, build checks into the scheme generator itself, or design scheme elements (such as primitives) that can be manipulated via a given syntax.
3. Completeness or unambiguousness. This property determines the ability of the scheme to support analysis and other engineering applications. A complete scheme must provide models with sufficient data for any geometric calculation to be performed on them.
4. Uniqueness. This property is useful to determine object equality. It is a custom in algebra to check for uniqueness but it is rare to do so in geometry. This is because it is difficult to develop algorithms to detect the equivalence of two objects and it is computationally expensive to implement these algorithms if they exist. Positional and permutational nonuniqueness are two simple cases shown in Fig. 7-10. Figure 7-10a shows a two-dimensional rectangular solid (of side lengths $a$ and $b$) in two different positions and orientations. The two-dimensional solid $S$ shown in Fig. 7-10b is divided into three blocks $A$, $B$, and $C$ that can be unioned in a different order.

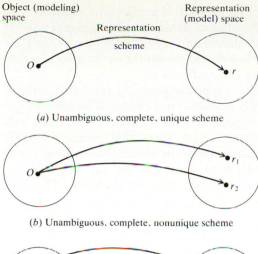

Object (modeling) space

Representation (model) space

Representation scheme

$O$

$r$

(a) Unambiguous, complete, unique scheme

$O$

$r_1$

$r_2$

(b) Unambiguous, complete, nonunique scheme

$O_1$

$O_2$

$r$

(c) Ambiguous and incomplete scheme

**FIGURE 7-9**
Classification of representation schemes.

There are other properties of representation schemes such as conciseness, ease of creation, and efficacy in the context of applications. These properties cannot be formalized and therefore are considered informal. Conciseness is a measure of the size of data a scheme requires to describe an object. Concise representation schemes generate compact databases that contain few redundant data, are convenient to store, and are efficient to transmit over data links (networks) from one system to another. Selectively imposed redundancy may save computational time and may increase the number of application algorithms that can utilize the stored data. Ease of creation of a representation is important to users and determines the user-friendliness of a scheme to a great extent. This is why most existing solid modelers have a CSG-compatible user input because it is

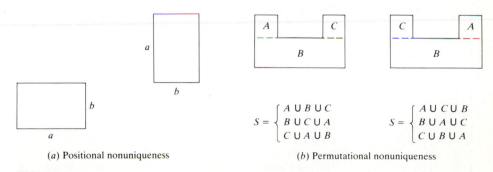

$a$

$b$

$b$

$a$

(a) Positional nonuniqueness

$A$  $C$

$B$

$$S = \begin{cases} A \cup B \cup C \\ B \cup C \cup A \\ C \cup A \cup B \end{cases}$$

$C$  $A$

$B$

$$S = \begin{cases} A \cup C \cup B \\ B \cup A \cup C \\ C \cup B \cup A \end{cases}$$

(b) Permutational nonuniqueness

**FIGURE 7-10**
Positional and permutational nonuniqueness.

concise and easy to create. Efficacy in the context of applications measures how accessible a representation is by downstream applications. Representations of objects in themselves are useless and should be viewed as sources of data for algorithms. Good representation schemes should permit the use of a wide variety of application algorithms for evaluating various functions.

Various representation schemes have been designed and developed, with the above properties in mind, to create solid models of real objects. Nine schemes can be identified. Some of them are more popular than the others. These are half-spaces, boundary representation (B-rep), constructive solid geometry (CSG), sweeping, analytic solid modeling, cell decomposition, spatial enumeration, octree encoding, and primitive instancing. Each of these schemes has its properties, advantages, and disadvantages that are discussed later in the chapter. The three most popular schemes are B-rep, CSG, and sweeping. Most existing solid modeling packages or systems use one or more of the known schemes. Table 7.1 lists some of the existing solid modelers with their core representation scheme. In most packages or systems, one scheme is considered the primary representation

**TABLE 7.1**
**Some available solid modelers**

| Modeler | Vendor | Primary representation scheme | | User modeling input based on | |
|---|---|---|---|---|---|
| | | B-rep | CSG | B-rep | CSG |
| BMOD | Auto-trol | × | | × | |
| CATIA | IBM | × | | × | × |
| CMOD | Auto-trol | | × | | × |
| DDM SOLIDS | GE Calma | × | | × | × |
| EUCLID | Matra Datavision | × | | × | × |
| GEMSMITH | Vulcan | × | | × | × |
| GEOMED | SDRC | × | | × | × |
| GEOMETRIC MODELING SYSTEM | Graftek | × | | × | |
| ICEM | CDC | | × | | × |
| ICM GMS | ICM | × | | × | × |
| INSIGHT | Phoenix Data Systems | × | | × | × |
| MEDUSA | Prime Computer | × | | × | × |
| PADL-2 | Cornell University | | × | | × |
| PATRAN-G | PDA Engineering | ASM† | | Hyper-patches | × |
| ROMULUS | Evans and Sutherland | × | | × | |
| SOLIDESIGN | Computervision | × | | × | × |
| SOLIDS MODELING II | Applicon | | × | × | × |
| SOLID MODELING SYSTEM | Intergraph | × | | × | × |
| SYNTHVISION | MAGI | | × | × | × |
| TIPS-1 | CAM-I | | × | | × |
| UNIS-CAD | Sperry Univac | | × | | × |
| UNISOLIDS | McDonell Douglas | | × | × | × |

† Analytic solid modeling (see Sec. 7.10 for details).

scheme that can be converted to others or other schemes can be converted to it. For example, a CSG-based system utilizes the CSG scheme as the core of its geometric modeling engine which, in turn, can be converted into a B-rep for application purposes, or sweeping can be converted into a B-rep or CSG by a package whose core representation is one of these two. Conversion between various schemes is not always possible and depends primarily on how data is stored. Conversion from CSG to B-rep, octree, or spatial enumeration is possible. However, converting B-rep to CSG is not well known (conversion in two dimensions is known). Simple sweep can be converted to B-rep, CSG, or cell decomposition.

The major geometric procedures needed for solid modeling, regardless of any representation scheme, are curve/curve, curve/surface, and surface/surface intersection calculations. Conceptually, any geometric entity (a primitive or a surface) could be added to any representation scheme to increase its modeling domain. However, unless such a scheme can support these intersections of the entity, its use in modeling and applications becomes useless. Support of sculptured surface geometries by solid modeling depends on developing efficient methods to perform the intersection calculations for these geometries. Once these methods are available, solid modeling systems would cater to both solid and surface modeling within the same conceptual and algorithmic framework.

Representations of solids are built and invoked via algorithms (sometimes called processors). Informally, an algorithm is a procedure that takes certain input and produces a desired output. Algorithms should be developed carefully and tested for a wide variety of input to ensure their generality, reliability, and consistency. Algorithms can be classified into three types according to their input and output. Some algorithms take data and produce representations; that is, a: data → rep (reads as algorithm a is defined as taking data and producing representation). These algorithms build, maintain, and manage representations. Representation schemes mentioned above fall into this type. The other type of algorithms compute property values by taking a representation and producing data; that is, a: rep → data. All application algorithms belong to this type. For example, a mass property algorithm takes a solid model representation and produces volume, mass, and inertial properties. Algorithms of the third type take representations and produce representations; that is, a: rep → rep. For example, an algorithm that converts CSG to B-rep or one that simulates (models) processes (such as motion or machining) on objects belongs to this type. An algorithm might take a piece of stock and end up with a machined part. Figure 7-11 illustrates the three types of algorithms.

Conversion of solid models into wireframe models or an edge representation is well understood and is used to generate orthographic views for display and drafting purposes. While the views are generated automatically, they are not dimensioned automatically, and a manual or semi-automatic dimensioning is required. However, the opposite problem of creating solid models from wireframe models or from existing orthographic views or drawings is largely unsolved. Mathematically, this is a problem of converting an edge representation into a solid representation. This problem is not complete or well defined due to two reasons. First, edges of curved solids (curved polyhedra) may not be easily found

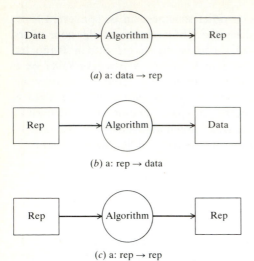

(a) a: data → rep

(b) a: rep → data

(c) a: rep → rep

**FIGURE 7-11**
Types of solid modeling algorithms.

from a finite number of projections. Second, the edge representation itself is ambiguous and can correspond to more than one object. Algorithms for dis-ambiguating wireframe models exist. These algorithms find all possible objects that correspond to one drawing. The main thrust to convert drawings to solid models stems from the large existing industrial base of wireframe models.

## 7.5  FUNDAMENTALS OF SOLID MODELING

Before covering the details of the various representation schemes, it is appropri-ate to discuss the details of some of the underlying fundamentals of solid model-ing theory. These are geometry, topology, geometric closure, set theory, regularization of set operations, set membership classification, and neighborhood. Geometry and topology have been covered in Sec. 7.2 and geometric closure is introduced in Sec. 7.4. This section covers set theory, regularization, classi-fication, and neighborhood. The significance of these topics to solid modeling stems from the definition of a solid model as a point set in $E^3$ as given in Eq. (7.1). They provide good rigorous mathematical foundations for developing and analyzing solids.

### 7.5.1  Set Theory

We begin the review of set theory by introducing some definitions followed by set algebra (operations on sets) and laws (properties) of the algebra of sets. At the end, the concept of ordered pairs and cartesian product is introduced. A set is defined as a collection or aggregate of objects. The objects that belong to the set are called the elements or members of the set. For example, the digits 0, 1, ..., 9 form a set (set of digits) $D$ whose elements are 0, 1, ..., 9. While the concept is relatively simple, the elements of a set must satisfy certain requirements. First, the elements must be well defined to determine unequivocally whether or not any object belongs to the set; that is, fuzzy sets are excluded. Second, the elements of a set must be distinct and no element may appear twice. Third, the order of the elements within the set must be immaterial. To realize the importance of these requirements in geometric modeling, the reader can apply them to a point set of eight elements which are the corner points of a block.

The elements of a set can be designated by one of two methods: the roster method or the descriptive method. The former involves listing within braces all the elements of the set and the latter involves describing the condition(s) that every element in the set must meet. The set of digits $D$ can be written using the roster and the descriptive methods respectively as

$$D = \{0, 1, 2, 3, 4, 5, 6, 7, 8, 9\} \tag{7.4}$$

and
$$D = \{x: x = 0, 1, 2, 3, 4, 5, 6, 7, 8, 9\} \tag{7.5}$$

Equation (7.4) reads as "$D$ is equal to the set of elements 0, 1, 2, 3, 4, 5, 6, 7, 8, 9." Equation (7.5) reads as "$D$ is equal to the set of elements $x$ such that $x$ equals 0, 1, 2, 3, 4, 5, 6, 7, 8, 9." The colon in Eq. (7.5) is sometimes replaced by a vertical bar, that is, $D = \{x \mid x = 0, 1, \ldots, 9\}$. Regardless of set designation, set membership and nonmembership is customarily indicated by $\in$ and $\notin$ respectively. If we write $9 \in D$, we mean 9 is an element (or member) of the set of digits $D$ or 9 belongs to $D$. Similarly, $-2 \notin D$ means that $-2$ is not an element of $D$.

Two sets $P$ and $Q$ are equal, written $P = Q$, if the two sets contain exactly the same elements. For example, the two sets $P = \{1, 3, 5, 7\}$ and $Q = \{1, 5, 7, 3\}$ are equal, since every element in $P$ is in $Q$ and every element in $Q$ is in $P$. The inequality is denoted by $\neq$ ($P \neq Q$ reads "$P$ does not equal $Q$").

A set $R$ is a subset of another set $S$ if every element in $R$ is in $S$. The notation for subset is $\subseteq$ and $R \subseteq S$ reads "$R$ is a subset of $S$." Analogous to $\in$ and $\notin$, the notation for not subset is $\nsubseteq$. If it happens that all elements in $R$ are in $S$ but all elements in $S$ are not in $R$, then $R$ is called a proper subset of $S$ and is written $R \subset S$. This means that for $R$ to be a proper subset of $S$, $S$ must have all elements of $R$ plus at least one element that is not in $R$. For example, given $S = \{1, 3, 5, 7\}$, then $R = \{1, 3, 5, 7\}$ is a subset of $S$ and $R = \{5, 7\}$ is a proper subset of $S$. Formally, $R \subset S \Leftrightarrow R \cap S = R$ and $R \neq S$ ($\Leftrightarrow$ reads "if and only if") or $R \subset S \Leftrightarrow R \cup S = S$ and $R \neq S$.

There are two sets that usually come to mind when discussing sets and subsets. The universal set $W$ is a set that contains all the elements that the analyst wishes to consider. It is problem-dependent. In solid modeling, $W$ contains $E^3$ and all points in $E^3$ are the elements of $W$. In contrast the null (sometimes referred to as the empty) set is defined as a set that has no elements or members. It is designated by the null set symbol $\varnothing$. The null set is analogous to zero in ordinary algebra.

Having introduced the required definitions, we now discuss set algebra. Set algebra consists of certain operations that can be performed on sets to produce other sets. These operations are simple in themselves but are powerful when combined with the laws of set algebra to solve geometric modeling problems. The operations are most easily illustrated through use of the Venn diagram named after the English logician John Venn. It consists of a rectangle that conceptually represents the universal set. Subsets of the universal set are represented by circles drawn within the rectangle or the universal set.

The three essential set operations are complement, union, and intersection. The complement of $P$, denoted by $cP$ (reads "$P$ complement"), is the subset of elements of $W$ that are not members of $P$, that is,

$$cP = \{x: x \notin P\} \tag{7.6}$$

The shaded portion of the Venn diagram in Fig. 7-12a shows the complement of $P$.

The union of two sets $P \cup Q$ (read "$P$ union $Q$") is the subset of elements of $W$ that are members of either $P$ or $Q$, that is,

$$P \cup Q = \{x: x \in P \text{ or } x \in Q\} \tag{7.7}$$

The union is shown in Fig. 7-12b as the shaded area.

The intersection of two sets $P \cap Q$ (read "$P$ intersect $Q$") is the subset of elements of $W$ that are simultaneously elements of both $P$ and $Q$, that is,

$$P \cap Q = \{x: x \in P \text{ and } x \in Q\} \tag{7.8}$$

The shaded portion in Fig. 7-12c shows the intersection of $P$ and $Q$. It is easy to realize that $P \cap W = P$ and $P \cap cP = \varnothing$. Sets that have no common elements are termed disjoint or mutually exclusive.

Two additional set operators that can be derived from the above set operations are difference and exclusive union. The difference of two sets $P - Q$ (read "$P$ minus $Q$") is the subset of elements of $W$ that belong to $P$ and not $Q$, that is,

$$P - Q = \{x: x \in P \text{ and } x \notin Q\} \tag{7.9}$$

or

$$Q - P = \{x: x \in Q \text{ and } x \notin P\} \tag{7.10}$$

Figure 7-12d and e shows the difference operator. The difference can also be expressed as

$$P - Q = P \cap cQ \tag{7.11}$$

(a) Complementation ($cP$)

(b) Union ($P \cup Q$)

(c) Intersection ($P \cap Q$)

(d) Difference ($P - Q$)

(e) Difference ($Q - P$)

(f) Exclusive union ($P \cup Q$) or symmetric difference ($P \Delta Q$)

**FIGURE 7-12**
Venn diagram of set algebra.

The exclusive union (also known as symmetric difference) of two sets $P \cup Q$ (also written as $P \Delta Q$) is the subset of elements of $W$ that are members of $P$ or $Q$ but not of both, that is,

$$P \cup Q = \{x : x \notin P \cap Q\} \tag{7.12}$$

Figure 7-12f shows the exclusive union. Using the Venn diagram it can be shown that $P \cup Q$ can also be expressed as $c(P \cap Q) \cap (P \cup Q)$, $(P \cap cQ) \cup (cP \cap Q)$, $(P - Q) \cup (Q - P)$, or $(P \cup Q) - (P \cap Q)$.

The laws of set algebra are in some cases similar to the laws of ordinary algebra. Just as the latter can be used to simplify algebraic equations and expressions, the former can be used to simplify sets. The laws of set algebra are stated here without any mathematical proofs. Interested readers can prove most of them using the Venn diagram. These laws are:

the commutative law (similar to ordinary algebra $p + q = q + p$ and $pq = qp$):

$$P \cup Q = Q \cup P \tag{7.13}$$

$$P \cap Q = Q \cap P \tag{7.14}$$

the associative law [similar to ordinary algebra $p + (q + r) = (p + q) + r$ and $p(qr) = (pq)r$]:

$$P \cup (Q \cup R) = (P \cup Q) \cup R \tag{7.15}$$

$$P \cap (Q \cap R) = (P \cap Q) \cap R \tag{7.16}$$

the distributive law [similar to $p(q + r) = pq + pr$]:

$$P \cup (Q \cap R) = (P \cup Q) \cap (P \cup R) \tag{7.17}$$

$$P \cap (Q \cup R) = (P \cap Q) \cup (P \cap R) \tag{7.18}$$

the idemoptence law:

$$P \cap P = P \tag{7.19}$$

$$P \cup P = P \tag{7.20}$$

the involution law:

$$c(cP) = P \tag{7.21}$$

and

$$P \cup \emptyset = P \tag{7.22}$$

$$P \cap W = P \tag{7.23}$$

$$P \cup cP = W \tag{7.24}$$

$$P \cap cP = \emptyset \tag{7.25}$$

$$c(P \cup Q) = cP \cap cQ \tag{7.26}$$

$$c(P \cap Q) = cP \cup cQ \tag{7.27}$$

where Eqs. (7.26) and (7.27) are DeMorgan's laws and Eqs. (7.13) to (7.26) provide the tools necessary to manipulate and simplify sets. For example, using Eqs. (7.13), (7.17), and (7.19) one can prove that the set $(P \cup Q) \cup (P \cap Q)$ is equal to the set $P \cup Q$. The Venn diagram can also be used as an informal method to reach the same conclusion. From a geometric modeling point of view, these equations, or the set theory in general, can operate on point sets that represent solids in $E^3$ or they can be used to classify other point sets in space against solids to determine which points in space are inside, on, or outside a given solid.

The concept of the cartesian product of two sets is useful to geometric modeling because it can be related to coordinates of points in space. The concept of an ordered pair must be introduced first. Let us assume that $a$ and $b$ are two elements. An ordered pair of $a$ and $b$ is denoted by $(a, b)$; $a$ is the first coordinate of the pair $(a, b)$ and $b$ is the second coordinate. This guarantees that $(a, b) \neq (b, a)$ if $a \neq b$. The ordered pair of $a$ and $b$ is a set and can be defined as

$$(a, b) = \{\{a\}, \{a, b\}\} \tag{7.28}$$

Equation (7.28) implies that the first coordinate of the ordered pair is the first element $\{a\}$ and the second coordinate is the second element $\{a, b\}$; both elements form the set of the ordered pair $(a, b)$. If $a = b$, then $(a, a) = \{\{a\}, \{a, a\}\}$ $= \{\{a\}, \{a\}\} = \{\{a\}\}$. Based on this definition, there is a theorem which states that two ordered pairs are equal if and only if their corresponding coordinates are equal, that is, $(a, b) = (c, d) \Leftrightarrow a = c$ and $b = d$.

The cartesian product is the concept that can be used to form ordered pairs. If $A$ and $B$ are two sets, the cartesian product of the sets, designated by $A \times B$, is the set containing all possible ordered pairs $(a, b)$ such that $a \in A$ and $b \in B$, that is,

$$A \times B = \{(a, b): a \in A \text{ and } b \in B\} \tag{7.29}$$

If, for example, $A = \{1, 2, 3\}$ and $B = \{1, 4\}$, then $A \times B = \{(1, 1), (1, 4), (2, 1), (2, 4), (3, 1), (3, 4)\}$. Note that $A \times B \neq B \times A$. We denote $A \times A$ by $A^2$. The cartesian product of three sets can now be introduced as

$$A \times B \times C = (A \times B) \times C = \{(a, b, c): a \in A, b \in B, c \in C\} \tag{7.30}$$

where $(a, b, c)$ is an ordered triple defined by $(a, b, c) = ((a, b), c)$. $A \times A \times A$ is usually denoted by $A^3$. In general, an $n$-tuple can be defined as the cartesian product of $n$ sets and takes the form $(a_1, a_2, \ldots, a_n)$. Ordered pairs and triples are considered 2-tuples and 3-tuples respectively.

Equations (7.29) and (7.30) can be used to define points and their coordinates in the context of set theory. If we consider a set of points (set of real numbers) $R^1$ in one-dimensional euclidean space $E^1$, then $R^2$ defines a set of points in $E^2$; each is defined by two numbers or an ordered pair. Similarly, $R^3$ defines a set of points in $E^3$; each is defined by three numbers or an ordered triple.

**Example 7.2.** A point set $S$ that defines a solid in $E^3$ is a set of ordered triples. Find the three sets whose cartesian product produces $S$.

**Solution.** The point set can be written as

$$S = \{P_1, P_2, \ldots, P_n\} \tag{7.31}$$

where $P_1, P_2, \ldots, P_n$ are points inside or on the solid. This set can also be written as

$$S = \{(x_1, y_1, z_1), (x_2, y_2, z_2), \ldots, (x_n, y_n, z_n)\} = \{(x_i, y_i, z_i): 1 \leq i \leq n\} \quad (7.32)$$

We can define three sets $A$, $B$, and $C$ such that

$$A = \{x_1, x_2, \ldots, x_n\} \quad (7.33)$$

$$B = \{y_1, y_2, \ldots, y_n\} \quad (7.34)$$

$$C = \{z_1, z_2, \ldots, z_n\} \quad (7.35)$$

Let us define the set $P$ as the cartesian product $A \times B \times C$, that is,

$$P = A \times B \times C = \{(x_i, y_j, z_k): 1 \leq i \leq n, 1 \leq j \leq n, i \leq k \leq n\} \quad (7.36)$$

The point set $S$ of the solid given by Eq. (7.32) is clearly a (proper) subset of the set $P$, that is, $S \subset P$. The elements of $S$ are equal to the elements of $P$ only when $i = j = k$.

Let us introduce a new notion called the *ordered* cartesian product. It is a more restricted special case of the cartesian product concept. It is applied only to sets that have the same number of elements. We denote it by " $\otimes$ " to differentiate it from " $\times$ " which is used for the cartesian product (not ordered). If we have two sets defined as $A = \{a_1, a_2, \ldots, a_n\}$ and $B = \{b_1, b_2, \ldots, b_n\}$, then

$$A \otimes B = \{(a_i, b_i): a_i \in A, b_i \in B, \text{ and } 1 \leq i \leq n\} \quad (7.37)$$

The ordered cartesian product of three sets is similarly given by

$$A \otimes B \otimes C = (A \otimes B) \otimes C = \{(a_i, b_i, c_i): a_i \in A, b_i \in B, c_i \in C, \text{ and } 1 \leq i \leq n)\} \quad (7.38)$$

Comparing Eqs. (7.32) and (7.38) shows that the *ordered* cartesian product of the three sets $A$, $B$, and $C$ given by Eqs. (7.33) to (7.35) gives the point set $S$ of a solid. This observation that $S$ can be related to $A$, $B$, and $C$ might be useful in classification problems.

## 7.5.2   Regularized Set Operations

The set operations (c, $\cup$, $\cap$, and $-$) covered in the previous section are also known as the set-theoretic operations. When we use these operations in geometric modeling to build complex objects from primitive ones, the complement operation is usually dropped because it might create unacceptable geometry. Furthermore, if we use the other operations ($\cup$, $\cap$, $-$) without regularization in solid modeling, they may cause user inconvenience (say, user must not have overlapping faces of objects or primitives). In addition, objects resulting from these operations may lack geometric closure, may be difficult to validate, or may be inadequate for application (e.g., interference analysis).

To avoid the above problems, the point sets that represent objects and the set operations that operate on them must be regularized. Regular sets and regularized set operations (boolean operations) are considered as boolean algebra.

A regular set is defined as a set that is geometrically closed [refer to Eq. (7.3)]. The notion of a regular set is introduced in geometric modeling to ensure the validity of objects they represent and therefore eliminate nonsense objects. Under geometric closure, a regular set has interior and boundary subsets. More importantly, the boundary contains the interior and any point on the boundary is in contact with a point in the interior. In other words, the boundary acts as a skin wrapped around the interior. The set $S$ shown in Fig. 7-7 is an example of a regular set while Fig. 7-8 shows a nonregular set because the dangling edge and face are not in contact with the interior of the set (in this case the box).

Mathematically, a set $S$ is regular if and only if

$$S = \text{ki}S \tag{7.39}$$

This equation states that if the closure of the interior of a given set yields that same given set, then the set is regular. Figure 7-13$a$ shows that set $S$ is not regular because $S' = \text{ki}S$ is not equal to $S$. Some modeling systems use regular sets that are open or do not have boundaries. A set $S$ is regular open if and only if

$$S = \text{ik}S \tag{7.40}$$

This equation states that a set is regular open if the interior of its closure is equal to the original set. Figure 7-13$b$ shows that $S$ is not regular open because $S' = \text{ik}S$ is not equal to $S$.

Set operations (known also as boolean operators) must be regularized to ensure that their outcomes are always regular sets. For geometric modeling, this means that solid models built from well-defined primitives are always valid and represent valid (no-nonsense) objects. Regularized set operators preserve homogeneity and spatial dimensionality. The former means that no dangling parts should result from using these operators and the latter means that if two three-dimensional objects are combined by one of the operators, the resulting object should not be of lower dimension (two or one dimension). Regularization of set operators is particularly useful when users deal with overlapping faces of different objects, or in other words when dealing with tangent objects, as will be seen shortly in an example.

Based on the above description, regularized set operators can be defined as follows:

$$P \cup^* Q = \text{ki}(P \cup Q) \tag{7.41}$$

$$P \cap^* Q = \text{ki}(P \cap Q) \tag{7.42}$$

$$P -^* Q = \text{ki}(P - Q) \tag{7.43}$$

$$\text{c}^* P = \text{ki}(\text{c}P) \tag{7.44}$$

where the superscript * to the right of each operator denotes regularization. The sets $P$ and $Q$ used in Eqs. (7.41) to (7.44) are assumed to be any arbitrary sets. However, if two sets $X$ and $Y$ are r-sets (regular sets), which is always the case for geometric modeling, then Eqs. (7.41) to (7.44) become

$$X \cup^* Y = X \cup Y \tag{7.45}$$

$$X \cap^* Y = X \cap Y \Leftrightarrow \text{b}X \text{ and } \text{b}Y \text{ do not overlap} \tag{7.46}$$

$$X -^* Y = \text{k}(X - Y) \tag{7.47}$$

$$\text{c}^* X = \text{k}(\text{c}X) \tag{7.48}$$

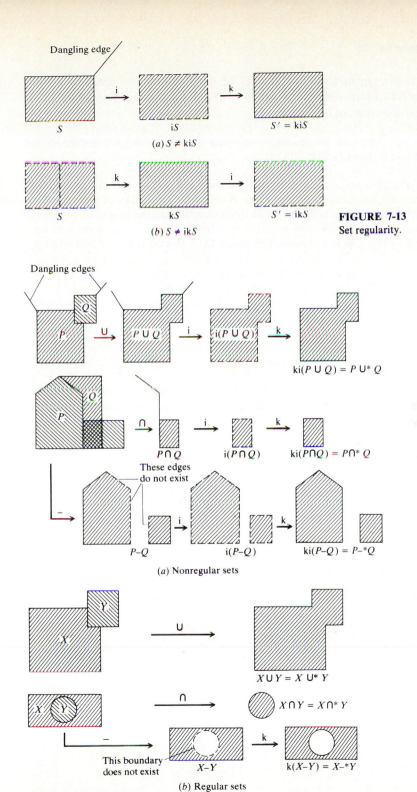

Dangling edge

$S$    $\xrightarrow{\ i\ }$    $iS$    $\xrightarrow{\ k\ }$    $S' = kiS$

$(a)\ S \neq kiS$

$S$    $\xrightarrow{\ k\ }$    $kS$    $\xrightarrow{\ i\ }$    $S' = ikS$

$(b)\ S \neq ikS$

**FIGURE 7-13**
Set regularity.

Dangling edges

$Q$
$P$   $\xrightarrow{\ \cup\ }$   $P \cup Q$   $\xrightarrow{\ i\ }$   $i(P \cup Q)$   $\xrightarrow{\ k\ }$

$ki(P \cup Q) = P \cup^* Q$

$Q$
$P$   $\xrightarrow{\ \cap\ }$   $P \cap Q$   $\xrightarrow{\ i\ }$   $i(P \cap Q)$   $\xrightarrow{\ k\ }$   $ki(P \cap Q) = P \cap^* Q$

These edges
do not exist

$P-Q$    $\xrightarrow{\ i\ }$    $i(P-Q)$    $\xrightarrow{\ k\ }$    $ki(P-Q) = P-^*Q$

$(a)$ Nonregular sets

$Y$
$X$    $\xrightarrow{\ \cup\ }$

$X \cup Y = X \cup^* Y$

$X$ $Y$    $\xrightarrow{\ \cap\ }$    $X \cap Y = X \cap^* Y$

This boundary
does not exist    $\xrightarrow{\ -\ }$    $X-Y$    $\xrightarrow{\ k\ }$    $k(X-Y) = X-^*Y$

$(b)$ Regular sets

**FIGURE 7-14**
Regularized set operators.

359

If b$X$ and b$Y$ overlap in Eq. (7.46), Eq. (7.42) is used and the result is a null object. Figure 7-14 illustrates Eqs. (7.41) to (7.48) geometrically. The figure does not include the complement operation.

> **Example 7.3.** What are the results of applying the regularized set operations to objects $A$ and $B$ shown in Fig. 7-15?
>
> **Solution.** The positions of objects $A$ and $B$ shown in Fig. 7-15 are chosen to illustrate some tangency cases of objects. $A$ and $B$ are r-sets. The results of applying Eqs. (7.45) to (7.47) are shown in Table 7.2 for each case. For all the cases, the results of the regularized union operations are obvious. However, the results of the intersection operations may be less obvious. For case 1, $A \cap B$ is the common face which is eliminated by the regularization process. For case 2, the intersection does not exist; therefore the result is an empty set or a null object. For case 3, $A \cap B$ is the common edge which is eliminated by the regularization process. For case 4, $A \cap B$ is the common block and the common face. The common face is eliminated after regularization. The results of the regularized difference operations are obvious. In cases 1, 2, and 3, $A -^* B$ is the object $A$ itself. For case 4, the difference is a disjoint object. Such an object should not be viewed as two objects. Any further set operation or rigid-body motion treats it as one object.

The reader is advised to carry the details of these results following the steps illustrated in Fig. 7-14. The reader should also try to use these cases to test any available solid modeling package.

### 7.5.3  Set Membership Classification

In various geometric problems involving solid models, we are often faced with the following question: given a particular solid, which point, line segment, or a portion of another solid intersects with such a solid? These are all geometric intersection problems. For a point/solid, line (curve)/solid, or solid/solid intersec-

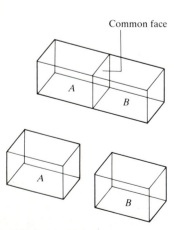

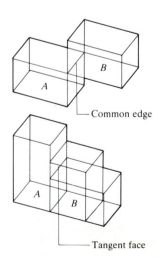

**FIGURE 7-15**
Sample objects.

**TABLE 7.2**
**Results of Example 7.3**

| Case | Objects | Set operation | Result |
|------|---------|---------------|--------|
| 1 | | $A \cup^* B$ | |
| | | $A \cap^* B$ | $\emptyset$ (null object) |
| | | $A -^* B$ | |
| 2 | | $A \cup^* B$ | |
| | | $A \cap^* B$ | $\emptyset$ (null object) |
| | | $A -^* B$ | |
| 3 | | $A \cup^* B$ | |
| | | $A \cap^* B$ | $\emptyset$ (null object) |
| | | $A -^* B$ | |
| 4 | | $A \cup^* B$ | |
| | | $A \cap^* B$ | |
| | | $A -^* B$ | |

tion, we need to know respectively which points, line segments, or solid portions are inside, outside, or on the boundary of a given solid. These geometric intersection problems have useful practical engineering applications. For example, line/solid intersection can be used to shade or calculate mass properties of given

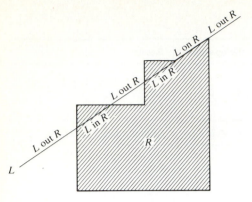

$M[L,R] = (L \text{ in } R, L \text{ on } R, L \text{ out } R)$

**FIGURE 7-16**
Line/polygon set membership classification.

solids via ray-tracing algorithms, while solid/solid intersection can be used for interference checking between two solids.

In each of the above problems, we are given two point sets: a reference set $S$ and a candidate set $X$. The reference set is usually the given solid whose inside (interior) and boundary are $iS$ and $bS$ respectively. The outside of $S$ is its complement $cS$. The candidate set is the geometric entity that must be classified against $S$. The process by which various parts of $X$ (points, line segments, or solid portions) are assigned to $iS$, $bS$, and/or $cS$ is called set membership classification.

A function called a set membership classification function exists which provides a unifying approach to study the behavior of the candidate set $X$ relative to the reference set $S$. The function is denoted by $M[.]$ and is defined as

$$M[X, S] = (X \text{ in } S, X \text{ on } S, X \text{ out } S) \qquad (7.49)$$

Equation (7.49) implies that the input to $M[.]$ is the two sets $X$ and $S$ and the output is the classification of $X$ relative to $S$ as in, on, or out $S$. Figure 7-16 shows an example of classifying a portion of a line $L$ against the polygon $R$.

The implementation of the classification function given by Eq. (7.49) depends to a great extent on the representations of both $X$ and $S$ and their data structures. Let us consider the line/polygon classification problem when the polygon (reference solid) is stored as a B-rep or a CSG. Figure 7-17 shows the B-rep case. The line $L$ is chosen such that no "on" segments result for simplicity.

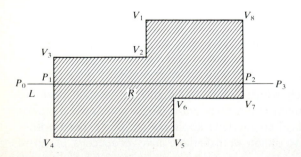

**FIGURE 7-17**
Line/polygon classification for B-rep.

The algorithm for this case can be described as follows:

1. Utilizing a line/edge intersection routine, find the boundary crossings $P_1$ and $P_2$.
2. Sort the boundary crossings according to any agreed direction for $L$. Let the sorted boundary crossing list be given by $(P_0, P_1, P_2, P_3)$.
3. Classify $L$ with respect to $R$. For this simple case, we know that the odd boundary crossings (such as $P_1$) flags "in" segments and the even boundary crossings (such as $P_2$) flags "out" segments. Therefore, the classification of $L$ with respect to $R$ becomes

$$[P_0, P_1] \subset L \text{ out } R$$

$$[P_1, P_2] \subset L \text{ in } R$$

$$[P_2, P_3] \subset L \text{ out } R$$

If the line $L$ contains an edge of the polygon, the above classification criterion of odd and even crossings would not work and another criterion should be found. In this case, a direction (clockwise or counterclockwise) to traverse the polygon boundaries is needed. Let us apply this idea to the problem at hand to see how it would work. If we choose the counterclockwise direction, polygon vertices would be numbered as shown in Fig. 7-17. Now we know that i$R$ is always to the left of any edge. The new classification criterion can be stated as follows. Let us assume that an edge is defined by the two vertices $V_i$ and $V_{i+1}$. Whenever there is a boundary crossing on an edge whose $V_i$ is above $L$ and $V_{i+1}$ is below $L$, this crossing is flagged as "in" and whenever $V_i$ is below $L$ and $V_{i+1}$ is above, it is flagged "out." This criterion obviously gives the same result as the previous criterion for this example.

Let us consider the same line/polygon classification problem when the polygon is stored as a CSG representation. The classification for this case is done at the primitive level and the algorithm becomes as follows:

1. Utilize a line/primitive intersection routine to find the intersection points of the line with each primitive of $R$.
2. Use these intersection points to classify the line against each primitive of $R$.
3. Combine the "in" and "on" line segments obtained in step 2 using the same boolean operators that combine the primitives. For example, if two primitives $A$ and $B$ are unioned, then the "in" and "on" line segments are added.
4. Find the "out" segments by taking the difference between the line (candidate set) and the "in" and "on" segments. Figure 7-18 shows the "classify" and "combine" strategy for the three boolean operations of two blocks $A$ and $B$. Notice that the polygon that results from the union operation is the same as the polygon $R$ used in the classification of the B-rep case. The classification of $L$ relative to $A$ and $B$ is straightforward. To combine these classifications, we first combine $L$ in $A$ and $L$ in $B$ to obtain $L$ in $R$, using the proper boolean operator. The $L$ on $R$ can result from combining three possibilities: $L$ in $A$ and

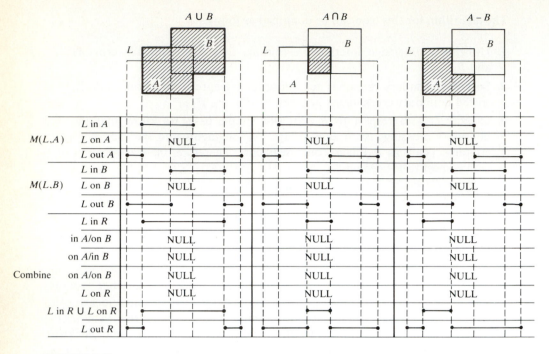

**FIGURE 7-18**
Line/polygon classification for CSG rep.

$L$ on $B$, $L$ on $A$ and $L$ in $B$, and $L$ on $A$ and $L$ on $B$. All these possibilities are obtained and then combined to give $L$ on $R$. The remaining classification $L$ out $R$ is obtained by adding $L$ in $R$ and $L$ on $R$, and subtracting the result from $L$ itself.

The above example has considered the polygon case. The example does not purposely include "on" segments because they are ambiguous and need more information (neighborhoods) to resolve their ambiguities for both B-rep and CSG (refer to Sec. 7.8 for details). Algorithms to classify candidate sets against three-dimensional solids can follow similar steps to those described in the above example but with more elaborate details.

## 7.6 HALF-SPACES

Half-spaces form a basic representation scheme for bounded solids. By combining half-spaces (using set operations) in a building block fashion, various solids can be constructed. Half-spaces are usually unbounded geometric entities; each one of them divides the representation space into two infinite portions, one filled with material and the other empty. Surfaces can be considered half-space boundaries and half-spaces can be considered directed surfaces.

A half-space is defined as a regular point set in $E^3$ as follows:

$$H = \{P : P \in E^3 \text{ and } f(P) < 0\} \tag{7.50}$$

where $P$ is a point in $E^3$ and $f(P) = 0$ defines the surface equation of the half-space boundaries. Half-spaces can be combined together using set operations to create complex objects.

## 7.6.1  Basic Elements

Various half-spaces can be described and created using Eq. (7.50). However, to make them useful for design and manufacturing applications, supporting algorithms and utility routines must be provided. For example, if one were to add a cylindrical half-space to a modeling package, intersecting routines that enable this half-space to intersect itself as well as other existing half-spaces must be developed and added as well.

The most widely used half-spaces (unbounded) are planar, cylindrical, spherical, conical, and toroidal half-spaces. They form the natural quadrics discussed earlier in Sec. 7.3 (with the exception of the torus which can be formed from the other half-spaces). The regular point set of each half-space is a set of ordered triplets $(x, y, z)$ given by

Planar half-space:  $\qquad H = \{(x, y, z): z < 0\}$  $\qquad$ (7.51)

Cylindrical half-space:  $\qquad H = \{(x, y, z): x^2 + y^2 < R^2\}$  $\qquad$ (7.52)

Spherical half-space:  $\qquad H = \{(x, y, z): x^2 + y^2 + z^2 < R^2\}$  $\qquad$ (7.53)

Conical half-space:  $\qquad H = \{(x, y, z): x^2 + y^2 < [(\tan \alpha/2)z]^2\}$  $\qquad$ (7.54)

Toroidal half-space:  $\qquad H = \{(x, y, z): (x^2 + y^2 + z^2 - R_2^2 - R_1^2)^2$

$$< 4R_2^2(R_1^2 - z^2)\} \quad (7.55)$$

Equations (7.51) to (7.55) are implicit equations and are expressed in terms of each half-space local coordinate system whose axes are $X_H$, $Y_H$, and $Z_H$. The implicit form is efficient to find surface intersections (refer to Chap. 6). The corresponding surface of each half-space is given by its equation when the right and left sides are equal. For the planar half-space, Eq. (7.51) is based on the vertical plane $z = 0$. Other definitions can be easily written. Figure 7-19 shows the various half-spaces with their local coordinate systems and the limits on their configuration parameters.

## 7.6.2  Building Operations

Complex objects can be modeled as half-spaces combined by the set operations. As a matter of fact, half-spaces are treated as lower level primitives and all the related construction techniques to CSG can be used here. As will be seen later, one form of CSG can be based on unbounded half-spaces. Regularized set operations can be used to combine half-spaces to form complex solids. Most often, half-spaces may have to undergo rigid motion via homogeneous transformations to be positioned properly before intersection.

Let us represent the solid $S$ shown in Fig. 7-20a using half-spaces. Parameters of the solid are shown. The hole is centered in the top face. The MCS of the

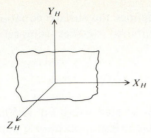

Planar half-space

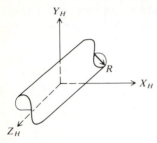

Cylindrical half-space ($R > 0$)

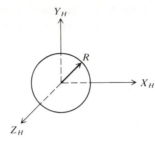

Spherical half-space ($R > 0$)

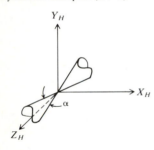

Two-sheet conical half-space
($0 < \alpha < \pi$)

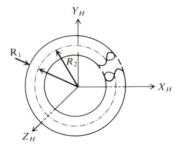

Toroidal half-space
($R_1 > 0,\ R_2 > 0,\ R_2 > R_1$)

**FIGURE 7-19**
Unbounded half-spaces.

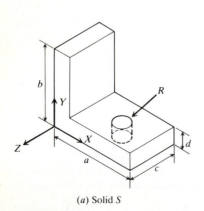

(a) Solid S

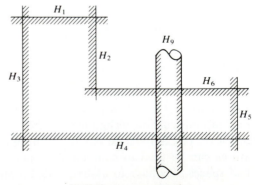

(b) Half-space representation

**FIGURE 7-20**
Half-space representation of solid S.

solid model is chosen as shown in the figure. Figure 7-20b shows that nine half-spaces (eight planes and one cylinder) $H_1$ to $H_9$ are needed to represent S. Half-spaces $H_7$ and $H_8$ that model the front and back faces of the model are not shown in the figure. Utilizing local coordinate systems shown in Fig. 7-19, some half-spaces have to be positioned first. For example, rotate H given by Eq. (7.51) an angle $-90°$ about the X axis and translate it up in the Y direction a distance b to obtain $H_1$. In a similar fashion, the other half-spaces can be positioned using the proper rigid motion. Only $H_7$ needs no positioning. The positioned half-spaces can be intersected and then boolean operations are used to combine them. $H_1$ to $H_8$ are unioned and $H_9$ is subtracted from the result.

**Example 7.4.** How can you create a solid fillet using unbounded half-spaces?

**Solution.** Surface fillet has been defined in Chap. 6 as a B-spline surface blending two given surfaces. Similarly, a solid fillet can be used to blend sharp edges of a solid as shown in Fig. 7-21a. The solid fillet is defined by its radius r and length d. Six half-spaces $H_1$ to $H_6$ (see Fig. 7-21b) are needed to construct the fillet. $H_1$, $H_2$, $H_3$, and $H_4$ represent the front, left, back, and bottom faces respectively. $H_6$ is the cylindrical face. $H_5$ is an auxiliary half-space positioned at distance r from the origin of the $X_L Y_L Z_L$ local coordinate system of the fillet and oriented at 45° as shown in

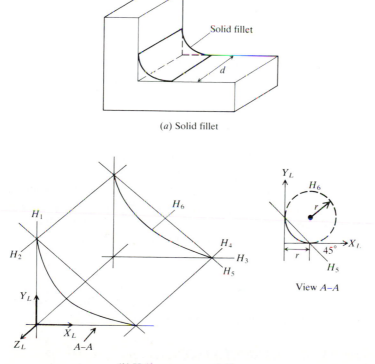

(a) Solid fillet

(b) Half-space representation

**FIGURE 7-21**
Solid fillet and its half-space representation.

view $A\text{-}A$ in Fig. 7-21$b$. $H_5$ is used to intersect $H_2$, $H_4$, and $H_6$ so that the boundaries of the fillet can be evaluated. This is because the cylindrical half-space is tangent to $H_2$ and $H_4$.

Except for $H_1$, the other half-spaces must be positioned before intersection and set operations are performed to create the fillet. Let us look at positioning $H_5$ and $H_6$ as an example. Using Eq. (7.51), $H$ has to rotate an angle of $90°$ about the $Y$ axis, followed by a $45°$ rotation about the $Z$ axis, and finally translated a distance $r$ in the positive $X$ direction to produce $H_5$. $H_6$ is obtained by translating the cylindrical half-space $H$ given by Eq. (7.52) an equal distance $r$ in both the positive $X$ and $Y$ directions. At this position, the complement of the cylindrical half-space, $cH$, is taken to obtain $H_6$. Theoretically, $cH$ is equal to $E^3$ minus the cylindrical half-space. For practical and implementation purposes, $E^3$ can be limited to a bounded volume, such as a box, enclosing the cylindrical half-space, or the complement process can be replaced by choosing a surface normal to be positive on one side of the half-space and negative on the other side.

The intersections of $H_1$ to $H_6$ with each other can now be performed and the results can be unioned to obtain the solid fillet. Notice that the solid fillet could have been created without the complement operation by subtracting the cylindrical half-space itself after its positioning from the intersection results of $H_1$ to $H_5$. However, the complement of a half-space is generally used to minimize the number of half-spaces used in modeling objects.

It should be noted from this example that using half-spaces and/or their complements or directed surface normals, any complex object can be modeled as the union of the intersection of half-spaces, that is,

$$S = \cup \left( \bigcap_{i=1}^{n} H_i \right) \qquad (7.56)$$

where $S$ is the solid and $n$ is the number of half-spaces and/or their complements. As an example, a box is the union of six intersected half-spaces.

### 7.6.3   Remarks

The half-space representation scheme is the lowest level available to represent a complex object as a solid model. The main advantage of half-spaces is its conciseness in representing objects compared to other schemes such as CSG. However, it has a few disadvantages. This representation can lead to unbounded solid models if the user is not careful. Such unboundedness can result in missing faces and abnormal shaded images. It can also lead to system crash or producing wrong results if application algorithms attempt to access databases of unbounded models. Another major disadvantage is that modeling with half-spaces is cumbersome for casual users and designers to use and may be difficult to understand. Therefore, half-space representation is probably useful only for research purposes. Modelers, such as SHAPES, TIPS, and PADL, attempt to shield users from dealing directly with the unbounded half-spaces.

### 7.7   BOUNDARY REPRESENTATION (B-rep)

Boundary representation is one of the two most popular and widely used schemes (the other is CSG discussed in Sec. 7.8) to create solid models of physical objects. A B-rep model or boundary model is based on the topological notion that a physical object is bounded by a set of faces. These faces are regions or subsets of closed and orientable surfaces. A closed surface is one that is continuous without breaks. An orientable surface is one in which it is possible to distinguish two sides by using the direction of the surface normal to point to the inside or outside of the solid model under construction. Each face is bounded by edges and each edge is bounded by vertices. Thus, topologically, a boundary model of an object is comprised of faces, edges, and vertices of the object linked together in such a way as to ensure the topological consistency of the model.

The database of a boundary model contains both its topology and geometry. Topology is created by performing Euler operations and geometry is created by performing euclidean calculations. Euler operations are used to create, manipulate, and edit the faces, edges, and vertices of a boundary model as the set (boolean) operations create, manipulate, and edit primitives of CSG models. Euler operators, as boolean operators, ensure the integrity (closeness, no dangling faces or edges, etc.) of boundary models. They offer a mechanism to check the validity of these models. Other validity checks may be used as well. Geometry includes coordinates of vertices, rigid motion and transformation (translation, rotation, etc.), and metric information such as distances, angles, areas, volumes, and inertia tensors. It should be noted that topology and geometry are interrelated and cannot be separated entirely. Both must be compatible otherwise nonsense objects may result. Figure 7-22 shows a square which, after dividing its top edges by introducing a new vertex, is still valid topologically but produces a nonsense object depending on the geometry of the new vertex.

In addition to ensuring the validity of B-rep models, Euler operators provide designers with drafting functionality. These allow solid models to be built up graphically by incrementally adding individual vertices, edges, and faces to the model in such a way as to always obey Euler's laws, as will be seen in Sec. 7.7.2. Euler operators are considered to be lower level operators than boolean operators in the sense that they combine faces, edges, and vertices to form B-rep models.

Boolean operations are not considered a part of the representation of a B-rep model, but they are often employed as one of the means of creating, manipulating, and editing the model as mentioned in Sec. 7.1 and shown in Table 7.1.

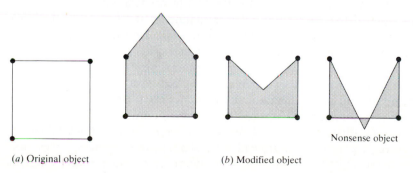

(a) Original object                    (b) Modified object

**FIGURE 7-22**
Effect of topology and geometry on boundary models.

The effect of a Boolean operation on a CSG model (see Sec. 7.8) is simply an addition to the CSG tree. However, since B-rep systems require an explicit representation of the boundary of the solid, they must evaluate the new boundary that is the result of the operation.

While B-rep systems store only the bounding surfaces of the solid, it is still possible to compute volumetric properties such as mass properties (assuming uniform density) by virtue of the Gauss divergence theorem which relates volume integrals to surface ones. The speed and accuracy of these calculations depend on the types of surfaces used by the models. More details are covered in Chap. 17.

The modeling domain (or the range of objects that can be modeled) of a B-rep scheme is potentially large and depends mainly on the primitive surfaces (planar, curved, or sculptured) that are admissible by the scheme to form the faces of various models. For example, given the modeling domain of a scheme based on half-spaces, a B-rep scheme with the same domain can be designed by using the boundary surfaces of the half-spaces as its primitive surfaces.

The desired properties of a representation scheme discussed in Sec. 7.4 apply to B-rep schemes. These schemes are unambiguous if faces are represented unambiguously, that is, as regions of closed orientable surfaces. This claim (unambiguous faces result in unambiguous B-rep) is based on the fact that an r-set is defined unambiguously by its boundary and that non-r-sets are not defined unambiguously by their boundaries. The validity of B-rep models is ensured via Euler operations which can be built into the syntax of a CAD/CAM system. However, these models are not unique because the boundary of any object can be divided into faces, edges, and vertices in many ways. Verification of uniqueness of boundary models is computationally expensive and is not performed in practice.

### 7.7.1   Basic Elements

If a solid modeling system is to be designed, the domain of its representation scheme (objects that can be modeled) must be defined, the basic elements (primitives) needed to cover such modeling domain must be identified, the proper operators that enable the system users to build complex objects by combining the primitives must be developed, and finally a suitable data structure must be designed to store all relevant data and information of the solid model. Other system and geometric utilities (such as intersection algorithms) may also need to be designed. Let us apply these ingredients to a B-rep system.

Objects that are often encountered in engineering applications can be classified as either polyhedral or curved objects. A polyhedral object (plane-faced polyhedron) consists of planar faces (or sides) connected at straight (linear) edges which, in turn, are connected at vertices. A cube or a tetrahedron is an obvious example. A curved object (curved polyhedron) is similar to a polyhedral object but with curved faces and edges instead. The identification of faces, edges, and vertices for curved closed objects such as a sphere or a cylinder needs careful attention, as will be seen later in this section. Polyhedral objects are simpler to deal with and are covered first.

The reader might have jumped intuitively to the conclusion that the primi-tives of a B-rep scheme are faces, edges, and vertices. This is true if we can answer the following two questions. First, what is a face, edge, or a vertex? Second, knowing the answer to the first question, how can we know that when we combine these primitives we would create valid objects? Answers to these ques-tions can help users to create B-rep solid models of objects successfully. To show that these answers are not always simple, consider the polyhedral objects shown in Fig. 7-23. Polyhedral objects can be classified into four classes. The first class (Fig. 7-23a) is the simple polyhedra. These do not have holes (through or not through) and each face is bounded by a single set of connected edges, that is, bounded by one loop of edges. The second class (Fig. 7-23b) is similar to the first

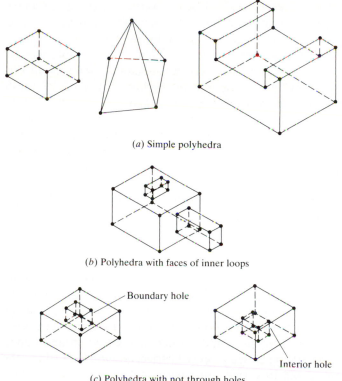

(a) Simple polyhedra

(b) Polyhedra with faces of inner loops

(c) Polyhedra with not through holes

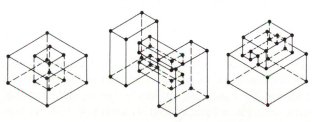

(d) Polyhedra with handles (through holes)

**FIGURE 7-23**
Types of polyhedral objects.

with the exception that a face may be bounded by more than one loop of edges (inner loops are sometimes called rings). The third class (Fig. 7-23c) includes objects with holes that do not go through the entire object. For this class, a hole may have a face coincident with the object boundary; in this case we call it a boundary hole. On the other hand, if it is an interior hole (as a void or crack inside the object), it has no faces on the boundary. The fourth and the last class (Fig. 7-23d) includes objects that have holes that go through the entire objects. Topologically, these through holes are called handles.

With the above physical insight, let us define the primitives of a B-rep scheme and other related topological items that enable a user to create the boundary model of an object. They apply to both polyhedral and curved objects. A vertex is a unique point (an ordered triplet) in space. An edge is a finite, non-self-intersecting, directed space curve bounded by two vertices that are not necessarily distinct. A face is defined as a finite connected, non-self-intersecting, region of a closed oriented surface bounded by one or more loops. A loop is an ordered alternating sequence of vertices and edges. A loop defines a non-self-intersecting, piecewise, closed space curve which, in turn, may be a boundary of a face. In Fig. 7-23a, each face has one loop while the top and the right side faces of the object shown in Fig. 7-23b have two loops each (one inner and one outer). A "not" through hole is defined as a depression in a face of an object. A handle (or through hole) is defined as a passageway that pierces the object completely. The topological name for the number of handles in an object is genus. The last item to be defined is a body (sometimes called a shell). It is a set of faces that bound a single connected closed volume. Thus a body is an entity that has faces, edges, and vertices. Such an entity may be a useful solid or an intermediate polyhedron. A minimum body is a point. Topologically this body has one face, one vertex, and no edges. It is called a seminal or singular body. It is initially attached as part of the world. The object on the right of Fig. 7-23c has two bodies (the exterior and interior cubes) and any other object in Fig. 7-23 has only one body.

Faces of boundary models possess certain essential properties and characteristics that ensure the regularity of the model; that is, the model has an interior and a boundary. The face of a solid is a subset of the solid boundary and the union of all faces of a solid defines such a boundary. Faces are two-dimensional homogeneous regions so they have areas and no dangling edges. In addition, a face is a subset of some underlying closed oriented surface. Figure 7-24 shows the relationship between a face and its surface. At each point on the face, there is a surface normal $\mathbf{N}$ that has a sign associated with it to indicate whether it points into or away from the solid interior. One convention is to assume $\mathbf{N}$ positive if it points away from the solid. It is desirable, but not required, that a face has a constant surface normal.

The representation of a face must ensure that both the face and solid interiors can be deduced from the representation. The direction of the face's surface normal can be used to indicate the inside or outside of the model. The surface equation must be consistent with the normal chosen convention. For example, if the face belongs to a Bezier or B-spline surface, the normal vector could be defined as $\partial\mathbf{P}/\partial v \times \partial\mathbf{P}/\partial u$ or $\partial\mathbf{P}/\partial u \times \partial\mathbf{P}/\partial v$ depending on the chosen normal convention and the directions of parametrizing the surface. Practically, some

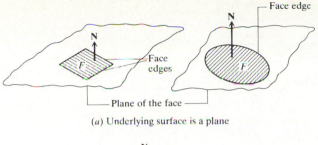

(a) Underlying surface is a plane

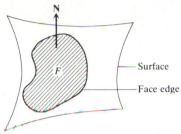

(b) A general underlying surface

**FIGURE 7-24**
Underlying surface of a face.

CAD/CAM systems store the surface normal and its sign as part of the face data (although it could be computed from the surface equation) since it is a useful parameter in many applications such as generating graphics displays or NC machining data. The face interior can be determined by traversing the face loops in a certain direction or assigning flags to them. In traversing loops, the edges of the face outer loop is traversed, say, in a counterclockwise direction and the edges of the inner loops are traversed in the opposite direction, say the clockwise direction. If one of the loops is a continuous or piecewise continuous curve, the parametrization direction is chosen to reflect the traversal direction. Figure 7-25 shows some traversal examples. The other alternative assigns one flag to outer loops and another one to inner loops.

Having defined the boundary model primitives, we now return to the question of how they can be combined to generate topologically valid models. The development of volume measure (valid models) based on faces, edges, and vertices is rigorous and not easy. Euler (in 1752) proved that polyhedra that are homomorphic to a sphere (i.e., their faces are non-self-intersecting and belong to closed orientable surfaces) are topologically valid if they satisfy the following equation:

$$F - E + V - L = 2(B - G) \qquad (7.57)$$

where $F$, $E$, $V$, $L$, $B$, and $G$ are the number of faces, edges, vertices, faces' inner loop, bodies, and genus (handles or through holes) respectively. Equation (7.57) is known as the Euler or Euler-Poincare law. The simplest version of this equation is $F - E + V = 2$ which applies to polyhedra shown in Fig. 7-23a. With Eq. (7.57) in hand, it has been easier to take it as the more primitive definition of a polyhedron on which to base its construction and data structure. From a user

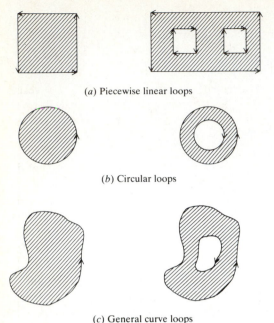

(a) Piecewise linear loops

(b) Circular loops

(c) General curve loops

**FIGURE 7-25**
Traversal of face's loops.

point of view, to create the boundary model of a given object, the user identifies the proper number for all the variables of Eq. (7.57) and substitutes them into the equation to ensure validity. Then system commands (Euler operations) are used to create the model and ensures the validity simultaneously. This is similar to identifying primitives and boolean operators in the case of a CSG-based user interface. Table 7.3 shows the counts of the various variables of Eq. (7.57) for polyhedra shown in Fig. 7-23. The numbering of these polyhedra in the table is taken from left to right and top to bottom with the top left cube being polyhedron number 1 and the bottom right object being number 9.

Euler's law given by Eq. (7.57) applies to closed polyhedral objects only. These are the valid solid models we like to deal with. However, open polyhedral objects do not satisfy Eq. (7.57). This class of objects includes open polyhedra

**TABLE 7.3**
**Counts of polyhedral values for objects of Fig. 7-23**

| Object number | F | E | V | L | B | G |
|---|---|---|---|---|---|---|
| 1 | 6 | 12 | 8 | 0 | 1 | 0 |
| 2 | 5 | 8 | 5 | 0 | 1 | 0 |
| 3 | 10 | 24 | 16 | 0 | 1 | 0 |
| 4 | 16 | 36 | 24 | 2 | 1 | 0 |
| 5 | 11 | 24 | 16 | 1 | 1 | 0 |
| 6 | 12 | 24 | 16 | 0 | 2 | 0 |
| 7 | 10 | 24 | 16 | 2 | 1 | 1 |
| 8 | 20 | 48 | 32 | 4 | 1 | 1 |
| 9 | 14 | 36 | 24 | 2 | 1 | 1 |

that may result during constructing boundary models of closed objects as well as all two-dimensional polygonal objects. Open objects satisfy the following Euler's law:

$$F - E + V - L = B - G \tag{7.58}$$

Figure 7-26 shows some examples of open objects. The reader can easily verify that they satisfy the above equation. In the above equation, $B$ refers to an open body which can be a wire, an area, or a volume. All the objects in Fig. 7-26 have one body and only bodies of Fig. 7-26c have one genus each. It might be interesting to mention that Eq. (7.58) can form the basis of creating a boundary model based on wireframe modeling. There are some systems such as MEDUSA that do that.

We now turn from polyhedral objects to curved objects such as cylinders and spheres. The same rules and guidelines for boundary modeling discussed thus far for the former objects apply to the latter. The major difference between the two types of objects results if closed curved edges or faces exist. Consider, for example, the closed cylinder and sphere shown in Fig. 7-27. As shown in Fig. 7-27, a closed cylindrical face (and alike) has one edge and two vertices and a spherical face (and alike) has one vertex and no edges. The boundary model of a cylinder has three faces (top, bottom, and cylindrical face itself), two vertices, and three edges connecting the two vertices. The other "edges" are for visualization purposes. They are called limbs, virtual edges, or silhouette edges. The problem of computing the silhouette curve of a solid object is covered in Chap. 10. The boundary model of a sphere, on the other hand, consists of one face, one vertex, and no edges. Notice that both models satisfy Euler laws $F - E + V = 2$ for simple polyhedra.

The representation of curved edges is more complex than representing piecewise linear edges. There are direct and indirect schemes. In direct schemes, an edge is represented by a curve equation and ordered endpoints. In indirect schemes, the edge is represented by the intersection of two surfaces. In practice,

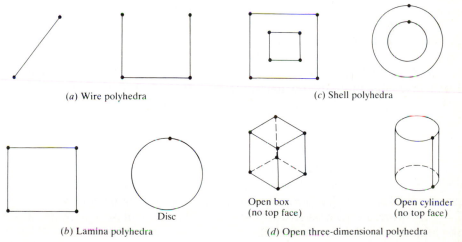

(a) Wire polyhedra

(c) Shell polyhedra

Disc

(b) Lamina polyhedra

Open box
(no top face)

Open cylinder
(no top face)

(d) Open three-dimensional polyhedra

**FIGURE 7-26**
Open polyhedral objects.

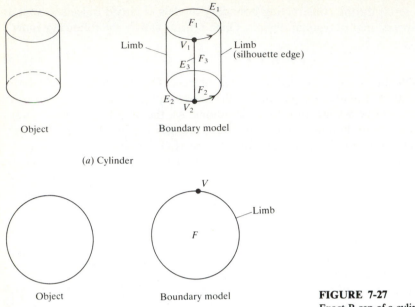

(a) Cylinder

Object          Boundary model

(b) Sphere

Object          Boundary model

**FIGURE 7-27**
Exact B-rep of a cylinder and a
sphere.

indirect schemes are probably preferred because the intersection of two underlying surfaces of two faces produces the curved edge of the two faces.

If the curved objects are represented by storing the equations of the underlying curves and surfaces of the object edges and faces respectively, the resulting
boundary scheme is known as an exact B-rep scheme. Another alternative is the
approximate or faceted B-rep (sometimes called tessellation rep). In this scheme,
any curved face is divided into planar facets—hence the name faceted B-rep.
Figure 7-28 shows a faceted B-rep of a cylinder and sphere. The faceted cylinder
is generated by rotating a line incrementally about the cylinder axis the desired
total number of facets. This is accomplished via a rotational sweep operator. A
faceted sphere is formed in a similar way by rotating $m$ connected line segments
(edges) about the sphere axis for a total of $n$ sides. MEDUSA, for example, is a
faceted B-rep package. The numbers $n$ and $m$ are user inputs. This representation,
although continuous, will no longer be smooth and as the number of facets
increases to give a more accurate representation, the computing time involved
increases dramatically.

A general data structure for a boundary model should have both topological and geometrical information. The structure shown in Fig. 7-29 is based on
Eq. (7.57). A relational database model is very effective to implement such a data
structure. Lists for bodies, faces, loops, edges, and vertices are generated and
stored in tables. Each line in Fig. 7-29 represents a pointer in the database.

The winged edge data structure is a particularly useful data structure which
has been adopted by several modeling systems such as GLIDE and BUILD. In
this structure, all the adjacency relations of each edge are described explicitly.
Since an edge is adjacent to exactly two faces, it is a component in two loops, one

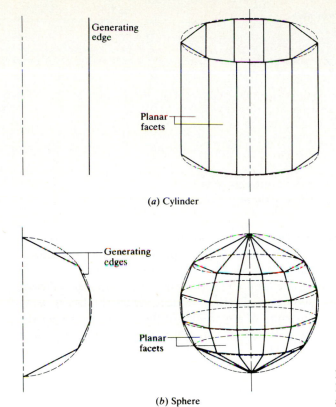

(a) Cylinder

(b) Sphere

**FIGURE 7-28**
Faceted B-rep of a cylinder and a sphere.

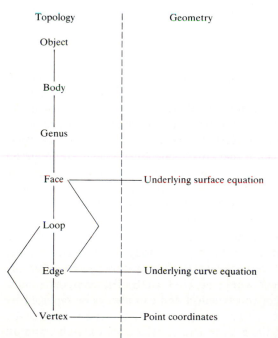

**FIGURE 7-29**
General data structure for boundary modeling.

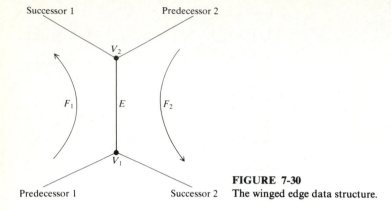

Successor 1          Predecessor 2

$V_2$

$F_1$      $E$      $F_2$

$V_1$

Predecessor 1          Successor 2

**FIGURE 7-30**
The winged edge data structure.

for each face. If these loops are oriented, that is, edges of a loop are traversed in a given direction, say counterclockwise, the edge has a predecessor and a successor in addition to the two bounding vertices (see Fig. 7-30). This edge structure together with its implication on the loop and face elements is extremely efficient for manipulation purposes (adding or deleting vertices, edges, or faces) using Euler's law. For example, the insertion of a new edge in the structure changes the predecessor/successor relationship of its adjacent edges, possibly splitting a face into two and therefore adding a new loop.

## 7.7.2  Building Operations

Equation (7.57) forms the basis to develop building operations to create boundary models of complex objects. Euler operators (or Euler primitives) are based on this equation. There are many variations on how these operators can be implemented. Sample operators are MBFV, MEV, MEF, and GLUE. In these operators, M and K stand for Make and Kill respectively and the other letters mean the same as in Eq. (7.57). Other operators are available to add convenience and flexibility to the construction process. Each operator usually has a complement that has the exact opposite effect on the construction process. Table 7.4 shows some Euler operators. The table shows that the user is not free to construct faces, edges, or vertices as with wireframe and surface modeling. There is no such operator as ME, MV, or MF only because they all violate Euler's law. To create an edge, for example, a new vertex or a new face must be created to preserve the topology. Thus, the two operators MEV and MEF are legitimate. The operator MBFV is usually used to begin constructing the boundary model and it returns a seminal or singular body. It could be thought of as creating the first vertex of the model. The gluing operator is used to glue bodies together at certain faces. The gluing can result in forming a genus or killing one body, as indicated by the operators KFEVMG and KFEVB respectively. Both operators can be called the GLUE operator whose complement would be UNGLUE. The composite commands are available for efficiency of construction and can always be replaced by a sequence of other basic operators. For example, ESPLIT can be replaced by KEV, MEV, and MEV. Similarly, the KVE operator which kills a vertex and all

**TABLE 7.4**
**Some Euler operations**

| Operation | Operator | Complement | Description of operator |
|---|---|---|---|
| Initialize database and begin creation | MBFV | KBFV | Make Body, Face, Vertex |
| Create edges and vertices | MEV | KEV | Make Edge, Vertex |
| Create edges and | MEKL | KEML | Make Edge, Kill Loop |
| faces | MEF | KEF | Make Edge, Face |
| | MEKBFL | KEMBFL | Make Edge, Kill Body, Face, Loop |
| | MFKLG | KFMLG | Make Face, Kill Loop, Genus |
| Glue | KFEVMG | MFEVKG | Kill Face, Edge, Vertex, Make Genus |
| | KFEVB | MFEVB | Kill Face, Edge, Vertex, Body |
| Composite operations | MME | KME | Make Multiple Edges |
| | ESPLIT | ESQUEEZE | Edge-Split |
| | KVE | | Kill Vertex, Edge |

attached edges to it can be replaced by an $(n-1)$ KEF followed by a single KEV if $n$ edges are attached to the vertex. Notice that some Euler operators do not tell directly their end result (see Table 7.5). For example, one would think the glue operator KFEVMG kills only one face. It is less confusing to write KFFEVMG, especially to new users of a system. The actual implementation and syntax of the above operators into modeling systems are not discussed here. Instead, Fig. 7-31 shows how they can be used conceptually in constructing boundary models.

Euler operators create changes in the number of the components in the topology under construction. The operators can be characterized by the transition status they make in the six-space defined by the parameters of Euler's law. Table 7.5 shows these changes for the operators listed in Table 7.4. Observe that the transition state of each operator satisfies Euler's law. This observation provides a general rule to design any new Euler operator. Moreover, if the operator

**TABLE 7.5**
**Transition states of some Euler operators**

| Operator | F | E | V | L | B | G |
|---|---|---|---|---|---|---|
| MBFV | 1 | 0 | 1 | 0 | 1 | 0 |
| MEV | 0 | 1 | 1 | 0 | 0 | 0 |
| MEKL | 0 | 1 | 0 | $-1$ | 0 | 0 |
| MEF | 1 | 1 | 0 | 0 | 0 | 0 |
| MEKBFL | $-1$ | 1 | 0 | $-1$ | $-1$ | 0 |
| MFKLG | 1 | 0 | 0 | $-1$ | 0 | $-1$ |
| KFEVMG | $-2$ | $-n$ | $-n$ | 0 | 0 | 1 |
| KFEVB | $-2$ | $-n$ | $-n$ | 0 | $-1$ | 0 |
| MME | 0 | $n$ | $n$ | 0 | 0 | 0 |
| ESPLIT | 0 | 1 | 1 | 0 | 0 | 0 |
| KVE | $-(n-1)$ | $-n$ | $-1$ | 0 | 0 | 0 |

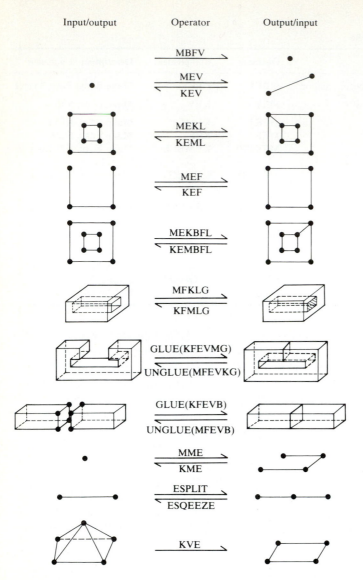

| Input/output | Operator | Output/input |
| --- | --- | --- |

**FIGURE 7-31**
Topology creation via Euler operators.

acts on valid topology and the state transition it generates is valid, then the resulting topology is a valid solid. Therefore, Euler's law is never verified explicitly by the modeling system and its software. It should also be noticed that intermediate topology during construction may not make geometrical sense or may not represent an acceptable, though valid, solid (see Fig. 7-31).

Higher-level Euler operators are possible to develop. Examples include MCUBE, MCYL, MSPH, SWEEPR, and SWEEPT to respectively create a cube, a cylinder, a sphere, axisymmetric (rotational sweep) objects, and uniform (translational) objects. Figure 7-32 shows a wire and a face that are rotated to

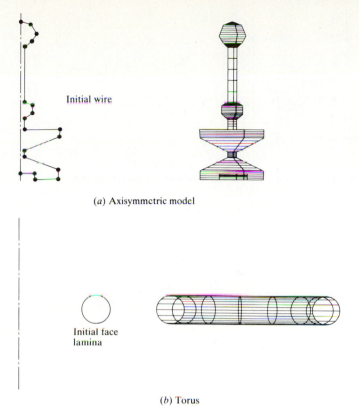

(a) Axisymmetric model

(b) Torus

**FIGURE 7-32**
Rotational sweep (approximate) boundary models.

create a symmetric object and a torus respectively. In addition, union, difference, and intersection operators can be developed.

The advantages of Euler operators are that they ensure creating valid topology, they provide full generality and reasonable simplicity, and they achieve a higher semantic level than that of manipulating faces, edges, and vertices directly. However, Euler operators do not provide any geometrical information to define a solid polyhedron. They do not impose any restriction on surface orientation, face planarity, or surface self-intersection. Nevertheless, in practice, Euler operators perform a useful role as a topological foundation for developing routines that embody more algebra and geometry.

**Example 7.5.** Create the boundary model of solid $S$ shown in Fig. 7-20a.

***Solution.*** First let us develop the boundary model of the solid $S$. For simplicity, we assume that an approximate B-rep scheme is used. Figure 7-33 shows the boundary model of the solid. Based on the figure, the model has 16 faces, 28 vertices, 42 edges, 2 loops, 1 body, and 1 genus. They all together satisfy Euler's law. A suggested sequence to create the model is shown in Fig. 7-34. The sequence matches the planning strategy reflected in Fig. 7-33 in which the cylindrical face of the hole has been approximated by eight facets. Figure 7-34 is shown in an isometric view although it

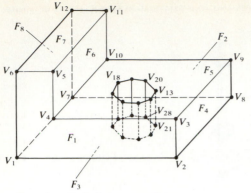

Faces $F_9$ to $F_{16}$ for hole are not shown

**FIGURE 7-33**
Boundary model of solid $S$.

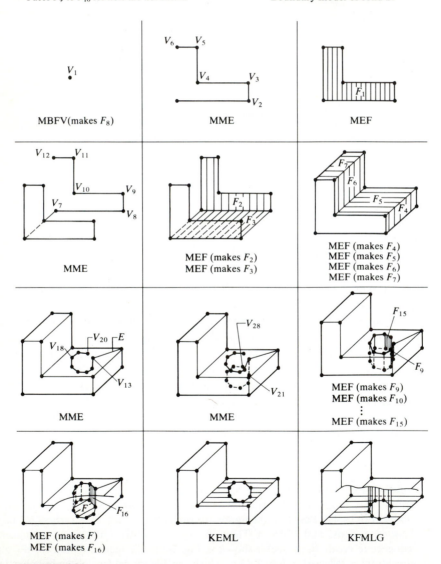

**FIGURE 7-34**
Creation of boundary model of solid $S$.

only reflects topology construction and not geometry. It is presented in this way for clarity and learning purposes. However, this topology could have been shown in one plane or one surface; that is, model $S$ is homomorphic to a sphere. Each step in the figure shows the operator(s) used and its result(s) and whatever is needed from the previous steps. It is observed that at the end of construction the final number of primitives (faces, edges, vertices, loops, bodies, and genus) created is equal to the number calculated from Fig. 7-33. Any intermediate topology (edge $E$ and face $F$) that is created for construction purposes has to be killed. Notice that the face created by the MBFV operator is chosen arbitrarily as $F_8$ for convenience. It could have been equally chosen to be $F_1$ or $F_3$. At this point, the face has no edges. However, its edges are formed automatically later on after creating the faces ($F_1$, $F_6$, and $F_7$) surrounding it. It would have been impossible to create $F_8$ otherwise and the topology would have been invalid if we had ignored the face created by MBFV.

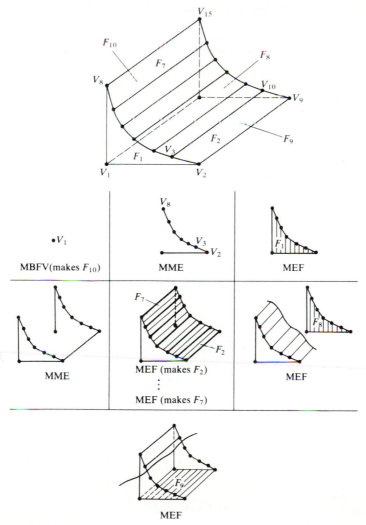

**FIGURE 7-35**
Creation of boundary model of solid fillet.

Similarly, the body created by this command is part of the topology and in fact is the body of solid $S$.

The reader can perhaps find a totally different set of steps than those shown in Fig. 7-34 to construct the boundary model, or these steps can change significantly depending on the available set of Euler operators. For example, if composite Euler operators for linear sweep and making cylinders are available, the model can easily be constructed in a smaller number of steps. The reader is encouraged to investigate this route.

**Example 7.6.** Create the boundary model of the solid fillet shown in Fig. 7-21.

**Solution.** Figure 7-35 shows the boundary model of the solid fillet and its creation. The curved face has been approximated by six facets. The construction steps follow the same general outline as in Example 7.5. It is obvious that the larger the number of facets, the more the CPU time and storage needed to create the model.

**Example 7.7.** Develop an algorithm that can enable the user to create and manipulate boundary models by using set operations.

**Solution.** We have mentioned in Sec. 7.2 and at the beginning of Sec. 7.7 that B-rep-based packages use set operations to create and manipulate boundary models. They seem to be more efficient than Euler operators. We have also mentioned in Sec. 7.7 that the effect of set operations on a B-rep model is different from that on a CSG model. In the former, the new boundary that results from the operation must be evaluated.

This example illustrates how to develop an algorithm to provide the user with the set operations union, difference, and intersection. The problem at hand can be stated as follows. Given two solids or primitives as boundary models, find their union, difference, and intersection. This problem is also known as boundary merging for B-reps and boundary evaluation for CSG. Set-operation algorithms are in general very complex programs. For simplicity purposes, let us assume that the two solids are not tangent to or touch each other. Thus, if two solids intersect

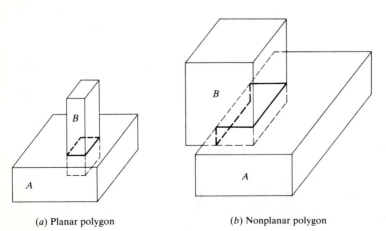

(a) Planar polygon

(b) Nonplanar polygon

**FIGURE 7-36**
Intersection polygons of two solids.

at all, they do so along one or more closed (maybe nonplanar) intersection polygons. Figure 7-36 shows a planar and a nonplanar intersection polygon. Such polygons result from the edges of each solid intersecting the faces of the other solid.

While set-operation algorithms could be written to manipulate the low-level data structures directly, basing them on Euler operators offers important advantages. They guarantee the topological validity of the result of the set operation and they hide the actual data structure during writing of the algorithms. The algorithm we develop here is therefore based on Euler operators. First, it finds the intersection polygon between the two solids and checks for its planarity. Then it splits each solid's boundary along this polygon. The two solids are then classified against each other. Finally, the proper parts of the two solids are glued together to give the desired result.

The detailed steps of the algorithm for the case of a planar intersection polygon (see Fig. 7-36a) are described as follows:

1. Compare each of the faces of $A$ with the edges of $B$. If an edge intersects a face in a point, we create a null edge as a loop into the face and split the edge by two coincident vertices to form a null edge. Figure 7-37a shows the result of this step. To create the null edges in the top face of $A$, the MME operator is used four times to create two edges (the dashed and the null edges) each time. This is followed by KEML four times to eliminate the undesired edges (dashed). To create the null edges in solid $B$, each of its intersecting edges is split twice by the ESPLIT operator. Each null edge in $A$ or $B$ has two coincident vertices, that is, geometrically identical vertices.

2. Repeat step 1 for faces of $B$ and edges of $A$. The result of this step is null for this example.

3. Connect neighbor null edges in each solid to create the intersection polygons. A "combine" algorithm is needed to achieve this step and is assumed to be available. Such an algorithm could be based on connecting the vertices of a null edge to the vertices of the nearest null edge on the same face by the MEF operator. To accomplish connecting the null edges, they are sorted in a given direction by the coordinates of their vertices. An intersection polygon is constructed for each solid and since these polygons are identical, the "combine" algorithm can construct both of them simultaneously.

   In order to construct the intersection polygons, null faces have to be created. Some of these faces will be killed with the null edges created in steps 1 and 2 to complete the construction process and some will remain to enable the two solids $A$ and $B$ to be split. To avoid any confusion that may result from this process, the following rule can be followed. Use the MEKL operator to create as many edges as there are loops created in steps 1 and 2 and then use the MEF operator to create the remaining edges. For solid $A$ in Fig. 7-37b, MEKL is applied four times to create four edges, say the top polygon, and kill the four loops created in step 1. Then MEF is used four times to create, say, the bottom polygon. The double-intersection polygon is needed to create the intersection faces. For solid $B$, MEF is applied eight times.

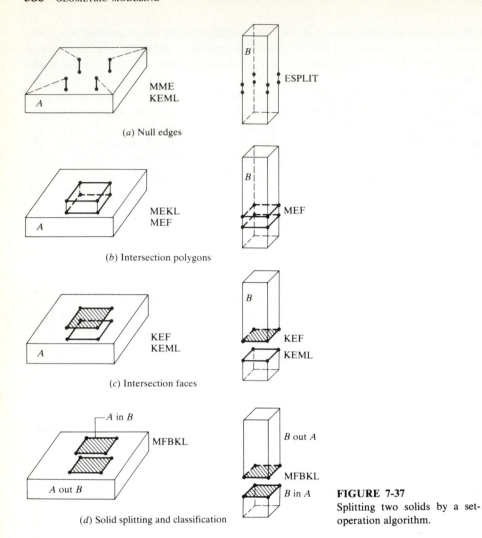

(a) Null edges

MME
KEML

ESPLIT

(b) Intersection polygons

MEKL
MEF

MEF

(c) Intersection faces

KEF
KEML

KEF
KEML

(d) Solid splitting and classification

—A in B

MFBKL

A out B

B out A

MFBKL

B in A

**FIGURE 7-37**
Splitting two solids by a set-operation algorithm.

4. Construct the intersection faces by deleting all the null edges in each solid. This is accomplished by using the KEF operator for all of the edges except the last null edge which is killed by KEML. In Fig. 7-37c, KEF is applied three times followed by one KEML.

5. Split each solid into two by using the MFBKL operator as shown in Fig. 7-37d.

   By following steps 1 to 5, the reader can easily find out that all the null edges are killed for each solid, eight edges and two faces are added to solid A, eight edges and six faces are added to solid B, and one body is added to each solid. While it is easy to interpret all these topological changes, faces need some clarification. For solid A, the two faces are the intersection faces. In this case, the top face of A is split into two faces; one of them is an intersection face. For solid B, two of the six faces are the intersection faces and the other four result from splitting each side face by the intersection polygon.

**6.** Each subsolid resulting from step 5 is classified against each original solid, that is, $A$ is classified as $A$ in $B$ and $A$ out $B$ and $B$ is classified as $B$ in $A$ and $B$ out $A$. $A$ in $B$, for example, means the parts of $A$'s boundary inside $B$ whereas $A$ out $B$ is outside $B$ (see Fig. 7-37d). Notice that any part of the original solid is by itself a valid solid.

**7.** Combine the proper parts of each solid to obtain the desired set operation as follows:

$$A \cup B = \text{Glue } (A \text{ out } B, B \text{ out } A)$$

$$A \cap B = \text{Glue } (A \text{ in } B, B \text{ in } A)$$

$$A - B = \text{Glue } (A \text{ out } B, B \text{ in } A)$$

or $$B - A = \text{Glue } (B \text{ out } A, A \text{ in } B)$$

The gluing operator for $A \cup B$ and $A - B$ is simple because all the subsolids involved are closed objects. The KFEVB operator described in Tables 7.4 and 7.5 can be used to glue the subsolids to give the proper results, that is, closed solids. However, if the same gluing operator is used in $A \cap B$ and $B - A$, open (unregularized open sets) objects would result. This is because one of its operands ($A$ in $B$) is an open object—in this example the two-dimensional intersection face. In such a case, the same previous gluing operator, KFEVB, can be used with the difference that it kills only one face instead of two. Therefore, the operator satisfies Eq. (7.58) and is used to kill the open object (intersection face).

The above described algorithm can be applied to any two boundary models whose classifications with each other do not yield $A$ on $B$ and/or $B$ on $A$ cases. This is why we mentioned in the beginning that $A$ and $B$ should not touch each other. The reader can extend solid $B$ to pierce through $A$ and apply the above steps.

The reader is also encouraged to apply this algorithm to the two solids shown in Fig. 7-36b. It can be assumed that an algorithm that sorts vertices by their planes is available. In this case, six null edges on each of $A$ and $B$, four loops for $A$, eight faces for $A$, two loops for $B$, and twelve faces for $B$ are created as intermediate results. After killing the null edges and splitting $A$ and $B$, four and eight faces are created to split $A$ and $B$ respectively to give the final result. The gluing process is exactly as above except that the number of faces the gluing operator has to kill is four instead of two.

### 7.7.3 Remarks

The B-rep scheme is very popular and has a strong history in computer graphics because it is closely related to traditional drafting. Its main advantage is that it is very appropriate to construct solid models of unusual shapes that are difficult to build using primitives. Examples are aircraft fuselage and automobile body styling. Another major advantage is that it is relatively simple to convert a B-rep model into a wireframe model because the model's boundary definition is similar to the wireframe definition. For engineering applications studied to date, algorithms based on B-rep are reliable and competitive with those based on CSG.

One of the major disadvantages of the boundary model is that it requires large amounts of storage because it stores the explicit definition of the model boundaries. It is also a verbose scheme—more verbose than CSG. The model is defined by its faces, edges, and vertices which tend to grow fairly fast for complex models. If B-rep systems do not have a CSG-compatible user interface, then it becomes slow and inconvenient to use Euler operators in a design and production environment. In addition, faceted B-rep is not suitable for many applications such as tool path generations.

## 7.8 CONSTRUCTIVE SOLID GEOMETRY (CSG)

CSG and B-rep schemes are the most popular schemes to create solid models of physical objects. This is apparent from the existing research and technological activities. They are the most popular because they are the best understood representations thus far. CSG offers representations that are succinct, easy to create and store, and easy to check for validity. Moreover, difference and intersection operations can respectively provide means for material removal processes and interference checking between objects. Interference checking is useful in many applications such as vision and robot path planning.

A CSG model is based on the topological notion that a physical object can be divided into a set of primitives (basic elements or shapes) that can be combined in a certain order following a set of rules (boolean operations) to form the object. Primitives themselves are considered valid CSG models. Each primitive is bounded by a set of surfaces; usually closed and orientable. The primitives' surfaces are combined via a boundary evaluation process to form the boundary of the object, that is, to find its faces, edges, and vertices. In addition to degenerating an object to a collection of primitives, a CSG model is fundamentally and topologically different from a B-rep model in that the former does not store explicitly the faces, edges, and vertices. Instead, it evaluates them whenever they are needed by applications' algorithms, e.g., generation of line drawings. The reader might then ask the question: if a CSG scheme has to evaluate faces, edges, and vertices, why not use a B-rep scheme from the beginning? The answer to this question entails close comparison between all aspects of both schemes including efficiency and performance. Such comparison is difficult to make due to all implementation and algorithmic details involved. However, one answer can be given. The concept of primitives offers a different conceptual way of thinking that may be extended to model engineering processes such as design and manufacturing. It also appears that CSG representations might be of considerable importance for manufacturing automation as in the study of process planning and rough machining operations.

There are two main types of CSG schemes. The most popular one, and the one we always mean when we talk about CSG models, is based on bounded solid primitives, that is, r-sets. The other one, less popular, is based on generally unbounded half-spaces, that is, non-r-sets. The latter scheme belongs more to half-space representation covered in Sec. 7.6. As a matter of fact, bounded solid primitives are considered composite half-spaces and the boundaries of these primitives are the surfaces of the corresponding half-spaces. CSG systems based on bounded primitives (e.g., PADL-2 and GMSOLID) allow their sophisticated users to use both their bounded primitives and/or half-spaces to create new primitives, typically called metaprimitives. It is also possible to extend the modeling domain of a system by implementing new half-spaces, and eventually new primitives, into its software. This implementation does not only require the trivial inclusion of the half-space equation into the software, but more importantly it requires developing supporting utilities such as intersecting the half-space with itself as well as other already existing half-spaces.

The modeling domain of a CSG scheme depends on the half-spaces that underlie its bounded solid primitives, on the available rigid motion and on the available set operators. For example, if two schemes have the same rigid motion and set operations but one has just a block and a cylinder primitive and the other has these two plus a tetrahedron, the two schemes are considered to have the same domain. Each has only planar and cylindrical half-spaces, and the tetrahedron primitive the other system offers is just a convenience to the user and does not extend its modeling domain. Similarly, the surfaces that a CSG scheme can represent directly depend on the bounding surfaces of its underlying half-spaces. The most widely represented surfaces are the quadric surfaces that bound most existing primitives. Extending the solid modeling domain to cover sculptured surfaces requires representing a "sculptured" half-space and its supporting utilities.

CSG schemes based on bounded primitives are usually more concise than those based on half-spaces because half-spaces are lower-level primitives. As an example, consider the solid shown in Fig. 7-38a. The model is represented by three bounded primitives (Fig. 7-38b) and seven half-spaces (Fig. 7-38c). Considering the half-spaces composing the three bounded primitives, it is obvious that 15 half-spaces (six for each block and three for the cylinder) have been used. Some of these half-spaces, such as the two at the bottom of blocks $A$ and $B$, are redundant. This redundancy is perfectly accepted by users in trade of the conveniences they gain from using bounded primitives. However, it raises the question of the minimal CSG representation of a solid.

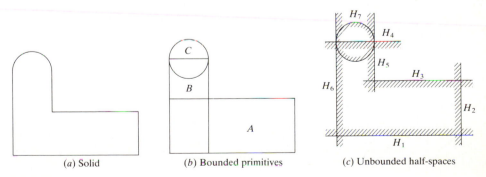

(a) Solid                (b) Bounded primitives                (c) Unbounded half-spaces

**FIGURE 7-38**
Bounded and unbounded primitives.

The database of a CSG model, similar to B-rep, stores its topology and geometry. Topology is created via the regularized set (boolean) operations that combine primitives. Therefore, the validity of the resulting model is reduced to the validity checks of the used primitives. For bounded primitives, these checks are usually simple (in the form of greater than zero) and the validity of the CSG model may be ensured essentially at the syntactical level. This means that in a CSG language a model is valid if it can be described syntactically correct using this language (user interface). The geometry stored in the database of a CSG model includes configuration parameters of its primitives and rigid motion and transformation. Geometry of faces, edges, and vertices are not stored but can be calculated via the boundary evaluation process.

While data structures of most boundary representations are based on the winged-edge structure developed by Baumgart in 1972, data structures of most CSG representations are based on the concept of graphs and trees. This concept is introduced here in enough depth to enable understanding of CSG data structures. The interested reader is referred to any standard textbook on Pascal or data structures for more details.

A graph is defined as a set of nodes connected by a set of branches or lines. Each branch in a graph is specified by a pair of nodes. Figure 7-39a illustrates a graph. The set of nodes is $\{A, B, C, D, E, F, G\}$ and the set of branches, or the set of pairs, is $\{\{A, B\}, \{A, C\}, \{B, C\}, \{B, E\}, \{B, F\}, \{B, G\}, \{C, D\}, \{C, E\}\}$. Notice that these pairs are unordered, that is, no relations exist between the elements of each pair. For example, the pair $\{A, B\}$ can also be $\{B, A\}$. If the pairs of nodes that make up the branches are ordered pairs, the graph is said to be a directed graph or digraph. This means that branches have directions in a digraph and become in a sense arrows going from one node to another, as shown in Fig. 7-39b. The tail of each arrow represents the first node in the pair and its head represents the second node. The set of ordered pairs for Fig. 7-39b is $\{(A, B), (A, C), (C, B), (B, E), (F, B), (B, G), (D, C), (E, C)\}$.

Each node in a digraph has an indegree and outdegree and has a path it belongs to. The indegree of a node is the number of arrow heads entering the node and its outdegree is the number of arrow tails leaving the node. For example, node $B$ in Fig. 7-39b has an indegree of 3 and an outdegree of 2 while node $D$ has a zero indegree and an outdegree of 1. Each node in a digraph

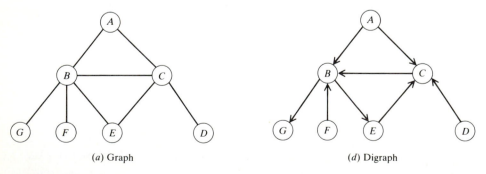

(a) Graph            (d) Digraph

**FIGURE 7-39**
Graphs and digraphs.

belongs to a path. A path from node $n$ to node $m$ is defined as a sequence of nodes $n_1, n_2, \ldots, n_k$ such that $n_1 = n$ and $n_k = m$ and any two subsequent nodes $(n_i, n_{i+1})$ in the sequence are adjacent to each other. For example, the path from node $A$ to node $G$ in Fig. 7-39b is $A, B, G$ or $A, C, B, G$. If the start and end nodes of a path are the same, the path is a cycle. If a graph contains a cycle, it is cyclic; otherwise it is acyclic.

We now turn our attention to trees. A tree is defined as an acyclic digraph in which only a single node, called the root, has a zero indegree and every other node has an indegree of 1. This implies that any node in the tree except the root has predecessors or ancestors. Based on this definition, a graph need not be a tree but a tree must be a graph. The digraph shown in Fig. 7-39b is not a tree. However, its modification shown in Fig. 7-40a is a tree. Node $A$ is the root of the tree and nodes $E$, $F$, and $G$, for example, have node $B$ as their ancestor or node $B$ has nodes $E$, $F$, and $G$ as its descendants. If the descendants of each node are in order, say, from left to right, then the tree is an ordered one. Moreover, when each node of an ordered tree has two descendants (left and right), the tree is called a binary tree (see Fig. 7-40b). Finally, if the arrow directions in a binary tree are reversed such that every node, except the root, in the tree has an out-degree of 1 and the root has a zero outdegree, the tree is called an inverted binary tree (see Fig. 7-40c). An inverted binary tree is very useful to understand the data structure of CSG models (sometimes called boolean models).

Any node in a tree that does not have descendants, that is, with an out-degree equal to zero, is called a leaf node and any node that does have descendants (outdegree greater than zero) is an interior node. In Fig. 7-40b, nodes $D$, $E$, $F$, and $G$ are leaf nodes and nodes $B$ and $C$ are interior nodes. Nodes $G$ and $D$ are called the leftmost leaf and the rightmost leaf of the tree respectively. Nodes

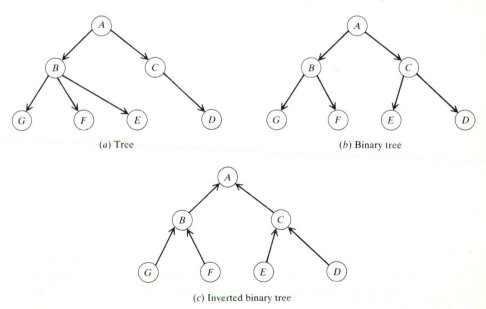

(a) Tree

(b) Binary tree

(c) Inverted binary tree

**FIGURE 7-40**
Types of trees.

in a tree can also be viewed from a different perspective as follows. Every node of a tree $T$ is a root of another tree, called a subtree of $T$, contained in the original tree $T$. A subtree is itself a binary tree. Any tree can be divided into two subtrees: left and right subtrees of the original tree. Considering Fig. 7-40*b*, the original tree consists of seven nodes with $A$ as its root. Its left subtree is rooted at $B$ and its right subtree is rooted at $C$. This is indicated by the two branches emanating from $A$ to $B$ on the left and to $C$ on the right. The absence of a branch indicates an empty subtree. The binary trees rooted at the leaves $D$, $E$, $F$, and $G$ have empty (nil) left and right subtrees.

Let us return back to the data structures of CSG representations and relate them to graphs and trees. Consider the solid shown in Fig. 7-41*a* with its MCS. A block and a cylinder primitive are enough to create the CSG model of the solid. Figure 7-41*b* shows one of the possible ways to decompose the solid into its primitives. Using the local coordinate systems of the primitives as shown in Fig. 7-4, and regardless of the user interface or command syntax offered by a particular CAD/CAM system, a user can construct the CSG model using the following steps:

$$
\left.
\begin{aligned}
B_1 &= \text{block positioned properly} \\
B_2 &= \text{block positioned properly} \\
B_3 &= \text{block} \\
B_4 &= B_3 \text{ moved properly in the } X \text{ direction} \\
C_1 &= \text{cylinder positioned properly} \\
C_2 &= C_1 \text{ moved properly in the } X \text{ direction} \\
C_3 &= \text{cylinder positioned properly} \\
C_4 &= C_3 \text{ moved properly in the } X \text{ direction}
\end{aligned}
\right\} \text{Primitives' definitions}
$$

$$
\left.
\begin{aligned}
S_1 &= B_1 \cup^* B_3 \\
S_2 &= S_1 \cup^* C_1 \\
S_3 &= S_2 \cup^* C_3
\end{aligned}
\right\} \text{Construct left half}
$$

$$
\left.
\begin{aligned}
S_4 &= B_2 \cup^* B_4 \\
S_5 &= C_2 \cup^* S_4 \\
S_6 &= C_4 \cup^* S_5
\end{aligned}
\right\} \text{Construct right half}
$$

$$
S = S_3 \cup^* S_6 \quad \} \text{ Model}
$$

To save the above steps in a data structure, such a structure must preserve the sequential order of the steps as well as the order of the boolean operations in any step; that is, the left and right operands of a given operator. The ideal solution is a digraph; call it a CSG graph. A CSG graph is a symbolic (unevaluated) representation and is intimately related to the modeling steps used by the user. This makes the CSG graph a very efficient data structure to define and edit a solid. The CSG graph representing the above steps is shown in Fig. 7-42. Each of the intermediate solids $S_1$ to $S_6$ is shown as the same node of its corresponding set operation node. Notice that the steps starting from $S_1$ and ending at $S$ can be replaced by

$$
S = B_1 \cup^* B_3 \cup^* C_1 \cup^* C_3 \cup^* B_2 \cup^* B_4 \cup^* C_2 \cup^* C_4
$$

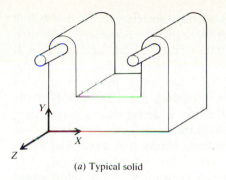

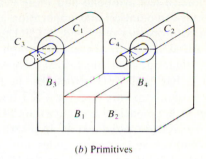

(a) Typical solid

(b) Primitives

**FIGURE 7-41**
A typical solid and its building primitives.

where set operations are evaluated from left to right unless otherwise indicated by parenthesis. In this case the intermediate solids $S_1$ to $S_6$ do not exist and should be removed from the CSG graph.

    While a CSG graph has a succinct data structure to represent a solid model and is suitable for convenient and efficient editing of the model, it is not suitable to use in geometric computation. This is mainly because of the cycles that the graph may have which, in turn, means graph nodes may be shared to reflect congruence relationships in the solid. This sharing means that useful information about the solid such as the locations of shared nodes is not explicitly stored by the graph structure. Another reason the CSG graph is not efficient in computations is its storage of real expressions that may be used in defining a solid (e.g., $c = b^2$, then use $c$ as a primitive parameter) as strings, that is, unevaluated.

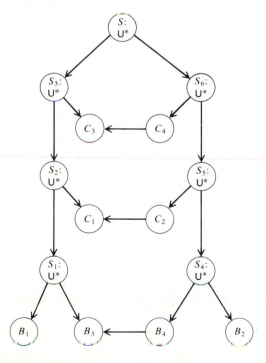

**FIGURE 7-42**
CSG graph of a typical solid.

Therefore, a less symbolic and more evaluated data structure is needed before involving computation and application algorithms such as boundary evaluation and mass properties of a solid. A CSG tree data structure is an ideal solution. It is a natural extension of the CSG graph and results from copying shared nodes and evaluating all strings (real expressions). Some solid modelers such as PADL-2 has both data structures. In these modelers, the CSG graph is the primary data structure and the CSG tree structure is derived from it whenever needed. Other modelers may have the CSG tree as their only data structure.

A CSG tree is defined as an inverted ordered binary tree whose leaf nodes are primitives and interior nodes are regularized set operations. Figure 7-43 shows the CSG tree derived from the CSG graph shown in Fig. 7-42. Notice that this CSG tree can be derived directly from the modeling steps without having to create the CSG graph. As a matter of fact, the tree can be created from the planning strategy shown in Fig. 7-41b. In Fig. 7-43, blocks $B_1$ to $B_4$, cylinders $C_1$ to $C_4$, and union operators are renamed as $P_1$ to $P_4$, $P_5$ to $P_8$, and $OP_1$ to $OP_7$ respectively to emphasize the fact that they are evaluated and stored explicitly compared to their counterparts used in the CSG graph (Fig. 7-42). The CSG tree is shown with its full details including arrows. In practice, the arrows are usually not shown, the leaf nodes are just shown as primitives' names without circles surrounding them, and a line extends from the tree root up to indicate the result of the final solid. Other styles of showing a CSG tree may replace primitive names by their sketches as well as showing each intermediate solid that results from an operator in the stream of the tree branches.

The total number of nodes in a CSG tree of a given solid is directly related to the number of primitives the solid is decomposed to. The number of primitives decides automatically the number of boolean operations required to construct the solid. If a solid has $n$ primitives, then there are $(n - 1)$ boolean operations for a total of $(2n - 1)$ nodes in its CSG tree. The balanced distribution of these nodes in the tree is a desired characteristic for various applications, especially those that use ray casting such as shading and mass properties. A balanced tree is defined as

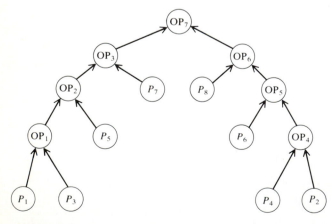

**FIGURE 7-43**
CSG tree of a typical solid.

a tree whose left and right subtrees have almost an equal number of nodes; that is, the absolute value of the difference $(n_L - n_R)$ is as minimal as possible where

$$n_L + n_R = 2n - 2 \qquad (7.59)$$

The root node is not included in this equation. $n_L$ and $n_R$ are the number of nodes of the left and right subtrees respectively. A perfect tree is one whose $|n_L - n_R|$ is equal to zero. A perfect tree results only if the number of primitives is even. For a perfect tree, the following equation applies:

$$n_L = n_R = n - 1 \qquad (7.60)$$

Each subtree has $n/2$ leaf nodes (primitives) and $(n - 2)/2$ interior nodes (boolean operations). Figure 7-43 shows a perfect tree.

The creation of a balanced, unbalanced, or a perfect CSG tree depends solely on the user and how he/she decomposes a solid into its primitives. The general rule to create balanced trees is to start to build the model from an almost central position and branch out in two opposite directions or vice versa; that is, start from two opposite positions and meet in a central one. The tree shown in Fig. 7-43 begins at the central blocks $B_1$ and $B_2$ and branches out. Another useful rule is that symmetric objects can lead to perfect trees if they are decomposed properly (see Figs. 7-41b and 7-42) starting from the plane(s) of symmetry. Figure 7-44 shows an unbalanced tree of the same solid shown in Fig. 7-41. This tree results if the user starts building the model from the left or right side. In this figure, primitives $P_1$ to $P_7$ correspond to primitives $C_1$, $C_3$, $B_3$, $B_1 + B_2$, $B_4$, $C_4$, and $C_2$ respectively, shown in Fig. 7-41b. In this tree $n_L = 11$ and $n_R = 1$. Reorganizing an unbalanced tree internally by a solid modeler is possible but is not practical to do.

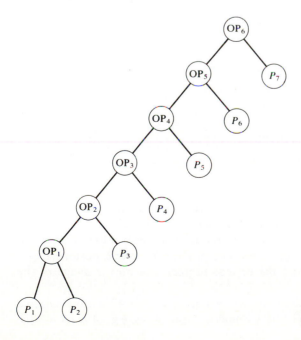

**FIGURE 7-44**
An unbalanced CSG tree.

Application algorithms must traverse a CSG tree, that is, pass through the tree and visit each of its nodes. Also traversing a tree in a certain order provides a way of storing a data structure. The order in which the nodes are visited in a traversal is clearly from the first node to the last one. However, there is no such natural linear order for the nodes of a tree. Thus different orderings are possible for different cases. There exist three main traversal methods. The methods are all defined recursively so that traversing a binary tree involves visiting the root and traversing its left and right subtrees. The only difference among the methods is the order in which these three operations are performed. The three methods are preorder, inorder, and postorder traversals. Sometimes, these methods are referred to as prefix, infix, and postfix traversals. Three other methods can be derived from these three main ones by reversing the order of the traversal to give reverse preorder, reverse inorder, and reverse postorder traversals.

To traverse a tree in preorder, we perform the following three operations in the order they are listed:

**1.** Visit the root.
**2.** Traverse the left subtree in preorder.
**3.** Traverse the right subtree in preorder.

In the reverse preorder method, the three operations are reversed to give the sequence of visiting the right subtree, then the left subtree, and then the root. Figure 7-45 shows the preorder, and its reverse, traversal of the tree shown in Fig. 7-43.

To traverse a tree in inorder (or symmetric order):

**1.** Traverse the left subtree in inorder.
**2.** Visit the root.
**3.** Traverse the right subtree in inorder.

In the reverse inorder method, the tree is traversed by visiting the right subtree, then the root, and then the left subtree (see Fig. 7-46).

To traverse a tree in postorder:

**1.** Traverse the left subtree in postorder.
**2.** Traverse the right subtree in postorder.
**3.** Visit the root.

In the reverse postorder method, the tree is traversed by visiting the root, then the right subtree, and then the left subtree, as shown in Fig. 7-47.

By comparing Figs. 7-45 to 7-47, the reader can easily observe that the reverse preorder is a mirror image of the postorder, the reverse postorder is a mirror image of the preorder, and the reverse inorder is a mirror image of the inorder.

Which of the traversal methods shown in Figs. 7-45 to 7-47 is more suitable to store a tree in a solid modeler? In arithmetic expressions, e.g., $A + (B + C)D$,

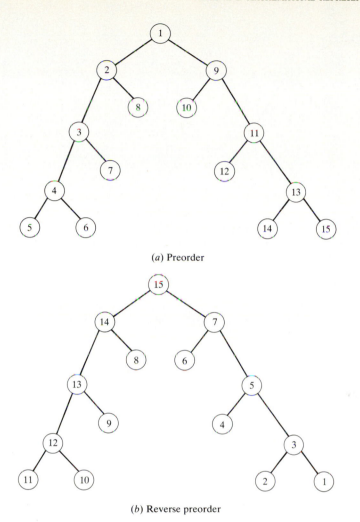

(*a*) Preorder

(*b*) Reverse preorder

**FIGURE 7-45**
Preorder and reverse preorder traversals of a tree.

the order of operations in an infix expression might require cumbersome parentheses while a prefix form requires scanning the expression from right to left.
Since most algebraic expressions are read from left to right, postfix is a more
natural choice. In addition, if the concept of a stack (refer to Pascal textbooks)
(last-in, first-out behavior) is used in an algorithm to evaluate an expression, the
postfix becomes the most efficient form. These same rationales can be extended to
binary trees. Trees are derived from steps that are commands input by a user to
create a solid. These commands are scanned from left to right by the software
and they might contain parentheses. In addition, if stacks are used in algorithms
that evaluate trees (PADL-2 does that), then the postorder is the ideal choice to
traverse a tree. However, the problem with the postorder traversal, as shown in
Fig. 7-47*a*, is that the root of the tree has the highest node number. It is more

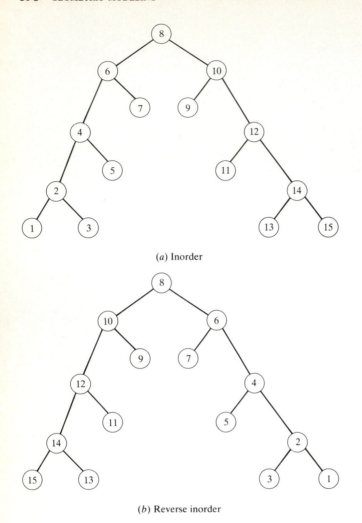

(a) Inorder

(b) Reverse inorder

**FIGURE 7-46**
Inorder and reverse inorder traversals of a tree.

natural to assign the root the number 1. Therefore, the reverse postorder seems the ideal traversal method of a CSG tree. PADL-2 solid modeler uses such a method. In this method also the leftmost leaf node of the tree has the highest node number in the tree.

### 7.8.1   Basic Elements

Bounded solid primitives, or primitives for short, are the basic elements or building blocks a CSG scheme utilizes to build a model. Primitives can be viewed as parametric solids which are defined by two sets of geometric data. The first set is called configuration parameters and the second is the rigid motion parameters. The most common primitives are shown in Fig. 7-4. Each one of these primitives

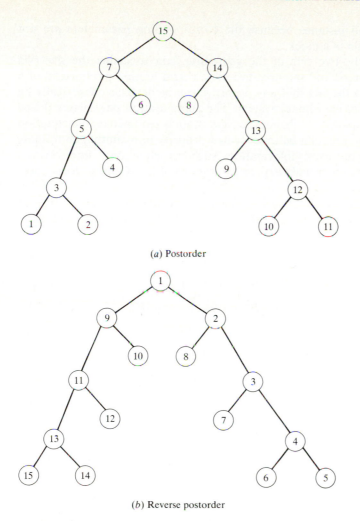

(a) Postorder

(b) Reverse postorder

**FIGURE 7-47**
Postorder and reverse postorder traversals of a tree.

is defined by its configuration and rigid motion parameters. For example, the configuration parameters of a block primitive is the triplet (ordered 3-tuple) $(W, H, D)$ and its rigid motion is given by the location of its origin $P$ relative to a reference coordinate system, say MCS or WCS, or by explicit rigid motion values (translation and/or rotation). The configuration parameters of the other primitives are shown in Fig. 7-4.

Each primitive, viewed as a parametric object, corresponds to a family of parts. Each given part of the family is called a primitive instance and corresponds to one and only one value set of the primitive configuration parameters. Each primitive has a valid configuration domain which is maintained by its solid modeler. User input values of any primitive parameters are usually checked against its valid domain. For example, a block primitive instance of the triplet

(0, 0, 0) is not a valid instance because the corresponding parameters are not within the valid domain of a block.

The choice of the two sets of the geometric data to define the size (via configuration parameters) and the orientation (via rigid motion parameters) of a primitive are based on the fact that any primitive can be described generically by an equation in its local coordinate system. The configuration parameters define such an equation completely. Utilizing the rigid motion parameters, the equation and, therefore, the primitive can be transformed properly into another coordinate system. Therefore, primitives' information such as equations, intersections, boundaries, and others are usually expressed in terms of the primitive local coordinate system $X_L Y_L Z_L$.

Mathematically, each primitive is defined as a regular point set of ordered triplets $(x, y, z)$. For the primitives shown in Fig. 7-4, these point sets are given by:

Block:    $\{(x, y, z): 0 < x < W, 0 < y < H, \text{ and } 0 < z < D\}$ (7.61)

Cylinder:    $\{(x, y, z): x^2 + y^2 < R^2, \text{ and } 0 < z < H\}$ (7.62)

Cone:    $\{(x, y, z): x^2 + y^2 < [(R/H)z]^2, \text{ and } 0 < z < H\}$ (7.63)

Sphere:    $\{(x, y, z): x^2 + y^2 + z^2 < R^2\}$ (7.64)

Wedge:    $\{(x, y, z): 0 < x < W, 0 < y < H, 0 < z < D,$

   and $yW + xH < HW\}$ (7.65)

Torus:    $\{(x, y, z): (x^2 + y^2 + z^2 - R_2^2 - R_1^2)^2 < 4R_2^2(R_1^2 - z^2)\}$ (7.66)

Comparing Eqs. (7.61) to (7.66) with the half-space equations (7.51) to (7.55), it is obvious that each of the above bounded primitives is a combination of a finite number of half-spaces. A block is the regularized union of six intersecting half-spaces. Each of these half-spaces is given by one limit of the three inequalities of Eq. (7.61). Similarly, a cylinder, cone, and a wedge are the union of three, three, and five half-spaces respectively. Figure 7-48 shows two-dimensional illustrations of the half-spaces of each primitive shown in Fig. 7-4. Some half-spaces for the block and wedge primitives are not shown in the figure for clarity purposes.

There are many representational alternatives for primitives. Some representations are terse and contain little or no redundant data. These are called input representations and are convenient for user input. Other representations are verbose, contain lots of redundant data, and are therefore convenient and efficient for computational purposes. They are called internal representations. Most CSG-based modelers use both and usually derive the internal representation from the input one. While one of these modelers may provide alternative input representation, mainly for user convenience, it usually has only one internal representation and all input alternatives are converted to it before storage. Consider, for example, the torus primitive shown in Fig. 7-4. A user can specify its $R_I$ and $R_O$ or $R_1$ and $R_2$ as input representations to create it.

What are the redundant data of a primitive that a solid modeler calculates based on user input representation and stores as its internal representation for computational purposes? An internal representation of a primitive that does not

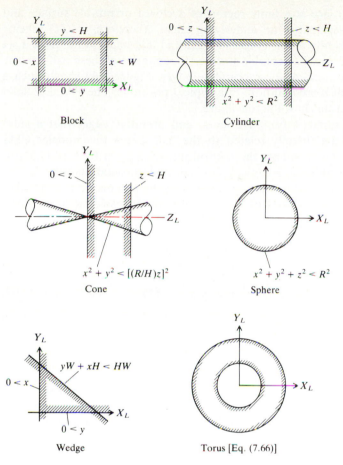

**FIGURE 7-48**
Half-spaces of bounded primitives.

have redundant data would only store the primitive's underlying half-spaces positioned and oriented properly in space, based on the user's configuration and rigid motion input parameters. Any other data such as primitive faces and edges that might be needed to evaluate the result of, say, a boolean operation must be derived by explicitly calculating the proper intersections of the underlying half-spaces. Such an approach would make application and computational algorithms totally inefficient. Therefore, underlying surfaces, faces, and edges, surface normals, and other data that are considered redundant are stored internally for each primitive in addition to its half-spaces. In essence the internal representation of each primitive is a CSG-rep plus a B-rep plus other information that is computationally useful. This "other information" could be engineering and design related in the case of implementing a new application into solid modeling.

Let us now look closely into how faces, edges, and other redundant data of a primitive are represented. Analogous to decomposing a solid into a combination of primitives, each primitive can be decomposed into a collection of faces and edges. Each face is a finite region of a closed orientable surface and each edge is a finite segment of an underlying curve. Therefore, a CSG scheme would have a set of primitives for its users to use and internally would have a set of half-spaces, a set of closed orientable surfaces (boundaries of these half-spaces), a set of primitive faces, and a set of primitive edges. Figure 7-49 shows such a data structure (internal representation) of a typical primitive. The PADL-2 solid modeler uses the structure shown in this figure.

The underlying surfaces, primitive faces, and primitive edges that a solid modeler can provide are directly related to the half-spaces the modeler CSG scheme utilizes. If a scheme utilizes the natural quadrics given by Eqs. (7.51) to (7.55), then planar, cylindrical, spherical, conical, and toroidal surfaces (called quadric surfaces) become the underlying surfaces of the scheme or the modeler. These surfaces are the boundaries of their corresponding half-spaces and their point sets are given by:

Planar surface: $\quad\quad\quad P = \{(x, y, z): z = 0\}$ (7.67)

Cylinder surface: $\quad\quad P = \{(x, y, z): x^2 + y^2 = R^2\}$ (7.68)

Spherical surface: $\quad\quad P = \{(x, y, z): x^2 + y^2 + z^2 = R^2\}$ (7.69)

Conical surface: $\quad\quad P = \{(x, y, z): x^2 + y^2 = [(R/H)z]^2\}$ (7.70)

Toroidal surface: $\quad\quad P = \{(x, y, z): (x^2 + y^2 + z^2 - R_2^2 - R_1^2)^2$

$$= 4R_2^2(R_1^2 - z^2)\} \quad (7.71)$$

These are infinite surfaces whose intersections yield infinite curves. These curves are usually classified against given primitives using set membership classification to determine which curve segments lie within these primitives and consequently within the solid.

Primitive faces are faces of primitives selected such that the boundary of any primitive may be represented as the union of a finite number of these faces after being positioned properly in space. The sufficient set of primitive faces to represent the boundary of any of the primitives shown in Fig. 7-4 consists of plate, triplate, disc, cylindrical, spherical, conical, and toroidal primitive faces. The equations of these primitive faces (Pfaces for short) are given by:

Plate Pface:: $\quad\quad F = \{(x, y, z): 0 < x < W, 0 < y < H, \text{ and } z = 0\}$ (7.72)

Triplate Pface: $\quad F = \{(x, y, z): 0 < x < W, 0 < y < H, \text{ and }$

$$yW + xH < HW\} \quad (7.73)$$

Disc Pface: $\quad\quad F = \{(x, y, z): x^2 + y^2 < R^2, \text{ and } z = 0\}$ (7.74)

Cylindrical Pface: $\quad F = \{(x, y, z): x^2 + y^2 = R^2, \text{ and } 0 < z < H\}$ (7.75)

Spherical Pface: $\quad F = \{(x, y, z): x^2 + y^2 + z^2 = R^2\}$ (7.76)

Conical Pface: $\quad\quad F = \{(x, y, z): x^2 + y^2 = [(R/H)z]^2, \text{ and } 0 < z < H\}$ (7.77)

Toroidal Pface: $\quad F = \{(x, y, z): (x^2 + y^2 + z^2 - R_2^2 - R_1^2)^2$

$$= 4R_2^2(R_1^2 - z^2)\} \quad (7.78)$$

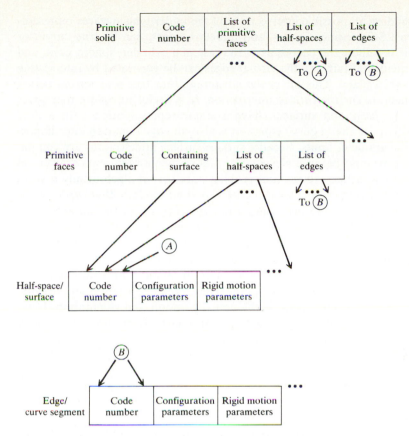

**FIGURE 7-49**
Data structure of a typical primitive solid.

Any of the above primitive faces is a subset of its underlying surface, that is, $F \subset P$. For example, the plate primitive face, Eq. (7.72), is a subset of the planar surface given by Eq. (7.67). The subset is bounded in the $x$ and $y$ directions. Similarly, a cylindrical primitive face is a finite (but not bounded) region (between $z = 0$ and $H$) of the cylindrical surface given in Eq. (7.68). In addition, the boundary of any primitive is a combination of these primitive faces. The boundary of a block primitive consists of six plate primitive faces positioned properly while that of a sphere or a torus consists of one spherical or toroidal primitive face respectively. The boundary of a cylinder consists of one cylindrical primitive face closed from each end by a disc primitive face. Lastly, a cone has a conical face closed by a disc and a wedge has three plates and two triplates.

Similar to primitive faces, primitive edges (sometimes called face bounding edges) are edges selected such that the boundary of any primitive face may be represented as the union of a finite number of these edges after being positioned properly in space. Each edge is a finite or bounded region of a corresponding underlying curve that may be possibly unbounded and disjointed (in which case curve segments make up the total curve). The underlying curves are usually obtained by finding all possible intersections of the underlying surfaces represented by a given CSG scheme. Curves, and therefore edges, can be true curves or virtual (profile or silhouette) curves, as in the case of a cylinder, sphere, cone, and a torus. For quadratic surfaces, the virtual edges can be generated by intersecting the surface with a plane. This offers the advantage that true and virtual curves can be obtained via surface/surface intersection. It is useful to realize that edges and curves, like faces and surfaces, have a single representation with a dual purpose. This implies that a curve equation is also an edge equation after imposing the proper parameter limits. Because surface/surface intersections are computationally intensive and because a particular CSG scheme represents a given set of surfaces, surface/surface intersection problems are solved algebraically *a priori* and stored in the corresponding solid modeler. A solid modeler that supports the primitives shown in Fig. 7-4 must contain the intersections of the surfaces given by Eqs. (7.67) to (7.71). For example, a cylinder/plane intersection may give a circle, an ellipse (could be very thin), two infinite parallel lines, or no intersection at all. What usually complicates the surface/surface intersection is the position and orientation of the two intersecting surfaces in space. One solution to this problem is to intersect the two surfaces in a given standard position and orientation and then transform the result to the actual position and orientation. Intersection curves are usually represented in parametric form because quadric surfaces can be parametrized conveniently as shown in Chap. 6. Equations of edges and curves are not given here. The reader is referred to Chap. 5 for some of these equations.

Surface/surface intersection is very crucial in geometric modeling in general and in solid modeling in particular. It is a decisive factor in determining and/or limiting the modeling domain of a solid modeler. As a matter of fact, it is the only factor in slowing down the implementations of sculptured surfaces into solid modeling. One might ask the following basic question. Why is surface/surface intersection so important while algebraic and numerical methods exist to solve virtually any two equations? The answer is not so much in the solution as it is in characterizing the solution. In order to perform boolean operations automatically, efficiently, and unambiguously, we need a precise description of the intersection curve. We need to know which surface pair results in which intersection curve, and we need to know when two different surface pairs give rise to the same curve. For example, if a boolean operation requires the intersection of a plane and a cylinder, the advance knowledge of their intersection curve provides a precise equation of the curve and consequently saves computation time that would otherwise be spent solving the two surface equations. In addition, two curves of a pair of surface/surface intersections can be checked out if they are identical, which can help eliminate many problems that may occur trying to differentiate between them. One may then conclude that if there was a universal description, say an equation, of any intersection curve in terms of the parameters of the two intersecting surfaces, the surface/surface intersection would have been solved once and for all.

While the full details of surface/surface intersections of quadric surfaces are beyond the scope of this book, the essence of the problem can be described as follows. Any of the surfaces described by Eqs. (7.67) to (7.71) can be rewritten in the following polynomial form:

$$Ax^2 + By^2 + Cz^2 + 2Dxy + 2Eyz + 2Fxz + 2Gx + 2Hy + 2Jz + K = 0$$

(7.79)

where $A, B, \ldots, K$ are arbitrary real constants. This equation can be expressed in a quadratic form as

$$F(x, y, z) = \mathbf{V}^T[Q]\mathbf{V} = 0 \qquad (7.80)$$

where $\mathbf{V}$ is a vector of homogeneous coordinates of a point on the surface and is given by $[x \quad y \quad z \quad 1]^T$. $[Q]$ is the coefficient matrix. It is symmetric and is given by

$$[Q] = \begin{bmatrix} A & D & F & G \\ D & B & E & H \\ F & E & C & J \\ G & H & J & K \end{bmatrix} \qquad (7.81)$$

The coefficient matrix $[Q]$ can be formed for any of the quadric surfaces [Eqs. (7.67) to (7.70)] by comparing the surface equation with Eq. (7.80). This gives

Planar surface: 
$$[Q] = \begin{bmatrix} 0 & 0 & 0 & 0 \\ 0 & 0 & 0 & 0 \\ 0 & 0 & 0 & \frac{1}{2} \\ 0 & 0 & \frac{1}{2} & 0 \end{bmatrix} \qquad (7.82)$$

Cylindrical surface: 
$$[Q] = \begin{bmatrix} 1 & 0 & 0 & 0 \\ 0 & 1 & 0 & 0 \\ 0 & 0 & 0 & 0 \\ 0 & 0 & 0 & -R^2 \end{bmatrix} \qquad (7.83)$$

Spherical surface: 
$$[Q] = \begin{bmatrix} 1 & 0 & 0 & 0 \\ 0 & 1 & 0 & 0 \\ 0 & 0 & 1 & 0 \\ 0 & 0 & 0 & -R^2 \end{bmatrix} \qquad (7.84)$$

Conical surface: 
$$[Q] = \begin{bmatrix} 1 & 0 & 0 & 0 \\ 0 & 1 & 0 & 0 \\ 0 & 0 & -(R/H) & 0 \\ 0 & 0 & 0 & 0 \end{bmatrix} \qquad (7.85)$$

The coefficient matrix $[Q]$ depends directly on the surface orientation and position. The above matrices are valid only if the local coordinate system of the primitive is identical to the MCS of the solid model to whom the primitive belongs. Otherwise, each matrix has to be transformed by the transformation matrix that results from the rigid motion (rotation and/or translation) of the primitive. This gives

$$[Q'] = [T]^T[Q][T] \qquad (7.86)$$

where $[Q']$ is the transformed coefficient matrix and $[T]$ is the transformation matrix given by Eq. (3.3). For example, if the origin of a sphere is located at point $(a, b, c)$ measured in the MCS, Eq. (7.84) is transformed by a translation vector to give

$$[Q'] = \begin{bmatrix} 1 & 0 & 0 & a \\ 0 & 1 & 0 & b \\ 0 & 0 & 1 & c \\ a & b & c & a^2 + b^2 + c^2 - R^2 \end{bmatrix} \tag{7.87}$$

This idea of transformation suggests that one can solve the intersection problem of two primitives in any convenient coordinate system and then transform the results as needed. Actually, there exists a set of algebraic constructs derived from $[Q]$ that remain invariant under rigid motions and completely specify the shape of the quadratic surface.

The intersection of two quadric surfaces can be solved as follows. If one surface has a coefficient matrix $[Q_1]$ and the other has $[Q_2]$, then the equation of the intersection curve is

$$\mathbf{V}^T([Q_1] - [Q_2])\mathbf{V} = 0 \tag{7.88}$$

This equation describes an infinite intersection curve. In order to determine finite segments of this curve (edges) which belong to the intersecting primitives, we must find the appropriate bounding points. One way of finding these points is if one of the intersecting surfaces can be parametrized in terms of two parameters $u$ and $v$, and Eq. (7.88) can be solved for one of the parameters in terms of the other. In this case, the equation of the intersecting curve can be written in a parametric form in terms of one parameter. To understand this approach, consider the cylinder/quadric intersection case. In this problem, we wish to find the intersection curve between a cylinder and any quadric surface. Assuming the cylinder is in a standard position, its parametric equation is known. In this case the vector $\mathbf{V}$ in Eq. (7.88) is $[R \cos u \quad R \sin u \quad v]^T$. Substituting Eqs. (7.81) and (7.83) for $[Q_1]$ and $[Q_2]$ respectively and simplifying the result we get

$$a(u)v^2 + b(u)v + c(u) = 0 \tag{7.89}$$

Equation (7.89) could also be obtained by substituting the parametric equation of a cylinder directly into Eq. (7.79). Equation (7.89) is a quadratic equation that can be solved for $v$ in terms of $u$. Moreover, the proper range of $u$ can be obtained by investigating the characteristics of the discriminant $b^2(u) - 4a(u)c(u)$. The same analysis can be done for the cone/quadric intersection. In this case the vector $\mathbf{V}$ (the cone parametric equation) is given by $[(R/H)v \cos u \quad (R/H)v \sin u \quad v]^T$.

What made it possible to reduce Eq. (7.88) to Eq. (7.89) in the case of a cylinder or a cone is the fact that either surface is a ruled quadric with $v$ being the parameter along the surface rulings. A sphere/quadric or torus/quadric intersection problem cannot be solved directly by following the above approach. A special transformation must be done first, which is not discussed here.

We have mentioned that surface normals are useful redundant data that is usually a part of the internal representations of solid primitives. A surface normal is usually useful in representing the direction of the unit normal vector of a face with respect to its solid so that the interior and exterior of the solid can be identified unambiguously. One convention is to choose the surface normal to be positive if it points away from its corresponding half-space. A positive surface normal of a cylindrical surface is one which points away from the cylindrical half-space given by Eq. (7.52) or simply points away from the cylinder axis. For a spherical surface, the positive surface normal points away from the center of the sphere. The surface normal can be calculated following the methods discussed in Chap. 6. Consider the cases of a plane, cylinder, and sphere (Fig. 7-19). The positive surface normal of a plane is simply the unit vector, $\hat{k}_L$, in the $Z_L$ direction. For a cylindrical surface, assuming that the position vector of a point on the surface is $\mathbf{P}$ (with respect to the local coordinate system), then the positive surface normal is given by $[\mathbf{P} - (\mathbf{P} \cdot \hat{k}_L)\hat{k}_L]/R$ where $\hat{k}_L$ is the unit vector in the $Z_L$ direction. For a sphere, it is $\mathbf{P}/R$.

The way the above convention of a surface normal is used to represent the direction of the unit normal of a face of a solid can be explained as follows. In some cases, the positive surface normal may point away from the solid interior or point into the interior, depending on the face position relative to its solid. In other cases, the negative surface normal may exhibit similar behavior. One can assign a variable to the face that may be 1 or $-1$. This variable always defines the outward pointing (away from the solid interiors) normal of the face of the solid. If the positive surface normal of the face underlying surface happens to point outward from the solid, then the variable is assigned the value 1. Otherwise, it is assigned the value $-1$ which indicates that the positive surface normal points inward into the solid. The value of the variable is predefined for each primitive face of all the primitive solids a solid modeler supports.

## 7.8.2 Building Operations

The main building operations in CSG schemes are achieved via the set operators or more specifically the regularized operators: union ($\cup^*$), intersection ($\cap^*$), and difference ($-^*$). Set operators are also known as boolean operators due to the close correspondence between the two. Union, intersection, and difference are equivalent to OR, AND, and NOT AND respectively. Due to the deep roots of CSG schemes in set theory or boolean algebra, CSG models are usually referred to as set-theoretic, boolean, or combinatorial models. Set-operation algorithms are amongst the most fundamental and delicate software components of solid modelers.

Unlike Euler operators, regularized set operators are not based on a given law or equation, but they derive their properties from the set theory and the concept of closure. They are considered higher-level operators than Euler operators. The validity checks for set operators are usually simple in the case of bounded primitives. They take the form of checking the user input of each primitive parameter. This is due to the fact that if primitives are valid and set operators are regularized, then the topology of the resulting solid is always valid.

Some solid modelers provide their users with other building operators that are less formal than set operators. ASSEMBLE and GLUE are two popular ones. Both of them operate on full solids and usually do not combine the solids.

They are merely assembled or glued. They only allow the user to refer to two solids as a single (usually) named entity. The two solids must be positioned properly first.

Almost all contemporary solid modelers provide their users with boolean operations. Boolean operators are mainly used to define solid models through proper combinations of solid primitives. Other important uses include modeling and simulation of manufacturing processes such as drilling and milling as well as detecting spatial interferences and collisions of positioned solid objects in space. In general, engineering processes that involve volumetric and spatial relationships are amenable to using boolean operations.

While boolean operators seem the same to users on all solid modelers, a set-operation algorithm depends primarily and solely on the solid representation scheme supported by each modeler. Example 7.7 in the previous section shows how set operations are performed for the B-rep scheme. In this section we see how they are performed for a CSG scheme. Regardless of the scheme the set operations depend on, an algorithm implementing them must evaluate the boundary of the resulting solid from a desired operation. For discussion purposes, let us write a typical set operation as $A\langle OP\rangle B$ where $A$ and $B$ are operands (primitives) and $\langle OP\rangle$ is any regularized set operator. The central question to implementing set operations can be asked as follows. Given the representations of two operands $A$ and $B$ of a given operator, evaluate the boundary of the resulting solid. The representations of $A$ and $B$ must contain representations of their boundaries and this is where the difference between schemes comes into play. In B-rep, boundaries are faces, edges, and vertices that are all stored explicitly. In CSG, boundaries are primitive faces and edges (no vertices) of primitives that are not stored explicitly and must be computed from their underlying surfaces/half-spaces.

Set operations are performed by so-called boundary merging in B-rep while in CSG they are performed by so-called boundary evaluation. In CSG, non-incremental and incremental boundary evaluations are available. In the non-incremental evaluator, only the boundary of the final solid $S$ is evaluated and not for subsolids of $S$. In the incremental evaluator, the boundaries of the intermediate subsolids are evaluated as the CSG tree is traversed to produce the boundary for the final solid. This latter evaluator is actually a boundary-merging procedure similar to what is used by B-rep schemes. Regardless of which evaluator is used in a modeler, performing set operations in CSG is equivalent to converting the CSG representation of a solid into its B-rep. The incremental boundary evaluator is more widely used than the nonincremental evaluator due to its computational efficiency and its speed in editing, displaying, and/or sectioning solids. GMSOLID and PADL-2 use incremental evaluators to perform set operations while PAD-1 uses a nonincremental boundary evaluator.

Set-operation algorithms based on a CSG scheme are quite similar in philosophy to those based on a B-rep scheme and discussed in Example 7.7. Both are based on edge/face intersection and classification. The differences arise mainly due to the amount and form of data stored by each scheme in its related data structure which, in turn, influences the classification process and how computations are organized. In Example 7.7, we have assumed that a classification algorithm exists to classify edges against a solid. A line of thinking for such an algorithm can follow the one described for the line/polygon classification problem covered in Sec. 7.5.3. While the line of thinking for the CSG representation of the same problem can be utilized in this section, we choose to investigate classification algorithms in more detail for better understanding.

To fully implement CSG-based set-operation algorithms, the required tools are an edge/solid intersection algorithm (not covered here), a classification algorithm to compute the set membership classification function $M[X, S]$ discussed in Sec. 7.5.3, and an algorithm to combine the resulting classifications. A classification algorithm can be based on the divide-and-conquer paradigm and combining classifications need the introduction of the concept of neighborhoods.

The divide-and-conquer paradigm is very similar in concept to the ray-casting algorithm (see Chap. 10). The main difference between the two is that the divide-and-conquer paradigm replaces the rays that are used by the ray-casting algorithm by the edges of a given solid. The line/polygon problem covered in Sec. 7.5.3 can be considered as a simple application of the paradigm. The paradigm is based on the fact that edge/solid classification is identically equivalent to classifying the edge against the left and right subtrees of the solid and then combining the two classifications using the same set operation that operates on the solid subtrees. If one of the subtrees is all primitives, then the edge is classified against all the primitives of the subtree. The paradigm, of course, requires procedures to classify edges with respect to primitives and to combine classifications (refer to Sec. 7.5.3 on how these procedures can be developed). The paradigm can also be extended to classify points and faces against a given solid.

The concept of neighborhoods is introduced to resolve the on/on ambiguities that result when combining "on" segments in a given classification. They are also used in converting a CSG representation into a B-rep. They are mainly useful with the divide-and-conquer paradigm to classify candidate sets and to combine classifications. Figure 7-18 does not show any "on" segments and also shows that combining "in" segments usually results in "in" segments. Figure 7-50 shows a case where a solid $S = A \cup B$. After classifying the edges of $A$ and $B$ against each other, combining the "on" segments of each primitive may result in "in" or "on" segments of $S$. The on/on ambiguities usually result when the two subsolids or primitives to be combined are tangent to each other along one or more faces. These ambiguities can be resolved using neighborhood information of any point on the "on" segments. The neighborhood of a point $P$ with respect to a solid $S$, denoted by $N(P, S)$, is the regularized intersection of a sphere with radius $R$ centered at $P$ with the solid. The value of the radius $R$ is arbitrary and should be chosen sufficiently small. We can generalize the set membership classification function $M[X, S]$ given by Eq. (7.49) to include neighborhood information by assuming that the candidate set $X$ has such information and therefore the resulting segment "$X$ on $S$" has it.

The representation of the neighborhood of a point is related to its position relative to the solid under investigation. Neighborhoods for points that are in the interior or outside of the solid are full or empty and can be represented easily. One alternative is to assign a variable, say $N$, to the neighborhood of the point such that it can take the values $-1$, $0$, and $1$ if the point is outside, on, or inside

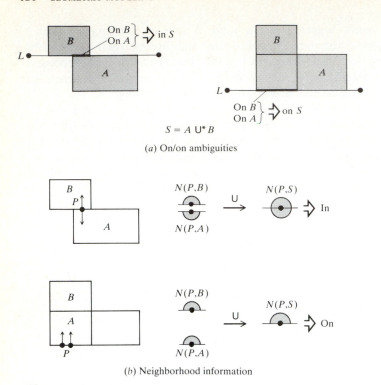

$S = A \cup^* B$

(a) On/on ambiguities

(b) Neighborhood information

**FIGURE 7-50**
On/on ambiguities and their resolutions using neighborhood information.

the solid respectively. If $N = -1$ or 1, no further information is needed. For points on the boundaries of the solid ($N = 0$), three cases can arise. A point may be in the interior of a solid's face. This becomes a case of the face neighborhoods, which can be represented using the face and surface normal signs described in Sec. 7.8.1. The second case is edge neighborhoods which result if the point lies on a solid's edge. Assuming that the edge is shared by two faces, the normal and tangent signs of the faces and their underlying surfaces serve to represent the neighborhood. The third case arises when the point is a vertex, thus resulting in vertex neighborhoods. A vertex is typically shared by three faces (or surfaces) and its neighborhood is complex and difficult to manipulate and is not needed in most algorithms. Figure 7-51 shows face and edge neighborhoods.

With all the tools in hand, we can now develop a CSG-based set-operation algorithm. The essence of an algorithm that can perform $S = A\langle OP\rangle B$ is to classify faces with respect to $S$ by the divide-and-conquer paradigm using face/solid classification, and then combine the classifications using $\langle OP\rangle$ to obtain the solid $S$. The resulting classifications produce portions of the faces of $A$ and $B$ that are on $S$ only; that is, that yields the boundary of $S$. The faces of $A$ and $B$ form a sufficient set of faces which include the boundary of $S$ and which can be used in the paradigm. This is based on the fact $b(A\langle OP\rangle B) \subset (bA \cup bB)$ where b means boundary. This fact is intuitively acceptable and can be proven mathematically. The faces of $A$ and $B$ are called tentative faces or t-faces.

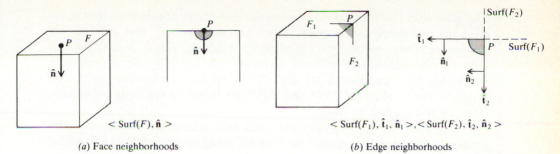

| | |
|---|---|
| < Surf($F$), $\hat{n}$ > | < Surf($F_1$), $\hat{t}_1$, $\hat{n}_1$ >, < Surf($F_2$), $\hat{t}_2$, $\hat{n}_2$ > |
| (a) Face neighborhoods | (b) Edge neighborhoods |

**FIGURE 7-51**
Face and edge neighborhood representations.

While face/solid classification is possible in theory using the divide-and-conquer paradigm, it is complex and not attractive because it must perform boolean operations on face subsets. Therefore, it is replaced by edge/solid classification which is much simpler. Thus a face classification is done indirectly by classifying its edges with respect to $S$. This, in turn, requires a set of tentative edges (t-edges) which if classified and combined result in the edges, and consequently the faces, of $S$. The two types of t-edges that may exist are self-edges, which are the edges of $A$ and $B$ themselves, and cross-edges, which result from intersections between faces of $A$ and $B$.

One can now devise a set-operation algorithm for a CSG scheme by using the divide-and-conquer paradigm which is based on edge/solid classification. The detailed steps of the algorithm can be envisaged as follows:

1. Generate a sufficient set of t-faces. If $A$ and $B$ are primitives, their primitive faces form such a set.
2. Classify self-edges of $A$ with respect to $A$ including neighborhoods. This is a trivial step because such classifications are already known. This step merely prepares information needed in step 4 below.
3. Classify self-edges of $A$ with respect to $B$ using the divide-and-conquer paradigm. The classification includes neighborhoods as well. If $A$ or $B$ is not a primitive, the paradigm becomes recursive. It is usually based on edge/primitive classifiers which are easy to write for primitives such as blocks, cylinders, and the like.
4. Combine classification results (via a "combine" algorithm) of steps 2 and 3 according to the desired boolean operation. Refer to Sec. 7.5.3 and Fig. 7-18 for an example on how to combine classifications. In following Fig. 7-18, the line $L$ is replaced by a given self(or cross)-edge. The result that is of interest to the set-operation algorithm is the segments that are on the boundary of $S$ (call them "on" segments). Thus, for any edge $E$ that is already classified via the divide-and-conquer paradigm, the combining classification is:

$\langle OP \rangle = \cup^*$: $E$ on $S = (E$ out $A$ INT* $E$ on $B$) UN* ($E$ on $A$ INT* $E$ out $B$)

$\langle OP \rangle = \cap^*$: $E$ on $S = (E$ in $A$ INT* $E$ on $B$) UN* ($E$ on $A$ INT* $E$ in $B$)

$\langle OP \rangle = -^*$: $E$ on $S = (E$ in $A$ INT* $E$ on $B$) UN* ($E$ on $A$ INT* $E$ out $B$)

The UN* and INT* operators are regularized union and intersection operators in one-dimensional space. They are not the $\cup^*$ and $\cap^*$ we are developing. They combine the classification results by simply comparing and merging the endpoints of different segments. Edges are usually expressed in parametric form and thus UN* and INT* find the appropriate parametric ranges of the result. Thus UN* and INT* are based on scanning parametric intervals and are easy to write.

The above combining rules are valid only if there are no on/on ambiguities. If there are, they should be resolved using neighborhoods and added accordingly to the results obtained from the above rules. This step gives the classification of self-edges of $A$ with respect to solid $S$.

5. Regularize the "on" segments that result from step 4 by discarding the segments that belong to only one face of $S$. This is done through testing neighborhoods of the segments.

6. Store the final "on" segments that result from step 5 as part of the boundary of $S$. Steps 2 to 6 are performed for each t-edge of a given t-face of $A$.

7. Utilize surface/surface intersection to find cross-edges that result from intersecting faces of $B$ (one face at a time) with the same t-face mentioned in step 6. This step results in "oversized" cross-edges which are reduced to "minimal" cross-edges by using step 8 below. Refer to Example 7.9 for more details.

8. Classify each cross-edge with respect to $S$ by repeating steps 2 to 4 with the replacement of self-edges of $A$ used in these steps by each cross-edge. Here, cross-edges are classified with respect to the faces of $A$ and $B$ they belong to. This is a two-dimensional classification and can be combined using the rule $E$ on $S = E$ in $A$ INT* $E$ in $B$ (see Example 7.9).

9. Repeat steps 5 and 6 for each cross-edge.

10. Repeat steps 2 to 9 for each t-face of $A$.

11. Repeat steps 2 to 6 for each t-face of $B$.

The above set-operation algorithm is not very efficient. For example, each self-edge is classified at least twice (each edge belongs to two t-faces). This can be easily avoided. Other shortcuts (such as spatial locality in geometric computations) can be used to avoid unnecessary calculations.

**Example 7.8.** Create the CSG model of the solid $S$ shown in Fig. 7-20a.

*Solution.* The creation of a CSG model of $S$ is by and large much simpler to create using solid primitives and boolean operators than using faces, edges, and vertices and Euler operators utilized in Example 7.5. That is not to say that creating a B-rep model of the same complexity is inefficient because the steps covered in this example can be used in B-rep if it supports boolean operations.

The first step in the creation procedure is the planning strategy. This model is simple and can be created by adding two blocks and subtracting one cylinder. The primitives and the CSG tree are shown in Fig. 7-52. Utilizing the local coordinate systems shown in Fig. 7-4, blocks $A$ and $B$ must be translated and cylinder $C$ translated and rotated relative to the MCS shown to be positioned and oriented properly. Points $P_A$, $P_B$, and $P_C$ show the origins of the local coordinate systems of these

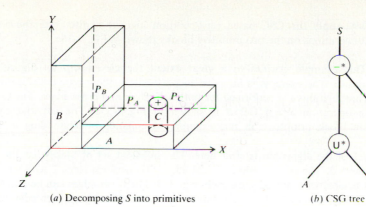

(a) Decomposing S into primitives          (b) CSG tree

**FIGURE 7-52**
CSG model of solid S.

primitives. The geometrical information of each of these primitives is:

Block $A$:        $x_L = a - d$, $y_L = d$, $z_L = c$, $P_A(x, y, z) = P_A(d, 0, -c)$

Block $B$:        $x_L = d$, $y_L = b$, $z_L = c$, $P_B(x, y, z) = P_B(0, 0, -c)$

Cylinder $C$:    $R = R$, $H = d$, $P_C(x, y, z) = P_C[(a+d)/2, d, -c/2]$, Rot about $X = 90°$

Assuming the user has created the primitives $A$, $B$, and $C$, the command $S = A \cup^* B -^* C$ creates the CSG tree shown in Fig. 7-52b.

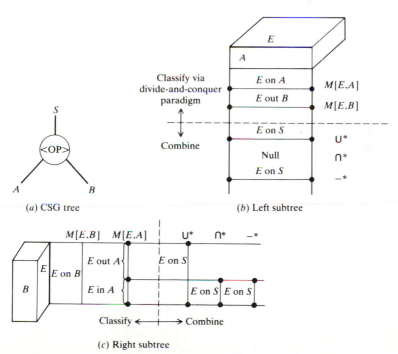

(a) CSG tree                    (b) Left subtree

(c) Right subtree

**FIGURE 7-53**
Self-edge classification and combination.

**Example 7.9.** Apply the CSG-based set-operation algorithm discussed above to perform set operations on the two primitive blocks shown in Fig. 7-36a.

*Solution.* This example clarifies to a great extent the eleven-step set algorithm described above. The example is intended to particularly show how classifications and their combinations are performed for both self-edges and cross-edges. The CSG tree of the operation $S = A\langle OP\rangle B$ is shown in Fig. 7-53a. The left and right subtrees are block primitives in this example to enhance understanding of the algorithm.

Step 1 of the algorithm is clear here. The sufficient set of t-faces has the 12 faces of $A$ and $B$. The set of t-edges has 24 edges. To perform steps 2 to 6 and 10 and 11, let us deal with one edge on each $A$ and $B$. The other edges can be handled in a similar way. Figure 7-53b and c shows the results. Classifying each edge with respect to its primitives (step 2) is trivial and produces "on" segments. Classifying the same edge with respect to the other primitive (step 3) requires an edge/primitive classifier. The classification results are combined according to the rules of step 4. This procedure is repeated for the other 23 edges.

Unlike self-edges, cross-edges must be found first by using step 7 as shown in Fig. 7-54. Consider faces $F_1$ of $A$ and $F_2$ of $B$. Their underlying surfaces are $S_1$ and

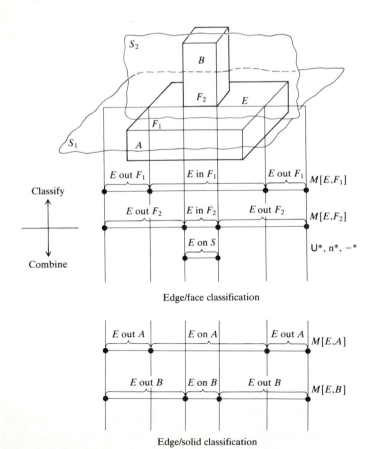

Edge/face classification

Edge/solid classification

**FIGURE 7-54**
Cross-edge classification and combination.

$S_2$ respectively. Intersecting $S_1$ and $S_2$ is easier than interesting $F_1$ and $F_2$ but produces an "oversized" edge that must be reduced down to the edge between $F_1$ and $F_2$; that is, the "minimal" edge. This is achieved via an edge/face classifier (step 8). The "minimal" edge is the "$E$ on $S$" segment shown in Fig. 7-54. This procedure (steps 7 and 8) is repeated for the other three cross-edges in this case (step 9).

The edge/face classification used in this example assumes that the B-rep of $A$ and $B$ is available in addition to their CSG. This assumption is acceptable for CSG modelers that use incremental boundary evaluations. Modelers that use nonincremental boundary evaluations can replace this classification by an edge/solid classification utilizing the divide-and-conquer paradigm. In this case, the classifier must eliminate the "on" segments, of the cross-edge, that belong to only one face, as mentioned in step 6 (see Fig. 7-54). This is normally done with the aid of neighborhoods and is not shown in Fig. 7-54 or discussed here because it requires new definitions of faces (maximum faces or m-faces) that may be confusing to the reader.

The reader is encouraged to apply the set-operation algorithms to other problems. However, the reader is advised that the algorithm needs refinements and details regarding neighborhoods to be universal.

The classification ideas used in this example and its related algorithm can be applied to algorithms based on B-rep (see Example 7.7). The major change to be done is to replace the divide-and-conquer paradigm that utilizes CSG tree structure by an algorithm that is based on a face/edge/vertex data structure.

## 7.8.3 Remarks

The CSG scheme is a very powerful representation scheme. It is not closely related to conventional drafting language and has many advantages. It is easy to construct out of primitives and boolean operations. It is concise and requires minimum storage to store solid definitions (the CSG graph). This is why it is slow to retrieve the model because it has to build a boundary from the CSG graph. It is also due to this fact that CSG is slow in generating wireframes, that is, line drawings. CSG must be converted internally into a B-rep (similar to the set-operation algorithm covered earlier) to display the model or generate its line drawings.

Application algorithms based on CSG schemes are very reliable and competitive with those based on B-rep schemes. However, the major disadvantage of CSG is in its inability to represent sculptured surfaces and half-spaces. This is an active area of research, and one would expect this limitation to go away with time.

## 7.9 SWEEP REPRESENTATION

Schemes based on sweep representation are useful in creating solid models of two-and-a-half-dimensional objects. The class of two-and-a-half-dimensional objects includes both solids of uniform thickness in a given direction and axisymmetric solids. The former are known as extruded solids and are created via linear or translational sweep; the latter are solids of revolution which can be created via rotational sweep. Sweeping is used in general as a means of entering object descriptions into B-rep or CSG-based modelers. There exists no sweeping-based modelers due to the limited modeling domain of sweep representations and the lack of a formal underlying theory of sweeping. For example, general validity and regularization conditions for sweep representations are not known and are usually left to the user.

Sweeping is based on the notion of moving a point, curve, or a surface along a given path. There are three types of sweep: linear, nonlinear, and hybrid sweeps. In linear sweep, the path is a linear or circular vector described by a linear, most often parametric, equation while in nonlinear sweep, the path is a curve described by a higher-order equation (quadratic, cubic, or higher). Hybrid sweep combines linear and/or nonlinear sweep via set operations and is, therefore, a means of increasing the modeling domain of sweep representations.

Linear sweep can be divided further into translational and rotational sweep. In translational sweep, a planar two-dimensional point set described by its boundary (or contour) can be moved a given distance in space in a perpendicular direction (called the directrix) to the plane of the set (see Fig. 7.55a). This is similar to entity projection and surface offsetting or translation in wireframe and surface representations respectively. The boundary of the point set must be closed otherwise invalid solids (open sets) result. In rotational sweep, the planar two-dimensional point set is rotated about an axis of rotation (axis of symmetry of the object to be created) by a given angle (see Fig. 7-55a). This is similar to entity rotation or a surface of revolution in wireframe and surface representations. Non-linear sweep is similar to linear sweep but with the directrix being a curve instead of a vector (Fig. 7-55b). Hybrid sweep tends to utilize some form of set operations. Figure 7-55c shows the same object shown in Fig. 7-55a but with a hole. In this case two point sets are swept in two different directions and the two resulting swept volumes are glued together to form the final object. Invalid solids or nonregular sets may result if the sweeping direction is not chosen properly, as shown in Fig. 7-55d.

Sweeping operations are useful in engineering applications that involve swept volumes in space. Two widely known applications are simulations of material removal due to machining operations and interference detection of moving objects in space. In the first application, the volume swept by a moving cutter along a specific direction is intersected with the raw stock of the part. The intersection volume represents the material removed from the part. In interference detection, a moving object collides with a fixed one if the swept volume due to the motion of the first intersects the fixed object.

## 7.9.1   Basic Elements

Wireframe curves, both analytic and synthetic, covered in Chap. 5 are valid basic elements, or primitives, to create two-dimensional contours for sweep operations. However, a solid modeler may not allow its users to use all wireframe entities it supports in sweep operations. Such a limitation usually stems from the modeling domain of the internal representation the modeler supports. It is this representation that user sweep operations are converted to before being stored in the data-

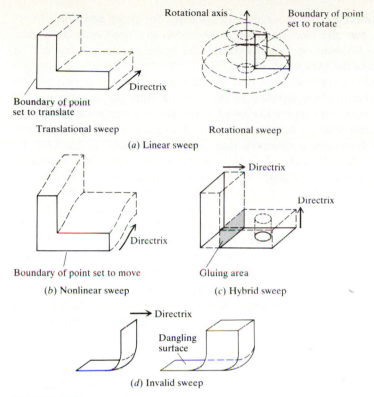

**FIGURE 7-55**
Types of sweep.

base of the model under construction. If the sweep operations are converted to a B-rep, the permissible wireframe entities are those that are the same as the underlying curves of edges that the B-rep supports. If they are converted to a CSG, the permissible entities are the ones that generate surfaces (boundaries) of supported half-spaces. Lines, arcs, circles, and B-splines are among the most widely used entities in sweep operations.

The boundary of a two-dimensional point set used in sweep operations can consist of nested contours up to one level only (one inner contour) within the outer contour. This is allowed to create holes in the resulting solid. There may also exist a maximum number of entities allowed to create any one contour. The number is usually adequate for practical uses. It is usually set for implementation purposes of the sweep algorithm.

### 7.9.2   Building Operations

The building operations of linear and nonlinear sweep models are simple: generate the boundary and sweep it. If hybrid sweep is available, these operations extend to include boolean operations. If there existed data structures designed only for sweep representation, algorithms to implement these boolean operations would have been different from those already discussed for B-rep and CSG. Practically, this is not the case and the sweep operations are used only as a user convenience and a boolean operation acts on the corresponding B-rep or the CSG data structure of the sweep operations.

How are sweep operations converted to a B-rep or CSG? It is well known that only linear sweep can be converted. In the case of a translational sweep, each entity in the swept boundary represents an edge and each corner point represents a vertex in the corresponding boundary model. For a one-contour boundary, each entity in the boundary indicates a face also. The number of faces of the model are equal to $(N + 2)$, where $N$ is the number of entities of the boundary and the 2 accounts for the front and back faces (see Fig. 7-55a). The number of edges is equal to $(2N + M)$, where $M$ is the number of corner points of the boundary and the number of vertices is equal to $2M$. These values satisfy the Euler equation (7.57), regardless of the number of entities of the boundary, if we notice that $N = M$ for a closed boundary. If the boundary has holes (nested contours), a similar relationship can be obtained. Rotational sweep operations can be converted in a similar fashion. In Sec. 7.7, we have discussed how a rotational sweep operator (see Fig. 7-32) based on approximate B-rep can be developed. The reverse of that discussion shows how a sweep model can be converted into an approximate B-rep. It is left to the reader to find the number of faces, edges, vertices, loops, bodies, and genus when the sweeping angle is a full 360° or a given range of it.

The linear sweep to CSG conversion must be based on unbounded CSG primitives, that is, half-spaces. Conversion to CSG based on bounded primitives might not be possible all the time and may be no better than the half-space alternative. In this case, each entity in the swept boundary represents a bounding surface of a corresponding half-space. A linear entity, for example, represents a planar surface and half-space and a circular entity or an arc represents a cylindrical surface and half-space. The CSG model is then composed of the union of these intersecting half-spaces.

In both conversions, the underlying surfaces must be oriented. This requires the direction of a surface normal. This can be achieved by choosing a direction (clockwise or counterclockwise) when inputting or creating the outer and inner contours of the given boundary. The interior of the boundary, which defines the interior of the solid, can be identified and the proper normal sign can be chosen accordingly. For example, one may choose to traverse the outer contour in a counterclockwise direction and then choose the positive normal sign to indicate the exterior of the solid. A vector that represents such a surface normal becomes very easy to create.

**Example 7.10.** Create the sweep model of the solid $S$ shown in Fig. 7-20a.

*Solution.* The solid $S$ is two-and-a-half dimensional, having uniform thickness in the negative $Z$ direction. If it was not the cylindrical hole, the model could have been created via translational sweep exactly as shown in Fig. 7-55a. Due to the hole presence, hybrid sweep is utilized as shown in Fig. 7-55c. The bottom face with the circular hole is created and swept vertically and the left face is created and swept to the right. The two subsolids are then glued along the hatched gluing area shown.

### 7.9.3   Remarks

Sweep representation is useful once it develops. Its modeling domain can be extended beyond two-and-a-half-dimensional objects if nonlinear (sometimes called general) sweep is available. Nonlinear sweep may be useful in creating nonrigid objects and studying their deformation as they travel in space. Complex mechanical parts such as screws, springs, and other components that require helical and special loci can be represented by sweeping. In any one of these parts, a two-dimensional polygon can form the basis of the desired boundary.

## 7.10   ANALYTICAL SOLID MODELING (ASM)

The historical development of ASM is closely related to finite element modeling. Those who are familiar with finite element analysis (FEA) can easily recognize that the mathematical foundations of ASM follow similar guidelines to three-dimensional isoparametric formulation of FEA for 8- to 20-node hexahedral elements. ASM is developed to aid designers and engineers in the arduous task of modeling complex geometry commonly found in design applications. ASM can be viewed as more of a representation scheme for design than for manufacturing purposes due to its formulation, as seen in this section, which does not involve orientable surfaces as does B-rep or CSG.

   While ASM originated from the need to solve the problem of finite element modeling, it has now a wide range of applications such as mass property calculations, composite material modeling, and computer animation. The widespread acceptance of ASM in the finite element and finite difference communities has been due to the efficiency and flexibility of mesh-generation algorithms that operate on hyperpatches (see the next section). A uniform transition, or nonuniform mesh, can be generated within a hyperpatch and, consequently, within the entire model. In addition, because a hyperpatch is a mapping of a unit cube, it is easy to subdivide it into hexahedral elements. The solid modeler PATRAN-G is based on ASM and interfaces to various FEA packages and other solid modelers.

### 7.10.1   Basic Elements

ASM is an extension of the well-established tensor product method, introduced to represent surfaces in Chap. 6, to three-dimensional parametric space with the parameters $u$, $v$, and $w$. Therefore, it involves only products of univariate basis polynomials and introduces no conceptual difficulty due to the higher dimensionality of a solid. The properties of tensor product solids can be easily deduced from properties of the underlying curve schemes. Thus, one can conceptually conceive and easily derive the representations of tricubic, Bezier, and B-spline solids analogous to bicubic, Bezier, and B-spline surfaces in two-dimensional parametric space $(u, v)$ and analogous to cubic, Bezier, and B-spline curves in one-dimensional parametric space $(u)$.

   The tensor product formulation in three-dimensional parametric space is a mapping of a cubical parametric domain described by $u$, $v$, and $w$ values into a solid described by $x$, $y$, and $z$ in the cartesian (modeling) space, as shown in Fig.

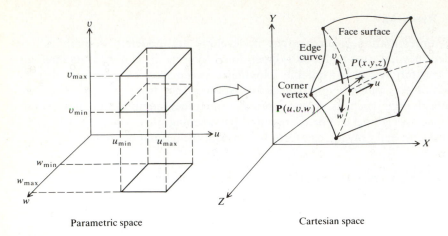

Parametric space                          Cartesian space

**FIGURE 7-56**
Hyperpatch representation.

7-56. The resulting solid is called a parametric solid or a hyperpatch (so called because hyperpatches are extensions of and bounded by surface patches) whose points in the interior or on the bounary are given by

$$P(u, v, w) = [x \quad y \quad z] = [x(u, v, w) \quad y(u, v, w) \quad z(u, v, w)],$$

$$u_{min} \leq u \leq u_{max}, v_{min} \leq v \leq v_{max}, w_{min} \leq w \leq w_{max} \qquad (7.90)$$

As with curves and surfaces, Eq. (7.90) gives the coordinates of a point inside or on the hyperpatch as the components of its position vector. The equation uniquely maps the parametric space ($E^3$ in $u$, $v$, and $w$ values) to the cartesian space. The parametric variables are constrained to intervals bounded by minimum and maximum values. In most hyperpatches, these intervals are [0, 1] which result in a unit cube in the parametric space.

Equation (7.90) suggests that ASM represents an object as an assembly of nonoverlapping hyperpatches. Each hyperpatch has six faces, each of which is any surface patch discussed in Chap. 6. Each face has four edge curves of the same type as the surface patches and four corner vertices (see Fig. 7-56). For example, a tricubic hyperpatch is bounded by six bicubic surface patches, each of which is bounded by four cubic splines. This hierarchy in topology, from a hyperpatch to patches, edges, and vertices, provides a means to construct a hyperpatch from given control points, curves, or patches. This feature makes ASM an extension of wireframe or surface modeling. For example, a user can create a Bezier hyperpatch by either entering its control points, Bezier curves, or Bezier patches.

While a hyperpatch can be described by a polynomial of any order, a cubic polynomial in each parameter is sufficient for practical design applications. As mentioned in Chap. 5, the higher the order, the more difficult it is to control the resulting shape. Analogous to Eqs. (5.74) and (6.39), a cubic hyperpatch can be given by the following equation:

$$P(u, v, w) = \sum_{i=0}^{3} \sum_{j=0}^{3} \sum_{k=0}^{3} C_{ijk} u^i v^j w^k, \quad 0 \leq u \leq 1, 0 \leq v \leq 1, 0 \leq w \leq 1 \qquad (7.91)$$

The face surfaces, edge curves, and corner vertices can be obtained by substituting the proper value(s) of the parameters into this equation. The face surfaces are given by $P(0, v, w)$, $P(1, v, w)$, $P(u, 0, w)$, $P(u, 1, w)$, $P(u, v, 0)$, and $P(u, v, 1)$. Similarly, the equation of any edge curve is obtained by fixing two of the three parametric variables and keeping the third one free. $P(u, 0, 0)$, $P(u, 1, 0)$, and $P(0, v, 0)$ are three of the available 12 edge curves, and the eight corner vertices are obviously $P(0, 0, 0)$, $P(1, 0, 0)$, $P(0, 1, 0)$, $P(1, 1, 0)$, $P(0, 0, 1)$, $P(1, 1, 0)$, $P(0, 1, 0)$, and $P(1, 1, 1)$.

There are 64 $C_{ijk}$ vector coefficients (called polynomial or algebraic coefficients, as they were called previously for curves and surfaces) that must be determined utilizing a given set of boundary conditions of a given hyperpatch. For a tricubic hyperpatch, these conditions are 8 position vectors (one $P$ at each corner vertex), 24 tangent vectors [three ($\partial P/\partial u$, $\partial P/\partial v$, and $\partial P/\partial w$) at each corner vertex], 24 twist vectors [three ($\partial^2 P/\partial u \, \partial v$, $\partial^2 P/\partial u \, \partial w$, and $\partial^2 P/\partial v \, \partial w$) at each corner vertex], and 8 triple mixed partial derivatives [one ($\partial^3 P/\partial u \, \partial v \, \partial w$) at each corner vertex]. For a cubic Bezier hyperpatch, there are 64 given control (data) points that form the characteristic (control) polyhedron of the hyperpatch. These points are arranged in a $4 \times 4 \times 4$ mesh. Each face surface of the hyperpatch uses 16 of these control points to define its control polygon and each edge curve uses four control points to form its control polygon. In the case of a cubic B-spline hyperpatch, an $n \times m \times q$ mesh can be used where the number of control points in the $u$, $v$, and $w$ directions are $n$, $m$, and $q$ respectively and are not necessarily equal. B-spline surfaces and curves at the boundary of the hyperpatch can be deduced in a similar fashion to the Bezier hyperpatch. If a hyperpatch that interpolates the data points is desired, an interpolating polynomial must be used.

The reduction of Eq. (7.91) to a matrix form as we have done to Eqs. (5.74) and (6.39) [see Eqs. (5.77) and (6.40)] is possible but produces an awkward form. In addition, converting Eq. (7.91) (sometimes called the algebraic form) to a form similar to Eqs. (5.83) and (6.42) (sometimes called the geometric form), in the case of a tricubic hyperpatch, requires relating the algebraic coefficients $C_{ijk}$ to the geometric coefficients, that is, the 64 boundary conditions. This conversion process is not different from the case of curves and surfaces and is very cumbersome for hyperpatches. Therefore, the developments of the geometric forms of Eq. (7.91) for a tricubic cubic Bezier and cubic B-spline hyperpatches are done here by extending (intuitively) the patterns we can identify in one- and two-dimensional parametric spaces into three-dimensional parametric space.

Let us look into developing the geometric form of a tricubic hyperpatch. If Eq. (5.83) is expanded and rearranged by collecting the terms of $u^3$, $u^2$, $u^1$, and $u^0$, we get the following equations:

$$P(u) = \sum_{i=0}^{3} \sum_{j=1}^{4} u^i M_{H_{ij}} V_j, \qquad 0 \le u \le 1 \qquad (7.92)$$

where $M_{H_{ij}}$ are the elements of the geometry matrix $[M_H]$ given by Eq. (5.84) and $V_j$ are the elements of the vector $V$ given by Eq. (5.85). It is more desirable to unify the summation limits of both the $i$ and $j$ indices in Eq. (7.92) by writing it as

$$P(u) = \sum_{i=1}^{4} \sum_{j=1}^{4} u^{i-1} M_{H_{ij}} V_j, \qquad 0 \le u \le 1 \qquad (7.93)$$

Equation (7.92) can also be written as

$$\mathbf{P}(u) = \sum_{i=0}^{3} \left( \sum_{l=1}^{4} M_{H_{il}} \mathbf{V}_l \right) u^i, \qquad 0 \le u \le 1 \tag{7.94}$$

Comparing Eqs. (7.94) and (5.74) gives

$$\mathbf{C}_i = \sum_{l=1}^{4} M_{H_{il}} \mathbf{V}_l \tag{7.95}$$

and, therefore, Eq. (7.93) becomes

$$\mathbf{P}(u) = \sum_{i=1}^{4} \mathbf{C}_i u^{i-1}, \qquad 0 \le u \le 1 \tag{7.96}$$

Equations (7.93) and (7.96) give the geometric and algebraic terms respectively of a Hermite cubic spline. Equation (7.95) relates the algebraic coefficients to the geometric ones.

Applying the same treatment to a bicubic surface, Eq. (6.42) can be rewritten as

$$\mathbf{P}(u, v) = \sum_{i=1}^{4} \sum_{j=1}^{4} \mathbf{C}_{ij} u^{i-1} v^{j-1}, \qquad 0 \le u \le 1, 0 \le v \le 1 \tag{7.97}$$

where

$$\mathbf{C}_{ij} = \sum_{l=1}^{4} \sum_{m=1}^{4} M_{H_{il}} M_{H_{jm}} \mathbf{b}_{lm} \tag{7.98}$$

and $\mathbf{b}_{lm}$ are the elements of the $[B]$ matrix given by Eq. (6.43).

Comparing Eqs. (7.96) and (7.97) shows that introducing an additional parametric variable amounts to adding an extra summation sign into the geometric form. Moreover, comparing Eqs. (7.95) and (7.98) shows the same effect, as well as the fact that the matrix $[M_H]$ always relates the algebraic coefficients to the geometric ones. On the basis of these observations, one can easily write the equation of a tricubic hyperpatch as

$$\mathbf{P}(u, v, w) = \sum_{i=1}^{4} \sum_{j=1}^{4} \sum_{k=1}^{4} \mathbf{C}_{ijk} u^{i-1} v^{j-1} w^{k-1}, \qquad 0 \le u \le 1, 0 \le v \le 1, 0 \le w \le 1 \tag{7.99}$$

where

$$\mathbf{C}_{ijk} = \sum_{l=1}^{4} \sum_{m=1}^{4} \sum_{n=1}^{4} M_{H_{il}} M_{H_{jm}} M_{H_{kn}} \mathbf{b}_{lmn} \tag{7.100}$$

Notice that Eq. (7.99) is the same as Eq. (7.91) with summation limits changed. The geometric coefficients $\mathbf{b}_{lmn}$ can be arranged in four $[B]$ matrices similar to the one given by Eq. (6.43). The first and second, say $[B_1]$ and $[B_2]$, are exactly like $[B]$ of Eq. (6.43) but for $w = 0$ and $w = 1$ respectively. For example, $\mathbf{b}_{122}$ and $\mathbf{b}_{211}$ are $\mathbf{P}_{110}$ and $\mathbf{P}_{001}$ respectively. The third and the fourth, say $[B_3]$ and $[B_4]$, are the derivatives of $[B_1]$ and $[B_2]$ respectively with respect to $w$. For example, $\mathbf{b}_{323}$ and $\mathbf{b}_{432}$ are $\mathbf{P}_{uvw000}$ and $\mathbf{P}_{uw011}$ respectively. If we choose the subscript $l$ in Eq. (7.100) to correspond to the four $[B]$ matrices, we can easily see that the four specific $\mathbf{b}$ elements we just mentioned become $\mathbf{b}_{122}$, $\mathbf{b}_{211}$, $\mathbf{b}_{323}$, and $\mathbf{b}_{432}$ respectively.

The tricubic hyperpatch suffers from all the disadvantages of the cubic curve and the bicubic surface. It even requires the input of $\partial^3 \mathbf{P}/\partial u\, \partial v\, \partial w$ as a boundary condition. Therefore, let us look into a cubic Bezier hyperpatch. Following the same approach we used for the tricubic patch, the cubic Bezier curve (see Prob. 5.10), the cubic Bezier surface [see Eq. (6.73)], and the cubic Bezier hyperpatch can also be described by Eqs. (7.96), (7.97), and (7.99) respectively. Equations (7.95), (7.98), and (7.100) then become respectively:

$$\mathbf{C}_i = \sum_{l=1}^{4} M_{B_{il}} \mathbf{P}_l \tag{7.101}$$

$$\mathbf{C}_{ij} = \sum_{l=1}^{4} \sum_{m=1}^{4} M_{B_{il}} M_{B_{jm}} \mathbf{P}_{lm} \tag{7.102}$$

$$\mathbf{C}_{ijk} = \sum_{l=1}^{4} \sum_{m=1}^{4} \sum_{n=1}^{4} M_{B_{il}} M_{B_{jm}} M_{B_{kn}} \mathbf{P}_{lmn} \tag{7.103}$$

where the elements of the matrix $[M_B]$, given by Eq. (6.74), are used in these three equations. The control points $\mathbf{P}_{lmn}$ for the hyperpatch are arranged in the $l \times m \times n$ mesh. The $m \times n$ mesh for the Bezier surface can easily be extended to form the $l \times m \times n$ mesh. Thus any point $\mathbf{P}_{lmn}$ is much easier to locate and understand than the $\mathbf{b}_{lmn}$ coefficients used in Eq. (7.100) for the tricubic hyperpatch.

The cubic B-spline hyperpatch would be more advantageous to use in design applications over the cubic Bezier hyperpatch due to the local control characteristics of the former. The cubic B-spline curve (see Prob. 5.15), the cubic B-spline surface [see Eq. (6.82)], and the cubic B-spline hyperpatch can be described by Eqs. (7.96), (7.97), and (7.99) respectively. The maximum value of any of the parameters may exceed the value of 1. Equations (7.101) to (7.103) can be extended to the cubic B-spline hyperpatch by replacing the matrix $[M_B]$ with the matrix $[M_S]$ given by Eq. (6.81).

### 7.10.2  Building Operations

The creation of an ASM model of an object simply involves dividing the object into the proper assembly of nonoverlapping hyperpatches. Each hyperpatch can be constructed from curves and/or surface patches. For example, a cubic Bezier or a B-spline hyperpatch can be constructed by creating Bezier or B-spline curves and connecting the curves by surfaces. This process reflects the natural nesting of curves, surfaces, and solids.

If the ASM model of an existing object is to be created, Bezier and B-spline hyperpatches introduced in Sec. 7.10.1 cannot be used because they extrapolate the given data points. Instead, we can use a B-spline hyperpatch that interpolates the data points or a cubic curve that interpolates through four data points at specified parametric locations can be used. If these points are at $u = 0$, $\frac{1}{3}$, $\frac{2}{3}$, and 1, they are sometimes referred to as the one-third points. Similarly 16 points and 64 points would be needed to create a bicubic surface and a tricubic hyperpatch respectively.

Other construction methods of ASM models can include ruled volumes and sweeping. A ruled volume can be created between two given surface patches by linearly interpolating between them as we did in developing ruled surfaces. Linear sweep creates hyperpatches that have uniform thickness normal to the surface patch. It is also possible to create hyperpatches that have thicknesses that vary bilinearly over the surface patch (see Sec. 6.6.6). Rotational and/or nonlinear sweep of surface patches can also create hyperpatches. The equations of the ruled or swept hyperpatches created by these construction methods are considered special cases of the general equations covered in Sec. 7.10.1 and can be obtained from them.

It is desirable, from a user point of view, for ASM schemes to support boolean operations. In such a case hyperpatches can be unioned, intersected, and differenced to model various objects. This would require developing intersecting algorithms of these hyperpatches as well as validity checks to ensure the creation of valid objects.

**Example 7.11.** Create the ASM model of the solid $S$ shown in Fig. 7-20$a$.

**Solution.** The ASM model of this solid $S$ consists mainly of hyperpatches that have planar face surfaces. Figure 7-57 shows the minimum number of hyperpatches that can be used to create the ASM model of the solid $S$.

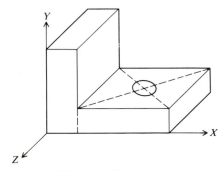

(a) Subdivision of $S$ into hyperpatches

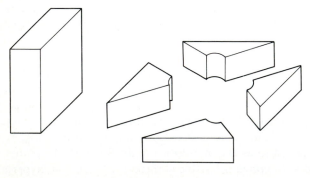

(b) Individual hyperpatches

**FIGURE 7-57**
ASM model of solid $S$.

### 7.10.3 Remarks

ASM has a strong history in finite element modeling. The fact that a hyperpatch is given by Eq. (7.99) makes it appealing in design and analysis applications that require information inside as well as on the boundary of a given object. This is desirable, for example, in modeling and studying composite materials and fracture mechanics problems. However, ASM is not adequate for manufacturing applications such as tool path generation because face surfaces of hyperpatches are not explicitly stored and are not orientable; that is, normals to face surfaces (surface patches) cannot indicate the interior or exterior of the object.

## 7.11   OTHER REPRESENTATIONS

We have covered the four most popular representations (B-rep, CSG, sweep, and ASM) used in solid modeling in Secs. 7.7 to 7.10. Other representations exist. However, they are less popular because their modeling domain is limited and/or they do not support a wide range of applications. These representations are primitive instancing, cell decomposition, spatial occupancy enumeration, and octree encoding.

Primitive instancing is based on the notion of families of objects or family of parts. All objects that have the same topology but different geometry can be grouped into a family called generic primitive. Each individual object within a family is called a primitive instance. Take, for example, a block primitive which can be represented by its length $L$, width $W$, and height $H$. Each block primitive instance is defined by specific values of $L$, $W$, and $H$. Primitive instancing is similar in philosophy to group technology used in manufacturing. It promotes standardization. It is also an unambiguous, unique, and easy to use and validate scheme. However, its main drawbacks are its limited domain of modeling unless we use an enormous number of generic primitives, and the lack of generality to develop any algorithms to compute properties of represented solids.

In a cell decomposition scheme, an object can be represented as the sum of cells into which it can be decomposed. Each cell in the decomposition can always be represented. Thus, cell decomposition may enable us to model objects, which may not otherwise be representable, by their cells. Take the case of a cup with a handle. It can be decomposed into two cells: a body and a handle. The body and/or handle can be decomposed further if needed. Cell decompositions are unambiguous, nonunique, and are computationally expensive to validate. They have been historically used in structural analysis. ASM and finite element modeling are forms of cell decomposition.

In a spatial enumeration scheme, a solid is represented by the sum of spatial cells that it occupies. These cells (sometimes called voxels for "volume elements") are cubes of a fixed size that lie in a fixed spatial grid. Each cell can be represented by its centroid coordinates in the grid. The smaller the size of the cube, the more accurate the scheme in representing curved objects. It is exact for boxlike objects. The scheme is unambiguous, unique, and easy to validate, but it is verbose when describing an object, especially curved ones.

The octree encoding (quadtree encoding in two-dimensions) scheme can be considered a generalization of the spatial enumeration scheme in that the cubes

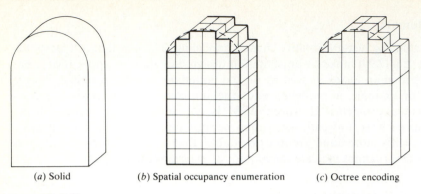

(a) Solid          (b) Spatial occupancy enumeration          (c) Octree encoding

**FIGURE 7-58**
Quasi-disjoint decomposition of a solid.

may have variable sizes. Octrees are hierarchical structures that reflect the recursive subdivision of objects into variably sized cubes. Figure 7-58 shows the difference between spatial occupancy enumeration and octree encoding. In octree encoding, we enclose the object to be modeled inside a cube. If the object does not uniformly cover the cube, then we subdivide the cube into eight octants. If any of the resulting octants is full (completely inside the object) or empty (completely outside the object), no further subdivision is made. If any of the octants is partially full, we subdivide it again into octants. We continue to subdivide the partially full octants until the resulting octants are either full or empty or until some predetermined level of subdivision is reached. Quadtree encoding is exactly the same as octree encoding but it begins with a square and recursively subdivides it into quadrants. Quadtree and octree encodings were originally developed for use in image representation. They have been adapted to finite element modeling (refer to Chap. 18 for more details).

## 7.12  ORGANIZATION OF SOLID MODELERS

Each of the solid modeling representations described above has its advantages and disadvantages. Some are more suitable for certain applications than others. This leads to the idea of developing solid modelers with more than one internal representation. While this idea might extend the geometric coverage (modeling domain) of a modeler, the development and maintenance of such a modeler requires more time and effort than a modeler based on only one representation.

The generic architecture of a solid modeler is shown in Fig. 7-59. This architecture applies also to most of the existing CAD/CAM systems. The figure shows that a solid modeler can be divided into four major systems. The input system consists primarily of the user interface and its related commands. Users can input commands (sometimes called a symbol structure) to define a new object or input application commands that invoke an application algorithm such as mass property calculations. The GMS (geometric modeling system) is the heart of the solid modeler. The GMS translates the symbol structure that defines an object into the internal representation the solid modeler supports. The applica-

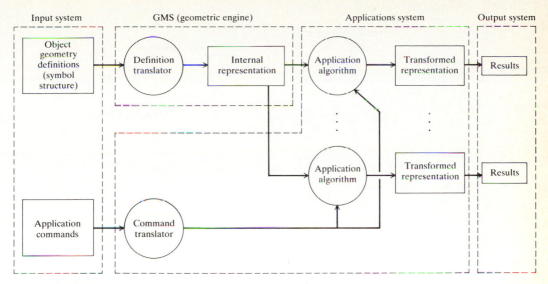

**FIGURE 7-59**
Architecture of a typical solid modeler.

tions system usually consists of various application algorithms. Each algorithm operates on the modeler's internal representation and transforms it into the proper geometric form (transformed representation) needed by the analysis procedure related to the application. In the case of finite element modeling, the mesh algorithm produces nodal (grid) points and elements. In general, the GMS is fixed once it is developed while the applications system is extending to accommodate new design and manufacturing applications. The engineering value of a solid modeler is usually assessed by the capabilities of its applications system. The output system displays the results in a graphical form. Its details have been discussed in Chapter 2. Typically, a solid modeler is a part of a total CAD/CAM system. In such a case, the architecture shown in Fig. 7-59 is an extension of that of the CAD/CAM system.

Solid modelers can be categorized into three types based on their GMSs as follows:

1. **Single representation modelers.** These modelers have only one internal representation they store. B-rep is usually such a representation. All modelers based on B-rep fall into this type. These modelers usually support CSG-like input and sweep operations to facilitate user input. These forms of input are not stored internally but are converted into the B-rep format before storing. Figure 7-60a shows the GMS of a single representation system.

2. **Dual representation modelers.** This type is very popular among modelers whose primary representation scheme is CSG (e.g., PADL-2). A modeler has both B-rep and CSG representations. However, B-rep is derived internally by the modeler from the CSG and the user has no control over it. In addition, the modeler does not usually store the B-rep. It only stores the CSG graph (and tree) and can always reevaluate the B-rep. The need for converting the CSG

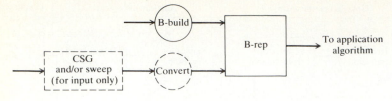

(a) Single representation modeler

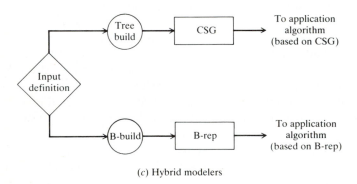

(b) Dual representation modeler

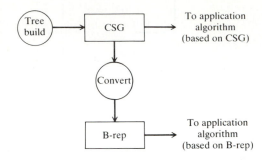

(c) Hybrid modelers

**FIGURE 7-60**
Types of solid modelers.

into B-rep is natural and is needed for display and other graphics purposes. Dual-rep modelers can support a wide range of applications. However, their developments and maintenance are usually more complex than single-rep modelers. Figure 7-60b shows the structure of the related GMS.

3. Hybrid modelers. Here the modeler uses two independent internal representations, usually B-rep and CSG (see Fig. 7-60c). GMSOLID is a true hybrid modeler. Hybrid modelers are different from dual-rep modelers in a few ways. Neither of the internal representations supported is derived from the other. The user can choose which representation and input form to use. Furthermore, the user can use both representations to solve a given problem and compare them. In such cases, it is the user's responsibility to ensure the consistency between input data. The need for a hybrid modeler is clearly to increase

the modeling domain of the modeler. For example, sculptured surfaces can be modeled on a hybrid modeler using its B-rep while this cannot be done on a dual-rep system if it does not support sculptured surfaces.

## 7.13  SOLID MANIPULATIONS

Solid manipulations are useful during the design phase of a given part or object. It is beneficial if the manipulation concepts (refer to Chaps. 5 and 6) utilized by the wireframe and surface modeling techniques can be extended to solid modeling. Most of these concepts can be foreseen as solving intersection problems and/or using set membership classification. In addition, any solid manipulation involves manipulation of both its geometry (as in curves and surfaces) and its topology to ensure that the resulting solids are valid.

### 7.13.1  Displaying

Displaying a solid can take two forms: wire display and shaded images. The wire display requires the B-rep of the solid, in which case the metric information of edges and vertices are used to generate the wireframe model of the solid. This wireframe model can be displayed, edited, or produce line drawings. Editing of the wireframe model derived from a solid database does not, of course, affect the solid itself. If the underlying surfaces of the solid faces are utilized, a mesh can be added to the display to help visualization.

Displaying solids as shaded images provides realistic visual feedbacks to users of solid modelers. Shading is perhaps considered the oldest and most popular application of solid modeling. In fact, there exists the mistaken notion that a shaded image is a solid model. As one expects, shading algorithms are directly related to the representation schemes of solids. Shaded images can be generated from B-reps by a variety of visible-surface algorithms. Most often, these exact B-reps are converted to faceted (approximate or polygonal) B-reps because such algorithms become simple for the latter. Special-purpose tiling engines based on algorithms for displaying approximate B-reps are available commercially for faster displays.

Shading can be performed directly from CSG by means of ray-casting (also called ray-tracing) algorithms or depth-buffer (also called z-buffer) algorithms. Many improvements and alterations have been introduced to ray casting to speed up the algorithms. Shading algorithms for both spatial enumeration and octree have been implemented in special-purpose hardware: a voxel machine for the former and an octree machine for the latter. Visible-surface algorithms can be utilized to shade ASM models. Details of some of these shading algorithms are covered in Chap. 10.

### 7.13.2  Evaluating Points, Curves, and Surfaces on Solids

Applications that require information about the boundary of a solid would need to evaluate points, curves, and surfaces on this boundary. Take the popular application of generating tool paths from solid modeling databases. To drive a tool along the solid boundary, proper curves are evaluated on the underlying surfaces of the solid faces. Points are then generated on each curve to generate the required tool locations.

Evaluating points and curves on solids can be viewed as intersection problems. Solutions of curve/solid and surface/solid intersection problems generate points and curves on solids respectively. Curves and surfaces, utilized in the intersection problems, must be of the type that the solid modeler supports. More specifically, it suffices for these intersecting curves and surfaces to be lines and boundaries of planar half-spaces (i.e., planes) respectively. A plane/solid intersection is also useful in sectioning a solid to generate desired cross-sections. An algorithm that evaluates points and curves on a solid is dependent on the representation of the solid, but in the most part follows similar outlines as described in Examples 7.7 and 7.9.

Evaluating surfaces on a solid can be regarded as extracting the underlying surfaces of the solid faces. These surfaces can be bounded by the proper solid edges or other user-defined boundaries. To keep the solid topology and geometry intact, the geometry and other related information of these surfaces must be copied. The parametric equations of the surfaces might have to be stored in the given solid modeler to facilitate editing the extracted surfaces.

The inverse problem involves checking whether given points, curves, and surfaces lie in, on, or outside a given solid. The solution of this problem is achieved by the set membership classification and neighborhoods. To classify a point against a solid, one passes a line through the point, intersects it with the solid, and classifies it with respect to the solid. A similar approach can be followed for curve/solid and surface/solid classification.

### 7.13.3   Segmentation

The segmentation concept introduced for curves and surfaces is applicable to solids. Segmenting a solid is equivalent to splitting it into two or four valid subsolids depending on whether it is to be split by a plane or a point respectively. Each resulting subsolid should have its own topology and geometry. In a B-rep model, new vertices, edges, and faces are created. Steps 1 to 5 described in Example 7.7 shows how to split a B-rep model into two subsolids. In a CSG model, splitting the solid would also require splitting its CSG graph and tree. Splitting an ASM model is an extension of segmenting curves and surfaces discussed in Chaps. 5 and 6.

### 7.13.4   Trimming and Intersection

Trimming a solid entails intersecting the solid with the trimming boundaries, say surfaces, followed by the removal of the solid portions outside these boundaries. In trimming a solid, it is split into three subsolids, two of which are removed. The trimming surfaces must be of the type supported by the given solid modeler to be able to solve the resulting surface/solid intersection problem.

The intersection problem involving solids is trivial to perform if boolean operations are supported by the solid modeler. All that needs to be done is to use the intersection operator. No additional development or programming is required.

### 7.13.5 Transformation

Homogeneous transformations, or rigid motion, of solids involve translating, rotating, or scaling them. These transformations can be used on two different occasions. When constructing a solid, its primitives are positioned and oriented properly before applying boolean operations by using these transformations. Here, the local coordinate system of each primitive is positioned and oriented relative to the MCS or a WCS of the solid under construction. If the solid is to be transformed later after its complete construction, the transformation operation must be applied to all, say, its faces, edges, and vertices for a B-rep solid or its primitives for a CSG solid.

### 7.13.6 Editing

Editing a solid model is an important feature for the design process. Most new designs are not totally new but rather alterations of existing ones. Editing a solid involves changing its existing topological and geometrical information. An efficient means of solids editing is to use its CSG graph which is only a symbolic structure, as discussed in Sec. 7.8. This is natural for CSG models but for other models such as a B-rep model a CSG graph and a tree can be created; otherwise editing would have to be done on the face/edge/vertex structure which may be slow.

Solid modelers must provide users with fast visual feedbacks when solids are edited. This implies that boundary representations must be updated rapidly, because displays are typically generated from face, edge, and vertex data. Thus, editing is faster if only the part of the boundary representation that is affected by the user changes is updated. Some updating algorithms are based on structural and spatial localities and are not covered here.

## 7.14  SOLID MODELING-BASED APPLICATIONS

Applications based on solid modeling have been increasing rapidly. The underlying characteristic of all these applications is full automation. Current applications can be divided into four groups:

1. Graphics. This is considered the most complete group. It includes generating line drawings with or without hidden line removal, shading, and animation.
2. Design. The most well-understood application in this group is the mass property calculations. Other applications include interference analysis, finite element modeling, and kinematic and mechanism analysis. Some of these applications are more developed than the others.

3. **Manufacturing.** The most active application in this group is tool path generation and verification. Other applications include process planning, dimension inspection, implementing form features needed for manufacturing into solid modelers, and representing geometric features such as tolerances and surface finish.
4. **Assembly.** This is a useful group of applications to robotics and flexible manufacturing. Applications include assembly planning, vision algorithms based on solid modeling, and robotic kinematics and dynamics driven by solid models.

Some of the above applications are covered in the appropriate chapters of Part V of the book while other applications are not covered because they are either beyond the scope of the book or at an early stage of research.

## 7.15 DESIGN AND ENGINEERING APPLICATIONS

Following are some examples to show how the solid modeling theory can be utilized in design and engineering applications. Readers can utilize available solid modelers to them to rework these examples or develop new ones along the same line of thinking.

**Example 7.12.** Figure 7-61 shows a $3 \times 3 \times 1$ inch block. A curved slot of 0.3 inch deep is to be milled in the block using a ball-end mill of 0.25 inch diameter. The equation of the centerline of the slot on the top face of the block is given by $z = -(1.5 - \sqrt[3]{x^2/3.5})$. Create the solid model of the block with the slot in it and show the swept volume of the tool if it moves perpendicular to the top plane of the block.

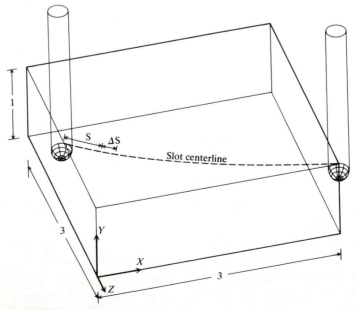

**FIGURE 7-61**
A block with a curved slot.

*Solution.* This example illustrates how complex shapes can be approximated to fit within the modeling domain of a given solid modeler. Here, we are assuming that the modeler supports boolean operations and has natural quadrics as its minimum set of primitives. If the block had a slot with a straight centerline in any orientation relative to it, its modeling would have been exact and trivial. In the case of a curved centerline, the tool motion is approximated by line segments along the centerline. The solid model of the block becomes a block primitive from which the tool, in its proper position and orientation, is subtracted. The swept volume of the tool is the union of the tool instances.

The tool is the union of a sphere and a cylinder both positioned at 0.175 inch below the top face. This position (0.175) assumes the slot is created in the block by removing all the material in one cut. The original position of the centerline of the tool is at the beginning of the slot centerline, as shown in Fig. 7-61. In this position, the tool is oriented vertically along the $Y$ axis. In order to obtain a fairly smooth slot, the tool is positioned every $d/4$, where $d$ is the tool diameter, that is, every 0.0625 inch. The top view of the profile of the tool swept volume is shown in Fig. 7-62. The curve length $S$ of the slot must be calculated to determine the required number of tool positions, $N$, to sweep the slot. This length is given by

$$S = \int \sqrt{1 + \left(\frac{dz}{dx}\right)^2}\, dx \qquad (7.104)$$

Profile of tool swept volume

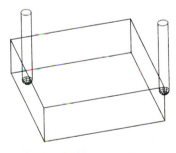

Initial and final tool positions

Swept volume of tool

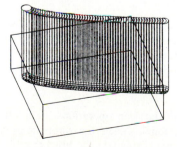

Swept volume superimposed on the block

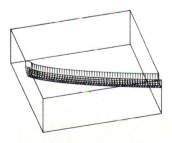

Solid model of the block

**FIGURE 7-62**
Block and tool swept volume.

where $dz/dx$ is calculated from the slot equation as

$$\frac{dz}{dx} = 0.439x^{-1/3} \tag{7.105}$$

Substituting Eq. (7.105) into (7.104), rearranging, and integrating, we obtain

$$S = \sqrt{(0.193 + x^{2/3})^3} \tag{7.106}$$

The curve length of the slot between $x = 0$ and $x = 3$ is $[S(x = 3) - S(x = 0)]$ 3.342 inch. Thus the number $N$ is given by

$$N = \text{INT}\left(\frac{S}{\Delta S}\right) + 1 \tag{7.107}$$

where INT is the INTEGER function. For $\Delta S = 0.0625$, 54 tool instances are required.

Rearranging Eq. (7.106) to give $x$ in terms of $S$ enables the calculation of the tool position, that is,

$$x = \sqrt{(S^{2/3} - 0.193)^3} \tag{7.108}$$

In a recursive form, this equation becomes

$$x_{i+1} = \sqrt{(S_{i+1}^{2/3} - 0.193)^3}, \qquad 0 \le i \le (N - 1) \tag{7.109}$$

and 
$$S_{i+1} = S_i + \Delta S \tag{7.110}$$

where the subscripts $i$ and $i + 1$ indicate the previous and current positions respectively. For the initial position, $i = 0$, $S_0 = 0.085$ inch, and $x_0 = 0$. Having the value of $x$ from Eq. (7.109), the $z$ coordinate of the tool position can be obtained from the equation of the slot centerline. The $y$ coordinate of the origins of the sphere and cylinder primitives that make up the tool (see Fig. 7.4) is 0.825.

The position of each primitive (sphere and cylinder) of the tool is given by $(x, y, z)$ as calculated above. The sphere and cylinder are unioned together to give the tool which, in turn, is subtracted from the block. For example, the initial and final positions of the sphere and the cylinder are given by $(0, 0.825, -1.5)$ and $(3.0, 0.825, -0.13)$ respectively. Repeating this process 54 times produces the block with the slot. The results are shown in Fig. 7-62.

The above solution is approximate: the more the number of tool instances, the more accurate the solid model of the block and, of course, the more expensive to operate on the model. The exact solution of this problem would require a solid modeler that supports nonlinear (or general) sweep of moving objects besides the natural quadrics primitives.

**Example 7.13.** Figure 7-63 shows a crankshaft mounted on a bracket. The given dimensions cause interference between the crankshaft and the bracket to occur. Find the maximum volume of interference. The interference can be removed by either creating a depression in the bracket or decreasing the length of the crankshaft. Find the amount of material removed for the latter solution.

*Solution.* This problem illustrates how to use boolean operations for interference detection between components of a solid model. Given the dimensions shown, create the solid model of the crankshaft system. Designate the bracket and crankshaft as solids $B$ and $C$ respectively. The interference volume is the intersection of the two; that is, $B \cap^* C$ and is shown in Fig. 7-63b. The mass properties of this volume can easily be calculated and are not discussed now (refer to Chap. 17).

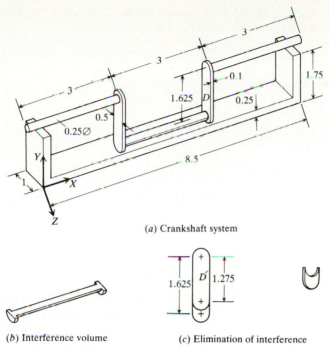

(a) Crankshaft system

(b) Interference volume

(c) Elimination of interference

**FIGURE 7-63**
Interference detection.

The crankshaft should be reduced by 0.35 inch (see Fig. 7-63c) so that the interference between the bracket and the crankshaft is removed, and to allow 0.1 inch clearance between the two. The amount of material removed is $2(D -* D')$, where $D -* D'$ is the difference between the original crankshaft $D$ (see Fig. 7-63a) and the new one $D'$ (Fig. 7-63c).

## PROBLEMS
## Part 1: Theory

**7.1.** A valid solid is defined as a point set that has an interior and a boundary as given by Eq. (7.1). A valid boundary must be in contact with the interior. Sketch a few two- and three-dimensional solids and identify $iS$ and $bS$ for each one. Is $iS$ always joint for any $S$? Can $bS$ be disjoint? What is your conclusion?

**7.2.** Three point sets in $E^2$ define three valid polygonal solids $S_1$, $S_2$, and $S_3$. The three solids are bounded by three boundary sets $bS_1$, $bS_2$, and $bS_3$ given by their corner points as: $bS_1 = \{(2, 2), (5, 2), (5, 5), (2, 5)\}$, $bS_2 = \{(3, 3), (7, 3), (7, 6), (3, 6)\}$, and $bS_3 = \{(4, 1), (6, 1), (4, 4), (6, 4)\}$. Find $S_1 \cup S_2 \cup S_3$, $S_1 \cap S_2 \cap S_3$, and $S_1 - S_2 - S_3$.

**7.3.** Reduce the following set expressions:
(a) $c(P \cap Q) \cup P$
(b) $c(P \cup Q) \cup (P \cap cQ)$
(c) $(P \cap Q) \cap c(P \cup Q)$
(d) $P - (P - Q)$
Use the set laws given by Eqs. (7.13) to (7.27) as well as the Venn diagram.

**7.4.** Using the set membership classification, classify the line $L$ shown in Fig. P7-4 with respect to the solid shown if the solid given is a B-rep and a CSG.

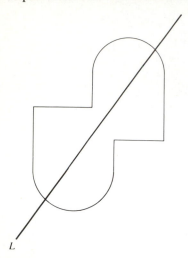

$L$                                          **FIGURE P7-4**

**7.5.** Implementing set operations given by Eqs. (7.41) to (7.48), or $S = A\langle OP \rangle B$ in general, involves finding intersections between $bA$ and $bB$, classifying the boundaries with respect to each operand, and combining classifications according to the rules:

$$\langle OP \rangle = \cup^*: bS = (A \text{ out } B \text{ UN}^* \ B \text{ out } A \text{ UN}^* \ A \text{ on } B+)$$

$$\langle OP \rangle = \cap^*: bS = (A \text{ in } B \text{ UN}^* \ B \text{ in } A \text{ UN}^* \ A \text{ on } B+)$$

$$\langle OP \rangle = -^*: bS = (A \text{ out } B \text{ UN}^* \ B \text{ in } A^{-1} \text{ UN}^* \ A \text{ on } B-)$$

where UN* is a regularized union operator in $E^1$. $A$ on $B+$ consists of those parts of $bA$ that lie on $bB$ so that the face normals of the respective faces are equal, whereas $A$ on $B-$ consists of the overlapping parts where the normals are opposite. $B$ in $A^{-1}$ denotes the complement of $B$ in $A$, that is, $B$ in $A$ with the orientation of all faces reversed. Notice that the $+$ and $-$, and the exponent $-1$, are a form of neighborhoods. Using the above equations:
(a) Explain the concept of closure used in Eqs. (7.41) to (7.48) and shown in Fig. 7-14. What should be done to implement the concept of interior to eliminate nonregular sets?
(b) Find $S = A\langle OP \rangle B$ for $A$ and $B$ shown in Fig. P7-5.

**7.6.** Solids in $E^2$ (i.e., two-dimensional solids) and their solid modelers are valuable in understanding many of the concepts needed to handle solids in $E^3$ (i.e., three-dimensional solids). A two-dimensional solid is a point set of ordered pairs $(x, y)$. A two-dimensional solid modeler based on half-spaces is to be developed utilizing linear and disc half-spaces. Find the equations of the two half-spaces.
    *Hint:* See Eqs. (7.51) and (7.52).
    Develop a parametric equation for their intersection.

**7.7.** Apply Euler laws, Eqs. (7.57) and/or (7.58), to models shown in Prob. 7.17 of Part 2. Using Euler operators, write the construction sequence needed to create them.

**7.8.** Following Example 7.7, write a procedure to split a solid with a plane.

**7.9.** Create an input sequence to create a CSG model of each object shown in Prob. 7.17 in Part 2. Based on this sequence, find the model CSG graph and tree. It is preferred to obtain balanced trees if possible.

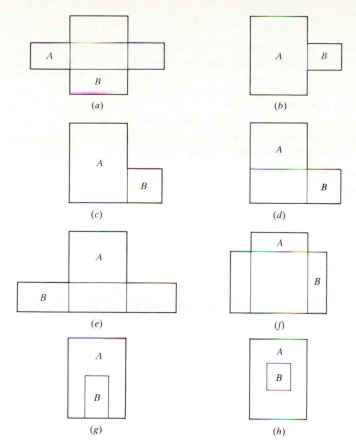

**FIGURE P7-5**

**7.10.** Apply the various available traversal methods to traverse each tree developed in Prob. 7.9.

**7.11.** It is desired to develop bounded primitives for a two-dimensional solid modeler based on the CSG scheme. A plate (rectangular plate and triplate) and disc primitives are to be developed. Find the mathematical definitions of these primitives.
 *Hint*: See Eqs. (7.61), (7.62), and (7.65).
 Develop intersection equations (refer to Prob. 7.6).

**7.12.** Apply the algorithms that perform boolean operations and described or used in Examples 7.7 (B-rep) and 7.9 (CSG) to the two primitive blocks shown in Fig. 7-36b.

**7.13.** How can you use a cylinder primitive to generate a sphere?

**7.14.** How can you generate a torus using other natural quadrics?

**7.15.** Repeat Prob. 7.8 but for a CSG model (follow Example 7.9).

**7.16.** Problems 7.6 and 7.11 introduced the mathematical foundations of a two-dimensional solid modeler. The further development of such a modeler requires representations of orientable surfaces, that is, represent surface normals, neighborhoods, classifications, and combining the classifications. Discuss the following for half-space, B-rep, and CSG schemes:
 (a) How to represent surface normals and neighborhoods
 (b) How to develop a classification algorithm
 (c) How to combine classifications

## Part 2: Laboratory

**7.17.** Create the solid models of the objects of Prob. 5-20. Similarly, create the solid models of the objects shown in Prob. 6.16. Use layers and colors provided by your CAD/CAM system to be able to manage the primitives and resulting subsolids. How do you compare the amount of effort it takes to create the wireframe, surface, and solid models of a given object?

**7.18.** Perform the set operations on the two objects shown in Fig. 7-15 on your CAD/CAM system. Compare the results with those of Example 7.3. Did your system fail to perform any of the operations? What is your conclusion?

## Part 3: Programming

Write a procedure to:

**7.19.** Classify lines with respect to two-dimensional solids for B-rep and CSG (see Figs. 7-17 and 7-18).

**7.20.** Combine the classifications that result from the above procedure.

**7.21.** Implement the results of Prob. 7.6.

**7.22.** Convert a user input into a CSG graph and a tree traversed in a reverse postorder.

**7.23.** Implement the results of Prob. 7.11.

**7.24.** Create two-dimensional primitives based on B-rep schemes (assume that low-level Euler operators exist).

**7.25.** Implement results of Prob. 7.16.

**7.26.** Implement results of Prob. 7.13 into an existing solid modeler.

**7.27.** Implement results of Prob. 7.14 into an existing solid modeler.

## BIBLIOGRAPHY

Baer, A., C. Eastman, and M. Henrion: "Geometric Modeling: A Survey," *Computer Aided Des. (CAD) J.*, vol. 11, no. 5, pp. 253–272, 1979.

Baumgart, B. G.: "Winged Edge Polyhedron Representation," AIM-179, report STAN-CS-320, Stanford University, 1972.

Baumgart, B. G.: "Geometric Modeling for Computer Vision," AIM-249, report CS-463, Artificial Intelligence Laboratory, Stanford University, 1974.

Berlin, E.: "Solid Modeling on a Microcomputer," *Computer Graphics World*, pp. 39–42, November 1984.

Bin, H.: "Inputting Constructive Solid Geometry Representations Directly from 2D Orthographic Engineering Drawings," *CAD J.*, vol. 18, no. 3, pp. 147–155, 1986.

Bobrow, J. E.: "NC Machine Tool Path Generation from CSG Part Representation," *CAD J.*, vol. 17, no. 2, pp. 69–76, 1985.

Bowerman, R. G.: "Drafting Links Up with Solid Modeling," *Mach. Des.*, pp. 46–49, September 12, 1985.

Boyse, J. W.: "Interference Detection among Solids and Surfaces," *Commun. ACM*, vol. 22, no. 1, pp. 3–9, 1979.

Braid, I. C.: "The Synthesis of Solids Bounded by Many Faces," *Commun. ACM*, vol. 18, no. 4, pp. 209–218, 1975.

Bronsvoort, W. F., J. J. Wijk, and F. W. Jansen: "Two Methods for Improving the Efficiency of Ray Casting in Solid Modeling," *CAD J.*, vol. 16, no. 1, pp. 51–55, 1984.

CAM-I, Inc.: "Design of an Experimental Boundary Representation and Management System for Solid Objects," report R-80-GM-02, CAM-I, Arlington, Tex., 1980.

CAM-I, Inc.: "Boundary File Design," report R-81-GM-02.1, CAM-I, Arlington, Tex., 1981.

CAM-I, Inc.: "Boundary Representation for Solid Objects," report R-82-GM-02.1, CAM-I, Arlington, Tex., 1982.

CAM-I, Inc.: "Extended Geometric Facilities for Surface and Solid Objects," report R-83-GM-02.1, CAM-I, Arlington, Tex., 1983.

CAM-I, Inc.: "Solid Modeling Applications; The Real Payback," *Proc. CAM-I's 3rd Geometric Modeling Seminar*, March 19–20, 1985, Nashville, Tenn.

Casale, M.: "Free-Form Solid Modeling with Trimmed Surface Patches," *IEEE CG&A*, pp. 33–43, January 1987.

Casale, M. S., and Stanton, E. L.: "An Overview of Analytic Solid Modeling," *IEEE CG&A*, pp. 45–56, February 1985.

Childress, R. L.: *Sets, Matrices, and Linear Programming*, Prentice-Hall, Englewood Cliffs, N.J., 1974.

Chiyakura, H., and F. Kimura: "A Method of Representing the Solid Design Process," *IEEE CA&A*, pp. 32–41, April 1985.

Choi, B. K., M. M. Barash, and D. C. Anderson: "Automatic Recognition of Machined Surfaces from a 3D Solid Model," *CAD J.*, vol. 16, no. 2, pp. 81–86, 1984.

Clark, A. L.: "Roughing It: Realistic Surface Types and Textures in Solid Modeling," *Computers In Mechanical Engineering (CIME) Mag.*, pp. 12–16, March 1985.

Congdon, R. M., and D. C. Gossard: "Interactive Graphic Input of Plane-Faced Solid Models," *Proc. Conf. on CAD/CAM Technology in Mechanical Engineering*, March 24–26, 1982, pp. 350–360, MIT, Cambridge, Mass.

Crocker, G. A.: "Screen-Area Coherence for Interactive Scanline Display Algorithms," *IEEE CA&G*, pp. 10–17, September 1987.

Doctor, L. J., and J. G. Torborg: "Display Techniques for Octree-Encoded Objects," *IEEE CG&A*, pp. 29–38, July 1981.

Eastman, C., and M. Henrion: "GLIDE: A Language for Design Information Systems," *Proc. First Annual Conf. on Computer Graphics in CAD/CAM Systems*, April 9–11, 1979, pp. 24–33, MIT, Cambridge, Mass.

Eastman, C., and K. Weiler: "Geometric Modeling Using the Euler Operators," *Proc. First Annual Conf. on Computer Graphics in CAD/CAM Systems*, April 9–11, 1979, pp. 248–259, MIT, Cambridge, Mass.

Farr, R., and G. Fredrickson: "Interactive Solid Modeling," *CAE Mag.*, pp. 46–48, November 1986.

Hakala, D. G., R. C. Hillyard, B. E. Nourse, and P. J. Malraison: "Natural Quadrics in Mechanical Design," in *AUTOFACT WEST*, pp. 363–378, Anaheim, Calif., November 17–20, 1980.

Halmos, P. R.: *Naive Set Theory*, Litton Educational Publishing, New York, 1960.

Hillyard, R.: "The Build Group of Solid Modelers," *IEEE CG&A*, pp. 43–52, March 1982.

Holt, M. G.: "Experiences with CAD Solids Modeling and Its Role in Engineering Design," *Computer-Aided Engng J.*, pp. 38–44, April 1985.

Hook, T. V.: "Advanced Techniques for Solid Modeling," *Computer Graphics World*, pp. 45–54, November 1984.

Hrbacek, K., and T. Jech.: *Introduction to Set Theory*, Marcel Dekker, New York, 1978.

Johnson, R. H.: "Product Data Management with Solid Modeling," *Computer-Aided Engng J.*, pp. 129–132, August 1986.

Kalay, Y. E.: "A Relational Database for Nonmanipulative Representation of Solid Objects," *CAD J.*, vol. 15, no. 5, pp. 271–276, 1983.

Kirk, D. B.: "Curved Surfaces in Solid Modeling: New Hardware Improves the View," *CIME Mag.*, pp. 10–14, May 1986.

Krouse, J. K.: "Solid Models for Computer Graphics," *Mach. Des.*, pp. 50–55, May 20, 1982.

Krouse, J. K.: "Sorting Out the Solid Modelers," *Mach. Des.*, pp. 94–101, February 10, 1983.

Krouse, J. K.: "Solid Modeling Catches On," *Mach. Des.*, pp. 60–64, February 7, 1985.

Lee, Y. C., and K. S. Fu: "Machine Understanding of CSG: Extraction and Unification of Manufacturing Features," *IEEE CG&A*, pp. 20–32, January 1987.

Levin, J.: "A Parametric Algorithm for Drawing Pictures of Solid Objects Composed of Quadratic Surfaces," *Commun. ACM*, vol. 19, no. 10, pp. 555–563, 1976.

Manty, M. la: "Boolean Operations of 2-Manifolds Through Vertex Neighborhood Classification," *ACM Trans. on Graphics*, vol. 5, no. 1, pp. 1–29, 1986.

Manty, M. la, and R. Sulonen: "GWB: A Solid Modeler with Euler Operators," *IEEE CG&A*, pp. 17–31, September 1982.

Manty, M. la, and M. Tamminen: "Localized Set Operations for Solid Modeling," *ACM Computer Graphics*, vol. 17, no. 3, pp. 279–288, 1983.

Miller, J. R., D. R. Starks, and M. D. Hastings: "An Evolving Volume Modeling-Based CAD/CAM System," *Proc. Conf. on CAD/CAM Technology in Mechanical Engineering*, March 24–26, 1982, pp. 33–53, MIT, Cambridge, Mass.

Mortenson, M. E.: *Geometric Modeling*, John Wiley, New York, 1985.

Myers, W.: "An Industrial Perspective on Solid Modeling," *IEEE CG&A*, pp. 86–97, March 1982.

Patnaik, L. M., R. S. Shenoy, and D. Krishnan: "Set Theoretic Operations on Polygons Using the Scan-Grid Approach," *CAD J.*, vol. 18, no. 5, pp. 275–279, 1986.

Pickett, M. S., and J. W. Boysl (Eds.): "Solid Modeling by Computers: From Theory to Applications," *Proc. Symposium on Solid Modeling*, September 25–27, 1983, General Motors Research Laboratories, Warren, Mich.

Post, F. H., and F. Klok: "Deformations of Sweep Objects in Solid Modeling," in *Eurographics '86* (Ed. A. A. G. Requicha), pp. 103–115, Elsevier Science, New York, 1986.

Pratt, M. J.: "Solid Modeling and the Interface Between Design and Manufacture," *IEEE CG&A*, pp. 52–59, July 1984.

Putnam, L. K., and P. A. Subrahmanyam: "Boolean Operations on *n*-Dimensional Objects," *IEEE CG&A*, pp. 43–51, June 1986.

Requicha, A. A. G. (Ed.): *Eurographics '86*, Elsevier Science, New York, 1986.

Requicha, A. A. G., and S. C. Chan: "Representation of Geometric Features, Tolerances, and Attributes in Solid Modelers Based on Constructive Geometry," *IEEE J. of Robotics and Automation*, vol. RA-2, no. 3, pp. 156–166, 1986.

Requicha, A. A. G., and H. B. Voelcker: "Solid Modeling: A Historical Summary and Contemporary Assessment," *IEEE CG&A*, pp. 9–24, March 1982.

Requicha, A. A. G., and H. B. Voelcker: "Solid Modeling: Current Status and Research Directions," *IEEE CG&A*, pp. 25–37, October 1983.

Requicha, A. A. G., and H. B. Voelcker: "Boolean Operations in Solid Modeling: Boundary Evaluation and Merging Algorithms," *Proc. IEEE*, vol. 73, no. 1, pp. 30–44, 1985.

Rossignac, J. R., and A. A. G. Requicha: "Offsetting Operations in Solid Modeling," *Computer Aided Geometric Des.*, vol. 3, pp. 129–148, 1986.

Rossignac, J. R., and A. A. G. Requicha: "Depth-Buffering Display Techniques for Constructive Solid Geometry," *IEEE CG&A*, pp. 29–39, September 1986.

Roth, S. D.: "Ray Casting as a Method for Solid Modeling," report GMR-3466, General Motors Research Laboratories, Warren, Mich., 1980.

Rouse, N. E.: "Linking Solids and Surfaces," *Mach. Des.*, pp. 82–86, May 7, 1987.

Sarraga, R. F.: "Algebraic Methods for Intersections of Quadric Surfaces in GMSOLID," *Computer Vision, Graphics, and Image Processing*, vol. 22, pp. 222–238, 1983.

Stewart, I. P.: "Quadtrees: Storage and Scan Conversion," *The Computer J.*, vol. 29, no. 1, pp. 60–75, 1986.

Tamminen, M., O. Karonen, and M. la Manty: "Ray-Casting and Block Model Conversion Using a Spatial Index," *CAD J.*, vol. 16, no. 4, pp. 203–208, 1984.

Tan, S. T., and M. M. F. Yuen: "Integrating Solid Modeling with Finite Element Analysis," *Computer-Aided Engng J.*, pp. 133–137, August 1986.

Tan, S. T., M. F. Yuen, and K. C. Hui: "Modeling Solids with Sweep Primitives," *Computers In Mechanical Engineering (CIME) Mag.*, pp. 60–73, September/October 1987.

The Merrit Company: *Solid Modeling Today*, vol. 1, no. 1–8, The Merrit Company, Santa Monica, Calif., 1986.

Tilove, R. B.: "Set Membership Classification: A Unified Approach to Geometric Intersection Problems," *IEEE Trans. on Computers*, vol. C-29, no. 10, pp. 874–883, 1980.

Tilove, R. B.: "A Null-Object Detection Algorithm for Constructive Solid Geometry," *Commun. ACM*, vol. 27, no. 7, pp. 684–694, 1984.

Tilove, R. B., and A. A. G. Requicha: "Closure of Boolean Operations on Geometric Entities," *CAD J.*, vol. 12, no. 5, pp. 219–220, 1980.

Tilove, R. B., A. A. G. Requicha, and M. R. Hopkins: "Efficient Editing of Solid Models by Exploiting Structural and Spatial Locality," *Computer Aided Geometric Des.*, vol. 1, pp. 227–239, 1984.

Toriya, H., T. Satoh, K. Ueda, and H. Chiyokura: "UNDO and REDO Operations for Solid Modeling," *IEEE CA&A*, pp. 35–42, April 1986.

Vandoni, C. E. (Ed.): *Eurographics '85*, Elsevier Science, New York, 1985.

Varady, T., and M. J. Pratt: "Design Techniques for the Definition of Solid Objects with Free-Form Geometry," *Computer Aided Geometric Des.*, vol. 1, pp. 207–225, 1984.

Voelcker, H. B., and A. A. G. Requicha: "Geometric Modeling of Mechanical Parts and Processes," *Computer*, pp. 48–57, December 1977.

Wagner, P. M.: "Solid Modeling for Mechanical Engineering," *Computer Graphics World*, pp. 10–24, September 1984.

Wang, W. P., and K. K. Wang: "Geometric Modeling for Swept Volume of Moving Solids," *IEEE CG&A*, pp. 8–17, December 1986.

Weiler, K.: "Edge-Based Data Structures for Solid Modeling in Curved-Surface Environment," *IEEE CG&A*, pp. 21–40, January 1985.

"What's Holding Back Solid Modeling?," *CAE Mag.*, pp. 46–52, December 1986.

Williams, N. H.: *Combinatorial Set Theory*, North-Holland, 1977.

Wilson, P. R.: "Euler Formulas and Geometric Modeling," *IEEE CG&A*, pp. 24–36, August 1985.

Woo, T. C.: "Computer Aided Recognition of Volumetric Designs," in *Advances in Computer-Aided Manufacture* (Ed. D. McPherson), pp. 121–136, North-Holland, 1977.

Woo, T. C.: "Feature Extraction by Volume Decomposition," *Proc. Conf. on CAD/CAM Technology in Mechanical Engineering*, March 24–26, 1983, pp. 76–94, MIT, Cambridge, Mass.

Woo, T. C.: "Interfacing Solid Modeling to CAD and CAM: Data Structures and Algorithms for Decomposing a Solid," *Computer*, pp. 44–49, December 1984.

Woo, T. C.: "A Combinatorial Analysis of Boundary Data Structure Schemata," *IEEE CG&A*, pp. 19–27, March 1985.

Woodwark, J. R.: "Generating Wireframes from Set-Theoretic Solid Models by Spatial Division," *CAD J.*, vol. 18, no. 6, pp. 307–315, 1986.

Wyvill, G., T. Kunii, and Y. Shirai: "Space Division for Ray Tracing in CSG," *IEEE CG&A*, pp. 28–34, April 1986.

Yamaguchi, F., and T. Tokieda: "A Solid Modeler with a $4 \times 4$ Determinant Processor," *IEEE CG&A*, pp. 51–59, April 1985.

Yerry, M. A., and M. S. Shephard: "A Modified Quadtree Approach to Finite Element Mesh Generation," *IEEE CG&A*, pp. 39–46, January/February 1983.

*1985 European Conference on Solid Modeling*, September 9–10, 1985, London.

# CHAPTER
# 8

## CAD/CAM
## DATA
## EXCHANGE

### 8.1  INTRODUCTION

Computer databases are now replacing paper blueprints in defining product geometry and nongeometry for all phases of product design and manufacturing. It becomes increasingly important to find effective procedures for exchanging these databases. Fundamental incompatibilities among entity representations greatly complicate exchanging modeling data among CAD/CAM systems. Even simple geometric entities such as circular arcs are represented by incompatible forms in many systems. The database exchange problem is complicated further by the complexity of CAD/CAM systems, the varying requirements of organizations using them, the restrictions on access to proprietary database information, and the rapid pace of technological change.

Transferring data between dissimilar CAD/CAM systems must embrace the complete product description stored in its database. Four types of modeling data make up this description. These are shape, nonshape, design, and manufacturing data. Shape data consists of both geometrical and topological information as well as part or form features. Entity attributes such as font, color, and layer as well as annotation are considered part of the entity geometrical information. Topological information applies only to products described via solid modeling. Features allow high-level concept communication about parts. Examples are hole, flange, web, pocket, chamfer, etc. Nonshape data includes graphics data such as shaded images, and model global data as measuring units of the database and the resolution of storing the database numerical values. Design data has to do with the information that designers generate from geometric models for analysis purposes. Mass property and finite element mesh data belong to this type of data. Manufacturing data is the fourth type. It consists of information as tooling, NC tool paths, tolerancing, process planning, tool design, and bill of materials.

Data formats that are designed to communicate product data among CAD/CAM systems must address-exchange these four types of data. Successful exchange cannot otherwise be achieved and the data format performing the exchange is viewed to be limited in scope. While it is desired for a data format to address the complete product description, it is not always feasible to design and implement such a format. Early attempts to design data formats, e.g., IGES, focused on CAD-to-CAD exchange where primarily shape and nonshape data were to be transferred from one system to another. Soon it became apparent that new data formats need to be designed or the scope of existing ones must be extended to include CAD-to-CAD and CAM-to-CAM exchanges, that is, exchange of complete product descriptions. PDES (see Sec. 8.4) is an example of a data format whose scope covers the four types of modeling data discussed earlier.

The need to exchange modeling data is directly motivated by the need to integrate and automate the design and manufacturing processes to obtain the maximum benefits from CAD/CAM systems. There is always the demand to be able to tie together two or more of these systems to form an application that shares common data. This demand exists either internally within a single organization or externally as in the case of subcontract manufacturers or component suppliers. Where similar CAD/CAM systems are operated by both parties, no difficulty of exchange exists as the files that store modeling data are compatible and therefore can be transferred directly. However, many dissimilar CAD/CAM systems are in existence, and here data communication problems arise as each system employs its own system-specific data structure to store product data; that is, each system stores drawings and modeling representation in its own way.

Having acknowledged the need to exchange modeling data, how can we meet such a need and solve the problem of data exchange? This problem has two solutions: direct and indirect. The direct solution entails translating the modeling data stored in a product database directly from one CAD/CAM system format to another, usually in one step. On the other hand, the indirect solution is more general and adopts the philosophy of creating a neutral database structure (also called a neutral file) which is independent of any existing or future CAD/CAM system. This structure acts as an intermediary and a focal point of communication among dissimilar database structures of CAD/CAM systems. The structure of the neutral database must be general, governed only by the minimum required definitions of any of the modeling data types, and be independent of any vendor format. Naturally, the structure of this database is influenced by structures of existing vendors' databases and can be viewed as the common denominator among them. In order to achieve such generality, efficiency and special enhancements to store and access the database suffer. Therefore, the size of the neutral database is larger and its access speed is slower in comparison to their counterparts created by various vendors' systems.

Figure 8-1 shows how both solutions work. Direct translators convert data directly in one step. They are typically written by computer service companies that specialize in CAD/CAM database conversion. Direct translators are considered to be dedicated translator programs, two of which link a system pair as indicated by the dual direction arrows shown in the figure. For example, two

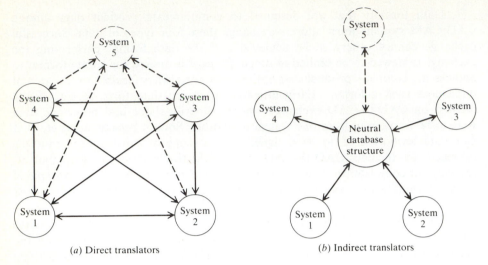

(a) Direct translators     (b) Indirect translators

**FIGURE 8-1**
Methods of exchanging modeling data among dissimilar CAD/CAM systems.

translators are needed to transfer data between systems 1 and 2: one from system 1 to system 2 and the other from system 2 to system 1. Indirect translators utilize some neutral file format, which reflects the neutral database structure, with each system having its own pair of processors to transfer data to and from this neutral format. The translator that transfers data from the database format of a given system to the neutral format is called a preprocessor while the translator that does the opposite transfer is known as a postprocessor.

Each type of translator has its advantages and disadvantages. Direct translators provide a satisfactory solution when only a small number of systems are involved, but as this number increases the number of translator programs that need to be written becomes prohibitive. In general, if modeling data is to be transferred between all possible pairs of $n$ CAD/CAM systems, then the total number of translators, $N$, that must be written are given by

$$N = 2\binom{n}{2} = n(n-1) \tag{8.1}$$

where
$$\binom{n}{2} = \frac{n!}{2!(n-2)!}$$

in which the symbol " ! " indicated factorial. The coefficient 2 of the left-hand side of the above equation reflects the fact that two translators must be written for each pair of systems (see Fig. 8-1a). Adding one system to the existing $n$ systems would require writing $2n$ additional translators. In Fig. 8-1a, eight additional translators (shown dashed in the figure) are needed to accommodate the addition of system 5. However, direct translators run more quickly than indirect ones and the data files they produce are smaller in size than the neutral files created by indirect translators.

On the other hand, indirect translators do not suffer from the increasing numbers of programs to be written as in the case of direct ones. For the case of $n$ systems, the total number of indirect translators that must be written is given by

$$N = 2n \qquad (8.2)$$

This equation applies for $n \geq 3$. For $n = 3$, Eqs. (8.1) and (8.2) give the same number of translators. For $n > 3$, the required number of translators based on the neutral format philosophy [Eq. (8.2)] is less than those based on direct conversion [Eq. (8.1)]. Adding one system (system 5 shown in Fig. 8-1b) to $n$ existing systems would only require writing two additional translators (shown dashed in the figure) regardless of the value of $n$. Moreover, indirect translator philosophy provides stable communication between CAD/CAM systems, protects against system obsolescence, and eliminates dependence on a single-system supplier. A side benefit of neutral files is that they can potentially be archived. Some companies, in the aerospace industry for example, need to keep CAD/CAM databases for 20 to 50 years. Indirect translators based on standard neutral file format are now the common practice while direct translators are seldom used. The remainder of this chapter presents material related to indirect translators only.

The design of a successful data exchange format or standard must meet a minimum set of requirements. The standard must address the four types of modeling data. It must be able to support the common entities of each type; that is, it tends to be a superset of modeling data found in existing CAD/CAM systems. The development of a compact form to store and retrieve data is crucial to the performance of the standard, that is, its speed to convert data to and from the neutral format and the size of the resulting neutral file. For example, latest versions of IGES utilize a binary form instead of ASCII for its neutral files. For its future survival, future versions of the standard must remain upward compatible with its old and/or existing versions. These requirements apply to standards such as IGES, PDES, and others discussed in this chapter.

The above design requirements of a standard to communicate modeling data between dissimilar CAD/CAM systems cannot be achieved without limitations and problems. There are problems of definition. The standard may use definitions and terminologies that do not correspond to that of vendors. Furthermore, not all systems may support all the entities supported by the standard, say IGES, thus creating a mismatch between systems that support a different list of entities. There are problems of implementation. As with any standard, different people may interpret the rules of a CAD/CAM standard (such as IGES) differently and therefore incompatible translators may result. Owing to the ongoing development of CAD/CAM software and the standard itself, the translators (or processors) will always tend to be outdated. Other problems may exist. For example, certain numerical errors can arise in converting certain entities such as free-form curves and surfaces into neutral format. Despite these problems, a data exchange standard is the best solution to integrate dissimilar CAD/CAM systems. Most of these problems are usually prominent in the early revisions of a standard and seem to go away in its late revisions.

This chapter describes the available data exchange standards and how they are driving towards developing an international standard. It then covers in more detail the IGES and PDES standards.

## 8.2  EVOLUTION OF DATA EXCHANGE FORMAT

The upsurge of interest in product data exchange has led various national and international groups and organizations to search for definitions of standards for this purpose. There exist few standards that have been adopted, implemented, and tested by various vendors and users. The evolution of these standards follows a similar path to the evolution of the CAD/CAM technology itself. The first tier of efforts concentrated on exchanging shape data only. Initially, the problem was to transfer mainly geometrical data. However, as users became more experienced, it was realized that topological data needed to be transferred as well for a complete definition of shape data.

It was soon realized that there is much more to a product definition than just shape data. For example, all the design and manufacturing information must be available in an exchange-sensible manner. Exchanging such information is a prerequisite to automate and integrate the various CAD/CAM functions. Therefore, the second tier of efforts extends the scope of the first tier. It emphasizes the definition and development of standards that are capable of exchanging the complete product description, that is, exchanging the four types of data (shape, nonshape, design, and manufacturing). These standards must be carefully defined to preserve valuable product information that may be lost if all data types are not considered. For example, it is necessary for a data format or standard to capture the numerical values of tolerances, and hence their meaning, rather than treating them as merely text strings placed in a drawing. Another example is that the representation of shape data may be unsatisfactory to define manufacturing data. Consider the NURBS (nonuniform rational B-spline) representation discussed in Chap. 5. Such a representation can define a variety of curves and surfaces. This may cause a loss in information. If a circular cylinder (a hole) is represented by NURBS, data indicating that the part feature was a circular cylinder may be lost unless it is carried along. Such data would be useful to manufacturing to identify it as a hole to be drilled or bored rather than a surface to be milled. Another case in point involves offset surfaces. Information indicating that a surface is offset from an original surface is not often carried along with the original surface. Offset dimensions and tolerance information, however, can ease manufacturing of a part.

With the above heritage and wealth of information and experience in hand, it became obvious that efforts to define exchange standards must be unified at the international level. This third tier of efforts is led by the ISO (International Standards Organization). These efforts are relatively new, being initiated in 1984, and are targeted to produce the first ISO standard in the 1990s.

Figure 8-2 shows the evolution of the data exchange format and the various existing standards. A brief description of the evolution and emphasis of each standard is given below.

### 8.2.1  Shape-Based Format

In the late 1970s the US Department of Defense recognized the need to transfer modeling data (mainly geometrical data at that time) between different types of

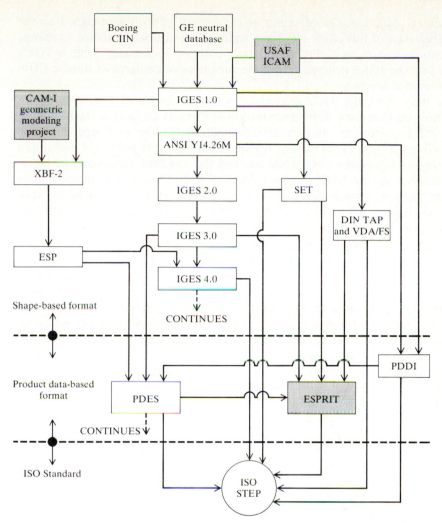

**FIGURE 8-2**
Evolution of data exchange format.

CAD/CAM systems. In September 1979, representatives of government and industry joined forces under the Air Force integrated computer aided manufacturing (ICAM) program (research programs are shown shaded in Fig. 8-2) to develop a method for data exchange. Funding for management and coordination was provided by the ICAM program while industrial users and CAD/CAM vendors provided resource material and personnel. A technical committee consisting of the Boeing Company, the General Electric Company, and the then National Bureau of Standards, or NBS (now National Institute of Standards and Technology, or NIST), was assigned to develop the data exchange method. Boeing and GE were chosen due to their prior experience in developing data exchange formats—Boeing with its CIIN (CAD/CAM Integrated Information Network) standard format it developed in the late 1960s and GE with its work on a neutral database. Coordination of the overall effort was assigned to the NBS. The result of this effort was the creation of the Initial Graphics Exchange Specification, known as IGES, which was published in January 1980 as IGES version 1.0. The IGES concept carries the considerable pedigree of Boeing CIIN and GE neutral databases.

In May 1980, the ANSI (American National Standards Institute) Y14.26 committees on Computer Aided Preparation of Product Definition Data voted to adopt IGES version 1.0 as part of the standard for the exchange of digital product data. A draft standard was released to the general public for evaluation and voting. In September 1981, ANSI adopted the Y14.26M document and therefore IGES became an ANSI standard. Since then, IGES has gone through three revisions: version 2.0 released in February 1983, version 3.0 released in April 1986, and version 4.0 released in June 1988. Technical details of IGES are covered in Sec. 8.3.

IGES has gained wide support among CAD/CAM users and vendors. Its public support is strong enough for its future development and enhancements to be most likely to continue. If a new standard replaces it, such a standard need not be directly upward compatible with IGES, but must accommodate a conversion path. IGES is implicitly targeted toward CAD/CAM systems of the 1970s and early 1980s. It addresses the exchange of mostly geometrical data and non-shape data. Recent revisions attempt to address design and topological data due to the lack of other existing standards. For example, version 2.0 and 3.0 support the exchange of finite element data and electronics printed wiring board product data. Version 4.0 supports topological data of the CSG modeling scheme.

Since its release, IGES has stimulated new efforts in a search for better standards that address the four types of modeling data or has directly or indirectly influenced other standards such as the French SET and German DIN TAP and VDA/FS that were developed in Europe for geometric data exchange (see Fig. 8-2). The SET (Standard d'Exchange et de Transfert) standard is developed by the French company, Aerospatiale. The range of entities transmitted is similar to that of IGES. However, SET utilizes a free format (IGES was a fixed format) to generate SET files to achieve greater efficiency. The first SET document was publicly available in 1983 and revision 1.1 was published in March 1984. SET translators are available for several widely used CAD/CAM systems.

With the conception, and later wide acceptance, of solid modeling as a viable modeling technique, efforts to search for standards to exchange shape data, both geometry and topology, were established by various organizations in anticipation of the need for such standards. The CAM-I (Computer Aided Manufacturing—International) organization funded the development of a solid modeling exchange standard called the experimental boundary file specification, or XBF. The first XBF document was published in 1981 and a revised version was published in 1982. This included both CSG and B-rep models. XBF extends IGES to solid modeling and the format of XBF files are closely similar to that of IGES files. While the XBF efforts were under way, the IGES committee was also considering including solids into the standard. Due to the many similarities between the two efforts, XBF was merged into the IGES Experimental Solids Proposal (ESP). The ideas of ESP are implemented in the IGES version 4.0.

## 8.2.2   Product Data-Based Format

The experience gained from standards based on exchanging shape and nonshape data, coupled with the need to automate CAD/CAM functions, has led to efforts being made to address the exchange of design and manufacturing data in addition to the other two types that have been already addressed. The emphasis in developing new standards is on exchanging the complete product data or description as opposed to exchanging shape data only, as discussed in the above section.

The product data definition interface (PDDI) was developed by the US Air Force ICAM program to address the need to exchange product data. This specification was released into the public domain in 1984 for testing and evaluation purposes. The PDDI project is targeted toward future replacement of the function of the engineering drawing. The PDDI system is intended to serve as the information interface between engineering and all manufacturing functions using blueprints, including process planning, NC programming, NC verification, quality assurance, tool design, robotics, and others. The scope of the PDDI system is therefore limited and has as its purpose the development of a product model and a data exchange format capable of conveying sufficient information to manufacture a part. It is not intended to be a complete database exchange format such as IGES and PDES.

The data elements that can be identified and exchanged by the PDDI system fit into five categories. These are geometry, topology, tolerances, form features, and part control information (such as material, process specifications, part identification, etc.). Other data such as analysis and/or manufacturing data are not included. The data defined by the PDDI system is applicable only to discrete mechanical parts. No assemblies are allowed. Four classes of parts can be handled: composite, machined, sheet metal, and turned.

The implementation of the PDDI system requires two processors. A preprocessor is needed to convert a part model database from a particular CAD/CAM system format to the exchange format required by the PDDI system. PDDI then takes the exchange format and converts it into what is called a "working form" which, in turn, can be accessed by manufacturing application programs (e.g., NC programming) of the CAD/CAM system via a postprocessor. The processors can interface to the PDDI system on an entity-by-entity "put" and "get" level to ease designing of the interfaces and speed up the conversion process. The link that the PDDI system provides between the part databases [mainly engineering (shape) information] and a manufacturing application program acts as, and therefore eliminates, the engineering drawing or the blueprint.

The significance of the PDDI system is that it went beyond just geometry and topology entities by identifying form features as an important part of product data. Features allow high-level concept communication about parts. The PDDI feature entities relate specific topology and geometry entities to a given feature so that identifying information for that feature can be explicit in the data, a necessary condition for the support of automation.

While the PDDI project tackles some product data, it can still be viewed as an advanced shape-based format. There is still the need to develop an exchange standard for product data in support of industrial automation. Product data must include the four types of data described at the beginning of this chapter. This data is relevant to the entire life cycle of a product: design, analysis, manufacturing, quality assurance, testing, support, etc. A long-term project has been initiated in 1985 to develop PDES (Product Data Exchange using STEP) to fulfill this need.

PDES is directly influenced by the existing data exchange heritage such as IGES, XBF-2, IGES ESP, and PDDI. PDES is being developed in coordination with IGES current and future efforts. PDES is, at a minimum, functionally equivalent to IGES version 4.0. When this functional equivalence is achieved, development efforts will be directed toward PDES, with the maintenance efforts being directed toward IGES.

PDES is designed to support any industrial application such as mechanical, electrical, plant design, and AEC (architecture and engineering construction). PDES defines a minimum set of conceptual data required by all applications. Then a data representation is used to store the information in a three-layer architecture comprised of the logical (conceptual) layer, the application layer, and the physical layer. PDES and its underlying philosophy are covered in more detail in Sec. 8.4.

Europe has also been involved in a search for standards to address the exchange of product data between the CAD/CAM system. A research program called ESPRIT began in 1984 (see Fig. 8-2) toward this goal. The ESPRIT project is funded by the European Commission and is led by West Germany. ESPRIT is intended to run for five years and to provide a focus for European developments in data exchange formats.

### 8.2.3 ISO Standard

This bewildering proliferation of data exchange "standards" is obviously of little advantage to CAD/CAM vendors and users as there is no accepted common format. Therefore, a Subcommittee, SC4, was formed in 1984 within the ISO Technical Committee TC184 (Industrial Automation Systems) to address the representation of product model data. There is agreement within the subcommittee that a single worldwide standard for the exchange of product data is needed. The name of the standard is to be STEP (Standard for the Transfer and Exchange of Product Model Data). The goal of the standard is to enable the exchange of a computerized product model with all its supporting types of data in a neutral format. STEP development is influenced by existing standards, in particular IGES, PDDI, SET, and DIN TAP and VDA/FS. STEP efforts are all also coordinated with other standards under development, mainly PDES and ESPRIT work.

### 8.3 IGES

IGES is the first standard exchange format developed to address the concept of communicating product data among dissimilar CAD/CAM systems. Many users of these systems have invested heavily in developing and/or acquiring special-purpose software for the design, analysis, manufacturing, and testing of their discrete products. As they seek to integrate this software into the total design and manufacturing environment, they use IGES to solve their problems of database communication. IGES has been used for a dual purpose: transfer of modeling data within corporate dissimilar systems and digital communication between the company and its suppliers and customers.

At its inception, IGES was viewed as a communication file for transmitting data between two systems or applications. The target data consisted of relatively simple mechanical models and drawings, with few internal relationships. This set of simple goals yielded processors that translated on an entity-by-entity basis and subscribed to the self-contained file limitation. Little attention was provided for logical associativities typically available in CAD/CAM databases. As the users' needs have grown, the view of the standard has evolved toward IGES functioning as a neural database structure. Such a structure can be accessed by various applications and systems via the proper interface processors. This view of a standard makes it dynamic and adjustable to accommodate future needs both for new applications and product data.

IGES has been revised a few times since its version 1.0 was released in 1980. The various IGES versions share some common characteristics. Each version must remain upwards compatible with the previous version for practical purposes. This protects any previous efforts and resources devoted by both vendors and users to develop and use IGES processors respectively. This also means that a processor that is fully conforming to the latest version can correctly interpret IGES files written in accordance with prior versions of IGES. In addition, each version usually represents both a refinement and extension of the earlier version. The wide public reviews and comments plus the feedback from implementing and testing the standard add dramatically to its clarity and precision. The extensions and enhancements are usually targeted toward extending the range of geometry covered and increasing the number of applications that can be supported.

Versions 2.0, 3.0, and 4.0 each contains refinements and extensions over its predecessor. Version 2.0 has tightened the description and added to the clarification of IGES. It has also included many technical extensions to augment IGES capabilities and expand it into new areas. Geometric entities have been enhanced in scope to be more generally applicable. This included the parametrization in the ruled surface entity, a more general form of the tabulated cylinder entity, and means of relating the surface of revolution entity to the common geometrical surfaces such as spheres and cones. Also, two new geometric entities have been added to provide a more general approach for surface and curve representations. These are a rational B-spline surface entity and a related rational B-spline curve entity. Algorithms were developed for an exact conversion between the rational B-spline method and the Bezier method of representation. Version 2.0 has improved the earlier work in the annotation area by specifying a larger set of text fonts, clarifying the intent for positioning and scaling of text material, and clearly defining the angular dimension entity. In the application area, version 2.0 has addressed the communication of finite element data and printed wiring product (printed circuit board) data. Finally version 2.0 has solved the problem of IGES large file lengths due primarily to the ASCII character representation by introducing an optional or alternate binary format representation which addresses the problems of file size and processing speed. Savings in file size of 50 to 68 percent have been estimated.

Version 3.0 has made considerable clarifications related to IGES entities; namely, in the view and drawing entities, in the global parameters including default values, in the unit flag, in the transformation matrix pointer, in the leader entity, and in the parametric spline curve and surface entities. In the geometric area, new entity capabilities have been added for offset curves, offset surfaces, curves on a parametric surface, and a trimmed surface entity (which allows the definition of a surface boundary). In the annotation area, new capabilities exist to represent a larger range of annotation style of the nominal (base) value and the tolerance limits. In addition, a more compact definition for cross-hatching is provided. Version 3.0 has enhanced the capabilities of user-defined MACROs essential for standard part libraries. This includes adding labels, branching, and calling arguments to the previous MACRO capability. Finally, version 3.0 introduces the compressed ASCII format that reduces the IGES file size to one third of its previous size to address the storage size and telecommunications costs.

Version 4.0 has extended IGES capabilities to solid modeling (CSG representation scheme only) for the first time. Entities to handle the CSG scheme have been introduced. Representations of regularized boolean operations (union, intersection, and difference) and primitives (block, wedge, cylinder, cone, sphere, torus, ellipsoid, solid of linear extrusion, and solid of revolution) have been established. Extension of IGES to the B-rep scheme will be in version 5.0. In addition, version 4.0 has concentrated on applications. The ability to attach predefined electrical attributes (e.g., maximum ratings and propagation delays) has been added to electrical/electronic applications. AEC applications have been extended to include the ability to define attribute data (e.g., pattern fill) and related graphic representations as well as piping information. Considering the finite element modeling and analysis application, version 4.0 has expanded IGES to describe nodal (e.g., temperature and displacement) and element (e.g., stress and strain) results. Table 8.1 shows the geometric entity types of IGES and their availabilities in its various versions. The numbers shown are those of the entity types as IGES assigns them. The definitions of some of these entities depend on the existence of subordinate entities. Such relationships can be easily identified based on the material covered in Chaps. 5, 6, and 7. Therefore, not only single entities but also structures of entities need to be considered when comparing IGES models.

### 8.3.1 Description

IGES defines a neutral database, in the form of a file format, which describes an "IGES model" of modeling data of a given product. The IGES model can be read and interpreted by dissimilar CAD/CAM systems. Therefore, the corresponding product data can be exchanged among these systems. IGES describes the possible information entities to be used in building an IGES model, the necessary parameters (data) for the definition of model entities, and the possible relationships and associativities between model entities.

**TABLE 8.1**
**IGES geometric entity types**

| Modelling type | IGES version | Code | Entity |
|---|---|---|---|
| Solids | IGES 4.0 | 150 | Block |
| | | 152 | Right angular wedge |
| | | 154 | Right circular cylinder |
| | | 156 | Right circular cone |
| | | 158 | Sphere |
| | | 160 | Torus |
| | | 162 | Solid of revolution |
| | | 164 | Solid of linear extrusion |
| | | 168 | Ellipsoid |
| | | 180 | Boolean tree |
| | | 184 | Solid assembly instance |
| | | 430 | Solid instance |
| Surfaces | IGES 2.0 | 108 | Plane |
| | | 114 | Parametric spline surface |
| | | 118 | Ruled surface |
| | | 120 | Surface of revolution |
| | | 122 | Tabulated cylinder |
| | | 128 | Rational B-spline surface |
| | IGES 3.0 | 140 | Offset surface |
| | | 144 | Trimmed parametric surface |
| Curves | IGES 2.0 | 100 | Circular arc |
| | | 102 | Composite curve |
| | | 104 | Conic arc |
| | | 110 | Line |
| | | 112 | Parametric spline curve |
| | | 116 | Point |
| | | 126 | Rational B-spline curve |
| | IGES 3.0 | 130 | Offset curve |
| | | 142 | Curve on a parametric surface |
| Others | IGES 2.0 | 106 | Copious data |
| | | 124 | Transformation matrix |
| | | 125 | Flash |
| | | 134 | Node |
| | | 136 | Finite element |
| | IGES 3.0 | 132 | Connect point |
| | | 138 | Nodal display and rotation |
| | | 146 | Nodal results |
| | | 148 | Element results |

Like most CAD/CAM systems, an IGES model is based on the concept of entities. The fundamental unit of information in the model, and consequently in the IGES file, is the entity; all product definition data are expressed as a list of predefined entities. Each entity defined by IGES is assigned a specific entity type number to refer to it in the IGES file. Entity numbers 1 through 599 and 700 through 5000 are allocated for specific assignments. Entity type numbers 600 through 699 and 10,000 through 99,999 are for implementor-defined entities (via MACRO definitions). These entities enable IGES to act as an archiving format where both the source (sending) and the target (receiving) systems are the same. Entity type number 5001 through 9999 are reserved for MACRO entities. Entities are categorized as geometric and nongeometric. Geometric entities represent the definition of the product shape and include curves and surfaces. Relations that may exist between various entities are included as parameters. For example, an IGES B-spline surface entity can point to a B-spline curve entity as part of its parameters. Nongeometric entities provide views and drawings of the model to enrich its representation and include annotation and structure entities. Annotation entities include various types of dimensions (linear, angular, ordinate), centerlines, notes, general labels, symbols, and cross-hatching. Structure entities include views, drawings, attributes (such as line and text fonts, colors, layers, etc.), properties (e.g., mass properties), subfigures and external reference entities (for assemblies), symbols (e.g., mechanical and electrical symbols), and macros (to define parametric parts).

IGES itself is just a document describing what should go into a data file. Interested developers (CAD/CAM vendors or companies specialized in database transfer) must write software to translate from their systems to the IGES format and vice versa. The software that translates from the native database format of a given CAD/CAM system to the IGES format is called a preprocessor. The software that translates in the opposite way (from IGES to a CAD/CAM system) is called a postprocessor. The preprocessors and postprocessors are also called translators and they determine the success of an IGES translation. Figure 8-3 shows the database exchange using IGES. The source system is the originating or sending CAD/CAM system and the target system is the receiving one. The archival database is a side benefit of IGES. Such archived databases could be

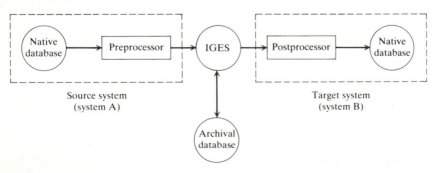

**FIGURE 8-3**
Database exchange using IGES.

kept for as long as needed. If system B in the figure becomes the sources and system A becomes the target, then the processors reverse positions.

### 8.3.2 Data Representation

While its fundamental unit of information is the entity, IGES has an information structure to describe geometric models similar to that found in typical CAD/CAM systems. For example, IGES defines a drawing entity and a view entity. The former allows a set of views to be identified and arranged and contains only the set of rules and parameters of extracting drawings from the geometric model. The latter provides information for view orientation, clipping, line removal, and other view characteristics. IGES also defines a property entity that enables relating attributes to entities and defines associativity entities to use when several entities must be related to one another.

Another part of the IGES information structure is the cross-reference between model entities. This is particularly useful in surface representation. A surface of revolution entity type, for example, is defined in terms of referencing (via pointers in the IGES file) a generative curve and an axis of revolution. The tabulated cylinder entity type has similar definitions.

With the information structure of geometric models captured, the remainder of data representation consists primarily of specifying the data and parameters of typical geometric entities such as curves, surfaces, and solids, and nongeometric entities such as annotation. The parameters required to represent geometric entities are directly related to the material covered in Chaps. 5, 6, and 7. The actual formats to store the parameters of each entity in an IGES file are described in Sec. 8.3.3. Due to the similarities between the IGES representation of various entities, only selected entity types are covered in some detail in this chapter. Readers who are interested in more information should refer to the IGES 4.0 report listed at the end of the chapter.

**8.3.2.1 GEOMETRY.** IGES uses two distinct but related cartesian coordinate systems to represent geometric entity types. These are the MCS and WCS introduced in Chap. 3. The MCS defines the model space and the WCS defines the working space (IGES refers to it as the definition space). The WCS plays a simplifying role in representing planar entities. In such a case, the $XY$ plane of the WCS is taken as the entity plane and therefore only $x$ and $y$ coordinates relative to the WCS are needed to represent the entity. To complete the representation, a transformation matrix is assigned (via a pointer) to the entity as one of its parameters to map its description from WCS to MCS. This matrix itself is defined in IGES as entity type 124. Each geometric entity type in IGES has such a matrix. If an entity is directly described relative to the MCS, then no transformation is required. This is achieved in IGES by setting the value of the matrix pointer to zero to prevent unnecessary processing. As a general rule, all geometric entity types in IGES are defined in terms of a WCS and a transformation matrix. The case when MCS and WCS are identical is triggered by a zero value of the matrix pointer.

Directionality is important when exchanging curves, especially parametric curves. Within IGES, all curves are directed. Each curve has a starting point, an ending point, and a parameter $u$ (see Chap. 5). The information may not be enough to uniquely define the curve as in the case of a circular or conic arc. Thus, some entity types refer to a "counterclockwise direction" with respect to a WCS. In IGES, the definition of this direction is based on an observer positioned along the positive $Z$ axis of the WCS and looking down upon the $XY$ plane of the WCS.

One of the main concerns in designing and/or using IGES processors is the accurate mapping of native entities of a given CAD/CAM system to IGES entities and/or mapping the latter to the former. If the mapping is not performed carefully, loss of shape, accuracy, and data of the entities may occur and consequently result in the failure of the exchange process. Two levels of concern can be identified. The first, and less serious, happens when the internal data representation of a native entity is different from that required by IGES. Consider the example of a circular arc entity. IGES defines this entity by an arc center, a starting point, and an ending point. If a system defines the arc using a center, a radius, and starting and ending angles, then a conversion utilizing the arc parametric equation must be performed by the designated IGES processor. Such conversion may be done twice (to and from IGES), and in each time the arc data is subject to truncation and round-off errors.

The second level of concern is more serious and occurs whenever an entity is not specifically supported and thus must be converted to the closest available entity. Consider the popular case of exchanging splines through IGES. There is a wide variety of spline entities and each CAD/CAM system has its own spline types. IGES itself supports seven types including the rational B-spline curve. In the case where the source system or the target system (Fig. 8-3) does not support directly one of these seven types, a conversion to the closest available spline type must be performed. In this case, the shape of the newly evaluated spline may be inadequate, especially in applications where tight tolerances are required. For example, if a CAD/CAM system has a native B-spline entity that could be of degree 1 to 7, then its IGES processor can be designed such that splines of degree 1 to 3 can be represented by the IGES parametric spline entity type 112 (see details below) and degrees 4 to 7 by the IGES rational B-spline entity type 126. The comparison between the shapes of the native and IGES splines is a matter of designing and testing IGES processors, covered in Sec. 8.3.4.

IGES reserves entity numbers 100 to 199 inclusive for its geometric entities. Entity type numbers that have been assigned thus far are shown in Table 8.1. There is no real scheme in assigning these numbers to their entities except that they are even numbers so far, with the exception of entity type number 125, and are assigned to entities as they become available in various revisions.

Specifications and descriptions of entities, including geometric entities, in IGES follow one pattern. Each entity has two main types of data: directory data and parameter data. The former is the entity type number and the latter are the parameters required to uniquely and completely define the entity. In addition, IGES specifies other parameters related to entity attributes and to the IGES file structure. These parameters are described in more detail in Sec. 8.3.3.

To gain a better understanding of how geometric entities are described in

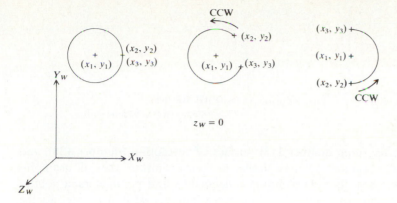

**FIGURE 8-4**
Examples of IGES circular arc entity.

IGES, let us consider sample wireframe and surface entities. The IGES circular arc entity has the type number 100 and its parameter data are a parallel $z_W$ displacement of arc from the $XY$ plane of the arc definition space (WCS), a center $(x_1, y_1)$, a starting point $(x_2, y_2)$, and an ending point $(x_3, y_3)$. The latter two points must be chosen to define a counterclockwise (CCW) direction, as shown in Fig. 8-4. The IGES line is a parametric line with the entity type 110, a starting point $(x_1, y_1, z_1)$, and an ending point $(x_2, y_2, z_2)$. The IGES parametric spline curve entity has the type number 112. This entity is a difficult entity to specify due to the many available methods that can be used to define a spline curve and which CAD/CAM systems use. The seven spline types available in IGES are linear, quadratic, cubic, Wilson-Flower, modified Wilson-Flower, B-spline, and rational B-spline. The first six splines fall under the entity type 112 and have types 1, 2, 3, 4, 5, and 6 respectively. The rational B-spline is treated as a separate entity with the type number 126. In addition to the spline type, other parameter data for the first six splines include degree of continuity, planarity/nonplanarity, number of segments, data points, and derivatives.

IGES surface entities are specified in a similar form to that of wireframe entities except for pointers that are used to refer to the latter whenever needed in surface specification. The IGES plane entity has the type number 108 and the implicit form:

$$Ax + By + Cz = D \tag{8.3}$$

Any other form of the plane must be converted by the proper IGES translator into Eq. (8.3). Thus, the parameter data for the plane are the plane coefficients ($A$, $B$, $C$, $D$), a pointer to a closed curve entity if the plane is bound, and a display symbol location $(x, y, z)$ and its size if the plane is unbounded. The symbol and its size are used to display unbounded planes as shown in Fig. 8-5. If the pointer value is zero, it indicates an unbounded plane. In this case, symbol location and a nonzero size is needed. If the pointer has a value, the symbol is set to zero to indicate that symbol information is not needed.

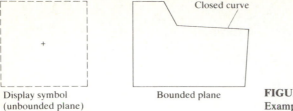

Display symbol  
(unbounded plane)

Bounded plane

**FIGURE 8-5**  
Examples of IGES plane entity.

Ruled surface (type number 118), surface of revolution (number 120), and tabulated cylinder (number 122) are similar in their definition; they all use pointers. For example, a surface of revolution is defined by four parameters: a pointer to the axis of revolution (must be a line entity number 110), a pointer to the generatrix (curve to be rotated), a starting angle, and an ending angle (both in radians). A parametric spline surface entity (type number 114) follows a similar format to that of the spline curve entity. Its parameter data includes the spline boundary type (1 to 6 spline curves as described previously), patch type (1 = cartesian product, 0 = unspecified), number of segments in both the $u$ and $v$ directions, and the coordinates of control points of the patch.

**8.3.2.2  ANNOTATION.** Drafting data are represented in IGES via its annotation entities. Many IGES annotation entities are constructed by using other basic entities that IGES defines such as copious data (centerline, section, and witness line), witness line, leader (arrow), and a general note. For example, the dimension entities may have 0, 1, or 2 pointers to witness line entities, 0, 1, or 2 pointers to leader entities, and a pointer to a general note entity. The number of 0, 1, or 2 pointers of a given entity depends on what the dimension entity is to look like. A dimension entity may have, for example, 0, 1, or 2 arrows, in which case a corresponding number of leader entities is required.

An annotation entity may be defined in definition space (WCS) or in two-dimensional space associated with a given drawing. This is analogous to the model and drawing modes introduced in Chap. 3. If a dimension is inserted by the user in the model mode, then it requires a transformation matrix pointer when translated into IGES.

As an example, a diameter dimension entity (type number 206) consists of a general note (the text itself), one or two leaders, and the centerpoint of the arc to be dimensioned to position the dimension line properly relative to the arc. Therefore, the parameter data required to define this dimension entity are a pointer to a general note, a pointer to the first leader, a pointer to the second leader or zero to eliminate it, and coordinates of the arc center $(x, y)$, as shown in Fig. 8-6. Table 8.2 shows the other IGES annotation entities.

**8.3.2.3  STRUCTURE.** The previous two sections show how geometric and drafting data can be represented in IGES. A typical product definition database includes much more information. IGES permits a valuable set of product data to be represented via its structure entities. These entities include associativity, drawing, view, external reference, property, subfigure, macro, and attribute

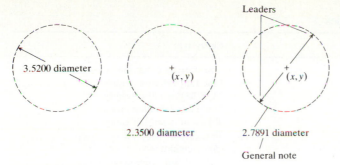

**FIGURE 8-6**
Examples of IGES diameter dimension entity.

entities. Attributes include line fonts, text fonts, and color definition. Table 8.3 shows IGES structure entities.

The associativity definition entity (type number 302) allows IGES to define a special relationship (called the associativity schema) between various entities of a given model. This entity specifies the syntax of the given relationship and not its semantics, that is, its validity. The collection of entities that are related to each other via the associativity schema is called a class. The existence of several classes implies an association among the classes as well as among the contents of each class. Two kinds of associativities are permitted within IGES. Predefined associativities have the (form) numbers 1 to 5000 and the second kind is implementor-defined and has the numbers 5001 to 9999. Each time an associativity relation is needed in the IGES file, an associativity instance entity (type number 402) is used. Consider, as an example, the group associativity. Most CAD/CAM systems allow a collection of a set of entities to be defined as a single, logical entity called a group. Groups are useful for animation (of linkages and mechanisms) and other purposes (see Chap. 13). IGES predefines group associativity (form number = 1) via the entity type number 302. The entity type number 402 is used in the IGES file every time a group instance in the model data arises.

**TABLE 8.2**
**IGES annotation entity types**

| Entity type number | Entity type |
| --- | --- |
| 106 | Copious data (centerline, section, witness line) |
| 202 | Angular dimension |
| 206 | Diameter dimension |
| 208 | Flag note |
| 210 | General label |
| 212 | General note |
| 214 | Leader (arrow) |
| 216 | Linear dimension |
| 218 | Ordinate dimension |
| 220 | Point dimension |
| 222 | Radius dimension |
| 228 | General symbol |
| 230 | Sectioned area |

**TABLE 8.3**
**IGES structure entity types**

| Entity type number | Entity type |
| --- | --- |
| 302 | Associativity definition entity |
| 304 | Line font definition entity |
| 306 | MACRO definition entity |
| 308 | Subfigure definition entity |
| 310 | Text font definition entity |
| 312 | Text display template |
| 314 | Color definition |
| 320 | Network subfigure definition |
| 402 | Associativity instance entity |
| 404 | Drawing entity |
| 406 | Property entity |
| 408 | Singular subfigure instance entity |
| 410 | View entity |
| 412 | Rectangular array subfigure instance entity |
| 414 | Circular array subfigure instance entity |
| 416 | External reference |
| 418 | Nodal load/constraint |
| 420 | Network subfigure instance |
| 600–699 or 10,000–99,999 (user-specified) | MACRO instance entity |

The external reference entity (type number 416) enables IGES files to relate to each other. This entity provides a link between an entity in one file and the definition or a logically related entity in another file. This concept is similar to referencing standard FORTRAN libraries (e.g., IMSL) in user-written programs. Three forms of external reference entity are defined. Form 0 is used when a single definition from the referenced file, which may contain a collection of definitions, is desired. Form 1 is used when the entire file is to be instanced, which is the case where the referenced file contains a complete subassembly. Form 3 is used when an entity in one file refers to another entity in a separate file. This is the case when each sheet of a drawing is a separate file and, say, a flange on one sheet mates with a flange on another sheet.

The property entity (type number 406) in IGES contains numerical and textural data. Due to the wide range of properties, each one is assigned a form number and each form number may contain different property types (p-types). For example, form number 11 contains tabular data that are organized under $n$ p-types. P-types 1, 2, 3, 4, as an example, refer to Young's modulus, Poisson's ratio, shear modulus, and material matrix respectively. There are 17 form numbers all together that can be specified with the property entity.

The MACRO capability in IGES (type number 306) allows the family of parts and/or entities grouped by the user for special purposes, which is a common practice on existing CAD/CAM systems to be exchanged. MACROS can only define a "new" entity in terms of the entities supported by IGES. This capability allows the extension of IGES beyond its common entity subset by utilizing a formal mechanism which is a part of IGES itself. A "new" entity can only be defined once in an IGES file but it can be referenced as many times in the file as needed by using the MACRO instance entity (type numbers 600 to 699 or 10,000 to 99,999). This number is referred to in the MACRO definition entity as the entity type ID.

The syntax for a MACRO definition entity takes the following form:

306, MACRO, ID, PAR1, PAR2, ..., PARN

MACRO Syntax

ENDM

The first line is the MACRO header. The first two variables are fixed. ID is the entity type that is used in the MACRO instance entity. The MACRO definition entity can be thought of as a subroutine and each time an instance is needed, a subroutine call is made using the ID and the macro parameters. PAR1, PAR2, ..., and PARN are the parameters that define the MACRO. They take specific values for specific instances.

As an example, consider a user who has two MACROs in a given model; one defines a set of concentric circles and the other defines a set of parallel lines. The two MACROs that an IGES translator must create have the headers:

306, MACRO, 605, XC, YC, ZC, R, N

and             306, MACRO, 610, X1, Y1, Z1, X2, Y2, Z2, D, N

The detailed syntax of each MACRO is not covered here. The ID numbers 605 and 610 are chosen arbitrarily. The first MACRO requires the center of the circle $(x_C, y_C, z_C)$, its radius $R$, and the number $N$ of concentric circles. The second MACRO requires the endpoints of the line $(x_1, y_1, z_1)$ and $(x_2, y_2, z_2)$, the offset distance $D$, and the number of lines $N$. Sample calls to these MACROs in the IGES file are:

605, 0.0, 0.0, 0.0, 3, 3
605, 2.0, 1.0, −4.0, 4, 2
610, 0.0, 0.0, 0.0, 1.0, 1.0, 0.0, 0.5, 3
610, 1.0, 0.0, 0.0, 3.0, 0.0, 0.0, 1.0, 4

### 8.3.3   File Structure and Format

A typical CAD/CAM system that supports IGES usually provides its users with two IGES commands. One command enables the user to create an IGES file of a given model residing in the system while the other allows the user to read an existing IGES file of a model into the system. The former command accesses the system IGES preprocessor (we refer to it as the "put IGES" command) and the latter accesses the system IGES postprocessor (we call it the "get IGES" command). Typically, each one of these commands engages the user in a dialogue to provide information to enable interpreting the IGES file (by future postprocessors) in the case of "put IGES" and to provide various file names in the case of "get IGES." For example, "put IGES" asks the user for an error log filename (optional), the name of the model (part) to be converted to the IGES format, the name of the IGES file that will be generated, and data related to the "start and global sections" of the IGES file described below. The error log file contains an error section that lists the entities that were not processed successfully and the related error messages. The file may contain a section that lists the numbers of entities and their types that were processed successfully and a section of information similar to that of the "start section" of the IGES file. Thus, the user can quickly review the error log file and determine the state of conversion. The "get IGES," on the other hand, asks the user for an error log filename that serves the same purpose as with "put IGES," the name of the IGES file to be processed by the postprocessor, and the name of the model (part) to be created.

Unless the user has the appropriate background and understanding regarding the IGES file structure and format, dealing with the above two commands and reading their related documentation become difficult. A knowledge of the IGES file structure and format serves this and other purposes. For example, a user may edit a particular IGES file although this is not always recommended.

An IGES file is a sequential file consisting of a sequence of records. The file formats treat the product definition to be exchanged as a file of entities, each entity being represented in a standard format, to and from which the native representation of a specific CAD/CAM system can be mapped. Depending on the chosen file format, the record length can be fixed or variable. There are two different formats to represent IGES data in a file: ASCII and binary. The ASCII form has two format types: a fixed 80-character record length format and a compressed format. In the fixed record length format, the entire file is of 80-character records (lines). The file is divided into sections (see the description below). Within each section, the records are labeled and numbered. IGES data is written in columns 1 through 72 inclusive of each record. Column 73 stores the section identification character. Columns 74 through 80 are specified for the section sequence number of each record. Before we continue describing the compressed ASCII and binary formats, let us describe the IGES file structure first to better understand these formats.

An IGES file consists of six sections which must appear in the following order: Flag section (optional), Start section, Global section, Directory Entry (DE) section, Parameter Data (PD) section, and Terminate section, as shown in Fig. 8-7. The identification character, also called the section code (column 73 of each record), for these sections respectively are S, G, D, P, and T (excluding the Flag section). The Flag section is used only with compressed ASCII and binary format. It is a single record (line) that precedes the Start section in the IGES file with the character "C" in column 73 to identify the file as compressed ASCII. The compressed ASCII form is intended to be simply converted to and from the regular ASCII form. The National Institute of Standards and Technology has software available to convert to and from both ASCII forms. In the binary file format, the Flag section is called the Binary information section and the first byte (8 bits) of this section has the ASCII letter "B" as the file identifier.

The Start section is a human-readable introduction to the file. It is commonly described as a "prologue" to the IGES file. As mentioned at the beginning of this section, the "put IGES" command can be designed to request, from the

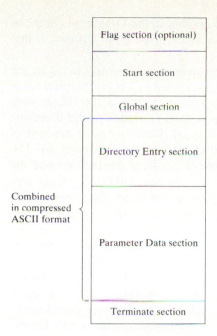

FIGURE 8-7
IGES file general structure.

user, relevant information such as the names of the sending (source) and receiving (target) CAD/CAM systems, and a brief description of the product being converted. IGES does not specify how this section could be used.

The Global section includes information describing the preprocessor and information needed by the postprocessor to interpret the file. Some of the parameters that are specified in this section are:

Characters used as delimiters between individual entries and between records (usually commas and semicolons respectively)
The name of the IGES file itself
Vendor and software version of sending (source) system
Number of significant digits in the representation of integers and single and double precision floating-point numbers on the sending systems
Date and time of file generation
Model space scale
Model units
Minimum resolution and maximum coordinate values
Name of the author of IGES file and his/her organization

Some of the above parameters can be implemented into the "put IGES" dialogues.

The DE section is a list of all the entities defined in the IGES file together with certain attributes associated with them. The entry for each entity occupies two 80-character records which are divided into a total of twenty 8-character fields. The first and the eleventh (beginning of the second record of any given entity) fields contain the entity type number (Tables 8.1 to 8.3). The second field contains a pointer to the parameter data entry for the entity in the PD section.

The pointer of an entity is simply its sequence number in the DE section. Some of the entity attributes specified in this section are line font, layer number, transformation matrix, line weight, and color.

The PD section contains the actual data defining each entity listed in the DE section. For example, a straight line entity is defined by the six coordinates of its two endpoints. While each entity has always two records in the DE section, the number of records needed for each entity in the PD section varies from one entity to another (the minimum is one record) and depends on the amount of data. Parameter data are placed in free format in columns 1 through 64. The parameter delimiter (usually a comma) is used to separate parameters and the record delimiter (usually a semicolon) is used to terminate the list of parameters. Both delimiters are specified in the Global section of the IGES file. Column 65 is left blank. Columns 66 through 72 on all PD records contain the entity pointer specified in the first record of the entity in the DE section.

The Terminate section contains a single record which specifies the number of records in each of the four preceding sections for checking purposes.

The fixed record length (80 characters) of the IGES ASCII file format coupled with fields that are not used in the DE section resulted in transmission of a high proportion of space characters which increase the field size significantly. The compressed ASCII and binary formats are introduced to address this file size problem. We now return to describe these formats.

In the compressed ASCII format, the Start, Global, and Terminate sections remain the same as the ASCII form, while the DE and the PD sections are combined into a single Data section. In addition, each record of the PD portion of the Data section is written in a free form similar to the ASCII PD section, but is of variable length (terminate before column 65) to eliminate storing space characters.

The binary file format is a bit stream binary representation of data. All entity parametrization and data organization are otherwise identical to the ASCII form. Each section in the file begins with the section code (8 bits), that is, S, G, D, P, or T, followed by a section byte count (32 bits) to specify the total number of bytes [excluding the 5 bytes (40 bits) required for the section code and section byte count] that belong to the section. Notice here that the section code is only specified once at the beginning of the section. Within the DE section, the entity DE variables remain exactly the same as the ASCII format. Similarly, the PD parameters remain the same within the PD section as the ASCII format.

Figure 8-8 shows a simple model and its IGES ASCII file. The model consists of a circular arc and a line. The arc has endpoints $P_2$ and $P_1$ (in this order) and a center as the midpoint of the line. The endpoints have coordinates of (8, 3, 0) and (3, 3, 0) respectively. The figure shows the various sections of a typical ASCII IGES file (see Fig. 8-7) and details of each section.

### 8.3.4  Processors

IGES in itself is just a format specification to exchange product data among dissimilar CAD/CAM systems. This format must be interpreted, understood, and implemented by CAD/CAM vendors into programs, often called processors or

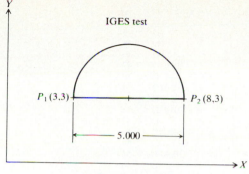

Y

IGES test

$P_1$ (3,3)     $P_2$ (8,3)

5.000

→ X

(*a*) Geometric model

This is a CADDS part
being tested for IGES conversion                                                    S    1
,,8HTESTIGES, 13Higes.TESTIGES, 49HCOMPUTERVISION  S    2
CADDStation REV 4.0 GRAPHIC SYSTEM, 28HIGES          G    1
VERSION 4.0 (06-OCT-88),32,38,6,308,14,8HTESTIGES,   G    2
1.0,1,4HINCH, 32767, 32.767, 13H891013.124852,0.000001,,  G    3
12HCENGIZ YEKER,2HNU,6,3;                                            G    4

| 124 | 1 | 1 | 0 | 0 | 0 | 0 | 0 | 1D | 1 |
|-----|---|---|---|---|----|----|---|-------|----|
| 124 | 0 | 0 | 1 | 0 |   |   |   | D | 2 |
| 124 | 2 | 1 | 0 | 0 | 0 | 0 | 0 | 1D | 3 |
| 124 | 0 | 0 | 1 | 0 |   |   |   | D | 4 |
| 124 | 3 | 1 | 0 | 0 | 0 | 0 | 0 | 1D | 5 |
| 124 | 0 | 0 | 1 | 0 |   |   |   | D | 6 |
| 124 | 4 | 1 | 0 | 0 | 0 | 0 | 0 | 1D | 7 |
| 124 | 0 | 0 | 1 | 0 |   |   |   | D | 8 |
| 124 | 5 | 1 | 0 | 0 | 0 | 0 | 0 | 1D | 9 |
| 124 | 0 | 0 | 1 | 0 |   |   |   | D | 10 |
| 124 | 6 | 1 | 0 | 0 | 0 | 0 | 0 | 1D | 11 |
| 124 | 0 | 0 | 1 | 0 |   |   |   | D | 12 |
| 406 | 7 | 1 | 0 | 0 | 0 | 0 | 0 | 10201D | 13 |
| 406 | 0 | 0 | 1 | 15 |   |   |   | D | 14 |
| 124 | 8 | 1 | 0 | 0 | 0 | 0 | 0 | 10101D | 15 |
| 124 | 0 | 0 | 1 | 0 |   |   |   | D | 16 |
| 108 | 9 | 1 | 0 | 0 | 0 | 0 | 0 | 10001D | 17 |
| 108 | 0 | 0 | 1 | 1 |   |   |   | D | 18 |
| 108 | 10 | 1 | 0 | 0 | 0 | 0 | 0 | 10001D | 19 |
| 108 | 0 | 0 | 1 | 1 |   |   |   | D | 20 |
| 108 | 11 | 1 | 0 | 0 | 0 | 0 | 0 | 10001D | 21 |
| 108 | 0 | 0 | 1 | 1 |   |   |   | D | 22 |
| 108 | 12 | 1 | 0 | 0 | 0 | 0 | 0 | 10001D | 23 |
| 108 | 0 | 0 | 1 | 1 |   |   |   | D | 24 |
| 410 | 13 | 1 | 0 | 0 | 0 | 15 | 0 | 10201D | 25 |
| 410 | 0 | 0 | 1 | 0 |   |   |   | D | 26 |
| 116 | 14 | 1 | 1 | 0 | 0 | 0 | 0 | 1D | 27 |
| 116 | 0 | 0 | 1 | 0 |   |   |   | D | 28 |
| 116 | 15 | 1 | 1 | 0 | 0 | 0 | 0 | 1D | 29 |
| 116 | 0 | 0 | 1 | 0 |   |   |   | D | 30 |
| 116 | 16 | 1 | 1 | 0 | 0 | 0 | 0 | 1D | 31 |
| 116 | 0 | 0 | 1 | 0 |   |   |   | D | 32 |
| 402 | 17 | 1 | 0 | 0 | 0 | 0 | 0 | 201D | 33 |
| 402 | 0 | 0 | 1 | 4 |   |   |   | D | 34 |
| 110 | 18 | 1 | 1 | 0 | 33 | 0 | 0 | 1D | 35 |
| 110 | 0 | 0 | 1 | 0 |   |   |   | D | 36 |
| 402 | 19 | 1 | 0 | 0 | 0 | 0 | 0 | 201D | 37 |
| 402 | 0 | 0 | 1 | 4 |   |   |   | D | 38 |
| 110 | 20 | 1 | 1 | 0 | 37 | 0 | 0 | 1D | 39 |
| 110 | 0 | 0 | 1 | 0 |   |   |   | D | 40 |
| 100 | 21 | 1 | 1 | 0 | 0 | 3 | 0 | 1D | 41 |

**FIGURE 8-8**

| | | | | | | | | | | |
|---|---|---|---|---|---|---|---|---|---|---|
| 100 | 0 | 0 | 1 | 0 | | | | | D | 42 |
| 110 | 22 | 1 | 1 | 0 | 0 | 0 | 0 | | 1D | 43 |
| 110 | 0 | 0 | 1 | 0 | | | | | D | 44 |
| 116 | 23 | 1 | 1 | 0 | 0 | 0 | 0 | | 1D | 45 |
| 116 | 0 | 0 | 1 | 0 | | | | | D | 46 |
| 212 | 24 | 1 | 0 | 0 | 0 | 0 | 0 | | 10101D | 47 |
| 212 | 0 | 0 | 1 | 0 | | | | | D | 48 |
| 214 | 25 | 1 | 0 | 0 | 0 | 0 | 0 | | 10101D | 49 |
| 214 | 0 | 0 | 1 | 2 | | | | | D | 50 |
| 214 | 26 | 1 | 0 | 0 | 0 | 0 | 0 | | 10101D | 51 |
| 214 | 0 | 0 | 1 | 2 | | | | | D | 52 |
| 106 | 27 | 1 | 0 | 0 | 0 | 0 | 0 | | 10001D | 53 |
| 106 | 0 | 0 | 1 | 40 | | | | | D | 54 |
| 106 | 28 | 1 | 0 | 0 | 0 | 0 | 0 | | 10001D | 55 |
| 106 | 0 | 0 | 1 | 40 | | | | | D | 56 |
| 216 | 29 | 1 | 0 | 0 | 0 | 0 | 0 | | 10101D | 57 |
| 216 | 0 | 0 | 1 | 0 | | | | | D | 58 |
| 212 | 30 | 1 | 0 | 0 | 0 | 0 | 0 | | 10101D | 59 |
| 212 | 0 | 0 | 2 | 0 | | | | | D | 60 |
| 212 | 32 | 1 | 0 | 0 | 0 | 0 | 0 | | 10101D | 61 |
| 212 | 0 | 0 | 1 | 0 | | | | | D | 62 |
| 212 | 33 | 1 | 0 | 0 | 0 | 0 | 0 | | 10101D | 63 |
| 212 | 0 | 0 | 1 | 0 | | | | | D | 64 |
| 212 | 34 | 1 | 0 | 0 | 0 | 0 | 0 | | 10101D | 65 |
| 212 | 0 | 0 | 1 | 0 | | | | | D | 66 |
| 212 | 35 | 1 | 0 | 0 | 0 | 0 | 0 | | 10101D | 67 |
| 212 | 0 | 0 | 2 | 0 | | | | | D | 68 |
| 212 | 37 | 1 | 0 | 0 | 0 | 0 | 0 | | 10101D | 69 |
| 212 | 0 | 0 | 1 | 0 | | | | | D | 70 |
| 212 | 38 | 1 | 0 | 0 | 0 | 0 | 0 | | 10101D | 71 |
| 212 | 0 | 0 | 1 | 0 | | | | | D | 72 |
| 212 | 39 | 1 | 0 | 0 | 0 | 0 | 0 | | 10101D | 73 |
| 212 | 0 | 0 | 2 | 0 | | | | | D | 74 |
| 406 | 41 | 1 | 0 | 0 | 0 | 0 | 0 | | 10201D | 75 |
| 406 | 0 | 0 | 1 | 15 | | | | | D | 76 |
| 406 | 42 | 1 | 0 | 0 | 0 | 0 | 0 | | 10201D | 77 |
| 406 | 0 | 0 | 1 | 17 | | | | | D | 78 |
| 406 | 43 | 1 | 0 | 0 | 0 | 0 | 0 | | 10201D | 79 |
| 406 | 0 | 0 | 1 | 16 | | | | | D | 80 |
| 404 | 44 | 1 | 0 | 0 | 0 | 0 | 0 | | 201D | 81 |
| 404 | 0 | 0 | 2 | 0 | | | | | D | 82 |

| | |
|---|---|
| 124,1.0,0.0,0.0,0.0,0.0,0.0,1.0.,0.0,0.0,0.0,0.0,0.0,1.0,0.,0.0; | 1P 1 |
| 124,1.0,0.0,0.0,0.0,0.0,0.0,0.0,0.0,.-1.0,0.0,0.0,0.1.0,0.0,0.0,0.0; | 3P 2 |
| 124,0.0,0.0,0.1.0,0.0,1.0,0.0,0.0,0.0,0.0,0.0,1.0,0.0,0.0; | 5P 3 |
| 124,1.0,0.0,0.0,0.0,0.0,0.0,.-1.0,0.0,0.0,0.0,0.0,0.0,.-1.0,0.0; | 7P 4 |
| 124,0.0,0.0,0.,-1.0,0.0,.-1.0,0.0,0.0,0.0,0.0,0.0,1.0,0.0,0.0; | 9P 5 |
| 124,-1.0,0.0,0.0,0.0,0.0,0.0,0.0,0.1.0,0.0,0.0,0.1.0,0.0,0.0; | 11P 6 |
| 406,1,5HFRONT; | 13P 7 |
| 124,1.0,0.0,0.0,0.0,0.0,0.0,0.0,0.1.0,0.0,0.0,0.0,.-1.0,0.0,0.0; | 15P 8 |
| 108,1.0,0.0,0.0,0.0,.-3.50103; | 17P 9 |
| 108,0.0,0.0,0.0,1.0,9.44148; | 19P 10 |
| 108,1.0,0.0,0.0,0.0,12.6263; | 21P 11 |
| 108,0.0,0.0,0.0,1.0,.-6.23409; | 23P 12 |
| 410,4,1.0,0,17,19,21,23,0,0,2,37,33,1,13; | 25P 13 |
| 116,3.0,0.0,0.3.0,0.0; | 27P 14 |
| 116,8.0,0.0,0.3.0,0.0; | 29P 15 |
| 116,0.0,0.0,0.0,0.0; | 31P 16 |
| 402,1,0.25,1,0,0,0; | 33P 17 |
| 110,0.0,0.0,0.0,0.0,10.5708,0.0,0.0; | 35P 18 |
| 402,1,0.25,1,0,0,0; | 37P 19 |
| 110,0.0,0.0,0.0,0.0,0.0,0.0,6.93429; | 39P 20 |
| 100,0.0,5.5,3.0,8.0,3.0,3.0,3.0; | 41P 21 |
| 110,3.0,0.0,0.3.0,8.0,0.0,3.0; | 43P 22 |
| 116,5.5,0.0,0.3.0,0.0; | 45P 23 |
| 212,1,5,0.78,0.156,1,1.5708,0.0,0.0,11.4683,8.422,0.0,5H5.000; | 47P 24 |
| 214,1,0.15,0.05,0.0,9.32444,8.5,11.3708,8.5; | 49P 25 |
| 214,1,0.15,0.05,0.0,14.3244,8.5,12.3458,8.5; | 51P 26 |

**FIGURE 8-8**

```
106,1,3,0.0,9.32444,10.1189,9.32444,10.1189,9.32444,8.37325;        53P     27
106,1,3,0.0,14.3244,10.1189,14.3244,10.1189,14.3244,8.37325;        55P     28
216,47,49,51,53,55;                                                 57P     29
212,1,34,8.95,0.25,1,1.5708,0.0,0,0,5.89528,3.01129,0.0,34HFIGUR     59P     30
E8.8 SAMPLE ASCII IGES FILE;                                        59P     31
212,1,1,0.216667,0.25,17,1.5708,0.0,0,0,16.9856,6.62526,0.0,1HX;    61P     32
212,1,1,0.216667,0.25,17,1.5708,0.0,0,0,5.94045,14.0113,0.0,1HY;    63P     33
212,1,7,1.5,0.25,1,1.5708,0.0,0,0,7.58932,9.923,0.0,7HP (3,3);      65P     34
212,1,7,1.525,0.25,1,1.5708,0.0,0,0,14,7495,9.94559,0.0,7HP (8,3    67P     35
);                                                                 67P     36
212,1,1,0.1309,0.187,1,1.5708,0.0,0,0,7.83778,9.83265,0.0,1H1;      69P     37
212,1,1,0.188,0.187,1,1.5708,0.0,0,0,15.0431,9.83265,0.0,1H2;       71P     38
212,1,9,2.45,0.25,1,1.5708,0.0,0,0,9.78029,13.7177,0.0,9HIGES TE    73P     39
ST;                                                                73P     40
406,1,3HEIN;                                                        75P     41
406,2,1,2HIN;                                                       77P     42
406,2,22.0,17.0;                                                   79P     43
404,1,25,6.32444,7.21253,9,57,59,61,63,65,67,69,71,73,0,3,75,77,    81P     44
79;                                                                81P     45
S 2G 4D 82P 45                                                       T       1
```

(*b*) IGES file

**FIGURE 8-8**
A sample ASCII IGES file.

translators. The user interface to access these processors usually takes the form of simple commands, such as "put IGES" and "get IGES," accompanied by proper dialogues. Primarily IGES processors provide:

1. Translation algorithms between the IGES formats of entities and their formats within specific CAD/CAM systems.
2. Read and write routines to access entities stored in IGES files and in the databases used by the specific CAD/CAM systems.

Two typical cases can arise when using IGES processors. The first case is encountered when both the source (sending) and target (receiving) systems support the same IGES entities but their corresponding processors may not. This case occurs when either processor implements only a subset of IGES entities. If, for example, the preprocessor of the source system supports an entity that the postprocessor of the target system does not, the data exchange between the two systems fails and error messages are usually reported by the latter processor. The second case happens, more often than the first one, when processors of both systems support the complete set of IGES entities but the systems themselves do not (such as the case with spline types). In this case, the data exchange is completed successfully, but the mapping of entities of the source system to their closest entities available on the target system may result in an unacceptable loss of geometric representation. If, for example, the source system supports the Wilson-Flower spline (type 4) while the target system supports only the B-spline (type 6), then the postprocessor of the target system must convert type 4 spline to type 6 and issue a message to the user as a warning.

**8.3.4.1  DESIGN.** The two cases discussed above reveal that designing IGES processors is a delicate task. In fact, IGES, or other standards, reduces the database exchange problem to the design of two processors: a preprocessor to map the source database into an IGES file and a postprocessor to map the IGES file to the target database. Writing IGES processors is still a significant challenge. A typical database might contain many instances of many entity types. Many of these entity types involve complex mathematics (e.g., sculptured surfaces) and complex data structures. Moreover, organizations using IGES processors can have widely different and conflicting objectives for database exchange. These objectives range from the easily fulfilled "identical graphics image" to the very difficult "minimum possible information loss."

Problems in writing an IGES processor relate to the definition and format of the IGES standard itself, and to the variability of implementations in IGES processors. Some of these problems are described below:

1. IGES entity set. IGES does not and cannot contain a real superset of entities which are found in all of today's CAD/CAM systems. IGES may contain an entity that has no equivalence on a specific CAD/CAM system, or the system may contain an entity for which no IGES entity exists. An IGES processor could either ignore translating the entity or translate it into a similar one destroying its original meaning. It is also possible that the definitions for the same type of entity vary between IGES and specific CAD/CAM systems. This can cause problems if one of the definitions is less complete than the other. Differences will occur between the representations of the entity in the source system, in IGES, and in the target system. Consider the example of a simple geometric entity: the planar circular arc. A circular arc is determined by four parameters: a center, radius, starting angle, and an ending angle. There is a variety of alternatives for choosing these parameters and conversion algorithms between these alternatives must be designed as part of the IGES processor.

2. IGES format. While IGES allows the exchange of complex structures and relationships, the IGES format must be processible by a wide range of different computer systems and therefore can only use simple data formats and management methods known to these systems and, in the meantime, independent of any system specifics.

3. Limitations of individual CAD/CAM systems. These limitations are based within a specific system and are related to things such as model size, data precision, or two-dimensional model space.

Designing IGES processors with all the above problems in mind divides into the following steps:

1. Analyzing and tabulating entity characteristics. This step involves the study of entity mathematical representations utilized by both IGES and the CAD/CAM system. In many cases, an entity can be represented by a number of nearly, but not completely, equivalent ways. A circular arc, for example, can be represented by various forms. These forms differ in their ability to handle special cases such as a zero-length arc (point), a line segment, and a full circle.

Entity characteristics of both representations of IGES and the CAD/CAM system are tabulated and compared.

2. **Defining conversion algorithms.** The above step clearly provides the necessary information required to design the proper conversion algorithms to convert an entity to and from IGES.

3. **Developing a complete specification of the processors.** The above two steps form the core of the design process of IGES processors. Once completed, other relevant specifications of the processors must be developed. These include the IGES revision the processor ought to support, the subset of IGES entities it can support, and the user interface of the processor.

4. **Designing verification procedures.** Careful verification of IGES processors is very important because IGES processors operate at the interface between different organizations and vendors. Direct inspection of a processor's software and its supporting documentation is not acceptable. IGES processors must be verified by constructing test data, running it through the processors, and comparing the actual results with those expected. Ideally, two sets of test data would be required: a set for implementations to use during processor development and a more comprehensive set for final processor verification. In addition, more customized tests for specific user requirements can also be developed in collaboration between users and implementors of IGES.

**8.3.4.2 IMPLEMENTATION.** Initially, many CAD/CAM system vendors were not strongly committed to the IGES concept and there were no guidelines for implementation. Therefore, each vendor implemented a subset of IGES entities that is the easiest to understand. The common core of entities implemented was consequently small, leading to many incompatibilities at first. Currently, this situation is greatly improved and most geometric entities can be accurately transferred between systems via IGES processors.

Since no subset structure is imposed on the IGES specification, vendors are faced with one of two options in implementing IGES. They can opt to implement the full set of IGES entities into their processors or to just implement a subset of entities. The former approach may delay the release of the processors longer than desired, unless vendors commit substantial resources to the implementation process, while the latter approach may result in user's dissatisfaction. In most cases, the former approach is preferred.

**8.3.4.3 TESTING AND VERIFICATION.** A newly developed IGES processor must be carefully tested before it is used in a production environment. There is an IGES Test, Evaluate, and Support Committee whose function is to provide test data. There is an IGES test library prepared by the Committee which allows testing of the basic implementation of an entity. However, the library does not allow variations to be checked that occur in production data due to numerical and computational errors. These variations must be tested by implementors and users themselves.

Verification of the results of a processor is a time-intensive task. In most cases, it is not sufficient to check the drawings visually and more comprehensive tests are needed. The common methods of testing are:

1. **Reflection test.** In this test, an IGES file created by a system's preprocessor is read by its own postprocessor to create another model. This test is used to establish that a system's processors could read and write common entities, making them symmetric.
2. **Transmission test.** Here an IGES file of a model created by a source system's preprocessor is transferred to a target system whose postprocessor is used to recreate the model on the target system. This test essentially determines the capabilities of the preprocessor and the postprocessor of the source and the target systems respectively.
3. **Loopback test.** In this test, an IGES file created by the source system is read by the target system which, in turn, creates another IGES file and then transfers this file back to the source system to read it. This test checks the pre- and postprocessors of both the source and the target systems.

Figure 8-9 shows these three tests.

Of the above three tests, the loopback test is the most comprehensive one. One method to implement this test can be described as follows:

A. On the source system:
    1. Generate an IGES input tape of a given IGES file.
    2. Generate a hard copy of the input image.
    3. Perform some specific modifications on the system database and note the resulting output expected.
    4. Generate an IGES output tape of step 3.
    5. Generate a hard copy of the output image of step 3.

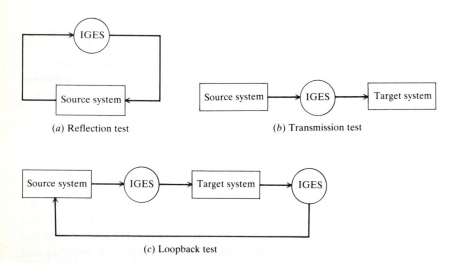

(a) Reflection test

(b) Transmission test

(c) Loopback test

**FIGURE 8-9**
Methods of verification of IGES processors.

B. On the target system:
6. Read the IGES input file and produce the specified input image.
7. Perform the same modifications as in step 3 above to the database created in step 6.
8. Verify that the target system responded to these modifications in precisely the same manner as it would to a database created on it.
9. Generate an IGES output tape of step 8.
10. Generate a hard copy of the output image of step 8 and compare it with that obtained in step 5.
11. Compare the two IGES tapes obtained in steps 4 and 9 using an external processor.

This method of implementation is based on using IGES files and graphics images because comparison of anything else, such as databases, is time-consuming and error-prone. If there are no external processors to compare the IGES files, they would be compared manually, which is not recommended.

**8.3.4.4   ERROR HANDLING.** Error handling and reporting when processing an IGES file is important. There are two major error sources when processing IGES files: program errors in the processor and misinterpretation of the IGES standard itself. These sources apply to both pre- and postprocessors. The way an IGES processor reports these errors and the information given with these reports determines whether the correction of an error becomes a laborious task or not. The preprocessor should report the entity type, number of unprocessed entries, reasons for unprocessing, and other relevant database information of these unprocessed entities. On the other hand, the postprocessor should report the number of unprocessed entities, their types, their forms, their record numbers in the DE and PD sections, and the reasons for unprocessing. It should also report any invalid or missing data encountered in reading IGES files, especially those that were edited.

### 8.3.5   Remarks

It is apparent that IGES provides a means of communicating data between dissimilar CAD/CAM systems, although it has some limitations and imperfections. The main virtue of IGES, however, is that it has provided a vehicle for discussion and a standard for comparison for the more advanced transmission formats that will emerge, such as PDES. These will certainly benefit from the experiences gained from IGES.

### 8.4   PDES

PDES is an exchange for product data in support of industrial automation. "Product data" is interpreted to be more general than "product definition data" which forms the core philosophy of IGES. "Product data" encompasses data relevant to the entire life cycle of a product such as design, manufacturing, quality assurance, testing, and support. In order to support industrial automation, PDES files are fully interpretable by computer. For example, tolerance information would be carried in a form directly interpretable by a computer rather than a computerized text form which requires human intervention to interpret. In addition, this information would be associated with those entities in the model affected by the tolerance. Thus, the general emphasis of PDES is to eliminate the human presence from the "product data" exchange, that is, to obviate the use of engineering drawings and other paper documents as a necessary means of passing information between different product phases that may be performed on similar or dissimilar CAD/CAM systems. The development of PDES as defined here requires selection of a set of logical structures to contain the product data information and also choice of a method to implement these structures in a computer form. Such structures and implementation are discussed in Sec. 8.4.1.

It is intended that PDES and STEP will be identical. Within ISO, the ISO Technical Committee TC184 (Industrial Automation Systems) and its Subcommittee SC4 (External Representation of Product Model Data) were formed in July 1984 to address the need for a single worldwide standard for the exchange of product data. Technical work for future versions of STEP will be accomplished by existing and future natural projects. Therefore, in June 1985, the IGES Steering Committee voted that PDES should represent US interests in the STEP effort. To emphasize the intention that PDES and STEP will be identical, the acronym PDES now stands for Product Data Exchange using STEP. Previously, it meant Product Data Exchange Standard.

PDES draws heavily on the existing data exchange heritage and the related experience. Thus far, two tiers of efforts can be identified. The first one is all strictly standards efforts, concerned with existing systems and techniques. This tier includes IGES efforts and others such as SET. The second tier of efforts are research and development projects concerned with finding the optimum methodologies of data exchange. The ICAM PDDI effort and the GMAP (Geometric Modeling Applications Interface Program) sponsored by the Air Force CIM Program are examples of the second tier. Efforts of both tiers have demonstrated that emphasis on solid modeling is needed because it offers a more complete definition of the shape of a part. They also indicated that the computerized part model must be a "complete" model to support automation. This means that the model contains all types of data discussed in Sec. 8.1.

PDES reflects the above dual heritage and extends it. PDES accommodates wireframes and surface models, as well as drawing representations, as in the efforts in the standards class. However, its contribution to this class is the accommodation of solid modeling. PDES provides a complete product model, as characterized by the research and development class. This model is computer-sensible, functional, and integrated with all types of information necessary to perform a given application.

## 8.4.1  Description

There is a fundamental difference in philosophy between exchanging data in IGES and in PDES. As discussed in Sec. 8.3.1, the central unit of data exchange in the "IGES model" is the entity due to the belief in the 1970s that it is the crucial unit of information exchange. As has been experienced, this belief made it impossible for IGES to support industrial automation which is an essential goal of the 1980s and years to come. On the contrary, the central unit of data exchange in the "PDES model" is the application that contains various types of entities. Therefore, when data is exchanged between systems, it is done in terms of "application" units. This approach maintains all the meaningful associativities and relationships between the application entities that make industrial automation possible.

To achieve the above PDES philosophy, product data is exchanged by PDES according to "discipline models" or "mental models." Both the sender who originates the discipline model and the receiver of the model must be aware of the meaning of the discipline model being exchanged in order to recover the correct meaning of data in the exchange. Discipline models are standardized and defined by PDES in order to be interpreted and used by another computer. This implies that the discipline model must be computer readable, must be able to be made explicit in the data, and must be able to be exchanged with its structure intact. Examples are a mechanical products discipline model, an electrical products discipline model, and AEC discipline model, etc. Thus, the concept of discipline models makes PDES flexible enough to accommodate any future models and application areas when they become available.

PDES characteristics reflect recent developments in database and information systems in general. The PDES methodology involves a three-layer architecture, reference models, formal languages, and coordination with other standards efforts. The three-layer architecture shown in Fig. 8-10 forms the core of PDES structure. This architecture is similar to the three-scheme framework for database management systems as identified by ANSI/X3/SPARC. Within PDES, three layers are identified: the application layer, the logical layer, and the physical layer.

The application layer is the interface between the user and PDES. It contains all the descriptions and information of various application areas. These

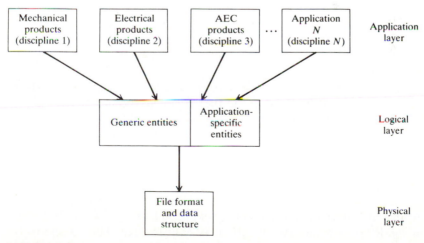

**FIGURE 8-10**
PDES three-layer architecture.

descriptions and information are expressed formally within PDES via information modeling techniques which are sometimes referred to as "reference models." Both ICAM IDEF1 and the Nijessen Information Analysis Model (NIAM) information modeling techniques have been used by PDES.

The purpose of the logical layer is to provide a consistent, computer-independent description of the data constructs that contain the information to be exchanged. Both generic and application-specific constructs are identified. A key objective here is to ensure that no redundancy occurs in PDES generic data structures and their relationships, and at the same time to ensure that such data structures are sufficient to support the wide range of applications.

The physical layer deals with the data structures and data format for the exchange file itself. The main goal here is to establish and maintain efficiency in the file size and processing time to avoid the related problems experienced with IGES.

PDES utilizes formal languages to define data structures and for the PDES file syntax. Languages with context-free grammars are chosen so that parsers can be built more simply.

The final characteristic of the PDES methodology is its coordination with other standard efforts. The goal here is to ensure compatibility and to minimize duplication.

To be able to understand the PDES methodology, the input to it, its overview, and its output are described here. The methodology is illustrated in Fig. 8-11. The input to the methodology is a set of discipline (reference) models. A discipline model represents an application expert's view of a discipline area such as flat-plate design or finite element modeling. The discipline model is created by the user or the expert. This is similar to creating a geometric model for IGES, but with more strict rules required by PDES.

The discipline model developed in the application layer is used to develop a logical layer model in the logical layer itself. The logical layer model (shown as the qualified discipline model in Fig. 8-11) is a generic binary model and is the summation of the resource models—geometry, topology, presentation, and geometry-topology associativities—and any cross-relationships among resource models. Resource models contain only generic entities and structures that are common to application areas. The logical layer model does not contain any discipline-specific entities.

In order to relate the logical layer model back to its discipline model, the set of correspondences (mappings) from the discipline-specific entities to the generic entities must be maintained. The global model is introduced for this purpose and can be viewed as a composite model that contains both discipline-specific and generic entities. Thus, the global model can be used to validate the correctness of the correspondence between the discipline model and the logical layer model.

Once the global model is developed, the data specification language (DSL) is used to generate an information model by expressing the global model (binary model) in the form of a text structure that is easy to read. DSL prepares the global model to be converted to a record model of specific exchange format which is used by the PDES processor to read PDES files.

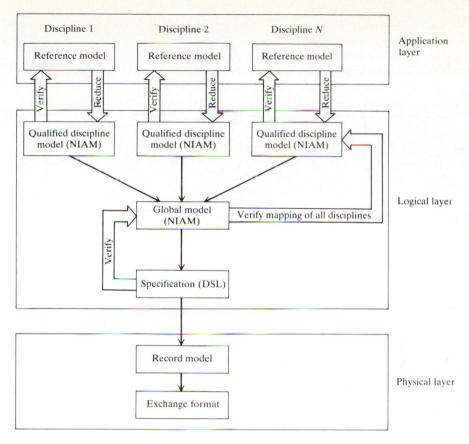

**FIGURE 8-11**
PDES methodology architecture.

In the context of the above PDES methodology, Fig. 8-12 shows the input/ output flow of PDES. The user, or expert, must define the product data, that is, application, at hand as a discipline model. The preprocessor then uses this model to produce a "data exchange unit." The data exchange unit is considered a three-part scheme consisting of the discipline model, the set of generic entities, and the set of correspondences (mappings) between the discipline-specific entities and the generic entities. The global model mentioned above is not equivalent to the data exchange unit. Instead, it can be viewed as a "smached" version of it used for verification purposes. The data exchange unit, as the name implies, is the basic unit of data exchange and the physical file data consists of instances of it. This unit is computer-interpretable and is made explicit in the data to be exchanged.

PDES is designed to have a proactive influence on both users and vendors. The demand of exchanging data in terms of units as stated above would require users to enter data into their own CAD/CAM systems in a certain way, and would require a new generation of translators or processors. Users must set up their own correspondences between PDES discipline models and their own set of generic entities provided by their CAD/CAM systems, and then must strictly

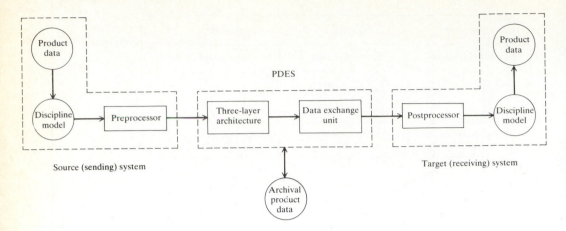

**FIGURE 8-12**
Product data exchange using PDES.

follow these correspondences when entering product data into their systems. Therefore, extensive analysis by users of their own application area in terms of the standard PDES discipline models is needed.

PDES preprocessors are designed to enable users to specify the processor actions from the generic entities of their own system to the generic entities of PDES based on their own correspondences between the discipline model and the generic entities of their system. Similarly, a postprocessor action can be based on information at the application layer as well as the logical layer by taking the correspondences from the PDES discipline-specific entities to the PDES generic entities into account. Both IGES processors have no correspondences to draw upon and therefore must determine their actions based solely on generic-type information. These actions are then fixed, either permanently or at least for an entire exchange set of data.

### 8.4.2 Data Representation

PDES version 1.0 was not available while this textbook was being written. Therefore, data representation of the discipline-specific model and generic entities of the logical layer as well as the file structure were not available. Future editions of the book will include the material as it becomes available.

### 8.4.3 Remarks

PDES is a much more comprehensive and complex standard than any of its predecessors. More user interaction with PDES is required. Therefore, the user interface is not as simple as "put IGES" and "get IGES." However, PDES reflects demands of the 1980s and 1990s from the CAD/CAM technology, that is, support of industrial automation.

# PROBLEMS
## Part 1: Theory

**8.1.** Discuss the requirements of product data exchange between dissimilar CAD/CAM systems.

**8.2.** Compare the shape-based and the product data-based exchange standards. Which has potential to support industrial automation? Why?

**8.3.** Describe the IGES methodology.

**8.4.** Compare the various testing methods of IGES processors. Which test is the most comprehensive? Why?

**8.5.** Describe the PDES methodology.

**8.6.** Compare IGES and PDES.

**8.7.** Refer to current literature and prepare a report on IDEF1 and NIAM.

**8.8.** Refer to current literature and prepare a report on DSL.

## Part 2: Laboratory

**8.9.** Generate IGES files for selected wireframe and surface geometric models created on your CAD/CAM system in Chaps. 5 and 6. Prepare files to demonstrate the exchange of geometric entities, annotation, and structure.

**8.10.** For the IGES files you generated in Prob. 8.9, perform the reflection test by reading the files back into your system. Did you get back the exact original models you created in Chaps. 5 and 6? What are your comments?

**8.11.** If you can use more than one CAD/CAM system, perform the other types of IGES tests. What are your observations?

# BIBLIOGRAPHY

Carringer, R. A.: "Product Definition Data Interface (PDDI)," *Proc. CIMTECH Conf.*, sponsored by CASA of SME, March 10–13, 1986, pp. 3.13–3.23, Chicago, Illinois.

Fletcher, S. K.: "Spline Transfer through IGES," report SAND-83-2131 (also DE84 002757), CAD/CAM Integration Division, Sandia National Laboratory, Albuquerque, N. Mex., October 1983.

Grabowski, H., and R. Glatz: "IGES Model Comparison System: A Tool for Testing and Validating IGES Processors," *IEEE CG&A*, vol. 7, no. 11, pp. 47–57, 1987.

"Initial Graphics Exchange Specification, Version 2.0," NBSIR 82-2631, NBS, Gaithersburg, Md., 1983. Published by NTIS, Springfield, Va., acquisition PB83-137448.

"Initial Graphics Exchange Specification, Version 3.0," NBSIR 86-3359, NBS, Gaithersburg, Md., 1986. Published by NTIS, Springfield, Va., acquisition PB86-199759.

"Initial Graphics Exchange Specification, Version 4.0," NBSIR 88-3813, NBS, Gaithersburg, Md., 1988. Published by NTIS, Springfield, Va., acquisition PB88-235452/LAA.

"Introducing IGES," *Engineering*, pp. 154–158, March 1985.

Lewis, J. W.: "Interchanging Spline Curves Using IGES," *CAD J.*, vol. 13, no. 6, pp. 359–364, 1981.

Lewis, J. W.: "Specifying and Verifying IGES Processors," *Proc. Conf. on CAD/CAM Technology in Mechanical Engineering*, March 24–26, 1982, pp. 377–398, MIT, Cambridge, Mass.

Liewald, M. H.: "Initial Graphics Exchange Specification: Successes and Evolution," *Computers and Graphics*, vol. 9, no. 1, pp. 47–50, 1985.

Mayer, R. J.: "IGES: One Answer to the Problems of CAD Database Exchange," *Byte*, vol. 12, no. 6, pp. 209–214, 1987.

Mayer, R. J.: "Experts Fight It Out Over IGES," *Mach. Des.*, vol. 59, no. 15, pp. 96–99, 1987.

Park, C.: "IGES Update: A Report from the Electrical Subcommittee," *Computer Graphics World*, vol. 7, no. 3, pp. 43–50, 1984.

"PDES Logical Layer Initiation Tasks," report SAND-86-1048C, Sandia National Laboratories, Albuquerque, N. Mex., 1980.

Pratt, M. J.: "IGES—The Present State and Future Trends," *Computer-Aided Engng J.*, pp. 130–133, August 1985.

"Product Data Exchange Standard (PDES)," report SAND-85-1248C, Sandia National Laboratories, Albuquerque, N. Mex., 1985.

Skinner, C. S., S. L. Cotter, and K. A. Gutmann: "Defining Product Data for Integrated CAD/CAM," *CAE Mag.*, vol. 4, no. 8, pp. 74–82.

Smith, B. M.: "IGES: A Key to CAD/CAM Systems Integrations," *IEEE CG&A*, pp. 78–83, November 1983.

Weissflag, U.: "Experience in Design and Implementation of an IGES Translator," *Computers and Graphics*, vol. 8, no. 3, pp. 269–273, 1984.

Wilkinson, D., and R. Hallam: "A Study of Product Data Transfer Using IGES," *Computer-Aided Engng J.*, vol. 4, no. 3, pp. 131–136, 1987.

Wilson, P. R.: "Communication Issues in Solid Modeling: IGES and All That," *Proc. European Conf. on Solid Modeling*, 1985, London, England.

Wilson, P. R., I. D. Faux, M. C. Ostrowski, and K. G. Pasquill: "Interfaces for Data Transfer between Solid Modeling Systems," *IEEE CG&A*, pp. 41–51, January 1985.

# PART III

TWO- AND
THREE-DIMENSIONAL
GRAPHICS
CONCEPTS

# CHAPTER
# 9

# GEOMETRIC
# TRANSFORMATIONS

## 9.1 INTRODUCTION

A crucial software module of a CAD/CAM system is its graphics package. Such a package contains many graphics concepts that produce the functionality and interactivity of the system. Some of these concepts are geometrical transformations, viewing in two and three dimensions, modeling and object hierarchy, algorithms for removing hidden edges and surfaces, shading and coloring, and clipping and windowing.

Geometric transformations play a central role in model construction and viewing. They are used in modeling to express locations of objects relative to others. In generating a view of an object, they are used to achieve the effect of different viewing positions and directions. Typical CAD/CAM construction commands to translate, rotate, zoom, and mirror entities are all based on geometric transformations covered in this chapter. Some of these commands have been utilized in Chaps. 5 and 6 to construct typical models. Once the model construction is complete, its viewing in its modeling space is achieved via geometric transformations again. Orthographic views for engineering drawings as well as perspective views of a geometric model can be obtained by projecting the model onto the proper plane. In addition, the model itself can be rotated or scaled up and down to view it in its three-dimensional space.

Geometric transformation can also be used to create animated files of geometric models to study their motion. For example, the motion of a spatial mechanism can be animated by first calculating its motion (displacements and/or rotations) using the proper kinematic and dynamic equations. The geometric model of the mechanism at the initial position is then constructed and transformed incrementally using the calculation results. The resulting configurations of the mechanism are grouped together and redisplayed to convey the continuous motion effect. Similarly, transformation can be applied to display vibrations and deformations of modeled objects.

Geometric transformations are ideally suited for computer graphics applications and object modeling because the utilized geometry is point-based. Chapters 5, 6, and 7 have shown that displaying and/or transforming a given entity require the transformation of its key points first. In applications where the viewpoint changes rapidly or where objects move fast in relation to each other, transformation of these points must be carried out rapidly and repeatedly. It is, therefore, necessary to find efficient ways of performing three-dimensional transformations. Most of these transformations are implemented at the hardware level, and firmware that perform them are commonly provided by CAD/CAM systems.

Geometric transformations is a well-established subject. However, this chapter covers them from a new perspective. A unified vector treatment of the subject is presented. This enables both two- and three-dimensional transformations to be handled at once. Sections 9.2 and 9.3 cover both transformations and mappings of geometric models respectively. Basic transformations such as translation, reflection, and rotation as well as their concatenations are covered. The recast of these transformations in terms of homogeneous coordinates is presented. In Sec. 9.3, it is shown how the same transformation equations are interpreted to map model representations from one coordinate system to another. Sections 9.4 and 9.5 describe the inverse operations and show how they are useful in a user's environment. Section 9.6 shows useful applications of geometric transformations.

## 9.2 TRANSFORMATIONS OF GEOMETRIC MODELS

By definition, geometric transformations are mappings from one coordinate system onto itself. In other words, the description of a geometric model of an object can change within its own MCS. This would imply that the geometric model must undergo motion relative to its MCS. The simplest motion is the rigid-body motion in which the relative distances between object particles remain constant; that is, the object does not deform during the motion. Geometric transformations that describe this motion are often referred to as rigid-body transformations and typically include translation, scaling, reflection, rotation, and any combination of them. These transformations can be applied directly to the parametric representations of objects such as points, curves, surfaces, and solids. Matrix notation provides a very expedient way of developing and implementing geometric transformations into graphic packages.

Transformation of a point represents the core problem in geometric transformation because it is the basic element of object representation. For example, a line is represented by its two endpoints, and a general curve, surface, or solid is represented by a collection of points as seen in the previous part of the book. The problem of transforming a point can be stated as follows. Given a point $P$ that belongs to a geometric model that undergoes a rigid-body motion, find the corresponding point $P*$ in the new position such that

$$\mathbf{P*} = f(\mathbf{P}, \text{transformation parameters}) \qquad (9.1)$$

that is, the new position vector $\mathbf{P*}$ should be expressed in terms of the old position vector $\mathbf{P}$ and the motion parameters. One of the characteristics of Eq. (9.1) that should be emphasized here is that geometric transformation should be unique. A given set of transformation parameters must yield one and only one new point for each old point. This characteristic is a direct outcome of the rigid-body motion requirement. Another characteristic is the concatenation, or combination, of transformations. Intuitively, two transformations can be concatenated to yield a single transformation which should have the same effect as the sequential application of the original two.

In order to implement Eq. (9.1) into graphics hardware or software, it is desirable to express it in terms of matrix notation as

$$\mathbf{P*} = [T]\mathbf{P} \qquad (9.2)$$

where $[T]$ is the transformation matrix. Its elements should be functions of the given transformation parameters. The matrix $[T]$ should have some important properties. It must apply to all rigid-body transformation (translation, scaling, reflection, and rotation) as well as clipping and windowing. It should also be applicable to both two- and three-dimensional graphics applications. As explained in the sections to follow, homogeneous representation of Eq. (9.2) is introduced in order to be able to recast translation in terms of this equation.

Applying Eq. (9.2) repeatedly to key points in a geometric model database or a particular entity enables the transformation of the model or the entity. For example, to transform a straight line, its two endpoints are transformed and then connected to produce the transformed line. Similarly, to transform a curve, points on the curve are generated utilizing its parametric equation, transformed, and then connected to give the transformed curve. Equation (9.2) may also be applied to the parametric equation of the entity as discussed in this chapter. The display of transformed entities entails displaying the line segments connecting the transformed key points as discussed in Part II of the book.

## 9.2.1 Translation

When every entity of a geometric model remains parallel to its initial position, the rigid-body transformation of the model is defined as translation. Translating a model implies that every point on it moves an equal given distance in a given direction. Translation can be specified by a vector, a unit vector and a distance, or two points that denote the initial and final positions of the model to be translated. Figure 9-1 shows a curve translated by a vector $\mathbf{d}$.

To relate the final position vector $\mathbf{P*}$ of a point $P$ to its initial position vector $\mathbf{P}$ after being translated by a vector $\mathbf{d}$, consider the triangle shown in Fig. 9-1. In this case Eq. (9.1) takes the form

$$\mathbf{P*} = \mathbf{P} + \mathbf{d} \qquad (9.3)$$

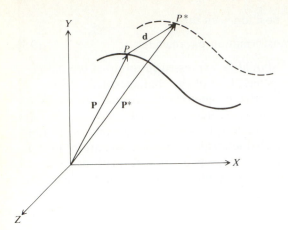

**FIGURE 9-1**
Translation of a curve.

This equation is applicable to both two- and three-dimensional points and can be written in a scalar form for the three-dimensional case as

$$x^* = x + x_d$$
$$y^* = y + y_d \qquad (9.4)$$
$$z^* = z + z_d$$

If Eq. (9.3) is applied to each point on a curve, that is, pointwise transformation, the curve is then translated by the vector **d**. However, it is more efficient and useful to relate the translation of an entity (curve, surface, or solid) to its geometric representation whether it is analytic or synthetic. For example, translating a circle or an ellipse requires translating its center only, and translating a parabola or hyperbola requires translating its vertex. Example 9.1 as well as problems at the end of the chapter provide more details.

As expected intuitively, translating a curve does not change its tangent vector at any of its points. This can be seen by differentiating Eq. (9.3) with respect to the parameter $u$ to obtain $\mathbf{P}^{*\prime} = \mathbf{P}'$ because the translation vector **d** is constant.

**Example 9.1.** Given a Hermite cubic spline, show that its pointwise translation and translating its geometric representation are identical.

*Solution.* As shown in Chap. 5, the geometric representation of a Hermite cubic spline is define by its two endpoints and two end slopes. Let us assume that the spline is defined in its initial and final positions by the geometry vectors $\mathbf{V} = [\mathbf{P}_0 \quad \mathbf{P}_1 \quad \mathbf{P}_0' \quad \mathbf{P}_1']^T$ and $\mathbf{V}^* = [\mathbf{P}_0^* \quad \mathbf{P}_1^* \quad \mathbf{P}_0^{*\prime} \quad \mathbf{P}_1^{*\prime}]^T$ respectively, as shown in Fig. 9-2. Because the spline undergoes translation only, $\mathbf{V}^*$ becomes

$$\mathbf{V}^* = [\mathbf{P}_0 + \mathbf{d} \quad \mathbf{P}_1 + \mathbf{d} \quad \mathbf{P}_0' \quad \mathbf{P}_1']^T = \mathbf{V} + \mathbf{D} \qquad (9.5)$$

where $\mathbf{D} = [\mathbf{d} \quad \mathbf{d} \quad 0 \quad 0]^T$.

Utilizing Eq. (5.83), the spline equation in the translated position is given by

$$\mathbf{P}^*(u) = \mathbf{U}^T[M_H]\mathbf{V}^*, \qquad 0 \le u \le 1 \qquad (9.6)$$

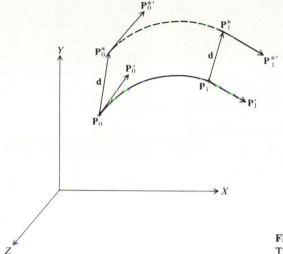

**FIGURE 9-2**
Translating a Hermite cubic spline.

Substituting Eq. (9.5) into Eq. (9.6) gives

$$\mathbf{P}^*(u) = \mathbf{U}^T[M_H](\mathbf{V} + \mathbf{D}) = \mathbf{U}^T[M_H]\mathbf{V} + \mathbf{U}^T[M_H]\mathbf{D}$$

or          $$\mathbf{P}^*(u) = \mathbf{P}(u) + \mathbf{U}^T[M_H]\mathbf{D} \qquad (9.7)$$

Substituting Eq. (5.84) into the second term of Eq. (9.7) and reducing the result gives

$$\mathbf{P}^*(u) = \mathbf{P}(u) + \mathbf{d} \qquad (9.8)$$

Equation (9.8) simply implies that each point on the translated spline is obtained by translating its corresponding point on the initial spline by the vector $\mathbf{d}$. Therefore, pointwise translation of a Hermite spline is identical to translating its geometry vector—more specifically its endpoints. This is beneficial because the geometric characteristics of the curve can be preserved and no intermediate points are needed to be calculated on the curve to translate it.

## 9.2.2   Scaling

Scaling is used to change, increase or decrease, the size of an entity or a model. Pointwise scaling can be performed if the matrix $[T]$ in Eq. (9.2) is diagonal, that is,

$$\mathbf{P}^* = [S]\mathbf{P} \qquad (9.9)$$

where $[S]$ is a diagonal matrix. In three dimensions, it is given by

$$[S] = \begin{bmatrix} s_x & 0 & 0 \\ 0 & s_y & 0 \\ 0 & 0 & s_z \end{bmatrix} \qquad (9.10)$$

Thus (9.9) can be expanded to give

$$x^* = s_x x \qquad y^* = s_y y \qquad z^* = s_z z \qquad (9.11)$$

The elements $s_x$, $s_y$, and $s_z$ of the scaling matrix $[S]$ are the scaling factors in the $X$, $Y$, $Z$ directions respectively. Scaling factors are always positive (negative factors produce reflection). If the scaling factors are smaller than 1, the geometric model or entity to which scaling is applied is compressed; if the factors are greater than 1, the model is stretched. If the scale factors are equal, that is, $s_x = s_y = s_z = s$, the model changes in size only and not in shape; this is the case of uniform scaling. For this case, Eq. (9.9) becomes

$$\mathbf{P^*} = s\mathbf{P} \tag{9.12}$$

Unlike translation, scaling proportionally changes tangent vectors by the factor $s$ as differentiating Eq. (9.12) with respect to $u$, for a curve, gives $\mathbf{P^{*\prime}} = s\mathbf{P}'$. However, uniform scaling does not change the slope, or direction cosines, at any point.

Differential scaling occurs when $s_x \neq s_y \neq s_z$; that is, different scaling factors are applied in different directions. Differential scaling changes both the size and the shape of a geometric model or curve. It also changes the direction cosines at any point. Differential scaling is seldom used in practical applications.

The scaling discussed above is said to be about the origin, that is, a model or a curve changes size and location with respect to the origin of the coordinate system, as shown in Fig. 9-3. The model or the curve gets closer to or further from the origin depending on whether the scaling factor is smaller or greater than one respectively (see Fig. 9-3). Scaling about any other point than the origin is possible, and its development is assigned in Prob. 9.3.

Uniform scaling is available on CAD/CAM systems in the form of a "zoom" command. The command requires users to input the scale factor $s$ and digitize the entity or the view to be zoomed. The zoom, or scaling, function is useful if a user needs to magnify a dense graphics area, on the screen, to be able

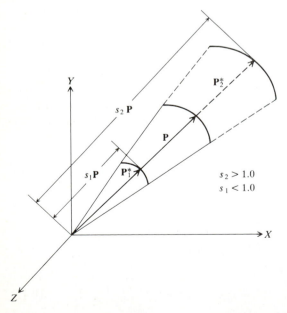

**FIGURE 9-3**
Scaling a curve relative to the origin.

to visually identify the geometry in the area for picking and selection purposes. If a view is zoomed, a "set view" or "reset view" command is usually required to make the view scaling permanent or to return the view to its original size respectively.

**Example 9.2.** Show that scaling the geometric representation of a Hermite cubic spline is identical to its pointwise scaling.

**Solution.** The solution is similar to that followed in Example 9.1. Here the geometry vector, based on Eq. (9.12), is given by

$$\mathbf{V}^* = [s\mathbf{P}_0 \quad s\mathbf{P}_1 \quad s\mathbf{P}_0' \quad s\mathbf{P}_1']^T = s\mathbf{V}$$

Substituting the above equation into Eq. (9.6) gives

$$\mathbf{P}^*(u) = s\mathbf{U}^T[M_H]\mathbf{V} = s\mathbf{P}(u)$$

Therefore, scaling the geometry vector $\mathbf{V}$ of a Hermite cubic spline is identical to scaling each point on it.

### 9.2.3 Reflection

Reflection (or mirror) transformation is useful in constructing symmetric models. If, for example, a model is symmetric with respect to a plane, then only half of its geometry is created which can be copied by reflection to generate the full model. A geometric entity can be reflected through a plane, a line, or a point in space, as illustrated in Fig. 9-4. Reflecting an entity through a principal plane $(x = 0, y = 0, z = 0$ plane) is equivalent to negating the corresponding coordinate of each point on the entity. Reflection through the $x = 0$, $y = 0$, or $z = 0$ plane can be achieved by negating the $x$, $y$, or $z$ coordinate respectively. Reflection through an axis is equivalent to reflection through two principal planes intersecting at the given axis. As shown in Fig. 9-4b, an entity is reflected through the $Y$ axis by reflection through the $z = 0$ plane followed by a reflection through the $x = 0$ plane. In this case, reflection is accomplished by negating the $x$ and $z$ coordinates of each point on the entity. Similarly, reflection through the $X$ and $Z$ axes requires negating the $y$ and $z$ and the $x$ and $y$ coordinates respectively. Reflection through the origin is equivalent to reflection through the three principal planes that intersect at the origin. Figure 9-4c shows reflection of an entity through the origin which is accomplished by negating the three coordinates of any point on the entity.

Equation (9.9) can be used to describe reflection if the diagonal elements of $[S]$ are chosen to be ones. Thus, the reflection transformation can be expressed by the following equation:

$$\mathbf{P}^* = [M]\mathbf{P} \tag{9.13}$$

where $[M]$ (mirror matrix) is a diagonal matrix with elements of $\pm 1$, that is,

$$[M] = \begin{bmatrix} m_{11} & 0 & 0 \\ 0 & m_{22} & 0 \\ 0 & 0 & m_{33} \end{bmatrix} = \begin{bmatrix} \pm 1 & 0 & 0 \\ 0 & \pm 1 & 0 \\ 0 & 0 & \pm 1 \end{bmatrix} \tag{9.14}$$

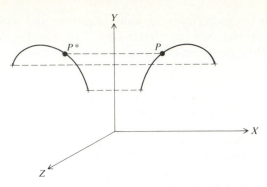

(*a*) Reflection through a principal plane (*x* = 0 plane)

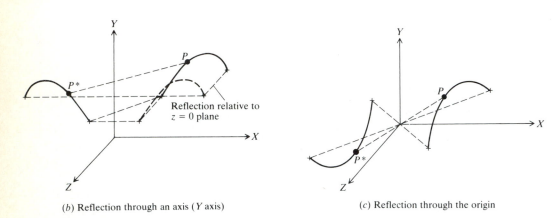

(*b*) Reflection through an axis (*Y* axis)

(*c*) Reflection through the origin

**FIGURE 9-4**
Reflecting a curve relative to a coordinate system.

The reflection (or mirror) matrix $[M]$ given by Eq. (9.14) applies only to reflections relative to planes, axes, or the origin of a coordinate system. For reflection through the $x = 0$ plane, $m_{11} = -1$ and $m_{22} = m_{33} = 1$. Similarly, setting $m_{11} = m_{33} = 1$ and $m_{22} = -1$, or $m_{11} = m_{22} = 1$ and $m_{33} = -1$, produces reflection through the $y = 0$ or $z = 0$ plane respectively. Reflection through the $X$ axis requires $m_{11} = 1$ and $m_{22} = m_{33} = -1$, through the $Y$ axis requires $m_{11} = m_{33} = -1$ and $m_{22} = 1$, and through the $Z$ axis requires $m_{11} = m_{22} = -1$ and $m_{33} = 1$. Selecting all the diagonal elements to be negative, that is, $m_{11} = m_{22} = m_{33} = -1$, produces reflection through the origin. For the latter case, Eq. (9.14) becomes $\mathbf{P}^* = -\mathbf{P}$ and, therefore, $\mathbf{P}^{*\prime} = -\mathbf{P}'$; that is, magnitudes of tangent vectors remain constant but their directions are reversed.

While reflections relative to a coordinate system have been discussed above, other reflections through general planes, lines, and points are possible and useful in practice. Figure 9-5 illustrates this general reflection of an entity. As seen from the figure, the common characteristic of the general reflection (the same as in Fig. 9-4) is that the distance from any point $P$ to be reflected to the reflection mirror (plane, line, or point) is equal to that from the mirror to the image

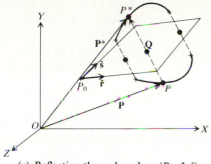

(a) Reflection through a plane $(P_0, \hat{r}, \hat{s})$

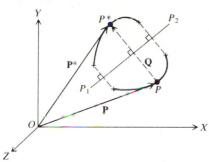

(b) Reflection through a line $(P_1, P_2)$

(c) Reflection through a point $P_r$

**FIGURE 9-5**
General reflection of a curve.

(reflected) point $P^*$. In the three cases, the triangle $OPP^*$ can be identified and can be used to relate the coordinates of $P^*$ to those of $P$ as follows:

$$P^* = P + Q \tag{9.15}$$

where $Q$ is the vector connecting $P$ and $P^*$.

In order to obtain $P^*$ from Eq. (9.15), the vector $Q$ must be evaluated first from the given geometry. This is when the difference between the three cases comes into play. Consider first the case shown in Fig. 9-5a. The plane is defined by a point $P_0$ and two unit vectors $\hat{r}$ and $\hat{s}$. The vector $Q$ is perpendicular to the plane and its magnitude is double the normal distance between $P$ and the plane. Utilizing Eq. (6.34), the normal distance $D$ can be obtained (compare Figs. 9-5a and 6-26) and we can write

$$Q = 2D\hat{n} \tag{9.16}$$

where $\hat{n}$ is the surface unit normal to the plane and can be calculated from Eq. (6.27) by using $\hat{r}$ and $\hat{s}$ in place of $(P_1 - P_0)$ and $(P_2 - P_0)$ respectively in the equation.

The vector $Q$ in the case of reflection through a general line shown in Fig. 9-5b can be evaluated by using the results of Example 5.9, case b. Comparing Fig. 5-20b and Fig. 9-5b, we can rewrite Eq. (5.22) to give

$$Q = -2\{(P - P_1) - [(P - P_1) \cdot \hat{n}_1]\hat{n}_1\} \tag{9.17}$$

The factor $-2$ is used in Eq. (9.17) because the magnitude of $\mathbf{Q}$ is twice the normal distance between $P$ and the line and its direction is opposite to $\hat{\mathbf{n}}_2$, shown in Fig. 5-20$b$.

In the case of reflection through a general point $P_r$, the vector $\mathbf{Q}$ can easily be written as (refer to Fig. 9-5$c$)

$$\mathbf{Q} = 2(\mathbf{P}_r - \mathbf{P}) \tag{9.18}$$

The effect of reflection on tangent vectors to a curve depends on each individual given case. Only for reflection through the origin or a general point can we write

$$\mathbf{P}^{*'} = -\mathbf{P}' \tag{9.19}$$

that is, tangent vectors reverse directions and their magnitudes remain unchanged. For other cases, only appropriate component(s) reverse directions.

**Example 9.3.** Show that reflecting the geometric representation of a Hermite cubic spline is identical to its pointwise reflection.

*Solution.* Let us consider the case of reflection through a general point. Utilizing Eqs. (9.15), (9.18), and (9.19), we can write the geometry vector $\mathbf{V}^*$ of the spline as

$$\mathbf{V}^* = [2\mathbf{P}_r - \mathbf{P}_0 \quad 2\mathbf{P}_r - \mathbf{P}_1 \quad -\mathbf{P}_0' \quad -\mathbf{P}_1']^T$$
$$= -[\mathbf{P}_0 \quad \mathbf{P}_1 \quad \mathbf{P}_0' \quad \mathbf{P}_1']^T + [2\mathbf{P}_r \quad 2\mathbf{P}_r \quad 0 \quad 0]^T = -\mathbf{V} + \mathbf{F} \tag{9.20}$$

where $\mathbf{F} = [2\mathbf{P}_r \quad 2\mathbf{P}_r \quad 0 \quad 0]$. Substituting Eq. (9.20) into Eq. (9.6), we get

$$\mathbf{P}^*(u) = \mathbf{U}^T[M_H](-\mathbf{V} + \mathbf{F}) = -\mathbf{U}^T[M_H]\mathbf{V} + \mathbf{U}^T[M_H]\mathbf{F} = -\mathbf{P} + \mathbf{U}^T[M_H]\mathbf{F} \tag{9.21}$$

Substituting Eq. (5.84) into the second term of Eq. (9.21) and reducing we obtain

$$\mathbf{P}^*(u) = -\mathbf{P} + 2\mathbf{P}_r = \mathbf{P} + 2(\mathbf{P}_r - \mathbf{P}) = \mathbf{P} + \mathbf{Q} \tag{9.22}$$

This equation is the same as combining Eqs. (9.15) and (9.18). Therefore, reflecting the end conditions of a Hermite cubic spline results in the pointwise reflection of the spline. Proofs of other cases of reflections can follow the same outlines discussed in this example and are left as exercises to the reader.

## 9.2.4   Rotation

Rotation is an important form of geometrical transformation. It enables users to view geometric models from different angles and also helps many geometric operations. For example, it can be used to create entities arranged in a circular pattern (circular arrays) by creating the entity once and then rotating/copying it to the desired positions on the circumference. In a similar fashion, rotation can be used to construct axisymmetric geometric models.

Rotation has a unique characteristic that is not shared by translation, scaling, or reflection—that is, noncommutativeness. The final position and orientation of an entity after going through two subsequent translations, scalings, or reflections are independent of the order of the operations, that is, commutative. On the contrary, two subsequent rotations of the entity about two different axes produce two different configurations of the entity depending on the order of the rotations. The reader can verify this by simply marking an edge of a box and then rotating it about two of its other edges and observe the final configuration of the marked edge. The same experiment can be performed on a CAD/CAM system. Interpretations of these experiments are covered in Sec. 9.2.6.

**9.2.4.1 ROTATION ABOUT COORDINATE SYSTEM AXES.** Rotating a point a given angle $\theta$ about the $X$, $Y$, or $Z$ axis is sometimes referred to as rotation about the origin. A convention for choosing signs of angles of rotations must be established. In this book, the right-hand convention is chosen. Therefore, a rotation angle about a given axis is positive in a counterclockwise sense when viewed from a point on the positive portion of the axis toward the origin.

To develop the rotational transformation of a point (or a vector) about one of the principal axes, let us consider the rotation of point $P$ a positive angle $\theta$ about the $Z$ axis, as shown in Fig. 9-6. This case is equivalent to two-dimensional rotation of a point in the $XY$ plane about the origin. The final position of $P$ after rotation is shown as point $P^*$. Equation (9.1) or (9.2) can be written here by relating the coordinates of $P^*$ to those of $P$ as follows:

$$x^* = r \cos(\theta + \alpha) = r \cos \alpha \cos \theta - r \sin \alpha \sin \theta$$

$$y^* = r \sin(\theta + \alpha) = r \sin \alpha \cos \theta + r \cos \alpha \sin \theta \qquad (9.23)$$

$$z^* = z$$

where $r = |\mathbf{P}| = |\mathbf{P^*}|$. To eliminate the angle $\theta$ from Eqs. (9.23), we can write (refer to the trigonometry in Fig. 9-6)

$$x = r \cos \alpha \qquad y = r \sin \alpha \qquad (9.24)$$

Substituting Eqs. (9.24) into (9.23) gives

$$x^* = x \cos \theta - y \sin \theta$$

$$y^* = x \sin \theta + y \cos \theta \qquad (9.25)$$

$$z^* = z$$

Rewriting Eqs. (9.25) in a matrix form gives

$$\begin{bmatrix} x^* \\ y^* \\ z^* \end{bmatrix} = \begin{bmatrix} \cos \theta & -\sin \theta & 0 \\ \sin \theta & \cos \theta & 0 \\ 0 & 0 & 1 \end{bmatrix} \begin{bmatrix} x \\ y \\ z \end{bmatrix} \qquad (9.26)$$

or

$$\mathbf{P^*} = [R_Z]\mathbf{P} \qquad (9.27)$$

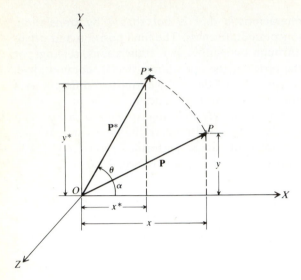

**FIGURE 9-6**
Rotation of a point about the $Z$ axis.

where

$$[R_Z] = \begin{bmatrix} \cos\theta & -\sin\theta & 0 \\ \sin\theta & \cos\theta & 0 \\ 0 & 0 & 1 \end{bmatrix} \tag{9.28}$$

Similarly, we can prove that matrices for rotations about $X$ and $Y$ axes are given by

$$[R_X] = \begin{bmatrix} 1 & 0 & 0 \\ 0 & \cos\theta & -\sin\theta \\ 0 & \sin\theta & \cos\theta \end{bmatrix} \tag{9.29}$$

$$[R_Y] = \begin{bmatrix} \cos\theta & 0 & \sin\theta \\ 0 & 1 & 0 \\ -\sin\theta & 0 & \cos\theta \end{bmatrix} \tag{9.30}$$

Thus, in general, we can write

$$\mathbf{P^*} = [R]\mathbf{P} \tag{9.31}$$

where $[R]$ is the appropriate rotational matrix. Equations (9.28) to (9.30) show some popular forms of $[R]$. Other forms are covered in the next two sections.

The columns of the rotation matrix $[R]$ have some useful characteristics. If we substitute the unit vector $[1 \quad 0 \quad 0]^T$ in the $X$ direction into Eq. (9.26), we obtain the first column of $[R_Z]$ as the components of the transformed unit vector. This implies that if we rotate the unit vector in the $X$ direction and angle $\theta$ about the $Z$ axis, the first column of $[R_Z]$ gives the coordinates of the transformed unit vector. Similarly, the second and third columns of $[R_Z]$ are the new coordinates of the unit vectors $[0 \quad 1 \quad 0]^T$ and $[0 \quad 0 \quad 1]^T$ in the $Y$ and $Z$ directions respectively after rotating them the same angle $\theta$. Therefore, the columns of a rotation matrix $[R]$ represent the unit vectors that are mutually orthogonal in a right-hand system, that is, $\mathbf{C}_1 \times \mathbf{C}_2 = \mathbf{C}_3$, $\mathbf{C}_2 \times \mathbf{C}_3 = \mathbf{C}_1$, and $\mathbf{C}_3 \times \mathbf{C}_1 = \mathbf{C}_2$, where $\mathbf{C}_1$, $\mathbf{C}_2$, and $\mathbf{C}_3$ are the first, second, and third columns of $[R]$ respectively. From linear algebra, a matrix with orthonormal columns is an orthogonal matrix and its inverse is equal to its transpose.

The effect of rotation on tangent vectors of a curve can be obtained from Eq. (9.31) as

$$\mathbf{P^{*\prime}} = [R]\mathbf{P'} \tag{9.32}$$

Thus, for a Hermite cubic spline, it is easily seen that $\mathbf{V^*} = [R]\mathbf{V}$. Therefore, the rotation of the spline about a given axis is equivalent to rotating its end conditions about the same axis.

### 9.2.4.2 TWO-DIMENSIONAL ROTATION ABOUT AN ARBITRARY AXIS.

The rotation of a point, or an entity in general, about an axis passing through an arbitrary point that is not the origin occurs when one point rotates about another one. In fact, the rotation of a point about the origin covered in the previous section is considered a special case of this problem we are about to solve. Rotation of a point or an entity about a point is useful in simulations of mechanisms, linkages, and robotics where links or members must rotate about their respective joints.

Figure 9-7 shows the rotation of point $P$ about point $P_1$ in the $XY$ plane. Figure 9-7a shows $P$ and its rotation, to its final position $P^*$, an angle $\theta$ about an axis parallel to the $Z$ axis and passing through $P_1$. In order to develop the rotation matrix correctly for this case, we can use Eq. (9.27) to rotate the vector $(\mathbf{P} - \mathbf{P}_1)$(not $\mathbf{P}$) about $P_1$ to obtain $(\mathbf{P^*} - \mathbf{P}_1)$ (not $\mathbf{P^*}$). Thus, we can write

$$\mathbf{P^*} - \mathbf{P}_1 = [R_Z](\mathbf{P} - \mathbf{P}_1) \tag{9.33}$$

Rearranging Eq. (9.33) gives

$$\mathbf{P^*} = [R_Z](\mathbf{P} - \mathbf{P}_1) + \mathbf{P}_1 \tag{9.34}$$

Equation (9.34) can also be obtained by considering the rotation of point $P$ about $P_1$ instead of considering the rotation of the vector $(\mathbf{P} - \mathbf{P}_1)$ about $P_1$ as we did. From this point of view, the rotation of $P$ about $P_1$ can be achieved in three steps as shown in Fig. 9.7b to d. In the first step, translate $P_1$ to the origin $O$. In this position, we refer to point $P_1$ as $P_{1t}$. Also, translate point $P$ to $P_t$ by the translation vector $-\mathbf{P}_1$ as shown in Fig. 9.7b. Therefore,

$$\mathbf{P}_t = \mathbf{P} - \mathbf{P}_1 \tag{9.35}$$

In the second step, rotate $P_t$, in the $XY$ plane, the angle $\theta$ about the origin, as shown in Fig. 9.7c. Consequently, Eq. (9.27) gives

$$\mathbf{P}_t^* = [R_Z]\mathbf{P}_t = [R_Z](\mathbf{P} - \mathbf{P}_1) \tag{9.36}$$

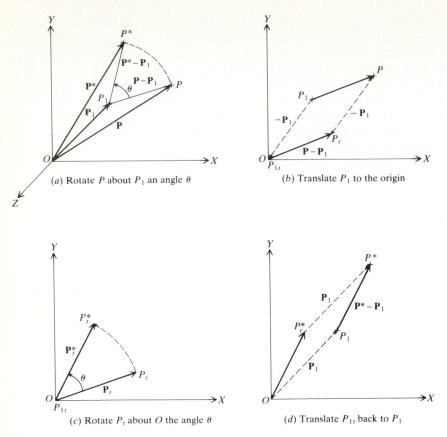

(a) Rotate $P$ about $P_1$ an angle $\theta$

(b) Translate $P_1$ to the origin

(c) Rotate $P_t$ about $O$ the angle $\theta$

(d) Translate $P_{1t}$ back to $P_1$

**FIGURE 9-7**
Two-dimensional rotation of a point about an arbitrary axis.

In the last step, translate points $P_{1t}$ and $P_t$ back to their original positions $P_1$ and $P$ respectively by the translation vector $\mathbf{P}_1$. This would require translating point $P_t^*$ by the same vector to the position $P^*$, as shown in Fig. 9.7$d$. Thus,

$$\mathbf{P}^* = \mathbf{P}_t^* + \mathbf{P}_1 = [R_Z](\mathbf{P} - \mathbf{P}_1) + \mathbf{P}_1 \qquad (9.37)$$

Equation (9.37) is the same as Eq. (9.34). Equation (9.37) applies to two-dimensional rotations in the $XZ$ or $YZ$ plane by replacing $[R_Z]$ by $[R_Y]$ or $[R_X]$ respectively.

**9.2.4.3  THREE-DIMENSIONAL ROTATION ABOUT AN ARBITRARY AXIS.**
Points undergoing rigid-body rotation describe arcs in a plane perpendicular to a fixed line, the axis of rotation. In planar two-dimensional rotation the axis is always perpendicular to the $XY$ plane; that is, parallel to the $Z$ axis. Consequently, the axis is completely defined by its intersection with the $XY$ plane (the origin $O$ in Fig. 9-6 and point $P_1$ in Fig. 9-7) and the orientation of the axis is implicitly defined (along the $Z$ axis) and does not appear as a parameter in the

rotation matrix $[R]$. As a result, the angle of rotation $\theta$ is the only transformation parameter required to completely define a two-dimensional rotation and, therefore, the corresponding $[R]$. It will be shown later in this section that two-dimensional rotation is a special case of three-dimensional rotation.

In the general spatial (three-dimensional) case, rotation is not constrained to the $XY$ plane and the axis of rotation may be oriented in any direction. Therefore, the orientation of the axis must be incorporated into the rotation matrix in addition to the angle of rotation. If we define the orientation by the unit vector $\hat{\mathbf{n}}$ (Fig. 9-8), Eq. (9.1) can be written as

$$\mathbf{P}^* = f(\mathbf{P}, \hat{\mathbf{n}}, \theta) \tag{9.38}$$

In this equation, it is assumed that the axis of rotation passes through the origin. If it does not, then a similar development to that presented in Sec. 9.2.4.2 should be followed. Equation (9.38) is derived below and recast in a matrix form for three-dimensional rotation about an arbitrary axis. Two cases are considered: the axis passes through the origin and the axis is in an arbitrary location.

Figure 9-8 shows the three-dimensional rotation of a point $P$ an angle $\theta$ about an arbitrary axis that passes through the origin. The positions of the point before and after rotation are $P$ and $P^*$ respectively. The orientation of the axis of rotation is defined by the unit vector $\hat{\mathbf{n}}$ such that

$$\hat{\mathbf{n}} = n_x \hat{\mathbf{i}} + n_y \hat{\mathbf{j}} + n_z \hat{\mathbf{k}} = \cos\alpha \hat{\mathbf{i}} + \cos\beta \hat{\mathbf{j}} + \cos\gamma \hat{\mathbf{k}} \tag{9.39}$$

where $n_x = \cos\alpha$, $n_y = \cos\beta$, and $n_z = \cos\gamma$ are the direction cosines of $\hat{\mathbf{n}}$. If the axis of rotation is defined as a line connecting the origin $O$ and any point, say $A$, on it (see Fig. 9-8), then $n_x = x_A/|\mathbf{A}|$, $n_y = y_A/|\mathbf{A}|$, and $n_z = z_A/|\mathbf{A}|$, where $x_A$, $y_A$, and $z_A$ are the coordinates of point $A$ and $|\mathbf{A}| = \sqrt{x_A^2 + y_A^2 + z_A^2}$.

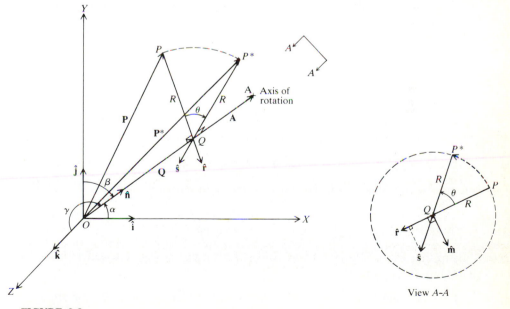

**FIGURE 9-8**
Three-dimensional rotation of a point about an arbitrary axis.

The rotation of $P$ about the axis $OA$ defines a circle whose plane is perpendicular to $OA$. Its center is point $Q$ which is the intersection between the axis and the plane. Its radius is $R$ which is the perpendicular distance between $P$ and $OA$ in any position, that is, $R = PQ = P*Q$. The angle of rotation $\theta$ is chosen in Fig. 9-8 to be positive according to the agreed-upon convention adopted for two-dimensional rotation. View $A$-$A$ shows $\theta$ counterclockwise, that is, positive, if the observer is placed at $A$-$A$, that is, on the positive portion of the axis.

In order to facilitate the development, let us define the directions of the lines $PQ$ and $P*Q$ by the unit vectors $\hat{r}$ and $\hat{s}$ respectively, as shown in Fig. 9-8. From the figure, it is obvious that the final position vector $\mathbf{P}*$ of point $P$ is the resultant of three vectors, that is,

$$\mathbf{P}* = \mathbf{P} + \mathbf{PQ} + \mathbf{QP}* \tag{9.40}$$

where the notation $\mathbf{PQ}$ indicates a vector going from point $P$ to point $Q$. Utilizing $\hat{r}$, $\hat{s}$, and $R$, Eq. (9.40) can be written as

$$\mathbf{P}* = \mathbf{P} + R\hat{r} - R\hat{s} \tag{9.41}$$

The remainder of the development that follows centers around expressing $\hat{r}$ and $\hat{s}$ in terms of $\mathbf{P}$, $\hat{n}$, $\theta$, that is, in terms of the desired rotation parameters. Utilizing the right triangle $OPQ$, we can write

$$\mathbf{PQ} = \mathbf{Q} - \mathbf{P} \tag{9.42}$$

Observing that $\mathbf{Q}$ is the component of $\mathbf{P}$ along the axis of rotation, we can write

$$\mathbf{Q} = (\mathbf{P} \cdot \hat{n})\hat{n} \tag{9.43}$$

Substituting Eq. (9.43) into (9.42) and dividing the result by $R$ (the magnitude of $\mathbf{PQ}$) gives

$$\hat{r} = \frac{(\mathbf{P} \cdot \hat{n})\hat{n} - \mathbf{P}}{R} \tag{9.44}$$

In order to express $\hat{s}$ in terms of $\mathbf{P}$ and $\hat{n}$, we need to introduce the intermediate unit vector $\hat{m}$ shown in view $A$-$A$ in Fig. 9-8. The vector is chosen to be perpendicular to $\hat{r}$ and lies in the plane of the circle; thus it is also perpendicular to $\hat{n}$. Utilizing the cross-product definition of two vectors, we can write

$$\hat{m} = \hat{n} \times \hat{r} \tag{9.45}$$

The unit vector $\hat{s}$ can now be written in terms of its components in the $\hat{r}$ and $\hat{m}$ directions as

$$\hat{s} = \cos\theta\hat{r} + \sin\theta\hat{m} \tag{9.46}$$

Substituting Eq. (9.45) into (9.46) and substituting the result together with Eq. (9.44) into (9.41), we obtain

$$\mathbf{P}* = (\mathbf{P} \cdot \hat{n})\hat{n} + [\mathbf{P} - (\mathbf{P} \cdot \hat{n})\hat{n}]\cos\theta + (\hat{n} \times \mathbf{P})\sin\theta - \hat{n} \times (\mathbf{P} \cdot \hat{n})\hat{n}\sin\theta$$

$$\tag{9.47}$$

The last term in the above equation is equal to zero because it represents the cross product of two collinear vectors $\hat{n}$ and $(\mathbf{P} \cdot \hat{n})\hat{n}$ (also $\mathbf{Q}$). Thus we have

$$\mathbf{P}^* = (\mathbf{P} \cdot \hat{n})\hat{n} + [\mathbf{P} - (\mathbf{P} \cdot \hat{n})\hat{n}]\cos\theta + (\hat{n} \times \mathbf{P})\sin\theta \qquad (9.48)$$

To write Eq. (9.48) in matrix form, we can write the following:

$$\mathbf{P} \cdot \hat{n} = xn_x + yn_y + zn_z = [n_x \quad n_y \quad n_z]\begin{bmatrix} x \\ y \\ z \end{bmatrix} \qquad (9.49)$$

$$\hat{n} \times \mathbf{P} = \begin{vmatrix} \hat{i} & \hat{j} & \hat{k} \\ n_x & n_y & n_z \\ x & y & z \end{vmatrix} = (n_y z - n_z y)\hat{i} + (n_z x - n_x z)\hat{j} + (n_x y - n_y x)\hat{k}$$

$$= \begin{bmatrix} 0 & -n_z & n_y \\ n_z & 0 & -n_x \\ -n_y & n_x & 0 \end{bmatrix}\begin{bmatrix} x \\ y \\ z \end{bmatrix} \qquad (9.50)$$

Substituting Eqs. (9.49) and (9.50) into Eq. (9.48) and rearranging, we get

$$\mathbf{P}^* = \left\{ (1 - \cos\theta)\begin{bmatrix} n_x \\ n_y \\ n_z \end{bmatrix}[n_x \quad n_y \quad n_z] + \cos\theta\begin{bmatrix} 1 & 0 & 0 \\ 0 & 1 & 0 \\ 0 & 0 & 1 \end{bmatrix} \right.$$

$$\left. + \sin\theta\begin{bmatrix} 0 & -n_z & n_y \\ n_z & 0 & -n_x \\ -n_y & n_x & 0 \end{bmatrix} \right\}\begin{bmatrix} x \\ y \\ z \end{bmatrix} \qquad (9.51)$$

or $$\mathbf{P}^* = [R]\mathbf{P} \qquad (9.52)$$

After reducing Eq. (9.51) further, $[R]$ becomes

$$[R] = \begin{bmatrix} n_x^2 \, v\theta + c\theta & n_x n_y \, v\theta - n_z \, s\theta & n_x n_z \, v\theta + n_y \, s\theta \\ n_x n_y \, v\theta + n_z \, s\theta & n_y^2 \, v\theta + c\theta & n_y n_z \, v\theta - n_x \, s\theta \\ n_x n_z \, v\theta - n_y \, s\theta & n_y n_z \, v\theta + n_x \, s\theta & n_z^2 \, v\theta + c\theta \end{bmatrix} \qquad (9.53)$$

where $c\theta = \cos\theta$, $s\theta = \sin\theta$, and $v\theta = \text{versine } \theta = 1 - \cos\theta$.

The general rotation matrix $[R]$ given by Eq. (9.53) has two important characteristics. First, it is skew symmetric because the third term in Eq. (9.51) is skew symmetric. Second, its determinant $|R|$ is equal to one. In general, $|R| = 1$ for any rotation matrix that describes rotation about the origin. The reader can verify this fact for Eqs. (9.28) to (9.30) and (9.53).

The rotation matrices given by Eqs. (9.28) to (9.30) can now be seen as special cases of Eq. (9.53) as follows. If the axis of rotation is the $Z$ axis, then its orientation is given by $\hat{n} = \hat{k}$, that is, $n_x = n_y = 0$ and $n_z = 1$. Substituting these values into Eq. (9.53) gives Eq. (9.28). Similarly, the $X$ and $Y$ axes of rotation are given by $\hat{n} = \hat{i}$ $(n_x = 1, n_y = n_z = 0)$ and $\hat{n} = \hat{j}$ $(n_x = n_z = 0, n_y = 1)$ respectively and Eqs. (9.29) and (9.30) can be easily obtained from Eq. (9.53).

We now return to the case of three-dimensional rotation about an arbitrary axis that does not pass through the origin. This case is conceptually similar to the two-dimensional case covered in Sec. 9.2.4.2 and its development follows exactly the same steps. Therefore, Eq. (9.37) is applicable for the three-dimensional case after replacing $[R_z]$ by $[R]$ given by Eq. (9.53), that is,

$$\mathbf{P}^* = [R](\mathbf{P} - \mathbf{P}_1) + \mathbf{P}_1 \tag{9.54}$$

Here the point $P_1$ can be any point in space and is not restricted to the $XY$ plane as in the two-dimensional case.

**Example 9.4.** Prove that if a point to be rotated about a given axis of rotation lies on the axis, the point does not change position in space and, therefore, its coordinates do not change.

*Solution.* Figure 9-9 shows this problem. It is expected that $P$ and $P^*$ are identical and, therefore, $\mathbf{P} = \mathbf{P}^*$ regardless of the angle of rotation $\theta$. Substituting Eq. (9.53) into Eq. (9.52) and expanding the result, we obtain for the $x$ coordinate:

$$x^* = (n_x^2\ v\theta + c\theta)x + (n_x n_y\ v\theta - n_z\ s\theta)y + (n_x n_z\ v\theta + n_y\ s\theta)z \tag{9.55}$$

If point $P$ lies on the axis of rotation, then we can write (see Fig. 9-9)

$$x = |\mathbf{P}|n_x \qquad y = |\mathbf{P}|n_y \qquad z = |\mathbf{P}|n_z \tag{9.56}$$

where $|\mathbf{P}|$ is the magnitude of $\mathbf{P}$. Substituting Eq. (9.56) into (9.55) and reducing the result, we obtain

$$x^* = |\mathbf{P}|[n_x\ v\theta(n_x^2 + n_y^2 + n_z^2) + n_x\ c\theta] \tag{9.57}$$

Using the identity $n_x^2 + n_y^2 + n_z^2 = 1$, Eq. (9.57) becomes

$$x^* = |\mathbf{p}|[n_x(1 - \cos\ \theta) + n_x\ \cos\ \theta] = |\mathbf{P}|n_x = x \tag{9.58}$$

Similarly, we can prove $y^* = y$ and $z^* = z$.

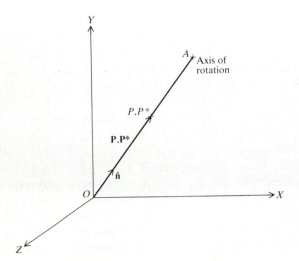

**FIGURE 9-9**

Rotation of a point about an axis passing through it and through the origin.

This example has useful practical implications. Typically, CAD/CAM systems let users define axes of rotations by inputting endpoints. If the axis of rotation happens to be an entity of a geometric model to be rotated, the coordinates of the endpoints of that entity should stay the same before and after rotation. The user can utilize the "verify entity" command available on the system before and after the rotation to display the coordinates and compare. If there are small differences, they usually result from the round-off errors. Coordinates should be the same within the given significant digits of the computer system used.

### 9.2.5 Homogeneous Representation

The various rigid-body geometric transformations have been developed in the previous section. Equations (9.3), (9.9), (9.13), and (9.52) represent translation, scaling, mirroring, and rotation respectively. While the last three equations are in the form of matrix multiplication, translation takes the form of vector addition. This makes it inconvenient to concatenate transformations involving translation. Equation (9.37) shows an example. It is desirable, therefore, to express all geometric transformations in the form of matrix multiplications only. Representing points by their homogeneous coordinates provides an effective way to unify the description of geometric transformations as matrix multiplications.

Homogeneous coordinates have been used in computer graphics and geometric modeling for a long time. With their aid, geometric transformations are customarily embedded into graphics hardware to speed their execution. Homogeneous coordinates are useful for other applications. They are useful to obtain perspective views of geometric models. The subjects of projective geometry, mechanism analysis and design, and robotics utilize them quite often in development and formulation. In addition, homogeneous coordinates remove many anomalous situations encountered in cartesian geometry such as representing points at infinity and the nonintersection of parallel lines. Also, they greatly simplify expressions defining rational parametric curves and surfaces.

In homogeneous coordinates, an $n$-dimensional space is mapped into $(n + 1)$-dimensional space; that is, a point (or a position vector) in $n$-dimensional space is represented by $(n + 1)$ coordinates (or components). In three-dimensional space, a point $P$ with cartesian coordinates $(x, y, z)$ has the homogeneous coordinates $(x^*, y^*, z^*, h)$ where $h$ is any scalar factor $\neq 0$. The two types of coordinates are related to each other by the following equations:

$$x = \frac{x^*}{h} \qquad y = \frac{y^*}{h} \qquad z = \frac{z^*}{h} \tag{9.59}$$

Equations (9.59) are based on the fact that if the cartesian coordinates of a given point $P$ are multiplied by a scalar factor $h$, $P$ is scaled to a new point $P^*$ and the coordinates of $P$ and $P^*$ are related by the above equations. Figure 9-10 shows point $P$ scaled by the two factors $h_1$ and $h_2$ to produce the two new points $P_1^*$ and $P_2^*$ respectively. These two points could be interpreted in two different

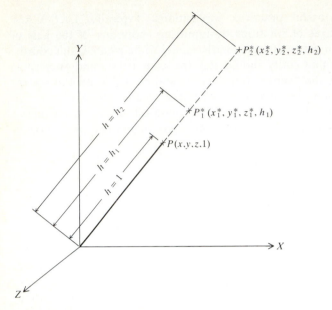

**FIGURE 9-10**
Homogeneous coordinates of point $P$.

ways. From a cartesian-coordinates point of view, Eq. (9.12) can be used with $s = h_1$ and $h_2$. Once the cartesian coordinates of $P_1^*$ and $P_2^*$ are calculated, their relationships to $P$ do not exist any more. Moreover, the three points still belong to the cartesian space. From a homogeneous-coordinates point of view, the original point $P$ is represented by $(x, y, z, 1)$ and $P_1^*$ and $P_2^*$ are represented by $(x_1^*, y_1^*, z_1^*, h_1)$ and $(x_2^*, y_2^*, z_2^*, h_2)$ respectively according to Eqs. (9.59). More importantly, the three points belong to the homogeneous space, with the cartesian coordinates obtained when $h = 1$, and the relationship between $P$ and $P_1^*$ or $P_2^*$ is maintained through the proper value of $h$. As a matter of fact, any two homogeneous-coordinates points $P_1^*$ and $P_2^*$ represent the same cartesian point if and only if $h_2 = ch_1$, for any nonzero constant $c$. Therefore, there is no unique homogeneous representation of a point. For the purpose of geometric transformations, the scalar factor $h$ used in Eqs. (9.59) is taken to be unity to avoid unnecessary division.

The translation transformation given by Eq. (9.3) can now be written as a matrix multiplication by adding the component of 1 to each vector in the equation and using a $4 \times 4$ matrix as follows:

$$[x^* \quad y^* \quad z^* \quad 1]^T = \begin{bmatrix} 1 & 0 & 0 & x_d \\ 0 & 1 & 0 & y_d \\ 0 & 0 & 1 & z_d \\ 0 & 0 & 0 & 1 \end{bmatrix} \begin{bmatrix} x \\ y \\ z \\ 1 \end{bmatrix} \qquad (9.60)$$

or
$$\mathbf{P^*} = [D]\mathbf{P} \qquad (9.61)$$

where $[D]$ is the translation matrix shown in Eq. (9.60). While scaling, reflection, and rotation are already expressed in terms of matrix multiplication, their corresponding matrices are changed from $3 \times 3$ into $4 \times 4$ by adding a column and a row of zero elements except the fourth, which is 1. Thus, the scaling matrix [Eq. (9.10)] becomes

$$[S] = \begin{bmatrix} s_x & 0 & 0 & 0 \\ 0 & s_y & 0 & 0 \\ 0 & 0 & s_z & 0 \\ 0 & 0 & 0 & 1 \end{bmatrix} \tag{9.62}$$

Similarly, the reflection matrix [Eq. (9.14)] becomes

$$[M] = \begin{bmatrix} \pm 1 & 0 & 0 & 0 \\ 0 & \pm 1 & 0 & 0 \\ 0 & 0 & \pm 1 & 0 \\ 0 & 0 & 0 & 1 \end{bmatrix} \tag{9.63}$$

and the rotation matrix [Eq. (9.53)] becomes

$$[R] = \begin{bmatrix} r_{11} & r_{12} & r_{13} & 0 \\ r_{21} & r_{22} & r_{23} & 0 \\ r_{31} & r_{32} & r_{33} & 0 \\ 0 & 0 & 0 & 1 \end{bmatrix} \tag{9.64}$$

Equations (9.28) to (9.30) can be rewritten in a similar fashion.

To illustrate the convenience gained from the homogeneous representation, Eq. (9.37) can be written as

$$\mathbf{P}^* = [D_1][R_Z][D_2]\mathbf{P} = [T]\mathbf{P} \tag{9.65}$$

where

$$[D_1] = \begin{bmatrix} 1 & 0 & 0 & -x_1 \\ 0 & 1 & 0 & -y_1 \\ 0 & 0 & 1 & -z_1 \\ 0 & 0 & 0 & 1 \end{bmatrix} \tag{9.66}$$

$$[R_Z] = \begin{bmatrix} \cos\theta & -\sin\theta & 0 & 0 \\ \sin\theta & \cos\theta & 0 & 0 \\ 0 & 0 & 1 & 0 \\ 0 & 0 & 0 & 1 \end{bmatrix} \tag{9.67}$$

$$[D_2] = \begin{bmatrix} 1 & 0 & 0 & x_1 \\ 0 & 1 & 0 & y_1 \\ 0 & 0 & 1 & z_1 \\ 0 & 0 & 0 & 1 \end{bmatrix} \tag{9.68}$$

and

$$[T] = [D_1][R_Z][D_2] = \begin{bmatrix} \cos\theta & -\sin\theta & x_1(\cos\theta - 1) - y_1\sin\theta & 0 \\ \sin\theta & \cos\theta & x_1\sin\theta + y_1(\cos\theta - 1) & 0 \\ 0 & 0 & 1 & 0 \\ 0 & 0 & 0 & 1 \end{bmatrix}$$

(9.69)

A closer look at the transformation matrices given in Eqs. (9.61) to (9.64) shows that they can all be embedded into one $4 \times 4$ matrix. This matrix takes the form:

$$[T] = \begin{bmatrix} t_{11} & t_{12} & t_{13} & t_{14} \\ t_{21} & t_{22} & t_{23} & t_{24} \\ t_{31} & t_{32} & t_{33} & t_{34} \\ t_{41} & t_{42} & t_{43} & t_{44} \end{bmatrix} = \begin{bmatrix} T_1 & T_2 \\ T_3 & 1 \end{bmatrix}$$

(9.70)

The $3 \times 3$ submatrix $[T_1]$ produces scaling, reflection, or rotation. The $3 \times 1$ column matrix $[T_2]$ generates translation. The $1 \times 3$ row matrix $[T_3]$ produces perspective projection, covered in Sec. 9.5.2. The fourth diagonal element is the homogeneous-coordinates scalar factor $h$ used in Eq. (9.59) and is chosen to be unity, as mentioned earlier.

Equation (9.70) gives the explicit form of the transformation matrix $[T]$ used in Eq. (9.2). It is usually written for one geometric transformation at a time by using any of Eqs. (9.60) to (9.64). If more than one transformation is desired, the resulting matrices are multiplied to produce the total transformation, as discussed in Sec. 9.2.6 that follows.

While the homogeneous representation and the resulting transformation matrix $[T]$ given by Eq. (9.70) are useful and convenient to think of and write compact equations, a computer program to implement them to transform entities should be carefully designed to avoid wasting time multiplying ones and zeros. As a matter of fact, they may not even be used at all to simplify the related programming logic.

### 9.2.6   Concatenated Transformations

So far we have concentrated on one-step transformations of points such as rotating or translating a point. However, in practice a series of transformations may be applied to a geometric model. Thus, combining or concatenating transformations are quite useful. Concatenated transformations are simply obtained by multiplying the $[T]$ matrices [Eq. (9.70)] of the corresponding individual transformations. However, because matrix multiplication may not be commutative in all cases, attention must be paid to the order in which transformations are applied to a given geometric model. In general, if we apply $n$ transformations to a point starting with transformation 1, with $[T_1]$, and ending with transformation

$n$, with $[T_n]$, then the concatenated transformation of the point is given by

$$\mathbf{P}^* = [T_n][T_{n-1}] \cdots [T_2][T_1]\mathbf{P} \tag{9.71}$$

As an example, consider rotating a point, or its position vector, in the fixed coordinate system $XYZ$, that is, MCS, by the following rotations in the following order: $\alpha$ about the $Z$ axis, $\beta$ about the $Y$ axis, and $\gamma$ about the $X$ axis. Substituting $\alpha$, $\beta$, and $\gamma$ in Eqs. (9.28), (9.30), and (9.29) respectively and multiplying, we obtain the concatenated transformation matrix as

$$[T] = [T_X][T_Y][T_Z] \tag{9.72}$$

or

$$\begin{bmatrix} [R] & \vdots & 0 \\ \hdashline 0 & \vdots & 1 \end{bmatrix} = \begin{bmatrix} [R_X] & \vdots & 0 \\ \hdashline 0 & \vdots & 1 \end{bmatrix} \begin{bmatrix} [R_Y] & \vdots & 0 \\ \hdashline 0 & \vdots & 1 \end{bmatrix} \begin{bmatrix} [R_Z] & \vdots & 0 \\ \hdashline 0 & \vdots & 1 \end{bmatrix} \tag{9.73}$$

or

$$[R] = [R_X][R_Y][R_Z] \tag{9.74}$$

Expanding the above equation gives

$$[R] = \begin{bmatrix} c\alpha \ c\beta & -s\alpha \ c\beta & s\beta \\ s\alpha \ c\gamma + c\alpha \ s\beta \ s\gamma & c\alpha \ c\gamma - s\alpha \ s\beta \ s\gamma & -c\beta \ s\gamma \\ s\alpha \ s\gamma - c\alpha \ s\beta \ c\gamma & c\alpha \ s\gamma + s\alpha \ s\beta \ c\gamma & c\beta \ c\gamma \end{bmatrix} \tag{9.75}$$

**Example 9.5.** Using the concatenated rotations about the axes of the coordinate system shown in Fig. 9-8, rederive Eq. (9.53).

*Solution.* The basic idea to solve this example is to rotate the axis of rotation $OA$ shown in Fig. 9-8 to coincide with one of the axes, rotate the point $P$ the angle $\theta$ about this coincident axis, and finally rotate $OA$ in the opposite direction to its original position. The rotation of $OA$ is achieved in two steps. In effect, this is equivalent to decomposing the rotation about the general axis into three rotations about the principal axes $X$, $Y$, and $Z$. Figure 9-11 shows one possible decomposition where point $B$ is the projection of point $A$ onto the $XZ$ plane. In this decomposition, the following sequence of rotations is followed:

1. Rotate $OA$ and point $P$ about the $Y$ axis an angle $-\phi$ so that $OB$ is collinear with the $Z$ axis, where

$$\tan \phi = \frac{x_A}{z_A} = \frac{x_A/|\mathbf{A}|}{z_A/|\mathbf{A}|} = \frac{n_x}{n_z} \tag{9.76}$$

This then gives

$$\cos \phi = \frac{n_z}{\sqrt{n_x^2 + n_z^2}} \qquad \sin \phi = \frac{n_x}{\sqrt{n_x^2 + n_z^2}} \tag{9.77}$$

Thus,

$$\mathbf{P}^* = [R_Y(-\phi)]\mathbf{P} \tag{9.78}$$

2. Following the above rotation, rotate $OA$ and $P$ about the $X$ axis an angle $\psi$ so that $OA$ is collinear with the $Z$ axis, where

$$\sin \psi = \frac{y_A}{|\mathbf{A}|} = n_y \qquad \cos \psi = \sqrt{1 - n_y^2} \tag{9.79}$$

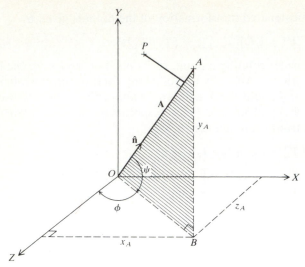

**FIGURE 9-11**
Decomposition of three-dimensional rotation of a point.

Equation (9.78) becomes

$$\mathbf{P}^* = [R_X(\psi)][R_Y(-\phi)]\mathbf{P} \tag{9.80}$$

2. Rotate point $P$ about the $Z$ axis an angle $\theta$. Note that $P$ is now given by Eq. (9.80). After rotating by the angle $\theta$, it then becomes

$$\mathbf{P}^* = [R_Z(\theta)][R_X(\psi)][R_Y(-\phi)]\mathbf{P} \tag{9.81}$$

4. Reverse step 2, that is, rotate about the $X$ axis an angle $-\psi$. This modifies Eq. (9.81) to

$$\mathbf{P}^* = [R_X(-\psi)][R_Z(\theta)][R_X(\psi)][R_Y(-\phi)]\mathbf{P} \tag{9.82}$$

5. Reverse step 1, that is, rotate about the $Y$ axis an angle $\phi$. This modifies Eq. (9.82) to

$$\mathbf{P}^* = [R_Y(\phi)][R_X(-\psi)][R_Z(\theta)][R_X(\psi)][R_Y(-\phi)]\mathbf{P} \tag{9.83}$$

Equation (9.83) should be the same as Eq. (9.51) but in a different form, and in comparing it with Eq. (9.52), we can write

$$[R] = [R_Y(\phi)][R_X(-\psi)][R_Z(\theta)][R_X(\psi)][R_Y(-\phi)] \tag{9.84}$$

If the matrix multiplications in the above equation are performed and the result is reduced, Eq. (9.53) will be obtained. The reader can carry out the details by using Eqs. (9.28) to (9.30) and using the identity $n_x^2 + n_y^2 + n_z^2 = 1$. If other sequences of decompositions of the rotation are used, the right-hand side of Eq. (9.84) changes but the final result, that is, $[R]$, stays the same. The reader is encouraged to try these sequences.

## 9.3 MAPPINGS OF GEOMETRIC MODELS

In the previous section, we concerned ourselves with rigid-body transformations of geometric models. Thus, we have discussed transforming a point (or a set of points) belonging to an object into another point (or another set of points), with both points (or sets) described in the same coordinate system. Thus, the model position and orientation change with respect to the origin of the coordinate system which stays unaltered in space. In this section, we think of rigid-body motion and its related matrices as mappings of geometric models between different coordinate systems. This is useful in geometric modeling as transformations (see Example 9.6). Mapping of a point (or a set of points) belonging to an object from one coordinate system to another is defined as changing the description of the point (or the set of points) from the first coordinate system to the second one. Thus, the model position and orientation stays unaltered in space with respect to the origins of both coordinate systems while only the description of such position and orientation changes. This is equivalent to transforming one coordinate system to another.

Mapping can be used in various applications. It is useful during model construction, as discussed in Chap. 3. When the user defines a WCS and creates geometry by inputting coordinates measured in this WCS, the software maps these coordinates to the MCS before storing them in the model database. Mapping is also useful in assemblies or model merging where one or more models, each defined in its own MCS, are combined or merged into a host model. The coordinates of each subassembly or merged model is expressed in terms of the MCS of the host assembly or model via mapping.

The same mathematical forms that we have developed for geometric transformations can be used to map points between coordinate systems. However, the interpretation of these forms is different. The problem of mapping a point from one coordinate system to another can be stated as follows. Given the coordinates of a point $P$ measured in a given $XYZ$ coordinate system, find the coordinates of the point measured in another coordinate system, say $X^*Y^*Z^*$, such that

$$\mathbf{P}^* = f(\mathbf{P}, \text{mapping parameters}) \qquad (9.85)$$

where $\mathbf{P}$ and $\mathbf{P}^*$ are the position vectors of point $P$ in the $XYZ$ and $X^*Y^*Z^*$ systems respectively. The mapping parameters describe the relationship between the two systems and consist of the position of the origin and orientation of the $X^*Y^*Z^*$ system relative to the $XYZ$ system. Equations (9.1) and (9.85) are the same, but their interpretations differ. Equation (9.85) can be expressed in the matrix form given by Eq. (9.2), where $[T]$ is referred to as the mapping matrix. This matrix describes the position of the origin and the orientation of one coordinate system relative to another one as expressed by Eq. (3.3). We now consider the three possible cases of mapping and develop their corresponding matrices.

## 9.3.1 Translational Mapping

When the axes of the two coordinate systems are parallel, the mapping is defined to be translational. In Fig. 9-12, the origins of the $XYZ$ and the $X^*Y^*Z^*$ systems are different but their orientations in space are the same. The point $P$ is described by the vectors $\mathbf{P}$ and $\mathbf{P}^*$ in the $XYZ$ and $X^*Y^*Z^*$ systems respectively. The vector $\mathbf{d}$ describes the position of the origin of the former system relative to the

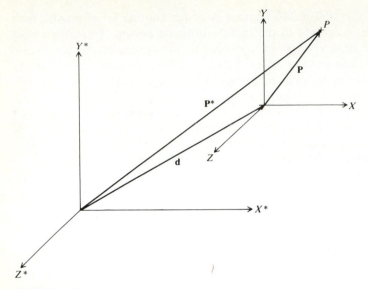

**FIGURE 9-12**
Translational mapping of a point.

latter. Equation (9.85) can be written exactly as Eq. (9.3) or (9.60) in the homogeneous form.

## 9.3.2   Rotational Mapping

Figure 9-13 shows rotational mapping between two coordinate systems. The two systems share the same origin and their orientations are different by the angle $\theta$. In this figure, we assume that the $XY$ and $X^*Y^*$ planes are coincident. Utilizing

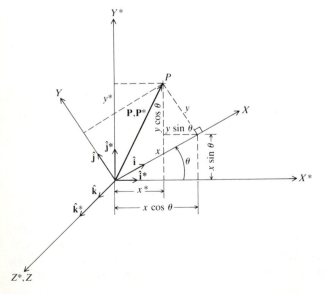

**FIGURE 9-13**
Rotational mapping of a point.

the trigonometric relationships shown in the figure, Eq. (9.25) can be derived. Therefore, the rotation matrix given by Eq. (9.28) or (9.67) is applicable for rotational mapping. Similarly, Eqs. (9.29), (9.30), (9.53), and (9.75) are applicable for their corresponding rotational mapping cases.

In rotational mapping, it is important to realize that the columns of a rotational matrix $[R]$ can be interpreted to describe the orientations of a given coordinate system in space. If we take the unit vectors $\hat{\imath}$, $\hat{\jmath}$, and $\hat{k}$ in the directions of the axes of the $XYZ$ system as shown in Fig. 9-13, these vectors can be expressed in terms of the $X*Y*Z*$ system as follows:

$$\hat{\imath} = \cos\theta\hat{\imath}* + \sin\theta\hat{\jmath}* + 0\hat{k}*$$

$$\hat{\jmath} = -\sin\theta\hat{\imath}* + \cos\theta\hat{\jmath}* + 0\hat{k}* \qquad (9.86)$$

$$\hat{k} = \hat{k}*$$

Rewriting Eqs. (9.86) in a matrix form, we obtain:

$$[\hat{\imath} \ \hat{\jmath} \ \hat{k}]^T = \begin{bmatrix} \cos\theta & -\sin\theta & 0 \\ \sin\theta & \cos\theta & 0 \\ 0 & 0 & 1 \end{bmatrix}^T \begin{bmatrix} \hat{\imath}* \\ \hat{\jmath}* \\ \hat{k}* \end{bmatrix} \qquad (9.87)$$

The matrix in the above equation is $[R_Z]^T$ and each of its untransposed columns represents the components of a unit vector. Comparing the columns of the matrix $[R_Z]$ with Eqs. (9.86) shows that the first column represents the direction cosines (components) of the unit vector $\hat{\imath}$. The second and the third columns represent the direction cosines of the unit vectors $\hat{\jmath}$ and $\hat{k}$ respectively. Therefore, the columns of any rotational matrix $[R]$ represent orthogonal unit vectors. This observation is useful in building $[R]$ from user input, as explained in Example 9.6. The reader can show that the columns of any rotation matrix [Eqs. (9.28) to (9.30) or (9.53) or (9.75)] all have unit magnitude and that they are orthonormal.

### 9.3.3  General Mapping

The general mapping combines both translational and rotational mappings as shown in Fig. 9-14. In this case, the general mapping matrix $[T]$ is given by Eq. (9.70) with the submatrix $[T_3]$ set to zero, that is,

$$[T] = \begin{bmatrix} r_{11} & r_{12} & r_{13} & x_d \\ r_{21} & r_{22} & r_{23} & y_d \\ r_{31} & r_{32} & r_{33} & z_d \\ 0 & 0 & 0 & 1 \end{bmatrix} = \begin{bmatrix} [R] & \mathbf{d} \\ 0 & 1 \end{bmatrix} \qquad (9.88)$$

where $[R]$ and $\mathbf{d}$ are the rotational and translational mapping parts of $[T]$ respectively.

Due to the involvement of two coordinate systems in mapping geometric models, the correct interpretation of Eqs. (9.2) and (9.88) can be explained as follows. In Eq. (9.2), the coordinates of point $P$ are measured in the $XYZ$ coordinate system while its coordinates measured in the $X*Y*Z*$ coordinate system are given by $\mathbf{P}*$ or $[T]\mathbf{P}$. Therefore, Eq. (9.88) gives the position vector of the origin of the $XYZ$ system as $\mathbf{d}$ and its orientation as $[R]$, both measured in the

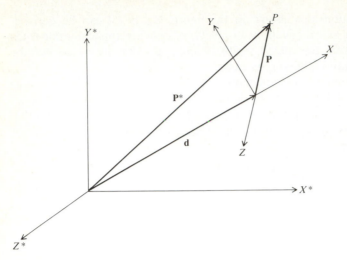

**FIGURE 9-14**
General mapping of a point.

$X*Y*Z*$ system. The columns of $[R]$ are the components of the unit vectors of the $XYZ$ system (along its axes) measured in the $X*Y*Z*$ system. If one reverses the descriptions, that is, given $\mathbf{P}$ measured in the $X*Y*Z*$ system, then the inverse of $[T]$ must be used (see Sec. 9.4). To emphasize this point, Eq. (9.2) can be written as $^B\mathbf{P} = {^B_A}[T]^A\mathbf{P}$, where $A$ and $B$ are the two coordinate systems involved in the mapping process. However, this notation is not used in the book to emphasize that we deal with one matrix $[T]$ for both mapping and transformation, but we interpret it differently.

### 9.3.4   Mappings as Changes of Coordinate System

In the previous section, 9.3.3, we have presented mappings as changing descriptions of points (or point sets) from one coordinate system to another. In this section we view the same mappings and their related matrices as changes of coordinate system. Let us assume that we are given a set of points described in a given coordinate system. Mapping this coordinate system to another is defined as changing the coordinate system so that the coordinates of the points in the transformed set with respect to the new coordinate system are the same as the coordinates of the points in the original set with respect to the original system.

Mappings as changes of coordinate system are useful in applications such as model merging or building solid models. The local coordinate system of each subassembly in model merging or of each primitive in solid modeling is positioned and oriented properly relative to a reference coordinate system. In both cases, the coordinates of the related point set stays the same with respect to its local coordinate system in its new configuration. The mapping matrix between the reference and the local coordinate systems can be derived as explained in the following section. This matrix can be used to find the coordinates of the point set in its final position, measured in the reference coordinate system as described in Sec. 9.3.3.

Figure 9-15 shows mapping as a change of coordinate system. Let us assume that the $X*Y*Z*$ reference coordinate system and point $P*$ are changed to the $XYZ$ system and point $P$ respectively such that the coordinates of $P*$ relative to the $X*Y*Z*$ system are the same as those of $P$ relative to the $XYZ$ system. In other words, the magnitudes and orientations of the vector $P*$ relative to $X*Y*Z*$ and of the vector $P$ relative to $XYZ$ are the same. The mapping matrix given by Eq. (9.88) can be used to describe the relationship between the $X*Y*Z*$ and $XYZ$ systems. In terms of Fig. 9-15, the columns of this matrix are precisely the coordinates of the unit vectors $\hat{i}$, $\hat{j}$, and $\hat{k}$, and the origin $O$ of the $XYZ$ systems, all measured relative to the reference system $X*Y*Z*$. Therefore, this matrix maps the reference coordinate system into a new system $(XYZ)$ whose origin, in reference system coordinates, is the point $(x_0, y_0, z_0)$ and whose $x$, $y$, $z$ direction cosines are given by $(i_x, i_y, i_z)$, $(j_x, j_y, j_z)$, and $(k_x, k_y, k_z)$.

Once the mapping matrix $[T]$ given by Eq. (9.88) is established, it can be used to map coordinates of points from the $XYZ$ system to the $X*Y*Z*$ as described in Sec. 9.3.3. Equation (9.2) can be used to find the coordinates of $P$ (vector $P**$ shown in Fig. 9-15) relative to the $X*Y*Z*$ system if its coordinates (vector $P$ in Fig. 9-15) relative to the $XYZ$ system are given. Figure 9-16 shows the various cases of mappings as changes of coordinate system with their corresponding mapping matrices.

**Example 9.6.** For the geometric model shown in Fig. 9-17, a user defines the $XYZ$ coordinate system shown as the MCS of the model database. The user later defines the two WCSs $\{W\}$ and $\{W_1\}$ shown to be able to construct the two circles whose centers are $C$ and $C_1$. Both $C$ and $C_1$ are the centerpoints of the respective faces of the model. Find:

(a) The mapping matrices $[T_W]$ and $[T_{W1}]$ that map points from any one of the two WCSs to the MCS.
(b) The coordinates of the centers $C$ and $C_1$ as they are stored in the model database.

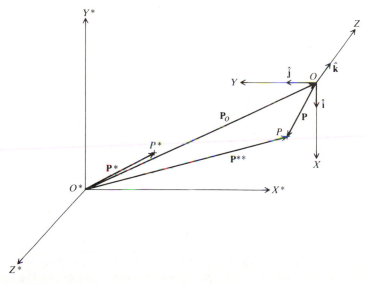

**FIGURE 9-15**
Mapping as a change of coordinate system.

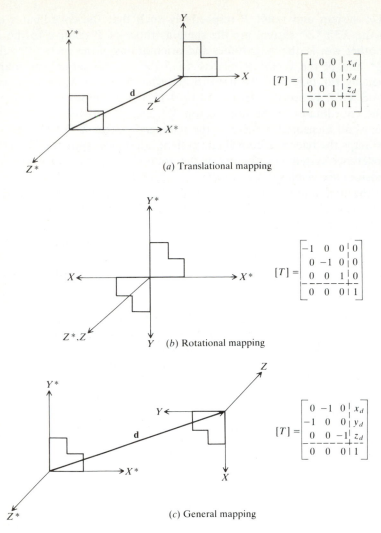

$$[T] = \begin{bmatrix} 1 & 0 & 0 & | & x_d \\ 0 & 1 & 0 & | & y_d \\ 0 & 0 & 1 & | & z_d \\ \hline 0 & 0 & 0 & | & 1 \end{bmatrix}$$

(a) Translational mapping

$$[T] = \begin{bmatrix} -1 & 0 & 0 & | & 0 \\ 0 & -1 & 0 & | & 0 \\ 0 & 0 & 1 & | & 0 \\ \hline 0 & 0 & 0 & | & 1 \end{bmatrix}$$

(b) Rotational mapping

$$[T] = \begin{bmatrix} 0 & -1 & 0 & | & x_d \\ -1 & 0 & 0 & | & y_d \\ 0 & 0 & -1 & | & z_d \\ \hline 0 & 0 & 0 & | & 1 \end{bmatrix}$$

(c) General mapping

**FIGURE 9-16**
Cases of mappings as changes of coordinate system.

### Solution

(a) Using a typical CAD/CAM system, the user creates all the lines of the model (assuming a wireframe construction). To create the circle with center $C$, the user defines the WCS $\{W\}$ by selecting the endpoints $P_1$, $P_2$, and $P_3$ of the existing lines shown such that $P_1$ defines the origin of the WCS, and the lines $P_1P_2$ and $P_1P_3$ define the $X_W$ and $Y_W$ axes respectively. The definition of this WCS is useful for two reasons. First, it defines the plane of the circle. Second, the coordinates of the center $C$ relative to this WCS are obviously (1.5, 2, 0). In order to calculate $[T_W]$, we need to calculate $[R]$ and $\mathbf{d}$ given in Eq. (9.88). The coordinates of points $P_1$, $P_2$, and $P_3$ relative to the MCS are (5, 0, 0), (5, 0, −3), and $(3, 2\sqrt{3}, 0)$ respectively. The unit vectors $\hat{\mathbf{i}}_W$, $\hat{\mathbf{j}}_W$, and $\hat{\mathbf{k}}_W$ can be calculated as

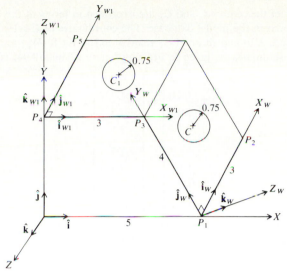

**FIGURE 9-17**
Utilizing general mapping in model construction.

follows:

$$\hat{i}_W = \frac{P_2 - P_1}{|P_2 - P_1|} = -\hat{k} \tag{9.89}$$

Notice that this relation between $\hat{i}_W$ and $\hat{k}$ is easily seen from Fig. 9-15 without substituting $P_1$ and $P_2$ into the above equation.

$$\hat{j}_W = \frac{P_3 - P_1}{|P_3 - P_1|} = \frac{1}{4}(-2\hat{i} + 2\sqrt{3}\hat{j}) = -0.5\hat{i} + 0.866\hat{j} \tag{9.90}$$

$$\hat{k}_W = \hat{i}_W \times \hat{j}_W = 0.866\hat{i} + 0.5\hat{j} \tag{9.91}$$

Writing Eqs. (9.89) to (9.91) in a matrix form, we obtain

$$[R_W] = \begin{bmatrix} 0 & -0.5 & 0.866 \\ 0 & 0.866 & 0.5 \\ -1 & 0 & 0 \end{bmatrix} \tag{9.92}$$

Substituting this equation into Eq. (9.88) and knowing that $d = P_1$, we obtain

$$[T_W] = \begin{bmatrix} 0 & -0.5 & 0.866 & \vdots & 5.0 \\ 0 & 0.866 & 0.5 & \vdots & 0 \\ -1 & 0 & 0 & \vdots & 0 \\ \cdots & \cdots & \cdots & & \cdots \\ 0 & 0 & 0 & \vdots & 1 \end{bmatrix} \tag{9.93}$$

A similar way can be followed to find $[T_{W1}]$ by using the points $P_3$ $(3, 2\sqrt{3}, 0)$, $P_4$ $(0, 2\sqrt{3}, 0)$, and $P_5$ $(0, 0, -3)$. However, by inspection we can see that $\hat{i}_{W1} = \hat{i}$, $\hat{j}_{W1} = -\hat{k}$, and $\hat{k}_{W1} = \hat{j}$. The vector $d$ is equal to $P_4$. Therefore,

$$[T_{W1}] = \begin{bmatrix} 1 & 0 & 0 & \vdots & 0 \\ 0 & 0 & 1 & \vdots & 3.464 \\ 0 & -1 & 0 & \vdots & 0 \\ \cdots & \cdots & \cdots & & \cdots \\ 0 & 0 & 0 & \vdots & 1 \end{bmatrix} \tag{9.94}$$

(b) The coordinates of the centers $C$ and $C_1$ are expressed in terms of the MCS before they are stored in the model database. The coordinates of $C$ relative to the WCS $\{W\}$ are (1.5, 2, 0) and those of $C_1$ relative to the WCS $\{W_1\}$ are (1.5, 1.5, 0). Their MCS coordinates are given by utilizing Eqs. (9.93) and (9.94) as follows:

$$C = [T_W] \begin{bmatrix} 1.5 \\ 2.0 \\ 0 \\ 1 \end{bmatrix} = \begin{bmatrix} 4.0 \\ 1.732 \\ -1.5 \\ 1 \end{bmatrix}$$

and

$$C_1 = [T_{W1}] \begin{bmatrix} 1.5 \\ 1.5 \\ 0 \\ 1 \end{bmatrix} = \begin{bmatrix} 1.5 \\ 3.464 \\ -1.5 \\ 1 \end{bmatrix}$$

The reader can verify these results by constructing the model on a CAD/CAM system and using the "verify entity" command or its equivalence.

## 9.4   INVERSE TRANSFORMATIONS AND MAPPINGS

Calculating the inverse of transformations and mappings is useful in both theoretical and practical aspects of geometric modeling. For example, using inverse mappings, some CAD/CAM systems (e.g., Computervision and GE Calma) provide their users with functions that can take an existing entity and return its coordinates relative to a given WCS. Normally, the "verify entity" command returns the coordinates relative to the MCS.

All the transformation and mapping matrices developed in Secs. 9.2 and 9.3 have inverses. These matrices have been collected into one general matrix given by Eq. (9.70). Thus, it is appropriate to find the inverse $[T]^{-1}$ of this matrix and then try to relate the result to the various matrices of the previous two sections. Because $[T]$ is partitioned into four submatrices, then $[T]^{-1} = [A]$ also has four submatrices such that

$$[T][A] = [I] \tag{9.95}$$

where $[I]$ is the identity matrix. This equation can be written as

$$\begin{bmatrix} T_1 & T_2 \\ T_3 & T_4 \end{bmatrix} \begin{bmatrix} A_1 & A_2 \\ A_3 & A_4 \end{bmatrix} = \begin{bmatrix} I & 0 \\ 0 & I \end{bmatrix} \tag{9.96}$$

Here we replaced the element $t_{44} = 1$ of $[T]$ given by Eq. (9.70) by the submatrix $T_4$ of size $1 \times 1$. The partitioned form implies four separate matrix equations, two of which are $T_1 A_1 + T_2 A_3 = I$ and $T_3 A_1 + T_4 A_3 = 0$. These can be solved simultaneously for $A_1$ and $A_3$. The remaining two equations give $A_2$ and $A_4$ and lead to the following result:

$$[T]^{-1} = \begin{bmatrix} (T_1 - T_2 T_4^{-1} T_3)^{-1} & -T_1^{-1} T_2 (T_4 - T_3 T_1^{-1} T_2)^{-1} \\ -T_4^{-1} T_3 (T_1 - T_2 T_4^{-1} T_3)^{-1} & (T_4 - T_3 T_1^{-1} T_2)^{-1} \end{bmatrix} \tag{9.97}$$

The inverse $[T]^{-1}$ given by the above equation does not take full advantage of the inherent structure of $[T]$ itself. First, $T_4$ is one element equal to 1.

Therefore, $T_4^{-1} = 1$ also. Second, in both transformation and mapping, $[T_3]$ has zero elements, that is, $[T_3] = [0 \quad 0 \quad 0]$. Moreover, $[T_1]$ and $[T_2]$ are the rotation matrix $[R]$ and the translational vector **d** in both transformation and mapping. Substituting all these properties into Eq. (9.97) gives

$$[T]^{-1} = \left[ \begin{array}{ccc|c} & [R]^{-1} & & -[R]^{-1}\,\mathbf{d} \\ \hline 0 & 0 & 0 & 1 \end{array} \right] \tag{9.98}$$

We have mentioned in Sec. 9.3.2 that the rotational matrix $[R]$ is orthogonal. Therefore, its inverse is equal to its transpose, that is,

$$[R]^{-1} = [R]^T \tag{9.99}$$

Substituting this equation into (9.98), we obtain the final form of $[T]^{-1}$ as

$$[T]^{-1} = \left[ \begin{array}{ccc|c} & [R]^T & & -[R]^T\,\mathbf{d} \\ \hline 0 & 0 & 0 & 1 \end{array} \right] \tag{9.100}$$

This equation is general and extremely useful to compute the inverse of a homogeneous transformation or mapping. The derivation we just followed to obtain Eq. (9.100) is quite general and other special derivatives of it may exist.

We may now ask ourselves the following questions. Does Eq. (9.100) agree with one's intuition for simple cases? And if it does, how? Take the example of translation. If a translational transformation or mapping is given by Eq. (9.60), the corresponding inverse is obtained by reversing the translational vector **d**, that is, by negating its components $x_d$, $y_d$, and $z_d$. Here, no rotation is involved and therefore $[R] = [R]^T = [I]$ and Eq. (9.100) yields exactly the same result. Similarly, for scaling $\mathbf{d} = 0$ and $[R]$ is a diagonal matrix, that is, $[S]$ given by Eq. (9.62). In this case, Eq. (9.98) must be used, because $[R]$ is not orthogonal, with the diagonal elements of $[R]^{-1}$ being the reciprocals of those of Eq. (9.62). For rotation, $\mathbf{d} = 0$ and $[R]^T$ to find the inverse is equivalent to negating the angle of rotation, as one would expect. The reader can check this for Eqs. (9.28) to (9.30), (9.53), and (9.75).

One last useful inverse transformation problem is that of determining the direction of the axis of rotation $\hat{n}$ and the angle of rotation $\theta$ from a given rotation matrix $[R]$. In general, the elements of $[R]$ are as shown by the top left submatrix of Eq. (9.88). They are also as shown by Eq. (9.53) in terms of $\hat{n}$ and $\theta$. To solve for $\hat{n}$ and $\theta$, we equate both forms, that is,

$$[R] = \begin{bmatrix} r_{11} & r_{12} & r_{13} \\ r_{21} & r_{22} & r_{23} \\ r_{31} & r_{32} & r_{33} \end{bmatrix} = \begin{bmatrix} n_x^2\, v\theta + c\theta & n_x n_y\, v\theta - n_z\, s\theta & n_x n_z\, v\theta + n_y\, s\theta \\ n_x n_y\, v\theta + n_z\, s\theta & n_y^2\, v\theta + c\theta & n_y n_z\, v\theta - n_x\, s\theta \\ n_x n_z\, v\theta - n_y\, s\theta & n_y n_z\, v\theta + n_x\, s\theta & n_z^2\, v\theta + c\theta \end{bmatrix}$$

$$\tag{9.101}$$

Adding elements (1, 1), (2, 2), and (3, 3) on both sides of this equation to each other and simplifying the result, we obtain

$$\theta = \cos^{-1}\left( \frac{r_{11} + r_{22} + r_{33} - 1}{2} \right) \tag{9.102}$$

Subtracting element (2, 3) from (3, 2) on both sides yields

$$n_x = \frac{r_{32} - r_{23}}{2 \sin \theta} \tag{9.103}$$

Similarly, we can find $n_y$ and $n_z$, and write

$$\hat{n} = \left(\frac{1}{2 \sin \theta}\right) \begin{bmatrix} r_{32} - r_{23} \\ r_{13} - r_{31} \\ r_{21} - r_{12} \end{bmatrix} \tag{9.104}$$

Equation (9.102) always computes a value of $\theta$ between 1 and 180°. Thus, for any axis-angle pair $(\hat{n}, \theta)$ there is another pair $(-\hat{n}, -\theta)$ that results in the same orientation in space, with the same $[R]$ describing it. Therefore, a choice has to always be made when converting from a rotation matrix into an axis-angle representation. It is also obvious from Eq. (9.101) that the smaller the angle of rotation $\theta$, the closer to zero the off-diagonal elements are, and consequently the more ill-defined the axis of rotation becomes as seen from Eq. (9.104). When $\theta = 0$ or 180°, the axis becomes completely undefined.

**Example 9.7.** An entity is rotated about the three principal axes of its MCS with equal angles of 45° each. Find the equivalent axis and angle of rotation.

*Solution.* Substituting $\alpha = \beta = \gamma = 45°$ into Eq. (9.75), we obtain

$$[R] = \begin{bmatrix} 0.5 & -0.5 & 0.707 \\ 0.854 & 0.146 & -0.5 \\ 0.146 & 0.854 & 0.5 \end{bmatrix}$$

Substituting these values into Eqs. (9.102) and (9.104), we obtain

$$\theta = 85.81°$$

and

$$\hat{n} = \begin{bmatrix} 0.679 \\ 0.281 \\ 0.679 \end{bmatrix}$$

## 9.5   PROJECTIONS OF GEOMETRIC MODELS

Databases of geometric models can only be viewed and examined if they can be displayed in various views on a display device or screen. Viewing a three-dimensional model is a rather complex process due to the fact that display devices can only display graphics on two-dimensional screens. This mismatch between three-dimensional models and two-dimensional screens can be resolved by utilizing projections that transform three-dimensional models onto a two-dimensional projection plane. Various views of a model can be generated using various projection planes.

To define a projection, a center of projection and a projection plane must be defined as shown in Fig. 9-18. To obtain the projection of an entity (a line connecting points $P_1$ and $P_2$ in the figure), projection rays (called projectors) are

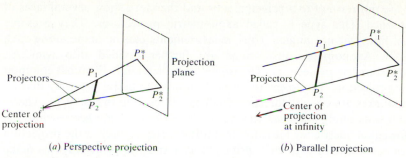

(a) Perspective projection        (b) Parallel projection

**FIGURE 9-18**
Projection definition.

constructed by connecting the center of projection with each point of the entity. The intersections of these projectors with the projection plane define the project- ed points which are connected to produce the projected entity. There are two different types of projections based on the location of the center of projection relative to the projection plane. If the center is at a finite distance from the plane, perspective projection results and all the projectors meet at the center. If, on the other hand, the center is at an infinite distance, all the projectors become parallel (meet at infinity) and parallel projection results. Perspective projection is usually a part of perspective, or projective, geometry. Such geometry does not preserve parallelism, that is, no two lines are parallel. Parallel projection is a part of affine geometry which is identical to euclidean geometry. In affine geometry, parallelism is an important concept and therefore is preserved.

Perspective projection creates an artistic effect that adds some realism to perspective views. As can be seen from Fig. 9-18a, the size of an entity is inversely proportional to its distance from the center of projection; that is, the closer the entity to the center, the larger its size is. Perspective views are not popular among engineers and draftsmen because actual dimensions and angles of objects, and therefore shapes, cannot be preserved, which implies that measurements cannot be taken from perspective views directly. In addition, perspective projection does not preserve parallelism.

Unlike perspective projection, parallel projection preserves actual dimen- sions and shapes of objects. It also preserves parallelism. Angles are preserved only on faces of the object which are parallel to the projection plane. There are two types of parallel projections based on the relation between the direction of projection and the projection plane. If this direction is normal to the projection plane, orthographic projection and views result. If the direction is not normal to the plane, oblique projection occurs.

There are two types of orthographic projections. The most common type is the one that uses projection planes that are perpendicular to the principal axes of the MCS of the model; that is, the direction of projection coincides with one of these axes. The front, top, and right views that are used customarily in engineer- ing drawings belong to this type. There are three other views that belong to this type and are typically provided by CAD/CAM systems. These are the bottom, rear, and left views. The other type of orthographic projection uses projection planes that are not normal to a principal axis and therefore show several faces of a model at once. This type is called axonometric projections. They preserve paralelism of lines but not angles. Thus, measurements can be made along each principal axis. Axonometric projections are further divided into trimetric, dimetric, and isometric projections. The isometric projection is the most common axonometric projection. The isometric projection has the useful property that all three principal axes are equally foreshortened, as will be seen in Sec. 9.5.1. Therefore measurements along the axes can be made with the same scale—thus the name: iso for equal, metric for measure. In addition, the normal to the projection plane makes equal angles with each principal axis and the principal axes make equal angles (120° each) with one another when projected onto the projection plane.

We may now ask the following question. How does the common practice of defining views, on CAD/CAM systems, of geometric models relate to both orthographic and isometric projections? Typically, a view definition requires a view origin, viewport (or view window), and a viewing direction, as shown in Fig. 9-19. The view origin defines the location of the origin of the MCS of the model (to be viewed) inside the view window. The viewing direction is the same as the projectors shown in Fig. 9-18b. The viewing plane is perpendicular to this direction and is the same as the projection plane. The viewport or view window defines the boundaries against which the view is clipped. Displayed graphics can always be zoomed in or out to scale within the viewport.

A view has a viewing coordinate system (VCS). It is a three-dimensional system with the $X_v$ axis horizontal pointing to the right and the $Y_v$ axis vertical pointing upward, as shown in Fig. 9-19. The $Z_v$ axis defines the viewing direction. The positive $Z_v$ axis has an opposite sense to the viewing direction to keep the VCS a right-handed coordinated system, even though a left-handed system may be more desirable here since its positive $Z_v$ axis is in the direction of the lines of sight emitting from the viewing eye. (This leads to the logical interpretation of

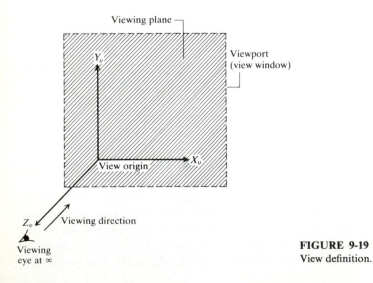

FIGURE 9-19
View definition.

larger $z$ values being further from the viewing eye.) To obtain views of a model, the viewing plane, the $X_v Y_v$ plane, is made coincident with the $XY$ plane of the MCS such that the VCS origin is the same as that of the MCS. Model views now become a matter of rotating the model with respect to the VCS axes until the desired model plane coincides with the viewing plane followed by projecting the model onto that plane. Thus, a view of a model is generated in two steps: rotate the model properly and then project it. These steps are usually performed when the user follows the view definition syntax on a given CAD/CAM system. Figure 9-20 shows the relationship between the MCS and VCS for typical views of a geometric model. We can apply the two-step procedure just described to the figure. For the front view, the $XY$ and $X_v Y_v$ plane are identical. To obtain this view, we simply project the geometry onto the viewing plane. For the top view,

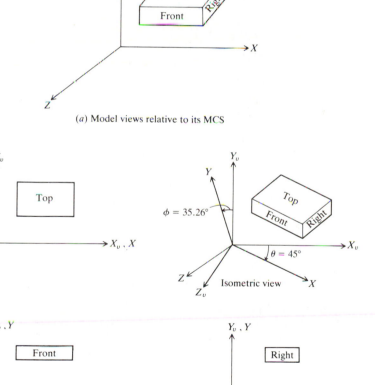

(a) Model views relative to its MCS

(b) MCS and VCS relationship

**FIGURE 9-20**
Relationship between MCS and VCS.

we must rotate the model about the $X_v$ axis by $90°$ so that the $XZ$ plane coincides with the $X_v Y_v$ plane. The other views can be obtained in a similar fashion. The MCS is shown in Fig. 9-20b as if it rotates in space with respect to the VCS exactly in the same way as the model rotates with respect to this VCS. This keeps the relationship between the model and its MCS unchanged in space. Another way to look at this observation is to say that views of a model are obtained by rotating the model and its MCS about the $X_v$ and $Y_v$ axes of the VCS.

The remainder of this section shows the underlying mathematics of projections and how views relate to geometric transformations.

### 9.5.1   Orthographic Projections

An orthographic projection (view) of a model is obtained by setting to zero the coordinate value corresponding to the MCS axis that coincides with the direction of projection (or viewing) after the model rotation. An orthographic view follows the definition shown in Fig. 9-19. To obtain the front view (see Fig. 9-20b), we only (no rotation is needed) need to set $z = 0$ for all the key points of the model. Thus, Eq. (9.70) becomes

$$[T] = \begin{bmatrix} 1 & 0 & 0 & 0 \\ 0 & 1 & 0 & 0 \\ 0 & 0 & 0 & 0 \\ 0 & 0 & 0 & 1 \end{bmatrix} \tag{9.105}$$

and Eq. (9.2) gives

$$\mathbf{P}_v = [T]\mathbf{P} \tag{9.106}$$

where $P_v$ is the point expressed in the VCS. For the front view, Eq. (9.106) gives $x_v = x$ and $y_v = y$. For the top view, the model and its MCS are rotated by $90°$ about the $X_v$ axis followed by setting the $y$ coordinate of the resulting points to zero. The $y$ coordinate is the one to set to zero because the $Y$ axis of the MCS coincides with the projection direction. In this case, $[T]$ becomes

$$[T] = \begin{bmatrix} 1 & 0 & 0 & 0 \\ 0 & 0 & -1 & 0 \\ 0 & 0 & 0 & 0 \\ 0 & 0 & 0 & 1 \end{bmatrix} \tag{9.107}$$

and Eq. (9.106) gives $x_v = x$ and $y_v = -z$. If we use the above equation to transform the MCS itself, the $X$ axis ($y = z = 0$) transforms to $x_v = x$ and the $Y$ axis ($x = z = 0$) transforms to $y_v = -z$. This result agrees with Fig. 9-20b. The right view shown in the figure can be obtained by rotating the model and its MCS about the $Y_v$ axis by $-90°$ and setting the $x$ coordinate to zero. Thus,

$$[T] = \begin{bmatrix} 0 & 0 & -1 & 0 \\ 0 & 1 & 0 & 0 \\ 0 & 0 & 0 & 0 \\ 0 & 0 & 0 & 1 \end{bmatrix} \tag{9.108}$$

which gives $x_v = -z$ and $y_v = y$.

Examining Eqs. (9.105), (9.107), and (9.108) shows that $[T]$ is a singular matrix with a column of zeros which corresponds to the MCS axis that coincides with the projection or viewing direction. These equations are obtained by rotation followed by setting a coordinate value to zero. They can also be obtained in the reverse order, that is, setting the coordinate value to zero followed by the rotation. Once the viewpoints $P_v$ are generated, they are clipped against the viewport boundaries, and then mapped into the physical device coordinate system (SCS discussed in Chap. 3) to display the view.

To obtain the isometric projection or view, the model and its MCS are customarily rotated an angle $\theta = \pm 45°$ about the $Y_v$ axis followed by a rotation $\phi = \pm 35.26°$ about the $X_v$ axis. These angles have been used for years in conventional manual drafting. In practice, the angle $\phi$ is taken as $\pm 30°$ to enable the drafting (plastic) triangles in manual construction of isometric views. The values of these angles are based on the fact that the three axes are foreshortened equally in the isometric view. This can be explained as follows. The two rotations give

$$\mathbf{P}_v = [T_x][T_y]\mathbf{P}$$

$$= \begin{bmatrix} 1 & 0 & 0 & 0 \\ 0 & \cos\phi & -\sin\phi & 0 \\ 0 & \sin\phi & \cos\phi & 0 \\ 0 & 0 & 0 & 1 \end{bmatrix} \begin{bmatrix} \cos\theta & 0 & \sin\theta & 0 \\ 0 & 1 & 0 & 0 \\ -\sin\theta & 0 & \cos\theta & 0 \\ 0 & 0 & 0 & 1 \end{bmatrix} \begin{bmatrix} x \\ y \\ z \\ 1 \end{bmatrix} \qquad (9.109)$$

Applying this equation to transform the unit vectors in the $X$ direction $[1 \ 0 \ 0 \ 1]^T$, in the $Y$ direction $[0 \ 1 \ 0 \ 1]^T$, and in the $Z$ direction $[0 \ 0 \ 1 \ 1]^T$, and ignoring the $Z$ component because we are projecting onto the $z_v = 0$ plane, we obtain respectively:

$$\begin{aligned} x_v &= \cos\phi & y_v &= \sin\phi \sin\theta \\ x_v &= 0 & y_v &= \cos\theta \\ x_v &= \sin\phi & y_v &= -\cos\phi \sin\theta \end{aligned} \qquad (9.110)$$

If the three axes are to be foreshortened equally, the magnitudes of the unit vectors given by the above equations must be equal. The first two equations give

$$\cos^2\phi + \sin^2\phi \sin^2\theta = \cos^2\theta \qquad (9.111)$$

and the last two equations give

$$\sin^2\phi + \cos^2\phi \sin^2\theta = \cos^2\theta \qquad (9.112)$$

Solving Eqs. (9.111) and (9.112) gives $\theta = \pm 45°$ and $\phi = \pm 35.26°$. The signs of the rotation angles $\theta$ and $\phi$ result in four possible orientations of isometric views. Figure 9-20$b$ shows the most common orientation where $\theta = -45°$ and $\phi = 35.26°$.

**Example 9.8.** Find the rotations that are necessary to define the front, top, right, rear, bottom, and left views of a model if the $XY$ plane of the MCS is ($a$) vertical, ($b$) horizontal.

***Solution.*** In Chap. 3, we have obtained these six views as two-dimensional and isometric views, as shown in Figs. 3-47 and 3-48. To generate these views, use the following axes and angles of rotations in the view definition command available on the CAD/CAM system being used.

| | Case (a) | | | | Case (b) | | | |
|---|---|---|---|---|---|---|---|---|
| | **Two dimensions** | | **Isometric** | | **Two dimensions** | | **Isometric** | |
| **View** | | | | | | | | |
| Front | — | — | $Y, X$ | $\theta, \phi$ | $X$ | $-90$ | $X, Y, X$ | $-90, \theta, \phi$ |
| Top | $X$ | 90 | $X, Y, X$ | $90, \theta, \phi$ | — | — | $Y, X$ | $\theta, \phi$ |
| Right | $Y$ | $-90$ | $Y, Y, X$ | $-90, \theta, \phi$ | $Y$ | $-90$ | $Y, Y, X$ | $-90, \theta, \phi$ |
| Rear | $X$ | 180 | $X, Y, X$ | $180, \theta, \phi$ | $X$ | 90 | $X, Y, X$ | $90, \theta, \phi$ |
| Bottom | $X$ | $-90$ | $X, Y, X$ | $-90, \theta, \phi$ | $X$ | 180 | $X, Y, X$ | $180, \theta, \phi$ |
| Left | $Y$ | 90 | $Y, Y, X$ | $90, \theta, \phi$ | $Y$ | 90 | $Y, Y, X$ | $90, \theta, \phi$ |

In this table, columns with $X$'s and $Y$'s show the order of rotation about the VCS axes and other columns show the corresponding angles of rotation. The angles $\theta$ and $\phi$ are $-45°$ and $35.26°$ as derived previously. The angles shown above are based on the assumption that the front and top views are the default views for cases (a) and (b) respectively. The reader is encouraged to test the proper case on the available CAD/CAM system.

## 9.5.2 Perspective Projections

One common way to obtain a perspective view is to place the center of projection along the $Z_v$ axis of the VCS and project onto the $z_v = 0$ or the $X_v Y_v$ plane. Figure 9-21 shows this case. The center of projection $C$ is placed at a distance $d$ (measured along the $Z_v$ axis) from the projection plane. In order to find the matrix $[T]$ for the case of perspective projection where the viewing eye lies on the $Z_v$ axis let us develop it from the trigonometry shown in Fig. 9-21. The viewing eye is located at the center $C$. A new coordinate system called the eye coordinate system (ECS) is introduced relative to the line of sight (see Fig. 9-21). The ECS has an origin located at the same position as the viewing eye. Its $X_e$ and $Y_e$ axes are parallel to the $X_v$ and $Y_v$ axes of the VCS. However, it is a left-handed system. The $Z_e$ axis is taken in the direction of the line of sight. Therefore, points with larger $Z_e$ values are taken to be further from the viewing eye. The ECS is useful in the hidden line and surface removal algorithms (see Chap. 10). The transformation matrix of coordinates of points from the VCS to the ECS or vice versa can be written as

$$[T] = \begin{bmatrix} 1 & 0 & 0 & \vdots & 0 \\ 0 & 1 & 0 & \vdots & 0 \\ 0 & 0 & -1 & \vdots & 0 \\ \cdots & \cdots & \cdots & & \cdots \\ 0 & 0 & 0 & \vdots & 1 \end{bmatrix} \qquad (9.113)$$

This matrix simply inverts the sign of the z coordinate. In the orthographic views, the ECS is located at infinity. It is obvious that the ECS can be replaced by the

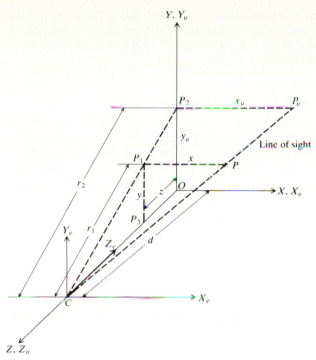

**FIGURE 9-21**
Perspective projection along the $Z_v$ axis.

VCS. In this case, points with smaller $z$ values are interpreted as being further from the viewing eye.

The figure shows the perspective projection of point $P$ as point $P_v$. To find the $y_v$ of $P_v$, the two similar triangles $COP_2$ and $CP_3P_1$ give

$$\frac{y_v}{y} = \frac{d}{d-z} = \frac{1}{1-z/d} \tag{9.114}$$

The two similar triangles $CP_vP_2$ and $CPP_1$ give $x_v$ of $P_v$ as

$$\frac{x_v}{x} = \frac{r_2}{r_1} = \frac{d}{d-z} = \frac{1}{1-z/d} \tag{9.115}$$

Rearranging Eqs. (9.114) and (9.115) to give $y_v$ and $x_v$ respectively and knowing $z_v = 0$, we can put the result in a homogeneous form as

$$\mathbf{P}_v = \begin{bmatrix} 1 & 0 & 0 & 0 \\ 0 & 1 & 0 & 0 \\ 0 & 0 & 0 & 0 \\ 0 & 0 & -1/d & 1 \end{bmatrix} \begin{bmatrix} x \\ y \\ z \\ 1 \end{bmatrix} \tag{9.116}$$

If this equation is expanded it gives $\mathbf{P}_v = [x \quad y \quad 0 \quad (1-z/d)]^T$. This would require the division of $x$ and $y$ by $(1-z/d)$ to obtain the corresponding cartesian coordinates of these homogeneous coordinates. Consequently, Eqs. (9.114) and (9.115) result. Thus, Eq. (9.116) gives the perspective projection onto the $z_v = 0$ plane when the center of projection is placed on the $Z_v$ axis at a distance $d$ from the origin. This result agrees with what was mentioned earlier that the matrix $[T_3]$ [Eq. (9.70)] produces perspective projection. If the center of projection is placed at a general point in space, the other elements of $[T_3]$ will be nonzero.

## 9.6 DESIGN AND ENGINEERING APPLICATIONS

Geometric transformations and mappings are useful in various design and engineering applications, especially those that are related to kinematics, mechanisms, linkages, and robotics. Most of these applications involve rotations and/or translations of various elements while maintaining the spatial and geometric constraints at the joints that connect these elements. The various configurations of these elements can be used to study their effects on the motion or kinematics of the corresponding geometric models, or they can be used to animate the motion of the models for visualization purposes.

When more than one degree-of-freedom model is to be transformed, such as in a robotics system, kinematic analysis is required to determine the relative motion between the elements. This motion can then be executed via geometric transformations. Due to the repetitive work to transform the elements (such as incremental rotation or translation), programming is usually more efficient than just typing commands. The following example shows how to use rotations in relation to a simple mechanism.

**Example 9.9.** Figure 9-22 shows a representation of a slider-crank mechanism. Find the locus of point $D$, the midpoint of the connecting rod $BC$. What is the angle $\theta$ at which the tangent to the locus at $D$ becomes horizontal?

*Solution.* The idea here is to find enough points on the locus of $D$ and then connect them with a closed B-spline. This is achieved by constructing the mechanism for $\theta = 0$. Then the crank $AB$ is rotated incrementally, say by $\Delta\theta = 10°$, and the mechanism is constructed for each $\theta$ (10, 20, ...). At each configuration the position of point $D$ is recorded by simply inserting a point at the origin of the line $BC$. The resulting positions are then connected with a closed B-spline command to produce the locus. To find the angle $\theta$ at which the tangent is horizontal, we construct a few tangents to the locus where they may be horizontal. Using the "measure angle" command, the angle that each tangent makes with the horizontal can be obtained

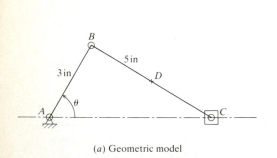

(*a*) Geometric model

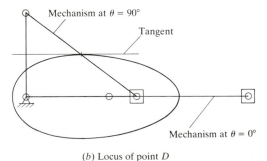

(*b*) Locus of point $D$

**FIGURE 9-22**
Slider-crank mechanism.

and compared to zero. The closest angle to zero with an allowable error gives the horizontal tangent and the corresponding angle is the solution. Figure 9-22b shows the locus and the tangent within an error of $2.6 \times 10^{-6}$°. The angle $\theta$ that corresponds to this tangent is 90°.

The reader can extend this method to study the effect of the lengths of the crank and the connecting rod on the locus and the tangent. Programming is useful in this study (refer to Chap. 15).

## PROBLEMS

### Part 1: Theory

**9.1.** A general curve such as a Bezier or B-spline is to be translated. Does translating the control points and then generating the curve give the same result as translating the original curve or not? Prove your answer.

**9.2.** Develop the translational transformation equation for a Hermite bicubic spline surface, a bicubic Bezier surface, and a bicubic B-spline surface. How can you extend the results to a cubic hyperpatch?

**9.3.** Derive the relationship between a point $P$ and its scaled counterpart $P^*$ if $P$ is scaled uniformly about a given point $Q$ which is not the origin.

**9.4.** How can a Bezier curve, B-spline curve, Hermite bicubic surface, Bezier surface, and B-spline surface be scaled uniformly?

**9.5.** Show that Eqs. (9.15) to (9.18) can reduce to Eqs. (9.13) and (9.14); that is, show that reflection relative to a coordinate system is a special case of general reflection.

**9.6.** Develop the reflection transformation equations for Bezier and B-spline curves and surfaces as well as a Hermite bicubic surface. Carry the developments for the case of reflection through a general point.

**9.7.** Figure P9-7 shows a cube of length 2 in. The cube is rotated an angle $\theta = 30$° about the cube diagonal $OD$. If point $B$ is the midpoint of side $AD$, find the coordinates of points $A$, $B$, and $C$ before and after rotation. Verify your answer by solving the problem on your CAD/CAM system.

**9.8.** Show how the homogeneous representation can help represent points at infinity and can also be used to force parallel lines to intersect.

**9.9.** Show that parallel and perpendicular lines transform to parallel and perpendicular lines.

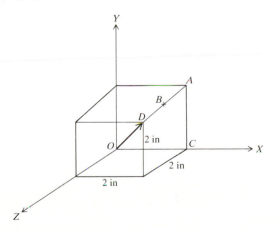

**FIGURE P9-7**

**9.10.** Show that the midpoint of a line transforms to the midpoint of the transformed lines.

**9.11.** A point is rotated about the $Z$ axis by two successive angles $\theta_1$ and $\theta_2$. Show that this is equivalent to rotating the point about the same axis once with an angle $\theta = \theta_1 + \theta_2$.

**9.12.** Show that:
(a) Translation is commutative.
(b) Mirror and two-dimensional rotation about the $Z$ axis are not commutative.
(c) Scaling and two-dimensional rotation about the $Z$ axis are commutative.
(d) Three-dimensional rotations are not commutative.

**9.13.** Given a point $P = (2, 4, 8)$ and using the homogeneous representation:
(a) Calculate the coordinates of the transformed point $P*$ if $P$ is rotated about the $X$, $Y$, and $Z$ axes by angles 30, 60, and 90° respectively.
(b) If the point $P*$ obtained in part (a) is to be rotated back to its original position, find the corresponding rotation matrix. Verify your answer.
(c) Calculate $P*$ if $P$ is translated by $\mathbf{d} = 3\hat{\mathbf{i}} - 4\hat{\mathbf{j}} - 5\hat{\mathbf{k}}$ and then scaled uniformly by $s = 1.5$.
(d) Calculate the orthographic projection $P_v$ of $P$.
(e) Calculate the perspective projection $P_v$ of $P$ if the center of projection is at a distance $d = 10$ in from the origin along the $Z_v$ axis.

**9.14.** Given three points $P_1$, $P_2$, and $P_3$ that belong to a geometric model and given three other points $Q_1$, $Q_2$, and $Q_3$, find the transformation matrix $[T]$ that:
(a) Transforms $P_1$ to $Q_1$.
(b) Transforms the direction of the vector $(\mathbf{P}_2 - \mathbf{P}_1)$ into the direction of the vector $(\mathbf{Q}_2 - \mathbf{Q}_1)$.
(c) Transforms the plane of the three points $P_1$, $P_2$, and $P_3$ into the plane of $Q_1$, $Q_2$, and $Q_3$.
This problem is sometimes called "three-point" transformation. It is useful to move two geometric models, mainly solids, to coincide with one another or to position entities in a geometric model.

**9.15.** Figure P9-15 shows the rotation of a point $P$ about an arbitrary axis of rotation that passes through the origin and lies in the $XZ$ plane. Derive the rotation matrix $[R]$ for this case. Verify your answer by substituting the proper values in the general matrix $[R]$ given by Eq. (9.53).

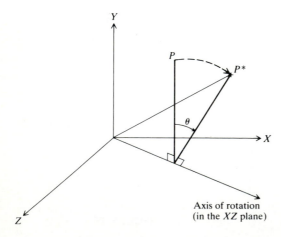

Axis of rotation
(in the $XZ$ plane)     **FIGURE P9-15**

## Part 2: Laboratory

**9.16.** Find the view definitions required by your CAD/CAM system to define the six two-dimensional views as well as the six isometric views described in Example 9.8.

**9.17.** Show that three-dimensional rotations are not commutative. Take a sequence of 90° rotations about the three principal axes and permutate them.

**9.18.** Redo Example 9.5 on your CAD/CAM system. Use $\hat{n} = 0.5\hat{i} + 0.707\hat{j} + 0.5\hat{k}$ and $P = (1, 3, 5)$.

**9.19.** A line is connecting the origin of the MCS of a model and a point $P(1, 2, 3)$. Find three different ways to rotate the line so that it coincides with the $Z$ axis of the MCS.

## Part 3: Programming

**9.20.** Write a program to implement Eq. (9.70). Use the form $\mathbf{P}^* = [T]\mathbf{P}$ where $\mathbf{P}$ is a vector of points, tangent vectors, or any other vectors of interest. For example, $\mathbf{P}$ could be four points, that is, $[\mathbf{P}_1 \quad \mathbf{P}_2 \quad \mathbf{P}_3 \quad \mathbf{P}_4]^T$, or it could be $[\mathbf{P}_0 \quad \mathbf{P}_1 \quad \mathbf{P}_0' \quad \mathbf{P}_1']^T$ as for the Hermite cubic spline. Write the program for rotation about the $Z$ axis, translation, and scaling uniformly.

**9.21.** Using the program developed in Prob. 9.20, write a program to translate a cubic spline curve, a Bezier curve, and a B-spline curve.

## BIBLIOGRAPHY

Demel, J. T., and M. J. Miller: *Introduction to Computer Graphics*, Brooks/Cole Engineering Division, Monterey, Calif., 1984.

Encarnacao, J., and E. G. Schlechtendahl: *Computer-Aided Design: Fundamental and System Architectures*, Springer-Verlag, New York, 1983.

Faux, I. D., and M. J. Pratt: *Computational Geometry for Design and Manufacture*, John Wiley, New York, 1981.

Foley, J. D., and A. van Dam: *Fundamentals of Interactive Computer Graphics*, Addison-Wesley, Reading, Mass., 1982.

Giloi, W. K.: *Interactive Computer Graphics: Data Structures, Algorithms, Languages*, Prentice-Hall, Englewood Cliffs, N.J., 1978.

Groover, M. P., and E. W. Zimmers: *CAD/CAM: Computer-Aided Design and Manufacturing*, Prentice-Hall, Englewood Cliffs, N.J., 1984.

Harrington, S.: *Computer Graphics: A Programming Approach*, McGraw-Hill, New York, 1983.

Harris, D.: *Computer Graphics and Application*, Chapman and Hall, New York, 1984.

Mortenson, M. E.: *Geometric Modeling*, John Wiley, New York, 1985.

Newman, W. M., and R. F. Sproull: *Principles of Interactive Computer Graphics*, 2d ed., McGraw-Hill, New York, 1979.

Park, C. S.: *Interactive Microcomputer Graphics*, Addison-Wesley, Reading, Mass., 1985.

Rogers, D. F., and J. A. Adams: *Mathematical Elements for Computer Graphics*, McGraw-Hill, New York, 1976.

# CHAPTER
# 10

# VISUAL
# REALISM

## 10.1 INTRODUCTION

One of the most recognized aids of CAD/CAM is its ability to provide its users with visual displays of the objects and scenes they model on CAD/CAM systems. Visualization has always been recognized as the most effective means of communicating new ideas and designs among designers, engineers, and others. There is always the saying that a picture is worth a thousand words. The truth of this saying may be related to the fact that an estimated 50 percent of the brain's neurons are associated with vision. Virtually all CAD/CAM-related application programs present their results to their users in a graphical form by coverting the corresponding numerical results to such forms.

Visualization embraces both image understanding and image synthesis; that is, visualization is a tool both for interpreting image data fed into a computer and for generating images from complex multidimensional data sets. Therefore, two types of visualization can be identified: visualization in geometric modeling and visualization in scientific computing. The former is related to displaying geometric models of objects while the latter is concerned with displaying results related to science and engineering. The most common methods of visualizing geometric models are projection and shading. Orthographic projections have been used in engineering drawings for years to visualize new designs. Isometric and perspective projections are, in addition, used in CAD/CAM due to the ease of (automatically) generating them and the rich visual information they provide about related designs and objects. Although orthographic projections are the oldest means of communicating engineering designs, they are difficult to read and require training for some time. They also do not convey any appearance characteristics (surface color and texture) of the related designs. Shaded images with a high degree of realism can make users believe that the images are of real

526

objects and not of synthetic ones (geometric models). High-quality shaded images, in many instances, provide an easy, more effective, and less expensive way of reviewing various design alternatives than building models and prototypes. Designs of many mechanical parts, automobile bodies, aircraft parts, ship hulls, shoes, and others are greatly enhanced by studying their corresponding shaded images.

Visualization in geometric modeling can be quite useful in determining spatial relationships in design applications. For example, it can be used for interference detection between mating parts in assemblies. By shading the parts with different colors and using shadowing and transparency, the designer can easily detect any undesired hidden interferences. Similarly, in the design of complex surfaces, such as those used in automobile bodies and aircraft frame structures, shading with various texture characteristics can help detect any undesired abrupt changes in surface curvature or smoothness which directly controls the aerodynamic characteristics of the bodies and structures.

Visualization in scientific computing is in itself viewed as a method of geometric modeling. It transforms the numerical data into image display, enabling users to observe their simulations and computations. Visualization offers a method of seeing the unseen. Visualization in scientific computing is of great interest to scientists and engineers. The need for such visualization is becoming more and more crucial with the increasing power and speed of supercomputers and parallel processing, which means more and more results to deal with. Interactive visual computing is an important process of visualization. It enables scientists and engineers to dynamically modify computations while they are occurring. This process and its related immediate visual feedback can help designers and researchers gain insight into design processes and anomalies, and can help them discover computational and modeling errors.

Many existing CAD/CAM applications utilize one form or another of visualization. Animation and simulation are considered a popular form. For example, flight and navigation simulators have been in existence for some time. The motion of various mechanisms can be simulated by generating and animating different frames of their corresponding geometric models. These frames are usually generated according to the kinematic and dynamic equations governing the mechanisms. Robot simulations and trajectory planning and generation on CAD/CAM systems can be achieved in a similar fashion. Verification of NC tool paths is another form of useful visualization where the numerical data describing the path of a cutting tool is used to generate and display the motion of the tool to verify its correctness. Other applications of visualization in scientific computing include the display of results of finite element analysis (displacements, deformed shaped, and stress contours), heat-transfer analysis (temperature and heat flux contours), computational fluid dynamics (simulation of the propagation of boundary layers and/or combustion processes), and structural dynamics and vibration (display and animate mode shapes). Another application is in the medical field in relation to hip-replacement operations. Custom hips can now be fabricated prior to surgery by accurate measurements using noninvasive three-dimensional imaging, thus reducing the number of postoperative body rejections from 30 to only 5 percent.

The main problem in visualization is the display of three-dimensional objects and scenes on two-dimensional screens. How can the third dimension, the depth, be displayed on the screen? How can the visual complexities of the real environment such as lighting, color, shadows, and texture be represented as image attributes? What complicates the display of three-dimensional objects even further is the centralized nature of the databases of their geometric models. If we project a complex three-dimensional model onto a screen, we get a complex maze of lines and curves. To interpret this maze, curves and surfaces that cannot be seen from the given viewpoint must be removed. Hidden line and surface removal eliminates the ambiguities of the displays of three-dimensional models and is considered the first step toward visual realism.

Various approaches to achieve visual realism exist. They are directly related to the types of geometric models utilized to represent three-dimensional objects. Thus, one would expect an upward trend in the efficiency and automation level of these approaches as the geometric modeling techniques have advanced from wireframes, to surfaces, to solids. Among the existing visualization approaches are parallel projections, perspective projections, hidden line removal, hidden surface removal, hidden solid removal, and the generation of shaded images of models and scenes. Parallel projections, especially orthographic projections, represent the oldest approach. They are easy to generate, as discussed in Chap. 9, take less computer time to generate compared to, say, shaded images, and are widely used in engineering drawings. Perspective projection is widely used in architecture engineering and construction (AEC). Shaded images can only be generated for surface and solid models. Shading is a two-step process. It begins by eliminating the hidden surfaces or solids first, and then shades the visible portions only. Shaded images represent the highest level of visual realism.

Increasing the level of realism in a model by, say, developing shaded images adds to its complexity. The creation and maintenance of such a model become complex as well. For example, generating a high-resolution shaded image of a fairly complex model with various types of lighting (source and ambient lights) can take a considerable number of CPU minutes and may degrade the CAD/CAM system's performance (slow down other concurrent users). Storing and/or retrieving models with these kinds of shaded images are usually slow. Generating real-time images, therefore, requires powerful computers with most of the shading algorithms embedded into the hardware. Shading firmware is commonly provided by various existing CAD/CAM systems to improve the performance of the shading algorithms.

The direct or brute force approach to hidden line/surface/solid removal requires large amounts of computing times. The position of each line or surface has to be compared with that of all the others, leading to a square-law combinatorial explosion. Therefore, extensive studies of existing algorithms have been pursued in an attempt to find the common underlying principles among these algorithms. Two main general principles can be identified. These are coherence and sorting. Objects and their related geometric models are more than a set of random discontinuities. They have some consistency both over the area of the image and over the time, from one frame to the next. Thus using this coherence between various frames can improve the efficiency of visualization algorithms.

Similarly, all these algorithms sort or search through collections of surfaces, edges, or objects according to various criteria, finally discovering the visible items and displaying them. Therefore, effective sorting is a key to good hidden line/ surface/solid removal.

The remainder of the chapter presents more details on the most common visualization approaches. Section 10.2 discusses the common practice that makes orthographic projection a realistic visualization approach. The other sections cover the other visualization algorithms.

## 10.2 MODEL CLEAN-UP

Model clean-up is perhaps considered the oldest most elementary method to achieve visual realism. As discussed in Chap. 3, the clean-up process begins by generating the proper orthographic views (projections). Visual realism is added to each view separately (by invoking the drafting mode) by eliminating the hidden lines in the view, changing their fonts from solid to dashed, and/or adding dimensions and texts. A major advantage of the manual model clean-up is the control of the user over which entities should be removed and which should be dashed. The complete elimination of hidden lines may result in loss of depth information. The major disadvantage is that it is a tedious, time-consuming, and error-prone process.

While the syntax details of the model clean-up differ from one CAD/CAM system to another, the basic core of the process is the same and can be described as follows. First, the user defines a drawing and orthographic views that make up the drawing. The user then creates the model database while in the model mode (refer to Chap. 3). For each view of the drawing, the user turns the drafting mode on followed by the clean-up process. The drafting mode ensures that the clean-up activities are applicable only to the view where the mode is on. This is necessary because hidden lines in one view may be visible in another. In addition, dimensions added to one view should not be projected and displayed in the other views. They would if added to the view in the model mode, that is, as model entities. Some CAD/CAM systems do not allow users to add dimensions if the model mode is on, that is, the drafting mode is off. Figures 10-1 and 10-2 show a typical model before and after clean-up to appreciate the added realism that results from this process.

The "repaint" command offered by all CAD/CAM systems is a useful tool in a typical model clean-up session. After each few commands of clean-up, the user can redraw the screen to update its display to check the results of these commands. The clean-up related commands consist mainly of the two commands "blank entity" and "change font." The first command is used to hide undesired entities and the second changes solid lines to other fonts (typically dashed). Other clean-up commands and procedures are not covered in this book because they are system dependent. Interested readers are referred to their system documentation.

In a typical model clean-up session, the most inconvenience a user might encounter results from dealing with overlapped and partially hidden lines. If, for example, $n$ hidden lines overlap, the "blank entity" command must be used $n$

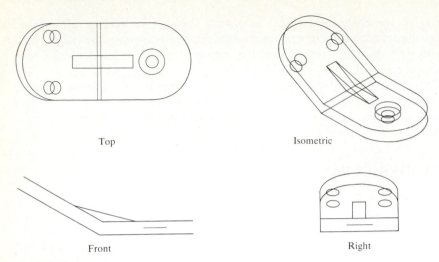

Top

Isometric

Front

Right

**FIGURE 10-1**
A typical model before the clean-up process.

times to hide them, or the command must be used $(n - 1)$ times followed by using the "change font" command once to dash the remaining line. The "repaint" command is usually used during the intermediate uses of these commands. If a line is mistakenly hidden, it can be made visible again in the view where clean-up

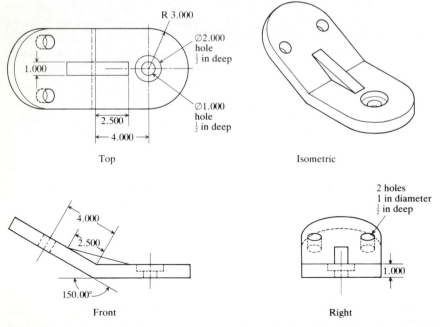

R 3.000

Ø2.000
hole
$\frac{1}{2}$ in deep

1.000

Ø1.000
hole
$\frac{1}{2}$ in deep

2.500

4.000

Top

Isometric

2 holes
1 in diameter
$\frac{1}{2}$ in deep

4.000

2.500

1.000

150.00°

Front

Right

**FIGURE 10-2**
Model of Fig. 10-1 after clean-up.

takes place by using the "unblank entity" command. To achieve this reverse process successfully, the entity must be visible in other views to be able to digitize it for the command unless it is referred to by name (tag name).

The inconvenience associated with the partially hidden lines stems from the fact that the "trim entity" command, which is expected to be used, affects the entity permanently in all views; that is, it has a global effect on the entity, not just a local effect in a given view. This would force the user to copy the partially hidden line (model entity) first while the drafting mode is on (to obtain a drafting entity) and then to trim the resulting line. Some CAD/CAM systems may have a special solution to this problem.

While the manual model clean-up as described above is applicable to wireframe models, its extension to surface models becomes practically useless. In some instances, the surface may have to be replaced by its boundaries (wireframe entities) before the clean-up process begins. Surface manual clean-up is usually not common because surfaces are seldom used in engineering drawings. Automatic hidden line and hidden surface removal algorithms, discussed in the following sections, are typically used instead.

## 10.3  HIDDEN LINE REMOVAL

Since the early development of computer graphics, there is always a demand for images (of objects) enhanced by removing the hidden parts that would not be seen if objects were constructed from opaque material in real life. Edges and surfaces of a given object may be hidden (invisible) or visible in a given image depending on the viewing direction. The determination of hidden edges and surfaces is considered one of the most challenging problems in computer graphics. Its solution typically demands large amounts of computer time and memory space. Techniques to reduce these demands and improve efficiencies of related algorithms exist and are discussed here.

The solution to the problem of removing hidden edges and surfaces draws on various concepts from computing, mainly sorting, and geometric modeling, mainly projection and intersection. This problem can also be viewed as a visibility problem. Therefore, a clear understanding of it and its solution is useful and can be extended to solve relevant engineering problems. Consider, for example, the vision and path planning problems in robotics applications. In the vision problem, the camera location and orientation provide the viewing direction which, in turn, can be used to determine the hidden edges and surfaces of objects encountered in the robot working environment. In the path planning problem, the knowledge of when a given surface changes from visible to hidden (via finding silhouette edges and curves as seen later in this section) can be utilized to find the minimum path of the robot end effector. Points on the surface where its status changes from visible to invisible or vice versa can be considered as critical points which the path planning algorithm can use as an input. Another example is the display of finite element meshes where the hidden elements are removed. In this case, each element is treated as a planar polygon and the collection of elements that forms the meshed object, from a finite element point of view, forms a polyhedron from a computer graphics viewpoint.

A wide variety of hidden line and hidden surface removing (visibility) algorithms is in existence today. The development of these algorithms is influenced by the types of graphics display devices they support (whether they are vector or raster) and by the type of data structure or geometric modeling they operate on (wireframe, surface, or solid modeling). Some algorithms utilize parallel, over the traditional serial, processing to speed up their execution. The formalization and generalization of these algorithms are useful and are required if one attempts to design and build special-purpose hardware to support hidden line and hidden surface removal, which is not restricted to a single algorithm. However, it is not a trivial task to convert the different algorithmic formulations into a form that allows them to be mapped onto a generalized scheme.

Algorithms that are applied to a set of objects to remove hidden parts to create a more realistic image are usually classified into hidden line and hidden surface algorithms. The former supports line-drawing devices such as vector displays and plotters, while the latter supports raster displays. Hidden line algorithms can, of course, be used with raster displays because they support line drawings. However, hidden surface algorithms are not applicable to vector displays. From a geometric modeling point of view, this classification is both confusing and deceiving. Hidden line removal does not mean (as the name may imply) that it is applicable to wireframe models only. Similarly, hidden surface removal is not only applicable to surface models. As a matter of fact, algorithms to remove hidden parts from an image cannot be applied to wireframe or surface models directly. They require an unambiguous data structure that represents an object as orientable faces. This means that each face has a surface normal with a consistent direction (say positive if face edges are input in a counterclockwise direction); that is, polyhedral objects are represented by orientable flat polygons. These polygons can be obtained from a wireframe, surface, or solid model. Users would have to input extra information to identify faces and orientation for wireframes or orientation for surface models. Solid models provide such information automatically. In spite of the above misleading classification, we have hidden line, hidden surface, and hidden solid removal sections in this chapter. This is merely done to reflect the historical order of the development of the related algorithms.

Hidden line and hidden surface algorithms have been classified as object-space methods, image-space methods, or a combination of both (hybrid methods). Image-space algorithms can be further divided into raster and vector algorithms. The raster algorithms use a pixel-matrix representation of the image and the vector algorithms use endpoint coordinates of line segments in representing the image. An object-space algorithm utilizes the spatial and geometrical relationships among the objects in the scene to determine hidden and visible parts of these objects. An image-space algorithm, on the other hand, concentrates on the final image to determine what is visible, say, within each raster pixel in the case of raster displays. Most hidden surface algorithms use raster image-space methods while most hidden line algorithms use object-space methods.

The two approaches (object-space and image-space) to achieve visual realism exhibit different characteristics. Object-space algorithms are more accurate than image-space algorithms. The former perform geometric calculations (such as intersections) using the floating-point precision of the computer hardware, while the latter perform calculations with accuracy equal to the resolution of the display screen used to present them. Therefore, enlargement of an object-space image does not degrade its quality of display as does the enlargement of an image-space image. As the complexity of the scene increases (large numbers of objects in the scene), the computation time grows faster for object-space algorithms than for image-space algorithms.

## 10.3.1    Visibility of Object Views

The visibility of parts of objects of a scene depends on the location of the viewing eye, the viewing direction, the type of projection (orthogonal or perspective), and the depth or the distance from various faces of various objects in the scene to the viewing eye. The hidden line removal of perspective views (see Fig. 9-21) is a fairly complex problem to solve. Many lines of sight (rays) from the viewing eye must be considered, and their points of intersection with objects' faces have to be calculated. The complexity of the problem is considerably reduced if orthographic views are utilized because no intersections would be necessary. Therefore, it is common practice to apply the perspective transformation given by Eq. (9.116) to the set of the points in the scene, and then apply orthographic hidden line visibility algorithms to the resulting (transformed) set of points. This is equivalent to saying that the orthographic viewing of the transformed (perspective) objects is identical to the perspective viewing of the original (untransformed) objects. Hence, only orthographic hidden line algorithms are discussed in this book.

The depth comparison is the central criterion utilized by hidden line algorithms to determine visibility. Depth comparisons are typically done after the proper view transformation given by Eqs. (9.105) and (9.116) for orthographic and perspective projections respectively. While these two equations destroy the depth information (the $z_v$ coordinate of projected points) to generate views, such information can be saved [by replacing the element $t_{33}$ of $[T]$ in both equations by 1 instead of the current 0; for Eqs. (9.107) and (9.108), set $t_{32}$ and $t_{31}$ respectively to 1] for depth comparisons by hidden line algorithms.

The depth comparison determines if a projected point $P_{1v}$ $(x_{1v}, y_{2v})$ in a given view obscures another point $P_{2v}$ $(x_{2v}, y_{2v})$. This is equivalent to determining if the two original corresponding points $P_1$ and $P_2$ lie on the same projector as shown in Fig. 10-3 (the MCS and VCS are shown as the $XYZ$ and the $X_v Y_v Z_v$ systems respectively). For orthographic projections, projectors are parallel. Therefore, two points $P_1$ and $P_2$ are on the same projector if $x_{1v} = x_{2v}$ and $y_{1v} = y_{2v}$. If they are, a comparison of $z_{1v}$ and $z_{2v}$ decides which point is closer to the viewing eye. Utilizing the VCS shown in Figs. 9-19 and 10-3, the point with the larger $z_v$ coordinate lies closer to the viewer. Applying this depth comparison to points $P_1$, $P_2$, and $P_3$ of Fig. 10-3 shows that point $P_1$ obscures $P_2$ (i.e., $P_1$ is visible and $P_2$ is hidden) and $P_3$ is visible. If the depth comparison is to be performed utilizing the ECS (see Fig. 9-21), the transformation matrix given by Eq. (9.113) must be applied to the projected points before the comparison. This simply reverses the comparison to say that points with smaller $z_v$ coordinates lie closer to the viewer.

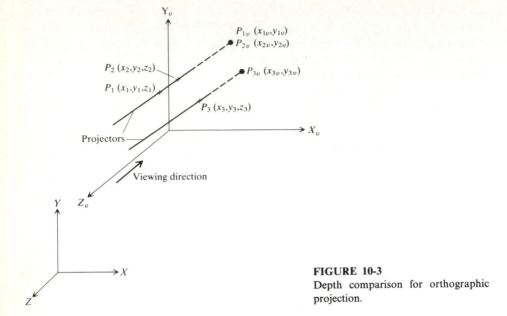

**FIGURE 10-3**
Depth comparison for orthographic projection.

## 10.3.2 Visibility Techniques

If the depth comparison criterion is used solely with no other enhancements, the number of comparisons grows rapidly [for $n$ points, $\binom{n}{2}$ tests are required] which leads to difficulties storing and managing the results by the corresponding hidden line algorithm. As a result, the algorithm might be slow in calculating the final image. Various visibility techniques exist to alleviate these problems. In general, these techniques attempt to establish relationships among polygons and edges in the viewing plane. The techniques normally check for overlapping of pairs of polygons (sometimes referred to as lateral comparisons) in the viewing plane (the screen). If overlapping occurs, depth comparisons are used to determine if part or all of one polygon is hidden by another. Both the lateral and depth comparisons are performed in the VCS.

**10.3.2.1 MINIMAX TEST.** This test (also called the overlap or bounding box test) checks if two polygons overlap. The test provides a quick method to determine if two polygons do not overlap. It surrounds each polygon with a box by finding its extents (minimum and maximum $x$ and $y$ coordinates) and then checks for the intersection for any two boxes in both the $X$ and $Y$ directions. If two boxes do not intersect, their corresponding polygons do not overlap (see Fig. 10-4). In such a case, no further testing of the edges of the polygons is required.

If the minimax test fails (two boxes intersect), the two polygons may or may not overlap, as shown in Fig. 10-4. Each edge of one polygon is compared against all the edges of the other polygon to detect intersections. The minimax test can be applied first to any two edges to speed up this process.

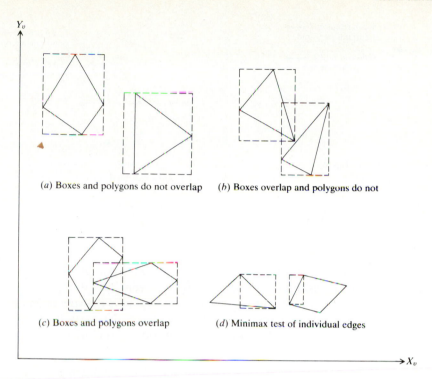

(a) Boxes and polygons do not overlap    (b) Boxes overlap and polygons do not

(c) Boxes and polygons overlap    (d) Minimax test of individual edges

**FIGURE 10-4**
Minimax tests for typical polygons and edges.

The minimax test can be applied in the $Z$ direction to check if there is no overlap in this direction. In all tests, finding the extents themselves is the most critical part of the test. Typically, this can be achieved by iterating through the list of vertex coordinates of each polygon and recording the largest and the smallest values for each coordinate.

**10.3.2.2 CONTAINMENT TEST.** Some hidden line algorithms depend on whether a polygon surrounds a point or another polygon. The containment test checks whether a given point lies inside a given polygon or polyhedron. There are three methods to compute containment or surroundedness. For a convex polygon, one can substitute the $x_v$ and $y_v$ coordinates of the point into the line equation of each edge. If all substitutions result in the same sign, the point is on the same side of each edge and is therefore surrounded. This test requires that the signs of the coefficients of the line equations be chosen correctly.

For nonconvex polygons, two other methods can be used. In the first method, we draw a line from the point under testing to infinity as shown in Fig. 10-5a. The semi-infinite line is intersected with the polygon edges. If the intersection count is even, the point is outside the polygon ($P_2$ in Fig. 10-5a). If it is odd, the point is inside ($P_1$ in the figure). If one of the polygon edges lies on the semi-infinite line, a singular case arises which needs special treatment to guarantee the consistency of the results.

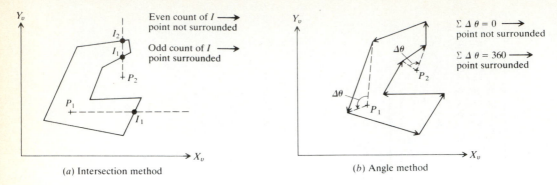

(a) Intersection method

(b) Angle method

**FIGURE 10-5**
Containment test for nonconvex polygons.

The second method for nonconvex polygons (Fig. 10-5b) computes the sum of the angles subtended by each of the oriented edges as seen from the test point ($P_1$ or $P_2$). If the sum is zero, the point is outside the polygon. If the sum is $2\pi$ or $-2\pi$, the point is inside. The minus sign reflects whether the vertices of the polygon are ordered in a clockwise direction instead of counterclockwise.

**10.3.2.3 SURFACE TEST.** This test (also called the back face or depth test) provides an efficient method for implementing the depth comparison. Figure 10-6a shows that face $B$ obscures part of face $A$. In this case the equation of a plane is used to perform the test. The plane equation is given by

$$ax_v + by_v + cz_v + d = 0 \qquad (10.1)$$

If a given point $(x_v, y_v, z_v)$ is not on the plane, the sign of the left-hand side of the above equation is positive if the point lies on one side of the plane and negative if it lies on the other side. The equation coefficients $a$, $b$, $c$, and $d$ can be arranged so that a positive value indicates a point outside the plane. The plane equation can also be used to compute the depth $z_v$ of a face at a given point $(x_v, y_v)$. The depths of two faces can, therefore, be computed at given points to decide which one is closer to the viewing eye.

Another important use of the plane equation in hidden line removal is achieved by using the normal vector to the plane. The first three coefficients $a$, $b$, and $c$ of Eq. (10.1) represent the normal to the plane and the vector $[a, b, c, d]$ represents the homogeneous coordinates of this normal. The coefficient $d$ is found by knowing a point on the plane. In Chap. 6, we have discussed the various ways of finding the plane equation. Figure 10-6b shows how the normal to a face can be used to decide its visibility. The basic idea of the test (Fig. 10-6b) is that faces whose outward normal vector points toward the viewing eye are visible (face $F_1$) while others are invisible (face $F_2$). This test is implemented by calculating the dot product of the normal vector $\mathbf{N}$ and the line-of-sight vector $\mathbf{S}$ (Fig. 10-6b) as

$$\mathbf{N} \cdot \mathbf{S} = |\mathbf{N}| \, |\mathbf{S}| \cos \theta \qquad (10.2)$$

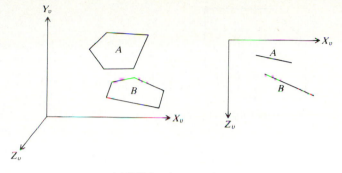

(a) Utilizing plane equation

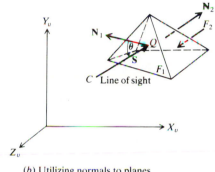

(b) Utilizing normals to planes

**FIGURE 10-6**
Surface test.

If, in this equation, we assume that **N** points away from the solid and $\theta$ is measured from **N** to **S**, the dot product is positive when **N** points toward the viewing eye or when the face and its edges are visible. The right-hand side of Eq. (10.2) gives the component of **N** along the direction of **S**. For orthographic projection, this direction coincides with the $Z_v$ axis. Thus the surface test can be stated as follows. Faces whose normal has a positive component in the $Z_v$ direction are visible and those whose normal has a negative $Z_v$ component are not visible. The surface test by itself cannot solve the hidden line problem except for single convex polyhedra. Even for convex polyhedra, the test may fail for perspective projection if more than one polyhedron exists in the scene.

**10.3.2.4  COMPUTING SILHOUETTES.** A set of edges that separates visible faces from invisible faces of an object with respect to a given viewing direction is called silhouette edges (or silhouettes). The signs of the $Z_v$ components of normal vectors of the object faces can be utilized to determine the silhouette. An edge that is part of the silhouette is characterized as the intersection of one visible face and one invisible face. An edge that is the intersection of two visible faces is visible, but does not contribute to the silhouette. The intersection of two invisible faces produces an invisible edge. Figure 10-7a shows the silhouette of a cube.

Figure 10-7b shows how to compute the silhouette edges. One cube is oriented at an angle with the axes of the VCS and the other is parallel to the axes. In the first case, the $Z_v$ component of each normal is calculated (shown

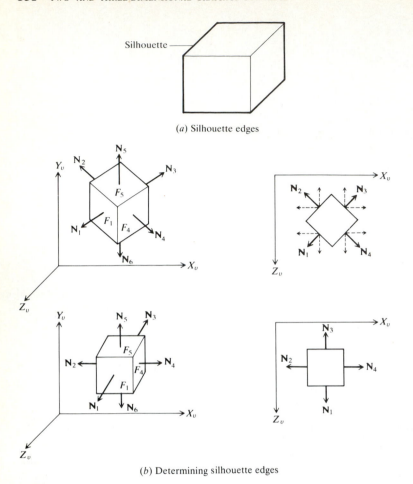

(a) Silhouette edges

(b) Determining silhouette edges

**FIGURE 10-7**
Silhouette edges of a polyhedral object.

dashed in the figure). The edges between faces $F_1$ and $F_2$, $F_1$ and $F_6$, $F_2$ and $F_5$, $F_3$ and $F_5$, $F_3$ and $F_4$, and $F_4$ and $F_6$ are silhouette edges according to the above criterion. If a normal does not have a $Z_v$ component, as in the second case of Fig. 10-7b, additional information is needed to compute the silhouette. This case implies that the corresponding face is parallel to the $Z_v$ axis and either the $X_v$ or $Y_v$ axis, and perpendicular to the remaining axis. For example, face $F_4$ is parallel to both the $Z_v$ and $Y_v$ axes and perpendicular to the $X_v$ axis. Therefore, the face normal is parallel to one of the VCS axes. If this normal points in the positive direction of the axis, the face is visible. Face $F_4$ is visible and $F_2$ is not. Similarly, face $F_5$ is visible while $F_6$ is not.

Determining silhouette curves for curved surfaces follows a similar approach, but is more involved. The silhouette curve of a surface is a curve on the surface along which the $Z_v$ component of the surface normal is zero, as shown in Fig. 10-8. To obtain this curve, the equation of the $Z_v$ component of the surface normal [Eq. (6.11)] is set to zero and solved for $u$ and $v$. This approach is

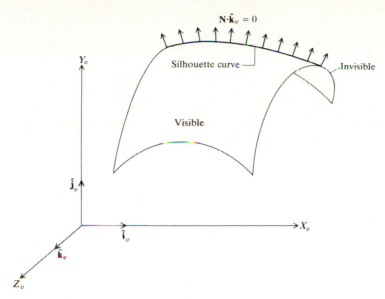

**FIGURE 10-8**
Silhouette curve of a curved surface.

usually inconvenient because the resulting equation is difficult to solve. For a bicubic surface, the equation is a quintic polynomial in $u$ and $v$. Other more efficient methods are available in the literature and are not discussed here.

**10.3.2.5 EDGE INTERSECTIONS.** All the visibility techniques discussed thus far cannot determine which parts are hidden and which are visible for partially hidden edges. To accomplish this, hidden line algorithms first calculate edge intersections in two-dimensions, that is, in the $X_v Y_v$ plane of the VCS. These intersections are used to determine edge visibility. Consider the edge $AB$ and the face $F$ shown in Fig. 10-9. The edges of the face $F$ are directed in a counterclockwise direction. Let us consider the intersection $I$ between $AB$ and edge $CD$ of face $F$. The visibility of $AB$ with respect to $F$ can fall into one of three cases: fully visible, $I$ indicates the disappearance of $AB$, or $I$ indicates the appearance of $AB$. In the first case, the depths $z_v$ at point $I$ are computed and compared. If it is considered a point on $F$, its $x_v$ and $y_v$ coordinates are substituted into the plane equation to find $z_v$. If it is a point on $AB$, the line equation is used instead to find the other depth. If the depth of the line is larger (we are dealing with a right-handed VCS) than the depth of the face, the line $AB$ is fully visible (Fig. 10-9a). Otherwise, if the directed edge $CD$ subtends a clockwise angle $\theta$ about $A$ (Fig. 10-9b), the edge disappears. If, on the other hand, the edge subtends a counterclockwise angle $\theta$ about $A$ (Fig. 10-9c), the edge appears. Notice that if the face edges are directed clockwise, the angle criterion reverses. The angle criterion is sometimes referred to as the vorticity of edge $CD$ with respect to point $A$.

**10.3.2.6 SEGMENT COMPARISONS.** This class of techniques is used to solve the hidden surface problem in the image (raster) space. It is covered in this section as

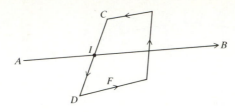

(a) Fully visible edge AB

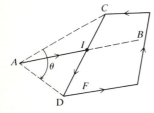

(b) CD marks the disappearance of partially hidden edge AB

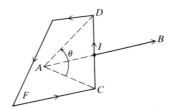

(c) CD marks the appearance of partially hidden edge AB

**FIGURE 10-9**
Computing visibility of edges.

another visibility technique. The techniques covered here are applicable to hidden surface and hidden solid algorithms as well. As discussed in Chap. 2, scan lines are arranged on a display screen from top to bottom, left to right. Therefore, instead of computing the whole correct image at once, it can be computed scan line by scan line, that is, in segments, and displayed in the same order as the scan lines. Computationally, the plane of the scan line defines segments where it intersects faces in the image (see Fig. 10-10). Computing the correct image for one scan line is considerably simpler.

The segment comparisons are performed in the $X_v Z_v$ plane (Fig. 10-10). The scan line is divided into spans (dashed lines shown in the bottom of Fig. 10-10 define the bounds of the spans). The visibility is determined within each span by comparing the depths of the edge segments that lie within the span. Plane equations are used to compute these depths. Segments with maximum depth are visible throughout the span.

The strategy to divide a scan line into spans is a distinctive feature of any hidden surface algorithm. One obvious strategy is to divide the scan line at each endpoint of each edge segment (lines $A$, $B$, $C$, and $D$ in Fig. 10-10). A better strategy is to choose fewer spans. In Fig. 10-10, it is optimum to divide the scan line via line $C$ into two spans only.

**10.3.2.7  HOMOGENEITY TEST.** The depth test described in Sec. 10.3.2.3 is concerned with comparing the depths of point sets (single points) to determine visibility. Computing homogeneity of point sets is another test to determine visibility. The notion of neighborhood (discussed in Chap. 7) must be used to determine homogeneity. The neighborhood of a point $P$, denoted here by $N(P)$, in

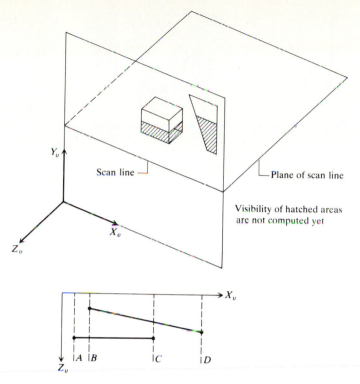

**FIGURE 10-10**
Computing visibility using scan lines.

a data set is all points in the set lying inside a sphere (or a circle for two-dimensional point sets) around it.

Three types of points can be identified based on computing homogeneity: homogeneously visible, homogeneously invisible, and inhomogeneously visible. If a neighborhood of a point $P$ can be bijectively projected onto a neighborhood of the projection of the point, then the neighborhood of $P$ is visible or invisible and $P$ is called homogeneously visible or invisible respectively. Otherwise, $P$ is inhomogeneously visible or invisible. If we denote the projection of $P$ by $\mathrm{pr}(P)$, $P$ is homogeneously visible or invisible if $\mathrm{pr}(N(P)) = N(\mathrm{pr}(P))$ and inhomogeneously visible or invisible if $\mathrm{pr}(N(P)) \neq N(\mathrm{pr}(P))$. Using this test, inner points of scenes are homogeneously visible (covering) or invisible (hidden) and contour (edge) points are inhomogeneously visible. Figure 10-11 shows an example. $N(P, F)$ denotes the neighborhood of a point $P$ that belongs to face $F$. It is obvious that contour points ($P_2$) are inhomogeneously visible (covering) and inner points are homogeneously visible ($P_1$ on face $F_2$) or invisible ($P_1$ on face $F_1$).

Homogeneity is important for both covering and hiding. No point needs to be tested against any homogeneously visible (covering) point and no homogeneously hidden point needs to be tested against any other point, since these points are homogeneously invisible in both cases. Moreover, homogeneously invisible point sets are of no interest to the visibility problem. Homogeneously

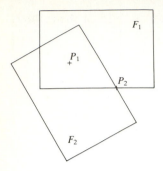

Two overlapping faces $F_1$ and $F_2$

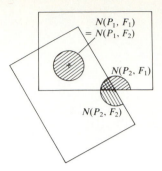

Neighborhoods

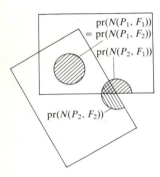

Projection of neighborhoods

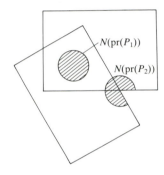

Neighborhoods of projected points

**FIGURE 10-11**
Homogeneity test.

visible point sets can be completely displayed without further tests against other points. Therefore, determination of homogeneous point sets reduces further computation. Some area-oriented algorithms (see Sec. 10.3.6.2) use this test to improve their efficiencies.

## 10.3.3   Sorting

Many visibility algorithms (hidden line, hidden surface, and hidden solid algorithms) make extensive use of sorting operations. For visibility algorithms, sorting and searching techniques operate on the records of the scene database. These records typically contain geometrical, topological, and viewing information about the polygons and faces that make the scene. Sorting is an operation that orders a given set of records according to a selected criterion. The time required to perform the sort depends on the number of records to be processed, the algorithm that performs the sort, and the statical properties of the initial ordering of the records (whether it is random or semi-ordered). Various sorting techniques are available. They are not covered here and the reader is referred to the bibliography at the end of the chapter.

### 10.3.4 Coherence

Naturally, the elements of a scene or its image have some interrelationships. This interrelationship is called coherence. Hidden line algorithms that utilize coherence in their sorting techniques are more effective than other algorithms that do not. Coherence is a measure of how rapid a scene or its image changes. It describes the extent to which a scene or its image is locally constant.

The coherence of a set of data can improve the speed of its sorting significantly. If, for example, an initially sorted deck of cards is slightly shuffled, the coherence remaining in the deck can be of great use in re-sorting it. Similarly, the gradual changes in the appearance of a scene or its image from one place to another can reduce the number of sorting operations greatly.

Several types of coherence can be identified both in the object space and image space:

1. *Edge coherence.* The visibility of an edge changes only when it crosses another edge.
2. *Face coherence.* If a part of a face is visible, the entire face is probably visible. Moreover, penetration of faces is a relatively rare occurrence, and therefore is not usually checked by hidden removal algorithms.
3. *Geometric coherence.* Edges that share the same vertex or faces that share the same edge have similar visibilities in most cases. For example, if three edges (the case of a box) share the same vertex, they may be all visible, all invisible, or two visible and one invisible. The proper combination depends on the angle between any two edges (less or greater than 180°) and on the location of any of the edges relative to the plane defined by the other two edges.
4. *Frame coherence.* A picture does not change very much from frame to frame.
5. *Scan-line coherence.* Segments of a scene visible on one scan line are most probably visible on the next line.
6. *Area coherence.* A particular element (area) of an image and its neighbors are likely to have the same visibility and to be influenced by the same face.
7. *Depth coherence.* The different surfaces at a given screen location are generally well separated in depth relative to the depth range of each.

The first three types of coherence are object-space based while the last four are image-space based. If an image exhibits a particular predominant coherence, the coherence would form the basis of the related hidden line removal algorithm.

### 10.3.5 Formulation and Implementation

The hidden line removal problem can be stated as follows. For a given three-dimensional scene, a given viewing point, and a given direction, eliminate from an appropriate two-dimensional projection of the scene all (parts of) edges and faces which the observer cannot see. For orthographic projections, the location of the viewing point is not needed. The two-dimensional projection is also known as the image (picture) space. It is the $X_v, Y_v$ plane of the VCS.

A set of generic steps that can implement a solution to the above problem is shown in Fig. 10-12. The three-dimensional scene is a set of three-dimensional objects. Each object is defined by its geometry and topology. A solid model is an ideal representation. The root of the tree represents the three-dimensional scene. Databases of wireframe and surface models have to be modified to be able to identity faces (polygons) and the order of their edges (clockwise or counter-clockwise). These modifications are typically achieved in an interactive way by requesting the user to digitize the edges of each face in a given order.

The second step is to apply the proper geometric transformations (Chap. 9) based on the viewing direction to the three-dimensional scene data to obtain the two-dimensional image "raw" data. At this stage, the image is a maze of all edges (visible and invisible). These transformations are modified as discussed earlier to also produce the depth information which is stored in the image database for depth-comparison purposes later.

The next steps (sorting and applying visibility techniques) shown in Fig. 10-12 are interrelated and may be difficult to identify as such. Nevertheless, we apply one or more of the visibility techniques covered earlier with the aid of a sorting technique. The surface test to eliminate the back faces is usually sufficient to solve the hidden line problem if the scene is only one convex polyhedron without holes. Otherwise a combination of techniques are usually required. It is this combination and sorting techniques that differentiate the existing algorithms. In order to apply the visibility techniques to the image data, the sorting of this data by either polygons or edges is required.

With the completion of the sorting according to the visibility criteria set by the visibility techniques, the hidden edges (or parts of edges) have been identified

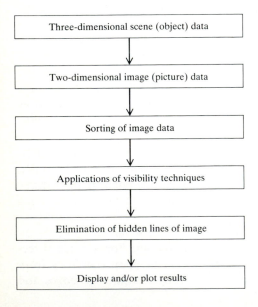

**FIGURE 10-12**
A generic hidden line algorithm.

and removed from the image data. The last step in the algorithm is to display and/or plot the final images.

### 10.3.6  Sample Hidden Line Algorithms

The basis of a visibility algorithm is to test whether a picture element is hidden by another or not. A visibility algorithm can be described as va($H$, $C$) where $H$ and $C$ are the sets of the hidden and the covering candidate edges of the scene consideration respectively. Since every edge in the $H$ set has to be tested against all the edges in the $C$ set, the algorithm requires $H * C$ visibility tests. The different visibility algorithms are mainly distinguished by the strategy used to apply the visibility test, which in turn determines the $H$ and $C$ sets.

The efficiency of a visibility algorithm depends on the choice of the $H$ and $C$ sets. Existing hidden line removal algorithms can take one of three approaches: the edge-oriented approach, silhouette (contour)-oriented approach or area-oriented approach. To examine the efficiency of each approach, let $E$ and $S$ be the number of the edges and silhouette edges of a picture respectively. The edge-oriented approach forms the basis of most existing hidden line algorithms. The underlying strategy is the explicit computation of the visible segments of all individual edges. The visibility calculation consists of the test of all edges against all surfaces. Therefore, $H = E$ and $C = E$. Thus $E * E$ visibility tests are required. This approach is inefficient because it tests all edges against each other, whether they intersect or not. It does not recognize the structural information of a picture. Sorting of all edges improves the efficiency of the approach.

In the silhouette (contour)-oriented approach, the silhouette edges are calculated first. The visibility calculation consists of testing all the edges against all the silhouette edges only. Thus, all the edges are hidden candidates but only the silhouette edges are covering candidates; that is, $H = E$, $C = S$. The calculation rate is estimated at $E * S$. This approach is more efficient than the edge-oriented approach. However, some edges (after sorting) will still be tested against some nonintersecting silhouettes. Moreover, intersections of homogeneously (fully) hidden edges against silhouette edges are unnecessarily computed. However, the tests of homogeneously covering (visible) edges against each other are avoided completely.

The area-oriented approach aims at the recognition of the visible areas of a picture. This approach calculates the silhouette edges and connects them to form closed polygons (areas). The connection can be done in $S * S$ steps. The visibility calculations begin with the test of all silhouette edges against each other. Thus $H = S$ and $C = S$, and the calculation rate is $S * S$. This approach works better than the silhouette-oriented approach. First, sorting is reduced to the silhouette edges only. Only silhouettes (contours) are tested against nonintersecting silhouettes. The computation of intersection points of both homogeneously hidden and homogeneously covering edges is avoided.

There exists a wide wealth of hidden line algorithms that utilize one or more of the visibility techniques discussed in Sec. 10.3.2 and follow one of the three approaches described above. These include the priority algorithm, the plane-sweep paradigm, algorithms for scenes of high complexities, algorithms for finite element models of planar elements, area-oriented algorithms, the overlay algorithm for surfaces defined by $u$–$v$ grids (Chap. 6), and algorithms for projected grid surfaces. In the remainder of this section, we discuss the priority algorithm and an area-oriented algorithm. Readers who are interested in details of other algorithms should consult the bibliography at the end of the chapter.

**10.3.6.1 THE PRIORITY ALGORITHM.** This algorithm is also known as the depth or $z$ algorithm. The algorithm is based on sorting all the faces (polygons) in the scene according to the largest $z$ coordinate value of each. This step is sometimes known as assignment of priorities. If a face intersects more than one face, other visibility tests besides the $z$ depth are needed to resolve any ambiguities. This step constitutes determination of coverings.

To illustrate how the priority algorithm can be implemented, let us consider a scene of two boxes as shown in Fig. 10-13. Figure 10-13 shows the scene in the standard VCS where the viewing eye is located at ∞—on the positive $Z_v$ direction. The following steps provide a guidance to implement the algorithm:

1. Utilize the proper orthographic projection to obtain the desired view (whose hidden lines are to be removed) of the scene. This results in a set of vertices with coordinates of $(x_v, y_v, z_v)$. To enable performing the depth test, the plane equation of any face (polygon) in the image can be obtained using Eq. (10.1). Given three points that lie in one face, Eq. (10.1) can be rewritten as

$$z_v = Ax_v + By_v + C \qquad (10.3)$$

where

$$A = \frac{(z_{v1} - z_{v3})(y_{v2} - y_{v3}) - (z_{v2} - z_{v3})(y_{v1} - y_{v3})}{D} \qquad (10.4)$$

$$B = -\frac{(z_{v1} - z_{v3})(x_{v2} - x_{v3}) - (z_{v2} - z_{v3})(x_{v1} - x_{v3})}{D} \qquad (10.5)$$

$$C = z_{v1} - Ax_{v1} - By_{v1} \qquad (10.6)$$

and $\qquad D = (x_{v1} - x_{v3})(y_{v2} - y_{v3}) - (x_{v2} - x_{v3})(y_{v1} - y_{v3}) \qquad (10.7)$

2. Utilize the surface test to remove back faces to improve the efficiency of the priority algorithm. Equation (10.2) can be used in this test. Any two edges of a given face can be used to calculate the face normal (refer to Chaps. 5 and 6). Steps 1 and 2 result in a face list which will be sorted to assign priorities. For the scene shown in Fig. 10-13a, six faces $F_1$ to $F_6$ form such a list. The order of the faces in the list is immaterial.

3. Assign priorities to the faces in the face list. The priority assignment is determined by comparing two faces at any one time. The priority list is continuously changed and the final list is obtained after few iterations. Here is how priorities can be assigned. The first face in the face list ($F_1$ in Fig. 10-13b) is assigned the highest priority 1. $F_1$ is intersected with the other faces in the list, that is, $F_2$ to $F_6$. The intersection between $F_1$ and another face may be an

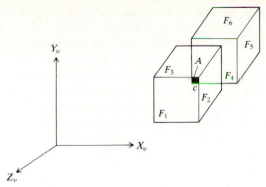

(a) Scene of two boxes

| Face list | Priority list |
|-----------|---------------|
| $F_1$ | 1 |
| $F_2$ | 1 |
| $F_3$ | 1 |
| $F_4$ | 2 |
| $F_5$ | |
| $F_6$ | |

Iteration 1

| Face list | Priority list |
|-----------|---------------|
| $F_2$ | 1 |
| $F_3$ | 1 |
| $F_4$ | 2 |
| $F_5$ | |
| $F_6$ | |
| $F_1$ | |

Iteration 2

| Face list | Priority list |
|-----------|---------------|
| $F_3$ | 1 |
| $F_4$ | 2 |
| $F_5$ | |
| $F_6$ | |
| $F_1$ | |
| $F_2$ | |

Iteration 3

| Face list | Priority list |
|-----------|---------------|
| $F_4$ | ~~1~~ 2 |
| $F_5$ | ~~1~~ 2 |
| $F_6$ | ~~1~~ 2 |
| $F_1$ | 1 |
| $F_2$ | 1 |
| $F_3$ | 1 |

Iteration 4

(b) Assignment of priorities

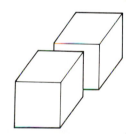

(c) Image with hidden lines removed

**FIGURE 10-13**
The priority algorithm.

area as in the case of $F_1$ and $F_4$, an edge as for faces $F_1$ and $F_2$, or an empty set (no intersection) as for faces $F_1$ and $F_6$. In the case of an area of intersection ($A$ in Fig. 10-13a), the $(x_v, y_v)$ coordinates of a point $c$ inside $A$ can be computed (notice the corner points of $A$ are known). Utilizing Eq. (10.3) for both faces $F_1$ and $F_4$, the two corresponding $z_v$ values of point $c$ can be calculated and compared. The face with the highest $z_v$ values is assigned the highest priority. In the case of an edge of intersection, both faces are assigned the same priority. They obviously do not obscure each other, especially after the removal of the back faces. In the case of no face intersection, no priority is assigned.

Let us apply the above stategy to the scene of Fig. 10-13. $F_1$ intersects $F_2$ and $F_3$ in edges. Therefore both faces are assigned priority 1. $F_1$ and $F_4$ intersect in an area. Using the depth test, and assuming the depth of $F_4$ is less than that of $F_1$, $F_4$ is assigned priority 2. When we intersect faces $F_1$ and $F_5$, we obtain an empty set, that is, no priority assignment is possible. In this case, the face $F_1$ is moved to the end of the face list and the sorting process to determine priorities starts all over again. In each iteration, the first face in the face list is assigned priority 1. The end of each iteration is detected by no intersection. Figure 10-13b shows four iterations before obtaining the final priority list. In iteration 4, faces $F_4$ to $F_6$ are assigned the priority 1 first. When $F_4$ is intersected with $F_1$, the depth test shows that $F_1$ has higher priority. Thus, $F_1$ is assigned priority 1 and priority of $F_4$ to $F_6$ is dropped to 2.

4. Reorder the face and priority lists so that the highest priority is on top of the list. In this case, the face and priority lists are $[F_1, F_2, F_3, F_4, F_5, F_6]$ and $[1, 1, 1, 2, 2, 2]$ respectively.

5. In the case of a raster display, hidden line removal is done by the hardware (frame buffer of the display). We simply display the faces in the reverse order of their priority. Any faces that would have to be hidden by others would thus be displayed first, but would be covered later either partially or entirely by faces of higher priority.

6. In the case of a vector display, the hidden line removal must be done by software by determining coverings. For this purpose, the edges of a face are compared with all other edges of higher priority. An edge list can be created that maintains a list of all line segments that will have to be drawn as visible. Visibility techniques such as the containment test (Sec. 10.3.2.2) and edge intersection (Sec. 10.3.2.5) are useful in this case. Figure 10-13c shows the scene with hidden lines removed.

In some scenes, ambiguities may result after applying the priority test. Figure 10-14 shows an example. The figure shows a case in which the order of faces is cyclic. Face $F_1$ covers $F_2$, $F_2$ covers $F_3$, and $F_3$ covers $F_1$. The reader is encouraged to find the priority list that produces this cyclic ordering and coverage. To rectify this ambiguity, additional software to determine coverings (similar to that of vector displays) must be added to the priority algorithm.

**10.3.6.2   AREA-ORIENTED   ALGORITHM.** The   area-oriented   algorithm   described here subdivides the data set of a given scene in a stepwise fashion until all visible areas in the scene are determined and displayed. The data structure that the algorithm operates on is of the winged edge structure discussed in Chap. 7 and shown in Figs. 7-29 and 7-30. In this algorithm as well as in the priority algorithm, no penetration of faces is allowed. Considering the same scene shown in Fig. 10-13a, the area-oriented algorithm can be described as follows:

1. Identify silhouette polygons. Silhouette polygons are polygons whose edges are silhouette edges. First, silhouette edges in the scene are recognized as described in Sec. 10.3.2.4. Second, the connection of silhouette edges to form closed silhouette polygons can be achieved by sorting all the edges for equal

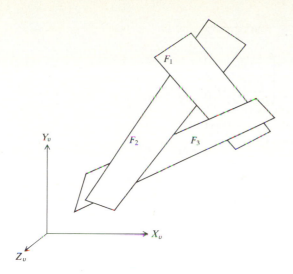

$Y_v$

$F_1$

$F_2$

$F_3$

$X_v$

$Z_v$

**FIGURE 10-14**
Ambiguous example of priority algorithm.

endpoints. For the scene shown in Fig. 10-15, two closed silhouette polygons $S_1$ and $S_2$ are identified.

2. Assign quantitative hiding (QH) values to edges of silhouette polygons. This is achieved by intersecting the polygons (the containment test can be utilized first as a quick test). The intersection points define the points where the value of QH may change. Applying the depth test to the points of intersection ($P_1$ and $P_2$ in Fig. 10-15), we determine the segments of the silhouette edges that are hidden. For example, if the depth test at $P_1$ shows that $z_v$ at $P_1$ of $S_1$ is smaller than that of $S_2$, edge $C_1 C_2$ is partially visible. Similarly, the depth test at $P_2$ shows that edge $C_2 C_3$ is also partially visible. To determine which segment of an edge is visible, the visibility test of Sec. 10.3.2.5 can be used. Determination of the values of QH at the various edges or edge segments of silhouette polygons is based on the depth test. Step 2 in Fig. 10-15 shows the values of QH. A value of 0 indicates that an edge or segment is visible and a value of 1 indicates that an edge or segment is invisible.

3. Determine the visible silhouette segments. From the values of QH, the visibility of silhouette segments can be determined with the following rules in mind. If a closed silhouette polygon is completely invisible, it has not to be considered any further. Otherwise, its segments with the least QH values are visible (step 3 in Fig. 10-15).

4. Intersect the visible silhouette segments with partially visible faces. This step is used to determine if the silhouette segments hide or partially hide non-silhouette edges in partially visible faces. In Fig. 10-15, edges $E_1$ to $E_6$ of $S_2$ are intersected with the internal edges (edges of the square in the face) of $F_1$ and the visible segments of the internal edges are determined. By accessing only the silhouette edges of the covering silhouette polygon only and the partially visible face only, the algorithm avoids any unnecessary calculations.

5. Display the interior of the visible or partially visible polygons. This step can be achieved using a stack and simply enumerates all faces lying inside a silhou-

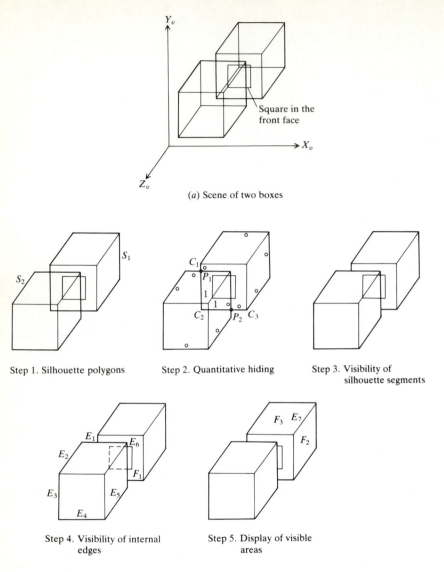

(a) Scene of two boxes

Step 1. Silhouette polygons

Step 2. Quantitative hiding

Step 3. Visibility of silhouette segments

Step 4. Visibility of internal edges

Step 5. Display of visible areas

(b) Steps of the alogrithm

**FIGURE 10-15**
Area-oriented algorithm.

ette polygon. The stack is initialized with a visible face which has a silhouette edge. We know this face belongs to a visible area. A loop begins with popping a face $(F_2)$ from the stack. We examine all the edges of the face. If an edge $(E_7)$ is not fully invisible, then the neighboring face $(F_3)$ has also visible edges and, therefore, is pushed into the stack if it has not already been pushed. The edge itself or its visible segments are displayed. The loop is repeated and the algorithm stops when the stack is empty.

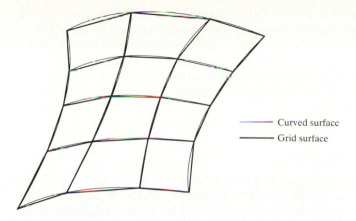

Curved surface

Grid surface

**FIGURE 10-16**
Grid surface approximation of a curved surface.

The two hidden line algorithms discussed in this section are sample algorithms to enable understanding of the basic nature of the hidden line removal problem. The area-oriented algorithm is more efficient than the priority algorithm because it hardly involves any unnecessary edge/face intersection.

### 10.3.7  Hidden Line Removal for Curved Surfaces

The hidden line algorithms described thus far are applicable to polyhedral objects with flat faces (planar polygons or surfaces). Fortunately, these algorithms are extendable to curved polyhedra by approximating them by planar polygons. The $u$–$v$ grid offered by parametric surface representation (Chap. 6) offers such an approximation. This grid can be utilized to create a "grid surface" consisting of straight-edged regions (see Fig. 10-16) by approximating the $u$–$v$ grid curves by line segments.

The overlay hidden line algorithm mentioned in Sec. 10.3.6 is suitable for curved surfaces. The algorithm begins by calculating the $u$–$v$ grid using the surface equation. It then creates the grid surface with linear edges. Various criteria discussed previously can be utilized to determine the visibility of the grid surface.

There is no best hidden line algorithm. Many algorithms exist and some are more efficient and faster in rendering images than others for certain applications. Firmware and parallel-processing computations of hidden line algorithms make it possible to render images in real time. This adds to the difficulty of deciding on a best algorithm.

### 10.4  HIDDEN SURFACE REMOVAL

The hidden surface removal and hidden line removal are identically one problem. Most of the concepts and algorithms described in Sec. 10.3 are applicable here and vice versa. While we limited ourselves to object-space algorithms for hidden line removal, we discuss image-space algorithms only for hidden surface removal. A wide variety of these algorithms exist. They include the $z$-buffer algorithm, Watkin's algorithm, Warnock's algorithm, and Painter's algorithm. Watkin's algorithm is based on scan-line coherence while Warnock's algorithm is an area-coherence algorithm. Painter's algorithm is a priority algorithm as described in the previous section for raster displays. Two sample algorithms are covered in this section.

### 10.4.1  The $z$-Buffer Algorithm

This is also known as the depth-buffer algorithm. In addition to the frame (refresh) buffer (see Chap. 2), this algorithm requires a $z$ buffer in which $z$ values can be sorted for each pixel. The $z$ buffer is initialized to the smallest $z$ value, while the frame buffer is initialized to the background pixel value. Both the frame and $z$ buffers are indexed by pixel coordinates $(x, y)$. These coordinates are actually screen coordinates. The $z$-buffer algorithm works as follows. For each polygon in the scene, find all the pixels $(x, y)$ that lie inside or on the boundaries of the polygon when projected onto the screen. For each of these pixels, calculate the depth $z$ of the polygon at $(x, y)$. If $z > $ depth $(x, y)$, the polygon is closer to the viewing eye than others already stored in the pixel. In this case, the $z$ buffer is updated by setting the depth $(x, y)$ to $z$. Similarly, the intensity of the frame buffer location corresponding to the pixel is updated to the intensity of the polygon at $(x, y)$. After all the polygons have been processed, the frame buffer contains the solution.

### 10.4.2  Warnock's Algorithm

This is one of the first area-coherence algorithms. Essentially, this algorithm solves the hidden surface problem by recursively subdividing the image into sub-images. It first attempts to solve the problem for a window that covers the entire image. Simple cases such as one polygon in the window or none at all are easily solved. If polygons overlap, the algorithm tries to analyze the relationship between the polygons and generates the display for the window.

If the algorithm cannot decide easily, it subdivides the window into four smaller windows and applies the same solution technique to every window. If one of the four windows is still complex, it is further subdivided into four smaller windows. The recursion terminates if the hidden surface problem can be solved for all the windows or if the window becomes as small as a single pixel on the screen. In this case, the intensity of the pixel is chosen equal to the polygon visible in the pixel. The subdivision process results in a window tree.

Figure 10-17 shows the application of Warnock's algorithm to the scene shown in Fig. 10-13a. One would devise a rule that any window is recursively subdivided unless it contains two polygons. In such a case, comparing the $z$ depth of the polygons determines which one hides the other.

While the subdivision of the original window is governed by the complexity of the scene in Fig. 10-17, the subdivision of any window into four equal

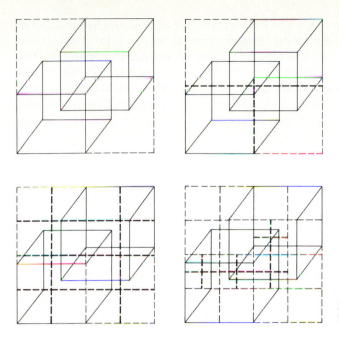

**FIGURE 10-17**
Warnock's algorithm.

windows makes the algorithm inefficient. A more efficient way would be to sub-divide a window according to the complexity of the scene in the window. This is equivalent to subdividing a window into four unequal subwindows.

## 10.5 HIDDEN SOLID REMOVAL

The hidden solid removal problem involves the display of solid models with hidden lines or surfaces removed. Due to the completeness and unambiguity of solid models as discussed in Chap. 7, the hidden solid removal is done fully auto-matically. Therefore, commands available on CAD/CAM systems for hidden solid removal, say a "hide solid" command, require minimum user input. In fact, all that is needed to execute the command is that the user identifies the solid to be hidden by digitizing it. The data structure of a solid model (see Fig. 7-29) has all the necessary information to solve the hidden line or hidden surface problem. All the algorithms discussed in Secs. 10.3 and 10.4 are applicable to hidden solid removal of B-rep models. Selected algorithms such as the z-buffer have been extended to CSG models.

For displaying CSG models, both the visibility problem and the problem of combining the primitive solids into one composite model have to be solved. There are three approaches to display CSG models. The first approach converts the CSG model into a boundary model that can be rendered with the standard hidden surface algorithms. The second approach utilizes a spatial subdivision strategy. To simplify the combinatorial problems, the CSG tree (half-spaces) is pruned simultaneously with the subdivision. This subdivision reduces the CSG evaluation to a simple preprocessing before the half-spaces are processed with standard rendering techniques.

The third approach uses a CSG hidden surface algorithm which combines the CSG evaluation with the hidden surface removal on the basis of ray classification. The CSG ray-tracing and scan-line algorithms utilize this approach. The attractiveness of the approach lies in the conversion of the complex three-dimensional solid/solid intersection problem into a one-dimensional ray/solid intersection calculation. Due to its popularity and generality, the remainder of this section covers in more detail the ray-tracing (also called ray-casting) algorithm.

## 10.5.1 Ray-Tracing Algorithm

The virtue of ray tracing is its simplicity, reliability, and extendability. The most complicated numerical problem of the algorithm is finding the points at which lines (rays) intersect surfaces. Therefore a wide variety of surfaces and primitives can be covered. Ray tracing has been utilized in visual realism of solids to generate line drawings with hidden solids removed, animation of solids, and shaded pictures. It has also been utilized in solid analysis, mainly calculating mass properties.

The idea of ray tracing originated in the early 1970s by MAGI (Mathematic Applications Group, Inc.) to generate shaded pictures of solids. To generate these pictures, the photographic process is simulated in reverse. For each pixel in the screen, a light ray is cast through it into the scene to identify the visible surface. The first surface intersected by the ray, found by "tracing" along it, is the visible one. At the ray/surface intersection point, the surface normal is computed, and knowing the position of the light source, the brightness of the pixel can be calculated.

Ray tracing is considered a brute force method for solving problems. The basic (straightforward) ray-tracing algorithm is very simple, but yet slow. The CPU usage of the algorithm increases with the complexity of the scene under consideration. Various alterations and refinements have been added to the algorithm to improve its efficiency. Moreover, the algorithm has been implemented into hardware (ray-tracing firmware) to speed its execution. In this book, we present the basic algorithm and some obvious refinements. More detailed work can be found in the bibliography at the end of the chapter.

**10.5.1.1 BASICS.** The geometric reasoning of tray tracing stems from light rays and camera models. The geometry of a simple camera model is analogous to that of projection of geometric models. Referring to Fig. 9-18, the center of projection, projectors, and the projection plane represent the focal point, light rays, and the screen of the camera model respectively. For convenience, we assume that the camera model uses the VCS as described in Chap. 9 and shown in Figs. 9-19 and 9-21. For each pixel of the screen, a straight light ray passes through it and connects the focal point with the scene.

When the focal length, the distance between the focal point and screen, is infinite, parallel views result (Fig. 9-19) and all light rays becomes parallel to the $Z_v$ axis and perpendicular to the screen (the $X_v Y_v$ plane). Figure 10-18 shows the geometry of a camera model as described here. The $XYZ$ coordinate system

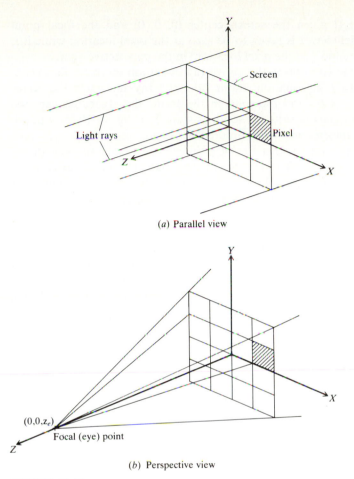

(a) Parallel view

(b) Perspective view

**FIGURE 10-18**
Camera model for ray tracing.

shown is the same as the VCS. We have dropped the subscript $v$ for simplicity. The origin of the $XYZ$ system is taken to be the center of the screen.

A ray is a straight line which is best defined in a parametric form as a point $(x_0, y_0, z_0)$ and a direction vector $(\Delta x, \Delta y, \Delta z)$. Thus, a ray is defined as $[(x_0, y_0, z_0) \quad (\Delta x, \Delta y, \Delta z)]$. For a parameter $t$, any point $(x, y, z)$ on the ray is given by

$$x = x_0 + t\Delta x$$

$$y = y_0 + t\Delta y \qquad (10.8)$$

$$z = z_0 + t\Delta z$$

This form allows points on a ray to be ordered and accessed via a single parameter $t$. Thus, a ray in a parallel view that passes through the pixel $(x, y)$ is defined as $[(x, y, 0) \quad (0, 0, 1)]$. In a perspective view, the ray is defined by

$[(0, 0, z_e) \ (x, y, -z_e)]$ given the screen center $(0, 0, 0)$ and the focal point $(0, 0, z_e)$. In the parallel view, $t$ is taken to be zero at the pixel location while it is zero at the focal point (and 1 at the pixel location) in the perspective view.

A ray-tracing algorithm takes the above ray definition given by Eqs. (10.8) as an input and output information about how the ray intersects the scene. Knowing the camera model and the solid in the scene, the algorithm can find where the given ray enters and exits the solid, as shown in Fig. 10-19 for a parallel view. The output information is an ordered list of ray parameters, $t_i$, which denotes the enter/exit points, and a list of pointers, $S_i$, to the surfaces (faces) through which the ray passes. The ray enters the solid at point $t_1$, exits at $t_2$, enters at $t_3$, and finally exits at $t_4$. Point $t_1$ is closest to the screen and point $t_4$ is furthest away. The lists of ray parameters and surface pointers suffice for various applications.

### 10.5.1.2  BASIC RAY-TRACING ALGORITHM.

While the basics of ray tracing is simple, their implementation into a solid modeler is more involved and depends largely on the representation scheme of the modeler. When boundary representation is used in the object definition, the ray-tracing algorithm is simple. For a given pixel, the first face of the object intersected by the ray is the visible face at that pixel.

When the object is defined as a CSG model, the algorithm is more complicated because CSG models are compositions of solid primitives. Intersecting the solid primitives with a ray yields a number of intersection points which requires additional calculations to determine which of these points are intersection points of the ray with the composite solid (object).

A ray-tracing algorithm for CSG models consists of three main modules: ray/primitive intersection, ray/primitive classification, and ray/solid (or ray/object) classification.

**Ray/primitive intersection.**  Utilizing the CSG tree structure, the general ray/solid intersection problem reduces to the ray/primitive intersection problem. The ray enters and exits the solid via the faces and the surfaces of the primitives.

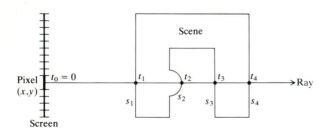

**FIGURE 10-19**
Output information from ray tracing.

For convex primitives (such as a block, cylinder, cone, and sphere), the ray/primitive intersection test has four possible outcomes: no intersection (the ray misses the primitives), the ray is tangent to (touches) the primitive at one point, the ray lies on a face of the primitive, or the ray intersects the primitive at two different points. In the case of a torus, the ray may be tangent to it at one or two points and may intersect it at as many as four points.

The ray/primitive intersection is evaluated in the local coordinate system of the primitive (see Fig. 7-4) because it is very easy to find. The equation of a primitive [Eqs. (7.61) to (7.66)] expressed in its local coordinate system is simple and independent of any rigid-body motion the primitive may undergo to be oriented properly in a scene. Arbitrary elliptic cylinders in the scene are all the same primitive in its local coordinate system: a right circular cylinder at the origin. Similarly, ellipsoids are the same sphere, elliptic cones are the same right circular cone, and parallelepipeds are all the same block.

Given a ray originating in the screen coordinate system (SCS), it must be transformed into the primitive (local) coordinate system (PCS) via the scene (model) coordinate system (MCS) in order to find the ray/primitive intersection points. Each primitive has its local-to-scene transform and inverse, but there is only one scene-to-screen transform and inverse. The local-to-scene $[T_{LS}]$ and scene-to-screen $[T_{SS}]$ transformation matrices are determined from user input to orient primitives or define views of the scene respectively. The transformation matrix $[T]$ that transforms a ray from a local-to-screen coordinate system is given by

$$[T] = [T_{SS}][T_{LS}] \tag{10.9}$$

Therefore, a ray can be transformed to the PCS of a primitive by transforming its fixed point and direction vector:

$$\begin{bmatrix} x_0 & \Delta x \\ y_0 & \Delta y \\ z_0 & \Delta z \\ 1 & 1 \end{bmatrix}_{\text{local}} = [T]^{-1} \begin{bmatrix} x_0 & \Delta x \\ y_0 & \Delta y \\ z_0 & \Delta z \\ 1 & 1 \end{bmatrix}_{\text{screen}} \tag{10.10}$$

As discussed in Chap. 9, geometric transformation is a one-to-one correspondence. Thus, the parameter $t$ that designates the ray/primitive intersection points need not be transformed once the intersection problem is solved in the PCS. Therefore, only rays need to be transformed between coordinate systems, not parameters.

The ray/plane intersection calculation is simple. For instance, to intersect the parameterized ray $[(x_0, y_0, z_0) \ \ (\Delta x, \Delta y, \Delta z)]$ with the $XY$ plane, we simultaneously solve $z = 0$ and $z = z_0 + t\Delta z$ for $t$ to get

$$t = -\frac{z_0}{\Delta z} \tag{10.11}$$

Having found $t$, the point of intersection is

$$[x_0 + (-z_0/\Delta z)\Delta x, \ y_0 + (-z_0/\Delta z)\Delta y, \ 0]$$

If $\Delta z$ is zero, the ray is parallel to the plane, so they do not intersect. If the point of intersection lies within the bounds of the primitive, then it is a good ray/primitive intersection point. The bounds test for this point on the $XY$ plane of a block given by Eq. (7.61) is

$$0 \le (x_0 + t\Delta x) \le W \qquad \text{and} \qquad 0 \le (y_0 + t\Delta y) \le H \qquad (10.12)$$

Finding ray/quadric intersection points is slightly more difficult. Consider a cylindrical surface given by Eq. (7.68). Substituting the $x$ and $y$ components of the ray's line equation into Eq. (7.68) yields

$$(x_0 + t\Delta x)^2 + (y_0 + t\Delta y)^2 = R^2 \qquad (10.13)$$

Rearranging gives

$$t^2[(\Delta x)^2 + (\Delta y)^2] + 2t(x_0 \Delta x + y_0 \Delta y) + x_0^2 + y_0^2 - R^2 = 0 \qquad (10.14)$$

Using the quadratic formula, we find $t$ as

$$t = \frac{-B \pm \sqrt{B^2 - 4AC}}{2A} \qquad (10.15)$$

where

$$A = (\Delta x)^2 + (\Delta y)^2$$
$$B = 2(x_0 \Delta x + y_0 \Delta y) \qquad (10.16)$$
$$C = x_0^2 + y_0^2 - R^2$$

Obviously, the ray will intersect the cylinder only if $A \ne 0$ and $(B^2 - 4AC) \ge 0$. Having found the one or two values of $t$, the bounds test for the cylindrical surface is

$$0 \le (z_0 - t\Delta z) \le H \qquad (10.17)$$

Intersecting rays with a torus is more complicated because it is a quartic surface. It is left as an exercise (see the problems at the end of the chapter) for the reader.

**Ray/primitive classification.** The classification of a ray with respect to a primitive is simple. Utilizing the set membership classification function introduced in Sec. 7.5.3 and the ray/primitive intersection points, the "in," "out," and "on" segments of the ray can be found. As shown in Fig. 10-19, the odd intersection points signify the beginning of "in" segments and the end of "out" segments.

For the convex primitives, if the ray misses or touches the primitive at one point, it is classified as completely "out." If the ray intersects the primitive in two different points, it is divided into three segments: "out-in-out." If the ray lies on a face of the primitive, it is classified as "out-on-out." With respect to a torus, a ray is classified as "out," "out-in-out," or "out-in-out-in-out."

**Ray/solid classification.** Combining ray/primitive classifications produces the ray/solid (or ray/object) classification. Ray/solid classification produces the "in," "on," and/or "out" segments of the ray with respect to the solid. It also reorders ray/primitive intersection points and gives the closest point and surface of the solid to the camera. To combine ray/primitive classifications, a ray-tracing algorithm starts at the top of the CSG tree, recursively descends to the bottom, classifies the ray with respect to the solid primitives, and then returns up the tree combining the classifications of the left and right subtrees. Combining the "on" segments requires the use of neighborhood information as discussed in Chap. 7. Figures 10-20 and 10-21 illustrate ray/solid classification. The solid lines in Fig. 10-21 are "in" segments.

    The combine operation is a three-step process. First, the ray/primitive intersection points from the left and right subtrees are merged in sorted order, forming a segmented composite ray. Second, the segments of the composite ray are classified according to the boolean operator and the classifications of the left and right rays along these segments. Third, the composite ray is simplified by merging contiguous segments with the same classification. Figure 10-22 illustrates these three steps for the union operator. Combining classifications involves boolean algebra where the complement operator is replaced by the difference operator. Table 10.1 defines the combine rules.

**The algorithm.** To draw the visible edges of a solid, a ray per pixel is generated moving top-down, left-right on the screen. Each ray is intersected with the solid and the visible surface in the pixel corresponding to this ray is identified. If the visible surface at pixel $(x, y)$ is different from the visible surface at pixel $(x - 1, y)$,

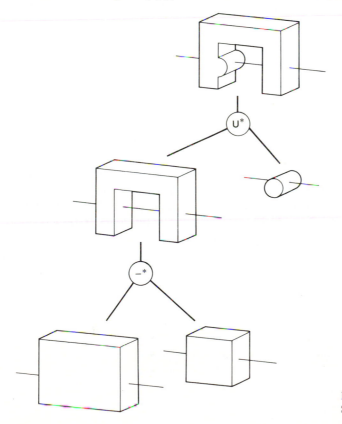

**FIGURE 10-20**
Sample ray and a CSG tree.

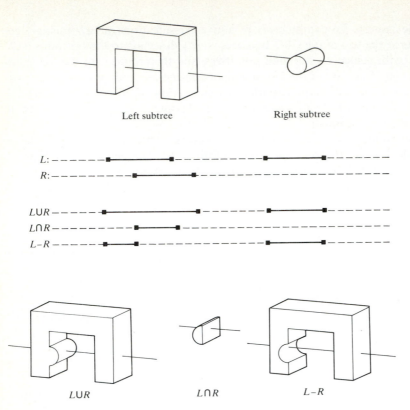

FIGURE 10-21
Combining ray classifications.

then display a vertical line one pixel long centered at $(x - 0.5, y)$. Similarly, if the visible surface at pixel $(x, y)$ is different from the visible surface at pixel $(x, y - 1)$, then display a horizontal line one pixel long centered at $(x, y - 0.5)$. The resulting line drawing with hidden solids removed will consist of horizontal and vertical edges only. Figure 10-23 shows a magnification of a drawing of a box with a hole. Figure 10-23a shows the pixel grid superimposed on the box and Fig. 10-23b shows the drawing only. As shown in Fig. 10-23a, the pixel-long horizontal and vertical lines may not coincide with the solid edges. However, the hidden solid still looks acceptable to the user's eyes because of the small size of each pixel.

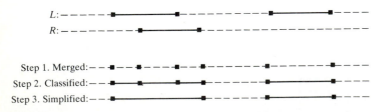

FIGURE 10-22
The three-step combine process.

**TABLE 10.1**
## The combine rules for boolean algebra

| Operator | Left subtree | Right subtree | Combine |
|---|---|---|---|
| Union | IN | IN | IN |
|  | IN | OUT | IN |
|  | OUT | IN | IN |
|  | OUT | OUT | OUT |
| Intersection | IN | IN | IN |
|  | IN | OUT | OUT |
|  | OUT | IN | OUT |
|  | OUT | OUT | OUT |
| Difference | IN | IN | OUT |
|  | IN | OUT | IN |
|  | OUT | IN | OUT |
|  | OUT | OUT | OUT |

In pseudo code, a ray-tracing algorithm may be expressed as follows:

```
Procedure RAY TRACE
  for each pixel (x,y) do
    generate a ray through pixel (x,y)
    if solid to be hidden is not a primitive {ray classification}
      then
        do {combine}
          classify ray against left subtree {L_classify}
          classify ray against right subtree {R_classify)
          combine (L_classify and R_classify)
          end {combine}
      else
        do {primitives}
          transform the ray equation from SCS to PCS
```

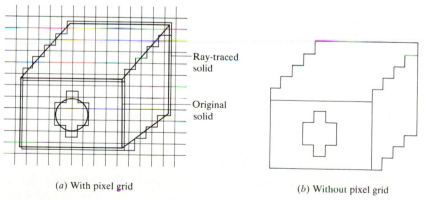

(a) With pixel grid                    (b) Without pixel grid

**FIGURE  10-23**
Solid appearance after applying ray-tracing algorithm.

```
        branch to the proper primitive case:
          Block:
              do 6 ray-plane intersection tests;
          Sphere:
              do 1 ray-quadric intersection test;
          Cylinder:
              do 2 ray-plane & 1 ray-quadric intersection tests;
          Cone:
              do 1 ray-plane & 1 ray-quadric intersection tests;
          Torus:
              do 1 ray-quartic intersection test;
          end {branch}
          classify ray against primitive
        end {primitives}
  end {ray classification}
  find the first visible surface S₁ in pixel (x,y)
  if S₁ in pixel (x,y) is different than S₁ in pixel (x-1,y)
      then
        display a pixel-long vertical line centered at (x-0.5,y)
      else
        if S₁ in pixel (x,y) is different than S₁ in pixel (x,y-1)
            then
                display a pixel-long horizontal line centered at (x,y-0.5)
        end {if}
    end {if}
end {pixel loop}
```

**10.5.1.3  IMPROVEMENTS OF THE BASIC ALGORITHM.** The basic ray-tracing algorithm as described in the above section is very slow and its memory and CPU usage is directly proportional to the scene complexity, that is, to the number of primitives in the solid. In practice, the use of memory is not as much a concern as how fast the algorithm is. To appreciate the cost of using ray tracing, consider the scenario of a scene of a solid composed of 300 primitives drawn on a raster display of 500 × 500 pixels. Since the solid is composed of 300 primitives, its CSG tree has 300 (actually 299) composite solids, making a total of 600 solids which a ray must visit via 600 calls of the algorithm to itself. Thus 600 × 500 × 500 calls of the ray-tracing algorithm are needed. At each composite solid in the tree, the left and right classifications must be combined, requiring 300 × 500 × 500 classification combines. In addition, 300 × 500 × 500 ray transformations from SCS to PCS are required. Finally, assuming an average of four surfaces per primitive, a total of 4 × 300 × 500 × 500 ray-intersection tests are performed. Therefore, the total cost of generating the hidden solid (or the shaded image) line drawing is the sum of these four costs.

The above high cost of the basic ray-tracing algorithm is primarily due to the multipliers 300 and 500 × 500. Many applications may require casting a ray from every other pixel (or more), thus reducing the latter multiplier to 250 × 250. This is equivalent to using a raster display of resolution 250 × 250 instead of

$500 \times 500$. The former multiplier can be reduced significantly for the large class of solids using box enclosures.

By using minimum bounding boxes around the solids in the CSG tree, the extensive search for the ray/solid intersection becomes an efficient binary search. These boxes enable the ray-tracing algorithm to detect the "clear miss" cases between the ray and the solid. The CSG tree can be viewed as a hierarchical representation of the space that the solid occupies. Thus, the tree nodes would have enclosure boxes that are positioned in space. Then, quick ray/box intersection tests guide the search in the hierarchy. When the test fails at an intermediate node in the tree, the ray is guaranteed to be classified as out of the composite; thus recursing down the solid's subtrees to investigate further is unnecessary.

Figure 10-24 shows the tree of box enclosures for the solid shown in Fig. 10-20. The ray/box intersection test is basically two dimensional because rays usually start at the screen and extend infinitely into the scene. When rays are bounded in depth (as in mass property applications), a ray/depth test can be added. Unlike the union and intersection operators, the subtraction operator does not obey the usual rules of algebra. The enclosure of $A - B$ is equal to the enclosure of $A$, regardless of $B$.

**10.5.1.4  REMARKS.** The ray-tracing algorithm to generate line drawings of hidden solids has few advantages. It eliminates finding, parameterizing, clas-

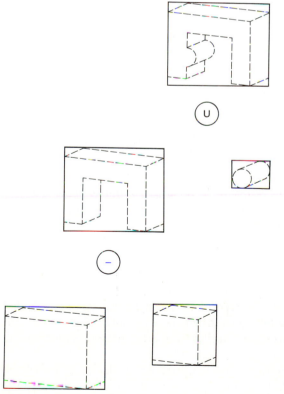

**FIGURE 10-24**
Tree of box enclosures.

sifying, and storing the curved edges formed by the intersection of surfaces. Finding the silhouettes of curved surface is a byproduct and can be found whenever the view changes.

The main drawbacks of the algorithm are speed and aliasing. Aliasing causes edges to be jagged and surface slivers may be overlooked. Speed is particularly important to display hidden solid line drawings in an interactive environment. If the user creates a balanced tree of the solid in the scene, the efficiency of ray tracing improves. Coherence of visible surfaces (surfaces visible at two neighboring pixels are more likely to be the same than different) can also speed up the algorithm. In addition, edges of the solid are only sought to generate line drawings. Thus, ray tracing should be concentrated around the edges and not in the open regions. This can be implemented by sparsely sampling the screen with rays and then locating (when neighboring rays identify different visible surfaces) the edges via binary searches. The sampling rate is under user control. As the sampling becomes sparser, the chance that solid edges and slivers may be overlooked becomes larger.

## 10.6   SHADING

Line drawings, still the most common means of communicating the geometry of mechanical parts, are limited in their ability to portray intricate shapes. Shaded color images convey shape information that cannot be represented in line drawings. Shaded images can also convey features other than shape such as surface finish or material type (plastic or metallic look).

Shaded-image-rendering algorithms filter information by displaying only the visible surface. Many spatial relationships that are unresolved in simple wireframe displays become clear with shaded displays. Shaded images are easier to interpret because they resemble the real objects. Shaded images also have viewing problems not present in wireframe displays. Objects of interest may be hidden or partially obstructed from view, in which case various shaded images may be obtained from various viewing points. Critical geometry such as lines, arcs, and vertices are not explicitly shown. Well-known techniques such as shaded-image/wireframe overlay (Fig. 10-25), transparency, and sectioning can be used to resolve these problems.

One of the most challenging problems in computer graphics is to generate images that appear realistic. The demand for shaded images began in the early 1970s when memory prices dropped enough to make the cost of raster technology attractive compared to the then-prevailing calligraphic displays. In shading a scene (rendering an image), a pinhole camera model is almost universally used. Rendering begins by solving the hidden surface removal problem to determine which objects and/or portions of objects are visible in the scene. As the visible surfaces are found, they must be broken down into pixels and shaded correctly. This process must take into account the position and color of the light sources and the position, orientation, and surface properties of the visible objects.

Careful shading calculations can be distorted by defects in the hardware to display the image. Some of the common defects are noise, the spot size of the deflection beam, and the fidelity of the display. The noise can occur either in the

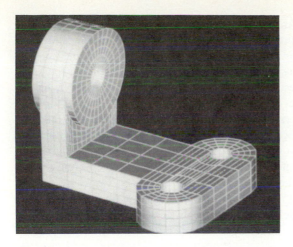

**FIGURE 10-25**
Shaded-image/wireframe overlay.

delivery of intensity (of pixels) information from the bit map of the display to its screen or in the deflection system that steers the beam over the pixel array. The spot size must be adjusted to minimize the user's perception of the raster array. An array of dots appears clearly if the spot size is too small but the sharpness of the image suffers if the spot size is too large. The fidelity of a display is a measure of how the light energy calculated by a shading model is reproduced on the screen. Nonlinearities in the intensity control circuits or in the phosphor response can distort the amount of energy actually emitted at a pixel.

### 10.6.1 Shading Models

Shading models simulate the way visible surfaces of objects reflect light. They determine the shade of a point of an object in terms of light sources, surface characteristics, and the positions and orientations of the surfaces and sources. Two types of light sources can be identified: point light source and ambient light. Objects illuminated with only point light source look harsh because objects are illuminated from one direction only. This produces a flashlight-like effect in a black room. Ambient light is a light of uniform brightness and is caused by the multiple reflections of light from the many surfaces present in real environments.

Shading models are simple. The input to a shading model is intensity and color of light source(s), surface characteristics at the point to be shaded, and the positions and orientations of surfaces and sources. The output from a shading model is an intensity value at the point. Shading models are applicable to points only. To shade an object, a shading model is applied many times to many points on the object. These points are the pixels for a raster display. To compute a shade for each point on a $1024 \times 1024$ raster display, the shading model must be calculated over one million times. These calculations can be reduced by taking advantage of shading coherence; that is, the intensity of adjacent pixels is either identical or very close.

Let us examine the interaction of light with matter to gain an insight into how to develop shading models. Particularly, we consider point light sources shining on surfaces of objects. (Ambient light adds a constant intensity value to the shade at every point.) The light reflected off a surface can be divided into two components: diffuse and specular. When light hits an ideal diffuse surface, it is reradiated equally in all directions, so that the surface appears to have the same brightness from all viewing angles. Dull surfaces exhibit diffuse reflection. Examples of real surfaces that radiate mostly diffuse light are chalk, paper, and flat paints. Ideal specular surfaces reradiate light in only one direction, the reflected light direction. Examples of specular surfaces are mirrors and shiny surfaces. Physically, the difference between these two components is that diffuse light penetrates the surface of an object and is scattered internally before emerging again while specular light bounces off the surface.

The light reflected from real objects contains both diffuse and specular components, and both must be modeled to create realistic images. A basic shading model that incorporates both a point light source and ambient light can be described as follows:

$$\mathbf{I}_P = \mathbf{I}_d + \mathbf{I}_s + \mathbf{I}_b \tag{10.18}$$

where $\mathbf{I}_P$, $\mathbf{I}_d$, $\mathbf{I}_s$, and $\mathbf{I}_b$ are respectively the resulting intensity (the amount of shade) at point $P$, the intensity due to the diffuse reflection component of the point light source, the intensity due to the specular reflection component, and the intensity due to ambient light. Equation (10.18) is written in a vector form to enable modeling of colored surfaces. For the common red, green, and blue color system, Eq. (10.18) represents three scalar equations, one for each color. For simplicity of presentation, we develop Eq. (10.18) for one color, and therefore refer to it as $I_P = I_d + I_s + I_b$ from now on (drop the vector notation).

To develop the intensity components in Eq. (10.18), consider the shading model shown in Fig. 10-26. The figure shows the geometry of shading a point $P$ on a surface $S$ due to a point light source. An incident ray falls from the source to $P$ at an angle $\theta$ (angle of incidence) measured from the surface unit normal $\hat{n}$ at $P$. The unit vector $\hat{I}$ points from the light source to $P$. The reflected ray leaves $P$ with an angle of reflection $\theta$ (equal to the angle of incidence) in the direction defined by the unit vector $\hat{r}$. The unit vector $\hat{v}$ defines the direction from $P$ to the viewing eye.

**10.6.1.1 DIFFUSE REFLECTION.** Lambert's cosine law governs the diffuse reflection. It relates the amount of reflected light to the cosine of the angle $\theta$

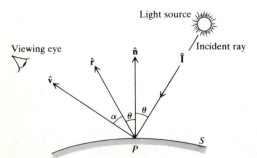

Light source

Viewing eye

Incident ray

**FIGURE 10-26**
The geometry of shading a point.

between $\hat{\mathbf{I}}$ and $\hat{\mathbf{n}}$. Lambert's law implies that the amount of reflected light seen by the viewer is independent of the viewer's position. The diffuse illumination is given by

$$I_d = I_L K_d \cos \theta \qquad (10.19)$$

where $I_L$ and $K_d$ are the intensity of the point light source and the diffuse-reflection coefficient respectively. $K_d$ is a constant between 0 and 1 and varies from one material to another. Replacing $\cos \theta$ by the dot product of $\hat{\mathbf{I}}$ and $\hat{\mathbf{n}}$, we can rewrite Eq. (10.19) as

$$I_d = I_L K_d (\hat{\mathbf{n}} \cdot \hat{\mathbf{I}}) \qquad (10.20)$$

Note that since diffuse light is radiated equally in all directions, the position of the viewing eye is not required by the computations, and the maximum intensity occurs when the surface is perpendicular to the light source. On the other hand, if the angle of incidence $\theta$ exceeds 90, the surface is hidden from the light source and $I_d$ must be set to zero. A sphere shaded with this model (diffuse reflection only) will be brightest at the point on the surface between the center of the sphere and the light source and will be completely dark on the far half of the sphere from the light.

Some shading models assume the point source of light to be coincident with the viewing eye, so no shadows can be cast. For parallel projection, this means that light rays striking a surface are all parallel. This means that $\hat{\mathbf{n}} \cdot \hat{\mathbf{I}}$ is constant for the entire surface, that is, the intensity $I_d$ is constant for the surface, as shown by Eq. (10.20).

**10.6.1.2 SPECULAR REFLECTION.** Specular reflection is a characteristic of shiny surfaces. Highlights visible on shiny surfaces are due to specular reflection while other light reflected from these surfaces is caused by diffuse reflection. The location of a highlight on a shiny surface depends on the directions of the light source and the viewing eye. If you illuminate an apple with a bright light, you can observe the effects of specular reflection. Note that at the highlight the apple appears to be white (not red), which is the color of the incident light.

The specular component is not as easy to compute as the diffuse component. Real objects are nonideal specular reflectors, and some light is also reflected slightly off axis from the ideal light direction (defined by vector $\hat{\mathbf{r}}$ in Fig. 10-26). This is because the surface is never perfectly flat but contains microscopic deformations.

For ideal (perfect) shiny surfaces (such as mirrors), the angles of reflection and incidence are equal. This means that the viewer can only see specular reflected light when the angle $\alpha$ (Fig. 10-26) is zero. For nonideal (nonperfect) reflectors, such as an apple, the intensity of the reflected light drops sharply as $\alpha$ increases. One of the reasonable approximations to the specular component is an empirical approximation and takes the form

$$I_s = I_L W(\theta) \cos^n \alpha \qquad (10.21)$$

For real objects, as the angle of incidence ($\theta$) changes, the ratio of incident light to reflected light also changes, and $W(\theta)$ is intended to model the change. In practice, however, $W(\theta)$ has been ignored by most implementors or very often is set to a constant $K_s$, which is selected experimentally to produce aesthetically pleasing results.

The value of $n$ is the shininess factor and typically varies from 1 to 200, depending on the surface. For a perfect reflector, $n$ would be infinite. $\cos^n \alpha$ reaches a maximum when the viewing eye is in the direction of $\hat{r}(\alpha = 0)$. As $n$ increases, the function dies off more quickly in the off-axis direction. Thus, a shiny surface with a concentrated highlight would have a large value of $n$, while a dull surface with the highlight covering a large area on the surface would have a low value of $n$, as shown in Fig. 10-27. Replacing $\cos \alpha$ by the dot product of $\hat{r}$ and $\hat{v}$, we can rewrite Eq. (10.21) as

$$I_s = I_L W(\theta)(\hat{r} \cdot \hat{v})^n \tag{10.22}$$

If both the viewing eye and the point source of light are coincident at infinity, $\hat{r} \cdot \hat{v}$ becomes constant for the entire surface. This is because $\hat{n} \cdot \hat{v}$, that is, $\cos(\theta + \alpha)$, and $\hat{n} \cdot \hat{l}$, that is, $\cos \theta$, become constant.

Other, and more accurate, shading models for specular reflection have been developed and are available but are not discussed in this book. Among these realistic models are the Blinn and Cook and Torrance models.

### 10.6.1.3 AMBIENT LIGHT. 
Ambient light is a light with uniform brightness. It therefore has a uniform or constant intensity $I_a$. The intensity at point $P$ due to ambient light can be written as:

$$I_b = I_a K_a \tag{10.23}$$

where $K_a$ is a constant which ranges from 0 to 1. It indicates how much of the ambient light is reflected from the surface to which point $P$ belongs.

Substituting Eqs. (10.20), (10.22), and (10.23) into Eq. (10.18), we obtain

$$I_P = I_a K_a + I_L[K_d(\hat{n} \cdot \hat{l}) + W(\theta)(\hat{r} \cdot \hat{v})^n] \tag{10.24}$$

If $W(\theta)$ is set to the constant $K_s$, this equation becomes

$$I_P = I_a K_a + I_L[K_d(\hat{n} \cdot \hat{l}) + K_s(\hat{r} \cdot \hat{v})^n] \tag{10.25}$$

All the unit vectors can be calculated from the geometry of the shading model while constants and intensities on the right-hand side of the above equation are

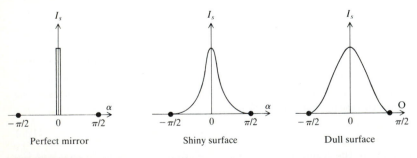

**FIGURE 10-27**
The reflectance of various surfaces as a function of $\alpha$.

assumed by the model. Additional intensity terms can be added to the equation if more shading effects such as shadowing, transparency, and texture are needed. Some of these effects are discussed later in this section.

## 10.6.2 Shading Surfaces

Once we know how to shade a point [Eq. (10.25)] we can consider how to shade a surface. To calculate shading precisely, Eq. (10.25) can be applied to each point on the surface. Relevant points on the surface have the same locations in screen coordinates as the pixels of the raster display. Determining these points is an outcome of hidden surface removal. The normal unit vector $\hat{n}$ used in Eq. (10.25) depends on the surface geometry and can be computed anew for each point of the display. This would require a large number of calculations. Sometimes they are evaluated incrementally. If a bicubic surface is to be shaded in this way, its parametric equation is used to subdivide it into patches whose sizes are equal to or less than the pixel size. After the visible surfaces are determined, the unit normal vector of each patch is calculated exactly and the patch is shaded using Eq. (10.25).

Most surfaces, including those that are curved, are described by polygonal meshes when the visible surface calculations are to be performed by the majority of rendering algorithms. The majority of shading techniques are therefore applicable to objects modeled as polyhedra. Among the many existing shading algorithms, we discuss three of them: constant shading, Gourand or first-derivative shading, and Phong or second-derivative shading.

**10.6.2.1 CONSTANT SHADING.** This the simplest and less realistic shading algorithm. Since the unit normal vector of a polygon never changes, polygons will have just one shade. An entire polygon has a single intensity value calculated from Eq. (10.25). Constant shading makes the polygonal representation obvious and produces unsmooth shaded images (intensity discontinuities). Actually, if the viewing eye or the light source is very close to the surface, the shade of the pixels within the polygon will differ significantly.

The choice of point $P$(Fig. 10-26) within the polygon becomes necessary if the light source and the viewing eye are not placed at infinity. Such a choice affects calculation of the vectors $\hat{l}$ and $\hat{v}$. $P$ can be chosen to be the center of the polygon. On the other hand, $\hat{l}$ and $\hat{v}$ can be calculated at the polygon corners and the average of these values can be used in Eq. (10.25).

**10.6.2.2 GOURAND SHADING.** Gourand shading is a popular form of intensity interpolation or first-derivative shading. Gourand proposed a technique to eliminate (not completely) intensity discontinuities caused by constant shading. The first step in the Gourand algorithm is to calculate surface normals. When a curved surface is being broken down into polygons, the true surface normals at the vertices of the polygons are retained. If more than one polygon shares the same vertex as shown in Fig. 10-28a, the surface normals are averaged to give the vertex normal. If smooth shading between the four polygons shown is required,

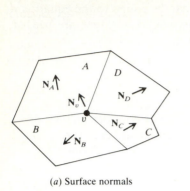

(a) Surface normals

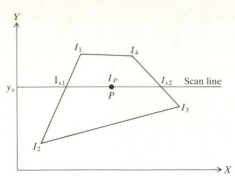

(b) Intensity interpolation along polygon edges

**FIGURE 10-28**
Gourand shading.

then

$$N_v = \tfrac{1}{4}(N_A + N_B + N_C + N_D) \tag{10.26}$$

If shading discontinuities are to be introduced deliberately across an edge to show a crease or a sharp edge in the object, the proper surface normals can be dropped from the above equation. For example, shading discontinuities occur along the $AD$ and $BC$ boundaries shown in Fig. 10-28a if we average only two face normals. $N_v = \tfrac{1}{2}(N_A + N_B)$ and $N_v = \tfrac{1}{2}(N_C + N_D)$ are used to interpolate shades between polygons $A$ and $B$ and $C$ and $D$ respectively. Thus, smooth shading occurs along the $AB$ and $CD$ boundaries while discontinuous shading occurs along the $AD$ and $BC$ boundaries.

The third step in the Gourand algorithm (after calculating surface and vertex normals) is to compute vertex intensities using the vertex normals and the desired shading model [Eq. (10.25)]. The fourth and the last step is to compute the shade of each polygon by linear interpolation of vertex intensities. If the Gourand algorithm is utilized with a scan-line hidden surface algorithm, the intensity at any point $P$ inside any polygon (Fig. 10-28b) is obtained by interpolating along each edge and then between edges along each scan line. This gives

$$I_P = \frac{x_{s2} - x_P}{x_{s2} - x_{s1}} I_{s1} + \frac{x_P - x_{s1}}{x_{s2} - x_{s1}} I_{s2} \tag{10.27}$$

$$I_{s1} = \frac{y_s - y_2}{y_1 - y_2} I_1 + \frac{y_1 - y_s}{y_1 - y_2} I_2 \tag{10.28}$$

$$I_{s2} = \frac{y_4 - y_s}{y_4 - y_3} I_3 + \frac{y_s - y_3}{y_4 - y_3} I_4 \tag{10.29}$$

Gourand shading takes longer than constant shading and requires more planes of memory to get the smooth shading for each color. How are the bits of each pixel divided between the shading grade and the shade color? For example, one may use the red color to obtain a light red shade, a dark red shade, or any variation in between. Let us consider a display with 12 bits of color output, that is, $2^{12}$ or 4096 simultaneous colors. If we decide that 64 shading grades per color are required to obtain fairly smooth shades, then $4096/64 = 64$ gross different colors are possible. Actually only 63 colors are possible as the remaining one is reserved for the background. Within each color, 64 different shading grades are possible. This means that six bits of each pixel are reserved for colors and the other six for shades. The lookup table of the display would reflect this subdivision, as shown in Fig. 10-29. Usually, the six least significant bits (LSB) correspond to the shade and the six most significant bits (MSB) correspond to the color, so that when interpolation is performed to obtain shading grades the six most significant bits (i.e., the color) remain the same.

**10.6.2.3  PHONG SHADING.** While Gourand shading produces smooth shades, it has some disadvantages. If it is used to produce shaded animation (motion sequence), shading changes in a strange way because interpolation is based on intensities and not surface normals that actually change with motion. In addition, Mach bands (a phenomenon related to how the human visual system perceives and processes visual information) are sometimes produced, and highlights are distorted due to the linear interpolation of vertex intensities.

Phong shading avoids all the problems associated with Gourand shading although it requires more computational time. The basic idea behind Phong shading is to interpolate normal vectors at the vertices instead of the shade intensities and to apply the shading model [Eq. (10.25)] at each point (pixel). To

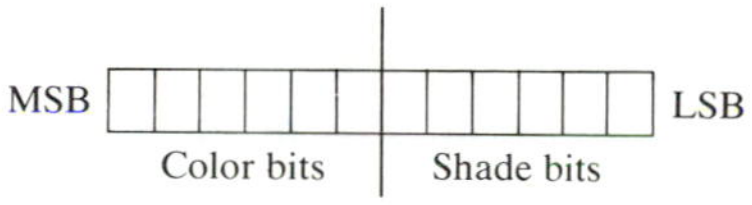

| Address | Color | Shade |
|---|---|---|
| 0 ⋮ 63 | Background | — |
| 64 ⋮ 127 | 1 | Light shade ⋮ Dark shade |
| 128 ⋮ 191 | 2 | Light ⋮ Dark |
| ⋮ | ⋮ | ⋮ |
| 4032 ⋮ 4095 | 63 | Light ⋮ Dark |

**FIGURE  10-29**
Split of pixel bits between colors and shades.

perform the interpolation, Eq. (10.26) can be used to obtain an average normal vector at each vertex. Phong shading is usually implemented with scan-line algorithms. In this case Fig. 10-28*b* is applicable if we replace the intensities by the average normal vectors, $N_v$, at the vertices. Similarly, Eqs. (10.27) to (10.29) are applicable if the intensity variables are replaced by the normal vectors.

### 10.6.3   Shading Enhancements

The basic shading model described in Sec. 10.6.1 is usually enhanced to produce special effects for both artistic value and realism purposes. These effects include transparency, shadows, surface details, and texture.

Transparency can be used to shade translucent material such as glass and plastics or to allow the user to see through the opaque material. Two shading techniques can be identified: opaque and translucent. In the opaque technique, hidden surfaces in every pixel are completely removed. In the translucent method, hidden surfaces are not completely removed. This allows some of the back pixels to show through, producing a screen-door effect.

Consider the box shown in Fig. 10-30. If the front face $F_1$ is made translucent, the back face $F_2$ can be seen through $F_1$. The intensity at a pixel coincident with the locations of points $P_1$ and $P_2$ can be calculated as a weighted sum of the intensities at these two points, that is,

$$I = KI_1 + (1 - K)I_2 \qquad (10.30)$$

where $I_1$ and $I_2$ are the intensities of the front and back faces respectively, calculated using, say, Eq. (10.25). $K$ is a constant that measures the transparency of the front face: when $K = 0$, the face is perfectly transparent and does not change the intensity of the pixel; when $K = 1$, the front face is opaque and transmits no light. Sometimes transparency is referred to as x-ray due to the similarity in effect.

Shadows are important in conveying realism to computer images. More importantly, they facilitate the comprehension of spatial relationships between objects of one image. The complexity of a shadow algorithm is related to the model of the light source. If it is a point source outside the field of view at

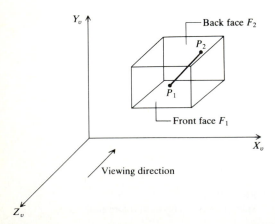

FIGURE 10-30
Transparency and visibility of back faces.

infinity, the problem is simplified. Finding which objects are in shadow is equivalent to solving the hidden surface problem as viewed from the light source. If several light sources exist in the scene, the hidden surface problem is solved several times—every time one of the light sources is considered as the viewing point. The surfaces that are visible to both the viewer and the light source are not shaded. Those that are visible to the viewer but not to the light source are shaded.

Surface details that are usually needed to add realism to the surface image are better treated as shading data than as geometrical data. Consider, say, adding a logo of an object to its image. The logo can be modeled using "surface-detail" polygons. Polygons of the object geometric model point to these surface-detail polygons. If one of the geometric model polygons is to be split for visibility reasons, its corresponding surface-detail polygon is split in the exact fashion. Surface-detail polygons obviously cover the surface polygons when both overlap. When the shaded image is generated, the surface details are guaranteed to be visible with their desired color attributes. Separating the polygons of geometric models and surface details speeds up the rendering of images significantly and reduces the possibility of generating erroneous images.

Texture is important to provide the illusion of reality. For example, modeling of a rough casting should include the rough texture nature of its surfaces. These objects, rich in high frequencies, could be modeled by many individual polygons, but as the number of polygons increases, they can easily overflow the modeling and display programs. Texture mapping (Fig. 10-31) is introduced to solve this problem and provide the illusion of complexity at a reasonable cost. It is a method of "wallpapering" the existing polygons. As each pixel is shaded, its corresponding texture coordinates are obtained from the texture map, and a lookup is performed in a two-dimensional array of colors containing the texture. The value in this array is used as the color of the polygon at this pixel, thus providing the "wallpaper."

Three-dimensional texture mapping is easier to use with three-dimensional objects than two-dimensional texture mapping. It is a function that maps the object's spatial coordinates into three-dimensional texture space and uses three-

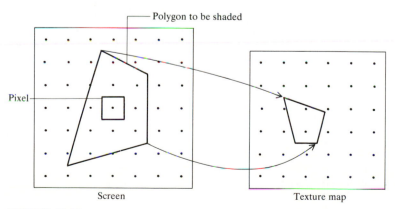

**FIGURE 10-31**
Texture mapping.

dimensional textures. Thus, no matter what the object's shape is, the texture on its surface is consistent. This is useful to model materials such as wood and marble. Due to the space required to store a three-dimensional array of pixels, procedural textures can be used. However, they are difficult to be antialiased. Texture mapping can contain other surface properties besides color to increase the illusion of complexity. For example, surface normal perturbations could be stored in the texture map (bump mapping) to enable the simulation of wrinkled surfaces.

**Example 10.1.** Apply the shading model given by Eq. (10.25) to the scene of the two boxes shown in Fig. 10-13a. Assume that the visible surfaces have been identified as shown in Fig. 10-13c. Use $n = 100$ for the specular reflection component.

*Solution.* Let us assume that the point light source is placed at infinity, that is, coincident with the viewing eye. Table 10.2 can be derived to enable use of Eq. (10.25). The table shows the components of the vectors of each face in the VCS shown in Fig. 10-13a. Substituting the values in Table 10.2 into Eq. (10.25) gives

Face $F_1$:         $I_P = I_a K_a + I_L[-K_d + (K_s)^{100}]$

Face $F_2$:         $I_P = I_a K_a + I_L(K_s)^{100}$

Face $F_3$:         $I_P = I_a K_a + I_L(K_s)^{100}$

Face $F_4$:         $I_P = I_a K_a + I_L[-K_d + (K_s)^{100}]$

Face $F_5$:         $I_P = I_a K_a + I_L(K_s)^{100}$

Face $F_6$:         $I_P = I_a K_a + I_L(K_s)^{100}$

The intensities $I_a$ and $I_L$ are chosen based on the maximum intensity a pixel may have. The coefficients $K_a$, $K_d$, and $K_s$ are chosen based on experimental measurements. The above equations assume constant shading. Notice that the shading of $F_1$ and $F_4$, and $F_2$, $F_3$, $F_5$, and $F_6$ are equal. This will make $F_2$ and $F_3$ and $F_5$ and $F_6$ indistinguishable in the shaded image. A solution to this problem would be to use Gourand or Phong shading. The reader is encouraged to calculate both of them as an exercise and compare intensities (see the problems at the end of the chapter).

## 10.6.4   Shading Solids

Shading is one of the most popular applications of solid modeling. In fact, shading and solid modeling are often erroneously equated. Shading algorithms of solids can be developed based on exact solid's representation schemes (B-rep and CSG) or on some approximations of these schemes (faceted B-rep). A consider-

**TABLE 10.2**
**Vectors for Example 10.1**

| Face | $\hat{n}$ | $\hat{l}$ | $\hat{r}$ | $\hat{v}$ |
|------|-----------|-----------|-----------|-----------|
| $F_1$ | (0, 0, 1) | (0, 0, −1) | (0, 0, 1) | (0, 0, 1) |
| $F_2$ | (1, 0, 0) | (0, 0, −1) | (0, 0, 1) | (0, 0, 1) |
| $F_3$ | (0, 1, 0) | (0, 0, −1) | (0, 0, 1) | (0, 0, 1) |
| $F_4$ | (0, 0, 1) | (0, 0, −1) | (0, 0, 1) | (0, 0, 1) |
| $F_5$ | (1, 0, 0) | (0, 0, −1) | (0, 0, 1) | (0, 0, 1) |
| $F_6$ | (0, 1, 0) | (0, 0, −1) | (0, 0, 1) | (0, 0, 1) |

able number of existing solid modelers utilize the latter schemes to speed up the rendering of a shaded image. The rationale behind this approach is that shaded images usually serve the purpose of visualization only. While the solid modeler maintains its exact representation scheme internally for analysis purposes (mass property calculations, NC tool path generation, and finite element modeling), an approximate representation (polygonal approximation of exact geometry) is derived for shading purposes. Sometimes, these exact and approximate (for visualization purposes) representations are referred to as analytic and visual solid modelers respectively.

A wide variety of shading algorithms of solids exist and are directly related to the representation scheme utilized. Three of these algorithms are described later in this section. The most popular algorithms are:

1. Shaded displays can be generated from B-reps (exact or faceted) by the algorithms of surface shading discussed in Secs. 10.6.2 and 10.6.3. These algorithms are particularly simple when B-reps are triangulations or other tessellations. Special-purpose tiling engines based on algorithms for displaying tessellations are commercially available. Other VLSI chips for high-speed display of tessellation are also available.

2. Algorithms to generate shaded images of octree and spatial enumeration representations exist and have been implemented in special-purpose hardware. Octree (for the first) and voxel (for the second) machines are available.

3. Shading of CSG models can be achieved directly by utilizing the CSG tree or indirectly by converting CSG into B-rep and then utilizing B-rep-based algorithms. Two of the popular shading algorithms that work directly on CSG are ray-racing and $z$-buffer (depth-buffer) methods. Ray tracing is general and can also be applied to all solid modeling representations.

**10.6.4.1  A RAY-TRACING ALGORITHM FOR CSG.** The ray-tracing algorithm described in Sec. 10.5.1 is considered a natural tool for making shaded images. To make a shaded image, cast one ray per pixel into the scene and identify the visible surface $S_1$ (Fig. 10-19) in this pixel. Compute the surface normal at the visible point $t_1$. Use Eq. (10.25) to calculate the pixel intensity value. Processing all pixels in this way produces a raster-type picture of the scene.

Special effects and additional realism are possible by adding transparency and shadowing. Transparent surfaces may be modeled with or without refraction. Nonrefractive transparency is easy because the ray remains as one straight line (no refraction by going through the solid). With transparency, more than one surface may be visible at the pixel. If surface $S_1$ (Fig. 10-19) transmits any light, then the surface normal at point $t_2$ must be computed to calculate the intensity at the point. The net intensity at the given pixel can be calculated using Eq. (10.30). If $S_2$ transmits light, then $S_3$ must be processed in a similar way, etc.

To model shadows, the following procedure is executed for each point light source:

1. Cast a second ray connecting the visible point $t_1$ with the point light source as shown in Fig. 10-32.

2. If the ray intersects any surface $(S_3)$ between the visible (relative to the viewing eye) point and the light source, then the point is in shadow.
3. If the surface that shadows $(S_3)$ is transparent, then attenuate the intensity of the light that passes through and add it to the intensity of the pixel at point $t_1$.

**10.6.4.2  A $z$-BUFFER ALGORITHM FOR B-REP AND CSG.** Depth-buffer or $z$-buffer algorithms that operate on B-reps, especially polygonal approximations, are simple and well known. The basic algorithm defines two arrays, the depth array $z(x, y)$ and the intensity array $I(x, y)$, to store the depth and intensity information of the screen pixels located by $(x, y)$ values. Considering Fig. 10-33, a $z$-buffer algorithm for B-rep can be described as follows. Initialize the $z(x, y)$ and $I(x, y)$ arrays with a large number and background intensity respectively. For each face of the solid, check the distance $d$ between each point on the face $(P_1$ and $P_2)$ and the viewing eye. If $d$ is smaller than the $z$ value in the proper element of the $z(x, y)$ array, write $d$ onto $z(x, y)$ and compute the intensity using Eq. (10.25). Update the intensity buffer by writing the intensity value onto $I(x, y)$. At the end of the scan, the intensity buffer contains the correct image values. Typically, the frame buffer of the graphics display is used to store the intensity array. Thus, the display can be updated incrementally, whenever a new intensity for a pixel is computed, by writing the new value onto the frame buffer.

To apply the above-described $z$-buffer algorithm for B-rep to CSG models, it must be modified to reflect the fact that the actual solid's faces are not available. Instead, the primitives' faces must be scanned to determine the elements of the $z(x, y)$ and $I(x, y)$ arrays. Scanning the primitives' faces yields a superset of the points needed by the $z$-buffer algorithm. Exploiting the generate-and-classify paradigm based on the point-membership classification, we can discard the

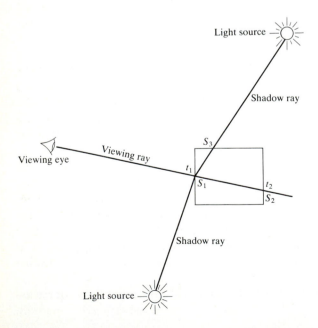

**FIGURE 10-32**
Modeling shadows.

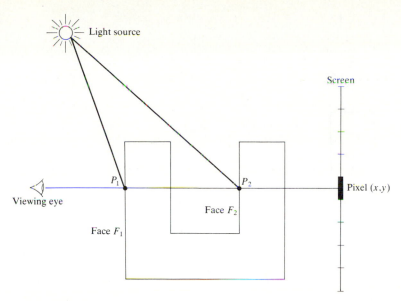

**FIGURE 10-33**
The z-buffer algorithm for B-rep models.

points, from the superset, that are not on the actual solid's faces. The z-buffer algorithm for CSG models is described as follows. Initialize the $z(x, y)$ and $I(x, y)$ arrays as in B-rep. For each face of each primitive of the solid, check the distance $d$ between each point on the face and the viewing eye. If $d$ is smaller than the $z$ value in the proper element of $z(x, y)$, then classify the point with respect to the solid (by traversing the CSG tree). If the point is classified as "on" the solid, then write $d$ onto $z(x, y)$ and compute the intensity and write it onto $I(x, y)$. Otherwise, if the point is classified as "in" or "out" the solid, discard it and move to the next point.

**10.6.4.3 AN ALGORITHM FOR OCTREE-ENCODED OBJECTS.** Display algorithms for octrees transform an octree representation of an object into a quadtree-encoded image. Quadtree represents an ideal structure of the shaded image because it is compact. Algorithms are applicable to parallel projections. A shading algorithm for octrees must first solve the hidden surface removal problem before applying a shading model. This problem is easy to solve for octree representation because the octree of the object can be traversed in a prescribed order. For example, if the octants of any node in the octree are processed in a depth-first manner and in an ascending order (0, 1, 2, ..., 7) (see Fig. 10-34), the front four octants of the universe (0, 1, 2, 3) are visited before the rear four (4, 5, 6, 7). Likewise, the front four octants (sons) of octant "0" are processed before its rear four. This continues recursively so that the frontmost octants of the universe are visited first and the rearmost octants are visited last. Similar observations can be made if octants are processed in a descending order.

The input to the hidden surface algorithm is an octant and the quadrant behind it, as shown in Fig. 10-35a. The result of the algorithm is a quadrant

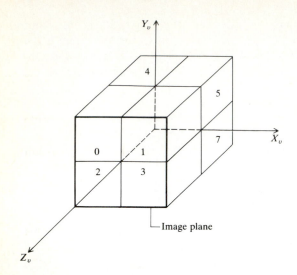

Image plane

**FIGURE 10-34**
Numbering and viewing octants of a
universe cube.

which represents the image area directly in front of the octant. The octant
remains unaffected during the hidden surface removal process. The hidden
surface algorithm proceeds recursively as follows. When the octant is homogeneous (completely full or completely empty), the corresponding quadrant is
always defined. If an octant is completely full (fully contained in the object), then
the old quadrant (back one) is obscured and overwritten by a new quadrant
(front one) of the octant. The new quadrant has a color that is dictated by the
material that overwrites it. If the octant is completely empty, the old quadrant is
not changed. When the octant is heterogeneous (partially full), the quadrant is
undefined and a new quadtree node in the quadtree structure must be created.

The value of each quadrant of the new quadtree node is determined by the
pair of octants aligned with that quadrant. Octants 0 and 4, 1 and 5, 2 and 6, and
3 and 7 determine the results of quadrants 0, 1, 2, and 3 respectively. The rear

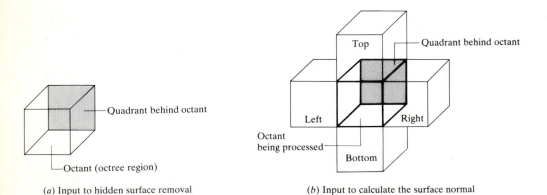

(a) Input to hidden surface removal

(b) Input to calculate the surface normal

**FIGURE 10-35**
Input parameters to an algorithm to shade octree-encoded objects.

octants (4, 5, 6, 7) are processed by the hidden surface algorithm before the front (0, 1, 2, 3) for all four quadrants. This is because the input to the algorithm is the rear quadrant to the octant being processed and the output is the front quadrant. After both the rear and front octants are processed, the quadtree holds the visible result from the octree.

The visible quadrants stored in the quadtree structure can be shaded via a suitable shading model, e.g., Eq. (10.25). Surface normals are required to compute the intensity of light reflected to the viewing eye. In the octree structure, no explicit information about surface orientation is stored. The shading algorithm for octree-encoded objects must therefore extract as much surface shading information as possible. To approximate the surface normal, the four octants representing the top, bottom, left, and right (Fig. 10-35b) neighbors of the octant being processed (call it the active octant) are utilized. If the active octant has a void (completely empty) octant on the left side, a surface normal exists in this direction. If any other neighbors are void, surface normal vectors must lie along the same area as well. The computations of these vectors are easy because the spatial orientations of the octants in the octree are known.

Transparency can be modeled with octree structures. For opaque materials (as described above), the quadtree is simply overwritten if an octant is found to obscure it. For transparent materials, however, the intensity of the old quadrant is used to add a contribution to the existing quadrant by using, say, Eq. (10.30).

## 10.7 COLORING

The use of colors in CAD/CAM has two main objectives: facilitate creating geometry and display images. Colors can be used in geometric construction. In this case various wireframe, surface, or solid entities can be assigned different colors to distinguish them. Color is one of the two main ingredients (the second being texture) of shaded images produced by shading algorithms. In some engineering applications such as finite element analysis, colors can be used effectively to display contour images such as stress or heat-flux contours.

Black and white raster displays provide achromatic colors while color displays (or television sets) provide chromatic color. Achromatic colors are described as black, various levels of gray (dark or light gray), and white. The only attribute of achromatic light is its intensity, or amount. A scalar value between 0 (as black) and 1 (as white) is usually associated with the intensity. Thus, a medium gray is assigned a value of 0.5. For multiple-plane displays, different levels (scale) of gray can be produced. For example, 256 ($2^8$) different levels of gray (intensities) per pixel can be produced for an eight-plane display. The pixel value $V_i$ (which is related to the voltage of the deflection beam) is related to the intensity level $I_i$ by the following equation:

$$V_i = \left(\frac{I_i}{C}\right)^{1/\gamma} \tag{10.31}$$

The values $C$ and $\gamma$ depends on the display in use. If the raster display has no lookup table, $V_i$ (e.g., 00010111 in an eight-plane display) is placed directly in the

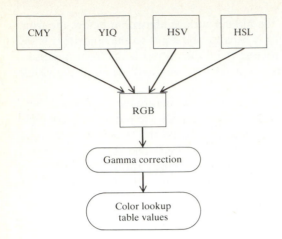

**FIGURE 10-36**
Transformation of a color model to RGB.

proper pixel. If there is a table, $i$ is placed in the pixel and $V_i$ is placed in entry $i$ of the table. Use of the lookup table in this manner is called gamma correction, after the exponent in Eq. (10.31).

Chromatic colors produce more pleasing effects on the human vision system than achromatic colors. However, they are more complex to study and generate. Color is created by taking advantage of the fundamental trichromacy of the human eye. Three different colored images are combined additively at photo-receptors in the eye to form a single perceived image whose color is a combination of the three prime colors. Each of the three images is created by an electron gun acting on a color phosphor. Using shadow-mask technology, it is possible to make the images intermingle on the screen, causing the colors to mix together because of spatial proximity. Well-saturated red, green, and blue colors are typically used to produce the wide range of desired colors.

Color descriptions and specifications generally include three properties: hue, saturation, and brightness. Hue associates a color with some position in the color spectrum. Red, green, and yellow are hue names. Saturation describes the vividness or purity of a color or it describes how diluted the color is by white light. Pure spectral colors are fully saturated colors and grays are desaturated colors. Brightness is related to the intensity, value, or lightness of the color.

There exists a rich wealth of studies and methods of how to specify and measure colors. Some methods are subjective such as Munsell and pigment-mixing methods. The Munsell method is widely used and is based on visually comparing unknown colors against a set of standard colors. The pigment-mixing method is used by artists. Other methods used in physics are objective and treat visible light with a given wavelength as an electromagnetic energy with a spectral energy distribution. Our primary interest in this section is not to review these studies and methods but to describe some existing color models, so that application programs can choose the desired colors properly. We will also show how some of these models can be converted to red, green, and blue since most of the commonly used CRTs demand three digital values, specifying an intensity for each of the colors.

## 10.7.1 Color Models

A color model or a space is a three-dimensional color coordinate system to allow specifications of colors within some color range. Each displayable color is represented by a point in a color model. There are quite a number of color models available. Some popular models are discussed here. These models are based on the red, green, and blue (RGB) primaries. For any one of these models, coordinates are translated into three voltage values in order to control the display. This process is shown in Fig. 10-36, which summarizes the sequence of transformation for some models. The gamma correction is performed to obtain a linear relationship between digital RGB values and the intensity of light emitted by the CRT.

**10.7.1.1  RGB MODEL.** The RGB color space uses a cartesian coordinate system as shown in Fig. 10-37a. Any point (color) in the space is obtained from the three RGB primaries; that is, the space is additive. The main diagonal of the cube is

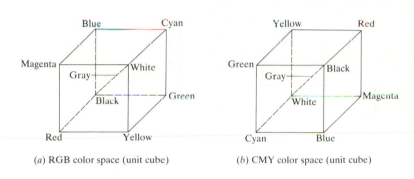

(a) RGB color space (unit cube)          (b) CMY color space (unit cube)

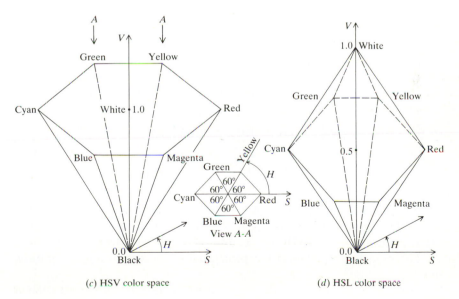

(c) HSV color space          (d) HSL color space

**FIGURE 10-37**
Some color models.

the locus of equal amounts of each primary and therefore represents the gray scale or levels. In the RGB model, black is at the origin and represented by (0, 0, 0) and white is represented by (1, 1, 1). Thus in the RGB model, the lowest intensity (0 for each color) produces the black color and the maximum intensity (1 for each color) produces the white color.

**10.7.1.2 CMY MODEL.** The CMY (cyan, magneta, yellow) model shown in Fig. 10-37b is the complement of the RGB model. The cyan, magneta, and yellow colors are the complements of the red, green, and blue respectively. The white is at the origin (0, 0, 0) of the model and the black is at point (1, 1, 1) which is opposite to the RGB model. The CMY model is considered a subtractive model because the model primary colors subtract some color from white light. For example, a red color is obtained by subtracting a cyan color from the white light (instead of adding magneta and yellow).

The conversion from CMY to RGB is achieved by the following equation:

$$
\begin{bmatrix} R \\ G \\ B \end{bmatrix} = \begin{bmatrix} 1 \\ 1 \\ 1 \end{bmatrix} - \begin{bmatrix} C \\ M \\ Y \end{bmatrix}
\tag{10.32}
$$

The unit column vector represents white in the RGB model or black in the CMY model.

**10.7.1.3 YIQ MODEL.** The YIQ space is used in raster color graphics. It has been in use as a television broadcast standard since 1953 when it was adopted by the National Television Standards Committee (NTSC) of the United States. It was designed to be compatible with black and white television broadcast. The $Y$ axis of the color model corresponds to the luminance (the total amount of light). The $I$ axis encodes chrominance information along a blue-green to orange vector and the $Q$ axis encodes chrominance information along a yellow-green to magneta vector.

The conversion from YIQ coordinates to RGB coordinates is defined by the following equation:

$$
\begin{bmatrix} R \\ G \\ B \end{bmatrix} = \begin{bmatrix} 1.0 & 0.95 & 0.62 \\ 1.0 & -0.28 & -0.64 \\ 1.0 & -1.11 & 1.73 \end{bmatrix} \begin{bmatrix} Y \\ I \\ Q \end{bmatrix}
\tag{10.33}
$$

**10.7.1.4 HSV MODEL.** This color model (shown in Fig. 10-37c) is user oriented because it is based on what artists use to produce colors (hue, saturation, and value). It is contrary to the RGB, CMY, and YIQ models which are hardware oriented. The model approximates the perceptual properties of hue, saturation, and value.

The conversion from HSV coordinates to RGB coordinates can be defined as follows. The hue value $H$ (range from 0° to 360°) defines the angle of any point on or inside the single hexacone shown in Fig. 10-37c. Each side of the cone bounds $60°$ as shown in view $A$-$A$. If we divide a given $H$ by 60, we obtain an integer part $i$ and a fractional part $f$. The integer part $i$ is between 0 and 5 ($H = 360$ is treated as if $H = 0$). Let us define the following quantities:

$$\begin{bmatrix} a \\ b \\ c \end{bmatrix} = V \begin{bmatrix} 1 - S \\ 1 - Sf \\ 1 - S(1 - f) \end{bmatrix} \tag{10.34}$$

Then we can write:

$$(R, G, B) = \begin{cases} (V, c, a) & i = 0 \\ (b, V, a) & i = 1 \\ (a, V, c) & i = 2 \\ (a, b, V) & i = 3 \\ (c, a, V) & i = 4 \\ (V, a, b) & i = 5 \end{cases} \tag{10.35}$$

**10.7.1.5  HSL MODEL.** The HSL (hue, saturation, lightness) color model shown in Fig. 10-37$d$ forms a double hexacone space. It is used by Tektronix. The saturation here occurs at $V = 0.5$ and not 1.0 as in the HSV model. The HSL model is as easy to use as the HSV model. The conversion from HSL to RGB is possible by using the geometry of the double hexacone as we did with the HSV model.

## 10.8  USER INTERFACE FOR SHADING AND COLORING

Major commercial CAD/CAM systems provide their users with shading packages that implement most of the shading and coloring concepts covered in Secs. 10.6 and 10.7. A shading command with the appropriate shading attributes and modifiers allows users to shade either surface or solid models. A generic syntax for a typical shading command may look as follows:

SHADE⟨geometric model type⟩⟨shading modifiers⟩⟨shading resolution⟩ ⟨geometric model entities⟩

The type of geometric model could be a "surface" or "solid." The "shade surface" command would require the user to digitize all the surfaces of the model that are to be shaded. The back invisible surfaces of a model need not be digitized. Digitizing surfaces could be very time-consuming and error-prone for complex and dense models. Users can use the "window" modifier offered by their respective systems to select all the entities displayed in one view by defining a window around them. The "shade solid" command, on the other hand, identifies the solid to be shaded by digitizing it only once. This reflects the completeness and unambiguities of solid models as discussed in Chap. 7.

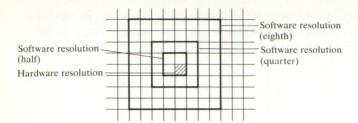

**FIGURE 10-38**
Relationship between hardware and software resolutions.

The shading modifiers of the "shade" command include all the input parameters and their values, required to define the desired shading model. These modifiers include a background color, a location of the point light source, and a desired intensity of an ambient light (between 0 and 1). The background color is the color of the picture (image) background. The location of the point light source is the coordinates $(x, y, z)$ of a point input relative to the MCS. Other modifiers to specify enhancements of shading models such as transparency and texture are also possible.

The shading resolution is an important input parameter because it controls the quality of the image to be generated. The higher the resolution of the image, the higher its quality is and the more CPU time and memory space it takes to generate it and store it. A low resolution is recommended when the user is experimenting in search for the best combination of colors and shading modifiers, especially the best location of the point light source to best illuminate the model. How can the resolution of an image be made different from the resolution of the graphics display? Let us define two types of resolution: hardware and software. A hardware resolution is the actual resolution of the display and is equal to the number of pixels in both the horizontal and vertical directions, as discussed in Chap. 2. A software resolution is a scale of hardware resolution and is used by shading software to control the quality of a shaded image. Consider, for example, a graphics display of hardware resolution of 1280 × 1024. If the shading software scales this resolution by a factor of 2, 4, and 8, half (640 × 512), quarter (320 × 256), and eighth (160 × 128) software resolutions result respectively. Figure 10-38 shows the relationship between hardware and software resolutions. The solid boundaries in the figure show the size of the software (virtual) pixel. Figure 10-39 shows the effect of resolution on the images of a surface model of a bracket.

The geometric model entities required by the "shade" command are the surfaces for surface models or the solids for solid models. The command usually allows the user to specify the desired colors of the entities at the time they are digitized. If different colors are required for different surfaces, the surfaces would have to be digitized explicitly and the "window" modifier described earlier would be useless. If a solid model is to be shaded with different colors, the user would digitize subsolids (intermediate solids) that make the solid instead of digitizing the final solid. Figure 10-40 shows the image of a solid model of a fly-cutter of a Bridgeport milling machine. The background shows views of the solid model

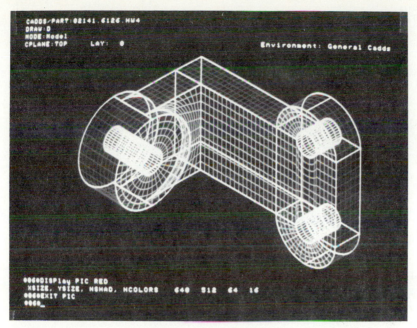

(a) Surface model of bracket

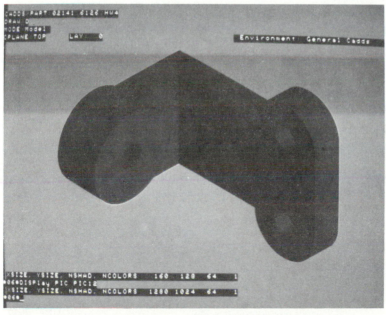

(b) Hardware resolution (1280 × 1024)

**FIGURE 10-39**

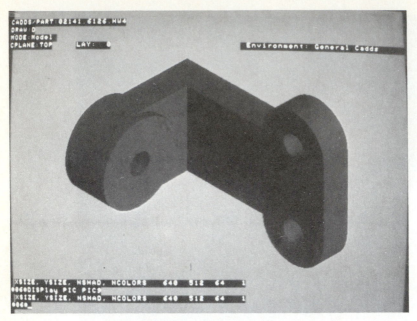

(c) Half resolution (640 × 512)

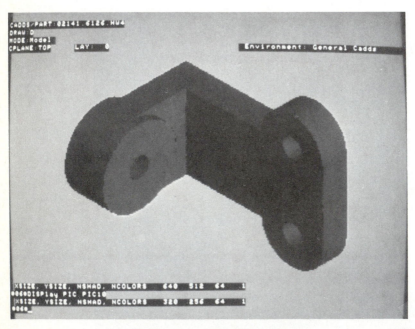

(d) Quarter resolution (320 × 256)

**FIGURE 10-39**
**(continued)**

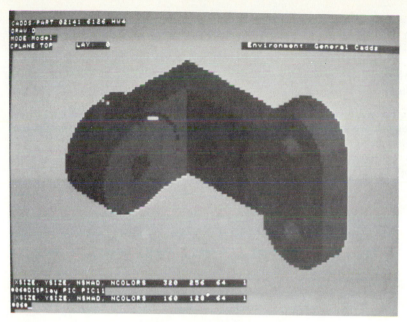

(e) Eighth resolution (160 × 128)

**FIGURE 10-39**
Effect of resolution on shaded images of a bracket.

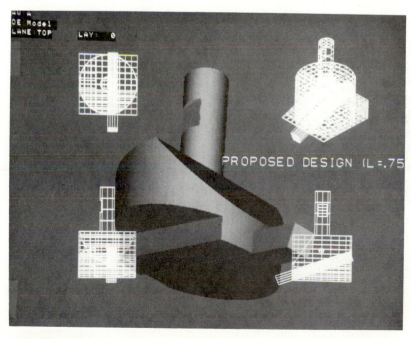

**FIGURE 10-40**
A shaded image of a solid model.

(with a surface mesh displayed) before shading. The software resolution of the image is $640 \times 512$.

## PROBLEMS
## Part 1: Theory

**10.1.** Hidden line removal algorithms can be applied to wireframe and surface models if additional input information is supplied by users. What is this information and how can it be input?

**10.2.** Sketch the minimax boxes for the tangent polygons shown in Fig. P10-2. What conclusions can you make?

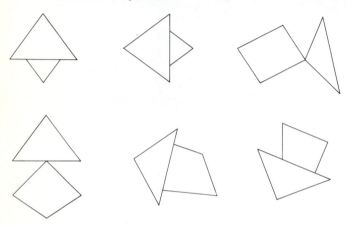

**FIGURE P10-2**

**10.3.** Find the sum $(\sum \Delta\theta)$ of the angles subtended by the edges of the polygons shown in Fig. P10-3 with respect to point $P$. What conclusions can you make?

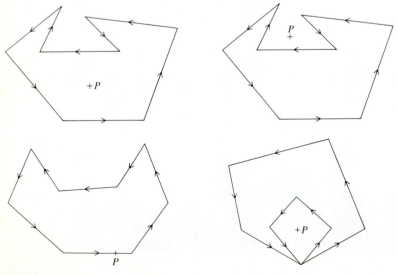

**FIGURE P10-3**

**10.4.** Find the face and priority lists of scenes in Fig. P10-4. Assume that back faces have been removed to simplify the problem.

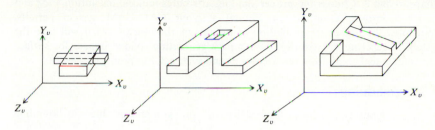

**FIGURE P10-4**

**10.5.** Apply the area-oriented algorithm to the scenes of Prob. 10.4.

**10.6.** Find the ray/sphere, ray/cone, and ray/torus intersections and the corresponding bounds tests. Use Eqs. (7.69) to (7.71). For the ray/torus intersection, a fourth-order polynomial in $t$ results which has a closed-form solution.

**10.7.** Solve Example 10.1 for both Gourand and Phong shading. How do you ensure that edges are clearly visible in the resulting shading images?

**10.8.** Apply shading solid techniques to the models shown in Fig. P10-8.

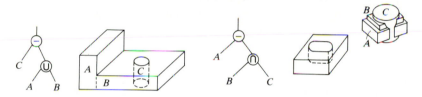

**FIGURE P10-8**

**10.9.** Generating and displaying contour images in engineering applications (e.g., stress contours in finite element analysis) provide designers with valuable information for sound design decisions. Propose a method and algorithm to develop these contours and their images.

**10.10.** Derive the equation that converts HSL coordinates into RGB coordinates.

## Part 2: Laboratory

**10.11.** For each of the geometric models of Prob. 5.20 define a drawing with the layout shown in Fig. P10-11. Perform the model clean-up on each drawing. Add dimensions as well.

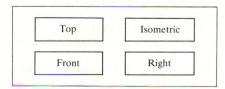

**FIGURE P10-11**

**10.12.** Create the shaded images of the surface and solid models of each of the geometric models of Prob. 5.20. These models should exist if you have solved lab problems in Chaps. 6 and 7. Choose the proper shading attributes (colors, resolution, etc.) and models (constant, Gourand, or Phong) as your CAD/CAM system permits. Remember the higher the resolution of the image, the more CPU it will take the system to generate it. How do you compare generating shaded images for surface and solid models both from the user input and system CPU points of view?

## Part 3: Programming

**10.13.** Write a program that sorts a list of vertices of a polygon and produces its bounding box.

**10.14.** Write a program to classify a ray for a ray-tracing algorithm:
(*a*) Against a primitive
(*b*) Against a solid

**10.15.** Write a program that would compute box enclosures around primitives in the SCS. Extend the program to combine the enclosures at tree nodes for the three operators $\cup^*$, $\cap^*$, and $-^*$.

**10.16.** Write a program to implement Eq. (10.25) for the case of a parallel (orthographic) view with a point light source coincident with the viewing eye. Implement it into a given display if you have access to its hardware details.

**10.17.** Write a program to implement ray tracing to generate shaded images.

**10.18.** Write a program to implement the $z$ buffer to generate shaded images.

**10.19.** Write a program to shade octree-encoded objects.

## BIBLIOGRAPHY

Amanatides, J.: "Realism in Computer Graphics: A Survey," *IEEE Computer Graphics and Applic.*, vol. 7, no. 1, pp. 44–56, January 1987.

Amanatides, J., and A. Woo: "A Fast Voxel Traversal Algorithm for Ray Tracing," in *Proc. Eurographics '87 Conf. and Exhibition*, August 24–28, 1987, pp. 3–10, Amsterdam, The Netherlands.

Andreson, D. P.: "Hidden Line Elimination in Projected Grid Surfaces," *ACM Trans. on Graphics*, vol. 1, no. 4, pp. 274–288, 1982.

Arvo, J., and D. Kirk: "Fast Ray Tracing by Ray Classification," *Computer Graphics*, vol. 21, no. 4, pp. 55–64, 1987 (Siggraph '87 Conference Proceedings, July 27–31, 1987, Anaheim, Calif.).

Barnell, J.: "More Realism in Workstation Graphics," *Mach. Des.*, vol. 57, no. 27, pp. 87–91, November 21, 1985.

Burnett, C.: "Understanding Image Quality in Computer Graphics," *Computers in Mech. Engng (CIME) Mag.*, vol. 3, no. 6, pp. 32–38, 1985.

Burton, R. P., and D. R. Smith: "A Hidden-Line Algorithm for Hyperspace," *SIAM J. Comput.*, vol. 11, no. 1, pp. 71–80, 1982.

Cabral, B., N. Max, and R. Springmeyer: "Bidirectional Reflection Functions from Surface Bump Maps," *Computer Graphics*, vol. 21, no. 4, pp. 273–282, 1987 (Siggraph '87 Conference Proceedings, July 27–31, 1987, Anaheim, Calif.).

Chen, L., G. T. Herman, R. A. Reynolds, and J. K. Udupa: "Surface Shading in the Cuberille Environment," *IEEE Computer Graphics and Applic.*, vol. 5, no. 12, pp. 33–43, 1985.

"Digital Color Halftone Reproduction," *IBM Tech. Disclosure Bull.*, vol. 28, no. 1, pp. 438–439, 1985.

Duff, T.: "Compositing 3-D Rendered Images," *Computer Graphics*, vol. 19, no. 3, pp. 41–44, 1985 (Siggraph '85 Conference Proceedings, July 22–26, 1985, San Francisco, Calif.).

Encarnacao, J., and E. G. Schlechtendahl: *Computer Aided Design: Fundamentals and System Architectures*, Springer-Verlag, Berlin, Germany, 1983.

Foley, J. D., and A. van Dam: *Fundamentals of Interactive Computer Graphics*, Addison-Wesley, Reading, Mass., 1982.

Giloi, W. K.: *Interactive Computer Graphics: Data Structures, Algorithms, Languages*, Prentice-Hall, Englewood Cliffs, N.J., 1978.

Hornung, C.: "An Approach to a Calculation-Minimized Hidden Line Algorithm," *Computers and Graphics*, vol. 6, no. 3, pp. 121–126, 1982.

Hu, M., and J. D. Foley: "Parallel Processing Approaches to Hidden-Surface Removal in Image Space," *Computers and Graphics*, vol. 9, no. 3, pp. 303–317, 1985.

Jansen, F. W.: "A Pixel-Parallel Hidden Surface Algorithm for Constructive Solid Geometry," in *Proc. Eurographics '86 Conference and Exhibition*, August 19-25, 1986, pp. 29–40, Lisbon, Portugal.

Kamada, T., and S. Kawai: "An Enhanced Treatment of Hidden Lines," *ACM Trans. on Graphics*, vol. 6, no. 4, pp. 308–323, 1987.

Klein, A.: "Modelling 3D Shaded Solids of Arbitrary Shape Using an Edge-Oriented Algorithm," *Computers and Graphics*, vol. 10, no. 4, pp. 327–331, 1986.

Laitano, E. E., and S. R. Idelsohn: "A Simple Hidden Line Algorithm for a Structural Model of Planar Elements," *Adv. Engng Software*, vol. 8, no. 1, pp. 2–7, 1986.

Lien, S., M. Shantz, and V. Pratt: "Adaptive Forward Differencing for Rendering Curves and Surfaces," *Computer Graphics*, vol. 21, no. 4, pp. 111–118, 1987 (Siggraph '87 Conference Proceedings, July 27–31, 1987, Anaheim, Calif.).

Mitchell, D. P.: "Generating Antialiased Images at Low Sampling Densities," *Computer Graphics*, vol. 21, no. 4, pp. 65–72, 1987 (Siggraph '87 Conference Proceedings, July 27–31, 1987, Anaheim, Calif.).

Mortenson, M. E.: *Geometric Modeling*, John Wiley, New York, 1985.

Newman, W. M., and R. F. Sproull: *Principles of Interactive Computer Graphics*, 2d ed., McGraw-Hill, New York, 1979.

Nishita, T., Y. Miyawaki, and E. Nakamae: "A Shading Model for Atmospheric Scattering Considering Luminous Intensity Distribution of Light Sources," vol. 21, no. 4, pp. 303–310, 1987 (Siggraph '87 Conference Proceedings, July 27–31, 1987, Anaheim, Calif.).

Nishita, T., I. Okamura, and E. Nakamae: "Shading Models for Point and Linear Sources," *ACM Trans. on Graphics*, vol. 4, no. 2, pp. 124–146, 1985.

Ohno, Y.: "A Hidden Line Elimination Method for Curved Surfaces," *CAD J.*, vol. 15, no. 4, pp. 209–216, 1983.

Okino, N., Y. Kakazu, and M. Morimoto: "Extended Depth-Buffer Algorithms for Hidden-Surface Visualization," *IEEE Computer Graphics and Applic.*, vol. 4, no. 5, pp. 79–88, 1984.

Ottmann, T., P. Widmayer, and D. Wood: "A Worst-Case Efficient Algorithm for Hidden-Line Elimination," *Int. J. Computer Math.*, vol. 18, no. 2, pp. 93–119, 1985.

Peng, Q., Y. Zhu, and Y. Liang: "A Fast Ray Tracing Algorithm Using Space Indexing Techniques," *Proc. Eurographics '87 Conf. and Exhibition*, August 24-28, 1987, pp. 11–23, Amsterdam, The Netherlands.

Rankin, J. R.: "A Geometric Hidden-Line Processing Algorithm," *Computers and Graphics*, vol. 11, no. 1, pp. 11–19, 1987.

Reeves, W. T., D. H. Salesin, and R. L. Cook: "Rendering Antialiased Shadows with Depth Maps," *Computer Graphics*, vol. 21, no. 4, pp. 283–292, 1987 (Siggraph '87 Conference Proceedings, July 27–31, 1987, Anaheim, Calif.).

Rossignac, J. R., and A. A. G. Requicha: "Depth-Buffering Display Techniques for Constructive Solid Geometry," *IEEE Computer Graphics and Applic.*, vol. 6, no. 9, pp. 29–39, September 1986.

Rushmeier, H. E., and K. E. Torrance: "The Zonal Method for Calculating Light Intensities in the Presence of a Participating Medium," *Computer Graphics*, vol. 21, no. 4, pp. 293–302, 1987 (Siggraph '87 Conference Proceedings, July 27–31, 1987, Anaheim, Calif.).

Schwarz, M. W., W. B. Cowan, and J. C. Beatty: "An Experimental Comparison of RGB, YIQ, LAB, HSV, and Opponent Color Models," *ACM Trans. on Graphics*, vol. 6, no. 2, pp. 123–158, 1987.

Smith, A. R.: "Planar 2-Pass Texture Mapping and Warping," *Computer Graphics*, vol. 21, no. 4, pp. 263–272, 1987 (Siggraph '87 Conference Proceedings, July 27–31, 1987, Anaheim, Calif.).

Snyder, J. M., and A. H. Barr: "Ray Tracing Complex Models Containing Surface Tessellations," *Computer Graphics*, vol. 21, no. 4, pp. 119–128, 1987 (Siggraph '87 Conference Proceedings, July 27–31, 1987, Anaheim, Calif.).

Sutherland, I. E., R. F. Sproull, and R. A. Schumacker: "A Characterization of Ten Hidden-Surface Algorithms," *ACM Comput. Surveys*, vol. 6, no. 1, pp. 1–55, 1974.

Tomlinson, D. J.: "An Aid to Hidden Surface Removal in Real Time CGI Systems," *The Computer J.*, vol. 25, no. 4, pp. 429–441, 1982.

Verroust, A.: "Visualization Algorithm for CSG Polyhedral Solids," *CAD J.*, vol. 19, no. 10, pp. 527–533, 1987.

Wallace, J. R., M. F. Cohen, and D. P. Greenberg: "A Two-Pass Solution to the Rendering Equation: A Synthesis of Ray Tracing and Radiosity Methods," *Computer Graphics*, vol. 21, no. 4, pp. 311–320, 1987 (Siggraph '87 Conference Proceedings, July 27–31, 1987, Anaheim, Calif.).

Willis, P. J.: "A Review of Recent Hidden Surface Removal Techniques," *Displays*, vol. 6, no. 1, pp. 11–20, 1985.

Wittram, M.: "Hidden-Line Algorithm for Scenes of High Complexity," *Computer Aided Des. (CAD) J.*, vol. 13, no. 4, pp. 187–192, 1981.

Wyvill, G., T. L. Kunii, and Y. Shirai: "Space Division for Ray Tracing in CSG," *IEEE Computer Graphics and Applications*, vol. 6, no. 4, pp. 28–34, 1986.

Yang, C-G.: "Illumination Models for Generating Images of Curved Surfaces," *CAD J.*, vol. 19, no. 10, pp. 544–554, 1987.

# INTERACTIVE TOOLS

# CHAPTER
# 11

## GRAPHICS AIDS

## 11.1  INTRODUCTION

The basic concepts of geometric modeling and computer graphics have been introduced in the first half of the book. The second half of the book discusses the influences of these concepts on the engineering design process, as well as presenting some relevant design applications. Some of the graphics aids and manipulations, such as geometric modifiers, described here are a direct outcome of the computational geometry theory, others, such as transformations, are related to computer graphics, and a third group, such as entity names, is based on management systems of CAD/CAM databases.

Although CAD/CAM systems may seem different in their capabilities due to the syntax and format of user interface offered by each one, they all provide a generic set of graphics aids and manipulations. It is the intent of this chapter and the next one to introduce some of the common aids and manipulations offered by many of these systems. To best utilize the material presented here, readers are encouraged to refer to system documentation and manuals of their respective CAD/CAM systems.

Awareness of the available graphics aids provides designers with the ability to think in terms of them when performing design tasks. These aids should help speed up the completion of design tasks utilizing CAD/CAM systems. To effectively utilize these aids during a design session on a CAD/CAM system, they should be made part of the planning strategy (refer to Chap. 3) of creating the design. For example, if a design consists of more than one component, the designer can utilize the layers concept to separate the geometry of these components, or create subassemblies of the individual components and then utilize the assembly concept to merge these components to obtain the total design.

In this chapter we present a set of common graphics aids useful for creating designs and geometric models on CAD/CAM systems. The aids covered here include geometric modifiers, entity names, colors, grids, layers, groups, and dragging and rubberbanding. The material is presented in a generic form with the goal of introducing the basic concepts. No syntax details are given. Readers can attempt to find the corresponding commands to these concepts on their particular CAD/CAM systems and study their syntax. In essence, we present only the semantics of these commands here.

## 11.2  GEOMETRIC MODIFIERS

Geometric modifiers are used by any typical commercial CAD/CAM system to facilitate entering and extracting information to and from the system. A geometric modifier is a word that changes the mode of input in a command. A major advantage of using geometric modifiers is to be able to deal with specific existing geometric information of a geometric model without having to calculate that information explicitly. These modifiers apply mostly to wireframe entities, that is, curves.

The three common geometric modifiers are the end, the origin (center), and the intersection modifiers, as shown in Fig. 11-1. The end modifier signifies an endpoint of an entity. Each entity has two endpoints. If the entity is a closed curve such as a circle, the two endpoints are coincident. The origin modifier

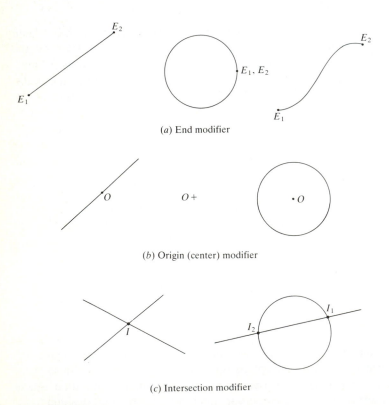

(a) End modifier

(b) Origin (center) modifier

(c) Intersection modifier

**FIGURE 11-1**
Typical geometric modifiers.

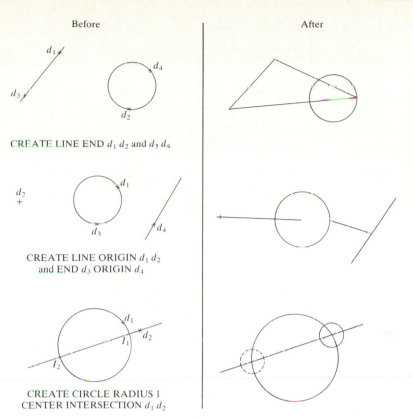

Before                                                                          After

CREATE LINE END $d_1$ $d_2$ and $d_3$ $d_4$

CREATE LINE ORIGIN $d_1$ $d_2$
and END $d_3$ ORIGIN $d_4$

CREATE CIRCLE RADIUS 1
CENTER INTERSECTION $d_1$ $d_2$

**FIGURE 11-2**
Typical uses of geometric modifiers.

identifies the center of an entity. The origin of a line is its midpoint, of a point is the point itself, and of a circle is its center. The origin of a general curve such as a B-spline is its center of curvature. The intersection modifier indicates the intersection point of two entities.

A command that uses a geometric modifier requires the user to type the modifier followed by digitizing the desired entity. In the case of the end or intersection modifier where more than one solution (point) exists, the user must digitize the entity close to the desired solution. In the case of an end modifier, the endpoints can be identified by their parametric values ($u$), as discussed in Chap. 5. The first input endpoint has a $u$ value of 0 and the second input endpoint has a $u$ value of 1. A modifier may remain in effect once used until another modifier is used. Figure 11-2 shows some examples of using these modifiers. The commands shown in the figure do not follow the syntax of a particular CAD/CAM system. The first example in the figure creates two lines between the endpoints of the line and the circle. The second example shows the use of the origin modifier and the interchange of modifiers. The third example shows creating the new circle with a center at $I_1$. If the digitizes $d_1$ and $d_2$ were close to $I_2$, the dashed circle would have been created instead.

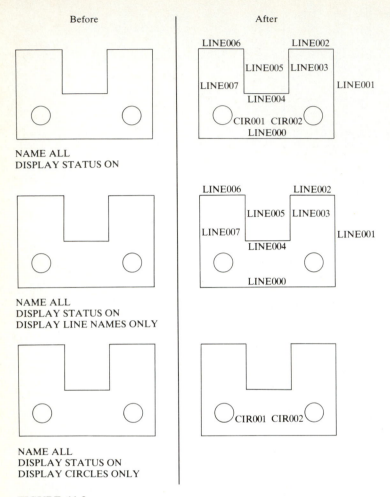

**FIGURE 11-3**
A typical system naming scheme.

## 11.3 NAMES

Entities (curves, surfaces, and solids) created by users and stored in a model database can be named or tagged for future reference purposes. Users can use names (tags) of entities to refer to them instead of digitizing them. This is very useful when writing interactive programs with minimum user interaction. Consider, for example, referencing the center of an existing circle. The user can use the origin modifier followed by digitizing the circle or specifying its name. Using names of entities enables the user to access entities that are not displayed but reside in the model database.

Naming entities usually involves either existing entities or entities to be created. For existing entities, the user requests the CAD/CAM system to name them. Each system has its own naming scheme which is usually self-explanatory. For example, lines are named LINXXX where XXX is a number. The number

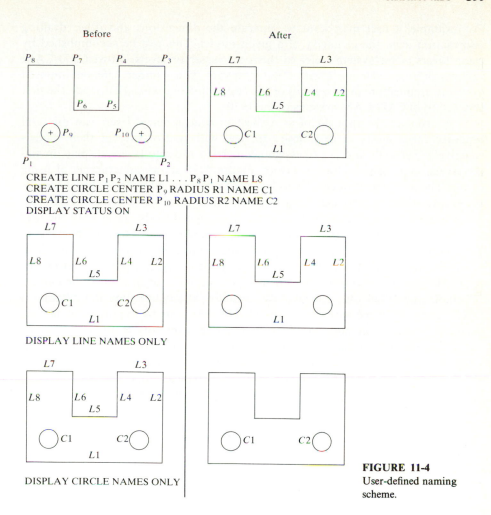

CREATE LINE $P_1 P_2$ NAME L1 ... $P_8 P_1$ NAME L8
CREATE CIRCLE CENTER $P_9$ RADIUS R1 NAME C1
CREATE CIRCLE CENTER $P_{10}$ RADIUS R2 NAME C2
DISPLAY STATUS ON

DISPLAY LINE NAMES ONLY

DISPLAY CIRCLE NAMES ONLY

**FIGURE 11-4**
User-defined naming
scheme.

usually reflects the order in which the entities of one kind, say lines, are created. An entity can be named at the time it is created by assigning a name to it. For example, the pseudo command "create line $P_1$ $P_2$ name L1" creates a line between points $P_1$ and $P_2$ with the name L1.

Names of entities are not usually displayed by the system unless the user requests so. To display entity names, the user must turn the status of name display on via the appropriate command. When this is done, names of only previously named entities are displayed. The status of name display can be turned on or off at the user control. Figures 11-3 and 11-4 show default and user-defined naming schemes respectively.

## 11.4 LAYERS

Users of CAD/CAM systems may often want to group and/or separate certain types of information related to the models or parts they create on these systems.

For example, a user may want to separate the dimensions and other drafting information from the geometry of a given model. This can be accomplished by using layers (some systems refer to them as classes or levels). Most CAD/CAM systems provide their users with 256 ($2^8$) layers. The number 256 was imposed from the architecture (8 bits/word) of the superminicomputers that made the first generation of CAD/CAM systems in the 1970s.

A layer can be thought of as a sheet of transparency. Users can mix and organize these sheets as they desire to deal with and/or present their models effectively. Typically, a user can assign geometric entities to layers, display on and off layers, and assign colors to layers. Figure 11-5 shows an example of using layers. The assignment of entities to layers can be done either globally or locally. If the user selects a particular layer number (make this layer active), any geometric or graphics entities created afterwards are assigned to this layer until the user selects another layer. Users can always change (modify) their layer assignments of entities by using the proper layer commands. Layer assignment can be done locally to an entity. When the user is creating the entity, a layer modifier can be used. The resulting entity resides on the specified layer regardless of the current selectable one. While only one layer can be active at any one time, more than one layer can be displayed simultaneously. CAD/CAM systems usually provide their users with layer commands that can display one layer, any combination, or all the layers. Users are usually advised to use the "repaint" command following the display commands of layers to see the final displayed result.

Various colors can be assigned to layers—typically one color per layer. If no color assignments are chosen, all layers have the same color (the default

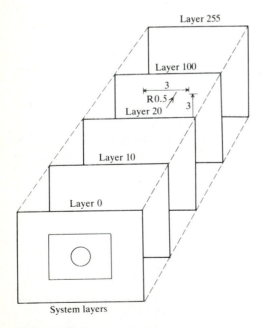

System layers

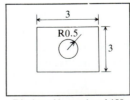

Display of layers 1 and 100

**FIGURE 11-5**
Layering concept.

color). Entities created on a given layer take the color attribute of such a layer. When more than one layer is displayed, the effect of different colors becomes obvious to users. Existing color assignments of layers can be modified by users. In such a case, the color of all existing entities that belong to the layer changes to the same color.

Users must be alert when dealing with layers. If, for example, a user attempts to digitize a displayed entity for manipulation or other purposes and the entity belongs to a different layer from the active one, the system always rejects the digitize and the entity selection. In this case making the entity layer active solves this problem.

How is the layering concept implemented into CAD/CAM software? Each entity is represented in its corresponding model database by a record of a certain length (number of words). The record is divided into subrecords. Each subrecord stores pertinent information about the entity such as geometry (coordinates), line font, and color attributes. One of the subrecords stores the layer attribute (layer number) of the entity. This subrecord is 8 bits long and, thus, layer numbers from 0 (00000000 in binary) to 255 (11111111 in binary) are possible to choose from. For example, if a user chooses layer number 10 for an entity, the layer subrecord of this entity is set to 00001010 in binary with the least significant bit being the rightmost bit. Thus, if a user decides to display layer 10, the software checks the number stored in the layer subrecord of each entity of the model against the number 10 and displays on the screen only those entities that pass the check.

Layers are especially useful in constructing assembly and architecture drawings. Different parts (subassemblies) of the assembly could be placed on certain layers and manipulated or studied. The entire assembly could also be seen and operated on by displaying all layers. In an architectural drawing, for example, of a room, the shell of the room could be placed on one layer with the structural, piping, wiring, and furniture arrangements placed on other layers.

In typical industrial CAD/CAM installations, standard layering schemes are usually developed by company management, and system users are required to adhere to these schemes. One example would be to use layer 1 for indexing, the next fifty layers for model geometry, the following fifty for manufacturing notes, another fifty for drafting (dimensioning and text), another fifty for analysis and technical illustrations, and the last fifty for construction aids and intermediate construction steps.

## 11.5   COLORS

Colors are helpful in distinguishing entities of geometric models from each other. A typical color command on a typical CAD/CAM system affects the color with which entities are displayed on the color graphics display screen. Most systems provide seven colors. These are red, green, blue, yellow, magneta, cyan, and gray. Other colors may be available directly or by using the RGB model described in Chap. 10 to create other colors. Some systems refer to colors by their names (e.g., red for the color red) while others refer to them by numbers (1 for red).

Color is assigned by layer; therefore, all entities assigned to a particular layer are displayed in the color assigned to that layer. If no color assignments are

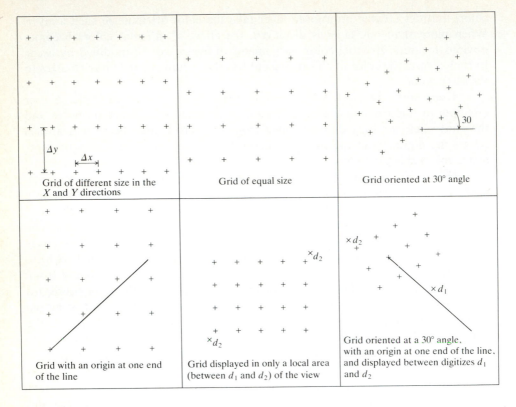

**FIGURE 11-6**
Some rectangular grids.

made, the default color is used for all layers. Users can assign a desirable color to a certain layer by using one command. For example, a command such as "select color 1 layer 10" assigns the red color to layer 10.

Color plots can be generated in a similar fashion. In the case of pen plotters, layer(s) can be assigned to a given pen number. For example "select pen 1 layer 10, 15–20" assigns layers 10 and 15 through 20 to pen 1. The choice of pen colors, that is, pen 1 is black, depends on the order of the pens in the plotter at the time of plotting.

## 11.6  GRIDS

A grid is a network of uniformly spaced points superimposed (overlayed) on the screen. Grids are useful for operations such as sketching, planning layouts, placing text at a specific location, and using freehand digitizing to indicate geometry location. When the grid is activated in a view, digitizes in that view are snapped (moved) to the nearest grid point.

There are two types of grids available, rectangular and radial. Grids are displayed as series of dots (in rectangular or radial patterns), one dot for each grid point. A grid can be selected for each view and can be turned on (activated)

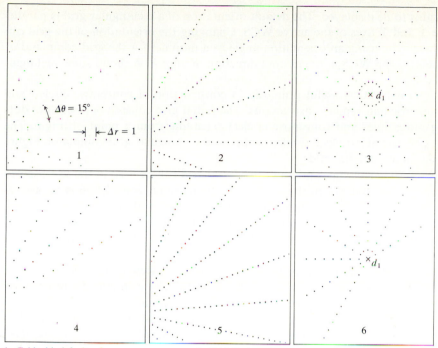

$\Delta\theta = 15°.$

$\rightarrow |\leftarrow \Delta r = 1$

× $d_1$

× $d_1$

1. Grid with default values
2. Grid with $\Delta r = 0.5$ and $\Delta\theta = 20°$
3. Grid with $d_1$ as an origin
4. Grid displayed between angles 30° and 120°
5. Grid with first radial line at 20°
6. Grid displayed between angles 30° and 245° with an origin at $d_1$

**FIGURE 11-7**
Some radial grids.

or off (deactivated) in any view at any time. The default origin for grids is usually the origin of the view. If a grid size (spacing between grid points) is specified as being too small for the grid points to be easily distinguished, the grid is not displayed on the screen; also, the grid is not visible if the grid points are further apart than the width and/or height of the view. Changing the magnification (zooming the view in or out) of the view will cause the grid points to become visible. Even if the grid is not visible, it is still active, and digitizes are snapped to the closest grid point.

A rectangular grid is particularly useful in constructing equally spaced entities, enabling the user to enter exact points by using digitizing rather than explicit coordinate entry. Use of rectangular grids also greatly facilitates sketching, moving entities from point to point, measuring between points, and trimming (using grid points as trimming boundaries). The user can specify the distance between grid points, the origin of the grid, and the angle (orientation) of the grid. Different views can contain different grids of different sizes.

A grid command typically activates or deactivates the rectangular grid in a specific view. A grid command must be followed by a "repaint" or "echo grid" command to display grid points. Various modifiers can be used with the command to control the origin, size, orientation, and/or the local area in which the grid is to be displayed. The default orientation of a rectangular grid is parallel to the $X$ and $Y$ axes of the active WCS. Changing the orientation of the grid can be achieved by using an orientation angle as a modifier for the grid command or by orienting the WCS in the desired direction. Figure 11-6 shows some rectangular grids.

A radial grid is useful primarily in construction of concentric circles, and axisymmetric and radial constructions. The radial grid consists of a series of dots (along radial lines and concentric circles) extending from a point of origin. The radial grid command has a similar syntax to a rectangular grid command. Figure 11-7 shows some radial grids.

Both the rectangular grid and the radial grid can be used in the same screen display (layout), but not in the same view, at the same time. There is no association between grid types; the parameters specified with the rectangular grid command do not apply to the radial grid command. It is not necessary to turn off a rectangular grid (or a radial grid) before a radial grid (or a rectangular grid) can be activated in the same view.

When changing grid type (whether in the same view or in different views), parameters are not retained from the previous grid. For example, if a rectangular grid exists in a particular view (with established parameters) and then a radial grid is placed in the same view, and then a rectangular grid is reestablished in that view, the new rectangular grid uses the default values if none is specified. Each time the grid type is changed, the default parameters are used, unless parameters are specifically set by the user.

## 11.7 GROUPS

Groups are useful when it is required to treat a selected number of entities temporarily as one (compound) entity. A group can be created by using the proper command which usually requires the user to digitize the individual entities that are to make up the group. A group can have a name and can be manipulated by using its name or digitizing an entity that belongs to it. A command executed on a group affects all the entities in the group. More than one group can be constructed if required.

Groups eliminate the need for continually constructing a window about entities that are to be manipulated several times, as in the case of animation. Consider the motion of a mechanism (or a robot) consisting of few links which may rotate and/or translate in space. It will be efficient to combine the entities of each link into a group which can be rotated or translated as one (compound) entity from one position to another.

Once a group is no longer needed, the user can delete (disassociate or degroup) it, that is, release its entities so that each can be treated individually again. A user may not be allowed to delete an entity that is a member of a group before deleting the group first.

## 11.8 DRAGGING AND RUBBERBANDING

Dragging is a technique of moving an object around with a locating device. Dragging can be used to position objects and symbols. Dragging is achieved by

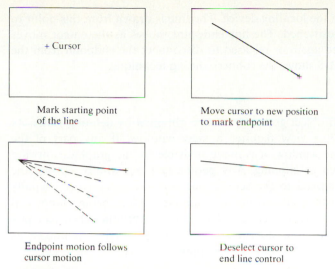

Mark starting point
of the line

Move cursor to new position
to mark endpoint

Endpoint motion follows
cursor motion

Deselect cursor to
end line control

**FIGURE 11-8**
Line construction with rubberbanding technique.

reading the locating device at least once each refresh cycle, and using the position of the device to move (translate) the object.

Rubberbanding is another technique usually used to construct lines. The push of a button marks the starting point of the line. The coordinates of the

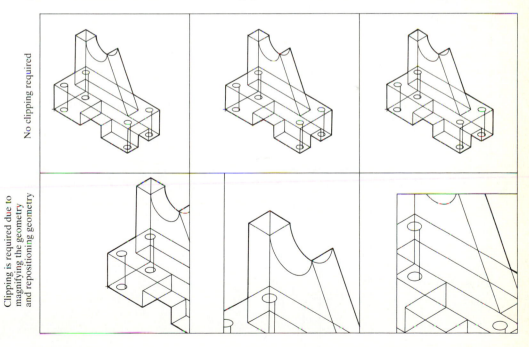

No clipping required

Clipping is required due to
magnifying the geometry
and repositioning geometry

**FIGURE 11-9**
Clipping.

point give the position of the locating device. The line is drawn from this point to wherever the cursor is positioned. The line endpoint moves as the cursor moves until the button is pushed again or released to disconnect the endpoint from the locating device. Figure 11-8 shows the rubberbanding technique.

## 11.9  CLIPPING

Various projections of an object geometry can be obtained by defining views. As we discussed in Chap. 3, a view requires a view window. If any part of the geometry is not inside the window, it is made invisible by the graphics software through a process known as clipping. Any geometry lying wholly outside the view boundary is not mapped to the screen, and any geometry lying partially inside and partially outside is cut off at the boundary before being mapped. If clipping is not done properly, a CAD/CAM system will produce incorrect pictures due to overflow of internal coordinate registers. This effect is known as wraparound. Figure 11-9 shows examples of clipping.

## PROBLEMS
## Part 1: Theory

**11.1.** A user used a line command and digitized close to endpoints $P_1$ and $P_2$ to close the polygon shown in Fig. P11-1. When the user utilized a later command to calculate the area enclosed by the polygon, the system issued an error message that the polygon is not closed. How is this possible and how can the user solve the problem?

FIGURE P11-1

**11.2.** How can you use the geometric modifiers to create the model shown in Fig. P11-2 without having to calculate any other dimensions explicitly?

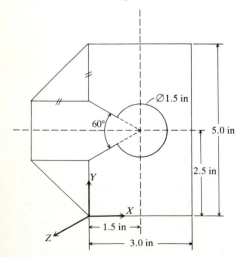

FIGURE P11-2

**11.3.** How can you know if a given entity is already named or not regardless of the name display status?

**11.4.** List some of the advantages of referring to entities by names instead of digitizing them.

**11.5.** Give some examples where the layering concept is useful to use.

**11.6.** If the layer subrecord is 8 bits, find the corresponding binary numbers to layers 50, 130, 200, 245, and 250.

## Part 2: Laboratory

**11.7.** Find the geometric modifiers available on your CAD/CAM system. What are their syntax? How do you use them? What is the default modifier?

**11.8.** To utilize names effectively on your system:
  (a) How can names (tags) be created with construction commands?
  (b) Find the command sequence to name and display the names of existing entities.
  (c) How can you modify names of already named entities?
  (d) How can you display names of selected entities?
  (e) How can you use entity names in commands such as "delete entity", "translate entity", "verify entity", etc.?

**11.9.** Find all the layer-related commands on your system—specifically how to select/deselect layers, assign entities to layers, assign layers to entities, assign colors to layers, modify layer colors, and modify layers of existing entities.

**11.10.** For the model shown in Prob. 11.2, create the straight lines on layer 10, the circle on layer 20, and the dimensions on layer 30.

**11.11.** For the models you created in Chaps. 5, 6, and 7 tag some entities and modify their current layer assignments.

**11.12.** Find all the color-related commands on your system—specifically how to assign and deassign colors to layers.

**11.13.** Find how you can generate color plots on your system plotters.

**11.14.** Find the commands related to rectangular and radial grids. How can you use them? Investigate their related modifiers.

**11.15.** With the aid of grids, sketch the geometry shown in Fig. P11-15.

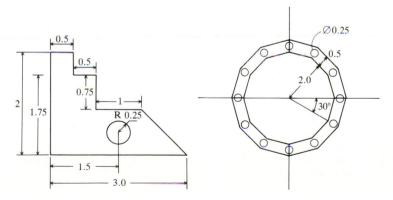

**FIGURE P11-15**

**11.16.** Find the commands related to groups on your system. How can you create and delete groups?

**11.17.** Using groups, rotate the slider-crank mechanism shown in Fig. P11-17 by the full 360° with an increment of 36. Assume any other missing dimensions.

Assume any other missing dimensions

**FIGURE P11-17**

**11.18.** If your system has dragging and rubberbanding capabilities, find out how to activate and use them.

## Part 3: Programming

The programming assignments in this chapter attempt to simulate the software routines that are involved when users use typical commands related to graphics aids discussed in the chapter. The digitize of an entity usually required by some of these commands is replaced (simulated) by the geometric data required to create the entity. For example, a program can simulate digitizing an existing line by reading its two endpoints from user input.

**11.19.** Write a program that can return the geometric information (coordinates) corresponding to the geometric modifiers (end, origin, intersection) when applied to lines and circles.

**11.20.** Write a program that reads text strings input by users as tags or names. Store these names and let the program print them back as text strings.

**11.21.** Write a program that converts any layer decimal number between 0 and 255 into its corresponding binary number. Let the program output the layer number both in decimal and binary. Use 8 bits to store any layer number where the least significant bit is the rightmost bit.

## BIBLIOGRAPHY

Foley, J. D., and A. van Dam: *Fundamentals of Interactive Computer Graphics*, Addison-Wesley, Reading, Mass., 1986.
Goetsch, D. L.: *MICROCADD: Computer-Aided Design and Drafting on Microcomputers*, Prentice-Hall, Englewood Cliffs, N.J., 1988.
Hordeski, M. F.: *CAD/CAM Techniques*, Reston, A Prentice-Hall Company, Reston, Va., 1986.
McKissick, M. L.: *Computer-Aided Drafting and Design*, Prentice-Hall, Englewood Cliffs, N.J., 1987.
System Documentation and Manuals of any CAD/CAM system.
Voisinet, D. D.: *Introduction to CAD*, 2d ed., McGraw-Hill, New York, 1986.
Zeid, I.: "CAD on Computervision System," unpublished document, CAD/CAM Laboratory, Northeastern University, Boston, Mass., 1984.
Zeid, I.: "CAD on GE Calma System," unpublished document, CAD/CAM Laboratory, Northeastern University, Boston, Mass., 1985.

# CHAPTER
# 12

## GRAPHICS
## MANIPULATIONS
## AND EDITINGS

## 12.1 INTRODUCTION

One of the most recognized benefits of using CAD/CAM systems in design and engineering applications is increasing productivity. The principal time-saving factor in CAD as compared with manual drafting and geometric construction comes in the manipulation and editing of geometrical data and entities. One of the most time-consuming tasks in manual drafting and construction is correcting and revising existing geometry. Because of the manipulation and editing capabilities offered by CAD/CAM systems, corrections and revisions can be made quicker and easier on these systems.

Good understanding and efficient use of manipulation and editing functions and commands on a CAD/CAM system enable users of the system to better correct errors and mistakes. While it is always the easiest way to delete a partially wrong entity and reconstruct it, it is also the most nonproductive way to deal with mistakes. It does not help the user advance in using the system. For example, if a correct entity is created in the wrong position, the user can move it to the desired one instead of deleting it and recreating it. It is important that users think in terms of manipulation and editing functions when correcting their mistakes.

An investigation of most existing CAD/CAM systems reveals that related software provides users with a generic set of manipulation and editing functions and commands. It is the purpose of this chapter to present these functions in a syntax-independent way. Readers of this chapter are encouraged to relate the chapter material to their particular system syntax and commands. This can be done by referring to the system documentation and manuals.

The mathematical and theoretical backgrounds of the material covered in this chapter have been covered in previous chapters of the book. Sections 5.7, 6.7, and 7.13 in Chaps. 5, 6, and 7 respectively present curve, surface, and solid manipulations and editings. Chapter 9 is devoted to transformations (rotations, translation, reflection or mirroring, and scaling). The material in this chapter can be considered an implementation of this background and the available related commands can be viewed as the user interface of such implementation.

The use of manipulations and editings can be made more efficient and productive if made in conjunction with the material covered in Chap. 11. For example, a user may group a set of entities before translating or rotating them or an entity can be trimmed to the endpoints of another entity by using these points as trimming boundaries in the trim command. Other graphics aids such as layers and names can also be used.

## 12.2   ENTITY SELECTION METHODS

Manipulation and editing commands usually act on existing entities. Each of these commands requires that the user identifies or selects the desired entities and inputs data. The most obvious method of identifying entities is digitizing them. This method, however, is inefficient if many entities are required or if the entities to be digitized happen to be in a dense area (closely displayed entities) of the display screen. This section discusses the most common methods of selecting entities. These methods can be mixed, that is, more than one method can be used simultaneously in a manipulation/editing command.

### 12.2.1   Individual Entity

If only one or few entities are to be selected, the user can simply digitize them or refer to them by names (tags) within the manipulation or editing command. In the case of digitizes, the user must digitize close to the entities. The CAD/CAM system usually highlights the successfully selected entities to provide the user with a visual feedback before hitting the return key for final acceptance and processing of the command. If no entity is highlighted, it means that the system did not recognize the digitizes and the user should digitize again. If the wrong entities are selected, the user can use the backspace key to deselect them and then digitize again.

### 12.2.2   All Displayed Entities

In some cases, all entities of a given view displayed on the screen are to be selected. Consider the example of construction of a two-and-a-half-dimensional model. The user first creates the geometry of the face of the model which is perpendicular to the direction of the uniform thickness. Next, the user must select all the displayed entities to project them to create the back face of the model. In these cases, there is a modifier that can be used with the desired manipulation/editing command or a "select entity" command to indicate that all entities in a displayed view are desired. This modifier is usually followed (as one would expect) with a digitize within the view boundary (window) to identify the desired view. In this case, all displayed entities in the selected (digitized) view are highlighted. If a substantial number of entities (not all of them) in a view are desired, these entities can be moved to a specific layer and the layer is selected and echoed before using the view modifier.

### 12.2.3 Groups

Groups discussed in Chap. 11 comprise another effective method of entity selection. Groups are easily managed and manipulated. Creating groups itself requires selecting the group entities. Thus, any of the methods described in this section can be used to create groups. Once groups are created, they can be used as a method of entity selection.

### 12.2.4 Enclosing Polygon or Window

Entities can be selected by enclosing them within the boundaries of a polygon. A polygon can be any shape, as shown in Fig. 12-1. A polygon is defined by its corner points which are input by the user as digitizes. A polygon can take any shape to selectively enclose the desired entities. If only two points are given to define a polygon, they are the endpoints of a diagonal of this polygon. The enclosing polygon becomes an enclosing window in this case. Entities or endpoints of entities that lie on the boundary of a polygon are considered to be

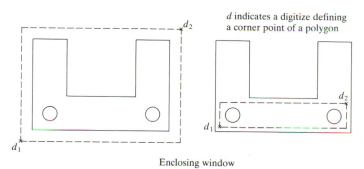

$d$ indicates a digitize defining a corner point of a polygon

Enclosing window

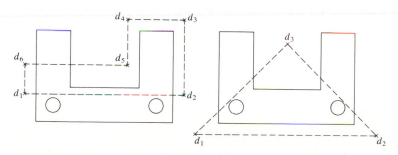

Enclosing polygon

**FIGURE 12-1**
Use of polygons to select entities.

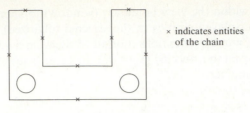

× indicates entities
of the chain

(a) Chain with single branch points

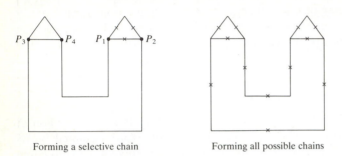

Forming a selective chain

Forming all possible chains

(b) Chain with multiple branch points
($P_1$ to $P_4$ are multiple branch points)

**FIGURE 12-2**
Chaining entities.

within the polygon. If both endpoints of an entity (such as a line or a circle) lie within the polygon, the entire entity is affected by the manipulation/editing commands. If only one endpoint is included (the entity is partially enclosed), different CAD/CAM systems give different results. Some systems do not allow the commands to affect the entity at all, while others apply the commands to the included endpoints only.

### 12.2.5   Chaining Contiguous Entities

An entity and all entities forming a contiguous path with it can be selected by chaining them. It is enough to digitize one entity of the contiguous entities to form a chain. Some CAD/CAM systems may require another digitize to indicate the direction of chaining. Multiple branch points, where several entities are contiguous at the same point, result in multiple chains. All these chains or a particular one of them can be selected. Chaining is extremely useful for selecting long series of contiguous entities that are surrounded by other geometry, making selection by a polygon impractical, and when selecting each entity in the chain individually would be tedious and time-consuming. Figure 12-2 shows some examples of forming chains.

### 12.2.6   Width

Entities can be selected on the basis of a given width. This method of selection is useful for three-dimensional model construction. A width table may have to be created first to use this method successfully.

## 12.3  MANIPULATION OPERATIONS

This section covers the common manipulation operations available on CAD/CAM systems. Users should always think in terms of these operations during contruction of geometric models. Here, importantly, they should extend them to aid solving their design and engineering problems when utilizing these systems.

### 12.3.1  Verification of Model and Database Parameters

A listing of the essential parameters of a model and its database can be obtained via a verification command. The listing provides the default values of these parameters at the time of listing. These parameters provide information about the active mode (model or drafting), the database units, drawings, views, dimensions, tolerancing, model extents, grid information, etc. The amount of information obtained from the verification command may differ from one CAD/CAM system to another.

Verification of model and database parameters is useful in two cases. First, the user may want to review the current values of these parameters before beginning geometric operations. This is particularly useful if the user has not worked with a particular model for a while. Second, if the results of the user's commands are unpredictable, reviewing these values is usually useful in an attempt to find the reason.

### 12.3.2  Entity Verification

Information about existing entities in a model database can be obtained by using an entity verification command. Such a command usually provides the entity type, the geometric information of the entity, its layer number, and its name. Other secondary information may be provided and may differ from one CAD/CAM system to another. For example, a verification of a line provides the coordinates of its start and endpoints, its length, its layer, and its name. Secondary information could include line type, weight, width, and date of creation and/or modification.

There is a number of modifiers related to the verification command. One can verify all the entities of a model without selecting them by using an "all" modifier. A useful set of modifiers is the one that enables the user to obtain the geometric information of an entity relative to MCS, WCS, or SCS. For example, the coordinates of the endpoints of a line can be obtained relative to one of these coordinate systems. Such a flexibility is helpful during geometric construction. Figure 12-3 shows an example.

### 12.3.3  Entity Copying (Duplication)

Existing entities can be copied or duplicated. Two cases may arise. An entity may be copied in its current location or may be copied and moved to a new location.

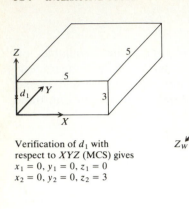

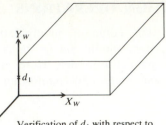

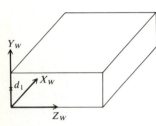

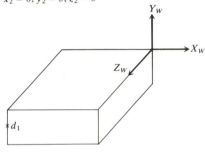

Verification of $d_1$ with
respect to $XYZ$ (MCS) gives
$x_1 = 0, y_1 = 0, z_1 = 0$
$x_2 = 0, y_2 = 0, z_2 = 3$

Verification of $d_1$ with respect to
$X_w Y_w Z_w$ (WCS) gives
$x_1 = 0, y_1 = 0, z_1 = 0$
$x_2 = 0, y_2 = 3, z_2 = 0$

Verification of $d_1$ with
respect to $X_w Y_w Z_w$ (WCS) gives
$x_1 = 0, y_1 = 0, z_1 = 0$
$x_2 = 0, y_2 = 3, z_2 = 0$

Verification of $d_1$ with respect to
$Z_w Y_w Z_w$ (WCS) gives
$x_1 = -5, y_1 = -3, z_1 = 5$
$x_2 = -5, y_2 = 0, z_2 = 5$

**FIGURE 12-3**
Entity verification.

The need for the first case arises when the user desires to copy an existing entity to a different layer than its current one. If the entity is complex and the user is about to experiment with it, it is important to back it up in case it is lost or damaged.

The second case is useful in constructing geometry that has repetitive patterns. The user creates one pattern and duplicates it in the proper locations. Consider, for example, constructing a wireframe model of a gear. The user creates the geometry of one tooth precisely and duplicates it around the gear blank. In this case, commands or modifiers based on geometric transformation (rotate, translate, and/or mirror) are typically used.

### 12.3.4 Geometric Arrays

The second case discussed in the previous section can also be viewed as constructing geometric arrays. A geometric array is a number of identical entities placed uniformly at specified locations. There are two types of geometric arrays: rectangular and circular. Entities in rectangular arrays are separated by increments along the $X$, $Y$, and/or $Z$ axes of the proper coodinate system. Entities in circular arrays are separated by increments in the radial and/or angular directions.

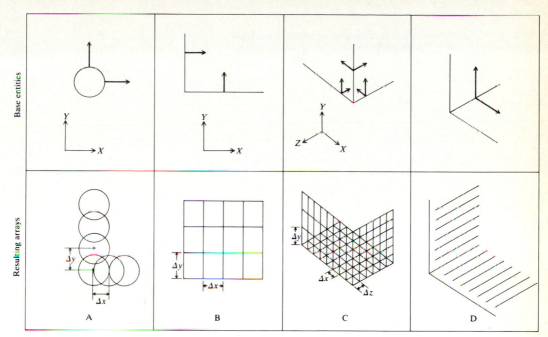

**FIGURE 12-4**
Rectangular arrays.

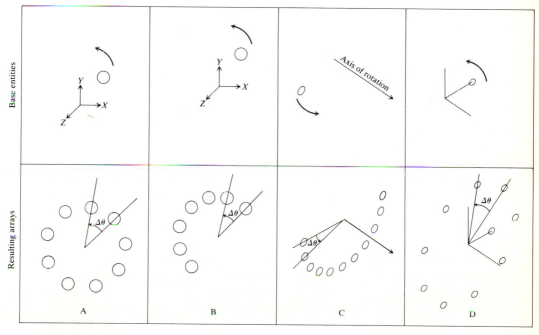

**FIGURE 12-5**
Circular arrays.

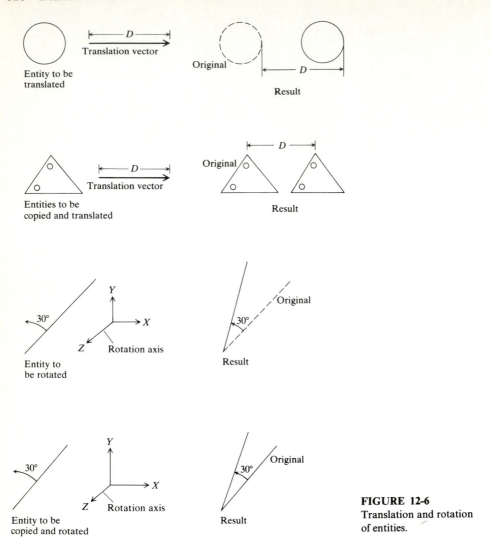

**FIGURE 12-6**
Translation and rotation of entities.

Figures 12-4 and 12-5 show some array examples. The base entities are the ones used to generate the arrays while the arrows show the directions of copying the base entities to form the arrays. The user can specify the number of copies and the spacing between them ($\Delta x$, $\Delta y$, and/or $\Delta z$ for rectangular arrays, or $\Delta r$ and/or $\Delta \theta$ for circular arrays) in each direction. Case D in both figures shows that the resulting arrays do not have to coincide with the base entity.

### 12.3.5   Transformation

Transformation commands are useful productivity tools to utilize on CAD/CAM systems in order to manipulate existing entities. They can be used to translate, rotate, mirror (reflect), and scale entities. These commands and their related

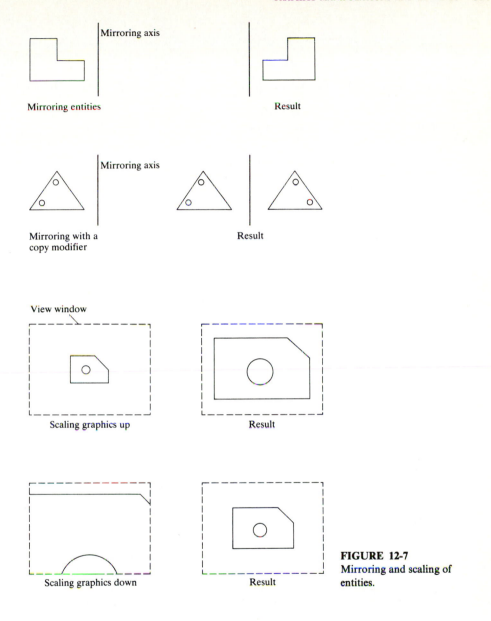

**FIGURE 12-7**
Mirroring and scaling of entities.

modifiers are based on the mathematical background covered in Chap. 9. A copy modifier is usually available with these commands in which case a copy of the existing entity is made and acted on by the command, thus keeping the existing entity unchanged.

The translation command requires a translation distance and a translation direction. If the copy modifier is used, a different layer for the copy may be chosen differently from the layer of the existing layer. The rotation command requires a rotational vector, that is, an axis of rotation and an angle of rotation. The axis of rotation could be a general axis or one of the MCS or WCS axes. The

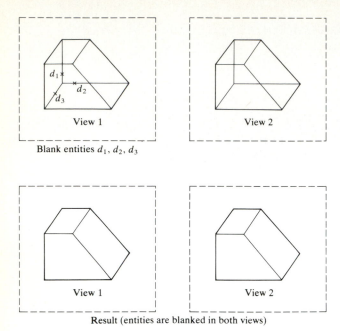

Blank entities $d_1, d_2, d_3$

Result (entities are blanked in both views)

**FIGURE 12-8**
Entity blanking.

angle of rotation could be positive or negative. A positive angle is an angle represented by a vector whose direction is in the positive direction of the axis of rotation. Figure 12-6 shows some examples of translating and rotating entities.

The "mirror" command requires a mirroring axis. The existing entity is always mirrored on the other side of the mirroring axis in reverse, that is, as if it is viewed in a mirror. On some CAD/CAM systems, the mirror command may move the existing entity to the other side of the axis while on others the command may move a copy of the entity. On the former systems, a copy modifier is available with the command. Figure 12-7 shows an example of mirroring.

The "scale" command requires a scale factor. Alternative names of this command which are also used by CAD/CAM systems are "zoom or magnify" command. Other names may exist. The command is usually used to adjust the size of displayed graphics within a view window. If the graphics size is too small, the command can enlarge it to fill the view window. If the size is too large and some entities are clipped against the window boundary, the "scale" command can be used to scale the size down. Figure 12-7 shows examples of scaling.

### 12.3.6   Entity Blanking/Unblanking

Blank/unblank operations are useful for creating auxiliary views for drawings and for temporarily removing entities from the display screen in crowded or complex models in order to make further construction simpler. Blank operation is also useful in model clean-up, as explained in Chap. 10 to hide entities.

The "blank entity" command is used to remove entities from the graphics screen and the "unblank entity" command is used to restore blanked entities back to the display screen. Blanking is a mode-dependent command. If used in

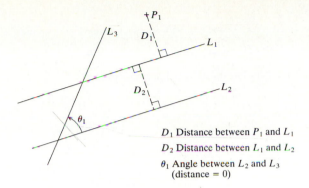

$D_1$ Distance between $P_1$ and $L_1$
$D_2$ Distance between $L_1$ and $L_2$
$\theta_1$ Angle between $L_2$ and $L_3$
  (distance = 0)

(*a*) Measurements between points and lines or lines and lines

| Open contour | Closed contour |
|---|---|

Digitize contour entities in a given order (clockwise or counterclockwise direction)

(*b*) Contour measurements

**FIGURE 12-9**
Examples of geometric measurements.

the model mode, entities are blanked in all views. In the drafting mode, entities are blanked only in the view they are selected in. The result of a blank command is different from a delete command. The latter removes entities permanently from a model database while blanked entities still remain the model database and are removed from the screen only. Figure 12-8 shows an example of entity blanking.

## 12.3.7 Geometric Measurements

A measure function or a command can be used to obtain basic geometric measurements such as:

1. The minimum distance between two entities, two points, or an entity and a point.
2. The angle between two lines.
3. The angle specified by three points.
4. The length of a contour of a connected set of entities.

If either two geometric entities, two points, or one of each are specified, the minimum distance is given. The minimum distance is defined as the measurement along a line mutually perpendicular to both entities or between the two closest

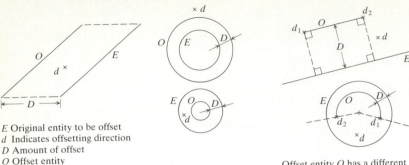

E Original entity to be offset
d Indicates offsetting direction
D Amount of offset
O Offset entity

Offset entity O has a different
length from original entity E.
$d_1$ Start point of O
$d_2$ Endpoint of O

(a) Uniform offsetting

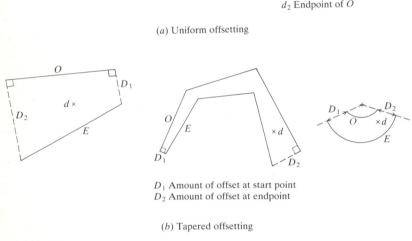

$D_1$ Amount of offset at start point
$D_2$ Amount of offset at endpoint

(b) Tapered offsetting

**FIGURE 12-10**
Examples of entity offsetting.

points of the entities. If two line entities are specified, the following results are obtained:

1. For two lines that intersect, the minimum distance is zero and the measurement of the angle (usually acute) between the lines is displayed on the screen.
2. For two parallel lines, the measurement of the distance between the two lines is displayed.
3. For two lines that do not intersect in space, the measurement of the distance between the lines at the two points of closest proximity is given. The measurement command has a modifier that can project the two lines temporarily onto a given plane (construction plane) of the WCS before any measurements are obtained. The resulting projected lines are treated as described in result 1 or 2.

If three points are specified, the measurement of the angle described is displayed on the screen. The angle used is usually the counterclockwise angle from the second point to the third point, using the first point as the origin point of the angle.

The "measure" command allows the user to measure the length of a contour. The curves that make up the contour may be any wireframe entity described in Chap. 5. Successive pairs of curves in the sequence, except for the first and the last, either match at endpoint or should intersect one another. The sequence of curves may form an open or a closed contour.

Figure 12-9 shows some examples of the "measure" command. Readers are advised to review Chap. 5 to relate this command to its mathematical background.

### 12.3.8 Entity Offsetting

Entity offsetting allows the user to construct an offset of a planar entity. The entity may be a wireframe or a surface. Uniform and tapered offsets can be obtained. In uniform offsetting, the offset entity (wireframe or surface) constructed is of the same type as the original entity. For example, offsetting a circular arc produces a circular arc. Tapered offsetting is usually applied to wireframe entities and the resulting entity may have a different type than the type of the original entity. Taper offsets of arcs and conics result in B-spline entities.

An "offset" command requires entities to be offset, an offsetting direction (the side on which the offset entity is to be constructed), and offsetting parameters. These parameters usually include the offset value (or values in tapered offset), and may include start and endpoints that may be different from those of the original entity.

An "offset" command finds applications in engineering design. It can be used to generate thicknesses of various geometric models such as thick and thin wall pressure vessels. The command can also be used in place of entity projection commands. Figure 12-10 shows some examples of entity offsetting.

## 12.4 EDITING OPERATIONS

This section covers the common editing operations available on CAD/CAM systems. Together with the manipulation operations covered in the previous section, the user can avoid many unnecessary calculations during geometric construction on CAD/CAM systems as compared to manual construction on drafting boards. Whenever encountering the need for geometric calculations during construction on these systems, the user must investigate whether commands or procedures can be used before thinking of manual calculations. Manipulation and editing operations provide useful tools for design and engineering applications.

### 12.4.1 Entity Trimming

Trimming is used to stretch or contract endpoints of geometric entities such as lines within particular boundaries specified by the user. Trimming can be performed by using a "trim" command or modifier. The trim modifier is useful when the correctly trimmed geometry is created simultaneously during construction.

A "trim" command requires the entity or entities to be trimmed and trimming boundaries. These boundaries can be specified as a given length or locations. Geometric modifiers are usually useful in defining trimming boundaries.

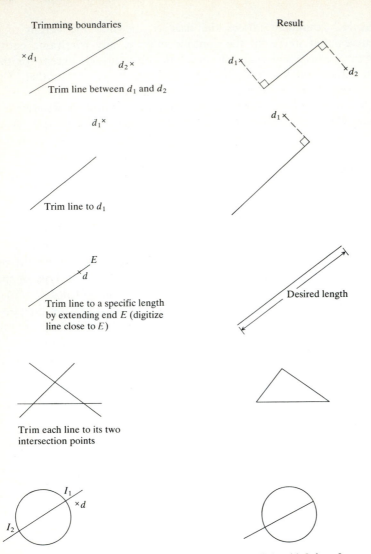

FIGURE 12-11
Trimming open entities.

Figure 12-11 shows examples of trimming open entities such as lines. It should be noticed that open splines can be trimmed within the original endpoints, but cannot be extended beyond those points (refer to Chap. 5 for an explanation).

Trimming closed entities such as circles follows the counterclockwise direction rule. According to this rule, the part of the closed entity that defines the counterclockwise direction from the first trimming boundary to the second boundary remains in the model database at the end of the trimming operation. Figure 12-12 shows examples of trimming closed entities such as circles. Notice that if a trimming operation results in keeping the undesired portion of an entity,

Trimming boundaries                           Result

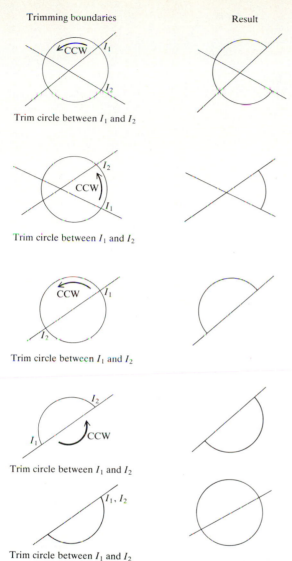

Trim circle between $I_1$ and $I_2$

Trim circle between $I_1$ and $I_2$

Trim circle between $I_1$ and $I_2$

Trim circle between $I_1$ and $I_2$

Trim circle between $I_1$ and $I_2$

**FIGURE 12-12**
Trimming closed entities.

the operation can be repeated on such a portion by using the same trimming boundaries used with the first operation but in a reversed order. If the original closed entity is to be obtained again, the two trimming boundaries can be chosen to be identical (same point).

## 12.4.2  Entity Division

An existing entity can be divided into separate smaller entities of the same type. The divisions of the entity are designated by explicit input of the number of divisions or by specifying the division boundary points. When a "divide entity"

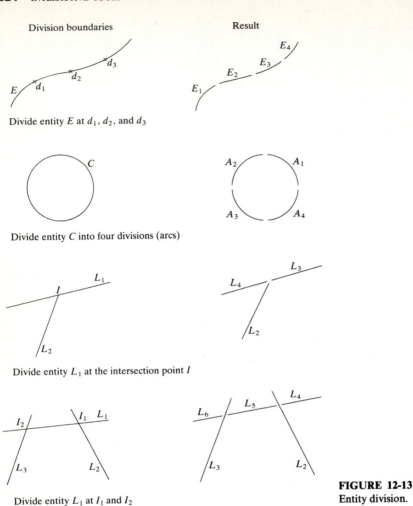

Division boundaries

Result

Divide entity $E$ at $d_1$, $d_2$, and $d_3$

Divide entity $C$ into four divisions (arcs)

Divide entity $L_1$ at the intersection point $I$

Divide entity $L_1$ at $I_1$ and $I_2$

**FIGURE 12-13**
Entity division.

command is applied to an entity, this original entity is deleted and is replaced by the newly formed entities. Figure 12-13 shows examples of entity division.

At the completion of a "divide entity" command, the user cannot visually see a difference. To overcome this uncertainty, the user can proceed with the next command to be executed, or a command, such as the "verify entity," that results in temporary highlighting can be used.

### 12.4.3 Entity Stretching

Sometimes an entity is drawn with a wrong endpoint, or the endpoint may have to move to a new location. It is always inefficient to delete the entity and recreate it, especially if the entity is defined by many points. CAD/CAM systems provide commands that can be used to stretch (modify) the undesired endpoint to the desired location. Figure 12-14 shows examples of entity stretching.

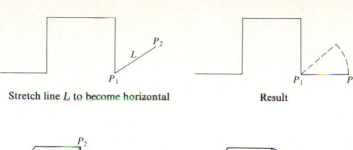

Stretch line *L* to become horizontal       Result

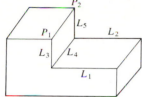

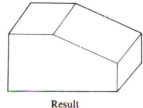

Delete line $L_3$, $L_4$, and $L_5$,       Result
and stretch lines $L_1$ and $L_2$
to the two endpoints $P_1$ and $P_2$

**FIGURE 12-14**
Entity stretching.

### 12.4.4 Entity Editing

Some CAD/CAM systems provide their users with specialized commands that enable them to edit existing entities in special ways. By using these commands, the user does not have to delete entities and recreate them. While some of these commands can be replaced by the already introduced editing operations, some can not. Some CAD/CAM systems, for example, provide their users with commands to edit existing fillets or change existing circles. A user can change the radius or diameter of an existing circle to a new value without having to delete the circle and recreate it with the desired value. These commands are useful and readers are encouraged to investigate their respective systems for some of these useful commands.

## 12.5 DESIGN AND ENGINEERING APPLICATIONS

The concepts covered in Chap. 11 and this chapter can be very helpful in engineering. While the concepts are always thought of as graphics tools to aid geometric construction, they can be extended innovatively to engineering. For example, if we think in terms of geometric modifiers, manipulation operations, and editing operations while solving design and engineering problems, we can avoid many explicit calculations, especially those related to analytic geometry and trigonometry. There are many graphical techniques and methods in various disciplines of engineering that lend themselves perfectly to using these concepts. In this section, we present two examples as an illustration.

> **Example 12.1.** The resultant **R** of the two forces **P** and **Q** shown in Fig. 12-15a is the 700 lb force. Determine the magnitudes of **P** and **Q**.

> *Solution.* To solve this problem on a CAD/CAM system, the user begins by using the part setup described in Chap. 3. Each force in Fig. 12-15a is treated as a line

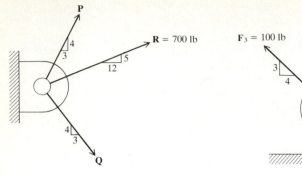

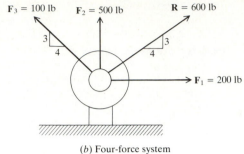

(a) Two-force system

(b) Four-force system

**FIGURE 12-15**
Force system acting on a pin.

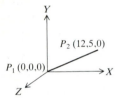

Create the **R** line

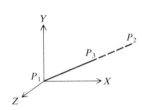

Trim line $P_1 P_2$ to a length of 7 inches

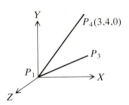

Create the **P** line

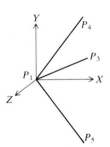

Mirror line $P_1 P_4$ to obtain line $P_1 P_5$

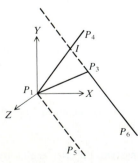

Translate line $P_1 P_5$ to $P_3$

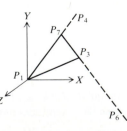

Trim lines $P_1 P_4$ and $P_3 P_6$ to their intersection point $I$

**FIGURE 12-16**
Solution to the
Fig. 12-15a problem.

that can be constructed. The resultant **R** can be created as a line connecting points $P_1(0, 0, 0)$ and $P_2(12, 5, 0)$, as shown in Fig. 12-16. We then trim the line to a length of 7 inches (assuming the database units are inches) using a "trim" command. This results in endpoint $P_3$ whose coordinates can be obtained by a "verify" command. We use a scale factor of 100. Next, create the force **P** as a line connecting points $P_1$ and $P_4(3, 4, 0)$. Point $P_1$ can be referenced in the line command as an endpoint of the line $P_1P_3$ by using the end modifier. Therefore, its coordinates are not necessary to be input again. The third step is to create the force **Q** by mirroring the line connecting $P_1$ and $P_4$ with respect to the $X$ axis because of the slopes of **P** and **Q** as shown in Fig. 12-15a. This produces the line $P_1P_5$.

To solve the problem, we need to create a force triangle. To achieve this, we translate line $P_1P_5$ to point $P_3$ using a translation command with a translation vector of $P_1P_3$. This produces line $P_3P_6$. We now trim lines $P_1P_4$ and $P_3P_6$ to their point of intersection $I$. Notice that the end modifier should be used to define the translation vector; the intersection modifier should be used to define $I$. If we use a "verify entity" or a "measure distance" command, we can find the lengths of the lines $P_1P_7$ and $P_3P_7$. Multiplying them by the scale factor 100, we obtain the magnitudes of **P** and **Q** respectively.

*Note:* one could avoid mirroring line $P_1P_4$ and translating line $P_1P_5$ by creating a line directly connecting $P_3$ and $P_6$. In this case, incremental coordinates are used to input $P_6$. The line command would have endpoint $P_3$ as one input and $\Delta x = 3$, $\Delta y = -4$, $\Delta z = 0$ as the input that defines $P_6$. The example is solved purposely the other way to illustrate more of the concepts introduced.

**Example 12.2.** Three forces of a four-force system acting on an eyebolt are shown in Fig. 12-15b. The resultant **R** of the system is also shown. Determine the fourth force.

*Solution.* Although we can follow similar steps to solve this example as we did in the previous one, let us perform some of the obvious basic calculations ourselves. Using the forces shown in Fig. 12-15b and their slopes, we can quickly determine the incremental coordinates of the endpoints of the lines representing the forces as shown in Fig. 12-17. We use one line command to create the piecewise line connecting $P_1$, $P_2$, $P_3$, and $P_4$. Using a scale of 100, lines $P_1P_2$, $P_2P_3$, and $P_3P_4$ represent

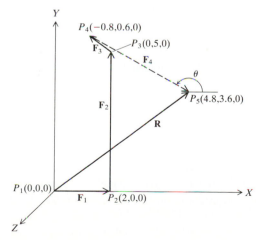

All dimensions shown are incremental except for $P_1$

**FIGURE 12-17**
Solution to the
Fig. 12-15b problem.

the forces 200, 500, and 100 lb respectively. Create line $P_1P_5$ to represent **R**. Create a line connecting $P_4$ and $P_5$ as endpoints of existing lines. This last line represents the fourth force $\mathbf{F}_4$ both in magnitude and direction. Using a "verify entity" or "measure distance" command gives the magnitude of $\mathbf{F}_4$ (after multiplying the length by the scale factor 100). A "measure angle" command gives the angle $\theta$ that $\mathbf{F}_4$ makes with the $X$ axis. The sense (shown by arrows) of $\mathbf{F}_4$ is easily determined from the concept of the equilibrium diagram. The magnitude of $\mathbf{F}_4$ is 412 lb and the angle $\theta$ is 150.95°.

# PROBLEMS
## Part 1: Theory

**12.1.** In addition to using the geometric modifiers, how can you use the manipulation and editing operations to easily create the model shown in Prob. 11.2 in Chap. 11 (we need a planning strategy more than a detailed sequence of commands)?

**12.2.** Why cannot B-spline and Bezier curves be trimmed beyond their existing endpoints?

**12.3.** Which of the manipulation and editing operations covered in this chapter are applicable to solid models? Explain your answers.

**12.4.** Identify some of the graphical techniques and methods in engineering that are in your area of interest (mechanics, fluid mechanics, heat transfer, design, etc.) and explain how you can implement them on your CAD/CAM system.

## Part 2: Laboratory

On your CAD/CAM system, find the following:

**12.5.** The available entity selection methods. What are their syntax and how can you use them?

**12.6.** How can you verify the model and database parameters? Obtain a listing of these parameters and their values.

**12.7.** How can you verify entities? What are the available modifiers?

**12.8.** How can you copy (duplicate) entities? What are the available modifiers?

**12.9.** How can you create geometric arrays?

**12.10.** The transformation commands, that is, the translation, rotation, mirroring, and scaling commands. How can you use each command with its various modifiers? Apply the commands to basic and simple graphics.

**12.11.** How can you blank/unblank entities? Investigate the mode dependence of the command.

**12.12.** The detailed syntax (with modifiers) and uses of the "measure" command.

**12.13.** The detailed syntax (with modifiers) and uses of the "offset" command.

**12.14.** The detailed syntax (with modifiers) and uses of the "trim" command.

**12.15.** The detailed syntax (with modifiers) and uses of the "divide" command.

**12.16.** Any commands for stretching, modifying, or editing entities. Can they be replaced by any of the commands covered in Probs. 12.8 to 12.15?

**12.17.** Reconstruct some of the geometric models introduced in Chaps. 5, 6, and 7 by stressing the use of the concepts and commands introduced in this chapter.

## Part 3:  Programming

The programming assignments in this chapter follow the same philosophy as those of Chap. 11.

**12.18.** Write a program that can measure the length of a line or a contour.

**12.19.** Write a program that can trim a line to a given point or to the intersection point with another line.

**12.20.** Write a program that can divide a line or a circle to $n$ equal or nonequal divisions.

## BIBLIOGRAPHY

Goetsch, D. L.: *MICROCADD: Computer-Aided Design and Drafting on Microcomputers*, Prentice-Hall, Englewood Cliffs, N.J., 1988.

McKissick, M. L.: *Computer-Aided Drafting and Design*, Prentice-Hall, Englewood Cliffs, N.J., 1987.

System Documentation and Manuals of any CAD/CAM System.

Voisinit, D. D.: *Introduction to CAD*, 2d ed., McGraw-Hill, New York, 1986.

Zeid, I.: "CAD on Computervision System," unpublished document, CAD/CAM Laboratory, Northeastern University, Boston, Mass., 1984.

Zeid, I.: "CAD on GE Calma System," unpublished document, CAD/CAM Laboratory, Northeastern University, Boston, Mass., 1985.

# CHAPTER
# 13

# COMPUTER
# ANIMATION

## 13.1 INTRODUCTION

Animation is the process in which the illusion of movement is achieved by creating and displaying a sequence of images with elements that appear to have motion. The principle of animation defined as such was conceived eighty years ago and is still valid today. The illusion of movement can be achieved in various ways. The most obvious way is to change the locations of various elements of various images in the sequence. Other ways include transforming an object to another (metamorphosis), changes in color of an object, or changes of light intensities.

The world of entertainment has used both conventional (manual) and computer animation in producing animated cartoons, movies, logos, and ads. The emphasis of conventional animation is usually on the artistic aspects and looks of the images in the animation sequence. The animator draws a sequence of pictures which, if animated, produces the illusion of the proper motion. No calculations or physical laws are involved in generating the pictures. The main criterion to generate these pictures is that the motion looks as real as possible to the observing eye. Consider, for example, the animation of a human body. The animator creates pictures that make the motions of the various parts of the body look real without utilizing any kinematic or dynamic analysis of articulated bodies.

The use of computer animation in entertainment has enabled more complicated motions and more realistic images to be introduced than manual animation can offer. Also, it has enabled physical laws to be incorporated into animation. Referring to the above example, animation software exists that can determine the human body motion based on kinematic or dynamic analysis. With more and more use of the computer in animation, we see a shift from conceiving animation as a mere art into a science. This shift does not suggest the elimination of art in animation. Rather, it suggests that animation should be based on a balance of art and science. At the present time, animation art is more advanced than animation science.

While animation may be viewed as having developed from puppetry in the entertainment world, it may be viewed as a valuable extension of modeling and simulation in the world of science and engineering. If the spatial or temporal scale of an engineering system is too large or too small to facilitate viewing in real-life modeling, then animation may be a viable alternative. In many engineering problems that involve simulation, animation may be considered as a discretization of the time domain, generating the sequence of appropriate images at discrete time values, and displaying the sequence to restore time continuity so that the development of events over time can be better understood.

Animation is a very useful visualization aid for many modeling and simulation applications, especially those where large amounts of scientific data and results are generated. Visualizing this data is important to its correct interpretation and understanding. With the increasing use of supercomputers in engineering, visualizing simulation results become more critical. A new visualization problem arises: volume visualization—the presentation of not just surface data so that it looks three dimensional, but of interior points in a structure. This problem is expected to require new languages and operating systems—entirely different computing environments.

Animation has already been used and has proved valuable in many engineering applications. To simulate the evolution of a thunderstorm, for example, software for a simulation model based on flow equations that describe thunderstorm dynamics is used to generate data at given time values. The amount of the data generated is too large to study in its numerical form. Its display in an animated form enables a thorough study of how gusts and tornadoes develop. Another example is the animation of facial expressions. These animations are based on models of muscles which can be pulled or squeezed to create the deformable topologies of faces. These animations could facilitate teaching lip-reading to the hearing-impaired and help determine preoperatively the mobility that would remain after facial surgery. Other useful animation applications are found in nuclear weapon simulations, car-crash simulations, fluid dynamics (flame propagation), meterology, chemical and molecular modeling, CT scans of parts of the human body (head, neck, spine, chest, liver, kidney, and stomach), ultrasound scans, trajectory planning and obstacle avoidance in robotics, verification of NC tool paths, and many others.

In the following sections, we cover conventional and computer animation, requirements of animation systems, types and techniques of animation, some animation-related problems, some animation applications, and how to do animation on microcomputers.

## 13.2   CONVENTIONAL ANIMATION

Conventional (also called traditional or manual) animation is the type used in various studies to produce cartoon animated films. It is used primarily to animate two-dimensional scenes. Its extension to three-dimensional animation is usually difficult and time-consuming. However, its study and understanding form the basis of computer animation (see next section). Most of the terminology and concepts utilized by animation software originates from conventional animation.

An animated film, as in ordinary films, tells a story. At its conception, the story is described by a synopsis or summary. The scenario of the story is developed next. The scenario is the detailed text of the story without any cinematographic references. A storyboard is then developed based on the scenario. The storyboard is a film in an outline form. It is a set of drawings resembling a comic strip which indicates the key sequences of the film scenes. These key sequences (also called keyframes) form the basis of animation to create the animated film.

In creating an animated film by conventional animation, the following steps are utilized:

1. Keyframe. Animators draw keyframes that correspond to the movement of the film characters and to the timing required. An animator is usually assigned to one specific character. Animators are skilled individuals who understand human and animal motions thoroughly and who have good imagination.

2. Inbetweening. Interpolation between any two keyframes must be made to produce smooth animation. This interpolation process, which is done manually, and the resulting frames are known as inbetweening and inbetweens respectively. Assistant animators draw the main artistry of inbetweens and inbetweeners draw the remaining figures. For the smoothest animation, 24 frames must be drawn for every second of animation. If the movement in a certain scene of the animation is to take $N$ seconds, the inbetweener must draw $24N$ frames of the same scene to complete this movement.

3. Line testing. The drawings (line drawings thus far) of the keyframes and inbetweens are photocopied onto transparent acetate (cel), and are filmed under a rostrum camera to test the quality of the movements produced.

4. Painting. After any modifications arising from the line tests, cells are painted to introduce color which results in a color film. Painting also gives the animated characters a sense of solidity. Static backgrounds have to be painted also.

5. Filming. The final photography under the rostrum camera is carried out on color films or videotapes. Soundtracks of voice, music, and effect (such as thunderstorming) are added to the film.

Figure 13-1 shows these steps with an illustration of a runner.

Special effects are used in conventional animation to improve the quality of the animated film. Pan and tilt are produced by moving the camera from one point to another horizontally and vertically respectively. A zoom is produced by moving the camera closer to or further away from a given object. Spin is an effect produced by rotating the camera. Fade-in, fade-out, cross-dissolve, and wipe are effects often used during transition between frames. Fade-in and fade-out are used at the beginning and the end of a frame or a scene respectively. Fade-in makes a frame gradually appear from the background and fade-out makes it gradually disappear onto the background. Cross-dissolve is a combination of

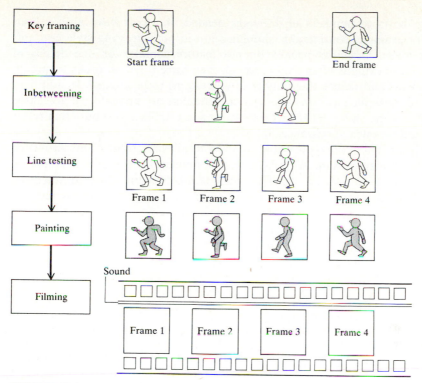

**FIGURE 13-1**
Steps of conventional animation.

both. It corresponds to a fade-out of one frame and a fade-in of the next frame over the same length of time. Wipe makes one frame appear to slide over the preceding frame. Sliding can occur horizontally, vertically, or diagonally, as shown in Fig. 13-2.

## 13.3  COMPUTER ANIMATION

Conventional animation, even for a small cartoon film, is a very labor-intensive process involving hundreds, perhaps thousands, of man-hours. The majority of the time is spent in making all the drawings of the characters and the backgrounds for both the keyframes and the inbetweens. Due to the high cost of

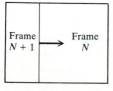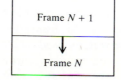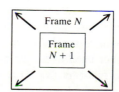

**FIGURE 13-2**
Various types of wipes.

producing the large numbers of drawings needed, shortcuts have always been sought. For example, considerable simplification in both drawing style and animation is made. Drawings usually do not feature complex or fine details of scenes, and a drawing is often shot twice, or even three times, in order to stretch the movements and reduce the number of inbetweens to be created. In addition, only limited parts such as arms and legs of characters are animated at all. These simplifications and shortcuts are acceptable for some applications but often result in a jerky and wooden appearance.

Computer animation is a viable solution to alleviate most of the problems of conventional animation. As expected, the animator's energy is no longer invested in repetitive drawing and the tedium of inbetweening. Instead, the focus is on creating the scenes and their related details and imagery. This has resulted in demanding new techniques and approaches for computer animation. The science of animation is therefore developing rapidly to meet the demands of computer animation. Such a trend has made animation a viable extension of modeling and simulation in engineering applications.

### 13.3.1 Entertainment Animation

Two classes of computer animation can be distinguished: entertainment and engineering. The role of the computer in the first class can be identified from Fig. 13-1. Computer graphics techniques can be used to generate the drawings of keyframes and inbetweens. The drawings of keyframes can be created with an interactive graphics editor, can be digitized, or can be produced by programming. The inbetweens can be completely calculated by the computer by interpolating along complex paths of motion. The use of layers in generating these drawings is very useful. Drawings that are shared by more than one frame are stored on separate layers and shared by all the frames by overlaying the proper layers together.

Shading techniques can be used to paint the drawings of the various frames. Not only do these techniques add great visual realism to the animation but they also provide a tenfold reduction in time taken for this task. Shading systems available for animation artists provide them with convenient user interfaces and special coloring and shading effects. For example, an artist can color an area of a picture by simply touching a point within the area with a lightpen. The palette of colors available to the artist is usually displayed on the screen and can be standardized throughout a film.

Filming an animated film can be assisted by computer. If filming is done by a camera, its movements can be controlled by the computer, or virtual cameras can be completely programmed. If a film is recorded via a video recorder, the computer can control the recorder. A still-frame video recorder, similar to a frame grabber used in image processing, is used to record the film. The average home video recorder (VCR) is not suitable for recording animation because it is designed to record long "takes" (a take is what is recorded between the start and end of recording). Animation usually require very short takes (snap shots) that must be laboriously edited.

The use of the computer in conventional animation as described above is known as computer assisted two-dimensional animation (sometimes called key-frame animation). Automatic inbetweening has been the primary focus of attention in moving from conventional animation to computer assisted animation. The techniques to assist in conventional animation do not fully exploit all the power of computer graphics and CAD. They can only automate processes that were previously carried out manually.

Modeled animation (sometimes called three-dimensional animation), on the other hand, is an entirely different medium. It opens up the possibility of utilizing the available computer graphics and CAD techniques to create scenes, movements, and images that are difficult to achieve by conventional means. In particular, accurate representations of objects, as well as smooth and complex three-dimensional movements, are facilitated. After the storyboard is prepared for a film, the following steps arise to achieve modeled animation:

1. **Geometric description.** In order to allow the complete and general three-dimensional animation of drawn objects, they should be described as geometric models utilizing wireframe, surface, or solid representations. A great degree of realism can be added to these models after their images are generated with shading attributed such as color, texture, reflectance, translucency, etc.

2. **Frame generation.** Cartoon animation has drawn objects frequently distorted by stretching, bending, and twisting for humorous effect. Apart from this effect, these objects can exhibit dynamic movements by applying various geometric transformations to their geometric models. After all movements of objects in a scene have been accomplished, the resulting set of geometry forms a frame in the animation sequence to be created. Applying another set of transformations yields another frame. The recording of these frames produces the animation sequence of the animated film. As with conventional animation, all movements can be "faired," that is, carefully accelerated from and decelerated to rest. Fairing can easily be incorporated into geometric transformations, which helps give computer animation its special characteristics.

3. **Line testing.** After all frames are generated, their corresponding images are generated by shading them. These images can be animated and displayed on the graphics display to test the movements in real time. Because of the complexity of the images to be generated, it is normal to create simplified images (low resolution and less fancy shading effects) to test the movements.

4. **Recording.** When all the frames and images are satisfactory, the images are recorded frame by frame onto video recorders or films. Sometimes images are produced on paper or cel for filming under a conventional rostrum camera, or full-size negatives are made of these images to enable optical tricks or even some hand coloring to be carried out.

Figure 13-3 shows the steps of modeled animation. The example of a slider-crank mechanism is used to illustrate these steps. As will be seen in the next section, modeled and engineering animations have the same basic concepts with differences in implementations.

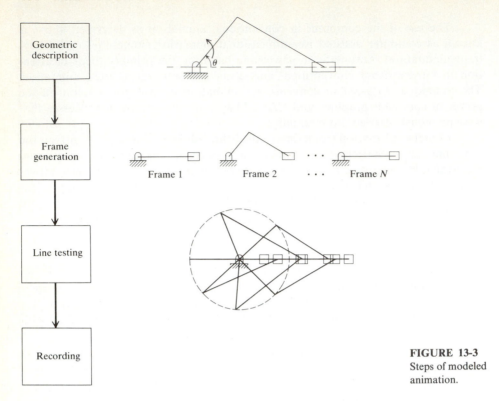

**FIGURE 13-3**
Steps of modeled animation.

### 13.3.2 Engineering Animation

Animation has been utilized in CAD/CAM-based applications primarily for visualization purposes. NC tool paths can be verified by displaying the tool as it moves on the surfaces to be machined. Design engineers can use animation to detect interferences during assembly processes. Similarly, mechanism engineers can display motions to verify kinematic contraints.

Animation as described above is mainly an extension of geometric modeling. A geometric model is usually created first. Some of the geometric parameters of the model are changed according to geometrical and/or analytical procedures. The various values of these parameters are used to generate the various frames that can be animated. These frames could be simply wireframe geometry or elaborated shaded images. Usually, wireframe geometry is used to test animation before generating the images for final display and recording. The slider-crank mechanism shown in Fig. 13-3 is an example. The angle $\theta$ of the crank is incrementally changed within its range (0 to 360°) to produce the various frames for animation.

Engineering animation can be thought of in another way, that is, as an extension of analytical modeling and simulation. In this context, animation becomes another effective way of analyzing the large number of numerical data that results from simulation. By visualizing the data in a continuous animated fashion (with "freeze" capabilities), engineers and scientists can track precisely the development of various phenomena.

Engineering animation is in itself a form of modeled animation, discussed in the previous section. The difference is that the former is science oriented while the latter is image oriented. Less emphasis on image realism and no fairing characterize engineering animation. In fact, wireframe animation can be sufficient in some applications. In addition, imagery in engineering animation could focus on the scientific contents of the frames instead of their realistic and scenic look.

An engineering animation system needs to meet the following requirements:

1. Exact representation and display of data. By data, we mean objects or numerical results of simulation. Data must be displayable in an image (shaded) form for better visualization and understanding. Raster displays are usually used to display animation (sometimes called raster animation) because of their ability to display solid areas of color and patterns.

2. High-speed and automatic production of animation. In order to use animation as an aid of communication between engineers and designers, animated frames must be produced fast and need a high-speed display. A real-time animation system is ideal. In addition, engineering animation must be performed automatically because engineers and designers who produce the animation frames are not professional animators.

3. Low host dependency. An engineering animation system must be available to engineers during their modeling or simulation. The system must not depend heavily on host computers that are usually overloaded. Implementing the system into workstations that share a host computer is ideal. In this case, geometric transformations can be performed locally on the workstation, thus producing a practical animation system even if the connection between the host and the workstation is not fast enough (a low-speed serial interface such as a local area network).

## 13.4 ANIMATION SYSTEMS

Two types of animation systems may be identified. The first type is the dedicated systems typically used by animation applications such as animated films, flight simulators, etc. These systems usually utilize high-speed dedicated host computers as well as high-speed graphics displays. Their software is primarily designed and developed for use in computer animation. The other type usually exists on some typical CAD/CAM systems as a special software module. This type is usually limited in its capabilities and does not require any special hardware. In this section we discuss some of the details of the dedicated animation systems.

### 13.4.1 Hardware Configuration

The ability of computers to generate pictorial displays opens the door to building animation systems around them. Theoretically, an animation system may consist of a computer to calculate and generate the required frames for animation and a graphics display (usually raster display) to display the frames successively as they are generated. However, the inability of computers to generate and display the frames fast enough makes it impossible to achieve animation. Most computers cannot achieve the animation quality of commercial movies or television. Television-quality animation stems from their video-display generation rates (30 frames/second). These rates prevent a computer from generating the successive frames quickly enough to create the illusion of continuous motion as perceived by the human eye.

One way to solve the above speed problem is to save the frames (computer visual output) as they are generated and then to redisplay them continuously later in order to create the animation effect. This approach allows a computer to take as much time as it needs to generate each frame. When the computer completes one frame, it is saved. The computer then begins to generate and display the next frame. When it finishes, it is saved. When the entire presentation has been saved one frame at a time in this manner, the entire sequence of frames is shown at sufficient speeds to be of interest to people.

Two of the common media to save frames are cameras and videotape recorders. The schematic of two animation systems based on them is shown in Fig. 13-4. The system with a movie camera (Fig. 13-4a) saves the frames photographically on a film. The camera is interfaced and therefore controlled by the computer. When the computer completes one frame, the movie camera, whose shutter is triggered electronically by the computer, transfers that image to a single frame and then advances the film by one frame. When the computer finishes the next frame, the camera records it and again advances its film one frame. When all the frames are photographed, the film is developed and shown to produce the animation sequence. Effective use of this filming technique requires an understanding of the factors that affect still photography of video images, such as exposure timing, ambient light, and the alignment of the camera with the screen. It is imperative, for these factors, to have the camera controlled by the computer.

The system with a videotape recorder (Fig. 13-4b) follows the same steps to save an animation sequence as described above, but with the frames saved on a videotape instead of a film. Figure 13-5 shows a detailed hardware configuration of the system. The main idea of this system is to convert the color graphics

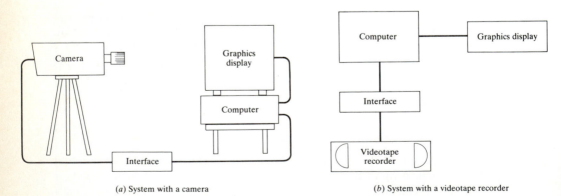

(a) System with a camera                    (b) System with a videotape recorder

**FIGURE 13-4**
Hardware configuration of animation systems.

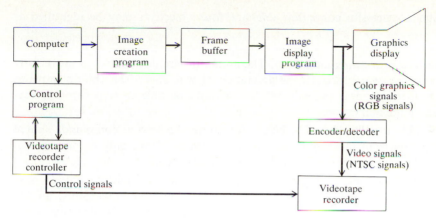

**FIGURE 13-5**
Animation system with a videotape recorder.

signals (usually RGB signals) into video signals (such as NTSC signals) via an encoder/decoder. The system's main components are a computer, a raster graphics display, and a videotape recorder. The computer has two main functions. It executes the programs that convert the animator's shading data and input into RGB signals, and it controls the recorder. The image-creation program shown in the figure calculates the intensities of the various pixels using the animator's data, and stores them in the frame buffer. The image-display program converts these intensities into RGB signals and displays them on the graphics screen.

The control program both translates and monitors the control codes that the animator inputs. As a translator, it converts commands from the computer into a form recognized by the videotape recorder controller shown in Fig. 13-5 and transmits status indications from the controller back to the computer. The control codes electronically transfer animations, frame by frame, directly from the image-display program to the videotape. With the assistance of the encoder/decoder, the image information (RGB signals) is assembled into a single frame which is fed to the videotape recorder for recording.

As a monitor, the control program detects and rejects invalid data sequences in both directions (from the computer to the recorder and vice versa). It also determines whether the videotape recorder responds in the correct manner to commands sent to it and, if not, takes remedial action. For example, if the recorder has not entered the record mode within, say, ten seconds after the appropriate command has been sent, the control program informs the computer and the command is tried again. If overnight recording sessions are used, the control program should be able to recover automatically from any errors that may occur.

## 13.4.2 Software Architecture

Animation software is the key component of animation systems. The software has two primary functions: produce the animation frames and control the hardware.

To design a successful animation software, four objectives may be identified as follows:

1. To have a previewing capability. The animator should be able to move forward or backward through the animation at will, should be able to move to any individual frame immediately, and should be able to step through keyframes of the animation.
2. To be able to edit animation before recording. Animation commands should be separate from image generation and routines so that only command files should have to be changed in order to edit animation.
3. To produce natural fairing of movement. Animators should be able to vary movement parameters in a way that simulates natural movements, and to allow smooth animation with gradual starts and stops. This objective is not acceptable in engineering animation where scientific data and results are usually animated.
4. To control the process of transferring computer-generated images (frames) to recording media such as film or videotape. The animation software should be able to control both the image generation and the videotape recorder, as well as recover from errors.

Figure 13-6 shows an architecture of an animation software that meets the above objectives. The graphics and animation editors are responsible for processing user commands. They are invoked by the corresponding user interfaces. In addition to managing the entire animation software, the global controller

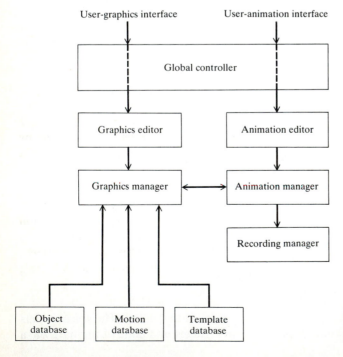

**FIGURE 13-6**
Software architecture of animation systems.

directs the user input to the proper editor. The graphics input data, received by the graphics manager through the graphics editor, is used to create the object database. Similarly, the animation input data is processed by the animation manager, together with the graphics manager, to create the motion database. A template database of standard parts or objects may be available to both managers. In an engineering environment, the three databases would also be shared by other systems such as a CAD/CAM system and a dynamic simulation system.

The graphics manager is also responsible for creating the images requested by the animation manager. These images, in turn, are processed to create the animation frames. The frames can be edited via the animation editor before recording.

The recording manager controls recording the frames and reports at the end of each frame. This manager can be written for cameras and/or videotape recorders.

The design and development of animation software involve four distinct types of input. Individuals who are responsible for the technical applications and plan to use the software discuss the outlines of these applications with a producer. The producer converts the outlines into a story. The graphics designer, who is concerned with the overall look of the graphics, decides which animation will be suitable for computer graphics and then develops a storyboard with the computer programmer. The programmer then develops the programs as described in Fig. 13-6. Most animations use FORTRAN or C programs that call standard graphics routines (based on GKS or PHIGS).

### 13.4.3   Classification

An animation system and a robot control system share a lot of basics. The main difference is what each system controls. From a control point of view, animation systems can be classified into three levels:

1. Guiding. Guiding animation systems are those with no mechanisms to specify motion algorithmically. These systems depend heavily on the animator's artistic background to define motion from one frame to another. They allow affine transformations of objects to create keyframes. Inbetweens are generated by interpolating the transformation parameters and transforming the objects. In guiding systems, the animator has complete control over the motion of an object and must specify in advance the details of this motion. This is reasonable only in simple as well as artistic motions.

2. Animator-level systems. These systems allow the animator to specify motion algorithmically. They provide significant improvements over guiding systems in terms of the ability to describe precisely, based on computational models, the required motions for animation. However, these systems are more difficult to develop and use than guiding systems.

3. Task-level systems. At the task level, the animation system acts as an expert animation system. The animator can only specify the broad outlines of a particular movement and the animation system fills in the details based on a

knowledge base of objects and figures in the environment. The knowledge base would contain information about positions, physical attributes, and functionality of these objects and figures. It would also contain the constraints that are imposed by the environment. Whether task-level animation is appropriate depends on the particular application.

Which of the above animation systems is more suitable for engineering simulation and animation? Unfortunately, there is no one system. A hybrid system may be the answer. For example, if the animation of an existing engineering system is required to study, say, its motion, an animator-level system is appropriate. If, on the other hand, a redesign of the system is required, guiding with algorithmic animation may be suitable. If guiding results in an optimal position on the motion curve, the solution of the inverse problem of equations of motions would produce the corresponding values of the system parameters.

## 13.5   ANIMATION TYPES

In the preceding section, animation systems are classified based on their role in the animation process. If we consider the time it takes to generate animation, three types of computer animation result. They are frame-buffer animation, real-time playback, and real-time animation. These types are controlled by the time it takes the computer to produce a frame. This time is, in turn, controlled by the complexity and realism of the frames as well as the speed and the scan rates of the computer and the graphics displays involved.

### 13.5.1   Frame-Buffer Animation

Hardware features widely available on current frame-buffer systems offer a surprising variety of techniques for limited animation. These features, if used properly, can provide the illusion of real-time animation for a large class of applications. They may not be suitable for engineering animation, though, because they are all based on static images, in the sense that the contents of the frame buffer (pixel memory) will never change during animation. Dynamics are added by modifying the pattern in which pixels are read from memory and the way in which their contents (bits) are interpreted to provide color information using a color lookup table. There are three types of frame-buffer animation.

**13.5.1.1   COLOR TABLE ANIMATION.** A color lookup table provides a limited form of frame-buffer animation. It is one technique for achieving dynamics with a static pixel memory. The value of a pixel is generally an index in a lookup table of colors. Lookup tables can be used to partition pixel values into displayed and nondisplayed colors. This partition can step a displayed image through an animation sequence.
    Color table animation is achieved by color cycling, alternate color, or bitplane extraction. In color cycling, the entries in the lookup table are rotated. This gives the illusion of motion as colors appear to flow across the screen. Alternate color animation paints more than one image into the frame-buffer memory using

different pixel values for each image. Setting the lookup table entries to background for all but one of the images allows a single image to be viewed. Changing the subset of visible colors alternately displays one image and then another. Bit-plane extraction partitions the frame buffer into bit planes and assigns each image to a separate set of these planes. Loading the lookup table to ignore all but the bit planes corresponding to a particular image allows images to be displayed separately. Using this technique, it is possible to display more than one image simultaneously, to mix them, or to overlay one with another using a priority scheme.

Figure 13-7 shows an example of displaying a pendulum motion using cycling animation. The figure shows how the frame buffer could be loaded; the numbers indicate the pixel values placed in each region of the buffer. Figure 13-8 shows how the lookup table is loaded at each step of animation to display the pendulum motion. The idea is to display all but one of the pendulums at the background color 1. The motion effects result by cycling the contents of the lookup table.

**13.5.1.2  ZOOM-PAN-SCROLL ANIMATION.** In this type of animation, the frame buffer (pixel memory) is logically divided into different regions, each containing a separate, low-resolution image. Figure 13-9 illustrates the cases in which four $256 \times 256$ images and sixteen $128 \times 128$ images are stored within a $512 \times 512$ frame buffer. By setting horizontal and vertical zoom factors of two with a $512 \times 512$ viewport, the display can cycle through the four images (Fig. 13-9a) by successively setting the window location to the upper left-hand pixel of each image. One advantage of this approach over lookup table animation is that each pixel maintains its full color range instead of sacrificing palette size for enhanced dynamics. However, it sacrifices the spatial resolution of the image. There is clearly a trade-off between spatial resolution and intensity resolution.

**13.5.1.3  CROSSBAR ANIMATION.** Crossbar animation is achieved by partitioning bits within pixels. In other words, crossbar animation consists of routing any of the bits from pixel memory to any of the input lines in the lookup table. The name crossbar stems from a crossbar switch, which is usually found in raster displays. The crossbar switch is responsible for routing pixel bits into the lookup table. In addition to routing the input, the crossbar switch is also capable of ignoring input and passing one bit through to any output.

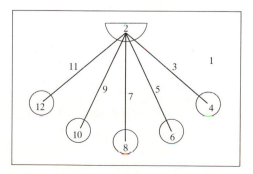

**FIGURE 13-7**
Contents of frame buffer for pendulum animation.

| Entry number | Animation steps with corresponding colors loaded in lookup table | | | | | | | | |
|---|---|---|---|---|---|---|---|---|---|
| | 1 | 2 | 3 | 4 | 5 | 6 | 7 | 8 | 9 |
| 1 | W | W | W | W | W | W | W | W | W |
| 2 | B | B | B | B | B | B | B | B | B |
| 3 | B | | | | | | | | B |
| 4 | R | | | | | | | | R |
| 5 | | B | | | | | | B | |
| 6 | | R | | | | | | R | |
| 7 | | | B | | | | B | | |
| 8 | | | R | | | | R | | |
| 9 | | | | B | | B | | | |
| 10 | | | | R | | R | | | |
| 11 | | | | | B | | | | |
| 12 | | | | | R | | | | |

W White
B Black
R Red

**FIGURE 13-8**
Lookup table for a pendulum motion (unspecified entries are white).

With a 32-bit frame buffer, we can store four 8-bit images, eight 4-bit images, sixteen 2-bit images, or even thirty-two 1-bit images. Figure 13-10 shows the partition of the buffer to obtain four images. Crossbar animation coupled with color table and zoom-pan-scroll animation is a particularly effective combination.

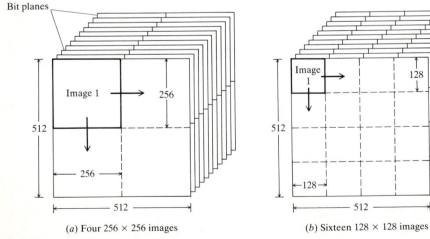

(a) Four 256 × 256 images          (b) Sixteen 128 × 128 images

**FIGURE 13-9**
Zoom-pan-scroll selection.

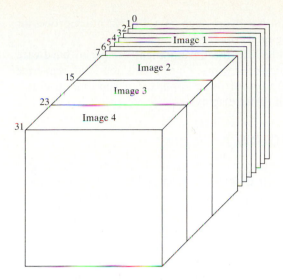

**FIGURE 13-10**
Crossbar animation with 32-bit frame buffer split into four 8-bit images.

### 13.5.2 Real-Time Playback

This is the most popular form of animation. Frames are generated frame by frame in advance at non-real-time rates, and then saved on a film or recorder. A real-time program can display the frames later to create the animation. In this case animation is in real time in the sense of playing back the sequence of frames.

Real-time playback is used where it is generally impossible to generate frames at the rates required for real-time presentation. This occurs in applications that require intensive calculations or in images that must display a certain degree of realism. Frames must be displayed at the rate of 24 frames per second for smooth animation, and at the rate of 30 frames per second for flicker-free display. Therefore, if the time it takes to calculate any frame in the animation sequence is greater than 1/24 s, the frames must be recorded and then played back at the rate of 24 frames/second.

### 13.5.3 Real-Time Animation

Real-time computer animation is limited by the capabilities of the computer and data-transfer rates. Few minutes may be required to calculate a single frame in number-crunching-oriented applications or in images with a great deal of realism. Even if a frame can be calculated in less than 1/30 second, the bandwidth of existing digital hardware has to transfer the frame data from the computer to the frame buffer of the graphics display. Let us consider an 8-bit plane display of $512 \times 512$ resolution. A 30 frame per second animation requires processing of 7.5 million bytes ($500 \times 500 \times 30$) of information per second. This amounts to a 60-MHz ($500 \times 500 \times 8 \times 30$) bandwidth. This is well beyond the capacity of most mini- and microcomputer disk controllers, I/O interfaces, and buses. To appreciate this data-transfer problem, color television is limited in a bandwidth of 3 to 4 MHz. Moreover, analog hardware has enough problems working at these high frequencies. Digitally creating animation by totally recomputing every frame in real time on existing digital hardware such as a 4-MHz microprocessor using 8-bit bytes is simply out of the question.

Real-time animation is possible with the development of special hardware such as array processors and graphics processors. Until this is achieved, real-time animation is not possible for calculation-intensive applications running on multi-user computers with high-resolution raster displays.

## 13.6 ANIMATION TECHNIQUES

The essence of the computer animation problem is in determining the proper sequence of frames based on a given animation model. The animation model could vary widely depending on the application at hand. In entertainment, the model is simply a given set of keyframes and a set of motion (spatial and temporal) constraints. The solution to the animation problem in this case is the automatic inbetweening (generation of the inbetweens). In engineering, the model is an equilibrium equation with the proper boundary and/or initial conditions. The solutions of this equation at certain time values are the desired sequence of frames.

The solution of the animation problem is directly related to the animation model. If the model is based on shape information only, the frames are generally fast to generate, but the quality of animation (how smooth and realistic the resulting motions are) usually suffers. If the model is based on physical laws in addition to shape information, more computer time is needed to generate the frames, but the quality of animation is better. Existing animation models or techniques offer various degrees of compromise between the exactness of the animation model and the time required to generate the frames.

### 13.6.1 Keyframe Technique

This technique is an extension of keyframing used in conventional animation. Given a set of keyframes, the technique automatically determines the correspondences between them and then generates all the frames (inbetweens) between them by some form of interpolation. The correspondence between the geometry (number of points and curves) of the keyframes is a simple process if they have the same number of points and curves. All that is needed to establish correspondences in this case is to request the animator to input the geometry of the keyframes in the same order. The other case, where the keyframes do not have the same number of points and curves, requires preprocessing to make the number equal for all the keyframes. Different preprocessing arises depending on whether only the number of points, only the number of curves, or both are different. Preprocessing methods are not covered in this book. We will always assume that all frames have the same number of points and curves.

There are two fundamental approaches to keyframe animation. The first is called image-based keyframe animation (also called shape interpolation). In this technique, the inbetweens are obtained by interpolating the keyframe images themselves. Linear and/or nonlinear interpolation algorithms exist. Linear algorithms, in general, produce undesirable effects such as lack of smooth motion,

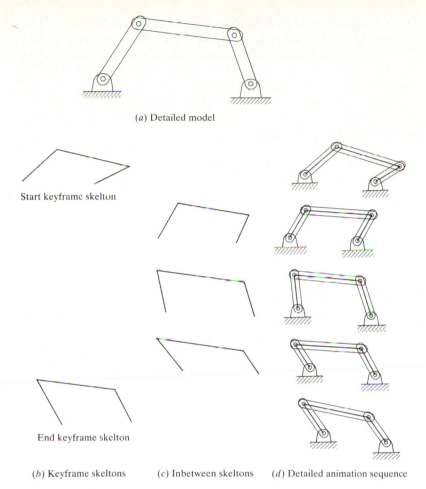

(*a*) Detailed model

Start keyframe skelton

End keyframe skelton

(*b*) Keyframe skeltons        (*c*) Inbetween skeltons        (*d*) Detailed animation sequence

**FIGURE 13-11**
Skelton animation.

discontinuities in motion speed, distortions in rotations, and contortions in the generated frames.

   The second type of keyframe animation is called parametric keyframe animation (or key-transformation animation). Here, better images can be produced by interpolating the parameters that describe the keyframe images, that is, the parameters of the geometric model itself. In a parameter model, the animator creates keyframes by specifying the parameter values. These values are interpolated and the inbetweens are individually constructed from the interpolated values. The remainder of this section describes some algorithms used for image-based keyframe animation.

**13.6.1.1  SKELTON ALGORITHM.** The idea behind the skelton algorithm stems from the manual inbetweening. Instead of using the images themselves as the basis for inbetweening, skeltons of the figures can be used. A skelton, or stick figure, is a simple image of the original one composed of only the key points and curves that describe the form of movement required. This allows the animator to create many keyframes consisting of skeltons only. These keyframe skeltons are then interpolated by the computer to create the inbetween skeltons. Details can be added to both keyframe and inbetween skeltons according to a single model. The inbetweens created in this way are much better because the keyframes are similar. Skeltons can be defined by curves or four-sided polygons. Figure 13-11 shows an example of skelton animation.

**13.6.1.2   THE PATH OF MOTION AND P-CURVES.** The keyframe technique is based solely on linear interpolation of the shape of the object to be animated. It does not consider the dynamics or movements of the object during generation of the inbetweens. The time is considered only while selecting the keyframes. If the motion is uniform, the positions of the keyframes in the time space become less critical in order to generate smooth animation. If abrupt motion or dynamics exist, unsmooth animation can result. The ideal solution is, of course, to develop and solve the dynamic equilibrium equations that describe the motion. This may be an expensive solution, especially for entertainment animation.

The use of path description and *P*-curves provides the animator with the information about the motion and its dynamics necessary to define the keyframes. A *P*-curve defines both the trajectory of a point in space and its location in time. Thus, the curve provides both spatial and temporal information about the motion. Figure 13-12 shows the *P*-curve for a person who walks along two adjacent walls to go from point *A* to point *C*. Both motion trajectory $[y = f(x)]$

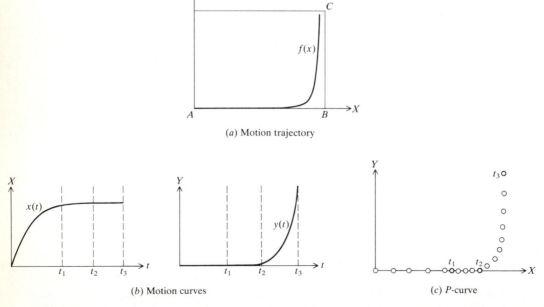

(*a*) Motion trajectory

(*b*) Motion curves

(*c*) *P*-curve

**FIGURE 13-12**
*P*-curve of a typical motion.

and motion curves $[x = x(t), y = y(t)]$ are necessary to describe the motion. The person accelerates in the $X$ direction until time $t_1$, continues with almost uniform motion (constant speed) until time $t_2$, and then accelerates in the $Y$ direction to reach point $C$. The $P$-curve combines Fig. 13-12$a$ and $b$. It has the shape of the trajectory, but a trail of symbols is used to indicate the path. These symbols are equally spaced in time.

The dynamics of a motion are represented on its $P$-curve by the local density of the symbols, as shown in Fig. 13-12$c$. The animator can use the density as a guidance to choose the locations of the keyframes and decide on the necessary number of inbetweens so that the resulting animation sequence is smooth and looks natural.

**13.6.1.3 INBETWEENING UTILIZING MOVING POINT CONSTRAINTS.** This technique allows the animator more control over the inbetweening process than the previous techniques. The animator can specify, in addition to a set of keyframes, a set of constraints called moving points. Moving points are curves varying in space and time which constrain both the trajectory and dynamics (i.e., path and speed) of certain points on the keyframes similar to $P$-curves. The set of keyframes and moving points form a constraint or patch specification of the desired dynamics.

Figure 13-13 shows an example of patch network. The network is formed from the animator's input. It consists of an ordered set of keyframes $\{\mathbf{K}_1, \mathbf{K}_2, \ldots, \mathbf{K}_{n-1}, \mathbf{K}_n\}$ which define the shape of the object to be animated at the animator-specified times $\{t_1, t_2, \ldots, t_{n-1}, t_n\}$ and a set of moving points $\{\mathbf{M}_1, \mathbf{M}_2, \ldots, \mathbf{M}_{q-1}, \mathbf{M}_q\}$. Each keyframe can be considered as a static shape positioned at a fixed point in time. Thus, each keyframe acts as a constraint in the motion sequence.

The patch network is subdivided into patches $P_1, P_2, \ldots, P_j$. Any patch $P_i$ is defined by four boundary curves; two are static boundaries derived from the two bounding keyframes and two are dynamic boundaries derived from the two

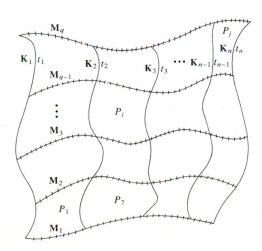

**FIGURE 13-13**
A patch network.

bounding moving points. The inbetweening of the patch network can thus be reduced to the sum of inbetweening the individual patches. A single patch geometry is shown in Fig. 13-14. The patch is described by the two parameters $u$ and $t$; $u$ is the parametric space variable and $t$ is the parametric time variable. The parameters are normalized to the interval $[0, 1]$. The two static boundaries are $\mathbf{P}(u, 0)$ and $\mathbf{P}(u, 1)$ which are parametric curves describing the geometry of keyframes. The two dynamic boundaries are $\mathbf{P}(0, t)$ and $\mathbf{P}(1, t)$. The corners of the patch are the endpoints of the keyframe curves. The subscript associated with each corner denotes its parametric values. For example, $P_{00}$ is the point $P(0, 0)$.

A betweening algorithm based on this technique is to find a parametric function $\mathbf{P}(u, t)$ for each patch in the patch network. A patch time interval $[t_s, t_e]$ ($t_s$ and $t_e$ are the start and end times of the patch respectively) should be normalized to the interval $[0, 1]$ to facilitate calculations. If an inbetween frame $\mathbf{P}(u, t_I)$ (refer to Fig. 13-14) is to be generated at time $t_I$, the corresponding normalized time is given by

$$t_n = \frac{t_I - t_s}{t_e - t_s} \tag{13.1}$$

If the functions $\mathbf{P}(u, t)$ are available for all the patches, the inbetween frame $\mathbf{P}(u, t_I)$ is obtained by evaluating all the patches overlapping $t_I$ by holding the time variable constant at $t_I$.

Three inbetween algorithms are available. They are linear space inbetweening, cubic metric space inbetweening, and Coons patch inbetweening. The linear space inbetweening algorithm is the Miura inbetweening algorithm originated from research at the Hitachi Research Laboratory in the mid 1960s. The algorithm first associates a linear time function called a basic curve with each keyframe $\mathbf{K}_0$ and $\mathbf{K}_1$ of a given patch, as shown in Fig. 13-15. Each time function

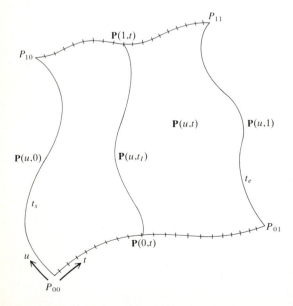

**FIGURE 13-14**
Patch specification.

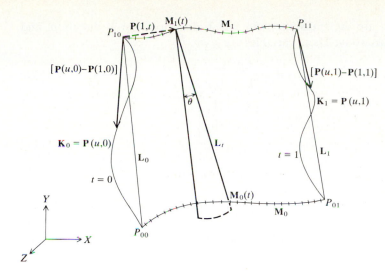

**FIGURE 13-15**
Linear space
inbetweening
algorithm.

passes through the two endpoints of its corresponding keyframe. To generate an inbetween frame $P(u, t)$ at time $t$, the basis curve $L_t$ (a line in this case) is obtained by connecting the two points $M_0(t)$ and $M_1(t)$. The inbetween frame is defined as

$$P(u, t) = (1 - t)A_0 + tA_1 \tag{13.2}$$

that is, the inbetween frame (or the static curve) interpolated for time $t$ is the time-weighted average of the term $A_0$ and $A_1$. $A_0$ is derived from the initial static shape $K_0$ at $t = 0$ and the other from the final static curve $K_1$ at $t = 1$. The term $A_0$ is the transformation of $K_0$ by a transformation matrix $[T_0]$, and $A_1$ is the transformation of $K_1$ by $[T_1]$. Thus,

$$A_0 = [T_0][P(u, 0) - P(1, 0)] \tag{13.3}$$

and
$$A_1 = [T_1][P(u, 1) - P(1, 1)] \tag{13.4}$$

The vector $P(1, 0)$ is subtracted from the keyframe $P(u, 0)$ in Eq. (13.3) to transform the origin of the $XYZ$ coordinate system to point $P_{10}$ to facilitate calculating $[T_0]$. Similarly, $P_{11}$ is used as the point of rotation of the keyframe $P(u, 1)$.

The matrix $[T_0]$ is found based on the argument that the matrix that transforms the basis curve $L_0$ at $t = 0$ to the basis curve $L_t$ at time $t$ should be the same matrix that transforms $K_0$ to $P(u, t)$. As seen in Fig. 13-15, $[T_0]$ involves rotation, translation, and scaling. Therefore, $[T_0]$ rotates, translates, and scales $K_0$ to a new orientation, position, and size, but with the same shape. Similarly $[T_1]$ maps $K_1$ to $A_1$.

To find $[T_0]$, the basis curve $L_0$ is rotated the angle $\theta$ shown in Fig. 13-15 about the point $P_{10}$, translated along the vector $P(1, t)$, and its length scaled from $L_0$ to $L_t$. Assuming the two-dimensional case for simplicity, we can write

$$[T_0] = s \begin{bmatrix} \cos \theta & -\sin \theta & l \\ \sin \theta & \cos \theta & m \\ 0 & 0 & 1 \end{bmatrix} \tag{13.5}$$

where $l$, $m$, and $s$ are the translations in the $X$ direction, the $Y$ direction, and scaling factor respectively. These variables are given by

$$\cos \theta = \frac{\mathbf{L}_0 \cdot \mathbf{L}_t}{|\mathbf{L}_0||\mathbf{L}_t|} \tag{13.6}$$

$$\sin \theta = \left| \frac{\mathbf{L}_0 \times \mathbf{L}_t}{||\mathbf{L}_0||\mathbf{L}_t||} \right| \tag{13.7}$$

$$l = x(1, t) \tag{13.8}$$

$$m = y(1, t)$$

$$s = \frac{|\mathbf{L}_t|}{|\mathbf{L}_0|} \tag{13.9}$$

Substituting the endpoints of $\mathbf{L}_0$ and $\mathbf{L}_t$ into Eqs. (13.6) to (13.9) and substituting the results into Eq. (13.5), we obtain

$$[T_0] = \frac{1}{|\mathbf{L}_0|^2} \left[ \begin{array}{cc|c} r_1 & -r_2 & l \\ r_2 & r_1 & m \\ \hline 0 & 0 & 1 \end{array} \right] \tag{13.10}$$

where

$$r_1 = (x_{00} - x_{10})(x_{0t} - x_{1t}) + (y_{00} - y_{10})(y_{0t} - y_{1t}) \tag{13.11}$$

$$r_2 = (y_{00} - y_{10})(x_{0t} - x_{1t}) - (x_{00} - x_{10})(y_{0t} - y_{1t}) \tag{13.12}$$

$$|\mathbf{L}_0|^2 = (x_{00} - x_{10})^2 + (y_{00} - y_{10})^2 \tag{13.13}$$

Similarly the matrix $[T_1]$ can be evaluated as

$$[T_1] = \frac{1}{|\mathbf{L}_1|^2} \left[ \begin{array}{cc|c} b_1 & -b_2 & n \\ b_2 & b_1 & q \\ \hline 0 & 0 & 1 \end{array} \right] \tag{13.14}$$

where

$$b_1 = (x_{01} - x_{11})(x_{0t} - x_{1t}) + (y_{01} - y_{11})(y_{0t} - y_{1t}) \tag{13.15}$$

$$b_2 = (y_{01} - y_{11})(x_{0t} - x_{1t}) - (x_{01} - x_{11})(y_{0t} - y_{1t}) \tag{13.16}$$

$$|\mathbf{L}_1|^2 = (x_{01} - x_{11})^2 + (y_{01} - y_{11})^2 \tag{13.17}$$

The interpolation algorithm is applied to each patch in the network. The linear space inbetweening algorithm suffers from slop discontinuity (i.e., speed discontinuity) along the boundaries of the patches.

The cubic metric space inbetweening algorithm is very similar to the Miura algorithm, but is designed to rectify its cross-boundary derivative discontinuities. In the cubic metric space algorithm, Hermite cubic splines are used as basis curves as shown in Fig. 13-16. The cubic splines $\mathbf{C}_0$ and $\mathbf{C}_1$ are the basis vectors of keyframes $\mathbf{K}_0$ and $\mathbf{K}_1$ respectively. Each spline is defined by two endpoints and two end slopes. At time $t$ which corresponds to an inbetween frame $\mathbf{P}(u, t)$, a similar spline curve $\mathbf{C}_t$ is defined. Its endpoints are the positions of the moving

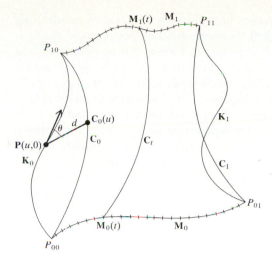

**FIGURE 13-16**
Cubic metric space inbetweening algorithm.

points at time $t$ and its end slopes are the time-weighted average of the slopes at the corresponding endpoints of $K_0$ and $K_1$.

In a cubic metric space, each point on a keyframe has a corresponding point on its basis curve, that is, both have the same value of the $u$ parameters as shown in Fig. 13-16. The distance $d$ between $P(u, 0)$ and $C_0(u)$ and the angle $\theta$ between the tangent to $K_0$ and the line joining the two points determine the transformation matrix $[T_0]$. Similarly, the matrix $[T_1]$ can be determined using $k_1$ and $C_1$. $[T_0]$ and $[T_1]$ transform $K_0$ and $K_1$ to $P(u, t)$ such that $P(u, t)$ matches the positions and slopes of $C_t$ at its endpoints. The details of developing $[T_0]$ and $[T_1]$ are not covered here.

The third algorithm is the Coons patch inbetweening algorithm. It is based on Coons patch representation as shown in Fig. 13-14 and has the advantage of controlling the normal derivatives across patch boundaries. This ensures slope continuities along boundaries between adjacent patches. A linear Coons patch is usually used for inbetweening. To find the inbetween frame $P(u, t)$ at time $t$, this time value is substituted into the patch equation (see Chap. 6).

The three inbetweening algorithms outlined above are general enough to handle keyframes made of any curves. However, the Coons patch inbetweening algorithm is the best of the three. It does not suffer from derivative discontinuities as the Miura algorithm; nor does it require the animator to input slope values at the endpoints of keyframes as the cubic metric space algorithm.

In addition, the average execution time required by the Coons algorithm for inbetweening is a little higher than that required by the Miura algorithm and much less than that required by the cubic metric space algorithm.

### 13.6.2   Simulation Approach

This approach (also called behavioral or algorithmic animation) is based on the physical laws that control the motion or the dynamic behavior of the object to be animated. The physical laws that describe the motion are developed and solved.

Thus, the motion is described algorithmically. A frame is obtained by substituting a given time value into the solution and the corresponding position and configuration of the object are calculated. A sequence of frames can be calculated and animated for a given time interval. The simulation approach of animation attempts to combine characteristics of objects and environment traditionally modeled in graphics (such as shape, shading, and illumination) with physical laws that require physical properties of objects (e.g., mass, inertia, etc.) and environment (e.g., friction, wind pressure, etc.). The simulation approach for animation is versatile, produces realistic animation quality, and eliminates the unnatural and jerky motion that may result from keyframe animation. However, finding the physical laws that control the motion may be difficult and solving them (in most cases numerical solutions are utilized) for a large sequence may be too expensive.

### 13.6.3 Hybrid Approach

While the simulation approach is attractive in describing the dynamic behavior of the animated object, it might be viewed as too restrictive for entertainment animation. Since motion is completely determined once the differential equations and their initial conditions are known, control that the animator needs to tailor motion is lost. In addition, the complexity of these differential equations denies the animator a simple relationship between the parameters that control the dynamics and the resulting motion.

Thus, on the one hand, the animator wishes to control the artistic look of motion (e.g., drama, expression, etc.) and, on the other hand, wishes to make the motion look as natural as possible. The hybrid approach solves this dilemma. It lets the animator specify keyframes, which convey the artistic look of animation, and then utilizes the physical laws of motion to interpolate between these keyframes to produce the inbetweens. These keyframes can be treated as constraints during the solution of the differential equations. In addition to the usual keyframe specifications (e.g., geometry and position), the animator is required to specify the dynamics (e.g., displacements and velocities) associated with the keyframes.

One way of formulating hybrid animation is possible by using the optimal control theory. Let us assume that the position, at time $t$, of any particle on a given rigid body is described by the vector $\mathbf{r}(t) = [x(t) \quad y(t) \quad z(t)]^T$ with respect to a fixed coordinate system in space. The equations of motion of the rigid body can be written as

$$\sum \mathbf{F} = m \frac{d^2 \mathbf{r}(t)}{dt^2} \qquad (13.18)$$

$$\sum \mathbf{M}_c = \frac{d\mathbf{H}(t)}{dt} \qquad (13.19)$$

where $\sum \mathbf{F}$ is the summation of forces acting on the body, $m$ is the mass, $\sum \mathbf{M}_c$ is the summation of moments of forces about the center of mass, and $\mathbf{H}$ is the angular momentum of the body about its center of mass. In many cases, the following equation is valid:

$$H(t) = [I] \frac{d\theta(t)}{dt} \tag{13.20}$$

where $\theta(t) = [\theta_x(t) \quad \theta_y(t) \quad \theta_z(t)]^T$, and $\theta_x(t)$, $\theta_y(t)$, $\theta_z(t)$ are the angles of rotation about the $X$, $Y$, and $Z$ axes respectively. $[I]$ is the inertia tensor given by

$$[I] = \begin{bmatrix} I_{xx} & -I_{xy} & -I_{xz} \\ -I_{xy} & I_{yy} & -I_{yz} \\ -I_{xz} & -I_{yz} & I_{zz} \end{bmatrix} \tag{13.21}$$

The keyframe constraints over a time interval $[0, t_n]$ can be described as follows:

$$\mathbf{r}(t_i) = \mathbf{r}_i \tag{13.22}$$

$$\theta(t_i) = \theta_i \tag{13.23}$$

$$\frac{d\mathbf{r}(t_i)}{dt} = \mathbf{v}_i \tag{13.24}$$

$$\frac{d\theta(t_i)}{dt} = \omega_i \tag{13.25}$$

where $0 \le t_i \le t_n$ are instants of time. The constraints given by Eqs. (13.22) to (13.25) may be partially or all specified at $t_i$.

The problem at hand can be stated as follows. Given the equations of motion (13.18) and (13.19) over the time interval $[0, t_n]$ and the keyframe constraints (13.22) to (13.25), find the trajectories $\mathbf{r}(t)$ and $\theta(t)$ that satisfy these equations and are natural and smooth. The state-space notation together with the optimum control theory can be used to solve the problem. The state $\mathbf{S}(t)$ can be defined as

$$\mathbf{S}(t) = \left[ \mathbf{r}(t) \quad \theta(t) \quad \frac{d\mathbf{r}(t)}{dt} \quad \frac{d\theta(t)}{dt} \right]^T \tag{13.26}$$

Using this state, and following the state-space formulation, a control vector function $\mathbf{u}(t)$ can be obtained by minimizing an appropriate control energy function (integral function $J$). The state $\mathbf{S}(t)$ and the function $\mathbf{u}(t)$ are related by rewriting Eqs. (13.18) and (13.19), using Eq. (13.26), as follows:

$$\frac{d\mathbf{S}(t)}{dt} = [F(t)]\mathbf{S}(t) + [G(t)]\mathbf{u}(t) \tag{13.27}$$

This equation can be used to solve $\mathbf{S}(t)$ once $\mathbf{u}(t)$ is known. The details of the solution are left as an exercise at the end of the chapter.

## 13.7  ANIMATION-RELATED PROBLEMS

The successful generation and display of an animation sequence depend on various factors such as the nature of objects being animated (whether they are fuzzy or not), the dynamic model used, the time interval between frames in the animation sequence (i.e., time sampling rate), interaction of the animated object with its environment (e.g., collision detection), the enhancement calculations used to generate the frames (such as shading, transparency, shadowing, etc.), and the hardware characteristics of graphics displays (resolution, refresh rate, etc.). This section describes briefly some of the common problems encountered in animation.

Two of the common problems in animation that control the frame rate (number of frames computed and displayed per second) calculations are frame-to-frame flicker and frame-to-frame discontinuity. Flicker is the blinking effect of a graphics display caused by the blank period between erasing and generating the contents of its pixels. It takes time to erase and redraw a screen image. To avoid frame flicker, a frame rate of 30 Hz must be used. This is the same rate used to scan graphics displays for flicker-free images. With such a high rate, only playback animation is possible if the animation sequence is to be flicker-free.

Frame-to-frame discontinuity results from motion sampling in the time domain. If the time sampling rate is not adequate, discontinuous and jerky motion results. Changes in position of image elements from one frame to the next should be gradual. Frame-to-frame discontinuities are unnoticeable and blend into smooth realistic motion at a time sampling rate of 24 frames/second.

Two other problems that control the quality of appearance of animation are spatial aliasing and temporal aliasing. Spatial aliasing is related to the resolution of the graphics display and results in the known staircase effect (jagged edges). Many spatial antialiasing techniques are available to solve this problem. Temporal aliasing is related to the discretization of the time domain into discrete time values at which frames are calculated. Temporal antialiasing is not as common as spatial antialiasing. However, it is desired when motion blur is to be modeled into computer animation. Algorithms for spatial antialiasing can be extended to temporal antialiasing. For example, supersampling algorithms with filtering can be used. In this case, each pixel intensity might be determined in four or more different locations, rather than just one. A filter may then be applied to derive the actual intensity of the output pixel. Supersampling is applied to each moving image (frame), and then filtering is applied to each resulting intensity function to "multiply expose" each output picture. Different filter types can be used to achieve different effects of motion blur in the image. For example, the standard box filter tends to create an image in which objects are fainter at their extremes in the direction of motion, where they cover pixels for a shorter duration. Gaussian and triangular filters exaggerate this effect. Refer to the problems at the end of the chapter for more details.

When several objects are animated at once in a scene, the problem of detecting and controlling object interactions is encountered. Most animation systems at present do not provide even minimal collision detection. Instead, they require the animator to visually inspect the scene for object interaction and respond accordingly. This is a time-consuming and difficult process even for key-frame systems where the user defines the motion explicitly. It is even more difficult for algorithmic animation where motion is obtained by solving the physical laws of motion. Collision detection has been studied extensively in the fields of CAD/CAM and robotics. Some algorithms are more general than what may be required by computer animation. An existing algorithm is designed to test the interpenetration of surfaces modeling flexible objects. If surfaces are modeled as triangular patches, collision between two surfaces is detected by testing for penetration of each vertex point through the planes of any triangle not including that vertex. Surfaces are assumed to be initially separate. For each time step of animation, the positions of the points at the beginning and the end of the time step are compared to see if any point went through a triangle during that time step. If so, collision has occurred. The mathematical details of the algorithm are not covered here. Another algorithm to test the collision between convex polyhedral solids is available. It is based on the Cyrus-Beck clipping algorithm. Interested readers are referred to the bibliography at the end of the chapter (see also the problems at the end of the chapter).

Once a collision is detected, a response to it is necessary. Keyframe animation systems can follow a predetermined set of rules about the motion of objects immediately following the collision. Animation systems using dynamic simulation inherently must respond to collisions automatically and realistically. The response is usually based on the conservation of linear and angular momentum. Surface friction and elasticity of the colliding objects can be considered in the momentum equations. Details of the collision response are not covered here (refer to the problems at the end of the chapter).

## 13.8 ANIMATION OF ARTICULATED BODIES

The animation of human and human-like characters is one of the major problems in computer animation. This is largely due to the dominance and importance of these characters in conventional animation which, in general, form the origin of computer animation. The key aspect of human animation is to achieve realistic motion with a minimal amount of input on the animator's part. Similar problems arise in animating articulated bodies and robots. There is a wealth of literature that addresses these problems; part of it is listed in the bibliography at the end of the chapter. Some of the published research work is listed as problems at the end of the chapter.

## 13.9 ANIMATION ON MICROCOMPUTERS

Microcomputer-based animation is usually limited to applications that do not require extensive shading or coloring, and where a small part of the screen graphics needs updating to reflect the effect of motion. This type of animation can be classified as real-time animation in the sense that moving graphics are displayed as soon as they are generated. However, resolution is not high, the color range is limited, and applications are simple.

There are two techniques used to create animation on microcomputers. The first technique is based on a simple idea. Moving objects can be animated by erasing an image and creating another in a slightly different position on the screen. In a system with a single buffer (one screen), the displayed image is first erased and then redisplayed in a new position. In a system with double buffering (two screens), two ways are possible: erase the first image and then display the second or display the second and then erase the first. Repeating this process in a loop achieves the effect of motion. The first method suffers from blinking (there will be times when nothing is displayed on the screen) while the second method may leave holes in the image (if the two images overlap). Programming languages available on microcomputers provide users with statements to allow them to erase and create images. For example, BASIC uses PUT and GET statements or FOR-NEXT loops (with erasing as equal to painting displayed graphics in a background color) to handle graphics animation. Filled (solid) graphics areas (e.g., a solid colored circle or rectangle) are created with the PAINT statement. The details of how to use the syntax of these languages in animation programs are not covered here (see programming problems at the end of the chapter).

A major flaw in the above color-switching animation technique is that screen flicker becomes unacceptable for images with complex graphics. Erasing an object from the screen takes as long as drawing it and a reasonable indicator of the time each frame is displayed is twice the time taken to draw it. Screen scrolling is another animation technique that avoids flicker. This technique is based on moving (scrolling) the bit map that corresponds to the frame window within the total bit map of the screen. This is equivalent to scrolling the screen. The display of the window bit map must be synchronized with the screen refresh cycle. A frame window is usually a rectangle. This makes this animation technique equivalent to scrolling a rectangle with pixels included inside it on the screen. Scrolling can be made in two directions: horizontally or vertically. The high speed of this technique allows the movements of many objects on the screen. An algorithm for animation using screen scrolling should first make all the calculations necessary to determine scrolling increments of all rectangles based on the number of animation frames and the total animation movements. Then, scrolling loops are executed in sequence to create the animation. Two scroll statements can be included in one FOR (or DO) loop, each dealing with a different rectangle, thus appearing to move concurrently. The speed achieved by this animation technique is adequate for the animation and simulation of a variety of PC-based applications.

The main disadvantage of screen scrolling is that it is not possible to produce trajectories other than either horizontal or vertical. Elaborate types of animation can be produced by combining screen scrolling with color switching. A variety of movement patterns can be simulated by color switching together with screen scrolling. For example, consider four frames of a runner placed side by side across the screen, each frame representing a different position of the runner. If only one frame is displayed in sequence, the runner would appear to move across the screen, but would also quickly disappear. However, if each frame window were scrolled to a fixed position on the screen and then its color is switched on, the runner would appear to be running on the spot. Diagonal movements on the screen can also be simulated with screen scrolling and color switching.

## 13.10  DESIGN AND ENGINEERING APPLICATIONS

Many design and engineering principles, concepts, and applications can benefit from computer animation. The visual contribution of animation enhances understanding the abstraction associated with them. For example, animating molecular interaction enables an understanding of chemical reaction. Animation can be applied to kinematic and dynamic analysis, wave propagation, vibration studies, simulation, and many other applications.

**Example 13.1.** Animate the four-bar linkage shown in Fig. 13-17.

*Solution.* This example can be solved in a few different ways. It can be animated on a microcomputer. It can also be animated on a dedicated animation system where the inbetweening and animation techniques discussed in this chapter can be used. It can also be animated using an animation procedure that may be available on a CAD/CAM system. The procedure requires the user to create the geometry at the various frames first. The user then (via animation commands) identifies the geometry of each frame with the procedure. These frames are stored by the procedure in order to create the animation sequence. The user saves the sequence for later playbacks on the graphics display. In this case, the type of animation is a real-time playback, and is usually performed by the CAD/CAM system and displayed on the graphics display.

The best approach to utilize this procedure is to create the geometry of all the frames first and then to animate them. This is useful in general if calculating these frames is time-intensive. It is also useful to place the geometry of each frame on a different layer so that the user can use the window option to define the geometry of each frame. Figure 13-18 shows a sequence of frames for this problem, each created at an angle increment $\Delta\theta$ of link $AB$ of 20° resulting in 18 frames to cover the full 360° angle of rotation of $AB$.

To construct the frame, the linkage can be constructed when $\theta = 0$. In this case, the links $AB$, $BC$, and $CD$ are horizontal, as shown in Fig. 13-19. For the next

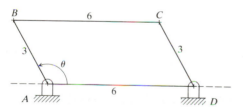

**FIGURE 13-17**
A four-bar linkage.

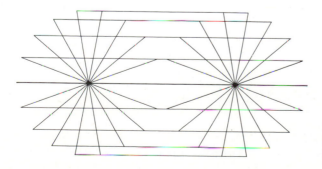

**FIGURE 13-18**
Frames for animating a four-bar linkage.

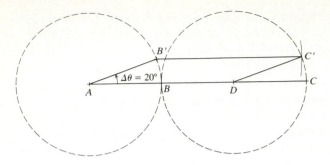

**FIGURE 13-19**
Constructing animation frames.

frame, the user chooses a new layer. Points $B$ and $C$ move in circular paths with centers at $A$ and $D$ and radii of $AB$ and $CD$ respectively. Thus, the user uses a "rotate" command with a copy modifier and an angle of rotation of $\theta + \Delta\theta$ to obtain $AB'$. The user then constructs a circle with center $B'$ and radius of $BC$. This circle intersects the circular path of $CD$ at point $C'$. Using the intersection, end, and origin modifiers, the user can construct the links $B'C'$ and $C'D$. Thus, the linkage $AB'C'D$ for the second frame of animation has been created. This process can be repeated for the other frames.

Once all the frames have been created on different layers, the animation procedure described above can be followed by displaying one layer at a time in the order of animation and creating a frame. The sequence of the frames can be saved and played back.

## PROBLEMS

### Part 1:  Theory

**13.1.** Each frame in an animation sequence takes 5 minutes of computer time to generate. Assuming 24 frames per second of animation, how long does it take to produce one minute of animation?

**13.2.** Figure P13-2 shows a four-bar linkage. Find the two extreme positions of the linkage. If these two positions are used as the start and end keyframes in keyframe animation:

(a) Find the motion trajectory, motion curves, and $P$-curves of points $B$ and $C$ if the $P$-curves animation method is used. Assume constant angular acceleration

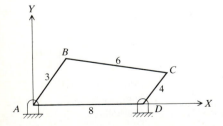

**FIGURE P13-2**

for links *AB* and *CD* of 4 and 6 rad/s$^2$ respectively. Also, find an appropriate number of inbetweens based on the *P*-curves. Generate these inbetweens.

(*b*) Using the same number and positions of the inbetweens as in (*a*), generate them using the linear space inbetweening algorithm.

(*c*) Repeat (*b*) but for the linear Coons patch inbetweening algorithm.

(*d*) Compare the inbetweens generated in (*a*), (*b*), and (*c*).

**13.3.** Develop the transformation matrices $[T_0]$ and $[T_1]$ for the cubic metric inbetweening algorithm. (Refer to A. S. Glassner, "Spacetime Ray Tracing for Animation," *IEEE Computer Graphics and Applic.*, vol. 8, no. 2, pp. 60–70, 1988.)

**13.4.** Study the algorithms for temporal antialiasing described in J. Korein and N. Badler, "Temporal Anti-Aliasing in Computer Generated Animation," *Computer Graphics*, *SIGGRAPH '83 Conf. Proc.*, pp. 377–388, 1983.

**13.5.** Derive an algorithm for collision detection of surfaces modeled as planar triangles. (Refer to M. Moore and J. Wilhelms, "Collision Detection and Response for Computer Animation," *Computer Graphics*, *SIGGRAPH '88 Conf. Proc.*, vol. 22, no. 4, pp. 289–298, 1988.)

**13.6.** Repeat Prob. 13.5 but for convex polyhedra.

**13.7.** Develop a collision response of two arbitrary articulated rigid bodies. (Refer to M. Moore and J. Wilhelms, "Collision Detection and Response for Computer Animation," *Computer Graphics*, *SIGGRAPH '88 Conf. Proc.*, vol. 22, no. 4, pp. 289–298, 1988.)

**13.8.** Study J. K. Hahn, "Realistic Animation of Rigid Bodies," *Computer Graphics*, *SIGGRAPH '88 Conf. Proc.*, vol. 22, no. 4, pp. 299–308, 1988, as an example of realistic animation of rigid bodies.

**13.9.** In Sec. 13.6.3 a formulation of a hybrid animation approach based on the optimal control theory is presented. Find the solution to the problem and apply it to some examples. (Refer to L. S. Brotman and A. N. Netravoli, "Motion Interpolation by Optimal Control," *Computer Graphics*, *SIGGRAPH '88 Conf. Proc.*, vol. 22, no. 4, pp. 309–315, 1988.)

**13.10.** Study M. Girard and A. A. Maciejewski, "Computational Modeling for the Computer Animation of Legged Figures," *Computer Graphics*, *SIGGRAPH '85 Conf. Proc.*, vol. 19, no. 3, pp. 263–270, 1985, as an example of animating legged figures based on forward and inverse kinematics as well as dynamic modeling and control.

**13.11.** Study W. W. Armstrong and M. W. Green, "The Dynamics of Articulated Rigid Bodies for the Purpose of Animation," *The Visual Computer*, vol. 3, no. 5, pp. 231–240, 1985, as an example of animating the human body as an articulated rigid body.

## Part 2: Laboratory

On your CAD/CAM system generate the real-time playback animation of the following objects:

**13.12.** A human heart. (Refer to D. Thalmann, O. Ratib, N. Magnenat-Thalmann, and A. Righetti, "A Model for the Three-Dimensional Reconstruction and Animation of the Human Heart," *The Visual Computer*, vol. 1, no. 4, pp. 241–248, 1985.)

**13.13.** Animate the systems shown in Fig. P13-13.

**13.14.** Choose your favorite object(s), system(s), or application(s) and animate them.

All dimensions in inches

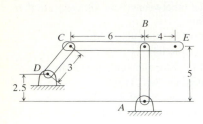

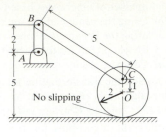

Linkage

Linkage

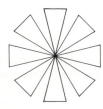

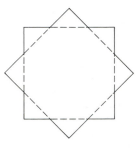

Windmill

Windmill

**FIGURE P13-13**

## Part 3: Programming

Write an animation program to animate:

**13.15.** Harmonic motion.

**13.16.** A bouncing ball.

**13.17.** A pendulum motion.

**13.18.** A yoyo motion.

**13.19.** Inserting a disc into a shaft.

**13.20** Your favorite application.

## BIBLIOGRAPHY

Armstrong, W. W., and M. W. Green: "The Dynamics of Articulated Rigid Bodies for the Purpose of Animation," *The Visual Computer*, vol. 3, no. 5, pp. 231–240, 1985.

Artwick, B. A.: *Microcomputer Displays, Graphics, and Animation*, Prentice-Hall, Englewood Cliffs, N.J., 1985.

Badler, N. I., J. D. Korein, J. U. Korein, G. M. Radack, and L. S. Brotman: "Positioning and Animating Human Figures in a Task-Oriented Environment," *The Visual Computer*, vol. 1, no. 4, pp. 212–220, 1985.

Badler, N. I., K. H. Manoochehri, and G. Walters: "Articulated Figure Positioning by Multiple Constraints," *IEEE Computer Graphics and Applic.*, vol. 7, no. 6, pp. 28–38, 1987.

Badler, N. I., and M. A. Morris: "Modeling Flexible Articulated Objects," *Computer Graphics '82, Proc. Online Conf.*, London, pp. 305–314, 1982.

Ball, H., R. R. Moore, and P. Quarendon: "Exploring New Worlds—Computer Animation in Education," *Perspectives in Computing*, vol. 6, no. 2, pp. 34–43, 1986.

Batty, M.: *Microcomputer Graphics*, Chapman and Hall Computing, London, 1987.

Booth, K. S., and S. A. Mackay: "Techniques for Frame Buffer Animation," *Proc. Graphics Interface '82*, pp. 213–220, 1982.

Brotman, L. S., and A. N. Netravoli: "Motion Interpolation by Optimal Control," *Computer Graphics, SIGGRAPH '88 Conf. Proc.*, vol. 22, no. 4, pp. 309–315, 1988.

Brown, P.: "Computer Animation in Australia," *Computer Graphics '86, Proc. Int. Conf.*, Pinner, London, pp. E17–E25, 1986.

Calkins, D. E., and J. Ishimaru: "Computer Graphics and Animation Come to Ship Designing," *Computers in Mech. Engng (CIME)*, vol. 3, no. 1, pp. 32–42, 1984.

Cann, P.: "Photographic Animation of Microcomputer Graphics," *Collegiate Microcomputer*, vol. 8, no. 10, pp. 350–362, 1983.

Cantwell, C.: "Animation on a Desktop," *Datamation*, vol. 27, no. 2, pp. 93–96, 1981.

Carlson, P. W.: "Easy IBM Full-Screen Animation," *Compute!*, vol. 8, no. 9, issue 76, pp. 61–63, 1986.

Chuang, R., and G. Entis: "3-D Shaded Computer Animation—Step by Step," *IEEE Computer Graphics and Applic.*, vol. 3, no. 9, pp. 18–25, 1983.

Comninos, P. P., and P. F. Hardie: "CGAL: The Soft Machine," *Computer Graphics '86, Proc. Int. Conf.*, Pinner, London, pp. E31–E48, 1986.

Dubis, J-E., S. Y. Yue, and J-P. Doucet: "Molecular Shape Embedding in a Grid Stage Modelling and Animation," *Proc. Computer Graphics Tokyo '86*, pp. 275–290, 1986.

Farrell, E. J., W. C. Yang, and R. A. Zappulla: "Animated 3D CT Imaging," *IEEE Computer Graphics and Applic.*, vol. 5, no. 12, pp. 26–32, 1985.

Finegold, L. S., and A. J. Asch: "Development of a Low-Cost 3-Dimensional Computer Graphics Training System," *Proc. Graphics Interface '82*, pp. 235–241, 1982.

Fishkin, K. P., and B. A. Barsky: "Algorithms for Brush Movement," *The Visual Computer*, vol. 1, no. 4, pp. 221–230, 1985.

Foley, J. D., and A. van Dam: *Fundamentals of Interactive Computer Graphics*, Addison-Wesley, Reading, Mass., 1982.

Forest, L., N. Magnenat-Thalmann, and D. Thalmann: "Integration Key-Frame Animation and Algorithmic Animation of Articulated Bodies," *Proc. Computer Graphics Tokyo '86*, pp. 263–274, 1986.

Forest, L., D. Ramband, and D. Magnenat-Thalmann: "Keyframe-Based Subactors," *Proc. Graphics Interface '86 and Vision Interface '86*, pp. 213–215, 1986.

Frenkel, K. A.: "The Art and Science of Visualizing Data," *Commun. ACM*, vol. 31, no. 2, pp. 111–112, 1988.

Gantz, J.: "The Quickening of Animation," *Computer Graphics World*, vol. 10, no. 2, pp. 21–23, 1987.

Geshwind, D. M.: "The NOVA Opening: A Case Study in Digital Computer Animation," *Computer Graphics '82, Proc. Online Conf.*, London, pp. 325–335, 1982.

Girard, M., and A. A. Maciejewski: "Computational Modeling for the Computer Animation of Legged Figures," *Computer Graphics, SIGGRAPH '85 Conf. Proc.*, vol. 19, no. 3, pp. 263–270, 1985.

Glassner, A. S.: "Spacetime Ray Tracing for Animation," *IEEE Computer Graphics and Applic.*, vol. 8, no. 2, pp. 60–70, 1988.

Grush, B.: *The Shoestring Animator*, Contemporary Books, Chicago, Ill., 1981.

Hahn, J. K.: "Realistic Animation of Rigid Bodies," *Computer Graphics, SIGGRAPH '88 Conf. Proc.*, vol. 22, no. 4, pp. 299–308, 1988.

Halas, J. (Ed.): *Computer Animation*, Hastings House, New York, 1974.

Hanrahan, P., and D. Sturman: "Interactive Animation of Parametric Models," *The Visual Computer*, vol. 1, no. 4, pp. 260–266, 1985.

Hayward, S.: *Computers for Animation*, Focal Press, London, 1984.

Ihnatowicz, E: "Solid Modelling of Sculptured Surfaces in Computer Animation," *Computer Graphics '86, Proc. Int. Conf.*, Pinner, London, pp. E27–E30, 1986.

Keith, S. R.: "A Transformation Structure for Animated 3-D Computer Graphics," *Computer Graphics (ACM)*, vol. 15, no. 1, pp. 72–91, 1981.

Kirchhof, C.: "Animation: Application to Tool," *Computer Graphics World*, vol. 10, no. 7, pp. 80–82, 1987.

Korein, J., and N. Badler: "Temporal Anti-Aliasing in Computer Generated Animation," *Computer Graphics, SIGGRAPH '83 Conf. Proc.*, pp. 377–388, 1983.

Lane, E. T., and V. A. Lane: "Animated Waves and Particles," *Collegiate Microcomputer*, vol. 5, no. 1, pp. 94–99, 1987.

Lansdown, R. J.: "Computer Aided Animation; A Concise Review," *Computer Graphics '82, Proc. Online Conf.*, London, pp. 279–290, 1982.

Lasseter, J.: "Principles of Traditional Animation Applied to 3D Computer Animation," *Computer Graphics, SIGGRAPH '87 Conf. Proc.*, vol. 21, no. 4, pp. 35–44, 1987.

Magnenat-Thalmann, N., and D. Thalmann: *Computer Animation: Theory and Practice*, Springer-Verlag, 1985.

Moore, M., and J. Wilhelms: "Collision Detection and Response for Computer Animation," *Computer Graphics, SIGGRAPH '88 Conf. Proc.*, vol. 22, no. 4, pp. 289–298, 1988.

Nahas, M., H. Huitric, and M. Saintourens, "Animation of a B-spline Figure," *The Visual Computer*, vol. 3, no. 5, pp. 272–276, 1988.

Neelamkavil, F., and L. Beare: "Techniques for Animation on Microcomputer," *Computer Graphics Forum*, vol. 7, no. 1, pp. 21–27, 1988.

Noma, T., and T. L. Kunii: "ANIMENGINE: An Engineering Animation System," *IEEE Computer Graphics and Applic.*, vol. 5, no. 10, pp. 24–33, 1985.

Person, R.: *Animation Magic with Your Apple IIe and IIc*, Osborne McGraw-Hill, Berkeley, Calif., 1985.

Reeves, W. T.: "Inbetweening for Computer Animation Utilizing Moving Point Constraints," *Computer Graphics, SIGGRAPH '81 Conf. Proc.*, vol. 15, no. 3, pp. 263–269, 1981.

Ressler, S. P.: "An Object Editor for a Real Time Animation Processor," *Proc. Graphics Interface '82*, pp. 221–225, 1982.

Reynolds, C. W.: "Computer Animation with Scripts and Actors," *Computer Graphics, SIGGRAPH '82 Conf. Proc.*, vol. 16, no. 3, pp. 289–296, 1982.

Reynolds, C. W.: "Flocks, Herds, and Schools: A Distributed Behavioral Model," *Computer Graphics, SIGGRAPH '87 Conf. Proc.*, vol. 21, no. 4, pp. 25–34, 1987.

Robertson, B.: "Animation for Engineering," *Computer Graphics World*, vol. 10, no. 2, pp. 46–50, 1987.

Rogers, D. F., and R. A. Earnshaw (Eds.): *Techniques for Computer Graphics*, Springer-Verlag, 1987.

Rosebush, J. G.: "Computer Animation in the 80's," *Proc. Graphics Interface '82*, p. 195, 1982.

Solomon, C., and R. Stark: *The Complete Kodak Animation Book*, Eastman Kodak Company, Rochester, N.Y., 1983.

Thalmann, D., and N. Magnenat-Thalmann: "Artificial Intelligence in Three-Dimensional Computer Animation," *Computer Graphics Forum*, vol. 5, no. 4, pp. 341–348, 1986.

Thalmann, D., O. Ratib, N. Magnenat-Thalmann, and A. Righetti: "A Model for the Three-Dimensional Reconstruction and Animation of the Human Heart," *The Visual Computer*, vol. 1, no. 4, pp. 241–248, 1985.

Thalmann, D., N. Thalmann, and P. Bergeron: "Dream Flight: A Fictional Film Produced by 3D Computer Animation," *Computer Graphics '82, Proc. Online Conf.*, London, pp. 353–367, 1982.

Traister, R. J.: *Graphics Programs for the IBM PC*, TAB Books, Blue Ridge Summit, Pa., 1983.

Vince, J. A.: "Having a Real Time," *Computer Graphics '86, Proc. Int. Conf.*, Pinner, London, pp. E3–E16, 1986.

Waters, K.: "Laugh, I Almost Cried; Expressive Three Dimensional Facial Animation," *Computer Graphics '86, Proc. Int. Conf.*, Pinner, London, pp. E49–E57, 1986.

Waters, K.: "A Muscle Model for Animating Three-Dimensional Facial Expression," *Computer Graphics, SIGGRAPH '87 Conf. Proc.*, vol. 21, no. 4, pp. 17–24, 1987.

Zeltzer, D.: "Presentation of Comolex Animated Figures," *Proc. Graphics Interface '82*, pp. 205–211, 1982.

Zeltzer, D.: "Towards an Integrated View of 3-D Computer Animation," *The Visual Computer*, vol. 1, no. 4, pp. 249–259, 1985.

# CHAPTER
# 14

# MECHANICAL ASSEMBLY

## 14.1  INTRODUCTION

In most engineering designs, the product of interest is a composition of parts, formed into an assembly. When the product is designed, consideration is generally given to the ease of manufacturing its individual parts and how the final product would look. Little attention is usually given to those aspects of design that will facilitate assembly of the parts, and great reliance is often placed on assembly or production engineers to solve any assembly-related problems. This approach worked well in the past because all mechanical assembly operations were performed manually, labor was inexpensive, and because products were not complex. As products are becoming more complex and labor is becoming more expensive, the demand to pay more attention to the assembly process during the design phase of a product is becoming increasingly high.

The most obvious way to facilitate the assembly process at the design phase is to simplify the product by reducing the number of different parts to a minimum. For example, designers may use welded parts or riveted joints instead of using screws, nuts, and washers. This would eliminate some assembly operations, but would result in a product that would be more difficult to repair. This might be acceptable if assembly costs are to be reduced, and customers should become more accustomed to the idea of replacing the complete product in the event of failure. The proper choice of manufacturing processes may enable complex parts to be produced; thus the designer can combine fewer simple parts into complex ones.

In addition to product simplification, the assembly process can be greatly facilitated by introducing guides and tapers into the design of various parts. Sharp corners usually hinder guiding parts into their correct positions during assembly. Figure 14-1 shows an example of assembling a pin into a groove. Here sharp corners are removed to improve assembly.

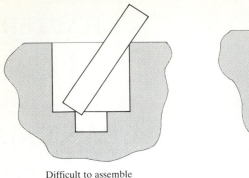
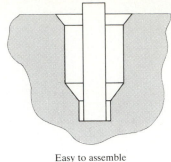

Difficult to assemble                    Easy to assemble

**FIGURE 14-1**
Redesign to facilitate assembly.


For automatic assembly, each product should have a base part (or a host part, or just a host) on which the assembly can be built. This base part must be designed with features that make it suitable for quick and accurate location on the assembly line or work carrier. The design of the base part decides the jigs and fixtures that may be used to support it during the assembly process.

Apart from the assembly considerations at the design phase, which are discussed above, modeling and representing assemblies, generating assembly sequencies, and analyzing assemblies are all relevant issues to geometric modelling and to the CAD/CAM technology. They form the focus of this chapter. Parts and/or subassemblies of a given product can be modeled separately, most often by different members of the design team, on a CAD/CAM system. Instances of these parts can then be merged into the base part or the host to generate the assembly database. The effectiveness of this database to study and analyze the assembly depends on whether it stores the hierarchical relationships between the assembly components and on whether efficient assembly sequences to assemble these components can be deduced from it. In addition, a link between the assembly database and the databases of its parts should exist so that when any part is modified by a member of the design team, the corresponding instance in the assembly is updated automatically. As a matter of fact, all assemblies that use instances of this part should be updated.

As in the case of engineering design, the primary focus in geometric modeling and CAD has been on the design and analysis of individual parts. However, it would be very useful if software packages (call them assembly modelers) would exist that would allow a designer to create individual parts, assemble them, and then perform the necessary analysis on the assembly. Consider, for example, the case of a designer who wants to study the kinematic and dynamic performance of an assembly. Conventionally, the designer generates a model of the assembly, runs the analysis, and evaluates the assembly design. Whenever any component is to be modified, the designer would have to regenerate the assembly model. This additional step tends to lead the designer either to consider fewer design alternatives or to skip the analysis.

The remainder of this chapter discusses the requirements for assembly modeling, the inference of a component configuration, the various schemes of representing assemblies, and how, for instance, individual parts or components can be assembled to generate an assembly model. Analysis of assemblies together with some applications are also covered.

## 14.2   ASSEMBLY MODELING

An assembly is a collection of independent parts. It is important to understand the nature and the structure of dependencies between parts in an assembly to be able to model the assembly properly. In order to determine, for example, whether a part can be moved and which other parts will move with it, the assembly model must include the spatial positions and hierarchical relationships among the parts, and the assembly or attachment relationships (or mating conditions) between parts. The modeling representation of hierarchical relationships and mating conditions are what distinguishes between modeling individual parts and assemblies, and consequently between geometric modelers and assembly modelers.

Most of the existing modeling packages offered by today's CAD/CAM systems that are in use in practice can be classified as geometric modelers. Their data structures are designed to store and manipulate geometric data of individual parts only. Most of the material covered in the book thus far apply to these parts. Assembly modelers can be thought of as more advanced geometric modelers where the data structure is extended to allow representation and manipulation of hierarchical relationships and mating conditions. Figure 14-2 shows how an assembly model can be created. The geometric modeler acts as a preprocessor to the assembly modeler. Designers first create all the shape information (both

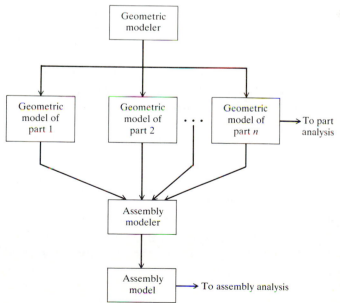

**FIGURE 14-2**

Generation of an assembly model.

Depth 0, hierarchy $n$

Depth 1, hierarchy $n-1$

Depth $n-1$, hierarchy 1

Depth $n$, hierarchy 0

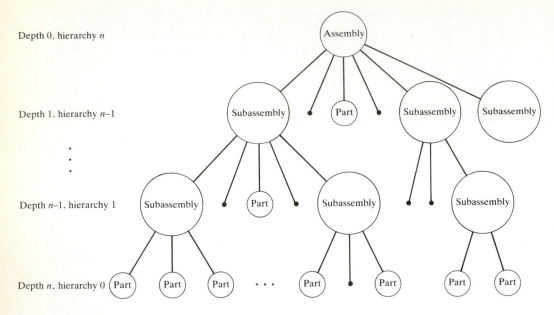

**FIGURE 14-3**
Assembly tree.

geometry and topology) of the individual parts. They can also analyze the parts separately. Part analysis may include mass property calculations and finite element analysis. Once the parts design is complete, designers can utilize the assembly modeler to create the assembly and analyze it. Creating the assembly from its parts requires specifying the mating and spatial relationships between the parts. Assembly analysis may include interference checking, mass properties, kinematic and dynamic analysis, and finite element analysis. The link between the geometric and assembly modelers is established such that designers need only to modify individual parts for design modification by using the geometric modeler, and the assembly model is updated automatically.

In the context of the foregoing, three requirements are necessary for assembly modeling: modeling of individual parts, specifying the hierarchical relationships between parts in the assembly, and specifying the mating conditions between parts or specifying the locations and orientations of the parts in their assembled positions.

### 14.2.1 Parts Modeling and Representation

This is the first step in creating an assembly model. Individual parts can be created using a geometric modeler with the proper representation scheme. Solid modeling, specifically boundary representation, is the appropriate scheme because the mating conditions are related to the faces, edges, and vertices of the assembled parts. In addition to the shape information, a part database can store assembly attributes such as the parts material type and properties, mass and inertial properties, frictional properties of the faces, and others. Some of these attributes such as the mass and inertial properties can be evaluated automatically by the geometric modeler after the part's database is created.

### 14.2.2   Hierarchical Relationships

The most natural way to represent the hierarchical relationships between the various parts of an assembly is an assembly tree as shown in Fig. 14-3. An assembly is divided into several subassemblies at different levels (shown in the figure as the tree depths), and each subassembly at depth $(n - 1)$ is composed of various parts. The leaves of the tree represent individual parts (a subassembly leaf can be decomposed into its individual parts), its nodes represent subassemblies, and its root represents the assembly itself. The assembly is located at the top of the tree at depth 0 or at the highest hierarchy $n$ of the assembly sequence.

Figure 14-4 shows an electric clutch assembly. The clutch consists of three main elements: the field, the rotor, and the armature. The field and the coil are held stationary, and the rotor is driven by the electric motor through the gear, pinion, and the rotor shaft. The armature is attached to the load shaft through the hub. The load shaft carries the load (not shown) to be overcome by the clutch. The assembly tree for this clutch is shown in Fig. 14-5. The tree represents an assembly sequence by which the clutch assembly can be produced. The assembly tree is not unique as it is possible to generate other valid assembly sequences.

### 14.2.3   Mating Conditions

Individual parts of an assembly are usually created separately using a CAD/CAM system and then merged (assembled) together, using a "merge" command, to form the assembly (see Fig. 14-2). Parts may have to be scaled up

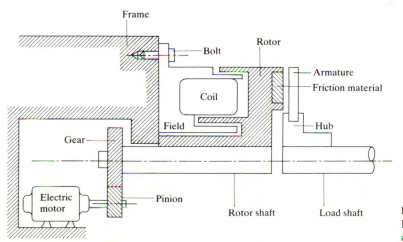

**FIGURE 14-4**
Electric clutch assembly.

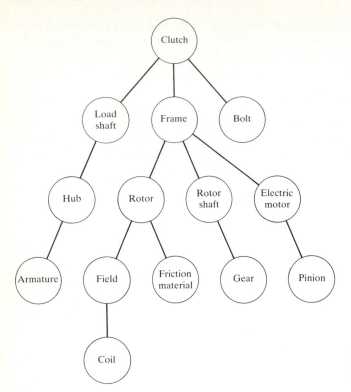

**FIGURE 14-5**
Assembly tree of the electric clutch assembly.

or down before merging to fit properly into the assembly. Each part has its own database with its own MCS (model coordinate system). Typically, the user selects one of the parts as a base part (host) and merges the other parts into it. The MCS of the host becomes the global coordinate system, that is, the MCS of the assembly and the MCS of each other part becomes a local coordinate system for this part. The final correct position of each part in the assembly is obtained by locating and orienting its corresponding MCS property with respect to the global coordinate system of the assembly. Figure 14-6 shows an example. The $XYZ$ is the global coordinate system of the database of the assembly model. Its origin $O$ is the (0, 0, 0) point. The $X_1Y_1Z_1$, $X_2Y_2Z_2$, $X_3Y_3Z_3$ and $X_4Y_4Z_4$ are local coordinate systems of four parts that make the assembly. Their origins $O_1$, $O_2$, $O_3$, and $O_4$ are located properly relative to the assembly origin $O$, and their orientations relative to the $XYZ$ coordinate system reflect the proper orientation of the parts in their assembly.

There are two alternatives of representing an assembly depending on how the locations and orientations of its various parts are provided by the user. The simplest alternative is to specify the location and orientation of each part in the assembly, together with the representation of the part itself, by providing a $4 \times 4$ homogeneous transformation matrix. This matrix transforms the coordinates of the geometric entities of the part from its local coordinate system to the global coordinate system of the assembly. One convenient way for the user to provide the transformation matrix interactively is by specifying the location of the local

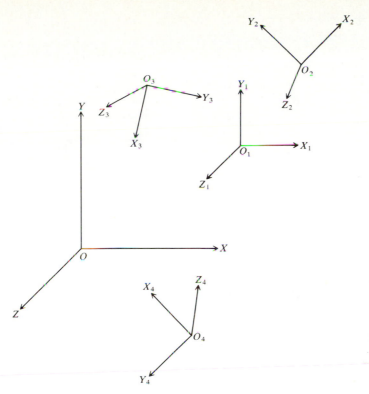

**FIGURE 14-6**
Positioning individual parts into their corresponding assembly.

coordinate system of a part relative to the assembly global coordinate system and by forcing the orientation of the local system to coincide with the orientation of the proper WCS (working coordinate system). This WCS is defined relative to the assembly global coordinate system. The transformation matrix is derived as discussed in Chap. 3 and given by Eq. (3.3).

As a WCS is completely defined by specifying its $X$ and $Y$ axes or its $XY$ plane, the proper WCS used to merge a part into its assembly can be defined such that its $XY$ plane coincides with the $XY$ plane of the part MCS. In other words, the WCS becomes the part local coordinate system after merging, and in effect the transformation matrix relates the part MCS to the host MCS (global coordinate system of the assembly). This alternative of merging parts into their assemblies is commonly used in various existing CAD/CAM systems. Figure 14-7 shows an example. The assembly consists of two parts, $A$ and $B$. Three instances of part $B$ are used in the assembly. The user first creates the databases of $A$ and $B$ with the MCS of each part as shown in Fig. 14-7a. To create the assembly, let us take part $A$ as the base part or the host and merge three instances of part $B$ into it. It is usually beneficial to assign a separate layer for each instance for ease of managing the assembly. To merge the instance of $B$ on top of $A$, the $X_1 Y_1 Z_1$ WCS is defined by the user as shown, and then the instance is merged. Similarly, the $X_2 Y_2 Z_2$ and $X_3 Y_3 Z_3$ WCSs are defined and shown in Fig. 14-7b. The transformation matrices to merge these instances into part $A$ are given by respectively:

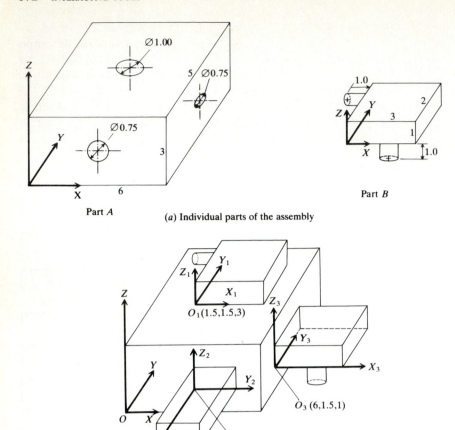

Part A

Part B

(a) Individual parts of the assembly

(b) Assembly

**FIGURE 14-7**
Creation of an assembly via the WCS alternative.

$$[T_1] = \begin{bmatrix} 1 & 0 & 0 & 1.5 \\ 0 & 1 & 0 & 1.5 \\ 0 & 0 & 1 & 3 \\ \hline 0 & 0 & 0 & 1 \end{bmatrix} \tag{14.1}$$

$$[T_2] = \begin{bmatrix} 0 & 1 & 0 & 2 \\ -1 & 0 & 0 & 0 \\ 0 & 0 & 1 & 1 \\ \hline 0 & 0 & 0 & 1 \end{bmatrix} \tag{14.2}$$

$$[T_3] = \begin{bmatrix} 1 & 0 & 0 & 6 \\ 0 & 1 & 0 & 1.5 \\ 0 & 0 & 1 & 1 \\ \hline 0 & 0 & 0 & 1 \end{bmatrix} \tag{14.3}$$

After these instances are merged into part $A$, they can be kept intact or they can be exploded by the user. Exploding a merged instance of a part means decomposing it back into its individual entities. If an instance is kept intact, the CAD/CAM system treats it as one complex entity. If the user manipulates it, all the entities that make up the instance are equally affected. This is useful in hiding or translating the instance. If the instance is exploded, the bond between its entities is lost and the user can treat each entity individually in the regular manner. This is useful when the user has to clean up the assembly to create an assembly drawing.

The other alternative to represent an assembly is based on specifying the spatial relationships between its individual parts as mating conditions. Mating feature information can be provided interactively with ease because mating features are simple graphics entities such as faces and centerlines. For example, a mating condition can consist of planar faces butting up against one another ("against" condition) or requiring centerlines of individual parts to be collinear ("fits" condition). Therefore, by simply selecting graphics entities the assembly data can be provided interactively. Providing the assembly data as mating features seems more natural than defining WCSs required by the first alternative.

By assigning the mating conditions, the transformation matrices that merge parts into their assembly can be automatically computed and stored for each part (see the next section). In addition, using mating conditions instead of providing transformation matrices determines whether or not the parts in an assembly can be assembled. If some parts cannot physically be assembled due to specifying inconsistent mating conditions, the transformation matrices satisfying all mating conditions do not exist. The computation algorithm for transformation matrices will diverge.

How can the mating conditions be identified, represented in the assembly data structure, and related to the actual positions of individual parts of an assembly? In most assemblies, the mating features between a pair of parts satisfy the conditions of "against," "fits," "tight fits," "contact," or "coplanar." As will be seen later in Sec. 14.4, a mating condition can be represented by its type ("against," "fits," etc.) and the two faces that mate. The relationship between the mating conditions and the actual positions of the various parts is discussed in the next section.

The "against" condition holds between two planar faces, or between a planar face and a cylindrical face (shaft). This condition is illustrated in Fig. 14-8. Parts 1 and 2 have the MCSs $X_1 Y_1 Z_1$ and $X_2 Y_2 Z_2$ respectively. The designated (shown dark in the figure) faces are the faces to be mated. Each face is specified by its unit normal vector and any one point on the face with respect to the part MCS. For example, the planar face of part 1 is specified by the unit normal $\hat{n}_1$, whose components are $n_{1x}$, $n_{1y}$, and $n_{1z}$, and by the point $P_1(x_1, y_1, z_1)$ with respect to the $X_1 Y_1 Z_1$ coordinate system. Similarly, the planar face of part 2 is specified by $\hat{n}_2 = [n_{2x} \quad n_{2y} \quad n_{2z}]^T$ and $P_2(x_2, y_2, z_2)$ with respect to the $X_2 Y_2 Z_2$ coordinate system. The "against" condition is satisfied by forcing $\hat{n}_1$ and $\hat{n}_2$ to be opposite to each other, and the two faces touch each other.

The "fits" condition holds between two cylindrical faces: a shaft cylindrical face and a hole cylindrical face as shown in Fig. 14-9. The "fits" condition is achieved by forcing the shaft and hole axes to be collinear. Each axis is specified

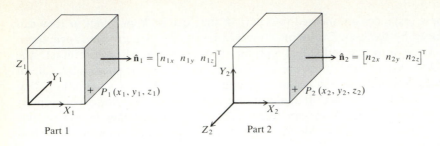

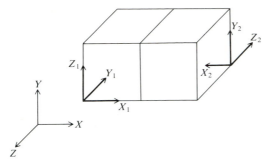

(a) "Against" condition between two planar faces

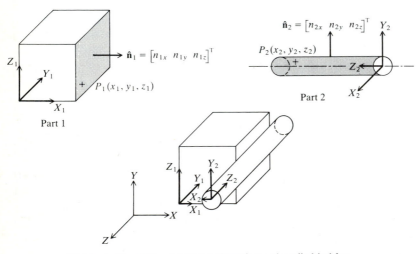

(b) "Against" condition between a planar face and a cylindrical face

**FIGURE 14-8**
"Against" condition.

by two points. The hole axis is specified by the two points $P_1(x_1, y_1, z_1)$ and $P_2(x_2, y_2, z_2)$ defined with respect to the $X_1 Y_1 Z_1$ coordinate system. Similarly, the shaft axis is specified by the two points $P_3(x_3, y_3, z_3)$ and $P_4(x_4, y_4, z_4)$ with respect to the $X_2 Y_2 Z_2$ coordinate system.

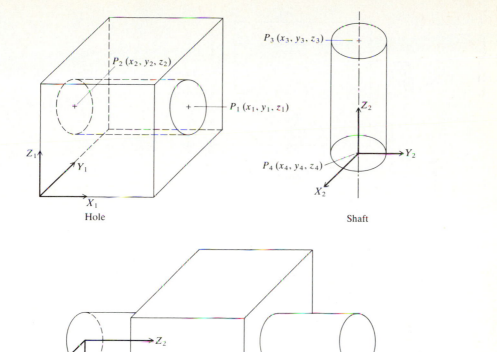

**FIGURE 14-9**
"Fits" condition.

The "against" and "fits" conditions as described above allow both rotational and translational freedom of movement between the mating parts. In the "against" condition shown in Fig. 14-8, part 2 can slide on part 1 or rotate relative to it after the two faces designated by the two normals $\hat{n}_1$ and $\hat{n}_2$ are mated together. Similarly, in the "fits" condition shown in Fig. 14.9, the shaft can slide and/or rotate inside the hole. There are some parts with the "against" or "fits" condition where rotational, translational, or both are not permitted. The "contact" and "tight fits" are introduced to handle these cases. The "contact" condition prevents the freedom of movement due to the "against" condition, and the "tight fits" prevents the movement due to the "fits" condition.

The "contact" condition is specified by requiring two points on the two mating parts to coincide. The "contact" condition does not exclude the "against" condition as the former can allow rotation about the contact point. Consider the example shown in Fig. 14-10. Faces are indicated by the letter $F$ with two subscripts. The first is the face number and the second is the part to which the face belongs. For example, $F_{2,1}$ and $F_{1,2}$ are the second face of part 1 and the first face of part 2 respectively. Points follow a similar convention as shown. The mating conditions between these two parts can be specified by three "against"

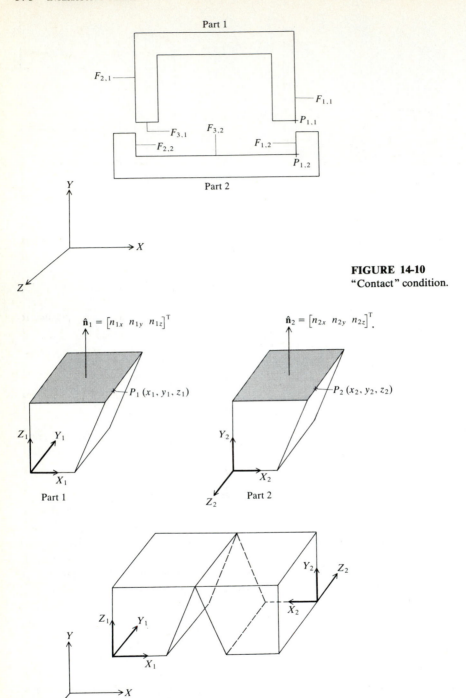

FIGURE 14-10
"Contact" condition.

FIGURE 14-11
"Coplanar" condition.

TABLE 14.1
**Mating conditions for electric clutch assembly of Fig. 14-4**

| Part 1 | Part 2 | Mating condition |
|---|---|---|
| Armature | Hub | "Tight fits" |
| Hub | Load shaft | "Tight fits" |
| Coil | Field | "Tight fits" |
| Field | Rotor | "Fits" |
| Friction material | Rotor | "Fits" |
| Rotor | Frame | "Against" |
| Gear | Rotor shaft | "Fits" |
| Rotor shaft | Frame | "Fits" |
| Pinion | Electric motor | "Fits" |
| Electric motor | Frame | "Against" and "contact" |
| Bolt | Frame | "Fits" |
| Gear | Pinion | "Against" |
| Load shaft | Rotor shaft | "Fits" |

conditions: between $F_{1,1}$ and $F_{1,2}$, between $F_{2,1}$ and $F_{2,2}$, and between $F_{3,1}$ and $F_{3,2}$. With these conditions specified only, part 1 should be free to move along the $Z$ direction against $F_{3,2}$. If we specify the "contact" condition that the points $P_{1,1}$ and $P_{1,2}$ are coincident, this undesired movement is eliminated. If only the "contact" condition is specified, part 1 can be tilted relative to part 2 and the proper faces may not be mated together.

The "tight fits" condition is introduced to prevent the rotational movement that may accompany the "fits" condition. The "tight fits" condition describes fits between parts where the force to rotate one part relative to another is too great to be called a rotational degree of freedom. One can think of "fits" as a clearance fit and of "tight fits" as an interference fit.

The "coplanar" condition holds between two planar faces when they lie in the same plane. This condition is illustrated in Fig. 14-11. It is similar to the "against" condition except that the points $P_1$ and $P_2$ are chosen to lie on the two edges to mate. The "coplanar" condition is the complement (opposite) of the "against" condition, and is satisfied by forcing the two normals $\hat{n}_1$ and $\hat{n}_2$ to be in the same direction.

As an example of how to apply the various mating conditions to assemblies, Table 14.1 shows these conditions for the electric clutch assembly shown in Fig. 14-4. The assembly tree shown in Fig. 14-5 is used to identify mating parts while developing the table.

## 14.3 INFERENCE OF POSITION FROM MATING CONDITIONS

Figures 14-8, 14-9, and 14-11 show that satisfying mating conditions requires repositioning of the parts of the assembly. The inference of the location and orientation of a part in an assembly from mating conditions requires computing its transformation matrix from these conditions. This matrix relates the part's local coordinate system to the global coordinate system of the assembly. With reference to Fig. 14-6, the location of part 1 is represented by the vector $\mathbf{OO}_1$ connecting the origin $O$ of the assembly global coordinate system to the origin $O_1$ of the $X_1 Y_1 Z_1$ local coordinate system of the part. The orientation is represented by the rotation matrix between the two systems. The transformation matrix can be written as [see Eq. (9.88)]

$$[T] = \left[\begin{array}{ccc|c} m_x & q_x & r_x & x \\ m_y & q_y & r_y & y \\ m_z & q_z & r_z & z \\ \hline 0 & 0 & 0 & 1 \end{array}\right] \tag{14.4}$$

This matrix has twelve variables (nine rotational and three translation elements) that must be determined from the mating conditions. For an assembly of $N$ parts, and choosing one of them as a host, $N - 1$ transformation matrices have to be computed. Therefore, the variables to solve for simultaneously are the $12 \times (N - 1)$ elements of these matrices.

In a typical assembly, the mating conditions between two components are not enough by themselves to completely constrain the two components. An intertwinement of mating conditions usually exists between all of the parts. In general, a group of parts must be solved simultaneously. The mating conditions along with the transformation matrix properties provide the constraint equations necessary to solve for the $12 \times (N - 1)$ variables. The number of equations is always equal to or greater than the number of variables. Therefore, the method of solution must account for the number of redundant equations, and eliminate these equations from the system of equations to be solved.

Before discussing the details of possible methods of solution, the development of constraint equations from mating conditions is presented. We discuss the three basic mating conditions: "against," "fits," and "coplanar." For the "against" condition shown in Fig. 14-8, each face where the two parts mate (butt up against one another) is specified by a unit normal and a point described in the local coordinate system of its corresponding part. Let $[T_1]$ and $[T_2]$ be the transformation matrices from the $X_1 Y_1 Z_1$ and $X_2 Y_2 Z_2$ coordinate systems respectively to the global coordinate system of the assembly. The unit normals and the two points specifying the mating conditions can be expressed in terms of the $XYZ$ system as follows:

$$\begin{bmatrix} n_{1x}^a \\ n_{1y}^a \\ n_{1z}^a \\ 0 \end{bmatrix} = [T_1] \begin{bmatrix} n_{1x} \\ n_{1y} \\ n_{1z} \\ 0 \end{bmatrix} \tag{14.5}$$

$$\begin{bmatrix} x_1^a \\ y_1^a \\ z_1^a \\ 1 \end{bmatrix} = [T_1] \begin{bmatrix} x_1 \\ y_1 \\ z_1 \\ 1 \end{bmatrix} \tag{14.6}$$

$$\begin{bmatrix} n^a_{2x} \\ n^a_{2y} \\ n^a_{2z} \\ 0 \end{bmatrix} = [T_2] \begin{bmatrix} n_{2x} \\ n_{2y} \\ n_{2z} \\ 0 \end{bmatrix} \tag{14.7}$$

and

$$\begin{bmatrix} x^a_2 \\ y^a_2 \\ z^a_2 \\ 1 \end{bmatrix} = [T_2] \begin{bmatrix} x_2 \\ y_2 \\ z_2 \\ 1 \end{bmatrix} \tag{14.8}$$

In the above equations, the superscript $a$ indicates assembly. The "against" condition requires the directions of the two unit normals to be equal and opposite and the two points to lie in the same plane at which the two faces mate. These can be expressed by the following four equations:

$$n^a_{1x} = -n^a_{2x} \tag{14.9}$$

$$n^a_{1y} = -n^a_{2y} \tag{14.10}$$

$$n^a_{1z} = -n^a_{2z} \tag{14.11}$$

$$[n^a_{1x} \quad n^a_{1y} \quad n^a_{1z} \quad 0] \left\{ \begin{bmatrix} x^a_1 \\ y^a_1 \\ z^a_1 \\ 1 \end{bmatrix} - \begin{bmatrix} x^a_2 \\ y^a_2 \\ z^a_2 \\ 1 \end{bmatrix} \right\} = 0 \tag{14.12}$$

Hence, four equations [(14.9) to (14.12)] are required for each "against" condition.

The "fits" condition requires that the centerlines of the shaft and the hole be collinear as shown in Fig. 14-9. The equation of the centerline of, say, the hole can be written as

$$\frac{x - x^a_1}{x^a_2 - x^a_1} = \frac{y - y^a_1}{y^a_2 - y^a_1} = \frac{z - z^a_1}{z^a_2 - z^a_1} \tag{14.13}$$

If the shaft axis is collinear with the hole centerline, points $P_3$ and $P_4$ defining the axis should satisfy Eq. (14.13). The points must first be transformed using $[T_2]$ to the assembly global coordinate system. The constraint equations required for each "fits" condition can be written as

$$\frac{x^a_3 - x^a_1}{x^a_2 - x^a_1} = \frac{y^a_3 - y^a_1}{y^a_2 - y^a_1} = \frac{z^a_3 - z^a_1}{z^a_2 - z^a_1} \tag{14.14}$$

and

$$\frac{x^a_4 - x^a_1}{x^a_2 - x^a_1} = \frac{y^a_4 - y^a_1}{y^a_2 - y^a_1} = \frac{z^a_4 - z^a_1}{z^a_2 - z^a_1} \tag{14.15}$$

Each of the above equations yields three combinations of equations resulting in a total of six equations for each "fits" condition. In general, two of these equations are redundant because Eqs. (14.14) and (14.15) each yields only two independent equations instead of three. However, it is necessary to carry all the three to cover the case where the centerline passing through points $P_1$ and $P_2$ is parallel to any of the global coordinate axes. For example, if the centerline is parallel to the $X$ axis as shown in Fig. 14-9, Eq. (14.14) becomes

$$\frac{x_3^a - x_1^a}{x_2^a - x_1^a} = \frac{y_3^a - y_1^a}{0} = \frac{z_3^a - z_1^a}{0} \tag{14.16}$$

which gives the following two equations only:

$$(y_3^a - y_1^a)(x_2^a - x_1^a) = 0 \tag{14.17}$$

and
$$(z_3^a - z_1^a)(x_2^a - x_1^a) = 0 \tag{14.18}$$

Hence, it can be seen that all three equations must be carried so that at least two independent equations can be written for all cases, although this introduces redundancy in the system of equations.

The constraint equations for the "coplanar" condition are the same as for the "against" condition except that the two unit normals are in the same direction as shown in Fig. 14-11. Thus Eqs. (14.9) to (14.12) can be used after replacing the minus sign in Eqs. (14.9) to (14.11) with a plus sign.

One last constraint equation can be written and applies to all free rotating parts in the assembly such as bolts, pins, shafts, etc. A free rotating part is defined here as a part that rotates freely about a centerline axis. The rotation of these parts usually does not alter the appearance of the assembly—thus the name "free rotation." Therefore, there can be infinitely possible orientations of a free rotating part. Figure 14-12 shows a bolt with its $X_1 Y_1 Z_1$ local coordinate system oriented in the $XYZ$ global coordinate system of the assembly. The bolt can rotate freely about its $X_1$ axis by any angle $\phi$. As seen from the figure, the orientation of the $Y_1$ and $Z_1$ axes are insignificant to the assembly. If this free rotation is not constrained, the calculations of the transformation matrix from the mating conditions will diverge. As long as the angle $\phi$ is arbitrary, it can be set to zero, that is,

$$\phi = 0 \tag{14.19}$$

This constraint equation represents the constraint equation associated with free rotating parts. With reference to Fig. 14-12, Eq. (14.19) is equivalent to the two equations:

$$a = b = 0 \tag{14.20}$$

where $a$ and $b$ are the components of the unit vectors (along the $Y_1'$ and $Z_1'$ axes) in the $Z_1$ and $Y_1$ directions respectively. If the $X_1$ axis is coincident with the $X$ axis of the global coordinate system of the assembly, Eq. (14.20) can be written in terms of the elements of the transformation matrix given by Eq. (14.4) as

$$q_z = r_y = 0 \tag{14.21}$$

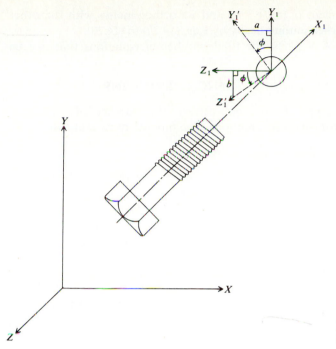

**FIGURE 14-12**
Free rotating part.

For a part rotating freely about its $Y_1$ axis or $Z_1$ axis, we can write respectively

$$m_z = r_x = 0 \qquad (14.22)$$

$$m_y = q_x = 0 \qquad (14.23)$$

With all the constraint equations derived for the various mating conditions, we now calculate the total number of equations and unknowns that can be used to infer the position of a part from mating conditions. For each "against" condition, 16 equations can be written: 12 are provided by Eqs. (14.5) to (14.8) and the other 4 are Eqs. (14.9) to (14.12). For each "fits" condition, 18 equations can be written: 12 are provided by Eqs. (14.5) to (14.8) and the other 6 are Eqs. (14.14) and (14.15). The "coplanar" condition provides the same number of equations as the "against." For each free rotating part, two equations [Eq. (14.21), (14.22), or (14.23)] are available. In addition, the properties of the transformation matrix [Eq. (14.4)] provides six equations: three from the unit vector length property and three from the orthogonality property. These can be written as

$$m_x^2 + m_y^2 + m_z^2 = 1 \qquad (14.24)$$

$$q_x^2 + q_y^2 + q_z^2 = 1 \qquad (14.25)$$

$$m_x q_x + m_y q_y + m_z q_z = 0 \qquad (14.26)$$

$$r_x = m_y q_z - m_z q_y \qquad (14.27)$$

$$r_y = m_z q_x - m_x q_z \qquad (14.28)$$

$$r_z = m_x q_y - m_y q_x \qquad (14.29)$$

The unit length requirement of $(r_x, r_y, r_z)$ and its orthogonality with the other two unit vectors are satisfied automatically by Eqs. (14.27) to (14.29).

For an assembly of $N$ parts, the total number of equations that can be written is given by

$$M = 6(N - 1) + 16NA + 16NC + 18NF + 2NR \qquad (14.30)$$

where NA, NC, NF, and NR are respectively the number of "against," "coplanar," "fits," and free rotation conditions. The number of variables are

$$V = 12(N - 1) + 12(NA + NC + NF) \qquad (14.31)$$

The first term of this equation represents the elements of the $(N - 1)$ transformation matrices and the second term is the number of variables introduced by the "against," "coplanar," and "fits" conditions. Each condition introduces 12 variables: 6 components of the unit normal vectors $\hat{n}_1$ and $\hat{n}_2$, and 6 coordinates of the two points $P_1$ and $P_2$ (or $P_3$ and $P_4$ for the "fits" condition). Equations (14.30) and (14.31) show that the number of equations and the number of variables are not equal. In general, the former is always equal to or greater than the latter.

The representation of the transformation matrix by Eq. (14.4) increases both the number of variables and the number of equations, which in turn makes finding a solution an expensive proposition. In an effort to reduce the number of variables, Eq. (14.4) may be rewritten using the following approach. The position of the part is still given by the elements $x$, $y$, and $z$. The orientation can be described by a sequence of rotation about the $X$, $Y$, or $Z$ axes instead of using the components of the unit vectors as given by the nine elements of the rotation matrix. Let us assume that the local coordinate system of a part can be oriented properly in its assembly by three rotations about the axes of the global coordinate system in the following order: $\alpha$ about the $Z$ axis, $\beta$ about the $Y$ axis, and $\gamma$ about the $X$ axis. Thus, we can rewrite Eq. (14.4) as

$$[T] = \begin{bmatrix} 1 & 0 & 0 & x \\ 0 & 1 & 0 & y \\ 0 & 0 & 1 & z \\ 0 & 0 & 0 & 1 \end{bmatrix} \begin{bmatrix} & & & 0 \\ & [R] & & 0 \\ & & & 0 \\ 0 & 0 & 0 & 1 \end{bmatrix} \qquad (14.32)$$

where $[R]$ is given by Eq. (9.75).

Using Eq. (14.32) instead of Eq. (14.4) reduces the number of variables per matrix by half: from 12 to 6 ($\alpha$, $\beta$, $\gamma$, $x$, $y$, $z$). Consequently, six equations [(14.24) to (14.29)] are eliminated. In addition, only one constraint equation [14.19)] is used for free rotation. To reduce the number of equations and variables even further, we can eliminate the 12 variables (the global description of unit normals and points) given by Eqs. (14.5) to (14.8) by considering them known in terms of the elements of the transformation matrix. Thus, Eqs. (14.30) and (14.31) become respectively

$$M = 4NA + 4NC + 6NF + NR \qquad (14.33)$$

and

$$V = 6(N - 1) \qquad (14.34)$$

To simplify the implementation of the free rotation condition, we can define the $Z$ axis as the axis of free rotation. Therefore the angle $\phi$ in Eq. (14.19) is about this axis. If the free rotating axis is the $Y$ axis of the local coordinate system, multiply the local coordinates of all the points and unit vectors by $[R_X(90)]$. This rotation matrix rotates the $Y$ axis by $90°$ about the $X$ axis (see Fig. 14-12). The new axis of rotation becomes the $Z$ axis. If the part rotates freely about its $X$ axis, then multiply all points and vectors by $[R_Y(-90)]$ to change the axis of rotation to the $Z$ axis. In an interactive system, before assigning mating conditions the user is prompted to select the free rotating parts displayed on the screen to determine the free rotating axis of each part.

We now discuss the solution of the system of equations that result from applying the mating conditions. This system is nonlinear due to the trigonometric functions that appear in the transformation matrix $[T]$. Since the number of equations is equal to or exceeds the number of variables, a method is needed to remove the redundant equations. The method discussed here utilizes the least squares technique to eliminate redundancy first, followed by using the Newton-Raphson iteration method to solve the resulting set of independent equations. The Newton-Raphson method for $n$ nonlinear equations of $n$ variables can be written as

$$X_{k+1} = X_k + [J(X_k)]^{-1}R_k \tag{14.35}$$

where $X_k$ is the solution vector at the $k$th iteration, $[J(X_k)]$ is the jacobian matrix, and $R_k$ is the residual vector, both of which are evaluated at the current solution vector $X_k$.

When redundancy exists, the inverse of the jacobian may not exist because the jacobian itself may not be square and/or it may be singular. The following procedure can be used to solve for $n$ variables ($X = [x_1 \; x_2 \; \cdots \; x_n]^T$) using the following $m$ equations:

$$f_1(x_1, x_2, \ldots, x_n) = 0$$
$$f_2(x_1, x_2, \ldots, x_n) = 0$$
$$\vdots \tag{14.36}$$
$$f_m(x_1, x_2, \ldots, x_n) = 0$$

In vector form, Eqs. (14.36) become

$$F(X) = 0 \tag{14.37}$$

To write Eqs. (14.36) in Newton-Raphson iterative form, let us assume that a solution $X_i$ exists at step $i$ and the solution at step $i + 1$ is $X_{i+1}$ such that

$$X_{i+1} = X_i + \Delta X_i \tag{14.38}$$

Linearizing Eqs. (14.36) about $X_i$ gives

$$F_{i+1} = F_i + \frac{\partial F(X_i)}{\partial X} \Delta X_i \tag{14.39}$$

If $X_{i+1}$ is the solution, then Eq. (14.37) holds, that is, $F_{i+1} = 0$. Thus Eq. (14.39) becomes

$$\frac{\partial F(X_i)}{\partial X} \Delta X_i = -F_i \tag{14.40}$$

Expanding this equation gives:

$$
\begin{bmatrix}
\dfrac{\partial f_1}{\partial x_1} & \dfrac{\partial f_1}{\partial x_2} & \cdots & \dfrac{\partial f_1}{\partial x_n} \\[2mm]
\dfrac{\partial f_2}{\partial x_1} & \dfrac{\partial f_2}{\partial x_2} & \cdots & \dfrac{\partial f_2}{\partial x_n} \\[2mm]
& \vdots & & \\[2mm]
\dfrac{\partial f_m}{\partial x_1} & \dfrac{\partial f_m}{\partial x_2} & \cdots & \dfrac{\partial f_m}{\partial x_n}
\end{bmatrix}_i
\begin{bmatrix}
\Delta x_1 \\[2mm]
\Delta x_2 \\[2mm]
\vdots \\[2mm]
\Delta x_n
\end{bmatrix}_i
= -
\begin{bmatrix}
f_1(x_1, x_2, \ldots, x_n) \\[2mm]
f_2(x_1, x_2, \ldots, x_n) \\[2mm]
\vdots \\[2mm]
f_m(x_1, x_2, \ldots, x_n)
\end{bmatrix}_i
\tag{14.41}
$$

or
$$
[J]_i \, \Delta \mathbf{X}_i = \mathbf{R}_i \tag{14.42}
$$

where $[J]_i = [J(\mathbf{X}_i)]$, $\Delta \mathbf{X}_i$, and $\mathbf{R}_i$ are the jacobian matrix, the incremental solution, and the residual vector at iteration $i$ respectively. The jacobian $[J(\mathbf{X}_i)]$ is a nonsquare matrix of size $m \times n$.

The least-squares method may be used to solve Eq. (14.42) for $\Delta \mathbf{X}_i$. The method is based on multiplying both sides of Eq. (14.42) by $[J]_i^T$, that is,

$$
[J]_i^T [J]_i \, \Delta \mathbf{X}_i = [J]_i^T \mathbf{R}_i \tag{14.43}
$$

Solving this equation for $\Delta \mathbf{X}_i$ gives

$$
\Delta \mathbf{X}_i = [J^T J]_i^{-1} [J]_i^T \mathbf{R}_i \tag{14.44}
$$

The algorithm to solve for $\Delta \mathbf{X}_i$ can be described as follows. An initial guess $\mathbf{X}_0$ is made. The jacobian $[J]_0$ and the residual vector $\mathbf{R}_0$ are computed. Next, Eq. (14.44) is used to calculate $\Delta \mathbf{X}_0$. Lastly, Eq. (14.38) is used to compute $\mathbf{X}_1$. These steps are repeated to obtain $\Delta \mathbf{X}_1$ and $\mathbf{X}_2$, $\Delta \mathbf{X}_2$ and $\mathbf{X}_3$, $\ldots$, and $\Delta \mathbf{X}_{n-1}$ and $\mathbf{X}_n$. Convergence is achieved when the elements of the residual vector $\mathbf{R}$ or the incremental solution $\Delta \mathbf{X}$ approaches zero.

**Example 14.1.** Figure 14-13 shows a pin and a block with their local coordinate systems. The pin is to be assembled into the hole in the block. Use the method of mating conditions to find the location and orientation of the pin in its assembly. Use the block as the host (base part).

*Solution.* The assembly consists of two parts: the pin and the block. The use of the block as the host implies that the global coordinate system is the local system of the block. The location and orientation of the $X_1 Y_1 Z_1$ local coordinate system of the pin must be found relative to the global $XYZ$ system. By inspection, it is obvious that the origin $P_3$ of the $X_1 Y_1 Z_1$ system must coincide with point $P_1$ in the assembled position, and its orientation is the same as that of the $XYZ$ system. We now see how to reach the same conclusion using the mating conditions.

As seen from the figure, there is one "against" condition, one "fits" condition, and one free rotation condition. In addition, the number of parts $N$ is equal to 2. Substituting this information into Eqs. (14.33) and (14.34) gives $M = 11$ and $V = 6$. The six variables are $\alpha$, $\beta$, $\gamma$, $x$, $y$, and $z$ where $x$, $y$, and $z$ are the coordinates of $P_3$

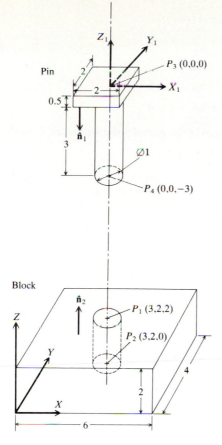

**FIGURE 14-13**
Pin/block assembly.

measured in the $XYZ$ system. From Fig. 14-13, we can write

$$\hat{\mathbf{n}}_1 = [0 \quad 0 \quad -1 \quad 0]^T \qquad \hat{\mathbf{n}}_2 = [0 \quad 0 \quad 1 \quad 0]^T$$

$$\mathbf{P}_1 = [3 \quad 2 \quad 2 \quad 1]^T \qquad \mathbf{P}_2 = [3 \quad 2 \quad 0 \quad 1]^T$$

$$\mathbf{P}_3 = [0 \quad 0 \quad 0 \quad 1]^T \qquad \mathbf{P}_4 = [0 \quad 0 \quad -3 \quad 1]^T$$

The normal $\hat{\mathbf{n}}_2$ and the points $P_1$ and $P_2$ are already expressed in terms of the global coordinate system of the assembly. To transform $\hat{\mathbf{n}}_1$, $P_3$, and $P_4$, use Eqs. (14.5) and (14.6) together with Eq. (14.32) to obtain

$$\hat{\mathbf{n}}_1^a = [-s\beta \quad c\beta s\gamma \quad -c\beta c\gamma \quad 0]^T$$

$$\mathbf{P}_3^a = [x \quad y \quad z \quad 1]^T$$

$$\mathbf{P}_4^a = [x - 3s\beta \quad y + 3c\beta s\gamma \quad z - 3c\beta c\gamma \quad 1]^T$$

The following constraint equations can be written:

The "against" condition [use Eqs. (14.9) to (14.12)]:

$$s\beta = 0$$

$$c\beta s\gamma = 0$$

$$c\beta c\gamma - 1 = 0$$

$$c\beta s\gamma(y - 2) - s\beta(x - 3) - c\beta c\gamma(z - 2) = 0$$

The "fits" condition [use Eqs. (14.14) and (14.15)]:

$$x - 3 = 0$$
$$y - 2 = 0$$
$$x - 3s\beta - 3 = 0$$
$$y + 3c\beta s\gamma - 2 = 0$$

Free rotation of the pin about the $Z$ axis of the global system gives the condition:

$$\alpha = 0$$

Note that the "fits" condition gives four constraint equations instead of six because the $X_1$ and the $Y_1$ axes are parallel to the $X$ and $Y$ axes of the global system respectively. This reduces $M$ from 11 to 9. Comparing the above nine constraint equations with Eq. (14.36), we find

$$\mathbf{X} = [\alpha \quad \beta \quad \gamma \quad x \quad y \quad z]^T$$

and

$$f_1 = s\beta$$
$$f_2 = c\beta s\gamma$$
$$f_3 = c\beta c\gamma - 1$$
$$f_4 = c\beta s\gamma(y - 2) - s\beta(x - 3) - c\beta c\gamma(z - 2)$$
$$f_5 = x - 3$$
$$f_6 = y - 2$$
$$f_7 = x - 3s\beta - 3$$
$$f_8 = y + 3c\beta s\gamma - 2$$
$$f_9 = \alpha$$

The jacobian matrix becomes

$$[J] = \begin{bmatrix}
0 & c\beta & 0 & 0 & 0 & 0 \\
0 & -s\beta s\gamma & c\beta c\gamma & 0 & 0 & 0 \\
0 & -s\beta c\gamma & -c\beta s\gamma & 0 & 0 & 0 \\
0 & \begin{matrix} -s\beta s\gamma(y-2) \\ -c\beta(x-3) \\ +s\beta c\gamma(z-2) \end{matrix} & \begin{matrix} c\beta c\gamma(y-2) \\ +c\beta s\gamma(z-2) \end{matrix} & -s\beta & c\beta s\gamma & -c\beta c\gamma \\
0 & 0 & 0 & 1 & 0 & 0 \\
0 & 0 & 0 & 0 & 1 & 0 \\
0 & -3c\beta & 0 & 1 & 0 & 0 \\
0 & -3s\beta s\gamma & 3c\beta c\gamma & 0 & 1 & 0 \\
1 & 0 & 0 & 0 & 0 & 0
\end{bmatrix}$$

To find the solution, let us assume the initial guess

$$\mathbf{X}_0 = [0 \quad 0 \quad 0 \quad 0 \quad 0 \quad 0 \quad 0 \quad 0 \quad 0]^T$$

Substituting this initial guess into the above matrix given $[J]_0$ as

$$[J]_0 = \begin{bmatrix} 0 & 1 & 0 & 0 & 0 & 0 \\ 0 & 0 & 1 & 0 & 0 & 0 \\ 0 & 0 & 0 & 0 & 0 & 0 \\ 0 & 3 & -2 & 0 & 0 & -1 \\ 0 & 0 & 0 & 1 & 0 & 0 \\ 0 & 0 & 0 & 0 & 1 & 0 \\ 0 & -3 & 0 & 1 & 0 & 0 \\ 0 & 0 & 3 & 0 & 1 & 0 \\ 1 & 0 & 0 & 0 & 0 & 0 \end{bmatrix}$$

Then

$$[J]_0^T = \begin{bmatrix} 0 & 0 & 0 & 0 & 0 & 0 & 0 & 0 & 1 \\ 1 & 0 & 0 & 3 & 0 & 0 & -3 & 0 & 0 \\ 0 & 1 & 0 & -2 & 0 & 0 & 0 & 3 & 0 \\ 0 & 0 & 0 & 0 & 1 & 0 & 1 & 0 & 0 \\ 0 & 0 & 0 & 0 & 0 & 1 & 0 & 1 & 0 \\ 0 & 0 & 0 & -1 & 0 & 0 & 0 & 0 & 0 \end{bmatrix}$$

$$[J^T J]_0 = \begin{bmatrix} 1 & 0 & 0 & 0 & 0 & 0 \\ 0 & 19 & -6 & -3 & 0 & -3 \\ 0 & -6 & 14 & 0 & 3 & 2 \\ 0 & -3 & 0 & 2 & 0 & 0 \\ 0 & 0 & 3 & 0 & 2 & 0 \\ 0 & -3 & 2 & 0 & 0 & 1 \end{bmatrix}$$

Notice that $[J^T J]$ is always symmetric. Also,

$$[J^T J]_0^{-1} = \begin{bmatrix} 1 & 0 & 0 & 0 & 0 & 0 \\ 0 & 0.1818 & 0 & 0.2727 & 0 & 0.5455 \\ 0 & 0 & 0.1818 & 0 & -0.2727 & -0.3636 \\ 0 & 0.2727 & 0 & 0.9091 & 0 & 0.8182 \\ 0 & 0 & -0.2727 & 0 & 0.9091 & 0.5455 \\ 0 & 0.5455 & -0.3636 & 0.8182 & 0.5455 & 3.3636 \end{bmatrix}$$

$[J^T J]_0^{-1} [J]_0^T =$

$$\begin{bmatrix} 0 & 0 & 0 & 0 & 0 & 0 & 0 & 0 & 1 \\ 0.1818 & 0 & 0 & 0 & 0.2727 & 0 & -0.2727 & 0 & 0 \\ 0 & 0.1818 & 0 & 0 & 0 & -0.2727 & 0 & 0.2727 & 0 \\ 0.2727 & 0 & 0 & 0 & 0.9091 & 0 & 0.0909 & 0 & 0 \\ 0 & -0.2727 & 0 & 0 & 0 & 0.9091 & 0 & 0.0909 & 0 \\ 0.5455 & -0.3636 & 0 & -1 & 0.8182 & 0.5455 & -0.8182 & -0.5455 & 0 \end{bmatrix}$$

To calculate the residual vector $\mathbf{R}_0$, substitute the initial guess $\mathbf{X}_0$ into the nine functions $\mathbf{f}$ to obtain

$$\mathbf{R}_0 = [0 \quad 0 \quad 0 \quad 2 \quad 3 \quad 2 \quad 3 \quad 2 \quad 0]^T$$

Using Eq. (14.44), we obtain

$$\Delta X_0 = \begin{bmatrix} 0 & 0 & 0 & 3 & 2 & 2 \end{bmatrix}^T$$

and Eq. (14.38) gives

$$X_1 = \begin{bmatrix} 0 & 0 & 0 & 3 & 2 & 2 \end{bmatrix}^T$$

We now use $X_1$ as an initial guess for the second iteration. Before proceeding, if we calculate $R_1$, we find it to be $0$. Therefore $\Delta X_1 = 0$ and $X_2 = X_1$. Thus, $X_1$ is the solution. Actually, it is the exact solution.

Substituting the solution $X_1$ into Eq. (14.32) gives the transformation matrix as

$$[T] = \left[\begin{array}{ccc:c} 1 & 0 & 0 & 3 \\ 0 & 1 & 0 & 2 \\ 0 & 0 & 1 & 2 \\ \hdashline 0 & 0 & 0 & 1 \end{array}\right]$$

Multiplying the local coordinates of each point of the pin transforms them to their global coordinates measured in the assembly coordinate system. Notice that this is the same transformation matrix that would be used if the WCS method were to be used to create the assembly.

**Example 14.2.** Calculate the locations and orientations of the instances of the assembly shown in Fig. 14-7. Use the method of mating conditions.

*Solution.* This assembly consists of four parts, that is, $N = 4$. Each instance of part B requires one "against" condition, one "fits" condition, and one free rotation condition to be assembled. Therefore, the number of equations is $M = 4 \times 3 + 6 \times 3 + 1 \times 3 = 33$, and $V = 6(4 - 1) = 18$. The 18 variables are the three rotations $(\alpha, \beta, \gamma)$ and the three coordinates $(x, y, z)$ of the origin of each of the $X_1 Y_1 Z_1$, $X_2 Y_2 Z_2$, and $X_3 Y_3 Z_3$ coordinate systems.

The constraint equations for the "against," "fits," and free rotation conditions can be obtained in a similar fashion to the previous example. However, the free rotation condition differs for the three instances. It is $\alpha = 0$, $\beta = 0$, and $\gamma = 0$ for the instance with origins $O_1$, $O_2$, and $O_3$ respectively. For the "fits" condition, four equations instead of six can be written for each instance. This brings $M$ down to 27 from 33.

The vector of variables to solve for are

$$X = \begin{bmatrix} \alpha_1 & \beta_1 & \gamma_1 & x_{01} & y_{01} & z_{01} & \alpha_2 & \beta_2 & \gamma_2 \\ & & x_{02} & y_{02} & z_{02} & \alpha_3 & \beta_3 & \gamma_3 & x_{03} & y_{03} & z_{03} \end{bmatrix}^T$$

Because the three instances are not interrelated, Eq. (14.42) can be written as

$$\left[\begin{array}{c:c:c} [J_{01}] & 0 & 0 \\ \hdashline 0 & [J_{02}] & 0 \\ \hdashline 0 & 0 & [J_{03}] \end{array}\right] \begin{bmatrix} \Delta X_{01} \\ \Delta X_{02} \\ \Delta X_{03} \end{bmatrix} = \begin{bmatrix} R_{01} \\ R_{02} \\ R_{03} \end{bmatrix}$$

This is a system of 27 equations in 18 variables ($[J_0]$ has 27 rows and 18 columns). However, the system is decoupled as shown, and the location and orientation for each instance can be solved for independently from the other two instances. Thus, we can write

$$[J_{01}] \, \Delta \mathbf{X}_{01} = \mathbf{R}_{01} \qquad \text{for instance with origin } O_1$$

$$[J_{02}] \, \Delta \mathbf{X}_{02} = \mathbf{R}_{02} \qquad \text{for instance with origin } O_2$$

$$[J_{03}] \, \Delta \mathbf{X}_{03} = \mathbf{R}_{03} \qquad \text{for instance with origin } O_3$$

From an algorithmic and software development point of view, solving the $27 \times 18$ system is preferred because the software may not have the intelligence to judge if the system of equations for an assembly is decoupled or not. The software may also be developed in such a way to allow the user to assemble two parts at a time instead of assembling all the parts at once by specifying all the mating conditions simultaneously.

The solution to this example is given by Eqs. (14.1) to (14.3). The reader is encouraged to work out the details to obtain the solution.

## 14.4  REPRESENTATION SCHEMES

An assembly of parts can be represented by the description of its individual components and their relationships in the assembly. An assembly database stores the geometric models of individual parts, the spatial positions and orientations of the parts in the assembly, and the assembly or attachment relationships between parts. The representation schemes of individual parts have been described in Chaps. 5 to 7 of this book. In this section, the assembly data structure to represent the relationships between parts is described. Few representation schemes and their related data structures exist. The inherent problem that all assembly data structures attempt to solve is how to assign assembly data interactively to build or develop the assembly. Some assembly representation schemes and their related data structures exist. The main difference between these schemes stems from the way the user provides the assembly data, that is, the locations and orientations of the various parts and their hierarchical relationships. Some schemes utilize the WCS method, while others utilize the mating conditions method. Four existing schemes are discussed below.

### 14.4.1  Graph Structure

In this scheme, an assembly model is represented by a graph structure in which each node represents an individual part or a subassembly. The branches of the graph represent relationships among parts. Four kinds of relationships exist: part-of (P), attachment (A), constraint (C), and subassembly (SA). The "part-of" relations represent the logical containment of one object in another. For example, the head and shaft of a screw are "part-of" the screw itself.

There are three types of "attachment" relationships: rigid, nonrigid, and conditional. Each "attachment" relation in an assembly indicates the type of attachment, the two parts related, and relative coordinate transformations between the two parts. In addition, nonrigid and conditional "attachment" relations contain information to qualify the particular relation. "Rigid attachment" occurs when no relative motion is possible between two parts. "Nonrigid attachment" occurs when parts cannot be separated by an arbitrarily large distance but relative motion between the two parts is possible. "Nonrigid attachment" is useful in mechanisms analysis. The popular types of these attachments are revolute and prismatic joints which allow translation and rotation respectively. Other types such as ball (spherical) joints also exist. The "conditional attachment" is related to parts supported by gravity, but not strictly attached. It describes the relationships of a part supporting another. In this case, the relationship defines the range of orientations over which the support relationships will hold, so that, for example, the supported part is not allowed to fall off. Conditional attachments form a conditional list, each element of which is a spatial position, orientation, or velocity condition expressed parametrically as a linear inequality. The intent is that the attachment holds only as long as the constraints are met.

"Constraint" relationships represent physical constraint of one part on another. A translational constraint, for example, specifies that the part cannot move in a given direction. A rotational constraint describes limits of rotation of an object about a given vector.

A "subassembly" relationship indicates that an assembly is merged into a higher assembly. The subassembly usually does not have a geometric representation of its own, but implicitly acquires the representations of its attached chain of parts. This allows an assembly to include groupings of assembled parts in its data structure.

A user can create a graph structure of an assembly by specifying the hierarchical structure defined by the "part-of" relation in the assembly model itself. In addition, the user must input the locations and orientations of the various parts relative to the global or assembly coordinate system. Thus, the graph structure scheme uses the WCS method. Figure 14-14 shows the graph structure of the electric clutch shown in Fig. 14-4. Each relationship in the figure is labeled with its kind.

### 14.4.2  Location Graph

Location of a part is a relative property. A coordinate system is the means used to specify location of one part relative to another. A location in one coordinate system also defines a new coordinate system for the located part, with its origin and axes. Other locations can be defined in terms of this second one. Thus, a chain of locations can be defined such that each location is defined in terms of another part's coordinate system. A set of these chains results in a graph referred to here as a location graph.

An assembly location is the root of the location graph and becomes the global coordinate system for the assembly. If two parts are located in coordinate systems not connected in the location graph, e.g., there are two or more disjoint graphs, then the relative locations of the parts in one graph are undefined relative to the parts in the other graph. If a location graph has any cycles (closed loops), then there are two alternative chains defining the location of one part relative to another, and the redundant definition encourages inconsistencies. Cycles in location graphs should be avoided. Moving any part's location in the location graph has the logical effect of also moving the parts that are located within its coordinate system, that is, its offsprings.

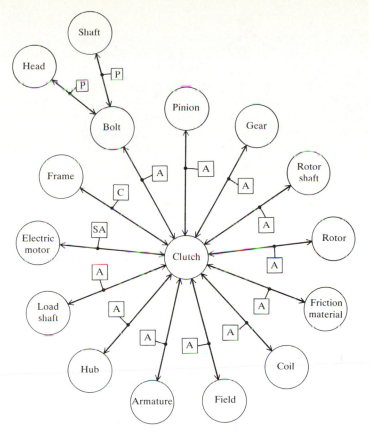

**FIGURE 14-14**
The graph structure of the electric clutch assembly shown in Fig. 14-4.

An example of the use of location information is shown in Fig. 14-15. The figure is the location graph of the clutch assembly shown in Fig. 14-4. In this example, the frame coordinate system is the root of the location graph, and it is taken as the global coordinate system of the assembly. There are at least 13 coordinate systems in which geometric entities are located. They are all related in the coordinate system of the frame. For example, the hub coordinate system is related to the armature coordinate system by the transformation (location) matrix $[T_2]$, and the latter is related to the frame coordinate system by the matrix $[T_1]$. These matrices must be input by the user via, say, the WCS method. Moving a branch in the location graph, such as the armature (relative to the frame), relocates those entities located relative to it, that is, the entities of the armature, of the hub, and of the load shaft.

The basic operations needed to insert part locations within a location graph, to move them within their current coordinate location, or to change their attachment on the tree are straightforward operations on the location matrix $[T]$. Two concatenated links in a location graph can be combined by multiplying the corresponding homogeneous transformation matrices together. This process

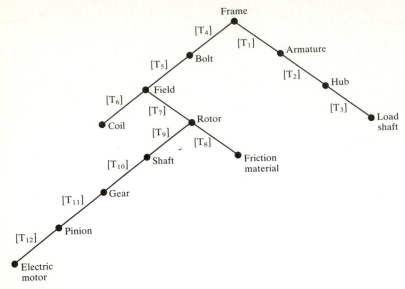

**FIGURE 14-15**
The location graph of the electric clutch assembly shown in Fig. 14-4.

can be used to relate the coordinate system of a given part in the assembly to its ancestors or offsprings. For example, the hub is located relative to the armature by the location matrix $[T_2]$, and is located relative to the load shaft by the inverse of $[T_3]$, that is, $[T_3]^{-1}$. Thus, to define a part location relative to any other part in the graph, its location matrix with regard to the root is computed and multiplied by the inverse of the location matrix of the coordinate system (relative to the root) within which it is to be defined. As a general rule, moving closer to the root of the graph utilizes the location matrix itself while moving away from it utilizes the inverse of the location matrix.

### 14.4.3  Virtual Link

The assembly representation schemes described in the above two sections require the user to input transformation matrices of the various parts of an assembly. The scheme discussed in the section requires more basic information such as mating conditions between the parts which in turn are used to calculate the transformation matrices as described in Sec. 14.3. The data structure of the scheme is based on the concept of virtual link. In the assembly data structure, any mating pair of two subassemblies, two parts, or one subassembly and one part is connected by a virtual link. If more than two parts are mutually related, several virtual links can be used so that every pair of mating parts is connected by one virtual link. A virtual link is defined as the complete set of information required to describe the type of attachment (as defined in the graph structure scheme in Sec. 14.4.1) and the mating conditions between the mating pair.

Figure 14-16 shows an assembly graph structure based on the concept of virtual link. The assembly is located at the top node. It is composed of one or

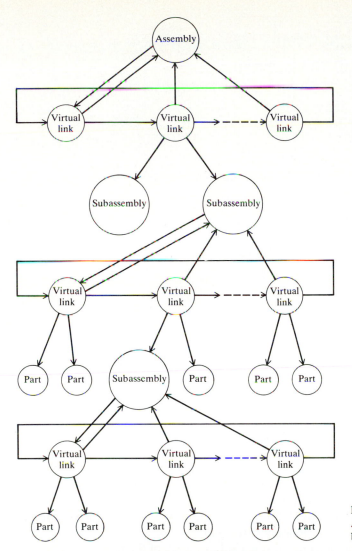

**FIGURE 14-16**
Assembly graph structure based on the virtual link.

more pairs of subassemblies, where every pair is connected by a virtual link. In turn, the subassemblies may be composed of subassemblies and parts which are connected into pairs by virtual links. The terminal nodes of the assembly graph are the parts of the assembly and geometric data for each one of these parts are connected to these terminal nodes via pointers in the data structures. If many identical parts appear in an assembly, data for only one part is stored, by using the concept of instance. In this case, virtual links point to the instances of the parts and not the parts themselves.

As shown in Fig. 14-16, the assembly and each subassembly point (via a pointer in the data structure) to the first virtual link of the immediate lower level. In addition, each virtual link points to the next virtual link and to the assembly or subassembly it belongs to in the immediate upper level. A data structure based on the concept of virtual link is shown in Fig. 14-17. Five main arrays are

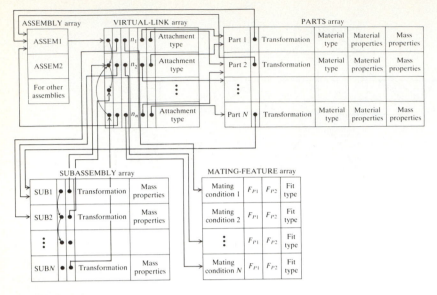

**FIGURE 14-17**
Data structure based on the virtual link.

required to define the data structure. Each array stores relevant information to the assembly as follows:

ASSEMBLY array:
   Name of the assembly
   Pointer to the first virtual link of the immediate low level
VIRTUAL-LINK array:
   Pointer to the next virtual link
   Pointer to an assembly or subassembly of the immediate upper level
   Pointer to the starting location in MATING-FEATURE array
   Number of records in MATING-FEATURE array related to the virtual link
   Pointer to a part or subassembly of the pair
   Pointer to the other part or the other subassembly of the pair
   Attachment type provided by the virtual link (rigid, nonrigid, or conditional)
SUBASSEMBLY array:
   Name of subassembly
   Pointer to subassembly or assembly of the immediate upper level
   Pointer to the first virtual link of the immediate lower level
   Transformation matrix from the assembly or subassembly of the immediate
      upper level
   Mass properties of the subassembly
MATING-FEATURE array:
   Type of mating condition ("against," "fits," etc.)
   Mating feature of the pair (planar face or centerline)
   The other mating feature of the pair (planar face or centerline)
   Fit type between mating feature (clearance, tight, or transition)

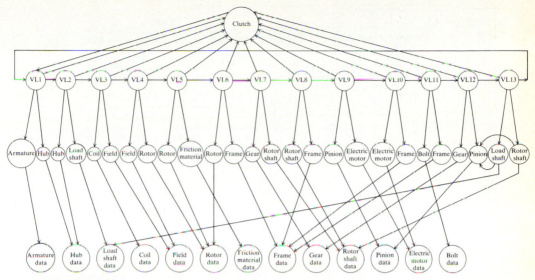

**FIGURE 14-18**
The graph structure based on the virtual link of the clutch assembly shown in Fig. 14-4.

PARTS array:
    Name of the part
    Pointer to subassembly or assembly of the immediate upper level
    Transformation matrix from the assembly or subassembly of the immediate
        upper level
    Material type of the part
    Material properties of the part
    Mass properties of the part

Figures 14-18 and 14-19 show respectively the assembly graph structure and the data structure based on the virtual link of the clutch assembly of Fig. 14-4. Both figures are based on the mating conditions listed in Table 14.1. In Fig. 14-18, a virtual link is designated by the two letters VL. In Fig. 14-19, all the pointers (from the virtual links to the assembly and to the mating feature or from the parts to the assembly) are not shown to facilitate understanding the figure. Also, the MATING-FEATURE array shows the parts and not the planes or the centerlines related to the mating conditions. The reader can easily identify these planes and centerlines using Fig. 14-4.

## 14.5  GENERATION OF ASSEMBLING SEQUENCES

It is always useful when studying the assembly of a product to identify the various ways in which the assembly process may be carried out. In most assemblies, there are more than one assembling sequence to generate assemblies from their respective parts. An assembly or production engineer must decide on the most optimum sequence. The sequence by which a set of parts is assembled plays an important role in determining important characteristics of the tasks of

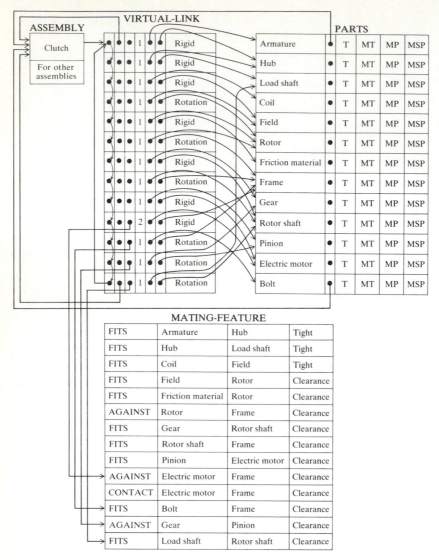

**FIGURE 14-19**
The data structure based on the virtual line of the clutch assembly shown in Fig. 14-4.

assembly and of the finished assembly. The choice of an assembly sequence affects matters such as the difficulty of assembly steps, the need for fixturing, the potential for parts damage during assembly and part mating, the ability to do in-process testing, the occurrence of the need for rework, time of assembly, assembly skill level, and the unit cost of assembly.

Exploring the choices of an assembly sequence by trial and error is practically impossible for two reasons. First, the number of valid sequences can be large even when the number of parts of the assembly is relatively small. The number can even be staggering when the number of parts increases. Second, minor design changes can drastically modify the available choices of assembly sequences.

Few techniques exist to generate all assembly sequences. Some of these techniques are manual while others are algorithmic. An obvious way of generating all assembly sequences is from the records of an exhaustive set of trials involving either all the ways of assembling the parts or all the ways of removing parts from an assembly and each of its subassemblies. This section discusses techniques based on the former means.

### 14.5.1  Precedence Diagram

The precedence diagram is designed to show all the possible assembly sequences of a product. To develop the precedence diagram for a product, each individual assembly operation is assigned a number and is represented by an appropriate circle with the number inscribed. The circles are connected by arrows showing the precedence relations. The precedence diagram is usually organized into columns. All the operations that can be carried out first are placed in the first column, and so on. Usually, one operation appears in the first column: the placing of the base part (the host) on the work carrier where assembling takes place.

To draw the precedence diagram, one places the base part in column I. Assembly operations that can only be performed when the operation in column I has been performed are placed in column II. Lines are then drawn to connect operations in columns I and II. Third-stage operations are placed in column III with appropriate connecting lines, and so on until the diagram is complete. Following all the lines from a given operation to the left indicates all the operations that must be completed before the operation under consideration can be performed.

Figure 14-20 shows the precedence diagram for the clutch assembly shown in Fig. 14-4. The frame is chosen as the base part. The precedence diagram shown in Fig. 14-20a is based on the individual parts of the clutch. This means that all these parts are assembled on a single assembling machine. The precedence diagram shown in Fig. 14-20 can be correlated to the assembly tree of the clutch shown in Fig. 14-5.

In the assembly of the clutch shown in Fig. 14-20a, there are 14 operations, and it is probably impractical to carry out all these on a single assembling machine for the simple reason that it would be difficult to design all the proper fixtures, and fixtures may be in the way of assembling some parts. Therefore, it is probably better to treat groups of parts as subassemblies. This is indicated in Fig. 14-20a by the dashed lines enclosing the proper operations. Thus, if a product is broken down into a number of subassemblies, all assembly sequences for each one can be easily studied and evaluated. In addition, if these subassemblies can be mechanically assembled, separate assembling machines may be used.

Figure 14-20b shows the precedence diagram for the subassemblies of the clutch assembly. The clutch is divided into four subassemblies. It can be seen that no flexibility exists in the ordering of operations 1, 4, 2, 6, and 7. Operations 3 and 5, however, can be carried out in any order except that they can not be

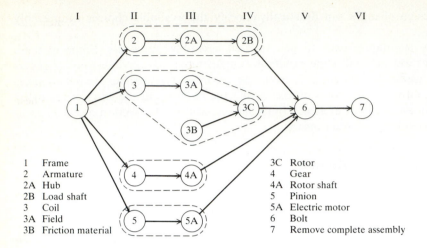

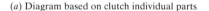

(*a*) Diagram based on clutch individual parts

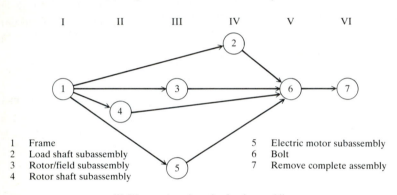

(*b*) Diagram based on clutch subassemblies

**FIGURE 14-20**
Precedence diagram for the clutch assembly shown in Fig. 14-4.

performed until operation 4 is complete. Thus, the precedence diagram shown in Fig. 14-20*b* represents two possible orderings (assembly sequences) of the various assembly operations.

## 14.5.2   Liaison-Sequence Analysis

Like the precedence diagram method, the liaison-sequence analysis method (liaison method for short) uses precedence relations to elaborate a complete set of assembly sequences. Both methods begin by using the assembly information contained in a parts list and an assembly drawing to develop all assembly sequences. However, while the production engineer generates possible assembly sequences directly from the precedence diagram (because it maintains precedence of assembly operations), the liaison method asks the engineer a series of questions about mating conditions between parts to gain knowledge about the assembly to generate the assembly sequences. These sequences can be generated either manually or algorithmically. When generated algorithmically, a series of rules that permit the generation of only valid sequences is derived from the answers of the production engineer.

The liaison method develops all possible assembly sequences in two steps. First, it characterizes the assembly by a network wherein nodes represent parts and lines between nodes represent any mating conditions between parts. These mating conditions are referred to in this method as liaisons. The network itself is known as the liaison diagram. Liaison diagrams are developed by production engineers from assembly drawings and parts lists. Part names or codes are attached to the nodes and liaison numbers are attached to the lines of the liaison diagram. It is useful, when developing a liaison diagram, to note that the liaison count (number of liaisons) $l$ is related to the part count (number of parts) $n$ by the following inequality:

$$(n - 1) \leq l \leq \frac{n^2 - n}{2} \tag{14.45}$$

Once the liaison diagram of an assembly is developed, the second step in the liaison method is to generate all possible assembly sequences and represent them in what is called the liaison-sequence graph. The graph is generated from the liaison diagram by asking the production engineer a series of questions as discussed above. The branches of the graph represent the assembly sequences with the initial disassembled state at the top of the graph. The bottom of the graph is the assembled state. The graph is arranged into levels or ranks. The zeroth rank corresponds to the disassembled state. Various transient states between the top and the bottom of the graph exist and represent intermediate subassemblies. Below and connected to each state in the graph at a given rank are all the states that can be reached from the reference state by adding a part. For example, if there are three ways to reach a particular state in the $n$th rank, three paths (branches) in the graph enter the state from rank $(n - 1)$. Thus the number of valid liaison sequences for completion of the assembly is equal to the number of paths from the top rank to the bottom one.

A liaison diagram is straightforward and simple to generate. It is a collection of liaisons among the assembly parts. An experienced production engineer is able to anticipate some possible assembly sequences from the diagram. The main difference between a liaison diagram and a precedence diagram is that the former does not maintain precedence relations between assembly operations as the latter does. This is why the liaison method asks the production engineer a series of questions to gain back this lost information due to the nature of the liaison diagram. As the reader can tell, there is no precise definition of "liaison" presented. The method does not fail if an obscure liaison is omitted or if conservatively too many are included. There are always two criteria that can be applied to find the minimum number of liaisons. The liaison count must satisfy Eq. (14.45), in particular the minimum liaison count $l = n - 1$. Additionally, a part is not part of an assembly unless it is connected, by a liaison, to another part or parts in the assembly.

The generation of assembly (liaison) sequences responds kindly to errors of omission or commission in the listing of liaisons. In the case of omission, fewer assembly sequences are generated, and if two parts do not share a liaison, they will not be assembled one to the other. In the case of commission, more assembly sequences are generated. It is therefore more conservative to include questionable liaisons instead of excluding them.

Once the liaison diagram is developed, the assembly engineer must answer a pair of questions for each liaison to generate the liaison-sequence graph which generates all the assembly sequences. The two questions to be answered are:

$$\text{for } i = 1 \text{ to } l:$$

Q1. What liaisons must be done prior to doing liaison $i$?
Q2. What liaisons must be left to be done after doing liaison $i$?

These two questions are to be answered for each liaison. Thus, $2l$ such questions are required per assembly. Answers are expressed in the form of precedence relationships between liaisons or a logical combination of them. Example answers to the two questions (respectively) may be as follows:

A1.  $(L_J \text{ or } (L_k \text{ and } L_m)) \rightarrow L_i$
A2.  $L_i \rightarrow (L_s \text{ or } (L_t \text{ and } L_u))$

where the symbol " $\rightarrow$ " reads "must precede." The production engineer must seek all the alternatives that permit each liaison to be done, and doing so results in a close knowledge of the design details of the assembly.

To generate assembly sequences, one scans the liaison list and the answers for those liaisons that are not precedented. Any of these may serve as the first liaison to be established. Next, line up representations of each first possible state across a rank and connect each with the starting state by a line. For each possible first liaison, one explores all possible subsequent states, by again scanning the liaison list and the precedence relations (answers), thus generating another rank. If there is more than one way to get to a state in a rank, an equal number of branches (lines) enters the state.

To illustrate how the liaison method works, let us consider the clutch assembly shown in Fig. 14-4. To establish the liaison diagram and list, the assembly drawing (Fig. 14-4) and the assembly tree shown in Fig. 14-5 are utilized. The liaison diagram is shown in Fig. 14-21. For this diagram, $l = 13$, $n = 13$, and $l > (n - 1)$. It is obvious that the liaison diagram and list are not unique. Even in the liaison diagram shown in Fig. 14-21 the reader may identify additional liaisons that may seem convincing but are not included. However, the diagram is valid because it satisfies the two criteria discussed earlier to establish a liaison diagram. Parts are indicated by their first and/or second letters, and liaisons are indicated by a number that begins with the letter L. Liaison L1, for example, connects the coil and the field. Similarly, the other liaisons can be interpreted. The liaisons numbers shown do not follow a strict order. Instead, they are numbered following more or less the order in which one would create the assembly from its individual parts.

With the liaison diagram established, the next step is to determine the precedence relationships between liaisons by asking and answering the two questions for each liaison. A total of 26 questions are asked and answered.

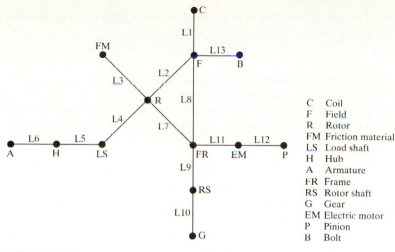

**FIGURE  14-21**
Liaison diagram for the clutch assembly shown in Fig. 14-4.

Q1. What liaisons must be done prior to doing liaison *i*?
*i* = 1: Nothing need precede the coil-to-field mate. Note that the question addresses what needs to be done, not necessarily what is convenient to do.
*i* = 2: Prior to putting the rotor in the field (liaison L2), the coil must be inserted into the field (L1), yielding the precedence relationship

$$L1 \rightarrow L2$$

*i* = 3: Prior to putting the rotor into the frame (L7), the friction material must be put into the rotor (L3):

$$L3 \rightarrow L7$$

*i* = 4: Prior to putting the load shaft close to the rotor (L4), the rotor must be assembled into the frame (L7), that is,

$$L7 \rightarrow L4$$

*i* = 5: Nothing need be done prior to doing L5.
*i* = 6: Nothing need be done prior to doing L6.
*i* = 7: Prior to putting the rotor into the frame (L7), the coil must be put into the field (L1), that is,

$$L1 \rightarrow L7$$

*i* = 8: Prior to putting the field into the frame (L8), the coil must be inserted into the field (L1), that is,

$$L1 \rightarrow L8$$

*i* = 9: Nothing need be done prior to doing L9.
*i* = 10: Prior to putting the gear into the rotor shaft (L10), something has to receive it and support it in place. Thus, the rotor shaft must be

assembled into the frame (L9) or the pinion is assembled into the electric motor (L12), that is,

$$L9 \text{ or } L12 \rightarrow L10$$

$i = 11$: Nothing need be done prior to doing L11.
$i = 12$: Nothing need be done prior to doing L12.
$i = 13$: Prior to putting the bolt into the frame (L13), the field must be assembled into the frame (L8) which implies that the coil would have been inserted into the field, that is,

$$L8 \rightarrow L13$$

This finishes answering half the required questions. The answers have resulted in precedence relationships that are unique. As a general rule, these relationships may have to be condensed by eliminating repeated ones or ones that are implied (weaker) in others. We now look into the answers to the second question. In some assemblies, no new precedence relationships or information may come out of the answers to Q2. In the case of the clutch assembly, the answers to the 13 questions of Q2, $i = 1$ to 13, are as follows:

Q2. What liaisons must be left to be done after doing liaison $i$?
$i = 1$: $L1 \rightarrow L2$. This may be read as "L2 and its successors, if any, must be left undone to allow doing L1." The phrase "and its successors, if any" may be appended to precedence relations. Note that this relation is implied in the relation (L1 and L3 $\rightarrow$ L2) obtained from answering question Q1 for $i = 2$.
$i = 2$: No liaison need be left undone.
$i = 3$: No liaison.
$i = 4$: No liaison.
$i = 5$: No liaison.
$i = 6$: No liaison.
$i = 7$: $L7 \rightarrow L4$ (included in the answer to Q1 for $i = 4$).
$i = 8$: No liaison.
$i = 9$: $L9 \rightarrow L10$ (included in the answer to Q1 for $i = 10$).
$i = 10$: No liaison need be left undone.
$i = 11$: No liaison.
$i = 12$: $L12 \rightarrow L11$ (included in the answer to Q1 for $i = 12$).
$i = 13$: No liaison.

Now the answers to questions Q1 and Q2 are complete. Note that different answers to Q1 and Q2 for $i = 1$ to 13 may exist. This nonuniqueness only affects the assembly sequences obtained from the liaison-sequence graph. Experienced production or assembly engineers may obtain better answers than others, especially if they are aware of actual assembling requirements on the shop floor. The precedence relationships are now summarized as follows:

$$L1 \rightarrow L2$$

$$L1 \rightarrow L7$$

$$L1 \rightarrow L8$$

$$L3 \rightarrow L7$$

$$L7 \rightarrow L4$$

$$L8 \rightarrow L13$$

$$L9 \text{ or } L12 \rightarrow L10$$

It remains now to algorithmically generate the assembly sequences (the liaison-sequence graph) of the 13 liaisons subject to the above 7 constraining (precedence) relations. Figure 14-22 shows the possible liaison (assembly) sequences of the clutch assembly. In Fig. 14-22, assembly states are represented by boxes, and each box has 13 cells in a two-row-by-four-column and one-row-

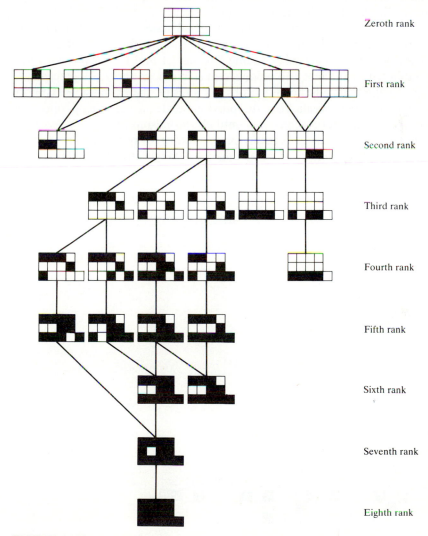

Zeroth rank

First rank

Second rank

Third rank

Fourth rank

Fifth rank

Sixth rank

Seventh rank

Eighth rank

**FIGURE 14-22**
Liaison-sequence graph for the clutch assembly shown in Fig. 14-4.

by-five-column array, corresponding to the 13 liaisons. The particular correspondence is liaisons one through four, left to right across the top row; five through eight, left to right across the middle row; and nine through thirteen, left to right across the bottom row. A blank cell implies that the corresponding liaison is not established while a marked cell implies that the corresponding liaison has been established. Lines connecting the boxes represent the possible state transitions. Each path through the graph starting at the top and moving along lines through succeeding ranks to the bottom represents a valid liaison (assembly) sequence.

To create a state in any rank in the liaison-sequence graph, one would first eliminate, from the 13 liaisons, all the liaisons that are already established prior to this rank, and which can only be obtained by moving upward in the graph to the first rank. Next, additional liaisons are eliminated by applying the precedence relationships. Figure 14-23 illustrates this process. Let us assume that the first two ranks are established as shown. To find the legal states in the third rank that will result from the state shown in the second rank, we list the available possible liaisons. These are the 11 liaisons excluding liaisons L1 and L3 because they are already established in the first and second ranks. We now apply the precedence relationships to eliminate additional liaisons. As a result, the permissible states in the third rank that may result from the previous state in the second rank are shown Fig. 14-23. This process continues until all the liaisons are established.

When an assembly has a relatively large number of parts, that is, a large number of liaisons, the number of states could be staggeringly large. To reduce

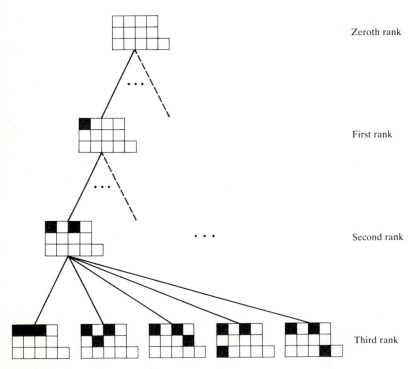

**FIGURE 14-23**
Establishing liaison states in the liaison-sequence graph.

the number, other conditions or constraints can easily be imposed in addition to the precedence relations. One useful constraint is to preclude more than a single subassembly. Precluding more than a subassembly can be interpreted in terms of the liaison diagram; each liaison established subsequent to the first liaison must associate with an already established liaison. This logic can be expressed as a table that lists prospective next liaisons as a function of already established liaisons. In addition to obeying the precedence relations, the assembly algorithm may choose liaisons only from those associated, by the table, with already established liaisons.

Table 14.2 shows the additional constraints on the clutch assembly that result from precluding more than one assembly. The liaisons in the second column of the table are easily obtained from the liaison diagram by listing all the liaisons connected to the corresponding liaisons in the first column. Each row in the table can be thought of as a subassembly. Therefore, two or more liaisons belonging to two or more rows can be added to a state simultaneously. For example, after applying the precedence relations and Table 14.2 to the second state from the left at the second rank in Fig. 14-22, the two liaisons L3 and L8 can be added to that state to produce the state in the third rank. Both liaisons are added simultaneously to the state in the second rank because each is obtained from a different row (subassembly) in Table 14.2—L3 from the second row and L8 from the first row.

Figure 14-22 shows four possible paths from the zeroth rank to the eighth rank. The states shown that are not continued or connected to the final state (assembly) imply that they are not extendable to other states after applying the precedence relations and Table 14.2. Each path through the graph starting at the top and moving along lines through succeeding ranks to the bottom represents a valid liaison sequence. To count how many sequences there are, we work upward from the next-to-the-last rank, answering (and recording) for each state in each rank the question, "From this state, how many paths to the last rank are there?"

**TABLE 14.2**

**Additional constraints on the clutch assembly**

| Established liaison | Prospective next liaisons |
| --- | --- |
| L1 | L2, L8, L13 |
| L2 | L1, L3, L4, L7, L8, L13 |
| L3 | L2, L4, L7 |
| L4 | L2, L3, L5, L7 |
| L5 | L4, L6 |
| L6 | L5 |
| L7 | L2, L3, L4, L8, L9, L11 |
| L8 | L1, L2, L7, L9, L11, L13 |
| L9 | L7, L8, L10, L11 |
| L10 | L9 |
| L11 | L7, L8, L9, L12 |
| L12 | L11 |
| L13 | L1, L2, L8 |

The answer to this question for the single state in the zeroth rank is the number of valid liaison (assembly) sequences.

### 14.5.3 Precedence Graph

Unlike the previous two methods, this method is fully automatic. It is based on the virtual-link data structure and requires the mating conditions as input to automatically generate assembly sequences for various assemblies. Once the mating conditions are provided, they are organized in the form of a mating graph. The parts in an assembly are then structured in a hierarchical assembly tree as shown in Fig. 14-3. Then an assembly sequence is generated with the aid of interference checking. In this method, the assembly sequence is referred to as a precedence graph.

To illustrate how the method works, let us apply it to the clutch assembly shown in Fig. 14-4. From Table 14.1, the mating graph is shown in Fig. 14-24. Notice that the branches (lines) of the mating graph are the virtual links. The assembly tree for the clutch is shown in Fig. 14-5. The assembly sequence is

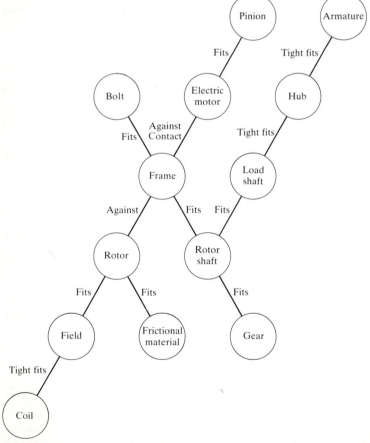

**FIGURE 14-24**
Mating graph for the clutch assembly shown in Fig. 14-4.

generated based on the mating graph as follows. First, identify the part that is connected to the largest number of parts by virtual links (frame for the clutch assembly). This is considered the base part. Next, gather all the parts directly connected to the base part by virtual links of the nonsubassembly type. A virtual link is classified as a subassembly type when the parts involved have movements relative to each other. The relative movement between two parts of a virtual link depends on the mating conditions specified by the link. For example, "against" and "fits" conditions, "against" and "contact," or "tight fits" result in virtual links of the nonsubassembly type. "Against" or "fits" alone allow relative movements. In the clutch assembly, based on Table 4.1, virtual links between the armature and the hub, the hub and load shaft, the coil and the field, and the frame and the electric motor are of the nonsubassembly type. The remaining virtual links are of the subassembly type.

The identification of the base part and the type of virtual links can be achieved automatically without user assistance utilizing the database which stores the virtual links and their information. Once the identification is complete, the automatic generation of an assembly sequence is achieved in two steps. In the first step, a part hierarchy is developed based on the type of virtual link, whether it is of the nonsubassembly or subassembly type. A gathering procedure develops such a hierarchy. In the second step, an assembling procedure utilizes the part hierarchy developed in the first step, together with interference checking, to develop the precedence graph which is effectively the desired assembly sequence.

The gathering procedure is a recursive algorithm which attempts to answer the following question. Given the geometric models of all parts of an assembly and the mating graph of the parts with virtual links, find the part hierarchy. The gathering algorithm begins by using the base part as a reference. It then scans all the parts of the assembly and gathers only those that are directly adjacent to the base part by a virtual link of the nonsubassembly type. Each of the newly gathered parts can be treated as a new base part and the gathering algorithm is applied again. Other parts, whose virtual links are of the subassembly type, not visited belong to different subassemblies. The gathering algorithm is applied to them in the same manner to group them. The result of this algorithm after applying it to the clutch assembly is shown in Fig. 14-25. In this figure a part at depth $(n - 1)$ serves as the base part for its children at depth $n$. For example, the hub is a base part for the armature and the load shaft is a base part for the hub.

The assembling procedure is a recursive algorithm. When traversing the part hierarchy to develop the precedence graph, two questions arise: how to assemble children parts (parts at different depths) into their base parts and how to assemble sibling parts (parts at the same depth). The order in which to assemble children is simple. If we assume that a base part should always be assembled after its children (descendants), then parts at depth $n$ must be assembled first. In Fig. 14-25, the armature is assembled to the hub first, then the hub into the load shaft. The assembling algorithm traverses the part hierarchy tree in postorder fashion in order to assemble the parts bottom-up.

The main criterion in assembling sibling is to ensure that they do not stand in the way of each other during the assembling sequence. In more technical terms, the space occupied by a part in its assembled state, in the assembly, should

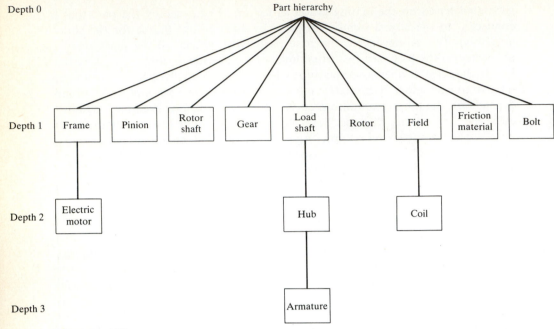

Depth 0     Part hierarchy

Depth 1     Frame    Pinion    Rotor shaft    Gear    Load shaft    Rotor    Field    Friction material    Bolt

Depth 2     Electric motor    Hub    Coil

Depth 3     Armature

**FIGURE 14-25**
Part hierarchy of the clutch assembly shown in Fig. 14-4.

not intersect the spaces (swept volumes) through which all other unassembled parts should traverse to be assembled. Therefore, the assembling algorithm must find the proper assembly sequence so that collision-free paths are possible. The approach that the assembling algorithm may use is similar to that of the find-path problem using the configuration space and discussed by R. A. Brooks and T. Lozano-Perez ("A Subdivision Algorithm in Configuration Space for Findpath with Rotation," *Proc. 8th Int. J. Artif. Intell.*, pp. 799–806, 1983). Interested readers should refer to the reference for more details. The paths are mostly straight translations in many assemblies. Figure 14-26 shows the collision between the bolt and the swept volume of the field if the former is assembled to the frame before the latter.

Siblings at any depth are considered separate subassemblies. To assemble siblings at depth 1, the root vertex at depth 0 (see Fig. 14-25) is added so that the subassemblies at depth 1 can be treated as siblings. If the swept volumes of two siblings do not intersect, no ordering is necessary to assemble the two sibling

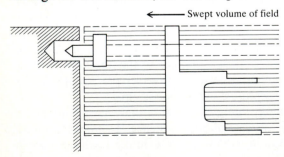

Swept volume of field

**FIGURE 14-26**
Interference between the field and the bolt.

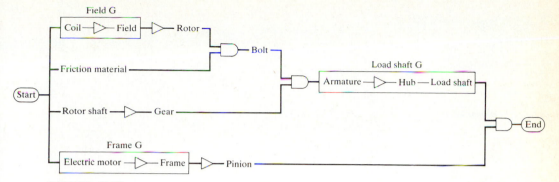

**FIGURE 14-27**
Precedence graph of the clutch assembly shown in Fig. 14-4.

parts. The calculations of the swept volumes can easily be made from the mating conditions. For example, the sweeping direction is the normal vector for the "against" condition or is the centerline in the "fits" condition. The length of the sweeping vector can be chosen arbitrarily from a reference point in space.

The assembling algorithm produces an ordering list of how parts should be assembled. The list for the clutch assembly is shown below with part $1 <$ part 2 means part 1 should be assembled before part 2. Note that the "$<$" relation is transitive, that is, P1 $<$ P2 and P2 $<$ P3 implies P1 $<$ P3. Also the letter G following a part name indicates grouping (a part and its children) of the part.

Armature $<$ hub
Hub $<$ load shaft
Electric motor $<$ frame
Coil $<$ field
Pinion $<$ frame G
Rotor shaft $<$ gear
Field G $<$ rotor
Rotor $<$ load shaft G
Field G $<$ bolt

Using these orderings, a precedence graph can be derived as shown in Fig. 14-27. The figure uses the AND gate and buffer in analogy with digital circuits. For the AND gate, all the inputs to the gate have to be present to produce its output, while the buffer is a special case of the AND gate where there is only one input.

## 14.6 ASSEMBLY ANALYSIS

Various CAD/CAM systems provide various analysis tools to analyze assemblies once they are created. Among the popular analysis tools are the generation of assembly drawings, exploded views of assemblies, shaded images of assemblies, cross-sectional views, mass property calculations, interference checking, kinematic and dynamic analyses, finite element analysis, and animation and simulation.

Applying most of the above analysis tools to assemblies is a simple extension of their applications to individual parts. Therefore, most of them are covered

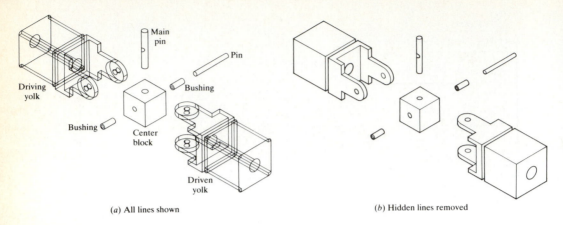

(a) All lines shown        (b) Hidden lines removed

**FIGURE 14-28**
Exploded view of a Cardan universal joint.

in detail in their respective chapters throughout the book. The user can analyze the individual parts separately or a group of them. This is the case with mass property calculations or finite element analysis. In the former, the user can merge (sum) the mass properties of individual parts to obtain the assembly mass properties. In finite element analysis, an individual part can be analyzed separately if the proper boundary conditions that describe its attachment to the assembly are used.

To generate assembly drawings and/or shaded images of assemblies, the user must create the assembly database first by merging the various parts by either using the WCS or the mating conditions method. Cleaning up the assembly drawings follows the same manual procedure discussed in Chap. 3. If parts of an assembly are created as wireframe models, shading them would require creating the corresponding surface models first. If, however, they are

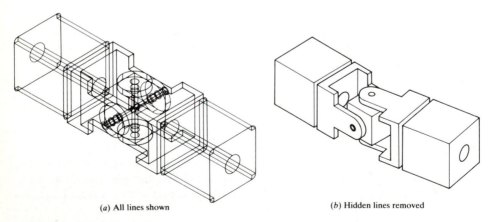

(a) All lines shown        (b) Hidden lines removed

**FIGURE 14-29**
Assembled view of Cardan universal joint shown in Fig. 14-28.

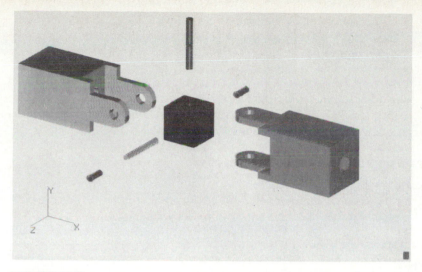

**FIGURE 14-30**
Shaded image of Cardan universal joint shown in Fig. 14-28.

created as solid models, they can be shaded automatically. In this case, automatic hidden line or surface removal is possible which can create a clear isometric view showing the positions of the parts of the assembly relative to each other. Figures 14-28 to 14-30 show drawings and images of a Cardan universal joint.

## PROBLEMS

### Part 1: Theory

Generate the assembly trees for the assemblies shown in:

**14.1.** Figure P14-1, see below.
**14.2.** Figure P14-2 ⎫
**14.3.** Figure P14-3 ⎬ see page 712.
**14.4.** Figure P14-4 ⎭
**14.5.** Figure P14-5 ⎫
**14.6.** Figure P14-6 ⎬ see page 713.
**14.7.** Figure 14-7 ⎭
**14.8.** Figure P14-8 ⎫ see page 714.
**14.9.** Figure P14-9 ⎭
**14.10.** Find the mating conditions in Probs. 14.1 to 14.9.
**14.11.** Use the methods of WCS and mating conditions to assemble the parts shown in Fig. P14-11, see page 715.

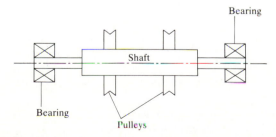

Bearing

Shaft

Bearing

Pulleys

**FIGURE P14-1**
Shaft system.

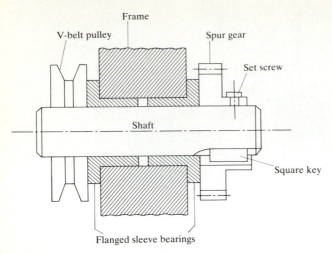

**FIGURE P14-2**
Driving unit of a small centrifugal fan.

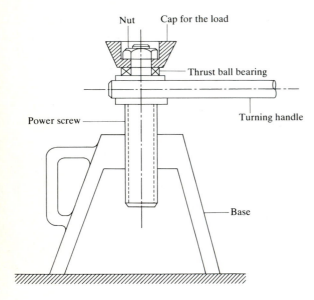

**FIGURE P14-3**
Screw jack.

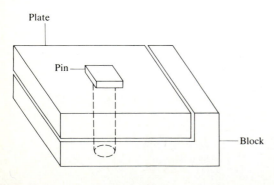

**FIGURE P14-4**
Block/plate assembly.

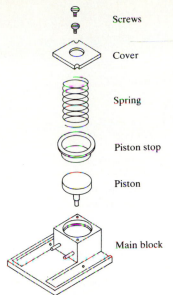

Screws

Cover

Spring

Piston stop

Piston

Main block

**FIGURE P14-5**
Pressure sensor.

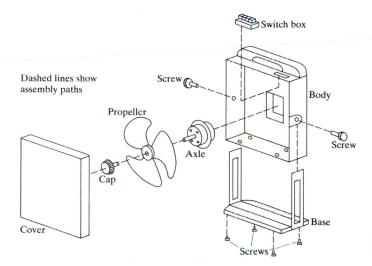

Switch box

Dashed lines show
assembly paths

Screw

Body

Propeller

Screw

Axle

Cap

Cover

Base

Screws

**FIGURE P14-6**
Household fan.

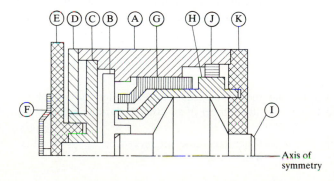

E D C B A G H J K

F

I

Axis of
symmetry

**FIGURE P14-7**
Mechanical assembly
(axisymmetric).

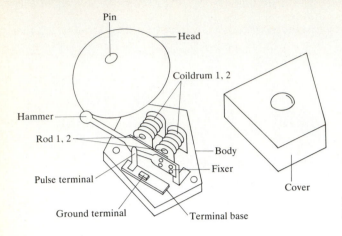

Pin

Head

Coildrum 1, 2

Hammer

Rod 1, 2

Body

Fixer

Pulse terminal

Cover

Ground terminal

Terminal base

**FIGURE P14-8**
Bell assembly.

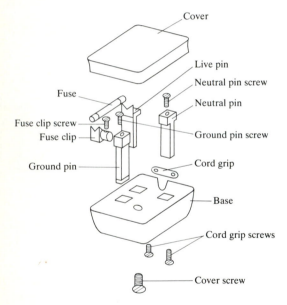

Cover

Live pin

Neutral pin screw

Fuse

Neutral pin

Fuse clip screw

Fuse clip

Ground pin screw

Ground pin

Cord grip

Base

Cord grip screws

Cover screw

**FIGURE P14-9**
Three-pin power plug.

**14.12.** Develop the graph structures for the assemblies shown in Probs. 14.1 to 14.9.

**14.13.** Develop the location graphs for the assemblies shown in Probs. 14.1 to 14.9.

**14.14.** Develop the graph structures and data structures, using the concept of virtual link, for the assemblies shown in Figs. P14-1 to P14-9.

**14.15.** Develop the precedence diagrams for the assemblies shown in Probs. 14.1 to 14.9. Count the number of all possible assembly sequences to create the assembly in each problem.

**14.16.** Develop the liaison diagrams and the liaison-sequence graphs for the assembles shown in Probs. 14.1 to 14.9. Count the number of all possible liaison sequences to create the assembly in each problem.

**14.17.** Develop the mating graphs, the part hierarchy, and the precedence graphs for the assemblies shown in Probs. 14.1 to 14.9.

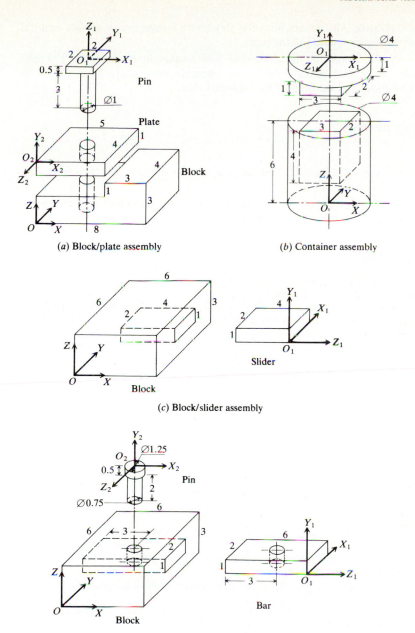

(a) Block/plate assembly

(b) Container assembly

(c) Block/slider assembly

(d) Block/bar assembly

**FIGURE P14-11**
Typical assemblies (all dimensions in inches).

## Part 2:  Laboratory

**14.18.** Create the databases of the individual parts shown in Probs. 14.1 to 14.9 and 14.11
using your CAD/CAM system. Use the merging method (most probably the WCS
method) to create the databases for the assemblies. Obtain drawings of individual
parts as well as assembly drawings. Obtain exploded views as well. Use solid

models if possible. Also obtain isometric views with hidden line or surface removal. Assume the dimensions for all the parts.

**14.19.** Obtain shaded images of the assemblies you created in Prob. 14.18. For each assembly obtain two images, one of the assembly itself where parts are in the correct positions and the other of the exploded view.

## Part 3: Programming

Write a program that:

**14.20.** Implements the WCS method. Test the program for simple assemblies (assemblies of two or three parts).

**14.21.** Can infer the location and orientation of parts of an assembly from the mating conditions using the least-squares method.

## BIBLIOGRAPHY

Ambler, A. P., and R. J. Popplestone: "Inferring the Positions of Bodies from Specified Spatial Relationships," *Artif. Intell.*, vol. 6, pp. 157–174, 1975.

Boothroyd, G., C. Poli, and L. E. Murch: *Automatic Assembly*, Marcel Dekker, New York, 1982.

Brooks, R. A., and T. Lozano-Perez: "A Subdivision Algorithm in Configuration Space for Findpath with Rotation," *Proc. 8th Int. J. Artif. Intell.*, pp. 799–806, 1983.

De Fazio, T. L., and D. E. Whitney: "Simplified Generation of all Mechanical Assembly Sequences," *IEEE J. Robotics and Automn*, vol. RA-3, no. 6, pp. 640–658, 1987.

Eastman, C. M.: "The Design of Assemblies," SAE Technical Paper Series, paper 810197, Society of Automotive Engineers, U.S.A., 1981.

Ko, H., and L. Lee: "Automatic Assembling Procedure Generation from Mating Conditions," *CAD J.*, vol. 19, no. 1, pp. 3–10, 1987.

Lee, K., and G. Andrews: "Inference of the Positions of Components in an Assembly: Part 2," *CAD J.*, vol. 17, no. 1, pp. 20–24, 1985.

Lee, K., and D. C. Gossard: "A Hierarchical Data Structure for Representing Assemblies: Part 1," *CAD J.*, vol. 17, no. 1, pp. 15–19, 1985.

Lieberman, L. I., and M. A. Wesley: "AUTOPASS: An Automatic Programming System for Computer Controlled Mechanical Assembly," *IBM J. Res. Dev.*, vol. 21, no. 4, pp. 321–333, 1977.

Liu, C. R., and I. Glassar: "The Construct of a High-Level Computer Language for Programmable Assembly," *Computers and Ind. Engng*, vol. 9, no. 3, pp. 203–214, 1985.

Mullineux, G.: "Optimization Scheme for Assembling Components," *CAD J.*, vol. 19, no. 1, pp. 35–40, 1987.

Poli, C., and F. Fenoglio: "Designing Parts for Automatic Assembly," *Mach. Des.*, vol. 59, no. 29, pp. 140–145, 1987.

Poppelstone, R. J., A. P. Ambler, and I. M. Bellos: "An Interpreter for a Language for Describing Assemblies," *Artif. Intell.*, vol. 14, pp. 79–107, 1980.

Prenting, T. O., and R. M. Battaglin: "The Precedence Diagram: A Tool for Analysis in Assembly Line Balancing," *J. Ind. Engng.*, vol. 15, no. 4, pp. 208–213, 1964.

Rocheleau, D. N., and K. Lee: "System for Interactive Assembly Modelling," *CAD J.*, vol. 19, no. 2, pp. 65–72, 1987.

Rush, C. L., and M. J. Bailey: "Computer-Aided Simulation of Solid Assemblies," *Computer in Mech. Engng (CIME)*, vol. 2, no. 3, pp. 40–43, 1983.

Tilove, R. B.: "Extending Solid Modeling Systems for Mechanism Design and Kinematic Simulation," *IEEE Computer Graphics and Applic.*, vol. 3, no. 3, pp. 9–19, 1983.

Wesley, M. A., T. Lozano-Perez, L. I. Lieberman, M. A. Lavin, and D. D. Grossman: "A Geometric Modeling System for Automated Mechanical Assembly," *IBM J. Res. Dev.*, vol. 24, no. 1, pp. 64–74, 1980.

# CHAPTER
# 15

# INTERACTIVE COMPUTER PROGRAMMING

## 15.1   INTRODUCTION

Most often users of CAD/CAM systems encounter situations that may require programming. For example, if a geometric model of the same topology, but of different geometry, is repetitively used, the user may create a program that accepts the geometric data and generates the model automatically. In many engineering applications, engineering analysis requires a geometric model (data). In such a case, programming is a useful tool to use.

The term "interactive computer programming" is used in this book to indicate programming that processes both graphics (entity data) and nongraphics (numbers and text strings) data on a CAD/CAM system, and that controls the operation of the system. In this sense, interactive computer programming (or interactive programming for short) is a powerful tool to customize CAD/CAM systems to perform certain design tasks and functions that cannot be achieved directly by using the standard application modules typically offered by these systems. Most available commercial systems supply their users with programming languages to enable them to extend their domains of applications.

The major advantage of interactive programming is to automate some phases of the design process, hence improving the user's productivity on the CAD/CAM system. As a matter of fact, some of the high productivity gains claimed by using the CAD/CAM technology can only be achieved via interactive programming. Following are some typical applications where interactive programming may be useful:

1. Automatic graphics creation. Repetitive graphics work can be performed automatically via interactive programming. Consider the example of creating families of parts. Parts in one family usually have the same shape with different

dimensions. To write a program to generate such a family, the first step is to parametrize the part that belongs to the family. This step involves describing the part's shape by a set of variables. When these variables are assigned certain values, the corresponding model of the part can be generated. The second step is to write a program in terms of these variables. When a user executes the program, it generates the part after the user supplies the desired values for all the variables.

2. Create customized menus. The ability to create customized menus helps tailor a CAD/CAM system to meet individual needs. For example, if a group of graphics commands is always used, the user can create a menu that includes these commands only, or a user may create a design or analysis menu. A menu, say, for beam analysis can include icons for various types of beams (simply supported, cantilever, etc.), of loads (concentrated, distributed, etc.) of material (steel, aluminum, etc.), and the type of analysis (linear, nonlinear, etc.).

3. Create user-defined geometry. If the user has a special graphics entity (a curve or a surface) described by a mathematical equation, a program can be written to create it. In addition, a command or an icon can be created to use the program in the same way as the rest of the system graphics commands. Consider, for example, creating a program and a command to generate the involute curve of the profile of a tooth of a spur gear. After a program to create the involute curve is written, the processor of the user interface can be used to develop the command or the icon.

4. Perform parametric design studies. By changing geometric and other input data for a given program, the designer can study the effect of certain parameters on a particular design. In addition to displaying the geometric models of various designs, the program can also display trends of how a given design parameter affects the design.

5. Automatic graphics functions and procedures. In many cases, various design groups in various organizations develop certain procedures that the members of these groups must use for standardization purposes. Consider, say, a title block for a drawing. A program may be written that can ask a member for the relevant data used by the block (such as name, data, scale, part number, etc.) and then generates the block automatically. Special geometric procedures can be automated in a similar way. For example, a program can be written that can change a curve or a surface into a set of points.

6. Perform engineering analysis and simulation. For example, the geometry of a mold created on a CAD/CAM system may be tested by writing an analysis program to check for tolerances of the mold, or programs for kinematic analysis may test mechanisms created on the system.

The reader may ask the following two questions. What is the difference between conventional and interactive programming and what are the advantages of the latter over the former? By "conventional programming," we mean programs that we write for number crunching only. The answer to the first question is covered in the following section of the chapter. The main advantage of interactive programming is that the user can generate and display numerical results simultaneously in one program. The reason that this advantage is achieved in interactive programming is because interactive programs are usually executed after a part is activated. Therefore, they have access to all the abilities associated with the centralized database of the part. Refer to examples covered in Sec. 15.3 for a better understanding. In conventional programming, the programmer stores the numerical results in a file and then plots them on a plotter or displays them on a graphics terminal.

Interactive programming is particularly useful because of the nature of part databases. In some instances, a part database may become corrupted for reasons beyond user control so that the user cannot access the part database. These instances usually result in loss of work worth many man-hours. If some of this work is created by programs, the programs can simply be reexecuted to create the geometry, thus saving the time of having to reconstruct the geometry from scratch. Interactive programming is also helpful when a CAD/CAM system crashes, resulting in possible loss of the entire part database.

The focus of this chapter is to present the concepts of interactive programming in a generic, syntax-independent, fashion. No specific programming languages are covered because their syntax differs widely from one CAD/CAM system to another. Even for a given system, existing languages may be outdated and replaced by better ones. However, reference to some of these languages is made throughout the chapter. Programming examples, on the other hand, are presented in a pseudocode (English-like statements) to present the new concepts covered in this chapter. Interested readers should refer to their system documentation and problems of Parts 2 and 3 at the end of the chapter to learn the syntax of the programming languages available on their respective systems.

## 15.2  REQUIREMENTS OF INTERACTIVE COMPUTER PROGRAMMING

Interactive programming processes both graphical and nongraphical information. To be useful in an engineering or design environment, interactive programming must be able to handle the related information—mainly analysis information and graphics information. Analysis information consists of numerical results and/or coordinate calculations of various points of the geometric model to which the analysis is applied. Graphics information includes either extracting existing information from a part database or adding new information to the database.

Due to the required graphics and analysis information, the following types of statements must exist in any interactive programming language:

1. **Declaration statements.** These statements declare the mode of variables to the program and specify dimensions of arrays. Variables can be numeric (real or integer), text strings, graphics location, and graphics entities. The latter two types of variables do not exist in conventional programming. A graphics-location variable holds three values. These are the coordinates of the point stored in this variable.

2. Arithmetic statements. Most arithmetic operations (single or double precision) provided by conventional programming are also available in interactive programming. Addition, subtraction, multiplication, division, and exponent operations are usually available. The precedence rules of these operations to evaluate expressions are the same as in conventional programming.

3. Control statements. These statements control the flow of the program and are identical to those used in conventional programming. They include IF, GO TO (BRANCH), and WHILE statements, as well as DO (FOR or REPEAT) loops.

4. I/O statements. Two types of I/O statements can be identified in interactive programming. Statements that are used to input analysis data and output analysis results are similar to conventional programming and include READ, WRITE, INPUT, and PRINT statements. File-handling statements are also available.

   The other type of I/O statements deals with inputting or extracting graphics information from a part database. A DIGITIZE statement, for example, can be used to input a location as a user digitize. A program to draw a circle may ask the user to input the value of the radius and digitize the location of its center on the screen. The value the user input as a radius is read by a READ statement and the user digitize is processed by the program via a DIGITIZE statement. A DIGITIZE statement may allow the users to input multiple digitizes. In such a case, the system syntax may allow the programmer to record the actual number of digitizes input by the user and/or recognized by the system. The programmer can compare this number in an IF statement with the original requested numbers of digitizes. If they are not equal, the programmer may force the user to redigitize again before continuing the program execution any further.

   The opposite to the DIGITIZE statement is a statement that extracts information about an existing entity. Let us refer to this statement as OBTAIN. If, for example, we need to construct a circle with a center as the midpoint of an existing line, the OBTAIN statement is used first to provide the location of the midpoint of the line. Obviously, the statement requires the user to identify the line by digitizing it, or the programmer can identify it within the program by using its name (tag), as discussed in Chap. 11.

5. Construction statements. When coordinate information is generated in a program, the construction statements of the interactive programming language are used to create the corresponding entities. The syntax of these statements may be very similar to the syntax of the graphics commands of the CAD/CAM system. This facilitates learning the syntax of the programming language. It is always natural that the user may attempt to extend the syntax of the graphics commands to the programming language. However, the user should expect some changes in the syntax of the graphics commands when they are transported to the programming language. In some other languages, the syntax of the construction statements may be totally different from that of graphics commands. In this case, it may seem that the user is learning two different syntax on the same CAD/CAM system.

There is another major difference between construction statements and graphics commands. When the user issues a graphics command, the resulting entity is displayed once the command is completed. On the other hand, the entity is not displayed at the end of executing the construction statement. The user must use a DISPLAY statement to display the resulting graphics. It is usually recommended that all the resulting entities be displayed at the end of the program using one DISPLAY command. This strategy usually speeds up the execution of the interactive program.

6. Graphics manipulation statements. These statements support manipulating entities (translate, rotate, mirror, and scale) that result in an interactive program. As in construction statements, the syntax of manipulation statements may or may not be drastically different from manipulation commands. In some interactive programming languages, the manipulation statements are a subset of the manipulation commands supported by the corresponding CAD/CAM system.

The reader can easily conclude that interactive programming supports the statement types (1 to 4 above) that exist in conventional programming. In addition, it supports graphics-related operations via statement types 5 and 6 above.

## 15.3 TYPES OF INTERACTIVE COMPUTER PROGRAMMING

The development of programming capabilities on CAD/CAM systems has followed the evolution of these systems. These systems have developed originally from a need for drafting automation. A large number of users have been draftsmen or designers with no programming background, yet the need for programming to automate some procedures and functions has been increasing. As a result, some elementary programming capabilities have been provided by CAD/CAM vendors. When users of more programming background began to use the CAD/CAM technology, the demand for better programming capabilities arose. As a result, a wide variety of interactive programming types exists on many CAD/CAM systems. The capabilities, characteristics, and limitations of each type are discussed in this section.

### 15.3.1 Elementary Level Programming

This is the simplest level of interactive programming available on a CAD/CAM system. It suits users without an extensive programming background. Writing a program using this level entails stacking graphics commands in a text file. Users can think of a program written at this level as if they have stored a particular command sequence in a file. No intelligence such as control or arithmetic statements are permitted in the program. The program is interpretive, that is, not compiled. Therefore, it tends to run slow.

To make a program written at this level more versatile, the syntax of graphics commands is slightly changed from its conventional syntax to allow for user variable input such as digitizes, scalars, and names. In addition, each command must be terminated explicitly in the program. For example, if a command is terminated by a carriage return, then a carriage return must be typed explicitly at the end of each command used in the program. In the examples to follow, we use the notation $D, $S, $P, $CR for variable digitize, variable scalar, prompt, and carriage return respectively.

Prompt statements are allowed in the program. They are a necessity if the user of the program is to communicate with it. When the program encounters a variable input in a command, it waits to allow the user to input the corresponding value(s). Thus, the programmer must precede a command with a variable input by a prompt statement so that the user can expect what to input. Comment statements may also be allowed in the program.

Elementary level programming is typically used to create a family of parts or to automate simple graphics procedures. This level of programming is known as "execute file" on the Computervision system or "macro" on the Calma system.

**Example 15.1.** Write an elementary level program that can create the geometric model shown in Fig. 15-1. The center of the circle coincides with the center of the rectangle.

*Solution.* Let us assume that the MCS of the model is located at the bottom left corner of the rectangle as shown in the figure. The model is created in the program by connecting the points $P_1$, $P_2$, $P_3$, $P_4$, and $P_5$ as shown. The coordinates of these points are easily calculated after the user decides on the values for $W$ and $H$. The center of the circle is $(W/2, H/2, 0)$. The pseudocode of the elementary level program to create this model may look as follows:

```
$P PLEASE INPUT THE WIDTH W FOLLOWED BY THE HEIGHT H OF THE
RECTANGLE $P
$P PLEASE HIT A CARRIAGE RETURN AFTER EACH INPUT #P
   CREATE MULTIPLE LINES CONNECTING POINTS P₁ TO P₅
$P PLEASE CALCULATE (W/2, H/2) AS (X,Y) COORDINATES OF THE
CENTER OF THE CIRCLE $P
$P ENTER THE RADIUS OF THE CIRCLE FOLLOWED BY ITS CENTER $P
   CREATE CIRCLE RADIUS $S CENTER X$S, Y$S, Z0, $CR
```

In the above line command, we assumed that if no new value for a particular coordinate is input, the previous value is used by the program. The notation X0Y0Z0 means a location with (0, 0, 0) for the $x$, $y$, $z$ coordinates. The notation X$S

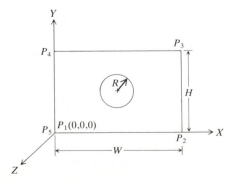

**FIGURE 15-1**
Typical geometric model.

implies a variable value for the $x$ coordinates of the corresponding point. During execution, the program waits for the user to input the desired values for a variable. After the user hits a carriage return at the end of an input, the program proceeds to the next variable and waits again for a value followed by a carriage return. Also, notice that a prompt statement may occupy more than one line, and an explicit carriage return ($CR) is only needed at the end of a graphics command.

After the above program is created using an appropriate text editor at the operating system level of the CAD/CAM system, the user needs to access the graphics level to execute the program. Because the program generates graphics entities, it must be executed after a part has been activated. More precisely, the user must type the part setup command sequence, discussed in Chap. 3, before executing the program. This sequence in itself can be another program that can be executed immediately after the user has activated the graphics level. To execute the program, the user may type the command RUN followed by the filename that contains the program. In this example, we type RUN RECT, assuming we have called the file RECT for rectangle.

**Example 15.2.** Find the locus in space of the midpoint $E$ of the bar $BC$ of the four-bar linkage shown in Fig. 15-2.

*Solution.* The previous example shows a program that applies to geometric construction. This example shows how to use elementary level programming in typical engineering applications. Let us assume that the MCS is chosen as shown in Fig. 15-2. The locus of point $E$ in space is defined as its path of motion, or the path it traverses in space during the motion of the linkage. The locus can be generated as follows:

1. Find the extreme right position of the linkage bars. Construct the linkage in this position. The extreme right position of the linkage occurs when the bar $AB$ is in the horizontal position to the right of the pin support at $A$. The location of point

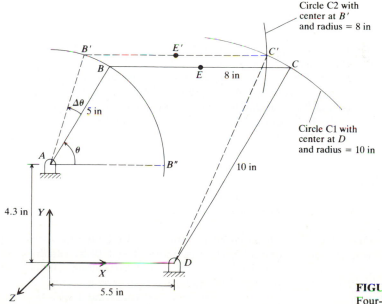

**FIGURE 15-2**
Four-bar linkage.

*C* in this position is obtained by intersecting two circles C1 and C2: C2 with center at *B″* and radius of 8 in (length of *BC*) and C1 with center at *D* and radius of 10 in.

2. Find the location of point *E* in the extreme right position. This is simply the midpoint of the line *B″ C″* (*C″* is not shown in the figure).

3. Rotate the bar *AB* an incremental angle $\Delta\theta$ in the counterclockwise direction. Repeat the procedure described in step 1 to find the midpoint of the bar *BC*. The dashed linkage of Fig. 15-2 shows point *E′* after *AB* is rotated by an angle $\Delta\theta$.

4. Repeat step 3 until no further motion of *AB* to the left is possible. This is detected when the intersection between circles C1 and C2 is no longer possible.

5. Connect the various points generated (from step 3) with a B-spline curve to obtain the locus. The smaller the angle $\Delta\theta$, the more points generated on the locus and the smoother the B-spline curve.

A program that would implement steps 2 to 5 to create the locus of point *E* is shown below:

```
$P PLEASE DIGITIZE NEAR END A, INPUT THE ANGLE OF ROTATION OF
AB, AND FINALLY DIGITIZE AB$P
    ROTATE ABOUT END $D ANGLE $S COPY $D $CR        {Create line AB'}
$P PLEASE DIGITIZE THE END B OF AB IN ITS NEW POSITION $P
    CREATE CIRCLE RADIUS 8 CENTER $D $CR            {Create circles C2}
$P PLEASE DIGITIZE THE END B OF AB IN ITS NEW POSITION
FOLLOWED BY DIGITIZING THE TWO CIRCLES $P
    CREATE LINE END $D INTOF $D $D $CR              {Create line B'C'}
$P PLEASE DIGITIZE END D OF DC FOLLOWED BY DIGITIZING THE TWO
CIRCLES $P
    CREATE LINE END $D INTOF $D $D $CR              {Create line DC'}
$P PLEASE DIGITIZE BC IN ITS NEW POSITION $P
    SELECT LAYER 200 $CR
    CREATE POINT ORIGIN $D $CR                      {Create point E'}
    SELECT LAYER 100 $CR
```

In the above program, END, INTOF, and ORIGIN are the end, intersection of, and origin modifiers introduced in Chap. 11. The program is based on few assumptions. The extreme right position of the linkage and the circle C1 have been constructed prior to executing the program. In addition, all lines and circles reside on layer 100 and the locus points reside on layer 200. Finally, the B-spline curve is created outside the program. Thus, the program generates the set of locus points only.

The program must be executed once to generate a point on the locus. The user runs the program *n* times to generate *n* points. The reason is that no DO (REPEAT) loops or control statements are permitted. This is why tasks that are performed only once such as constructing the extreme right position of the linkage or the B-spline curve are not included in the program. Figure 15-3 shows the locus of point *E*.

This program can be modified by replacing all the fixed numbers (lengths of the linkage bars and distances between supports *A* and *D*) by variables using $S. The resulting program can be used to perform parametric design studies to investigate the effect of the linkage parameters on the shape and length of the locus. The locus length can be evaluated using the "measure" command discussed in Chap. 12.

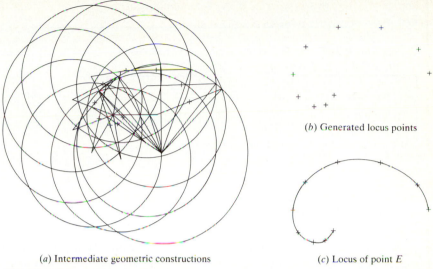

(b) Generated locus points

(a) Intermediate geometric constructions          (c) Locus of point E

**FIGURE 15-3**
Locus of four-bar linkage shown in Fig. 15-2.

In this context, the user chooses a set of values for the parameters and runs the program repetitively to generate the corresponding locus. The user can choose another set and generate the new locus on, perhaps, a new layer. After enough loci are generated, they are studied to enable making the proper design decisions.

### 15.3.2  On-Screen Menu Programming

An on-screen menu is a collection of screen buttons or icons that are displayed on the screen. These icons are accessible with the mouse of the graphics display by overlapping the crosshair into the icon area and then clicking one of the mouse buttons. With the widespread use of the Unix operating system by many CAD/CAM systems, developing on-screen menus would require knowledge of the programming tools and toolkits provided by UIMS (user-interface management systems) or SUIMS (syntactic UIMS). One can think of a user interface toolkit as a set of facilities that a programmer can use to develop pop-up menus or screen buttons. Details of developing on-screen menus under Unix are not discussed in the book. The interested reader is referred to the bibliography at the end of the chapter. On-screen menus must be activated to use them and deactivated when finished.

Figure 15-4 shows a sample design of an engineering menu. Behind each menu icon is a main program which is executed when the user selects this particular icon. The communication between the user icon selection and the execution of the programs is achieved under the control of the UIMS or SUIMS. The details of developing this menu are not discussed here.

### 15.3.3  High Level Programming

This is perhaps the most popular type of programming among users of CAD/CAM systems. Most commercial systems provide programming languages

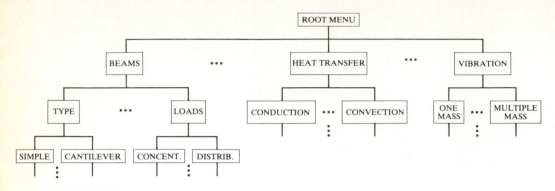

**FIGURE 15-4**
Sample engineering menu.

that support this level of programming. Computervision, Calma, and McDonnell Douglas offer CVMAC, DAL, and GRIP languages respectively, to name a few. These languages and others are similar in nature and philosophy to conventional languages such as BASIC, FORTRAN, Pascal, C, etc. A program written in one of these languages, that is, CVMAC, DAL, or GRIP, is usually compiled before it is executed. Therefore, the program runs much faster than a corresponding program written using the elementary level programming. If the main program calls other programs or procedures, the latter ones must be linked to the former before it is executed. As in the elementary level programming, source files that store the high level programs are developed and compiled at the operating system level, and the executables are run at the graphics level. Figure 15-5 shows the steps of developing, compiling, and executing a high level program.

High level programming languages support the six statement types described in Sec. 15.2. Therefore, these languages are system-dependent and programs written for one system cannot be transported to another system. These languages have also the ability to interface with conventional languages, in particular FORTRAN and C. The benefits of the interface are twofold. First, it makes accessible to a high level program all the libraries supported by the conventional language. For example, if a program calls a FORTRAN subroutine, routines from standard libraries such as IMSL can be called within the FORTRAN routine. Second, most of the number-crunching operations can be programmed in, say, FORTRAN, and only the main routine can be written in the system-dependent language. In this case, the main routine has all the I/O, construction, and manipulation statements, as well as the procedural calls to the FORTRAN routines. In this case, to transport all the routines to another CAD/CAM system, all that is needed is to write the main routine using the system-dependent language of the new system as long as it supports the FORTRAN interface.

High level programming is useful for almost all engineering and design applications that involve processing both graphics and analysis information. It is useful for parametric design studies, automating various engineering analyses, and automating geometric and graphics procedures.

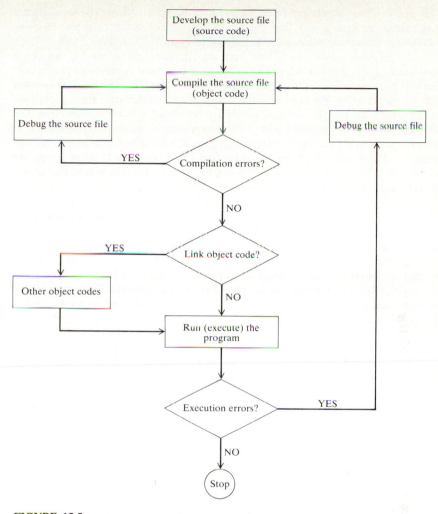

**FIGURE 15-5**
Developing and executing high level programs on CAD/CAM systems.

While Examples 15.1 and 15.2 can be rewritten using this level of program-ming, other examples are covered below. Readers are encouraged to use high level programming languages available on their respective CAD/CAM systems to resolve Examples 15.1 and 15.2.

**Example 15.3.** Write a program that can evaluate and display the saddle surface given by the equation:

$$Cz = \frac{x^2}{A^2} - \frac{y^2}{B^2}$$

where $A$, $B$, and $C$ are constants.

***Solution.*** The above equation is an implicit nonparametric equation. The surface can be evaluated by generating a set of points, using the above equation, for a given

*A*, *B*, and *C*. These points can be connected by a B-spline surface. The set of points is calculated in such a way that the points can be connected to create B-spline curves first. Then these curves are interpolated to create the B-spline surface.

To calculate point coordinates from the above equation, we fix, say, the *y* coordinate to an arbitrary value. The equation is then used to calculate the *z* coordinate for various values of *x* of our choice. The resulting set of points is connected with a B-spline curve. We increment the *y* coordinate and calculate the *z* coordinate for the same *x* values. The resulting points are connected with a B-spline curve. The resulting curves are connected with a B-spline surface. Fixing the *y* coordinate in these calculations is equivalent to intersecting the saddle surface with planes perpendicular to the *Y* axis at various *y* values (see Fig. 15-6). The B-spline curves we calculate are the intersection curves. We can fix the *x* coordinate instead and repeat the same calculations to evaluate the surface.

The B-spline curves and surface are scaled and displayed as soon as their respective commands are executed in the program. The program that generates this surface is executed at the graphics level within an active part. The user does not have to deal with managing the resulting coordinate values to scale and plot them as in the case of conventional programming. The surface is considered as a part (model) by the CAD/CAM system used to run the program and display the surface. The user can apply all the proper commands to better understand the surface. For example, various views of the surface can be obtained and the surface can be verified as an entity.

The following program is written in pseudocode to generate the surface. It is based on the following data. The values for *x* and *y* coordinates are $-5 \leq x \leq 4$ and $-5 \leq y \leq 5$ respectively. The *y* value is held constant at six values, that is, $-5$, $-3$, $-1$, 1, 3, and 5. For each *y* value, ten *x* values are used, that is, $-5$, $-4$, $-3$, $-2$, $-1$, 0, 1, 2, 3, and 4. Thus the surface is generated by creating six B-spline curves and each curve is generated by creating ten points.

```
PROCEDURE SADDLE
    DECLARE LOCATION P(10)
    DECLARE NUMERIC A, B, C, X, Y, Z, I, N, M
```

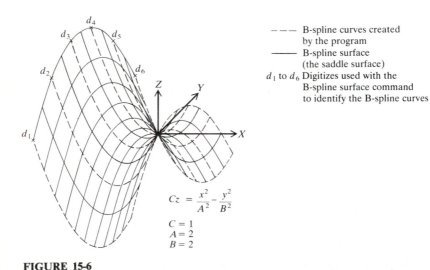

— — — B-spline curves created
by the program
——— B-spline surface
(the saddle surface)
$d_1$ to $d_6$ Digitizes used with the
B-spline surface command
to identify the B-spline curves

$$Cz = \frac{x^2}{A^2} - \frac{y^2}{B^2}$$

$C = 1$
$A = 2$
$B = 2$

**FIGURE 15-6**
Saddle surface.

```
     DECLARE ENTITY S(6), BS
"PLEASE INPUT VALUES FOR A, B, AND C"
     READ A, B, C
     N = 1                                    {Initialize N}
     M = 0                                    {Initialize M}
DOG                                           {Label}
     M = M + 1
     Y = -6 + N                               {Set values for y coordinate}
     FOR I = 1 TO 10, INCREMENT BY 1
        X = -6 + I
        Z = (X*X/A*A - Y*Y/B*B)/C
        P(I) = (X, Y, Z)                      {Store values into location P(I)}
     END                                      {FOR loop}

     S(M) = BSPLINE CURVE P                   {Connect points in the P array
                                              with the B-spline curve and store it
                                              as entity S(M)}
     N = N + 2                                {Increment N}
     IF (N LESS THAN 12) GOTO DOG
     BS = BSPLINE SURFACE S                   {Create B-spline surface using the
                                              B-spline curve entities}

     DISPLAY S, BS                            {So the user can see the graphics
                                              on the screen}
END
```

This program allows the user to input different values for $A$, $B$, and $C$. It can be used to study the influence of these parameters on the surface shape. Figure 15-6 shows the surface for $A = B = 2$ and $C = 1$. The surface and its B-spline curves pass through one common point. This is the origin $(0, 0, 0)$ which satisfies the surface equation.

**Example 15.4.** The leaves of two roses can be described by the following equations:

$$r = a \sin n\theta \tag{15.1}$$

$$r = a \cos n\theta \tag{15.2}$$

where $a$ is a constant that decides the length of the leaf and $n \geq 2$. The variable $n$ decides the number of leaves of any rose. If $n$ is an even integer, the corresponding rose has $2n$ leaves. If $n$ is an odd integer, the rose has $n$ leaves. Write a program that evaluates the leaves and displays the rose for various values of $a$ and $n$.

*Solution.* Figure 15-7 shows a sketch of Eqs. (15.1) and (15.2) for $n = 2$ and 3. The roses generated from both equations are identical except that their orientations with respect to the $X$ and $Y$ axes are different. A point $P$ on a leaf is described by $(r, \theta)$ or $(x, y)$ coordinates. A leaf has a maximum radius equal to $a$ and begins and ends with $r$ values equal to zero. For Eq. (15.1), $r$ is maximum at $n\theta = \pi/2$ or $\theta = \pi/2n$. Thus the first leaf has an angle $\phi = \pi/2n$ with the $X$ axis as shown in the figure. The number of leaves per rose is equal to $n$ if $n$ is odd or $2n$ if $n$ is even. Thus the angle $\Delta\phi$ that separates leaves from each other is equal to $360/n$ or $360/2n$. The limits of a leaf, using Eq. (15.1), occurs when $0 \leq n\theta \leq \pi$ or $0 \leq \theta \leq \pi/n$. Similarly, Eq. (15.2) shows that the first leaf has an angle $\phi = 0$ with the $X$ axis, as shown in the figure, and limits of $-\pi/2n \leq \theta \leq \pi/2n$.

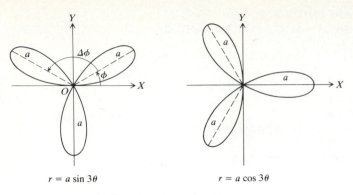

$r = a \sin 3\theta$ $\qquad\qquad\qquad$ $r = a \cos 3\theta$

(a) Three-leaved roses ($n = 3$)

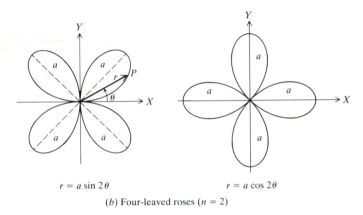

$r = a \sin 2\theta$ $\qquad\qquad\qquad$ $r = a \cos 2\theta$

(b) Four-leaved roses ($n = 2$)

**FIGURE 15-7**
Schematic three- and four-leaved roses.

Having found all the necessary geometric parameters of a rose, the strategy to write a program to generate and display a rose can be described as follows. The program reads the values for $a$ and $n$ from user input. The program must check that the user input for $n$ is equal to or greater than two. Next, the user is prompted to identify the type of rose—whether it is a sine [Eq. (15.1)] or cosine [Eq. (15.2)] function. With the user input complete, points on a leaf profile can be generated and connected by a B-spline curve to create the profile. The coordinates of any point are given by $x = r \cos \theta$, $y = r \sin \theta$, and $z = 0$, where $r$ is given by Eq. (15.1) or (15.2) above. The range for the angle $\theta$ for one leaf can be divided into small enough increments to generate a smooth B-spline curve. In the program below, 19 points are used.

With the geometry of one leaf complete, the "rotate entity" command with the copy modifier can be used to generate the remaining leaves. The rotation occurs about the $Z$ axis an angle equal to $360/n$ for odd $n$ or $360/2n$ for even $n$.

The following program implements the above ideas. It also shows how the FORTRAN interface can be achieved. A FORTRAN subroutine called LEAF is shown below in actual FORTRAN syntax.

```
PROCEDURE ROSE
    DECLARE NUMERIC A,X(20),Y(20),Z(20),N,THETA,I,R,ANG,K
```

```
DECLARE LOCATION P(20)
DECLARE TEXTSTRING T1
DECLARE ENTITY S
"PLEASE INPUT A VALUE FOR LEAF LENGTH"
READ A
"NOW INPUT A INTEGER N GREATER THAN OR EQUAL TO 2"
CAT                                                        {Label}
READ N
IF (N LESS THAN 2) "N MUST BE GREATER OR EQUAL TO 2. INPUT N
AGAIN"
  GO TO CAT
ELSE
  "PLEASE INPUT E FOR EVEN N OR O FOR ODD N"
  READ T1
  "PLEASE INPUT 1 FOR A SINE ROSE OR 2 FOR COSINE ROSE"
  READ K
ENDIF
CALL LEAF (A,X,Y,Z,N,THETA,I,R,K)        {Call FORTRAN subroutine to
                                          calculate points on the leaf}

FOR I = 1 TO 20, INCREMENT BY 1          {Assigns values in the X, Y and Z
                                          arrays to array P}

  P(I) = (X(I),Y(I),Z(I))
END                                      {FOR loop}
S = BSPLINE CURVE P
DISPLAY S
IF (T1 EQUAL TO "E")
  ANG = 360/2*N
  THETA = -90.0/N
  N = 2*N - 1
ELSE
  ANG = 360/N
  THETA = 0.0
  N = N - 1
ENDIF
FOR I = 1 TO N, INCREMENT BY 1
  "PLEASE DIGITIZE THE LEAF DISPLAYED LAST"
  S = ROTATE ENTITY COPY ANGLE ANG $D X0Y0Z0
                              {The X0Y0Z0 is the axis of rotation}
  DISPLAY S
END                                      {FOR loop}
END                                      {Program}
  SUBROUTINE LEAF (A,X,Y,Z,N,THETA,I,R,K)
  DOUBLE PRECISION A,X(20),Y(20),Z(20),N,THETA,R,INC,FAC
  INTEGER I,K
C CALCULATE INCREMENT INC IN ANGLE THETA
  FAC = 3.14159/180.0              {Factor to change degrees to radians}
  INC = (180.0/19*N)*FAC
  THETA = THETA*FAC
  DO 10 I = 1,20
    IF (K.EQ.1)
    R = A*SIN(N*THETA)
```

```
            ELSE
               R = A*COS(N*THETA)
            ENDIF
            X(I) = R*COS(THETA)
            Y(I) = R*SIN(THETA)
            Z(I) = 0.0
            THETA = THETA + INC
      10    CONTINUE
            RETURN
            END
```

The above program and the FORTRAN subroutine are based on generating 20 points to create the first leaf. Nineteen increments result from the 20 points. Thus, the increment in angle between any two points on a leaf is calculated as $180/19N$. Note that the angles in the main program are assumed to be in degrees, and are changed to radians in the subroutine. Also, note that in the "rotate entity" command, the variable digitize input ($D) is used to facilitate creating the rose. It can be replaced by using names (tags). The reader is encouraged to change the program to implement this possibility. Figure 15-8 shows the results of the program for certain values of $a$ and $n$.

### 15.3.4   Database Level Programming

The three types of interactive programming that have already been discussed have one common goal, that is, to shield the user from having to get deeply involved in the details of database structure and management of a particular CAD/CAM system. However, programs developed based on these types may not be very efficient and may require redundant user input due to the limitations imposed by the capabilities of the corresponding programming language.

The database level programming is the highest level of programming a CAD/CAM system can offer to its users. However, it requires considerable knowledge of the system graphics database structure and management at the record and subrecord levels. System routines that manipulate (read, write, and modify) records and subrecords are usually called within a program at this level. Also a knowledge of how to modify the system user interface to accept new commands is required. Programs developed at this level are usually more efficient than if they are developed at the other levels.

Programming at the database level is a useful tool to adapt and extend CAD/CAM systems to accommodate new engineering, design, and manufacturing applications. A user can design a data structure that can store analysis data together with graphics and geometric data. Thus, the corresponding database stores analysis attributes besides the graphics and geometric attributes. Let us consider the simple example of cantilever beam analysis. The geometric attributes of the beam include its length and the dimensions of its cross section. The graphics attributes include the beam display data such as line font, layers, colors, etc. The analysis attributes include the beam material, the type of cross section (rectangular, circular, etc.), the load value, and the load type (concentrated, distributed, etc.).

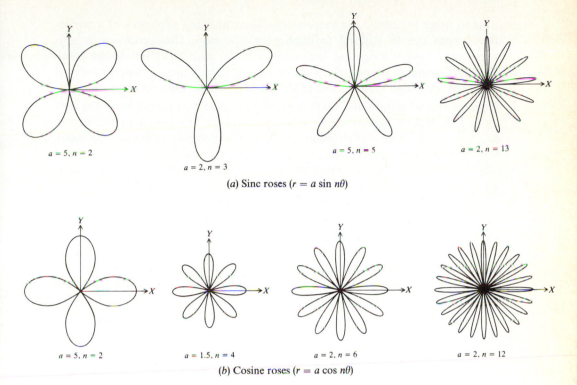

$a = 5, n = 2$

$a = 2, n = 3$

$a = 5, n = 5$

$a = 2, n = 13$

(a) Sinc roses ($r = a \sin n\theta$)

$a = 5, n = 2$

$a = 1.5, n = 4$

$a = 2, n = 6$

$a = 2, n = 12$

(b) Cosine roses ($r = a \cos n\theta$)

**FIGURE 15-8**
Samples of generated roses.

Programs developed at the database level are usually executed on a CAD/CAM system via commands implemented into the system user interface. For example, a "generate deflection" command can execute a program that performs cantilever beam analysis. Analysis modifiers can be used with the command in a similar way to the graphics and geometric modifiers. If the user interface is menu-driven, an icon called "beam analysis" can be developed and used to execute the program. In this case, the beam modifiers can be organized into menus and submenus.

The process to implement a program written at the database level of a CAD/CAM system depends greatly on the internal structure of the system database, software, and hardware. Regardless of the specific details related to such a structure, the following generic steps are required to generate a command that executes the program:

1. Generate user interface. Following the existing command syntax, the new command structure and modifiers are added to the software. The user interface that includes the new command must be loaded at the runtime otherwise the system would not be able to recognize the new command and would issue an error message. If we think of the user interface as a command table, the new command must be added to the table and the new table must be activated

prior to using the command successfully. The command processor of a particular system can be utilized to implement this step. This processor is also responsible for ensuring that the commands the user types online are legal. For every command typed, the processor scans the command table to check if the typed command matches one in the table. If not, it issues an error message.

2. Supply required input data. Data is input in the form of modifiers of the new command. Values associated with these modifiers are stored in the model database as attributes. Analysis, graphics, and/or geometric data can be input. For the beam analysis, modifiers for material type, cross-section type, and load type and value can be associated with the beam command. The system modifier processor checks the validity of the user-typed modifiers by comparing them with a stored modifier table of the command.

3. Extract data. Data already input and stored in a model database may be accessed to perform geometrical and/or analysis calculations. A get-data processor accepts input taken from graphics entities displayed by digitizing them or the input of explicit coordinates or real values to be used as input to the data processor that performs the analysis calculations. A read from a model database may be performed by this processor.

4. Perform calculations. Once all the input and extracted data are complete from steps 2 and 3 respectively, the data processor (consisting of the analysis programs) makes the calculations. The processor also writes the proper results to the model database. These results are stored in records and subrecords as analysis attributes similar to storing graphical and geometrical attributes.

5. Display results. A graphics processor takes the results from the data processor after being written to the model database and generates their corresponding graphics display. This step typically utilizes the extensive graphics library and its routines provided by the CAD/CAM system in use.

To illustrate the use of the above steps to develop database level programs, let us consider the cantilever beam example mentioned earlier in this section. A cantilever beam loaded at its free end with a point load $P$ is shown in Fig. 15-9. The MCS of the beam database is shown in the figure. The deflection curve of the beam is given by the equation

$$\delta = \frac{Px^2(x - 3L)}{6EI} \tag{15.3}$$

A main program (call it CANT) is written to accept as input the load $P$, the beam length $L$, the modulus of elasticity $E$, and the cross-section moment of inertia $I$. The program uses this input and calculates the deflection $\delta$ at various points along the beam (at different $x$ values) by substituting into Eq. (15.3). If a point on the undeflected beam is defined by the coordinates $(x, y, z)$, the corresponding point on the deflected beam has the coordinates $(x, y + \delta, z)$. The deflected points are connected by a B-spline curve to generate the deflected beam graphics, as shown in Fig. 15-9.

A new command called "generate deflection" is created and added to the command table of the CAD/CAM system in use. To supply data required by

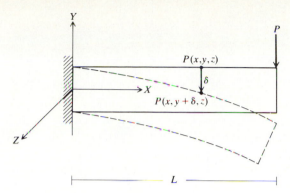

**FIGURE 15-9**
MCS for cantilever beam deflection.

steps 2, 3, and 4, the command with its modifiers and digitizes *may* take the following form:

GENERATE DEFLECTION MATERIAL STEEL LOAD 1000
TYPE 1 XSECTION RECTANGULAR $D$D$D

The modifiers are MATERIAL, LOAD, TYPE, and XSECTION. In the above case the material is steel, which provides $E$ as $30 \times 10^6$ lb/in$^2$, the load is concentrated (TYPE 1) with a value of 1000 lb, and the cross section is rectangular. Other permissible values of the modifiers can be stored in the appropriate modifier tables which are not discussed here. The three digitizes shown in the command require the user to digitize the beam length and the width and the height of the cross section. If the beam cross section is circular, the command would require only two digitizes (the beam length and the circle). When the results are complete, they are scaled properly and displayed by the graphics processor.

When the user types the above command, the command processor would invoke the routine—in this case CANT. In a sense, the user types the command (or chooses the icon in menu-driven systems), instead of typing RUN CANT, to execute the CANT routine. Results of this command are shown in Fig. 15-10. The detailed syntax and structures of the routines written to implement this command are not included here in the discussion. Refer to questions in Part 2 of the problems at the end of the chapter.

## 15.3.5   Device Level Programming

In all graphics-oriented programming and applications, results are displayed on a graphics display. The need to interface application programs with these displays is essential for the success of these programs. Programs, usually referred to as device drivers or handlers, are written to display the results on a specific display. These programs form the heart of the graphics processor discussed in the previous section. Device drivers are hardware-dependent and require a knowledge of hardware details. Drivers can be written for either input devices (lightpens, electronic pens, mice, etc.) or output devices (displays, plotters, printers, etc.). Device drivers generate device-dependent output and handle device-dependent interaction.

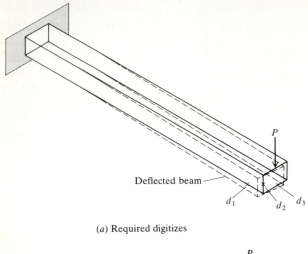

(a) Required digitizes

(b) Front view

**FIGURE 15-10**
Beam deflection and required
digitizes.

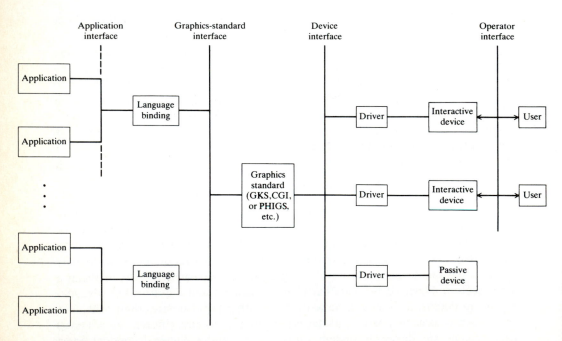

**FIGURE 15-11**
Interfaces using graphics standards.

Graphics standards are available as discussed in Chap. 3 to protect device drivers against obsolescence of hardware. Using these standards, very few hardware-dependent routines need to be changed to adapt an existing driver to a new device. These standards allow exchanging drivers among various CAD/CAM systems with minimum reprogramming effort.

Figure 15-11 shows the interface between an application program and a device using graphics standards. The figure shows that the graphics standard is situated between the application interface and the device interface. In most applications, the programmer only interfaces the application program with the standard's interface via language bindings. These bindings are subroutines stored in a library that comes with the device and is accessible by the programmer. If the device is an interactive one such as a workstation, communication with the application is achieved via the operator interface. Details of the device-dependent programming are not covered in this book. Interested readers can refer to the documentation of their respective devices.

## 15.4 OBJECT-ORIENTED PROGRAMMING

Despite the need to use very high level objects and operations in computer graphics, low level data structures such as arrays or files are often utilized to write programs and implement algorithms. One often finds a gap between sophisticated algorithms, using very high level data structures, and poor methods of program and object construction. Particular programming languages are too readily adopted for solving problems before a precise specification of the problem to be solved has been formulated.

Object-oriented programming can be defined as a technique or paradigm for writing "good" programs for a set of problems. In traditional programming, the languages that support them cannot distinguish between the general properties of any shape or object (a shape has a color, it can be drawn, etc.) and the properties of a specific shape (a circle is a shape that has a radius and center, is drawn by a circle-drawing function, etc.). Object-oriented programming has the ability to express this distinction and take advantage of it. Benefits of object-oriented programming include easier program design, as the objects correspond closely to the behavior of items being simulated or calculated; fewer program errors, as objects promote modularity and encapsulation; and easier program extension, as new kinds of objects can be added easily.

Object-oriented languages support user-defined types of data. They provide users with constructs that let them express the distinction discussed in the above paragraph. In object-oriented languages, programs are based on objects, which are record-like data structures. Each type, or "class," of object is associated with a particular set of procedure-like operations called "methods," and methods are performed when objects are invoked by "messages." Each item of data within a program is regarded as an attribute of some object and only accessed by invoking one of the methods defined for that object's class.

Object-oriented languages have been developing since the introduction of ALGOL. The first language was Simula. Other languages are Smalltalk, Alphard, CLU, C++, ADA, Loops, and APL. Object-oriented designs of software are possible in languages such as C, Pascal, Modula-2, CHILL, and even FORTRAN. "Object-oriented" has become a synonym for "good;" thus each language would attempt to support it. However, some languages would be more appropriate than others. In general, a good object-oriented language should have features that support the desired programming style in the desired application area. Specifically, it is important that:

1. All features are efficiently integrated into the language.
2. It is possible to combine features to achieve solutions that would otherwise require extra, separate, features.
3. Very few spurious and special-purpose features exist in the language.
4. Implementing a feature does not impose significant overhead on programs that do not have it.
5. A user need only know about the language subset used explicitly to write a program.

The last two principles imply that changing or modifying already-written object-oriented programs should not represent a formidable task on the user's part which would require significant changes to a program or rewriting it all over again.

While syntax of specific object-oriented languages is not covered here, an example may be helpful to understand the essence of object-oriented programs. Let us consider the analysis of beams. Considering the types of beam supports, one can identify two classes of beams: cantilever and simply supported. Let us refer to the two classes as CANT-BEAM and SIMP-BEAM respectively. Within each class, various objects may exist. Objects are considered instances of classes. For example, a cantilever beam with specific loading, geometric specifications, and material properties is an object that belongs to the class CANT-BEAM. To calculate, say, the deflection of this specific beam, we may use a statement like "CANT-BEAM deflection." In this statement, the object "CANT-BEAM" receives the message "deflection," calculates the deflection, and sends back the resulting deflection as another object that belongs to another class, call it DEF. The message "deflection" invokes the "methods" by which the beam deflection can be calculated. These "methods" account for the various types of beam loading, geometric specifications, and material properties.

From this beam example, we conclude that an object-oriented program is designed on the basis of classes, objects, methods, and messages. Thus, the design methodology and the thinking process to develop such a program are almost identical to those that users follow to analyze and solve the application to be programmed. This is the strongest advantage of object-oriented programming.

## 15.5   DEVELOPMENT OF INTERACTIVE PROGRAMS

To develop a program that performs number-crunching operations, users typically use flowcharts to organize their thinking to generate the most efficient programs. The development of interactive programs follows a slightly different approach. These programs usually deal with two types of information: graphics and analysis as discussed in Sec. 15.2. While the flowchart concept can still be used to help develop interactive programs, it must be augmented by the planning strategy that creates new geometry or manipulates existing entities. The interface between analysis data/results and graphics entities/databases is crucial to avoid unnecessary calculations and programming steps. To write an efficient interactive program that performs a given task, the user should develop the strategy to achieve the task without programming with the goal of avoiding excessive calculations, and then program the resulting strategy.

Another factor to consider in developing interactive programs is the user/ program communication. The program should be written in such a way as to minimize the amount of interaction between the program and the user. This results in speeding program execution, relieving the user from intensive communication requirements and reducing the possibilities of crashing or aborting the program during execution. User/communication interaction can be minimized using graphics aids and manipulations discussed in Chaps. 11 and 12 such as using names (tags) and geometric modifiers.

A third consideration is related to displaying results for user evaluation. Results should be displayed in a graphical form with the option that the user can verify desired graphics results. In addition, results can be displayed on various layers or in various windows for clarity purposes. Colors, contour displays, and superposition of layer displays are among the enhancement techniques that enable users to reach a decision fairly quickly.

A last and important factor is to develop "idiot-proof" programs. This characteristic is important to guard the program against any intentional or unintentional user abuse during execution. It also serves as a mechanism to ensure that the program interprets the user input properly. Consider, for example, user digitizes as input. A program may request four digitizes from the user. Let us assume that the program has a loop that checks how many digitizes the user inputs and does not let the program continue execution until four digitizes are input by the user. If the user inputs less than four digitizes or if one or more of four digitizes input by the user are not accepted as legal digitizes, the loop that checks the number of legal digitizes from the user should prompt the user to try to input four digitizes again. Refer to the program of Example 15.4 for a similar loop which checks the user input for the variable $N$.

## 15.6  APPLICATIONS OF INTERACTIVE COMPUTER PROGRAMMING

Interactive computer programming can be applied to many engineering and design applications. Some applications, especially those that require graphical and geometrical solutions, are better suited for interactive programming than others. When interactive programming is applied to these applications, the user must not think of only conventional programming tools, but should also think of the added tools that CAD/CAM centralized databases with their geometrical and graphical capabilities (discussed in Parts II, III, and IV of this book) can typically provide. It is possible to extend many of the graphics and geometrical concepts learned from CAD/CAM to solve engineering problems. The enrichment of the user's mind by these concepts is sometimes considered one of the most valuable gains of CAD/CAM.

Many engineering and design disciplines can utilize interactive programming at various levels—educational, applied, or research. In machine design, for example, various interactive programs can be written to analyze and design various elements of machine design. Typically, a program requests a set of input parameters from the user and displays the element's geometry, performance curves (deflection, stress, etc.), and/or display maximum/minimum values (stresses, temperatures, etc.). Examples include the pairing of various types of gears, the design of cam profiles, coil springs, transmission elements (belts, chains, clutches, etc.), etc.

In the area of strength of materials, beam analysis, determining principal stresses using Mohr's circle, buckling analysis, etc., can be investigated. Kinematics of various mechanisms (slider-crack, four-bar linkage, etc.) can be studied and effectively visualized via animation techniques on graphics displays.

In heat transfer, various conduction and convection problems can be analyzed via interactive programming. Temperature distribution in fins and related parametric studies can be performed. Designing heat exchangers is also possible. In fluid mechanics, the analysis and design of air foils, studies of various types of flows and simulation of stream functions, etc., can all be performed.

Vibration analysis and control of mechanical and nonmechanical systems can be investigated via interactive programming. Temporal responses of lumped mass damped/undamped systems can be evaluated, displayed, and animated. Designs of various controls can be achieved and their effect on a system response can be studied.

In conclusion, interactive computer programming is a viable tool for analysis, design, and simulation of many engineering systems and applications. The utilization of interactive programming in these applications is limited by the user's imagination.

## 15.7   DESIGN AND ENGINEERING APPLICATIONS

This section demonstrates the utilization of interactive programming in selective applications. The high level programming type discussed in Sec. 15.3.3 is used to solve the examples. To solve each example, necessary formulation is performed first to determine the suitable form of the equations to program. In all these examples, the reader may conclude that using conventional programming to solve them is not efficient. Readers are encouraged to program these examples or variants of them on their CAD/CAM systems.

**Example 15.5.** A clock manufacturer wants to make a Valentine cuckoo clock with a heart-shaped (cardioid shape) pendulum. The heart is to be hung sidewise with the tip facing left and the curves facing right. The heart should fit exactly in a rectangle of 5-inch length and 4-inch height. The shape is symmetric about a horizontal axis.

The heart must be hung on the connecting rod of the clock so that the part of the heart on the left side would reach a length twice the length on the right side.

The manufacturer calculated that the area of the shape must lie between 9 and 11 in², so that the manufacturer can adjust the thickness to obtain the desired weight. For manufacturing purposes, the manufacturer wants the shape to consist of curves and straight lines, with the curves to be not more than two different types of different radii.

Design the clock to meet the above design specifications.

*Solution.* The heart must be hung on the connecting rod of the clock at its centroid so that its weight does not exert any moment on the rod. Thus, the centroid must be located at a distance from the tip that is twice the distance from the curves. The design contraints may be listed as follows:

1. The cardioid shape should fit in a 5 × 4 in frame.
2. It must be symmetric about a horizontal axis.
3. The centroid must lie on the horizontal axis at 3.3333 in from the tip.
4. The area $A$ must be $10 \pm 1$ in².
5. Two different types of curves are allowed in the design.

Figure 15-12 shows a possible design with lines and arcs. The circle on the right has a fixed radius and center $A_C$ while the arc at the tip has a center $B$ that can move on the $Y$ axis; its radius has a value RAD which can change. The radius of the circles on the right is taken as 1 inch to meet the first design constraint listed above. Arcs at the tip are connected to circles at the right by straight lines tangent to both. Due to symmetry of the cardioid, half of its shape is created and then mirror-copied with respect to the $X$ axis to complete the shape.

In order to find the radius RAD, the tangent lines, and for the centroid $C$ to be at the desired coordinates ($H = 3.3333$) with the area requirement, a program is written using the elementary level programming to enable changing RAD and calculating $H$ and $A$. The program is based on the construction strategy shown in Fig. 15-13. The program begins by constructing the two circles C1 and C2 with centers $A_C(4, 1, 0)$ and $B(0, \text{RAD}, 0)$ respectively. A tangent line $L_1$ is constructed between the two circles which are then trimmed to produce half of the cardioid

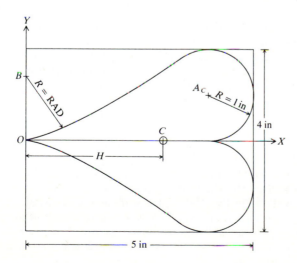

**FIGURE 15-12**
Possible design of the cardioid clock.

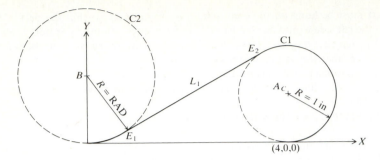

**FIGURE 15-13**
Construction of the cardioid clock.

shape. The trimmed parts of the circles are shown dashed. The trimming boundaries of C1 are the end $E_2$ of the tangent $L_1$ and point (4, 0, 0). Circle C2 is trimmed between the end $E_1$ of $L_1$ and point (0, 0, 0). A "mirror" command with the copy modifier is used to complete the cardioid shape. A command to calculate the centroid and the area of the shape is invoked. The program listing is shown below and assumes that the user has completed the part setup.

```
CREATE CIRCLE TAG C1 RADIUS 1 CENTER X4Y1Z0 $CR
$P PLEASE INPUT THE RADIUS OF THE CIRCLE AT THE TIP OF THE CAR-
    DIOID $P
CREATE CIRCLE TAG C2 RADIUS X0Y$SZ0 X0Y0Z0 $CR
CREATE LINE TAG L1 TANGENT TO TAG C1 TAG C2 $CR
MIRROR COPY TAG C1 TAG C2 TAG L1 WITH RESPECT TO X0Y0Z0,
    X1Y0Z0$CR
$P PLEASE DIGITIZE THE BOUNDARY OF THE CARDIOID SHAPE IN
    ORDER $P
CALCULATE AREA $D $CR
$P PLEASE INPUT LAYER NUMBER FOR THE NEXT DESIGN $P
SELECT LAYER $S $CR
```

The layer command is added so that various designs are stored in various layers for display and management purposes. The line command that creates the tangent $L_1$ typically requires additional modifiers (such as near modifier) to choose the proper tangent (four tangent possibilities usually exist). If the program does not use tags and asks the user to digitize C1 and C2 instead, the CAD/CAM system usually creates the tangent closest to the digitizes. The command that calculates the area provides the location of the centroid as well.

To find the solution, the above program is executed several times, each with a different value of RAD, starting from 1 to 4 in in increments of 0.5 in. The values of $H$ and $A$ for each RAD are tabulated in Table 15.1. The cardioids made out of the different values of RAD are shown in Fig. 15-14. To find the value of RAD that would produce a value of 3.3333 in for $H$, the $H$ values in Table 15.1 are plotted versus the RAD values in a graph. The resulting points are connected with a B-spline curve. The curve is shown in Fig. 15-15. Each point on the curve has coordinates (RAD, $H$) for the $x$ and $y$ coordinates respectively. Due to the use of the part centralized database, the points are scaled properly and the user does not have to spend any additional effort to scale the results. Note that the $X$ and $Y$ axes shown

**TABLE 15.1**
**Centroid and area**

| RAD | H | Area |
|-----|-----|------|
| 1.0 | 3.2182936 | 11.1415929 |
| 1.5 | 3.2606937 | 10.879692 |
| 2.0 | 3.3043787 | 10.6036205 |
| 2.5 | 3.3494915 | 10.3100204 |
| 3.0 | 3.3962168 | 9.9935846 |
| 3.5 | 3.4448239 | 9.6444244 |
| 4.0 | 3.495990 | 9.2321634 |

in Fig. 15-15 are line entities with arrowheads, points at the tec marks, and text as values and labels.

Working backwards on the graph, a horizontal line is created with an arbitrary length and passing through point (0, 3.3333, 0). The $y$ coordinate is the desired location ($H$ value) of the centroid. At the point where this line intersects the B-spline curve, a vertical line is constructed. The $x$ coordinate of the point of intersection between the vertical line and the $X$ axis is the solution for the RAD value. Using the "verify entity" command, the coordinates of the point of intersection are obtained. The $x$ coordinate is found to be 2.322857 in.

A final heart is created, but this time the value of RAD equals the above $x$ coordinate. Thus, the desired parameters of the heart are:

Radius of circles on the right = 1 inch

Radius of circles on the left RAD = 2.322857 inch

Area of the cardioid = 10.41632 in$^2$

Centroid of the cardioid at $x = 3.3333$, $y = 0.0$, $z = 0.0$

The final heart design is shown in Fig. 15-16.

**Example 15.6.** The thin-walled pressure vessel shown in Fig. 15-17 is made of steel and has the shape of a paraboloid. The equation of the generating parabola is $x^2 = cy$, where $x$ and $y$ are in inches and $c$ is a constant. The existing design of the vessel has a thickness $t$ and a constant $c$ of 0.25 and 4 in respectively.

The vessel is closed by a thick flat plate on its top. The internal pressure in the vessel is set to 250 lb/in$^2$ gage. The axial and hoop stresses in the shell at a point 16 in above the bottom due to the internal pressure have been calculated and found to be $\sigma_a = 4120$ lb/in$^2$ and $\sigma_h = 8000$ lb/in$^2$. It is observed from field service data that the high value of $\sigma_h$ causes frequent failure of the vessel. After careful analysis of the failure data, it is decided that the vessel must be redesigned such that $\sigma_a \leq 4000$ and $\sigma_h \leq 4000$ at the location 16 in above the bottom. In addition, the volume of the shell must stay unchanged for cost and weight purposes. However, the paraboloid volume can change.

Find the thickness $t$ and the constant $c$ of the new vessel that meet the above-stated design specifications and constraints.

*Solution.* To formulate this problem correctly, we express $\sigma_a$, $\sigma_h$, and the shell volume $V$ in terms of $t$ and $c$. Let us consider the free-body diagram of the vessel at

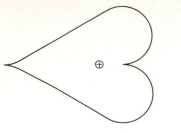

RAD = 1.0, $H$ = 3.22, $A$ = 11.14

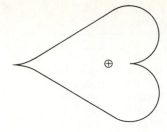

RAD = 1.5, $H$ = 3.26, $A$ = 10.88

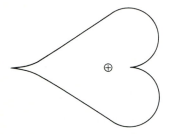

RAD = 2.0, $H$ = 3.30, $A$ = 10.60

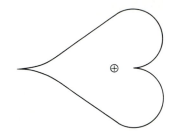

RAD = 2.5, $H$ = 3.35, $A$ = 10.31

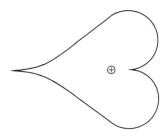

RAD = 3.0, $H$ = 3.40, $A$ = 9.99

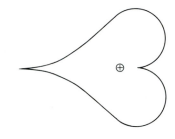

RAD = 3.5, $H$ = 3.44, $A$ = 9.64

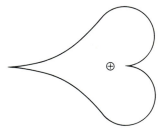

RAD = 4.0, $H$ = 3.50, $A$ = 9.23

**FIGURE 15-14**
Various cardioid clock designs.

a cross section at a distance $y$ from the $X$ axis, as shown in Fig. 15-18. Considering the force equilibrium in the $Y$ direction, we can write

$$\int dP = \int dF \cos \alpha \qquad (15.4)$$

or

$$P\pi x^2 = \sigma_a(2\pi x t \cos \alpha) \qquad (15.5)$$

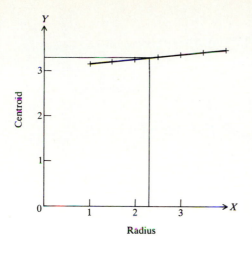

**FIGURE 15-15**
Centroid location $H$ versus radius RAD.

where $P$ is the internal pressure. This equation gives

$$\sigma_a = \frac{Px}{2t \cos \alpha} \tag{15.6}$$

Observing that $\tan \alpha = dx/dy$ and utilizing the parabola equation, Eq. (15.6) becomes

$$\sigma_a = \frac{P\sqrt{cy(1 + c/4y)}}{2t} \tag{15.7}$$

The axial and hoop stresses are related by the equation:

$$\frac{P}{t} = \frac{\sigma_a}{r_1} + \frac{\sigma_h}{r_2} \tag{15.8}$$

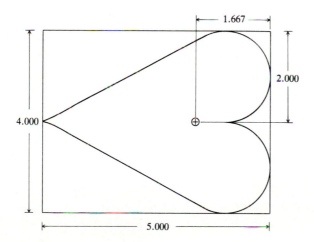

Centroid at $H = 3.33333$ in from tip
Area $= 10.4163188$ in$^2$

**FIGURE 15-16**
Final pendulum design.

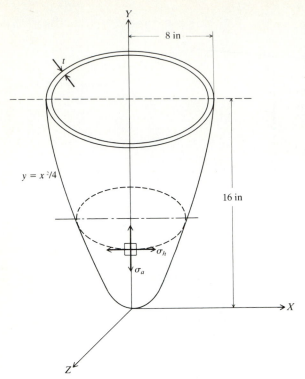

**FIGURE 15-17**
Thin-walled pressure vessel.

where $r_1$ is the radius of curvature of the shell in the $XY$ plane and $r_2$ is the radius of curvature of the shell in the perpendicular plane ($XZ$ plane). They are given by

$$r_1 = \frac{[1 + (dy/dx)^2]^{1.5}}{d^2y/dx^2} \tag{15.9}$$

$$r_2 = \frac{\cos\alpha}{\sqrt{cy}} = \frac{1}{\sqrt{cy(1 + c/4y)}} \tag{15.10}$$

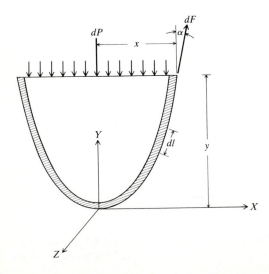

**FIGURE 15-18**
Free-body diagram of vessel.

Substituting Eqs. (15.9) and (15.10) into (15.8) and rearranging we obtain

$$t = \frac{250}{2\sigma_a/c[1 + 4(16)/c]^{1.5} + \sigma_h/\sqrt{c(16 + c/4)}}$$ (15.11)

The volume of the shell is given by

$$V = At$$ (15.12)

where $A$ is the shell surface area. If we assume an incremental shell length $dl$ (see Fig. 15-18), we can write

$$dA = 2\pi x \, dl$$ (15.13)

Hence

$$A = 2\pi \int_0^{12} x \sqrt{1 + \left(\frac{dy}{dx}\right)^2} \, dx$$ (15.14)

This equation can be integrated by substitution to give

$$A = \frac{\pi}{6} c^2 \left\{ \left[ 1 + \frac{4(16)}{c} \right]^{1.5} - 1 \right\}$$ (15.15)

Substituting Eq. (15.15) into (15.12) and rearranging we obtain

$$t = \frac{V}{(\pi/6)c^2 \{[1 + 4(16)/c]^{1.5} - 1\}}$$ (15.16)

For the existing design, the volume $V$ is calculated to be 144.7 in³. Then Eq. (15.16) becomes

$$t = \frac{144.7}{(\pi/6)c^2 \{[1 + 4(16)/c]^{1.5} - 1\}}$$ (15.17)

We now have the three equations (15.7), (15.11), and (15.17) that any new design of the vessel, that is, $t$ and $c$ values, must satisfy. Assuming that $\sigma_a$ and $\sigma_h$ are known, then we have three equations in two unknowns. We can use them to solve for $t$ and $c$ utilizing the inequality constraints on $\sigma_a$ and $\sigma_h$ as follows. We can substitute the highest values of $\sigma_a$ and $\sigma_h$ (4000 lb/in²) into Eq. (15.11) and solve it with Eq. (15.17) for $t$ and $c$. Equations (15.11) and (15.17) can be rewritten respectively as

$$t = f(c) \qquad \text{and} \qquad t = g(c)$$ (15.18)

If the functions $f(c)$ and $g(c)$ are plotted for various values of $c$, their intersection point is the solution.

To plot these functions, various values of $c$ are assumed and the corresponding $t$ values are calculated using Eqs. (15.11) and (15.17). The points on each function are plotted with $(c, t)$ values as their $(x, y, 0)$ coordinates. Each set that results from one equation are connected with a B-spline curve. The coordinates of the intersection point of the two spline curves, when verified, are the $c$ and $t$ values that meet the design specifications. Figure 15-19 shows the $c$–$t$ graph. The solution is $c = 2.1$ inch and $t = 0.36$ inch.

We now must substitute the new values for $c$ and $t$ into Eq. (15.7), the one we have not used so far, and check if the resulting $\sigma_a$ satisfies the inequality $\sigma_a \leq 4000$. These values produce a normal stress $\sigma_a = 2045.45$ lb/in² which is acceptable. We

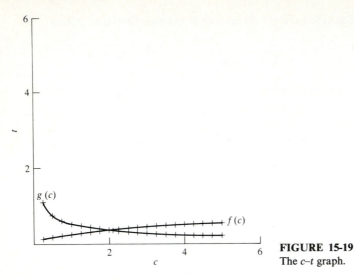

**FIGURE 15-19**
The $c$–$t$ graph.

now must recalculate $\sigma_h$ using Eq. (15.8) and ensure that it satisfies the inequality $\sigma_h \leq 40,000$. Equations (15.9) and (15.10) give $r_1 = 185.4$ and $r_2 = 5.891$ respectively. Substituting these values into Eq. (15.8) together with $\sigma_a = 2045.45$, $P = 250$, and $t = 0.36$, we obtain $\sigma_h = 4025.98$ lb/in$^2$, which is 0.6 percent higher than 4000. Therefore $c = 2.1$ inch and $t = 0.36$ inch are the desired redesign values. Figures 15-20 to 15-22 show the surface models of the old and new vessels.

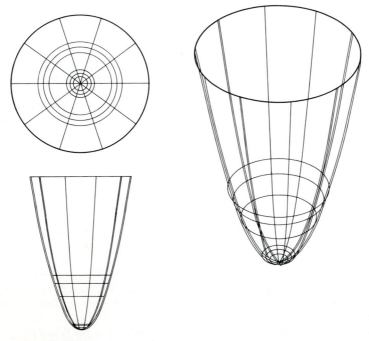

**FIGURE 15-20**
Original design of pressure vessel.

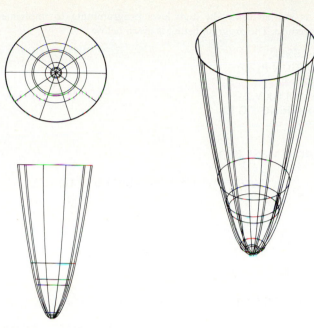

**FIGURE 15-21**
Final design of pressure vessel.

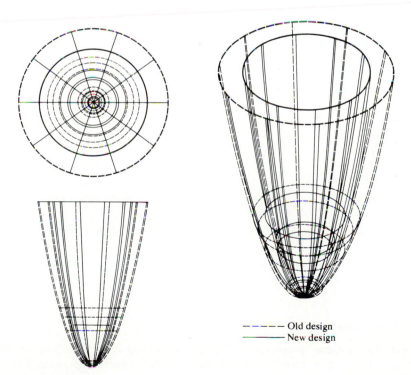

– – – – – Old design
———— New design

**FIGURE 15-22**
Superposition of original and final designs of pressure vessel.

A program is written using the high level programming to implement the above solution procedure. The program listing is given below:

```
PROCEDURE VESSEL
DECLARE LOCATION P(20)
DECLARE ENTITY S
DECLARE REAL X,I,DX,Y,Z
DECLARE TEXTSTRING F,G,P1
X = 0
DX = 0.25
FOR I = 1,20, INCREMENT BY 1
   X = X + DX
   Y = (144.7*6)/(3.14159*(X**2)*(((1 + (64/X))**1.5) − 1))
   Z = 0
   P(I) = (X,Y,Z)
END                                                    {FOR loop}
S = BSPLINE CURVE TAG G P
DISPLAY S
SELECT LAYER 100
X = 0
FOR I = 1,20, INCREMENT 1
   X = X + DX
   Y = 250.0/((8000/X)/((1 + (64 + X))**1.5) + (4000/(X*(16 + (X/4)))**0.5))
   Z = 0
   P(I) = (X,Y,Z)
END                                                    {FOR loop}
S = BSPLINE CURVE TAG F P
DISPLAY S
S = POINT TAG P1 INTOF TAG F TAG G
VERIFY ENTITY TAG P1
END PROGRAM
```

## PROBLEMS

### Part 1: Theory

**15.1.** What are the advantages of interactive programming over conventional programming?

**15.2.** Why are the requirements of interactive programming different from conventional programming?

**15.3.** Refer to the bibliography at the end of the chapter to find more about UIM and SUIM. What are the differences between the two types of user interfaces?

### Part 2: Laboratory

**15.4.** Find the available programming languages and capabilities offered by your CAD/CAM system. Identify them in terms of the classification covered in this chapter.

**15.5.** Study and learn the syntax of each language.

## Part 3: Programming

Use the appropriate interactive programming level (refer to Sec. 15.3) to solve the following problems:

**15.6.** The bar $AB$ shown in Fig. P15-6 moves with its ends in contact with the horizontal and vertical walls. Find the locus of point $G$ on the bar. If point $G$ becomes the center of $AB$, what will its locus be? Prove your answer.

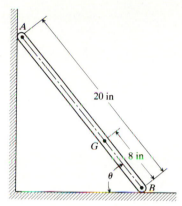

FIGURE P15-6

**15.7.** The cross section shown in Fig. P15-7 must be redesigned under the following conditions:

Design goal: Increase the area of the cross section by 30 percent.
Design constraints:
1. The centroid should not change location relative to the shown $XYZ$ coordinate system.
2. The ratio $t/L$ must satisfy the inequality $1/3 \leq t/L \leq 1/2$.

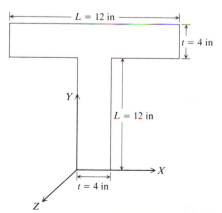

FIGURE P15-7

**15.8.** The area shown in Fig. P15-8 represents a cross section. It has to be redesigned to meet certain strength requirements. The final design should meet the following specifications:

Design goal: The $y$ coordinate of the centroid of the cross section should be zero.

Design constraints:
1. Only $\alpha$ and/or $b$ can change.
2. $A$ must be a minimum.

Find the values of $\alpha$, $b$, or both for the final design.

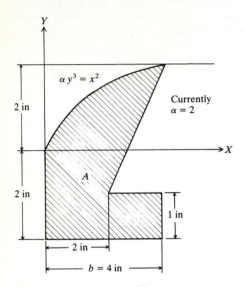

**FIGURE P15-8**

**15.9.** Write programs to create the quadratic surfaces shown in Prob. 6.17 of Chap 6.

**15.10.** Write programs to create the following surfaces:

$$z = \frac{x^2}{A^2} + \frac{y^2}{B^2}$$

$$z = \frac{y^2 - xy + 1}{x + 1.5}$$

$$z = y^2 - x$$

$$z^2 - x^2 - y^2 = K$$

$$r^2 = 2A^2 \cos 2\theta$$

$$r = 1 + \cos \theta$$

$$z = xy\left(\frac{x^2 - y^2}{x^2 + y^2}\right)$$

**15.11.** Write programs to create the following two curves:

$$x = 2a \cos \theta + a \cos 2\theta$$

$$y = 2a \sin \theta - a \sin 2\theta$$

and

$$x = \tfrac{1}{2}a(3 \cos \theta - \cos 3\theta)$$

$$y = \tfrac{1}{2}a(3 \sin \theta - \sin 3\theta)$$

**15.12.** Write a program that can perform cantilever beam analysis.

**15.13.** Write a program that animates the mechanism shown in Fig. P15-13.

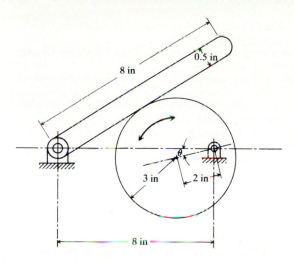

FIGURE P15-13

**15.14.** Write a program to study the kinematics of a slider-crank mechanism.

**15.15.** Write a program to study the kinematics of the quick-return mechanism shown in Fig. P15-15. Assume the lengths of the rods $A$, $B$, $C$, and $E$ to be 1.5, 3, 2, and 3.5 inches respectively. In addition, assume the distance $D$ to be 1.75 inches and the thickness of any rod to be 0.25 inches.

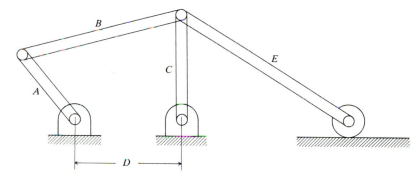

FIGURE P15-15

**15.16.** Write a program to study projectile motion.

**15.17.** Write a program to perform buckling analysis of beams.

**15.18.** Write a program to study and animate the damped vibration of a two-degree-of-freedom system.

**15.19.** Write a program that can generate the teeth of a spur gear.

**15.20.** Write a program that constructs Mohr's circle.

**15.21.** Write a program that calculates the temperature distribution in a fin.

**15.22.** Write a program that generates profiles of different cams.

# BIBLIOGRAPHY

Barris, W. C., and D. R. Riley: "Programmer-Friendly Graphics Libraries," *CIME*, vol. 5, no. 1, pp. 21–26, 1986.

Conner, M. S.: "Device-Independent Interfaces Enhance Graphics Compatibility," *EDN*, pp. 166–172, July 9, 1987.

Dufourd, J. F.: "Construction of Interactive Programs in Computer Graphics," *Computer Graphics Forum*, vol. 7, no. 3, pp. 161–176, 1988.

Goldberg, A., and D. Robson: *Smalltalk80: The Language and Its Implementation*, Addison-Wesley, Reading, Mass., 1983.

Green, M.: "The University of Alberta User Interface Management System," *Computer Graphics, SIGGRAPH '85 Conf. Proc.*, vol. 19, no. 3, pp. 205–213, 1985.

Haeberli, P. E.: "ConMan: A Visual Programming Language for Interactive Graphics," *Computer Graphics*, vol. 22, no. 4, pp. 103–111, 1988.

Jacky, J. P., and I. J. Kalet: "An Object-Oriented Programming Discipline for Standard Pascal," *Commun. ACM*, vol. 30, no. 9, pp. 772–776, 1987.

Lyons, T. G. L.: "The Public Tool Interface in Software Engineering Environments," *Software Engng J.*, vol. 1, no. 6, pp. 254–258, 1986.

Olsen, D. R., E. P. Dempsey, and R. Rogge: "Input/Output Linkage in a User Interface Management System," *Computer Graphics, SIGGRAPH '85 Conf. Proc.*, vol. 19, no. 3, pp. 191–197, 1985.

Sibert, J. L., W. D. Hurley, and T. W. Bleser: "An Object-Oriented User Interface Management System," *Computer Graphics, SIGGRAPH '86 Conf. Proc.*, vol. 20, no. 4, pp. 259–268, 1986.

Stroustrup, B.: "What Is Object-Oriented Programming?" *IEEE Software*, vol. 5, no. 3, pp. 10–20, 1988.

Valliere, D.: "Using CAD Macros and Languages for Productivity on a Unigraphics System," *CAD J.*, vol. 18, no. 8, pp. 250–252, 1986.

Zeid, I., and T. Bardasz: "The Role of Turnkey CAD/CAM Systems in the Development of the 'GRAPHYSIS' Concept," *Proc. ASME 1985 Int. Computers in Engineering Conf.*, pp. 1–10, 1985.

Zygmont, A.: "Object-Oriented Programming and CACSD," *CoED*, vol. VIII, no. 3, pp. 7–11, 1988.

# PART
# V

## DESIGN
## APPLICATIONS

# CHAPTER
# 16

# MECHANICAL TOLERANCING

## 16.1 INTRODUCTION

The design of a product includes many factors besides determining the loads, stresses, and material selection. Before construction or manufacturing can begin, design engineers specify the shape of the product in terms of an assembly and detailed drawings to convey all the necessary information to the shop floor to manufacture the product. Drawings are usually checked to ensure that dimensioning is done in a manner that is most convenient and understable to production engineers. It is essential that a drawing should be made in such a way that it has one and only one interpretation. No involved calculations should be made by shop personnel to calculate missing dimensions before the production machines can be set up.

Drawing annotations, specifically dimensions and tolerances, are among other important information that drawings contain. The annotations must be chosen and disposed of in such a way that neither over- nor underdimensioned drawings result. Various available dimensioning schemes (such as linear—standard, baseline, ordinate, or isometric dimensions—angular, and circular dimensions) must meet this requirement. The proper dimensioning of a drawing is a separate subject and is not covered here.[1]

Manufacturing parts of exactly equal dimensions is known to be impossible from practical experience. A variety of physical limitations on manufacturing processes (such as cutting conditions, hardware accuracy, software accuracy, skills of machine operators, etc.) and assembly processes, as well as material

---

[1] See F. E. Giesecke, A. Mitchell, H. C. Spencer, I. L. Hill, and J. T. Dygdon, *Technical Drawing*, 8th ed., Macmillan, 1986, and T. E. French, C. J. Vierck, and R. J. Foster, *Graphics Science and Design*, 4th ed., McGraw-Hill, New York, 1984.

properties, contribute to limiting the precision with which we can manufacture parts. To account for this variability of dimensions (due to manufacturing) at the design phase, we assign a tolerance or a range of acceptable values to each suitable (not every dimension requires a tolerance) dimension of the part. If a part size and shape are not within the maximum and minimum limits defined by the part tolerances, the part is not acceptable. The assignment of actual values to the tolerance limits has a major influence on the overall cost and quality of an assembly or a product. If the tolerances are too small (tight), the individual parts will cost more to make. If the tolerances are too large (loose), an unacceptable percentage of assemblies may be scrapped (rejected) or require rework.

In addition to manufacturing cost considerations, tolerances are usually specified to meet functional requirements of assemblies. In order for mating features (faces) of mating parts to fit together and operate properly, each part must be manufactured within these tolerance limits. For example, sliding parts such as journals and pistons must be made so that they are capable of moving relative to other parts but without so much freedom that they will not function properly. On the other hand, keys, gears on shafts, and other similar members mounted by press or shrink-fit are toleranced so that the desired interference is maintained without being so large as to make the assembly impossible or the resulting stresses too high.

Tolerancing is an essential element of mass production and interchangeable manufacturing, by which parts can be made in widely separated locations and then brought together for assembly. For example, an automobile manufacturer subcontracts the manufacturing of many parts of a design to other companies. Tolerancing makes it also possible for spare parts to replace broken or worn ones in existing assemblies successfully. In essence, without interchangeable manufacturing, modern industry could not exist, and without effective size control by the engineer, interchangeable manufacturing could not be achieved.

Tolerancing information is essential for part process planning, assembly operations, part inspection, and for other design and production activities. Design engineers need tolerance analysis to distribute allowances among related design dimensions, to check design results, or to design assemblies. Production engineers need tolerance analysis to transform design coordinates into manufacturing coordinates and to perform tolerance calculations and distributions in process planning.

Automatic tolerance analysis and good tolerance software are important ingredients of CAD/CAM systems. They are important to change the current belief among many design engineers that production engineers should produce whatever they have designed and specified. Design engineers usually do not give serious consideration to production costs and feasibility. However, often there is a natural reluctance to change a proven design to reduce manufacturing cost. Therefore, the availability of good tolerance software is important to determine tolerances and their optimal values of a product, especially when manual tolerance analysis is often tedious, cumbersome, and error-prone. This is always the case when the number of related dimensions to be analyzed is large.

This chapter presents relationships between tolerance and manufacturing; existing tolerance concepts and theories; tolerance representations in CAD/CAM systems; and methods of tolerance synthesis, analysis, distribution, and integration. Some examples and applications are covered to illustrate the chapter material.

## 16.2 TOLERANCE CONCEPTS

This section introduces the reader to the basic concepts, standards, and common practice of tolerancing of mechanical parts. These issues are essential to the understanding, representation, and analysis of tolerances. We begin by introducing the ANSI (American National Standards Institute) definitions of some related terms.

"Nominal size" is the designation used for the purpose of general identification. It is usually expressed in common fractions. For example, a pipe designated as a $3\frac{1}{2}$-inch diameter pipe may have an actual diameter of $3\frac{1}{2}$ inches (in some designations the diameter may be different from $3\frac{1}{2}$, for example, 3.625). "Basic size" is the theoretical size from which limits of size are derived by the application of allowances and tolerances. The basic size is the decimal equivalence of the nominal size. The number of decimal places determines the precision or accuracy required. For example, if the nominal width of a slot is $1\frac{3}{4}$ in and we require accuracy to three decimal places, its basic size becomes 1.750 in. "Actual size" is the measured size of the finished part. "Tolerance" is the total amount by which a dimension may vary. It is used to determine the permissible limits (maximum and minimum) of the dimensions. Tolerance can be expressed in either of two ways. A "bilateral tolerance" is specified as plus or minus deviation from the basic size, for example, $1.750 \pm 0.002$ in. A "unilateral tolerance" is a tolerance in which variation is permitted only in one direction from the basic size, for example, $1.750^{+0.004}_{-0.000}$ or $1.750^{-0.003}_{-0.004}$. "Allowance" is the difference between the maximum material limits of mating parts. It is the minimum clearance (positive allowance) or maximum interference (negative allowance) between mating parts. It is also known as the tightest fit between mating parts.

The foregoing definitions are usually applied by design engineers (designers) to determine the proper dimensions of mating features so that parts can be manufactured and can be interchangeable. Let us consider, for example, a shaft transmitting a load through a gear mounted on it. After the designer has performed the stress calculations, a nominal size of the shaft diameter is obtained, say 2 inches. This is also the nominal size of the hole in the gear as shown in Fig. 16-1a. Assuming that allowance and tolerances are specified to four decimal places, the precision required is then the same and the basic size is expressed as 2.0000 as shown in Fig. 16-1b. From the functional and assembly requirements between the gear and the shaft (power is transmitted between the two via a key), a clearance fit between the two parts is adequate. Let us assume an allowance $a$ of 0.0030 in, a tolerance $h$ for the hole in the gear of value 0.0030 in, and a shaft tolerance $s$ of 0.0030 in.

In order to determine the diameters of the gear hole and the shaft, given the above values for $a$, $h$, and $s$, let us consider all the possible variations in the dimensions of these diameters relative to the basic size. Ten possibilities exist.

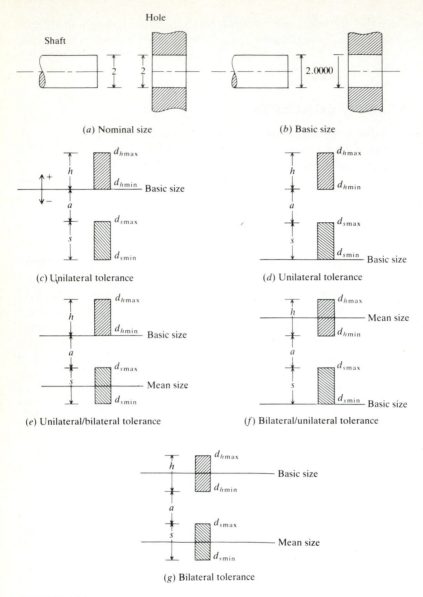

**FIGURE 16-1**
Clearance fit calculations.

Five of them are shown as bar diagrams in Fig. 16-1c to g. The other five possibilities are complements (reverse the location of the bar adjacent to the "basic size" datum) of those shown in the figure. The hatched bars are the tolerance zones. The relative locations of these zones with respect to the basic size and the mean size determine the type of tolerance, and consequently the hole and shaft diameters. To determine the maximum and minimum diameters for the shaft ($d_{s\max}$ and $d_{s\min}$) and the hole ($d_{h\max}$ and $d_{h\min}$), a tolerance zone ($s$ or $h$) is located relative to the basic size, followed by the allowance $a$ and the other tolerance zone. From the definition of allowance (minimum clearance), $h$, $s$, and $a$ do not overlap in the case of clearance fit. In Fig. 16-1e, the hole has unilateral tolerance while the shaft has a bilateral tolerance. Figure 16-1f shows the opposite case.

The calculations of the toleranced dimensions from the bar diagrams are simple. For the five possibilities shown in Fig. 16-1c to $g$, the hole dimensions are respectively $2.0000^{+0.0030}_{-0.0000}$, $2.0060^{+0.0030}_{-0.0000}$, $2.0000^{+0.0030}_{-0.0000}$, $2.0060^{+0.0015}_{-0.0015}$, and $2.0000^{+0.0015}_{-0.0015}$. The shaft dimensions are respectively $1.9970^{+0.0000}_{-0.0030}$, $2.0000^{+0.0030}_{-0.0000}$, $1.9955^{+0.0015}_{-0.0015}$, $2.000^{+0.0030}_{-0.0000}$, and $1.9955^{+0.0015}_{-0.0015}$. Using these dimensions for both the shaft and the hole, the reader can easily verify that the minimum clearance is equal to the allowance $a$. The other five cases that are not shown in Fig. 16-1 are obtained from the cases shown in the figure by flipping the tolerance zone adjacent to the basic-size datum to the other side of the datum; except for the case shown in Fig. 16-1g, the basic-size and mean-size datums are interchanged. The toleranced dimensions of the hole for the complements of possibilities of Fig. 16-1c to $g$ are respectively $2.0000^{+0.0000}_{-0.0030}$, $2.0030^{+0.0030}_{-0.0000}$, $2.0000^{+0.0000}_{-0.0030}$, $2.0045^{+0.0015}_{-0.0015}$, and $2.0045^{+0.0030}_{-0.0000}$, and the toleranced dimensions of the shaft are respectively $1.9955^{+0.0000}_{-0.0030}$, $2.0000^{+0.0000}_{-0.0030}$, $1.9925^{+0.0015}_{-0.0015}$, $2.0000^{+0.0000}_{-0.0030}$, and $2.0000^{+0.0015}_{-0.0015}$.

Of the ten possibilities of the toleranced dimensions of both the shaft and the hole, what is the best possibility? Practice shows that drawings made with unilateral tolerances are usually easier to check than those made with bilateral tolerances. Thus, the six possibilities that result from Fig. 16-1e to $g$ and their complements can be eliminated. In addition, it is usually easier to machine shafts to any desired size. This eliminates the possibility shown in Fig. 16-1d and its complement. The complement possibility of Fig. 16-1c is also eliminated as it is practically easier for the machinist or workman to aim at minimum hole diameter equal to basic size (basic size is usually a rounded number such as 2.000, 2.500, etc.) instead of a diameter equal to basic size minus $h$. The clearance fit for the possibility of Fig. 16-1c is shown in Fig. 16-2.

In some situations, the bilateral method of tolerancing is very appropriate. Examples include the location of holes when the variation from the basic size is equally critical in both directions, welded assemblies, and loosely toleranced

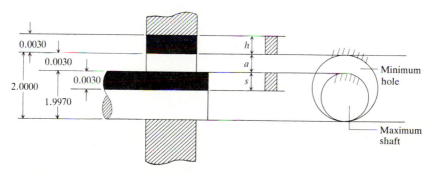

**FIGURE 16-2**
Clearance fit of Fig. 16-1c.

dimensions (for large tolerances, it is sometimes more convenient to give the mean dimension and the variation each way).

Figures 16-3 and 16-4 show how to apply the clearance fit calculations shown in Figs. 16-1 and 16-2 to an interference fit. The key difference is that $h$, $a$, and $s$ overlap, based on the definition of allowance $a$ (maximum interference) in the case of interference fit. In this interference example we use $h = 0.0008$ in, $s = 0.0008$ in, and $a = 0.0020$ in. For the five possibilities shown in Fig. 16-3c to g,

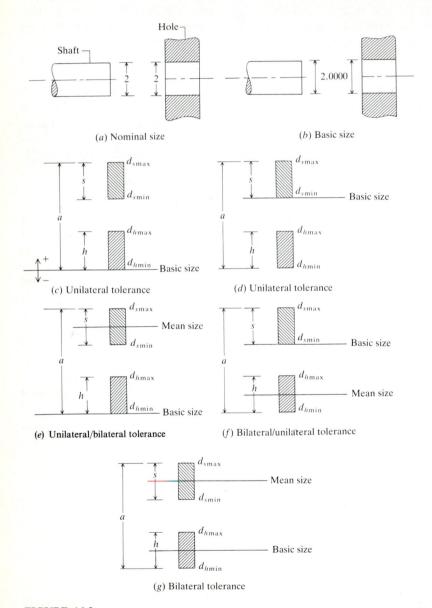

(a) Nominal size    (b) Basic size

(c) Unilateral tolerance    (d) Unilateral tolerance

(e) Unilateral/bilateral tolerance    (f) Bilateral/unilateral tolerance

(g) Bilateral tolerance

**FIGURE 16-3**
Interference fit calculations.

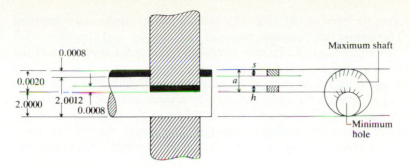

**FIGURE 16-4**
Interference fit of Fig. 16-3c.

the hole dimensions are respectively $2.0000^{+0.0008}_{-0.0000}$, $1.9996^{+0.0000}_{-0.0008}$, $2.0000^{+0.0008}_{-0.0000}$, $1.9992^{+0.0004}_{-0.0004}$, and $2.0000^{+0.0004}_{-0.0004}$. The shaft dimensions are respectively $2.0012^{+0.0008}_{-0.0000}$, $2.0000^{+0.0008}_{-0.0000}$, $2.0016^{+0.0004}_{-0.0004}$, $2.0000^{+0.0008}_{-0.0000}$, and $2.0012^{+0.0004}_{-0.0004}$. Using these dimensions for both the shaft and the hole, the reader can easily verify that the maximum interference is equal to the allowance $a$. The preferred interference fit of Fig. 16-3c is shown in Fig. 16-4.

### 16.2.1 Limits and Fits

Fits between mating parts signify the range of tightness or looseness that may result from the application of a specific combination of allowances and tolerances in mating parts. In the previous section, the limits (the maximum and minimum diameters) of the shaft and the hole are determined on the basis of the fit type (clearance or interference), given an allowance $a$, a hole tolerance $h$, and a shaft tolerance $s$. As seen, the type of fit is crucial in interpreting the allowance $a$ (minimum clearance or maximum interference), which in turn determines the limits. The differences between the hole and the shaft limits determine the range of tightness or looseness.

Fits between mating parts can be identified as cylindrical or location fits. Cylindrical fits apply when an internal member fits in an external member as a shaft in a hole. Location fits are intended to determine only the locations of mating parts. Both cylindrical and location fits can be divided into three types: clearance fits, interference fits, and transition fits. In a clearance fit, one part is always loose relative to the other, e.g., a shaft is loose in a hole or two stationary parts can be freely assembled or disassembled. In an interference fit, one part is forced tight into the other during assembly and an internal pressure between the two results. Interference fits which can transmit torques or forces between mating parts (e.g., a shaft and a pulley) are usually referred to as force fits. Location interference fits are used when accuracy of location is important and for parts requiring alignment. A transition fit is a fit that may result in either a clearance or interference condition.

When utilizing a given fit to calculate the limits and toleranced dimensions of mating parts, the design engineer must assign the basic size found from design calculations to either the shaft or the hole, as seen in the previous section. The choice determines the basis or the system for calculating the limits and toleranced dimensions. There exist two systems: the basic hole system and the basic shaft system. Both systems assume unilateral tolerances *h* and *s* for the hole and the shaft. In the basic hole system, the minimum hole is taken as the basic size, and the allowance and tolerances are applied accordingly. Figure 16-1*c* shows an example of using the basic hole system to apply tolerances and calculate limits. In the basic shaft system, the maximum shaft is taken as the basic size, and the allowance and tolerances are applied accordingly. The complement of Fig. 16-1*d* would show an example of using the basic shaft system to apply tolerances and calculate limits.

The basic hole system is the most widely used and recommended system in practice. This is due to manufacturing considerations. It is usually easier to machine shafts to any desired size than holes. Holes are often produced by using standard reamers, broaches, and other standard tools; standard plug gages are used to check the actual sizes. The basic shaft system should not be used unless there is a reason for it. For example, it is advantageous when several parts having different fits, but one basic size, are mounted on a single shaft. Typically, the textile industry uses the basic shaft system.

Having calculated the limits of hole and shaft sizes for a given fit, how can we show them on an engineering drawing? There are two types of dimensioning a drawing: one is based on the maximum material condition (MMC) and the other is based on the least material condition (LMC). The MMC is defined as the condition where the maximum amount of material is contained. The allowance *a* for clearance and interference fits has been defined to maintain the MMC. The LMC is the opposite of the MMC. It is the condition where the least amount of material is contained. Figure 16-5 shows how the MMC and LMC are used to show limits or dimensions for the clearance fit shown in Fig. 16-2 and the interference fit shown in Fig. 16-4.

Figure 16-5 shows that the dimensions of the maximum shaft and minimum hole represent the MMC and consequently the tightest fit, while the dimensions of the minimum shaft and maximum hole represent the LMC and, therefore, the loosest fit. Thus, the MMC produces the least possibility of assembly because it produces the most dangerous condition (tightest fit). However, once assembled, the resulting assembly best meets its functional requirement. The LMC, on the other hand, produces the best possibility of assembly because it produces the less dangerous condition (loosest fit). However, once assembled, the resulting assembly least meets its functional requirement. The MMC is usually preferred over the LMC.

Dimensioning based on MMC has another advantage. During manufacturing, the machinist or worker aims at the principal dimension, which is the one shown above the dimensioning line, that is, maximum shaft or minimum hole in the MMC as shown in Fig. 16-5*a*. Should the machinist, through error, produce an oversized hole or an undersized shaft, the parts might still be acceptable providing the dimensions are within the limits specified by the drawing. In effect, the MMC reduces the amount of scrap which is a valuable economic gain.

In order to achieve interchangeable manufacturing, calculations of fits, tolerances, and limits have to be standardized. Starting with the functional require-

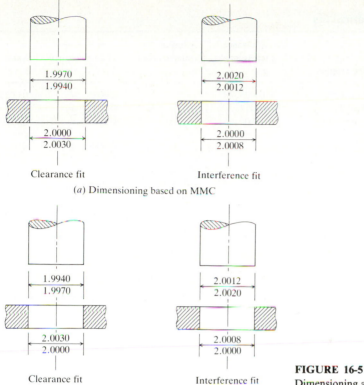

1.9970
1.9940

2.0020
2.0012

2.0000
2.0030

2.0000
2.0008

Clearance fit

Interference fit

(a) Dimensioning based on MMC

1.9940
1.9970

2.0012
2.0020

2.0030
2.0000

2.0008
2.0000

Clearance fit

Interference fit

(b) Dimensioning based on LMC

**FIGURE 16-5**
Dimensioning according to MMC and LMC.

ments of the mating parts, the designer can choose a suitable fit. Allowances and tolerance zones and limits can then be evaluated, and the dimensioning limits are calculated. ANSI has established eight classes (types) of cylindrical fit that specify the amount of allowance $a$, the hole tolerance $h$, and the shaft tolerance $s$ as functions of the basic size (diameter) $d$ (see Table 16.1). With a basic size (diameter) $d$ and a desired fit, the designer can use Table 16.1 to calculate $a$, $h$, and $s$, which in turn can be used to calculate the toleranced dimensions and limits as discussed in the previous section and shown in Figs. 16-1 to 16-5.

Given $a$, $h$, and $s$ and assuming unilateral tolerances and using the basic hole system, the limits of a cylindrical fit can be written as (see the previous section and Figs. 16-1 to 16-5)

$$d_{hmin} = d \tag{16.1}$$

$$d_{hmax} = d + h \tag{16.2}$$

$$d_{smax} = d - a \tag{16.3}$$

$$d_{smin} = d - a - s \tag{16.4}$$

Equations (16.1) to (16.4) apply for clearance, transition, or interference fit. In Eqs. (16.3) and (16.4), the allowance $a$ is an algebraic value; it is positive for

**TABLE 16.1**
**ANSI classification of cylindrical fits**

| Class of fit | Type | Description | Allowance (clearance) $a$ | Allowance (interference) $a$ | Hole tolerance $h$ | Shaft tolerance $s$ |
|---|---|---|---|---|---|---|
| 1 | Clearance | Loose fit | $0.0025\sqrt[3]{d^2}$ | | $0.0025\sqrt[3]{d}$† | $0.0025\sqrt[3]{d}$ |
| 2 | fit | Free fit | $0.0014\sqrt[3]{d^2}$ | | $0.0013\sqrt[3]{d}$ | $0.0013\sqrt[3]{d}$ |
| 3 | | Medium fit | $0.0009\sqrt[3]{d^2}$ | | $0.0008\sqrt[3]{d}$ | $0.0008\sqrt[3]{d}$ |
| 4 | Transition | Snug fit | 0 | | $0.0006\sqrt[3]{d}$ | $0.0004\sqrt[3]{d}$ |
| 5 | fit | Wringing fit | | 0 | $0.0006\sqrt[3]{d}$ | $0.0004\sqrt[3]{d}$ |
| 6 | Interference | Tight fit | | $0.00025d$ | $0.0006\sqrt[3]{d}$ | $0.0006\sqrt[3]{d}$ |
| 7 | fit | Medium force fit | | $0.0005d$ | $0.0006\sqrt[3]{d}$ | $0.0006\sqrt[3]{d}$ |
| 8 | | Heavy force (shrink) fit | | $0.001d$ | $0.0006\sqrt[3]{d}$ | $0.0006\sqrt[3]{d}$ |

† $d$ = basic size (diameter).

clearance fits and negative for interference fits. The tolerances $h$ and $s$ are always positive values. Subtracting Eq. (16.3) from (16.1) gives

$$a = d_{hmin} - d_{smax} \qquad (16.5)$$

This equation agrees with the definition of allowance $a$ introduced in the previous section. It also shows that $a$ is positive for clearance fits and negative for interference fits.

While Table 16.1 and Eqs. (16.1) to (16.4) can be used to calculate hole and shaft sizes, it is useful in practice to transform this information into tables that designers can look up to determine limits and toleranced dimensions. ANSI and ISO (International Standards Organization) systems exist and are equivalent.

ANSI, to incorporate the ISO system, rearranged Table 16.1 and extended it to include location fits. As a result, there exist five standard classes of fits (with their ISO equivalences) with several grades (a grade of a fit is determined by the permissible variation in its relative looseness or tightness; it depends on the quality of machining or workmanship) under each class. The five classes are running or sliding fits RC, location fits (location clearance fits LC, location transition fits LT, and location interference fits LN), and force or shrink fits FN. The classes, grades, and their equivalent ISO symbols are shown in Table 16.2.

In ANSI standards, maximum and minimum allowance, hole tolerance limits (0, $h$), and shaft tolerance limits ($a$, $a + s$) are calculated (using equations similar to those in Table 16.1) for a practical range of basic sizes. The results are available in tables (see Table 16.3 for a sample).[2] If the tolerance limits are applied algebraically to the basic size, the limits of a given dimension (see Fig. 16-5a) that produce a given fit can be obtained. The results listed in these tables can be

---

[2] See A. D. Deutschman, W. J. Michels, and C. E. Wilson, *Machine Design: Theory and Practice*, Macmillan, New York, 1975.

**TABLE 16.2**
**ANSI fits and their ISO equivalents**

| Class | Grade | ISO symbol Hole | Shaft | Class | Grade | ISO symbol Hole | Shaft |
|-------|-------|------|-------|-------|-------|------|-------|
| Running | RC1 | H5 | g4 | Location | LT1 | H7 | js6 |
| or | RC2 | H6 | g5 | transition | LT2 | H8 | js7 |
| sliding | RC3 | H7 | f6 | fits | LT3 | H7 | k6 |
| fits | RC4 | H8 | f7 | LT | LT4 | H8 | k7 |
| RC | RC5 | H8 | e7 | | LT5 | H7 | n6 |
| | RC6 | H9 | e8 | | LT6 | H7 | n7 |
| | RC7 | H9 | d8 | | | | |
| | RC8 | H10 | C9 | Location | LN1 | H6 | n5 |
| | RC9 | H11 | C11 | interference | LN2 | H7 | p6 |
| | | | | fits LN | LN3 | H7 | r6 |
| Location | LC1 | H6 | h5 | | | | |
| clearance | LC2 | H7 | h6 | Force or | FN1 | H6 | n6 |
| fits | LC3 | H8 | h7 | shrink | FN2 | H7 | s6 |
| LC | LC4 | H10 | h9 | fits | FN3 | H7 | t6 |
| | LC5 | H7 | g6 | FN | FN4 | H7 | u6 |
| | LC6 | H9 | f8 | | FN5 | H8 | x7 |
| | LC7 | H10 | e9 | | | | |
| | LC8 | H10 | d9 | | | | |
| | LC9 | H11 | C10 | | | | |
| | LC10 | H12 | C12 | | | | |
| | LC11 | H13 | C13 | | | | |

related back to $a$, $h$, and $s$ as follows. The lowest number in the clearance column (or the highest number in the interference column) is the allowance $a$. The largest tolerance limit (the other is always zero due to use of basic hole system) in the hole column is $h$. The shaft tolerance $s$ can be obtained from the column that shows its tolerance limits using the following equation:

$$s = |L_{min} + a| \tag{16.6}$$

where $L_{min}$ is the algebraic minimum limit and $a$ is the algebraic allowance (positive for clearance and negative for interference); $|\cdot|$ denotes the absolute value.

ISO symbols can also be utilized to determine tolerances and limits as follows. The capital letter H is used to describe the classes of holes and small letters are used to describe the classes of shafts. The grade within a class is described by a number called the tolerance grade number or IT number. For example, H7/f6 (see Table 16.2) represents a clearance fit where the hole has class H and IT number 7, and the shaft has class f and IT number 6. For each shaft class, there is an amount of tolerance called the fundamental deviation $\delta_F$, and for each IT number, there is another amount of tolerance, call it $\Delta d$. There are tables[3] that list $\delta_F$ and $\Delta d$ for various classes and IT numbers, for a practical

---

[3] See J. E. Shigley and C. R. Mischke, *Mechanical Engineering Design*, 5th ed., McGraw-Hill, New York, 1989.

range of basic sizes. The allowance $a$, the hole tolerance zone $h$, and the shaft tolerance zone $s$ can be related to $\delta_F$ and $\Delta d$ as follows:

$$h = \Delta d_h \tag{16.7}$$

$$s = \Delta d_s \tag{16.8}$$

$$a = \begin{cases} \delta_F & \text{clearance} \\ \delta_F + \Delta d_s & \text{interference} \end{cases} \tag{16.9}$$

where the subscripts $h$ and $s$ denote the hole and shaft respectively.

**Example 16.1.** Calculate the tolerance zones $h$ and $s$, the allowance $a$, and the limits for two fits of classes RC2 and FN4 if the basic size is 3.0000 inches.

*Solution.* Table 16.2 shows that RC2 and FN4 are clearance and interference fits respectively. They also have the ISO symbols of H6/g5 and H7/u6. The clearance fit has tolerance grades (IT numbers) of 6 and 5 for the hole and shaft respectively, and a shaft class g. Using the proper ANSI tables (Table 16.3 shows the information pertinent to this example and found in ANSI and ISO tables), we obtain $a = 0.0004$ in, $h = 0.0007$ in, and $s = |-0.0009 + 0.0004| = 0.0005$ in. The tables give tolerance limits of (0.0007, 0) for the hole and $(-0.0004, -0.0009)$ for the shaft. These limits

**TABLE 16.3**
**Extracted information for Example 16.1**

**ANSI information**

| Basic size range, in | | RC2 | | | FN4 | | |
|---|---|---|---|---|---|---|---|
| Over | To | Limit clearance | Hole H6 | Shaft g5 | Limit interference | Hole H7 | Shaft u6 |
| 2.56 | 3.15 | 0.0004 0.0016 | +0.0007 0.0000 | −0.0004 −0.0009 | 0.0028 0.0047 | 0.0012 0.0000 | +0.0047 +0.0040 |

**ISO information**

| Basic size range, in | | Tolerance grade (IT number) | | |
|---|---|---|---|---|
| Over | To | IT5 | IT6 | IT7 |
| 2.00 | 3.20 | 0.0005 | 0.0007 | 0.0012 |

| Basic size range, in | | Fundamental deviation ($\delta_F$) | |
|---|---|---|---|
| Over | To | g | u |
| 2.60 | 3.20 | −0.0004 | 0.0040 |

give $d_{h\max} = 3.0007$ in, $d_{h\min} = 3.0000$ in, $d_{s\max} = 2.9996$ in, and $d_{s\min} = 2.9991$ in. These values agree with Eqs. (16.1) to (16.4).

Using the proper ISO tables gives $\Delta d_h = 0.0007$ in for IT number 6, $\Delta d_s = 0.0005$ in for IT number 5, and $\delta_F = -0.0004$ in for class g. Using Eqs. (16.7) to (16.9) gives $h = 0.0007$ in, $s = 0.0005$ in, and $a = 0.0004$ in, which agrees with ANSI calculations.

For the interference fit FN4, similar calculations can be performed. Using the proper ANSI tables, we obtain $a = -0.0047$ in, $h = 0.0012$ in, and $s = |0.004 - 0.0047| = 0.0007$ in. The tables gives tolerance limits of (0.0012, 0) for the hole and (0.0047, 0.0040) for the shaft. These limits give $d_{h\max} = 3.0012$ in, $d_{h\min} = 3.0000$ in, $d_{s\max} = 3.0047$ in, and $d_{s\min} = 3.0040$ in. These values agree with Eqs. (16.1) to (16.4).

Using the proper ISO tables gives $\Delta d_h = 0.0012$ in for IT number 7, $\Delta d_s = 0.0007$ in for IT number 6, and $\delta_F = 0.0040$ in for class u. Using Eqs. (16.7) to (16.9) gives $h = 0.0012$ in, $s = 0.0007$ in, and $a = 0.0047$ in, which agrees with ANSI calculations.

## 16.2.2  Tolerance Accumulation

Placing toleranced dimensions on an engineering drawing is not a random process and requires careful consideration of the effect of one tolerance on another. Whenever more than one tolerance in a given direction (e.g., horizontal or vertical) affect the location of a given surface of a part, tolerances are cumulative. Noncumulative tolerances and dimensioning are always preferred because cumulative ones may result in a higher percentage of rejection (scrap) during inspection. Accumulation (stack-up) of tolerances usually occurs statistically.

Figure 16-6 shows three possibilities of dimensioning the same drawing. Figure 16-6a shows each dimension in the horizontal direction with its associated tolerance. This practice of dimensioning should be discouraged and is not recommended. It is confusing to the machinist that will produce the part. The machinist cannot produce the four toleranced dimensions; only any three can be controlled and the fourth is an outcome. Therefore, the machinist must make a decision on which three dimensions to aim for. If the left-out dimension is crucial to the function of the part, the rejection rate of the part during inspection will most likely be high.

Figure 16-6b shows a possible solution to the above problem of over-tolerancing (or superfluous dimensioning). Here, only the three necessary dimensions are given with their associated tolerances. The overall length is just a reference and, therefore, shown without tolerance as well as being marked REF. With this solution the designer assumes that the overall length is not important to the functional requirements (assembly) of the part.

While Fig. 16-6b may seem an acceptable solution, it has two disadvantages. First, tolerances are cumulative. For example, the middle dimension is controlled by two tolerances and any of the side dimensions is controlled by three tolerances. Second, three surfaces ($A$, $B$, and $C$) are reference surfaces (datums) from which dimensions are measured. Thus, the machinist must carefully machine these surfaces even if the machining quality of these surfaces is immaterial to the part functions. This may increase the machining cost of the part.

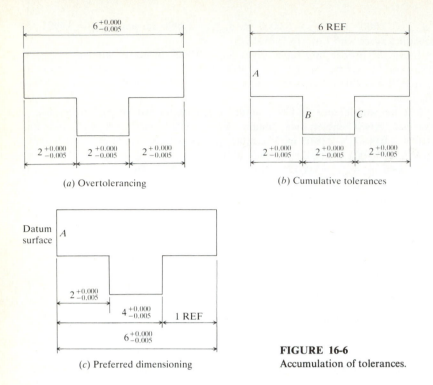

(a) Overtolerancing

(b) Cumulative tolerances

(c) Preferred dimensioning

**FIGURE 16-6**
Accumulation of tolerances.

Figure 16-6c shows the preferred solution of dimensioning the part. Here only one reference surface ($A$) is used as a datum and only three toleranced dimensions are shown. Therefore, as a general rule, it is best to dimension surfaces so that each surface is affected by only one tolerance. This can be achieved by referring all dimensions to a single datum surface, as shown in Fig. 16-6c.

### 16.2.3 Tolerance/Cost Relationship

Tolerance is a key factor in determining the cost of a part. The relationship between tolerances and manufacturing cost is shown in Fig. 16-7. The manufacturing (total) cost is divided into machining and scrap cost. The machining cost is the cost of first producing the part. This cost consists of labor, overhead, gages, tools, jigs and fixtures, inspection, etc. The scrap cost is the cost encountered due to rejecting some parts that fall outside the specified tolerance range, and/or due to repairing some of these parts. The cost of parts which are produced by multiple manufacturing processes (Fig. 16-7b) is the sum of costs encountered in each process. An example of multiple processes is a cylinder that may need rough turning, finish turning, and grinding.

As expected, and shown in Fig. 16-7, the tighter the tolerance, the more expensive it is to manufacture a part. This trend provides the fundamental rule in selecting tolerances by designers at the design phase; that is, tolerances should be chosen as large as possible as long as they meet the functional and assembly requirements of the part. It may be worth while to change designs to relax tolerance requirements for cost purposes. Larger tolerances result in using less skilled

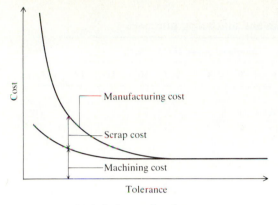

Manufacturing cost

Scrap cost

Machining cost

Tolerance

(*a*) A single manufacturing process

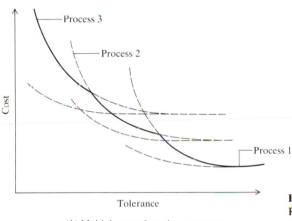

Process 3

Process 2

Process 1

Tolerance

(*b*) Multiple manufacturing processes

**FIGURE 16-7**
Relationship between tolerance and manufacturing cost.

machinists (less expensive labor), lower inspection costs, and reduced scrapping of material.

It is important for designers to be aware of manufacturing accuracies attainable by various manufacturing processes. To assist designers in relating tolerances to machining processes, Table 16.4, developed by ANSI, relates tolerance grades (IT number) to machining processes. The range of grades for each process accounts for the conditions (old, new, well maintained, etc.) of the machine, and the level of skills possessed by the machine operator. Having selected a tolerance based on functional requirements of a part, the designer can determine the tolerance grade for a given size using tolerance grade tables (see Table 16.3 as an example). With the tolerance grade known, the designer can determine the proper machining process by using Table 16.4. With the knowledge presented in Table 16.4 and the tolerance grade table, the designer can effectively judge the manufacturing method to produce the design.

It should be noted here that choosing tolerances is often difficult, even with the aid of available tables and Table 16.4. Not only is the accuracy of a given

**TABLE 16.4**
**Relationship between tolerance grades and machining processes**

| Machining process | Tolerance grade (IT number) | | | | | | | | | |
|---|---|---|---|---|---|---|---|---|---|---|
| | 4 | 5 | 6 | 7 | 8 | 9 | 10 | 11 | 12 | 13 |
| Lapping and honing | × | × | | | | | | | | |
| Cylindrical grinding | | × | × | × | | | | | | |
| Surface grinding | | × | × | × | × | | | | | |
| Diamond turning | | × | × | × | | | | | | |
| Diamond boring | | × | × | × | | | | | | |
| Broaching | | × | × | × | × | | | | | |
| Powder metal—sizes | | | × | × | × | | | | | |
| Reaming | | | × | × | × | × | × | | | |
| Turning | | | | × | × | × | × | × | × | × |
| Powder metal—sintered | | | | × | × | × | × | | | |
| Boring | | | | | × | × | × | × | × | × |
| Milling | | | | | | | × | × | × | × |
| Planing and Shaping | | | | | | | × | × | × | × |
| Drilling | | | | | | | × | × | × | × |
| Punching | | | | | | | × | × | × | × |
| Die casting | | | | | | | | × | × | × |

machine tool difficult to determine but also the relationship between accuracy and cost is not exactly known. Skilled operators produce smaller tolerances using the same machine as unskilled operators. Tolerance/cost relationships are usually more deterministic for accepted manufacturing methods and for batch products.

### 16.2.4  Surface Quality

Tolerance specifications are imposed on dimensions to ensure functional and assembly requirements of mating parts. Tolerances determine, to a large extent, the manufacturing processes required to produce the parts. Surface quality is another important factor that affects the performance of mating parts relative to each other, as well as the choice of the manufacturing processes. Tolerances and surface quality are interrelated in the sense that they are both a direct outcome of manufacturing processes. A manufacturing process, such as lapping and honing, that produces small tolerances (small tolerance grades) also produces smooth surfaces. Therefore, in specifying tolerances, a designer should consider the requirements of surface finish in addition to functional and assembly requirements. For example, an interference fit made on rough surfaces may have a reduced contact area which results in a subsequent reduction of the interference force between the mating parts. Higher surface quality results in higher production costs. Thus, designers would normally leave a surface as rough as is feasible.

Surface finish can be evaluated quantitatively by using various measures. The most popular measures are surface roughness and waviness. Figure 16-8a shows that a close look at a surface texture shows that it is not absolutely smooth. Instead, it consists of fine irregularities (Fig. 16-8b) superposed on larger

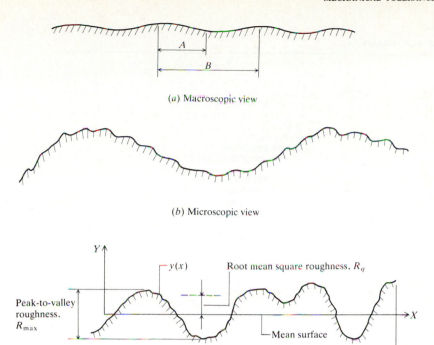

(a) Macroscopic view

(b) Microscopic view

(c) Roughness measures

**FIGURE 16-8**
Surface roughness.

or wavelike variations. The measure of the irregularities over a sampling length $A$ (Fig. 16-8a) is defined as the surface roughness, whereas the measure of the large variations over a wavelength $B$ defines the waviness of the surface.

There are three methods to calculate the surface roughness $R$ of a surface. Let us define an imaginary mean surface (Fig. 16-8c) such that the total variations (measured by the sum of the areas between the mean surface and the profile of the actual surface) above the mean surface are equal to the total variations beneath it. The roughness average $R_a$ measures the average of the absolute displacement (variation) relative to the mean surface:

$$R_a = \frac{1}{L} \int_0^L |y| \, dx \qquad (16.10)$$

where $|y|$ is the absolute value of the roughness function $y(x)$. The roughness average $R_a$ is also known as the arithmetic average (AA). It is usually measured using a planimeter to calculate the area above and below the mean surface.

$R_a$ values are usually expressed as micrometers ($\mu$m) or microinches ($\mu$ in). The value of $R_a$ can vary quite considerably without affecting the surface function. Table 16.5 gives the ANSI ranges of $R_a$ produced by common manufacturing processes. The ISO roughness code is also shown in the table.

**TABLE 16.5**
**ANSI surface roughness**

| | Roughness average, $R_a$ | | | | | | | | | | | |
| --- | --- | --- | --- | --- | --- | --- | --- | --- | --- | --- | --- | --- |
| Micrometer ($\mu$m) | 50 | 25 | 12.5 | 6.3 | 3.2 | 1.6 | 0.8 | 0.4 | 0.2 | 0.1 | 0.05 | 0.025 | 0.012 |
| Microinches ($\mu$ in) | 2000 | 1000 | 500 | 250 | 125 | 63 | 32 | 16 | 8 | 4 | 2 | 1 | 0.5 |
| ISO roughness code (N) | N12 | N11 | N10 | N9 | N8 | N7 | N6 | N5 | N4 | N3 | N2 | N1 | |

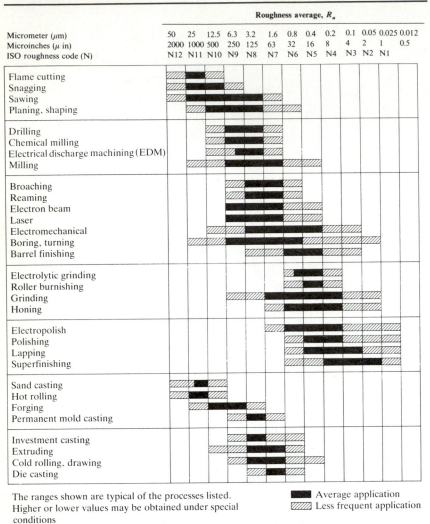

Flame cutting
Snagging
Sawing
Planing, shaping

Drilling
Chemical milling
Electrical discharge machining (EDM)
Milling

Broaching
Reaming
Electron beam
Laser
Electromechanical
Boring, turning
Barrel finishing

Electrolytic grinding
Roller burnishing
Grinding
Honing

Electropolish
Polishing
Lapping
Superfinishing

Sand casting
Hot rolling
Forging
Permanent mold casting

Investment casting
Extruding
Cold rolling, drawing
Die casting

The ranges shown are typical of the processes listed.
Higher or lower values may be obtained under special
conditions

▬ Average application
▨ Less frequent application

Another measure of surface roughness is given by the rms (root mean square) value $R_q$ (Fig. 16-8c) given by

$$R_q^2 = \frac{1}{L} \int_0^L y^2 \, dx \tag{16.11}$$

The rms method is still an averaging method.

The third measure of roughness is given by the maximum peak-to-valley height $R_{max}$ (Fig. 16-8c). Sometimes $R_{max}$ is evaluated at various locations over the length of the surface and an average value is calculated.

Various methods exist to determine the wavelength and consequently the waviness of a surface. These methods concentrate on counting the frequency rather than the amptitude of the surface profile. Some of these methods use concepts from Fourier analysis to determine average wavelengths. These methods are not covered here.

## 16.3   GEOMETRIC TOLERANCING

The tolerance concepts covered in the previous section are referred to as traditional, conventional, or coordinate tolerancing methods. They typically control the variability of linear dimensions that describe location, size, and angle as shown in Fig. 16-9. Conventional tolerancing is also known as tolerancing of perfect form because variability of shape is not addressed. For example, the tolerances $\Delta C$ and $\Delta E$ shown in Fig. 16-9 do not convey any information about the perpendicularity of the horizontal and vertical planes shown in the figure. Conventional tolerancing methods have three main shortcomings. First, they are incapable of controlling all aspects of the shape of a part. In addition to the control of location, size, and angle, control is also needed for the form (shape) of features, such as straightness, flatness, parallelism, or angularity of specific portions of the part.

Second, conventional tolerancing does not use the concept of datum: an important concept to manufacture and inspect the part. It does not explicitly specify datums or their precedence. Datums are usually implied from the way the part drawings are dimensioned. For example, Fig. 16-9 implies that the bottom horizontal and the left vertical planes are used as datums. However, which plane is more important than the other (precedence of datums) cannot be determined. In some cases, implied datums are too ambiguous to identify easily.

Third, extending the conventional tolerancing methods to control locations (i.e., two- or three-dimensional control) described in rectangular coordinates or angular dimensions introduces awkward situations. Consider, for example, the

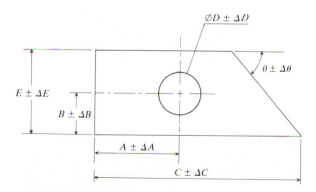

$\Delta A$ and $\Delta B$ control location
$\Delta C$, $\Delta D$, and $\Delta E$ control size
$\Delta\theta$ controls angle

**FIGURE 16-9**
Control of location, size, and angle
via conventional tolerancing.

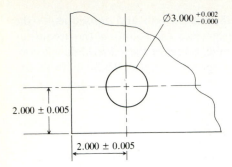

$\varnothing 3.000 \, {}^{+0.002}_{-0.000}$

$2.000 \pm 0.005$

$2.000 \pm 0.005$

(a) Toleranced dimensions of the center

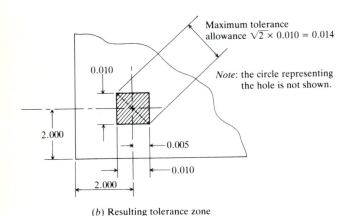

Maximum tolerance
allowance $\sqrt{2} \times 0.010 = 0.014$

0.010

*Note*: the circle representing
the hole is not shown.

2.000

0.005

0.010

2.000

(b) Resulting tolerance zone

**FIGURE 16-10**
Positional control of the
center of a hole.

control of the location of the center of the hole shown in Fig. 16-10. Specifying a tolerance of $\pm 0.005$ in on the coordinates of the center (2.000, 2.000) results in the $0.010 \times 0.010$ in square tolerance zone shown in Fig. 16-10b. While the designer might think that he/she is controlling the location of the center of the 2.000-in hole within a 0.010-in boundary, the center could actually vary across the diagonal of the square tolerance zone, yielding a maximum tolerance of 0.014 in instead of 0.010 in. Figure 16-11 shows similar results for other tolerance zones. Thus, conventional tolerancing gives more freedom in the diagonal directions and may unnecessarily constrain errors in the horizontal (or radial) and vertical (angular) directions.

Determining tolerance zones of locations, say of holes, that are interrelated are more complex due to the combined effects of individual tolerance zones. This discrepancy indicates that a better tolerancing system to control locations is needed. In the new system, it seems natural to force the tolerance zone of the hole center to become circular instead of square, rectangular, or polar. This new tolerancing system is known as geometric tolerancing. It provides a more functional and better-controlled location, size, angle, and form of parts.

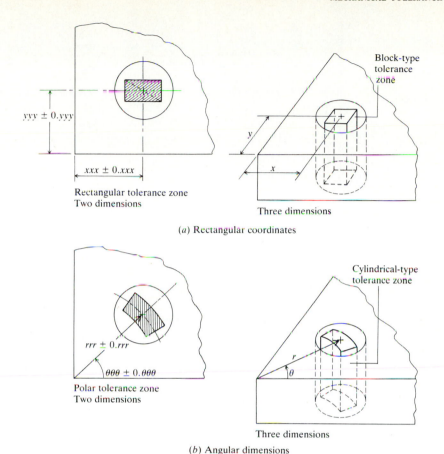

$yyy \pm 0.yyy$

$xxx \pm 0.xxx$

Rectangular tolerance zone
Two dimensions

Block-type
tolerance
zone

Three dimensions

(a) Rectangular coordinates

$rrr \pm 0.rrr$

$\theta\theta\theta \pm 0.\theta\theta\theta$

Polar tolerance zone
Two dimensions

Cylindrical-type
tolerance zone

Three dimensions

(b) Angular dimensions

**FIGURE 16-11**
Influence of coordinate systems on positional control of the center of a hole.

### 16.3.1 Datums

Geometry of a geometric model (potential design) of a part is usually described in three-dimensional euclidean space relative to an *XYZ* coordinate system. The *X*, *Y*, and *Z* axes are mutually perpendicular and form the principal directions along which all dimensions of the model are measured. The manufacture of such a model requires mapping the (*x*, *y*, *z*) coordinates and dimensions from the engineering drawing to the workpiece used to produce the part. This mapping should enable accurate measurements on the workpiece for manufacturing and inspection purposes.

Mapping is achieved via using the concept of datums. A datum is a "theoretical plane" which acts as a master reference to locate features (surfaces) of a part during manufacturing and inspection. A datum may be a point, a line, or a plane. A datum plane may be created along each of the three (*X*, *Y*, and *Z*) axes, as shown in Fig. 16-12. The three mutually orthogonal datum planes form what is sometimes called the datum reference system (other names are the datum

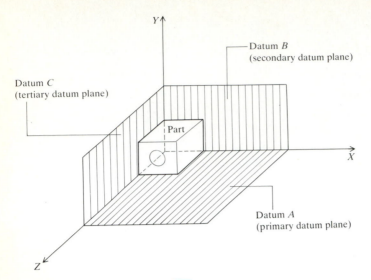

**FIGURE 16-12**
Datum reference system
using planar datums.

reference frame or master datum system). The most important plane to part measurements is termed datum $A$ (the horizontal plane in Fig. 16-12) and is known as the primary datum plane. The second most important is datum $B$ (known as the secondary plane or datum), and the least important is datum $C$ (the tertiary plane or datum). The designer usually decides the priority of importance of the three datum planes based on the functional and assembly requirements of the part.

A distinction must be made between the datum plane and the actual feature or surface of the part in contact with the plane. Such a feature or surface is known as the datum feature or surface. For example, the bottom surface of the part shown in Fig. 16-12 which is in contact with the primary datum plane is called the primary datum feature or the primary datum surface. Similarly, the rear and left surfaces of the part are called the secondary datum feature and the tertiary datum feature respectively.

While the datum features are actual physical surfaces of a part, the datum planes represent surfaces of machine tables, fixtures, inspection devices, and gages. From this point of view, we can explain the physical interpretation and implication of datum priority described above as follows. In reference to Fig. 16-12, the part is pushed onto datum plane $A$ until contact is established. Then, the part is pushed onto datum plane $B$ while in contact with datum $A$ (i.e., slide it on $A$) until contact with $B$ is achieved. Finally, contact with datum $C$ is established in a similar way while maintaining the contact with $A$ and $B$ simultaneously. This order of establishing contact with the datums defines datum priority during inspection.

Figure 16-12 shows datum planes and datum features as perfect planes. In practice, these planes are not perfect and have surface irregularities because they are manufactured. If we assume that datum planes (surfaces of inspection devices) are manufactured with much higher precision than datum features (actual surfaces of parts), one can assume a perfect datum plane and an irregular datum

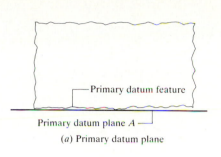

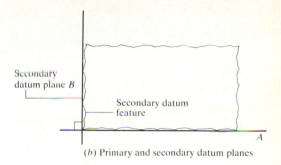

Primary datum feature

Primary datum plane $A$

(a) Primary datum plane

Secondary datum plane $B$

Secondary datum feature

$A$

(b) Primary and secondary datum planes

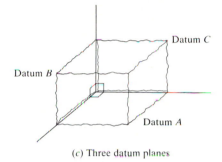

Datum $C$

Datum $B$

Datum $A$

(c) Three datum planes

**FIGURE 16-13**

Contact between datum planes and datum features.

feature as shown in Fig. 16-13. The primary datum plane $A$ contacts the actual primary datum feature of the part at three or more noncollinear points, the secondary datum plane $B$ contacts the secondary datum feature of the part at two or more distinct points not on the same normal to $A$, and the tertiary datum plane contacts the tertiary datum feature of the part at one or more points. It is clear now that the order or precedence of the datums is important, since the actual datum feature may sit differently against the datum plane depending on the number of points that are required to make contact.

In defining the datum reference system above, the contact points between the datum features and the datum planes are not precisely located. It is assumed that the surface finish of the parts is so good that the precise contact points are unimportant. For less regular surfaces such as those of castings and forgings, spot and line datums are required for precise datum definition. Functionally, a primary datum feature might be replaced by a three-point contact or by a contact of one point and one line, as shown in Fig. 16-14. These replacements are known as datum targets. Figure 16-14b shows how the primary datum $A$ is specified using datum targets. The letter $A$ in the circular symbols refers to the datum name. The digit on the right (2 in the figure) specifies the number of targets used to define the datum while the number on the left (1 or 2 in the figure) reflects the sequence of the corresponding target.

Figure 16-15 shows the interpretation of datum targets in terms of gages and placement procedure for inspection. Datum $A$ is the primary datum and is defined by two target points (two points are sufficient for two dimensions while three points are needed for three dimensions). Datum $B$ is the secondary datum, thus requiring one target point. Figures 16-15b and c show how the part is placed

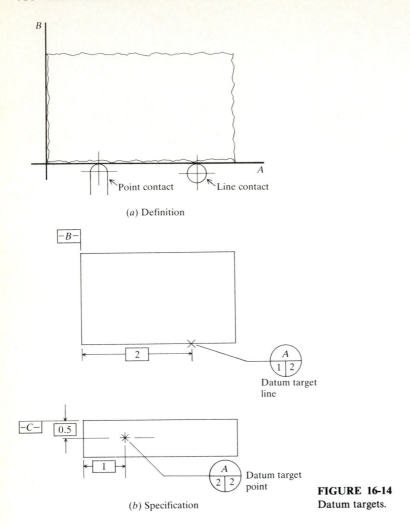

(a) Definition

Point contact

Line contact

Datum target line

Datum target point

(b) Specification

**FIGURE 16-14**
Datum targets.

and inspected. The part is pushed onto the gage in such a way that surface $F_1$ is in contact first with the target points of $A$ and then surface $F_2$ is brought in contact with the target point of $B$. If $F_1$ and $F_2$ touch the gage in the three target points simultaneously, it passes inspection; otherwise it is rejected.

Other datum reference systems may be established using cylindrical datums. Datum reference systems may also be formed by combinations of cylindrical and planar datums. Precedence is again important. Figure 16-16 shows examples.

The specification of datum features includes definition of the criteria for applying material conditions (MMC, LMC, or RFS—regardless of feature size) to the datum. These conditions specify how datums are established during inspection of the part. Material conditions are used with features that have size (features of size) and may be used as datums such as holes. If the surface of a hole is specified as a datum at the MMC, then the surface of the minimum hole should be used during inspection. If, on the other hand, the RFS is used, then the effect

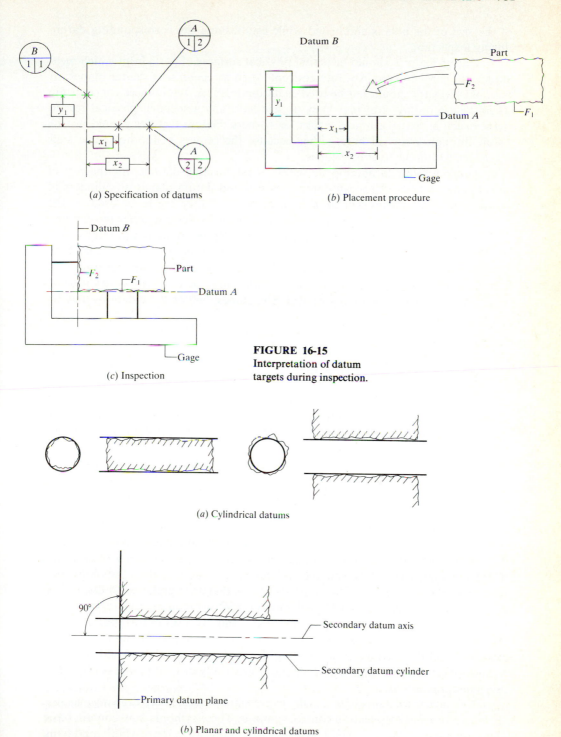

(a) Specification of datums

(b) Placement procedure

(c) Inspection

**FIGURE 16-15**
Interpretation of datum targets during inspection.

(a) Cylindrical datums

(b) Planar and cylindrical datums

**FIGURE 16-16**
Datum reference system using cylindrical datums.

of the size of the hole is eliminated while establishing the corresponding datum during inspection.

Datum features should be actual physical surfaces of parts from which measurements can be made. Abstract concepts such as centerlines, center planes, or axes of symmetry should not be used as datum features. While these concepts are useful to use on line drawings, they do not have any associated physical resemblance such as actual surfaces, edges, or corners. Datum features should not be taken through a row of holes either because the centers of the holes can shift anywhere within the tolerance zones.

Designers should develop the habit of establishing datums (for the rest of the chapter, datums will mean datum features and datum planes or cylinders of datum reference systems will be written explicitly if they are to be used) at the design phase of a part. Proper choices of datums make designs easier to read and interpret, and easier to manufacture and inspect. Proper datums can reduce production errors which in turn can reduce the number of scrap parts, and, therefore, can be very cost-effective. The number of datums needed to control positions on a part depends on the function of the part and how the designer wishes it to be made. A single datum is usually needed. Sometimes, two or three datums may be required.

### 16.3.2    Types of Geometric Tolerances

Three types of geometric tolerances exist: size, location, and form tolerances. Size tolerance (as shown in conventional tolerancing) controls the size (lengths and/or diameters) of part features. Location tolerance controls position and concentricity of various features. Form tolerance controls the shape of individual features (e.g., flatness of a single surface) or related features (e.g., parallelism or perpendicularity of two surfaces). Geometric tolerancing permits an explicit definition of datums, with a clear specification of the datum precedence in relation to each tolerance specification.

In order to communicate geometric tolerances with their related information effectively and concisely on engineering drawings, ANSI and ISO have developed a set of standards and symbology for dimensioning and tolerancing (D&T). ANSI Y14.5M-1982 standards incorporates most of the symbology and features of the international ISO 1101 standards that are significant to D&T. The 1982 version of ANSI Y14.5-M includes the ISO symbology related to datum targets as well as methods for specifying limits and fits. ISO in turn has accepted the projected tolerance zone, the three-plane datum concept, the multiple datum principle, and the method of designating datum targets from the ANSI Y14.5M. In spite of this mutual exchange, differences (especially in symbology) between the two standards still exist.

Geometric tolerances are usually expressed on engineering drawings via displaying symbols called feature control symbols. These symbols may contain basic sizes, tolerances, datums, ANSI symbols (see Table 16.6), and ANSI modifying symbols (Table 16.7). Samples are shown in Fig. 16-17. Applications and use of these symbols are explained in more detail in the following sections.

**TABLE 16.6**
## ANSI symbols for geometric tolerancing

| Feature | Type of tolerance | Characteristic | Symbol |
|---|---|---|---|
| For individual features | Form | Straightness | — |
| | | Flatness | ▱ |
| | | Circularity (roundness) | ○ |
| | | Cylindricity | ⌭ |
| For individual or related features | Profile | Profile of a line | ⌒ |
| | | Profile of a surface | ⌓ |
| For related features | Orientation | Angularity | ∠ |
| | | Perpendicularity | ⊥ |
| | | Parallelism | // |
| | Location | Position | ⊕ |
| | | Concentricity | ◎ |
| | Runout | Circular runout | ↗ |
| | | Total runout | ⌰ |

**TABLE 16.7**
## ANSI modifying symbols

| Term | Symbol |
|---|---|
| At maximum material condition | Ⓜ |
| Regardless of feature size | Ⓢ |
| At least material condition | Ⓛ |
| Projected tolerance zone | Ⓟ |
| Diameter | ⌀ |
| Spherical diameter | S⌀ |
| Radius | R |
| Spherical radius | SR |
| Reference | ( ) |
| Arc length | ⌒ |

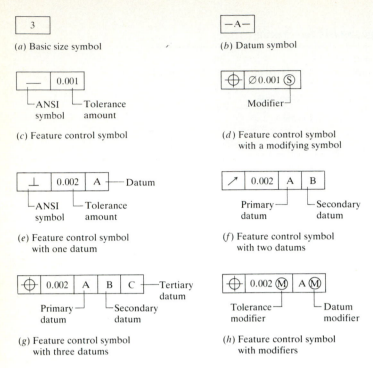

**FIGURE 16-17**
Sample feature control symbols for geometric tolerancing.

**16.3.2.1 SIZE TOLERANCES.** A feature of size, such as a hole or a slot, may be toleranced in a similar way to conventional tolerancing: by direct attachment of the size tolerance ($T_s$) to the basic (nominal) size or by specifying the upper and lower limits of size as shown in Fig. 16-18.

**16.3.2.2 LOCATION TOLERANCES.** This type of geometric tolerancing is designed to eliminate the square or rectangular tolerance zone (see Figs. 16-10 and 16-11) that results from using conventional tolerancing methods to control locations of features. There are two types of location tolerances (see Table 16.6): position and concentricity. Positional tolerancing is the most widely used and is known as the true position tolerancing. In positional tolerancing, tolerances are specified on the actual position and not on its coordinates—thus the name true position tolerances. Figure 16-19 converts Fig. 16-10 from conventional tolerances to true position tolerances. The resulting circular tolerance zone shows that the tolerance on the center of the hole is indeed 0.010 in, and the axis of the hole must lie within the cylindrical tolerance zone which has a circle of diameter equal to the tolerance value (0.010 in) and a height equal to the hole depth.

In true position tolerancing, the true position of a feature is the basic (nominal) location of a point, line, or a plane of a feature with respect to a datum or other feature. It is specified using basic dimensions as shown in Fig. 16-19. For cylindrical features, the true position tolerance is the diameter of the cylindrical

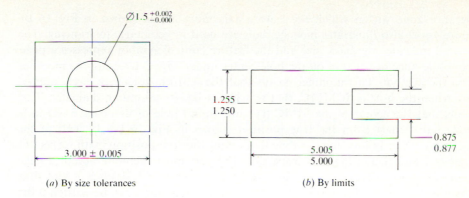

(a) By size tolerances

(b) By limits

**FIGURE 16-18**
Specification of size tolerances.

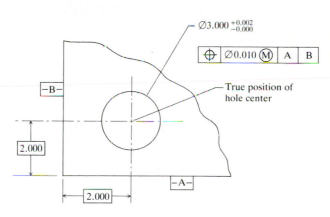

(a) Size and position tolerances

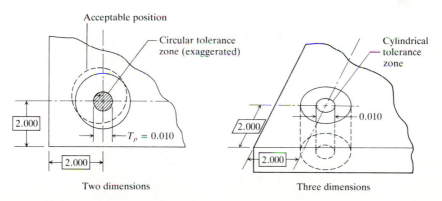

(b) Resulting tolerance zone

**FIGURE 16-19**
Positional tolerancing for Fig. 16-10.

tolerance zone within which the feature axis must lie, as shown in Fig. 16-19. Figure 16-19 also illustrates how symbols are used in geometric tolerancing. The boxed 2 indicates the basic size and the feature control symbol expresses a positional tolerance of 0.010 in on the hole center under MMC by using the modifier Ⓜ in the symbol. This modifier means that the 0.010-in tolerance circle applies only when the hole is at MMC, that is, when it has its minimum diameter. This means that the center of the MMC (i.e., minimum) hole of diameter 3.000 in is allowed to depart from the true position shown in Fig. 16-19a by half of the amount 0.010 in in any angular direction about the true position, with respect to datums $A$ and $B$. During part inspection, the part must therefore accept a gage pin (i.e., a shaft) of diameter $d_s$ and make contact with both datums $A$ and $B$ as shown in Fig. 16-20. From the figure, the following equation can be written if the part must pass inspection:

$$d_s = d_{hmin} - T_p \qquad (16.12)$$

where $T_p$ is the true position tolerance (0.010 in in Fig. 16-19b).

Figure 16-19a shows that two values of tolerances are needed to fully describe the hole. A size tolerance on the hole diameter is given as 0.002 in and a true position tolerance of 0.010 in on the position of its center is specified. True position and feature size tolerances ($T_p$ and $T_s$) may be interdependent or independent depending on the material condition attached to the former. The MMC or LMC makes $T_p$ and $T_s$ interdependent while the RFS condition makes them independent of each other. Consider the example shown in Fig. 16-19a which uses the Ⓜ modifier in the feature control symbol that specifies $T_p$. If the hole is at the MMC (smallest size), $T_p$ is not affected, but if the hole is larger, the available $T_p$ is larger. As shown in Fig. 16-21a, if the hole is exactly 3.000 in in diameter (MMC, or smallest size), its center location may vary from 1.995 to 2.005

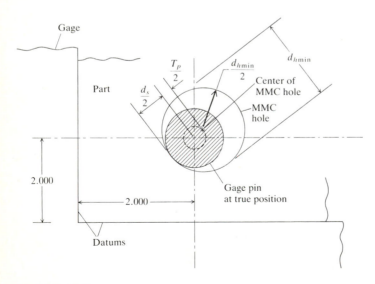

**FIGURE 16-20**
Interpretation of positional tolerance of Fig. 16-19a.

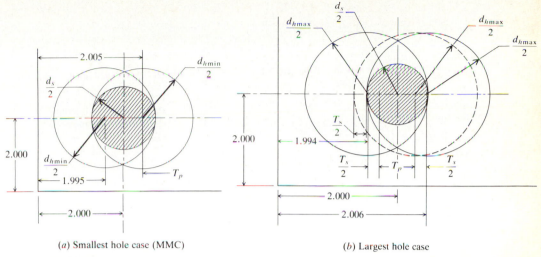

(*a*) Smallest hole case (MMC)                    (*b*) Largest hole case

**FIGURE 16-21**
Interdependence of $T_s$ and $T_p$.

in, that is, within the $T_p$ zone of 0.010. The MMC hole can receive a gage pin of 2.990 in diameter [Eq. (16.12)] located at the true position (2.000, 2.000). At the extreme position of the hole to the right, the outer side of the hole contacts the outer side of the pin. At the extreme position of the hole to the left, the inner side of the hole contacts the inner side of the pin.

 If the hole is 3.002 in in diameter, that is, maximum size, it will be acceptable by the gage pin in both extreme positions (right or left) if the proper sides (inner or outer) of both the hole and the pin are in contact, as shown by the solid circles in Fig. 16-21*b*. As seen from the figure, the hole center location may vary from 1.994 to 2.006 in, which is outside the specified positional tolerance permitted by $T_p$ alone. In effect, when the hole is not at MMC, the positional tolerance is greater than $T_p$ and reaches $T_p + T_s$ at the maximum hole diameter. Thus, when the hole is not at the MMC, a greater positional tolerance becomes available. Therefore, it has become common practice for both manufacturing and inspection to assume that positional tolerance applies to the MMC and that greater positional tolerance becomes permissible when the part is not at the MMC.

 If, for any reason, the hole center must be arbitrarily held within the tolerance zone specified by $T_p$, the modifying symbol Ⓢ (see Table 16.7) should replace the symbol Ⓜ in the feature control symbol shown in Fig. 16-19*a*. For this case (RFS), the hole size and its center location is shown by the dashed circle in Fig. 16-21*b*. The oversize tolerance zone of $T_p + T_s$ diameter is no longer permissible since the modifier Ⓢ restricts its diameter to $T_p$. In this case, the ordinary fixed gage (shown in Fig. 16-20) is no longer able to inspect the part for the location of the hole. Other methods (perhaps more expensive) must be used to inspect the hole if its center is to be restricted to the tolerance zone of the diameter $T_p$.

 In addition to solving the problem of square or rectangular tolerance zones, positional tolerancing also eliminates the accumulation of tolerances, even in a

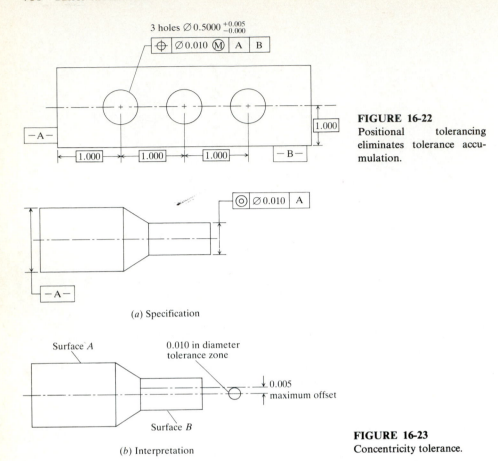

3 holes ∅ 0.5000 $^{+0.005}_{-0.000}$

⊕ | ∅ 0.010 Ⓜ | A | B

1.000

−A−

1.000    1.000    1.000    −B−

**FIGURE 16-22**
Positional tolerancing eliminates tolerance accumulation.

◎ | ∅ 0.010 | A

−A−

(a) Specification

Surface A

0.010 in diameter tolerance zone

0.005 maximum offset

Surface B

(b) Interpretation

**FIGURE 16-23**
Concentricity tolerance.

chain of dimensions as shown in Fig. 16-22. This is because positional tolerances are usually measured relative to the true positions which are specified by basic sizes only. For example, the true position of the middle hole shown in Fig. 16-22 is measured from datums *A* and *B* by 2.000 and 1.000 in respectively. This true position is independent of the tolerances of the left hole.

The second type of location tolerances is concentricity tolerance. It is defined as a relationship between the axis of the toleranced feature(s) and the axis of a datum feature(s). It specifies the collinearity of axes of adjacent features as shown in Fig. 16-23. Concentricity tolerance is less used than positioned tolerance because it is difficult to locate actual centers. It is usually preferred to replace this type of tolerance with the runout tolerance (Table 16.6). The axis of the concentricity cylindrical tolerance zone is taken to be the axis of the datum feature(s), as shown in Fig. 16-23.

**Example 16.2.** A designer has designed the plate shown in Fig. 16-24 with its nominal dimensions. A tolerance of 0.005 in is acceptable from drilling the holes. For successful assembling of the plate, a tolerance of 0.002 in is allowed on the position of the center of any hole, and another tolerance of 0.001 in is allowed on

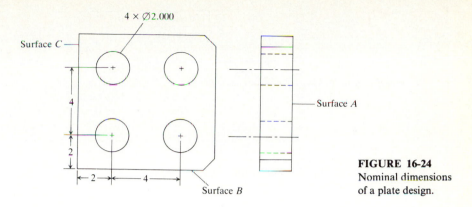

4 × ∅2.000

Surface C

Surface A

Surface B

**FIGURE 16-24**
Nominal dimensions
of a plate design.

the axis of any hole. Re-sketch Fig. 16-24 to show these tolerances. Sketch the resulting tolerance zones.

*Solution.* The tolerance 0.005 in resulting from drilling is a size tolerance on any hole size; that is, $T_s = 0.005$ in. There is a positional tolerance $T_p = 0.002$ in on the location of any hole center. This tolerance requires datums. From the nominal dimensions shown in Fig. 16-24, surfaces $A$ (bottom of the plate), $B$, and $C$ are proper datums. These surfaces would have to be machined and prepared properly before the holes are drilled. The third tolerance on the axis of any hole can be thought of as a form tolerance (see next section); that is, $T_F = 0.001$ in. If we measure this tolerance with respect to surface $A$ of the plate, $T_F$ would become a perpendicularity tolerance between the axis of any hole and the bottom surface of the plate. Figure 16-25a shows the specifications of the three types of tolerances ($T_s$, $T_p$, and $T_F$), assuming the MMC and unilateral tolerance for $T_s$. Figure 16-25b shows the two tolerance zones in two and three dimensions that result from $T_p$ and $T_F$. Figure 16-25b shows one of the many possible tolerance zone patterns.

**Example 16.3.** Two parts such as the one shown in Fig. 16-26 are to be assembled by a bolt whose diameter ranges from 0.250 to 0.260 in. The designer, using positional tolerancing, specifies tolerances on the hole where the bolt is to be assembled as shown in the figure. Assuming the parts to be in contact with the datums during assembly, is it possible to assemble the two parts with the bolt? Why? How can the problem be solved?

*Solution.* Using Eq. (16.12) with the MMC, we write

$$d_{smax} = d_{hmin} - T_p$$

substituting $d_{smax} = 2.260$ in and $d_{hmin} = 0.270$ in into this equation gives $T_p = 0.010$ in. The specified tolerance $T_p$ is 0.015, which means that it is not possible to assemble the two parts with the bolt with the given tolerances. For successful assembling, Eq. (16.12) implies that $T_p < d_{hmin} - d_{smax}$. In this example, $T_p$ should be less than 0.010 in.

**16.3.2.3   FORM TOLERANCES.** Form tolerances ($T_F$), together with location tolerances, provide the designer with complete tolerancing theory capable of controlling most aspects of part geometry and, therefore, its manufacturing. Form

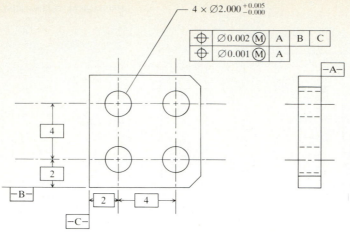

$4 \times \varnothing 2.000 ^{+0.005}_{-0.000}$

| $\oplus$ | $\varnothing 0.002$ Ⓜ | A | B | C |
|---|---|---|---|---|
| $\oplus$ | $\varnothing 0.001$ Ⓜ | A | | |

−A−

4

2

−B−

2 ← 4 →

−C−

(a) Tolerance specifications

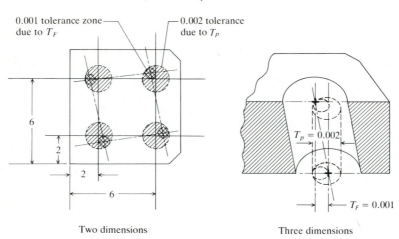

0.001 tolerance zone due to $T_F$

0.002 tolerance due to $T_P$

6

2

2

6

Two dimensions

$T_P = 0.002$

$T_F = 0.001$

Three dimensions

(b) Tolerance interpretation

**FIGURE 16-25**
Tolerance specifications and interpretations for plate design shown in Fig. 16-24.

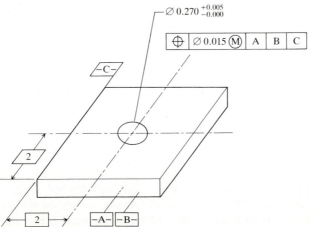

$\varnothing 0.270 ^{+0.005}_{-0.000}$

| $\oplus$ | $\varnothing 0.015$ Ⓜ | A | B | C |
|---|---|---|---|---|

−C−

2

2

−A−−B−

**FIGURE 16-26**
Toleranced dimensions for an assembly.

790

**TABLE 16.8**
**ANSI form tolerances**

| Form tolerance | Specification on a drawing | Interpretation |
|---|---|---|
| Straightness | $\varnothing 1.00^{+0.02}_{-0.00}$    $-$   $0.05$ | 1.02 MMC (boundary of perfect form)<br>0.05 wide tolerance zone<br>Control surface straightness |
| | $\varnothing 1.00^{+0.02}_{-0.00}$    $-$   $\varnothing 0.05$ (M) | $\varnothing 0.07$ diameter tolerance zone<br>$\varnothing 1.07$<br>Control axis straightness |
| Flatness | $\square$ $0.08$ | 0.08 wide tolerance zone |
| Roundness (circularity) | $A$   $\bigcirc$ $0.05$ | 0.05 wide tolerance zone<br>Actual contour<br>Section $A$–$A$ |
| Cylindricity | $\diamondslash$ $0.06$ | 0.06 wide tolerance zone |
| Profile | $\frown$ $0.08$ $A$ $B$    $-A-$    $-B-$ | 0.08 wide tolerance zone<br>Datum $A$<br>Datum $B$<br>Control profile of a line |
| | $\cap$ $0.05$ $A$    $-A-$ | 0.05 wide tolerance zone   True profile relative to datum $A$<br>Datum $A$   Actual profile<br>Control profile of a surface |

**TABLE 16.8** (*continued*)

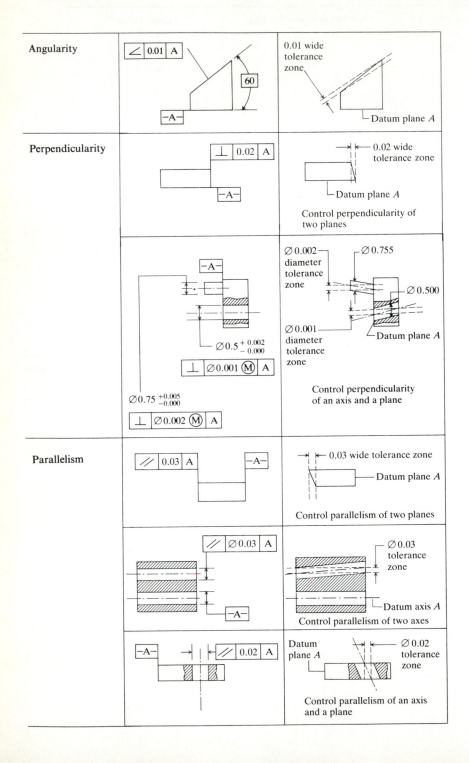

**TABLE 16.8** (*continued*)

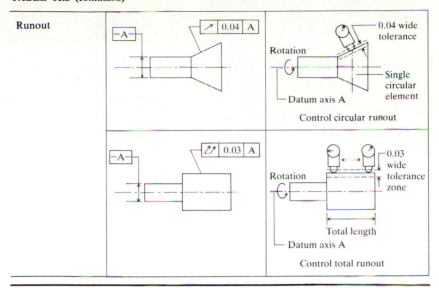

| Runout | | |
|---|---|---|

Control circular runout

Control total runout

tolerances control the variabilities in the shape of parts or features. Table 16.6 shows the various form tolerances. Some of them apply to individual features. They are straightness, flatness, roundness, cylindricity, and profile. Other form tolerances are applicable to related features. They control the variability of one feature relative to another. They are angularity, perpendicularity, parallelism, runout, and profile. The specification on a drawing and the meaning of each form tolerance are summarized in Table 16.8.

The table is self-explanatory for all forms. The runout tolerance deserves some comments. It is a measure of deviation from perfect form. It is defined by a tolerance zone contained between two surfaces of revolution with respect to a specified datum axis or between two planes normal to the axis. It is determined by rotating the form 360° about its axis. It can be found by reading the net change on a dial as shown in the table. For circular runout, the dial is fixed in a certain location; that is, it measures a circular element only. For total runout, the dial moves along the entire surface while the form is rotating. The runout is measured in FIM (full indicator movement) units of the dial. Runout tolerance is a composite tolerance that incorporates variations in straightness, roundness, and parallelism.

The tolerance zone of the runout tolerance is used to express the zone within which the feature surface must remain when the part is rotated about the given axis. The axis is established by one long cylinder, two widely separated coaxial cylindrical surfaces, or one cylindrical surface and a plane face at right angles to it, as shown in Fig. 16-27.

Figure 16-28 shows an example to illustrate the use of geometric tolerancing. The interpretation of the various tolerances shown is left as an exercise to the reader (refer to the Problems at the end of the chapter).

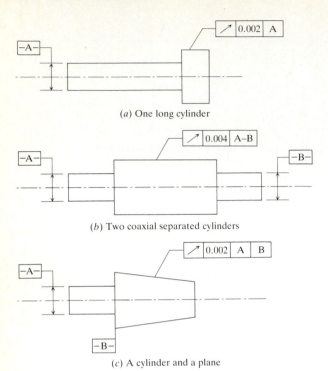

(a) One long cylinder

(b) Two coaxial separated cylinders

(c) A cylinder and a plane

**FIGURE 16-27**
Axis establishment for runout tolerance.

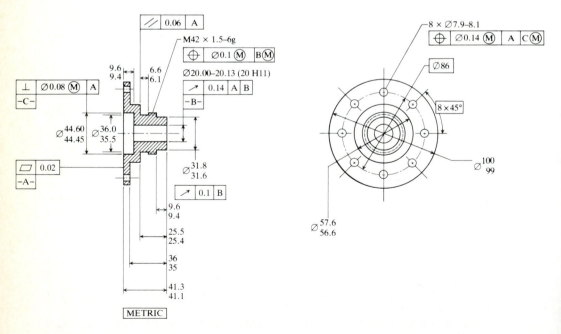

METRIC

**FIGURE 16-28**
Use of geometric tolerancing (ANSI metric example).

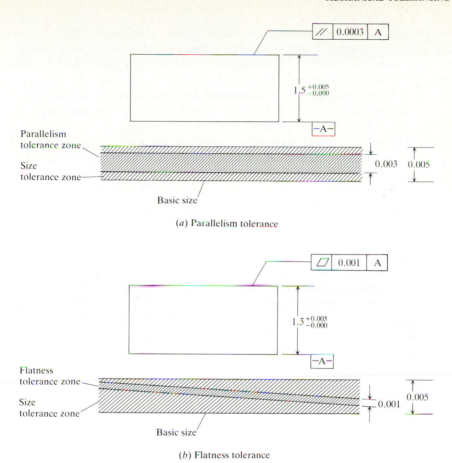

(a) Parallelism tolerance

(b) Flatness tolerance

**FIGURE 16-29**
Interrelationship between size and form tolerances.

Several form tolerances may be applied to a feature. The resulting tolerance zones may be interrelated and careful interpretation of the interrelationship is required. For example, one tolerance form may imply the other. In addition, the location of form tolerance zones is usually not specified in the form tolerance itself, but in the related positional or size tolerance. In general, all form tolerance zones must lie within the size or positional tolerance zones that determine their location. In Fig. 16-29, for example, the parallelism tolerance zone lies somewhere within the size tolerance zone. Similarly, the flatness tolerance zone must lie within the size tolerance zone, but neither its location nor orientation is precisely fixed.

## 16.4 DRAFTING PRACTICES IN DIMENSIONING AND TOLERANCING

Both conventional and geometric tolerancing methods have been covered in Secs. 16.2 and 16.3. While conventional tolerances are easy to understand and use, they are ambiguous in most cases and do not allow control of form (shape).

On the contrary, geometric tolerances are more specific and control both location and form, but assigning and interpreting them usually require more care. One would expect the wide use of geometric tolerances in industry and no or little use of conventional tolerances. Practices in D&T show a reasonable mixture of using conventional and geometric tolerances. In spite of the distinct advantages of geometric tolerances, conventional tolerancing has been retained in many companies where tolerance requirements are less stringent.

For many parts in which assembly conditions are not critical, or where known process capability ensures satisfactory results, it is accepted practice to indicate linear and angular dimensions from feature to feature without preferring either as a datum from which the other is measured. This results in greater flexibility to production and inspection (quality) departments which may reduce costs. In practice, geometric tolerancing is applied only when real advantages result or when specific needs and functional requirements are demanded.

It should be emphasized here that geometric tolerancing is quite crucial to achieve the automation of design and manufacturing demanded from the CAD/CAM technology. Unlike conventional tolerances, geometric tolerances are unambiguous and can be modeled geometrically and used for part manufacturing and inspection, as discussed in Sec. 16.6. Thus, geometric tolerancing provides full automation of D&T, while conventional tolerancing provides computer assisted D&T because human interpretation is always required.

## 16.5 DIMENSIONING AND TOLERANCING IN MANUFACTURING

D&T practices are defined for use in engineering drawings (blueprints) that are subsequently used for manufacturing a part or a product. In manufacturing practice, the engineering drawing is the legal document that transmits the product geometry to the various groups and departments involved in producing it. D&T is used to provide precise size, location, orientation, and form information. In defining functional dimensions, mating faces between individual components play a central role, and it is generally recommended that these faces be used in establishing dimensioning datums.

When the general assembly drawing and the detail drawings are complete, the dimensional information is used by the checker (a person who checks the validity and compatibility of dimensions) to ensure correct assembly of the parts and compliance with functional requirements. Such checking is initially applied to the ideal (nominal) part. Then the functional dimensions of the assembly can be computed from the individual tolerances on part dimensions. This amounts to performing tolerance analysis (see Sec. 16.7) in which the statistical properties of the manufacturing processes are taken into account.

In addition to using drawings for tolerance analysis, they are widely used to plan manufacturing operations (process planning); order and process material; design toolings and jigs and fixtures (clamps); machine parts; inspect product quality; assemble and test finished products; and for postmanufacturing tasks such as shipping, installation, operation, and maintenance.

The process planner (see Chap. 20) has the task of breaking down the manufacturing of components into individual process steps. The planner specifies the machines on which the components are to be made, and the functional requirements of tools, jigs and fixtures, and inspection gages. For each process step, the planner identifies the process datum(s) which determines the way in which the part is to be set up. Process dimensions and tolerances are also defined. In the early stages of machining of castings, forgings, or sheet metal parts, spot or line datums are often used since the surfaces of these parts are usually irregular. In general, faces, edges, and vertices of a part may be used as datums.

The process plan also includes a description of in-process inspection requirements for each process step. These requirements are deduced from the process D&T. From known information about process capabilities, the process planner decides on which dimensions are to be checked, and the frequency and nature of the checks.

The clamping method, besides the machining process, must be taken into account in assessing the process capability to achieve a given tolerance. In the design of clamps, the resulting distortion and its effect on critical dimensions must be considered. Where distortion is unavoidable, general tolerances are specified, or the maximum permissible loads which may be applied to produce the correct shape are given. These loads should be acceptable by the designer as safe loads to be applied on the assembly.

If the process planner decides that the design requirements are too expensive or impossible to meet with existing equipment, the planner may notify the responsible designer with suggested changes. This usually involves a need to relax close tolerances or a recommendation to use standard sizes. In many companies, process planning is brought earlier into the design cycle to reduce the need for this iterative process. Therefore, tentative process planning would have already taken place before the formal planning process begins.

Process plans must also include instructions for assembly. In planning a subassembly, dimensional allowances may be required to provide for final machining of the subassembly to true size to remove assembly-induced distortion. Both the function of the assembly and the resulting distortion have to be considered in determining assembly forces for interference fits, applied torques for bolted assemblies, etc.

Product inspection consists of in-process inspection and final inspection. The former is usually carried out using simple gaging techniques involving go/no-go gages as shown in Fig. 16-30. The use of these gages is usually more effective if

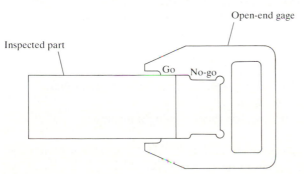

**FIGURE 16-30**
Go/no-go inspection gage.

the MMC is utilized in D&T. The final inspection of components usually requires checking every dimension and note on the drawing. The final inspection of a product consists of checks of overall dimensioning, inspection of functional dimensions and tolerances, and testing for required functions. In addition to other methods of inspection, a coordinate measuring machine (CMM) may be used. A control tape is produced from a part program similar to that used in NC, and the results of the measurements may be used in analysis programs.

Ideally, the required function of the whole product and present/future process capability and cost should determine the dimensions and tolerances of every component. However, breaking down a product D&T first into components and then into individual processes limits the size of optimization tasks to be undertaken, at the expense of requiring an iterative design cycle.

## 16.6   TOLERANCE MODELING AND REPRESENTATION

The activities that require specification of D&T have been discussed in the previous section and are summarized below:

1. Functional performance, specification, and checking.
2. Tolerance analysis, control, and distribution.
3. Assembly checking.
4. Process selection (steps, machines, fixtures, gages, cost, etc.).
5. In-process and final inspection.
6. Robot assembly.

In meeting the above requirements, modeling and representing tolerances are constrained by:

1. Allowing a mixture of conventional and geometric tolerancing. Conventional tolerances are subsumed into geometric tolerances that contain no implicit datums or unstated constraints. Some of them can be treated as defaults when assigning tolerances to a geometric model.
2. The need to avoid explicit definition of D&T unless essential. Default tolerances must be assigned automatically. This is equivalent to the general tolerance notes typically found on engineering drawings.
3. Allowing for feature modeling and representation as features, positions, and orientations form the core of the geometric tolerancing practice as discussed in Sec. 16.3.
4. Allowing for combining or nesting a group of features to create composite (or compound) features. The latter may be required to define subassemblies.

Thus, data structure for D&T must include the following information:

1. Assembly data structure as discussed in Chap. 14. That structure may be expanded to include fits and classes as described in Sec. 16.2.

2. Feature structure.

3. Datum reference structure including definitions and precedence.

4. Material conditions (MMC and RFS). Global default conditions should be possible.

5. Conventional and geometric tolerances.

## 16.6.1  Modeling

Geometric modeling as described in Part II of this book defines the nominal (ideal) shape of a mechanical part. Geometric modelers that lack tolerancing facilities cannot support fully automatically the manufacturing and production activities described in the previous section. The first attempt to add tolerancing information to nominal geometric models is to treat it as a mixture of textural and graphical data which users can insert as modifiers of dimensioning commands. This approach has been implemented by some wireframe-based modelers. Linear and angular dimensioning commands have a tolerance modifier. For example, the command:

CREATE LINEAR DIMENSION TOLERANCE POSITIVE 0.001 NEGATIVE 0.001 $d_1 d_2$

adds a tolerance of $\pm 0.001$ to entities identified by digitizes $d_1$ and $d_2$. This approach usually applies to conventional tolerances and can be used for tolerance stack analysis (see Sec. 16.7). However, the primary emphasis of this approach is to add tolerances as texts to engineering drawings in the same fashion of the manual generation of these drawings.

If full automation of design and manufacturing is to be achieved, the solid modeling theory should be utilized in tolerance modeling and representation. Thus, nominal parts should be modeled as solid models if viable and effective tolerancing facilities are to be developed. Tolerance modeling and representation discussed here are based on having solid models of nominal parts. Industrial users of solid modelers that do not support tolerancing information are forced to download nominal geometric data from these modelers into their wireframe systems to add the information manually to produce standard engineering drawings.

Before we define a D&T model, let us discuss the mathematical and geometric implications of tolerance specifications. Tolerances define a class of parts that are topologically identical to their corresponding nominal part and that are interchangeable in assembly operations and functionally equivalent. Such classes are referred to as variational classes. Mathematically, a variational class is modeled by a set of solid models that contain the solid model of the nominal part. Thus, variational classes may be modeled as a collection of r-sets (see Chap. 7). The precise nature of such a collection is largely unknown.

Geometrically, tolerance specifications are representations of variational classes. They can be thought of as entities of a computational nature; that is, they yield valid solid models if computed. Consider the part shown in Fig. 16-31. Only

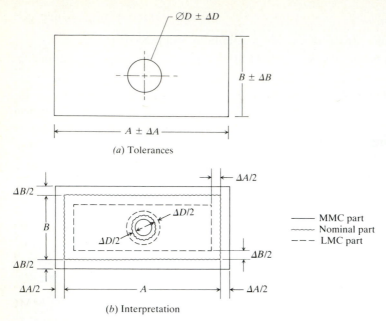

(a) Tolerances

(b) Interpretation

**FIGURE 16-31**
Geometric interpretation of tolerance specifications.

size conventional tolerances are shown for simplicity. Three entities are needed to store the tolerances $\pm\Delta A$, $\pm\Delta B$, and $\pm\Delta D$. When these entities are evaluated relative to the nominal part, two extreme or limiting models can be defined. One extreme model corresponds to MMC and is called the MMC model, while the other is the LMC model and corresponds to LMC. These are shown by the solid and dashed boundaries respectively in Fig. 16-31b. In this example, the variational class is all valid solid models (i.e., r-sets) contained between, and including, the MMC and LMC parts.

Readers should be cautious when extending the interpretation shown in Fig. 16-31 to more complex parts or to include geometric tolerances. The difficulty usually arises when tolerance zones overlap or when datums are implicitly assumed as in conventional tolerances. In Fig. 16-31, we divide the tolerances $\Delta A$ and $\Delta B$ equally around the nominal part. If, however, we assume that the bottom and left faces of the part are implicit datums, all $\Delta A$ and $\Delta B$ would be applied to the right and the top faces respectively. This would result in a different variational class. Readers are encouraged to change the part shown in Fig. 16-31 into an L shape, assign dimensions and tolerances to it (without overdimensioning), and attempt to interpret the tolerances geometrically.

There are two approaches to generate variational classes from tolerance specifications: parametrization and nonparametrization. Parametrization is similar to instancing or generating a family of parts. As shown in Fig. 16-31a, one would parametrize the nominal part into parameters $A$, $B$, and $D$. These parameters represent a family of parts. The limiting parts correspond to the tolerance limits as shown in Fig. 16-31b. Assigning these parameters specific values results in a specific part of the family. In B-rep, parametrization is achieved by assigning parameters to the vertices of the part. In CSG, it is achieved by assigning variables to the parameters of the primitives that make the part.

Variational geometry is a parametric approach to generate variational classes. In conventional solid modelers, dimensions are subordinate to geometry. The designer creates the exact geometry of a nominal part and dimensions are derived from the geometry. Variational geometry enables the reversal of this relationship. The designer creates the part topology and a set of dimensions (parameters) from which exact geometry is derived. Therefore, dimensions can be changed directly and the new model is created automatically. (In conventional solid modelers, a change of dimensions requires erasing part or all of the geometric model and creating it again.) Thus conventional tolerances can be represented as allowable ranges (limits) of the explicitly represented dimensions. A disadvantage of the variational geometry approach is that there is no obvious way to deal with geometric tolerances. The details of variational geometry is not covered here.[4]

Offsetting is a nonparametric approach to generate variational classes. In this approach, the boundary of the nominal part is offset by the amount of the specified tolerances to generate the limiting parts. One can think of Fig. 16-31*b* as offsetting the nominal part once outward and once inward to generate the MMC and LMC parts respectively. Offsetting seems to be more appealing than parametrization because it catches the spirit of tolerance zones. It is not always easy or straightforward to parametrize the part in such a way to correspond to specified tolerances. In offsetting, each tolerance zone corresponds to an offsetting zone.

The offsetting theory in geometric modeling is not covered here,[5] but it should be mentioned that offsetting supports bilateral size tolerances of equal value (e.g., $\pm 0.005$). Both of these values must not be null. Unilateral tolerance specification or unequal bilateral limits must be redefined to have equal values. For example, a size tolerance of $^{+0.0015}_{-0.0005}$ must be redefined to become $\pm 0.001$ for the offsetting theory to apply. However, this may not be acceptable in practice.

How can a variational class be displayed on a graphics display? This question may seem trivial at first. To appreciate the question, let us assume that displaying the three parts as shown in Fig. 16-31*b* is an acceptable display. It provides users with the nominal part display together with the limiting acceptable parts. Even with this simplification, the dimensions of the two limiting parts relative to the nominal part form an ill-conditioned display problem. Tolerances are usually in the order of hundredths of thousandths of the nominal dimensions. Therefore, the display of the three parts appear coincident on the screen. Magnifying local zones of the display provides a partial solution.

---

[4] Interested readers should refer to R. Light and D. Gossard, "Modification of Geometric Models through Variational Geometry," *Computer Aided Design*, vol. 14, no. 2, pp. 209–214, 1982, and V. C. Lin, D. Gossard, and R. Light, "Variational Geometry in Computer-Aided Design," *Computer Graphics*, vol. 15, no. 3, pp. 171–177, 1981.

[5] Interested readers should refer to J. R. Rossignac and A. A. G. Requicha, "Offsetting Operations in Solid Modelling," *Computer Aided Geometric Design*, vol. 3, pp. 129–148, 1986.

We now turn to the definition of a D&T model. A geometric (solid) model defines the nominal shape of a part. The D&T model is an extension of this model which defines the dimensions of this part and the tolerances (conventional and/or geometric) on these dimensions. This model and its data structure must support the activities, meet the requirements, and include the information discussed at the beginning of Sec. 16.6. The D&T model should be able to detect overdimensioning and, therefore, overtolerancing. Assignment of dimensions and tolerances should be separate from creating the geometric model of the nominal part, but yet should be considered an integral part of the design process. This enables designers to create their conceptual designs without having to worry about tolerances. As a result, the D&T model must provide independent D&T data, and it must simultaneously define the relations between this data and the geometric model of the nominal part. Figure 16-32 shows the relationship between the geometric and D&T models. The geometric modeler acts as a preprocessor to the D&T modeler. After designers have created and analyzed geometric models of individual nominal parts, they can utilize the D&T modeler to specify tolerances and analyze them as discussed in Sec. 16.7. The link between the geometric and D&T modelers is established such that the basic sizes (nominal dimensions) of the D&T model is updated automatically when designers change dimensions of the corresponding geometric models of the nominal parts.

In practice, it would be beneficial to have a CAD/CAM system that can provide designers with geometric models of nominal parts, assembly models, and D&T models in a modular way. This would provide designers with the capability of using one or more model at any one time. For example, a designer may want to analyze assemblies first before assigning tolerances. In such a case only geometric and assembly models are needed. In other cases, the three (geometric, assembly, and D&T) models may be needed. Figure 16-33 shows the relationship

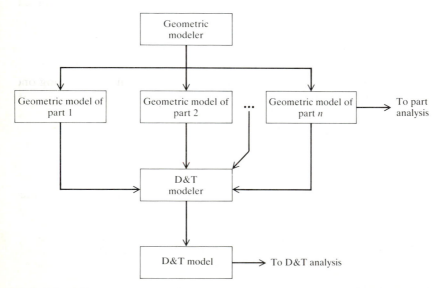

**FIGURE 16-32**
Generation of a D&T model.

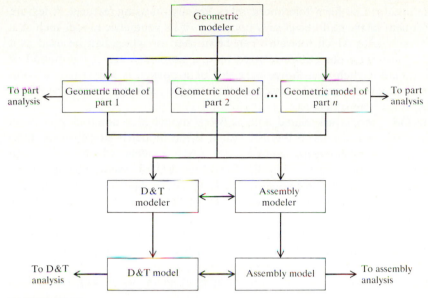

**FIGURE 16-33**
Generation and interaction of geometric, assembly, and D&T models.

between the three models. The figure is a combination of Figs. 14-2 (Chap. 14) and 16-32. Figure 16-33 shows that the geometric modeler is the backbone modeler which supports both the assembly and D&T modelers. The figure also shows that the latter modelers and their models are interrelated and can share information together. If geometric, assembly, and D&T models share information via pointers in the corresponding databases, redundancy in information between the three models and databases can be eliminated. One major advantage of the structure shown in Fig. 16-33 is its modularity. Users can begin by buying and using the geometric modeler. Later, they can choose to add none, either, or both assembly and D&T modelers. From the point of view of developing the software, the modularity shown in Fig. 16-33 helps CAD/CAM developers to organize and concentrate their efforts. The coupling of the three modelers, that is, creating one data structure that contains geometry, assembly, and D&T information, usually leads to software that is complex to develop and inefficient to use.

### 16.6.2   Representation

The D&T model as described in the previous section has two main functions: generating variational classes and performing tolerance analysis. Due to the practical importance of the latter function over the former, the data structure of the D&T model is usually designed to support tolerance analysis. This section describes representations of the D&T model for both B-rep and CSG models of nominal parts.

**16.6.2.1   B-REP-BASED D&T MODEL.** Before getting into the details of the model, it is useful to define what a feature is. Geometric tolerancing, especially datum definitions and form tolerances, relies heavily on using features. A feature is a component of the real object or a boundary of its geometric model such as a surface or a hole to which tolerance can be applied or which can be used as a datum. This feature is sometimes referred to as a simple feature. A compound or composite feature, such as a flange or a slot or a pocket, is one that is a combination of simple features. Throughout this chapter, a feature means a simple one unless it is stated otherwise.

The D&T model associated with a B-rep model of a nominal part is an evaluated model analogous to the fact that a B-rep model is stored in its database in an evaluated form (contrary to CSG models which are stored unevaluated). The model is called an EDT (evaluated D&T) model. Primarily, the EDT model consists of a set of nodes called EDT nodes. The nodes are related to each other as well as to the geometric model of the nominal part as shown in Fig. 16-34.

The rationale for this EDT model is based on the D&T concepts embodied in ANSI Y14.5-M standards. There are four types of nodes linked to each other and to the B-rep model of the nominal part. The primary node is the D/T (D&T) node. It corresponds to a feature control symbol as shown in Fig. 16-17. It is

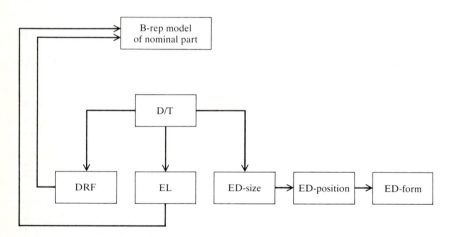

(a) One feature control symbol

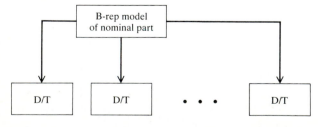

(b) Overall representation

**FIGURE 16-34**
Representation of the EDT model.

linked to the other nodes of the model as shown in Fig. 16-34*a*. The datum reference frame (DRF) node defines the datums (Fig. 16-12) used in the corresponding D/T node. The DRF node is linked to a set of three entities on the B-rep model which are used to define a DRF. Many D/T nodes may share, and therefore are linked to, the same DRF node.

The entity linking (EL) node is used to define the feature to which the D/T node is attached. The sole purpose of the EL node is to link a set of entities on the B-rep model to a D/T node. This set of entities makes up the feature associated with the D/T node. The evaluated data (ED) node stores a specific value of a certain type of geometric tolerance. Three types of ED nodes may exist and may be linked to a D/T node as shown in Fig. 16-34*a*. They are size, position, and form nodes which correspond to the three types of geometric tolerancing. Obviously, the ED-form node needs to be further classified to reflect the various types of form tolerances.

A typical EDT model contains multiple D/T nodes to represent the feature control symbols representing the various geometric tolerancings on a given B-rep model. All these nodes are linked to the B-rep model as shown in Fig. 16-34*b* through their corresponding DRF and EL nodes.

The EDT model as described above can be used to represent conventional tolerancing by setting the DRF, ED-position, and ED-form nodes to null, and using the ED-size node to store conventional tolerances. If one needs to sort out nodes corresponding to conventional or geometric tolerancing, a conventional/geometric (CG) node may be added to the EDT model shown in Fig. 16-34*a*. The CG node could be assigned a value of 1 (or C) for conventional tolerances and a value of 2 (or G) for geometric tolerances.

Let us apply the EDT model to some of the sample feature control symbols shown in Fig. 16-17. The D/T node corresponding to the case (*c*) symbol has only one ED-form node in addition to the EL node. The rest of the nodes are null. The D/T node for case (*g*) has a DRF (with three datums defined) node, an EL node, and one ED-position node. No ED-size or ED-form nodes are required.

**Example 16.4.** Sketch the corresponding EDT model to the ANSI example part shown in Fig. 16-28. Assume that the nominal part is a B-rep solid model.

***Solution.*** Figure 16-35*a* shows the toleranced faces of the solid model of Fig. 16-28 labeled as $F_1$, $F_2$, etc. These faces are also considered the features to which tolerances apply. Applying the generic structure shown in Fig. 16-34 to the solid model at hand results in the EDT model shown in Fig. 16-35*b*. Only the faces that are needed to define the EDT model are shown for simplicity (no edges or vertices are shown either).

Fifteen D/T nodes are needed to define all the geometric tolerances shown in Fig. 16-28. Only one D/T node (7) is shown out of the eight nodes (7 to 14) needed to define the positional tolerances on the eight holes (see Fig. 16-28). For each D/T node, the corresponding EL and DRF nodes are shown as pointers to the proper faces of the B-rep model. The symbol (e.g., $\perp$) for each form tolerance is shown in this way for simplicity. It can be replaced by a code number if needed. Each ED node shows the type and value of tolerance together with the material condition if applicable. Node $(D/T)_{15}$ shows an example of defining conventional tolerances.

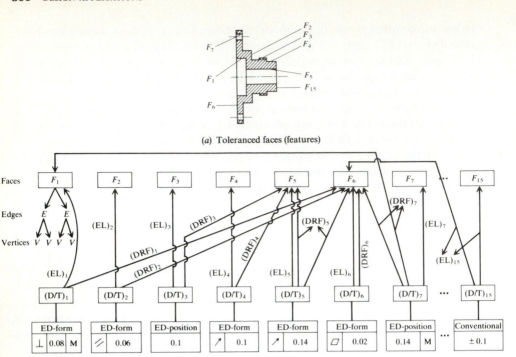

(a) Toleranced faces (features)

(b) EDT model

**FIGURE 16-35**
EDT model of the part shown in Fig. 16-28.

**16.6.2.2 CSG-BASED D&T MODEL.** In the EDT model, the faces of the B-rep model are directly associated as features using the EL node because the B-rep model has all faces of the nominal part explicitly defined. Unlike B-rep models, a CSG model is an unevaluated model and is usually stored as a CSG graph or tree. To explicitly define faces, the CSG model must be evaluated into a form like the B-rep model to yield these explicit faces (refer to Chap. 7).

The problem of associating a feature with a CSG model can be illustrated by the simple example shown in Fig. 16-36. In this example, the CSG model is defined as the union of a cylinder and a block (Fig. 16-36a). Before evaluating the model, there is one cylindrical primitive face (Pface). When the model is evaluated (see Fig. 16-36b), Pface is divided into faces $F_1$ and $F_2$ and two additional edges

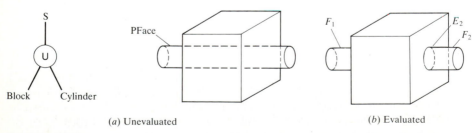

(a) Unevaluated

(b) Evaluated

**FIGURE 16-36**
Associating features with a CSG model.

$E_1$ and $E_2$ are created. In tolerancing practice, each of these two faces may be a different feature with different size, position, and/or form tolerances.

The above simple example shows that the extension of the EDT model to the CSG model requires one of two alternatives. The first and most obvious is to evaluate the CSG model and store it. This simply changes the model into a B-rep model and, therefore, the advantages of CSG representation are lost. The other alternative is to force the user to think in terms of features. For the example shown in Fig. 16-36, the user would have to model the part as the union of two cylinders and a block instead of a cylinder and a block. It is impractical to force designers to think in terms of tolerances early enough in the design cycle.

We therefore need to develop a D&T model that is specifically applicable to CSG models. We refer to this model as the UDT (unevaluated D&T) model. This name is chosen in analogy to the fact that CSG models are stored in an unevaluated form. The UDT model is not based on solid primitives (bounded solids) as one would expect (similar to using them together with boolean operations to create solid models of nominal parts). The reasons for the inadequacy of this approach is that some geometric tolerancing concepts such as datums cannot be defined in terms of solid primitives. A datum that is a surface is of lower dimensionality than a solid (see Chap. 7) and, therefore, cannot be obtained by manipulating primitives. Another reason is the possible division of a primitive into completely disjoint or disconnected parts. As shown in Fig. 16-36, the primitive cylindrical face is divided into the two separate faces $F_1$ and $F_2$ after the union operation.

We therefore need lower-level primitives to be able to define a UDT model. Thus, the UDT model described here is based on using half-spaces (see Chap. 7) to describe solid models of nominal parts and for associating dimensions of the parts to the parameters of these half-spaces. Tolerances are assigned to the half-spaces that make the boundary of the nominal solid model. The bounded portions of half-spaces that may be the solid's boundary are sometimes known as boundary-component (or b-component) half-spaces.

Parameters that define instances of half-spaces are divided into configuration parameters and position parameters. The former usually define the size of the instance half-space. For example, the configuration parameter of a cylindrical half-space is its diameter. The latter (position) parameters are the homogeneous transformations or rigid-body motions (translation, rotation, etc.) of the half-space. These parameters usually define the dimensions of the related solid. For example, if an instance half-space is displaced a distance $A$ in the $X$ direction of the MCS, then most probably there is a dimension of the related model associated with $A$.

Like the EDT model, the UDT model defines simple features by simple (individual) primitive half-spaces and compound features by composite (compound) half-spaces. Composite half-spaces are primitive half-spaces combined via boolean operations. For example, a composite half-space representing a slot can be defined as the intersection of three planar half-spaces. In addition to defining features, the UDT model defines datums by specifying proper half-spaces as such. It also defines tolerances as attributes assigned to proper half-spaces.

Comparisons between the EDT and UDT models reveal that they are similar in philosophy. Both are based on ASNI Y14.5-M standards. Therefore, they both define features and datums. Both also define tolerances as attributed to nominal faces. However, because feature identification in B-rep is easier than in CSG, the major difference between the two is their implementation into the solid modeler that model nominal parts. Because boundary faces of a B-rep model are explicitly defined in the model, datum definitions and tolerance assignments to these faces are done in a direct way. Conversely, they are done in an indirect way for a CSG model (by using primitive half-spaces) because boundary faces are represented by primitive half-spaces.

Tolerancing by constraining configuration and position parameters of primitive half-spaces is ideal for conventional tolerancing. Conventional D&T can be interpreted as a variancing scheme based on specification of limit ($\pm$) constraints on independent configuration and position parameters of primitive half-spaces. While configuration parameters are easy to deal with (just assign them), position parameters must be carefully assigned because such assignment determines the dimensioning scheme of the part.

For example, consider the part shown in Fig. 16-37$a$. The part is a two-dimensional solid (rectangle with a hole) consisting of a block (rectangle) minus a cylinder (circle). The block is defined by the intersection of four primitive planar

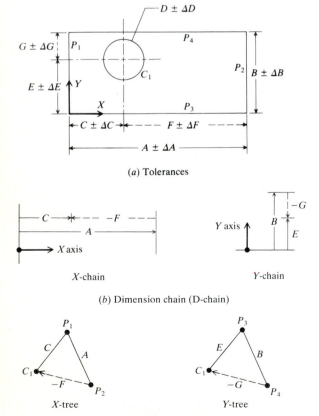

(a) Tolerances

(b) Dimension chain (D-chain)

(c) Dimension tree (D-tree)

**FIGURE 16-37**
Conventional tolerancing via parameters of primitive half-spaces.

half-spaces $P_1$ to $P_4$ and the cylinder is defined by a cylindrical half-space $C_1$. Planar half-spaces $P_1$ to $P_4$ do not have configuration parameters while half-space $C_1$ has the diameter $D$ as its configuration parameter. If the dimension scheme is to be as shown, then half-spaces $P_1$ and $P_3$ have zero position (translation is zero) parameters, $P_2$ has an $X$ translation of $A$, $P_4$ has a $Y$ translation of $B$, and $C_1$ has $X$ and $Y$ translations of $C$ and $E$ respectively. Thus, this dimensioning scheme requires the position parameters $A$, $B$, $C$, and $E$.

Two other dimensioning schemes are possible for this part. In one of them, we can locate the hole by the distances $F$ and $G$ (shown dashed) instead of $C$ and $E$ respectively. This scheme would then require the position parameters $A$, $B$, $F$, and $G$. In the other scheme, the hole is located by $C$ and $E$ while $P_2$ and $P_4$ are located by $F$ and $G$ respectively, thus requiring the position parameters $C$, $E$, $F$, and $G$. Notice that $A$, $C$, and $F$ cannot exist simultaneously otherwise over-dimensioning would result; only any two parameters are allowed. Similarly, any two parameters of $B$, $E$, and $G$ are permitted to avoid overdimensioning in the $Y$ direction.

The position parameters that determine the locations of primitive half-spaces (part features) can be defined systematically and algorithmically by introducing the concept of dimension chain. A dimensional chain (D-chain) defines a location relative to another location. Thus, a D-chain is a directed chain with positive and negative dimensions. A positive dimension is one that is measured in the positive direction of the corresponding axis and a negative dimension is measured in the negative direction of an axis. There is one D-chain per axis of measurement. Figure 16-37b shows the two D-chains corresponding to the $X$ and $Y$ directions. The $X$-chains for the three dimensioning schemes discussed above are $A$ and $C$, $A$ and $-F$, and $C$ and $F$, and the $Y$-chains are $B$ and $E$, $B$ and $-G$, and $E$ and $G$.

D-chains could be used to further develop the concept of dimension trees (D-trees). D-trees are more useful algorithmically than D-chains. Their traversal can be used to detect existing overdimensioning problems of a given dimensioning scheme, as well as to calculate tolerance stack analysis as seen in the following section. A D-tree is a tree whose nodes (interior or exterior) are half-spaces and whose branches are directed (positive or negative) dimensions. Like D-chains, there is one D-tree per axis of dimensioning. Figure 16-37c shows the $X$-tree and the $Y$-tree for the dimensioning scheme shown in Fig. 16-37a. Two branches per tree exist. The trees of the other two dimensioning schemes are also shown. The trees for both schemes still have the nodes $P_1$, $P_2$, and $C_1$ for the $X$-tree, and $P_3$, $P_4$, and $C_1$ for the $Y$-tree. However, the branches correspond to the D-chains mentioned in the previous paragraph.

Two important rules for manipulating D-chains and D-trees exist. The first rule is related to null dimensions if they occur. Null dimensions have no effect on a D-chain or a D-tree. Thus, the D-chain $A + \text{null} - F - \text{null}$ is equivalent to the chain $A - F$. The second rule is related to condensing a D-chain or a D-tree. A component dimension and its immediate successor cancel one another if and only if they are identical and opposite in direction. The D-chain $A - A$ is null while the D-chain $A + B - B$ is equivalent to $A$. However, the D-chain $A + B - A$ is not equivalent to $B$.

How can D-trees be used to detect overdimensioning? If the dimensioning scheme of a part results in a dimensioning graph (see Chap. 7) and not a tree, then the scheme is inappropriate. Thus, for properly dimensioned parts there is a unique path between any two nodes of D-trees. For example, if we specify the dimensions $A$, $C$, and $F$ simultaneously in the $X$ direction, and $B$, $E$, and $G$ in the $Y$ direction, the $X$-tree and $Y$-tree shown in Fig. 16-37$c$ become closed (have cycles or loops), thus changing from trees to graphs. Thus, overdimensioning occurs if any of the part dimensioning results in graphs and not trees.

While the above discussion is limited to translations of half-spaces, it can be extended in a similar fashion to rotations of the half-spaces as well, resulting in angular D-chains and D-trees. May these chains and trees be decomposed into several independent chains and trees (one per "direction"), or are they able to detect overdimensioning? The answers to these questions are left as an exercise for the reader.

In the context of D-chains or D-trees, conventional tolerances are applied to the branches (dimensions) of these chains or trees. The nominal value of a D-chain can be found by applying "+" or "−" as arithmetic operators to the nominal values of the component dimensions. The tolerance values, however, would require tolerance stack analysis to evaluate.

**Example 16.5.** Sketch the corresponding D-chains and D-tree to the ANSI example part shown in Fig. 16-28.

**Solution.** The part shown in Fig. 16-28 is axisymmetric. Two dimensioning axes can be identified: the horizontal and the radial. Dimensions in the radial axis are the diameters of the various cylindrical faces shown in the figure. A D-chain or D-tree

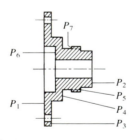

(a) Half-spaces dimensioned horizontally

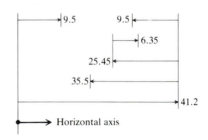

(b) D-chain (horizontal axis)

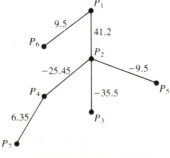

(c) D-tree (horizontal axis)

**FIGURE 16-38**
The D-chain and D-tree (horizontal axis) of the part shown in Fig. 16-28.

for this axis is useless (no chain effect exists) because the diameters themselves are not interrelated although they collectively affect the various radial thicknesses of the part.

Figure 16-38a shows the half-spaces dimensioned in the horizontal direction labeled as $P_1$, $P_2$, etc. Assuming bilateral tolerances, the nominal dimensions of the limits shown in Fig. 16-28 can be easily calculated. Figure 16-38b shows the D-chain and Fig. 16-38c shows the corresponding D-tree, assuming that the horizontal axis originates from the location of half-space $P_1$ (datum $A$) and is positive to the right. It is obvious that no overdimensioning problems exist because the dimensioning scheme shown in Fig. 16-28 results in a tree and not a graph.

Having discussed how the configuration and position parameters of half-spaces are used in conventional tolerances together with D-chains and D-trees, we now turn to their use in geometric tolerancing. The UDT model must support size, position, and form tolerances. Size tolerances can be represented as variations on the configuration parameters of the half-spaces. For example, the size of a hole is equivalent to the diameter of the corresponding cylindrical half-space, and a size tolerance on the hole can be attached to the diameter. Similarly, position tolerances can be represented as variations on the position parameters of the half-spaces. Geometric tolerances are represented as attributes assigned to the proper half-spaces (in the EDT mode, they are attributes of the model faces). Datums are defined by assigning proper half-spaces as such. The UDT model should also be able to define composite half-spaces from primitive ones to allow for defining composite features such as slots.

The UDT model associated with a CSG model of a nominal part is represented by a graph called VGraph (variational graph) whose structure is shown in Fig. 16-39. The VGraph consists of various nodes. The lowest-level entities in the VGraph are the PFace (primitive face) nodes. PFaces are used to evaluate the boundary faces of a CSG model as discussed in Chap. 7. These PFaces are the primitive half-spaces with the proper configuration and position parameters. NFace nodes are the nominal faces of the part. They are sometimes called BFaces (boundary faces). If the VGraph structure shown in Fig. 16-39 is detached from the CSG modeler (by eliminating the PFaces), it can be attached to another modeler by establishing the correlation between NFaces of the VGraph and the nominal faces used by the modeler. For example, one can simply attach the VGraph to the B-rep modeler by identifying NFaces with the faces in the B-rep.

A VFace (variational face) node represents a subset of NFace. It is a user-defined portion of the part's boundary. In almost all cases, VFaces and NFaces are identical. The concept of VFaces is useful if it is necessary to assign tolerance attributes to features that are smaller than those in a nominal representation. As an example, a large plane face (NFace) of a part may be loosely toleranced but have a small region (VFace) that must be tightly toleranced for mating purposes with another part. In another example, NFaces in some solid modelers may be disconnected (cylindrical face in Fig. 16-36b), and one may need to isolate the various connected components as VFaces (faces $F_1$ and $F_2$ in the figure) to assign distinct tolerances to them.

In addition to pointing to an NFace node, a VFace points to three nodes: a solid it belongs to, an MCO (membership classification operation) that may be

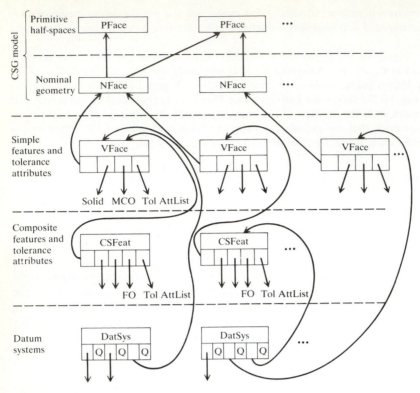

**FIGURE 16-39**
VGraph of the UDT model.

"in," "on," or "out," and a tolerance attribute list which corresponds to the feature control symbol assigned to the VFace. VFaces represent simple features or simple half-spaces.

CSFeat (compound surface feature) nodes represent collections of VFaces (simple features) to form compound features. A CSFeat node points to one or more VFace node, an FO (face operation) node which is always a union (of VFaces), and a tolerance attribute list.

DatSys (datum system) nodes represent datum systems and contain an ordered set of datums (one, two, or three). Each datum is represented by pointing to a VFace nodes and/or CSFeat nodes. Each datum has an associated qualifier ($Q$), typically an MMC or RFS.

In addition to the nodes shown in Fig. 16-39, nodes (called global default nodes) that define global default tolerances that apply to the entire part can be added to the graph. The most important of these is the global size node that specifies the default size tolerance.

Can we relate the UDT model shown in Fig. 16-39 to the EDT model shown in Fig. 16-34? First, let us introduce two simplifications to Fig. 16-39. Assume that NFaces and VFaces are identical and eliminate CSFeat nodes for now. One can then easily realize that NFaces in the VGraph correspond to the faces of the B-rep model in the EDT model. VFace nodes relate to D/T nodes.

The pointer from a VFace to an NFace corresponds to the EL node of the EDT model, and the tolerance attribute list corresponds to the ED node. Finally, the DatSys node corresponds to the DRF node.

> **Example 16.6.** Sketch the VGraph of the UDT model corresponding to the ANSI example part of Fig. 16-28. Assume that the nominal part is a CSG model.

> **Solution.** In this example, let us assume that NFaces of the CSG model exist (after being evaluated from PFaces) and they are as shown in Fig. 16-35a. These NFaces are the same as the VFaces. In addition, no CSFeat nodes exist because given tolerances apply to simple features only. In the VGraph we are about to construct, only geometric tolerances are shown and no global default nodes are considered.
>
> Figure 16-40 shows the resulting VGraph. Field(s) in a node marked with "×" means "not used." In all VFace nodes, the second field is not used because we only have one solid model in this example. (The second field is used to point to more than one solid model.) Also, the third field is not used because all VFaces are identical to NFaces. One datum system is defined. The first field indicates datum $A$ with no qualifier. The second and third fields indicate datums $B$ and $C$ respectively with M (MMC) as a qualifier. Various tolerance attribute lists are shown and correspond to the feature control symbols shown in Fig. 16-28. The M qualifier is removed from datum specifications in these lists because it is already included in the definition of the datum system.

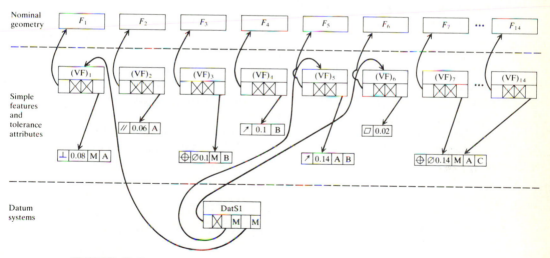

**FIGURE 16-40**
VGraph of the UDT model of the part shown in Fig. 16-28.

## 16.7 TOLERANCE ANALYSIS

Thus far, we have discussed the tolerancing concepts and how tolerances are represented in geometric modelers. We have also seen that tolerance assignments to various components of an assembly must meet its functional requirements and must reduce manufacturing cost. With all tolerances assigned to the various components (parts) of an assembly, the designer must check that the combined effect or accumulation of all these tolerances (tolerance stackup) does not cause an inoperable or malfunctioning assembly. Analysis of tolerances and their stackup is important because tolerance assignments are usually done on a part-by-part basis. Thus, tolerance analysis is defined as the process of checking the tolerances to verify that all the design constraints are met. Tolerance analysis is sometimes known as design assurance.

Tolerance analysis requires two steps (assuming tolerances have already been assigned). First, all dimensions that affect the analysis must be identified. Since dimensions in one direction are often related, the relations among them must be found. All required dimensional relationships and information of a part including all of the individual nominal dimensions and their attached tolerances can be automatically extracted from a solid modeling database using schemes such as D-chains and D-trees, discussed in the previous section. When the extraction of the tolerance information is complete, the second step is to use it to perform tolerance analysis via one of the analysis methods discussed in this section. These methods can be implemented into a D&T modeler to create an automatic tolerance analysis. Manual methods for tolerance analysis are time-consuming and error-prone. For complicated problems, they are usually infeasible.

The objective of tolerance analysis is to determine the variability of any quantity that is a function of product dimensions. Most often, these quantities are also dimensions, and are called design functions. Product dimensions and variables that control the behavior of a design function are called design function variables. The variability of the design functions is used to assess the suitability of a particular tolerance specification. Figure 16-41 shows an example of a design function for the case of two blocks assembled into a slot. The design function $F$ is the clearance between the two blocks, and is a function of the dimensions of the slot and the two blocks. A tolerance specification for these dimensions is satisfactory if it prevents $F$ from being less than zero.

The formulation of tolerance analysis can be stated as follows. Given a set of tolerances $\{T\} = \{T_1, T_2, \ldots, T_n\}$ on a set of dimensions $\{d\} = \{d_1, d_2, \ldots, d_n\}$, and given a set of design constraints $\{C\} = \{C_1, C_2, \ldots, C_m\}$, is $\{T\}$ satisfactory?

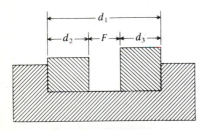

$d_1, d_2, d_3$ = product dimensions
(design function variables)
$F$ = design function
$F = f(d_1, d_2, d_3) = d_1 - d_2 - d_3$
$F \geq 0$, accept assembly
$F < 0$, reject assembly

**FIGURE 16-41**
Formulating a design function for tolerance analysis.

Constraints could be functional requirements of an assembly, manufacturing costs, etc. The dimensions in the set $\{d\}$ include both the nominal dimensions $\{d_N\}$ and their tolerances $\{T\}$, that is, $\{d\} = \{d_N\} + \{T\}$. To assess tolerance suitability, we formulate a design function in terms of $\{d\}$, that is,

$$F = f(\{d\}) = f(d_1, d_2, \ldots, d_n) \tag{16.13}$$

The variability of $F$ due to variability in $\{d\}$ is determined (using methods described below). If $F$ satisfies $\{C\}$ all the time, $\{T\}$ is satisfactory and assembly is accepted. If not, $\{T\}$ is unsatisfactory and assembly is rejected. Design functions are often complex and their formulation form the hardest part of tolerance analysis and can be time-consuming.

   Tolerance analysis methods can be divided into two types. In the simpler type, dimensions have conventional tolerances, and the result of tolerance analysis is the nominal value of the design function ($F_N$) and its upper ($F_{max}$) and lower ($F_{min}$) limits. This type of analysis is sometimes called worst-case analysis. This means that all possible combinations of in-tolerance parts must result in an assembly that satisfies the design constraints. The upper and lower limits of the design function represent the worst possible combination of the tolerances of the design function variables. However, the likelihood of worst-case combination of these tolerances in any particular product is very low. Therefore, worst-case tolerance analysis is very conservative.

   The other type of tolerance analysis is performed on a statistical basis. Tolerance analysis methods of this type allow statistical tolerances and output a statistical distribution for the design function. This allows for more realistic analysis. Manufacturing costs are reduced by loosening up the tolerances, and accepting a calculated risk that the design constraints $\{C\}$ may not be satisfied 100 percent of the time. By assuming a probability distribution for each toleranced dimension, it is possible to determine the likelihood that the specified design limits will be exceeded. Effectively, a reject rate is determined for the assembly. A nonzero reject rate may be preferable to an increase in individual part manufacturing costs due to tight tolerances. Both the worst-case and statistical approaches are important in practice.

## 16.7.1   Worst-Case Arithmetic Method

The arithmetic tolerance method is the worst-case analysis method. It uses the limits of dimensions to carry out the tolerance calculations. The actual or expected distribution of dimensions is not taken into account. All manufactured parts are interchangeable since the maximum values are used. Arithmetic tolerances require greater manufacturing accuracy. It is used in job shop production (very few parts are produced) and in cases where totally or 100 percent interchangeable assembly is required.

   Let us assume a closed-loop (meaning the resultant dimension is obtained by adding and/or subtracting the given dimensions) dimension set $\{d\}$ of $n$ elements such that the design function (resultant dimension) $F$ is obtained by adding the first $m$ elements (called increasing dimensions) and subtracting the last $(n - m)$ elements (called decreasing dimensions). Using this method, all tolerance information about $F$ is obtained by adding and/or subtracting the corresponding information of the individual dimensions. Thus, we can write:

Nominal dimension:

$$F_N = \sum_{i=1}^{m} d_{iN} - \sum_{i=m+1}^{n} d_{iN} \tag{16.14}$$

Maximum dimension:

$$F_{max} = \sum_{i=1}^{m} d_{imax} - \sum_{i=m+1}^{n} d_{imin} \tag{16.15}$$

Minimum dimension:

$$F_{min} = \sum_{i=1}^{m} d_{imin} - \sum_{i=m+1}^{n} d_{imax} \tag{16.16}$$

Tolerance on $F$:

$$T_F = F_{max} - F_{min} = \sum_{i=1}^{m} T_i + \sum_{i=m+1}^{n} T_i = \sum_{i=1}^{n} T_i \tag{16.17}$$

Upper tolerance on $F$:

$$T_{uF} = F_{max} - F_N = \sum_{i=1}^{m} (d_{imax} - d_{iN}) - \sum_{i=m+1}^{n} (d_{imin} - d_{iN})$$

$$= \sum_{i=1}^{m} T_{ui} - \sum_{i=m+1}^{n} T_{Li} \tag{16.18}$$

Lower tolerance on $F$:

$$T_{LF} = F_{min} - F_N = \sum_{i=1}^{m} (d_{imin} - d_{iN}) - \sum_{i=m+1}^{n} (d_{imax} - d_{iN})$$

$$= \sum_{i=1}^{m} T_{Li} - \sum_{i=m+1}^{n} T_{ui} \tag{16.19}$$

where $T_{ui}$ and $T_{Li}$ are the upper and lower tolerances on dimension $d_{iN}$ respectively. For unilateral tolerances, one of these variables is zero.

**Example 16.7.** Figure 16-42 shows a part design with assigned tolerances. Use the arithmetic method to calculate the tolerance information for the axial dimension $F$ of the outside surface shown.

*Solution.* Figure 16-42b and c shows the D-chain and D-tree of the dimensions of the part design. It is obvious that the design function $F$ is affected by the dimensions in its chain. The dimension $d_6$ is independent of the chain and, therefore, is not expected to affect $F$ in the tolerance analysis. There are five dimensions in the chain ($d_1$ to $d_5$) excluding $F$; two of them are increasing (positive) dimensions ($d_1$ and $d_3$) and three are decreasing (negative) dimensions ($d_2$, $d_4$, and $d_5$).

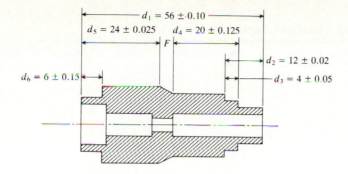

(a) Tolerance information

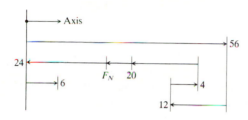

(b) D-chain

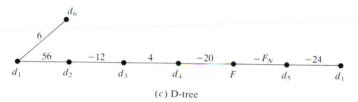

(c) D-tree

**FIGURE 16-42**
Tolerance analysis
of a part design.

Substituting the tolerance information shown in Fig. 16-42a into Eqs. (16.14) to (16.19), we obtain

$$F_N = (56 + 4) - (12 + 20 + 24) = 4 \text{ in}$$

$$F_{max} = (56.1 + 4.05) - (11.98 + 19.875 + 23.975) = 4.32 \text{ in}$$

$$F_{min} = (55.9 + 3.95) - (12.02 + 20.125 + 24.025) = 3.68 \text{ in}$$

$$T_F = 4.32 - 3.68 = 0.64 \text{ in}$$

$$T_{uF} = (0.10 + 0.05) - (-0.02 - 0.125 - 0.025) = 0.32 \text{ in}$$

$$T_{LF} = (-0.10 - 0.05) - (0.02 + 0.125 + 0.025) = -0.32 \text{ in}$$

## 16.7.2   Worst-Case Statistical Method

This method, like the arithmetic method, uses the limits of dimensions to perform tolerance analysis. However, unlike the arithmetic method, it takes into consideration the fact that dimensions of parts of an assembly follow a probabilistic distribution curve. Consequently, the frequency distribution curve of the dimensions of the final assembly follow a probabilistic distribution curve. Typically, the

**TABLE 16.9**
**$\alpha$ and $K$ values of typical distributions**

| Distribution | Normal | Uniform | Quasi-uniform | Triangle | Left skew | Right skew |
|---|---|---|---|---|---|---|
| Shape | | | | | | |
| $\alpha$ | 0 | 0 | 0 | 0 | −0.26 | 0.26 |
| $K$ | 1.0 | 1.73 | 1–1.5 | 1.22 | 1.17 | 1.17 |

probabilistic distribution curve is assumed to be a normal distribution curve. This method is used in both batch and mass production. It allows for variabilities in manufacturing conditions such as tool wear, machine conditions, random errors, etc. It increases the manufacturing efficiency by increasing tolerance limits and, therefore, reducing the required accuracy of manufacturing.

   This method is applied to a closed-loop dimension set $\{d\}$ with each element $d_i$ of the set having a probability distribution curve. The design function $F$ is obtained in the same way as in the arithmetic method. The tolerance information about $F$ [similar to Eqs. (16.14) to (16.19)] can be obtained statistically as follows. Normal distribution is considered the basis of the analysis. Parameters relating other distributions to the normal distribution are shown in Table 16.9. Figure 16-43 shows the parameters of a distribution curve for one of the elements of the dimension set. When the elements in the dimension set become large enough, the distribution of the design function $F$ (the resulting dimension) will be

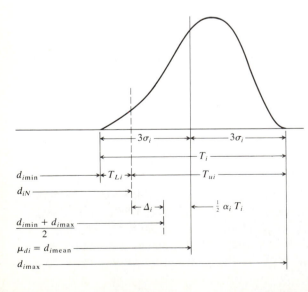

**FIGURE 16-43**
Tolerance information and dimension distribution.

asymptotically normal and independent of the distributions of the individual dimension. Thus we can write

$$F_N = \sum_{i=1}^{m} d_{iN} - \sum_{i=m+1}^{n} d_{iN} \tag{16.20}$$

$$T_F = \sqrt{\sum_{i=1}^{n} K^2 T_i^2} \tag{16.21}$$

$$\sigma_F = \sqrt{\sum_{i=1}^{n} \sigma_i^2} = \frac{T_F}{6} \tag{16.22}$$

$$T_{uF} = \Delta + \tfrac{1}{2} T_F \tag{16.23}$$

$$T_{LF} = \Delta - \tfrac{1}{2} T_F \tag{16.24}$$

$$\Delta = \sum_{i=1}^{m} (\Delta_i + \tfrac{1}{2}\alpha_i T_i) - \sum_{i=m+1}^{n} (\Delta_i + \tfrac{1}{2}\alpha_i T_i) \tag{16.25}$$

The $\sigma = T_F/6$ used in Eq. (16.22) is based on assuming a range of $6\sigma$ for the distribution curve ($3\sigma$ on each side of the mean tolerance), as shown in Fig. 16-43. Equations (16.23) to (16.25) can be viewed as dividing $T_F$ into upper and lower limits $T_{uF}$ and $T_{LF}$ respectively. If bilateral equal tolerance limits are assumed, then $T_{uF} = \tfrac{1}{2} T_F$ and $T_{LF} = -\tfrac{1}{2} T_F$.

**Example 16.8.**  Use the worst-case statistical method to calculate the tolerance information of $F$ for Example 16.7. Assume a normal distribution curve.

**Solution.**  The use of Eqs. (16.20) to (16.22) is straightforward and gives $F_N = 4$ in, $T_F = 0.341$ in, and $\sigma_F = 0.0568$ in. To calculate $T_{uF}$ and $T_{LF}$, notice that $\alpha_i = 0$ for a normal distribution curve. For worst-case analysis, $d_{iN}$ shown in Fig. 16-43 is either $d_{i\min}$ or $d_{i\max}$. In either case, it is shown from the figure that $\Delta_i = (d_{i\max} - d_{i\min})/2 = T_i/2$. Thus Eq. (16.25) becomes

$$\Delta = \sum_{i=1}^{m} \frac{T_i}{2} - \sum_{i=m+1}^{n} \frac{T_i}{2} = \tfrac{1}{2}(0.2 + 0.1) - \tfrac{1}{2}(0.04 + 0.25 + 0.05) = -0.02 \text{ in}$$

Substituting into Eqs. (16.23) and (16.24) gives $T_{uF}$ and $T_{LF}$ as 0.151 and $-0.190$ in respectively.

## 16.7.3  Monte Carlo Simulation Method

The previous two methods are only applicable to conventional tolerances—mainly for closed-loop dimensional sets with linear design functions. When these functions become more complex or nonlinear, applying these methods becomes less obvious, if not impossible. Consider the simple example of a box that has sides of lengths $a$, $b$, and $c$ with 1 percent tolerances on each dimension. To calculate the resulting tolerances on a diagonal of the box, the design function $F$ is the length of the diagonal and given by the relation:

$$F = \sqrt{a^2 + b^2 + c^2} \tag{16.26}$$

To calculate the tolerances on the diagonal using the worst-case arithmetic method, we reduce (or increase) each dimension by 1 percent, which gives a tolerance on the diagonal of $\pm 0.01 \sqrt{a^2 + b^2 + c^2}$. To use the worst-case statistical method is less obvious and may require linearizing Eq. (16.26).

One may conclude that it is not always possible to find a design function to perform tolerance analysis, or even if one is found, it may be difficult to use. In addition, geometric tolerances must be considered in the tolerance analysis. To bring these tolerances into the analysis, two-dimensional (such as area) or three-dimensional (such as volume) design functions may have to be formulated instead of the one-dimensional (dimension) functions used in the previous methods. These two- and three-dimensional functions can be written first in terms of nominal dimensions and then perturbed using the geometric tolerances. While this approach enables geometric tolerances to be included in the tolerance analysis, it still requires a design function which may not be possible to find.

One of the methods that seems useful in performing tolerance analysis using geometric tolerances without a need for an explicit design function is the Monte Carlo simulation method. The idea of the Monte Carlo method stems from the manufacturing practice where a prototype of a given assembly is built and assembled to test its tolerances and functionality. The Monte Carlo method achieves the same result (within the reliability limits of the method) without the time and cost of part manufacturing.

The Monte Carlo method is applicable whether the design variables (these are the design functions $F$ used in the previous two methods) are linear or not. Either a worst-case statistical analysis (using tolerance limits that give $d_{i\max}$ and $d_{i\min}$ as shown in Fig. 16-43) or just a statistical analysis (using tolerance values between tolerance limits that give $d_{iN}$ as shown in Fig. 16-43) can be performed. The method operates by generating (computer random generation) a large sample of assembly instances of the assembly model to be analyzed. Each assembly instance consists of parts instances, each of which corresponds to a set of dimensions that are generated randomly using the statistical distributions assumed for nominal dimensions (Fig. 16-43). Each generated assembly instance is checked to determine whether it meets the specified design constraints or not. Thus a statistical distribution can be generated for each design constraint. Probabilities of accepting or rejecting assemblies can, therefore, be estimated.

An algorithm based on the Monte Carlo method and implemented into a solid modeler with the EDT or UDT model can be described as follows:

1. Generate a candidate instance of an as-manufactured part using a normal (or other) distribution random number generator to perturb the vertices of the part within the specified size tolerance zone.

2. Check if the part instance meets the specified form tolerances. This is needed because form tolerances may be tighter than size tolerances, and because normal distribution may, in rare cases, generate perturbations with standard deviations beyond the size tolerance zone.

3. If one or more of the vertices of the part instance are found out of zones of form tolerances, the part instance is rejected. If all vertices are inside the zone, the part instance may be accepted.

4. Repeat steps 1 to 3 for all other parts in the assembly.
5. Use the solid modeler to create the assembly instance using all the instances created in step 4. These instances are positioned relative to datums established by part features.
6. Check if the assembly instance from step 5 satisfies the design constraints. If yes, the assembly is accepted. If not, the assembly is rejected.
7. Repeat steps 1 to 6 as many times as the desired sample size (number of assembly instances) is used to calculate the statistics. The larger the sample size the better, and the more confidence we have in the results.

In theory, the potential accuracy of the Monte Carlo method is unlimited. In practice, the sample size required to obtain reasonable estimates can be prohibitive. For a large complex assembly, the dimensionality of the sample space is so great that a Monte Carlo analysis, even one based on a large number of assembly instances, does not carry a high degree of reliability. Nevertheless, a Monte Carlo analysis may act to focus attention on potential problem areas.

To generate an instance of a part (step 1 above) depends primarily on whether the solid modeler is a B-rep or CSG. For B-rep models, where faces, edges, and vertices are stored, size and form tolerances are applied as variations to the surface equations of the part faces. Position tolerances are applied to features of position. New edges and vertices are computed at the intersections of the varianced faces. For CSG models, since part geometry is stored in an unevaluated form as an ordered sequence of features (primitive or composite half-spaces), each feature is perturbed first using proper tolerances and then combined to generate a unique instance of the part. This maintains features as the main building block of the assembly and reduces the instancing task to that of simply perturbing features individually.

The random generation of vertices is achieved by random perturbation of the coordinates of the vertices as follows. Given a nominal location $d_N$ of a vertex with a size tolerance $T_s(d_N \pm T_s/2)$, the vertex is moved an amount $x$ according to the size tolerance. Figure 16-43 can be used to interpret this movement. Assuming a normal distribution for $x$ and using $\sigma = T_s/6$, the density distribution function of the variational component $x$ is given by

$$f(x) = \frac{1}{\sigma\sqrt{2\pi}} e^{-x^2/2\sigma^2}, \qquad \sigma > 0 \qquad (16.27)$$

Statistical routines (such as IMSL library routines) can be used to randomly generate a series of normally distributed $x$ values for each vertex.

If vertices are perturbed as mentioned in step 1 of the above algorithm, the solid model must be triangulated before the algorithm begins. This may be unacceptable. In this case, tolerances may be applied directly to equations of surfaces in B-rep or of half-spaces in CSG.

### 16.7.4 Other Methods

Other methods for tolerance analysis exist. Optimization methods do treat the design function and the design constraints as an optimization problem while the design variables are viewed as the decision variables for optimization. The design function may be nonlinear. Linear programming methods linearize the design function and design constraint equations, and solve tolerance analysis using linear programming. After the linear programming problem is solved, a sensitivity analysis can be performed to determine the relative contribution of each of the tolerances.

The Taylor series method and the quadrature method are statistical methods, and they approximate the probability density function of the design function without generating the large number of samples required by the Monte Carlo method.[6]

Tolerance charts are a semi-graphical, spreadsheetlike, method which can be used to formulate design functions and perform tolerance analysis on the resulting function. This method deals only with one-dimensional problems.[7]

Bjorke's method formulates design functions and performs their statistical analysis. The method uses the concept of tolerance chain and can deal with general three-dimensional problems. However, it cannot accommodate design functions that are not dimensions. Bjorke assumes that the design function and the dimensions have a beta ($\beta$) distribution.[8]

## 16.8  TOLERANCE SYNTHESIS

In tolerance analysis, dimension tolerances are known and the tolerance on a resultant dimension is to be determined. In terms of a design function, the variability of the function is to be obtained. Tolerance synthesis (also called tolerance distribution) is the inverse problem; the allowable tolerance, or allowable variations in the design function, on a dimension is known and the dimension tolerances are to be determined. In terms of Fig. 16-44, given the tolerance on $A$, $\pm \Delta A$, find the tolerances on $B$, $C$, and $D$ such that their tolerance analysis produces back the tolerance $\pm \Delta A$ on $A$.

There are two criteria to distribute a tolerance: the equal tolerance criterion and the equal precision criterion. The former simply distribute the tolerance equally among dimensions that affect the given tolerance. While it is easy to

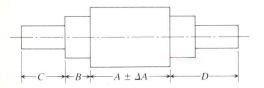

**FIGURE 16-44**
Tolerance synthesis.

---

[6] Interested readers are referred to D. H. Evans, "Statistical Tolerancing: The State of the Art, Part I: Background," *J. Quality Technology*, vol. 6, no. 4, pp. 188–195, 1974, and "Statistical Tolerancing: The State of the Art, Part II: Methods for Estimating Movements," *J. Quality Technology*, vol. 7, no. 1, pp. 1–12, 1975.

[7] See O. R. Wade, *Tolerance Control in Design and Manufacturing*, Industrial Press, New York, 1967.

[8] Refer to O. Bjorke, *Computer-Aided Tolerancing*, Tapir Publisher, Trondheim, Norway, 1978.

compute manually, it is not accurate based on design and manufacturing experience. The latter is more suitable to real designs, but it involves more complicated calculations.

When the nominal dimensions are less than or equal to 500 mm, there is a relation between the standard tolerance and precision in ISO:[9]

$$T_i = aI_i = a(C\sqrt[3]{d_{iav}} + 0.001d_{iav}) \qquad (16.28)$$

where $C$ is a constant, $a$ is the precision coefficient for different IT grades, $d_{iav}$ is the sectioned standard length value, and $I_i$ is the fundamental tolerance unit. The first term of $I_i$ in the above equation accounts for manufacturing errors, and is a statistical result of machine shop data. The second term includes temperature change error and measurement error caused by dimension change. It is very small and is neglected. The length $d_{iav}$ is used instead of the actual dimension $d_i$ to reduce the number of constants in the equation as shown. There are tables[10] that relate $d_{iav}$ to $d_i$. Equation (16.28) forms the basis of two methods for tolerance synthesis discussed below.

### 16.8.1   Arithmetic Method

Neglecting the second term in Eq. (16.28) gives

$$T_i = aI_i = aC\sqrt[3]{d_{iav}} \qquad (16.29)$$

Substituting this equation into Eq. (16.17) gives

$$a = \frac{T_F}{C\sum_{i=1}^{n}\sqrt[3]{d_{iav}}} \qquad (16.30)$$

Substituting $a$ given by Eq. (16.30) back into (16.29) gives the tolerance $T_i$ on each dimension in the dimension set.

The tolerance $T_i$ can be divided into upper and lower limits in various ways. For a shaft-based dimension, $T_{ui} = 0$, $T_{Li} = -T_i$, for a hole-based dimension, $T_{ui} = T_i$, $T_{Li} = 0$. For any other less critical dimension, we have:

For an increasing dimension:

$$T_{uadj} = T_{uF} + \sum_{m+1}^{n} T_{Li} - \sum_{i=1, i\neq adj}^{m} T_{ui} \qquad (16.31)$$

$$T_{Ladj} = T_{LF} + \sum_{m+1}^{n} T_{ui} - \sum_{i=1, i\neq adj}^{m} T_{Li} \qquad (16.32)$$

---

[9] See Z. Dong and A. Soom, "Automatic Tolerance Analysis from a CAD Database," Design Engineering Technical Conference paper 86-DET-36, pp. 1–8, 1986, and E. T. Fortini, *Dimensioning for Interchangeable Manufacture*, Industrial Press, New York, 1967.

[10] See Z. Dong and A. Soom, "Automatic Tolerance Analysis from a CAD Database," Design Engineering Technical Conference paper 86-DET-36, pp. 1–8, 1986.

For a decreasing dimension:

$$T_{uadj} = -T_{uF} + \sum_{i=1}^{m} T_{Li} - \sum_{i=m+1,\, i \neq adj}^{n} T_{ui} \tag{16.33}$$

$$T_{Ladj} = -T_{LF} + \sum_{i=1}^{m} T_{ui} - \sum_{i=m+1,\, i \neq adj}^{n} T_{Li} \tag{16.34}$$

## 16.8.2 Statistical Method

Substituting Eq. (16.29) into Eq. (16.21) gives

$$a = \frac{T_F}{\sqrt{\displaystyle\sum_{i=1}^{n} K_i^2 (C \sqrt[3]{d_{iav}})^2}} \tag{16.35}$$

Substituting this equation into Eq. (16.29) produces the tolerance $T_i$ on each dimension in the dimension set.

The division of $T_i$ into upper and lower limits for a shaft- or hole-based dimension is the same as in the arithmetic method. Modifications of the tolerance of other dimensions are given by (assuming $\alpha_F = 0$)

$$T_{adj} = \frac{1}{K_{adj}} \sqrt{T_F^2 - \sum_{i=1}^{n} K_i^2 T_i^2} \tag{16.36}$$

For an increasing dimension:

$$\Delta_{adj} = \Delta_F + \sum_{m+1}^{n} (\Delta_i + \tfrac{1}{2}\alpha_i T_i) - \sum_{i=1,\, i \neq adj}^{m} (\Delta_i + \tfrac{1}{2}\alpha_i T_i) - \tfrac{1}{2}\alpha_{adj} T_{adj} \tag{16.37}$$

For a decreasing dimension:

$$\Delta_{adj} = \Delta_F + \sum_{i=1}^{m} (\Delta_i + \tfrac{1}{2}\alpha_i T_i) - \sum_{i=m+1,\, i \neq adj}^{n} (\Delta_i + \tfrac{1}{2}\alpha_i T_i) - \tfrac{1}{2}\alpha_{adj} T_{adj} \tag{16.38}$$

where

$$\Delta_i = \tfrac{1}{2}(\Delta_{ui} + \Delta_{Li}) \tag{16.39}$$

Then the adjusted dimension is

$$d_{adj} + \Delta_{adj} \pm \frac{T_{adj}}{2} \tag{16.40}$$

where $d_{adj}$ is $d_i$ to be adjusted.

## 16.8.3 Taguchi and Other Methods

As in the case of tolerance analysis, the above two methods for tolerance synthesis are limited in their application. Some other methods are based on mathematical programming.[11] Another method is the Taguchi method which is based on the principles of experimental design.[12]

The Taguchi method not only determines tolerances but also determines the ideal nominal values for the dimensions. This is referred to as dimension centering. The method finds the nominal dimensions that allow the largest, lowest-cost tolerances to be assigned. It selects dimensions and tolerances with regard to their effect on a single design function. The method uses fractional factorial experiments to find the nominal dimensions and tolerances that maximize the so-called signal-to-noise (S/N) ratio. The signal is a measure of how close the design function is to its desired nominal value. The noise is a measure of the variability of the design function caused by tolerances.

The main disadvantage of the Taguchi method is its inability to handle more than one design function. Finding one design function for a product may not be at all practical.

## 16.9  DESIGN AND ENGINEERING APPLICATIONS

The section demonstrates the utilization of the material covered in the chapter in some design and engineering applications. Readers are encouraged to check their CAD/CAM systems to find what they offer in tolerance modeling, representation, analysis, and synthesis.

**Example 16.9.** Figure 16-45 shows a part with a hole and slot. Required datums and tolerances to manufacture the part are shown. Write a command sequence in English-like syntax to create the nominal solid model and its associated D&T model.

*Solution.* This example shows the reader how to input tolerance information into a solid modeler at the user level. The command sequence about to be written assumes that the solid modeler supports boolean operations. We have mentioned in Chap. 7 that most, if not all, solid modelers support these operations, due to their efficiency, regardless of the modeler's internal representation.

The command sequence is divided into two main blocks: a nominal definition block and a tolerance definition block. The former defines the nominal untolerated part as boolean combinations of primitives. The latter specifies both conventional and geometric tolerances. The advantage of having two separate blocks is that it enables the designer to concentrate on creating the (nominal) design first.

There are four geometric tolerances shown in Fig. 16-45: two size tolerances on both the hole and the slot, and two true position tolerances on the hole and the slot. The remaining shown tolerances are conventional tolerances (these may also be

[11] See P. Martino and G. A. Gabriele, "A Review of Tolerance Design Techniques for Computer Integrated Manufacturing," in *Proceedings of the ASME International Computers in Engineering Conference and Exhibition*, vol. 2, pp. 343–350, 1987.

[12] See G. Taguchi, E. A. Elsayed, and T. Hsiang, *Quality Engineering in Production Systems*, McGraw-Hill, New York, 1989.

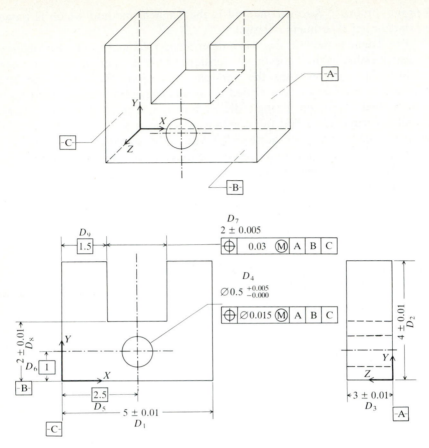

**FIGURE 16-45**
A block with a hole and a slot.

viewed as size tolerances). The hole is treated as a simple feature while the slot is treated as a composite feature with regard to geometric tolerances.

The command sequence may look as follows:

```
C BEGIN PART DEFINITION
   BEGIN PDEF
C BEGIN NOMINAL DEFINITION
   BEGIN NDEF
C DEFINE DIMENSIONS AS VARIABLES
   D1 = 5, D2 = 4, D3 = 3, D4 = 0.5, D5 = 2.5
   D6 = 1, D7 = 2, D8 = 2, D9 = 1.5
C DEFINE THE BLOCK
   B = BLOCK (X = D1, Y = D2, Z = D3)
C DEFINE THE HOLE
   H = CYLINDER (D = D4, L = D3) LOCATED AT (X = D5, Y = D6, Z = 0)
C DEFINE THE SLOT
   S = BLOCK (X = D7, Y = D2-D8, Z = D3)
   LOCATED AT (X = D9, Y = D8, Z = 0)
```

```
C DEFINE THE PART
   P = B-H-S
C END NOMINAL DEFINITION
   END NDEF
C BEGIN TOLERANCE DEFINITION
   BEGIN TDEF
C DEFINE DATUMS A, B, AND C
   A = ZFACE LOCATED AT (X = 0, Y = 0, Z = 0)
   B = YFACE LOCATED AT (X = 0, Y = 0, Z = 0)
   C = XFACE LOCATED AT (X = 0, Y = 0, Z = 0)
C DEFINE DATUM REFERENCE SYSTEM AS A = PRIMARY DATUM,
C B = SECONDARY DATUM, AND C = TERTIARY DATUM
   D = DATUM SYSTEM (A, B, C)
C DEFINE TRUE POSITION DIMENSIONS AS SUCH
   D5 = TRUE POSITION DIMENSION
   D6 = TRUE POSITION DIMENSION
   D9 = TRUE POSITION DIMENSION
C DEFINE TOLERANCES ON THE HOLE
   T1(D4) = SIZE TOLERANCE (0.005, 0)
C IDENTIFY THE HOLE SURFACE AS A FEATURE AND ASSIGN POSITION
C TOLERANCE TO IT
   F1 = ZSURFACE (H)
   T2 (F1, D) = POSITION TOLERANCE (0.015(MMC))
C DEFINE TOLERANCES ON THE SLOT
   T1(D7) = SIZE TOLERANCE (0.005, −0.005)
C IDENTIFY THE SLOT AS A COMPOSITE FEATURE AND ASSIGN
C POSITION TOLERANCE TO IT
   F2 = YZSURFACE (S)
   T2 (F2, D) = POSITION TOLERANCE (0.03(MMC))
C DEFINE A DEFAULT TOLERANCE FOR UNTOLERANCED
C DIMENSIONS
   T3 = DEFAULT SIZE TOLERANCE (0.01, −0.01)
C END TOLERANCE DEFINITION
   END TDEF
C END PART DEFINITION
   END PDEF
```

Some observations regarding the above command sequence are worth noting:

1. Dimensions are defined as variables at the beginning of the nominal definition block to be able to assign tolerances to them later in the tolerance definition block.

2. The width of the slot in the $Y$ direction is used as $D_2 - D_8$ in the slot definition command to reflect the effect of their tolerances on the slot width in the $Y$ direction. This is a good example to show that if no consideration is given to tolerances or manufacturing, the designer may just use $2(D_2 - D_8)$ as the $y$ dimension for the block.

3. The faces that define the datums are identified by the directions of their normals. For example, datum $A$ is defined as ZFACE because its normal is in the $Z$ direction.

4. The dimensions $D_5$, $D_6$, and $D_9$ are identified as true position dimensions so that default size tolerance ($\pm 0.01$) is not applied to them.

5. When a size tolerance is defined, the related dimension is used as an argument for the tolerance variable, e.g., $T_1(D_4)$ is the tolerance on dimension $D_4$.

6. The hole feature $F_1$ is defined as the $Z$ surface (the hole is parallel to the $Z$ axis) of the hole. $F_1$ is considered a simple feature.

7. $T_2 (F_1, D)$ indicates a tolerance on $F_1$ relative to the datum system $D$. The MMC qualifier is used in the related position tolerance command.

8. $F_2$ is the slot (composite) feature. It is defined as the two faces parallel to the $YZ$ plane via the function $YZSURFACE$. This means that the faces move rigidly together, that is, their relative location does not change due to position tolerance. However, this location can change due to the slot size location. In this way, the effects of the size and position tolerances on the slot size and position are decoupled as the two tolerances are intended to imply.

9. Comment statements are included in the command sequence to facilitate understanding it. A comment statement begins with the letter "C."

Once the above command sequence is entered via a suitable user interface, either an EDT or UDT model is created and stored in the model database. Such a model can be used for tolerance analysis using the Monte Carlo simulation method.

**Example 16.10.** Figure 16-46 shows an assembly consisting of an inverted U-shape $A$ with two parts $B$ and $C$ that must "just" fit into it when put end to end. The three parts have a 0.1 percent tolerance of their basic dimensions. The designer responsible for this design must answer the following questions:

1. Is interference likely to happen in the assembly?
2. If it does, what is the percentage of assemblies that must be rejected due to interference?

**Solution.** Figure 16-46 shows the three dimensions of $d_1$, $d_2$, and $d_3$ that control the assembly behavior. Let us consider $d_1$ as the design function $F$ and $d_2$ and $d_3$ as the design variables. To check for interference, consider the given tolerance on $d_1$ ($^{+0.006}_{-0.000}$) as a design constraint. Thus, the tolerance analysis problem is formulated as follows. Given $\{T\} = \{\pm 0.002, \pm 0.004\}$ on $\{d_N\} = \{2, 4\}$, and given $\{C\} = \{^{+0.006}_{-0.000}\}$ and assuming $F = d_1 = d_2 + d_3$, will interference occur?

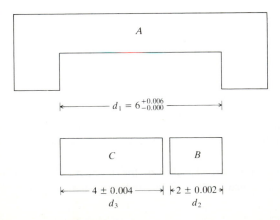

$$d_1 = 6^{+0.006}_{-0.000}$$

$C$          $B$

$4 \pm 0.004$    $2 \pm 0.002$
$d_3$          $d_2$

**FIGURE 16-46**
A three-parts assembly.

The answer to the above question entails finding the tolerance $T_F$ on $F$ and comparing it with $\{C\}$. Using Eq. (16.21) with $K_i = 1$ gives

$$T_F = \sqrt{(0.004)^2 + (0.008)^2} = 0.0089 \text{ in}$$

This means that $\{T_F\} > \{C\}$, which means that interference is bound to happen.

To find the rejection percentage due to interference, let us find the mean $\mu_F$ and the standard deviation $\sigma_F$ of the sample $F$. The means $\mu_B$ and $\mu_C$ of the samples $B$ and $C$ are 2 and 4 respectively. The mean $\mu_F$ is calculated as $\mu_B + \mu_C = 6$. Using Eq. (16.22) gives $\sigma_F = 0.0015$. Thus the probability density function $f(z)$ of the design function $F$, assuming normal distribution, is given as

$$f(z) = \frac{1}{\sqrt{2\pi}} e^{-z^2/2}$$

This is a standard normal distribution with $z$ as the normalized variable, that is, $z = (x - \mu_F)/\sigma_F$, where $x$ is any dimension. To find the probability of rejection we substitute the mean value of $d_1$ which is $(d_{1\max} + d_{1\min})/2 = 6.003$ for $x$ to obtain $z = 2$. Using probability tables for a standard normal distribution gives a probability of 0.0228 for $z = 2$. Thus the rejection percentage of the assembly at hand due to interference is 2.28 percent.

*Note:* readers should be familiar with population combinations in statistics. For a population $R = X_1 \pm X_2 \pm \cdots \pm X_n$, the mean and standard deviation are

$$\mu_R = \mu_1 \pm \mu_2 \pm \cdots \pm \mu_n$$

$$\sigma_R = \sqrt{\sigma_1^2 + \sigma_2^2 + \cdots + \sigma_n^2}$$

# PROBLEMS

## Part 1: Theory

**16.1.** Describe in detail the various available dimensioning schemes.

**16.2.** Find the basic rules that ANSI requires in dimensioning any drawing.

**16.3.** Cylindrical fits are fits between shafts and holes as shown in Figs. 16-1 to 16-4. For a nominal diameter $d = 1.75$ inches, calculate the toleranced and preferred dimensions for:

(a) Clearance fit with class fit 2 (free fit) where $h = 0.0013\sqrt[3]{d}$, $s = 0.0013\sqrt[3]{d}$, and $a = 0.0014\sqrt[3]{d}$.

(b) Transition fit with class fit 4 (snug fit) where $h = 0.0006\sqrt[3]{d}$, $s = 0.0004\sqrt[3]{d}$, and $a = 0$.

(c) Interference fit with class fit 6 (tight fit) where $h = 0.0006\sqrt[3]{d}$, $s = 0.0006\sqrt[3]{d}$, and $a = 0.00025d$.

(d) Interference fit with class fit 7 (medium force fit) where $h = 0.0006\sqrt[3]{d}$, $s = 0.0006\sqrt[3]{d}$, and $a = 0.0005d$.

Sketch the bar diagrams for each fit taking into consideration the actual values of $h$, $s$, and $a$.

**16.4.** Calculate the tolerance zones $h$ and $s$, the allowance $a$, and the limits for the following fits:

RC8: basic size = 2.5000 in
LC3: basic size = 7.0000 in
LT4: basic size = 10.0000 in
LN2: basic size = 12.0000 in
FN2: basic size = 3.0000 in

**16.5.** Interpret the positional tolerances shown in Fig. P16-5. Sketch the tolerance zone(s) for each case.

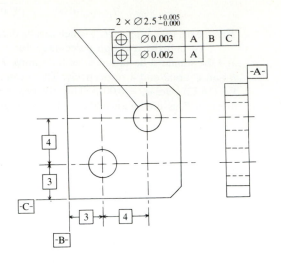

$2 \times \varnothing 2.5^{+0.005}_{-0.000}$

| ⊕ | $\varnothing 0.003$ | A | B | C |
|---|---|---|---|---|
| ⊕ | $\varnothing 0.002$ | A | | |

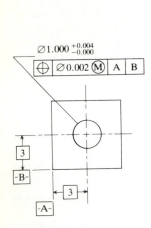

$\varnothing 1.000^{+0.004}_{-0.000}$

| ⊕ | $\varnothing 0.002$ Ⓜ | A | B |
|---|---|---|---|

**FIGURE P16-5**

**16.6.** Two parts of the one shown in Fig. P16-6 are to be assembled with pins or bolts. Derive the positional tolerance $T_p$ on the hole center to satisfy the following fits:
(a) Clearance fit 2 H8/f7.
(b) Transition fit 2 H7/k6.
(c) Interference fit 2 H7/s6.
Sketch the tolerance zone for each fit and compare their sizes. Write the tolerance specifications for each fit.

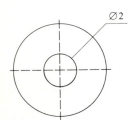

$\varnothing 2$

**FIGURE P16-6**

**16.7.** Interpret the various geometric tolerances used in the example shown in Fig. 16-19, as well as those shown in Fig. P16-7 on pages 831 and 832.

**16.8.** Develop the EDT models for tolerances shown in Probs. 16.5 and 16.7.

**16.9.** Sketch the D-chains and D-trees for the models shown in Probs. 16.5 and 16.7. Are there any overdimensioning cases? Why?

**16.10.** Sketch the VGraphs of the UDT models for tolerances shown in Probs. 16.5 and 16.7.

**16.11.** Identify the design function(s) $F$ in the part designs shown in Fig. P16-11. Then use the worst-case arithmetic method of tolerance analysis to calculate the tolerance information for $F$.

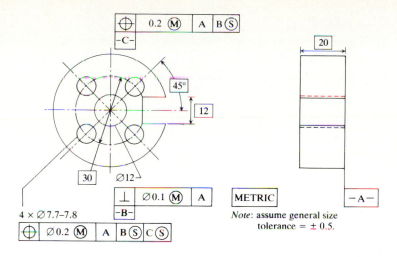

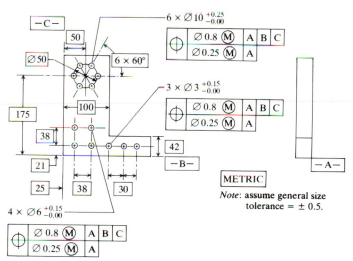

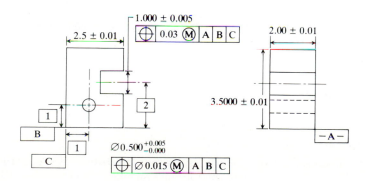

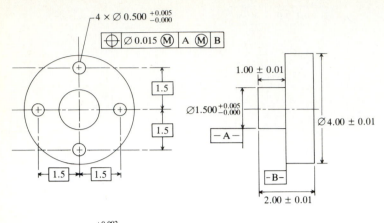

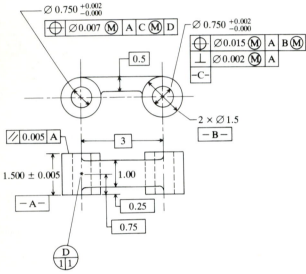

**FIGURE P16-7**

**16.12.** Use the worst-case statistical method of tolerance analysis to calculate the tolerance information for *F* for parts shown in Prob. 6.11.

**16.13.** A shaft and a hole of nominal diameters of 1 inch are shown in Fig. P16-13.

    (*a*) Will the assembly have a clearance of less than 0.006 in?

    (*b*) What is the percentage of assembly rejection due to a clearance of less than 0.006 in?

## Part 2: Laboratory

**16.14.** Find the dimension commands and their respective modifiers that are supported by your CAD/CAM systems. Are there any tolerance modifiers? What types of tolerances (conventional or geometric) do they support?

**16.15.** Does your system have tolerance analysis or synthesis commands? If it does what type has it (arithmetic, statistical, Monte Carlo, etc.)?

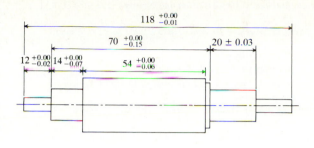

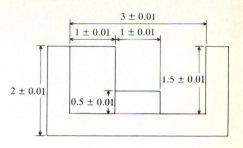

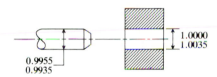

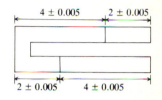

**FIGURE P16-11**

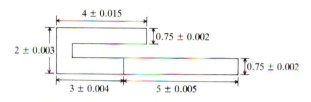

**FIGURE P16-13**

**16.16.** Write an English-like command sequence to create and tolerance the parts shown in Probs. 16.5 and 16.7. Use solid modeling theory.

## Part 3: Programming

**16.17.** Write a program that can read a text string of a dimension and its tolerance, and then separate the basic dimension and the tolerance into two numerical values.

**16.18.** Write a program that implements the worst-case arithmetic method for tolerance analysis.

**16.19.** Repeat Prob. 16.18 but for the worst-case statistical method.

**16.20.** Write a program that can randomly generate coordinates of a point (vertex) using IMSL library routines. Assume a size tolerance zone $T_s$ on the point and study the statistical behavior of the point.

## BIBLIOGRAPHY

Ahluwalia, R. S., and A. V. Karolin: "CAT-A Computer Aided Tolerance Control System," *J. of Manufacturing Systems*, vol. 3, no. 2, pp. 153–160, 1984.

Bjorke, O.: *Computer-Aided Tolerancing*, Tapir Publisher, Trondheim, Norway, 1978.

Burkett, W. C.: "PDDI Approach to Dimensioning and Tolerancing a Solid Model," Report P-85-ASPP-02, Computer Aided Manufacturing—International, Inc., Arlington, Texas, 1985.

Chang, T., and E. Wysk: *An Introduction to Automated Process Planning Systems*, Prentice-Hall, Englewood Cliffs, N.J., 1985.

Deutschman, A. D., W. J. Michels, and C. E. Wilson: *Machine Design: Theory and Practice*, Macmillan, New York, 1975.

Dieter, G.: *Engineering Design*, McGraw-Hill, New York, 1983.

"Dimensioning and Tolerancing," ANSI Y14.5-1982, 1983.

Dong, Z., and A. Soom: "Automatic Tolerance Analysis from a CAD Database," Design Engineering Technical Conference paper 86-DET-36, pp. 1–8, 1986.

Doughtie, V. L., A. Vallance, and L. F. Kreisle: *Design of Machine Members*, 4th ed., McGraw-Hill, New York, 1964.

Emanuel, R.: "Establishing Mechanical Tolerances with CAD/CAM," *Computer Aided Engng*, vol. 4, no. 7, pp. 76–80, 1985.

Evans, D. H.: "Statistical Tolerancing: The State of the Art, Part I: Background," *J. Quality Technol.*, vol. 6, no. 4, pp. 188–195, 1974.

Evans, D. H.: "Statistical Tolerancing: The State of the Art, Part II: Methods for Estimating Movements," *J. Quality Technol.*, vol. 7, no. 1, pp. 1–12, 1975.

Faux, I. D.: "Preliminary Study of the Requirements for the Incorporation of Dimensions and Tolerances in a Geometric Model," Report R-81-GM-03, Computer Aided Manufacturing—International, Inc., Arlington, Texas, 1981.

Fortini, E. T.: *Dimensioning for Interchangeable Manufacture*, Industrial Press, New York, 1967.

French, T. E., C. J. Vierck, and R. J. Foster: *Graphics Science and Design*, 4th ed., McGraw-Hill, New York, 1984.

Giesecke, F. E., A. Mitchell, H. C. Spencer, I. L. Hill, and J. T. Dygdon: *Technical Drawing*, 8th ed., Macmillan, 1986.

Hillyard, R. C., and I. C. Braid: "Analysis of Dimensions and Tolerances in Computer-Aided Mechanical Design," *Computer Aided Des.*, vol. 10, no. 3, pp. 161–166, 1978.

Hoffmann, P.: "Analysis of Tolerances and Process Inaccuracies in Discrete Part Manufacturing," *Computer Aided Des.*, vol. 14, no. 2, pp. 83–88, 1982.

Johnson, R. H.: "Dimensioning and Tolerancing Final Report," Report R-84-GM-02.2, Computer Aided Manufacturing—International, Inc., Arlington, Texas, 1985.

Lagodimos, A. G., and A. J. Scarr: "Computer-Aided Selection of Interference Fits," *Computer in Mech. Engng.*, vol. 2, no. 2, pp. 49–55, 1983.

Lagodimos, A. G., and A. J. Scarr: "Interactive Computer Program for the Selection of Interference Fits," *Computer Aided Des.*, vol. 16, no. 5, pp. 272–278, 1984.

Light, R., and D. Gossard: "Modification of Geometric Models through Variational Geometry," *Computer Aided Des.*, vol. 14, no. 2, pp. 209–214, 1982.

Lin, V. C., D. Gossard, and R. Light: "Variational Geometry in Computer-Aided Design," *Computer Graphics*, vol. 15, no. 3, pp. 171–177, 1981.

Marrelli, R. S.: "Clear Part Specifications with Geometric Tolerancing," *Mach. Des.*, vol. 57, no. 19, pp. 119–122, 1985.

Martino, P., and G. A. Gabriele: "A Review of Tolerance Design Techniques for Computer Integrated Manufacturing," in *Proceedings of the ASME International Computers in Engineering Conference and Exhibition*, vol. 2, pp. 343–350, 1987.

Pandit, V., and J. M. Starkey: "Mechanical Tolerance Analysis Using a Statistical Generate-and-Test Procedure," in *Proceedings of the ASME International Computers in Engineering Conference and Exhibition*, vol. 2, pp. 29–34, 1988.

Parkinson, D. B.: "The Application of Reliability Methods by Tolerancing," *ASME J. Mech. Des.*, vol. 104, no. 3, pp. 612–618, 1982.

Parkinson, D. B.: "Tolerancing of Component Dimensions in CAD," *Computer Aided Des.*, vol. 16, no. 1, pp. 25–32, 1984.

Parkinson, D. B.: "Assessment and Optimization of Dimensional Tolerances," *Computer Aided Des.*, vol. 17, no. 4, pp. 191–199, 1985.

Requicha, A. A. G.: "Toward a Theory of Geometric Tolerancing," *Int. J. Robotics Res.*, vol. 2, no. 4, pp. 45–60, 1983.

Requicha, A. A. G.: "Representation of Tolerances in Solid Modeling: Issues and Alternative Approaches," in *Solid Modeling by Computers* (Eds. M. S. Pickett and J. W. Boyse), Plenum Publishing Corporation, pp. 3–22, 1984.

Requicha, A. A. G., and S. C. Chan: "Representation of Geometric Features, Tolerances, and Attributes in Solid Modelers Based on Constructive Geometry," *IEEE J. Robotics and Automn*, vol. RA-2, no. 3, pp. 156–166, 1986.

Rossignac, J. R., and A. A. G. Requicha: "Offsetting Operations in Solid Modeling," *Computer Aided Geometric Des.*, vol. 3, pp. 129–148, 1986.

Shepherd, D.: "Geometric Tolerancing: Key to Assembly and Interchangeability," *Mach. Des.*, vol. 58, no. 18, pp. 65–67, 1986.

Shepherd, D.: "Improved Productivity with Geometric Tolerancing," *Mach. Des.*, vol. 58, no. 29, pp. 131–133, 1986.

Shigley, J. E., and C. R. Mischke: *Mechanical Engineering Design*, 5th ed., McGraw-Hill, New York, 1989.

Spotts, M. F.: *Design of Machine Elements*, 6th ed., Prentice-Hall, Englewood Cliffs, N.J., 1985.

Taguchi, G., E. A. Elsayed, and T. Hsiang: *Quality Engineering in Production Systems*, McGraw-Hill, New York, 1989.

Turner, J. U., and M. J. Wozny: "Tolerances in Computer-Aided Geometric Design," *Visual Computer*, vol. 3, no. 4, pp. 214–226, 1987.

Turner, J. U., M. J. Wozny, and D. D. Hoh: "Tolerance Analysis in Solid Modeling Environment," in *Proceedings of the ASME International Computers in Engineering Conference and Exhibition*, vol. 2, pp. 169–175, 1987.

Varghese, M., and J. Atkinson: "Automated Dimensional Tolerancing on a Turnkey CAD System," 2nd International Conference on Computer-Aided Production Engineering, pp. 43–48, 1987.

Wade, O. R.: *Tolerance Control in Design and Manufacturing*, Industrial Press, New York, 1967.

# CHAPTER
# 17

## MASS PROPERTY CALCULATIONS

### 17.1 INTRODUCTION

Mass property calculations is one of the earliest engineering applications implemented into CAD/CAM systems. This is, perhaps, due to the strong dependence of these calculations on geometry and topology of objects. These calculations typically involve masses, centroids (centers of gravity), and inertial properties (moments of inertia). They form the basis for the study and analysis of rigid as well as deformable body mechanics (statics and dynamics). For various shapes, one can create their geometric models first, and then use them to calculate their mass properties which can later be used for analysis.

Mass property calculations usually involve evaluating various integrals. Exact evaluation of these integrals is only possible for simple shapes. For complex shapes, approximate methods are usually used to evaluate these integrals. These methods have the important property that they monotonically converge to the exact solution which is, of course, not known.

Mass property algorithms are directly influenced by the type of geometric model used. Algorithms based on wireframe and surface models are not automatic and require a substantial amount of user input and preparation. Algorithms based on solid modeling are, as expected, fully automatic and require no additional input except mass attributes such as the density of the model.

This chapter covers the formulation and evaluation of mass properties (mass, centroid, and inertia). Geometrical properties, including length, area, and volume, are covered also to provide a foundation for mass property calculations. The chapter also covers the concept of composite masses and how it is used to enable calculating mass properties of complex objects. The generic features of

software and user input required by CAD/CAM systems for mass property calculations are included in the chapter together with some design and engineering
applications.

## 17.2  GEOMETRICAL PROPERTY FORMULATION

In this section, we develop the equations needed to calculate geometrical properties, specifically length, area, surface area, and volume. These properties form the
basis for mass property calculations. As seen in this section, length, area, and
volume are formulated as single, double, and triple integrals respectively.

### 17.2.1  Curve Length

Calculating the length of a given curve between two endpoints is useful in many
applications. In mechanism analysis, we might be interested in calculating the
length of a given locus in space. In mass property calculations, an area integral
can be evaluated by changing it to a line integral using Green's theorem. In other
general applications, a curve length may be used in various geometrical and
spatial design problems.

   To calculate the length of a spatial curve between two given points $P_1$ and
$P_2$, consider the curve shown in Fig. 17-1. Given an incremental length $\Delta L$ of the
curve, the curve total length between $P_1$ and $P_2$ can be given by the following

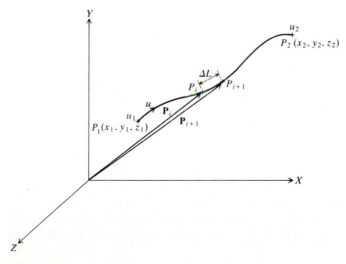

**FIGURE 17-1**
Length of a spatial curve.

integral:

$$L = \int_{P_1}^{P_2} dL \tag{17.1}$$

Curves are usually represented in a parametric form as discussed in Chap. 5 and given by Eq. (5.3). If the length element $\Delta L$ is bounded by the points $P_i$ and $P_{i+1}$ as shown in Fig. 17-1, $\Delta L$ can be approximated by the length of the vector connecting the two points, that is,

$$\Delta L = |\mathbf{P}_{i+1} - \mathbf{P}_i| \tag{17.2}$$

or

$$\Delta L = \sqrt{(x_{i+1} - x_i)^2 + (y_{i+1} - y_i)^2 + (z_{i+1} - z_i)^2} \tag{17.3}$$

or

$$\Delta L = \sqrt{(\Delta x)^2 + (\Delta y)^2 + (\Delta z)^2} \tag{17.4}$$

Dividing both sides by $\Delta u$ and taking the limit when $u$ approaches zero, we get

$$\lim_{u \to 0} \frac{\Delta L}{\Delta u} = \lim_{u \to 0} \sqrt{\left(\frac{\Delta x}{\Delta u}\right)^2 + \left(\frac{\Delta y}{\Delta u}\right)^2 + \left(\frac{\Delta z}{\Delta u}\right)^2} \tag{17.5}$$

or

$$\frac{dL}{du} = \sqrt{\left(\frac{dx}{du}\right)^2 + \left(\frac{dy}{du}\right)^2 + \left(\frac{dz}{du}\right)^2} \tag{17.6}$$

or

$$dL = \sqrt{x'^2 + y'^2 + z'^2}\ du \tag{17.7}$$

Substituting Eq. (17.7) into Eq. (17.1) gives

$$L = \int_{u_1}^{u_2} \sqrt{x'^2 + y'^2 + z'^2}\ du \tag{17.8}$$

or

$$L = \int_{u_1}^{u_2} \sqrt{\mathbf{P}' \cdot \mathbf{P}'}\ du \tag{17.9}$$

Equation (17.9) gives the exact length of a curve segment bounded by the parametric values $u_1$ and $u_2$ as the integral, with respect to $u$, of the square root of the dot product of the tangent vector of the curve. Equation (17.9) requires that the curve is $C^1$ continuous. It applies to both open and closed curves.

## 17.2.2  Cross-Sectional Area

A cross-sectional area is a planar region bounded by a closed boundary. The boundary consists of a set of $C^1$ continuous curves connected together. Thus, the boundary is piecewise continuous. There are three properties associated with a planar region: the length of its contour (boundary), its area, and its centroid.

Figure 17-2a shows a region $R$ that is oriented generally in space. The region's plane coincides with the $X_L Y_L$ plane of the $X_L Y_L Z_L$ local coordinate system. The mapping between this local system and the $XYZ$ coordinate system (MCS) shown in the figure can be achieved by using the mapping matrix given by Eq. (9.88).

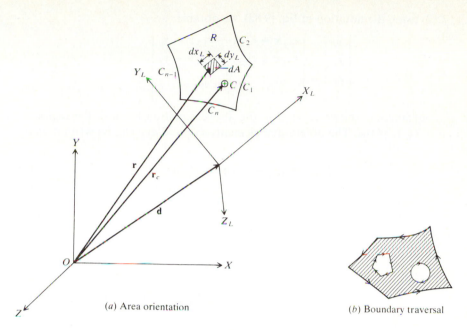

(a) Area orientation

(b) Boundary traversal

**FIGURE 17-2**
A cross-sectional area.

The region is bounded by the curves $C_1, C_2, \ldots, C_{n-1}$, and $C_n$. The length of the contour is given by the sum of the lengths of $C_1, C_2, \ldots$, and $C_n$, that is,

$$L = \sum_{i=1}^{n} L_i \tag{17.10}$$

where $L_i$ is the length of curve $C_i$ which can be calculated using Eq. (17.9).

To calculate the area $A$ of the region $R$, consider an area element $dA$ of sides $dx_L$ and $dy_L$, as shown in Fig. 17-2a. Integrating over the region gives

$$A = \iint_R dA = \iint_R dx_L\, dy_L \tag{17.11}$$

The evaluation of the integral in the above equation requires mapping this equation from the local coordinate system of the region $R$ to the global system (MCS) because the parametric equations describing the boundary curves are stored in the region's geometric database with respect to the latter. Utilizing the matrix $[T]$ given by Eq. (9.88), we can write:

$$\begin{bmatrix} x_L \\ y_L \\ z_L \\ 1 \end{bmatrix} = [T] \begin{bmatrix} x \\ y \\ z \\ 1 \end{bmatrix} \tag{17.12}$$

The matrix $[T]$ can be evaluated using three noncollinear points as discussed in Chap. 3. Endpoints of curves $C_i$ can be used for this purpose. Expanding

Eq. (17.12) using the notation of Eq. (9.88), we obtain:

$$\begin{bmatrix} x_L \\ y_L \\ z_L \\ 1 \end{bmatrix} = \begin{bmatrix} r_{11}x + r_{12}y + r_{13}z + x_d \\ r_{21}x + r_{22}y + r_{23}z + y_d \\ 0 \\ 1 \end{bmatrix} \tag{17.13}$$

The $z_L$ coordinate is shown as zero in the above equation because the region $R$ lies in the $X_L Y_L$ plane. The differential elements $dx_L$ and $dy_L$ can be written using the above equation as

$$dx_L = r_{11}dx + r_{12}\,dy + r_{13}\,dz \tag{17.14}$$

$$dy_L = r_{21}dx + r_{22}\,dy + r_{23}\,dz \tag{17.15}$$

Substituting Eqs. (17.14) and (17.15) into Eq. (17.11) and reducing, the area $A$ becomes

$$A = \alpha_1 \iint_R dx\,dx + \alpha_2 \iint_R dy\,dy + \alpha_3 \iint_R dz\,dz$$

$$+ \alpha_4 \iint_R dx\,dy + \alpha_5 \iint_R dx\,dz + \alpha_6 \iint_R dy\,dz \tag{17.16}$$

where

$$\alpha_1 = r_{11}r_{21} \tag{17.17}$$

$$\alpha_2 = r_{12}r_{22} \tag{17.18}$$

$$\alpha_3 = r_{13}r_{23} \tag{17.19}$$

$$\alpha_4 = r_{11}r_{22} + r_{12}r_{21} \tag{17.20}$$

$$\alpha_5 = r_{11}r_{23} + r_{13}r_{21} \tag{17.21}$$

$$\alpha_6 = r_{12}r_{23} + r_{13}r_{22} \tag{17.22}$$

It is beneficial to change the double integrals shown in Eq. (17.16) into line integrals that can be evaluated over the boundary of the region $R$. This is achieved by using Green's theorem which can be stated as follows. Given two functions $f(x, y)$ and $g(x, y)$ over a closed region $R$, we can write

$$\iint_R \left( \frac{\partial f}{\partial x} + \frac{\partial g}{\partial y} \right) dx\,dy = \oint_B (f\,dy - g\,dx) \tag{17.23}$$

where $f$ and $g$ are continuous and single-valued functions of $x$ and $y$. The line integral is taken in the positive direction around the boundary $B$. The functions $f$ and $g$ are usually not unique. There are three choices for both $f$ and $g$ for each integral of Eq. (17.16). For example, Green's theorem can be used for the $\iint_R dx\,dy$ with $f = x$ and $g = 0, f = 0$ and $g = y$, or $f = x/2$ and $g = y/2$. Applying Green's theorem to Eq. (17.16) using $f = x$ and $g = 0, f = y$ and $g = 0, f = z$ and $g = 0, f = x$ and $g = 0, f = 0$ and $g = z$, and $f = y$ and $g = 0$ for the six integrals respectively, we obtain

$$A = \oint_B (\alpha_1 x - \alpha_5 z)\,dx + \oint_B (\alpha_2 y + \alpha_4 x)\,dy + \oint_B (\alpha_3 z + \alpha_6 y)\,dz \tag{17.24}$$

Considering the curves $C_i$ of the boundary $B$, Eq. (17.24) becomes

$$A = \sum_{i=1}^{n} \left[ \int_{C_i} (\alpha_1 x - \alpha_5 z) \, dx + \int_{C_i} (\alpha_2 y + \alpha_4 x) \, dy + \int_{C_i} (\alpha_3 z + \alpha_6 y) \, dz \right] \quad (17.25)$$

The boundary $B$ must be traversed such that the interior lies to the left, that is, traverses it in the counterclockwise direction for the exterior boundary and the clockwise direction for hole boundaries (Fig. 17-2b). The traversed direction affects the limits of the integrals (see Example 17.2). Assuming that each curve $C_i$ is represented parametrically, then $x = x(u)$, $y = y(u)$, and $z = z(u)$. This gives

$$dx = x' \, du$$

$$dy = y' \, du \quad (17.26)$$

$$dz = z' \, du$$

Substituting this equation into Eq. (17.25) gives

$$A = \sum_{i=1}^{n} \left[ \int_{u_{i1}}^{u_{i2}} (\alpha_1 x - \alpha_5 z) x' \, du + \int_{u_{i1}}^{u_{i2}} (\alpha_2 y + \alpha_4 x) y' \, du + \int_{u_{i1}}^{u_{i2}} (\alpha_3 z + \alpha_6 y) z' \, du \right]$$

$$(17.27)$$

where $u_{i1}$ and $u_{i2}$ are the limits of the parameter $u$ for the $i$th curve.

The terms in Eq. (17.27) are reduced significantly depending on the relative orientation of the $X_L Y_L Z_L$ coordinate system with respect to the MCS. For example, if both systems are identical, then $[T]$ becomes an identity matrix. As a result, $\alpha_1 = \alpha_2 = \alpha_3 = \alpha_5 = \alpha_6 = 0$, $\alpha_4 = 1$, and Eq. (17.27) reduces to

$$A = \sum_{i=1}^{n} \int_{u_{i1}}^{u_{i2}} xy' \, du \quad (17.28)$$

The centroid of the region $R$ is obtained by equating the moment of the entire area $A$ lumped at its centroid $C$ (see Fig. 17-2a) to the sum of the moments of the element areas of the region. The moments are taken with respect to the origin $O$ of the MCS. Figure 17-2a shows an element area $dA$ located by the position vector $\mathbf{r}$ from the origin $O$. The centroid of the region is located by the vector $\mathbf{r}_c$. Equating the two moments, we get

$$\mathbf{r}_c = \frac{\iint_R \mathbf{r} \, dA}{\iint_R dA} = \frac{\iint_R \mathbf{r} \, dx_L \, dy_L}{A} \quad (17.29)$$

which, in scalar form, gives

$$x_c = \frac{\iint_R x \, dx_L \, dy_L}{A} \qquad y_c = \frac{\iint_R y \, dx_L \, dy_L}{A} \qquad z_c = \frac{\iint_R z \, dx_L \, dy_L}{A} \quad (17.30)$$

where $x_c$, $y_c$, and $z_c$ are the MCS coordinates of the centroid. The integrals $\iint_R x \, dx_L \, dy_L$, $\iint_R y \, dx_L \, dy_L$, and $\iint_R z \, dx_L \, dy_L$ are sometimes called the first moments of the area with respect to the $YZ$ plane, the $XZ$ plane, and the $XY$ plane respectively.

Substituting Eqs. (17.14) and (17.15) into Eqs. (17.30), using Green's theorem to change the resulting double integrals into line integrals and reducing the results in a similar way to the area calculations, Eqs. (17.30) become

$$x_c = \frac{1}{A} \sum_{i=1}^{n} \left[ \int_{u_{i1}}^{u_{i2}} \left( \frac{\alpha_1 x^2}{2} - \alpha_5\, xz \right) x'\, du + \int_{u_{i1}}^{u_{i2}} \left( \alpha_2\, xy + \frac{\alpha_4\, x^2}{2} \right) y'\, du \right.$$
$$\left. + \int_{u_{i1}}^{u_{i2}} (\alpha_3\, xz + \alpha_6\, xy)z'\, du \right]$$

$$y_c = \frac{1}{A} \sum_{i=1}^{n} \left[ \int_{u_{i1}}^{u_{i2}} (\alpha_1 xy - \alpha_5\, yz)x'\, du + \int_{u_{i1}}^{u_{i2}} \left( \frac{\alpha_2\, y^2}{2} + \alpha_4\, xy \right) y'\, du \right.$$
$$\left. + \int_{u_{i1}}^{u_{i2}} \left( \alpha_3\, yz + \frac{\alpha_6\, y^2}{2} \right) z'\, du \right] \qquad (17.31)$$

$$z_c = \frac{1}{A} \sum_{i=1}^{n} \left[ \int_{u_{i1}}^{u_{i2}} \left( \alpha_1 xz - \frac{\alpha_5\, z^2}{2} \right) x'\, du + \int_{u_{i1}}^{u_{i2}} (\alpha_2\, yz + \alpha_4\, xz)y'\, du \right.$$
$$\left. + \int_{u_{i1}}^{u_{i2}} \left( \frac{\alpha_3\, z^2}{2} + \alpha_6\, yz \right) z'\, du \right]$$

It is worth noting that $x$, $y$, and $z$ variables are used in Eqs. (17.30) and not $X_L$, $Y_L$, and 0, to facilitate the formulation of the centroid calculations. If the coordinates of the centroid $C$ are needed relative to the $X_L Y_L Z_L$ coordinate system, they can be mapped from the MCS to the local coordinate system as explained in Sec. 17.3.5.

The above formulations apply to a singly connected region, that is, a region with only one outside closed boundary. They can be extended to multiply connected regions. A multiply connected region is a region with holes inside it; that is, it has one outside boundary with more than one inside boundary. In this case, the net cross-sectional area of a multiply connected region is given by

$$A_m = A - \sum_{j=1}^{m} A_{hj} \qquad (17.32)$$

where $A_m$ is the net area, $A$ is the area of the singly connected (i.e., excluding the holes) region bounded by the outside boundary, and $A_{hj}$ is the area of the $j$th hole. Each of the areas $A$ and any $A_{hj}$ can be calculated using Eq. (17.27). The method by which to obtain the centroid of a multiply connected region is covered in Sec. 17.6.

### 17.2.3   Surface Area

The surface area $A_s$ of a bounded surface (Fig. 17-3) can be formulated in a similar way to the cross-sectional area. The major difference is that $A_s$ is not planar in general as in the case of a cylindrical, spherical, B-spline, or Bezier surface.

Figure 17-3 shows a surface $S$ and a surface area element $dA_s$. The surface area $A_s$ of the surface is given by

$$A_s = \int_S dA_s \qquad (17.33)$$

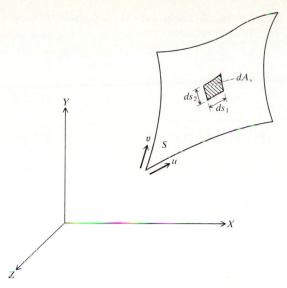

**FIGURE 17-3**
Surface area calculation.

To evaluate the integral in Eq. (17.33), $dA_s$ which is a rectangle must be written as the product of the lengths of two perpendicular sides. These sides are taken along the $u$ and $v$ directions as $ds_1$ and $ds_2$ (Fig. 17-3). This is a convenient choice because surfaces are represented in parametric form in CAD/CAM systems. Thus, Eq. (17.33) can be written as

$$A_s = \iint_S ds_1 \, ds_2 \tag{17.34}$$

Using the first fundamental quadratic form of a surface given by Eq. (6.13) and observing that $ds_1$ is taken along the $u$ direction (that is, $dv = 0$) and $ds_2$ is taken along the $v$ direction (that is, $du = 0$), we can write

$$ds_1 = \sqrt{\mathbf{P}_u \cdot \mathbf{P}_u} \, du \tag{17.35}$$

and

$$ds_2 = \sqrt{\mathbf{P}_v \cdot \mathbf{P}_v} \, dv \tag{17.36}$$

Substituting Eqs. (17.35) and (17.36) into Eq. (17.34), we get

$$A_s = \int_{v_1}^{v_2} \int_{u_1}^{u_2} \sqrt{(\mathbf{P}_u \cdot \mathbf{P}_u)(\mathbf{P}_v \cdot \mathbf{P}_v)} \, du \, dv \tag{17.37}$$

Equation (17.37) can also be written in terms of the jacobian $[J] = \|\mathbf{P}_u \times \mathbf{P}_v\|$ (see Prob. 17.5 at the end of the chapter).

For an object consisting of multiple surfaces, its total surface area is equal to the sum of its individual surface areas, that is,

$$A_s = \sum_{i=1}^{n} A_{si} \tag{17.38}$$

Notice that the contribution of holes in an object to its surface area is positive as shown by this equation. (Their contribution to the object volume is negative.)

The centroid of a surface can be found using Eq. (17.29). Substituting $dA_s = ds_1 ds_2$ for $dA$ in Eq. (17.29), Eqs. (17.30) become

$$x_c = \frac{\int_{v_1}^{v_2} \int_{u_1}^{u_2} xK \, du \, dv}{A_s} \qquad y_c = \frac{\int_{v_1}^{v_2} \int_{u_1}^{u_2} yK \, du \, dv}{A_s} \qquad z_c = \frac{\int_{v_1}^{v_2} \int_{u_1}^{u_2} zK \, du \, dv}{A_s}$$

$$\text{(17.39)}$$

where

$$K = \sqrt{(\mathbf{P}_u \cdot \mathbf{P}_u)(\mathbf{P}_v \cdot \mathbf{P}_v)} \qquad \text{(17.40)}$$

### 17.2.4 Volume

Figure 17-4 shows an object whose volume is $V$. The volume $V$ can be expressed as a triple integral by integrating the volume element $dV$, that is,

$$V = \iiint_V dV = \iiint_V dx \, dy \, dz \qquad \text{(17.41)}$$

This volume integral can be changed into a surface integral using the Gauss divergence theorem which can be expressed as follows:

$$\iiint_V \mathbf{V} \cdot \mathbf{F} \, dV = \iint_S \mathbf{F} \cdot \hat{\mathbf{n}} \, dS \qquad \text{(17.42)}$$

where

$$\mathbf{V} = \frac{\partial}{\partial x} \hat{\mathbf{i}} + \frac{\partial}{\partial y} \hat{\mathbf{j}} + \frac{\partial}{\partial z} \hat{\mathbf{k}} \qquad \text{(17.43)}$$

and $\hat{\mathbf{n}}$ is the unit normal vector of the surface of the body. The vector function $\mathbf{F}$ must be $C^1$ continuous over the closed regular surface $S$ and its interior $V$. To convert Eq. (17.41) to a surface integral, we have to find a vector function $\mathbf{F}$ that

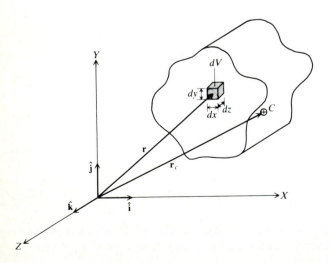

**FIGURE 17-4**
Volume and mass calculations.

satisfies the condition $\mathbf{V} \cdot \mathbf{F} = 1$. As in Green's theorem, the vector function $\mathbf{F}$ is not unique. Some candidate functions are $x\hat{\mathbf{i}}$, $y\hat{\mathbf{j}}$, $z\hat{\mathbf{k}}$, or $(x/3)\hat{\mathbf{i}} + (y/3)\hat{\mathbf{j}} + (z/3)\hat{\mathbf{k}}$. Choosing one of the first three candidates is better than the last one because it can eliminate unnecessary surface integrals. For example, choosing $\mathbf{F} = z\hat{\mathbf{k}}$ eliminates integrals over surfaces with normals perpendicular to the $Z$ axis. Using this choice, the volume can be written as

$$V = \iiint_V dV = \iint_S z(\hat{\mathbf{k}} \cdot \hat{\mathbf{n}}) \, dA_s \qquad (17.44)$$

Substituting Eqs. (17.35) and (17.36) for $dA_s$ in the above equation, we obtain

$$V = \iiint_V dV = \int_{v_1}^{v_2} \int_{u_1}^{u_2} z(\hat{\mathbf{k}} \cdot \hat{\mathbf{n}}) \sqrt{(\mathbf{P}_u \cdot \mathbf{P}_u)(\mathbf{P}_v \cdot \mathbf{P}_v)} \, du \, dv \qquad (17.45)$$

If the object is closed by multiple surfaces, its volume becomes

$$V = \sum_{i=1}^{n} \int_{v_{i1}}^{v_{i2}} \int_{u_{i1}}^{u_{i2}} z(\hat{\mathbf{k}} \cdot \hat{\mathbf{n}}_i) \sqrt{(\mathbf{P}_{ui} \cdot \mathbf{P}_{ui})(\mathbf{P}_{vi} \cdot \mathbf{P}_{vi})} \, du \, dv \qquad (17.46)$$

Equation (17.46) applies to a singly connected object (no holes in its interior). The volume $V_m$ of a multiply connected object (with holes) is given by

$$V_m = V - \sum_{j=1}^{m} V_{hj} \qquad (17.47)$$

where $V_{hj}$ is the volume of hole $j$.

The centroid of the object is located by the vector $\mathbf{r}_c$ shown in Fig. 17-4. This vector is given by [similar to Eq. (17.29)]

$$\mathbf{r}_c = \frac{\iiint_V \mathbf{r} \, dV}{\iiint_V dV} = \frac{\iiint_V \mathbf{r} \, dV}{V} \qquad (17.48)$$

or

$$x_c = \frac{1}{V} \iiint_V x \, dV \qquad y_c = \frac{1}{V} \iiint_V y \, dV \qquad z_c = \frac{1}{V} \iiint_V z \, dV \qquad (17.49)$$

Using the Gauss divergence theorem with $\mathbf{F} = xz\hat{\mathbf{k}}$, $\mathbf{F} = yz\hat{\mathbf{k}}$, and $\mathbf{F} = (z^2/2)\hat{\mathbf{k}}$ for $x_c$, $y_c$, and $z_c$ respectively, Eqs. (17.49) become

$$x_c = \frac{1}{V} \iint_S xz(\hat{\mathbf{k}} \cdot \hat{\mathbf{n}}) \, dA_s$$

$$y_c = \frac{1}{V} \iint_S yz(\hat{\mathbf{k}} \cdot \hat{\mathbf{n}}) \, dA_s \qquad (17.50)$$

$$z_c = \frac{1}{V} \iint_S \frac{z^2}{2} (\hat{\mathbf{k}} \cdot \hat{\mathbf{n}}) \, dA_s$$

Using $dA_s = ds_1 \, ds_2$ together with Eqs. (17.35), (17.36), and (17.40), Eqs. (17.50) become

$$x_c = \frac{1}{V} \int_{v_1}^{v_2} \int_{u_1}^{u_2} xz K(\hat{\mathbf{k}} \cdot \hat{\mathbf{n}}) \, du \, dv$$

$$y_c = \frac{1}{V} \int_{v_1}^{v_2} \int_{u_1}^{u_2} yz K(\hat{\mathbf{k}} \cdot \hat{\mathbf{n}}) \, du \, dv \qquad (17.51)$$

$$z_c = \frac{1}{V} \int_{v_1}^{v_2} \int_{u_1}^{u_2} \frac{z^2}{2} K(\hat{\mathbf{k}} \cdot \hat{\mathbf{n}}) \, du \, dv$$

If the object is closed by multiple surfaces, the integrals in Eqs. (17.51) are summed over these surfaces to give

$$x_c = \frac{1}{V} \sum_{i=1}^{n} \int_{v_{i1}}^{v_{i2}} \int_{u_{i1}}^{u_{i2}} xz K_i(\hat{\mathbf{k}} \cdot \hat{\mathbf{n}}_i) \, du \, dv$$

$$y_c = \frac{1}{V} \sum_{i=1}^{n} \int_{v_{i1}}^{v_{i2}} \int_{u_{i1}}^{u_{i2}} yz K_i(\hat{\mathbf{k}} \cdot \hat{\mathbf{n}}_i) \, du \, dv \qquad (17.52)$$

$$z_c = \frac{1}{V} \sum_{i=1}^{n} \int_{v_{i1}}^{v_{i2}} \int_{u_{i1}}^{u_{i2}} \frac{z^2}{2} K_i(\hat{\mathbf{k}} \cdot \hat{\mathbf{n}}_i) \, du \, dv$$

## 17.3  MASS PROPERTY FORMULATION

Mass properties of an object is a set of useful properties used in various engineering applications. These properties include mass, centroid, first moments, and second moments of inertia. The main difference between mass and geometrical properties is the inclusion of the density of the object material in the former. Formally an object can have a centroid (of its volume), a center of mass (of its mass), and a center of gravity (of its weight) that may be different if the acceleration of gravity $g$ and/or the density $\rho$ of the object material is not constant. In this chapter, we assume that $g$ and $\rho$ are constants. Therefore the three centers (of volume, of mass, and of weight) coincide and are equal to the centroid (of the volume) of the object. This assumption implies that objects of interest are homogeneous and are always close to the surface of the earth.

### 17.3.1  Mass

The mass of an object can be formulated in a similar way to formulating its volume. If we replace the element volume $dV$ shown in Fig. 17-4 by an element mass, we can write

$$dm = \rho \, dV \qquad (17.53)$$

Integrating this equation over the distributed mass of the object gives

$$m = \iiint_m \rho \, dV \qquad (17.54)$$

Assuming the density $\rho$ to be uniform (constant), Eq. (17.54) becomes

$$m = \rho \iiint_V dV = \rho V \qquad (17.55)$$

Thus, once the volume of an object is calculated, it is multiplied by its density to obtain its mass.

### 17.3.2  Centroid

Equation (17.48) can be used to find the center of mass of an object by replacing the volume $V$ by the mass $m$, that is,

$$\mathbf{r}_c = \frac{\iiint_m \mathbf{r}\, dm}{m} \qquad (17.56)$$

If we substitute $\rho V$ for $m$ in the above equation, Eq. (17.48) results and the center of mass is coincident with the center of volume. Therefore, Eqs. (17.49) to (17.52) apply to the center of mass (centroid).

### 17.3.3  First Moments of Inertia

The first moment of an area, volume, or a mass is a mathematical property that appears in various calculations. It is defined as the moment of an object property (area, volume, or mass) with respect to a given plane. For a lumped mass, the first moment of the mass about a given plane is equal to the product of the mass and its perpendicular distance from the plane. Using this definition, the first moments of a distributed mass of an object with respect to the $XY$, $XZ$, and $YZ$ planes are given by respectively

$$M_{xy} = \iiint_m z\, dm = \rho \iiint_V z\, dV \qquad (17.57)$$

$$M_{xz} = \iiint_m y\, dm = \rho \iiint_V y\, dV \qquad (17.58)$$

$$M_{yz} = \iiint_m x\, dm = \rho \iiint_V x\, dV \qquad (17.59)$$

We can easily recognize that the above expressions for first moments appear in the centroid equations (17.49). Substituting Eqs. (17.49) into the above three equations gives

$$M_{xy} = \rho V z_c = m z_c \qquad (17.60)$$

$$M_{xz} = \rho V y_c = m y_c \qquad (17.61)$$

$$M_{yz} = \rho V x_c = m x_c \qquad (17.62)$$

Thus, the first moments are byproducts of the volume and centroid calculations.

### 17.3.4   Second Moments and Products of Inertia

The second moment of inertia of a lumped mass about a given axis is the product of the mass and the square of the perpendicular distance between the mass and the axis. The second moments of inertia of a distributed mass about the $X$, $Y$, and $Z$ axes can be written as

$$I_{xx} = \iiint_m (y^2 + z^2)\, dm = \rho \iiint_V (y^2 + z^2)\, dV \tag{17.63}$$

$$I_{yy} = \iiint_m (x^2 + z^2)\, dm = \rho \iiint_V (x^2 + z^2)\, dV \tag{17.64}$$

$$I_{zz} = \iiint_m (x^2 + y^2)\, dm = \rho \iiint_V (x^2 + y^2)\, dV \tag{17.65}$$

The physical interpretation of a second moment of inertia of an object about an axis is that it represents the resistance of the object to any rotation about the axis (examples: torque equation $T = I\alpha$ and stress equation $\sigma = Mc/I$).

The inertia integrals can be changed to surface integrals via the Gauss divergence theorem. Using $\mathbf{F} = (y^2 z + z^3/3)\hat{\mathbf{k}}$, $\mathbf{F} = (x^2 z + z^3/3)\hat{\mathbf{k}}$, and $\mathbf{F} = (x^2 z + y^2 z)\hat{\mathbf{k}}$ for $I_{xx}$, $I_{yy}$, and $I_{zz}$ respectively, Eqs. (17.63) to (17.65) become

$$I_{xx} = \rho \iint_S z\left(y^2 + \frac{z^2}{3}\right)(\hat{\mathbf{k}} \cdot \hat{\mathbf{n}})\, dA_s \tag{17.66}$$

$$I_{yy} = \rho \iint_S z\left(x^2 + \frac{z^2}{3}\right)(\hat{\mathbf{k}} \cdot \hat{\mathbf{n}})\, dA_s \tag{17.67}$$

$$I_{zz} = \rho \iint_S z(x^2 + y^2)(\hat{\mathbf{k}} \cdot \hat{\mathbf{n}})\, dA_s \tag{17.68}$$

Using $dA_s = ds_1\, ds_2$ together with Eqs. (17.35), (17.36), and (17.40), Eqs. (17.66) to (17.68) become

$$I_{xx} = \rho \int_{v_1}^{v_2} \int_{u_1}^{u_2} zK\left(y^2 + \frac{z^2}{3}\right)(\hat{\mathbf{k}} \cdot \hat{\mathbf{n}})\, du\, dv \tag{17.69}$$

$$I_{yy} = \rho \int_{v_1}^{v_2} \int_{u_1}^{u_2} zK\left(x^2 + \frac{z^2}{3}\right)(\hat{\mathbf{k}} \cdot \hat{\mathbf{n}})\, du\, dv \tag{17.70}$$

$$I_{zz} = \rho \int_{v_1}^{v_2} \int_{u_1}^{u_2} zK(x^2 + y^2)(\hat{\mathbf{k}} \cdot \hat{\mathbf{n}})\, du\, dv \tag{17.71}$$

If the object is closed by multiple surfaces, the integrals in Eqs. (17.69) to (17.71) are summed over these surfaces to give

$$I_{xx} = \rho \sum_{i=1}^{n} \int_{v_{i1}}^{v_{i2}} \int_{u_{i1}}^{u_{i2}} z K_i \left( y^2 + \frac{z^2}{3} \right) (\hat{\mathbf{k}} \cdot \hat{\mathbf{n}}_i) \, du \, dv \tag{17.72}$$

$$I_{yy} = \rho \sum_{i=1}^{n} \int_{v_{i1}}^{v_{i2}} \int_{u_{i1}}^{u_{i2}} z K_i \left( x^2 + \frac{z^2}{3} \right) (\hat{\mathbf{k}} \cdot \hat{\mathbf{n}}) \, du \, dv \tag{17.73}$$

$$I_{zz} = \rho \sum_{i=1}^{n} \int_{v_{i1}}^{v_{i2}} \int_{u_{i1}}^{u_{i2}} z K_i (x^2 + y^2)(\hat{\mathbf{k}} \cdot \hat{\mathbf{n}}) \, du \, dv \tag{17.74}$$

Like second moments of inertia, products of inertia are useful and are defined by the following equations:

$$I_{xy} = \iiint_m xy \, dm = \rho \iiint_V xy \, dV \tag{17.75}$$

$$I_{xz} = \iiint_m xz \, dm = \rho \iiint_V xz \, dV \tag{17.76}$$

$$I_{yz} = \iiint_m yz \, dm = \rho \iiint_V yz \, dV \tag{17.77}$$

Following a similar approach to that used in the second moments of inertia, Eqs. (17.75) to (17.77) can be rewritten as

$$I_{xy} = \rho \int_{v_1}^{v_2} \int_{u_1}^{u_2} xyz K(\hat{\mathbf{k}} \cdot \hat{\mathbf{n}}) \, du \, dv \tag{17.78}$$

$$I_{xz} = \rho \int_{v_1}^{v_2} \int_{u_1}^{u_2} \frac{xz^2}{2} K(\hat{\mathbf{k}} \cdot \hat{\mathbf{n}}) \, du \, dv \tag{17.79}$$

$$I_{yz} = \rho \int_{v_1}^{v_2} \int_{u_1}^{u_2} \frac{yz^2}{2} K(\hat{\mathbf{k}} \cdot \hat{\mathbf{n}}) \, du \, dv \tag{17.80}$$

and for multiple surface objects:

$$I_{xy} = \rho \sum_{i=1}^{n} \int_{v_{i1}}^{v_{i2}} \int_{u_{i1}}^{u_{i2}} xyz K_i(\hat{\mathbf{k}} \cdot \hat{\mathbf{n}}_i) \, du \, dv \tag{17.81}$$

$$I_{xz} = \rho \sum_{i=1}^{n} \int_{v_{i1}}^{v_{i2}} \int_{u_{i1}}^{u_{i2}} \frac{xz^2}{2} K_i(\hat{\mathbf{k}} \cdot \hat{\mathbf{n}}_i) \, du \, dv \tag{17.82}$$

$$I_{yz} = \rho \sum_{i=1}^{n} \int_{v_{i1}}^{v_{i2}} \int_{u_{i1}}^{u_{i2}} \frac{yz^2}{2} K_i(\hat{\mathbf{k}} \cdot \hat{\mathbf{n}}_i) \, du \, dv \tag{17.83}$$

## 17.3.5 Property Mapping

The volume and mass properties already presented are formulated with respect to a given $XYZ$ coordinate system. For a geometric model, this system is the MCS (model coordinate system) of the model database. The MCS is a convenient system because all the curve, surface, or solid equations and other geometric information are stored with respect to this system. If mass properties are to be calculated with respect to other systems, these properties must be mapped from the MCS to other systems. In the following discussions, let us assume that mass properties are already available in the MCS $(XYZ)$ and we need to map them to a given WCS $(X_W Y_W Z_W)$.

The coordinates of the centroid $(x_c, y_c, z_c)$ can be mapped from the MCS to the WCS using the following equation:

$$r_{cW} = [T]r_c \tag{17.84}$$

or

$$
\begin{bmatrix} x_{cW} \\ y_{cW} \\ z_{cW} \\ 1 \end{bmatrix} = [T] \begin{bmatrix} x_c \\ y_c \\ z_c \\ 1 \end{bmatrix} \tag{17.85}
$$

where $[T]$ is the general mapping matrix given by Eq. (9.88).

Once the centroid is mapped, the first moments $M_{xy}$, $M_{xz}$, and $M_{yz}$ are automatically mapped, as seen from Eqs. (17.60) to (17.62). These equations become

$$
\begin{bmatrix} M_{xyW} \\ M_{xzW} \\ M_{yzW} \\ 1 \end{bmatrix} = m \begin{bmatrix} x_{cW} \\ y_{cW} \\ z_{cW} \\ 1 \end{bmatrix} = m[T] \begin{bmatrix} x_c \\ y_c \\ z_c \\ 1 \end{bmatrix} \tag{17.86}
$$

The second and product moments of inertia of an object with respect to its MCS can be mapped to compute the moment of inertia of the object about any arbitrary axis. The axis may or may not pass through the origin of the MCS. Figure 17-5 shows some examples. The axis $AA$ passes through the origin $O$ and its direction in space is defined by the unit vector $\hat{n}$. By definition, $I_{aa} = \iiint_m b^2 \, dm$, where $b$ is the perpendicular distance from $dm$ to $AA$. If the position of $dm$ is located using $\mathbf{r}$, then $b = r \sin \theta = |\hat{n} \times \mathbf{r}|$. Hence, $I_{aa}$ can be expressed as

$$I_{aa} = \iiint_m |\hat{n} \times \mathbf{r}|^2 \, dm = \iiint_m (\hat{n} \times \mathbf{r}) \cdot (\hat{n} \times \mathbf{r}) \, dm \tag{17.87}$$

Using $\hat{n} = n_x \hat{i} + n_y \hat{j} + n_z \hat{k}$, $\mathbf{r} = x\hat{i} + y\hat{j} + z\hat{k}$, and Eqs. (17.63) to (17.65) and (17.75) to (17.77), Eq. (17.87) becomes

$$I_{aa} = I_{xx} n_x^2 + I_{yy} n_y^2 + I_{zz} n_z^2 - 2I_{xy} n_x n_y - 2I_{xz} n_x n_z - 2I_{yz} n_y n_z \tag{17.88}$$

If the moment of inertia about a centroidal axis $BB$ (Fig. 17-5) parallel to axis $AA$ is to be computed, the parallel axis theorem can be used to give

$$I_{bb} = I_{aa} - mb_1^2 \tag{17.89}$$

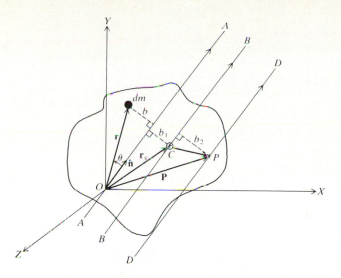

**FIGURE 17-5**
Moment of inertia about an arbitrary axis.

Substituting $|\hat{\mathbf{n}} \times \mathbf{r}_c|$ for $b_1$, Eq. (17.89) becomes

$$I_{bb} = I_{aa} - m(\hat{\mathbf{n}} \times \mathbf{r}_c) \cdot (\hat{\mathbf{n}} \times \mathbf{r}_c) \tag{17.90}$$

which can be reduced to

$$I_{bb} = I_{aa} - m[(n_y z_c - n_z y_c)^2 + (n_z x_c - n_x z_c)^2 + (n_x y_c - n_y x_c)^2] \tag{17.91}$$

The moment of inertia about any axis $DD$ (Fig. 17-5) that is parallel to $AA$ but passes through a general point $P$ can be computed using the parallel axis theorem again, that is,

$$I_{dd} = I_{bb} + mb_2^2 \tag{17.92}$$

The distance $b_2$ is equal to $|\hat{\mathbf{n}} \times \mathbf{CP}| = |\hat{\mathbf{n}} \times (\mathbf{P} - \mathbf{r}_c)|$. Hence Eq. (17.92) becomes

$$I_{dd} = I_{bb} + m\{[n_y(z - z_c) - n_z(y - y_c)]^2$$

$$+ [n_z(x - x_c) - n_x(z - z_c)]^2 + [n_x(y - y_c) - n_y(x - x_c)]^2\} \tag{17.93}$$

The moment of inertia of an object about axes parallel to the MCS but passing through a given point in space can be determined using the parallel axis theorem. Figure 17-6 shows three parallel coordinate systems: the MCS, a centroidal system $(X_C Y_C Z_C)$, and a WCS that has an origin at point $P$. Let us assume that the moments and products of inertia have been calculated with respect to the MCS. Applying the parallel axis theorem twice, we can write:

$$(I_{xx})_C = I_{xx} - m(y_c^2 + z_c^2)$$

$$(I_{yy})_C = I_{yy} - m(x_c^2 + z_c^2)$$

$$(I_{zz})_C = I_{zz} - (x_c^2 + y_c^2)$$

$$(I_{xy})_C = I_{xy} - mx_c y_c$$

$$(I_{xz})_C = I_{xz} - mx_c z_c$$

$$(I_{yz})_C = I_{yz} - my_c z_c$$

$$\tag{17.94}$$

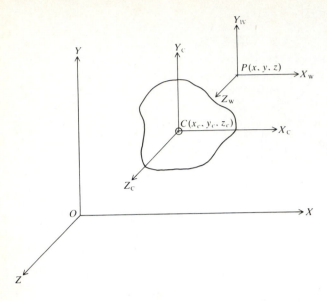

**FIGURE 17-6**
Moments of inertia about axes parallel to the MCS.

and

$$(I_{xx})_W = (I_{xx})_C + m[(y - y_c)^2 + (z - z_c)^2]$$
$$(I_{yy})_W = (I_{yy})_C + m[(x - x_c)^2 + (z - z_c)^2]$$
$$(I_{zz})_W = (I_{zz})_C + m[(x - x_c)^2 + (y - y_c)^2]$$
$$(I_{xy})_W = (I_{xy})_C + m(x - x_c)(y - y_c)$$
$$(I_{xz})_W = (I_{xz})_C + m(x - x_c)(z - z_c)$$
$$(I_{yz})_W = (I_{yz})_C + m(y - y_c)(z - z_c)$$

$$(17.95)$$

Principal moments of inertia are useful properties to determine for an object. Their corresponding axes are the principal axes of inertia. These axes are characterized so that products of inertia with respect to them are zeros. Thus, the principal inertia tensor is a diagonal matrix whose diagonal elements are the principal moments of inertia $I_x$, $I_y$, and $I_z$. These moments are the roots of the following cubic equation:

$$I^3 - (I_{xx} + I_{yy} + I_{zz})I^2 + (I_{xx}I_{yy} + I_{yy}I_{zz} + I_{zz}I_{xx} - I_{xy}^2 - I_{yz}^2 - I_{xz}^2)I$$
$$- (I_{xx}I_{yy}I_{zz} - 2I_{xy}I_{yz}I_{xz} - I_{xx}I_{yz}^2 - I_{yy}I_{xz}^2 - I_{zz}I_{xy}^2) = 0 \quad (17.96)$$

The determination of the directions of the principal axes of inertia are not discussed here. It involves angular momentum equations. Readers interested on how to derive Eq. (17.96) or how to find these directions should refer to books on statics and dynamics subjects.

The radius of gyration $R_g$ of an object with respect to an axis is one of the inertial properties of the object. It is defined by the following equation:

$$R_g = \sqrt{\frac{I}{m}}$$

$$(17.97)$$

where $I$ is the moment of inertia with respect to the same axis of $R_g$.

## 17.4  PROPERTY EVALUATION

Geometric and mass properties of an object have been formulated in the previous two sections. These properties and the final equations that describe them are listed in Table 17.1. All these final equations are integral equations in the parametric space. Some are line integrals to be evaluated over curves or edges of the object. In this case, the integrand is a function of the parameter $u$. Others are surface integrals to be evaluated over surfaces or faces of the object, in which case the integrand is a function of the parameters $u$ and $v$. These surface integrals can be further reduced to line integrals for analytic surfaces.

The equations listed in Table 17.1 are applicable to all curves, surfaces, and solids covered in Chaps. 5, 6, and 7 respectively. For many curves such as lines, circles, and conics, and for many surfaces such as spherical, conic, and bicubic surfaces, the integrals shown in the equations can be integrated exactly to generate the final expressions of geometric and mass properties in closed form.

Over some surfaces, like B-spline surfaces or curved surfaces with an irregular boundary, the surface integrals for geometric and mass properties cannot be evaluated exactly for various reasons. First, the parametric equation of the surface may not allow the conversion of the double integration over a parametric domain into a set of line integrals directly. In this case, numerical integration such as Gauss quadrature can be used to approximately evaluate the integral. Second, the analytic parametric equation of the boundary over which the line integral is to be evaluated may not be available. In this case, an interpolation function for the boundary of the parametric domain must be chosen even when simple surfaces such as arbitrarily oriented cylinders intersect each other. Third, the final line integral may only be possible to evaluate numerically.

Numerical integration (sometimes called numerical quadrature) is a major subject in itself. Interested readers should refer to books on numerical analysis. In

**TABLE 17.1**
**Summary of geometric and mass properties**

| Type | Property | Equation number |
|---|---|---|
| Geometric | Curve length | (17.9) |
| | Cross-sectional area: | |
| |   Area | (17.27) |
| |   Centroid | (17.31) |
| | Surface area: | |
| |   Area | (17.37) |
| |   Centroid | (17.39) |
| | Volume: | |
| |   Volume | (17.46) |
| |   Centroid | (17.52) |
| Mass | Mass: | |
| |   Mass | (17.55) |
| |   Center of mass | (17.52) |
| | First moments of inertia | (17.60) to (17.62) |
| | Second moments of inertia | (17.72) to (17.74) |
| | Products of inertia | (17.81) to (17.83) |

this section, we review two methods: Newton-Cotes quadrature and Gauss quadrature. These methods are numerically efficient. Mass property equations whose numbers are listed in Table 17.1 involve either one-dimensional (in $u$) or two-dimensional (in $u$ and $v$) integrals. These two types of integrals take the following general forms:

$$I = \int_{u_1}^{u_2} f(u) \, du \tag{17.98}$$

$$I = \int_{v_1}^{v_2} \int_{u_1}^{u_2} f(u, v) \, du \, dv \tag{17.99}$$

First, let us apply the two methods to Eq. (17.98). To evaluate the integral in the equation numerically, the integral is approximated by a polynomial and a remainder, that is,

$$I = I_a + R \tag{17.100}$$

where $I_a$ is the approximate value of $I$ and $R$ is the remainder. $I_a$ usually takes the form of a polynomial and $R$ is the source of error or approximation in the numerical evaluation.

In Newton-Cotes integration, it is assumed that the sampling points of $f(u)$ are spaced at equal distances in the interval $[u_1, u_2]$. If we use $(n + 1)$ sampling points, we can define

$$u_0 = u_1 \qquad u_n = u_2 \qquad h = \frac{u_2 - u_1}{n} \tag{17.101}$$

and the integration formula for the Newton-Cotes method can be written as

$$I = \int_{u_1}^{u_2} f(u) \, du = (u_2 - u_1) \sum_{i=0}^{n} C_i^n f_i + R \tag{17.102}$$

where $C_i^n$ are the Newton-Cotes constants and $f_i = f(u_i) = f(u_0 + ih)$. The cases $n = 1$ and $n = 2$ are the well-known trapezoidal and Simpson rules shown in Fig. 17-7. The constants for these two rules are given by

$$n = 1: \qquad C_0^1 = C_1^1 = \tfrac{1}{2} \tag{17.103}$$

$$n = 2: \qquad C_0^2 = \tfrac{1}{6} \qquad C_1^2 = \tfrac{4}{6} \qquad C_2^2 = \tfrac{1}{6} \tag{17.104}$$

Therefore, Eq. (17.102) becomes, for the trapezoidal rule (Fig. 17.7a),

$$I = \int_{u_1}^{u_2} f(u) \, du \approx \frac{u_2 - u_1}{2} (f_0 + f_1) = \frac{h}{2} (f_0 + f_1) \tag{17.105}$$

and for the Simpson rule (Fig. 17.7b),

$$I = \int_{u_1}^{u_2} f(u) \, du \approx \frac{u_2 - u_1}{6} (f_0 + 4f_1 + f_2)$$

$$= \frac{h}{3} (f_0 + 4f_1 + f_2) \tag{17.106}$$

The minimum number of sampling points in the interval $[u_1, u_2]$ are two and three for the trapezoidal and Simpson rules respectively. The two methods

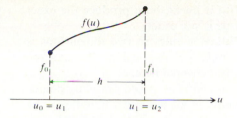

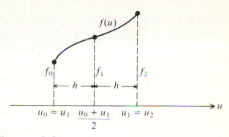

(a) Trapezoidal rule {two points at $u = u_1$ and $u_2$}

(b) Simpson rule {three points at $u = u_1$, $(u_1 + u_2)/2$, and $u_2$}

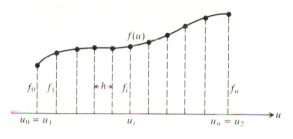

(c) Using $(n + 1)$ sampling points in $[u_1, u_2]$

**FIGURE 17-7**
Sampling points for trapezoidal and Simpson rules.

can still be applied if $(n + 1)$ sampling points are used in the interval $[u_1, u_2]$ as shown in Fig. 17-7c. In this case, each method can be applied repetitively using two sampling points for the former or three for the latter, and then add the results to obtain:

Trapezoidal rule:

$$I = \int_{u_1}^{u_2} f(u)\, du \approx h(\tfrac{1}{2}f_0 + f_1 + f_2 + \cdots + f_{n-1} + \tfrac{1}{2}f_n) \qquad (17.107)$$

Simpson rule:

$$I = \int_{u_1}^{u_2} f(u)\, du \approx \frac{h}{3}(f_0 + 4f_1 + 2f_2 + 4f_3 + \cdots + 2f_{n-2} + 4f_{n-1} + f_n) \quad (17.108)$$

For the Simpson rule, $n$ must be an even number.

The error $E$ introduced by Newton-Cotes integration is equal to the remainder $R$ [Eq. (17.100)]. An upper bound on the error $E$ is given by:

Trapezoidal rule:
$$E < \frac{(u_2 - u_1)^3}{10}\left(\frac{d^2f}{du^2}\right)_{u_2} \qquad (17.109)$$

Simpson rule:
$$E < \frac{(u_2 - u_1)^5}{1000}\left(\frac{d^4f}{du^4}\right)_{u_2} \qquad (17.110)$$

The actual errors can be significantly less than these upper bounds. They mainly depend on the function $f(u)$ which is integrated. It should be emphasized here that actual errors include, in addition to the remainder, the truncation and round-off errors that result during computations. The Simpson rule is usually more accurate than the trapezoidal rule. If $f(u)$ is linear, both produce the exact integration. If $f(u)$ is parabolic, the Simpson rule is still exact.

The extension of the trapezoidal and Simpson rules to two dimensions is straightforward. In this case the sampling points are obtained by dividing the intervals $[u_1, u_2]$ and $[v_1, v_2]$. The details are not covered here.

Both the trapezoidal and Simpson rules use equally spaced sampling points. These methods are effective when measurements of an unknown function to be integrated have been taken at equal intervals. However, in the integration of geometric-related equations such as mass property calculations, a subroutine is called to evaluate the function $f(u)$ at given points. These points may be chosen anywhere in the interval $[u_1, u_2]$. Therefore, it seems natural to optimize the positions of the sampling points to improve the accuracy of numerical integration. Gauss quadrature is a numerical integration method in which both the positions of the sampling points and associated weights have been optimized.

The basic formula in Gauss numerical integration is

$$I = \int_{u_1}^{u_2} f(u) \, du = \sum_{i=1}^{n} V_i \, f_i + R \tag{17.111}$$

where $V_i$ are the weighting factors. $V_i$ and $u_i$ (the sampling points) are weights and zeros associated with a Legendre polynomial of order $n + 1$, and are given by

$$V_i = \frac{u_2 - u_1}{2} W_i \tag{17.112}$$

$$u_i = \frac{u_2 - u_1}{2} C_i + \frac{u_1 + u_2}{2} \tag{17.113}$$

where $C_i$ and $W_i$ are abscissas of sampling points and weighting factors that have been published for values of $n = 1$ to $n = 16$. Table 17.2 shows values for $n = 1$ to 6 for $C_i$ and $W_i$ (to 15 decimal places). Substituting Eqs. (17.112) and (17.113)

**TABLE 17.2**
**Gauss quadrature data**

| Number of sampling points $n$ | Locations $C_i$ | Weights $W_i$ |
|---|---|---|
| 1 | 0.000000000000000 | 2.000000000000000 |
| 2 | $\pm$ 0.577350269189626 | 1.000000000000000 |
| 3 | $\pm$0.774596669241483 | 0.555555555555556 |
|   | 0.000000000000000 | 0.888888888888889 |
| 4 | $\pm$0.861136311594053 | 0.347854845137454 |
|   | $\pm$0.339981043584856 | 0.652145154862546 |
| 5 | $\pm$0.906179845938664 | 0.236926885056189 |
|   | $\pm$0.538469310105683 | 0.478628670499366 |
|   | 0.000000000000000 | 0.568888888888889 |
| 6 | $\pm$0.932469514203152 | 0.171324492379170 |
|   | $\pm$0.661209386466265 | 0.360761573048139 |
|   | $\pm$0.238619186083197 | 0.467913934572691 |

into Eq. (17.111) and eliminating $R$, we obtain

$$I = \int_{u_1}^{u_2} f(u)\, du \approx \frac{u_2 - u_1}{2} \sum_{i=1}^{n} W_i f\left(\frac{u_2 - u_1}{2} C_i + \frac{u_1 + u_2}{2}\right) \quad (17.114)$$

It must be emphasized that Gauss quadrature produces exact results for integrating polynomials according to the following rule: $n$ sampling points yield exact results for polynomials of degree $\leq 2n - 1$. Thus, three-point ($n = 3$) Gauss quadrature (Fig. 17-8a) seems adequate for mass property calculations.

If the interval $[u_1, u_2]$ over which the integral is to be evaluated is large and the function $f(u)$ changes significantly over the interval, the interval can be divided into subintervals as shown in Fig. 17-8b. Each subinterval could be chosen to have a width $\Delta u$ of 1 or 2. Gauss quadrature is applied to each subinterval and the results are added. This is similar to applying the Simpson rule to Fig. 17-7c.

In summary, in Newton-Cotes formulas, we use $(n + 1)$ equally spaced sampling points (as shown in Fig. 17-7a and b) and polynomials of order at most $n$ are integrated exactly. In Gauss quadrature $n$ unequally spaced sampling points (Fig. 17-8a) are required to integrate a polynomial of order at most $(2n - 1)$. Polynomials of orders less than $n$ and $(2n - 1)$ for the two methods respectively

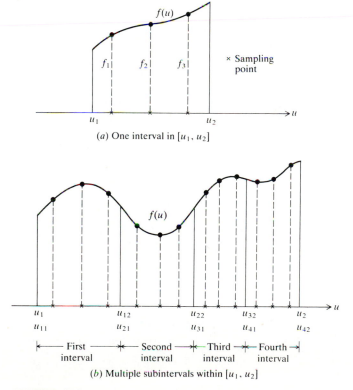

(a) One interval in $[u_1, u_2]$

(b) Multiple subintervals within $[u_1, u_2]$

**FIGURE 17-8**
Three-point Gauss quadrature.

would also be integrated exactly. For both methods, if a nonpolynomial function is integrated, an error will result, the magnitude of which depends on how well the polynomial matches the function. In general, Gauss quadrature is usually more accurate than Newton-Cotes methods.

The extension of Gauss quadrature to evaluate double integrals can be accomplished without much difficulty. Analogous to Eq. (17.111), the integral in Eq. (17.99) can be written as

$$
I = \int_{v_1}^{v_2} \int_{u_1}^{u_2} f(u, v) \, du \, dv = \sum_{i=1}^{n} \sum_{j=1}^{n} V_i V_j f_{ij} + R \tag{17.115}
$$

where

$$
V_j = \frac{v_2 - v_1}{2} W_j \tag{17.116}
$$

$$
v_j = \frac{v_2 - v_1}{2} C_j + \frac{v_1 + v_2}{2} \tag{17.117}
$$

and $V_i$ and $u_i$ are given by Eqs. (17.112) and (17.113) respectively. Thus, Eq. (17.115) can be written as

$$
I = \int_{v_1}^{v_2} \int_{u_1}^{u_2} f(u, v) \, du \, dv \approx \frac{(u_2 - u_1)(v_2 - v_1)}{4}
$$

$$
\times \sum_{i=1}^{n} \sum_{j=1}^{n} W_i W_j f\left( \frac{u_2 - u_1}{2} C_i + \frac{u_1 + u_2}{2}, \frac{v_2 - v_1}{2} C_j + \frac{v_1 + v_2}{2} \right) \tag{17.118}
$$

The sampling points for two-dimensional Gauss quadrature is shown in Fig. 17-9.

Other integration schemes are available. One particular scheme is the Monte Carlo method discussed in Chap. 16. What kind of integration scheme should be used? It seems that Gauss quadrature with three sampling points is attractive for mass property calculations based on the formulation presented in this chapter.

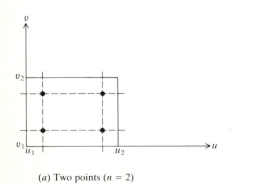

(a) Two points ($n = 2$)

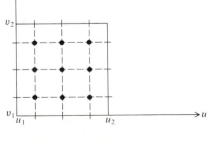

(b) Three points ($n = 3$)

**FIGURE 17-9**
Sampling points for two-dimensional Gauss quadrature.

The details of applying the equations listed in Table 17.1 to specific curves, surfaces, and solids as well as reducing these equations to closed forms or numerical integrations are not covered here. They are assigned to problems at the end of the chapter. Only sample examples are discussed here.

**Example 17.1.** Find the length of the B-spline curve developed in Example 5.21 in Chap. 5. The curve and its control points are shown in Fig. 17-10.

*Solution.* The B-spline curve equation is given by

$$\mathbf{P}(u) = \mathbf{P}_0(1 - u)^3 + 3\mathbf{P}_1 u(1 - u)^2 + 3\mathbf{P}_2 u^2(1 - u) + \mathbf{P}_3 u^3, \qquad 0 \le u \le 1$$

$$(17.119)$$

where $\mathbf{P}_0$, $\mathbf{P}_1$, $\mathbf{P}_2$, and $\mathbf{P}_3$ are given in Example 5.19. Using this equation, the tangent vector is

$$\mathbf{P}' = (3\mathbf{P}_3 - 9\mathbf{P}_2 + 9\mathbf{P}_1 - 3\mathbf{P}_0)u^2 + (6\mathbf{P}_2 - 12\mathbf{P}_1 + 6\mathbf{P}_0)u + 3(\mathbf{P}_1 - \mathbf{P}_0) \quad (17.120)$$

which gives

$$\begin{bmatrix} x' \\ y' \\ z' \end{bmatrix} = \begin{bmatrix} 6u - 6u^2 \\ 3 - 6u \\ 0 \end{bmatrix}$$

$$(17.121)$$

Substituting into Eq. (17.8) and reducing, the length of the B-spline is given by

$$L = 3 \int_0^1 \sqrt{4u^4 - 8u^3 + 8u^2 - 4u + 1} \; du$$

$$(17.122)$$

This integral can be evaluated numerically using Gauss quadrature with $f(u)$ being the square root function. Using two-point quadrature, we can write

$$L \approx \frac{3}{2} \sum_{i=1}^{2} W_i f_i = \frac{3}{2}(W_1 f_1 + W_2 f_2)$$

$$(17.123)$$

Using Eq. (7.113) and Table 17.2, we calculate $u_1$ and $u_2$ as 0.211324865 and 0.788675134. Calculating $f_1$ and $f_2$ and knowing that $W_1 = W_2 = 1$, Eq. (17.123)

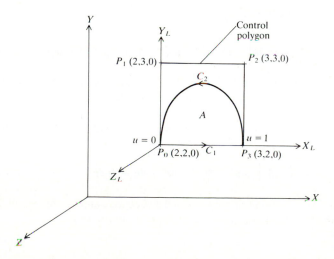

**FIGURE 17-10**
A cubic B-spline curve.

gives

$$L \approx \tfrac{3}{2}(0.666666667 + 0.666666665) = 1.999999998$$

Using three-point quadrature, the length $L$ is given by

$$L \approx \tfrac{3}{2} \sum_{i=1}^{3} W_i f_i = \tfrac{3}{2}(W_1 f_1 + W_2 f_2 + W_3 f_3) \qquad (17.124)$$

Note that $u_1$, $u_2$, and $u_3$ are calculated to be 0.112701665, 0.5, and 0.887298334 respectively. Thus,

$$L \approx \tfrac{3}{2}(0.444444445 + 0.444444444 + 0.444444444) = 2.0$$

Notice that because the B-spline is symmetric about an axis passing through the point $u = 0.5$ and the weighting factors are symmetric, the resulting $f_i$ are also symmetric. The reader is encouraged to apply the trapezoidal and Simpson rules and compare their accuracies with that of Gauss quadrature.

**Example 17.2.** Calculate the area (and its centroid) bounded by the B-spline curve of Example 5.21 and the $X_L$ axis shown in Fig. 17-10.

**Solution.** Let us assume that the $X_L Y_L Z_L$ coordinate system has the same orientation as the MCS, as shown in Fig. 17-10. Thus, Eq. (17.28) applies and gives

$$A = \sum_{i=1}^{2} \int_{u_{i1}}^{u_{i2}} xy' \, du \qquad (17.125)$$

The boundary consists of two curves: $C_1$ which is a straight line and $C_2$ which is the B-spline. Using Eqs. (5.11) and (5.13), we can write for $C_1$:

$$P(u) = P_0 + u(P_3 - P_0), \qquad 0 \le u \le 1 \qquad (17.126)$$

$$P' = P_3 - P_0 \qquad (17.127)$$

Substituting Eqs. (17.119) and (17.120) and Eqs. (17.126) and (17.127) into Eq. (17.125) and reducing we obtain

$$A = \int_0^1 [x_0 + u(x_3 - x_0)](y_3 - y_0) \, du$$

$$+ \int_1^0 x_0(1 - u)^3 + 3x_1 u(1 - u)^2 + 3x_2 u^2(1 - u) + x_3 u^2](3 - 6u) \, du \qquad (17.128)$$

The first term on the right-hand side of Eq. (7.128) is zero because the coordinates $y_3$ and $y_0$ are equal. The limits of the second integral reflect the fact that the curves $C_i$ of the boundary must be traversed in a counterclockwise order so that the area usually lies to the left of any boundary curve. Substituting the values for the $x$ coordinates and reducing, Eq. (17.128) becomes

$$A = 3 \int_1^0 (4u^4 - 8u^3 + 3u^2 - 4u + 2) \, du \qquad (17.129)$$

Exact integration of this equation gives $A = 0.6$. Let us compare this result with two-point and three-point Gauss quadrature. For two points, we get

$$A \approx -\tfrac{1}{2}(W_1 f_1 + W_2 f_2) \qquad (17.130)$$

Note that $u_1$ and $u_2$ are given by 0.211324865 and 0.788675134 respectively. Calculating $f_1$ and $f_2$ and substituting into Eq. (17.130), we obtain

$$A \approx -\tfrac{3}{2}(1.221153452 - 1.665597891) = 0.666666659$$

For three-point integration, the area is

$$A \approx -\tfrac{3}{2}(W_1 f_1 + W_2 f_2 + W_3 f_3) \qquad (17.131)$$

Note that $u_1$, $u_2$, and $u_3$ are 0.112701665, 0.5, and 0.887298334 respectively. Equation (17.131) gives

$$A \approx -\tfrac{3}{2}(0.875828708 + 0 - 1.275828689) = 0.599999971$$

The above area should be exact, that is, 0.6, because, as mentioned earlier, using $n = 3$ should yield exact results for polynomials of degree $\leq 5$. The polynomial in this example [Eq. (17.129)] is of degree 4. The difference is due to truncation and round-off errors. On the other hand, the accuracy of the two-point integration is not as good as expected.

The centroid of the area is given by Eq. (17.31), which reduces to

$$x_c = \frac{1}{A} \sum_{i=1}^{n} \int_{u_{i1}}^{u_{i2}} \frac{x^2}{2} y' \, du$$

$$y_c = \frac{1}{A} \sum_{i=1}^{n} \int_{u_{i1}}^{u_{i2}} xyy' \, du \qquad (17.132)$$

$$z_c = \frac{1}{A} \sum_{i=1}^{n} \int_{u_{i1}}^{u_{i2}} xzy' \, du$$

Because the area lies in the $XY$ plane, then $z_c = 0$. Similar to reducing Eq. (17.128), the above equations reduce to

$$x_c = \frac{3}{2A} \int_1^0 (-8u^7 + 28u^6 - 30u^5 + 25u^4 - 32u^3 + 12u^2 - 8u + 4) \, du \quad (17.133)$$

$$y_c = \frac{3}{A} \int_1^0 (-12u^6 + 36u^5 - 25u^4 + 5u^3 - 12u^2 - 2u + 4) \, du \qquad (17.134)$$

Exact integration of the above two equations gives $x_c = 2.5$ and $y_c = 2.321428571$. Notice that due to symmetry of the area with respect to the axis given by the equation $x_L = 0.5$ (or $x = 2.5$), the centroid lies on this axis. Thus $x_c = 2.5$ is correct as expected.

Let us compare the above exact results with three-point Gauss quadrature. Applying the same Gauss data we needed to calculating $A$ in Eqs. (17.133) and (17.134), we obtain

$$x_c = -\frac{3}{4A}(1.782523371 + 0 - 3.782523365)$$

$$= -\frac{3 \times -1.999999994}{4 \times 0.599999971} = 2.500000113$$

$$y_c = -\frac{3}{2A}(2.014406029 + 0 - 2.934406026)$$

$$= -\frac{3 \times -0.919999997}{2 \times 0.599999971} = 2.300000104$$

In summary, the exact area $A$ is 0.6 and the exact centroid is (2.5, 2.321428571, 0) with respect to the MCS ($XYZ$ system). Three-point Gauss quadrature gives very good results.

**Example 17.3.** A plane passes through the three points $P_0(1, 2, 3)$, $P_1(2, 4, 5)$, and $P_2(4, 2, 3)$. Find the surface area that is bounded by the parametric domain $u = [0, 1]$ and $v = [0, 1]$.

*Solution.* The plane equation is given by Eq. (6.25), that is,

$$\mathbf{P}(u, v) = \mathbf{P}_0 + u(\mathbf{P}_1 - \mathbf{P}_0) + v(\mathbf{P}_2 - \mathbf{P}_0), \qquad \begin{cases} 0 \le u \le 1 \\ 0 \le v \le 1 \end{cases} \qquad (17.135)$$

This equation gives

$$\mathbf{P}_u = \mathbf{P}_1 - \mathbf{P}_0 \qquad \mathbf{P}_v = \mathbf{P}_2 - \mathbf{P}_0 \qquad (17.136)$$

Substituting into Eq. (17.37), we get

$$A_s = C \int_0^1 \int_0^1 du \, dv = C \qquad (17.137)$$

where

$$C = \sqrt{[(x_1 - x_0)^2 + (y_1 - y_0)^2 + (z_1 - z_0)^2][(x_2 - x_0)^2 + (y_2 - y_0)^2 + (z_2 - z_0)^2]}$$

$$= 9$$

Thus, $A_s = 9$.

Let us use Gauss quadrature with two points of integration. Equation (17.118) gives

$$A_s = \frac{C}{4} \sum_{i=1}^2 \sum_{j=1}^2 W_i W_j f_{ij}$$

$$= \frac{C}{4} (W_1 W_1 f_{11} + W_1 W_2 f_{12} + W_2 W_1 f_{21} + W_2 W_2 f_{22}) \qquad (17.138)$$

From Eq. (17.137), $f(u, v) = 1$. Thus $f_{11} = f_{12} = f_{21} = f_{22} = 1$. Therefore, Eq. (17.138) gives

$$A_s = \frac{C}{4} (1.000000000 + 1.000000000 + 1.000000000 + 1.000000000) = C = 9$$

which is the exact answer as expected.

## 17.5  PROPERTY EVALUATION BASED ON SOLID MODELING

The geometric and mass property formulation and evaluation presented in the preceding sections are directly applicable to wireframe and surface models that have parametric representations ($u$ space for curves and $u$–$v$ space for surfaces). As a result, the resulting property integrals can be evaluated exactly in closed form or numerically with good accuracy.

While this approach can be extended to various solid representation schemes as discussed shortly here, other methods to calculate mass properties of solid models exist and are discussed here as well. It will be seen that the scheme used to represent a solid dominates the method and consequently the algorithm that can be used to evaluate its mass properties. Errors attributed to approximations in representing solids usually dominate errors from more traditional and well-studied resources such as truncations, round-off, and numerical integration. In dealing with multiple integrations over a solid domain, we are faced with integrating simple functions (integrands) over geometrically complicated domains (geometry of the solid). In traditional computational studies, the converse problem occurs: the integrand may be complex but the domain is geometrically simple (e.g. cube, cylinder, or sphere).

The known methods for computing integral properties of solid models fall into three broad categories:

1. Methods that use a given solid representation scheme directly without a need to convert it to another one.
2. Methods that involve representation conversion.
3. Monte Carlo and related methods.

In the following discussion, we briefly present an integral property computational method for each solid representation scheme introduced in Chap. 7.

In the primitive instancing representation scheme, a certain finite number of object families are defined. These families are used to model other objects that belong to them. For example, we can define a family of prisms with parameters $N$(number of sides), $R$(radius of containing cylinder), and $H$(prism height). Instantiating the family with various values of $N$, $R$, and $H$ allows various objects to be modeled. To compute the mass properties of objects represented by primitive instancing, a special formula is needed for each primitive. For example, the volume of any instance of the prism primitive described above is

$$V = \pi R^2 H \frac{\sin (2\pi/N)}{2\pi/N} \tag{17.139}$$

The volumes of all instances of various primitives that make a given object are calculated and added together to yield the volume of the object. Similarly, other material properties are calculated by adding the contributions from the instances.

Cell decomposition, spatial enumeration, and octree encoding schemes all represent a solid as the sum of quasi-disjoint (non-overlapping) cells. Thus, any integral over a solid is decomposed into a sum of integrals over the cells that make the solid, that is,

$$\int_{\text{solid}} f \, dV = \sum_{i=1}^{n} \int_{\text{cell } i} f \, dV \tag{17.140}$$

The integrals on the right-hand side of the above equation are easy to evaluate for cells with simple shapes. Errors of approximating a solid with cells is carried over to mass property calculations in addition to the traditional errors of computations. There is a trade-off between the shape of a cell and the difficulty to evaluate integrals. Simple-shaped cells make integrals easy to evaluate, but they may not be adequate to represent complex solids. The use of more complicated cells leads to more precise and concise representations, but integration over such cells becomes more difficult.

Analytic solid modeling (ASM) is another method of dividing a solid into quasi-disjoint cells called hyperpatches. Each hyperpatch is described by Eq. (7.91) and is a cube in the parametric space $(u, v, w)$. The volume of a solid is given by the sum of volumes of the hyperpatches that make the solid. Calculating any integral is achieved by mapping it from the cartesian space to the parametric space as follows:

$$\iiint_{\text{hyper.}} f(x, y, z) \, dx \, dy \, dz = \iiint_{\text{cube}} f(u, v, w)|J| \, du \, dv \, dw \qquad (17.141)$$

where $|J|$ is the determinant of the jacobian $[J]$ which is given by (see Prob. 17.13 at the end of the chapter)

$$[J] = \begin{bmatrix} \dfrac{\partial x}{\partial u} & \dfrac{\partial y}{\partial u} & \dfrac{\partial z}{\partial u} \\[2mm] \dfrac{\partial x}{\partial v} & \dfrac{\partial y}{\partial v} & \dfrac{\partial z}{\partial v} \\[2mm] \dfrac{\partial x}{\partial w} & \dfrac{\partial y}{\partial w} & \dfrac{\partial z}{\partial w} \end{bmatrix} \qquad (17.142)$$

The jacobian $[J]$ can be evaluated using Eq. (7.91). For mass properties, the function $f(x, y, z)$ is readily expressed in terms of $u$, $v$, and $w$, as can also be seen from Eq. (7.91). When $f(x, y, z) = 1$, Eq. (17.141) gives the volume of the hyperpatch. It should be mentioned here that Eq. (17.46) is not convenient to use with hyperpatches because their surfaces are byproducts of their equations (7.91).

Mass property calculations of swept solids take advantage of the fact that these are two-and-a-half-dimensional solids. For linear sweep, the volume of a solid is given by

$$V = AL \qquad (17.143)$$

where $A$ is the planar area that is swept to create the solid and $L$ is the length of the solid in the sweeping direction. For translational sweep, $L$ is the depth of the solid, and for rotational sweep it is the length of the circumference of the solid. We can then use the approach and equations described in Sec. 17.2.2 to calculate $A$. The centroid of the solid is the same as the area centroid but located halfway in the sweeping direction, that is, at $L/2$. If the area centroid is given by $(x_c, y_c, z_c)$, the solid centroid is

$$x_{cs} = x_c + \frac{L}{2} n_x$$

$$y_{cs} = y_c + \frac{L}{2} n_y \qquad (17.144)$$

$$z_{cs} = z_c + \frac{L}{2} n_z$$

where $(n_x, n_y, n_z)$ is the unit vector $\hat{n}$, which defines the sweeping direction for translational sweep or the axis of revolution for rotational sweep. To evaluate the inertial properties, the approach described in Secs. 17.3.3 and 17.3.4 can be used. The related equations may be simplified by taking advantage of the swept solid geometric characteristics. Readers can easily convince themselves with simple cases such as a cube or a cylinder. Mass property calculations for nonlinear sweep is more involved than linear sweep and need further investigations.

Integral properties of solids represented by their boundaries (B-rep scheme) may be evaluated by surface integration summarized in Table 17.1. Because of the directed curves and surfaces used to represent B-rep models, the traversal of the solid edges and faces is achieved automatically (without the user's help). These surface integrals can be derived using direct integration as presented in typical statics books or by using the Gauss divergence theorem as described in the previous section. Surface integrals can be changed to line integrals during their evaluation. Surface integrals are very attractive to use for polyhedral objects. The faces of these objects can be decomposed into triangular facets, and the objects can be divided into tetrahedra. Curved objects can be approximated in this way (the approximate B-rep scheme). A few algorithms that calculate integrals of polyhedral objects exist and are not discussed here.[1]

Few methods exist to evaluate the integral properties of solids based on the CSG schemes. A natural method to compute these properties is to exploit the "divide-and-conquer" paradigm by applying recursively the following formulas:

$$\iiint_{A \cup B} f \, dV = \iiint_{A} f \, dV + \iiint_{B} f \, dV - \iiint_{A \cap B} f \, dV \qquad (17.145)$$

$$\iiint_{A - B} f \, dV = \iiint_{A} f \, dV - \iiint_{A \cap B} f \, dV \qquad (17.146)$$

However, one must first solve the basic problem of evaluating an integral over the intersection of a number of primitive solids. A major drawback of this method is the possible out-of-hand growth of the interactions between primitives of a given solid which, in turn, would require intensive computations to evaluate all the intersection integrals.

Two methods are based on representation conversion. A B-rep of a solid can always be obtained from its CSG representation using boundary evaluation as discussed in Chap. 7. Then B-rep-based algorithms to evaluate integrals can be used. This method is usually not efficient. Another method uses recursive subdivision to convert CSG representation of a solid into an octree representation (see Fig. 7-58). This conversion is at best an approximation of the CSG model. An

---

[1] As an example see J. Cohen and T. Hickey, "Two Algorithms for Determining Volumes of Convex Polyhedra," *J. Association for Computing Machinery*, vol 26, no. 3, pp. 401–414, 1979.

algorithm to evaluate the integral properties that is based on this method is available.[2]

One attractive method that has been widely used to evaluate integral properties for CSG and other representations is ray tracing (or ray casting). This method is also used to generate shaded images. The basics of the method are introduced in Chap. 10. The method calculates the integral properties by partitioning a solid into volume elements (rectangular parallelepipeds), evaluating the properties for each element, and then summing all the results. The volume elements are generated by casting rays through the solids, classifying the rays with respect to the solid, and then enclosing each "in" segment of each ray with a rectangular parallelepiped. Rays are evenly cast in the $XY$ plane (the screen) of the camera model (refer to Chap. 10), but have varying lengths along the $Z$ axis. A random number generator can be used to position the rays in the pixels. The density of the rays determines the accuracy.

Let us consider, as an example, a volume algorithm based on ray tracing. The algorithm casts rays through the screen into the solid. The two-dimensional spacing of rays in the screen defines two dimensions of the volume elements (the rectangular parallelepipeds). The third dimension is defined by the enter-exit points returned by classifying the rays. Specifically, if the horizontal and vertical distance in the screen is $D$, then the volume detected by each ray is

$$V_R = D^2 L[(t_2 - t_1) + (t_4 - t_3) + \cdots + (t_n - t_{n-1})] \tag{17.147}$$

where $L = \sqrt{(\Delta x)^2 + (\Delta y)^2 + (\Delta z)^2}$, the length of the ray's direction vector. Each $(t_i - t_{i-1})L$ is the length of a ray segment that is inside the solid, that is, an "in" segment. The volume of the solid is the sum of $V_R$ for all the rays. It is obvious that the exact volume of a solid is the limit as the number of rays goes to infinity.

The last method to be reviewed here is the Monte Carlo method. This method applies to all representation schemes of solids. The method provides an estimate of the integral $\iiint_V f\, dV$ by averaging values of $f$ computed at a large number of randomly selected points, that is,

$$\iiint_V f\, dV \approx \frac{V}{N} \sum_{i=1}^{N} f(\mathbf{P}_i) \tag{17.148}$$

where $\mathbf{P}_i$ are random points independently and uniformly distributed inside the solid, $V$ is the volume of the solid, and $N$ is the number of sampling points. Equation (17.148) cannot be applied directly to a solid because its volume is not known *a priori*, and techniques for generating uniformly distributed points over complex regions are not available. To solve these problems, the solid is enclosed in a simpler virtual solid such as a box. Let us define the following function:

$$f^*(\mathbf{P}) = f(\mathbf{P})g(\mathbf{P}) \tag{17.149}$$

---

[2] See Y. T. Lee and A. A. G. Requicha, "Algorithms for Computing the Volume and Other Integral Properties of Solids. II. A Family of Algorithms Based on Representation Conversion and Cellular Approximation," *Communications of the ACM*, vol. 25, no. 9, pp. 642–650, 1982.

where $g(\mathbf{P})$ is a characteristic function of the true solid such that

$$g(\mathbf{P}) = \begin{cases} 1, & \text{if } \mathbf{P} \text{ belongs to the true solid} \\ 0, & \text{otherwise} \end{cases} \tag{17.150}$$

Thus, $g(\mathbf{P})$ is a step function. Therefore, the integral of $f^*$ over the box equals the desired integral of $f$ over the true solid. Utilizing Eqs. (17.149) and (17.150), Eq. (17.148) becomes

$$\iiint\limits_{V} f \, dV \approx \frac{V_b}{N_s} \sum_{i=1}^{N_s} f^*(\mathbf{P}_i) \tag{17.151}$$

where $V_b$ is the volume of the box and the $s$ indicates that the sum ranges only over points $\mathbf{P}_i$ that belong to the true solid, that is, for points $\mathbf{P}_i$ that have $g(\mathbf{P}_i) = 1$. The membership classification function can be used to determine if $\mathbf{P}_i$ belongs to a solid ("in" or "on") or not ("out"). Good error bounds and the general applicability of the Monte Carlo method are its advantages. Its major drawback is its slow convergence which implies that large number of points are required for accurate results.

## 17.6 PROPERTIES OF COMPOSITE OBJECTS

Frequently an object can be divided into two or more simple subobjects whose integral properties can be readily or easily obtained. For example, a three-dimensional object may be decomposed into a few two-and-a-half-dimensional subobjects. The original object is referred to as a composite object.

When an object can be divided into a number of simple subobjects, the volume of the object is the sum of the separate subobjects (if subobjects are removed such as holes, their corresponding volumes are subtracted), that is,

$$V = \sum_{i=1}^{n} V_i \tag{17.152}$$

where $V$, $V_i$, and $n$ are the volume of the object, the volume of the $i$th subobject, and the number of subobjects respectively. The centroid of the object is given by

$$\mathbf{r}_c = \frac{\sum\limits_{i=1}^{n} \mathbf{r}_{ci} V_i}{\sum\limits_{i=1}^{n} V_i} = \frac{\sum\limits_{i=1}^{n} \mathbf{r}_{ci} V_i}{V} \tag{17.153}$$

or

$$x_c = \frac{\sum\limits_{i=1}^{n} x_{ci} V_i}{V} \qquad y_c = \frac{\sum\limits_{i=1}^{n} y_{ci} V_i}{V} \qquad z_c = \frac{\sum\limits_{i=1}^{n} z_{ci} V_i}{V} \tag{17.154}$$

where $r_{ci}$ is the position vector of the centroid of the $i$th subobject with respect to the MCS. The first moments, second moments, and products of inertia are given by

$$M_{xy} = \sum_{i=1}^{n} (M_{xy})_i \qquad M_{xz} = \sum_{i=1}^{n} (M_{xz})_i \qquad M_{yz} = \sum_{i=1}^{n} (M_{yz})_i \qquad (17.155)$$

$$I_{xx} = \sum_{i=1}^{n} (I_{xx})_i \qquad I_{yy} = \sum_{i=1}^{n} (I_{yy})_i \qquad I_{zz} = \sum_{i=1}^{n} (I_{zz})_i \qquad (17.156)$$

$$I_{xy} = \sum_{i=1}^{n} (I_{xy})_i \qquad I_{xz} = \sum_{i=1}^{n} (I_{xz})_i \qquad I_{yz} = \sum_{i=1}^{n} (I_{yz})_i \qquad (17.157)$$

In Eqs. (17.155) and (17.157), the sums are algebraic sums because first moments and products of inertia may be positive or negative.

## 17.7 MASS PROPERTY CALCULATIONS ON CAD/CAM SYSTEMS

Mass property calculations is one of the earliest applications supported by many CAD/CAM systems. Mass property packages available on these systems provide properties for the three geometric modeling techniques: wireframes, surfaces, and solids. The properties formulated in Secs. 17.2 and 17.3 are usually calculated.

Due to the significant reductions in algorithm development, execution time, and user input, separate mass property commands and/or modifiers exist on most CAD/CAM systems to support calculations for two-dimensional planar areas ("calculate area" command), two-and-a-half-dimensional objects (uniform thickness and surface of revolution), and three-dimensional objects ("calculate mass properties" command).

The "calculate area" command usually requires the user to digitize the boundary curves of the area in a given order, followed by digitizing the boundary curves of the holes. The command then utilizes the composite area calculation similar to what is described in Sec. 17.6 to calculate the area properties. A common error associated with this command arises when the ends of adjacent boundary curves are not identical (within the floating-point accuracy) due to mistakes in geometric constructions.

The "calculate mass properties" command can have various modifiers. For two-and-a-half-dimensional objects with uniform thickness, the user is required to digitize the boundary (with holes if any) of the planar area, define the projection direction, and define the object thickness. Equations (17.143) and (17.144) are utilized. For axisymmetric objects, the user must define the planar area, the axis of rotation, and the angle of rotation. For three-dimensional objects, a surface model of the object of interest must be used. The user must first cut (slice) the model along a given axis, say the $Z$ axis, by a perpendicular plane to the axis at, say, $\Delta z = h$ intervals. The slicing of the model generates boundaries of cross-sectional areas. Once these boundaries are generated, the user inputs the number of cross sections, the length $h$ of the interval, and the material density, and digitizes the boundaries of each cross section so that the system can evaluate the mass properties using either the trapezoidal or Simpson rule. This is a very labor-intensive and error-prone process.

Solid models are much easier to use to calculate mass properties from a user's point of view. Once the solid model is available, all that the user needs to do is to identify the solid by one digitize and input the density. The mass properties algorithm calculates these properties automatically. In this case of solid models, there are no advantages of two-and-a-half-dimensional objects over three-dimensional objects at the user's input level.

Another useful mass property command is "merge mass point." This command implements the concept of composite objects described in the previous section. To utilize this command, a nongraphics entity (call it mass point) may be created to store all the mass properties of the various subobjects. This entity can be displayed on the screen as a symbol. The "merge mass point" command would require the user to digitize the individual mass points to be added algebraically. In dividing an object into subobjects, the user may create and/or delete geometric entities to create closed boundaries of various cross sections.

In using mass property commands, it is the user's responsibility, as is the case with any computer program, to check the correct units of the density so that the proper units for the mass and inertial properties are produced. It is also the user's responsibility to be aware of the errors involved with the calculations, as discussed in this chapter, and attempt to reduce these errors as much as the error analysis permits.

## 17.8 DESIGN AND ENGINEERING APPLICATIONS

In this section, some examples are presented to show how the material in this chapter can be applied to various applications. In some cases, especially in open-ended design problems, it may be beneficial to implement mass property commands into interactive programs as illustrated in Chap. 15.

**Example 17.4.** Calculate the mass properties of the object shown in Fig. 17-11. Use the density of 1.0 slug/in$^3$.

*Solution.* The object shown in Fig. 17-11a is a two-and-a-half-dimensional object with a uniform thickness of 2 in. Let us assume that the user creates this model as a wireframe one. A "calculate mass property" command with a density of 1, thickness $t = 2$ in, and a projection direction in the $Y$ direction can be used to evaluate the object properties. The boundary of the area is digitized in the counterclockwise direction as shown in Fig. 17-11a. The calculated mass properties of the object are:

$V = 30.7854022$ in$^3$, $m = 30.7854022$ slug

$x_c = 2.5510241$, $y_c = 1$, $z_c = 1.5$ in

$M_{xy} = 46.1781005$, $M_{xz} = 30.7854022$, $M_{yz} = 78.5343017$ slug $\cdot$ in

$I_{xx} = 135.6674194$, $I_{yy} = 360.6104736$, $I_{zz} = 307.037445$ slug $\cdot$ in$^2$

$I_{xy} = 78.5343094$, $I_{xz} = 117.8014602$, $I_{yz} = 46.1781043$ slug $\cdot$ in$^2$

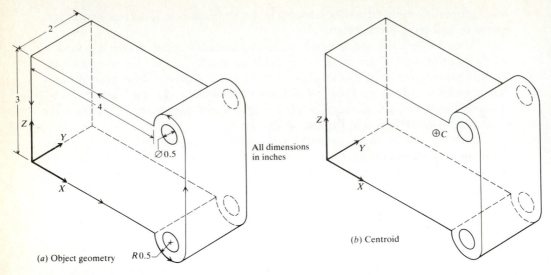

(a) Object geometry    R0.5

All dimensions
in inches

(b) Centroid

**FIGURE 17-11**
Mass properties of an object with uniform thickness.

Figure 17-11b shows the centroid of the object.

**Example 17.5.** Figure 17-12a shows a hollow cross section (lies in the $XZ$ plane)
that is rotated about an axis parallel to the $Z$ axis (in the $XZ$ plane) at a distance
$x = -2$. The cross section is rotated an angle of 120°. The beginning and ending
angles of rotation are 30° and 150° degrees respectively (see Fig. 17-12a). Calculate
the mass properties of the resulting axisymmetric object assuming a density of 1.0
slug/in³.

**Solution.** A surface of revolution command is used to rotate the cross section shown
in Fig. 17-12a about the shown axis of revolution. Figure 17-12b shows the resulting
surface relative to the original cross section. A mass property command with a
density of 1, an axis of revolution as shown, and beginning and ending angles of 30°
and 150° respectively is used. The outer circle is digitized as an outside boundary
and the inner circle is digitized as a hole. The resulting mass properties of the object
are:

$$V = 9.8696031 \text{ in}^3, \ m = 9.8696031 \text{ slug}$$

$$x_c = -2.0, \ y_c = 1.7832046, \ z_c = 0 \text{ in}$$

$$M_{xy} = 0, \ M_{xz} = 17.5995216, \ M_{yz} = -19.7392043 \text{ slug} \cdot \text{in}$$

$$I_{xx} = 37.524929, \ I_{yy} = 56.8531493, \ I_{zz} = 88.2095718 \text{ slug} \cdot \text{in}^2$$

$$I_{xy} = -35.1990432, \ I_{xz} = 0, \ I_{yz} = 0 \text{ slug} \cdot \text{in}^2$$

Figure 17-12c shows the centroid of the object. Notice that the zero values in the
above properties are due to symmetry of the cross section with respect to the $X$ and
$Z$ axes, as shown in Fig. 17-12a.

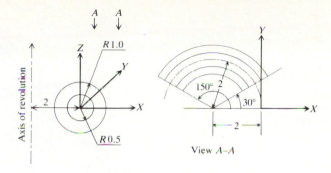

(*a*) Cross section and axis of rotation

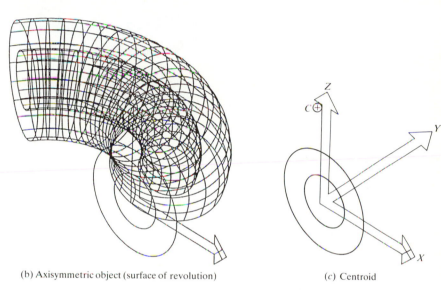

(b) Axisymmetric object (surface of revolution)          (*c*) Centroid

**FIGURE 17-12**
Mass properties of an axisymmetric object.


**Example 17.6.** Figure 17-13*a* shows a three-dimensional object of density 1.0 slug/in³. Calculate the mass properties of the object.

*Solution.* Let us create a surface model of the object. Ruled surfaces are adequate. Create a circle and a rectangle five inches apart along the *Y* axis. Divide the circle into four divisions and create a ruled surface between each division and the corresponding side of the rectangle. Figure 17-13*b* shows the resulting surface model.

To calculate the mass properties, the surface model is sliced at eleven locations along the *Y* axis. Thus, eleven cross sections perpendicular to the *Y* axis are generated 0.5 in apart. The model is cut by a series of planes perpendicular to the *Y* axis and positioned properly. The four ruled surfaces of the model must be digitized to intersect each plane. An interactive program to create the cutting plane and generate the cross sections (with possibly placing each one on a separate layer) is very

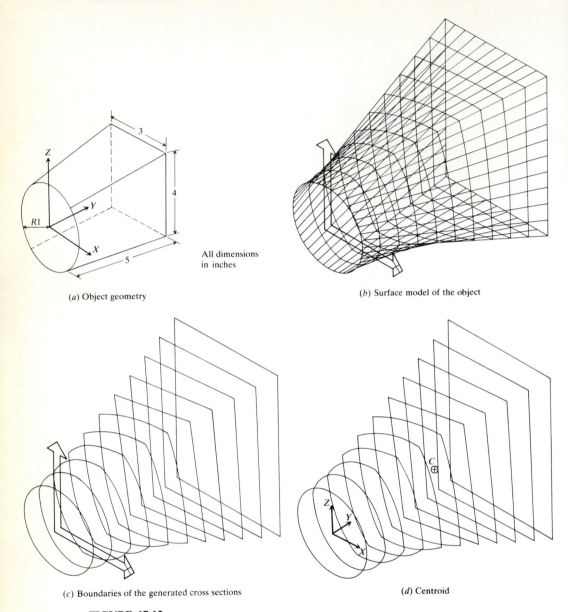

(a) Object geometry

All dimensions
in inches

(b) Surface model of the object

(c) Boundaries of the generated cross sections

(d) Centroid

**FIGURE 17-13**
Mass properties of a three-dimensional object.

helpful in this labor-intensive process. Figure 17-13c shows the resulting cross sections.

A mass property command utilizing the trapezoidal rule is used to calculate the mass properties. The boundary of each cross section must be digitized in order. The resulting values are listed below:

$$V = 31.8458938 \text{ in}^3, \ m = 31.8458938 \text{ slug}$$

$$x_c = 0, \ y_c = 3.0518979, \ z_c = 0 \text{ in}$$

$M_{xy} = 0$, $M_{xz} = 97.190422$, $M_{yz} = 0.0000015$ slug $\cdot$ in

$I_{xx} = 382.1100463$, $I_{yy} = 40.107357$, $I_{zz} = 372.731903$ slug $\cdot$ in$^2$

$I_{xy} = 0.0000015$, $I_{xz} = -0.424032$, $I_{yz} = 0.0000023$ slug $\cdot$ in$^2$

Notice the effect of the object symmetry as well as the effect of errors on the results. Figure 17-13d shows the centroid.

**Example 17.7.** Figure 17-14a shows an object geometry. Calculate the mass properties of the object assuming a density of 1 slug/in$^3$.

**Solution.** The object shown in Fig. 17-14a is a composite object which can be decomposed into three two-and-a-half-dimensional subobjects with uniform thickness: the subobjects on the right and left of thickness 1 inch and the center subobject. Using the mass property command for two-and-a-half-dimensional objects, the mass points $M_1$ and $M_2$ are calculated. The mass point $M_3$ is obtained by mirroring the mass point $M_2$ about the $YZ$ plane. The three mass points $M_1$, $M_2$, and $M_3$ are then merged (added) to yield the total mass point $M$ of the object. Figure 17-14b shows these mass points. The results attached with the final mass point $M$ are:

$V = 94.1514587$ in$^3$, $m = 94.1514587$ slug

$x_c = -0.0003542$, $y_c = 2.0$, $z_c = 2.1884136$ in

$M_{xy} = 206.0423278$, $M_{xz} = 188.3029174$, $M_{yz} = -0.0333519$ slug $\cdot$ in

$I_{xx} = 1192.638916$, $I_{yy} = 1208.4245605$, $I_{zz} = 1015.6827392$ slug $\cdot$ in$^2$

$I_{xy} = -0.0667038$, $I_{xz} = -0.0333943$, $I_{yz} = 412.0846557$ slug $\cdot$ in$^2$

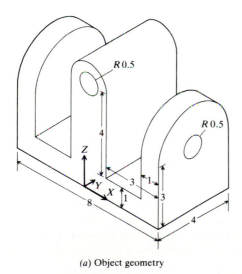

(a) Object geometry

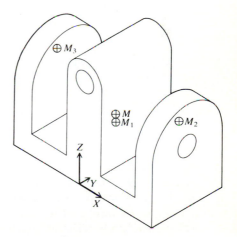

(b) Mass points of the object

**FIGURE 17-14**
A composite object.

Readers are encouraged to resolve the above examples on their respective CAD/CAM systems. Solid models might be created for these examples and comparisons between the amount of labor needed by various types of modeling techniques may be performed.

## PROBLEMS

### Part 1: Theory

**17.1.** Derive Eq. (17.31) in detail.

**17.2.** Derive Eq. (17.88) in detail.

**17.3.** How would Eqs. (17.88), (17.91), and (17.93) change if the arbitrary axis is defined by two points instead of a point and a unit vector $\hat{\mathbf{n}}$?

**17.4.** How does Eq. (17.31) reduce for the cases shown in Fig. P17-4?

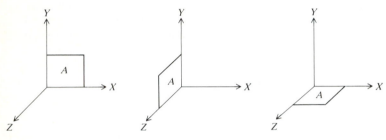

**FIGURE P17-4**

**17.5.** Prove that Eq. (17.37) can be rewritten as:

$$A_s = \int_{v_1}^{v_2} \int_{u_1}^{u_2} |J|\, du\, dv$$

where $|J|$ is the determinant of the jacobian $[J]$ which is given by the equation

$$[J] = \|\mathbf{P}_u \times \mathbf{P}_v\| = \begin{bmatrix} \hat{\mathbf{i}} & \hat{\mathbf{j}} & \hat{\mathbf{k}} \\ \dfrac{\partial x}{\partial u} & \dfrac{\partial y}{\partial u} & \dfrac{\partial z}{\partial u} \\ \dfrac{\partial x}{\partial v} & \dfrac{\partial y}{\partial v} & \dfrac{\partial z}{\partial v} \end{bmatrix}$$

*Hint:* Expand $(\mathbf{P}_u \cdot \mathbf{P}_u)(\mathbf{P}_v \cdot \mathbf{P}_v)$ in Eq. (17.37) and expand $|J|$. Then use the fact that $\mathbf{P}_u \cdot \mathbf{P}_v = 0$ to prove the equality.

**17.6.** Derive the principal moments of inertia of an object given its moments about a coordinate system.

**17.7.** Use Eq. (17.9) to find the curve length for all analytic and synthetic curves covered in Chap. 5. Which are exact and which are approximate?

**17.8.** Use equations listed in Table 7.1 to find properties of objects made of surfaces covered in Chap. 6. Which are exact and which are approximate?

**17.9.** Evaluate the following integrals using trapezoidal rule, Simpson rule, and Gauss quadrature (with one, two, and three sampling points). Solve the integrals exactly and compare.

(a) $\displaystyle\int_0^1 (4 + 5u^2)\, du$

(b) $\displaystyle\int_0^{\pi/2} \cos u\, du$

(c) $\displaystyle\int_{-1}^1 (8u^3 + 5u^2)\, du$

(d) $\displaystyle\int_0^2 ue^{2u}\, du$

(e) $\displaystyle\int_0^6 \left(\frac{u^2}{4} - 4u + 16\right) du$

(f) $\displaystyle\int_0^1 u^5\, du$

(g) $\displaystyle\int_{-1}^2 (u + 2 - u^2)\, du$

**17.10.** Evaluate the following integrals using Gauss quadrature with two and three sampling points. Solve the integrals exactly and compare the results.

(a) $\displaystyle\int_0^1 \int_0^1 (3u^3v + 4u^2v^3)\, du\, dv$

(b) $\displaystyle\int_0^\pi \int_0^2 r^2 \sin\theta\, dr\, d\theta$

(This is the first moment of half a circle of radius 2 about the $X$ axis.)

**17.11.** A cubic spline curve in the $XY$ plane of the MCS passes through the origin and is described by the following equation:

$$\mathbf{P}(u) = \begin{bmatrix} 0.03u^3 + 0.3u^2 + 1.5u \\ -0.05u^3 + 1.5u \\ 0 \end{bmatrix}, \qquad 0 \le u \le 1$$

Calculate the curve length, the area under the spline, and its centroid. Compare the exact results with the three-point Gauss quadrature.

**17.12.** A sphere with a radius $R$ and a center at $(x_0, y_0, z_0)$ is described by the following equation:

$$\mathbf{P}(u, v) = \begin{bmatrix} x_0 + R\cos u \cos v \\ y_0 + R\cos u \sin v \\ z_0 + R\sin u \end{bmatrix}, \qquad \begin{cases} -\pi/2 \le u \le \pi/2 \\ 0 \le v \le 2\pi \end{cases}$$

For $R = 1$ and center at $(1, 1, 1)$, calculate the surface area and the centroid exactly. Use the three-point Gauss quadrature. Compare the results.

**17.13.** Given a volume element $dV = dx\, dy\, dz$, prove that

$$dV = dx\, dy\, dz = |J|\, du\, dv\, dw$$

where $[J]$ is given by Eq. (17.142).

## Part 2: Laboratory

**17.14.** Use your CAD/CAM system to calculate the mass properties of:
   (a) Wireframe models created in Chap. 5 problems
   (b) Surface models created in Chap. 6 problems
   (c) Solid models created in Chap. 7 problems
   Compare with exact results if possible.

## Part 3: Programming

**17.15.** Write various computer programs that implement the equations listed in Table 17.1. Use exact solutions if possible or use Gauss quadrature with three-point integration.

## BIBLIOGRAPHY

Boyse, J. W., and J. E. Gilchrist: "GM Solid: Interactive Modeling for Design and Analysis of Solids," *IEEE Computer Graphics and Applic.*, vol. 2, no. 2, pp. 27–40, 1982.

Cohen, J., and T. Hickey: "Two Algorithms for Determining Volumes of Convex Polyhedra," *J. Ass. for Computing Machinery*, vol. 26, no. 3, pp. 401–414, 1979.

Cook, P. N., L. T. Cook, S. Batnitzky, K. R. Lee, W. H. Anderson, and S. J. Dwyer III: "Volume and Surface Area Estimates Using Tomographic Data," *IEEE Trans. Pattern Anal. Mach. Intell.*, vol. PAMI-2, no. 5, pp. 478–479, 1980.

Davis, P. J., and P. Rabinowitz: *Methods of Numerical Integration*, Academic Press, New York, 1975.

Eberhardt, A. C., and G. H. Williard: "Calculating Precise Cross-Sectional Properties for Complex Geometries," *Computers in Mech. Engng (CIME)*, pp. 32–37, September/October 1987.

Fabrikant, V. I., V. Latinovic, and T. S. Sankar: "Contour Integration on the Graphics Screen and Its Application in CAD/CAM," *Computer Aided Des.*, vol. 17, no. 2, pp. 60–68, 1985.

Haber, S.: "Numerical Evaluation of Multiple Integrals," *SIAM Rev.*, vol. 12, no. 4, pp. 481–526, 1970.

Hamming, R. W.: *Numerical Methods for Scientists and Engineers*, McGraw-Hill, New York, 1962.

Henrici, P.: *Applied and Computational Complex Analysis*, vol. 1, John Wiley, New York, 1974.

Kalos, M. V., and P. A. Whitlock: *Monte Carlo Methods*, vol. 1, John Wiley, New York, 1986.

Lee, Y. T., and A. A. G. Requicha: "Algorithms for Computing the Volume and Other Integral Properties of Solids. I. Known Methods and Open Issues," *Commun. ACM*, vol. 25, no. 9, pp. 635–641, 1982.

Lee, Y. T., and A. A. G. Requicha: "Algorithms for Computing the Volume and Other Integral Properties of Solids. II. A Family of Algorithms Based on Representation Conversion and Cellular Approximation," *Commun. ACM*, vol. 25, no. 9, pp. 642–650, 1982.

Lien, S. L., and J. T. Kajiya: "A Symbolic Method for Calculating the Integral Properties of Arbitrary Nonconvex Polyhedra," *IEEE Computer Graphics and Applic.*, vol. 4, no. 10, pp. 35–41, 1984.

Messner, A. M., and G. Q. Taylor: "Algorithm 550: Solid Polyhedron Measures," *ACM Trans. on Math. Software*, vol. 6, no. 1, pp. 121–130, 1980.

Miles, R. G., and J. G. Tough: "A Method for the Computation of Inertial Properties for General Areas," *Computer Aided Des.*, vol. 15, no. 4, pp. 196–200, 1983.

Roth, S. D.: "Ray Casting for Modeling Solids," *Computer Graphics and Image Processing*, vol. 18, no. 2, pp. 109–144, 1982.

Stolk, R., and G. Ettershank: "Calculating the Area of an Irregular Shape," *BYTE*, vol. 12, no. 2, pp. 135–136, 1987.

Timmer, H. G., and J. M. Stern: "Computation of Global Geometric Properties of Solid Objects," *Computer Aided Des.*, vol. 12, no. 6, pp. 301–304, 1980.

Wilson, Jr., H. B., and D. S. Farrior: "Computation of Geometrical and Inertial Properties of General Areas and Volumes of Revolution," *Computer Aided Des.* vol. 8, no. 4, pp. 257–263, 1976.

Wylie, C. E.: *Advanced Engineering Mathematics*, 4th ed., McGraw-Hill, New York, 1975.

# CHAPTER
# 18

# FINITE ELEMENT MODELING AND ANALYSIS

## 18.1 INTRODUCTION

Finite element modeling (FEM) and analysis (FEA) are two of the most popular mechanical engineering applications offered by existing CAD/CAM systems. This is attributed to the fact that the finite element method is perhaps the most popular numerical technique for solving engineering problems. The method is general enough to handle any complex shape or geometry (problem domain), any material properties, any boundary conditions, and any loading conditions. The generality of the finite element method fits the analysis requirements of today's complex engineering systems and designs where closed-form solutions of governing equilibrium equations are usually not available. In addition, it is an efficient design tool by which designers can perform parametric design studies by considering various design cases (different shapes, materials, loads, etc.), analyzing them, and choosing the optimum design.

The finite element method is a numerical analysis technique for obtaining approximate solutions to a wide variety of engineering problems. The method originated in the aerospace industry as a tool to study stresses in complex airframe structures. It grew out of what was called the matrix analysis method used in aircraft design. The method has gained increased popularity among both researchers and practitioners.

Since its birth in the 1950s the finite element method has gone through various development stages. One may perhaps identify four phases of development. From the 1950s to the 1970s the main thrust of research was devoted to formulate the finite element theory. Variational and Galerkin's approaches for two- and three-dimensional problems were formulated. Isoparametric formulations for both linear and nonlinear, static and dynamic analyses were developed. Also, computational methods required by the finite element method were established. These methods included numerical integration of element equations, solution of the global system of algebraic equations, and study of numerical stability and error analysis of solution algorithms.

Once the method gained popularity and most of the underlying theory and concepts were developed, various institutions and firms developed finite element analysis codes during the 1970s and 1980s. There is a wide variety of analysis codes which support a wide range of applications. Like CAD/CAM software, FEA codes may seem different, but they all share the same underlying theory. Any of these codes has six distinct modules: node module which generates nodal coordinates and degrees of freedom; boundary conditions module which applies given boundary conditions to the appropriate nodes; material properties module which usually stores various material types in a materials library; element module which stores various element types in an element library, evaluates element matrices, and assembles these matrices; loading module which applies the externally applied loads to the proper nodes and elements; and solution module which solves the system of equations that result from the assembly process of the element module for the unknown variables.

The third phase of development of the finite element method was the creation of pre- and postprocessors beginning in the mid 1970s. With the increased popularity of the finite element method and its use by aerospace, automative, and other industries, it became clear that a large percentage of the total computational cost of a typical FEA goes into the manual construction and preparation of the finite element mesh. Pre- and postprocessors are classified as such relative to the FEA process. Preprocessing precedes this process and helps a user to automate and/or facilitate the input required by the first five modules described in the previous paragraph. Postprocessing follows the analysis process and enables users to display the analysis results in various graphical forms to better understand and interpret them.

The fourth phase of the finite element method development is porting existing commercial FEA codes to the microcomputer environment with the benefit of the related low costs. There are usually limits on the maximum number of nodes and elements that can be processed on microcomputers due to their limited storage and processing speeds. Microcomputer-based FEA is adequate for first-cut quick evaluations of various design alternatives.

The finite element method is a major subject of study and there are numerous textbooks devoted to it. We cannot possibly cover all the facets of the method in one chapter in this book. It is often needed, during teaching of a CAD course, to provide students with a quick, but adequate, background of FEM and FEA so that they understand and utilize the method as an effective design tool, be able to use the related pre- and postprocessors, and more importantly interpret the FEA results correctly. Therefore, this chapter covers the basics of both FEA and FEM. In the first part of the chapter, we cover the various methods of finite element formulation, isoparametric elements and their evaluations, and sample applications. Isoparametric elements are chosen over other elements

because they produce more accurate results and their underlying theory is close
to the parametric representations discussed in Part II of the book. Readers who
are already familiar with the finite element method can consider the material
covered here as a review.

The second part of the chapter is devoted to FEM. In particular, various
methods, concepts, and algorithms of mesh generation are covered. Mesh generation methods based on wireframe, surface, and solid modeling techniques are
discussed. Their classification as semi-automatic and fully automatic methods is
presented.

## 18.2  GENERAL PROCEDURE OF THE FINITE ELEMENT METHOD

The solution of a continuum problem by the finite element method usually
follows an orderly step-by-step process. The following steps show in general how
the finite element method works. These steps will become more understandable
when the FEA is covered in more detail later.

1.  Discretize the given continuum. The essence of the finite element method is to
    divide a continuum, that is, problem domain, into quasi-disjoint non-
    overlapping elements. This is achieved by replacing the continuum by a set of
    key points, called nodes, which when connected properly produce the ele-
    ments. The collection of nodes and elements forms the finite element mesh. A
    variety of element shapes and types are available. The analyst or designer can
    mix element types to solve one problem. The number of nodes and elements
    that can be used in a problem is a matter of engineering judgment. As a
    general rule, the larger the number of nodes and elements, the more accurate
    the finite element solution, but also the more expensive the solution is; more
    memory space is needed to store the finite element model, and more computer
    time is needed to obtain the solution. Figure 18-1 shows an example of dis-
    cretizing a cantilever beam made of steel and supporting a concentrated load
    at its free end. Figure 18-1c shows two types (four-node and six-node) of a
    quadrilateral (element shape) element.

2.  Select the solution approximation. The variation of the unknown (called field
    variable) in the problem is approximated within each element by a poly-
    nomial. The field variable may be a scalar (e.g., temperature) or a vector (e.g.,
    horizontal and vertical displacements). Polynomials are usually used to
    approximate the solution over an element domain because they are easy to
    integrate and differentiate. The degree of the polynomial depends on the
    number of nodes per element, the number of unknowns (components of field
    variable) at each node, and certain continuity requirements along element
    boundaries.

3.  Develop element matrices and equations. The finite element formulation pre-
    sented in the next section involves transformation of the governing equi-
    librium equations from the continuum domain to the element domain. Once
    the nodes and material properties of a given element are defined, its corre-
    sponding matrices (stiffness matrix, mass matrix, etc.) and equations can be

Material:
Steel $E = 3 \times 10^7$ lb/in$^2$
$v = 0.3$

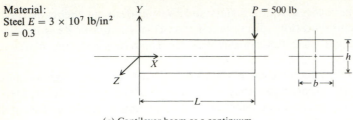

(a) Cantilever beam as a continuum

(b) Node generation and numbering

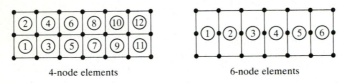

4-node elements                    6-node elements

(c) Element generation and numbering

**FIGURE 18-1**
Continuum discretization for FEA.

derived. Four methods are available to derive element matrices and equations: the direct method, the variational method, the weighted residual method, and the energy method. In this chapter, we cover the second (suitable for solid mechanics problems) and the third (suitable for thermalfluids problems) methods.

4. Assemble the element equations. The individual element matrices are added together by summing the equilibrium equations of the elements to obtain the global matrices and system of algebraic equations. Before solving this system, it must be modified by applying the boundary conditions. If boundary conditions are not applied, wrong results are obtained, or a singular system of equations may result.

5. Solve for the unknowns at the nodes. The global system of algebraic equations is solved via Gauss elimination methods to provide the values of the field variables at the nodes of the finite element mesh. Values of field variables and their derivatives at the nodes form the complete finite element solution of the original continuum problem before discretization. Values at other points inside the continuum other than the nodes are possible to obtain, although it is not customarily done.

6. Interpret the results. The final step is to analyze the solution and results obtained from the previous step to make design decisions. The correct inter-

pretation of these results requires a sound background in both engineering and FEA. This is why treating FEA and FEM codes as a black box is usually not recommended and, in fact, is considered dangerous.

In the context of the above step-by-step procedure, it is clear that there are various critical decisions that practitioners of the finite element analysis have to make, e.g., the type of analysis, the number of nodes, the degrees of freedom (components of the field variable) at each node, the element shape and type, the material type, and finally the interpretation of the results. Making these decisions becomes more obvious after we discuss the above steps in more detail while covering FEA.


## 18.3   FINITE ELEMENT ANALYSIS

There are two distinct approaches to model physical systems: the discrete lumped approach and the continuum approach. In the former, the actual system is idealized as an assemblage of elements, an equilibrium equation is written for each element, and the resulting set of equations is solved to yield the solution. In the latter approach, the actual system is treated as one continuum, and one or more equilibrium equation is written and solved for the system response.

Two different methods exist to develop the governing differential (equilibrium) equations for a continuum: the differential formulation and the variational formulation. In the former, the equations are derived by considering the equilibrium of a differential (infinitesimal) element. In the latter formulation, equilibrium equations are expressed as integral equations by considering energy and work balance in the system under study. Both formulations are equivalent and one can be derived from the other. Equilibrium equations are usually accompanied by boundary conditions (called boundary value problems) and/or initial conditions (called initial value problems).

There are many alternatives to solve linear and nonlinear boundary and initial value problems, ranging from completely analytical to completely numerical. Exact solutions are usually available for a few problems, usually with simple domains (e.g., rectangular domains). These solutions can be obtained by direct integration of the differential equations. This can be achieved by techniques such as separation of variables, and Fourier and Laplace transformations—to name only a few. Approximate solutions are usually sought for equations whose solutions cannot be obtained in a closed form. A score of approximate techniques exist and include perturbation, power series, probability schemes, the method of weighted residuals, the finite difference method, the Rayleigh-Ritz method, and the finite element method.

The finite element analysis is based on the following premise. Instead of solving the governing differential (equilibrium) equations directly, the finite element method solves an integral form of these equations. The solution of such an integral form is approximate. In obtaining such a solution, the finite element method leaves the differential operator intact and approximates the solution space (by using interpolation functions). In contrast, the finite difference method amounts to a finite difference approximation of the differential operator while keeping the solution space intact.

### 18.3.1   Development of Integral Equations

In this section, we cover two popular approaches to derive integral equations which form the basis of the finite element solution. These approaches are the variational principle and the method of weighted residuals. In both approaches, we begin with the governing differential equations. We might point out that the variational principle can also use work and energy concepts to derive the integral equations.

**18.3.1.1   VARIATIONAL APPROACH.** In this approach, a variational statement (called a functional $\Pi$) is derived from the differential equation of the problem. Let us assume that the continuum problem at hand can be described in general by the following equation:

$$L(\phi) - f = 0 \tag{18.1}$$

over the continuum domain $D$ and subject to the boundary conditions (b.c.):

$$\phi = \Phi \qquad \text{on } B \tag{18.2}$$

where $\phi$ is the dependent variable to solve for, $f$ is a known function of the independent variable, $L$ is a linear or nonlinear differential (ordinary or partial) operator, and $B$ is the boundary of the domain $D$.

To derive the functional $\Pi$ corresponding to Eqs. (18.1) and (18.2), we seek an integral $I(\phi)$ whose first variation with respect to $\phi$ vanishes. The rationale behind this step is well explained and documented in books on variational calculus (calculus of variations) and on the finite element method. Readers should refer to these books to convince themselves. Multiplying Eq. (18.1) by the first variation of $\phi$, $\delta\phi$, and integrating over the domain $D$, we get

$$\int_D \delta\phi[L(\phi) - f]\,dD = 0 \tag{18.3}$$

The objective now is to manipulate Eq. (18.3) in such a way that allows the variational operator $\delta$ to be moved outside the integral sign; that is, Eq. (18.3) becomes

$$\delta \int_D [L^*(\phi) - f^*]\,dD = 0 \tag{18.4}$$

where $L^*(\phi)$ is the new operator resulting from the original operator $L(\phi)$. When Eq. (18.4) is reached, the integral [call it $I(\phi)$] in the equation is the desired functional $\Pi$ because its first variation [as given by Eq. (18.4)] is zero; that is, Eq. (18.4) can be written as

$$\delta[I(\phi)] = \delta\Pi = 0 \tag{18.5}$$

where

$$\Pi = I(\phi) = \int_D [L^*(\phi) - f^*] \, dD \qquad (18.6)$$

There are no general consistent set of rules that can be followed during manipulation of Eq. (18.1) to obtain Eq. (18.6). However, it is always the case that the boundary conditions, given Eq. (18.2), must be invoked during the manipulation process. Integration by parts is usually useful for one- and two-dimensional problems, as seen from the examples below. In some cases, the Gauss divergence [Eq. (17.42)] theorem may be useful. While Eq. (18.1) is a good start to derive Eq. (18.6) for field problems (heat transfer, etc.), it is efficient for solid mechanics problems to write the functional $\Pi$ directly using the variational principle (equivalent to minimum potential energy) or the Hamilton principle.

**Example 18.1.** Derive the functional $\Pi$ for the following systems:

(a) A bar of length $L$ and variable cross-sectional area $A$ is under a concentrated axial load $P$ at its free end, as shown in Fig. 18-2a. The axial displacement at any point along the bar is governed by the following differential equation:

$$AEu'' = 0 \qquad (18.7)$$

where $E$ is the modulus of elasticity of the bar material and $u'' = d^2u/dx^2$.

(b) A cantilever beam of length $L$ and cross-sectional moment of inertia $I$ has a transverse load $P$ at its free end, as shown in Fig. 18-2b. The beam lateral deflection $y$ is given by

$$EIy'''' = 0 \qquad (18.8)$$

where $y'''' = d^4y/dx^4$.

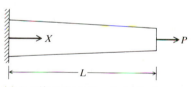

(a) A uniform bar with uniaxial loading

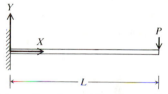

(b) A cantilever beam with concentrated load

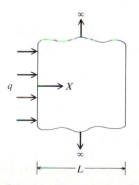

(c) An infinite slab with constant heat flux input

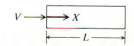

(d) A permeable media with constant fluid flow

**FIGURE 18-2**
Examples of one-dimensional continuum problems.

(c) An infinite slab of width $L$ is subjected to a constant uniform heat flux input $q$, as shown in Fig. 18-2c. The temperature $T$ inside the slab is governed by the following heat conduction equation:

$$KT'' = 0 \tag{18.9}$$

where $K$ is the thermal conductivity and $T'' = d^2T/dx^2$.

(d) A permeable media with a cross-sectional area $A$ has a constant fluid flow $V$ across the end at $x = 0$ (Fig. 18-2d). The pressure $P$ inside the media is given by the following differential equation:

$$AgP'' = 0 \tag{18.10}$$

where $g$ is the coefficient of permeability and $P'' = d^2P/dx^2$.

**Solution.** This example illustrates how a functional $\Pi$ is derived from a differential equation by using integration by parts. Care must be given to the boundary conditions during the derivation, as seen below. In terms of Eq. (18.1), Eq. (18.7) shows that $\phi = u$, $L = AE\, d^2/dx^2$, and $f = 0$. Similarly, Eqs. (18.8), (18.9), and (18.10) give respectively $\phi = y$, $L = EI\, d^4/dx^4$, $f = 0$; $\phi = T$, $L = K\, d^2/dx^2$, $f = 0$; and $\phi = P$, $L = Ag\, d^2/dx^2$, $f = 0$. The domain $D$ in the four systems of this example is one dimensional. It is the length $L$, that is, $D = \{x: x \in [0, L]\}$. Utilizing Eq. (18.3), the following shows how Eqs. (18.4) and (18.6) are derived for each system.

(a) Multiplying Eq. (18.7) by $\delta u$ and integrating over the bar length, we can write

$$\int_0^L AEu'' \,\delta u \, dx = 0 \tag{18.11}$$

Using the equality $u'' \, dx = du'$ and integrating by parts, we get

$$AEu' \,\delta u \Big|_0^L - \int_0^L AEu' \, d(\delta u) = 0 \tag{18.12}$$

where we used the integration by parts rule:

$$\int_0^L U \, dV = UV \Big|_0^L - \int_0^L V \, dU \tag{18.13}$$

where $U = EA\,\delta u$ and $V = u'$. The first term in Eq. (18.12) can be reduced using the boundary conditions. At $x = 0$, $u = 0$; thus $\delta u = 0$. At $x = L$, $AEu' = P$. This condition is derived from axial loading where the stress $\sigma = P/A = E\varepsilon = Eu'$. The second term in Eq. (18.12) can be reduced if we interchange $\delta$ and $d/dx$ in order to write:

$$d(\delta u) = \frac{d(\delta u)}{dx} \, dx = \delta\!\left(\frac{du}{dx}\right) dx = \delta(u') \, dx \tag{18.14}$$

Substituting the boundary conditions and Eq. (18.14) into Eq. (18.12), we get

$$\int_0^L AEu' \,\delta u' \, dx - P\,\delta u_L = 0 \tag{18.15}$$

where $u_L$ is the axial displacement at $x = L$. Rewriting $u' \,\delta u'$ as $\frac{1}{2}\delta u'^2$ (similar to $du^2 = 2u\, du$), Eq. (18.15) becomes

$$\int_0^L \delta\!\left(\frac{AE}{2} u'^2 \, dx\right) - \delta(Pu_L) = 0 \tag{18.16}$$

Interchanging $\delta$ and $\int$, we get

$$\delta\left(\int_0^L \frac{AE}{2} u'^2 \, dx - Pu_L\right) = 0 \tag{18.17}$$

Comparing Eq. (18.4) and (18.17), the functional $\Pi$ corresponding to the differential equation (18.7) is

$$\Pi = \tfrac{1}{2}\int_0^L AEu'^2 \, dx - Pu_L \tag{18.18}$$

(b) Multiplying Eq. (18.8) by $\delta y$ and integrating over the beam length, we get

$$\int_0^L EIy'''' \, \delta y \, dx = 0 \tag{18.19}$$

Integrating by parts ($U = EI \, \delta y$ and $V = y'''$), the above equation becomes

$$EIy''' \, \delta y \Big|_0^L - \int_0^L EIy''' \, d(\delta y) = 0 \tag{18.20}$$

Let us use the boundary conditions of the beam to reduce the first term in the above equation. At $x = 0$, $y = 0$; thus $\delta y = 0$. At $x = L$, $EIy''' = V = -P$. This condition states that the shear force $V$ at the free end (given by $EIy'''$ from the beam theory) is equal to minus the applied load $P$. Similar to Eq. (18.14), $d(\delta y) = \delta(y') \, dx$. Thus, Eq. (18.20) becomes

$$\int_0^L EIy''' \, \delta y' \, dx + P \, \delta y_L = 0 \tag{18.21}$$

where $y_L$ is the beam deflection at the free end. Using $dy'' = y''' \, dx$ and integrating by parts ($U = EI \, \delta y'$ and $V = y''$), the first term of the above equation reduces to

$$\int_0^L EIy''' \, \delta y' \, dx = EIy'' \, \delta y' \Big|_0^L - \int_0^L EIy'' \, d(\delta y') \tag{18.22}$$

At $x = 0$, $y' = 0$; thus $\delta y' = 0$ (slope at fixed end $= 0$). At $x = L$, $EIy'' = 0$. This condition states that the bending moment (given by $EIy''$ from the beam theory) at the free end is zero. Using these boundary conditions, and rewriting $y'' \, d(\delta y')$ as $\tfrac{1}{2}\delta y''^2 \, dx$, Eq. (18.22) becomes

$$\int_0^L EIy''' \, \delta y' \, dx = -\int_0^L \frac{EI}{2} \delta y''^2 \, dx \tag{18.23}$$

Substituting Eq. (18.23) into Eq. (18.21), we get

$$\delta\left(\int_0^L \frac{EI}{2} y''^2 \, dx - Py_L\right) = 0 \tag{18.24}$$

which gives

$$\Pi = \tfrac{1}{2}\int_0^L EIy''^2 \, dx - Py_L \tag{18.25}$$

(c) Multiplying Eq. (18.9) by $\delta T$ and integrating over the slab width gives

$$\int_0^L KT'' \, \delta T \, dx = 0 \tag{18.26}$$

This equation is identical in form to Eq. (18.11). The same mathematical manipulation used to obtain Eq. (18.18) can be followed here. Thus, Eq. (18.26) gives

$$KT' \, \delta T \Big|_0^L - \int_0^L KT' \, d(\delta T) = 0 \tag{18.27}$$

At $x = 0$, $KT' = -q$ (which is Fourier's law). At $x = L$, $T = T_L$ (assuming that the slab is insulated at $x = L$ to keep $T_L$ constant); thus $\delta T = 0$. After manipulating Eq. (18.27), the functional $\Pi$ is

$$\Pi = \tfrac{1}{2} \int_0^L KT'^2 \, dx - qT_0 \tag{18.28}$$

where $T_0$ is the temperature at $x = 0$.

(d) Multiplying Eq. (18.10) by $\delta P$ and integrating over the media length, we get

$$\int_0^L AgP'' \, \delta P \, dx = 0 \tag{18.29}$$

Again, this equation is identical in form to Eq. (18.11). Integrating it gives

$$AgP' \, \delta P \Big|_0^L - \int_0^L AgP' \, d(\delta P) = 0 \tag{18.30}$$

At $x = 0$, $AgP' = -V$ (fluid flow is proportional to the pressure gradient $P'$). At $x = L$, $P = P_L$ (assuming that pressure at $x = L$ is kept constant). Thus, Eq. (18.30) gives

$$\Pi = \tfrac{1}{2} \int_0^L AgP'^2 \, dx - VP_0 \tag{18.31}$$

The development of the functionals of the above systems show that applying the boundary conditions correctly is crucial to their correct development. Boundary conditions that are functions of derivatives of the dependent variable (e.g., $AEu'$, $EIy'''$, $EIy''$, $KT'$, and $AgP'$ in this example) are so-called natural boundary conditions. Conditions that are specified directly on the dependent variable (e.g., at $x = 0$, $u = 0$, $y = 0$, $T = T_0$, and $P = P_0$ in this example) are referred to as geometric or rigid boundary conditions. It is only the geometric boundary conditions that are applied to the global system of equations mentioned in step 4 in Sec. 18.2.

**Example 18.2.** Derive the functional $\Pi$ for the following systems:

(a) Plane stress problem over a rectangular domain, as shown in Fig. 18-3a. The concentrated loads $P_1$ and $P_2$ are applied at the center distances in the $X$ and $Y$ directions respectively.

(b) A steady-state heat conduction over a rectangular domain (Fig. 18-3b) subject to heat fluxes $q_1$ and $q_2$ in the $X$ and $Y$ directions respectively. The governing differential equations is given by

$$K \frac{\partial^2 T}{\partial x^2} + K \frac{\partial^2 T}{\partial y^2} = 0 \tag{18.32}$$

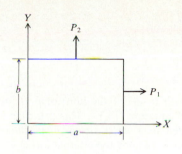

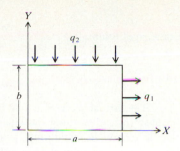

(*a*) Plane stress problem                (*b*) Heat conduction over a rectangle

**FIGURE 18-3**
Examples of two-dimensional continuum problems.

assuming the material is isotropic and homogeneous with a thermal conductivity of $K$.

*Solution*

(*a*) For plane stress continuum problems, the following equations can be written:

Displacement vector: $\qquad\qquad \mathbf{d} = [u \quad v]^T$ $\qquad\qquad$ (18.33)

where $u$ and $v$ are the displacement components in the $X$ and $Y$ directions respectively.

Strain vector: $\qquad \boldsymbol{\varepsilon} = [\varepsilon_x \quad \varepsilon_y \quad \gamma_{xy}]^T = \left[\dfrac{\partial u}{\partial x} \quad \dfrac{\partial v}{\partial y} \quad \dfrac{\partial u}{\partial y} + \dfrac{\partial v}{\partial x}\right]^T$ $\qquad$ (18.34)

where $\varepsilon_x$, $\varepsilon_y$, and $\gamma_{xy}$ are the normal strain in the $X$ direction, the normal strain in the $Y$ direction, and the shear strain respectively. In addition,

$$\varepsilon_z = \frac{v}{1 - v}(\varepsilon_x + \varepsilon_y) \qquad\qquad (18.35)$$

where $v$ is Poisson's ratio. Here, we are assuming isotropic material.

Stress vector: $\qquad\qquad \boldsymbol{\sigma} = [\sigma_x \quad \sigma_y \quad \tau_{xy}]^T$ $\qquad\qquad$ (18.36)

where $\sigma_x$, $\sigma_y$, $\tau_{xy}$ are the normal stress in the $X$ direction, the normal stress in the $Y$ direction, and the shear stress respectively. For plane stress, $\sigma_z = \tau_{xz} = \tau_{yz} = 0$.

Hook's law: $\qquad\qquad\qquad \boldsymbol{\sigma} = [D]\boldsymbol{\varepsilon}$ $\qquad\qquad\qquad$ (18.37)

where $[D]$ is the material stiffness matrix and is given by

$$[D] = \frac{E}{1 - v^2}\begin{bmatrix} 1 & v & 0 \\ v & 1 & 0 \\ 0 & 0 & \dfrac{1 - v}{2} \end{bmatrix} \qquad\qquad (18.38)$$

where $E$ is the modulus of elasticity.

Static equilibrium:

$$\frac{\partial \sigma_x}{\partial x} + \frac{\partial \tau_{xy}}{\partial y} = 0$$

(18.39)

$$\frac{\partial \tau_{xy}}{\partial x} + \frac{\partial \sigma_y}{\partial y} = 0$$

The virtual work principle is utilized to derive the functional $\Pi$ for this problem instead of using the differential equation (18.39). This principle states that if a body in equilibrium is subjected to a small virtual displacement, it remains in equilibrium and the virtual work done by the internal forces is equal to the virtual work done by the externally applied forces, that is,

$$W_{int} = W_{ext}$$

(18.40)

For an elastic body $W_{int}$ results from the internal stresses and any other internal and body forces. For our problem, Eq. (18.40) becomes

$$\int_V \delta\varepsilon^T \sigma \, dV = P_1 \, \delta u_1 + P_2 \, \delta v_2$$

(18.41)

where $u_1$ and $v_2$ are the horizontal and vertical displacement components under $P_1$ and $P_2$ respectively. We may mention that the virtual work expression given by the left-hand side of Eq. (18.41) is general and applies to any solid mechanics problem. Equation (18.41) can be recast as

$$\int_V \delta\varepsilon^T \sigma \, dV - P_1 \, \delta u_1 - P_2 \, \delta v_2 = 0$$

(18.42)

Our goal now is to find the functional $\Pi$ from the above equation by reducing it to the form given by Eq. (18.5). Substituting Eq. (18.37) into the first term in Eq. (18.42) gives

$$\int_V \delta\varepsilon^T \sigma \, dV = \int_V \delta\varepsilon^T [D]\varepsilon \, dV$$

(18.43)

For linear elastic material the matrix $[D]$ is symmetric, and $\delta\varepsilon^T[D]\varepsilon$ can be reduced to $\frac{1}{2}\delta(\varepsilon^T[D]\varepsilon)$. To verify this result for our problem, we expand $\delta\varepsilon^T[D]\varepsilon$ to obtain

$$\delta\varepsilon^T[D]\varepsilon = [\delta\varepsilon_x \quad \delta\varepsilon_y \quad \delta\gamma_{xy}][D]\begin{bmatrix} \varepsilon_x \\ \varepsilon_y \\ \gamma_{xy} \end{bmatrix}$$

(18.44)

Expanding this equation and reducing, we obtain

$$\delta\varepsilon^T[D]\varepsilon = \frac{E}{1-v^2}\left[ \varepsilon_x \, \delta\varepsilon_x + v(\varepsilon_x \, \delta\varepsilon_y + \varepsilon_y \, \delta\varepsilon_x) + \varepsilon_y \, \delta\varepsilon_y + \frac{1-v}{2} \gamma_{xy} \, \delta\gamma_{xy} \right]$$

$$= \frac{E}{2(1-v^2)}\left[ \delta\varepsilon_x^2 + 2v \, \delta(\varepsilon_x \varepsilon_y) + \delta\varepsilon_y^2 + \frac{1-v}{2} \delta\gamma_{xy}^2 \right]$$

$$= \frac{E}{2(1-v^2)} \delta\left( \varepsilon_x^2 + 2v\varepsilon_x \varepsilon_y + \varepsilon_y^2 + \frac{1-v}{2} \gamma_{xy}^2 \right)$$

$$= \tfrac{1}{2}\delta(\varepsilon^T[D]\varepsilon)$$

(18.45)

Substituting Eq. (18.45) into Eq. (18.42) gives

$$\tfrac{1}{2}\delta \int_V \boldsymbol{\varepsilon}^T [D]\boldsymbol{\varepsilon}\ dV - \delta(P_1 u_1) - \delta(P_2 v_2) = 0 \tag{18.46}$$

where $P_1\ \delta u_1$ and $P_2\ \delta v_2$ in Eq. (18.42) are written as shown above because $P_1$ and $P_2$ are constants (or independent of displacements). Factoring the operator $\delta$, Eq. (18.46) becomes

$$\delta\left(\frac{1}{2}\int_V \boldsymbol{\varepsilon}^T [D]\boldsymbol{\varepsilon}\ dV - P_1 u_1 - P_2 v_2\right) = 0 \tag{18.47}$$

Therefore:

$$\Pi = \tfrac{1}{2}\int_V \boldsymbol{\varepsilon}^T [D]\boldsymbol{\varepsilon}\ dV - P_1 u_1 - P_2 v_2 \tag{18.48}$$

or

$$\Pi = \tfrac{1}{2}\int_V \boldsymbol{\varepsilon}^T \boldsymbol{\sigma}\ dV - P_1 u_1 - P_2 v_2 \tag{18.49}$$

Equation (18.48) or (18.49) is also recognized as the potential (strain) energy equation of the continuum solid under study. The integral term is the strain energy due to deformation, and the other two terms are the work done by the external loads $P_1$ and $P_2$. For plane problems $dV = t\ dA$, where $t$ is the thickness of the solid. Thus Eq. (18.49) becomes

$$\Pi = \tfrac{1}{2}t\int_A \boldsymbol{\varepsilon}^T \boldsymbol{\sigma}\ dA - P_1 u_1 - P_2 v_2 \tag{18.50}$$

Notice that the above approach to derive $\Pi$ is general and applicable to a wide variety of two- and three-dimensional solid mechanics problems.

(b) Equation (18.32) is the familiar Laplace equation. Many physical problems are governed by the Laplace equation. In this example, we follow a general procedure to derive $\Pi$ for systems governed by the Laplace equation. This equation can be written as

$$\nabla^2 \phi = 0 \tag{18.51}$$

where the two-dimensional Laplace operator is

$$\nabla^2 = \frac{\partial^2}{\partial x^2} + \frac{\partial^2}{\partial y^2} \tag{18.52}$$

For Eq. (18.32), $\phi$ is obviously the temperature $T$. Substituting Eq. (18.51) into Eq. (18.3), we obtain

$$\int_A \nabla^2 \phi\ \delta\phi\ dA = 0 \tag{18.53}$$

Recognizing that $\nabla^2 \phi = \nabla \cdot \nabla \phi$, where $\nabla = (\partial/\partial x)\hat{\mathbf{i}} + (\partial/\partial y)\hat{\mathbf{j}}$, we can write

$$\nabla \cdot (\delta\phi \nabla\phi) = \nabla(\delta\phi) \cdot \nabla\phi + \delta\phi \nabla \cdot \nabla\phi$$

$$= \nabla(\delta\phi) \cdot \nabla\phi + \delta\phi \nabla^2 \phi \tag{18.54}$$

or

$$\delta\phi \nabla^2 \phi = \nabla \cdot (\delta\phi \nabla\phi) - \nabla(\delta\phi) \cdot \nabla\phi \tag{18.55}$$

Substituting this equation into the integral equation (18.53) gives

$$\int_A \nabla \cdot (\delta\phi \nabla\phi)\ dA - \int_A \nabla(\delta\phi) \cdot \nabla\phi\ dA = 0 \tag{18.56}$$

Applying the Gauss divergence theorem to the first term to invoke the boundary conditions, we get

$$\oint_B \hat{n} \cdot \delta\phi\nabla\phi \, dS - \int_A \nabla(\delta\phi) \cdot \nabla\phi \, dA = 0 \qquad (18.57)$$

The first term can be reduced further if we recognize that $\delta\phi$ is scalar and $\hat{n} \cdot \nabla\phi = \partial\phi/\partial n$ where $\partial\phi/\partial n$ is the derivative normal to the boundary. For the second term, we realize that $\nabla(\delta\phi) = \delta\nabla\phi$ and $\nabla(\delta\phi) \cdot \nabla\phi = \delta\nabla\phi \cdot \nabla\phi = \frac{1}{2}\delta(\nabla\phi \cdot \nabla\phi)$. Thus, Eq. (18.57) becomes

$$\frac{1}{2}\delta \int_A \nabla\phi \cdot \nabla\phi \, dA - \oint_B \delta\phi \frac{\partial\phi}{\partial n} \, dS = 0 \qquad (18.58)$$

The second term is the natural boundary conditions. Once this term is known for a given problem, the functional $\Pi$ is easily obtained from Eq. (18.58). For example, if all natural boundary conditions are zero, Eq. (18.58) gives

$$\Pi = \int_A \nabla\phi \cdot \nabla\phi \, dA = \int_A \left[ \left( \frac{\partial\phi}{\partial x} \right)^2 + \left( \frac{\partial\phi}{\partial y} \right)^2 \right] dA \qquad (18.59)$$

Let us apply Eq. (18.58) to the heat conduction problem described by Eq. (18.32) and boundary conditions shown in Fig. 18-3b. Assuming that the sides $x = 0$ and $y = 0$ are insulated, then the temperatures at these two sides are constants, or $\delta T = 0$. Thus, we can write

$$\oint_B \delta T \frac{\partial T}{\partial n} \, dS = \int_0^b \delta T \frac{\partial T}{\partial x} \, dy + \int_0^a \delta T \frac{\partial T}{\partial y} \, dx \qquad (18.60)$$

Using the Fourier law $K(\partial T/\partial n) = -q_n$, the above equation reduces to

$$\oint_B \delta T \frac{\partial T}{\partial n} \, dS = \frac{1}{K} \int_0^b \delta T(-q_1) \, dy + \frac{1}{K} \int_0^a \delta T(q_2) \, dx$$

$$= -\frac{1}{K} \int_0^b \delta(q_1 T) \, dy + \frac{1}{K} \int_0^a \delta(q_2 T) \, dx \qquad (18.61)$$

Substituting the above equation into Eq. (18.58) gives

$$\delta\left( \frac{1}{2} \int_A K\nabla\phi \cdot \nabla\phi \, dA + \int_0^b q_1 T \, dy - \int_0^a q_2 T \, dx \right) = 0 \qquad (18.62)$$

which gives

$$\Pi = \frac{1}{2} \int_0^b \int_0^a K\left[ \left( \frac{\partial T}{\partial x} \right)^2 + \left( \frac{\partial T}{\partial y} \right)^2 \right] dx \, dy + \int_0^b q_1 T \, dy - \int_0^a q_2 T \, dx \quad (18.63)$$

Notice that the above development can easily be extended to three-dimensional domains. In this case, the three-dimensional Laplace operator is

$$\nabla^2 = \frac{\partial^2}{\partial x^2} + \frac{\partial^2}{\partial y^2} + \frac{\partial^2}{\partial z^2}$$

### 18.3.1.2 METHOD OF WEIGHTED RESIDUALS.

In some applications such as transient (time-dependent) heat conduction and some types of flow in fluid mechanics, variational principles may not exist or are unknown. However, the governing differential equations and the boundary conditions (natural and geometric) are known. In these applications, the method of weighted residuals is used instead of the variational principle to derive the integral equation which is equivalent to the equation $\delta\Pi = 0$ when the variational principle is used.

The method of weighted residuals is a numerical technique for obtaining approximate solutions to differential equations. Its application involves two steps. First, an approximate solution which satisfies the differential equation and its geometric boundary conditions is chosen. This approximate solution may be, say, a polynomial with unknown coefficients. When this solution is substituted into the differential equation and boundary conditions, an error or residual results. This residual is chosen to vanish in some average sense over the entire solution domain. This leads to the integral equation which corresponds to the original differential equation. In the second step, the integral equation is solved for the unknown coefficients, thus producing the solution.

To solve Eq. (18.1) subject to Eq. (18.2) using the method of weighted residuals, we begin by assuming an approximate solution $\phi_a$ such that

$$\phi \approx \phi_a = \sum_{i=1}^{n} c_i g_i \tag{18.64}$$

where $c_i$ are unknown coefficients and $g_i$ are known assumed functions of the independent variable(s).

Substituting Eq. (18.64) into Eqs. (18.1) and (18.2) gives

$$L(\phi_a) - f \neq 0 \tag{18.65}$$

or

$$R = L(\phi_a) - f \tag{18.66}$$

where $R$ is the residual or error. To minimize the residual $R$ over the entire domain, we form a weighted average which should vanish over the domain; then

$$\int_D R W_i \, dD = \int_D [L(\phi_a) - f] W_i \, dD = 0, \qquad i = 1, 2, \ldots, n \tag{18.67}$$

There is a weighting factor $W_i$ associated with each term of Eq. (18.64). These factors may be chosen according to different criteria. The Galerkin method, for example, uses the known functions $g_i$ in Eq. (18.64) as $W_i$. Thus, Eq. (18.67) for the Galerkin method becomes

$$\int_D R g_i \, dD = \int_D [L(\phi_a) - f] g_i \, dD = 0, \qquad i = 1, 2, \ldots, n \tag{18.68}$$

This equation can be solved for the coefficients $c_i$.

Equation (18.68) is the integral equation for the Galerkin method, as Eq. (18.5) is the integral equation for the variational principle. Either equation forms the core of the finite element formulation.

**Example 18.3.** Solve the following differential equation using Galerkin's method:

$$y'' + y = -x, \qquad 0 \leq x \leq 1$$

using the boundary condition

$$y(0) = y(1) = 0$$

*Solution.* The above boundary conditions are only geometric. Let us assume the following approximate solution (which satisfies the above geometric boundary conditions):

$$y(x) \approx y_a(x) = c_1 x(1 - x^2) + c_2 x^2 (1 - x)$$
$$= c_1 g_1(x) + c_2 g_2(x)$$
$$= \sum_{i=1}^{2} c_i g_i(x)$$

Notice that $y_a(x)$ satisfies the b.c. Using the above differential equation, we can write

$$R = y_a'' + y_a + x$$
$$= (-5x - x^3)c_1 + (2 - 6x + x^2 - x^3)c_2 + x$$

Using Eq. (18.68), we can write

$$\int_0^1 R g_1 \, dx = 0 \qquad \int_0^1 R g_2 \, dx = 0$$

or

$$\int_0^1 x(1 - x^2) R \, dx = 0$$

and

$$\int_0^1 x^2 (1 - x) R \, dx = 0$$

Substituting $R$ into these two equations, performing the integrations, and reducing give

$$0.7238c_1 + 0.2738c_2 = 0.1333$$
$$0.2738c_1 + 0.1238c_2 = 0.05$$

Solving these two equations together gives $c_1 = 0.19211$ and $c_2 = -0.0210$. Thus

$$y(x) \approx y_a(x) = 0.19211x(1 - x^2) - 0.0210x^2(1 - x), \qquad 0 \le x \le 1$$

*Note:* compare this approximate solution with the exact solution:

$$y(x) = \frac{\sin x}{\sin 1} - x$$

**Example 18.4.** Use the Galerkin method to derive the integral equation of the cantilever beam described by Eq. (18.8) in Example 18.1.

**Solution.** Assuming an approximate beam deflection in the form given by Eq. (18.64), we can write

$$y_a(x) = \sum_{i=1}^{n} c_i g_i(x)$$

Using the above equation together with Eqs. (18.8) and (18.67), the beam integral equation is

$$\int_0^L EI y_a'''' g_i(x) \, dx = 0, \qquad i = 1, 2, \ldots, n$$

Integrating by parts, this equation gives

$$EIy_a''' g_i(x) \Big|_0^L - \int_0^L EIy_a''' g_i'(x) \, dx = 0, \qquad i = 1, 2, \ldots, n$$

At $x = L$, $EIy_a''' = V = -P$. At $x = 0$, $y_a(0) = 0$; thus $g_i(0) = 0$ because $c_i \neq 0$. Substituting and integrating by parts again give

$$-Pg_i(L) - EIy_a'' g_i'(x) \Big|_0^L + \int_0^L EIy_a'' g_i''(x) \, dx = 0, \qquad i = 1, 2, \ldots, n$$

The second term in the above equation vanishes because the slope at the fixed end equals 0; that is, $y_a'(0) = 0$ which gives $g_i'(0) = 0$, and because $EIy_a''(L) = 0$. Thus, the above integral equation becomes

$$\int_0^L EIy_a'' g_i''(x) \, dx - Pg_i(L) = 0, \qquad i = 1, 2, \ldots, n$$

This equation and the functional $\Pi$ given by Eq. (18.25) gives identical finite element equations, as presented later in Sec. 18.3.2.6.

## 18.3.2  Continuum Discretization

As presented in Sec. 18.2, in the FEA the continuum under study is approximated as an assemblage of discrete finite elements with the elements being interconnected at the nodal points (nodes) on the element boundaries. Therefore, the integral Eq. (18.5) or (18.68) which is written for the continuum domain is changed into a sum, over the number of elements, of integral equations, each of which is written for the element domain. Equation (18.5) or (18.68) takes the following general form:

$$\int_D H(\phi) \, dD = 0 \tag{18.69}$$

where $H(\phi)$ is a general function of the unknown variable $\phi$. If the continuum domain $D$ is divided into element domains $D^e$, then

$$D \approx \sum_{e=1}^m D^e \tag{18.70}$$

$$\phi \approx \sum_{e=1}^m \phi^e \tag{18.71}$$

where $m$ is the total number of elements that form the continuum domain and $\phi^e$ is the unknown variable within the element domain. Equation (18.69) can be rewritten at the element level as

$$\int_D H(\phi) \, dD = \sum_{e=1}^m \int_{D^e} H^e(\phi^e) \, dD^e = 0 \tag{18.72}$$

This equation forms the basis to derive element equations and matrices, as explained later in this section.

Finite elements are characterized by several features. An element is completely described by its shape, the number and type (interior or exterior) of nodes, nodal variables, and the type of shape function. These element features control the behavior of the element in solving problems; that is, they control the accuracy and convergence of the finite element solution to the exact solution (which in many problems does not exist).

**18.3.2.1 ELEMENT SHAPE.** Most finite elements are geometrically simple to meet the fundamental premise of the finite element method that a continuum of arbitrary shape can be accurately modeled by an assemblage of elements. This fundamental premise also implies that element dimensionality is the same as the continuum dimensionality. For one-dimensional elements, there is one independent variable and elements are line segments, as shown in Fig. 18-4a. The number of nodes per element depends on the nodal variables (degrees of freedom) and the continuity requirements between the elements. At first glance, one-dimensional elements may not seem necessary because one-dimensional problems are usually governed by linear or nonlinear ordinary differential equations whose solutions can be obtained via other analytic or numeric techniques. However, these elements are useful in modeling two- and three-dimensional problems where part of the problem is one dimensional. Consider, for example, a cantilever beam with a spring attached to its free end. While the beam can be modeled by two-dimensional elements, the spring is modeled as a one-dimensional element. Two popular one-dimensional elements in solid mechanics are the truss and beam elements. The truss element has two to four nodes with one variable (axial displacement) per node. This element can only support tension or compression (no bending). The beam element has also two, three, or four nodes with two variables (beam deflection and slope) per node.

Figure 18-4b shows common two-dimensional elements. Historically, triangular elements were developed first because they were easy to develop and formulate by hand. The three-node flat triangular element is the simplest two-dimensional element. The ten-node triangular element has nine nodes on its boundary (called exterior nodes) and one node inside the boundary (called interior node). A quadrilateral element has a minimum of four nodes and as many as twelve nodes. In addition to modeling plane stress and strain problems, two-dimensional elements can be used to model axisymmetric problems. In this case, an element represents the cross section of an axisymmetric element whose thickness is given by the length of its arc segment. In general, two-dimensional elements can model two- and two-and-a-half-dimensional objects (continuums).

Three-dimensional elements (Fig. 18-4c) are usually three-dimensional counterparts of two-dimensional elements. These elements can be used to discretize three-dimensional objects (continuums). Creating (and visualizing) three-dimensional finite element meshes is usually labor-intensive and an error-prone process. Thus, using preprocessors and automatic mesh generation algorithms are beneficial in discretizing three-dimensional objects.

The exterior nodes of any element shown in Fig. 18-4 can be divided into two types: corner and mid-side nodes. Corner nodes are the minimum required nodes to define the element shape and, as the name implies, are located at the

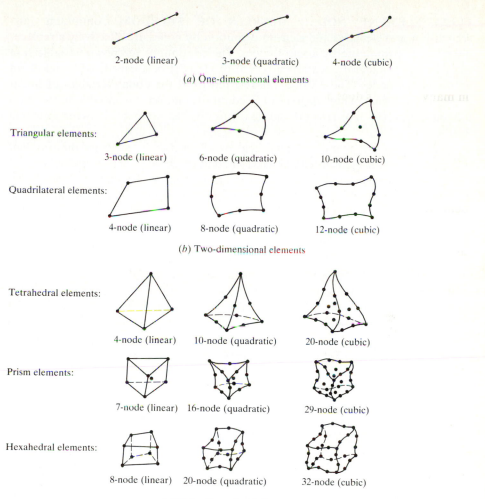

2-node (linear)    3-node (quadratic)    4-node (cubic)

(a) One-dimensional elements

Triangular elements:

3-node (linear)    6-node (quadratic)    10-node (cubic)

Quadrilateral elements:

4-node (linear)    8-node (quadratic)    12-node (cubic)

(b) Two-dimensional elements

Tetrahedral elements:

4-node (linear)    10-node (quadratic)    20-node (cubic)

Prism elements:

7-node (linear)    16-node (quadratic)    29-node (cubic)

Hexahedral elements:

8-node (linear)    20-node (quadratic)    32-node (cubic)

(c) Three-dimensional elements

**FIGURE 18-4**
Element types and dimensionality.

corners of the element. Mid-side nodes are added to improve the element accuracy or to meet continuity requirements between elements; they are located along the sides of the element. There is usually a specific order of defining element nodes (see Examples 18.5 and 18.6 and Sec. 18.4).

The various elements supported by a particular FEA code are sometimes known as the element library. The more the number of elements in the library, the more versatile the code, which means the larger the number of problems it can handle. In addition, many of these codes provide their users with the ability to interface customized (special) elements to the code. This is useful in a research and development environment. The element library should form an important criterion in evaluating existing FEA codes.

**18.3.2.2 ELEMENT NODAL DEGREES OF FREEDOM.** Continuum discretization into elements and element definition by nodes are effectively a replacement of the continuum by a set of nodes. The continuous unknown variable(s) in the governing integral equation(s) [(18.5) or (18.68)] becomes discrete unknown values at the nodes. These values are referred to as the nodal variables or nodal degrees of freedom. These degrees of freedom depend on the variable in the integral equation [(18.5) or (18.68)] and on continuity requirements between elements (see Sec. 18.3.2.3).

The specific meaning of nodal degrees of freedom stems from the problem at hand. A specific element can be used in solid mechanics, heat transfer, or fluid mechanics applications as long as the associated integral equation is the same and has the same characteristics. For example, one degree of freedom at a node can represent an axial displacement for the uniaxial loading problem (Fig. 18-2a), temperature for the heat conduction problem (Fig. 18-2c), or pressure for the fluid flow problem (Fig. 18-2d).

**18.3.2.3 ELEMENT DESIGN.** One of the most important decisions in FEA is the selection of particular elements and the definition of the appropriate interpolation (approximating) polynomial within each element. For practitioners and users of the finite element method, this means choosing the proper element(s) with the proper number of nodes from the available element library (see Fig. 18-4). The correlation between the integral equation governing the problem and the polynomial, and between the polynomial and the number of element nodes enables a proper decision to be made.

The ability to formulate the individual element equations from the integral equation [(18.5) or (18.68)] and the privilege to assemble these equations to obtain the global system of equations rely on the assumption that the interpolation polynomial within each element ($\phi^e$) must satisfy certain requirements. These requirements stem from the need to ensure that Eq. (18.72) holds and that the approximate solution converges to the correct solution when we use an increasing number of small elements, that is, when we refine the element mesh. This type of convergence (the finer the mesh, the better the solution) is known as monotonic convergence.

The assurance of monotonic convergence as the element size decreases is controlled primarily by the highest derivative that appears in the element integral equation which is identical in form to the continuum integral equation [see Eq. (18.72)]. Let us assume that the element integral equation contains derivatives up to the $r$th order. For monotonic convergence, the assumed element interpolation polynomial must meet the following two requirements:

1. Compatibility requirement. $C^{r-1}$ continuity must hold at element interfaces (boundaries); that is, the field variable and any of its derivatives up to one order less than the highest-order derivative appearing in the integral equation [(18.5) or (18.68)] must be continuous at the interelement boundary. Elements that satisfy this requirement are desirable and are known as compatible or conforming elements. If elements are nonconforming, a patch test can be used to test their validity. Nonconforming elements are not discussed here. Physi-

cally, compatibility assures that no gaps occur between elements when the assemblage is loaded.

2. Completeness requirement. $C^r$ continuity must hold within an element; that is, all uniform states of $\phi$ and its derivatives up to the highest order appearing in the integral equation should be represented in $\phi^e$ when, in the limit, the element size shrinks to zero. Physically, when the finite element mesh has a very large number of elements, the element domain $D^e$ becomes so small that $\phi$ and its related derivatives are approximately constant in the domain. This condition is achieved by having a constant in $\phi^e$ and its derivatives. For example, if $\phi^e = \alpha_1 + \alpha_2\, x + \alpha_3\, x^2$, then $\alpha_1$ and $\alpha_2$ are the uniform states of $\phi^e$ and its first derivative respectively. As the element size goes to zero, that is, as $x$ goes to zero, $\phi^e$ and $d\phi^e/dx$ go to $\alpha_1$ and $\alpha_2$ respectively, that is, constants (uniform states). Thus, the completeness requirement is achieved by having all the terms in the polynomial $\phi^e$ up to the order that provides a constant (at least) value of the $r$th derivative. Therefore, the polynomial $\phi^e$ should be of at least order $r$. In one dimension, any polynomial is complete, and a polynomial of order $r = n$ may be written as

$$\phi^e(x) = \sum_{i=1}^{L} \alpha_i\, x^{i-1} \tag{18.73}$$

where $L = n + 1$.

In two dimensions, a complete $n$th-order polynomial is given by

$$\phi^e(x,\, y) = \sum_{k=1}^{L} \alpha_k\, x^i y^j, \qquad i + j \le n \tag{18.74}$$

where $L = (n + 1)(n + 2)/2$. The well-known Pascal triangle (Fig. 18-5a) helps identify the term of a complete polynomial.

In three dimensions, a complete $n$th-order polynomial is given by

$$\phi^e(x,\, y,\, z) = \sum_{l=1}^{L} \alpha_l\, x^i y^j z^k, \qquad i + j + k \le n \tag{18.75}$$

where $L = (n + 1)(n + 2)(n + 3)/6$. The coefficients $\alpha_i$ in Eqs. (18.73), (18.74), or (18.75) are unknown constants called the generalized coordinates of the element.

If an interpolation polynomial is incomplete (has less terms than $L$ shown above) or if it has additional terms (more terms than $L$), then it is desirable but not necessary that the polynomial contains the appropriate terms to preserve symmetry as shown in Fig. 18-5. When symmetry is preserved, the polynomial is said to have geometric invariance, that is, it is independent of the origin and orientation of the coordinate system. In addition to selecting the symmetric terms, geometric invariance requires that an element shape be unbiased, i.e. does not have a preferred direction. For example, if we use triangular elements other than isosceles, rectangular elements of aspect ratio (ratio between the two dimensions of the rectangle) not equal to one, or an element with a different number of nodes for each of its sides, a preferred direction already exists. This is why geometric invariance is fortunately a desirable but not necessary requirement.

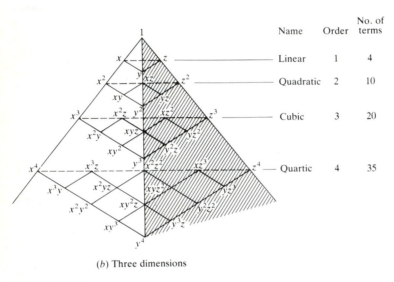

| | Name | Order | No. of terms |
|---|---|---|---|
| 1 | | | |
| $x \mid y$ | Linear | 1 | 3 |
| $x^2 \quad xy \quad y^2$ | Quadratic | 2 | 6 |
| $x^3 \quad x^2y \mid xy^2 \quad y^3$ | Cubic | 3 | 10 |
| $x^4 \quad x^3y \quad x^2y^2 \quad xy^3 \quad y^4$ | Quartic | 4 | 15 |
| $x^5 \quad x^4y \quad x^3y^2 \mid x^2y^3 \quad xy^4 \quad y^5$ | Quintic | 5 | 21 |
| $x^6 \quad x^5y \quad x^4y^2 \quad x^3y^3 \quad x^2y^4 \quad xy^5 \quad y^6$ | Hexadic | 6 | 28 |
| $x^7 \quad x^6y \quad x^5y^2 \quad x^4y^3 \mid x^3y^4 \quad x^2y^5 \quad xy^6 \quad y^7$ | Septic | 7 | 36 |

Axis of symmetry

(*a*) Two dimensions

| Name | Order | No. of terms |
|---|---|---|
| Linear | 1 | 4 |
| Quadratic | 2 | 10 |
| Cubic | 3 | 20 |
| Quartic | 4 | 35 |

(*b*) Three dimensions

**FIGURE 18-5**
Array of terms in complete polynomials.

If the interpolation polynomial of a finite element satisfies both the compatibility and the completeness requirements, how can we investigate the monotonic convergence of the FEA to the exact solution which may not be known? This question can also be asked as follows. How can we refine a finite element mesh to study the monotonic convergence of FEA? There are three conditions by which a mesh can be refined in a regular fashion: (1) elements can be made as small as possible without any gaps or other discontinuities between them; (2) all coarser (previous) meshes must be contained in the refined ones; and (3) the types of element (the number of nodes per element) must remain the same for all meshes. This means that the interpolation polynomial remains unchanged during the process of mesh refinement.

Having decided on the order and the number of terms of the interpolation polynomial of a finite element, how can we use this information to design (or choose an element from a library) an element, that is, determine its number of nodes and the degrees of freedom per node? Applying the completeness requirement, we can determine the order of the interpolation polynomial and its number of terms, that is, the number of the generalized coordinates $n_G$. Applying the compatibility requirement, we can determine the degrees of freedom per node ($n_F$) which equal the field variable $\phi$ and any of its derivatives up to the $r$th derivative. At last, the number of nodes per element is equal to $n_G/n_F$, that is, the total degrees of freedom at the nodes of an element must be equal to $n_G$ so that we can have enough equations to determine the generalized coordinates $\alpha_i$, as seen in the next section.

The above procedure to design or choose a finite element is iterative in nature. For example, $n_G/n_F$ may have to be rounded to the nearest digit or the geometry of the element may impose a minimum number of nodes. A linear triangular element requires three corner nodes to define its geometry while a linear quadrilateral element requires four corner nodes. Once the number of nodes $n_G/n_F$ is changed, then the interpolation polynomial (that is, the number of $\alpha_i$) must be adjusted accordingly. The excess of the number of nodes from the corner nodes is distributed between the element sides in such a way as to satisfy compatibility at the interelement boundary. At the first iteration of the procedure, emphasis is usually put on satisfying the compatibility requirement at the element corner nodes.

The design of finite elements becomes more complex as the compatibility requirements increase. $C^0$ continuity elements, that is, elements that require continuity of the field variable only, are much easier to design than $C^1$ continuity (continuity of $\phi$ and its first derivative) elements. This is why nonconforming elements are sometimes used. While these nonconforming elements preserve continuity of $\phi$, they may violate its slope (or higher derivatives) continuity between elements, though naturally not at the nodes where such continuity is imposed.

**Example 18.5.** Design finite elements for the systems whose integral equations are derived in Example 18.1.

*Solution.* One-dimensional elements are used for all the domains described in Example 18.1 because these domains are one dimensional. The functionals given by Eqs. (18.18), (18.28), and (18.31) are identical in form. Therefore, the same finite element can be used to discretize their corresponding domains. The highest derivative in any of the three equations is a first derivative. Thus, a linear interpolation polynomial is sufficient to satisfy the completeness requirement, and can be written, using Eq. (18.73), as

$$\phi^e = \alpha_1 + \alpha_2 x$$

The compatibility requirement is satisfied by requiring continuity of $\phi^e$ at the interelement boundary. Because one-dimensional elements are connected at nodes, then specified nodal values of $\phi^e$ cause elements to conform. This also implies that one degree of freedom per node ($n_F = 1$) is sufficient. With $n_G = 2$ (unknown $\alpha_1$ and $\alpha_2$) and $n_F = 1$, a two-node element is sufficient to provide monotonic convergence of the FEA analysis related to the three systems under discussion. Figure 18-6a shows the element and its degrees of freedom. Notice that $\phi = u$, $T$, or $P$. In solid mechanics applications, this element is known as a truss or bar element. The two-node

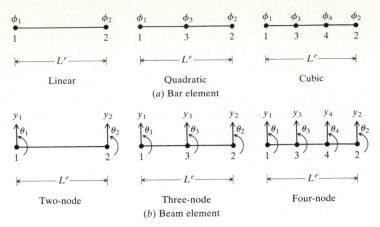

**FIGURE 18-6**
Various one-dimensional elements.

element is known as a linear element because $\phi^e$ is linear as written above. It is also known as a constant-strain element because the strain within the element domain is $\varepsilon_x = d\phi/dx = \alpha_2$. In using this element to obtain a problem solution, the element length $L^e$ must be sufficiently small to yield good results. For faster converging elements (which means using larger size elements), we can use quadratic or cubic elements. The quadratic element requires three nodes (two corner nodes and one mid-side node) because it uses the second-order polynomial $\phi^e = \alpha_1 + \alpha_2 x + \alpha_3 x^2$. This polynomial provides a linear function $(\alpha_2 + 2\alpha_3 x)$ of the slope $d\phi^e/dx$. Similarly, the cubic element requires four nodes (two corner and two mid-side), uses the third-order polynomial $\phi^e = \alpha_1 + \alpha_2 x + \alpha_3 x^2 + \alpha_4 x^3$, and provides a quadratic function $(\alpha_2 + 2\alpha_3 x + 3\alpha_4 x^2)$ of the slope $d\phi^e/dx$.

The functional $\Pi$ given by Eq. (18.25) has a second derivative as its highest derivative. Thus, interelement compatibility is achieved by using deflection $y$ and slope $y' = \theta$ as degrees of freedom at each node. Completeness requires that the interpolation polynomial must be at least quadratic. Realizing that a minimum of two nodes is required to define an element geometrically, and with two degrees of freedom per node, a cubic polynomial $(y^e = \alpha_1 + \alpha_2 x + \alpha_3 x^2 + \alpha_4 x^3)$ of the beam deflection meets the compatibility, completeness, and geometric requirements of the element. Figure 18-6b shows the two-node beam element. Within the element, the polynomial provides a quadratic function of the slope $y'$ and a linear function of the curvature $y''$. Three-node and four-node beam elements which use quintic and septic polynomials respectively are also shown in Fig. 18-6b.

The numbering system of element nodes shown in Fig. 18-6 is worth noting. The corner nodes are numbered first followed by the mid-side nodes. This system is useful in particular when we cover the isoparametric evaluation in Sec. 18.4.

**Example 18.6.** Design finite elements for the systems whose integral equations are derived in Example 18.2.

*Solution.* The two functionals given by Eqs. (18.48) or (18.63) have the same form. Therefore, the element requirements for both the plane stress and heat conduction problems are the same. Finite elements for these problems are two-dimensional and so are their interpolation polynomials. Completeness requires at least a linear function which results in a constant strain element for the plane stress problem, and a constant heat flux element for the steady-state heat conduction problem. Compatibility requires the continuity of $\phi$ at the nodes and the element sides. Thus, a one degree of freedom per node is sufficient.

Let us now consider a triangular and a quadrilateral element. A triangular element requires a minimum of three nodes to define its geometry. Using the Pascal triangle shown in Fig. 18-5$a$, the interpolation polynomial $\phi^e = \alpha_1 + \alpha_2 x + \alpha_3 y$ is the required one. For plane stress, the displacement has two components $u$ and $v$ which must be applied to each node. Thus, two interpolation polynomials, $u^e = \alpha_1 + \alpha_2 x + \alpha_2 y$ and $v^e = \beta_1 + \beta_2 x + \beta_3 y$, are required for each node. For the heat conduction problem, the polynomial is $T^e = \alpha_1 + \alpha_2 x + \alpha_3 y$. Figure 18-7$a$ shows the three-node linear triangular element. Higher-order elements such as quadratic and cubic elements (Fig. 18-7$a$) can be used to improve the convergence of the FEA. The

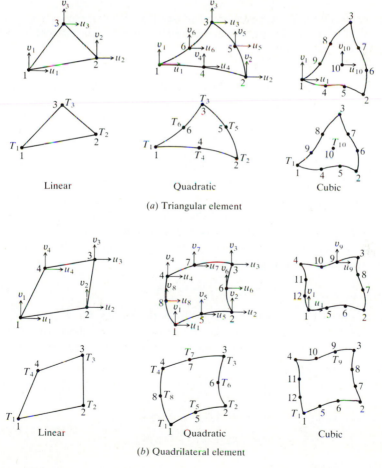

Linear                    Quadratic                    Cubic

(a) Triangular element

(b) Quadrilateral element

**FIGURE 18-7**
Various two-dimensional elements.

complete quadratic polynomial is given by $\phi^e = \alpha_1 + \alpha_2 x + \alpha_3 y + \alpha_4 x^2 + \alpha_5 xy + \alpha_6 y^2$. With six unknown $\alpha$ values, a six-node element is used. Similarly, a ten-node element is required to support the complete cubic polynomial $\phi^e = \alpha_1 + \alpha_2 x + \alpha_3 y + \alpha_4 x^2 + \alpha_5 xy + \alpha_6 y^2 + \alpha_7 x^3 + \alpha_8 x^2 y + \alpha_9 xy^2 + \alpha_{10} y^3$.

A quadrilateral element requires a minimum of four nodes to define its geometry as shown in Fig. 18-7b. With three generalized coordinates ($\alpha_1$, $\alpha_2$, and $\alpha_3$) in the linear polynomial, a minimum of four nodes, and one degree of freedom per node, we have three unknowns ($\alpha_1$, $\alpha_2$, and $\alpha_3$) and four equations (one for each node). To resolve this discrepancy, we must add an additional term to the linear interpolation polynomial. This term must maintain the linearity of the polynomial in both $x$ and $y$, as well as its geometric invariance (symmetry). The $xy$ term in the quadratic row in the Pascal triangle is the one that meets these two criteria. Therefore, the interpolation polynomial for the linear quadrilateral element is given by $\phi^e = \alpha_1 + \alpha_2 x + \alpha_3 y + \alpha_4 xy$. This argument can be extended to the quadratic and cubic elements. The polynomial for the eight-node quadrilateral element is given by $\phi^e = \alpha_1 + \alpha_2 x + \alpha_3 y + \alpha_4 x^2 + \alpha_5 xy + \alpha_6 y^2 + \alpha_7 x^2 y + \alpha_8 xy^2$ and for the twelve-node cubic element by $\phi^e = \alpha_1 + \alpha_2 x + \alpha_3 y + \alpha_4 x^2 + \alpha_5 xy + \alpha_6 y^2 + \alpha_7 x^3 + \alpha_8 x^2 y + \alpha_9 xy^2 + \alpha_{10} y^3 + \alpha_{11} x^3 y + \alpha_{12} xy^3$. The cubic terms $\alpha_{11} x^3 y$ and $\alpha_{12} xy^3$ are chosen to maintain the symmetry of $\phi^e$ (see Fig. 18-5a). If the term $x^2 y^2$ were to be chosen from the Pascal triangle and maintain symmetry, the element would have to have eleven nodes only.

### 18.3.2.4 ELEMENT SHAPE FUNCTION.

Thus far we have seen how a finite element can be designed to meet the monotonic convergence criteria (compatibility and completeness) and how a field variable can be represented within an element as a polynomial whose coefficients are the generalized coordinates of the element. In this section, we present how these coefficients are expressed in terms of the nodal values of the field variable as well as the coordinates of the nodes. This enables us later to express the element and global equations in terms of the nodal values that we need to solve for. It also enables us to express the element matrices in terms of the element geometry.

To illustrate the procedure, let us consider a linear triangular and a linear quadrilateral element with one degree of freedom per node. The field variable within a triangular element is given by

$$\phi^e(x, y) = \alpha_1 + \alpha_2 x + \alpha_3 y \tag{18.76}$$

Let us assume that the field variable $\phi$ assumes the values $\phi_1$, $\phi_2$, and $\phi_3$ at the element nodes 1, 2, and 3 respectively, as shown in Fig. 18-8a. Because these nodal values must satisfy Eq. (18.76), substituting the values with the nodal coordinates into the equation gives

$$\phi_1 = \alpha_1 + \alpha_2 x_1 + \alpha_3 y_1$$
$$\phi_2 = \alpha_1 + \alpha_2 x_2 + \alpha_3 y_2 \tag{18.77}$$
$$\phi_3 = \alpha_1 + \alpha_2 x_3 + \alpha_3 y_3$$

or, in matrix notation,

$$\boldsymbol{\phi} = [G]\boldsymbol{\alpha} \tag{18.78}$$

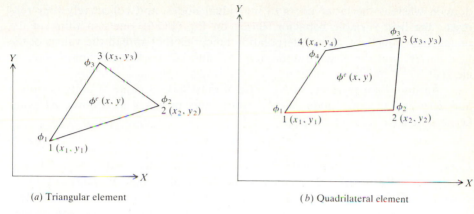

(a) Triangular element

(b) Quadrilateral element

**FIGURE 18-8**
Nodal values and geometry of linear two-dimensional elements.

where

$$\boldsymbol{\phi} = [\phi_1 \quad \phi_2 \quad \phi_3]^T \tag{18.79}$$

$$[G] = \begin{bmatrix} 1 & x_1 & y_1 \\ 1 & x_2 & y_2 \\ 1 & x_3 & y_3 \end{bmatrix} \tag{18.80}$$

$$\boldsymbol{\alpha} = [\alpha_1 \quad \alpha_2 \quad \alpha_3]^T \tag{18.81}$$

Solving Eq. (18.78) for the generalized coordinates gives

$$\boldsymbol{\alpha} = [G]^{-1}\boldsymbol{\phi} \tag{18.82}$$

Equation (18.76) can be written in vector form as

$$\phi^e = [P]\boldsymbol{\alpha} \tag{18.83}$$

where $[P] = [1 \quad x \quad y]$. Substituting Eq. (18.82) into (18.83), we obtain

$$\phi^e = [P][G]^{-1}\boldsymbol{\phi} = [N]\boldsymbol{\phi} \tag{18.84}$$

where

$$[N] = [P][G]^{-1} = [N_1 \quad N_2 \quad N_3] \tag{18.85}$$

Thus, Eq. (18.84) can be expanded to give

$$\phi^e = N_1\phi_1 + N_2\phi_2 + N_3\phi_3 = \sum_{i=1}^{3} N_i\phi_i \tag{18.86}$$

where $N_i = N_i(x, y)$.

The matrix $[N]$ is known as the element shape function and $N_i$ is the nodal shape function associated with node $i$. The procedure followed above to obtain Eq. (18.86) from Eq. (18.76) is general and applicable to other elements. The original interpolation polynomial as given by Eq. (18.76) should not be confused with the nodal shape functions $N_i$. The former applies to the whole element domain whereas the latter refers to individual nodes, and collectively they represent the field variable behavior. Based on Eq. (18.86), one can think of the nodal shape functions $N_i$ as interpolation functions that spread the values of the field variables at the element nodes ($\phi_1$, $\phi_2$, and $\phi_3$) back to the interior of the element.

An important property of $N_i$ is that it must have a value of unity at node $i$ and zero at any other node. We can prove this property as follows. At point $(x_1, y_1)$, we have

$$\phi^e(x_1, y_1) = \phi_1 \tag{18.87}$$

Comparing this equation with Eq. (18.86) shows that $N_1(x_1, y_1) = 1$ and $N_2(x_1, y_1) = N_3(x_1, y_1) = 0$. Another property is that $N_i$ has an identical form to $\phi$ with the difference in the coefficients (see Prob. 18.9 at the end of the chapter). These two properties are useful in deriving shape functions by inspection, as seen later in the isoparametric evaluation (Sec. 18.4).

Let us apply the above procedure to the linear quadrilateral element shown in Fig. 18-8b. The interpolation polynomial is given by

$$\phi(x, y) = \alpha_1 + \alpha_2 x + \alpha_3 y + \alpha_4 xy \tag{18.88}$$

Substituting the element nodal values into this equation, similar equations to (18.77) to (18.86) can be developed. In this case:

$$\boldsymbol{\Phi} = [\phi_1 \quad \phi_2 \quad \phi_3 \quad \phi_4]^T \tag{18.89}$$

$$[G] = \begin{bmatrix} 1 & x_1 & y_1 & x_1 y_1 \\ 1 & x_2 & y_2 & x_2 y_2 \\ 1 & x_3 & y_3 & x_3 y_3 \\ 1 & x_4 & y_4 & x_4 y_4 \end{bmatrix} \tag{18.90}$$

$$\boldsymbol{\alpha} = [\alpha_1 \quad \alpha_2 \quad \alpha_3 \quad \alpha_4]^T \tag{18.91}$$

$$[P] = [1 \quad x \quad y \quad xy] \tag{18.92}$$

$$[N] = [N_1 \quad N_2 \quad N_3 \quad N_4] \tag{18.93}$$

$$\phi^e = N_1 \phi_1 + N_2 \phi_2 + N_3 \phi_3 + N_4 \phi_4 = \sum_{i=1}^{4} N_i \phi_i \tag{18.94}$$

The above equations can easily be generalized for an $n$-node element. For example, the shape function and the interpolation polynomial can be written as

$$[N] = [N_1 \quad N_2 \quad \cdots \quad N_n] \tag{18.95}$$

$$\phi^e = N_1 \phi_1 + N_2 \phi_2 + \cdots + N_n \phi_n = \sum_{i=1}^{n} N_i \phi_i \tag{18.96}$$

The procedure described above to derive element shape functions is straightforward. However, its main disadvantages are: (1) the inverse of the matrix $[G]$ may not exist for all element orientations in the global coordinate system (the $XYZ$ system) and (2) even if $[G]^{-1}$ exists, the computational effort to obtain it is not trivial, especially for higher-order elements such as cubic ones. This led to finding $N_i$ by inspection, which relies on the use of element local coordinate systems called natural coordinates. In this book, we refer to them as the parametric coordinates which are an extension of the parametric spaces used in geometric modeling.

**18.3.2.5 ELEMENT EQUATIONS USING THE VARIATIONAL APPROACH.** The variational finite element formulation is equivalent to the Ritz method used to solve differential equations with some differences not discussed here. The finite element solution to a problem involves choosing the unknown nodal values ($\phi_i$) of the field variable $\phi$ to make the functional $\Pi$ stationary or minimum. This is equivalent to minimizing the potential energy of a system in solid mechanics problems. To make $\Pi$ as given by Eq. (18.6) stationary with respect to the nodal values of $\phi$, we require (from variational calculus) that

$$\delta\Pi = \delta I(\phi) = 0 \tag{18.97}$$

Applying the discretization equation (18.72), the above equation becomes

$$\delta\Pi = \sum_{i=1}^{m} \delta\Pi^e = \sum_{i=1}^{m} \delta I^e(\phi^e) = 0 \tag{18.98}$$

where the variation of $I^e$ is taken only with respect to the nodal values associated with the element $e$. Equation (18.98) implies

$$
\begin{bmatrix}
\dfrac{\partial I^e}{\partial \phi_1} \\[2mm]
\dfrac{\partial I^e}{\partial \phi_2} \\[2mm]
\vdots \\[2mm]
\dfrac{\partial I^e}{\partial \phi_n}
\end{bmatrix}
=
\begin{bmatrix}
0 \\
0 \\
\vdots \\
0
\end{bmatrix}
\tag{18.99}
$$

where $n$ is the element number of nodes. In vector form, the above equation is

$$\left\{\frac{\partial I^e}{\partial \phi_i}\right\} = \{0\}, \qquad i = 1, 2, \ldots, n \tag{18.100}$$

Equation (18.99) or (18.100) provides the necessary element equations that characterize the behavior of the element $e$. The development of the element matrices involves substituting Eq. (18.96) into the element functional $\Pi^e$ or $I^e$, and then differentiating the result to obtain Eq. (18.99). This procedure is illustrated in the following examples.

**Example 18.7.** Use the variational approach to derive the element equations for the systems whose integral equations are derived in Example 18.1.

*Solution.* We have designed two-node, three-node, and four-node elements for these systems. To derive the element equations and matrices, let us use a two-node element to reduce the amount of manipulations. Three-node and four-node elements follow the same path. Using Eq. (18.96) with $n = 2$, we get

$$\phi^e = N_1\phi_1 + N_2\phi_2 \tag{18.101}$$

Applying Eq. (18.72) to the functional $\Pi$ given by Eq. (18.18), we write

$$\Pi^e = I^e = \frac{1}{2}\int_0^{L^e} AEu^{e'2}\, dx - Pu_L^e \tag{18.102}$$

where $L^e$ is the element length. Substituting Eq. (18.101) into the above equation and realizing that $\phi^e = u^e$, we obtain

$$I^e = \frac{1}{2}\int_0^{L^e} AE(N_1'u_1 + N_2'u_2)^2\, dx - P(N_1u_1 + N_2u_2)_L \tag{18.103}$$

Applying the stationarity condition given by Eq. (18.99), the above equation gives

$$\int_0^{L^e} AE(N_1'u_1 + N_2'u_2)N_1'\, dx - PN_{1L} = 0 \tag{18.104}$$

and

$$\int_0^{L^e} AE(N_1'u_1 + N_2'u_2)N_2'\, dx - PN_{2L} = 0 \tag{18.105}$$

Because $N_i = 1$ at node $i$ and zero elsewhere, the second term in the above two equations is equal to zero except for the boundary element adjacent to the bar end at $x = L$. Thus, for any other element, Eqs. (18.104) and (18.105) can be written in a matrix form as

$$\begin{bmatrix} k_{11} & k_{12} \\ k_{21} & k_{22} \end{bmatrix}^e \begin{bmatrix} u_1 \\ u_2 \end{bmatrix}^e = \begin{bmatrix} 0 \\ 0 \end{bmatrix} \qquad \text{or} \qquad [K]^e \mathbf{U}^e = \mathbf{0} \tag{18.106}$$

where any element $k_{ij}$ is given by

$$k_{ij} = \int_0^{L^e} AEN_i'N_j'\, dx \tag{18.107}$$

Equation (18.106) is the element equation. $[K]^e$ is known as the element stiffness matrix and $\mathbf{U}^e$ is the element displacement vector. Notice that $[K]^e$ is symmetric, as can easily be seen from Eq. (18.107). In general, many engineering physical applications result in a symmetric positive define stiffness matrix. If we know the exact form of $N_i$ and perform the integral of Eq. (18.107), $[K]^e$ would be completely evaluated. Isoparametric evaluation of $[K]^e$ is discussed in Sec. 18.4.

For the boundary element adjacent to the end $x = L$, $N_{1L} = 0$ and $N_{2L} = 1$. Thus, this element equation is

$$\begin{bmatrix} k_{11} & k_{12} \\ k_{21} & k_{22} \end{bmatrix}^e \begin{bmatrix} u_1 \\ u_2 \end{bmatrix}^e = \begin{bmatrix} 0 \\ P \end{bmatrix} \tag{18.108}$$

The element equations and matrices can be obtained for the heat conduction and fluid flow systems following the same approach. Using $\phi^e = T^e$ in Eq. (18.101) together with Eq. (18.28), the element equation is identical to Eq. (18.106), but with replacing $u$ by $T$. In this case, the element matrix $[K]^e$ is known as the element conductivity matrix, and any element $k_{ij}$ is given by

$$k_{ij} = \int_0^{L^e} KN_i'N_j'\, dx \tag{18.109}$$

The equation of the boundary element at $x = 0$ is

$$\begin{bmatrix} k_{11} & k_{12} \\ k_{21} & k_{22} \end{bmatrix}^e \begin{bmatrix} T_1 \\ T_2 \end{bmatrix}^e = \begin{bmatrix} q \\ 0 \end{bmatrix} \tag{18.110}$$

Similarly for the fluid flow system, $\phi^e = P^e$ and the element matrix is the flow matrix whose elements are given by

$$k_{ij} = \int_0^{L^e} AqN_i'N_j' \, dx \tag{18.111}$$

The equation of the boundary element at $x = 0$ is

$$\begin{bmatrix} k_{11} & k_{12} \\ k_{21} & k_{22} \end{bmatrix}^e \begin{bmatrix} P_1 \\ P_2 \end{bmatrix}^e = \begin{bmatrix} V \\ 0 \end{bmatrix} \tag{18.112}$$

We now develop the element equations and matrices for beam bending. Here, we have two degrees of freedom ($y$ and $\theta$) per node. From Example 18.5, the beam deflection for a two-node element is given by

$$y^e = \alpha_1 + \alpha_2 x + \alpha_3 x^2 + \alpha_4 x^3 \tag{18.113}$$

Differentiating this equation, the beam slope is

$$\theta^e = \frac{dy^e}{dx} = \alpha_2 + 2\alpha_3 x + 3\alpha_4 x^2 \tag{18.114}$$

Substituting the nodal values ($y_1$, $\theta_1$, $y_2$, $\theta_2$) into these two equations and reducing, as we showed in Sec. 18.3.2.4, Eq. (18.113) can be written as

$$y^e = N_1 y_1 + N_2 \theta_1 + N_3 y_2 + N_4 \theta_2 \tag{18.115}$$

Although we can continue with Eq. (18.115) to develop the element equations and matrices, we stop the development here and leave it to interested readers. The beam element that utilizes the polynomial given by Eq. (18.113) is $C^1$ continuous and is sometimes known as the standard-type beam element. Its nodal shape functions cannot easily be obtained by inspection for isoparametric evaluation. We here present another approach that treats both $y$ and $\theta$ as independent variables. The merit of this approach is that it leads to beam elements that are $C^0$ continuous and whose shape functions are easier to find by inspection.

Using the beam slope $y' = \theta$, Eq. (18.25) can be written as

$$\Pi = \tfrac{1}{2} \int_0^L EI\theta'^2 \, dx - Py_L \tag{18.116}$$

This equation is identical in form to the equations of the other systems we have discussed. We can write an interpolation polynomial (linear for two-node elements) for both $y$ and $\theta$ as follows:

$$y^e = N_1 y_1 + N_2 y_2 \tag{18.117}$$

$$\theta^e = N_1 \theta_1 + N_2 \theta_2 \tag{18.118}$$

Notice that $N_1$ and $N_2$ in both equations are the same because they are functions of the element geometry only.

Applying Eq. (18.72) to Eq. (18.116) and using Eqs. (18.117) and (18.118) in the result, we obtain

$$\Pi^e = I^e = \tfrac{1}{2} \int_0^{L^e} EI(N_1'\theta_1 + N_2'\theta_2)^2 \, dx - P(N_1 y_1 + N_2 y_2)_L \tag{18.119}$$

Following the same procedure we used to obtain Eqs. (18.106) to (18.108), Eq. (18.119) gives

$$\begin{bmatrix} k_{11} & k_{12} \\ k_{21} & k_{22} \end{bmatrix}^e \begin{bmatrix} \theta_1 \\ \theta_2 \end{bmatrix}^e = \begin{bmatrix} 0 \\ 0 \end{bmatrix} \tag{18.120}$$

where

$$k_{ij} = \int_0^{L^e} EI N_i' N_j' \, dx \tag{18.121}$$

Equation (18.108) applies to the boundary beam element at $x = L$ after replacing $u_1$ and $u_2$ by $\theta_1$ and $\theta_2$ respectively.

Equation (18.120) does not include the displacements $y_1$ and $y_2$ at the element nodes. This is because Eq. (18.116) does not include them. Terms including $y_1$ and $y_2$ appear in Eq. (18.116) if the strain energy due to the beam shear deformation is included, and/or if external distributed loads and moments are added (see Prob. 18.11 at the end of the chapter). In this case, the matrix equation (18.120) changes from $2 \times 2$ to $4 \times 4$, and the element displacement vector $U^e$ becomes $[y_1 \quad \theta_1 \quad y_2 \quad \theta_2]^{eT}$. Using Eq. (18.120), the finite element solution gives the slopes $\theta$ at the nodes. To obtain the nodal displacements, we use $\theta = dy/dx$, or $\Delta y = \theta \, \Delta x$, or $y_{i+1} = y_i + \theta_i L^e$.

Readers can easily extend the procedure presented in this example to three-node and four-node elements. Refer to Prob. 18.10 at the end of the chapter.

**Example 18.8.** Use the variational approach to derive the element equations and matrices for the systems whose integral equations are derived in Example 18.2.

**Solution.** We have designed triangular and quadrilateral elements of various numbers of nodes for these systems. Let us derive the element equations and matrices for a three-node triangular element. The two independent displacement components $u$ and $v$ for the plane stress problem can be written using Eq. (18.86) as follows:

$$u^e = N_1 u_1 + N_2 u_2 + N_3 u_3 \tag{18.122}$$

$$v^e = N_1 v_1 + N_2 v_2 + N_3 v_3 \tag{18.123}$$

which can be written in vector form as:

$$\begin{bmatrix} u \\ v \end{bmatrix}^e = \begin{bmatrix} N_1 & 0 & N_2 & 0 & N_3 & 0 \\ 0 & N_1 & 0 & N_2 & 0 & N_3 \end{bmatrix} \begin{bmatrix} u_1 \\ v_1 \\ u_2 \\ v_2 \\ u_3 \\ v_3 \end{bmatrix}^e = [N]U^e \tag{18.124}$$

Substituting Eqs. (18.122) and (18.123) into Eq. (18.34) and rearranging, we obtain

$$\varepsilon = [B]U^e \tag{18.125}$$

where

$$[B] = \begin{bmatrix} N_{1,x} & 0 & N_{2,x} & 0 & N_{3,x} & 0 \\ 0 & N_{1,y} & 0 & N_{2,y} & 0 & N_{3,y} \\ N_{1,y} & N_{1,x} & N_{2,y} & N_{2,x} & N_{3,y} & N_{3,x} \end{bmatrix} \tag{18.126}$$

and $\mathbf{U}^e$ is the element displacement vector shown in Eq. (18.124). $N_{i,x}$ and $N_{i,y}$ means the partial derivative of $N_i$ of node $i$ with respect to $x$ and $y$ respectively. Applying Eq. (18.72) to Eq. (18.48) and using Eq. (18.125), we obtain:

$$\Pi^e = I^e = \tfrac{1}{2} \int_{V^e} \mathbf{U}^{eT}[B]^T[D][B]\mathbf{U}^e \, dV$$

$$-P_1[N_1 \quad 0 \quad N_2 \quad 0 \quad N_3]_1 \mathbf{U}^e$$

$$-P_2[0 \quad N_1 \quad 0 \quad N_2 \quad 0 \quad N_3]_2 \mathbf{U}^e \tag{18.127}$$

where $[N_1 \quad 0 \quad \cdots]_1$ and $[0 \quad N_1 \quad 0 \quad \cdots]_2$ are evaluated at the points where the loads $P_1$ and $P_2$ are applied respectively. These terms contribute as loads in the appropriate element equation as shown in Example 18.7. Applying the stationarity condition given by Eq. (18.99), six independent element equations can be written in a matrix form as

$$[K]^e \mathbf{U}^e = \mathbf{0} \tag{18.128}$$

where the element stiffness matrix $[K]^e$ is given by

$$[K]^e = \int_{V^e} [B]^T[D][B] \, dV$$

$$= t \int_{A^e} [B]^T[D][B] \, dA \tag{18.129}$$

This matrix is symmetric and positive definite. Readers are encouraged to expand Eq. (18.129) to obtain the individual elements of $[K]^e$. Elements whose nodes carry the applied loads $P_1$ and $P_2$ have a nonzero right-hand side in Eq. (18.128). For two-dimensional problems, more than one element may share, say, the node that carries $P_1$. In such a case, $P_1$ can be divided equally between the elements. There is a better approach to handling applied loads than this, given when we discuss the assembly process and the application of external loads.

For the heat conduction problem, we assume the temperature polynomial to be

$$T^e = N_1 T_1 + N_2 T_2 + N_3 T_3 = [N]\mathbf{T}^e \tag{18.130}$$

where $[N] = [N_1 \quad N_2 \quad N_3]$ and $\mathbf{T}^e = [T_1 \quad T_2 \quad T_3]^T$. To simplify manipulations, Eq. (18.63) can be rewritten as

$$\Pi = \tfrac{1}{2} \int_0^b \int_0^a \mathbf{T}'^T[K_T]\mathbf{T}' \, dx \, dy + \int_0^b q_1 T \, dy - \int_0^a q_2 T \, dx \tag{18.131}$$

where

$$\mathbf{T}' = \left[ \frac{\partial T}{\partial x} \quad \frac{\partial T}{\partial y} \right]^T$$

and $[K_T]$ is given by

$$[K_T] = \begin{bmatrix} K & 0 \\ 0 & K \end{bmatrix} \tag{18.132}$$

The first term in Eq. (18.131) is identical in form to that of Eq. (18.48). Using Eqs. (18.72) and (18.130), Eq. (18.131) gives

$$\Pi^e = I^e = \tfrac{1}{2} \int_0^{b^e} \int_0^{a^e} \mathbf{T}^{eT}[B]^T[K_T][B]\mathbf{T}^e \, dx \, dy + \int_0^{b^e} q_1[N]\mathbf{T}^e \, dy - \int_0^{a^e} q_2[N]\mathbf{T}^e \, dx$$

$$\tag{18.133}$$

where

$$[B] = \begin{bmatrix} N_{1,x} & N_{2,x} & N_{3,x} \\ N_{1,y} & N_{2,y} & N_{3,y} \end{bmatrix} \tag{18.134}$$

Similar to the plane stress problem, applying the stationarity condition given by Eq. (18.99) to Eq. (18.133), we obtain the element equations as

$$[K]^e \mathbf{T}^e - \mathbf{Q}_1 + \mathbf{Q}_2 = \mathbf{0} \tag{18.135}$$

where $[K]^e$ is the element conductivity matrix:

$$[K]^e = \int_0^{b^e} \int_0^{a^e} [B]^T [K][B] \, dx \, dy \tag{18.136}$$

and $\mathbf{Q}_1$ and $\mathbf{Q}_2$ are heat flow vectors:

$$\mathbf{Q}_1 = \int_0^{b^e} q_1 [N]^T \, dy \tag{18.137}$$

$$\mathbf{Q}_2 = \int_0^{a^e} q_2 [N]^T \, dx \tag{18.138}$$

**18.3.2.6 ELEMENT EQUATIONS USING THE METHOD OF WEIGHTED RESIDUALS.** Galerkin's method described by Eqs. (18.64) and (18.68) forms the basis of the Galerkin finite element method. Using the interpolation polynomial given by Eq. (18.96) in place of Eq. (18.64) and applying the discretization equation (18.72) to Eq. (18.68), the element equations based on the Galerkin finite element method are given by

$$\int_{D^e} [L(\phi^e) - f^e] N_i \, dD = 0, \qquad i = 1, 2, \ldots, n \tag{18.139}$$

where $n$ is the number of element nodes. Before deciding on the explicit form of the shape function $N_i$ to meet the compatibility and completeness requirements, Eq. (18.139) is always integrated by parts (or by using the Gauss divergence theorem), as we did in deriving a functional $\Pi$ to lower the highest-order derivative appearing in this equation. This, in turn, reduces the order of the element continuity ($C^0$, $C^1$, etc.) and, therefore, the order of its interpolation polynomial. There is close similarity between manipulating Eq. (18.139) to develop the element equations and the variational approach as illustrated in the following example.

**Example 18.9.** Use the Galerkin method to derive the element equations and matrices of the cantilever beam described by Eq. (18.8) in Example 18.1.

*Solution.* We have derived the integral equation of the beam using the Galerkin method in Example 18.4. That equation can be utilized to develop the standard-type beam element. In this example, as we did in Example 18.7, we use the beam slope $\theta$. From Example 18.4, the approximate solution and the beam integral equation are given in terms of the beam deflection $y$ as follows:

$$y_a = \sum_{i=1}^{n} c_i g_i(x) \tag{18.140}$$

and

$$\int_0^L EI y_a'' g_i''(x)\, dx - P g_i(L) = 0, \qquad i = 1, 2, \ldots, n \tag{18.141}$$

Differentiating Eq. (18.141), the beam approximate slope is

$$y_a' = \theta_a = \sum_{i=1}^{n} c_i g_i'(x) = \sum_{i=1}^{n} c_i g_{i\theta}(x) \tag{18.142}$$

Using the above equation in Eq. (18.141) by substituting $y_a'' = \theta_a'$ and $g_i''(x) = g_{i\theta}'(x)$, we obtain

$$\int_0^L EI \theta_a' g_{i\theta}'(x)\, dx - P g_i(L) = 0, \qquad i = 1, 2, \ldots, n \tag{18.143}$$

Replacing Eqs. (18.140) and (18.142) by Eqs. (18.117) and (18.118) respectively and writing Eq. (18.143) for an element domain, we obtain

$$\int_0^{L^e} EI(N_1' \theta_1 + N_2' \theta_2) N_i' - P N_{iL} = 0, \qquad i = 1, 2, \ldots, n \tag{18.144}$$

The element equations given by this equation are identical to those obtained by applying the stationarity condition to Eq. (18.119). Therefore, the element matrices are identical to those obtained via the variational approach.

## 18.3.3 Assembly of Element Equations

The discretization process given by Eq. (18.72) has resulted in a matrix equation for each element in the continuum finite element mesh. To obtain the finite element solution, we must combine these equations into one master matrix equation. This combination process of element equations is known as the assembly process and the master algebraic system of equations is known as the global or total equations. The assembly process and its procedure for constructing the system (continuum) equations from the element equations are the same regardless of the type of problem being analyzed, the complexity of the system of elements, or the mixture of element types that make the mesh.

An element matrix equation cannot be solved by itself for the unknown nodal values at the element nodes because it does not represent element equilibrium. No internal reactions are included in the element equation, and because these reactions always appear in pairs of equal and opposite magnitudes, assembly of the elements eliminates these reactions in principle (they are not included in element equations) and makes the global system of equations a valid system to solve for the nodal values. In addition, the assembly process is valid because of the compatibility requirement at the nodes, which is reinforced during the element design and the choice of the interpolation polynomial.

To be able to discuss the general assembly procedure and algorithm to execute it, the matrix equation of any element can be written in the following form:

$$[K]^e\phi^e = \mathbf{R}^e \tag{18.145}$$

where $[K]^e$ is the element matrix, $\phi^e$ is the vector of the element nodal values, and $\mathbf{R}^e$ is the element load vector. From the previous examples, $[K]^e$ may represent stiffness, conductivity, or the flow matrix; $\phi^e$ may represent displacements, temperatures, or pressures; $\mathbf{R}^e$ includes the effect of external and internal (body forces, heat sources, etc.) loads. Realizing that $[K]^e\phi^e - \mathbf{R}^e$ is equivalent to the element integral in Eq. (18.72), this equation can be rewritten as

$$\int_D H(\phi)\, dD = \sum_{e=1}^m ([K]^e\phi^e - \mathbf{R}^e) = 0 \tag{18.146}$$

or

$$\sum_{e=1}^m [K]^e\phi^e = \sum_{e=1}^m \mathbf{R}^e \tag{18.147}$$

or

$$[K]\phi = \mathbf{R} \tag{18.148}$$

where

$$[K] = \sum_{e=1}^m [K]^e \tag{18.149}$$

$$\phi = [\phi_1 \quad \phi_2 \quad \cdots \quad \phi_n]^T \tag{18.150}$$

$$\mathbf{R} = \sum_{e=1}^m \mathbf{R}^e \tag{18.151}$$

$[K]$, $\phi$, and $\mathbf{R}$ are the global (assembled) system matrix, global nodal values, and global load vector respectively. The size of $[K]$, $\phi$, and $\mathbf{R}$ is determined from the number of nodes $N$ multiplied by the number of degrees of freedom per node $n_F$. The matrix $[K]$ is $n \times n$ where $n = N \times n_F$, and both $\phi$ and $\mathbf{R}$ have $n$ elements.

The assembly procedure and its related algorithm is based on accumulating (adding) the contributions of elements attached to a given node to the equations, in the global system, that correspond to the degrees of freedom of this node. To illustrate this procedure, let us consider the three-node, two-element mesh shown in Fig. 18-9. There are two numbering schemes. An arbitrary global numbering scheme (Fig. 18-9a) is established to identify the mesh nodes and elements. This numbering scheme can be created manually or algorithmically. A local numbering scheme is used to generate the element equations before assembly. The relationship between the two schemes is established at the user input level of mesh data. For example, when a user inputs nodes 2 and 3 as the nodes of element ② in Fig. 18-9a, node 2 becomes its local node 1 and node 3 becomes its local node 2.

Utilizing the global numbering scheme, the element equations of the two elements shown in Fig. 18-9a can be written as follows:

$$\begin{bmatrix} k_{11}^{①} & k_{12}^{①} \\ k_{21}^{①} & k_{22}^{①} \end{bmatrix} \begin{bmatrix} \phi_1 \\ \phi_2 \end{bmatrix} = \begin{bmatrix} R_1^{①} \\ R_2^{①} \end{bmatrix} \tag{18.152}$$

Global numbering scheme

Local numbering scheme

(a) Two-element mesh

$$[K]\boldsymbol{\phi} = \mathbf{R}, \text{ where } [K] = [K]^{①} + [K]^{②}, \boldsymbol{\phi} = \boldsymbol{\phi}^{①} \cup \boldsymbol{\phi}^{②}, \text{ and } \mathbf{R} = \mathbf{R}^{①} + \mathbf{R}^{②}$$

or

$$\begin{bmatrix} K_{11} & K_{12} & K_{13} \\ K_{21} & K_{22} & K_{23} \\ K_{31} & K_{32} & K_{33} \end{bmatrix} \begin{bmatrix} \phi_1 \\ \phi_2 \\ \phi_3 \end{bmatrix} = \begin{bmatrix} R_1 \\ R_2 \\ R_1 \end{bmatrix}$$

that is,

(b) Global system of equations

**FIGURE 18-9**
Assembly of element equations.

$$\begin{bmatrix} k_{11}^{②} & k_{12}^{②} \\ k_{21}^{②} & k_{22}^{②} \end{bmatrix} \begin{bmatrix} \phi_2 \\ \phi_3 \end{bmatrix} = \begin{bmatrix} R_1^{②} \\ R_2^{②} \end{bmatrix} \tag{18.153}$$

Each row in an element equation represents an equation in a given direction, that is, an equation for a given degree of freedom. For example, the equation $k_{11}^{①}\phi_1 + k_{12}^{①}\phi_2 = R_1^{①}$ from the first element matrix equation (18.152) is associated with the degree of freedom of global node 1, that is, $\phi_1$. Global node 2 has two contributions, an equation from each element. When these two equations are added algebraically, the total node equation results, which can be placed in the global system of equations as shown in Fig. 18-9b. In addition, each element in the element matrix is identified by a row and a column. Each corresponds with a nodal degree of freedom. For example, the element $K_{12}^{②}$ is identified by the row of $\phi_2$ and the column of $\phi_3$. This row and column must be preserved while $K_{12}^{②}$ is placed in the global matrix. Therefore, the assembly procedure is equivalent to stretching and placing the element equations in the global system of equations and adding the overlapping elements in $[K]$ and $\mathbf{R}$ as illustrated in Fig. 18-9b.

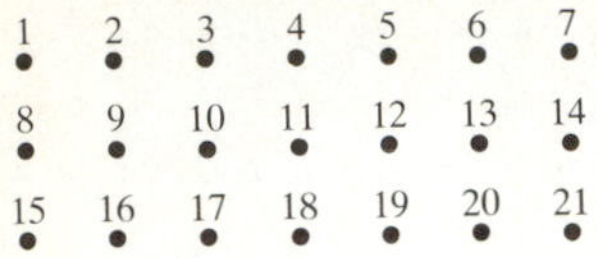

```
1    2    3    4    5    6    7
•    •    •    •    •    •    •

8    9    10   11   12   13   14
•    •    •    •    •    •    •

15   16   17   18   19   20   21
•    •    •    •    •    •    •
```

**FIGURE 18-10**
Nonoptimal global numbering scheme of mesh shown in Fig. 18-1b.

Due to the stretching effect during the assembly process, the global numbering scheme controls directly the locations of nonzero elements of the global matrix (the matrix bandwidth). A bandwidth is the width of nonzero elements of a matrix measured from the diagonal. Nodal numbering must be done to reduce the bandwidth. There are various algorithms for optimal nodal numbering for bandwidth reduction. The general rule is that the smaller the difference between the maximum and minimum global node numbers of any element, the smaller the bandwidth of the resulting global matrix and consequently the more optimum the global numbering scheme. On this basis, the numbering scheme shown in Fig. 18-1b is optimum while the numbering scheme of the same mesh shown in Fig. 18-10 is not. In Fig. 18-1b, the maximum difference between the numbers of any two nodes of any element is 4 for a four-node element and 5 for a six-node element. In Fig. 18-10, the difference jumps to 8 and 15 respectively.

### 18.3.4  Imposing Boundary Conditions

The global system of equations resulting from the assembly process is given by Eq. (18.148). As these equations stand, they cannot be solved for the nodal values $\phi$ of the field variable until they have been modified to account for the geometric boundary conditions of the problem at hand. The natural boundary conditions are already included in the element equations. Element matrices $[K]^e$ and the global (assembled) matrix $[K]$ are always singular, that is, their inverse cannot be found because their determinants are zero.

Before applying the geometric boundary conditions, the system of equations is not completely defined. This is analogous to being unable to find constants in solutions of differential equations if enough boundary conditions are not available. In solid mechanics applications, having enough fixed nodal displacements prevents the structure (continuum) from moving in space as a rigid body when external loads are applied.

Applying zero boundary conditions is relatively simple. It amounts to eliminating the row and the column from the global system of equations that correspond to the zero degree of freedom. If, for example, a global system of equations is given by

$$
\begin{bmatrix}
k_{11} & k_{12} & \cdots & & k_{1n} \\
 & k_{22} & \cdots & & k_{2n} \\
 & & & & \vdots \\
\text{Symmetric} & k_{mm} & \cdots & & k_{mn} \\
 & & & & \vdots \\
 & & & & k_{nn}
\end{bmatrix}
\begin{bmatrix}
\phi_1 \\ \phi_2 \\ \vdots \\ \phi_m \\ \vdots \\ \phi_n
\end{bmatrix}
=
\begin{bmatrix}
R_1 \\ R_2 \\ \vdots \\ R_m \\ \vdots \\ R_n
\end{bmatrix}
\tag{18.154}
$$

and $\phi_1 = \phi_m = 0$ as boundary conditions, applying these conditions to the above equation eliminates the first row and column (for $\phi_1 = 0$) and the $m$th row and column (for $\phi_m = 0$). Thus, the number of simultaneous equations reduces to $(n-2)$ from $n$, and Eq. (18.154) becomes

$$
\begin{bmatrix}
k_{22} & k_{23} & \cdots & & k_{2n} \\
 & k_{33} & \cdots & & k_{3n} \\
 & & & & \vdots \\
\text{Symmetric} & k_{m-1} & \cdots & & k_{m-1,n} \\
 & & k_{m+1} & \cdots & k_{m+1,n} \\
 & & & & \vdots \\
 & & & & k_{n-1}
\end{bmatrix}
\begin{bmatrix}
\phi_2 \\
\phi_3 \\
\vdots \\
\phi_{m-1} \\
\phi_{m+1} \\
\vdots \\
\phi_n
\end{bmatrix}
=
\begin{bmatrix}
R_2 \\
R_3 \\
\vdots \\
R_{m-1} \\
R_{m+1} \\
\vdots \\
R_n
\end{bmatrix}
\qquad (18.155)
$$

Applying nonzero geometric boundary conditions is possible and usually requires rearranging Eq. (18.154), and is not discussed here.

### 18.3.5   Lumping External Applied Loads

Equation (18.155) represents the original continuum problem after applying the finite element method to it. The continuum is replaced by a set of nodes. Therefore, any continuum-related property (such as mass, damping, internal heat source, etc.) or external applied loads must be lumped at the nodes. Two types of external loads exist: concentrated or distributed. Examples of the former include concentrated forces, moments, etc. Examples of the latter include distributed force, moments, heat fluxes, fluid fluxes, etc. To apply concentrated loads to a finite element model, all that is needed is to create a node at the point of application of each load. In the cantilever beam example shown in Fig. 18-1, node number 21 carries the load $P$ of 500 lb, or $P$ is applied to the model through node 21. In some cases, the load might be split between more than one node. In other cases, distributed loads may be lumped at various nodes. This is a matter of judgment left to the FEA analyst.

Distributed loads are automatically applied through the proper equations of boundary elements. Equations (18.137) and (18.138) show an example. In this case, the user would have to input the intensity of the distributed load and the nodes they are to be applied to or lumped at. As mentioned while developing element equations and assembling them, external applied (distributed or concentrated) loads may be neglected until the development of the global load vector **R** is performed. This makes all element equations similar and eliminates the need to identify boundary and interior elements. The development of the global load vector can be based on the nodes and the boundary segments they are applied to.

### 18.3.6   Solution of Global Equations

Thus far we have discussed in detail the derivation of the finite element equations given by the global system equation (18.155). For practical design and FEA problems, this system may have hundreds or thousands of degrees of freedom. The effective solution of such a large number of simultaneous algebraic equations not only determines the cost of the analysis but also controls the accuracy of the solution. In nonlinear analysis, the convergence to the correct solution is also controlled by the method of solution.

For linear static or steady-state analysis, Gauss elimination is the common numerical method of solution. After applying the boundary conditions and the external applied loads to Eq. (18.148), the nodal values can be obtained as

$$\phi = [K]^{-1}\mathbf{R} \tag{18.156}$$

where $[K]^{-1}$ is the inverse of the global matrix $[K]$. Many implementation factors are usually considered to speed up the inversion process. For example, a skyline (a piecewise line that separates the zero elements from the nonzero elements of $[K]$) is sometimes used to avoid wasting time dealing with zero elements in $[K]$.

For linear dynamic or transient analysis, numerical time integration is used. There are various integration schemes that keep the cost of the solution down and converge. Experienced FEA analysts realize that the time increment (time step) is the most sensitive factor that controls the convergence of dynamic analysis.

In nonlinear static analysis, iterative solution methods are utilized to solve for the system response. The Newton-Raphson method is the method commonly used in the solution. The solution is incremental in nature. The external load is applied to the continuum in increments (load increments or steps), and the continuum response is calculated incrementally. The load step is very crucial for the solution to converge and requires experience to choose. Two types of nonlinearities may exist in a problem. A geometrically nonlinear problem is one in which a large deformation occurs and a materially nonlinear problem is one in which material properties are nonlinear.

Nonlinear dynamic analysis is the most complex and sensitive analysis an FEA analyst may face. Here the interaction between the load step and the time step depends to a great deal on the nonlinearity at hand, and changes from one problem to another. The solution is usually obtained by trial and error. Readers who are interested in knowing more about the solution methods are referred to the bibliography at the end of the chapter.

### 18.3.7 Other Finite Element Analyses

The material presented in this chapter thus far has been related to linear static FEA because this is the simplest analysis. However, other analyses can be viewed as (nontrivial) extensions of all the concepts presented here. The steps of the finite element method presented here are essentially the same. The major difference comes in finding the integral equation and the numerical methods to solve the global systems of equations.

Equation (18.148) represents linear static problems. Linear dynamic problems can be represented by

$$[M]\ddot{\phi} + [C]\dot{\phi} + [K]\phi = \mathbf{R} \tag{18.157}$$

where $[M]$, $[C]$, $\ddot{\phi}$, and $\dot{\phi}$ are respectively the system mass matrix, damping matrix, acceleration vector, and velocity vector. In solid mechanics applications, the first term is the system inertia forces, the second term is the dissipation or frictional forces, and the third term is the spring forces. $[M]$ and $[C]$ are, for many applications, symmetric and positive definite. They can be formulated in a similar way to the stiffness matrix $[K]$.

When numerical integration schemes are applied to solve Eq. (18.157), the equation takes the following form:

$$[\bar{K}]\phi_{n+1} = \mathbf{R}_{n+1} \qquad (18.158)$$

where $[\bar{K}]$ is an effective matrix given by

$$[\bar{K}] = \alpha[M] + \beta[C] + \gamma[K] \qquad (18.159)$$

where $\alpha$, $\beta$, and $\gamma$ are constants determined by the time numerical integration scheme. $\mathbf{R}_{n+1}$ is the load vector at step $(n+1)$. It is also a function of the original load vector $\mathbf{R}$ and other terms that result from the integration scheme.

For nonlinear static analysis, Eq. (18.148) becomes

$$[K(\phi)]\phi = \mathbf{R} \qquad (18.160)$$

and for nonlinear dynamic analysis, the equation is

$$[M]\ddot{\phi} + [C(\phi)]\dot{\phi} + [K(\phi)]\phi = \mathbf{R} \qquad (18.161)$$

Utilizing the Newton-Raphson method to solve these two equations, an initial guess of $\phi$ (usually $\mathbf{0}$) for Eq. (18.160) and of $\dot{\phi}$ (also $\mathbf{0}$) and $\ddot{\phi}$ (also $\mathbf{0}$) additionally for Eq. (18.161) is required.

### 18.3.8   Convergence of Finite Element Solutions

The finite element method is an approximate method and thus introduces errors to problem solutions. There are two sources of errors of approximations: the discretization process and the element shape functions. The type and number of elements employed in a finite element mesh as well as the order of element shape functions control the quality of finite element solutions.

To reduce the errors of approximation and therefore improve the solution quality, we can either refine the mesh or increase the order of the interpolation polynomial. In conventional FEA, the convergence criteria discussed in Sec. 18.3.2.3 are used, and the approximation error is controlled by refining the mesh; as the continuum is broken into smaller elements for each analysis iteration, the error decreases. This method of controlling the quality of approximation is called the *h*-version (because the size of the elements is generally denoted by *h*) of the finite element method. In the *h*-version, solution improvement is achieved by an orderly sequence of uniform mesh refinements, and the shape functions remain constant and constructed from polynomials of low order (linear, quadratic, or cubic).

The other method of improving the quality of a finite element solution is called the *p*-version (because the polynomial degree of shape functions is generally denoted by *p*). Instead of changing the mesh, the degree of the polynomials describing the elements is changed. In the *p*-version, the finite element mesh is fixed, and the solution improvement is achieved by increasing the polynomial order of the shape functions. Polynomial order can change between one and eight.

Although it has been established theoretically and by examples that, for a given problem, *h*- or *p*-versions alone converge to the same solution, it has also been shown that the *p*-version is more efficient. With the proper mesh design, the *p*-version can achieve the near-optimum rates of convergence. Meshes for the *p*-version are usually different from those for the *h*-version. Meshes for the *p*-version are often simpler than those for the *h*-version, partly because *p*-version meshes can be graded so that the element size decreases in geometric progression toward points of interest such as stress concentration in elasticity. In some cases, the meshes in the *p*-version are so simple that mesh generators and, therefore, expensive graphics displays are not needed. Because *p*-version elements are generally large, mapping techniques based on the blending function method are used so that curves such as circles, ellipses, hyperbolas, and parabolas are represented exactly.

## 18.4   ISOPARAMETRIC EVALUATION OF ELEMENT MATRICES

Element matrices have been expressed as integrals, over the element domain, of the shape functions and other element properties. The evaluation of these matrices is a crucial step in FEA because it directly affects the cost and accuracy of the FEA. While shape functions and element integrals can be evaluated in the global coordinate system, such an approach is not very efficient. It is not always possible to find $[G]^{-1}$ as discussed in Sec. 18.3.2.4; nor is it efficient to compute $[G]^{-1}$ if it exists. Isoparametric evaluation of element matrices offers a very attractive alternative. In addition to avoiding the inversion problem, it enables curve-sided elements to be created which means that smaller numbers of elements could be used to represent curved boundaries. Consequently, the size and cost of the finite element model are reduced.

The basic idea behind the isoparametric evaluation of finite elements is similar to that of parametric representation in Part II of the book. Element geometry is mapped from the cartesian space to the parametric space where element geometric shape is simple. In the element parametric space, one-dimensional elements are always line segments of length 2; two-dimensional elements are squares of $2 \times 2$ sides; and three-dimensional elements are cubes of $2 \times 2 \times 2$ sides. Figure 18-11 shows some isoparametric elements. Due to the simple shapes of elements in the parametric space, element shape functions can be easily constructed in this space by inspection, and numerical integration to evaluate element matrices can be easily performed.

### 18.4.1   Element Mapping

The isoparametric evaluation of a finite element requires mapping both its geometric shape and interpolation polynomial from the cartesian space to the para-

(a) One-dimensional element

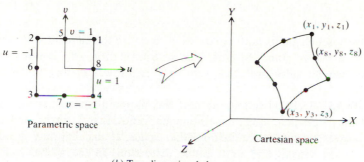

(b) Two-dimensional element

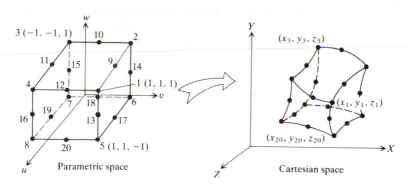

(c) Three-dimensional element

**FIGURE 18-11**
Sample isoparametric elements.

metric space. Instead of interpreting Eq. (18.96) to be over the element cartesian domain, we use it over the element parametric domain. Thus $N_i$ changes from functions of the cartesian space to functions of the element parametric space. For one, two, and three dimensions, $N_i$ becomes $N_i(u)$, $N_i(u, v)$, and $N_i(u, v, w)$ respectively. To map the element geometric shape, we can think of the coordinates $(x, y, z)$ of any point within the element in a similar way to the field variable $\phi$. Thus, Eq. (18.96) can be rewritten to interpolate the coordinates as follows:

$$x = \sum_{i=1}^{m} F_i x_i \qquad y = \sum_{i=1}^{m} F_i y_i \qquad z = \sum_{i=1}^{m} F_i z_i \qquad (18.162)$$

where $F_i$ is the shape function that interpolates the element geometry, $(x_i, y_i, z_i)$ are the cartesian coordinates of node $i$, and $m$ is the number of nodes that interpolates the geometry. In comparison to the number of nodes $n$ that interpolates the field variable, superparametric elements occur when $m > n$, subparametric elements occur when $m < n$, and isoparametric elements occur when $m = n$ (consequently $F_i = N_i$). In this section, we consider only isoparametric elements for which Eq. (18.162) changes to

$$x = \sum_{i=1}^{n} N_i x_i \qquad y = \sum_{i=1}^{n} N_i y_i \qquad z = \sum_{i=1}^{n} N_i z_i \qquad (18.163)$$

Equation (18.163) allows elements with curved sides because the element sides are fitted between the nodes. In standard-type element formulation, this is not possible because the element sides are always straight regardless of any mid-side nodes.

When writing Eq. (18.163), we assume that the mapping between the parametric and cartesian spaces is unique (one-to-one mapping); that is, each point in one space has a corresponding point in the other space. If the mapping is not unique, we may expect violent and undesirable distortions in the global space that may fold the curved element back upon itself. A method for checking for nonuniqueness and violent distortion is by evaluating the jacobian $[J]$ of the element (see below) and checking the sign of its determinant $|J|$. If the sign of $|J|$ does not change for all the elements in the solution domain (mesh), acceptable mapping is guaranteed.

An important consideration in isoparametric evaluation is to ensure that element shape functions established normally in the parametric space preserve the convergence conditions (compatibility and completeness) in the global space. Without proof, it turns out that if $N_i$ meets these conditions in the parametric space, they are automatically met in the global space. Therefore, the element design discussed in Sec. 18.3.2.3 for the global space applies for the parametric one.

## 18.4.2   Shape Functions by Inspection

Element shape functions for isoparametric elements can be established in the element parametric space by inspection based on the insight gained about them in Sec. 18.3.2.4. The fundamental property of a shape function $N_i$ is that its value is unity at node $i$ and zero at all other nodes. Using this property, $N_i$ corresponding to a specific nodal layout could be obtained in a systematic manner by inspection as follows. First, we construct the shape functions $N_i$ corresponding to the corner (basic) nodes of the element. The addition of another node (interior or mid-side) results in an additional shape function and a correction to be applied to the already existing shape functions.

Let us illustrate the above concept for one-dimensional elements (see Fig. 18-11a). The origin of the parametric coordinate system is taken as the midpoint of the element as shown. If the element has only the two corner nodes 1 and 2, their shape functions are linear and can be written as

$$N_1 = \tfrac{1}{2}(1 - u) \qquad N_2 = \tfrac{1}{2}(1 + u) \qquad (18.164)$$

Notice that $N_1$ and $N_2$ satisfy the fundamental property of a shape function. If we add node 3 at the center ($u = 0$), its shape function is (by inspection)

$$N_3 = 1 - u^2 \qquad (18.165)$$

At node 3, $u = 0$ which gives $N_1 = N_2 = \frac{1}{2}$. This violates the fundamental property. To make $N_1$ or $N_2$ equal to zero at the node 3 location, we simply modify Eq. (18.164) to give $N_1 = \frac{1}{2}(1 - u) - \frac{1}{2}$ and $N_2 = \frac{1}{2}(1 + u) - \frac{1}{2}$. These new $N_1$ and $N_2$ still violate the fundamental property at (now) nodes 1 and 2. The proper solution is

$$N_1 = \tfrac{1}{2}(1 - u) - \tfrac{1}{2}N_3 \qquad N_2 = \tfrac{1}{2}(1 + u) - \tfrac{1}{2}N_3 \qquad (18.166)$$

For a four-node element, nodes 3 and 4 are located at $u = -\frac{1}{3}$ and $u = \frac{1}{3}$ respectively. Following the same concept, we can write

$$N_3 = \tfrac{9}{16}(1 - u^2)(1 - 3u) \qquad (18.167)$$

$$N_4 = \tfrac{9}{16}(1 - u^2)(1 + 3u) \qquad (18.168)$$

The correct $N_1$ and $N_2$ can be written as

$$N_1 = \tfrac{1}{2}(1 - u) - \tfrac{2}{3}N_3 - \tfrac{1}{3}N_4 \qquad (18.169)$$

$$N_2 = \tfrac{1}{2}(1 + u) - \tfrac{1}{3}N_3 - \tfrac{2}{3}N_4 \qquad (18.170)$$

This systematic generation of shape functions for one-dimensional elements is illustrated in Fig. 18-12.

The above concept can be extended to two-dimensional elements. Considering the quadrilateral element shown in Fig. 18-11$b$, we can write for the corner

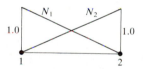

($a$) Linear element

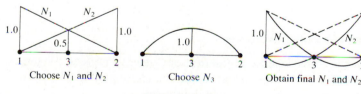

($b$) Quadratic element

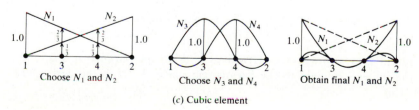

($c$) Cubic element

**FIGURE 18-12**
By-inspection generation of shape functions for one-dimensional elements.

nodes:

$$N_1 = \tfrac{1}{4}(1 + u)(1 + v) \tag{18.171}$$

$$N_2 = \tfrac{1}{4}(1 - u)(1 + v) \tag{18.172}$$

$$N_3 = \tfrac{1}{4}(1 - u)(1 - v) \tag{18.173}$$

$$N_4 = \tfrac{1}{4}(1 + u)(1 - v) \tag{18.174}$$

and for the mid-side nodes, we have

$$N_5 = \tfrac{1}{4}(1 - u^2)(1 + v) \tag{18.175}$$

$$N_6 = \tfrac{1}{4}(1 - v^2)(1 - u) \tag{18.176}$$

$$N_7 = \tfrac{1}{4}(1 - u^2)(1 - v) \tag{18.177}$$

$$N_8 = \tfrac{1}{4}(1 - v^2)(1 + u) \tag{18.178}$$

For each present mid-side node, the shape function of each surrounding corner node must by corrected by subtracting half of the shape function of this mid-side node. For example, if node 5 is present, then $\tfrac{1}{2}N_5$ is subtracted from $N_1$ and $N_2$ given in the above equations, that is, node 1 has $N_1 - \tfrac{1}{2}N_5$ and node 2 has $N_2 - \tfrac{1}{2}N_5$. If, in addition, node 8 is present, then node 1 has $N_1 - \tfrac{1}{2}N_5 - \tfrac{1}{2}N_8$ and node 4 has $N_4 - \tfrac{1}{2}N_8$.

Similarly, we can write the following equations for the three-dimensional element shown in Fig. 18-11c:

$$N_1 = \tfrac{1}{8}(1 + u)(1 + v)(1 + w) \tag{18.179}$$

$$N_9 = \tfrac{1}{4}(1 - u^2)(1 + v)(1 + w) \tag{18.180}$$

Equations for $N_2$ to $N_8$ are similar to Eq. (18.179) with the proper sign change and equations for $N_{10}$ to $N_{20}$ are similar to Eq. (18.180).

### 18.4.3   Evaluation of Element Matrices

In order to perform the isoparametric evaluation of element matrices, we should be able to evaluate the integrals that appear in the element equations. In general, for two-dimensional problems these integrals take the form

$$\int_{A^e} f\left(\phi^e, \frac{d\phi^e}{dx}, \frac{d\phi^e}{dy}, \ldots\right) dx\, dy \tag{18.181}$$

Thus, we should express the above derivatives of $\phi^e$ in the parametric space and transform the integral domain to the parametric space. To map the derivatives, Eq. (18.96) gives

$$\frac{\partial \phi^e}{\partial x} = \sum_{i=1}^{n} \frac{\partial N_i}{\partial x} \phi_i \qquad \frac{\partial \phi^e}{\partial y} = \sum_{i=1}^{n} \frac{\partial N_i}{\partial y} \phi_i \tag{18.182}$$

To express $\partial N_i/\partial x$ and $\partial N_i/\partial y$ in terms of $u$ and $v$, we can write $[N_i = N_i(u, v)]$

$$\frac{\partial N_i}{\partial u} = \frac{\partial N_i}{\partial x}\frac{\partial x}{\partial u} + \frac{\partial N_i}{\partial y}\frac{\partial y}{\partial u} \tag{18.183}$$

$$\frac{\partial N_i}{\partial v} = \frac{\partial N_i}{\partial x}\frac{\partial x}{\partial v} + \frac{\partial N_i}{\partial y}\frac{\partial y}{\partial v} \tag{18.184}$$

or

$$\begin{bmatrix} \dfrac{\partial N_i}{\partial u} \\[2mm] \dfrac{\partial N_i}{\partial v} \end{bmatrix} = \begin{bmatrix} \dfrac{\partial x}{\partial u} & \dfrac{\partial y}{\partial u} \\[2mm] \dfrac{\partial x}{\partial v} & \dfrac{\partial y}{\partial v} \end{bmatrix} \begin{bmatrix} \dfrac{\partial N_i}{\partial x} \\[2mm] \dfrac{\partial N_i}{\partial y} \end{bmatrix} \tag{18.185}$$

Using Eq. (18.163), the above equation becomes

$$\begin{bmatrix} \dfrac{\partial N_i}{\partial u} \\[2mm] \dfrac{\partial N_i}{\partial v} \end{bmatrix} = \begin{bmatrix} \displaystyle\sum_{i=1}^{n} \dfrac{\partial N_i}{\partial u} x_i & \displaystyle\sum_{i=1}^{n} \dfrac{\partial N_i}{\partial u} y_i \\[4mm] \displaystyle\sum_{i=1}^{n} \dfrac{\partial N_i}{\partial v} x_i & \displaystyle\sum_{i=1}^{n} \dfrac{\partial N_i}{\partial v} y_i \end{bmatrix} \begin{bmatrix} \dfrac{\partial N_i}{\partial x} \\[2mm] \dfrac{\partial N_i}{\partial y} \end{bmatrix} = [J] \begin{bmatrix} \dfrac{\partial N_i}{\partial x} \\[2mm] \dfrac{\partial N_i}{\partial y} \end{bmatrix} \tag{18.186}$$

where $[J]$ is known as the jacobian, that is,

$$[J] = \begin{bmatrix} \displaystyle\sum_{i=1}^{n} \dfrac{\partial N_i}{\partial u} x_i & \displaystyle\sum_{i=1}^{n} \dfrac{\partial N_i}{\partial u} y_i \\[4mm] \displaystyle\sum_{i=1}^{n} \dfrac{\partial N_i}{\partial v} x_i & \displaystyle\sum_{i=1}^{n} \dfrac{\partial N_i}{\partial v} y_i \end{bmatrix} = \begin{bmatrix} \dfrac{\partial N_1}{\partial u} & \dfrac{\partial N_2}{\partial u} & \cdots & \dfrac{\partial N_n}{\partial u} \\[3mm] \dfrac{\partial N_1}{\partial v} & \dfrac{\partial N_2}{\partial v} & \cdots & \dfrac{\partial N_n}{\partial v} \end{bmatrix} \begin{bmatrix} x_1 & y_1 \\ x_2 & y_2 \\ \vdots & \vdots \\ x_n & y_n \end{bmatrix} \tag{18.187}$$

Inverting Eq. (18.186), the global derivatives are given by

$$\begin{bmatrix} \dfrac{\partial N_i}{\partial x} \\[2mm] \dfrac{\partial N_i}{\partial y} \end{bmatrix} = [J]^{-1} \begin{bmatrix} \dfrac{\partial N_i}{\partial u} \\[2mm] \dfrac{\partial N_i}{\partial v} \end{bmatrix} \tag{18.188}$$

With this equation, $\partial\phi^e/\partial x$ and $\partial\phi^e/\partial y$ can be expressed in the parametric spaces as follows. Equation (18.182) can be rewritten as

$$\begin{bmatrix} \dfrac{\partial \phi^e}{\partial x} \\[2mm] \dfrac{\partial \phi^e}{\partial y} \end{bmatrix} = \begin{bmatrix} \dfrac{\partial N_1}{\partial x} & \dfrac{\partial N_2}{\partial x} & \cdots & \dfrac{\partial N_n}{\partial x} \\[3mm] \dfrac{\partial N_1}{\partial y} & \dfrac{\partial N_2}{\partial y} & \cdots & \dfrac{\partial N_n}{\partial y} \end{bmatrix} \begin{bmatrix} \phi_1 \\ \phi_2 \\ \vdots \\ \phi_n \end{bmatrix} \tag{18.189}$$

Using Eq. (18.188), this equation becomes

$$\begin{bmatrix} \dfrac{\partial \phi^e}{\partial x} \\[2mm] \dfrac{\partial \phi^e}{\partial y} \end{bmatrix} = [J]^{-1} \begin{bmatrix} \dfrac{\partial N_1}{\partial u} & \dfrac{\partial N_2}{\partial u} & \cdots & \dfrac{\partial N_n}{\partial u} \\[3mm] \dfrac{\partial N_1}{\partial v} & \dfrac{\partial N_2}{\partial v} & \cdots & \dfrac{\partial N_n}{\partial v} \end{bmatrix} \begin{bmatrix} \phi_1 \\ \phi_2 \\ \vdots \\ \phi_n \end{bmatrix} \tag{18.190}$$

**TABLE 18.1**
# Reliable and reduced Gauss quadrature order

| Dimension | Number of nodes | Element shape | Reliable Gauss quadrature order | Reduced order used in practice |
|---|---|---|---|---|
| One | 2 | | 1 | Same |
| | 3 | | 2 | Same |
| | 4 | | 3 | Same |
| Two | 4 | | $2 \times 2$ | Same |
| | 8 | | $3 \times 3$ | $2 \times 2$ |
| | 12 | | $4 \times 4$ | $3 \times 3$ |
| Three | 8 | | $2 \times 2 \times 2$ | Same |
| | 20 | | $3 \times 3 \times 3$ | $2 \times 2 \times 2$ |
| | 32 | | $4 \times 4 \times 4$ | $3 \times 3 \times 3$ |

To complete the evaluation of the integral, we need to express the area element $dx\, dy$ in terms of $du\, dv$. As shown in Chap. 17, we write

$$dx\, dy = |J|\, du\, dv \qquad (18.191)$$

For one-dimensional elements,

$$dx = \frac{dx}{du}\, du = \sum_{i=1}^{n} \frac{dN_i}{du}\, du = |J|\, du$$

For three-dimensional elements readers can follow a similar development ($[J]$ becomes $3 \times 3$) [see Eq. (17.141)].

With these mappings, the integral in Eq. (18.181) reduces to

$$\int_{-1}^{1} \int_{-1}^{1} g(u,\, v)\, du\, dv \qquad (18.192)$$

This integral can be evaluated numerically using Gauss quadrature as explained in Chap. 17. Because the integration limits always go from $-1$ to $1$ in the element parametric space, Eqs. (17.112) and (17.113) become

$$V_i = W_t \qquad (18.193)$$

$$u_i = C_i \qquad (18.194)$$

Equations (17.114) and (17.118) can be adjusted accordingly.

As discussed in Chap. 17, sampling points are required to use Gauss quadrature to evaluate integrals numerically. The number of sampling points used to evaluate element integrals is known as the order of quadrature in the finite element literature. The practical rule is that it is desirable to keep the order of quadrature as low as possible to reduce the cost of the FEA and minimize computational errors. There is a lower limit on the number of sampling points because as the element size decreases, the integrand becomes constant and the integral becomes the area or the volume of the element. Thus, all the discussions presented in Chap. 17 regarding the choice of the number of sampling points apply here. Table 18.1 shows the reliable integration order and the reduced order used in practice.

## 18.5  FINITE ELEMENT MODELING

In the above four sections, we have presented the finite element theory in order to provide readers with a clear understanding of the requirements of FEA (which FEM must meet), the information typically needed to perform one, and how to interpret the results. While much of the theory presented has been codified in commercial codes available for engineers and designers to use, most of the burden of finite element modeling (as with any engineering modeling) and of analyzing the results lies on the engineers and designers themselves. A typical finite element model is comprised of nodes, degrees of freedom, boundary conditions, elements, material properties, externally applied loads, and analysis type. Engineers and designers must carefully create the model to ensure the relevance of the results of the corresponding FEA. These results depend solely on the model.

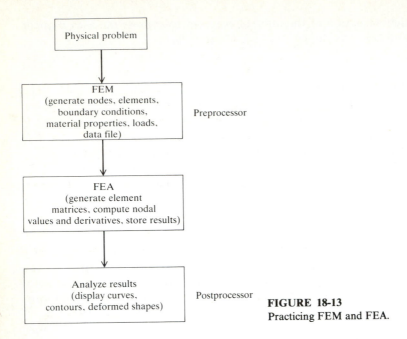

**FIGURE 18-13**
Practicing FEM and FEA.

In practice, practitioners must first decide on the mesh layout, that is, the number of nodes and elements. Zones of expected abrupt changes in the field variable (such as stress concentration around holes) require a denser number of nodes and elements than zones where gradual changes occur. This is known as mesh gradation. After the mesh layout is chosen, practitioners must choose the type of analysis (static/dynamic, linear/nonlinear, beam theory, plane stress, three-dimensional, etc.), the type (deflection, rotation, temperature, flux, etc.) and number of degrees of freedom at each node, the boundary conditions, the element information (type, number of nodes per element, and Gauss quadrature order), material properties, and lumping external loads at the nodes.

Once the finite element model is defined by choosing all of the above parameters of the corresponding mesh, it must be input to the code that performs the FEA. A review of most existing FEA commercial codes reveals that they require their users to provide the data of the finite element model in a data file with a specific format. A data file consists of records (these used to be known as cards); each record consists of 80 spaces (columns on cards). Each record is divided into various fields (a field consists of a certain number of spaces), and each field holds a specific variable. Apart from the specific detailed format of a certain data file, such a file usually has five major sections: control, nodal, element, material, and loading sections. While the latter four sections are self-explanatory, the control section includes information such as the problem description (heading), the total number of nodes, the type of analysis, etc.

The output from FEA codes is primarily in numerical form. It usually consists of the nodal values of the field variable and its derivatives. For example, in solid mechanics problems the output is nodal displacements and element stresses. In heat-transfer problems, the output is nodal temperatures and element heat fluxes. Graphical outputs are usually more informative in providing trends of continuum behavior. Thus, curves and contours of the field variable can be plotted and displayed. Also deformed shapes can be displayed superposed on undeformed shapes.

As the finite element modeling is a very labor-intensive task, it has been the target of automation utilizing the computer graphics and CAD/CAM technology. Powerful pre- and postprocessors exist that can perform most of the functions described above. The classification pre and post is relative to the FEA phase as shown in Fig. 18-13. Preprocessors, with the help of the user, can generate data files automatically. Users no longer need to know the exact format of the data file required by an FEA code. After the user generates a finite element model using a certain preprocessor, a single command can produce the data file with the required format. Most preprocessors support most of the existing commercial FEA codes. If not, a preprocessor can be interfaced with a given FEA code. This task usually requires a good knowledge of the database structure of the preprocessor and the format of the data file of the FEA code.

Postprocessors are usually automatic and do not require user assistance. They usually process the numerical results and display them in the desired form requested by the user. User postprocessor commands are usually simple to use.

## 18.6   MESH GENERATION

Mesh generation forms the backbone of the FEA. Mesh generation refers to the generation of nodal coordinates and elements. It also includes the automatic numbering of nodes and elements based on a minimal amount of user-supplied data. Automatic mesh generation reduces errors and saves a great deal of user time, therefore reducing the FEA cost.

Before the existence of preprocessors, finite element meshes were generated manually. In manual mesh generation, the analyst discretizes the simplified geometry of the object to be studied, that is, the geometric model of the object, into nodes and elements. Nodes are defined by specifying their coordinates while element connectivity (connecting nodes) defines the elements. Manual meshing is inefficient, error-prone, and meshing data can grow rapidly and become confusing for complex objects—especially three-dimensional ones.

The early development of computer-based finite element mesh generation methods began by attempting to extract the nodal coordinates and element nodes from the manually prepared (by the analyst) finite element data file and then plotting this mesh to ensure its correctness. With the advent of digitizers, analysts could prepare the finite element grid layout and strategy, define the grid origin and its coordinate system, and then digitize the nodal locations, which are stored in a file for later use to plot the mesh, to verify it, and to prepare the finite element data file.

With the widespread use of computer graphics and CAD technology, mesh generation has been a target for automation. There is a wide variety of algorithms, schemes, and methods for mesh generation. They have various levels of automation and different user-input requirements. This section classifies mesh generation into semi-automatic and fully automatic. A fully automatic mesh generation is taken to mean a method in which only the shape (both geometry and topology) of the object to be meshed and the mesh attributes (mesh density, element type, boundary conditions, loads, etc.) are required as input. Any other method that may require additional input such as subdividing the object into subdomains or regions is a semi-automatic one.

### 18.6.1  Mesh Requirements

Before we describe the various existing mesh generation methods, it is important to list the requirements that make a mesh valid, that is, produces the correct FEA results, and economical. Some of the requirements are listed below (some are necessary while others are desirable):

1. **Nodal locations.** Nodes must lie inside or on the boundaries of the geometric model to be meshed. Nodes that are very close to the boundaries must be pulled to lie on them to accurately mesh the model. Some generation methods offset (shrink) the model boundary by a small amount $\varepsilon$, generate the nodes based on the offset boundary, and then pull the boundary nodes to the original boundary of the model.

2. **Element type and shape.** It is desirable if various elements (large element library) can be generated to provide users with the required flexibility to meet the compatibility and completeness requirements.

3. **Mesh gradation.** This usually refers to mesh grading and density control. Most often, objects on which FEA is performed may have holes or sharp corners. It is usually required that mesh density (number of nodes and elements) is increased around these regions to capture the rapid change (e.g., stress variation around holes and sharp corners) of the field variable.

    Some generation methods allow users to specify various mesh densities for various regions. Mesh gradation is usually encountered in transition regions. A transition region is one that connects two neighboring regions with either different types of elements or the same type but with different numbers of nodes. Transition elements are usually employed in the transition region to merge (connect) the meshes in the other two regions.

4. **Mesh conversion.** It may be desirable to convert a mesh of a given type of element to another mesh of a different element type. In two-dimensional meshes, for example, it is always possible to convert a triangular element into three quadrilateral elements (a tetrahedron can be subdivided into four hexahedra) or combine two triangular elements to produce a quadrilateral element. A quadrilateral element mesh may be converted into a triangular element mesh by splitting each quadrilateral into two triangles. Mesh conversion must be done with care as poorly formed elements (especially in three dimensions) may result.

5. **Element aspect ratio.** For geometric invariance, as discussed earlier, it is important to keep the aspect ratio of any element close to 1, that is, all sides of an element are equal in length.

6. **Mesh geometry and topology.** As the object to be meshed has geometry and topology, so does its mesh. Mesh geometry refers to the coordinates of nodal

points and the connectivity information of elements. Mesh topology refers to the mesh orientation relative to object topology. Object topology always determines the mesh topology, as shown in Fig. 18-14.

7. Compatibility with representation schemes. A mesh generation method is inherently related to the type of geometric model to be meshed—whether it is a wireframe, surface, or solid model. For example, generation methods based on wireframes or surfaces can never be fully automatic due to their lack of topological information.

8. Cost effectiveness. The time it takes to generate the mesh and the time it takes to perform the FEA are crucial. To reduce both, it is important that the mesh generation method optimizes the mesh and minimizes the number of nodes and elements that comprise the mesh and yet meets the conversion requirements.

## 18.6.2  Semi-Automatic Methods

These methods are sometimes referred to as "interactive mesh generation methods" to emphasize their essential property that they require the analyst interaction with the mesh generator to create the mesh. The semi-automatic mesh generation methods can be divided into two groups: the wireframe- and surface-based group and the solid-modeling-based group. Within each group, methods can be classified further into the node-based approach and the region-based approach. Some of the prominent methods are discussed here.

**18.6.2.1  WIREFRAME- AND SURFACE-BASED METHODS.** Some of these methods represent early attempts to automate manual generation. Others are developed to utilize the parametric representation of curves and surfaces. Typi-

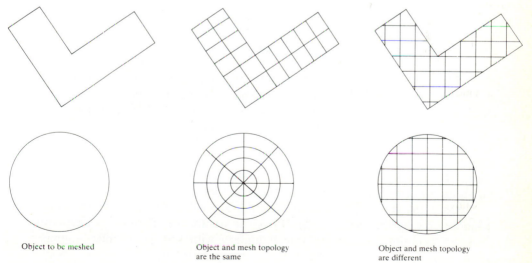

Object to be meshed                    Object and mesh topology                    Object and mesh topology
                                       are the same                                are different

**FIGURE 18-14**
Geometry and topology of finite element meshes.

cally, these methods require the user to aid the generation process and therefore are not fully error-proof. Graphics aids such as colors, layers, and blanking are usually helpful for the user to deal with the enormous amount of information involved.

**Node-based approach.** Methods following this approach consider the nodes to be the primary control factor of a mesh and therefore generate them first. Then these nodes can be connected to form elements. Various types of elements (e.g., triangular or quadrilateral for two-dimensional objects) can be generated. Element generation may be followed by a mesh refinement step, possibly with element rearrangements, to produce a reasonable mesh. The mesh refinement step can be avoided if all the nodes in the final mesh are generated in compliance with mesh density specifications. In such a case, the algorithm has only to consider how to connect the nodes to form the best possible elements.

   **Triangulation methods.** Triangulation methods have received a great deal of attention when considering mesh generation. This is due in part to the historical development of the finite element method itself and in another part to the nature of the triangular element. Historically, the triangular element has always been the easiest to formulate and understand for the FEA. Moreover, triangular elements were available before the isoparametric formulation was discovered, which is centered around quadrilateral elements. From a mesh generation point of view, triangular meshes are easier to generate than quadrilateral meshes. This is true because a triangle is a simplex while a quadrilateral is not.

   There is a wide variety of triangulation methods available to generate meshes. The majority of them are two-dimensional, that is, apply to polygonal domains. Some of these methods cannot be extended to three dimensions and neither can they be implemented within the current geometric modeling theory. In this section we discuss only two triangulation algorithms: one for two-dimensional domains and one for three-dimensional domains.

   The two-dimensional algorithm is modular with two main modules: point generation and mesh construction; that is, nodal points are specified and/or automatically generated in the first module, while the final triangulation is accomplished in the second module.

   The basic strategy of the two-dimensional triangulation algorithm is described as follows. The user must first specify the locations of the nodes on the boundaries of the planar structure to be meshed, that is, discretize the boundaries. The user must order the boundary nodes in a counterclockwise (or clockwise) direction including the nodes on interior boundaries, that is, holes as shown in Fig. 18-15b. The user is also required to input a node density $r$ and a shrinking factor $\varepsilon$. To vary node density, the user must decompose the region into a disjoint ensemble of subregions (Fig. 18-15c).

   Once the user input is complete, the algorithm first shrinks the region boundary $bR$ by the factor $\varepsilon$ (Fig. 18-15d) to ensure that the final triangulation does not contain triangles near the boundary with very acute interior angles. Then, starting with zone 1, the zone is circumscribed by the smallest possible rectangle. A square rectilinear grid is superimposed over the circumscribing rectangle (Fig. 18-15e). Then a random generator is used to randomly generate one

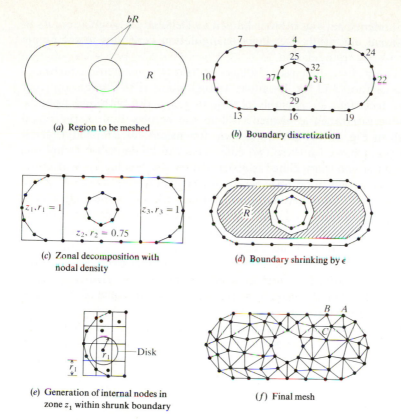

(a) Region to be meshed

(b) Boundary discretization

(c) Zonal decomposition with nodal density

(d) Boundary shrinking by $\epsilon$

(e) Generation of internal nodes in zone $z_1$ within shrunk boundary

(f) Final mesh

**FIGURE 18-15**
A two-dimensional triangulation algorithm.

interior node in each square. A disk of radius $r_i$ centered at each node is used to test that no other surrounding nodes are enclosed in the disk. If so, the node in question is regenerated.

The final set of nodes is then used to generate the triangular elements. This is achieved by first connecting any two nodes $A$ and $B$ on the boundary. Then a node $C$ is found to form a triangle such that $C$ lies to the left of the directed segment $AB$ (left indicates the interior of the boundary as boundary nodes are input in a counterclockwise direction) and such that the triangle $ABC$ is in some sense optimum. This process is continued until all elements are generated (Fig. 18-15f).

A smoothing process is then applied to the elements so that they are more closely equilateral triangles. If node $i$ is connected to $K$ surrounding nodes, then the following equation is used:

$$(x_i, y_i) = \frac{\sum\limits_{j=1}^{K} (x_j, y_j)}{K} \qquad (18.195)$$

Hence, each interior node $(x_i, y_i)$ of the triangulation is replaced by the centroid of the polygon composed of the triangles that surround that node.

The three-dimensional algorithm is known as Delaunay triangulation. As in the two-dimensional algorithm, Delaunay triangulation generates nodal points first and then performs triangulation.

Triangulating a three-dimensional object utilizes two geometric constructs: the Dirichlet tessellation and the Delaunay triangulation. The former construct decomposes an object into an assembly of disjoint polyhedra with nodal points. The latter produces tetrahedral elements filling the convex hull of the nodal points as shown in Fig. 18-16a (shown in two dimensions for simplicity). A triangulation algorithm based on these two constructs can be described as follows. First a set of points to be triangulated is generated on the boundaries and inside the object. This set is then used to generate adjacent polygons (called Voronoi polygons) according to a minimum euclidean distance criterion. The collection of these Voronoi polygons is called the Dirichlet tessellation. The three points within each of three neighboring polygons are used to create a triangle. The set of the resulting triangles is called Delaunay triangulation (Fig. 18-16a). In the three-dimensional case, Voronoi polyhedra are generated using a given set of points and each four points within each four adjacent polyhedra are used to create a tetrahedron. The collection of Delaunay tetrahedra can be thought of as a three-dimensional triangulation of the given set of points.

Delaunay triangulation can be generated directly from the given set of points (nodes) without constructing the Dirichlet tessellation first. In this case, the triangulation is based on the observation that each of three noncollinear points define a triangle and a circle (called the circumcircle of the triangle). In two dimensions, three points form a Delaunay triangle if and only if the circumcircle

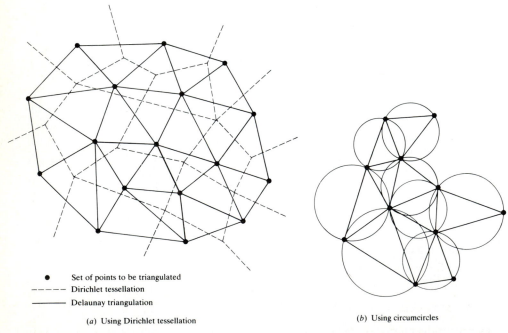

•       Set of points to be triangulated
- - - - -  Dirichlet tessellation
————  Delaunay triangulation

(a) Using Dirichlet tessellation

(b) Using circumcircles

**FIGURE 18-16**
Delaunay triangulation.

defined by these points does not contain any other nodal points in its interior. The set of triangles with empty circumcircles form the Delaunay triangulation (Fig. 18-16b).

**Recursive subdivision.** This method is based on recursively dividing the object to be meshed into subdivisions and generating nodes along their boundaries. The method can be used to generate triangular or quadrilateral elements within any general, planner, $n$-sided simply-connected regions, that is, without holes. The mesh density (number of elements) along the region boundaries can be specified. Multiply-connected regions, that is, with holes, must be reduced to simply-connected ones by connecting the hole boundaries to outer boundaries. Figure 18-17 shows how the method works. The method can be described as follows for two-dimensional regions. First, specify the mesh density required for the various curves that make up the region boundary. Second, connect the holes—if there are any—to the outer boundary of the region. Utilizing a one-dimensional node generator, create nodes on the connected boundary, by marching along it, based on the mesh density, thus resulting in a polygon of nodes. Split the polygon along the "best splitting line" and generate nodes along this new line, based on the mesh density, resulting in two subpolygons. This process is continued until all subpolygons have been reduced to a trivially simple polygon; these are the finite elements.

For three-dimensional regions, the boundary surfaces are meshed first. This results in a polyhedron. Then the "best splitting" plane is determined and nodes

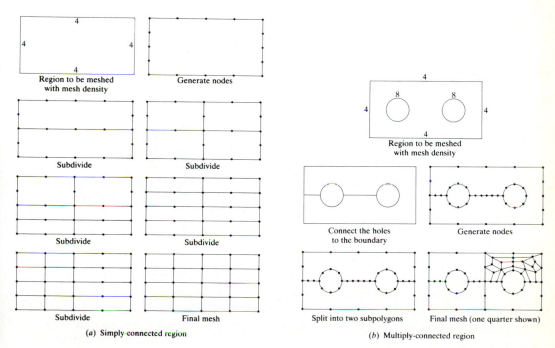

(a) Simply-connected region

(b) Multiply-connected region

**FIGURE 18-17**
Recursive subdivision.

and faces are generated on this plane. The resulting faces are added to both halves of the region, resulting in two subpolyhedra. This process is continued until all subpolyhedra have been reduced to simple polyhedra; these are used to create the solid finite elements by dividing them into tetrahedra.

**Region-based approach.** Most commercial mesh generators rely on this approach. This approach requires subdividing the object to be meshed into patches or regions with specific topologies (mostly four-sided regions). Within each region, the mesh is generated automatically by mapping the region into a regularized domain (normally a square). Various mapping methods exist; two of them are described below. The total mesh is obtained by merging or piecing the individual meshed regions together. The common sides that neighboring regions share must have the same number of nodes. This requirement can be enforced manually or algorithmically while generating the meshes of the neighboring regions.

**Isoparametric mapping.** This method utilizes the same idea of the isoparametric evaluation discussed in Sec. 18.4. It requires that the object (in two or three dimensions) to be meshed is divided into a collection of four-sided (two-dimensional) or boxlike (three-dimensional) regions. To generate the mesh within each region, the analyst inputs, at a minimum, the coordinates of the four (two-dimensional) or eight (three-dimensional) corner points of the region. Depending on the order of the curves describing the boundaries of the region, coordinates of additional points on these curves may be required. For example, in two dimensions, four, eight, and twelve input points per region are required to define linear, quadratic, and cubic mappings respectively. Once these key points are input, nodes within and on the boundary of each region are automatically positioned and referenced to the global cartesian coordinate system of the object.

The isoparametric mapping transforms a given region from the global cartesian coordinate system describing the object into a local parametric coordinate system via shape functions (polynomial interpolation functions). In the parametric space, any four-sided region becomes a square. The isoparametric mapping between the cartesian and the parametric spaces can be defined by Eq. (18.163). This equation can be utilized to generate a mesh in the region by simply incrementing the $u$ and $v$ values for two-dimensional regions (or $u$, $v$, and $w$ for three-dimensional regions) and generating cartesian coordinates of the resulting nodes. Typically, the nodes are located at the intersections of the constant $u$ and $v$ lines, as shown in Fig. 18-18. Finite elements are automatically generated by connecting the nodes along the constant $u$ and $v$ curves. The resulting elements are quadrilateral and can easily be diagonalized to form triangular elements.

**Transfinite mapping.** Transfinite mapping enables objects to be meshed with complex boundary curves and surfaces without introducing geometric errors while modeling these curves and surfaces by the mapping. In isoparametric mapping, boundaries are matched out at only a finite number of points (the input points). In contrast, transfinite mapping can exactly model the boundaries of a given region. In other words, it approximates the boundaries (curves or surfaces) at a nondenumerable (infinite) number of points—thus the name transfinite mapping.

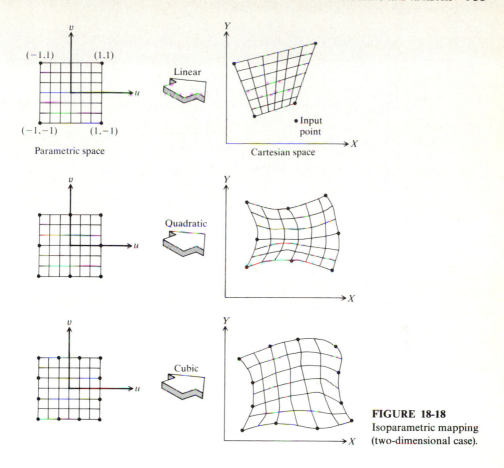

**FIGURE 18-18**
Isoparametric mapping (two-dimensional case).

To generate an FE mesh using transfinite mapping, the analyst divides the objects into mappable regions (bounded by four sides in two dimensions or eight faces in three dimensions), and may specify the required number of nodes on the proper sides (or faces) of these regions. Sometimes, the analyst may not have to specify such numbers of nodes and the mesh generator can interpolate the boundaries. However, when connecting (gluing) regions together, the analyst must ensure the validity of the mesh at the interface boundaries.

Transfinite mapping creates a surface or volume within a closed boundary. In two dimensions, it blends four closed boundary curves to create a planar surface. In three dimensions, it blends six closed boundary surfaces to create a volume. The blending functions can be linear, quadratic, or cubic. For FE mesh generation, linear blending is sufficient. Transfinite mapping is a generalization of the theory underlying the Coons surface described in Chap. 6.

In order to describe transfinite mapping, the concept of a projector is utilized. A linear projector is defined as a function that creates a ruled surface $\mathbf{F}$ between two parametric curves $\mathbf{C}_1$ and $\mathbf{C}_2$ or a ruled volume between two parametric surfaces $\mathbf{S}_1$ and $\mathbf{S}_2$. For example, $\mathbf{P}_1(u, v)$ and $\mathbf{P}_2(u, v)$ given by Eqs. (6.84) and (6.85) respectively are simple linear projectors. Substituting $\mathbf{C}_1$ and $\mathbf{C}_2$,

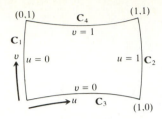

(a) Closed boundary

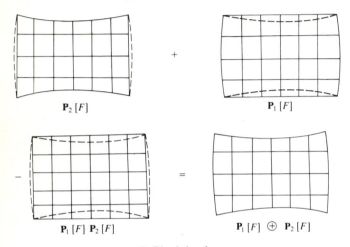

(b) Blended surface

FIGURE 18-19
Linear transfinite mapping
(two-dimensional case).

shown in Fig. 18-19, into Eq. (6.84) and $C_3$ and $C_4$ into Eq. (6.85), the following two linear projectors are obtained:

$$P_1[F] = P_1(u, v) = (1 - u)C_1(v) + uC_2(v) \qquad (18.196)$$

$$P_2[F] = P_2(u, v) = (1 - v)C_3(u) + vC_4(u) \qquad (18.197)$$

The boolean sum of these two projectors as given by Eq. (6.91) defines a bidirectional linear projector $P[F]$, that is,

$$P[F] = P_1[F] \oplus P_2[F] = P_1[F] + P_2[F] - P_1[F]P_2[F] \qquad (18.198)$$

where $P_1[F]P_2[F]$ is given by Eq. (6.89).

The three-dimensional case is a straightforward extension of the above two-dimensional case. Three linear projectors connecting the boundary surfaces $S_1$ and $S_2$, $S_3$ and $S_4$, $S_5$ and $S_6$ can be defined as

$$P_1[F] = P_1(u, v, w) = (1 - v)S_1(u, w) + vS_2(u, w) \qquad (18.199)$$

$$P_2[F] = P_2(u, v, w) = (1 - u)S_3(v, w) + uS_4(v, w) \qquad (18.200)$$

$$P_3[F] = P_3(u, v, w) = (1 - w)S_5(u, v) + wS_6(u, v) \qquad (18.201)$$

The tridirectional linear projector which describes the blending volume $F$ is given as the boolean sum of the linear projectors $P_1[F]$, $P_2[F]$, and $P_3[F]$. It is written as

$$\mathbf{P}[F] = \mathbf{P}(u, v, w) = \mathbf{P}_1[F] \oplus \mathbf{P}_2[F] \oplus \mathbf{P}_3[F]$$
$$= \mathbf{P}_1[F] + \mathbf{P}_2[F] + \mathbf{P}_3[F] - \mathbf{P}_1[F]\mathbf{P}_2[F] - \mathbf{P}_2[F]\mathbf{P}_3[F]$$
$$- \mathbf{P}_1[F]\mathbf{P}_3[F] + \mathbf{P}_1[F]\mathbf{P}_2[F]\mathbf{P}_3[F] \qquad (18.202)$$

To generate a mesh for a two-dimensional (Fig. 18-19) or three-dimensional region, a rectangular grid in the $u$, $v$ or $u$, $v$, $w$ space can be specified and Eq. (18.198) or (18.202) is used to evaluate the nodal coordinates. Generating the element connectivity becomes an automatic process as in isoparametric mapping. The grid does not have to adhere to a uniform distribution in any of the three coordinate directions. The values of $u$, $v$, and $w$ can be biased so that nodes at the grid locations are more closely spaced in some areas than others. Mesh biasing can be accomplished using free-form methods or by transforming the coordinates $u$, $v$, and $w$ from a linear domain between 0 and 1 to a distorted parabolic domain between 0 and 1. The sensitivity of the distortion in the new domain is controlled by a biasing factor supplied by the analyst.

Transfinite mapping can be applied to regions of three sides. Topologically, a three-sided region is equivalent to a four-sided region, one of which is degenerate. At the user interface, the analyst is required to digitize the region boundary in a given order, that is, clockwise or counterclockwise direction. In some mesh generators, this order is not required as the generator sorts the curves of the digitized boundary and reorders them properly.

**18.6.2.2 SOLID-MODELING-BASED METHODS.** Some of the above described meshing methods can be extended to solid models. However, due to their original designs, these methods can only work on the faces and edges of the solid analogous to using surfaces and curves in surface and wireframe models. Therefore, they do not fully utilize the informationally complete characteristic of a solid (specifically its topology). Users would have to extract faces and edges from the solid database before using the meshing method.

Both node-based and region-based approaches are applicable. For example, Delaunay triangulation is an appropriate method to use with solid models. However, pure Delaunay triangulation does not ensure the correct representation of the boundary of the solid because it, in effect, approximates a curved face by a set of planar polygons. Similarly, transfinite mapping can be applied to solids. First, faces of the solid must be extracted, divided, and grouped into valid regions. Second, Eq. (18.202) is used to generate the mesh inside each region. Also, recursive subdivision can be utilized in meshing solids.

Another region-based method which is solely developed for solid models is hyperpatch decomposition. The basis of this method has been covered in Sec. 7.10 as the analytic solid modeling (ASM). To mesh a solid with this method, the solid must be decomposed into hyperpatches (Fig. 18-20a and b) and its database is created as such. Then Eq. (7.91) is utilized to generate the mesh in each hyperpatch (Fig. 18-20c) in the same way as the transfinite mapping creates its mesh in each region. Equation (7.91) maps a $u$, $v$, $w$ grid in a unit cube into nodal coordinates in the global $XYZ$ system of the solid.

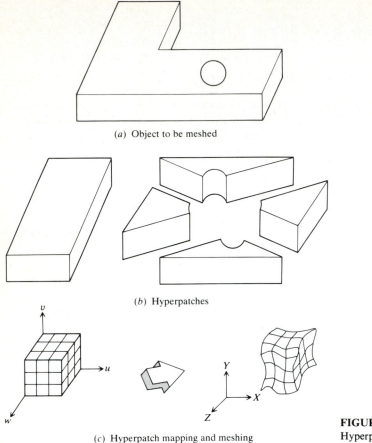

(a) Object to be meshed

(b) Hyperpatches

(c) Hyperpatch mapping and meshing

**FIGURE 18-20**
Hyperpatch decomposition.

### 18.6.3   Fully Automatic Methods

These methods are primarily designed based on the solid modeling theory to achieve full automation, and operate on solid models only. A fully automatic mesh generator can be invoked at the user's level by using a command such as "mesh solid att:d" where att is the mesh attributes and d is a digitize that identifies the solid to be meshed. This implies that mesh automation limits user interaction to defining the solid and specifying mesh density parameters.

**18.6.3.1   NODE-BASED APPROACH.** One of the methods that is based on this approach is the boolean-based method. This method utilizes the CSG representation of a solid. The method utilizes the CSG tree and boolean operations to generate a mesh. It requires two steps for mesh generation. First, it places well-distributed nodal points inside each primitive. It then traverses the CSG tree to combine these nodal points. Based on the particular boolean operator ($\cup$, $\cap$, $-$), the method follows a predefined set of rules to decide how to merge the nodal points in the overlapped zones. The method utilizes the divide-and-conquer algorithm and point membership classification technique in the merging process.

Once the nodal points are merged, they are then connected to create the elements. This method is capable of generating predominantly "good" quadrilateral elements. Triangular elements are generated only when quadrilateral elements are not feasible. The method has been applied to two-dimensional objects only. Figure 18-21 illustrates the method.

**18.6.3.2  ELEMENT-BASED APPROACH.** This approach is the solids' counterpart to the region-based approach. However, methods utilizing the element-based approach discretize the solids' domain into a set of valid elements directly without having to subdivide the domain into regions first.

**Modified quadtree and octree method.** Spatial decomposition schemes (e.g., spatial enumeration, cell decomposition, etc.) discussed in Chap. 7 are similar in philosophy to mesh generation methods. Both decompose a given solid into cells. If these cells are processed properly, they can represent finite elements. Quadtree and octree schemes have been modified to adapt to mesh generation. We first review the two schemes.

In the quadtree encoding, the object of interest is placed inside a square with an integer coordinate system attached to it at one of its corners, as shown in Fig. 18-22a. This square is called the universe square. The length of the side of this square in the integer space is $2^n$ where $n$ is the tree (subdivision) level. To

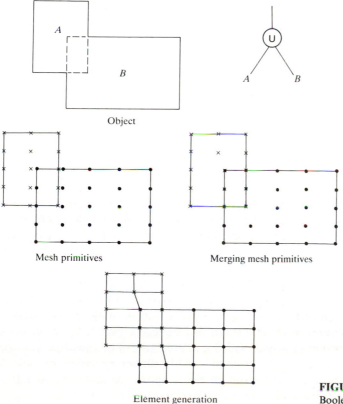

Object

Mesh primitives

Merging mesh primitives

Element generation

**FIGURE 18-21**
Boolean-based mesh generation.

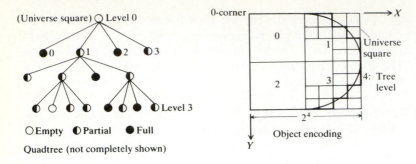

(a) Quadtree encoding

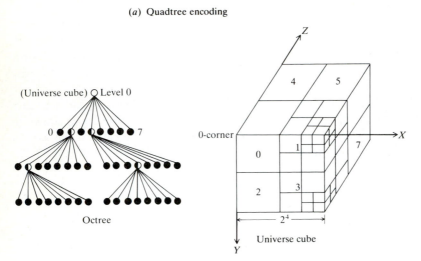

(b) Octree encoding

**FIGURE 18-22**
Quadtree and octree encoding schemes.

decompose the object, the universe square is divided into four squares or quadrants. Each quadrant is then checked against the object (via calculating and classifying the intersections between the square and the object) to see if it is inside the object (full), outside the object (empty), or if part of it is inside and part is outside (partial). Homogeneous quadrants (full or empty) require no further subdivisions while each partial quadrant (parent) is further divided into four quadrants (siblings). These new four quadrants are tested as above, and this process of subdivision continues until all the quadrants have the status of "full" or, practically, until a specified resolution level (tree level) is reached. Figure 18-22a shows a quadtree representation of an object with its related tree. The quadtree storage array (integer tree) shown is based on the quadrant numbering system (0, 1, 2, 3) shown for any four quadrant siblings and on choosing the numbers 0, 1, and 2 to represent "empty," "partial," and "full" respectively.

In the octree encoding, the object of interest is placed inside a universe cube with its attached integer coordinate system as shown in Fig. 18.22b. Here, the cube is divided into eight subdivisions or octants. Each octant is classified against the object, and each "partial" octant (parent) is subdivided into eight suboctants (siblings) which, in turn, are classified. Once a given resolution is achieved, the octree encoding of the given object is complete and stored in an octree and its corresponding array. The octant numbering system (0, 1, 2, 3, 4, 5, 6, 7) for any eight octant siblings is shown in Fig. 18-22b.

The mapping between the integer space and the cartesian space is established if the solid modeler utilizing the encoding provides the size (length of the side) of the universe unit (square or cube) and the coordinates of either its center or one of its corners.

The above encoding schemes can be modified to meet the mesh requirements; namely, element size must not be drastically different and the number of elements must be small enough to keep the FEA cost down. The modification includes the introduction of "cut" quadrants and octants to the encoding schemes. A cut quadrant or octant is treated as a terminal node in the encoding tree (similar to "full" or "empty" status) and is represented in the tree storage array by the number 3 (in addition to 0, 1, and 2, introduced previously). This modification enables the "close" representation of the boundaries of the object within the tree level specified by the analyst. Figure 18-23 shows the modified quadtree representation of the object shown in Fig. 18-22a.

As an example of mesh generation, a quadtree-based algorithm works as follows (see Fig. 18-24). The boundary representation of the object to be meshed is provided by the solid modeler. The algorithm requires an ordered list of curves to define a region. The analyst provides such information during material assignment to the various regions. The analyst can also specify mesh control parameters (mesh tolerance values or tree level) along the object boundaries. Control parameters can also be specified for a given location at the object interior. The

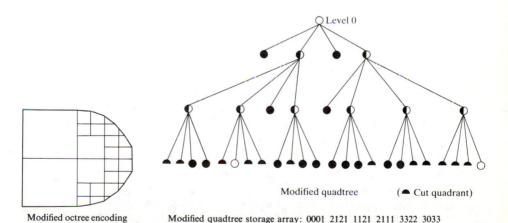

Level 0

Modified quadtree  ( ▲ Cut quadrant)

Modified octree encoding  Modified quadtree storage array: 0001 2121 1121 2111 3322 3033
2322 2223 2233 3330

**FIGURE 18-23**
Modified quadtree of object shown in Fig. 18-22a.

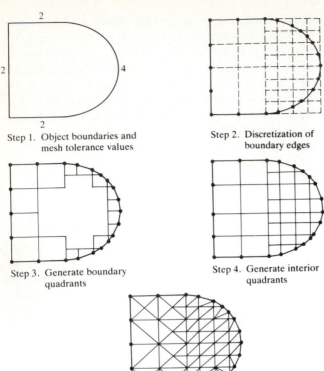

Step 1.  Object boundaries and
mesh tolerance values

Step 2.  Discretization of
boundary edges

Step 3.  Generate boundary
quadrants

Step 4.  Generate interior
quadrants

Step 5.  Generate elements

**FIGURE 18-24**
Mesh generation by modified
quadtree scheme.

algorithm then utilizes the finest control parameter (4 in this example) to create the space in the data structure of the tree. No geometric information is stored at this step. In the second step, the boundary edges are discretized using the mesh control parameters. After the discretization process is complete, the boundary quadrants are generated and stored with their proper information in the quadtree data structure. The next step is to create the quadrants in the interior domain. This is achieved by searching to find out which sides of the boundary quadrants are within the domain and then utilizing a rightward search to create the interior quadrants. If part of the right side of the quadrant is within the interior of the object, its adjacent right neighbors are found by traversing the tree. If these neighbors exist, the rightward search stops and another quadrant is examined. This search continues and the interior quadrants are completely generated when all of the boundary quadrants have been used as starters in the rightward search. The final step is then to generate the mesh by creating either valid triangular or quadrilateral elements.

**Tetrahedral decomposition.** This method is sometimes referred to as a topological-based method. We here describe an algorithm for dividing an arbitrary polyhedron, with or without holes, into tetrahedral elements. The algorithm first decomposes a polyhedral object into tetrahedral elements without adding new vertices to the geometric model. Then it refines the resulting elements by

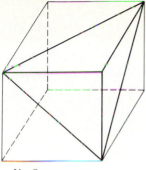

Use first operator

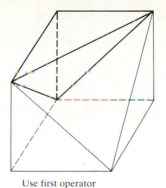

Use first operator

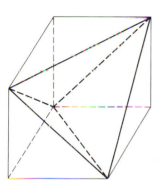

Use first operator

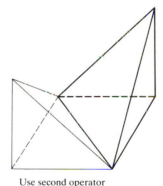

Use second operator

**FIGURE 18-25**
Tetrahedral decomposition
of a block.

further subdividing them. Two mesh operators have been developed to achieve
the first step. The first operator is used to "slice" a corner, from the solid, that
must be convex and trivalent, that is, has three edges. The second operator is
used to "dig out" a tetrahedron from a convex edge if the remaining polyhedron
(after applying the first operator) does not have a convex trivalent vertex.

The algorithm first finds a vertex that has three edges. It then determines if
the vertex is convex. If so, a tetrahedron is constructed from the vertex by
making a slice in the polyhedron, that is, cutting it with one plane using the first
operator. Once all the trivalent convex vertices are consumed, the second operator is used to carve a tetrahedron out of the polyhedron by cutting it with two
planes. The second operator is used repeatedly until the decomposition process is
complete. For polyhedra with holes (genus), the polyhedra must be opened (cut)
through the hole. This is achieved by changing it into a nonmanifold object (by
creating a singular edge and then a singular vertex) which can be changed to a
simple polyhedron, that is, without the hole. After decomposing the polyhedron,
the algorithm further subdivides the resulting tetrahedra into finer ones.

The above two mesh operators change the topology of the polyhedron
under decomposition. Thus, its related data structure must be updated after
applying any of the two operators. Therefore, the two operators are based on
Euler's law. Thus the change (by either operator) in vertices, $V$, edges, $E$, faces, $F$,

of a simple polyhedron satisfies the equation

$$\Delta V - \Delta E + \Delta F = 0 \qquad (18.203)$$

and for a polyhedron with a hole, the equation is

$$\Delta V - \Delta E + \Delta F = 2 \, \Delta G \qquad (18.204)$$

where $G$ is the number of holes (genus) in the polyhedron. Figure 18-25 illustrates this algorithm.

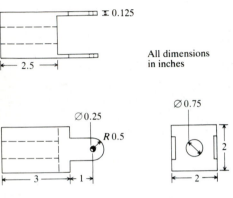

☐ 0.125

All dimensions
in inches

2.5

∅ 0.75

∅ 0.25

R 0.5

2

3

1

2

(a) Geometric model

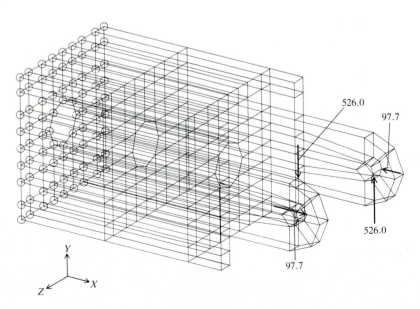

526.0

97.7

526.0

97.7

Y

Z

X

(b) Finite element mesh with loads

**FIGURE 18-26**

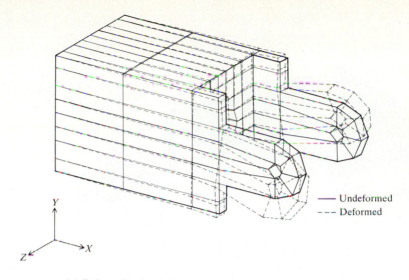

Y

Z ← →X

—— Undeformed
--- Deformed

(*c*) Deformed and undeformed shapes superimposed

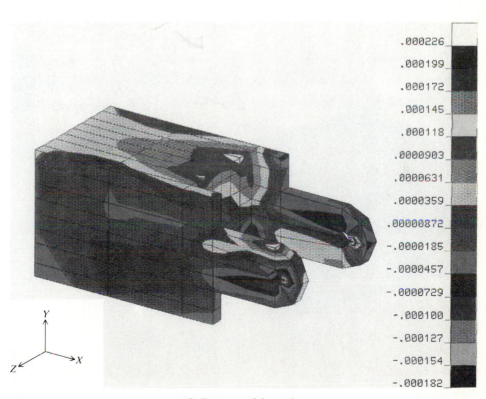

.000226
.000199
.000172
.000145
.000118
.0000903
.0000631
.0000359
.00000872
-.0000185
-.0000457
-.0000729
-.000100
-.000127
-.000154
-.000182

Y

Z ← →X

(*d*) Contours of the strain $\varepsilon_{xz}$

**FIGURE 18-26**

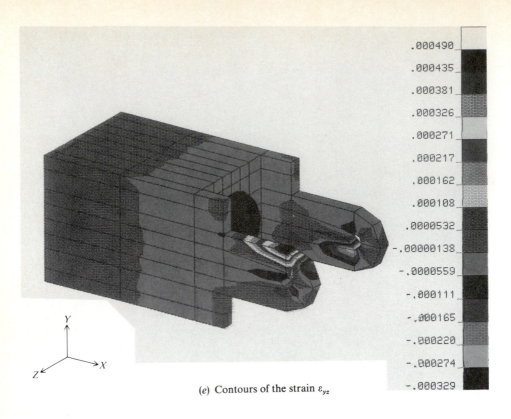

.000490
.000435
.000381
.000326
.000271
.000217
.000162
.000108
.0000532
-.00000138
-.0000559
-.000111
-.000165
-.000220
-.000274
-.000329

$Y$

$Z$  $X$

(e) Contours of the strain $\varepsilon_{yz}$

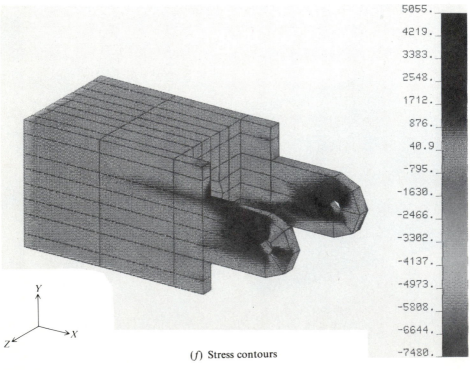

5055.
4219.
3383.
2548.
1712.
876.
40.9
-795.
-1630.
-2466.
-3302.
-4137.
-4973.
-5808.
-6644.
-7480.

$Y$

$Z$  $X$

(f) Stress contours

**FIGURE 18-26**
Driving yoke of a universal joint.

**946**

## 18.7 DESIGN AND ENGINEERING APPLICATIONS

Finite element modeling and analysis provide a valuable tool in engineering problems and projects. To utilize the finite element method in the design environment, designers must have access to good pre- and postprocessors that are also interfaced with FEA codes. In this section, we present an example that illustrates the use of the finite element method in engineering problems. More examples are discussed in Chap. 19.

> **Example 18.10.** Figure 18-26 shows the simplified geometry of the driving yoke of a universal joint. The force analysis shows that the input torque generates the forces shown in the figure. Find the maximum stresses in the yoke.
>
> *Solution.* The geometric model shown in Fig. 18-26a is created as a solid model using the PATRAN-G modeler. The model is created as a sum of hyperpatches shown in Fig. 18-26b. The figure shows the pair of forces of 526 lb, which creates a couple in the vertical plane and the pair of forces of 97.7 lb, which creates a couple in the horizontal plane.
>
> Utilizing PATRAN-G, each hyperpatch is automatically discretized into eight-node solid elements. Such an element is sufficient to produce the correct FEA results because it satisfies the compatibility and completeness requirements of this three-dimensional solid mechanics problem as discussed in the chapter. The nodal degrees of freedom are the $u$, $v$, $w$ displacements in the $X$, $Y$, and $Z$ directions respectively. The geometric boundary conditions applied are zero displacements of all the nodes of the face on the left end of the yoke as shown with the circular symbols in Fig. 18-26b.
>
> Utilizing steel for the yoke material, the FEA data file is generated and the FEA is performed. An output file is generated and processed by PATRAN. Figure 18-26c shows the deformed yoke superimposed on the undeformed one. As shown, the deformation shape (maximum at the free ends) agrees with the intuition, given the force directions shown in Fig. 18-26b. In addition, the strain contours for $\varepsilon_{xz}$ and $\varepsilon_{yz}$ are shown in Fig. 18-26d and e respectively. The resulting principal stress contours are shown in Fig. 18-26f.

## PROBLEMS

### Part 1: Theory

**18.1.** Derive the functional $\Pi$ for the following one-dimensional systems:

(a) A bar under a distributed axial load $P = P(x)$ with the following differential equation:

$$AEu'' + P(x) = 0$$

and the boundary conditions shown in Fig. P18-1a.

(b) A cantilever beam under a distributed load of density $q(x)$ per unit length. The differential equation is

$$EIy'''' - q(x) = 0$$

with the boundary conditions shown in Fig. P18-1b.

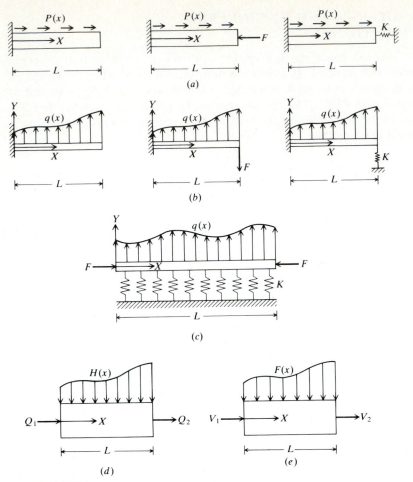

**FIGURE P18-1**

(c) A beam on elastic foundation of stiffness density $K(x)$ per unit length and subjected to a distributed load of density $q(x)$ per unit length. The differential equation is

$$EIy'''' + K(x)y - q(x) = 0$$

with the boundary conditions shown in Fig. P18-1c.

(d) A region of a cross-sectional area $A$ with a steady-state uniaxial heat flow $Q$ and a source heat flux $H(x)$. The differential equation is

$$AKT'' + H(x) = 0$$

with the boundary conditions shown in Fig. P18-1d.

(e) A steady-state flow through a region with a cross-sectional area $A$, a fluid flow $V$ across the ends of the region, and a fluid flux $F(x)$ per unit length distributed along the length of the region as shown in Fig. P18-1e. The differential equation is

$$AgP'' + F(x) = 0$$

**18.2.** Derive the functional $\Pi$ for the following two dimensional problems:

(a) Plane stress problem over a rectangular domain as shown in Fig. P18-2a, where $q_1$ and $q_2$ are the distributed load densities.

(b) A steady-state heat conduction over a rectangular domain as shown in Fig. P18-2b, where $Q_1$ and $Q_2$ are heat flows, $H(x)$ is a heat flux, and $Q$ is the internal heat generation (heat source). The differential equation of this system is given by

$$K \frac{\partial^2 T}{\partial x^2} + K \frac{\partial^2 T}{\partial y^2} + Q = 0$$

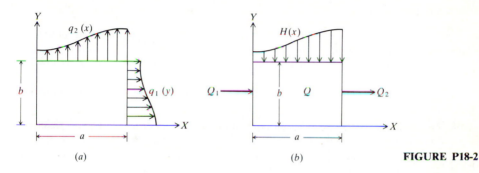

(a)  (b)  **FIGURE P18-2**

**18.3.** Starting with Eq. (18.32) in Example 18.2, rederive Eq. (18.63) using integration by parts.

**18.4.** Use the Galerkin method to derive the integral equation of the systems described in Example 18.1(a) and (c), as well as Example 18.2(b).

**18.5.** Equation (18.49) in Example 18.2 applies to three-dimensional solid mechanics problems. Knowing that the displacement components are $u$, $v$, and $w$ in the $X$, $Y$, and $Z$ directions respectively, design three-dimensional linear, quadratic, and cubic elements. Write the interpolation polynomial for each element.

**18.6.** The three-dimensional counterpart of Eq. (18.32) in Example 18.2 is given as

$$K \frac{\partial^2 T}{\partial x^2} + K \frac{\partial^2 T}{\partial y^2} + K \frac{\partial^2 T}{\partial z^2} = 0$$

Derive the corresponding functional $\Pi$. [*Hint:* modify Eq. (18.63).] Design finite elements for the problem and write the interpolation polynomial for each element.

**18.7.** The differential equilibrium equation for bending of an isotropic thin plate of constant thickness $t$ and subject to a distributed load of intensity $q$ per unit area is given by:

$$\frac{\partial^4 w}{\partial x^4} + 2 \frac{\partial^4 w}{\partial x^2 \, \partial y^2} + \frac{\partial^4 w}{\partial y^4} + \frac{12(1 - v^2)}{Et^3} q = 0$$

where $w$ is the plate lateral deflection in the $Z$ direction and the plate lies in the $XY$ plane (Fig. P18-7). Derive the following corresponding functional $\Pi$:

$$\Pi = \frac{1}{2} \iint_A \frac{Et^3}{12(1 - v^2)} \left\{ \left( \frac{\partial^2 w}{\partial x^2} + \frac{\partial^2 w}{\partial y^2} \right)^2 - 2(1 - v) \left[ \frac{\partial^2 w}{\partial x^2} \frac{\partial^2 w}{\partial y^2} - \left( \frac{\partial^2 w}{\partial x \, \partial y} \right)^2 \right] + qw \right\} dx \, dy$$

**18.8.** The eight-node rectangular element shown in Fig. P18-8 has been designed for a given problem. Write the corresponding interpolation polynomial for the field vari-

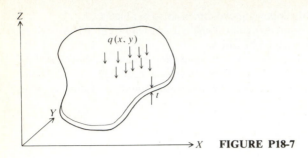

FIGURE P18-7

able $\phi$ assuming the element is a $C^0$ element [i.e., one degree of freedom ($\phi$) per node]. Investigating the order and the terms of the polynomial, how does $\phi$ change in the problem domain? Does the polynomial enjoy geometric invariance? Explain your answer.

    *Note:* the node numbering follows the isoparametric element numbering scheme.

**18.9.** Derive the explicit form of $N_i$ (in terms of nodal coordinates) used in Eq. (18.86) for the linear triangular element and Eq. (18.94) for the linear quadrilateral element.

**18.10.** Following Example 18.7, develop the element equations and matrices for three-node and four-node elements for the systems of Example 18.1.

**18.11.** The functional $\Pi$ for a beam loaded with a distributed load of intensity $p(x)$ and a distributed moment of intensity $m(x)$, and including the shear deformation, is given by

$$\Pi = \tfrac{1}{2}\int_0^L EI\theta'^2 \, dx + \tfrac{1}{2}\int_0^L GAS(y' - \theta')^2 \, dx - \int_0^L Py \, dx - \int_0^L m\theta \, dx$$

Derive the element equations and matrices for two-node, three-node, and four-node elements. $G$, $A$, and $S$ are the shear modulus, the cross-sectional area, and the shear correction factor respectively.

**18.12.** Write the polynomials and all relevant equations for quadratic and cubic triangular elements. Follow Example 18.8. Repeat for linear, quadratic, and cubic quadrilateral elements.

**18.13.** Derive the element equations and matrices for differential equations of Probs. 18.5, 18.6, and 18.7 using the variational approach. Use three-dimensional linear, quadratic, and cubic tetrahedral and hexahedral elements.

**18.14.** Utilizing the integral equations developed in Prob. 18.4, derive the corresponding element equations and matrices. Compare the results with those of Example 18.7.

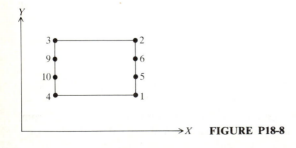

FIGURE P18-8

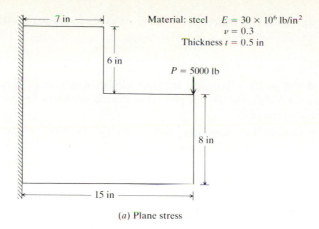

Material: steel   $E = 30 \times 10^6$ lb/in$^2$
$\nu = 0.3$
Thickness $t = 0.5$ in

7 in

6 in

$P = 5000$ lb

8 in

15 in

(a) Plane stress

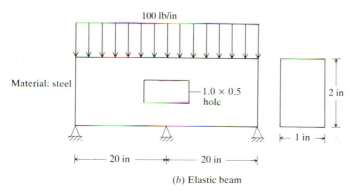

100 lb/in

Material: steel

1.0 × 0.5
hole

2 in

1 in

20 in          20 in

(b) Elastic beam

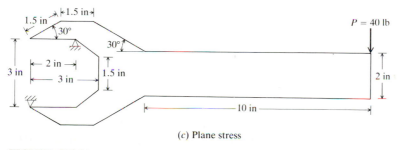

1.5 in   1.5 in
30°
30°
$P = 40$ lb
2 in
3 in   2 in
1.5 in
3 in
10 in

(c) Plane stress

**FIGURE P18-21**

**18.15.** Assuming each continuum in Example 18.1 is modelled by three two-node elements, write the element equations for each one and assemble them into a global system of equations.

**18.16.** Assuming each continuum in Example 18.2 is modeled by two four-node rectangular elements, write the element equations for each one and assemble them into a global system of equations.

**18.17.** Apply the geometric boundary conditions to each of the global systems of equations that results in Probs. 18.15 and 18.16. Write the final systems of equations.

**18.18.** Apply the external loads to the global systems of equations that resulted from Prob. 18.17.

**18.19.** Develop the jacobian [J] for three-dimensional elements for isoparametric evaluation.

## Part 2: Laboratory

**18.20.** Utilizing the FEM package provided by your CAD/CAM system, generate the meshes and the FEA data files for the geometric models shown in the problems at the end of Chaps. 5, 6, and 7.

**18.21.** Perform finite element modeling and analysis on the problems shown in Fig. P18-21.

## Part 3: Programming

**18.22.** Write a program that generates a finite element mesh for a four-sided two-dimensional region using the transfinite mapping method.

**18.23.** This is the same as Prob. 18.22 but for a three-dimensional region with six surfaces.

## BIBLIOGRAPHY

Ang, K. K., and S. Valliappan: "Mesh Grading Technique Using Modified Isoparametric Shape Functions and Its Applications to Wave Propagation Problems," *Int. J. Num. Meth. Engng,* vol. 23, no. 2, pp. 331–348, 1986.

Baker, A. J.: *Finite Element Computational Fluid Mechanics,* McGraw-Hill, New York, 1983.

Baker, T. J.: "Mesh Generation by a Sequence of Transformations," *Appl. Num. Math.,* vol. 2, no. 2, pp. 515–528, 1986.

Bathe, K. J.: *Finite Element Procedures in Engineering Analysis,* Prentice-Hall, Englewood Cliffs, N.J., 1982.

Brown, R. P., and D. S. Hayhurst: "Using the Schwarz-Christoffel Transformation in Mesh Generation for the Solution of Two-Dimensional Problems," *Computer in Mech. Engng (CIME) Mag.,* pp. 73–79, 1982.

Bryant, C. F.: "Two Dimensional Automatic Triangular Mesh Generation," *IEEE Trans. Magnetics,* vol. MAG-21, no. 6, pp. 2547–2550, 1985.

Buell, W. R., and B. A. Bush: "Mesh Generation—A Survey," *J. Engng for Industry, Trans. ASME,* vol. 95, no. 1, pp. 332–338, 1973.

Bykat, A.: "Automatic Generation of Triangular Grid: I—Subdivision of a General Polygon into Convex Subregions. II—Triangulation of Convex Polygons," *Int. J. Num. Meth. Engng,* vol. 10, pp. 1329–1342, 1976.

Cavendish, J. C.: "Automatic Triangulation of Arbitrary Planar Domains for the Finite Element Method," *Int. J. Num. Meth. Engng,* vol. 8, pp. 679–696, 1974.

Cavendish, J. C., D. A. Field, and W. H. Frey: "An Approach to Automatic Three-Dimensional Finite Element Mesh Generation," *Int. J. Num. Meth. Engng,* vol. 21, no. 2, pp. 329–347, 1985.

Cendes, Z. J., and D. N. Shenton: "Adaptive Mesh Refinement Computation of Magnetic Fields," *IEEE Trans. Magnetics,* vol. MAG-21, no. 5, pp. 1811–1816, 1985.

Cheng, J.: "Automatic Adaptive Remeshing for Finite Element Simulation of Forming Processes," *Int. J. Num. Meth. Engng,* vol. MAG-26, no. 1, pp. 1–18, 1988.

Cook, R. D.: *Concepts and Applications of Finite Element Analysis,* 2d ed., John Wiley, New York, 1981.

Cook, W. A.: "Body Oriented (Natural) Coordinates for Generating Three-Dimensional Meshes," *Int. J. Num. Meth. Engng,* vol. 8, pp. 28–43, 1974.

Cook, W. A., and W. R. Oakes: "Mapping Methods for Generating Three-Dimensional Meshes," *Computers in Mech. Engng (CIME) Mag.,* pp. 67–72, August 1982.

Frey, W. H.: "Selective Refinement: A New Strategy for Automatic Nodal Placement in Graded Triangular Meshes," *Int. J. Num. Meth. Engng,* vol. 24, pp. 2183–2200, 1987.

Fukuda, J., and J. Suhara: "Automatic Mesh Generation for Finite Element Analysis," in *Advances in Computational Methods in Structural Mechanics and Design* (Eds. J. T. Oden, R. W. Clough, and Y. Yamamoto), UAH Press, Huntsville, Ala., 1972.

Ghassemi, F.: "Automatic Mesh Generation Scheme for a Two- or Three-Dimensional Triangular Curved Surface," *Computers and Structs*, vol. 15, no. 6, pp. 613–626, 1982.

Gold, V., and C. Wei: "An Algorithm for Discretizing General Parametric Surfaces for Finite Element Mesh Generation," *Proceedings of the 1987 ASME International Computers in Engineering Conference and Exhibit*, New York City, August 9–13, 1987, vol. 3, pp. 195–198, 1987.

Gordon, W. J., and C. A. Hall: "Construction of Curvilinear Coordinate Systems and Applications to Mesh Generation," *Int. J. Num. Meth. Engng*, vol. 7, no. 4, pp. 461–477, 1973.

Grandin, Jr., H.: *Fundamentals of the Finite Element Method*, Macmillan, New York, 1986.

Haber, R., and J. Abel: "Discrete Transfinite Mappings for the Description and Meshing of Three-Dimensional Surfaces Using Interactive Computer Graphics," *Int. J. Num. Meth. Engng*, vol. 18, no. 1, pp. 41–66, 1982.

Haber, R., M. S. Shephard, J. F. Abel, R. H. Gallagher, and D. P. Greenberg: "A General Two-Dimensional Graphical Finite Element Preprocessor Utilizing Discrete Transfinite Mappings," *Int. J. Num. Meth. Engng*, vol. 17, no. 7, pp. 1015–1044, 1981.

Heighway, E. A., and C. S. Biddlecombe: "Two Dimensional Automatic Triangular Mesh Generation for the Finite Element Electromagnetics Package PE2D," *IEEE Trans. Magnetics*, vol. MAG-18, no. 2, pp. 594–598, 1982.

He-Le, K.: "Finite Element Mesh Generation Methods: A Review and Classification," *Computer Aided Des.*, vol. 20, no. 1, pp. 27–38, 1988.

Huebner, K. H., and E. A. Thornton: *The Finite Element Method for Engineers*, 2d ed., John Wiley, New York, 1982.

Imafuku, I., Y. Kodera, M. Sayawaki, and M. Kono: "A Generalized Automatic Mesh Generation Scheme for Finite Element Method," *Int. J. Num. Meth. Engng*, vol. 15, pp. 713–731, 1980.

Kela, A., R. Perucchio, and H. B. Voelcker: "Toward Automatic Finite Element Analysis," *ASME Computers in Mech. Engng (CIME)*, vol. 5, no. 1, pp. 57–71, 1986.

Lee, Y. T., A. De Pennington, and N. K. Shaw: "Automatic Finite-Element Mesh Generation from Geometric Models—A Point-Based Approach," *ACM Trans. Graphics*, vol. 3, no. 4, pp. 287–311, 1984.

Lo, S. H.,: "A New Generation Scheme for Arbitrary Planar Domains," *Int. J. Num. Meth. Engng*, vol. 21, no. 8, pp. 1403–1426, 1985.

Logan, D. L.: *A First Course in the Finite Element Method*, PWS Publishers, Boston, Mass., 1986.

Moscardini, A. O., M. Cross, and B. A. Lewis: "Assessment of Three Automatic Triangular Mesh Generators for Planar Regions," *Adv. Engng Software*, vol. 3, no. 3, pp. 108–114, 1981.

Moscardini, A. O., B. A. Lewis, and A. M. Cross: "AGTHOM_ Automatic Generation of Triangular and Higher Order Meshes," *Int. J. Num. Meth. Engng*, vol. 19, pp. 1331–1353, 1983.

Ngyen-Van-Phai: "Automatic Mesh Generation with Tetrahedron Elements," *Int. J. Num. Meth. Engng*, vol. 18, pp. 273–289, 1982.

Rivara, M.: "Design and Data Structure of Fully Adaptive, Multigrid, Finite-Element Software," *ACM Trans. Math. Software*, vol. 10, no. 3, pp. 242–264, 1984.

Sadek, E.: "A Scheme for the Automatic Generation of Triangular Finite Elements," *Int. J. Num. Meth. Engng.*, vol. 15, pp. 1813–1822, 1980.

Sapidis, N., and R. Perucchio: "Advanced Techniques for Automatic Finite Element Meshing from Solid Models," *Computer Aided Des.*, vol. 21, no. 4, pp. 248–253, 1989.

Schoofs, A. J. G., L. H. Th. Van Beukering, and M. L. C. Sluiter: "A General Purpose Two-Dimensional Mesh Generator," *Adv. Engng Software*, vol. 1, no. 3, pp. 131–136, 1979.

Shaw, R. D., and R. G. Pitchen: "Modifications to the Suhara-Fukuda Method of Network Generation," *Int. J. Num. Meth. Engng*, vol. 123, pp. 93–99, 1978.

Shenton, D. N., and Z. J. Cendes: "Three-Dimensional Finite Element Mesh Generation Using Delaunay Tesselation," *IEEE Trans. Magnetics*, vol. MAG-21, no. 6, pp. 2523–2538, 1985.

Sluiter, M. L. C., and D. L. Hansen: "A General Purpose Automatic Mesh Generator for Shell and Solid Finite Elements," in *Computers in Engineering* (Ed. L. E. Hulbert), vol. 3, book G00217, pp. 29–34, ASME, 1982.

Stefanou, G. D., and K. Syrmakezis: "Automatic Triangular Mesh Generation in Flat Plates for Finite Elements," *Computers and Structs*, vol. 11, pp. 439–464, 1980.

Taniguchi, T.: "Flexible Mesh Generator for Triangular and Quadrilateral Areas," *Adv. Engng Software*, vol. 9, no. 3, pp. 142–149, 1987.

Watson, D. F.: "Computing the $n$-Dimensional Delaunay Tessellation with Application to Voronoi Polytopes," *Computer J.*, vol. 24, no. 2, 1981.

Woo, T. C., and T. Thomasma: "An Algorithm for Generating Solid Elements in Objects with Holes," *Computers and Structs*, vol. 18, no. 2, pp. 333–342, 1984.

Wordenweber, B.: "Finite Element Mesh Generation," *Computer Aided Des.*, vol. 16, no. 5, pp. 285–291, 1984.

Yerry, M. A., and M. S. Shephard: "A Modified Quadtree Approach to Finite Element Mesh Generation," *IEEE Computer Graphics and Applic.*, vol. 3, no. 1, pp. 39–46, 1983.

Yerry, M. A., and M. S. Shephard: "Automatic Three-Dimensional Mesh Generation by the Modified-Octree Technique," *Int. J. Num. Meth. Engng*, vol. 20, no. 11, pp. 1965–1990, 1984.

Zienkiewicz, O. C.: *The Finite Element Method*, 3d ed., McGraw-Hill, New York, 1977.

Zienkiewicz, O. C., and D. V. Phillips: "An Automatic Mesh Generation Scheme for Plane and Curved Surfaces by Isoparametric Co-ordinates," *Int. J. Num. Meth. Engng*, vol. 3, pp. 519–528, 1971.

# CHAPTER 19

# DESIGN PROJECTS WITH CAE FOCUS

## 19.1 INTRODUCTION

Thus far, we have presented and illustrated the basic CAD/CAM concepts and demonstrated their use in various examples and applications. In the last three chapters, in particular, we have discussed three of the most popular design applications offered by most existing commercial CAD/CAM systems; namely, mechanical tolerancing, mass property calculations, and finite element modeling and analysis. The utilization of these concepts and systems is customary in both industrial and academic environments. Although each environment has its own specific needs, two common needs can be identified. First, a specific application may arise which may not be supported by a given system and, therefore, requires customization of the system to support such an application. In such a case, interactive programming, discussed in Chap. 15, provides the best way to customize a system.

The second need typically arises when CAD/CAM systems are utilized to tackle design projects in search for optimal solutions and answers to design questions. Readers with experience may have already concluded that not every design project can be best solved using CAD/CAM systems. In some projects, the justification of using a system, even if it is available, may not be practical. This leads to the question of what type of design projects are more suitable than others?

In this chapter, we address this question by discussing the requirements and formulation of design projects with CAE (computer aided engineering) focus. The discussion is particularly useful when assigning projects to students as a requirement of teaching design courses. Sample examples are presented to illustrate some of the chapter ideas.


## 19.2  PROJECT REQUIREMENTS

The development of successful design projects with CAE focus requires these projects to meet two distinct sets of conditions. The first set stems from the design contents of the projects. This set is common among CAE and non-CAE design projects. Some of the requirements that belong to this set are:

1. Open-ended nature. The design content versus the science content of a project is determined by how loosely the problem is defined. Typically, if a problem is well defined and has one solution, it does not qualify as a design project. By "loosely," we do not mean that the problem is vague, but rather we mean a problem that has more than one alternative to solve it. For example, instead of specifying certain values for certain variables, one would specify ranges of values instead which would force the designer into trying various options and alternatives.

2. Design goal and constraints. Due to the open-ended nature, a problem statement is usually accompanied by a design goal. To limit the infinite number of solution options and to make the project more challenging and realistic (resources are always limited in real life), design constraints are usually specified. These constraints can vary widely depending on the nature of the problem.

3. Interdisciplinary nature. It is nonexistent that the design and manufacturing of a worthwhile product do not require multidisciplinary background and expertise. Various design groups are usually involved in product design in industry.

4. Span of full design cycle. We have discussed the design process and product cycle in Chap. 1. A useful design project is one that forces students to go through all or most of the steps of a product cycle from conception to production.

5. Limiting available resources. During carrying a project idea and designing the product, it is useful to limit available resources during the design process. Examples include time limits, amount of consultation a design team may have with experts, financial limits, limit on computer time and accessibility, etc.

6. Defining project deliverables. Some projects may require submitting a final report while more interesting ones may require, in addition, a prototype that students must build.

The second set of requirements stem from the CAE focus of the projects. Unless a project requires an effective use of the various capabilities of the CAD/CAM technology, it becomes superficial. It is usually nonconvincing and confusing to attempt to justify using CAD/CAM systems to carry the ideas of such projects. Some of the requirements of a successful project with CAE focus are:

1. Real need for CAD/CAM capabilities. Some projects may only require geometric modeling and computer graphics while others may require using application software such as mass property calculations and finite element modeling and analysis. A third class of projects may require interactive programming.

2. Project modulation. It is important to structure the project in modules, each requiring a different capability of the CAD/CAM system in use. This is useful when students have to complete design projects while learning new CAD/CAM concepts and topics.

3. Terminal time allocation. It is always the case that the ratio of the number of graphics terminals to students is small, and access to CAD/CAM systems is limited. This fact must seriously be considered when defining the project scope. In addition to limiting the scope, students may be encouraged to work in groups.

4. Capabilities of existing software. Successful implementation of project ideas is only possible if accessible software to designers provides them with the capabilities they need to complete the projects. A project scope is usually a function of what accessible software can offer.

5. Integration of CAD/CAM functions. To illustrate the real gain in productivity, it is beneficial if projects require the majority of these functions. For example, a project may require geometric modeling to create the design model, mass properties to calculate inertial properties, drafting to document the design, FEM/FEA to analyze the model, process planning to plan its manufacturing, and NC to produce an NC program to manufacture the design.

## 19.3  PROJECT FORMULATION

Subject to the above requirements, projects of various foci can be formulated. A project focus depends primarily on the goal of the course within which the project is assigned. While a given project is, by nature, engineering-oriented, various CAE foci may exist. Some of them are discussed below.

1. Computer graphics focus. Projects of this focus would concentrate on utilizing concepts of interactive graphics and geometric modeling to solve the design problem. Project ideas of this nature are numerous. Projects related to locus finding and design of various mechanisms, to simulation of mechanical systems, and to satisfying space or geometric constraints belong to this group. In these projects, the solution strategy is primarily manual, and interactive tools and aids discussed in Chaps. 11 and 12 are employed to facilitate obtaining the solution.

2. Interactive programming focus. Projects based on this concept may be similar to or more advanced than the projects of the above focus in the engineering content. However, they require programming knowledge in addition to graphics background. The bulk of engineering projects fits into this group. A list of project ideas is endless and may include the design of typical machine elements (cams, springs, gears, etc.), solid-mechanics-based projects (beam

analysis and design, vibration analysis, etc.), heat-transfer-based projects (fin design, heat-exchanger design, etc.), fluid-mechanics-based projects, thermodynamics-based projects, materials-based projects, etc.

3. **Database level programming.** This focus may be suitable for graduate level projects where students may design their own databases and data structures. Projects with this focus could aim at implementing engineering ideas into existing or new software, or aim at programming computer graphics and geometric modeling concepts.

4. **Application focus.** Projects with this focus would tend to utilize existing application software to complete the projects. For example, a project may be designed around using a mass property calculation package or a finite element modeling and analysis package, or both.

## 19.4   DESIGN AND ENGINEERING APPLICATIONS

This section illustrates the use of some ideas of this chapter in problem solving and project design. These ideas and problems have been implemented and solved utilizing CAD/CAM systems. Readers are encouraged to solve them on their systems. For more project ideas, readers can refer to the bibliography listed at the end of the chapter.

**Example 19.1.** A packaging company has a need for stamping boxes automatically at the end of an assembly line. The ink pad is located at the initial linkage position while the boxes travel along on the assembly line and stop at the final position in order to be stamped as shown in Fig. 19-1a. Design the linkage such that:

Design goal. The linkage has a straight-line motion into the box so that the stamp imprint is not smudged. In addition, the linkage should hit the ink pad such that it applies uniform pressure on the pad to load the stamping end of the linkage uniformly with ink.
Design constraints. Any box to be stamped must be located from the bottom of the ink pad as shown in Fig. 19-1a so that the synchronization of the assembly line is maintained.

A proposed design idea which has not been fully investigated yet is shown in Fig. 19-1b.

**Solution.** This example can be considered as a design project with computer graphics focus as described in Sec. 19.3. The strategy to attack this project is to choose a prospective linkage first. Then we construct the linkage in various positions beginning from the initial (stamper in contact with the ink pad) position to the final (stamper in contact with the box) position. The locus of the stamping face in space is obtained by connecting its various positions by a B-spline curve. The locus is checked against the design goal. If it has horizontal (when approaching the ink pad) and vertical (when approaching the box) segments near its ends, it satisfies the goal, and the linkage is acceptable. Of course, the design constraints would have been maintained during the generation of the locus. The test for horizontal and vertical segments can easily be achieved by checking the $x$ and $y$ coordinates of the generated points on the locus.

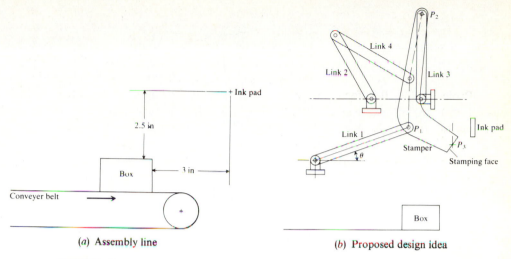

**FIGURE 19-1**
Stamping mechanism.

As can be seen, and purposely done, from the project formulation, there is an infinite number of solutions or designs that can achieve the given design goal and still meet the stated design constraints. We now discuss three design alternatives. We naturally begin with the proposed design to gain some insight into the problem which may lead to better alternatives.

**Design 1.** Figure 19-2 shows the detailed dimensions of the first design shown in Fig. 19-1b. The analysis of the design begins by obtaining the stick figure (skeleton or wireframe) of the linkage by considering only the centerlines of the links. The analysis consisted of tracing the full range of motion of link number "3" and also the full range of the stamper. An analysis of the motion of links "2" and "4" is not performed because it is determined prior to the analysis that these links do not affect the path of the stamper; they only constrict the range of this motion. The driving or "input" link is to be link "1." The analysis started with $\theta$ of the driving link at 20°. A brief procedure is presented below.

1. The path of motion of the stamper is determined by rotating the stamper about its connection to the driving link. The motion traced is that of point $P_2$ where link "3" and the stamper are connected. This step provides a circular locus of motion for $P_2$ with center at $P_1$ and radius equal to the distance between $P_1$ and $P_2$.

2. The path of link "3" is determined in a similar fashion. The origin of rotation is the pivot connected to the link. This step provides another circular locus for $P_2$.

3. The two circular loci created intersected at two points. A point at the proper intersection is created using an "insert point" command with the intersection modifier.

4. The input link is then incremented by two degrees and steps 1 through 3 are repeated. This provides the locations of point $P_1$ on its circular path.

5. It is determined that the two circles created only intersect when $\theta$ of the driving link is between 0 and 21.62°. The useful range of motion of the input link is now determined.

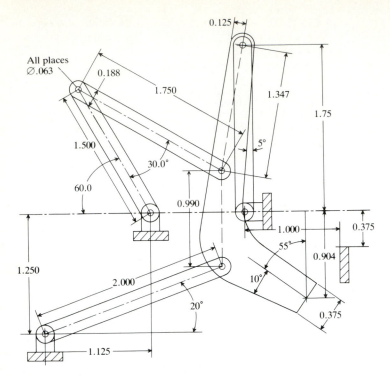

**FIGURE 19-2**
Detailed dimensions of design 1.

6. By limiting the range of motion of the driving link and having determined how link "3" and the stamper link interact, an analysis of the motion of the stamping face of the stamper could be performed.

7. The path of motion of the stamping face could be determined with a similar procedure to the one used to determine the path of point $P_2$. The midpoint $P_3$ on the stamping face is assumed to represent the face. A circle is then created with center at $P_2$ and radius equal to the distance between $P_2$ and $P_3$.

8. Another circular locus of $P_3$ with center at $P_1$ and radius equal to the distance between $P_1$ and $P_3$ is created.

9. The intersection of these two loci defines the location of $P_3$ in space.

10. The driving link is then again incremented in two-degree intervals between 0 and 21.62°, and steps 7 through 9 are repeated. The path for the stamping face is thereby determined.

After determining the path of motion of the input link and the stamper, it is found that straight line motion existed when the stamper was stamping the box; however, this is not the case for the straight line motion onto the inking pad. After several attempts to alter the linkage design to create straight line motion in both cases it was determined that a solution was too complicated. The design changes were focused on the orientation of the inking pad. Several options were considered. These are listed below.

1. Have the ink pad come to the stamper by the implementation of a slider mechanism which would be activated by a cam. The cam would be timed with the rotation of the driving link.
2. Have the ink pad closer to the stamper and positioned on a pivot so that when the stamper reaches its horizontal position it would hit the pad. The pivot would allow the pad to come into full contact with the face of the stamper.
3. Have an ink roller instead of the ink pad. This roller would be positioned along the path of the stamping face. The stamper face would pass the roller twice, once in the upward motion and once in the downward, which would ensure a good application of ink.

The option found to be most efficient and practical is that of the roller. The implementation of extra components such as cams, sliders, and pivots would complicate the design and create room for errors and mechanical failure. For the presented design a solenoid driving link "1" should be set so as to restrict the range of motion of the driving link from 0 to 22. This is the useful range of motion determined in the analysis. Several orientations of the assembly are provided in Fig. 19-3 to show how the assembly travels through its full cycle.

The range of motion of the various links is shown in Fig. 19-4. Notice that the locus of the stamper center achieves the design goal. Figure 19-5 shows the inking motion $(12° \le \theta \le 21.62°)$ and Fig. 19-6 shows the stamping motion $(0° \le \theta \le 21.62°)$.

**Design 2.** This design simplifies the proposed idea and is based on simplicity; the simpler a design can be in a case like this where the machine has a very simple task to perform, the better the design. This design is called the Z-stamper due to its shape. The Z-stamper is a simple stamper with two links to control its range of motion. The two controlling links are exactly the same, thus simplifying the manufacturing process and reducing the overall cost of the design.

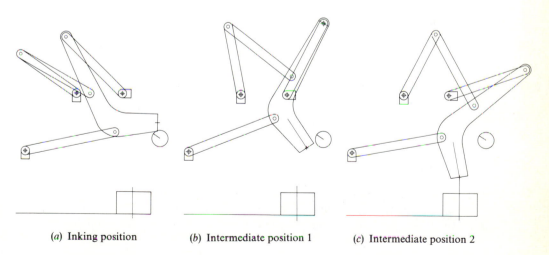

(a) Inking position      (b) Intermediate position 1      (c) Intermediate position 2

**FIGURE 19-3**
Full travel cycle of design 1.

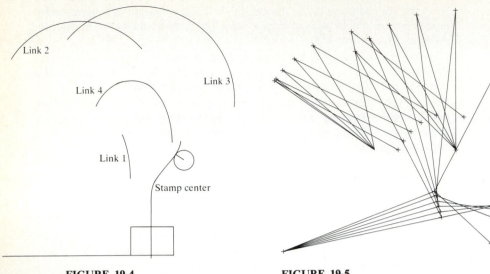

**FIGURE 19-4**
Linkage range of
motion of design 1.

**FIGURE 19-5**
Inking motion of design 1.

The range of motion of each link is 90°, as shown in Fig. 19-7. The detailed dimensions of any of the links and the stamper are shown in Figs. 19-8 and 19-9 respectively. The stamper could be driven by a straight line motion actuator, such as a solenoid, that alternately pulls the stamper through its first half of motion and then pushes it into its final position. The locus of the stamping face is shown in Fig. 19-10. It confirms the straight line motion of the stamping face at the end positions.

**Design 3.** This design ignores the proposed design idea shown in Fig. 19-1*b* completely. It uses only one link and the stamper as shown in Figs. 19-11 and 19-12.

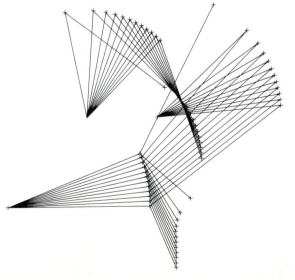

**FIGURE 19-6**
Stamping motion of design 1.

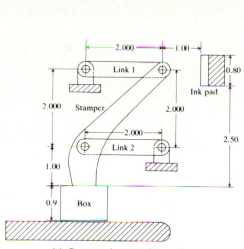

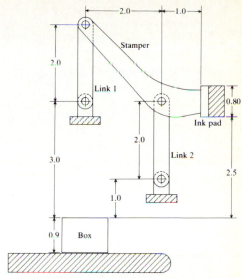

(a) Stamper in stamping position

(b) Stamper in inking position

**FIGURE 19-7**
Details of design 2.

*R* 0.2 TYP.

Ø0.2 TYP.

**FIGURE 19-8**
Details of links of design 2.

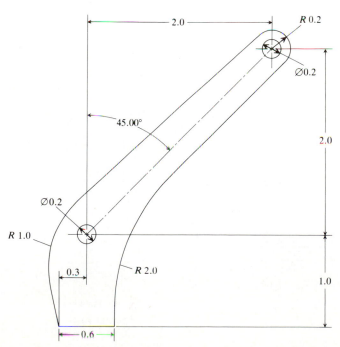

**FIGURE 19-9**
Details of the stamper of
design 2.

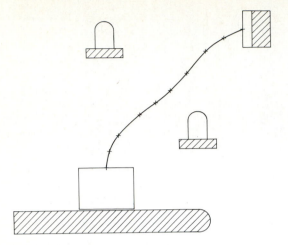

**FIGURE 19-10**
Locus of the stamping face of design 2.

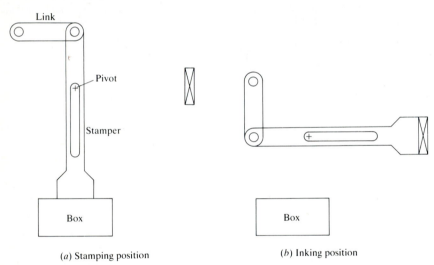

(a) Stamping position

(b) Inking position

**FIGURE 19-11**
Linkage of design 3.

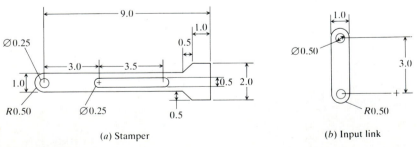

(a) Stamper

(b) Input link

**FIGURE 19-12**
Details of design 3.

The link is the input link. The range of rotation for the input link is 270°. It moves in a counterclockwise rotation when translating to the ink pad and clockwise when stamping the box. The stamper is slotted and rotates 90° while translating linearly in the horizontal and vertical directions. There is a slot in the stamper to allow its sliding about a fixed pivot (Fig. 19-11a) which is located at equal distances from the top of the box and the inked face of the ink pad.

To obtain the locus of the stamping face (its midpoint), the linkage is moved incrementally every 15° and constructed as shown in Fig. 19-13. The resulting points are connected with a B-spline curve. As shown in Fig. 9-14, the stamper approaches and leaves the box and the ink pad in straight line motion.

The investigation of this design project and its solution require the use of most of the interactive design tools and aids discussed in Part IV of the book. The use of geometric modifiers helps find intersections between various loci. Layers are most helpful in managing all the graphics. For example, the geometry of each position of a linkage is stored in a different layer. This geometry is created using "rotate entity" and "translate entity" commands. Animation techniques are also used to stimulate the stamping motion of the linkage of each of the three designs.

**Example 19.2.** Figure 19-15 shows the details of a current design of a simplified cantilever beam (called a turnover casting) used in bottle-cleaning machines. The beam material is cast aluminum. The casting as shown in the figure frequently breaks, causing delays and inconvenience to the bottling process. After investigating the details of fracture, it became obvious that the load-carrying capacity of the turnover casting must increase by 30 percent of its current value. As a member of the design team responsible for the redesign of the casting:

(a) Find the current load-carrying capacity of the turnover casting. Consider the weight of the casting.

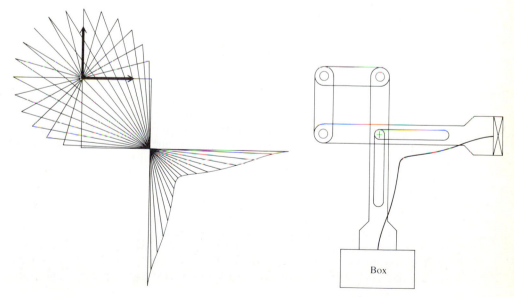

**FIGURE 19-13**
Linkage motion of design 3

**FIGURE 19-14**
Locus of the stamping face of design 3.

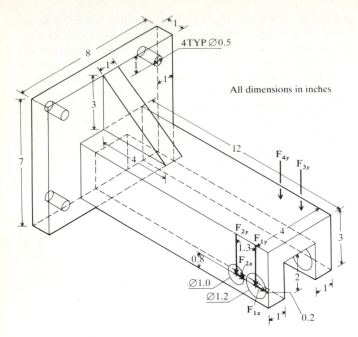

4TYP ∅0.5

All dimensions in inches

**FIGURE 19-15**
Details of turnover casting.

(b) Redesign the casting to increase its load-carrying capacity by 30 percent (design goal) subject to the following design constraints: the overall length of the casting cannot change, due to space constraints of the machine; the number of holes must stay as two, and the first hole must remain 0.2 in away from the free end of the casting. In addition, the casting material should not change.

*Solution.* This design project fits the project category with the application focus discussed in Sec. 19.3. Investigating Fig. 19-15 shows that the casting geometry does not allow any closed-form solutions or the use of the beam theory to solve the problem. Thus, numerical solutions must be sought. The finite element method is utilized here. The applied loads on the casting are shown as concentrated forces in the horizontal and vertical directions at the centers of the two holes. Because the weight of the casting must be considered in the analysis, the mass property calculations provided by the CAD/CAM systems must be utilized. While the calculations of the weight and the centroid are possible manually due to the simplified geometry of the casting, it is tedious to do. To calculate the mass properties with minimal user input, a solid model of the casting is created.

Once the magnitude and the location of the weight is determined, an FEM/FEA is performed to determine the current load-carrying capacity. This is achieved in a trial-and-error fashion. We first assume values of the applied loads (the beam theory can be used if we ignore the two holes and the triangular web). We then run the FEA and determine the maximum stresses. We compare these stresses with the casting yield strength after applying a reasonable factor of safety. This cycle of calculations is repeated until the maximum loads the casting can support are determined.

The next step is to increase these loads by 30 percent. Of course, the current casting design would fail under these loads. Thus, the casting design must be altered to increase its strength without changing its material. Some design changes are discussed here. To determine the strength of the new proposed designs, the FEM/FEA cycle described above is used here.

To evaluate the mass properties of the current casting, its solid model is created first. Figure 19-16 shows its CSG tree. The mass properties of the solid model is obtained using a specific weight of 0.10079 lb/in$^3$ for the cast aluminum. The weight of the casting is 15.46 lb and the centroid is located at $x = -8.64$ in, $y = 2$ in, and $z = 2.13$ in measured in the MCS shown in Fig. 19-17.

Having determined the weight, the finite element model of the current design of the casting is created using hyperpatches. The complete finite element model and the data file are created on the corresponding preprocessor. Figures 19-18 and 19-19 show the hyperpatches and loads utilized to analyze the current design. After performing the FEA and using an allowable stress of 20,000 lb/in$^2$ (using a factor of safety of 1.5), this stress would correspond to a total force of 9055 lb which rep-

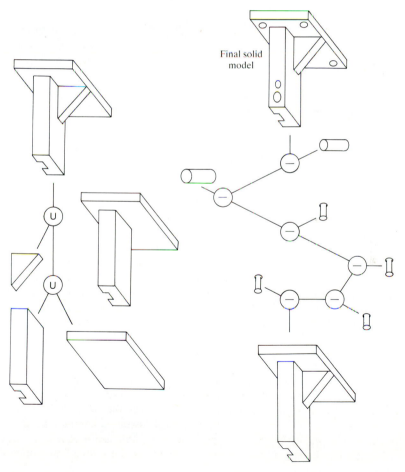

Final solid
model

**FIGURE 19-16**
CSG tree of the casting.

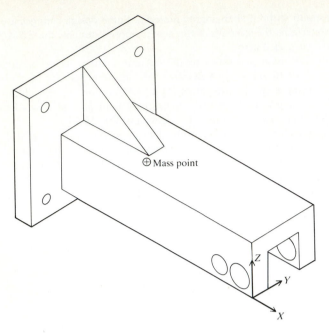

**FIGURE 19-17**
The casting solid model
and its mass point.

resents the current load-carrying capacity. The corresponding stress distribution and deformed shape of the current casting are shown in Figs. 19-20 and 19-21.

The modification of the current design to increase its load-carrying capacity by 30 percent (1.3 × 9055 = 11,771.5 lb) can be accomplished by changing the thickness of the sides of the casting, adding thickness in the web area, adding thickness to the top portion of the casting, or changing the cross-section dimension. The latter

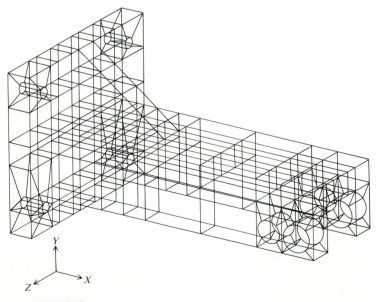

**FIGURE 19-18**
Hyperpatches of current design.

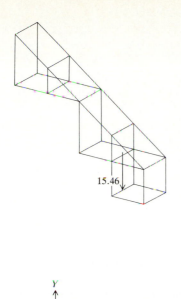

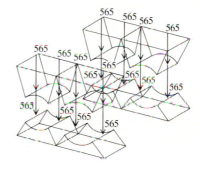

**FIGURE 19-19**
Loads and weight acting on current design.

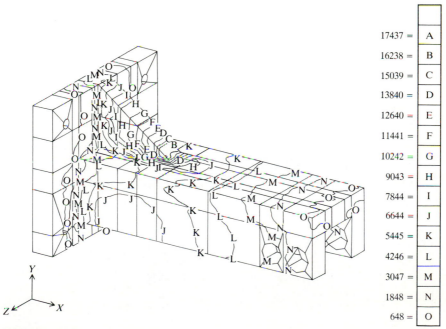

| | |
|---|---|
| 17437 = | A |
| 16238 = | B |
| 15039 = | C |
| 13840 = | D |
| 12640 = | E |
| 11441 = | F |
| 10242 = | G |
| 9043 = | H |
| 7844 = | I |
| 6644 = | J |
| 5445 = | K |
| 4246 = | L |
| 3047 = | M |
| 1848 = | N |
| 648 = | O |

**FIGURE 19-20**
Stress contours in current design.

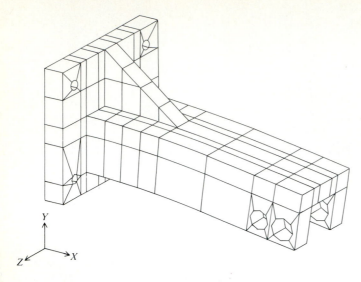

**FIGURE 19-21**
Deformation of current
design.

design change is chosen by changing the dimensions of the C-channel from
3 × 4 × 1 to 4 × 4 × 1. The finite element model due to this change is created by
modifying the model of the current design. The FEA is performed on the new design
which has proved that the new design is acceptable. Of course, the mass properties
of the new design have to be evaluated first. Figures 19-22 to 19-24 show the hyperpatches, stresses, and deformation of the new design respectively.

The two projects discussed in this section represent two of the foci listed in
Sec. 19.3. The other two foci (interactive programming and database level
programming) have been illustrated in Chap. 15. Readers are encouraged to formulate projects of direct interest to them in a similar fashion to what has been discussed here.

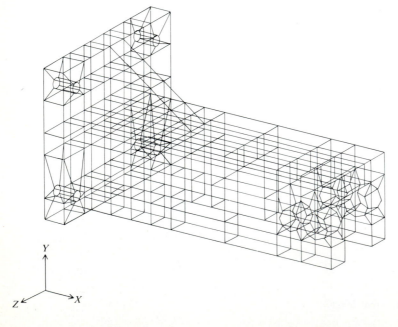

**FIGURE 19-22**
Hyperpatches of
new design.

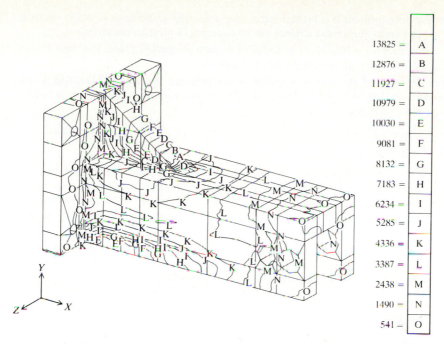

| | |
|---|---|
| 13825 = | A |
| 12876 = | B |
| 11927 = | C |
| 10979 = | D |
| 10030 = | E |
| 9081 = | F |
| 8132 = | G |
| 7183 = | H |
| 6234 = | I |
| 5285 = | J |
| 4336 = | K |
| 3387 = | L |
| 2438 = | M |
| 1490 = | N |
| 541 − | O |

**FIGURE 19-23**
Stress contours in new design.

# PROBLEMS

## Part 1: Theory

**19.1.** What are the characteristics of a successful designer?

**19.2.** Creativity in design has been cited as a major criterion for innovation. Discuss this statement as well as the personality of a creative person.

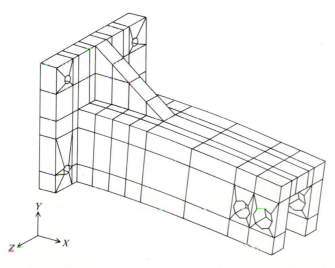

**FIGURE 19-24**
Deformation of new design.

**19.3.** Project management is a crucial factor in completing projects successfully on time. Use your library and write a paper on the concepts of product management.

**19.4.** Product liability is important in today's competitive market place. Use your library and write a paper on the subject and its legalities.

**19.5.** Discuss some of the common problems encountered in using CAD/CAM laboratories and CAE software in design courses and environments. Suggest ways to alleviate these problems.

## Part 2: Laboratory

**19.6.** Using Example 19.1 as a basis, design a linkage with the same design goal, but with new design constraints. These are (1) the relative location between the box in the stamping location and the ink pad is the same as in Example 19.1 and (2) the path of the stamper from the ink pad to the box must be as minimum as possible to increase the capacity of the stamping station.

**19.7.** Using Example 19.2 as a basis, redesign the turnover casting with the design goal of increasing its current load-carrying capacity by 50 percent. The design constraint in this case is only the hole sizes, and the web size can change.

## Part 3: Programming

**19.8.** Write a program that takes a set of data points and classify them by value for contour plotting.

**19.9.** Write a program to animate the linkage that results from Prob. 19.6 above.

## BIBLIOGRAPHY

Ardafio, D. D., J. Watson, and G. T. Eberhadt: "Short Communication: Computer Aided Design of Bevel and Worm Gears," *Adv. Engng Software*, vol. 7, no. 4, pp. 204–207, 1985.

Barker, C. R.: "Analyzing Mechanisms with the IBM PC," *CIME*, vol. 2, no. 1, pp. 37–47, 1983.

Dawson, G.: "The Dynamic Duo: DRAM and ADAMS," *CIME*, vol. 3, no. 5, pp. 20–24, 1985.

Erdman, A. G., and G. N. Sandor: *Mechanism Design: Analysis and Synthesis*, vol. I, Prentice-Hall, Englewood Cliffs, N.J., 1984.

Lai, H. Y.: "Computer-Aided Design and Manufacturing of Curved Shapes," *ASME Computers in Engng Conf.*, vol. 2, pp. 183–188, 1987.

Meyer, J., and R. Jauaraman: "Simulating Robotic Applications on a Personal Computer," *Computers in Mech. Engng (CIME)*, vol. 2, no. 1, pp. 14–19, 1983.

Nakazawa, M., and K. Kikuchi: "CAE System in Mechanical Engineering," *Computers and Graphics*, vol. 7, no. 3–4, pp. 315–325, 1983.

Orthwein, W. C.: "Designing Disk Springs with a Personal Computer," *CIME*, vol. 2, no. 3, pp. 45–52, 1983.

Orthwein, W. C.: "Designing and Selecting Screw Threads Using a Personal Computer," *CIME*, vol. 2, no. 5, pp. 49–56, 1984.

Parkinson, A. R., R. J. Balling, and J. C. Free: "Exploring Design Space: Optimization as Synthesizer of Design and Analysis," *CIME*, vol. 3, no. 5, pp. 28–36, 1985.

Reynier, M., and J. Fouet: "Automated Design of Crankcases: Carter System," *Computer Aided Des.*, vol. 16, no. 6, pp. 308–313, 1984.

Spotts, M. F.: *Design Engineering Projects*, Prentice-Hall, Englewood Cliffs, N.J., 1968.

Yang, R. J., and M. J. Fiedler: "Design Modeling for Large-Scale Three-Dimensional Shape Optimization Problems," *ASME Computers in Engng Conf.*, vol. 2, pp. 177–182, 1987.

# PART
# VI

## CAD AND
## CAM
## INTEGRATION

# CHAPTER 20

# PART
# PROGRAMMING AND
# MANUFACTURING

## 20.1  INTRODUCTION

The primary goal of engineering is to transform ideas into products that are economical and reliable. The process of designing and introducing a part to manufacturing often involves a sizable investment and draws on various disciplines and resources. Engineering is an important key to product design, product manufacturing flow, and the ability of a company to produce good products. Product design determines the function, appearance, cost of production, and the ability to plan and control manufacturing operations. It is known that about 80 percent of the resources and cost required to produce a part are committed at its design phase. The further the part in its production, the more costly it is for any design change, as shown in Fig. 20-1.

Engineering design has been influenced heavily by the CAD technology and tools available to designers. Similarly, manufacturing has undergone major changes with the introduction of numerically controlled (NC) and computer numerically controlled (CNC) machine tools. These replace conventional machines, thus offering increased flexibility, superior accuracy, and shorter production cycles. Machining of complex (sculptured) surfaces with conventional machining is neither economical nor accurate. These surfaces are found in a wide range of components including those for aircrafts, automobiles, construction and agricultural equipments, machine tools themselves, appliances, cameras, and instrument cases.

The potential benefits of integrating engineering and manufacturing are well recognized. More specifically, full integration of CAD and CAM is an important aspect of factory automation. Factory automation is a vast and complex subject. Achieving automation in a given company depends on the company corporate

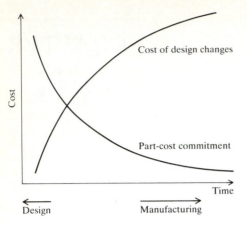

FIGURE 20-1
Influence of part design on its cost.

Cost

Cost of design changes

Part-cost commitment

Time

Design    Manufacturing

strategy, which includes such aspects as product improvements, manufacturing efficiency, quality improvements, market development, the organization, and the people.

Historically, CAD/CAM integration began with the development of the NC technology. NC machine tools have been improving steadily in both areas of hardware control and software developments. NC part programming and interactive computer graphics have contributed heavily to these developments. The integration of CAD and CAM places increasing emphasis on tools and paths for NC machines. It is interesting to note how independent developments (CAD and CAM) which began at completely opposite ends of the CAD/CAM spectrum during evolution of CAD/CAM systems have gradually approached each other.

In an attempt to show the influence of engineering design on the manufacturing process (as shown in Fig. 20-1), to help designers overcome their isolation from the manufacturing process, and to appreciate the subtleties facing CAD/CAM integration, this chapter is organized in two parts. Sections 20.2 to 20.4 provide a quick overview of manufacturing—specifically part production cycles and manufacturing systems and processes. The remaining sections in the chapter focus on process planning, NC part programming, and tool path generation and verification from geometric models.

## 20.2 PART PRODUCTION CYCLE

The main engineering functions involved in the creation and subsequent conversion of design data into a product are discussed in Chap. 1 and shown in Fig. 1-1. Clearly the extent, or even the formal existence, of each of these functions depends on a variety of company- and product-related factors. However, with the possible exception of NC functions where no NC machines are in use, all the activities have to be undertaken as part of the overall process of design and manufacturing.

In this section, we expand on a subset of the functions shown in Fig. 1-1, especially those related to manufacturing a part. Assuming that a part design has been completed, the following activities are utilized to produce it:

1. Part classification and identification. Parts are classified, numbered, and coded to enable them to be identified, permit easy storage and retrieval, and facilitate their production (if they are similar to other existing parts). Group technology (GT) plays an important role in producing a part. If a part has been classified using a GT part coding system (such as OPTIZ, CODE, KK-3, MICLASS, and DCLASS systems) and its process plan has been coded, new parts with similar features can be produced by using and/or modifying the existing process plan. Process planning typically consumes considerable engineering time.

2. Process planning. This is the planning strategy for manufacturing the part. This step involves identifying the sequence of processes to manufacture the part, identifying proper jigs and fixtures, planning and/or ordering of material, ordering and/or design of tools, and scheduling part production and inspection. In short, process planning is a complete time schedule that begins with a part design on paper and ends with a real product.

3. Design and procurement of new tools. A part process plan may reveal that new tools are needed to produce the part. If this is the case, the process planner (a person who develops the process plan) initiates a request for design or purchase of new tools so that they become available at the time of manufacturing.

4. Planning and ordering of material. The process planner ensures that all the materials shown on the bill of material, developed by the part designer at the design phase, are in stock for part production. Other accepted uses of a bill of material include product definition, service-parts control, liability and warranty protection, and part costing. Material handling systems such as MRPII [material resource (or requirement) planning] helps a great deal in this step.

5. NC programming. If NC machine tools are used to machine the part, an NC programmer must develop the program needed to machine the part. The integration between CAD and CAM facilitates the program generation and verification a great deal.

6. Production scheduling. Part production must be scheduled within the capacity and schedule of shops involved in producing the part as well as the availability of materials, tools, jigs or fixtures, and other items needed to produce the part. In addition, existing inventory, anticipated orders and shipments, and future market needs play a decisive factor in the additional production of existing parts.

7. Manufacturing. The culmination of all the above activities is the manufacturing of the part. Various manufacturing processes exist and one or more of them would have been identified during process planning to manufacture the part. This identification is based on the tolerance and surface finish requirements specified by the designer, as discussed in Chap. 16.

8. Inspection. Inspections of produced parts are required by internal (the company producing the part) quality control and outside customers such as the government. Quality control inspects parts for dimensional accuracy (tolerance), finish, material, and other physical properties. Typically, statistical

concepts of quality control are used to eliminate the need for inspection of every individual part.

9. Other activities. Parts that survive inspection and quality control requirements may follow different paths depending on various factors. If the company that produces a given part also produces the product the part belongs to, the next step after inspection is to assemble the part with others to form the product. If not, the part is packaged and shipped to prospective customers who assemble the parts into their products.

In the context of the foregoing steps or activities of a part production cycle, it is evident that the various aspects of part design affect, to various degrees, all of them. For example, if part design requires special material or material processing, this would result in raising the final cost of the part and delaying its production. Similarly, unnecessarily tight tolerances result in a high rejection rate, and consequently a high scrap rate. This would also raise the cost per unit. From a quality and reliability point of view, the improper choice of materials and/or tolerances result in degrading the part performance in field service which leads to customer dissatisfaction and ultimate rejection of the part from the market place.

## 20.3 MANUFACTURING SYSTEMS

Manufacturing of products draws on different resources such as people, machinery, and equipment. A manufacturing system can be defined as a combination of people, machinery, and equipment which is constrained by material and information flow. Manufacturing systems can be classified into discrete part manufacturing and continuous process manufacturing. The former refers to manufacturing a product where the product undergoes a finite number of production or assembly operations. The latter refers to the production of a product that undergoes continuous changes such as chemical reactions which transform raw materials into final products. Here, we concern ourselves with discrete part manufacturing and, more specifically, the manufacturing of machined parts.

Manufacturing systems have undergone radical changes since the industrial revolutions began around the 1900s. These changes can be viewed as an outcome of customer demands for higher performance and yet less expensive products. Changes in materials and society combined with fierce competition have shortened product life expectancy. The general trend is that manufacturing systems are becoming more flexible, to adapt to rapid product changes, and more automated to meet accuracy and cost requirements.

Manufacturing procedures vary from transfer line techniques for high-volume production to job shop procedures for low-volume items. Figure 20-2 classifies manufacturing systems into the following systems:

1. Transfer line. This represents the oldest type of manufacturing systems. Transfer lines are very efficient for mass production (large volumes at a high output rate). They represent what is sometimes called hard automation. They are suitable for manufacturing identical parts. Thus, they are inflexible and cannot tolerate variations in part design. Any change in part design requires the line

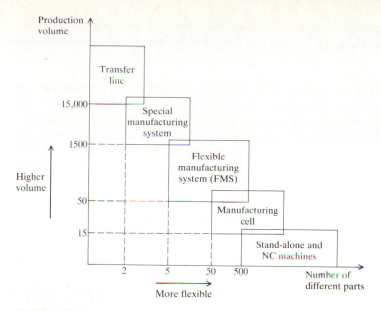

**FIGURE 20-2**
Types of manufacturing systems.

to shut down and be retooled. Moreover, if drastic changes in part design occur, the line becomes obsolete.

2. Special manufacturing system. This system together with the next two systems represent the various types of CIM (computer integrated manufacturing) systems. This system is the least flexible while a manufacturing cell is the most flexible. The system is suitable to produce a very limited number of different parts and a medium production rate per part. The system is configured in a similar way to transfer lines; thus only limited changes in the system are possible.

3. Flexible manufacturing system (FMS). A mid-volume, mid-variety production range is covered by this system. Most of the system activities and coordination are done automatically under computer control. Work parts are automatically loaded at central locations on to the handling system (pallets) and are routed to the proper machine tools. The computer job in an FMS includes the control of machine tools and the material handling system, monitoring the performance of the system, and scheduling production. FMSs are not totally under computer control. Humans and human labor are needed to set up machine tools for production, that is, load and unload workparts, prepare and change tools, and perform initial settings of machine tools.

4. Manufacturing cell. This is the most flexible CIM system. It has the lowest production rate of the three types (systems 2, 3, and 4 presented here). A manufacturing cell typically contains many stand-alone machine tools and robots.

5. Stand-alone and NC machines. These machines are highly flexible. Their production rates are too slow due to tool setting-up time and tool changes. They

are highly programmable and can deal with product changeovers and part design changes. They are appropriate for job shop and small batch manufacturing.

It is important to mention that although flexibility and automation are desired characteristics of manufacturing systems, these systems must be carefully designed to handle the expected production volume from them on a continuous basis. If a system is designed to handle a given production volume that is higher than the expected market projections or consumer needs, the system would have to stay idle for certain periods of time. Therefore, productivity gains during the operation of the system is lost by its periods of shutdown.

## 20.4 MANUFACTURING PROCESSES

Quite often design engineers are accused by their fellow manufacturing engineers that they pay too much attention to analysis and safety aspects of new designs and little, if no, attention to the manufacturing aspects of their designs. While we do not intend to debate this issue here, we do present a review of some common manufacturing processes in this section. This presentation is aimed at helping to achieve the goal of integrating design and manufacturing. Consider, for example, the two designs shown in Fig. 20-3. If the part design requires a hole regardless of the shape of the hole bottom, then the design option shown in Fig. 20-3b is easier to manufacture than that shown in Fig. 20-3a. The flat-bottom hole requires a milling operation while the conical-bottom hole requires a drilling operation to create. While little difference between the two design alternatives occurs at the design level, two different significant problems arise for the manufacturing engineer. The notion of considering manufacturing issues at the design phase is known as "design for manufacturing."

The better that design engineers understand the nature of materials and the rationale for their physical and mechanical properties, the more wisely they are able to predict the most favorable shaping, forming, cutting, and finishing processes to transform their designs into products of high quality and least cost. The materials used to produce the majority of discrete goods can be classified into ferrous, nonferrous, plastics, ceramics, powdered metals, and composites. The

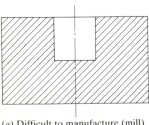

(a) Difficult to manufacture (mill)

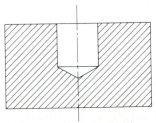

(b) Easy to manufacture (drill)

**FIGURE 20-3**
Design for manufacturing.

detailed discussions of the properties of these materials are beyond the scope of this book. Interested readers should refer to relevant books in the open literature.

It is fundamental that design engineers should know the competitive manufacturing processes that can be specified to transform raw materials into finished products. Knowledge about a manufacturing process should include details of its dimensional capacity (tolerances), rates of output, cost of tooling, and skills required. Refer to Chap. 16 regarding the tolerance limits of the various manufacturing processes. The other details of each process (output rate, tooling cost, skills, etc.) are left to the readers' investigations (see Prob. 20.3 at the end of the chapter).

In this section, we use the term "manufacturing process" to mean the transformation of a material's shape and form. (The other meaning, not used here, encompasses all the production activities, from process planning to packaging and shipping needed to produce the product.) Manufacturing processes, as defined here, can be classified into four categories: removing, forming, deforming, and joining processes. As a general guideline to relating materials and manufacturing processes, we can use the following recommendation. Metals can be processed via removing, deforming, forming, and joining processes; polymers (plastics) can be processed via forming (in some seldom cases, removing, deforming, and joining can be used); forming is widely used to process ceramics (removing is seldom used); composites are widely processed via forming also (removing and joining may be used on occasion).

## 20.4.1 Removing Processes

Removing processes are very popular and are widely used to machine metals such as steel, cast iron, aluminum, brass, bronze, and magnesium. As the name indicates, removal of material being processed occurs during the process itself. Various mechanisms for material removal exist: mechanical, thermal, chemical, and electrochemical. In mechanical removing, material is removed via a cutting tool which overcomes the strength of the material being cut. Thermal removing is achieved by melting or vaporizing the volume of the material to be removed. Chemical reactions with the material to be removed form the basis of chemical removing. In electrochemical removing, an electric field is induced into an electrolyte to enable chemical reactions to remove material.

Machining (mechanical removing process) is the most popular form of removing processes. Machining processes include turning, milling, drilling, grinding, facing, shaping, planning, boring, counterboring, countersinking, reaming, broaching, and sawing. These processes involve a relative motion between the workpiece and the cutting tool. In some processes such as drilling, the tool moves (translates and/or rotates) while the workpiece is held stationary. In other processes such as turning, the workpiece itself moves while the tool is only moved at the feed rate. Other processes such as grinding require moving both the workpiece and the cutting tool.

Machinability is a term used to indicate the relative ease of material removal with consideration to the surface finish, the tool life, and the rate of metal removal. Aluminum and magnesium alloys are known to have the highest machinability followed by brass and bronze, and then by steel. Cast irons are generally machined easily and a good surface finish at low cost is obtained. In the machining of metals, different surface finish results depending on the metal being cut, the cutting tool (amount of tool wear), the coolant, the process and type of machine tool, and the cutting conditions. There are always tables that accompany machine tools to specify the optimum machining parameters for a given material (see Prob. 20.4 at the end of the chapter).

The machinability of materials is a major factor in the cost of a product. The best machining conditions and lowest costs can be obtained by selecting the proper size, strength, composition, and the heat treatment of the material being cut. Machinability is an important item to be considered by design engineers.

### 20.4.2 Forming Processes

These are the processes in which a desired defined shape is created from an undefined shape of material. The state of material of the undefined shape may be molten, gaseous, or solid particles. Among forming processes are casting, powder metallurgy, and molding.

Metal casting processes are usually used when parts with intricate shapes and internal cavities are needed. Castings usually do not have a good surface quality. The most traditional casting process is sand casting in which the mold is made of sand. Other types of casting exist such as plaster-mold casting and ceramic-mold casting.

In powder metallurgy, the raw material is powder which is compacted in a suitable die and then heated (without melting). Parts produced by this process include gears, cams, bushings, and cutting tools.

Molding processes are usually used to produce plastic parts. There is a wide variety of molding processes such as injection molding, compression molding, and extrusion. Injection molding is the most popular process. In this process, molten material is injected into a mold (called resin) which is firmly kept shut during injection. Mold design is very crucial to obtain products with no cavities (due to air being trapped in the mold). There are various CAD/CAM systems that support the designing of molds interactively.

### 20.4.3 Deforming Processes

These processes convert a given shape into another via permanent deformation. They include rolling, forging, extrusion, and sheet metal forming. In rolling, the material is deformed by feeding it between two rolls that rotate. Hot or cold rolling is possible. Hot rolling refines the grain structure of the material, whereas cold rolling distorts it.

Forging is shaping metal under impact or pressure and improving its mechanical properties through controlled plastic deformation. Die forgings are made in steel dies constructed of hard, tough, and wear- and heat-resisting tool steel. As the metal halves of the die close, the material is shaped into the form of the die. Successive operations may be required to produce forged parts. Die forgings are usually formed hot. After the forging operation, the part must be trimmed to remove the flash.

Sheet metal forming techniques include drawing, bending, and shearing. Applying a combination of tensile and compressive deformations, drawing can change a sheet to form a hollow body of smaller size without a change in sheet thickness. In bending, the plastic deformation of the sheet is achieved by means of a bending load. Similarly, shearing is achieved by applying shearing loads to the sheet metal.

### 20.4.4   Joining Processes

These processes unite individual workpieces to make final products. The major industrial joining processes are welding, brazing, soldering, adhesive bonding, and mechanical fastening. Various types of welding exist and include arc welding, resistance welding, gas welding, and solid-state welding. Acceptable welds are those that are free from defects such as cracks, hard spots, porosity, or nonmetallic inclusions and are able to perform satisfactorily in their intended services. Producing acceptable welds depends on the melting point, thermal conductivity, thermal expansion, surface finish, and change in microstructure of the workpieces to be welded.

Brazing is the joining of metallic parts by filler metals placed between the facing surfaces to be joined. Filler metals usually have a melting point (greater than 800°F) below that of the base metal (metal of the parts to be welded). After cooling and solidifying the filler metal, a strong joint is obtained.

Soldering is a similar process to brazing. In soldering, however, the filler metal (solder) has low strength, a melting temperature below 800°F, and results in less alloying between the base metal and the filler. Soldering is used extensively in the electronics industry. Soldered joints cannot be used in applications that require high temperatures. They also do not have much strength in general, and therefore are not used in structural load-carrying applications.

Adhesive bonding is becoming more important in manufacturing and construction. It is replacing rivets, bolts, screws, nails, and other types of mechanical fasteners. In adhesive bonding, an adhesive material is applied to the two parts to be joined. Adhesive applicators include brushes, sprayers, rollers, scrappers, pressure guns, and tapes. Some machines can apply the correct bonding pressure and heating to cause adhesion quickly. The principal advantages of adhesive bonding is the ability to join dissimilar materials, bonding very thin sections to heavy sections without distortion, joining heat-sensitive alloys, and the low cost of the bonding process itself. Joints created by adhesive bonding cannot sustain impact and/or shear loads in general, and have a limited temperature range (below 500°F).

Mechanical fasteners are well known to all design engineers and include nuts, bolts, screws, rivets, and metal stitching. In metal stitching, the joint is created by stitching two parts with a wire. Metal stitching is used to join thin-section metals and nonmetals. It is a very low-cost process and does not require any precleaning, drilling, punching, or hole alignment.


## 20.5   INTEGRATION REQUIREMENTS

The need for increasing productivity and surviving in an increasingly competitive market has fueled the importance of integrating CAD and CAM, and has prompted the notion of CIM. The original tenet of CIM thinking is to establish a common database to hold the great bulk of data necessary to run a company. Based on a typical product cycle, such as the one shown in Fig. 1-1, a CIM database must support all of the design and manufacturing activities, and related information, shown in the figure. Thus, the CIM database acts as a central databank for the company. The information in the database would be accessed by any element in the CIM environment, to either extract or enter data. Therefore, the database would always contain the most recent information.

The technologies that would take part in the two-way communication with the central database include CAD, CAM, FMSs, production management, and automated assembly. The achievement of a true CIM system is doubtful to achieve with today's hardware and software technologies. The response time would not meet the numerous queries that a CIM database, with all the data it stores, would be expected to handle. In addition, the heterogeneous types of data stored in various elements of a CIM database may hinder its justification in practice. For example, the geometry data stored in the CAD database would be of no use to production management, and vice versa; data related to production management is of no use to the design department. Moreover, the restricted individual databases currently used, which make CIM, are of different types. For example, a relational database is required for a production management package, whereas a hierarchical database is needed for CAD. A probable solution to this latter problem is to use a distributed database system.

Another difficulty regarding the viability of CIM is that individual packages are designed and developed using specific programming languages on a restricted range of computers. For example, packages related to production management are written in data processing languages such as COBOL or BASIC, while CAD is traditionally written in FORTRAN or C. Each of these packages runs more efficiently on specific types of computer architecture. Thus, the integration of various packages into a full CIM system would involve much effort by hardware and software suppliers. However, these suppliers, being in constant competition, normally protect their products to a level that makes their integration practically impossible.

In the context of the above formidable problems to have a central CIM database, it might seem attractive and practical to create what is called here "a linked software CIM system." The linked CIM system would involve linking together all the existing elements and packages that are needed for CIM in a coherent way which provides a quick and cost-effective means of achieving integration between CAD and CAM. The linked CIM system would rely heavily on communication software to link the various involved hardware (mainly computers). There are many problems to be faced in terms of compatibility of different computer systems and hardware peripherals, compatibility of computer languages, compatibility of software packages in terms of their input requirements (structure and logic), and the compatibility of all the elements of a linked CIM system with the widely differing requirements of the various sectors of industry. The Ethernet local area network is showing reasonable success in attempting to network various computers. Communication protocols at the software level [such as IGES, PDES, MAP (manufacturing automation protocol), etc.] are achieving similar success. However, there are currently no hardware and software that are specifically designed to carry out the complex tasks involved in running a CIM system.

In conclusion, while CIM as now perceived is considered a future goal, considerable benefits can be gained from adopting the concepts behind CIM. One such concept is DNC (direct numerical control) technology which represents the formation of an "island of automation." The rest of this section and some other sections of this chapter are devoted to covering the details of NC which forms the backbone of CNC (computer numerical control) and DNC.

A crucial step in NC programming is the effective use of CAD information by the NC programming activity. This is sometimes referred to as linking CAD to NC. To appreciate the problems associated with this linkage, we present the following discussion. The CAD/CAM technology has been used extensively to generate all kinds of drawings. For the purpose of our discussion, we can identify three types of engineering drawings. These are design drawings, detail drawings, and NC programming drawings. A design drawing is generally composed of a large number of components in a general arrangement with few or no dimensions. Design drawings are commonly known as assembly drawings. Detail drawings are invariably of one component, plus a complete set of dimensions together with all the other engineering attributes (tolerance, surface finish, hardness, etc.) that need to be specified for the purpose of manufacturing. A programming drawing is usually extracted from a detail drawing. It is composed of an ordered set of the geometrical entities that make the part to be produced, plus a complete set of machining requirements and instructions such as to tolerances, etc.

To assure the successful linking between CAD and NC, certain requirements must be imposed on how the detail drawings are generated to enable automatic extraction of NC programming drawings from them, and consequently the generation of the NC programs themselves. These requirements form a CAM constraint that must be considered at the CAD phase to avoid scrapping detail drawings entirely and creating new ones to develop NC part programs. Most of these requirements are organizational in nature and involve ordering the geometric entities of the part and separating its detail drawing information on various layers.

Sometimes, a detail drawing that looks fine from a drafting or CAD point of view is a garbled mess when transferred to an NC software package. Although some NC packages include "translation" routines that automatically format detail drawings, some do not, and others require the NC programmer to prepare the detail drawing for programming. For a successful electronic linking of CAD to CAM, or more specifically detail drawings to NC programming, designers (or design engineers) should keep the following set of guidelines in mind when entering CAD data that will be used for NC programming:

1. Part numerical data must be accurate. The coordinates must be accurate before the detail drawing is sent to the NC package. The accuracy of the data is directly related to the way the host computer of the CAD system stores and manipulates real numbers which occurs because of the inevitable variances that arise with real number to binary number conversion, manipulation in binary, and reconversion to real number form. These variances are the known round-off and truncation errors. The host computer normally stores data in binary format according to the floating-point arithmetic system used— whether it is single or double precision. In addition, any subsequent manipulation of graphics, such as translating, rotating, copying, mirroring, etc., could result in further cumulative errors. As a result, graphics that may visually seem acceptable and satisfactory on either a graphics screen or paper may be totally unacceptable at the database level for NC programming.

   The importance of the variances (errors) present between the stored and displayed numbers (sizes) is entirely related to the tolerance associated with the dimension. If the variances are of the same order of magnitude as the tolerances, then the latter are useless. Consider, for example, storing the dimension $10 \pm 0.0005$ in a database. If the nominal size is stored as 10.0003 instead of 10.0000, then the tolerance becomes meaningless. Many CAD/CAM systems let their users specify the accuracy of the database (single or double precision). It is also possible to write programs to check the accuracy of numbers stored in databases.

2. Part data should be toleranced properly. If parts are drawn using nominal dimensions, they must be edited to actual sizes before data are sent to NC programming.

3. Drawing tolerances should match manufacturing practice. Certain dimensions and tolerances are easier to program than others because of the limits of NC programs and machine tools. Bilateral tolerances are usually easier to program. Preferred dimensioning tolerance notation usually specifies the nominal size plus or minus a tolerance since compensation can be made for cutter wear, material deflection, and tooling setup inaccuracies.

4. Separate drawing annotation (notes and dimensions) from drawing data (geometrical entities). If such information is on the same layer as part geometry, the NC package might reject the part data file.

5. Establish standards for layer assignments. The layer facility available with most CAD/CAM systems can be employed to separate information within a detail drawing. By adopting a layer structure which controls the placement of graphics, subsequent isolation of geometry for NC part programming purposes can be effected in an easy and rapid manner. Assigning specific layers for data such as dimensions, notes, drawing border, and material lists makes it easier for different programmers and designers to work on the same files.

6. Profile entities must connect. A typical CNC machine tool will cut continuously only within 0.0001 in. It is, therefore, imperative that CAD databases meet or exceed machine tool accuracy.

7. Avoid overlapping drawing entities. If an arc or line is copied over an existing one, the resulting tool path will be unusable.

8. Dimensionality requirement must be specified. If a design calls for milling operations which require Z-axis motion, the CAD package used to create the part database must have three-dimensional capabilities. However, two-dimensional only packages would be sufficient for some applications such as routing.

9. Input part geometry in manufacturing order. Geometric entities should be input in the order in which they are connected to each other so that they form a continuous unidirectional (clockwise or counterclockwise) path that the cutting tool might follow. However, requiring designers to enter geometry in manufacturing order could severely curtail creativity and design productivity. Therefore, most existing NC packages reorder geometry automatically or semi-automatically. Part profile (geometric entities) is usually reordered according to coincident endpoints. Reordering on the basis of the endpoints usually generates imperfect results and requires editing. For instance, automatic reordering may default to the tool starting point which may or may not be optimum. To alleviate imperfect results, NC packages allow their users to input NC data before reordering, such as tool park position, direction of cut, beginning and ending of cutting, and whether to cut inside or outside the geometry. NC packages that accept IGES files generated on different CAD systems follow this approach. They may also give the user the option to order data in design or to have the NC programmer resequence it after IGES transfer. In addition, they may convert points and full circles to base points which are used as centerpoints for drilling, boring, tapping, and reaming.

## 20.6 PROCESS PLANNING

Process planning is a common task in discrete part manufacturing. It is the manufacturing activity responsible for the conversion of design data to manufacturing or work instructions. More specifically, process planning is defined as the activity that translates part design specifications from an engineering drawing into the manufacturing operation instructions required to convert a part from a rough to a finished state. Process planners must first evaluate design data and specifications such as geometric features, dimensional sizes, tolerances, surface finish, and materials of the part in order to select an appropriate sequence of processing operations and specific machines. Operation details such as setup and cut planning, stock preparation, jigs and fixtures planning, speeds, feeds, tooling, and assembly steps are then determined, and standard times and costs are calculated. The outcome of the design evaluation and the preparation of operation details is a process plan to produce the part. This plan is then documented as either a job routing (or operation) sheet, a cost estimate, or as coded instructions for NC machine tools.

Process planning, as viewed by some, represents the link between engineering design and shop-floor manufacturing. It is a major determinant of manufacturing cost and profitability. The gap between CAD and CAM can be shortened considerably by developing better systems for process planning. The difficulty in developing a reliable process planning system is attributed to the qualitative nature of many of the decisions involved in devising a process plan, in addition to the many variables that interact and which the process planner must consider. Actually, we can make the analogy between the nature of the conceptual design phase in engineering design and the nature of the process planning phase in manufacturing.

Before we discuss the various approaches to process planning, it is useful to examine the set of functions included in process planning. These are design input, material selection, process selection, process sequencing, machine and tool selection, intermediate surface determination, fixture selection, machining parameter selection, cost/time estimation, process plan preparation, and NC tool path generation. Many of these functions are interrelated, which makes a single universally applicable process planning for all manufacturing processes unlikely to be attained, or even attempted.

There are essentially three approaches to accomplish the task of process planning: the traditional manual approach; the computer assisted variant approach; and the computerized generative approach. The latter two approaches are known as computer aided process planning (CAPP). Each approach is appropriate under certain conditions. The superiority of any one approach can be evaluated in terms of a set of specific requirements. Knowledge of the nature, advantages, and limitations of each approach is therefore important.

## 20.6.1   Manual Approach

The traditional manual approach involves examining an engineering part drawing and developing manufacturing process plans and instructions based on knowledge of manufacturing such as process and machine capabilities, tooling, materials, related costs, and shop practices. This approach depends heavily on the personal experience and background of the process planner to develop process plans that are feasible, low cost, and consistent with other plans for similar parts. If the part to be produced belongs to an existing family of parts, the process planning usually involves recalling existing process plans of a similar part and modifying them to create a routing for the new part. Process plans are manually classified, stored in workbooks, and retrieved. If the part to be produced is new, the planner may have to generate a routing as a unique plan. The manual approach is highly subjective, labor-intensive, time-consuming, tedious, and often boring. Furthermore, it requires personnel well trained and experienced in manufacturing shop-floor practices.

Two levels of process planning (these also apply to the CAPP approaches) can be identified: high-level planning and low-level planning. During the high-level planning, the planner identifies the machinable features (surfaces) of the part, groups them into setups, and orders these setups. Each setup is listed in the order in which it is to be done, the features to be cut in each of the setups, and the tools for cutting each feature. The low-level planning includes specifying the details of performing each step that results from the first level such as choosing

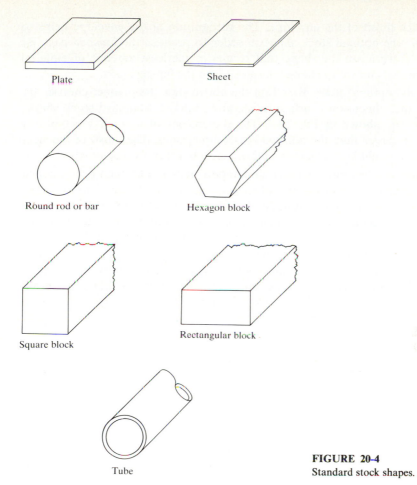

Plate

Sheet

Round rod or bar

Hexagon block

Square block

Rectangular block

Tube

**FIGURE 20-4**
Standard stock shapes.

machines, cutting conditions (speed and feed), type of fixturing, cost and time estimates, etc.

Most often, human process planners follow more or less a consistent set of steps to develop process plans for new products. These steps involve primarily stock preparation, plan generation, and specification of manufacturing parameters. The detailed steps can be listed as follows:

1. Get oriented. The process planner investigates the engineering drawing to identify the basic structure and the potential difficulties of the part to be produced. The planner reads the design data and specifications on the drawing and checks for any major problems: "Can the part be clamped or does it fit between the jaws of a vice?", "Is the part too long and thin in certain directions that it will bend when clamped?", and so on.

2. Recognize the outer envelope of the finished part. With the help of the engineering drawing, the planner can easily recognize the outer, or bounding, envelope of the part. The recognition includes both the geometric shape and

the surface finish of the envelope. The recognition of the envelope helps to determine the optimal stock shape in order to produce the finished part. By "optimal," we mean the shape that results in the least amount of material waste and/or removal to change the stock into the finished part.

3. Choose the optimal stock. Based on the above step, the planner chooses the stock shape, dimensions, surface finish, and material. Standard stock shapes exist and are shown in Fig. 20-4. The dimensions of a stock are typically about $\frac{1}{4}$ in larger than the finished part's dimensions. The finish of the sides of the stock could be rough (saw-cut) or smooth (rolled or machined).

4. Recognize part features. The second component of the finished part recognition (besides its outer envelope) is listing its features and subfeatures that are subtracted from its outer envelope. Sample examples of typical features and subfeatures encountered in manufacturing are shown in Figs. 20-5 and 20-6 respectively. Features are the individual geometric shapes that are cut into the stock to form the part. Although shapes of features are simple by themselves, they can be combined in many ways to form more complex shapes. Shapes that make useful features reflect the shapes of the tools, the move-

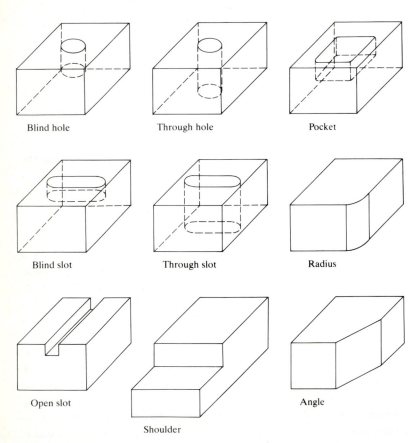

Blind hole      Through hole      Pocket

Blind slot      Through slot      Radius

Open slot      Angle

Shoulder

**FIGURE 20-5**
Sample manufacturing features.

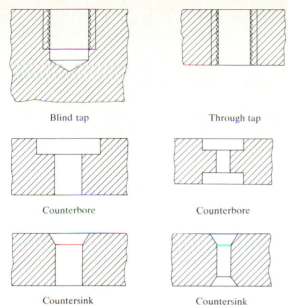

Blind tap                    Through tap

Counterbore                  Counterbore

Countersink                  Countersink

**FIGURE 20-6**
Sample manufacturing subfeatures.

ments of the machines, and the paths the tools can cut through the workpiece (stock).

5. Choose a stock preparation plan. The next step is for the planner to outline all methods for getting the raw material (stock) into an accurate shape with the minimum waste of material. This step serves as a framework to which the part features can be added. For example, sides of a stock normally need to be "squared off." The stock preparation plan represents the constraints on the order in which each of the sides may be squared off. This order is stored in what is sometimes called the squaring graph. Because there are only a limited number of ways in which a block can be squared up and still maintain high accuracy, the majority of stock shapes and sizes can be squared up using a limited number of squaring graphs. Some of the squaring rules utilized to develop squaring graphs are: machine the largest area side first; machine a medium area side second; machine the remaining sides in any order. Figure 20-7 shows the squaring graph of a rectangular block. Besides following the above squaring rules, this squaring plan takes advantage of the length to obtain squaring accuracy and compensates for the difficulty that all sides are rough at first. Also, when two arrows are branching at the same level as for setups C and D, the corresponding setups can be done in any order.

6. Consider alternative methods for producing each feature. There may be many ways of making each part feature recognized in step 4. For example, to make an angle, a sine table, sine bar, angle plate, or an angle cutter might be used. Some ways of producing a feature are better than others, so the alternatives generated must be ordered according to preference. The preference is usually determined based partly on judgment and partly on strategy. Judgment is related to the particular characteristics of the part. The planner may

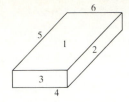

(a) Stock with labeled sides

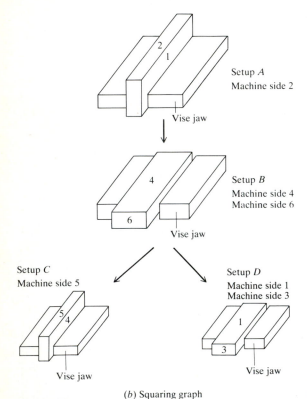

Setup A
Machine side 2

Vise jaw

Setup B
Machine side 4
Machine side 6

Vise jaw

Setup C
Machine side 5

Vise jaw

Setup D
Machine side 1
Machine side 3

Vise jaw

(b) Squaring graph

**FIGURE 20-7**
Squaring graph of a rectangular stock.

ask questions such as: "If I make an angle using the sine bar, can the part be clamped firmly enough?" Strategy, on the other hand, determines how trade-offs are to be made during manufacturing. For example, the planner may ask questions such as: "Which is more important for the part: cost or accuracy?", "Is it worth risking breaking a tool to discover a faster way to make the part?", "Is the part a prototype or is it used in mass production?"

7. Generate a plan by exploring feature interactions. Utilizing the outcomes of step 6, the planner is able to generate a plan to produce the part. In generating such a plan, one can view the collection of features as subgoals to be achieved in the machining plan. The difficulty in making a plan is to find an order to machine the features in which no subgoal interferes too seriously

with achieving the others. Feature interaction happens when machining operations of one collection of features affect or destroy the way in which other features can be machined. Feature interactions have several different causes. Most commonly they result from clamping or reference problems. For example, producing one feature may destroy the clamping surfaces needed to grip the workpiece while cutting another feature. In more general terms, interactions arise when the result (postcondition) of making one feature destroys a requirement (precondition) needed for making other features. The result of cutting a feature is the removal of a volume from the part. The requirements are things such as reference surfaces (datums or clamping surfaces), features on the part, tools, and fixtures. Consider, for example, drilling a through hole while clamping the part in a vise. We would have two reference (datum) machined sides from which to measure the hole position, two parallel machined sides on which to clamp (with sufficient surface area to provide rigid clamping), and a drill of the appropriate diameter. The final generation plan of the part may be expressed graphically in what may be called a feature interaction graph (see the example at the end of this section).

8. Integrate the squaring graph with the feature interaction graph. The planner now has a graph showing the order in which the sides may be cut (from step 5) and a graph showing the orders in which the features can be produced (from step 7). Each graph represents a separate set of constraints on the process plan. The two graphs must be merged with as much overlap between the steps as possible so that we can get a compact sequence. The more overlap, the better and more concise the final plan. The planner uses the requirements of each setup in the interaction graph and looks for the earliest similar setup in the squaring graph that meets those requirements. If one is found, the interaction setup is combined with the squaring setup. If a setup that meets the requirements is not found, then a new setup must be created and added to the squaring graph. The process continues until all setups in the interaction graph have been merged with or added to setups in the squaring graph. The resulting graph is the final plan outline and represents the final ordering for the desired process plan. We refer to the resulting graph as a plan graph.

9. Make a final check on the plan. At this point, the planner typically verifies the plan by checking that the setups are actually feasible, that the clamps are not in the way of the tools, etc.

10. Elaborate the process plan. The above nine steps represent the high-level planning of the process plan. This step represents the low-level planning. Now the planner generates more details for producing each individual feature, choosing feeds and speed, estimating costs and standard times, etc. The planner verifies all these details and releases the process plan to the various departments for execution.

> **Example 20.1.** Figure 20-8a shows a typical part. The part has a slot and two tapped holes with countersinks (on side 4). Develop the plan graph that can successfully produce the part. Show the squaring graph and the interaction graph as well.

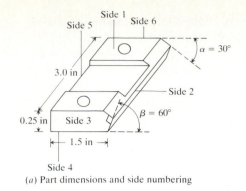

(a) Part dimensions and side numbering

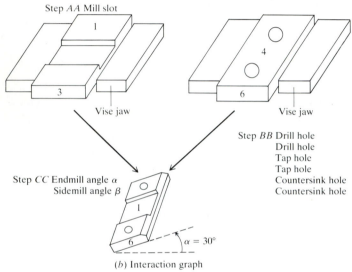

(b) Interaction graph

**FIGURE 20-8**
Typical part with its interaction plan.

***Solution.*** The above ten steps are followed here to develop the plan graph. Figure 20-8a shows a part with five features: a slot in the center, two holes, and two angles α and β. The part also has four subfeatures: two taps and two countersinks. The appropriate stock to produce this part is a block of dimensions 1.75 × 0.5 × 3.25 in (following the rule of a quarter inch larger than the finished part's dimensions). Assuming the stock block to be saw-cut, Fig. 20-7 shows the correct squaring graph. Studying the interactions of the part five features reveals some interesting observations.

The two holes interact with the angle α and the slot interacts with the angle β. In both cases, making the angle first will cause difficulty in gripping the part to make the other feature. The restriction that this interaction puts on the plan is that the two holes must be made before angle α and the slot before angle β. Thus, one would cut the slot and drill the two holes before cutting the two angles.

Because there are no tolerance specifications on the part dimensions, the process planner must assume them to be tight and recommend the machining methods accordingly. In situations like this, the planner usually requests a single roughing and finishing pass on the machine. If tolerances were specified, the planner could take advantage of this information to make the plan more efficient by reducing the number of machining steps (by requesting, say, a single roughing pass only on the machine and eliminating the finishing pass).

Figure 20-8*b* shows the resulting interaction graph for this part. The graph shows the machining sequence in which features are eliminated from the stock to produce the finished part. Each step in the sequence shows the machining activities that must be achieved. As the figure shows, the order of performing steps *AA* and *BB* is unimportant.

We now merge the squaring graph (Fig. 20-7*b*) with the interaction graph (Fig. 20-8*b*) to obtain the final plan graph. Investigating both graphs reveals that setup *D* of the squaring graph overlaps with the setup requirement of step *AA* in the interaction graph. Therefore, these two setups can be combined. This implies that while setting up the stock to machine its sides 1 and 3, we mill the slot afterwards. This results in saving one unnecessary setup. Figure 20-9 shows the plan graph to produce the part from the rectangular saw-cut stock.

Figure 20-9 can be translated into the following process plan (shown at the high-level planning only):

```
SET-UP SIDE 2
ENDMILL SIDE 2
SET-UP SIDES 4 AND 6
ENDMILL SIDE 4
SIDEMILL SIDE 6
SET-UP SIDE 5
ENDMILL SIDE 5
SET-UP SIDES 1 AND 3
ENDMILL SIDE 1
SIDEMILL SIDE 3
ENDMILL SLOT
SET-UP SIDE 4
DRILL HOLES
TAP HOLES
COUNTERSINK HOLES
SET-UP ANGLES α AND β
ENDMILL ANGLE α
SIDEMILL ANGLE β
```

The last step to fully develop the process plan is to perform the low-level planning by adding the required details for each operation and rewriting the plan utilizing customary process planning codes known to process planners. To be able to specify the proper cutting conditions (feeds and speeds), the planner must know the material of the part. Aluminum or steel can be used. As part of the low-level planning the planner can add the cost and time estimates as well as the final inspection of the finished part.

## 20.6.2 Variant Approach

The manual approach for process planning has some benefits and drawbacks. It is often the best approach for small companies with few process plans to generate and with good process planners on board. The approach is flexible and has low

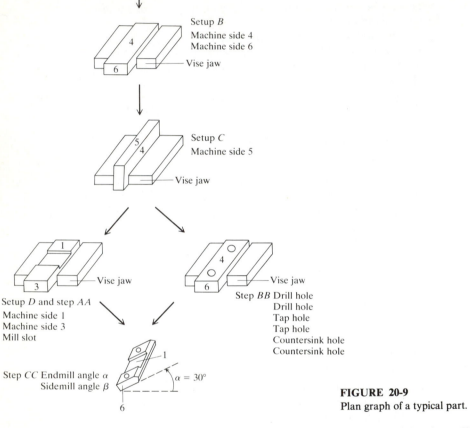

Setup *A*
Machine side 2
Vise jaw

Setup *B*
Machine side 4
Machine side 6
Vise jaw

Setup *C*
Machine side 5
Vise jaw

Vise jaw

Setup *D* and step *AA*
Machine side 1
Machine side 3
Mill slot

Step *BB* Drill hole
Drill hole
Tap hole
Tap hole
Countersink hole
Countersink hole

Vise jaw

Step *CC* Endmill angle $\alpha$
Sidemill angle $\beta$

$\alpha = 30°$

**FIGURE 20-9**
Plan graph of a typical part.

investment costs (no computer hardware or software is needed and neither is staff training). However, the drawbacks of the manual approach are manyfold. The approach becomes rapidly inefficient and unmanageable when the number of process plans and revisions to those plans increases. Consequently, inconsistent plans and large time requirements for planning often result. Furthermore, manu-ally generated process plans always reflect the personal experiences and prefer-ences of the process planner. The manual approach is also considered a poor use of engineering skills because of the high clerical content in most of its functions.

Due to the above limitations and drawbacks of the manual approach, the computerized approach to process planning is only logical. With the computer speed and consistency, better and faster process plans can be generated. This is crucial in an environment where the rate of change of design is increasing, the product life cycle is decreasing, the need to satisfy customer requirements more quickly is increasing, and manufacturing technology itself is subject to an increasing rate of change. With the computerized process planning approach, the following benefits can be attained: clerical effort can be reduced; lead times can be cut; consistency of application of manufacturing data can be improved; the accuracy with which data is used can be improved; a standardized method of manufacturing can be introduced; records can be better maintained; and manufacturing logic can be captured and used by anyone.

The variant approach (also called the retrieval approach), one of the two existing CAPP approaches to process planning, is a computer assisted extension of the manual approach. A process plan for a new part can be created by recalling, identifying, and retrieving an existing plan for a similar part and making the necessary modifications required by the new part. In other words, the variant approach creates new process plans by editing existing ones. The computer assists by providing an efficient system for data management, retrieval, editing, and high-speed printing of process plans.

The two major functions of a variant CAPP system are the creation of a new plan and the modification of an existing one. The creation of a plan requires a detailed knowledge of the manual process planning itself, that is, knowledge of machine tool capabilities, the types of tools, and the gage requirements. Also required is a knowledge of manufacturing processes, how parts are to be made, how they can be held, what should be done first, and so on. Another area of knowledge is in the detailed manufacturing data (speeds, feeds, etc.) and how to use it to generate time and cost estimates.

The modification of an existing process plan is the most common function performed on a variant CAPP system. The variant system typically has a catalog or menu of standard process plans. All plans are classified, coded, and stored based upon a coding scheme that codes parts into their families of parts. There is a master process plan, stored by the system, for each family of parts that can be retrieved and modified for any part that belongs to the family. When master plans are retrieved from the system, planners rely on their experience and manufacturing knowledge to either accept or edit these plans. A variant system also has the ability to examine the coding scheme of the part being planned in order to identify and retrieve any existing master plan that matches this part.

There are five main components to any variant CAPP system, as follows:

1. Coding and classification. This is an essential component that makes the system efficient or not. Similar parts are classified and coded using group technology techniques. Parts are usually grouped for process planning purposes. Thus, all parts in a family must require similar plans. The master plan for the family is therefore shared by the entire family.

2. Database creation and maintenance. The design of the data structure of the CAPP system and the creation and maintenance of its database depends on the needs of potential users of the system. However, the likely contents of the database are synthetic data, work patterns, documentation text, and tabular data. Synthetic data covers the elements of time and cost that a company uses in building up its work values. Some companies use a predetermined motion time system; others use time-studied values. Work patterns refer to the large building blocks in the database. These blocks define the relations between

smaller data elements (blocks). Cross-referencing (via pointers) between the smaller and larger blocks ensures the consistency of the data and information stored in the database. The documentation text is the text used in generating a process plan. This text forces all planners to use the same text and the same descriptions. The tabular data consists of the database records that describe the various manufacturing capabilities such as tooling, gaging, etc.

3. Logic processor. This is the heart of any CAPP system. It ensures that different people plan the same manufacturing method in the same way. As in the manual approach, such processors fall into one of two categories: either low level or high level. A low-level logic processor is one that stores the detailed method of manufacturing at the operation level. The high-level logic processor is one that is capable of specifying the sequence of the process plan itself. In fact, the high-level logic processor is the key factor in integration CAD and CAM.

4. Documentation production. Obviously, if the manufacturing sequence can be determined within the CAPP system, it is sensible that documents used at the shop-floor level for controlling manufacturing should be produced by the computer. Typical documents that are in fact produced from a CAPP system are method sheets, routing sheets, and toolkit sheets. Method sheets are typical manufacturing instructions at an operation level (low level) and give details of the exact method an operation must follow, including details such as speeds, feeds, tooling, etc. Routing sheets give details of the work centers through which the part passes and may include details of time value, tooling, etc. (high-level information). Toolkit sheets act as a listing from which tools can be ordered and/or designed prior to the beginning of production.

5. File maintenance. This covers the storage and control of records, and the ability to retrieve and edit data as required.

The variant approach to process planning has several advantages. It removes the clerical work from process planning; all data management, retrieval, and text editing are done on the computer. In addition, it forces companies to classify and code their parts into families to ensure the consistency of master process plans. However, the variant CAPP system is substantial. When new machines or processes become available, the existing database and files must be manually reviewed, revised, and tested by the process planner. Second, a significant financial investment on the company part may be required to cover the cost of the system, classifying and coding the parts and creating and storing the database. Third, the biggest disadvantage is that an experienced process planner is required to use a variant CAPP system. The knowledge and experience of the user are the key factors in determining the quality of the resulting plans. In effect, the computer is just a tool to assist in manual process planning activities.

### 20.6.3 Generative Approach

The generative process planning approach is viewed as the automated approach to process planning. Unlike the variant approach, the generative approach does not require assistance from the user to generate a process plan. It usually accepts part geometrical and manufacturing data from the user, and utilizes computerized searches and decision logics to develop the part process plan automatically. Generative process planning systems do not require or store predefined master process plans. Instead, the system automatically generates a unique plan for a part every time the part is ordered and released for manufacturing.

By definition, the generative approach generates a plan. In generating such a plan, an initial state of the part (stock) must be defined in order to reach the final state (goal or the finished part). The path taken between the two states defines the process plan or the sequence of processes. Two types of planning are available: forward and backward planning. Using forward planning, we begin with the stock as the initial state and remove part features until the finished part, the final state, is obtained. Backward planning uses the reverse procedure. We begin with the finished part as the initial state, and our goal is to fill it with part features until the stock is obtained. In backward planning, a drilling process fills a hole instead of its conventional meaning in forward planning. Forward and backward planning are not similar, as they may seem, and they affect the programming of a corresponding system significantly. The requirements (preconditions) and the results (postconditions) of a setup in forward planning are the results and requirements respectively of the setup in backward planning. While forward planning seems natural, why does one need backward planning? Forward planning suffers from conditioning problems; the result of a setup affects the next setup. In backward planning, conditioning problems are eliminated because setups are selected to satisfy the initial requirements only.

A generative process planning system consists of two major components: coding scheme and decision logic. The coding scheme is geometry based and defines all geometric features, feature sizes and locations, and feature tolerances for all process-related surfaces. Furthermore, the coding scheme must relate these features to their manufacturing requirements; it must relate them to the various manufacturing processes, the individual machine in each process, the available tooling on each machine, and the equipment for part positioning and clamping. The coding scheme can therefore be viewed as a universal language that relates all elements of manufacturing.

The decision logic involves the structuring of manufacturing planning logic and data into formats that will facilitate program coding and documentation. The decision logic results in determining the appropriate processing operations, selecting the machine for each operation, determining cut planning or other operation details subject to available tooling and fixturing, and calculating the setup and cycle time and cost for each operation.

There are various difficulties encountered when developing a decision logic. The decision logic software usually attempts to capture all aspects of manual process planning. In doing so, it is not usually straightforward to synthesize the manual steps. Furthermore, it is difficult to capture and program all the heuristics used in process planning. It is also a challenging task to develop and code the rules that can be used to judge and determine feature interactions. Typically, not all features of a part interact with each other. In the manual approach, the process planner can quickly filter the unnecessary interactions. However, this is a difficult task to achieve algorithmically.

There are various approaches used to build and develop decision logic. One approach that seems quite acceptable is the flowchart approach. This approach attempts to synthesize and computerize the manual process planning approach described earlier. It requires exhaustive collection and analysis of many existing process plans to extract the proper decision logic. Another approach uses decision trees. A decision tree is a natural way to represent information. Conditional statements are usually used to traverse the tree to make the proper decisions. Using decision tables is another approach to implement decision logic. Decision trees and tables based on various criteria can be built. They can be created based on structuring manufacturing processes and data by families of parts. They can be based on defining the manufacturing capabilities of each process and each piece of equipment in terms of a universally common geometry-based language. A third way is to base them on the manufacturing capabilities of each machine.

Expert systems techniques are becoming a viable option to develop decision logic for process planning. In a knowledge-based generative system, knowledge is represented by rules. Rule-based deduction is frequently used to make decisions. Two types of knowledge are needed in process planning: component knowledge and process knowledge. The former refers to the state of the part at any time while the latter defines how a component can be changed by processes. The details of knowledge-based generative systems are not covered here.

The generative approach has all the advantages of the variant approach. It also has the additional advantages that it is fully automatic and that an up-to-date process plan is generated each time a part is ordered. However, a generative system would require major revisions in decision logic if new equipment or processing capabilities were to become available. In addition, the development of the system in the first place is a formidable task.

## 20.6.4   Hybrid Approach

One can combine some of the characteristics of the variant and generative approaches to develop useful and successful hybrid process planning systems. For example, some built-in decision logic can be embedded in generative systems. In addition, some generic plans can even be stored and used as modules by decision logic in generative systems. The stock preparation plan is an example. Having standard stock shapes and knowing all the possible ways to prepare a given stock, such information can be made available to the decision logic in order to shorten it. Other similar techniques may import useful canned information from the variant approach to use with the generative one.

## 20.6.5   Geometric Modeling for Process Planning

In the context of process planning covered in this chapter and geometric modeling covered in the book, it is obvious that geometric information as required by process planning (identifying machinable surfaces or features and tolerances) is not readily available in the CAD database. Additional algorithms to identify machined surfaces from a CAD model are needed and may not exist. This leaves the option of user intervention and the human-computer interactive system as the solution to identify these surfaces. Solid modeling theory has the promise to meet this need. Feature-based modeling offers an attractive geometric modeling approach for process planning and other machining activities.

## 20.7 PART PROGRAMMING

One of the outcomes of process planning is the method sheet which lists the manufacturing instructions for various machines to produce the finished part. By producing these sheets, we imply that machine operators are humans who read the instructions and operate the machines accordingly. In an attempt to automate part production, the information in these sheets can be utilized directly to program machine tools to machine the part. In this section, we cover the basics of part programming. Readers should realize that the topic of part programming deserves a book by itself, and they should refer to existing books for further information.

We can view the various types of manufacturing systems presented in Sec. 20.3 as various attempts to automate manufacturing to meet the increasing demand for accuracy and production. Modern manufacturing systems utilize computers as an integral part of their control. The concept of an automatically controlled factory represents the central goal of the automation revolution. Simple production machines and mechanization that were introduced in the late 1700s were replaced by fixed automatic mechanisms and transfer lines around the 1900s. Next came machine tools with simple automatic control.

The introduction of machine tools was not enough to enable discrete part manufacturing to cope with the requirements facing it. Highly skilled operators were required to operate these machine tools to produce parts. Manual operations were not fast or accurate enough, and therefore prompted the need to automate part machining. The premise was that if the machining steps of a part could be captured in a program (called part program) and the control system of the machine tool could read it, then the part can be machined automatically without the need for human operators. Part programming is considered the turning point of the metal-cutting industry to meet the demands of intricate designs, with tight tolerances, which could never have been machined using the manual approach.

The premise of part programming was seriously considered during the Second World War. The aircraft and missile demands of the U.S. Airforce, combined with the demands for commercial jets, were not possible to meet by conventional manufacturing. A U.S. government study in 1947 showed that the entire U.S. metal-cutting industry could not produce the parts needed by the Airforce alone. As a result, the U.S. Airforce has contracted the Parsons Corporation to develop a dynamic and accurate manufacturing system that both meets the accuracy requirements and allows design changes easily and inexpensively. The Parsons Corporation subcontracted the development of the control system to the MIT in 1951. In 1952, the MIT developed and demonstrated a machine tool (Cincinnati Hydrotel Milling Machine) with the new technology which was named numerical control (NC). Ever since, machines that use this technology are known as NC machine tools. NC opened a new era in automation.

The NC technology is based on controlling the motion of the drives of the machine tool as well as the motion of the cutting tool via an NC part program. The program is a set of statements which can be interpreted by the machine control system and converted into signals that move the machine spindles and drives. NC has been defined by EIA (Electronic Industries Association) as "A system in which actions are controlled by the direct insertion of numerical data at some point. The system must automatically interpret at least some portion of the data." The numerical data required to produce a part is known as the part program.

The first generation of NC machine tools used vacuum tube technology, and the second generation utilized improved electronic tubes and later solid-state circuits. Both generations would read part programs prepared on certain input medium known as punched tapes. A part program is stored on a punched tape which can be read by the machine controller. The third generation of machine tools used much improved integrated circuits and NC control builders introduced ROM (read only memory) technology from computer hardware into the controllers of this generation. This led to the appearance of the CNC (computer numerical control) machine tools. Display monitors were later added to the machine for visual editing of part programs. CNC machines allowed the storage of part programs directly in the memory of the machine computer. The CNC concept made it possible to operate machine tools directly from memory. Programs that are always used in routine operation were programmed as subroutines (called canned cycles) and stored directly into the ROMs. The CNC concept facilitated editing part programs and eliminated all the problems related to reading punched tapes. The fourth generation of CNC machines has better storage capabilities (using bubble memory technology), faster memory access, and macro capabilities. This generation sets the stage for DNC (direct numerical control) machines.

While the NC hardware technology has been improving steadily, the NC software has been advancing as well. NC programming has benefited a great deal from the CAD technology and the advancements in geometric modeling. NC software permits NC programmers to generate tool paths for complex parts using part geometry and to verify these paths visually on graphics displays. NC tool path generation and verification have eliminated manual programming and have allowed machining sculptured surfaces which were not possible to machine before.

## 20.7.1  Fundamentals of NC

In a typical NC system, the part program is stored on a punched tape, while it is stored in ROM in a CNC system. For the rest of the chapter, NC will imply CNC (for ease of reference) unless otherwise stated explicitly. A part program is a combination of machine tool code and machine specific instructions. It contains geometric information about part geometry and motion information to move the cutting tool. Cutting speed, feedrate, and auxiliary functions (coolant on and off, spindle direction, etc.) are programmed according to surface finish and tolerance requirements.

In conventional machining, the machine operator produces a part by moving a cutting tool along a workpiece by means of the handwheels of the machine tool. Expert operators are needed to perform contour cutting by sight. The NC system replaces the manual actions of the operator typically required in conventional machine tools. Thus, operators of NC machine tools do not have to be skilled machinists as they are only required to monitor the operation of the machine and set up workpieces. Part programs contain all the thinking operations. However, alert intelligent operators with good judgment are usually required to operate NC machine tools since they are sophisticated and expensive.

Part programmers usually prepare part programs for NC machine tools. They must have knowledge of manufacturing requirements such as tools, cutting fluids, fixture design techniques, use of machinability data, and process engineering. In addition, they must be familiar with the function of NC machine tools and machining processes.

**20.7.1.1  DESCRIPTION OF MACHINE TOOLS.** Figure 20-10 shows a schematic of an NC machine tool. A typical NC machine tool contains the MCU (machine control unit) and the machine tool itself. The MCU (also known as the controller unit) is considered the brain of the machine. It reads the part program and controls the machine tool operations. After reading the part program, the MCU decodes it to provide commands and decoded instructions to the various control loops of the machine axes of motion.

The MCU performs two functions: it reads the part program and controls the machine tool. It consists of two units, one for each function. The DPU (data processing unit) reads and decodes the part program statements, processes the decoded information, and provides data to the CLU (control loop unit), the other unit. The CLU receives the data from the DPU and converts it to control signals.

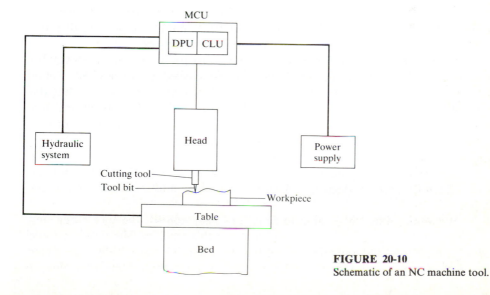

**FIGURE 20-10**
Schematic of an NC machine tool.

The data usually provides the control information such as the new required position of each axis, its direction of motion and velocity, and auxiliary control signals to relays. The CLU also instructs the DPU to read new instructions from the part program when needed, controls the drives attached to the machine leadscrews, and receives feedback signals on the actual position and velocity of the axes of the machine. Each axis of motion of the machine tool has its own leadscrew, control signals, and feedback signals.

In the first- and second-generation NC machine tools, the MCU has a tape drive that is able to mount and read punched tapes that contain part programs. The tape drive is part of the DPU of the machine. In CNC and DNC machine tools, the DPU is replaced by the ROM and the tape drive is eliminated all together.

**20.7.1.2   MOTIONS OF MACHINE TOOLS.** A workpiece is machined to the finished shape by allowing a relative motion between the workpiece and the cutting tool. Such relative motion can be provided by holding the workpiece stationary and moving the cutting tool as in milling and drilling, or by moving the cutting tool and the workpiece simultaneously as in turning. Regardless of whether the tool or the workpiece moves, each motion adds to the versatility of the machine tool and requires its own axis of motion. A machine tool may have more than one head to provide additional axes of motion.

In NC machine tools, each axis of motion is equipped with a driving device to replace the handwheel of the conventional machine. The type of driving device is selected mainly by the power requirements of the machine. A driving device may be a dc motor, a hydraulic actuator, or a stepper motor.

An axis of motion is defined as an axis where relative motion between the cutting tool and workpiece occurs. This movement is achieved by the motion of the machine tool table (slides). The primary three axes of motion are referred to as the $X$, $Y$, and $Z$ axes and form the machine tool coordinate system. The $XYZ$ system is a right-hand system and the location of its origin may be fixed or adjustable. The positive directions of the axes are usually defined by the manufacturer of the machine tool. However, it is a convention that the positive direction of the $Z$ axis moves the cutting tool away from the workpiece.

In addition to the primary slide motions in the $X$, $Y$, and $Z$ directions, secondary slide motions may exist and may be labeled $U$, $V$, and $W$. Rotary motions around axes parallel to $X$, $Y$, and $Z$ may also exist and are designated $a$, $b$, and $c$ respectively. These notations are EIA standards. It is conventional that machine tools are designated by the number of axes of motion they can provide to control the tool position and orientation. Most often, we hear terminology such as $2\frac{1}{2}$-axis, 3-axis, or 5-axis milling machines. These axes refer to the possible axes of motion the machine tool can control simultaneously. We should not correlate them with spatial degrees of freedom to avoid confusion and misinterpretation.

If the machine tool is able to simultaneously control the tool along only two axes, it is classified as a 2-axis machine. In this machine, the tool is parallel to and independently controlled along the third axis. Geometrically, this means that the machine tool control system is able to guide the cutting tool along a

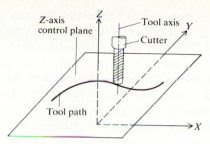

(a) 2-axis machine tool
  (Z-axis control plane is parallel to XY plane;
  tool axis is parallel to Z axis)

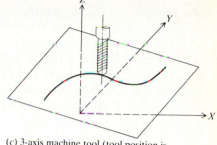

(c) 3-axis machine tool (tool position is
  simultaneously controlled in X, Y, and Z
  directions; orientation of tool axis does not
  change with tool motion)

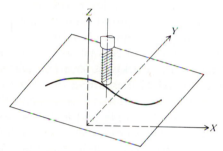

(b) $2\frac{1}{2}$-axis machine tool
  (Z-axis control plane is arbitrarily oriented in
  space; tool axis is parallel to Z axis)

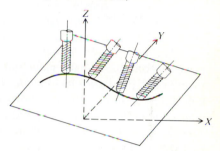

(d) Multiple-axis machine tool
  (tool position is simultaneously controlled
  in X, Y, and Z directions; orientation of tool
  axis changes simultaneously with tool motion)

**FIGURE 20-11**
Classification of machine tools according to simultaneous control of motion axes.

two-dimensional contour with only independent movement specified along the third axis. Figure 20-11a shows an example of a 2-axis machine. Here the Z-axis control plane is parallel to the XY plane. If the tool can be controlled to follow an inclined Z-axis control plane, we have a $2\frac{1}{2}$-axis machine as shown in Fig. 20-11b. In a 3-axis machine, the tool is controlled along the three axes (X, Y, and Z) simultaneously, but the tool orientation does not change with the tool motion, as shown in Fig. 20-11c. If the tool axis orientation varies with the tool motion in three dimensions, we have a multiaxis orientation machine (4-, 5-, or 6-axis). A 6-axis machine, for example, is capable of moving the tool simultaneously along each primary axis and, in the meantime, simultaneously rotate it about each primary axis. Figure 20-11d shows a multiaxis machine tool. Up to 9-axis machine tools may be commercially available.

**20.7.1.3  CLASSIFICATION OF MACHINE TOOLS.** The classification of NC machine tools can be done according to two main criteria. According to the type of machining, we can divide them into point-to-point (PTP) and continuous-path machining. According to the structure of the controller unit (MCU), there are NC, CNC, DNC, and adaptive control machines. The main features of each classification is discussed below.

PTP NC machining is considered the simplest type of machining. An obvious example is a drilling operation. In PTP machining, the cutting tool performs operations on the workpiece at specific points. The tool is not always in contact with the workpiece throughout its motion or its path. The exact path the tool takes in moving from point to point is immaterial assuming, of course, that the time required is reasonable and the tool does not collide with either the workpiece or the holding fixture. Straight line tool paths from one point to another are common. Drilling holes shown in Fig. 20-12a provides an example of PTP machining. In a drilling machine, the drill is moved, according to NC instructions, to a position directly over the hole to be drilled. Once in position, the drill is lowered at a predetermined speed and feed. After drilling a hole, the drill moves out of the hole at a rapid retract rate. The drill moves to another point and the cycle is repeated. Figure 20-12a shows the tool path for a typical drilling operation. The PTP concept is also used in other applications such as punching holes using a turret punch press and bending tubes using a tube bender.

Unlike PTP machining, the cutting tool is always in contact with the workpiece in continuous-path machining. Therefore, the workpiece is being affected throughout the entire movement. The entire travel of the cutting tool must be controlled to close accuracy both as to position and velocity. As a result, the control unit for a continuous path machine is more complicated and expensive than for a PTP machine. Milling operations provide an example of continuous machining. Figure 20-12b shows contour milling. NC milling machines (or centers as they are sometimes called) are very popular. This may be attributed to their versatility. NC milling machines can be used for PTP machining as well as for continuous-path machining. Obviously PTP machine tools cannot be used for continuous-path machining. NC lathes provide another example of continuous-path machining for NC turning.

Considering the structure of the controller unit, machine tools can be divided into NC, CNC, and DNC machines. In an NC machine tool, the DPU is

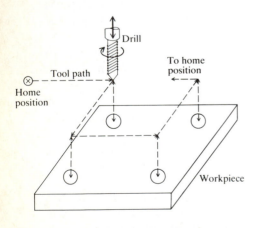

(a) PTP machining (drilling holes)

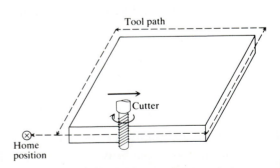

(b) Continuous-path machining (milling a contour)

**FIGURE 20-12**
Classifying machine tools according to types of machining.

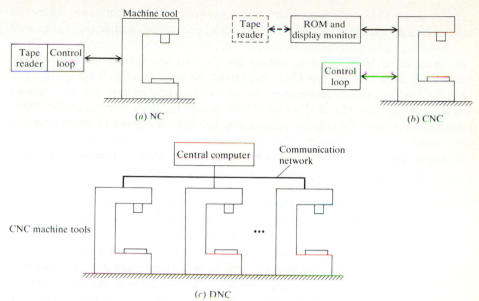

**FIGURE 20-13**
Classifying machine tools according to controller structure.

a tape reader, as shown in Fig. 20-13a. Each time the NC program needs to be executed to machine a new part, the punched tape must be read via the tape reader. In a CNC machine tool, the DPU is a ROM installed aboard the machine with a display monitor as shown in Fig. 20-13b. ROMs and display monitors have been enhanced enough that they are sometimes called computers aboard the CNC machine tools.

In some CNC machine tools, tape readers still exist. In the context of CNC, the reader is usually used to read an NC program only once to originally load the program to the machine ROM. In addition to editing and storing NC programs, CNC controllers perform other important functions such as dynamic corrections of machine tool motions for changes or errors that occur during processing, and diagnostic capabilities to assist in maintaining and repairing the machine tool system.

The CNC concept has few drawbacks. If the same NC program is used on various machine tools, it has to be loaded separately into each machine, thus unnecessarily repeating nonproductive tasks. More importantly, CNC systems are limited in the area of feedback. It has become necessary for some sort of feedback of information to be generated by the CNC system to track such variables as machine downtimes, work-in-process, production rates, scrap rates, and other production data. The DNC structure (Fig. 20-13c) has been created for this purpose. A DNC system consists of a central computer to which a group of CNC machine tools are connected via a communication network. The communication in a DNC system is usually achieved using the defacto standard protocol MAP. The central computer has various functions. It stores the NC programs. It also downloads these programs to any number of CNC machine tools in the network, thus avoiding the reentrance of the same program. It also performs the entire feedback task described above (data collection, processing, and reporting) as well as providing communications between the various components of the DCN system. In some DNC systems, various hierarchical levels of computers and networks may exist between the CNC machine tools and the central computer. The intermediate computers (sometimes called satellite computers) provide various levels of local controls of different CNC machines that belong to the DNC system. As a result of this philosophy, DNC has been redefined by many to mean distributed numerical control. The major advantage of a DNC system is that it centralizes the system monitoring. This is important when dealing with different operators, on different shifts, working with different machines.

**20.7.1.4   NC PROGRAMMING PROCEDURE.** Few steps are required to produce a part using NC programming. The first step is to produce a process plan for the part using its engineering drawings. This step identifies which NC operations are needed. The second step is to get an NC programmer to program the part. Using their machining and NC knowledge, part programmers plan the required tool path(s) with the proper cutting parameters. The tool path is the path that the cutting tool must follow from its home (park) position to machine the part and back to the home position. The path is usually repeated more than once (each time is called a pass) to finish the part. Each pass, the tool is fed further into the stock to eliminate more material and eventually change the stock into the finished part.

Once the path is planned, NC programmers utilize their NC knowledge to generate the details of the path. This includes calculating the $x$, $y$, and $z$ coordinates of the necessary points of the path, writing the NC program using the syntax of one of the programming languages such as APT, COMPACT II, etc., verifying the generated tool path, postprocessing the program, and finally using the program to produce the actual part. The calculations of the point coordinates of the path involve the geometry of the part profile (boundaries) and requires trigonometric and geometric skills of the programmer if the tool path is generated manually.

Programming languages such as APT are considered high level languages. They provide the programmer with geometric statements that utilize the calculated coordinates and machining statements that control the motion of the cutting tool as well as the cutting conditions. NC controllers are usually designed to accept low level languages utilizing G-code and M-code. Therefore, APT programs are usually postprocessed to generate the low level programs. Before using the NC program in production, it is usually necessary to verify it for possible programming mistakes. Therefore, the tool path is either plotted or displayed on a monitor to check visually. Once verified, the NC program is ready for use to produce the part.

If NC programming is used in an integrated CAD/CAM environment, the tool path generation, verification, and postprocessing is performed by the NC package which is part of the overall software. Utilizing the CAD database with all its geometrical information, the required coordinates to define the path are automatically evaluated. The NC software usually provides programmers with tool libraries that they can use to define various tools with the proper tool geometric parameters. For example, the programmer can define a drill or a cutting mill. Once a tool is defined, a tool path can be generated using the defined tool and the part geometry. When a tool path is generated, it can be verified on the graphics display. This usually involves the display of an animation sequence that shows the tool moving along its generated path which is superimposed on the part geometry. Colors and shaded images are usually used to enhance the visualization process. If mistakes are spotted, the tool path can be modified accordingly and reverified. Once accepted, time and cost estimates can be calculated. In addition, the NC package can generate the APT program automatically from the tool path information stored in the part database. Finally, the NC package can postprocess the APT program to translate it to the proper format (G- and M-codes) that a particular NC controller requires.

## 20.7.2 Basics of NC Programming

NC programming is quite different from high level programming languages we are accustomed to such as FORTRAN, Pascal, C, etc. This is attributed to the fact that NC programming deals with and controls machine tools with all their related components such as the machine table and coolant system, cutting tools, jigs and fixtures, workpieces, etc. Of course, NC programming languages have semantics and syntax to follow. Before we discuss the details and syntax of some of these languages, we shall review some of the concepts and basics typically employed in writing NC programs.

### 20.7.2.1 MACHINE TOOL COORDINATE SYSTEM. Most of the existing machine tools use cartesian coordinate systems. The origin and orientation of these systems are usually provided by the manufacturers with the machines' documentations. Typically, the $Z$ axis is always the tool axis. It is important for the NC programmer to ensure that the orientation of the coordinate system where the tool path is described geometrically is identical to that of the machine tool that will read and execute the corresponding NC program. If the tool path is generated from a CAD database, the MCS, or a chosen WCS, must be of an identical orientation to that of the machine tool.

### 20.7.2.2 MATHEMATICS OF TOOL PATHS. in order for the cutting tool to produce the required finished part, its path must guide it properly. This is especially important when different machinable surfaces meet or intersect. Transitional points must be calculated properly. Figure 20-14 shows some examples. If programming is done manually, the programmer must calculate the location $(\Delta x, \Delta y)$ of the center of the cutter at the intersecting edge of the two surfaces $A$ and $B$. The calculations of $\Delta x$ and $\Delta y$ for various cases require various mathematical backgrounds including trigonometry, angular relationships, analytic geometry, and cutter geometry. Actually, many common cases for $\Delta x$, $\Delta y$ have been calculated and tabulated for use by NC programmers.

   If the tool path is generated from a CAD database, most of these calculations have been incorporated into the NC software and are performed without user intervention.

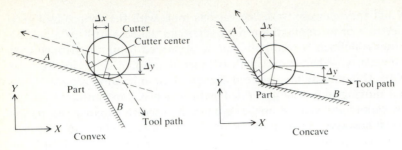

(a) Intersection of two planar surfaces

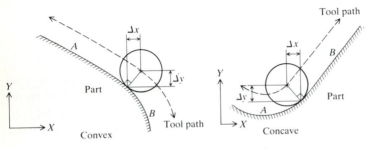

(b) Tangency of planar and cylindrical surfaces

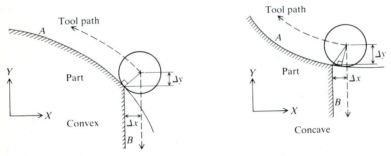

(c) Intersection of planar and cylindrical surfaces

**FIGURE 20-14**
Sample key points of tool paths.

**20.7.2.3   MACHINING FORCES.** It is the responsibility of the part programmers to choose the proper feeds and speeds for the various machining processes. As the dimensional accuracy of the workpiece relates to the programming accuracy, the efficiency of the machining process relates to the feeds and speeds. The feeds and speeds control the surface finish of the resulting part. They also control the machining forces that impact the cutting tool during machining. Higher feeds and speeds result in rough finish of the part surface and in excessive machining forces that may result in damaging the tool bit or breaking the cutting tool.

There are two common types of materials for cutting tools: high-speed steel (HSS) and carbide. Carbide tool bits usually stand higher machining forces than

HSS tool bits; thus they can stand higher feeds and speeds. However, they are more expensive. There are formulas and tables that help the programmer calculate and/or choose the cutting speed of the cutting tool, the rate of metal removal, and the horsepower for each machining process (drilling, milling, turning, etc.)

The most important consideration in the selection of feeds and speeds is the available horsepower of the machine tool. The average unit power is usually used to express the power required for drilling, milling, or turning. The average unit power is defined as the power required to remove one cubic inch of metal in one minute.

**20.7.2.4  CUTTER PROGRAMMING.** There are two methods to generate and program a tool path. The first method (less common now) is called the cutter center programming or "programming the tool." It considers the actual diameter of the cutter during the tool path generation. In this method, the program guides the cutter around the part contour. The cutter will have to follow the path at a set distance away from the part, at each point on the path, corresponding to the cutter radius. This method is error prone and the resulting program is valid for only one cutter diameter.

The second method (more popular) is used with NC machine tools that have advanced cutter compensation features. This method is called the zero radius programming or "programming the part." In this method, the program still guides the zero radius cutter around the part contour. By specifying that a tool offset equals the cutter diameter, the machine tool compensates for the diameter during execution of the program by using the proper tool offset (the cutter radius in this case). This method of programming is easier than the above one. It is also more flexible; it allows the use of cutters of different diameters.

Regardless of the method used in programming the cutter, a part program uses some or all of the following principles:

1. Absolute programming. In this mode of programming, an origin for each axis of motion of the machine tool has to be selected prior to starting the program. The intersection of the $X$, $Y$, and $Z$ axes of motion defines the machine origin which must be coincident with the part program origin. Once the origin is selected, all locations are defined with respect to this origin and all motions have to be started with respect to this origin.

2. Incremental programming. Unlike absolute programming, it is not necessary to select an origin in incremental programming. All motions are specified from the immediate last position of the tool. Thus, motions are described in increments from a given position. The distinct advantage of incremental programming is that the programmed motion matches directly the actual motion of the tool, as both occur from the last position of the cutter.

3. Rapid positioning. In PTP machining, rapid positioning of the programmed machine slides to a required location is desired to increase the machine productivity. Such motion is useful in drilling and boring when the tool is approaching the part, or moving between holes. The tool or the machine table always travels at the highest machine speed (e.g., 10 in/s). It should be mentioned here that regardless of the actual element of the machine that provides

the motion, it is always the tool that moves relative to the part in any NC program.

4. Linear interpolation. This feature is programmed using a certain programming code. It produces straight line motions by allowing the corresponding machine slides to move at different speeds independent of each other.

5. Circular interpolation. To achieve this motion, the relative positions and velocities of two machine slides are initiated and maintained on a constantly changing basis, but starting and stopping at the same time. A special code is used in the NC program to indicate circular interpolation. Both linear and circular interpolations are used to guide the tool along linear and circular paths respectively.

6. Canned cycles. These can be defined as standard subroutines stored in the "library" of the machine tool. Any NC program can call and use any of these subroutines using a special code. Examples include subroutines for drilling, tapping, boring, turning, and threading. The word "canned" indicates that the subroutine is prewritten and stored while the word "cycle" indicates that the subroutine is used over and over in a repetitive (cyclic) fashion. There are two types of canned cycles: fixed and variable canned cycles. The former implies that subroutines are not adaptable to specific user's needs while the latter allows the programmer to write canned cycles to suit particular applications.

**20.7.2.5  TOOL OFFSETS.** Tool offsets are useful features provided by NC machines to make part programs more flexible and easier to create. In the previous section, we have discussed the cutter diameter compensation. Tool length compensation is also possible, and it allows a program to be written to perform milling, drilling, tapping, or boring without presetting the tool to a specific length.

Other useful tool offsets are those used for flexible positioning of holding fixtures or parts and multiple part machining. Generally, in addition to the machine and program origins discussed previously, we have the fixture origin and the part origin. If a fixture holding the part is located in a certain position on the machine table, the part must be located properly relative to the fixture origin. The part origin (identical to the program origin) is therefore offset from the machine origin. Tool offset can be used to compensate for this offset during program execution. Similarly, if more than one part (similar or different) are set up on a machine table, they can be machined one after the other without stopping or interrupting the machine tool by using offsets to compensate for the various origins of the various parts. Offset values are usually stored by the programmer in specific offset registers programmed for various tools by using a special code.

### 20.7.3  NC Programming Languages

A part programming language allows a part programmer to take the geometry of a part from an engineering drawing (in manual programming) or from a CAD database (when using an NC package) and describe the path to be taken by a cutter to produce the part. Most of these languages have Englishlike syntax and statements. The Englishlike statements are composed of a combination of vocabulary words, symbolic names, and data (normally obtained from a drawing or a CAD database), these being separated by punctuation. These statements are designed to require the minimal amount of data input from the programmer, leaving the computer to generate the vast amount of tool offsets and other required information.

There is a handful of part programming languages. Most of them handle PTP and continuous-path machining, particularly milling. In many languages, turning is treated as a subset of milling with few deletions or additions. Some languages are developed specifically for turning operations such as CINTURN II. Some of the popular and powerful part programming languages are APT (automatically programmed tools) and APT-like (ADAPT, EXAPT, UNIAPT, MINIAPT, etc.), COMPACT II, SPLIT, PROMPT, and CINTURN II. Most of these languages were developed first for mainframe computers. They were later modified and adapted to run on minicomputers and microcomputers.

The statements of a part programming language can be divided into eight groups or types:

1. Language features. These features are similar to those found in programming languages such as FORTRAN or C. Part programmers can define variables (symbolic names) and subscripted scalar variables and geometric entities in a part program.

2. Geometric statements. These statements follow the same description of curves and surfaces covered in Chaps. 5 and 6. Analytic representations (points, lines, conics, planes, quadric surfaces, etc.), Z-surface representation for coordinate assignments (used for synthetic surfaces), and nesting statements are available. In a CAD environment, geometric information required by these statements are generated interactively when the user digitizes the surface to be machined when using NC software. Geometric statements must precede the motion statements in a part program.

3. Tool statements. Tool shape (geometry), tool axis orientation, and tool-to-part tolerance can be defined via these statements. Tool shapes are usually stored in a tool library maintained by NC software. A user can choose a specific tool and enter specific values for the tool variables, or the user can define a new tool and add it to the library.

4. Motion statements. These statements guide the tool in its motion. They provide information for type of machining (PTP or continuous path), direction of cutting, speed and feedrate, etc.

5. Arithmetic statements. As in FORTRAN, addition, subtraction, etc., as well as functions (square root, sine, etc.) are available.

6. Repetitive programming. Statements that provide looping, branching, coordinate copying, and coordinate transformations are provided. Moreover, macro facilities enable programmers to deal with repetitive programming more effectively.

7. Output facilities. A part program forms what is called the CL-data, CL meaning cutter location. The CL-data is stored in a file called CL-file. A computer printout of this file is known as the CL-printout. The CL-file is usually stored in ASCII format. In an attempt to reduce the file size and speed its postprocessing, the part program can be stored in a binary format. In this case, the file is known as the BCL-file, with the "B" meaning binary. Other output facilities include listing the part program with syntax error diagnostics.

8. Postprocessor statements. The CL-file (or BCL-file) is written in the programming language syntax. This syntax is a high level form which the machine tool controller is incapable of reading and interpreting. Therefore, the CL-file must be postprocessed via a postprocessor. Postprocessors are hardware-dependent. A postprocessor is written for a specific controller and machine combination. It converts the CL-file data into instructions and codes that the controller can accept. Many controllers accept G-codes and M-codes, which describe types of movements of the machine tool, as well as F-codes, which describe feedrate. These codes can be considered as low level programming machine languages (similar to the assembler for example). Interested readers can find listings and instructions of these codes in NC programming textbooks and manuals.

**20.7.3.1 PROGRAMMING IN APT.** In this section, we review the APT syntax and show examples of how to develop a part program to illustrate the many concepts discussed above. In this review, we present the minimum set of syntax. Not all of the modifiers that accompany many of the statements are shown.

A geometric statement in APT takes the following form:

symbol = geometric entity/geometric data to define the entity

"Symbol" is a user-chosen variable; "geometric entity" is an APT reserved word. Examples are shown below:

```
P1 = POINT/3.0, 2.0, −1.0
P2 = POINT/6.0, 5.0, 3.0
L1 = LINE/P1, P2
C1 = CIRCLE/CENTER, P1, RADIUS, 1.5
```

The first two statements define two points $P_1$ and $P_2$ with coordinates $(3, 2, -1)$ and $(6, 5, 3)$ respectively, the third statement creates a line $L_1$ between $P_1$ and $P_2$, and the last statement creates a circle with $P_1$ as a center and a radius of 1.5. Other modifiers exist to define points (e.g., intersection modifier) and lines (e.g., parallel, perpendicular, tangent, etc.).

A motion statement in APT takes the form:

motion command/motion data

Three of the useful motion statements are:

```
FROM/P1
GO TO/3.0, 4.0, −2.0
GODLTA/2.0, −1.0, 0.0
```

The first statement instructs the tool to move from its current position to point $P_1$. The second statement is an absolute programming statement where the tool must move to the point defined by $x = 3.0$, $y = 4.0$, and $z = -2.0$ relative to the machine tool origin. The third statement represents incremental programming where the tool is to move from its current position to a new position defined by $\Delta x = 2.0$, $\Delta y = -1.0$, and $\Delta z = 0$.

The above three statements are useful in PTP machining in particular. A tool needs more guidance, and therefore instructions, for continuous-path machining. A motion statement for continuous-path machining requires three surfaces to guide the tool motion: part, drive, and check surfaces as shown in Fig. 20-15. The part surface is defined as the surface that controls the tool motion along the tool axis. It controls the depth of the machining operation. The drive surface controls the tool motion perpendicular to the tool path, that is, the direction in which the tool is moving. Alternately, the drive surface is always tangent to the tool path or the direction of the tool motion. The check surface terminates the tool motion. The part surface usually intersects both the drive and check surfaces. However, the drive and check surfaces may or may not intersect. APT provides a GO command using these three surfaces with the following syntax.

GO/TO, drive surface, TO, part surface, TO, check surface

The "TO" in the above statement is a modifier. Other modifiers are "ON" and "PAST." These modifiers control the relative position of the cutter with respect to the surface following the modifier. Figure 20-16 shows these meanings graphically.

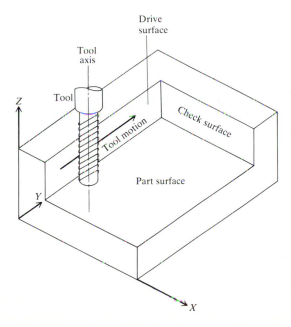

**FIGURE 20-15**
Tool guiding surfaces in APT continuous-path machining.

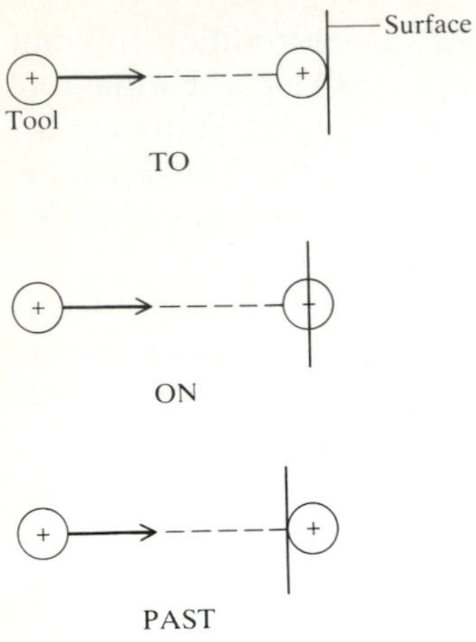

**FIGURE 20-16**
Meaning of TO, ON, and PAST modifiers in APT.

A tool statement describes the tool to be used in machining. It takes the form:

CUTTER/d

where $d$ is the cutter diameter. Tool description is required so that the APT program can use the tool offset ($d/2$) to compute the coordinates of the tool path.

Postprocessor commands are usually included in the APT program itself because there is normally only one input to the postprocessor, the CL-file. Sample commands are MCHIN, PARTNO, COOLNT, RAPID, RETRACT, STOP, and FINI. Use of some of these statements is shown in the following two examples.

**Example 20.2.** Figure 20-17a shows a two-and-a-half-dimensional part. The part thickness is 1.000 in. Two holes in the positions shown are to be drilled in the part. Write an APT program to drill the two holes. Use the tool home position shown, a drilling speed of 500 r/min and a feedrate of 3.55 in/min. The machine that performs the drilling is number 5 and its controller is coded as DRILL.

*Solution.* This example illustrates using APT in PTP machining. Utilizing the home position shown, the tool path is shown in Fig. 20-17b with the order in which the holes are drilled. In this example, the machine coordinate system ($X_m$, $Y_m$, $Z_m$) is coincident with the part origin ($X_p$, $Y_p$, $Z_p$). In addition, the machine origin, the part origin, and the program origin are identical and located at the left bottom corner of the part. Using this information, the APT program may look as follows:

```
PARTNO PTP-EXAMPLE
MACHIN/DRILL, 5
CLPRINT
CUTTER/0.750
P0 = POINT/−2.0, 0.0, 0.0
P1 = POINT/1.25, 1.0, 0.0
P2 = POINT/4.75, 1.0, 0.0
SPINDL/500
```

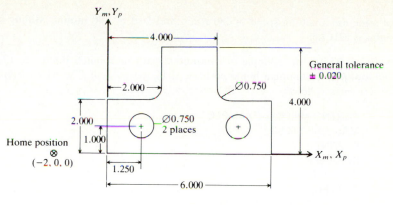

(a) Part geometry

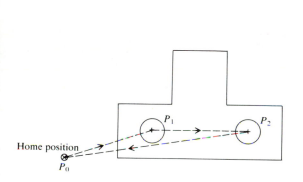

(b) Drilling tool path

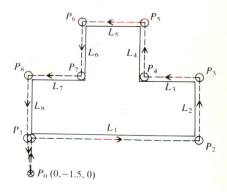

(c) Milling (contouring) tool path

**FIGURE 20-17**
Machining a typical part.

```
FEDRAT/3.55
COOLNT/ON
FROM/P0
GO TO/P1
GODLTA/0, 0, −1.5
GODLTA/0, 0, 1.5
GO TO/P2
GODLTA/0, 0, −1.5
GODLTA/0, 0, 1.5
GO TO/P0
COOLNT/OFF
FINI
```

This program is the CL-file to drill the two holes. It must be postprocessed to generate the G-code, M-code, and F-code before downloading into the controller of the machine tool.

**Example 20.3.** Write an APT program to mill the contour of the part shown in Fig. 20-17a. Use an end-mill cutter with a diameter of 0.75 in. Assuming the part is made of steel, use a cutting speed of 580 r/min and feed of 2.30 in/min. Milling machine name is MILL5.

**Solution.** This is an example of continuous path machining. Figure 20-17c shows the tool path assuming a home position at $P_0$ (0, $-1.5$, 0). The APT program may be written as follows:

```
PARTNO MILLING-EXAMPLE
MACHINE/MILL5
CLPRNT
INTOL/0.002
OUTOL/0.002
CUTTER/0.750
P0 = POINT/0, −1.5, 0
P1 = POINT/0, 0, 0
P2 = POINT/6.0, 0, 0
P3 = POINT/6.0, 2.0, 0
P4 = POINT/4.0, 2.0, 0
P5 = POINT/4.0, 4.0, 0
P6 = POINT/2.0, 4.0, 0
P7 = POINT/2.0, 2.0, 0
P8 = POINT/0, 2.0, 0
L1 = LINE/P1, P2
L2 = LINE/P2, P3
L3 = LINE/P3, P4
L4 = LINE/P4, P5
L5 = LINE/P5, P6
L6 = LINE/P6, P7
L7 = LINE/P7, P8
L8 = LINE/P8, P1
PL1 = PLANE/0, 0, −1.5, 6.0, 0, −1.5, 0, 2.0, −1.5
SPINDL/580
FEDRAT/2.30
COOLNT/ON
FROM/P0
GO/TO, L1, TO, PL1, TO, L8
GORGT/L1, PAST, L2
GOLFT/L2, PAST, L3
GOLFT/L3, TO, L4
GORGT/L4, PAST, L5
GOLFT/L5, PAST, L6
GOLFT/L6, TO, L7
GORGT/L7, PAST, L8
GOLFT/L8, PAST, L1
RAPID
GO TO/P0
COOLNT/OFF
FINI
```

To properly guide the tool on its path in continuous-path machining, one could imagine guiding the tool with a remote control around the part profile, thus making the proper left and right turns.

## 20.8 TOOL PATH GENERATION AND VERIFICATION

Four steps can be identified for successful NC machining of parts utilizing CAD databases. These are recognition of machined surfaces (feature recognition), tool path generation, tool path verification, and collision detection. Automatic feature recognition for manufacturing is a crucial step in integrating CAD and CAM. We have touched on it in process planning. Collision detection is concerned with finding if the cutting tool and its assembly (tool, shank, etc.) may collide with other components of the manufacturing environment such as the workpiece itself, jigs, fixtures, spindles, etc. One method to solve the collision problem, using solid modeling, is to find the swept volume generated by the tool motion in space and intersect it with any of the models of the components (jigs, fixtures, etc.). If the intersection is not null, collision occurs and the components would have to be rearranged. This section is devoted primarily to generating and verifying tool paths. Readers interested in obtaining more detail on feature recognition and collision detection and avoidance should consult the bibliography at the end of the chapter.

### 20.8.1  Generation of Tool Paths

The more advanced NC programming languages such as APT employ surface geometry techniques to generate tool paths. APT has the capability to describe, and therefore to generate, tool paths for all analytic curves and surfaces covered in Chaps. 5 and 6. APT utilizes the nonparametric representation to describe these geometric entities. Each entity has a corresponding geometric statement that takes the required input parameters to define it as described in the previous section.

NC programming based on APT uses only unoriented, unbounded, single surfaces. APT does not recognize connected surfaces. This is why APT cannot be used directly with solid models that use orientable and bounded surfaces (faces). In an attempt to enhance NC programming capabilities, machining free-form and sculptured surfaces have received wide attention. These capabilities enable machining synthetic surfaces which are usually represented in a parametric form.

In cutting free-form surfaces, one is faced with the problems of cutter offset, accuracy, and cutter interference with the workpiece (tool gouging). Customarily, ball-end (or ball-nosed) cutters are used in these circumstances and the calculation of cutter offsets is achieved by finding the directions of normals on the surface. Having determined the directions of normals, the tool is then offset by the radius of the cutter along the normal vectors over the surface.

Before calculating cutter offsets, we must generate the tool path. To mill a surface parametrically, cutter location points are generated using the surface definition and stated machining parameters. These points are stored in the CL-file. The ball-end cutter tip is directed to move in straight line segments to each point. The fewer the points, the more rough the resulting machined surface. However, large point files should be avoided to minimize storage, computation, and milling time. For example, if we choose a step size of 0.1 in both the $u$ and $v$ directions of a parametric surface, and assuming both $u$ and $v$ have a range from 0 to 1, 11 points are created along each of the $u$ and $v$ directions. Thus, 121 points are needed to mill the surface. The problem with this step definition is that it does not take into account areas of local flattening, where fewer points are required. In addition, it should take tolerance specification on the surface into account.

Out of the many existing algorithms to generate tool paths for sculptured surfaces, we present the following algorithms. The algorithm applies to 3-axis machining. The algorithm subdivides the three-dimensional sculptured surface into three-dimensional parametric space curves (in the $u$ or $v$ direction). Each curve is further approximated by a sequence of line segments. The tool paths for these segments are calculated for ball-end cutters. The algorithm attempts to minimize the number of points evaluated on the surface to generate the tool path, and yet stay within the specified tolerance. The key issues in the algorithm is how to approximate the parametric space curves and how to determine the tool paths.

Figure 20-18a shows a parametric surface patch. In this figure, we assume the machining direction of the surface to be the $u$ direction. Thus, the tool moves along curves of $v = \text{constant}$ on the surface, that is, $\mathbf{P}(u, c)$. Figure 20-18b shows

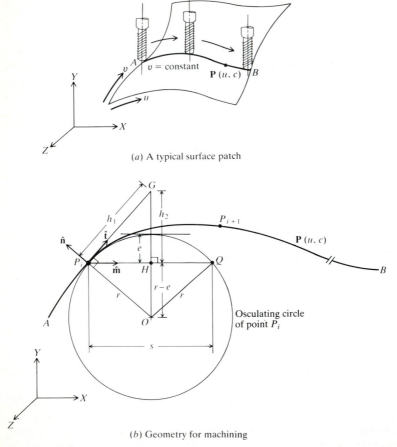

(a) A typical surface patch

(b) Geometry for machining

**FIGURE 20-18**
Geometry needed to machine sculptured surfaces.

the geometry required to divide one of these curves into line segments. Assume that a point $P_i$ on the curve is given. The unit normal and unit tangent vectors at this point are $\hat{\mathbf{n}}$ and $\hat{\mathbf{t}}$ respectively. The point and the unit vectors are obtained from the surface representation. The tolerance $e$ is known from the surface machining data. The step length $s$ of a line segment that a tool traverses during machining must be consistent with the tolerance requirement.

For a small enough step, it is reasonable to approximate the curve around $P_i$ by its osculating circle (circle with the greatest contact with the curve at $P_i$; its center $O$ is on the normal to the curve and its radius $r$ is equal to the radius of curvature at $P_i$). If the tool moves the step length $s$, point $Q$ on the osculating circle results which may not exactly coincide with point $P_{i+1}$ on the curve $\mathbf{P}(u, c)$. However, for a small step size $s$, point $Q$ is very close to point $P_{i+1}$. In this algorithm, they are both assumed to be the same.

The step length $s$ can be written in terms of the specified tolerance by considering, say, the triangle $P_iOH$, that is,

$$r^2 = \left(\frac{s}{2}\right)^2 + (r - e)^2 \tag{20.1}$$

or

$$s^2 = 4e(2r - e) \tag{20.2}$$

The unit vector $\hat{\mathbf{m}}$ in the direction of the step length is given by

$$\frac{s}{2}\hat{\mathbf{m}} = h_1\hat{\mathbf{t}} - \frac{h_2}{h_2 + r - e}(r\hat{\mathbf{n}} + h_1\hat{\mathbf{t}}) \tag{20.3}$$

or

$$s\hat{\mathbf{m}} = \frac{2h_1(r - e)}{h_2 + r - e}\hat{\mathbf{t}} - \frac{2h_2 r}{h_2 + r - e}\hat{\mathbf{n}} \tag{20.4}$$

Using the right triangle $P_iOG$ and the perpendicular line $P_iH$, we can write

$$h_1 = \frac{rs}{2(r - e)} \tag{20.5}$$

and

$$h_2 = \frac{s^2}{4(r - e)} \tag{20.6}$$

Substituting Eqs. (20.5) and (20.6) into Eq. (20.4), and using Eq. (20.2) to reduce the result, we obtain

$$\hat{\mathbf{m}} = \frac{r - e}{r}\hat{\mathbf{t}} - \frac{s}{2r}\hat{\mathbf{n}} \tag{20.7}$$

The position vector $\mathbf{Q}$ of point $Q$ on the osculating circle is given by

$$\mathbf{Q} = \mathbf{P}_i + s\hat{\mathbf{m}} \tag{20.8}$$

or

$$\mathbf{Q} = \mathbf{P}_i + \frac{s(r - e)}{r}\hat{\mathbf{t}} - \frac{s^2}{2r}\hat{\mathbf{n}} \tag{20.9}$$

The parametric value $u_{i+1}$ which corresponds to point $P_{i+1}$ must now be evaluated so that the unit normal and tangent vectors as well as the radius of curvature $r$ at point $P_{i+1}$ can be evaluated using the surface equation.

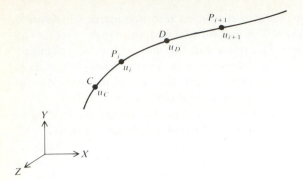

**FIGURE 20-19**
Points needed to calculate $u_{i+1}$ at point $P_{i+1}$.

Figure 20-19 shows the points needed for the calculation. We use the second divided difference method to approximate $u_{i+1}$. We choose an intermediate point $D$ between point $P_i$ and $P_{i+1}$ (e.g., choose the midpoint) and calculate its corresponding parameter $u_D$. Let us assume that point $C$, and its $u_C$, is the previous intermediate point used to calculate $u_i$ at point $P_i$.

Using the four points $C$, $P_i$, $D$, and $P_{i+1}$ and their corresponding parameter values, first calculate the distance between the two points $C$ and $D$ in each of the coordinate directions and find the maximum absolute value and its direction. Then select the corresponding value in that direction (could be $x$, $y$, or $z$) as variable $w$. Using the second divided difference method, we can write

$$u_{i+1} = u_C + (w_{i+1} - w_C)[K_1 + (w_{i+1} - w_i)K_3] \tag{20.10}$$

where

$$K_1 = \frac{u_i - u_C}{w_i - w_C} \tag{20.11}$$

$$K_2 = \frac{u_D - u_i}{w_D - w_i} \tag{20.12}$$

and

$$K_3 = \frac{K_2 - K_1}{w_D - w_C} \tag{20.13}$$

Having $u_{i+1}$, we can now correct the point $P_{i+1}$, which was assumed to be the same as point $Q$, by calculating its coordinates from the surface equations using $u_{i+1}$. In addition, the unit tangent vector $\hat{\mathbf{t}}$, the unit normal vector $\hat{\mathbf{n}}$, and the radius of curvature $r$ can be evaluated at $P_{i+1}$ to compute the next point $P_{i+2}$ and its parameter value $u_{i+2}$.

To machine a part contour, the points as calculated above must be offset by the radius of the cutting tool to produce the correct tool path. Assuming a ball-end cutter with diameter $d$, the offset geometry is shown in Fig. 20-20. To calculate the endpoints $E_i$ of the tool path line segments, we can write the following equations using Fig. 20-20:

$$\cos \theta = \hat{\mathbf{t}} \cdot \hat{\mathbf{m}} \tag{20.14}$$

$$\frac{d}{2} = |\mathbf{E}_i - \mathbf{P}_i| \cos \theta \tag{20.15}$$

$$\mathbf{P}_i = \mathbf{E}_i - |\mathbf{E}_i - \mathbf{P}_i|\hat{\mathbf{n}} \tag{20.16}$$

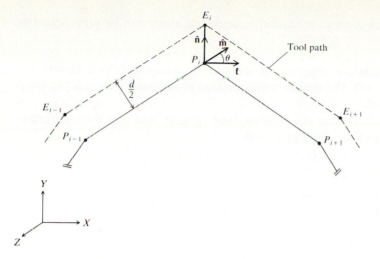

**FIGURE 20-20**
Offset geometry for ball-end cutter.

Reducing Eq. (20.16) using Eqs. (20.14) and (20.15), we get

$$\mathbf{E}_i = \mathbf{P}_i + \frac{d}{2\hat{\mathbf{t}} \cdot \hat{\mathbf{m}}} \, \hat{\mathbf{n}} \tag{20.17}$$

Equation (20.17) gives the cutter offset points $E_i$ which determine the line segments (tool path) of the cutter center path. Figure 20-21 shows sample tool paths that may be obtained using Eq. (20.17). The tool path takes the same form as the curve. The tool path may not be a simple curve for relatively large cutter radius and complex spatial curves (Fig. 20-21b). A tool path with loops and cusps results when the radius of curvature at a point on the curve is less than the cutter radius and the normal vector points away from the center of curvature. In practice, this problem can be rectified by selecting cutters with proper radii.

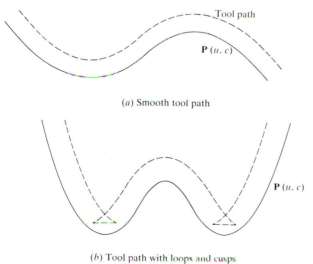

(a) Smooth tool path

(b) Tool path with loops and cusps

**FIGURE 20-21**
Sample tool paths.

The tool path generation algorithm described above can be summarized as follows:

1. Select the machining direction of the surface. Assuming the $u$ direction is selected, set $v = 0$. The surface representation is assumed to include the tolerance on the surface.
2. Set $u_C = 0$ and evaluate the position, unit tangent, and unit normal vectors and the radius of curvature at $u_A = 0$.
3. Evaluate an extra point $D$ at a parameter value $u_D$.
4. Select the variable $w$.
5. Update the parameter value $u$ using Eq. (20.10).
6. If $u = u_{max}$, then let $u = u_{max}$; otherwise go to step 7.
7. Evaluate the position, unit tangent, and unit normal vectors and the radius of curvature at the new $u$ value.
8. Calculate the cutter path point at $u$ using Eq. (20.17).
9. If $u < u_{max}$, set $u_C = u_D$ and go to step 3; otherwise go to step 10.
10. If $v < v_{max}$, set $v = v + D_v$ and go to step 2; otherwise go to step 11.
11. Stop.

Further enhancements and extensions to the above algorithm are possible. A collision detection mechanism between the tool, the workpiece, and the machine can be developed. The algorithm can also be extended to calculate multiple passes where a number of cutting passes may be required to machine an object. In addition, the algorithm can be extended to calculate paths using other types of milling cutters.

The above algorithm is based on the fact that the surface patch to be machined occupies the range described by $0 \le u \le u_{max}$ and $0 \le v \le v_{max}$. If the patch range is not the same as this range, the algorithm must be modified significantly. This case is shown in Fig. 20-22. In practice, we refer to this problem as machining trimmed or untrimmed surfaces. To use the same algorithm would require the programmer to subdivide the trimmed surface into few untrimmed ones.

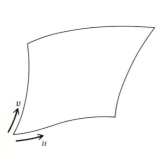

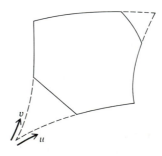

(a) Untrimmed surface                     (b) Trimmed surface

**FIGURE 20-22**
Types of sculptured surfaces for machining.

## 20.8.2    Verification of Tool Paths

NC part programs and the tool paths that are supposed to guide tools during actual machining usually include a lot of coordinate values that are impossible to verify manually. Graphical and visual display of the tool paths are beneficial in the verification process. NC verification software can simulate the actual machining process by displaying the cutting tool moving, following its tool path, relative to the stock and jigs and fixtures. Shaded images of the tool and stock are used to enhance the visualization process. This would enable an NC programmer to spot any potential errors in NC programs.

The animated tool path is generated by displaying the tool position and orientation, using the NC program data, at various points on the tool path, creating frames, and storing the frames for playback as discussed in Chap. 13. Other methods of tool path verification, especially for solid models, have been discussed earlier.

The advantages of tool path verification are numerous. Check whether:

1. The cutter removes the necessary material from the stock.
2. The cutter hits any clamps or fixtures on approach.
3. The cutter passes through the floor or side of a pocket, or through a rib.
4. The tool paths are as efficient as they could be.

Other advantages include rapid turnaround of program development, rapid training for potential NC programmers without danger, and freeing machine tools to only cut real parts (no testing of programs or training individuals). In addition, tool path verification by software reduces the wear on machine tools. When polyurethane foam is cut, for verification purposes, it rises in the air and then falls into the machine gears where it acts as a grinding paste.

Even with tool path verification, some machining problems go undetected. For example, problems such as tool chatter and stock warpage (due to heat stresses resulting during machining) are simply beyond the scope of verification. However, tool verification detects the majority of errors in general.

## 20.9    DESIGN AND ENGINEERING APPLICATIONS

The material in this chapter has provided background in manufacturing and part programming. The utilization of such material in engineering applications would require readers to have the proper software packages to practice the concepts covered in the chapter. We here discuss an example of NC programming that is carried using an NC package.

**Example 20.4.** Figure 20-23 shows a two-and-a-half-dimensional part. The part is to be machined from a 1.2 in thick low-carbon steel. Develop an APT program to mill the top face of the part, mill the pocket out, and drill the four holes. The speed and feed for milling is 400 r/min and 0.1 in/min respectively and for drilling they are 900 r/min and 3.0 in/min respectively.

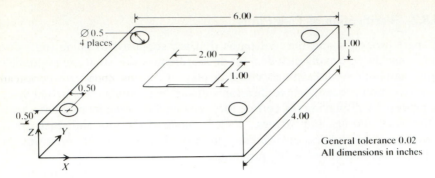

**FIGURE 20-23**
Geometry of a typical part.

*Solution.* We have solved this example using an NC software package. We here present only the strategy that has gone into the solution to avoid boring the reader with the detailed user interface of the package. The strategy is as follows:

1. Tools required for machining
    (*a*) 0.2 in of material will be milled off the top of the stock. This will be done with a 1 in diameter flat-end mill.
    (*b*) The center pocket will be milled out with a $\frac{1}{4}$ in diameter flat-end mill.
    (*c*) The four holes will be drilled out with a $\frac{1}{2}$ in diameter drill.

2. Define tools as follows:

| Name | Tool |
|---|---|
| MILL 0.5 | 0.5 in radius flat-end mill |
| MILL 0.125 | 0.125 in radius flat-end mill |
| Drill 0.25 | 0.25 in radius carbide drill |

3. Procedure for generating tool paths
    (*a*) Milling off 0.2 in from stock: in order to get square corners on the edge of the stock, the milling tool must go on the outer edge boundary as shown in Fig. 20-24. Therefore a boundary line was inserted as shown in order to achieve this result with the package used. A large 1 in diameter flat-end mill will be used to quickly cut the material.

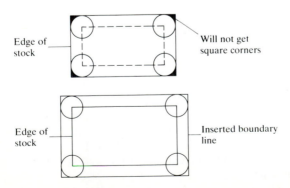

**FIGURE 20-24**
Intermediate boundary to mill top face.

(b) Use the proper command with the $\frac{1}{4}$ in diameter flat-end mill to cut out the pocket. The corners of the pocket must be filleted in order to use a standard milling tool. A 0.125 in radius fillet was selected.

(c) Use a 0.5 in diameter drill bit with a drilling command for the four holes.

(d) Use a "display tool" command to check tool paths. Zoom in on critical areas for visual verification.

(e) Generate the CL-file (APT program) with the proper command.

(f) Use a "calculate cost" command to provide some initial estimates on the job cost and time.

*Note:* in generating tool paths, the location of part clamps may need to be considered to ensure that the tool(s) does not hit the clamps. They were not considered in this solution.

In executing the above strategy on the CAD/CAM system, the user is able to verify tool paths as geometric entities, list information about tool specifications, and list cost information. This information is not listed here. Figures 20-25 and 20-26 show the stock of 1.2 in thickness and the tool path to mill the top face. Figure 20-27 shows the tool path to mill the pocket out. Figure 20-28 shows the tool path to drill the holes and Fig. 20-29 shows the finished part. The listing of the APT file generated by the NC package is shown below.

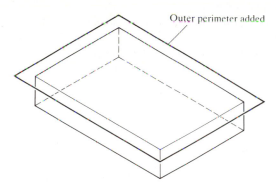

Outer perimeter added

**FIGURE 20-25**
Stock model.

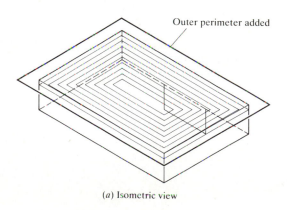

Outer perimeter added

(a) Isometric view

(b) Top view

**FIGURE 20-26**
Tool path to mill the top face.

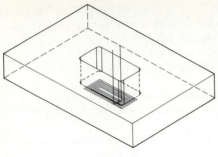

(*a*) Isometric view

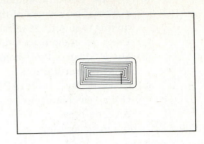

(*b*) Top view

**FIGURE 20-27**
Tool path to mill the pocket out.

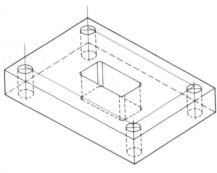

(*a*) Isometric view

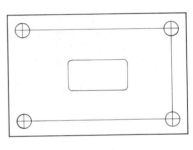

(*b*) Top view

**FIGURE 20-28**
Tool path to drill the holes.

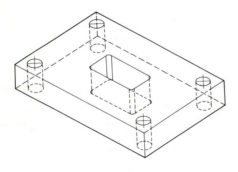

**FIGURE 20-29**
The finished part.

```
APTSOURCE.NC.&BCD.MILL1
3-10-87  16:33:31  FUTIL 6.21

PARTNO *V3.MODD.CHRIS.NC
MACHIN/BGBG02
CUTTER/1
OUTTOL/0
PPRINT LAB #7:ME3500
FROM/     0.00000,     0.00000,     0.00000
LOADTL/     1
```

MILLING OF ⌀0.2″ (Top face milling) (This is a comment. It is not an APT statement)
FEDRAT/     400.00000.IP M
COOLNT/ON
SPINDL/     400.00000,CL W
GOTO/     4.00000,     2.00000,     2.00000
GOTO/     4.00000,     2.00000,     1.50000
FEDRAT/     0.10000,IP M
GOTO/     4.00000,     2.00000,     1.00000
INTOL/     0.00500
GOTO/     4.22222,     2.00000,     1.00000
GOTO/     4.22222,     1.77778,     1.00000
GOTO/     1.77778,     1.77778,     1.00000
GOTO/     1.77778,     2.22222,     1.00000
GOTO/     4.22222,     2.22222,     1.00000
GOTO/     4.22222,     2.00000,     1.00000
GOTO/     4.44444,     2.00000,     1.00000
GOTO/     4.44444,     1.55556,     1.00000
GOTO/     1.55556,     1.55556,     1.00000
GOTO/     1.55556,     2.44444,     1.00000
GOTO/     4.44444,     2.44444,     1.00000
GOTO/     4.44444,     2.00000,     1.00000
GOTO/     4.66667,     2.00000,     1.00000
GOTO/     4.66667,     1.33333,     1.00000
GOTO/     1.33333,     1.33333,     1.00000
GOTO/     1.33333,     2.66667,     1.00000
GOTO/     4.66667,     2.66667,     1.00000
GOTO/     4.66667,     2.00000,     1.00000
GOTO/     4.88889,     2.00000,     1.00000
GOTO/     4.88889,     1.11111,     1.00000
GOTO/     1.11111,     1.11111,     1.00000
GOTO/     1.11111,     2.88889,     1.00000
GOTO/     4.88889,     2.88889,     1.00000
GOTO/     4.88889,     2.00000,     1.00000
GOTO/     5.11111,     2.00000,     1.00000
GOTO/     5.11111,     0.88889,     1.00000
GOTO/     0.88889,     0.88889,     1.00000
GOTO/     0.88889,     3.11111,     1.00000
GOTO/     5.11111,     3.11111,     1.00000
GOTO/     5.11111,     2.00000,     1.00000
GOTO/     5.33333,     2.00000,     1.00000
GOTO/     5.33333,     0.66667,     1.00000
GOTO/     0.66667,     0.66667,     1.00000
GOTO/     0.66667,     3.33333,     1.00000
GOTO/     5.33333,     3.33333,     1.00000
GOTO/     5.33333,     2.00000,     1.00000
GOTO/     5.55556,     2.00000,     1.00000
GOTO/     5.55556,     0.44444,     1.00000
GOTO/     0.44444,     0.44444,     1.00000
GOTO/     0.44444,     3.55556,     1.00000
GOTO/     5.55556,     3.55556,     1.00000
GOTO/     5.55556,     2.00000,     1.00000
GOTO/     5.77778,     2.00000,     1.00000
GOTO/     5.77778,     0.22222,     1.00000
GOTO/     0.22222,     0.22222,     1.00000
GOTO/     0.22222,     3.77778,     1.00000
GOTO/     5.77778,     3.77778,     1.00000
GOTO/     5.77778,     2.00000,     1.00000

```
GOTO/      6.00000,     2.00000,     1.00000
GOTO/      6.00000,     4.00000,     1.00000
GOTO/      0.00000,     4.00000,     1.00000
GOTO/      0.00000,     0.00000,     1.00000
GOTO/      6.00000,     0.00000,     1.00000
GOTO/      6.00000,     2.00000,     1.00000
FEDRAT/    1.00000,IPMM
GOTO/      6.00000,     2.00000,     1.50000
FEDRAT/    400.00000,IPM
GOTO/      6.00000,     2.00000,     2.00000
COOLNT/OF F
GOTO/      0.00000,     0.00000,     0.00000
STOP
LOADTL/    2
MILLING OF POCKET (This is a comment. It is not an ATP statement)
COOLNT/ON
SPINDL/    400.00000,CLW
GOTO/      3.50000,     2.00000,     2.00000
GOTO/      3.50000,     2.00000,     1.10000
FEDRAT/    0.10000,IP M
GOTO/      3.50000,     2.00000,    − 0.25000
GOTO/      3.50000,     1.94643,    − 0.25000
GOTO/      2.44643,     1.94643,    − 0.25000
GOTO/      2.44643,     2.05357,    − 0.25000
GOTO/      3.55357,     2.05357,    − 0.25000
GOTO/      3.55357,     1.94643,    − 0.25000
GOTO/      3.50000,     1.94643,    − 0.25000
GOTO/      3.50000,     1.89286,    − 0.25000
GOTO/      2.39286,     1.89286,    − 0.25000
GOTO/      2.39286,     2.10714,    − 0.25000
GOTO/      3.60714,     2.10714,    − 0.25000
GOTO/      3.60714,     1.89286,    − 0.25000
GOTO/      3.50000,     1.89286,    − 0.25000
GOTO/      3.50000,     1.83929,    − 0.25000
GOTO/      2.33929,     1.83929,    − 0.25000
GOTO/      2.33929,     2.16071,    − 0.25000
GOTO/      3.66071,     2.16071,    − 0.25000
GOTO/      3.66071,     1.83929,    − 0.25000
GOTO/      3.50000,     1.83929,    − 0.25000
GOTO/      3.50000,     1.78571,    − 0.25000
GOTO/      2.28571,     1.78571,    − 0.25000
GOTO/      2.28571,     2.21429,    − 0.25000
GOTO/      3.71429,     2.21429,    − 0.25000
GOTO/      3.71429,     1.78571,    − 0.25000
GOTO/      3.50000,     1.78571,    − 0.25000
GOTO/      3.50000,     1.73214,    − 0.25000
GOTO/      2.23214,     1.73214,    − 0.25000
GOTO/      2.23214,     2.26786,    − 0.25000
GOTO/      3.76786,     2.26786,    − 0.25000
GOTO/      3.76786,     1.73214,    − 0.25000
GOTO/      3.50000,     1.73214,    − 0.25000
GOTO/      3.50000,     1.67857,    − 0.25000
GOTO/      2.17857,     1.67857,    − 0.25000
GOTO/      2.17857,     2.32143,    − 0.25000
GOTO/      3.82143,     2.32143,    − 0.25000
GOTO/      3.82143,     1.67857,    − 0.25000
GOTO/      3.50000,     1.67857,    − 0.25000
```

```
GOTO/      3.50000,     1.62500,     -0.25000
GOTO/      3.87500,     1.62500,     -0.25000
GOTO/      3.87500,     2.37500,     -0.25000
GOTO/      2.12500,     2.37500,     -0.25000
GOTO/      2.12500,     1.62500,     -0.25000
GOTO/      3.50000,     1.62500,     -0.25000
GOTO/      3.50000,     1.62500,     1.10000
FEDRAT/    400.00000,IPM
GOTO/      3.50000,     1.62500,     2.00000
PRINT DRILL HOLES
COOLNT/OFF
GOTO/      0.00000,     0.00000,     0.00000
STOP
LOADTL/      3
DRILL 4 HOLES (This is a comment. It is not an APT statement)
COOLNT/ON
SPINDL/    900.00000,CLW
GOTO/      0.50000,     3.50000,     2.00000
GOTO/      0.50000,     3.50000,     1.10000
FEDRAT/    3.00000,IPM
GOTO/      0.50000,     3.50000,     -0.25000
FEDRAT/    1.00000,IPM
GOTO/      0.50000,     3.50000,     1.10000
FEDRAT/    400.00000,IPM
GOTO/      5.50000,     3.50000,     1.10000
FEDRAT/    3.00000,IPM
GOTO/      5.50000,     3.50000,     -0.25000
FEDRAT/    1.00000,IPM
GOTO/      5.50000,     3.50000,     1.10000
FEDRAT/    400.00000,IPM
GOTO/      5.50000,     0.50000,     1.10000
FEDRAT/    3.00000,IPM
GOTO/      5.50000,     0.50000,     -0.25000
FEDRAT/    1.00000,IPM
GOTO/      5.50000,     0.50000,     1.10000
FEDTAT/    400.00000,IPM
GOTO/      0.50000,     0.50000,     1.10000
FEDRAT/    3.00000,IPM
GOTO/      0.50000,     0.50000,     -0.25000
FEDRAT/    1.00000,IPM
GOTO/      0.50000,     0.50000,     1.10000
FEDRAT/    400.00000,IPM
GOTO/      0.50000,     0.50000,     2.00000
COOLNT/OFF
GOTO/      0.00000,     0.00000,     0.00000
STOP
END
FINI
```

# PROBLEMS

## Part 1: Theory

**20.1.** Choose a part or a few parts and apply the part production cycle described in Sec. 20.2 to them. To be more specific, contact and visit the company(s) that make them if possible.

**20.2.** Using your local library and visiting local companies, write a report on each type of manufacturing system presented briefly in Sec. 20.3. Describe the details of each system.

**20.3.** For each manufacturing process described in Sec. 20.4, compile details regarding the dimensional capacity, rate of output, cost of tooling, cost of initial setup, required skills of operator, etc. Generate a table of such details.

**20.4.** For the major machining processes (drilling, milling, etc.), compile details regarding the best machining conditions for various metals, e.g., speed, feedrate, level of tool wear, etc. Generate a table of such details.

**20.5.** Develop the squaring graph, the interaction graph, and the plan graph of the parts whose geometries are shown in the problems in Chaps. 5, 6, and 7. Write the sequence of steps (using a high level planning format) of the resulting process plan for each part.

**20.6.** This problem illustrates how to handle process planning with incomplete information due to lack of information itself or due to lack of experience. Consider yourself as a process planner with very little hands-on experience with titanium. How can you develop process plans for the part made of titanium shown in Fig. P20-6?

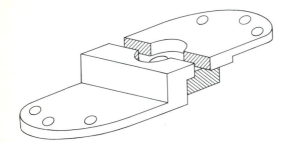

**FIGURE P20-6**

**20.7.** Search the various resources and collect the existing commercial CAPP systems. Classify them into variant and generative systems. What are the documents each one produces (method sheets, routing sheets, toolkit sheets, etc.)? For the variant systems, what is the classifying and code system used (group technology or not)? For the generative system, what method and decision logic are used (AI techniques or not)?

**20.8.** Using the APT language, generate the CL-files (part programs) for the parts shown in Fig. P20-8. For each part, write two programs, one to drill the holes and the other to perform contour milling. The part material is low-carbon steel and the cutters are HSS. Part thicknesses are 0.500 in.

## Part 2: Laboratory

**20.9.** Using a process planning package, generate process plans for the parts shown in Prob. 20.8.

**20.10.** Same as Prob. 20.9 but for the models shown at the end of Chaps. 5, 6, and 7.

**20.11.** Using an NC package, generate the CL-files (use APT) for the parts shown in Prob. 20.8. Compare with the manual procedure you followed in Prob. 20.8.

**20.12.** Same as Prob. 20.11 but for the models shown at the end of Chaps. 5, 6, and 7.

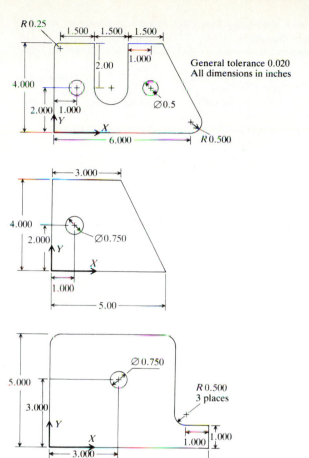

General tolerance 0.020
All dimensions in inches

**FIGURE P20-8**

## Part 3: Programming

**20.13.** Write a computer program to implement the algorithm described in Sec. 20.8.1 to generate tool paths for sculptured surfaces.

## BIBLIOGRAPHY

Anderson, R. O.: "Detecting and Eliminating Collisions in NC Machining," *Computer Aided Des. J.*, vol. 10, no. 4, pp. 231–238, 1978.

Armstrong, G. T., G. C. Carey, and A. De Pennington: "Numerical Code Generation from Geometric Modeling System," in *Proceedings of a Symposium on Solid Modeling*, 1983, pp. 139–158, Warren, Mich.

Bard, J. F., and T. A. Feo: "The Cutting Path and Tool Selection Problem in Computer Aided Process Planning," *J. Manufacturing Systems*, vol. 8, no. 1, pp. 17–26, 1989.

Bariani, P., and F. Vallese: "An Integrated Computer-Aided Process Planning Procedure for Fine Blanking," in *Proceedings of the Second International Conference on Computer-Aided Production Engineering*, 1987, pp. 179–184, Edinburgh.

Black, I.: "Assuring Confident Re-use of Production Drawing Information by the NC Programming Activity," *Computer-Aided Engng J.*, vol. 3, no. 4, pp. 159–163, 1986.

Blore, D.: "Computer Aided Process Planning," *Chartered Mech. Engr*, vol. 31, no. 5, pp. 31–34, 1984.

Bobrow, J. E.: "NC Machine Tool Path Generation from CSG Part Representations," *Computer Aided Des. J.*, vol. 17, no. 2, pp. 69–76, 1985.

Bussmann, J., R. Granow, and H. Hammer: "Economics of CNC Lathes," *J. Manufacturing Systems*, vol. 2, no. 1, pp. 1–14, 1983.

Chang, T., and R. A. Wysk: "CAD/Generative Process Planning with TIPPS," *J. Manufacturing Systems*, vol. 2, no. 2, pp. 127–135, 1983.

Chang, T., and R. A. Wysk: *An Introduction to Automated Process Planning Systems*, Prentice-Hall, Englewood Cliffs, N.J., 1985.

Chen, S. J., and S. Hinduja: "Checking for Tool Collision in Turning," *Computer Aided Des. J.*, vol. 20, no. 5, pp. 281–289, 1988.

Chilfs, J. J.: *Principles of Numerical Control*, Industrial Press, New York, 1969.

Choi, B. K., M. M. Barash, and D. C. Anderson: "Automatic Recognition of Machined Surfaces from a 3D Solid Model," *Computer Aided Des. J.*, vol. 16, no. 2, pp. 81–86, 1984.

Descotte, Y., and J. Latombe: "GARI: An Expert System for Process Planning," in *Proceedings of a Symposium on Solid Modeling*, 1983, pp. 329–346, Warren, Mich.

Donmez, A., C. R. Liu, M. Barash, and F. Mirski: "Statistical Analysis of Positioning Error of a CNC Milling Machine," *J. Manufacturing System*, vol. 1, pp. 33–41, 1982.

Earl, C. R.: "NC Verification Comes to CAD/CAM," *Automation*, vol. 36, no. 8, pp. 46–49, August 1988.

Eichner, D., C. Anderson, and D. Sly: "An Enhanced Micro Computer Program for CNC Mathematics with Graphics," *Computers and Ind. Engng*, vol. 11, nos. 1–4, pp. 454–458, 1986.

Groover, M. P.: *Automation, Production Systems, and Computer-Aided Manufacturing*, Prentice-Hall, Englewood Cliffs, N.J., 1980.

Hayes, C., and P. Wright: "Automating Process Planning: Using Feature Interaction to Guide Search," *J. Manufacturing Systems*, vol. 8, no. 1, pp. 1–15, 1989.

Hook, T. V.: "Real-Time Shaded NC Milling Display," *Computer Graphics*, vol. 20, no. 4, pp. 15–20, 1986.

Husbands, P., F. G. Mill, and S. W. Warrington: "Representation of Components for Automated Process Planning," in *Proceedings of the Second International Conference on Computer-Aided Production Engineering*, 1987, pp. 359–360, Edinburgh.

Kim, K., and J. E. Biegel: "A Path Generation Method for Sculptured Surface Manufacture," *Computers and Ind. Engng*, vol. 14, no. 2, pp. 95–101, 1988.

Kochan, D. (Ed.): *CAM; Developments in Computer-Integrated Manufacturing*, Springer-Verlag, New York, 1986.

Koren, Y.: *Computer Control of Manufacturing Systems*, McGraw-Hill, New York, 1983.

Kral, I. H.: *Numerical Control Programming in APT*, Prentice-Hall, Englewood Cliffs, N.J., 1986.

Krouse, J. K.: "Sculptured Surfaces for CAD/CAM," *Mach. Des.*, vol. 53, no. 5, pp. 115–120, 1981.

Liu, Y. S., and R. Allen: "Interactive Symbolic Description of Components as a Basis for a Computer-Assisted Process Planning System," in *Proceedings of the Second International Conference on Computer-Aided Production Engineering*, 1987, pp. 315–320, Edinburgh.

Loney, G. C., and T. M. Ozsoy: "NC Machining of Free Form Surfaces," *Computer Aided Des. J.*, vol. 19, no. 2, pp. 85–90, 1987.

Marciniak, K., J. I. Wojciechowski, A. Mazurek, D. Zaborowska, and S. Borucki: "An Interactive NC Milling Machine Programming System," in *Proceedings of the Second International Conference on Computer-Aided Production Engineering*, 1987, pp. 167–170, Edinburgh.

Marshall, P.: "Computer-Aided Process Planning and Estimating as Part of an Integrated CAD/CAM System," *Computer-Aided Engng J.*, vol. 2, no. 5, pp. 167–172, 1985.

Marsland, D. W.: "A Review of Computer-Aided Part Programming System Developments," *Computer-Aided Engng J.*, vol. 1, no. 6, pp. 193–197, 1984.

Miller, D. A., and K. H. Ng: "Computer Aided Design of Form Rolls for NC Manufacture," *J. Manufacturing Systems*, vol. 3, no. 1, pp. 91–98, 1984.

Mills, R. B.: "Linking CAD and CAM," *Computer-Aided Engng J.*, vol. 6, no. 9, pp. 66–77, 1987.

Mills, R. B.: "NC Keeps Pace," *Computer-Aided Engng J.*, vol. 8, no. 2, pp. 28–39, 1989.

Morris, H. M.: "Should You Connect CNC to CIM Without First Converting to DNC?," *Control Engng*, vol. 33, no. 5, pp. 56–57, 1986.

Niebel, B. W., A. B. Draper, and R. A. Wysk: *Modern Manufacturing Process Engineering*, McGraw-Hill, New York, 1989.

Persson, H.: "NC Machining of Arbitrary Shaped Pockets," *Computer Aided Des. J.*, vol. 10, no. 3, pp. 169–174, 1978.

Plossl, K. R.: *Engineering for the Control of Manufacturing*, Prentice-Hall, Englewood Cliffs, N.J., 1987.

Pottorf, D. L.: "CAM Cuts NC Programming Time by 60%," *Manufacturing Engng*, vol. 99, no. 6, pp. 63–66, December 1987.

Pratt, M. J.: "Solid Modeling and the Interface between Design and Manufacture," *IEEE Computer Graphics and Applic.*, vol. 4, no. 7, pp. 52–59, 1984.

Pusztai, J., and M. Sava: *Computer Numerical Control*, Reston Publishing Company, Reston, Va., 1983.

Ranky, P. G.: *Computer Integrated Manufacturing*, Prentice-Hall, Englewood Cliffs, N.J., 1986.

Razavi, S. E., and D. A. Milner: "Design and Manufacture of Free-Form Surfaces by Cross-Sectional Approach," *J. Manufacturing Systems*, vol. 2, no. 1, pp. 69–77, 1983.

Ross, D. T., and J. E. Ward: "Investigations in Computer-Aided Design for Numerically Controlled Production," MIT report ESL-IR-320, Project DSR 79442, 1967.

Rouse, N. E.: "NC Systems Close CAD/CAM Gap," *Mach. Des.*, vol. 58, no. 29, pp. 88–100, 1986.

Roy, G., D. Norrie, and R. Fauvel: "3-D Simulation of CNC Machine Tool Operation on a Super-minicomputer and on a PC," *J. Manufacturing Systems*, vol. 7, no. 3, pp. 267–271, 1988.

Satyanarayana, B., P. N. Rao, and N. K. Tewari: "Machining of Plate Cam Profiles on CNC Machine Tools Using a Highly Integrated Part Programming System," *Int. J. Adv. Manufacturing Technol.*, vol. 3, no. 4, pp. 105–125, 1988.

Stacey, T. W., and A. E. Middleditch: "The Geometry of Machining for Computer-Aided Manufacture," *Robotica*, vol. 4, pt. 2, pp. 83–91, 1986.

Steudel, H. J.: "Computer-Aided Process Planning: Past, Present, and Future," *Int. J. Prod. Res.*, vol. 22, no. 2, pp. 253–266, 1984.

Stout, P., and R. Leonard: "The Introduction of DNC Technology as a Partial Approach to Achieving the Objectives of CIM," *Computer-Aided Engng J.*, vol. 6, no. 1, pp. 16–20, 1989.

Subramani, G., S. G. Kapoor, R. E. Devor, and G. B. Hayashida: "An Enhanced Model for the Simulation of Face Milling Operations," in *Proceedings of the Second International Conference on Computer-Aided Production Engineering*, 1987, pp. 171–178, Edinburgh.

Thien-My, D., and F. Lamy: "A Computerized Approach for Classification of Parts and Optimization of Machines Selection in the Design of (GT) Manufacturing Cells," in *Proceedings of the Second International Conference on Computer-Aided Production Engineering*, 1987, pp. 311–314, Edinburgh.

Voelcker, H. B., and W. A. Hunt: "The Role of Solid Modelling in Machining-Process Modelling and NC Verification," SAE technical paper 810195, pp. 1–8, 1981.

Wang, H., and C. Lin: "Automated Generation of NC Part Programs for Turned Parts Based on 2-D Drawing Files," *Int. J. Adv. Manufacturing Technol.*, vol. 2, no. 3, pp. 23–35, 1987.

Wang, W. P.: "Solid Modeling for Optimizing Metal Removal of Three-dimensional NC End Milling," *J. Manufacturing Systems*, vol. 7, no. 1, pp. 57–65, 1988.

Weill, R., G. Spur, and W. Eversheim: "Survey of Computer Aided Process Planning Systems," *Ann. CIRP*, vol. 31, no. 2, pp. 539–550, 1982.

Zafar, A. W., and I. A. Chaudhry: "A Computer-Aided Process Planning Application of a Microprocessor in a Rice Mill," in *Proceedings of the Second International Conference on Computer-Aided Production Engineering*, 1987, pp. 343–350, Edinburgh.

# INDEX